"This is an amazing collection! Not only is it the first book of its kind, defining the territory of the new and rapidly developing field of philosophy of engineering, it contains chapters by a truly international and multidisciplinary group of scholars. The compilation is rich and exciting, and very timely."

Deborah G. Johnson, *Anne Shirley Carter Olsson Professor of Applied Ethics Emeritus, University of Virginia*

"Neelke Doorn and Diane Michelfelder have curated an impressive body of works that turn the clarifying and critical lens of philosophy upon engineering. This volume begins to reveal the depths of an essential human enterprise, one that philosophers for too long treated as a superficial craft rather than what it is: a creative endeavor of social imagination in action."

Shannon Vallor, *Baillie Gifford Chair in the Ethics of Data and Artificial Intelligence, University of Edinburgh*

"Traditional philosophy of technology largely ignores engineers and engineering, but *The Routledge Handbook of the Philosophy of Engineering* takes engineers, their methods, their responsibilities, and their future seriously with a world-class collection of spot-on papers sure to stimulate your reflection. Beg, borrow, or steal this volume and start treating the humans and human activity of engineering in a philosophically serious way, today."

David E. Goldberg, *Professor Emeritus, University of Illinois*

THE ROUTLEDGE HANDBOOK OF
THE PHILOSOPHY OF ENGINEERING

Engineering has always been a part of human life but has only recently become the subject matter of systematic philosophical inquiry. *The Routledge Handbook of the Philosophy of Engineering* presents the state-of-the-art of this field and lays a foundation for shaping future conversations within it. With a broad scholarly scope and 55 chapters contributed by both established experts and fresh voices in the field, the *Handbook* provides valuable insights into this dynamic and fast-growing field. The volume focuses on central issues and debates, established themes, and new developments in:

- Foundational perspectives
- Engineering reasoning
- Ontology
- Engineering design processes
- Engineering activities and methods
- Values in engineering
- Responsibilities in engineering practice
- Reimagining engineering

The Routledge Handbook of the Philosophy of Engineering will be of value for both students and active researchers in philosophy of engineering and in cognate fields (philosophy of technology, philosophy of design). It is also intended for engineers working both inside and outside of academia who would like to gain a more fundamental understanding of their particular professional field.

The increasing development of new technologies, such as autonomous vehicles, and new interdisciplinary fields, such as human-computer interaction, calls not only for philosophical inquiry but also for engineers and philosophers to work in collaboration with one another. At the same time, the demands on engineers to respond to the challenges of world health, climate change, poverty, and other so-called "wicked problems" have also been on the rise. These factors, together with the fact that a host of questions concerning the processes by which technologies are developed have arisen, make the current *Handbook* a timely and valuable publication.

Diane P. Michelfelder is Professor of Philosophy at Macalester College, USA. Along with philosopher Natasha McCarthy and engineer David E. Goldberg, she edited *Philosophy and Engineering: Reflections on Practice, Principles, and Process* (2013). Her most recent book is *Philosophy and Engineering: Exploring Boundaries, Expanding Connections,* edited with Byron Newberry and Qin Zhu (2016).

Neelke Doorn is Distinguished Antoni van Leeuwenhoek Professor of "Ethics of Water Engineering" at Delft University of Technology, the Netherlands. Recent book publications include co-editing the volumes *Responsible Innovation: Innovative Solutions for Global Issues* (2014) and *Early Engagement and New Technologies: Opening up the Laboratory* (2013). She is also the author of *Water Ethics: An Introduction* (2020).

ROUTLEDGE HANDBOOKS IN PHILOSOPHY

Routledge Handbooks in Philosophy are state-of-the-art surveys of emerging, newly refreshed, and important fields in philosophy, providing accessible yet thorough assessments of key problems, themes, thinkers, and recent developments in research.

All chapters for each volume are specially commissioned, and written by leading scholars in the field. Carefully edited and organized, *Routledge Handbooks in Philosophy* provide indispensable reference tools for students and researchers seeking a comprehensive overview of new and exciting topics in philosophy. They are also valuable teaching resources as accompaniments to textbooks, anthologies, and research-orientated publications.

Also available:

THE ROUTLEDGE HANDBOOK OF LINGUISTIC REFERENCE
Edited by Stephen Biggs and Heimir Geirsson

THE ROUTLEDGE HANDBOOK OF DEHUMANIZATION
Edited by Maria Kronfeldner

THE ROUTLEDGE HANDBOOK OF ANARCHY AND ANARCHIST THOUGHT
Edited by Gary Chartier and Chad Van Schoelandt

THE ROUTLEDGE HANDBOOK OF THE PHILOSOPHY OF ENGINEERING
Edited by Diane P. Michelfelder and Neelke Doorn

THE ROUTLEDGE HANDBOOK OF MODALITY
Edited by Otávio Bueno and Scott A. Shalkowski

THE ROUTLEDGE HANDBOOK OF PRACTICAL REASON
Edited by Kurt Sylvan and Ruth Chang

For more information about this series, please visit: www.routledge.com/Routledge-Handbooks-in-Philosophy/book-series/RHP

THE ROUTLEDGE HANDBOOK OF THE PHILOSOPHY OF ENGINEERING

Edited by Diane P. Michelfelder and Neelke Doorn

Routledge
Taylor & Francis Group

NEW YORK AND LONDON

First published 2021
by Routledge
52 Vanderbilt Avenue, New York, NY 10017

and by Routledge
2 Park Square, Milton Park, Abingdon, Oxon, OX14 4RN

Routledge is an imprint of the Taylor & Francis Group, an informa business

Library of Congress Cataloging-in-Publication Data
A catalog record for this book has been requested

ISBN: 978-1-138-24495-5 (hbk)
ISBN: 978-1-315-27650-2 (ebk)

Typeset in Bembo
by Apex CoVantage, LLC

CONTENTS

List of Figures   *xii*
List of Tables   *xiv*
List of Contributors   *xv*
Acknowledgments   *xxiii*

    Introduction   1
    Diane P. Michelfelder and Neelke Doorn

PART I
Foundational Perspectives   **9**

1 What Is Engineering?   11
   Carl Mitcham

2 A Brief History of Engineering   25
   Jennifer Karns Alexander

3 Western Philosophical Approaches and Engineering   38
   Glen Miller

4 Eastern Philosophical Approaches and Engineering   50
   Glen Miller, Xiaowei (Tom) Wang, Satya Sundar Sethy, Fujiki Atsushi

5 What Is Engineering Science?   66
   Sven Ove Hansson

6 Scientific Methodology in the Engineering Sciences   80
   Mieke Boon

PART II
Engineering Reasoning   **95**

7   Engineering Design and the Quest for Optimality   97
Maarten Franssen

8   Prescriptive Engineering Knowledge   111
Sjoerd Zwart

9   Engineering as Art and the Art of Engineering   127
Lara Schrijver

10   Creativity and Discovery in Engineering   138
David H. Cropley

11   Uncertainty   149
William M. Bulleit

12   Scenarios   160
Christian Dieckhoff and Armin Grunwald

13   Systems Engineering as Engineering Philosophy   176
Usman Akeel and Sarah Bell

14   Assessing Provenance and Bias in Big Data   191
Brent Mittelstadt and Jan Kwakkel

PART III
Ontology   **207**

15   Artifacts   209
Beth Preston

16   Engineering Objects   222
Wybo Houkes

17   Use Plans   233
Auke Pols

18   Function in Engineering   245
Boris Eisenbart and Kilian Gericke

19   Emergence in Engineering   263
Peter Simons

20 Towards an Ontology of Innovation: On the New, the Political-Economic
   Dimension and the Intrinsic Risks Involved in Innovation Processes   273
   Vincent Blok

PART IV
Engineering Design Processes   **287**

21 Engineering Design   289
   Peter Kroes

22 Values and Design   300
   Ibo Van de Poel

23 Design Methods and Validation   315
   Sabine Ammon

24 Human-Centred Design and its Inherent Ethical Qualities   328
   Marc Steen

25 Sustainable Design   342
   Steven A. Moore

26 Maintenance   356
   Mark Thomas Young

PART V
Engineering Activities and Methods   **369**

27 Measurement   371
   Lara Huber

28 Models in Engineering and Design: Modeling Relations and
   Directions of Fit   383
   Michael Poznic

29 Scale Modeling   394
   Susan G. Sterrett

30 Computer Simulations   408
   Hildrun Lampe

31 Experimentation   421
   Viola Schiaffonati

32 On Verification and Validation in Engineering   435
    Francien Dechesne and Tijn Borghuis

PART VI
Values in Engineering   **447**

33 Values in Risk and Safety Assessment   449
    Niklas Möller

34 Engineering and Sustainability: Control and Care in Unfoldings
    of Modernity   461
    Andy Stirling

35 The Role of Resilience in Engineering   482
    Neelke Doorn

36 Trust in Engineering   494
    Philip J. Nickel

37 Aesthetics   506
    Stefan Koller

38 Health   521
    Marianne Boenink

39 Philosophy of Security Engineering   533
    Wolter Pieters

PART VII
Responsibilities in Engineering Practice   **545**

40 Ethical Considerations in Engineering   547
    Wade L. Robison

41 Autonomy in Engineering   558
    Eugene Schlossberger

42 Standards in Engineering   569
    Paul B. Thompson

43 Professional Codes of Ethics   580
    Michael Davis

44 Responsibilities to the Public—Professional Engineering Societies 592
    Joseph Herkert and Jason Borenstein

45 Engineering as a Political Practice   607
    Govert Valkenburg

46 Global Engineering Ethics   620
    Pak-Hang Wong

47 Engineering Practice and Engineering Policy: The Narrative Form
    of Engineering Policy Advice   630
    Natasha McCarthy

PART VIII
Reimagining Engineering   **645**

48 Feminist Engineering and Gender   647
    Donna Riley

49 Socially Responsible Engineering   661
    Jessica M. Smith and Juan C. Lucena

50 Engineering and Social Justice   674
    Caroline Baillie

51 Engineering and Environmental Justice   687
    Benjamin R. Cohen

52 Beyond Traditional Engineering: Green, Humanitarian, Social Justice,
    and *Omnium* Approaches   700
    George D. Catalano

53 Engineering and Contemporary Continental Philosophy of Technology   710
    Diane P. Michelfelder

54 Engineering Practice from the Perspective of Methodical Constructivism
    and Culturalism   722
    Michael Funk and Albrecht Fritzsche

55 Reimagining the Future of Engineering   736
    *Neelke Doorn, Diane P. Michelfelder, Elise Barrella, Terry Bristol,
    Francien Dechesne, Albrecht Fritzsche, Gearold Johnson, Michael Poznic,
    Wade L. Robison, Barbara Sain, Taylor Stone, Tonatiuh Rodriguez-Nikl,
    Steven Umbrello, Pieter E. Vermaas, Richard L. Wilson*

Index   745

FIGURES

6.1	Hypothetical Deductive method in natural sciences. Schema based on Hempel (1966)	88
6.2	The B&K method	89
7.1	Eggert's belt-and-pulley example	102
10.1	Change as a driver of problems and solutions	140
10.2	Five pathways in engineering problem-solving	141
10.3	Diminishing returns and engineering solutions	143
10.4	Change and the cycle of replication–incrementation–disruption	144
12.1	Simplified argumentative structure of a model-based scenario	164
13.1	Generic systems engineering methodology	180
13.2	An open system	182
13.3	Continuum of system complexity	183
14.1	The data chain	192
18.1	Black box that causally links an input to a (desired) output	246
18.2	General process model for system development	247
18.3	Iterative synthesis and evaluation steps during conceptual design	248
18.4	Example of a functional structure of a tensile testing machine	248
18.5	Function decomposition and mapping to function carriers	249
18.6	Comparison by Stone and Wood (2000) of their Functional Basis with the function taxonomies proposed by Pahl et al. (2007), Hundal (1990), and Altschuller (1999)	250
18.7	The FBS framework	253
18.8	Chromosome model	254
18.9	Schema of reasoning from user goals to required function carriers	256
18.10	The 5-key-term approach	257
18.11	Classification of (a) FBS framework, (b) Pahl and Beitz approach, (c) Theory of Domains, and (d) Use case-based reasoning	258
20.1	Schumpeter's business cycles	275
22.1	The operationalization of the values environmental sustainability, safety and health for the case of refrigerant coolants	305
22.2	Possible values hierarchy for biofuels	307
23.1	General procedure of systematic design according to VDI 2221	321

23.2	Testing as verification and validation in design processes	322
25.1	The three-legged stool of sustainability	343
25.2	The Planners' Triangle	348
25.3	Example of a fitness landscape	351
29.1	Scale modeling in an educational setting	396
29.2	Experimental Model Basin, Washington Navy Yard, Washington, DC—interior view, c. 1900	398
29.3	Postcard of Waterways Experiment Station	399
29.4	A worksheet designed for use with middle school students showing how to use your own body to determine the height of a tree, from shadow measurements	401
32.1	Verification and validation for the elementary case	436
32.2	Verification and validation for multi-stage development	438
32.3	Verification and validation for systems of systems	439
33.1	The traditional two-stage picture of risk analysis	450
33.2	Hansson and Aven's model of the risk assessment process	456
36.1	Diffusion of linear-but-indirect trust in engineering	496
36.2	Trust in automation	499
50.1	Human rights defenders interviewed by our team	683
52.1	Expanding wave from traditional to green engineering	701
52.2	Expanding wave from green to humanitarian engineering	703
52.3	Expanding wave from humanitarian to engineering for social justice	705
52.4	Expanding wave from social justice to *omnium* engineering	706
54.1	Illustration of a family recipe for a common German bean dish	728
54.2	Lecture script illustrating genetic editing with scissors	729

TABLES

7.1	Different types of specifications generated by two basic distinctions	99
7.2	Overview of acceptable physical values for the design parameters and their assessment values	106
8.1	Differences between descriptive and perspective knowledge	115
10.1	Convergent (critical) thinking versus divergent thinking	145
10.2	The stages of creative problem-solving	146
10.3	Mapping phases to engineering design models	146
20.1	Differences between the contemporary, Schumpeterian and ontological concepts of innovation	283
24.1	Ethical qualities inherent in human-centred design practices	337
29.1	Generalization of similarity in Euclidean geometry to similarity in physics (mechanics, including heat, fluids, etc.)	403
29.2	Using dimensionless ratios in scale modeling	404
32.1	System life cycle processes according to ISO/IEC 15288	436

CONTRIBUTORS

Usman Akeel is Researcher/Lecturer at the Air Force Institute of Technology, Nigeria. He is a registered civil engineer in Nigeria and holds a PhD in engineering sustainability from University College London (UCL), UK. His current research focuses on engineering philosophy and systems thinking.

Jennifer Karns Alexander is Director of Graduate Studies in the Program in History of Science, Technology, and Medicine at the University of Minnesota, USA, where she is also Associate Professor in the Department of Mechanical Engineering. She is the author of *The Mantra of Efficiency: From Waterwheel to Social Control* (2008), and is completing a manuscript on religion, engineering, and technology since World War II.

Sabine Ammon is Professor of Knowledge Dynamics and Sustainability in the Technological Sciences at TU Berlin, Germany, with a focus on philosophy and ethics of design and technology. Recent publications include *The Active Image. Architecture and Engineering in the Age of Modeling* (co-edited with Remei Capdevila Werning, 2017).

Fujiki Atsushi is Associate Professor at Kobe City College of Nursing, Japan. His current research interests are engineering ethics, environmental ethics, ethics of environmental health, and engineering education.

Caroline Baillie is Professor of Praxis in Engineering and Social Justice at the University of San Diego, USA, and co-founder of 'Waste for Life' (http://wasteforlife.org/spt/) and the ESJP network (esjp.org). She has over 200 publications including *Green Composites: Waste and Nature-Based Materials for a Sustainable Future* (with Randika Jayasinghe, 2017).

Elise Barrella is Assistant Professor and Founding Faculty member of the Department of Engineering at Wake Forest University, USA. Her scholarly interests and recent publications focus on sustainability and design education, urban development, and community-engaged projects.

Sarah Bell is Professor of Environmental Engineering at the Institute for Environmental Design and Director of the Engineering Exchange at University College London (UCL), UK. Her most recent book was *Urban Water Sustainability* (2018).

Vincent Blok is Associate Professor in the Philosophy Group and the Management Studies Group, Wageningen University, the Netherlands. His books include *Ernst Jünger's Philosophy of Technology. Heidegger and the Poetics of the Anthropocene* (2017).

Marianne Boenink is Professor in Ethics of Healthcare at the Department of IQ Healthcare, Radboud University Medical Centre, Nijmegen, the Netherlands. Her research focuses on philosophical and ethical challenges posed by the development and use of technology in healthcare, including the visions guiding such innovation (like 'personalized medicine' or 'smart healthcare'). She co-edited the volume *Emerging Technologies for Diagnosing Alzheimer's Disease. Innovating with Care* (2017).

Mieke Boon is Professor in Philosophy of Science in Practice at the University of Twente, the Netherlands, where she is dean of the University College ATLAS. She is a core teacher in the philosophy of science and technology and society programs and in the University College ATLAS.

Jason Borenstein is Director of Graduate Research Ethics Programs and Associate Director of the Center for Ethics and Technology at the Georgia Institute of Technology, USA. His appointment is divided between the School of Public Policy and the Office of Graduate Studies.

Tijn Borghuis is Project Manager and Lecturer at the Philosophy and Ethics group of Eindhoven University of Technology, the Netherlands. He teaches modelling to engineering students of several BSc and MSc programmes.

Terry Bristol is Professor and President, Institute for Science, Engineering and Public Policy, affiliated with Portland State University, Portland, Oregon, USA.

William M. Bulleit is Professor of Structural Engineering at Michigan Tech, USA, a position he has held for 37 years. Prior to coming to Michigan Tech, he designed submarines in Florida and bridges in Washington. Much of his research and teaching has revolved around the need for designers to make decisions under uncertainty.

George D. Catalano is Emeritus State University of New York (SUNY) Distinguished Service Professor, Emeritus Professor of Biomedical Engineering, and Emeritus SUNY Bartle Professor, Binghamton University, USA. His most recent books include *Engineering Ethics: Peace, Justice, and the Earth* (second edition, 2014) and *Tragedy in the Gulf: A Call for a New Engineering Ethic* (2011).

Benjamin R. Cohen is Associate Professor of Engineering Studies and Environmental Studies at Lafayette College, USA. Among other works, he is the co-editor (with Gwen Ottinger) of *Technoscience and Environmental Justice: Expert Cultures in a Grassroots Movement* (2011) and author of *Pure Adulteration: Cheating on Nature in the Age of Manufactured Food* (2019).

David H. Cropley is Professor of Engineering Innovation at the University of South Australia. He is the author or co-author of nine books, including *The Psychology of Innovation in Organizations* (2015) and *Homo Problematis Solvendis—Problem Solving Man: A History of Human Creativity* (2019).

Michael Davis is Senior Fellow at the Center for the Study of Ethics in the Professions and Professor of Philosophy, Illinois Institute of Technology, USA. Among his recent books are *Engineering Ethics* (2005) and *Engineering as a Global Profession* (2019).

Francien Dechesne is Assistant Professor at the eLaw Center for Law and Digital Technologies at Leiden University, the Netherlands. She has a background in mathematics, computer science and philosophy, and her research is on responsible innovation in the field of digital technologies.

Christian Dieckhoff works at the NOW GmbH in Berlin, Germany, which supports the German government in implementing green transport technologies. He holds a diploma degree in mechanical engineering and a PhD in philosophy of science for a work on energy scenarios in policy advice.

Neelke Doorn is Distinguished Antoni van Leeuwenhoek Professor "Ethics of Water Engineering" at Delft University of Technology, the Netherlands. Recent book publications include co-editing the volumes *Responsible Innovation: Innovative Solutions for Global Issues* (2014) and *Early Engagement and New Technologies: Opening Up the Laboratory* (2013). She is the author of *Water Ethics: An Introduction* (2019).

Boris Eisenbart is Associate Professor and Course Director for Product Design Engineering at Swinburne University of Technology, Australia. He is also the Research Director of Swinburne's Industry 4.0 Testlab focusing on the integration of design, engineering and manufacturing processes through advancing digitalization.

Maarten Franssen is Associate Professor at Delft University of Technology, the Netherlands, where he engages with foundational and methodological issues in the philosophy of science and technology. He trained as a physicist and a historian and obtained a PhD in philosophy.

Albrecht Fritzsche is Acting Chair of Technology and Process Management at Ulm University, Germany. He holds doctoral degrees in philosophy and management and worked for many years in the manufacturing industry. His research is focused on innovation studies.

Michael Funk is a researcher and lecturer at the Chair of Philosophy of Media and Technology, Institute of Philosophy, and the Research Group Cooperative Systems, Faculty of Computer Science, at the University of Vienna, Austria. His latest publications include *Roboter- und Drohnenethik. Eine methodische Einführung* (2021).

Kilian Gericke is Professor for Product Development at the University of Rostock, Germany. He is an elected member of the advisory board of the Design Society. He is co-editor of the book *Pahl/Beitz: Konstruktionslehre* and co-author of the revised VDI 2221 guideline.

Armin Grunwald is Professor of Philosophy and Ethics of Technology, and Director of the Institute for Technology Assessment and Systems Analysis (ITAS), at Karlsruhe Institute of Technology (KIT), Germany. His recent monograph is *Technology Assessment in Practice and Theory* (2019).

Sven Ove Hansson is Professor in the Philosophy of Technology at the Royal Institute of Technology, Stockholm, Sweden. He is member of the Royal Swedish Academy of Engineering Sciences and past president of the Society for Philosophy and Technology. His recent books include *The Role of Technology in Science* (2015), *The Ethics of Technology. Methods and Approaches* (2017) and *Technology and Mathematics* (2018).

Joseph Herkert is Associate Professor Emeritus of Science, Technology and Society at North Carolina State University, USA. He is co-editor of *The Growing Gap Between Emerging Technologies and Legal-Ethical Oversight* (2011) and Editor of *Social, Ethical, and Policy Implications of Engineering: Selected Readings* (1999).

Wybo Houkes is Professor in Philosophy of Science and Technology at Eindhoven University of Technology, the Netherlands. He is the author of *Technical Functions* (with Pieter Vermaas, 2010) and numerous publications on the nature of artefacts, technological knowledge, scientific modelling, and evolutionary theories of technology.

Lara Huber is Senior Researcher and Lecturer in Philosophy at the Universities of Kiel and Wuppertal, Germany. She is the co-editor of *Standardization in Measurement: Philosophical, Historical and Sociological Issues* (with Oliver Schlaudt, 2015), and author of books on standardization in science from an epistemological point of view, on relevance and scientific research, on scientific concepts and societal perceptions of normality.

Gearold Johnson is Senior Research Scientist at Colorado State University, USA. He is also the Emeritus George T Abell Chair and Professor Emeritus in the Mechanical Engineering Department. His most recent book, co-authored with Thomas J. Siller and Gearold Johnson, is *Just Technology: The Quest for Cultural, Economic, Environmental and Technical Sustainability* (2018).

Stefan Koller is Associate Attorney at Gibson, Dunn & Crutcher LLP, USA. He holds doctoral degrees in law and philosophy, and co-edits *Architecture Philosophy*, a peer-reviewed interdisciplinary journal. He has received the Gibbs Prize in Philosophy, Duns Scotus Prize in Medieval Philosophy, Exhibitioner Award, Scatcherd European Award, and more, all from the University of Oxford.

Peter Kroes is Emeritus Professor in the Philosophy of Technology of Delft University of Technology, the Netherlands. His most recent book publication is *Philosophy of Technology after the Empirical Turn*, co-edited with Maarten Franssen, Pieter E. Vermaas, Peter Kroes and Anthonie W.M. Meijers (2016).

Jan Kwakkel is Associate Professor in Policy Analysis at Delft University of Technology, the Netherlands. He is interested in quantitative approaches for supporting public decision-making. In his research he uses a broad range of data analytic and simulation modelling techniques.

Hildrun Lampe is a PhD student at the High Performance Computing Center (HRLS) in Stuttgart, Germany. In her dissertation, she is concerned with a philosophical reflection of computer simulations in the environmental sciences.

Juan C. Lucena is Professor and Director of Humanitarian Engineering at the Colorado School of Mines, USA. His recent books include *Engineering Education for Social Justice: Critical Explorations and Opportunities* (2013), and *Engineering Justice: Transforming Engineering Education and Practice* (with Jon Leydens, 2017).

Natasha McCarthy is a head of policy at the Royal Society, UK, where she leads the Society's work on data and digital technology. She is author of *Engineering: A Beginner's Guide* (2009), a tour of the social, cultural and historical impact of engineering; and a co-editor of *Philosophy and Engineering: Reflections on Practice, Principles and Process* (2014).

Diane P. Michelfelder is Professor of Philosophy at Macalester College, USA. She is a past president of the Society for Philosophy and Technology, and a co-founder of the Forum for Philosophy, Engineering, and Technology (fPET). Her most recent book publication is *Philosophy and Engineering: Exploring Boundaries, Expanding Connections*, edited with Byron Newberry and Qin Zhu (2017).

Glen Miller is Instructional Associate Professor of Philosophy at Texas A&M University, USA, where he regularly teaches engineering ethics and cyberethics. His publications include *Reimagining Philosophy and Technology, Reinventing Ihde*, edited with Ashley Shew (2020).

Carl Mitcham is International Distinguished Professor of Philosophy of Technology at Renmin University of China and Emeritus Professor of Humanities, Arts, and Social Sciences at the Colorado School of Mines, USA. His publications include *Thinking Through Technology* (1994) and *Steps Toward a Philosophy of Engineering* (2020).

Brent Mittelstadt is a Senior Research Fellow at the Oxford Internet Institute, University of Oxford, UK. He is a philosopher specialising in data ethics in relation to machine learning, automated decision-making, and medical expert systems.

Niklas Möller is Professor of Practical Philosophy at Stockholm University, Sweden. His research interest lies in value questions in the philosophy of risk and moral philosophy. He is co-author of *The Practical Turn in Political Theory* (with Eva Erman, 2018), and Editor-in-Chief of the *Handbook of Safety Principles* (2018) and has published numerous articles in international peer review journals.

Steven A. Moore is Bartlett Cocke Regents Professor Emeritus in Architecture at the University of Texas at Austin, USA. His most recent books include *Pragmatic Sustainability: Dispositions for Critical Adaptation* (2nd edition, 2016) and, with Barbara B. Wilson, *Questioning Architectural Judgment: The Problem of Codes in the United States* (2014).

Philip J. Nickel is Associate Professor in the Department of Philosophy and Ethics at Eindhoven University of Technology, the Netherlands. He is the author of a number of articles on the philosophy of trust, testimonial reliance, and other phenomena involving human dependence.

Wolter Pieters is Professor of Work, Organisations and Digital Technology at the Faculty of Social Sciences, Radboud University Nijmegen, the Netherlands. His research focuses on cyber risk management, human factors in cybersecurity, and philosophy and ethics of security.

Auke Pols is Postdoctoral Researcher in Responsible Research and Innovation at Wageningen University & Research, the Netherlands. His interests include the conceptualization of actions performed with technical artefacts and ethical aspects of new energy technologies.

Michael Poznic is Postdoctoral Researcher at Karlsruhe Institute of Technology, Germany, at the Institute of Technology Assessment and Systems Analysis. His most recent publication is, "Thin versus Thick Accounts of Scientific Representation," in *Synthese*.

Beth Preston is Professor Emerita of Philosophy at the University of Georgia, USA. She is the author of *A Philosophy of Material Culture: Action, Function, and Mind* (2013), and of the 'Artifact' entry in the *Stanford Encyclopedia of Philosophy* (2018).

Donna Riley is Kamyar Haghighi Head and Professor in the School of Engineering Education and at Purdue University. She is the author of *Engineering and Social Justice* and *Engineering Thermodynamics and 21st Century Energy Problems* (2011).

Wade L. Robison is the Ezra A. Hale Professor of Applied Ethics at the Rochester Institute of Technology. He has published extensively in philosophy of law, David Hume, and practical and professional ethics. Among his books is *Decisions in Doubt: The Environment and Public Policy* (1994), which won the Nelson A. Rockefeller Prize in Social Science and Public Policy. His latest is *Ethics Within Engineering* (2016).

Tonatiuh Rodriguez-Nikl is Associate Professor of Civil Engineering at California State University, Los Angeles, USA, specializing in earthquake engineering and infrastructure resilience and sustainability. His philosophic interests concern technology, uncertainty, and the good life. He is vice chair of the Engineering Philosophy Committee of the Structural Engineering Institute.

Barbara Sain is Associate Professor of Systematic Theology at the University of St. Thomas, St. Paul, Minnesota, USA. Her teaching and research focus on Christian theology and on the interdisciplinary connections between theology and engineering.

Viola Schiaffonati is Associate Professor of Logic and Philosophy of Science at Dipartimento di Elettronica, Informazione e Bioingegneria of Politecnico di Milano, Italy. Her main research interests include the philosophical foundations of artificial intelligence and robotics and the philosophy of computing sciences, with particular attention to the epistemology of experiments in the engineering sciences.

Eugene Schlossberger is Professor Emeritus of Philosophy at Purdue University Northwest, USA. His publications include *A Holistic Approach to Rights* (2008), *The Ethical Engineer* (1993), and *Moral Responsibility and Persons* (1992).

Lara Schrijver is Professor in Architecture at the University of Antwerp Faculty of Design Sciences, Belgium. She is editor for *KNOB Bulletin*. In 2016 she co-edited the volume *Autonomous Architecture in Flanders*, and she was co-editor for the annual review *Architecture in the Netherlands* from 2016–2019.

Satya Sundar Sethy is Associate Professor of Philosophy in the Department of Humanities and Social Sciences at Indian Institute of Technology Madras, Chennai (India). His publications include *Contemporary Ethical Issues in Engineering* (edited, 2015), *Meaning and Language* (2016), and *Higher Education and Professional Ethics* (edited, 2017).

Peter Simons is Emeritus Professor of Philosophy at Trinity College Dublin, Ireland, specialising in metaphysics and Central European philosophy. His most recent co-authored books are *Joint Ventures in Philosophy* (with Edgar Morscher, 2014), and *Reflections on Free Logic* (with Karel Lambert and Edgar Morscher, 2017).

Jessica M. Smith is Associate Professor of Engineering, Design and Society and Director of Humanitarian Engineering Graduate Programs at the Colorado School of Mines, USA. An anthropologist, she is author of *Extracting Accountability: Engineers and Corporate Social Responsibility* (2021) and *Mining Coal and Undermining Gender* (2014).

Marc Steen is a senior research scientist at TNO, a Dutch research and technology organization. He earned MSc, PDEng and PhD degrees in industrial design engineering at Delft University of Technology. His mission is to support organizations to use technologies in ways that help to create a just society and to promote people's flourishing.

Susan G. Sterrett is the Curtis D. Gridley Distinguished Professor of the History and Philosophy of Science at Wichita State University, USA, and author of *Wittgenstein Flies a Kite: A Story of Models of Wings and Models of the World* (2005).

Andy Stirling is Professor of Science and Technology Policy and co-directs the ESRC STEPS Centre at the Science Policy Research Unit at Sussex University. He researches and engages on issues of uncertainty, sustainability, power and democracy around science and technology.

Taylor Stone is cross-appointed as a lecturer and postdoctoral researcher in the Department of Values, Technology and Innovation at Delft University of Technology, the Netherlands. His current research focuses on engineering ethics education and the philosophy of the city. He is co-editor of *Technology and the City: Towards a Philosophy of Urban Technologies* (2021).

Paul B. Thompson is the W.K. Kellogg Professor of Agricultural, Food and Community Ethics at Michigan State University. He has served in advisory capacity to for-profit, non-profit and governmental organizations including the United Egg Producers, the Food and Agriculture Organization (FAO) of the United Nations, and the U.S. National Academy of Engineering. His book *The Spirit of the Soil: Agriculture and Environmental Ethics* was reissued in an updated edition in 2017.

Steven Umbrello currently serves as the Managing Director at the Institute for Ethics and Emerging Technologies, Turin, Italy. His primary research interests are on the ethics and design of emerging technologies, primary artificial intelligence systems and advanced nanotechnology.

Govert Valkenburg is a researcher at the Department of Interdisciplinary Studies of Culture of the Norwegian University of Science and Technology, Trondheim. He trained in electrical engineering and philosophy and now works within the multidisciplinary field of science and technology studies. He has contributed to many books, including the recent *Cambridge Handbook of Social Problems* (2018).

Ibo Van de Poel is Antoni van Leeuwenhoek Professor in Ethics and Technology at the Technical University Delft in the Netherlands. Recent books he co-authored or co-edited include *New Perspectives on Technology in Society: Experimentation Beyond the Laboratory* (2018) and *Moral Responsibility and the Problem of Many Hands* (2015).

Pieter E. Vermaas is Associate Professor in Philosophy at Delft University of Technology, the Netherlands. His research focuses on the methodology and epistemology of design. Recent publications are *Handbook of Ethics, Values and Technological Design* (2015) with Jeroen van den Hoven and Ibo Van de Poel, and *Advancements in the Philosophy of Design* (2018) with Stéphane Vial.

Xiaowei (Tom) Wang is Associate Professor in the School of Philosophy at Renmin University of China. His research focuses on the philosophy and ethics of technology. He is the author of *Human Right and Internet Access: A Philosophical Investigation* (2016) and, together with Pak-Hang Wong, editor of *Harmonious Technology: A Confucian Ethics of Technology* (2021).

Richard L. Wilson is Professor at Towson University and Research Fellow in the Hoffberger Center for Professional Ethics at the University of Baltimore, USA. His latest book is, *Glossary of Cyber Warfare, Cyber Crime and Cyber Security* (coauthored with Dan Remenyi, 2018).

Pak-Hang Wong is Research Associate at the Department of Informatics, Universität Hamburg, Germany. He has published articles on philosophy of technology, ethics of emerging technologies, and Confucian philosophy, and is the co-editor of *Well-Being in Contemporary Society* (2015).

Mark Thomas Young is Associate Professor of Philosophy at the University of Bergen, Norway. His research covers two fields: the philosophy of technology, where he focuses on technologies in use; and the history and philosophy of science, where he explores instruments, craft practices and tacit knowledge in the early modern period.

Sjoerd Zwart is Assistant Professor at the Delft University of Technology, The Netherlands. His current research interests include methodology of and values in engineering and design. He co-edited, with Léna Soler, Vincent Israel-Jost, and Michael Lynch, *Science After the Practice Turn* (2014); and co-authored, with Ibo Van de Poel and Lambèr Royakkers, *Moral Responsibility and the Problem of Many Hands* (2015).

ACKNOWLEDGMENTS

Wherever we find a machine, the pioneering philosopher of technology Ernst Kapp once observed, we find an extension of the human hand. This *Handbook* has benefited from many who have lent their hands to its development. We are particularly grateful to our editor at Routledge, Andrew Beck, for his interest, enthusiasm, and support throughout this project. It was truly a pleasure to work with him. We are also grateful to the three anonymous reviewers of our proposal whose valuable suggestions and comments bettered this work as a whole. For giving us the opportunity to "jump-start" the final chapter of this volume by organizing a pre-fPET 2018 workshop, our thanks go to the meeting's organizers: Zach Pirtle, Guru Madhaven, and David Tomblin. A Wallace Scholarly Activities from Macalester College allowed Diane to come to the Netherlands so we could bypass Skype and work together on the *Handbook* at a critical point. Our heartfelt thanks also goes out to all the authors represented in this volume for participating in this endeavor, sharing their insights, and helping to move forward the field of philosophy of engineering. Lastly, we want to extend our appreciation to the readers of this book. As you look through its pages, we hope it will prove handy in inspiring your own reflection and research.

INTRODUCTION

Diane P. Michelfelder and Neelke Doorn

As recently as twenty years ago, this *Handbook* would have been unimaginable. From an academic standpoint, sustained philosophical inquiry into engineering in the form of a philosophy *of* engineering, as Ibo Van de Poel pointed out in his introduction to *Philosophy and Engineering: An Emerging Agenda* (2010), did not yet even exist. The reasons for this surge and maturation of interest can be tied directly to the myriad of ways in which engineering innovations are directly responsible for transforming the social fabric and established institutions of everyday life. These ways are not simply a matter of having more electric vehicles on the highways or relying on social media to get in touch with friends instead of talking to them on the phone. They are more a matter of simultaneous systemic shifts in agency, as decision-making is transferred from humans to algorithmic processes; in responsibility for these decisions; and in how individual artifacts are linked and networked together. The electric vehicle is not just a means of transportation, but a means of data collection and energy storage. Social media are not just a means of staying in touch with others but also ways of "curating" the news and other information that reach the user. These widespread shifts have prompted the need not simply for additional reflection on engineering, but also for an intensive conceptual and methodological understanding of the engineering-related activities and practices leading to them, as well as their implications for the future.

The purpose of *The Routledge Handbook of the Philosophy of Engineering* aligns directly with this state of affairs. In one sense our purpose in this *Handbook* is to present an in-depth look at what has now been imagined under the name of the philosophy of engineering. It intends to show the state of the art of a field of philosophical inquiry focused on an endeavor which is entangled with the sciences, mathematics, social sciences, design, and art-making activities. In another sense our aim is to bring out the thematic "infrastructure" associated with the philosophy of engineering, a spectrum of questions that surround these themes, and theoretical debates and differences of perspectives that have arisen in the course of addressing them. Our aim is decidedly not, though, to resolve any of these particular contestations. We want to show their richness as well as their urgency, an urgency driven not simply by intellectual curiosity but also, given the recent scandals from Volkswagen, Cambridge Analytica, and other engineering companies, by an increasing unease and questioning of engineering not only on the part of the general public but also by some engineers themselves (cf. Floridi et al. 2018; Hecht et al. 2018).

In thinking about how to organize the contents of this *Handbook*, we were guided by each of the aspects of its overall purpose as just described. The *Handbook* is divided into eight parts: (1) foundational perspectives; (2) engineering reasoning; (3) ontology; (4) engineering design processes; (5) engineering activities and methods; (6) values in engineering; (7) responsibilities in engineering practice; and (8) reimagining engineering. As this structure indicates, the volume moves from the consideration of more methodological issues to ones that are more value-laden. While this style of organizing the volume was intentional, it also carries with it the appearance of being quite traditional. But, one sign of just how rapidly the field of philosophy of engineering has developed is that some subjects within it have already reached almost a canonical status, while others are in more preliminary stages. In each of the sections of this *Handbook*, we aimed to include treatments of both.

The *Handbook* begins with coverage of foundational perspectives, starting with a chapter on what might be considered the most basic question within the philosophy of engineering, namely, what is engineering itself? (Chapter 1, Carl Mitcham). As Mitcham shows, although the question "What is engineering?" might seem to be open to a single answer, the question itself is not so simple, and so a single answer is not possible given the variety of perspectives on this question. He explores a representation of answers to this question from both engineering and philosophical domains, with an eye to identifying their individual strengths and weaknesses. The section moves from this question to a succinct historical overview of engineering that looks at its global development, the beginnings of engineering as a profession, and contemporary engineering practices in the context of different nation-states (Chapter 2, Jennifer Karns Alexander). Glen Miller (Chapter 3) aims to put Western philosophical traditions and the philosophy of engineering into dialogue with one another for the enrichment of both. Chapter 4 (Glen Miller, Xiaowei (Tom) Wang, Satya Sundar Sethy, and Fujiki Atsushi) explores the relationships among some Eastern philosophical traditions and the development of engineering within the context of the state. From the beginning, a key issue within the philosophy of engineering has been to show how the engineering sciences can be distinguished from other branches of science. Sven Ove Hansson (Chapter 5) addresses this issue by showing how the unique epistemological and ontological perspectives within the engineering sciences put these sciences on a par with natural, social, behavioral, and medical science. Based on a comparison between the phenomena in engineering sciences and natural sciences, Mieke Boon (Chapter 6) proposes a methodology for how scientific models of phenomena are constructed in engineering, which is on par with the well-known hypothetical-deductive methodology in the natural sciences.

Part II takes up the topic of engineering reasoning, taken in a broad sense to be the steps involved in conceptualizing and framing an engineering problem, including the role of engineering knowledge in this process. This section begins with two foundational chapters, "Engineering Design and the Quest for Optimality" (Chapter 7, Maarten Franssen) and "Prescriptive Engineering Knowledge" (Chapter 8, Sjoerd Zwart), that describe as well as critically engage with some conventional assumptions about engineering reasoning. Franssen questions the claim that optimality can be a goal of engineering design, while Zwart argues that prescriptive knowledge, defined as a justified and effective belief, is always involved in the solution of an engineering problem. Other chapters in Part II serve to disrupt the conventional view of engineering reasoning as a step-by-step process of thinking by highlighting the place of art, creativity, heuristics, and uncertainty as key dimensions of engineering reasoning. In "Engineering as Art and the Art of Engineering" (Chapter 9) Lara Schrijver takes a historical perspective to show key ties between art and engineering, questioning the wisdom of seeing art and engineering as discrete disciplines. In "Creativity and Discovery in Engineering" (Chapter 10), David H. Cropley shows how, in a world of rapid and ubiquitous change, engineering reasoning requires a combination of critical thinking—i.e. logic, analysis, and evaluation—and also creative thinking—i.e. novelty, unconventionality, and flexibility. Like Schrijver, he argues for a re-emphasis on creative thinking in engineering education.

William Bulleit ("Uncertainty", Chapter 11) implicitly underscores the importance of imagination in engineering reasoning by stressing that a new engineering "mindset" is required for the design of complex and complex adaptive systems. Continuing with the theme of uncertainty, Christian Dieckhoff and Armin Grunwald (Chapter 12) offer an overview of scenarios as ways of developing actions to shape the future and also to cope with the uncertainties that the future poses. In their contribution, Usman Akeel and Sarah Bell (Chapter 13) also call attention to uncertainty as one of the considerations to take into account in planning and decision-making. The primary point of their chapter, though, is to build a case for thinking of systems engineering as engineering philosophy and to show the advantages that such an identification brings. As contemporary engineering reasoning is increasingly reliant on data analytics, Part II ends with an examination of the challenges involved in a variety of strategies of dealing with bias in big data analytics (Chapter 14, Brent Mittelstadt and Jan Kwakkel).

With Part III, the volume turns to consider issues within the ontology of engineering: What comes out of a design process? Part III begins with Beth Preston's "Artifacts" (Chapter 15) which brings insights from contemporary philosophical discussion of artifacts into the discussion of artifacts in engineering, while also bringing out what the philosophy of engineering might offer to the wider discussion of artifacts in philosophy. In a similar vein, the next chapter, "Engineering Objects" (Chapter 16, Wybo Houkes), argues that a consideration of the full spectrum of the objects of engineering requires going beyond the conventional lens of analytic ontology, which looks at these objects as de-contextualized from their practical contexts of use. How this contextualization takes place is the topic of the next two chapters. Auke Pols (Chapter 17) provides an ontological perspective on use plans as the objects of engineering design. While Chapter 15 focuses on the artifacts that come out of a design process and Chapter 17 on what users do when they use these artifacts, Chapter 18 ("Function in Engineering", Boris Eisenbart and Kilian Gericke) focuses on the functions of an artifact as a means to represent and thus comprehend a system under development at an abstract level. Stressing that there is no single concept of emergence, "Emergence in Engineering" (Chapter 19, Peter Simons) explains how emergence can best be understood as the occurrence of certain novel and initially unexpected properties of complex engineering systems that are unpredicted, unexplained, or irreducible on the basis of the simpler parts of those systems. In keeping with the theme of complexity, Part III's concluding chapter, "Towards an Ontology of Innovation" (Chapter 20, Vincent Blok) draws on the history of philosophy to reveal innovation to be both a constructive and deconstructive concept, a disclosure which serves as a basis for questioning the "techno-economic" paradigm that has come to dominate how innovation is frequently thought of today.

With Part IV, the focus shifts to engineering design *processes*. The first chapter, "Engineering Design" (Chapter 21, Peter Kroes), looks at a variety of fundamental issues within engineering design while presenting a challenge to two accepted ideas. Looking at how engineers talk about their designs challenges the conventional distinction between facts and values; and investigating the nature of technical artifacts challenges the idea that they can be understood as the "embodiment" of a design itself. Ibo Van de Poel, in his chapter "Values and Design" (Chapter 22) notes that while engineers and designers have obligations to address values in a systematic way in the design process, there are a number of challenges that arise. Van de Poel turns his attention to several of these challenges and proposes how they might be addressed, including the challenges of how to deal with conflicting values and with value change. Subsequent chapters in Part IV also point out the complexity involved in design methods in general as well as with respect to particular methods such as human-centered design (Chapter 24, Marc Steen). "Design Methods and Validation" (Chapter 23, Sabine Ammon), for instance, shows the limitations of a decontextualized, "one-size-fits-all" approach to design methods, underscoring how different phases and different domains of design call for different methods. The chapter "Sustainable Design" (Chapter 25, Steven A. Moore) makes a strong case for questioning the "modernist" concept of sustainability, defending the idea that this theoretical approach is inextricably interconnected with knowledge from other disciplines and so must be receptive to them. In Chapter 26 ("Maintenance"), Mark Thomas Young argues that maintenance should get a more prominent place in the study of engineering. Recognizing the temporal nature of technologies would allow us to appreciate the true character of maintenance as a creative and transformative technical activity upon which the efficiency and reliability of technologies often depend, and give us a richer understanding of engineering practice itself.

Part V is devoted to engineering activities and methods, starting off with Lara Huber's chapter on the activity and method of measurement (Chapter 27). Huber takes up both the goals of measurement as well as uncertainty and other challenges to it. While not focused on modeling itself, this chapter introduces modeling as a specific activity within engineering and so sets the stage for the other chapters in Part V, which are all implicitly or explicitly related to modeling and ways to improve the accuracy of engineering models to represent aspects of reality. Building on the notion of "direction of fit", the chapter "Models in Engineering and Design: Modeling Relations and Directions of Fit" (Chapter 28, Michael Poznic) shows the richness of different uses of models and explicitly questions the very assumption that models are intended only to provide an accurate representation of reality. The next two chapters look at two specific types of models: scale models (Chapter 29, Susan G. Sterrett) and various kinds of computer models (Chapter 30, Hildrun Lampe). Sterrett illustrates the complexity of scale modeling, which goes much further than simply shrinking the object to a smaller size while keeping geometrical similarity. In a similar vein, Lampe challenges the lay understanding of computer simulations as mere calculations, bringing out their complexity, for instance, by showing how computer simulations are both similar to and different from thought experiments and laboratory experiments. Experimentation is the main focus of Chapter 31, by Viola Schiaffonati. Schiaffonati considers the plurality of approaches to experimentation as revealed in computer science and engineering, and presents the idea of an "explorative experiment" as a way of showing the difference between experiments in engineering and the natural sciences. The last chapter of Part V discusses the epistemic activities of verification and validation as well as contemporary challenges, especially in relation to the growing importance of formal verification methods in digital technologies (Chapter 32, Francien Dechesne and Tijn Borghuis).

While Part V ended with a discussion of methods to ensure epistemic values in engineering, Part VI picks up with a consideration of a variety of non-epistemic values central to engineering practices. Chapter 33, by Niklas Möller, focuses on risk assessment, showing how it is unavoidably value-laden, and so drawing a contrast to the received view which considers risk assessment to be a purely scientific activity. In showing how non-epistemic values are an integral part of risk assessment, Möller brings out how their presence ultimately calls for *transparency*. Transparency also features prominently in the chapters "Trust in Engineering" (Chapter 36, Philip J. Nickel) and "Philosophy of Security Engineering" (Chapter 39, Wolter Pieters). Whereas engineers are traditionally regarded as trustworthy professionals who "stand behind" their products, the rise of automated systems, which *invite* the trust of those who rely upon them, brings added complexity. Nickel proposes grounds for when trust in automated systems is justified, including hypothetical transparency. A defining characteristic of security engineering (Chapter 39, Wolter Pieters) is that it requires dealing with adversaries and adversarial risk. This immediately raises questions for transparency, as the transparency needed to create trust may make technologies thereby also more vulnerable to malicious attacks. One of the key debates in security engineering is therefore on whether security measures should be obscure ("secret security") or made explicit ("transparent security").

Other non-epistemic values considered in Part VI are sustainability, resilience, health, and aesthetics. Andy Stirling (Chapter 34) analyzes sustainability in the context of modernity, showing that while important to pursue due to the failure to control environmental degradation, sustainability itself has become another countervailing effort at control. Stirling argues that control should be replaced by a more caring approach that recognizes the inherent political dimensions of sustainability itself. Chapters 35 and 38 both ask the fundamental question what it means to function well, be it for sociotechnical systems or for human beings. Neelke Doorn (Chapter 35) focuses on resilience engineering as a new approach to managing risks. The chapter shows that, though based on a fundamentally different view of how systems function, measures taken in the context of safety management and those taken in the context of resilience engineering may not be so different after all. Most strikingly, resilience engineering puts greater emphasis on how well a system is able to deal with and adapt to changing circumstances. Where resilience and safety are closely linked to their opposite (i.e. risk), the meaning of health (Chapter 38) is also closely intertwined with the meaning of three different concepts that can be interpreted as the opposite of health: disease, illness, and sickness. Marianne Boenink shows how the emergence of technologies in healthcare has narrowed health to "absence of disease", marginalizing experiences of illness and sickness. Stefan Koller (Chapter 37) discusses how the fact that recent aesthetics has widened its scope of investigation beyond the traditional study of artworks and their beauty opens possibilities for it to contribute to the philosophy of engineering, through offering perspectives on whether engineered objects are also works of art and whether they can and ought to be experienced aesthetically.

Part VII takes an in-depth look at the "public-facing" responsibilities of engineers. "Ethical considerations in engineering" (Chapter 40, Wade L. Robison) shows how stages in the engineering design process are implicitly value-laden and argues that this should be made more explicit, particularly in the training of future engineers, so that they could better anticipate possible "downstream" adverse consequences of their work. "Autonomy in Engineering" (Chapter 41, Eugene Schlossberger) shows the broad extent to which this value plays a role in engineering decision-making, ranging from the need for engineers to have and to respect autonomy, to a host of issues raised by autonomous machines, including the question of when to give them rights. Paul B. Thompson (Chapter 42) shows how technical standards, although clearly having positive ethical valence in that they significantly lower the social costs of engineering systems, may bias engineering practice toward certain powerful individuals, firms, and even nations. Standard setting for Thompson is thus not simply a technical but also a morally related practice. Where Chapter 42 focuses on norms that apply to products, work processes, or designs, Michael Davis in Chapter 43 discusses the professional codes of ethics that apply to the people and organizations working on these products and designs and in these work processes. He provides an overview of what role these codes play in professional engineering today and sketches how they might evolve.

From these chapters, Part VII goes on to look at these "public facing" responsibilities in a more collective light, including the responsibilities of engineering societies and the responsibilities of engineering within a political context. Joseph Herkert and Jason Borenstein (Chapter 44) survey the history of professional and quasi-professional engineering societies, and argue that professional societies need to continue to take leadership in articulating the social responsibilities of engineers with regard to the development of AI and other emerging technologies. In Chapter 45, Govert Valkenburg analyzes engineering as a political practice and the implications this has for engineering and innovation practices. It is critical, Valkenburg proposes, for engineers to recognize that these practices are rooted in particular epistemologies, themselves not value-free but political, which serve to privilege certain points of view in agenda setting. The question of engineering responsibilities is opened up in an even wider aperture than in the previous chapters in Part VII by Pak-Hang Wong (Chapter 46) and Natasha McCarthy (Chapter 47). Wong finds the conventional approach to "global engineering ethics" as rooted in cross-cultural norms to be insufficient. He argues for a more ambitious understanding that would look to promote well-being via engineering, involving increased collaboration among engineers, philosophers, and social scientists. In her contribution, McCarthy draws upon accounts of narrative explanation in order to highlight the distinctive role that engineers, as contrasted with other professionals, can play in policymaking. She backs up her claim that engineering policy advice has a narrative form with several illustrations, including the Internet of Things and greenhouse gas removal.

The *Handbook* concludes with Part VIII, on reimagining engineering itself. In some sense, engineering is always a work in progress, but in Part VIII, how that work might be taken up so as to make sure that all can benefit from it becomes an intentional subject of questioning and reflection.

"Feminist Engineering and Gender" (Chapter 48, Donna Riley) looks toward a dramatic re-shaping of engineering, grounded in critiques of masculinist engineering, which uses feminist perspectives that take account of lived experiences to reconfigure engineering ontologies, epistemologies, and ethics. In Chapter 49, Jessica M. Smith and Juan Lucena similarly critique the established framework of socially responsible engineering within the corporate realm, arguing that it supports businesses more than communities, and propose a set of five criteria that a corporate engineer could draw upon to better support the communities within which they work and so to practice engineering that is truly socially responsible. For Caroline Baillie as well ("Engineering and Social Justice", Chapter 50), the first step in creating more socially just engineering is to acknowledge and undo established "paradigms of thought" found within much of current engineering practice so as to make social transformations that would lead to more equitable distributions of wealth and political power and create more pathways to healthy and fulfilling lives. This chapter is followed by Benjamin R. Cohen's consideration of the relationship between engineering and environmental justice (Chapter 51). Cohen looks at how engineering practitioners can benefit the ecological health of communities using a four-fold framework of different approaches to justice: procedural, recognition, distributive, and capabilities. All three of these chapters put special weight on the importance of changing engineering education programs so that the approaches they discuss become more embedded in how future engineers will think about the work that they do. Building on green, humanitarian, and social justice approaches to engineering, George D. Catalano (Chapter 52) argues for what he calls "*omnium engineering*", that is, an engineering profession that is ethically responsible toward the wants and needs of all life forms, not only that of the human species.

The next two chapters also take up this section's theme of reimagining engineering, based on philosophical developments whose relations to the philosophy of engineering have been underexplored. From the perspective of how it might contribute to the field of contemporary engineering, Diane P. Michelfelder (Chapter 53) looks at the resurgence of interest in contemporary Continental philosophy of technology in investigating technology as a whole. She highlights how work resulting from this interest can disclose alternatives to key understandings that engineering increasingly relies upon of what it is to be a human subject. Michael Funk and Albrecht Fritzsche (Chapter 54) map out key concepts, ideas, and argumentative patterns within methodical constructivism and culturalism and show how they might bear upon engineering practices. Of special note is the concept of transsubjectivity, which they see as holding potential for the development of robotics and other new technologies. Part VIII as well as the *Handbook* comes to a close on an exploratory note with Chapter 55 (Neelke Doorn, Diane P. Michelfelder, et al.) on "Reimagining the Future of Engineering". Focusing on designing, action, problem framing, professional and disciplinary identity, and the training of future engineers, the ultimate aim of this concluding chapter is to inspire future dialogue between philosophers and engineers.

Just as some themes have emerged within the individual parts of this volume, some emphases have also emerged within the *Handbook* as a whole, networking the chapters together in interesting and unexpected ways. Using the philosophy of engineering as a lens, we can see the depth to which engineering thinking involves creativity and so bears a closer relationship to the arts than is reflected in the dominant view of engineering as an activity. This lens also brings out the complexity and open-endedness of engineering design as an activity and, through this, also underscores the fact that the knowledge one gains through engineering is always going to be incomplete and subject to revision. Last but far from least, it also shows just how normative engineering is as an activity to begin with and how, because of this, maintaining the conventional idea that there is a sharp demarcation between the methodological and the normative side of engineering becomes increasingly difficult to sustain.

These same characteristics—creativity, complexity, and open-endedness—can also be affirmed of the subject of this *Handbook* itself, the philosophy of engineering. To go back to the sentence with which this introduction started, as recently as twenty years ago this *Handbook* would have been unimaginable, and now it no longer is. The dimensions of creativity, complexity, and open-endedness give rise to the idea that no matter at what point in the future, philosophy of engineering will always be in a dynamic and beta-state, marked by imagining and reimagining, and making important contributions not only to the understanding of engineering, but also to philosophy itself.

References

Floridi, L., Cowls, J., Beltrametti, M., Chatila, R., Chazerand, P., Dignum, V., Luetge, C., Madelin, R., Pagallo, U., Rossi, F., Schafer, B., Valcke, P. and Vayena, E. (2018). AI4People—An Ethical Framework for a Good AI Society: Opportunities, Risks, Principles, and Recommendations. *Minds and Machines,* 28(4), 689–707. doi:10.1007/s11023-018-9482-5.

Hecht, B., Wilcox, L., Bigham, J.P., Schöning, J., Hoque, E., Ernst, J., Bisk, Y., De Russis, L., Yarosh, L., Anjum, B., Contractor, D. and Wu, C. (2018). It's Time to Do Something: Mitigating the Negative Impacts of Computing Through a Change to the Peer Review Process. *ACM Future of Computing Blog.* https://acm-fca.org/2018/03/29/negativeimpacts/.

Van de Poel, I. (2010). Philosophy and Engineering: Setting the Stage. In I. Van de Poel and D. Goldberg (eds.), *Philosophy and Engineering: An Emerging Agenda.* Dordrecht: Springer, pp. 1–11.

PART I

Foundational Perspectives

1

WHAT IS ENGINEERING?

Carl Mitcham

Any regionalized philosophy must make some effort to identify what it wants to think about, its subject matter. "What is science?" is a key question for philosophy of science. "What is religion?" is a key issue in philosophy of religion. Such questions are sometimes called demarcation problems. Philosophy of engineering, likewise, must at some point seek to identify engineering and mark it off from near neighbors, that is, other dimensions of experience from which it can be distinguished.

This is not as easy as may initially appear, because engineering today is a contested concept—indeed, a contested activity. "Engineering" is not a rigid designator. There is no simple answer to what is truly not a simple question. Speaking generally, different answers tend to be given by engineers and by non-engineers. (In like manner, scientists do not always agree with the definitions of science given by non-scientists, nor do those who describe themselves as religious always like identities constructed by others.) Additionally, engineers and non-engineers do not always fully agree among themselves. In the present case, the question concerning engineering will be explored by considering a selective spectrum of responses along with their strengths and weaknesses.

1.1. Engineering Accounts

One engineering response to the question is represented by the British Royal Academy of Engineering web site: "Engineering covers many different types of activity. Engineers make things, they make things work and they make things work better" (www.raeng.org.uk/publications/reports/engineering-and-economic-growth-a-global-view). Interestingly, few national engineering academies—not even the International Council of Academies of Engineering and Technological Sciences—provide a definition of engineering. However, expanding on the Royal Academy description, a scholarly apologia for engineering authored by two engineers and a sociologist argues at length that engineering is the making of "ingenious devices" and is coeval with the origin of the human species.

> In its earliest form, engineering involved the making of stone tools and other artifacts to aid in human survival [see also Chapter 2, "A Brief History of Engineering" by Jennifer Karns Alexander, this volume]. During the ensuing millennia, the manufacture of ingenious devices expanded and contributed to the shaping of civilizations, to the establishment of human institutions, and to the enhancement of standards of living. Now, in the 21st century, engineering may be viewed as a profession which involves creative thought and skilled actions related to conceptualizing, planning, designing, developing, making, testing,

implementing, using, improving, and disposing of a variety of devices, invariably seeking to meet a perceived societal interest.

(Harms et al. 2004: v)

Nuclear engineer and philosopher Billy V. Koen goes even further in an extended, detailed, systematic argument regarding the essential human activity. According to Koen, engineering is defined by its method, which is "the use of heuristics to cause the best change in a poorly understood situation within the available resources" (Koen 2003: 59). Expanding, "a *heuristic* is anything that provides a plausible aid or direction in the solution of a problem but is in the final analysis unjustified, incapable of justification, and potentially fallible" (Koen 2003: 28). In effect, every decision that humans make to change things is engineering. "*To be human is to be an engineer*" (Koen 2003: 7, 58, italics in the original both times).

One strength of this view is that it presents engineering as crucial to being human and thereby justifies the importance of engineers. It is the basis of a strong *philosophy for engineers*, helping them explain and justify themselves to non-engineers (who, on Koen's account, are actually engineers without realizing it). A weakness is virtually the same: it turns so much into engineering that there doesn't seem to be anything left out. It fails, for instance, to distinguish between the editing of a poem (deleting or replacing words for better ones) and the engineering of a roadway. Even if all engineering is heuristics, does that mean all heuristics is engineering? Additionally, it fails to account for the very specific curriculum that engineering schools teach to engineers. A strong *philosophy for engineering* is not necessarily the same as a well-developed *philosophy of engineering*.

1.2. The Word "Engineering"

As has often been noted, the term "engineering" is of a distinctly modern provenance that challenges an expansive definition. In the Harms, Baetz, and Volti quotation in the previous section, "engineering" is derived from the Latin *ingeniosus*. From the earliest times, some humans have been described as ingenious—but not all humans. In fact, it was a rare person who was termed ingenious. Vitruvius's *De architectura* (1st century BCE) distinguishes the ingenious (hand-working) artisan (Greek *tekton*) from the *architekton* or architect, who supervised building. In Shakespeare's *Troiles and Cressida* (Act II, Scene 3), it is the great warrior Achilles who is referred to as a "rare enginer."

The first group of people to whom the word "engineer" was consistently applied were designers and operators of "engines of war." Leonardo da Vinci (1452–1519), for instance, was often referred to differentially as engineer, architect, or painter in order to emphasize, respectively, his military, civilian, or aesthetic activities. Later in the Leonardo century, Tommaso Garzoni (1549–1589), in an eccentric compendia of professions, further differentiated the engineer from the mechanic: the former working primarily with the mind, the latter more with hands.

Although etymology cannot resolve the philosophical issue of what engineering is, it does indicate that the concept emerged at a particular time, suggesting that the activity to which it refers was something historically new. The newness of the word and, by implication, activity is further endorsed in the founding of the British Institution of Civil Engineers (ICE), the oldest professional engineering society. When in 1828 the ICE applied for a Royal Charter, King George IV requested an identification of that to which this new institution was to be dedicated; he needed a legal solution to the demarcation problem. He was given, and used in the charter, what has become a classic definition: Civil (meaning all non-military) engineering is "the art of directing the great sources of power in nature for the use and convenience of man" (or, we would now say, "humans").

Three observations can be made about this definition: First, the qualifier "civil" was necessary because, as already noted, at that time engineering was a military profession in the Corps of Royal Engineers. Engineering was undergoing a shift from military to civilian affairs precisely during that historical period in which civil society as a whole was both expanding and being transformed by the rise of the bourgeois class and capitalist political economy. Engineering participated in and contributed to that transformation. Second, the "great sources of power in nature" were assumed best revealed by the new natural philosophy that conceived the physical world in mechanistic terms. Third, the aim of "use and convenience" echoes a theory of morals associated with the thought of John Locke (1632–1704) and David Hume (1711–1776). More consideration of the second and third points can help further reflection on the demarcation question.

There are a number of strengths and weaknesses of this classic ICE definition. It has certainly functioned well in promoting a legitimating professional self-understanding of English-speaking engineering. However, because it makes engineering virtually synonymous with human benefit, it has also served to short-circuit critical reflection.

1.3. Engineering, Applied Science, and Technology

Engineering has often been described as involving modern natural science—a view that is implicit in the ICE definition. Indeed, an ICE-commissioned concept paper by Thomas Tredgold from which the classic definition is drawn makes this explicit by describing engineering as "that *practical application of the most important principles of natural philosophy* which has in a considerable degree realized the anticipations of [Francis] Bacon, and changed the aspect and state of affairs in the whole world" (italics added). (The complete text, from the minutes of an 1828 ICE meeting, is available in Mitcham 2020: 368–369.) Bacon (1561–1626) had argued for creation of a new approach to knowledge production that stressed its practical utility. "Human knowledge and human power meet in one [but nature] to be commanded must be obeyed" (*Novum organum* I, 3). Tredgold's paper built on this view and maintained that it was through the new mathematicized science of hydraulics that the Dutch had separated engineering from hydraulic architecture. Similar claims regarding mathematicization are often made for conceiving of modern technology as (at least in part) applied science, that is, putting science to work. In considering what engineering is, then, it is appropriate to dig deeper into the extensive discussion of possible relationships between science, engineering, and technology.

Consider the polysemic term "technology." In an extensive historical examination of this "odd concept," Eric Schatzberg (2018) notes a plethora of closely related terms—"mechanical arts," "applied arts," "useful arts," "industrial arts," "industrial techniques"—with different usages among, for example, social scientists, engineers, and humanists. For historians, anthropologists, and other social scientists, "technology" tends to be co-extensive with material culture. Like historian Lewis Mumford's preferred term "technics" (Mumford 1934), it covers not just tools and machines but everything from clothing, shelter, utensils, utilities, and decorative objects of craft and art to mega-artifacts such as monuments, transport infrastructures, and communication networks—in short, all physical things made by humans. As social scientists have long argued, there are a host of underappreciated interactions (mediations) both ancient and modern between social orders and technologies.

Carl Mitcham (1994) has likewise argued for a typology that recognizes technology not just as objects but also as distinctive forms of knowledge, activity, and even volition. In their conceptualization of technology as craft object, for instance, anthropologists often describe artisanship knowledge as manual skill. In the words of French physical anthropologist André Leroi-Gourhan (1993: 254–255), "in preindustrial societies the individual level of technicity was relatively high [because of lives] filled with manual activities of many kinds," whereas contemporary technicity has been "demanualized." British cultural anthropologist Tim Ingold, taking Mitcham's typology as a starting point (Ingold 2000: 295ff.), provides further phenomenological descriptions of craft making (Ingold 2013) as the background against which engineering comes into relief as a unique form of technology as knowledge, activity, and even intention. Simplifying, hand-craft artisans are less explicitly motivated than engineers to transform; instead they seek to live in expressive harmony with the world.

Yanagi Sōetsu's classic account of *The Unknown Craftsman* (1972) richly describes how *mingei*, or the "hand-crafted art of ordinary people," is based in aesthetic appreciation of the material qualities of utilitarian objects such as bowls and cups.

For engineers, however, "technology" can paradoxically serve both as an umbrella term including all forms of engineering and to name something less scientifically based but still related. An "institute of technology" (such as MIT) teaches multiple types of engineering; yet engineers simultaneously contrast their more intellectual work to the manual skills of technologists or technicians who install, operate, and maintain what has been engineered. A bachelor of engineering degree requires greater knowledge of science than an associate degree in engineering technology; the former is more likely to be awarded by a university, the latter by what in the U.S. is called a "community college." (Mixing things up, American technicians are in Great Britain often called "engineers"; in Germany and Austria, *Techniker* can sometimes be translated as "engineer"; and even in the United States, the operators of railroad locomotives and other large mechanical devices such as heating plants can be called "engineers.")

According to a widely quoted statement attributed to the pioneering aeronautical engineer Theodore von Kármán (1881–1963), "The scientist describes what is; the engineer creates what never was" (Allibone 1980: 110). For von Kármán, engineering creation nevertheless utilizes scientific knowledge. In his words,

> In thermodynamics, . . . theoretical discoveries preceded by many years the actual production of engines and similar hardware. Similarly, the development of the theory of electromagnetism occurred long before some engineers saw how to apply it to create the electrical industry. And, of course, atomic theory preceded practical applications by several decades. In aerodynamics, . . . the discovery of the fundamental laws of lift [gave] us the first real understanding of what makes flight possible, and . . . set the stage for the amazingly swift progress to follow.
>
> *(von Kármán 1967: 59)*

Despite its specialized accuracy, von Kármán overlooks how thermodynamics also arose through work by the French military engineer Sadi Carnot (1796–1832) to improve the efficiency of the steam engine, well before it was used in designing the internal combustion engine.

In a more detailed conceptualizing of the science-engineering relationship, philosopher Mario Bunge (1967) argues that engineering can complement if not replace some practices of "prescientific arts and crafts" with what he calls "grounded rules." Scientific laws describe relationships and patterns in what is; rules (both social and technological) prescribe how to use these laws to create what is not yet. Trial and error or empirical craft rules of thumb become grounded when they can be accounted by scientific laws. Scientific accounting takes place through the formulation of "technological theories" (as found in the "technological" or "engineering sciences").

Technological or engineering theories are of two types: substantive and operative. Substantive theories deploy scientific laws in the engineering analysis of concrete situations, the way models from the physics of fluid dynamics assist in the aerodynamic engineering of airfoils. Operative theories arise when scientific methods are used to examine the workings of technologies or human-technology interactions. Thermodynamics, as noted, arose from systematic or scientific examination of steam engines; industrial engineering focuses on how to improve human-technological interactions in production or utilization processes.

Although often criticized, Bunge's "scientific philosophy" of technology remains one of the earliest to recognize the distinctiveness of engineering. Additionally, his description of how engineering produces a special type of knowledge accords well, for instance, with engineer Walter Vincenti's account of the growth of knowledge about airfoil design (see also Chapter 8, "Prescriptive Engineering Knowledge," Sjoerd Zwart, this volume). Initially, aeronautical engineers chose their airfoils from catalogs . . . of profiles devised by research engineers or designers to achieve various categories of performance. Most of these profiles were developed by modifying previously successful forms, using rules learned from experience, such theoretical understanding and methods as were available, and a vast amount of wind-tunnel testing. The mix of methods changed with time, with fluid-dynamic theory slowly replacing rules of experience.

(Vincenti 1990: 34)

Nathan Rosenberg and Edward Steinmueller (2013) make a similar case for knowledge production in chemical as well as aeronautical engineering.

By contrast, historians of technology such as Edwin Layton Jr. (1971b), Edward Constant II (1980), and Eugene Ferguson (1992) have argued against any easy description of engineering as applied science. Layton's catchy formulation is that the two forms of knowledge are "mirror-image twins." Making use of ideas from Thomas Kuhn (1962), Constant develops a model of technological change from normal through "presumptive anomaly" ("when assumptions derived from science indicate either that under some future conditions the conventional system will fail . . . or that a radically different system will do a better job") to technological revolutions distinct from scientific ones (Constant 1980: 15). Ferguson calls attention to Renaissance books called "theaters of machines" filled with graphic illustrations for a welter of technological mechanisms on which engineers can draw when designing complex useful machines. Other books provided graphical narrations of complex engineering projects, enabling them to be imitated or transferred from one place to another, as in Agricola's *De re metalica* (1556). Such instruments of communication were later supplemented with graphical means for analysis encoded in the standards of engineering drawing and physical tools of analysis (such as the slide rule) as well as in visual algebra and geometric calculus. All such technologies exist alongside rather than as direct applications of science.

Despite its limitations, the concept of engineering as applied science remains attractive insofar as engineering is identified with engineering sciences such as mechanics, thermodynamics, electronics, aerodynamics, and materials science. However, as even von Kármán recognizes, engineering is more than engineering science: it involves creation or what engineers regularly call design (see also Chapter 21, "Engineering Design," Peter Kroes, this volume).

1.4. Engineering as Designing

"Design" is another polysemic word. In the contemporary world, designers are all over the place. There are fashion designers, graphic designers, interior designers, landscape designers, industrial designers, and web designers. In both architecture and engineering design is referenced as a central, even defining, feature. As another approach to the question concerning engineering, is it possible to distinguish different types of design?

Historically, engineering arose as an activity distinguished from skilled artisanship and architecture. Arnold Pacey (1990) wants to place subsistence agriculture even earlier in this path. The career of Thomas Telford (1757–1834), the founding president of ICE, nicely illustrates a trajectory across and thus suggests distinctions among these four human engagements with the world.

Telford was born into a subsistence farming family on the English-Scottish border and began his own career as a stone mason working under the direction of architects to construct civic structures (houses and buildings) in the New Town of Edinburgh, Scotland (1770s and after). Having worked as an artisan for a few years while simultaneously educating himself—as he records in his "Architectural Commonplace Book," by copying and commenting on texts from Colen Campbell's *Vitruvius Britannicus* (3 volumes, 1715–1725), Bernard de Montfaucon's *Antiquity Explained and Represented in Diagrams* (15 volumes, 1721–1725), along with many others—he began to design and construct homes and other urban structures, and so called himself an architect. Then a decade later, as he turned from civic structures to designing and constructing the canal transport infrastructure of England for commercial utility, he assumed the title of "engineer."

As the ICE charter had elaborated, "use and convenience" is synonymous with enhancing

> the means of production and of traffic in states both for external and internal trade, as applied in the construction of roads, bridges, aqueducts, canals, river navigation and docks, for internal intercourse and exchange, and in the construction of ports, harbors, moles, breakwaters and lighthouses, and in the art of navigation by artificial power for the purposes of commerce, and in the construction and adaptation of machinery, and in the drainage of cities and towns.

This type of commercially useful design work to facilitate transport via roads and canals may be contrasted with that of architecture, which aims to humanize a particular place. Architecture is more concerned with ordering space to accommodate repetitive movement in a sedentary life than with movement across great distances for power or commercial gain. From its historical beginnings, engineering in England emerged as a handmaid of those economic institutions promoted by Adam Smith's *Wealth of Nations* (1776) that became known as capitalism.

In one of the original expositions of a Western (Greek and Roman) architectural tradition, Vitruvius identified three criteria for good building: *firmatas* (durability), *utilitas* (utility), and *venustas* (beauty) (*De architectura* I, 3, 2). Although engineering "use and convenience" echoes *utilitas*, and engineering to some degree aims for durability (although capitalist creative destruction has been known to require engineers to design-in obsolescence), what stands out is the fact that engineering design *qua* engineering does not aim for beauty—especially beauty as classically understood, which cultivates delight and the raising of spirits among those who inhabit its carpentered spaces (see also Chapter 9, "Engineering as Art and the Art of Engineering," Lara Schrijver, this volume). As one philosopher of technology puts it, "For the architect, function and aesthetics take center stage [while for] engineers, the design of an artifact . . . is approached with questions of utility and efficiency" (Pitt 2011: 133). In another's succinct phrasing, "What engineers tend to emphasize in place of beauty is cost, safety, and efficiency" (Davis 2010: 21).

For engineering "use and convenience" functions, however unconsciously, as another distinguishing technical term based in the proto-utilitarian moral theory of Locke and Hume and thereby biased toward cost and efficiency. Locke's labor theory of value can be read as domesticating Bacon's technoscientific aspiration to "conquer nature" for "the relief of man's estate." As further developed by Hume's this-worldly philosophy of possessive individualism, the foundation of moral judgment is explicitly denominated as "utility." Whereas premodern philosophers such as Aristotle had understood the virtues as practices that realize or perfect innate human potentials, Hume "ascribe[s] to their utility the praise, which we bestow on the . . . virtues." And with human artifacts, all goodness is said to rest on their "use and conveniency" (*Enquiry Concerning the Principles of Morals*, Section II, Part II).

Over the course of the 20th century, this ideal-type distinction between architectural and engineering design in terms of end or goal orientation has been complicated by the emergence of a bewildering diversity of design professions and the development of design studies focused as well on subject matters and methods (see, e.g., Cross 1986). As Richard Buchanan, a leading contributor to this interdisciplinary field, observes, "Design studies—the history, criticism, and theory of design, as well as design research through empirical investigation and philosophical speculation—[has exerted ever] greater influence on the study of design methods and methodology" (Buchanan 2009: 414). Insofar as these studies generalize a conception of design as grounded in processes of composition or synthesis (in contrast to division or analysis) and have sought to identify methods common to architecture, engineering, and other design fields, they weaken any appeal to designing as engineering-specific.

At the same time, as Buchanan also notes, design studies commonly recognize two broad traditions: arts and crafts versus engineering. The craft method, as already mentioned, is based on intuitive judgments combined with "traditional practices of trial and error in the making of artifacts and the gradual evolution of product forms adapted to particular circumstances." Engineering methods focus "on drawing and draftsmanship [in which] the designer sketches possible product forms that satisfy the needs of manufacturers and the marketplace and then develops detailed scale drawings that can be used as instructions or specifications to guide manufacture and construction" (Buchanan 2009: 414–415). Although drawing occurs in the arts in the form of preliminary sketching, engineering graphics becomes a method for representing information from "the natural sciences and mathematics [that facilitates] calculation and forethought about materials, energy, and how the parts of a product may be fabricated, combined, and assembled to bear loads, distribute stress and heat, and so forth" (Buchanan 2009: 413).

Although designing is probably the most common response to a question about the essence of engineering, what again undermines it as an unambiguous circumscription is that such calculative designing has migrated from material construction into the fields of economic analysis, social organization and bureaucratic processes, services delivery, commercial and political communication via symbolic environments, and deliberative decision-making in general. Multiple activities have been colonized by engineering or adopted methods from engineering. Further challenges for taking engineering design as its defining feature include the fact that engineers perform many more activities than designing, disagreements over its history about the proper role of design in engineering education (Grayson 1993), debates about design methods, methodology, and the computer transformation of engineering design (see, e.g., Kroes 2009), and arguments for engineering methods as analogous to the dynamic process of biological evolution occurs (see Wimsatt 2007).

1.5. Engineering as Profession

Still another response to the question concerning engineering considers its character as a profession (see also Chapter 43, "Professional Codes of Ethics," by Michael Davis, this volume). This approach prompts further ethical interrogation of the original ICE description.

According to Michael Davis, there are four increasingly specific concepts of a profession: as vocation or work, as occupation by which one earns a living, as honest or respected occupation, and as specific type of honest occupation. In philosophical dialogue with diverse practicing professionals, he has developed a concept of the necessary and sufficient conditions for a profession in the fourth sense: "A profession is a number of individuals in the same occupation voluntarily organized to earn a living by openly serving a certain moral ideal in a morally-permissible way beyond what law, market, and [public opinion] would otherwise require" (Davis 1997: 417, 2014: 490). Professions can thus be distinguished in regard to "underlying occupation," "moral ideal," or the ways by which "they seek to achieve [an] ideal" (Davis 1997: 420).

Adopting this framework, the ICE definition can be restated as

> The engineering profession is a voluntary organization of people (e.g., the ICE) who earn their living through the art of directing the great sources of power in nature (occupation) for the use and convenience of humans (moral ideal) in a morally-permissible way beyond what law, market, and public opinion would otherwise require.

"Art" here can be expanded as "the methods of engineering design" with the "sources of power in nature" coming into play through applied science. Such an expansion would revive the contentious issue of the science-engineering relationship. Moreover, Davis is skeptical about relying on some art of engineering as a distinguishing feature. In his view, engineering cannot really be grasped by means of some specific function such as design, since "engineers do a great many things: design, discover, inspect, invent, manage, teach, test, testify, and so on." Engineers are instead best identified by a common education and "a shared way of doing certain things" the precise contours of which are "not a matter of abstract definition, logic, or the like, but of history" (Davis 2009: 336, 337). This places education, moral ideal, and shared way of doing things at the center of the what-is-engineering question.

The alleged commonality of educational experience presents something of a challenge, since what is included in any engineering curriculum can reflect how the engineer is to be employed and the cultural context. Engineering education in the military includes a history of warfare. Historically, the engineering curriculum at the École Polytechnique has been substantially more mathematical than in British polytechnics. American corporations that hire engineers now want engineers to be taught economics and entrepreneurship.

Prescinding from education and the vague "way of doing things," the moral ideal of use and convenience and its history is central to understanding the engineering profession. As suggested, the original English-speaking engineer interpretation presumed that use and convenience was best pursued by capitalism. This effectively transferred the military way of doing things, where obedience to authority played a prominent role, into the civilian sector, as loyalty to a corporate employer or (in cases where engineers incorporated their own consulting firms) engineering clients. This had the effect of making engineers accessories for many of the negative social effects of the Industrial Revolution to which they made prominent if often not fully appreciated contributions.

In the United Kingdom, the problems engendered by capitalist engineering were addressed primarily by political reform through the emergence of a strong labor movement and ideals of socialism. One contributor to this development was Robert Owen (1771–1858), a self-made industrialist and proto-engineer. Disturbed by the way workers were treated in his own textile mills, Owen set out to create socialist alternative cooperatives in which use and convenience was oriented more directly to worker welfare. His engineering-like design of utopian socialist communities in both England and America failed to be successful as practical endeavors (in part because of corporate power opposition) but nevertheless helped inspire important political reforms especially in the UK.

In the United States, by contrast, the moral ideal of engineering became subject to extended internal debate, as analyzed by Edwin Layton's now classic *The Revolt of the Engineers: Social Responsibility and the American Engineering Profession* (1971a). There Layton highlighted the tension between the moral ideals of the engineer as both applied scientists and business persons. In his words, "Engineering is a scientific profession, yet the test of the engineer's work lies not in the laboratory, but in the marketplace" (Layton 1971a: 1).

In this social context, a number of historical tensions developed. One was reflected in the thought of social scientist Thorstein Veblen. His *Engineers and the Price System* (1921) contrasted the engineering notion of technical efficiency as a moral ideal with the capitalist ideal of economic or market efficiency as, in effect, alternative interpretations of use and convenience. Indeed, there has been within the engineering profession a persistent attraction to a moral-political ideal of technocracy in which engineers and engineering expertise would play a prominent role in the state. For example, in late-20th- and early-21st-century China, engineers came to play leading roles in government (see Kirby 2000; Liu 2016).

In an ostensibly democratic social context, however, a technocratic moral ideal was at a philosophical disadvantage. Instead, over the course of the 20th century in the American engineering community, the explicit articulation of moral ideals socially evolved from explicit commitments to company loyalty to holding paramount obligations to protect public safety, health, and welfare, even though this ideal is not as integral to the profession as, for example, the ideal of health is to medicine (Mitcham 2020, ch. 9; for some comparison with other countries, see ch. 12).

1.6. Engineering as Modernity

Looking back on and extending his own earlier analysis, Layton too has noticed a persistent American engineering uneasiness with use and convenience.

> Some of the worst horrors of the Industrial Revolution, after all, took place in factories, mines, and mills that were useful and convenient in some sense. Trusts and monopolies similarly might be considered useful and convenient for some, but most Americans regarded them as evil. Beginning in the late 19th century, American engineers were increasingly beset with a feeling of guilt and anxiety about the negative social effects of technology—for whose accomplishments, up to that time, they had been taking full credit. . . . American engineers reacted . . . by insisting that its true end is the good of humanity. This involved redefining the goals of engineering. Tredgold's "use and convenience of man" no longer seemed strong enough, and many engineers sought to raise the sights of engineering to a stronger notion of social benefit and moral improvement.
>
> *(Layton 1989: 60–61)*

The Revolt of the Engineers had examined the effort within the engineering community to construct a professional self-understanding as obligated to protect public safety, health, and welfare in a way that would give engineers sufficient autonomy to resist excessive control by distorting commercial interests. Now almost two decades later, Layton wanted to examine the ways engineers have appealed to the concept of engineering as applied science as an external ideological justification addressed to society at large.

The engineering as applied science ideology had a twofold social function. On the one hand, this "ideological portrait of engineering and engineers as agents of progress through science" raised the social status of engineers. It appropriated for engineering the good name of science as the source of modern progress. On the other, scientists did not object, and they readily adopted the same ideology as a means to justify the value of science and persuade the public to fund scientific research. As the linear model of Vannevar Bush (1945) had it, basic science creates a reservoir of knowledge from which engineers and entrepreneurs can draw. In the course of critically analyzing the inadequacy of this interpretation, Layton draws on other important studies in the history of technology that argue for a more subtle interaction of scientific and engineering (or technological) knowledge and reviews a history of alternative definitions of engineering as design or systems construction and management.

One alternative definition to which he gives extended consideration is the argument of Martin Heidegger ([1954] 1977) for the essence of technology as a kind of truth or revealing, insofar as it might reveal something about engineering. Because of his rejection of any strictly instrumental conception of technology, Heidegger provides a "radical challenge to Tredgold's definition" (Layton 1989: 64). Without fully endorsing Heidegger's concept of modern technology as a "challenging" of nature, Layton argues that there is considerable resonance with the historical reality created by engineering:

> We are challenged by the long term, unexpected, and often undesirable effects of rapid technological change in the modern era. The things that challenge us are the erosion of cherished values and institutions, the reduction of humans to mere cogs in giant social machines, environmental pollution, the psychological and moral dangers of modern mass media, . . . and the excessive concentration of power and wealth in a few hands. We are particularly challenged by threats to safety and even the continuation of the human species implicit in modern weapons systems, or by unforeseen dangers to the environment and to humans arising from technological activities.
>
> *(Layton 1989: 65)*

Moreover, the way Heidegger turns around the science-engineering relationship—describing science as theoretical engineering instead of engineering as applied science—points toward the historical ways in which "technology created the intellectual atmosphere out of which physics [as the archetypical modern science] arose" (p. 66). Layton suggests that engineers archetypically function as those who implement what Heidegger calls the *Gestell* or enframing of the world as *Bestand* or resource.

One strength of this radical description is that it places engineering squarely at the center of modernity and thus rightly corrects more common beliefs that the modern world is defined by, for example, science or democracy. It is instead defined by a kind of science founded in engineering which is also the source of industrial technology. The world in which humans now live is an engineered and continuously engineering world.

There are, however, two glaring weaknesses in this conception of engineering as the apotheosis of modernity. First, at least in the version attributed to Heidegger, it highlights a dark side of techno-scientific progress at the expense of appreciating its promise and brightness. Agreeing with the idea that engineering is the defining feature of the modern historical period but deeply disagreeing with an interpretation that stresses disruptive consequences, at the turn of the century the U.S. National Academy of Engineering established a project to identify and publicize how 20 major engineering achievements from electrification to high-performance materials have positively transformed the 20th century. In his foreword to a collaborative coffee table book promotion of this view, engineer Neil Armstrong, the first human to walk on the Moon, summarized the thesis: A century of engineering expansions, from civil and mechanical, through chemical and metallurgical, to electrical, automotive, electronic, and aeronautical, "deepened with the development of new methods, powerful computational tools, and dependable testing techniques," has turned the "explanations and understandings [of science] into new or improved machines, technologies, and processes—to bring reality to ideas and to provide solutions to societal needs" without which "our world would be a very different and less hospitable place" (Constable and Somerville 2003: vi, vii).

In a gloss on this grand achievements interpretation, the NAE President William Wulf stated an optimistic belief that "2100 will be 'more different' from 2000 than 2000 was from 1900 [and] that the differences will bring further improvements in our quality of life, and . . . be extended to many more of the people of the planet" (Wulf 2000: 7). To assist this happening, Wulf promoted a second project, building on the fact "that the Earth is already a humanly engineered artifact" (Wulf 2000: 9), to identify 14 "grand challenges" for 21st-century engineering that have been endorsed by a number of engineering schools, which also use them as recruiting tools (see engineeringchallenges.org). Reiterating aspirations to create beneficial transformations in the human condition, on the occasion of its 50th anniversary, the NAE issued another publication proclaiming how engineering was "making a world of difference" by "providing ever expanding services to people" (NAE 2014)—reaffirming the aim of use and convenience.

Second, as the contest between dark and bright interpretations of modernity indicates, modernity itself is a contested concept at least as much as engineering. To define engineering as modernity is thus to use one contested phenomenon to explain another. Modernity has been variously understood in terms of cosmology (the decentered universe of Copernicus), theology (Protestant Reformation), new political modes and orders (democracy), secularization (via the Enlightenment criticisms of Western religion), economy (capitalist industrialization), and culture (individualism). The conception of an emergent Anthropocene is another highly engineering-dependent version of modernity with both dark and bright sides. Is any one of these useful or more useful than another in pursuit of a philosophical understanding of engineering?

1.7. The Meaning of Engineering

To consider a possible connection between engineering and modernity deepens the original question. "What is engineering?" can become a question not just about demarcation but also about meaning: not just "What is *the definition* of engineering?" but "What is *the meaning* of engineering?" Beyond the isolating function of definitions, meaning considers relationships (ethical, political, ontological, epistemological, aesthetic) with other aspects of reality. Ultimately, any attempt to circumscribe engineering necessarily enters into a dialectic between what is inside and what is outside; the original question becomes coextensive with the philosophy of engineering as a whole.

What, then, is the meaning of engineering? One interpretation of this question would focus on the meaning it might give to engineers themselves thus reiterating experience within the profession. Samuel Florman (1976), for instance, contends that the practice of engineering brings with it existential pleasure in constructing the world. Another, more general version could ask about its historico-philosophical meaning after the manner of G.W.F. Hegel (1770–1831). What is the meaning of engineering as a central if often overlooked feature of modernity? Is there some reason why engineering as we experience it arose only once at a particular time and place in history and from there has spread into the whole world?

Full engagement with such a Hegelian question is quite beyond the scope of the present chapter. Hegel is rather marginalized in English-speaking analytic philosophy. Nevertheless, as a speculative exercise, consider the following. Robert Pippin, a contemporary interpreter of Hegel, in a provocative reading of modernity, places the rise of subjectivity and autonomy—both the assertion and the questioning of the primacy of individual, personal life—at its epistemological, ethical, political, ontological, and aesthetic core.

> [T]he philosophic idea of liberation and the achievement of autonomy began with and was largely defined by, an attack on scholasticism, religious authority, and feudal power in the name of method, skeptical inquiry and universal, disinterested reason [as with Descartes]. It progressed to a deeper, critical investigation of the "possibility" of such reason itself [with Kant], a kind of question that broadened into a social inquiry and so a critique of social self-consciousness or eventually of ideology [as with Nietzsche et al.].
>
> *(Pippin 1999: 112)*

> [T]he most persistent kind of question continually asked within modernity [is] whether a specific form of collective or individual independence, true self-determination, is possible; and if not, how to understand and state the nature of our "dependence" on tradition, nature, biology, history, in general how to understand our finitude.
>
> *(Pippin 1999: 115)*

According to Pippin's controversial but plausible reading, it was Hegel who most tried "to formulate some view of the internal logic and phenomenology of an 'eternally' self-determining, collective subjectivity, of Spirit" (p. 116)—and whose effort deserves reconsideration in the wake of modernists' dissatisfactions, radical challenges (from Heidegger), or weak defenses (from Jürgen Habermas).

What might also be considered is how the dialectic of independence and dependence in engineering mirrors the problematic of a more general individualist autonomy as well as the necessarily central role of the engineering enterprise in any conceptualization and defense of individualism (e.g., as manifested in declarations of human rights and libertarian economics) as a viable project. How can the human subject possibly realize its absolute worth, or practice even a measured autonomy, in the face of the limitations presented by nature or tradition without the practice of engineering? A social ontology in which individuals are metaphysically prior to groups would seem to depend on replacement of family and bonded community with an engineered shopping mall where producer-consumer exchanges trump personal interrelationships. The meaning of modernity and engineering are intertwined.

Modernism, on Pippin's interpretation, constitutes an ongoing series of "dissatisfactions with the affirmative, normative claims essential to European modernization." Although the individualist, consumer society has become a dominant social ideal, philosophers at least since Rousseau have challenged its rationality and philosophically reproduced its features in efforts to develop their creatively unique personal philosophies. Such is the case even with postmodernism which, although based in an experience "of modernization as a kind of spiritual failure [or] loss," nevertheless repeats many of the "ideas of high culture, criticism, skepticism or enlightenment" on which modernity depends (Pippin 1999: xi). For Pippin, the repetitive character of modernism invites a reconsideration of the Hegelian synthesis of a broadly positive and calming response to criticisms from the initial wave of challenges to modernity. Many of the issues at the core of those challenges and the Hegelian response can be found echoed in the philosophy of technology and engineering in ways that can contribute to critical reflection on the meaning of engineering.

Take the dense German philosophical dissatisfaction with Kant's epistemology: The delimitation of positive knowledge to perceptual appearance (what appears to us as phenomena) while affirming the reality of an unknowable something beyond appearance (noumena) was for a post-Kantian such as Hegel a scandal. Hegel's *Phenomenology of Spirit* is a sustained argument against such a bifurcation. In a parallel although admittedly less philosophically profound manner, the German engineer-philosopher Friedrich Dessauer (1927) also wanted to affirm that humans can know noumena through technological invention. In Dessauer's simplified Platonism, engineer inventors acquire positive knowledge of reality that transcends perceptual appearance through invention by bringing this supernatural reality into physical existence.

Hegel, however, anticipates Dessauer by arguing that something like a social engineering does the same thing by bringing an ideal state into real-world existence. Hegel defends (against its critics) the essential rationality of bourgeois society in a dialectic that finds satisfaction in recognizing, upholding, and working within traditional social role determinations, in a kind of synthesis of the classical heroic ideal with socioeconomic progress. Is it not possible to see in the engineering experience of design creativity within constraints, of the quotidian practice of trade-offs, a concrete manifestation of the dynamic of civil society as a mediation between family and state? Finally, take the effort to see and thereby introduce rationality into public affairs: Is this not precisely what is present in the various versions of social engineering defended by figures as diverse as John Dewey, Karl Popper, Daniel Sarewitz, and Braden Allenby, as well as Andrew Feenberg's neo-Marxist critical theory of technology? Engineering is the great hidden-in-plain-sight of modernity that deserves to be thought through if we are to understand ourselves philosophically.

1.8. Conclusion

Insofar as it is a philosophical question, there is more than one possible answer to the interrogation, "What is engineering?" Philosophically, it is possible to distinguish a number of different types of definitions in response to "what is" questions. One incomplete but to some degree necessary first step is simply to point out the relevant something ("That is engineering") or list (all?) the members of a class ("Engineering is civil engineering, mechanical engineering, electrical engineering, etc."). Such definitions are termed ostensive and extensional, respectively. Closely associated are prescriptive and persuasive definitions: "Let engineering be taken as including all forms of making," as some engineering apologists have argued. Since this is what many engineers themselves think, we probably ought to let engineers themselves have a say about what engineering is. Persuasion can be buttressed by lexical and linguistic definitions that depend on careful analyses of semantic usage.

Efforts to provide more penetrating responses to the whatness question have involved etymological and essential definitions. Genetic definitions aim in the same direction but replace words with phenomena. Essential definitions are classically illustrated in Aristotle's analyses of concepts and in biological taxonomies that attempt to grasp whatness by means of genus (general category) and species (differentia). The ideas of engineering as applied science or some specific form of designing or as a type of profession all exemplify this approach. However, Saul Kripke's argument for "natural kinds" as real but limited to scientific determination (e.g., chemical elements) deprives engineering of any essentialness, although one could still argue in an Aristotelian manner for an essence by analogy.

In conclusion, any answer to the whatness question with regard to engineering will have strengths and weaknesses. There is no knock-down argument in favor or one or the other because definitions always implicate meanings, which are always (in philosophy) subject to argumentative reflection. In the end, the best we can do is seek some degree of reflective equilibrium that takes into account both weaknesses and strengths, minimizing the one while protecting the other.

Related Chapters

Chapter 2: A Brief History of Engineering (Jennifer Karns Alexander)
Chapter 8: Prescriptive Engineering Knowledge (Sjoerd Zwart)
Chapter 9: Engineering as Art and the Art of Engineering (Lara Schrijver)
Chapter 21: Engineering Design (Peter Kroes)
Chapter 43: Professional Codes of Ethics (Michael Davis)

Further Reading

Blockley, David. (2012). *Engineering: A Very Short Introduction.* Oxford: Oxford University Press. (A senior British civil engineer's deft interpretation in terms of history, achievements, failures, and relationships to art, craft, science, and technology.)

Dias, Priyan. (2019). *Philosophy for Engineering: Practice, Context, Ethics, Models, Failure.* Singapore: Springer Nature. (By a Sri Lankan engineer who studied with David Blockley and has done an independent synthesis of much western philosophical reflection.)

Madhavan, Guru. (2015). *Applied Minds: How Engineers Think.* New York: W.W. Norton. (Historical and contemporary cases illustrating successful engineering practice by a biomedical engineer at the U.S. National Academy of Engineering.)

McCarthy, Natasha. (2009). *Engineering: A Beginner's Guide.* Oxford, UK: One World. (A philosopher's sympathetic interpretation.)

Moriarty, Gene. (2008). *The Engineering Project: Its Nature, Ethics, and Promise.* University Park, PA: Pennsylvania State University Press. (An electrical engineer's effort to rethink his profession under the influence of Albert Borgmann's philosophy.)

Petroski, Henry. (1985). *To Engineer Is Human: The Role of Failure in Successful Design.* New York: St. Martin's Press. (Popular and influential explanation for the general public.)

Reynolds, Terry S. (ed.) (1991). *The Engineer in America: A Historical Anthology from Technology and Culture.* Chicago: University of Chicago Press. (Historical analyses of key moments in the American experience.)

Van de Poel, Ibo and David Goldberg (eds.) (2010). *Philosophy and Engineering: An Emerging Agenda.* (A multi-authored broad introduction.)

References

Allibone, Thomas Edward (1980). Dennis Gabor. *Bibliographical Memoirs of Fellows of the Royal Society*, 26, 106–147.

Buchanan, Richard (2009). Thinking about Design: An Historical Perspective. In Anthonie Meijers (ed.), *Handbook of the Philosophy of Science*, Vol. 9: *Philosophy of Technology and Engineering Sciences*. Amsterdam: Elsevier, Part III, pp. 409–453.

Bunge, Mario (1967). *Scientific Research II: The Search for Truth.* New York: Springer.

Bush, Vannevar (1945). *Science: The Endless Frontier.* Washington, DC: United States Government Printing Office.

Constable, George and Somerville, Bob (2003). *A Century of Innovation: Twenty Engineering Achievements that Transformed Our Lives.* Washington, DC: Joseph Henry Press.

Constant, Edward II (1980). *The Origins of the Turbojet Revolution.* Baltimore: Johns Hopkins University Press.

Cross, Nigel (1986). The Development of Design Methodology in Architecture, Urban Planning and Industrial Design. In Robert Trappl (ed.), *Cybernetics and Systems '86: Proceedings of the Eighth European Meeting on Cybernetics and Systems Research.* Dordrecht: Riedel, pp. 173–180.

Davis, Michael (1997). Is There a Profession of Engineering? *Science and Engineering Ethics*, 3(4), 407–428.

Davis, Michael (2009). Defining Engineering from Chicago to Shantou. *Monist*, 92(3), 325–338.

Davis, Michael (2010). Distinguishing Architects from Engineers: A Pilot Study in Differences Between Engineers and Other Technologists. In Ibo Van de Poel and David Goldberg (eds.), *Philosophy and Engineering: An Emerging Agenda*. Dordrecht: Springer, pp. 15–30.

Davis, Michael (2014). Profession and Professionalism. In J. Britt Holbrook and Carl Mitcham (eds.), *Ethics, Science, Technology, and Engineering: A Global Resource*, Vol. 4. Macmillan Reference, pp. 489–493.

Dessauer, Friedrich (1927). *Philosophie der Technik: Das Problem der Realisierung* [*Philosophy of Technique: The Problem of Realization*]. Bonn: F. Cohen.

Ferguson, Eugene S. (1992). *Engineering and the Mind's Eye*. Cambridge, MA: MIT Press.

Florman, Samuel C. (1976). *The Existential Pleasures of Engineering*. New York: St. Martin's Press.

Grayson, Lawrence P. (1993). *The Making of an Engineer: An Illustrated History of Engineering Education in the United States and Canada*. New York: John Willey.

Harms, A.A., Baetz, B.W. and Volti, R.R. (2004). *Engineering in Time: The Systematics of Engineering History and Its Contemporary Context*. London: Imperial College Press.

Heidegger, Martin (1954). Die Frage nach der Technik. In *Vorträge und Aufsätze*. Pfullingen: Neske, pp. 9–40. English trans. William Lovitt. The Question Concerning Technology. In *The Question Concerning Technology and Other Essays*. New York: Harper and Row, 1977, pp. 3–35.

Ingold, Tim (2000). *The Perception of the Environment: Essays in Livelihood, Dwelling and Skill*. London: Routledge.

Ingold, Tim (2013). *Making: Anthropology, Archeology, Art and Architecture*. London: Routledge.

Kirby, William C. (2000). Engineering China: The Origins of the Chinese Developmental State. In Yeh Wen-hsin (ed.), *Becoming Chinese: Passages to Modernity and Beyond*. Berkeley, CA: University of California Press, pp. 137–160.

Koen, Billy Vaughn (2003). *Discussion of the Method: Conducting the Engineer's Approach to Problem Solving*. New York: Oxford University Press.

Kroes, Peter (ed.) (2009). Philosophy of Engineering Design. In Anthonie Meijers (ed.), *Handbook of the Philosophy of Science*, Vol. 9: *Philosophy of Technology and Engineering Sciences*. Amsterdam: Elsevier, Part III, pp. 405–630.

Kuhn, Thomas (1962). *The Structure of Scientific Revolutions*. Chicago: University of Chicago Press.

Layton, Edwin T. Jr. (1971a). *The Revolt of the Engineers: Social Responsibility and the American Engineering Profession*. Cleveland, OH: Case Western Reserve Press.

Layton, Edwin T. Jr. (1971b). Mirror-Image Twins: The Communities of Science and Technology in 19th-Century America. *Technology and Culture*, 12(4), 562–580, October.

Layton, Edwin T. Jr. (1989). A Historical Definition of Engineering. In Paul T. Durbin (ed.), *Critical Perspectives on Nonacademic Science and Engineering*. Bethlehem, PA: Lehigh University Press, pp. 60–79.

Leroi-Gourhan, André (1993). *Gesture and Speech*. Trans. Anna Bostock Berger. Cambridge, MA: MIT Press.

Liu, Yongmou (2016). The Benefits of Technocracy in China. *Issues in Science and Technology*, 33(1), 25–28, Fall.

Mitcham, Carl (1994). *Thinking through Technology: The Path between Engineering and Philosophy*. Chicago: University of Chicago Press.

Mitcham, Carl (2020). *Steps toward a Philosophy of Engineering: Historico-Philosophical and Critical Essays*. London: Rowman Littlefield International.

Mumford, Lewis (1934). *Technics and Civilization*. New York: Harcourt Brace.

National Academy of Engineering (2014). *Making a World of Difference: Engineering Ideas into Reality*. Washington, DC: National Academy of Science.

Pacey, Arnold (1990). *Technology in World Civilization: A Thousand-Year History*. Cambridge, MA: MIT Press.

Pippin, Robert B. (1999). *Modernism as a Philosophical Problem: On the Dissatisfactions of European High Culture*. Malden, MA: Blackwell.

Pitt, Joseph C. (2011). *Doing Philosophy of Technology: Essays in a Pragmatist Spirit*. Dordrecht: Springer.

Rosenberg, Nathan and Edward Steinmueller, W. (2013). Engineering Knowledge. *Industrial and Corporate Change*, 22(5), 1129–1158, October.

Schatzberg, Eric (2018). *Technology: History of a Concept*. Chicago: University of Chicago Press.

Veblen, Thorstein (1921). *Engineers and the Price System*. New York: B.W. Huebsch.

Vincenti, Walter (1990). *What Engineers Know and How They Know It*. Baltimore: Johns Hopkins University Press.

von Kármán, Theodore (1967). *The Wind and Beyond*. Boston: Little Brown and Company.

Wimsatt, William C. (2007). *Re-Engineering Philosophy for Limited Beings: Piecewise Approximations to Reality*. Cambridge, MA: Harvard University Press.

Wulf, William A. (2000). Great Achievements and Grand Challenges. *The Bridge*, 20(3–4), 5–10, Fall–Winter.

Yanagi, Sōetsu (1972). *The Unknown Craftsman: A Japanese Insight into Beauty*. New York: Kodansha International.

2

A BRIEF HISTORY OF ENGINEERING

Jennifer Karns Alexander

Knowledge of the history of engineering is necessary to an understanding of engineering's power and authority in the present world (Alexander 2008). A survey of that history reveals ancient associations between state power and engineering, and between engineering and military might. For many centuries after its earliest appearance in irrigation and walled city defense, engineering centered on projects: discreet constructions that required design and management and brought together the skills and labor of multiple people. Its authority derived from the kings and pharaohs who were wealthy enough, and whose societies were stable enough, for them to deploy it. Added to the project orientation, over time, were traditions of codifying and transmitting technical knowledge, and processes for transforming materials. In the modern era, engineering became a recognizable profession and began to draw its authority from additional sources, such as its own developing professional associations and increasing connections with commerce. In the contemporary era, of which we are part, engineering has proliferated into a bewildering variety of forms, united by an emphasis on technical training in what have become known as the engineering sciences, and in the physical, and increasingly biological and medical, sciences. Although much of engineering remains organized around projects, in the contemporary era, engineering is characterized by systems, of resource extraction and materials creation, and of supply, coordination, and distribution, that overlap and transcend discreet projects. Authority in the contemporary era derives from engineering's claims to expert scientific knowledge and increasingly from its ties to global commerce and the great fortunes generated by information technologies.[1]

This survey of the history of engineering centers on three eras: the project and process era, running from the earliest historical times to the brink of modernity; the modern era, here including the eighteenth and nineteenth centuries, during which engineering took on a professional form; and the contemporary era, from the late nineteenth century into the present, as engineering has proliferated into a nearly unfathomable array of specialties and possibilities.

Although the term "engineering" as we know it developed in the eighteenth century and is thus relatively modern, it carries valences connecting it to earlier traditions. Etymologically, it is connected to engines: devices to generate and control motion, early examples of which are levers, ramps, and pulleys. Engines also include engines of war, the battering rams, catapults, and, later, cannon that were the province of military engineers. It also overlaps with architecture, and many early people identified as architects could as easily be termed engineers: important was the responsibility to build as well as design structures. The term engineering also shares history with notions of the ingenious and of ingenuity, highlighting engineering's connection not only to mechanisms with powerful effects in the world, but to intellectual creativity and prowess. Engineering is thus a broader category than skill, and a more focused category than technology. It requires design and management, in addition to construction. Engineering draws on traditions of both material and intellectual power.

So what is to come? Projects: ziggurats, pyramids, and temples, in ancient Mesopotamia, Egypt, and Mexico, evidence of surprisingly precise construction and vast powers of organizing and administration; roads and bridges and aqueducts and walls, from Rome to Peru to China, that protected and served peoples and their cities; and metals and chemical and incendiary devices, from Arabia and Turkey, combining projects and systems people used to better understand the world and to defend the part of it in which they lived. From the modern era come the great engineering schools of France, and the engineering associations that founded a self-aware profession in Britain, Germany, and the United States, and the spread of both in an era of colonialism. From the contemporary world come chemical and electrical engineering, with roots in the sciences and not in traditional engineering practices, and a host of other engineering fields that followed. From them, too, came the increasing association of engineering with industry and private commerce.

2.1. Engineering, Power, and Coordination

Engineering, in the form of packed-earth dikes and walls of early irrigation and defensive systems, has generally been thought to date to roughly the same time as writing, approximately 3500 BCE. The chronological proximity of engineering to writing is suggestive of its roots in coordinated action; writing originated as an administrative tool, connected to activities of the earliest states. Both engineering and writing arose out of the Neolithic context of epochal change in human affairs, with the domestication of wild creatures as livestock and the beginning of settled agriculture. The remains of pottery found in many excavations in Mesopotamia and the eastern Mediterranean support the argument for the transition from a nomadic lifestyle into settlements; because it is heavy to move and fragile, pottery flourished only after settlement.

Settled societies often left evidence of administrative practices and material tools of coordination. Writing was used first for the keeping of records of things such as the paying of tribute and taxes, and the boundaries of land; it was not first used for philosophical speculation or personal expression. The earliest cuneiform tablets from Sumer are lists: accounts of temple income, drawings of ears of corn, and circular impressions taken to be numerals. Early civilizations were characterized not only by writing and engineering, but by the development of coordinated institutions we still recognize: permanent government; people recognized as citizens; social classes, often including a military class and other classes of specialized skill, such as priests and scribes; and organized bodies of knowledge, passed on within skilled classes and, in time, increasingly committed to writing. The advent of writing distinguishes historical time, for which deliberately created records exist, from prehistoric time.

Most of engineering's history thus falls within historic time, but not all of it. Recent investigations have revealed complex projects that predate writing, and historians have long known of projects built by civilizations that did not develop writing systems. An example is the Budj Bim of Australia, in southwest Victoria, an extensive system of water channels and ponds cut into and alongside basalt lava flows and designed to trap and hold eels until they grew large enough for harvest. Parts of the Budj Bim date back to 4600 BCE, arising both before writing and in a society that did not develop writing, although otherwise experiencing the Neolithic transition into settlements (Jordan 2012). Recent work on the Paracas and Nasca civilizations of what is now Peru suggest an extensive hydraulic system of dug open wells that channeled wind to drive water for settlements and agriculture, dating to the early first millennium CE and developed within a civilization without writing (Lasaponara 2017). South American civilizations are well-known for extensive engineering works in the absence of writing, and several are discussed in this chapter: thousands of miles of Inca highways and the pyramids and temples of Mesoamerica.

Engineering has an intimate connection with discipline, government, and the state. Since ancient times, massive and extended projects have been engineering's most recognizable form, associated with governmental authority, and, because material works of any great size require coordinated effort, such structures evidence ancient connections between engineering and administration. The ancient city wall of Zhengzhou, in the Henan province of central China, is an example. Remains of the rammed earth wall, built in 1600 BCE, still exist; at its greatest it extended seven kilometers and rose ten meters high; estimates are that it required 875,000 cubic meters of earth, and, at the rate of 8 people to lay one cubic meter of earth per day, would have taken a team of 3,500 laborers ten years to build (Shaughnessy 2009: 85). The wall was designed, its site chosen and leveled, and the dimensions and the slope of its sides designated; sources of earth and methods for its transport devised; and forms built to contain each layer as the wall mounted. Laborers were mustered and fed, and supervised at work. Engineering the wall was a feat of design and administration, in addition to construction (on engineering and political practice, see Chapter 45, "Engineering as a Political Practice" by Govert Valkenburg, this volume).

2.2. Pre-professional Engineering

Engineering developed by accretion, as early forms persisted alongside newer processes, methods, and materials, although later periods did not always build on earlier work. Engineering's history is filled with lost methods and processes: the building techniques of the great Egyptian pyramids are still debated, and the secret of hydraulic concrete died with ancient Rome, resuscitated in the eighteenth century by British engineer John Smeaton. What follows is a description of pre-modern engineering, beginning with its earliest forms in land-bound projects and tracing additional forms of engineering that added to its complexity before it became professional in the eighteenth and nineteenth centuries.

2.2.1 Landscape Defining Projects: From 6000 BCE

The earliest engineering projects became features of the landscape and bore close relation to what would in later millennia be called civil engineering. They were structures, cut into or built upon the land and fashioned out of local materials, and archaeological remains exist for many. The very earliest were most likely hydraulic projects, such as the eel-cultivating system of Banj Bim in Australia mentioned above. The forms of engineering discussed here were all further pursued at later times although sometimes by quite different peoples; despite irrigation's antiquity, for example, canal locks were not built until the fourteenth century, yet even at that late date they were part of municipal works and thus connected to authority and government.

Irrigation structures were among the earliest engineering works, built alongside large river systems. Irrigation along the Tigris and Euphrates of Mesopotamia, and the Nile of Egypt, dates to the fourth millennium BCE. The Harappan civilization built systems along the Indus River in the Indus Valley early in the third millennium BCE, and in approximately 2100 BCE, Chinese peoples began building irrigation systems along the Yellow River and the great hydraulic works that would long identify their culture. Irrigation is labor intensive, and in its most ancient form consisted of dikes of rammed earth and reeds, without sluices or gates, in which people dug and refilled openings as they needed them. Systems of any size required dedicated planning, labor, and coordination, and an early form of public works in Mesopotamian cities was the maintaining of nearby marshes to produce reeds needed for local dikes.

Irrigation systems were forms of environmental engineering, and they required both deep knowledge of the character and geography of the rivers and a suite of associated technologies. Basin irrigation systems in ancient Mesopotamia and Egypt were built to aid in both agriculture and flood control, and took advantage of periodic flooding by capturing floodwaters in fields enclosed by dikes and holding it while the water saturated the soil and dropped its valuable sediment. Floodwaters either overtopped the dikes, inundating the fields, or, if canals carried floodwaters alongside a field, ran in or out through specially dug openings in the dikes. Systems depended on long and coordinated observation of seasons and river behavior. In Egypt, for example, the Nile flooded each fall, carrying waters to the dry regions downstream. A series of gauges called Nilometers were cut into the river's rock walls as far upriver as Egyptian power extended, to judge the flood's height and determine its date of arrival. Egyptians also built masonry and stone dams, most famously at Wadi Garawi outside Cairo.

Ancient Egypt also illustrates engineering's association with power. Egypt's system of basin irrigation dates to the earliest of the great Egyptian dynasties, as represented in a bas-relief on a mace head dated to approximately 3100 BCE, depicting the first of the true dynastic kings holding a hoe alongside a canal (Miller 1990, 1991). Although farmers were responsible for maintaining their own dikes, the Egyptian system relied on official surveyors who reestablished boundaries when the packed-earth dikes deteriorated. These surveyors were the fabled cord-stretchers, who also laid out the sites of the pyramids and are depicted in tomb murals wearing the pleated garments of high officials.

The most spectacular of ancient projects were the ziggurats and pyramids that dominated the horizon wherever they were built. In ancient Babylon, the ziggurat Etemenanki rose 300 feet above the plain, faced with enameled brick and the inspiration for the biblical story of the Tower of Babel, where God confused the languages of the laborers so they could no longer complete what was described as humans' attempt to become gods themselves by reaching the heavens. The Great Pyramid Cheops, or Khufu, the largest of all pyramids and set at Giza, west of modern Cairo, was a project of the consolidated power of Old Kingdom Egypt; it stood more than 480 feet high, measured 756 feet along one side of its square foundation, and required the placing of more than 2 million blocks of two-ton-plus stone, raised by the muscle power of thousands of laborers along enormous ramps. The bulk was made from local limestone, supplemented with finer limestone for the face. These ancient projects serve as evidence of design and management on a vast scale. The Great Pyramid required people to design it, survey the site and the stone quarries, build roads and ramps for transporting and placing the stone, and provide for the daily needs of a very large workforce. It was the work of a consolidated and wealthy power.

Landscape-defining projects were undertaken around the globe. The massive earthen Ma'rib dam in present-day Yemen dates from the eighth century BCE, and Ceylon began its irrigation system in 543 BCE. Construction on the Great Wall of China and the Grand Canal began in the third century BCE; the Great Wall and the Great Pyramid were the two largest structures of the ancient world, and the Grand Canal came to run 1,200 miles north to south. The Buddhist stupa of Duthagamini, an enormous domed temple of brick at Anuradhapura in current Sri Lanka, was built in the second century BCE (legend has it that its rubble mix of earth and broken brick was compacted by elephants wearing boots). The great projects of Mesoamerica were built in its Formative and Classic periods, from the first millennium BCE through 900 CE, including the city of Teotihuacan, the Sun Pyramid, and a host of other temples and pyramids, and the dug Paracas irrigation system of what is now Peru. Other such pre-modern projects include the gothic cathedrals of the High Middle Ages in Europe, China's millennia-long struggle that continues today to contain the Hangzhou Bay, and the elaborate multi-starred fortifications of early modern Europe.

These landscape-defining projects illustrate the long-time association of engineering with stable systems of authority and power. The great engineering projects of both Egypt and China can be mapped alongside periods of central stability and power. In Egypt the irrigation system, the great and lesser pyramids, and the famous networks of tombs and temples all date from periods of secure dynastic authority; such projects did not appear during chaotic intermediary periods between the effective dynasties. In China, the great hydraulic works and the canals similarly date from times of strong imperial authority.

2.2.2 *Extending Influence: 1000 BCE+*

Place-bound engineering projects were supplemented after about 1000 BCE by projects for moving across land and water. These projects extended the influence of a powerful center by controlling movement into and away from that center. The roads, bridges, and aqueducts for which ancient Rome became famous are examples of such influence-extending projects, although they were also set-in-place constructions. Aqueducts totaling 260 miles deployed stone and hydraulic concrete to carry water into Rome from surrounding mountains, and thus extended Roman control of the territory surrounding the city. The aqueducts were also charismatic projects, especially the relatively few miles of channels carried above valleys on arcades supported by famous Roman arches. Roman armies also built standardized roads around the Mediterranean, into the Middle East, and north into Britain, and the Inca road system of more than 2,000 miles is an example from pre-Columbian America of engineering projects extending the reach of a powerful and stable empire. Along their roads Romans marched armies and transported war materiel, including catapults, which combined structural or civil with mechanical engineering. The catapult's origins were Greek, and it is attributed to Dionysius in his defense of Syracuse against the Romans; Alexander the Great deployed it to conquer his vast ancient empire, and it was deployed, later and in advanced form, to even greater effect by Imperial Rome. The catapult required both a structural frame, solid enough to withstand strong forces of tension and release, and precision mechanisms, to hold the propelling force of the rope in even tension on both sides, and then smoothly to release the rope in firing a projectile. Roman armies transported catapults long distances in pieces, and reassembled and re-tuned them on the spot as an essential feature of siege craft. In response developed something of an arms race, as people built higher and stronger city walls to withstand ever-stronger catapults (Hacker 1997). Catapults were expensive engineering projects both to build and to maintain. They required expert structural knowledge to design, assemble, and load, as any asymmetry would cause a firing catapult to shake apart or send its missile dangerously off-target.

The engineering of ships carried influence across water. The most celebrated was the trireme of ancient Phoenician and Greek warfare, so-named for having three rows of men at oar, vastly increasing the power of a relatively modestly sized vessel. Triremes were made in quantity, and their design required geometrical analysis of the arrangement of rowers and their oars. In the east, by the turn of the first millennium CE, the Chinese were building and sailing as far as the Indian Ocean in large and compartmented ships able to withstand being holed. Merchant galleys of Venetian shipwrights of the fifteenth century illustrate the growing influence of commerce on engineering with the approach of the modern world, as do the ships of maritime Spain and Portugal that carried Europeans to what became the Americas. The influence of government and crown on the engineering of early exploration, even for commercial purposes, can be seen in Portugal's financing of the voyage of Columbus, much as the British government would later finance the vessels of the British East India Company. Not until railroads in the nineteenth century would mechanisms of overland travel take on the characteristics of engineering that marked ship design and construction.

Early processes of transmitting technical knowledge can also be described as extending engineering's influence, and historians have identified schools and centers of knowledge, and textbook traditions, from the classical period on. The earliest were tied to military engineering, the first school of siege craft founded in the fourth century BCE by Philip of Macedonia at Thessaly, and a series of textbooks, including treatises on engines of war by both Philo of Byzantium and Hero of Alexandria. Within this tradition lies Archimedes of Syracuse, the revered mathematician whose mechanical writings have not survived, who worked at the great Library of Alexandria, and who gained fame for engineering the defenses of Syracuse against the Roman general Marcellus in the third century BCE. Roman soldiers were said to quiver in fright when one of Archimedes's war engines appeared above Syracuse's city walls.

2.2.3 *Transforming Materials: 1000 CE+*

Efforts to transform materials and processes date back to the ancient world. Iron and steel production, early chemical work, and the development of sophisticated mechanisms may be seen as forerunners of contemporary forms of engineering, although they hold an attenuated connection to the projects and governmental authority that defined the iconic forms of civil, military, and mechanical engineering. Their connection to authority lies in their often having been performed under patronage relationships, and in their positions in large-scale systems of production and advance, placing them within organizations and often under management.

Evidence for connections between iron and authority are stronger than for steel. Iron smelting has been dated back to the second millennium BCE in China and, more recently, the north of what is now Nigeria (Darling 2013). Evidence indicates intensive experimentation in a contained geographic region; early iron was used in both military and agricultural artifacts, and in Nigeria, the working of iron had religious significance. By the fourth century BCE, China was exporting cast iron to Rome and, by the eleventh century, had developed an iron industry centered on smelting households with a variety of official roles in local governments (Arnoux 2004; Wagner 2001). Both Austria and India produced steel by the late first millennium BCE; in the case of India, the famous Wootz steel from which Damascus steel was made. Ingots were carried from India to Damascus to be forged into blades highly prized for distinctive patterns that emerged as the steel was worked. Wootz steel was produced in clay crucibles under high temperatures, and although it required a supporting system of supply of raw materials, fuel, and apparatus, it may well have been produced as a craft rather than engineering tradition. By the early modern era, cast iron had been put to use in cannon in Europe, where it stimulated the development of star-shaped fortifications designed to prevent the new weapons from ever getting a straight shot at a wall.

Two lines of development may be seen as early examples of chemical engineering: alchemy and incendiaries. The tradition of alchemy dates from the ancient world; its object was the study and transformation of materials, particularly metals, by subjecting them to heat and other processes. Alchemy remained influential into the early modern world and was an important court practice under the Muslim caliphate, where its practitioners developed many of the chemical apparatus still in use in laboratories today. Incendiary mixtures date from China at the turn of the second millennium CE, in recipes for mixing saltpeter with charcoal or sulfur, and shortly thereafter they were enclosed in containers to make explosive devices. Incendiary work took on the form of engineering in the legendary Greek fire deployed to protect the Byzantine Empire against Arab attack after the seventh century CE; Greek fire was a liquid incendiary, attributed to the architect and engineer Kallinikos, which had the power of burning when wet or under water. Its composition was a tightly held secret that has been lost, and deploying it required an elaborately designed system.

2.2.4 *Increasing Mechanical Sophistication: 300 BCE+*

Increasingly sophisticated mechanisms have been developed since the earliest uses of ramps and ropes. Ancient Egypt and Mesopotamia did not know the pulley, but by the time of the great school at Alexandria, the concept of mechanical advantage and the simple machines was known: the lever and its incarnations as ramp, pulley, inclined wedge, and screw; and the wheel and axle. Gears appeared in archaeological remains and in writings from Greece, Rome, and China from the third century BCE through the fourth century CE, but not until joined with the precise metal-working of Arab engineers after the seventh century did they become truly functional; the great mechanist al-Jazari made use of them, working in the courts of the Muslim Artuqid dynasty where he produced one of the most important pre-modern treatises of mechanical engineering: *A Book of the Knowledge of Ingenious Mechanical Devices*. Also crucially important was the development of the escapement in China in the eighth century CE; an escapement checks and releases motion, controlling it through a tripping mechanism, and was necessary for development of the mechanical clocks from which the European tradition of mechanisms evolved. The printing press with movable type designed by Gutenberg in the fifteenth century was one of the most significant achievements in mechanisms before the modern era.

2.3. Engineering as a Profession

Engineering became a profession during the eighteenth and nineteenth centuries. Although the term "engineering" had been used to denote military engineering, until the eighteenth and nineteenth centuries there was little recognition of a distinct technical profession for the design and management of projects. Two institutional developments marked engineering's appearance as a self-conscious profession and undergirded a new sense of its authority: specialized institutions of engineering education, and self-governing engineering societies. Academic training affirmed that engineers possessed true expertise in the form of specialized knowledge, and professional societies gave engineers platforms through which to exercise their expert authority, for instance in the development and control of licensure.

Allied with these institutional measures were two additional features of engineering professionalization: the increasing association of engineering practices with the sciences—in some fields but not all—and the increasing alliance of engineering with commerce. The context of these developments was industrialization and developing forms of capitalism, an Enlightenment emphasis on applying human knowledge and sciences, and burgeoning European imperialism. This was also the period of the most significant European influence on engineering, carried to colonies and territories by European imperial expansion and by intellectual networks formed during the Enlightenment.

The most famous institution of engineering education was France's École Polytechnique, founded by the revolutionary government in 1794 as an elite technical preparatory school with a heavily mathematical and theoretical curriculum. It prepared students to learn applied engineering at one of the *Grandes écoles*, such as the military engineering academy at Mézières or the School of Roads and Bridges, both founded earlier in the century. The École Polytechnique was much imitated; the United States' military academy at West Point was modeled on it, as was early engineering education in Egypt under the Napoleonic incursion at the turn of the nineteenth century. The École Polytechnique embodied the association of professional engineering with authority and power (see also Strunz 1970; Olesko 2009).

In its exclusivity, the model of the polished and highly theoretic French polytechnician fails to capture whole vistas of engineering in the professionalization phase. France itself offers other examples, in the technical practice of mechanicians who only later in the nineteenth century took on the title "engineer" following vocational and industrial training for which both public and private institutions were founded in the early nineteenth century. As academic credentials replaced workshop and job-site training, mechanicians also came to master tools that gave them managerial authority, such as mechanical drawing, which provided a new language in which to express mechanical knowledge and, by providing templates against which work could be judged, a new method of oversight and control of mechanical production (Edmonson 1987: 17; Alexander 1999). The *Société des ingénieurs civil*, France's first engineering association, was established in 1848 and required academic credentials of its members. They came to be heavily drawn from mechanical engineering in private industry, thus representing the increasing authority of commerce.

Professionalization in Britain relied on the self-made engineer. John Smeaton coined the term "civil" engineer in the eighteenth century to distinguish his work from military engineering. He founded the first British engineering society, known as the Smeatonians, which was followed by the Institution of Civil Engineers; it began as a dining club to polish the manners of engineers who needed parliamentary approval for projects that crossed county lines. Until the twentieth century, membership in an engineering society was a more common credential for British engineers than formal academic training. Mechanical engineers were similarly self-made; the best example is James Watt, who built his developed practice through observation and workshop experience (Hills 1996). Watt's development of a separate condenser for steam engines, which greatly increased their efficiency, illustrates the significance of design as a necessary component of engineering: he conceived of and patented the device long before he was able to produce it. Watt's business association with the manufacturer Matthew Boulton enabled his commercial success outside the government patronage or project model. The Boulton and Watt connection illustrates the importance of commerce in the development of engineering, even for what we would now consider civil engineering projects. Thomas Telford, first president of the Institution of Civil Engineers, built roads, bridges, and harbors that enabled the transport of goods and materials in industrializing Britain, and the Institution took the occasion of Isambard Kingdom Brunel's 200th birthday, in 2006, to call for more engineers to maintain the infrastructure that underlay Britain's economy.

Large projects were critically important in engineering professionalization, regardless of a country's educational model. In France, railroad engineering brought together mathematically trained engineers from the School of Roads and Bridges and graduates of the newer vocational and technical schools in ways that strengthened the claims of those less elite graduates to use the title "engineer". Railroads were also important for British engineering, although the crucial ones were often in India; British imperialism stimulated an entire "hydraulics of empire", in the building of water delivery systems, docks, harbors, lighthouses, and shipping facilities that served as training grounds for engineering design and management (Cannadine 2004: 171). Smaller industrial projects were also significant, as millwrights designing and erecting water and then steam mills for burgeoning manufactories increasingly came to see themselves as engineers. Important in the United States was construction of the Erie Canal, on which engineers could move up through different levels of design and management responsibility. Independent societies underscored the self-consciousness of a new profession, especially as they came to control professional licensing: the American Society of Civil Engineers was founded in 1852; closely modeled on it was the American Society of Mechanical Engineers, founded in 1880 (on engineering societies, see Chapter 44, "Responsibilities to the Public—Professional Engineering Societies" by Joseph Herkert and Jason Borenstein, this volume).

2.4. Proliferation Within Contemporary Engineering

Contemporary engineering dates from the late nineteenth century to the present. It is defined by new engineering disciplines born out of the sciences, in contrast to the new uses of science in traditional forms of engineering that helped to characterize the period of professionalization (Kline 1995). It is further defined by the authority it derives from commercial success within the consumer marketplace. It is characterized by the proliferation of varied forms of engineering and allied professions; a wider set of relations with systems of government and authority; and increasingly global connections.

Electrical and chemical engineering are the iconic forms of engineering for the contemporary era. Both were rooted in nineteenth-century scientific work: in electrical engineering in work on electricity and signals; in chemical engineering in the chemistry of bleaches and dyes as developed by German industrial chemists. Both have achieved enormous prestige for their contributions to the consumer marketplace, electrical engineering most recently through computer engineering and computer science, and chemical engineering through pharmaceuticals and materials engineering (Duvall and Johnston 2000). The two fields differed in their early relationships to traditional engineering projects. Electrical engineers were involved in projects such as the laying of England's underwater telegraph to India, and in a famous episode from British technological history, it was physicists trained in electrical sciences and not practically trained telegraph engineers who were able successfully to interpret and design against the attenuating effect of water on submarine cable signals.

Traditional project connections were more attenuated in chemical engineering, which was close to the textile and pharmaceutical industries; the far-reaching significance of textiles and chemical engineering can be seen in the twentieth-century pharmaceutical work of Paul Ehrlich, who developed Salvarsan, a targeted syphilis treatment developed through the study of aniline dyes, and the early chemotherapy drug prontosil red, whose name identifies its origins as a dye.

University training in the fields grew quickly, as did professional societies. Darmstadt's University of Technology appointed its first electrical engineering faculty in 1882, and the Massachusetts Institute of Technology shortly thereafter began offering coursework. One of the earliest chemical engineering degrees was offered at MIT, and by the turn of the twentieth century, textbooks and degrees were widely available. Britain's Society of Telegraph Engineers was founded in 1871 by members of the military signal corps and later became the Institution of Electrical Engineers. The American Institute of Electrical Engineers was founded in 1884 and merged in 1963 with the Institute of Radio Engineers to form the Institute of Electrical and Electronics Engineers (IEEE), currently the largest professional society in the world. The American Institute of Chemical Engineers was founded in 1908, in a statement of the field's independence from chemistry, its parent science.

Computer and nuclear engineering are examples of the proliferation of engineering disciplines in the contemporary world. Computer engineering is rooted in programmable digital computers built during the Second World War, and it remained a relatively small field within electrical engineering until the explosion of the consumer market for computing technologies late in the twentieth century. Within computer engineering there are now a proliferation of sub-specialties, in systems engineering, network engineering, security and operations engineering, algorithm engineering, etc.

Nuclear engineering similarly grew out of the Second World War, from the United States' Manhattan Project, and was from the outset a heavily scientific discipline. Postwar naval nuclear programs were the forerunner of nuclear engineering programs at universities, which began to offer nuclear engineering degrees in the 1950s and 1960s. Nuclear engineering, in turn, had close ties to the development of programs in applied physics, which developed in the 1960s and 1970s and provided many faculty for traditional disciplines such as mechanical engineering.

Bioengineering, biomedical engineering, and geoengineering are even more recent examples of the explicit connections between scientific and engineering disciplines that characterize contemporary engineering (Lucena and Schneider 2008). Unlike computer and nuclear engineering, these fields did not emerge to serve new technologies. Instead, they grew from increasing recognition of the overlap between engineering projects and scientific research being carried out in other fields. Ventilation systems have long been at the core of mechanical engineering expertise; combined with work on human physiology and new attention to indoor spaces as environments, ventilation can be seen as an exercise in bioengineering. Biomedical engineering is an outgrowth of medical electronics, biomechanics and the mechanics of motion, and materials engineering. Geoengineering has emerged alongside proliferating evidence of humans' influence on the global environment and has moved from methods of controlling local weather, through cloud-seeding for example, to proposed projects to reshape the global climate.

The contemporary proliferation of forms of engineering developed in the globalizing context of the twentieth century, which in turn had its roots in nineteenth-century imperialism. This context had implications for the historically important relationship between engineering and authority or government, and for the increasing role of formal training as a requirement for entry into the engineering professions (*European Journal of Engineering Education* 2017). The high colonialism that marked the early twentieth century, and which in fact turned a European squabble into the First World War, gave way after the Second World War to a period of postcolonialism, and with it a plethora of new governments and different governmental forms.

Engineering in India and Peru illustrate different trajectories following colonial rule. Under colonialism in India, superintending engineers at the project management level had largely been expatriates recruited from Britain, with indigenous engineers serving at intermediate and local levels. Indian engineers who served in local private industry or as officers of the colonial state helped India develop a technical indigenous identity, and they became important as a technocratic cadre after India achieved independence in 1947. Under colonial rule India had developed both engineering education and professional engineering societies, and so it had in place a modern set of engineering institutions at the time of independence. A continuing link with a more informal engineering past is India's recognition of training through practice, enabling engineers to become associate members in the Institution of Engineers (AMIE) by exam rather than through verification of educational credentials (Ramnath 2013). Peru similarly established engineering education later in the nineteenth century, with the founding of the School of Civil Construction and Mining Engineers (now the National University of Engineering) in 1876 in Lima, designed to support Peru's developing mining industry and closely modeled on Europe (Palma et al. 2012). Dissimilar, however, was Peru's early postcolonial status, for it had established itself as an independent republic in 1821. Such early post-colonial status is an important characteristic of the Latin American regions, and its bearing on the history of engineering has yet to be explored.

China illustrates another structure within which contemporary engineering has developed. Although not under colonial rule, China too established engineering in the late nineteenth century along Western lines, in engineering curricula adopted at Shanghai Nanyang University and the Imperial Chinese Railway College. China underwent considerable governmental upheaval in the early twentieth century, culminating in the 1949 creation of the People's Republic of China under Mao Zedong. The PRC increased the number of universities and technical institutes offering engineering curricula, and since the Cultural Revolution, Chinese engineering education has met both ideological opposition to its creation of a technocratic elite and growing demands for technical support for a nation moving rapidly through industrialization and, in some areas, into post-industrialization (Dong and Liu 2017; Andreas 2009).

Large parts of Africa have seen the more recent establishment of engineering institutions. Morocco founded a School of Engineering in 1959; Zambia, at the University of Zambia, more recently, in 1969, Analysis of engineering practice in Zambia shows that the country continues to import construction and mining equipment rather than designing and producing it domestically, and engineers trained internally remain mid-level maintenance engineers without developmental or supervisory roles. UNESCO reports that access to engineering education remains limited in much of Africa and the ratio of engineers to population quite low; in 2014, one engineer per 6,000 people in Zimbabwe, Namibia, and Tanzania, in contrast to China's ratio of one per 200 (Onyancha et al. 2016; UNESCO 2010, 2014). Exceptions include South Africa and Egypt. South Africa's recent democratic transition and continuing political conservatism pose challenges for remaking engineering education previously built along European lines; in Egypt, French theoretical and mathematical influence is strong, supplemented by practically oriented British models from the period of British occupation in the mid-nineteenth century.

Nationalism has become a force in contemporary engineering, in contrast to the universal models of engineering knowledge espoused in the Enlightened French tradition. Greece developed a distinctive identity for engineering, through engineering societies, educational institutions, and periodicals that emphasized continuity with Greece's ancient engineering past. Projects to improve water supply to Athens and to expand the port of Piraeus were viewed as evidence of the modernity of the Greek nation-state and as extensions of the aqueducts and seafaring of ancient Athens (Tzokas 2017). In Japan in the 1920s, in contrast to its earlier welcoming of western scientific and technological models under the Meiji Restoration, engineers worked to develop a Japanese style of engineering and to establish themselves as technocrats with decision-making power within the Japanese government; a tool of this movement was the early engineering association the Kojin Club (Mizuno 2010).[2]

Engineering's ancient authority remains visible in contemporary projects, many of great scale and scope, with charismatic appeal. Construction of the Panama Canal, a colonial project of the United

States billed as a strategic interest and necessary for trade and commerce, drew observers from around the world. Later in the century, space engineering projects in the United States and Soviet Russia captivated the world and drew on extensive networks of materials and expertise. Nuclear weapons developed during and after World War II drew anxiety as well as admiration; the United States' Manhattan Project was an engineering achievement as well as a work of science, and it required massive sub-projects in industrial and materials engineering. Restoration of the Hanford Nuclear Reservation on the Columbia River in Washington State, the site of essential nuclear materials engineering, is itself a massive engineering project. The Three Gorges Dam on China's Yangtze River, the world's largest hydroelectric dam at nearly 600 feet tall, has drawn worldwide attention and criticism; because of their immense environmental, social, and ecological effects, both upstream and down, large dams remain among the most contested of engineering projects. The former need by projects of such magnitude for official sources of support may be changing, as private enterprise captures ever-increasing resources and develops the will to take on such projects itself.

2.5. Conclusion

In the contemporary era, the power of engineering is no longer as visible as it has been until very recently. People are used to seeing massive engineering projects all around them, and those projects are increasingly evidence of private commercial power in addition to governmental, or public, authority. Furthermore, many of the most significant engineering projects of our times are hardly visible at all; consider the Internet itself, wireless technologies, and the very notion of data stored in a cloud. The history of engineering can help us to recognize engineering's power, and its deep historical alliance with the powers that be.

Related Chapters

Chapter 1: What Is Engineering? (Carl Mitcham)
Chapter 5: What Is Engineering Science? (Sven Ove Hansson)
Chapter 9: Engineering as Art and the Art of Engineering (Lara Schrijver)
Chapter 44: Responsibilities to the Public—Professional Engineering Societies (Joseph Herkert and Jason Borenstein)

Further Reading

Andersen, Casper (2011). *British Engineers and Africa, 1975–1914*. London: Pickering & Chatto. (Engineers as active proponents of British Empire, their workings in London at the center of imperial power, their work in Africa, and their reputations at home.)

Camprubi, Lino (2014). *Engineers and the Making of the Francoist Regime*. Cambridge, MA: MIT Press. (Engineers as active in the development of Francoist Spain. Along with Andersen, another fine work expanding our understanding of engineering in geo-political contexts beyond Britain, France, and the United States.)

de Camp, L. Sprague (1960). *The Ancient Engineers*. New York: Ballantine Books. (A classic survey of ancient technologies, stuffed with excellent descriptions of surprising structures and machines; de Camp argues that civilization is based on engineering. Read it with Finch [1960] for a view of classic engineering history as understood in the 1950s and 1960s.)

Finch, James Kip (1960). *The Story of Engineering*. New York: Anchor Books. (A classic from the former Dean of Columbia's School of Engineering, the evolution of engineering from early craft and architecture, especially good on the last five centuries. See de Camp [1960] for the earlier period.)

Hall, Carl W. (2008). *A Biographical Dictionary of People in Engineering: From the Earliest Records Until 2000*. West Lafayette, IN: Purdue University Press. (A very useful reference of engineers, institutions, and firms.)

Headrick, Daniel R. (2009). *Technology: A World History*. Oxford: Oxford University Press. (Accessible, short survey of technological development around the globe, with good suggestions for further reading.)

Oleson, John Peter (ed.) (2008). *Oxford Handbook of Engineering and Technology in the Ancient World*. New York: Oxford University Press. (An extraordinary collection of essays on all aspects of ancient engineering and technology, with excellent bibliographies.)

Singer, Charles, Holmyard, E.J., Hall, A.R. and Williams, Trevor I. (1954–1958). *A History of Technology*, 5 vols., from early times to the late nineteenth century. New York and London: Oxford University Press. (Exhaustive and technical, the five volumes of Singer are the classic survey of the history of technology. Main categories are materials, methods of production, types of machinery, and processes; engineering and engineers appear throughout. Covers only through the nineteenth century.)

Notes

1. This brief history considers as engineering the building of complex structures and devices requiring management and coordination of resources, labor, and skill. This use of engineering encompasses both ancient projects and contemporary design teams driven to develop and improve consumer goods. This use distinguishes engineering from technology more generally; technology is a larger category and may include tools created and used by individuals, and it refers most specifically to discreet objects, structures, and devices and less to the process of their creation. This use is in contrast to Carl Mitcham's presentation at the 2014 Forum on Philosophy, Engineering, and Technology, "Why Ancient Egyptian Engineering Is Not Engineering".
2. Mizuno, passim.

References

Alexander, Jennifer Karns (1999). The Line Between Potential and Working Machines: Cesar Nicolas Leblanc and Patent Engravings, 1811–1835. *History and Technology*, 15, 175–212.

Alexander, Jennifer Karns (2008). *The Mantra of Efficiency: From Waterwheel to Social Control*. Baltimore: Johns Hopkins.

Andreas, Jack (2009). *Rise of the Red Engineers: The Cultural Revolution and the Origins of China's New Class*. Stanford: Stanford University Press.

Arnoux, Mathieu (2014). European Steel vs. Chinese Cast-Iron: From Technological Change to Social and Political Choices (Fourth Century BC to Eighteenth Century AD). *History of Technology*, 32, 297–312.

Cannadine, David (2004). Engineering History, or the History of Engineering? Re-Writing the Technological Past. *Transactions of the Newcomen Society*, 74, 163–180.

Darling, Patrick (2013). The World's Earliest Iron Smelting? Its Inception, Evolution, and Impact in Northern Nigeria. In Jame Humphris et al. (eds.), *The World of Iron*. London: Archetype Publishing, pp. 158–150.

Dong, Xisong and Liu, Xiwei (2017). A Review of Engineering Education in China: History, Present and Future. Paper presented to the *American Society for Engineering Education*, Columbus, OH, June.

Duvall, Colin and Johnston, Sean F. (2000). *Scaling Up: The Institution of Chemical Engineers and the Rise of a New Profession*. Dordrecht: Kluwer.

Edmonson, James (1987). *From mécanicien to ingénieur: Technical Education and the Machine-building Industry in Nineteenth Century France*. New York: Garland.

European Journal of Engineering Education (2017). Issue, European Models of Engineering Education: Evolution and Challenges, 42, passim.

Hacker, Barton C. (1997). Greek Catapults and Catapult Technology: Science, Technology, and War in the Ancient World. In Terry S. Reynolds et al. (eds.), *Technology and the West: A Historical Anthology from Technology and Culture*. Chicago: University of Chicago Press.

Hills, R.L. (1996). James Watt, Mechanical Engineer. *History of Technology*, 18, 59–79.

Jordan, J.W. (2012). The Engineering of Budj Bim and the Evolution of a Societal Structure in Aboriginal Australia. *Australian Journal of Multi-Disciplinary Engineering*, 9, 63–68.

Kline, Ronald (1995). Construing Technology as 'Applied Science': Public Rhetoric of Scientists and Engineers in the United States, 1880–1945. *Isis*, 86, 194–221.

Lasaponara, Rosa (2017). Puquios, the Nasca Response to Water Shortage. In Lasaponara et al. (eds.), *The Ancient Nasca World: New Insights from Science and Archaeology*. Springer Verlag, pp. 279–327.

Lucena, J. and Schneider, J. (2008). Engineers, Development, and Engineering Education: From National to Sustainable Community Development. *European Journal of Engineering Education*, 33, 247–257.

Miller, Nicholas B. (1990, 1991). The Narmer Macehead and Related Objects. *Journal of the American Research Center in Europe*, 27, 53–59; corr. 28, 223–225.

Mizuno, Hiromi (2010). *Science for the Empire: Scientific Nationalism in Modern Japan*. Stanford: Stanford University Press.

Olesko, Kathryn M. (2009). Geopolitics and Prussian Technical Education in the Late Eighteenth Century. *Actes d'Història de la Ciència I de la Tècnica Nova Època*, 2, 11–44.

Onyancha, Richard, Siame, John, Musando, Kabaso, Sampa Ng'andu, Harrison (2016). *A Review of Engineering Education in Zambia for the 21st Century: Historical, Current and Future Trends Part I*. Conference paper, Engineers Institution of Zambia, Livingstone, Zambia, April.

Palma, Martín, Ríos, Ignacio de los and Guerrero, Dante (2012). Higher Education in Industrial Engineering in Peru: Towards a New Model Based on Skills. *Procedia—Social and Behavioral Sciences*, 46, 1570–1580.

Ramnath, Aparajith (2013). *Engineers in India: Industrialisation, Indianisation, and the State, 1900–1947*. Ph.D. Thesis, Imperial College London.

Shaughnessy, Edward L. (2009). *Exploring the Life, Myth, and Art of Ancient China*. N.p.: Rosen Group.

Strunz, Hugo (1970). *Von Bergakademie zur Technischen Universität Berlin, 1770–1970*. Essen: Glückauf.

Tzokas, Spyros (2017). Greek Engineers, Institutions, Periodicals and Ideology: Late Nineteenth and Early Twentieth Century. *History of Technology*, 33, 157–178.

UNESCO (2014). *Africa Engineering Week 1–5 September 2014*. www.unesco.org/new/en/unesco/events/prizes-and-celebrations/celebrations/international-weeks/unesco-africa-engineering-week/. Accessed May 24, 2018.

United Nations Educational, Scientific, and Cultural Organization (2010). *Engineering: Issues, Challenges, and Opportunities for Development*. Paris: UNESCO.

Wagner, Donald B. (2001). The Administration of the Iron Industry in Eleventh-Century China. *Journal of the Economic and Social History of the Orient*, 44, 175–197.

3

WESTERN PHILOSOPHICAL APPROACHES AND ENGINEERING

Glen Miller

The relationship between philosophical traditions and engineering is complex. At any one time in a single culture, a number of philosophical schools vie for attention—even in ancient Greece, the Neoplatonists, Aristotelians, and Stoics offered different ways to understand the world—and various perspectives arise within each school. These variations multiply over time, and the influence and purchase of different philosophical perspectives ebbs and flows.

Engineering is also multiple: the term can refer to engineering practice, its social institutions, or a way of thinking about the world. The practice of engineering is the design, development, and maintenance of sociotechnical systems for the production of services, systems, and artifacts (collectively, technology), and, for the latter, their decommissioning, guided by mathematics and the physical and engineering sciences (see Chapter 1, "What Is Engineering?" by Carl Mitcham, and Chapter 5, "What Is Engineering Science?" by Sven Ove Hansson, this volume). Engineering practice today normally takes place as part of a technically challenging project that is part of a collective or joint venture involving a myriad of agents. The practice is explicitly defined through the creation and publication of standards, codes, and best-practices; it is implicitly determined by customs and traditions that are passed along to new engineers as they learn how to contribute to projects (see Chapter 54, "Engineering Practice From the Perspective of Methodical Constructivism and Culturalism" by Michael Funk and Albrecht Fritzsche, this volume). Engineers are often asked to assess uncertainty and manage risk, weigh various values, and take into account the interests of a variety of stakeholders affected, either directly or indirectly, by their projects. The social institutions of engineering are shaped by the roles that engineers occupy within their organizations and the socioeconomic structures in which they work, characteristics of the educational system that produces these engineers, and, in some cultures, their self-organization in professional organizations. More generally, engineering can be thought of as a way of thinking or a way of life, i.e., one in which human activity is always a form of problem-solving, especially as it employs quantitative methods and technological innovations. Engineering, in each of these forms, also advances over time.

Engineering practice and its social institutions, much like practice and social institutions in general, are products of a myriad of influences, including historical traditions, education, ethical beliefs, politics, religion, available technologies, and media. They have been in dialogue, as it were, with influential philosophical ideas that have varied between cultures, and even within one culture, over time.

While engineering is not yoked to one philosophical school, contemporary engineering practice is closely connected with modern technology and modern science, on which it depends and to which it contributes (see Chapter 5, "What Is Engineering Science?" by Sven Ove Hansson, and Chapter 6, "Scientific Methodology in the Engineering Sciences" by Mieke Boon, this volume). Modern technology and modern science provide common practical and epistemic elements regardless of where practice takes place today, though assessments of their importance and fundamental nature vary. At the extreme, in the 1950s and 1960s, Martin Heidegger (1993) finds these elements determinant in his influential analysis of the reductive nature of modern technology, which draws a direct lineage from ancient Greek foundations to its present culmination. More commonly, analyses of engineering practice show that much of it is sensitive to local values, though modern technology and science offer a critical inflection point in the historical development of artisanship. In spite of more than two thousand years of Greek and Roman influences, the technology developed from engineering practices, its adoption, and its cultural impact varies across the "West" in their development and their acceptance. Differences are further brought into relief in contrast with non-Western cultures, where the development of modern engineering can be traced to an infusion of modern science and technology into traditional artisanship and epistemic practices. As an example, based on a historical, anthropological, and linguistic analyses of this transformation in China, Qin Zhu (2010) shows that important residual elements of earlier technological practices and an orientation toward social benefit differentiate Chinese engineering from its Western permutations, which he sees as more closely yoked to its military heritage and capitalist goals (see also Chapter 4, "Eastern Philosophical Approaches and Engineering" by Glen Miller et al., this volume).

Perhaps these complexities have been impediments to the development of a more thorough understanding of engineering and various philosophical traditions. Technology has been called "the path between philosophy and engineering" (Mitcham 1994), yet the path has rarely been fully traversed. The voluminous literature about philosophy and technology rarely extends to engineering; most philosophical investigations of engineering are narrow, often emphasizing technical know-how, processes, and decision-making techniques, at the expense of considering its place in broader philosophical traditions.

In this chapter, I sketch out some connections between engineering and several well-known Western philosophical traditions. My approach overlaps slightly with Carl Mitcham and Robert Mackey's (2009) foray into the space: they briefly describe six different approaches to philosophy of engineering—phenomenological philosophy, postmodernist philosophy, analytic philosophy, linguistic philosophy, pragmatist philosophy, and Thomist philosophy—before focusing on the linguistic approach. I proceed in roughly chronological order starting with Aristotelian philosophy and the ethical framework developed from it, virtue ethics. Next I turn to Immanuel Kant's systematic philosophy, which yields important epistemological insights and a different approach to ethics based on duty, which I supplement with the *prima facie* duties of W.D. Ross. Then I consider consequentialist theories that are more empirically oriented, especially the utilitarian theories of Jeremy Bentham and John Stuart Mill, along with the proto-utilitarianism of David Hume. At the end of the section, I turn to the pragmatist tradition, which has many familial resemblances with utilitarianism and an engineering view of the world.

3.1. Aristotelian Philosophy and Virtue Ethics

Using an Aristotelian framework to analyze contemporary engineering indulges in an anachronism: the natures of craft and knowledge known to the ancient Greek and medieval philosophers differ significantly from what is taught at engineering schools today. Both precede the development of modern experimental science, which takes its form in René Descartes's method, described in his 1637 book *Discourse on Method* (2007); Francis Bacon's instrument-based epistemology, described at length in *The New Organon* (2000), originally published in 1620; and modern mathematics, including calculus and differential equations. The elements of modern technology that Hans Jonas (1982) thought made it a suitable topic for ethics—its transformation of human activity, the subject of ethics, in kind, not just in degree, by working at a more fundamental level, with dramatically increased power, with effects that last far longer, and with a much higher level of uncertainty—were not present in earlier periods. Moreover, the influence of Aristotelian and related philosophical schools waned in philosophical circles during the period in which modern engineering and its societies came into being.

Yet important aspects of Western culture, technology, and engineering can be traced to the ancient Greeks. The Scholastic influence, most importantly Thomas Aquinas (1225–1274), provided a ground for cultural development that can integrate religious revelation with Platonic and Aristotelian thought, and its influence is shot through Western culture today. More recently, the renewed interest in virtue ethics in philosophical circles beginning in the 1960s spread to professional ethics, especially bioethics, whose problems arose largely because of technical interventions, and recently has been extended to engineering ethics.[1] C. E. (Ed) Harris (2008) has argued that virtue ethics provides a framework in which dispositions toward nature, attunement toward risk and the social context of technological change, and commitment to the public good can more easily be expressed; it also emphasizes the discretion, judgment, and commitment that recede into the background in other theories. Along similar lines, Jon A. Schmidt (2014) has argued that that ethics in engineering is expressed in the accepted practices and common habits of engineers, which aligns with a virtue ethics framework.

On account of its comprehensiveness and its foundational influence on later philosophers, Aristotle's philosophical system, which includes ontology, epistemology, politics, logic, and ethics, is the best starting point to investigate engineering, and, of these areas, the most fitting departure point is ethics, understood as the study of free human behavior. Working off a definition of virtue (*arete*) as whatever makes a thing an excellent specimen of its kind, Aristotle (384–322 BCE) offered a new definition for the term *eudaimonia* (often translated as happiness or flourishing) as the end that all humans ought to seek by their nature. *Eudaimonia* is being in accordance with articulate speech and reason (*logos*), i.e., acting in accordance with virtue (Aristotle 2002). Human virtues can be divided into two categories. Intellectual virtues (or virtues of thought) are gained through instruction and experience, whereas moral virtues (or virtues of character) are gained through repeated and intentional action that "ingrains" particular habits into our being. A virtue is a mean between deficiency and excess. For intellectual virtues, one can err in excess by affirming truths that are not present and by deficiency by denying truths, i.e., through incoherence between what is and what one thinks. For moral virtues the error is found in disposition and act, e.g., the deficiency of courage is cowardice and its excess is rashness, although some dispositions, such as envy, and some actions, such as murder, are always vicious. While all humans by their nature should seek the moral virtues, the way they express themselves differ for individuals, and individuals have tendencies to err toward one extreme or the other.

The intellectual virtue most directly related to the practice of engineering is *techne*. It is craft know-how or knowledge of the art of making, a stable capability that a person can possess over and above particular understanding of how to make a specific object. Those who hold it know how to best bring about a desired outcome, which depends on an appropriate assessment of contingent conditions. *Techne* operates on contingent beings, especially the material stuff of the world: in this way, it differs from other virtues, because its activity results in the modification of the world outside the agent, whereas the other virtues primarily result in a modification or a shaping of the agents themselves. Aristotle considered *techne* a productive art, subordinate to higher ends, which was ultimately political, i.e., having to do with the welfare of the *polis*.[2]

Like *techne*, the intellectual virtue *phronesis* (prudence, practical judgment, or sometimes practical wisdom) is also concerned with contingent beings, though in this case the virtue is properly assessing a particular situation to determine the proper or good (i.e., beautiful, useful, and pleasant) act for a virtuous agent. It is called the queen of the virtues because all of the virtues depend on it. It serves as a bridge between virtues of thought and virtues of character. As one of the cardinal virtues, it directs courage, the ability to work against one's fear; temperance, the ability to moderate one's inclination or desires; and justice, the fundamental political virtue that is most simply defined as giving each his or her due.[3]

The remaining intellectual virtues have as their object that which cannot be otherwise; they are purely intellectual. *Episteme* (intelligence) is the process of deductive reasoning, which can be hypothetical, from principles to necessary conclusions, an act that is often connected to the kind of reasoning found in sciences. It is always employed in *techne*.[4] *Nous* (understanding) is the grasping, immediately or inductively, of fundamental truths or principles.[5] *Sophia* (wisdom) is understood as *episteme* and *nous* ordered toward their highest and best ends.

The virtues align with many aspects of the practice of engineering. The physical processes of engineering, which operate over contingent objects, are connected most closely with *techne*, though the dependence of modern engineering on the natural and engineering sciences, on mathematics, and on abstract reasoning, in particular in the processes of modeling, emphasizes its interactive relationship with *episteme*. The actions of an engineer in bringing about these processes depends on *phronesis*, which includes cleverness, or reasoning well from ends to means. These three virtues come together in engineering judgment, which includes self-assessment of competence, risk assessment and management, and how to appropriately respond to a myriad of concerns and interests. *Phronesis* allows an engineer to recognize conflicts of interest, and the virtue of justice is necessary to respond appropriately in these situations. Working well within the complex organizations in which engineering is usually practiced frequently depends on the moral virtues, where fear can prevent one from speaking out against unfair or dangerous conditions, and one's self-interest must be balanced against the interests of other stakeholders, proper aims of the organization, and the engineering profession. For the secular philosopher-engineer, the final goal of the processes, the individual engineer and the profession, and the employing organizations is social *eudaimonia*.

In addition to being a virtue, *techne* is also important in Aristotle's study of material beings (Aristotle 1999). It refers to the principle of beings whose formal, efficient, and final causes depend at least in part on an external agent. It is described in contrast to the principle of *physis* (nature), for which these causes are inherent to the being. A natural being such as an acorn does not depend on an external agent to take its form, in contrast to a wooden bed frame, which, if it were to sprout, would yield a tree—the bed frame comes to be through the crafter. To use another of Aristotle's examples, a medical doctor whose body automatically heals itself operates according to *physis*, whereas the doctor prescribing a medicine is an act of *techne*. The principle of engineered objects is *techne*. For premoderns, nature offered a normative orientation toward harmony and proportionality, though sometimes human participation made possible a perfection. While Heidegger traced the perniciousness of modern technology back to Aristotle's *techne*, Aristotle did not think of it as *the* way of thinking. He subordinated practical concerns, which were ephemeral, to intellectual virtues, especially contemplation (*theoria*).

Aristotle offers a few resources that can be used to understand social institutions, which can be applied to the professional societies of engineering. Professional societies often aim to provide the three kinds of friendships that Aristotle differentiated. Social activities of the society aim to provide some pleasure for their members. The connections forged in them regularly offer mutual professional benefit. Societies seem to flourish, though, when their members have a friendship of the highest kind, which arises between virtuous individuals and that inspires further development of intellectual and moral virtue. This form of friendship results in camaraderie between members; its emphasis on virtue and social eudaimonia also directs these societies toward the common good. The orientation toward the common good is especially important for engineers, whose work affects all people, and most engineering societies explicitly mention human welfare or well-being in their codes of ethics.

Aristotelian systems offer a robust framework for thinking about engineering. They have philosophical breadth, give ethical concerns an expansive scope that includes intellectual development and judgment, and take contingencies found in a world that is continually remade, in part by engineering, seriously. Yet they can be criticized for, among other weaknesses, (a) being overly relative, so, e.g., the virtue of courage does little to guide action because it differs among individuals; (b) placing too much burden on the agent's judgment, who must determine the beautiful act in each situation without recourse to rules or methods; and (c) retaining residual elements based on a pre-modern view of the world separated by two thousand years from the origins of modern engineering, ranging from remnants of geocentrism to his flawed understanding of the capabilities of women. These issues are addressed, at least to some degree, by more recent philosophers.

3.2. Immanuel Kant's Philosophy and Duty Ethics

Immanuel Kant's focus on necessary and transcendental truths and the limits of reason may appear to make him weakly relevant to engineering practice, which, as outlined in the previous section, deals with material and contingent beings and acts. To be sure, his approach and goals are radically different from those of Aristotelian philosophy, where ethics included the task of sketching, as best as possible but necessarily imperfectly, the elements of human life, including the moral and intellectual practices that one can develop over the course of time, that lead to flourishing. Kant (1724–1804), commonly considered the greatest of the modern philosophers, aimed to purify epistemology and metaphysics of their errors, whether they be excessive dogmatic affirmations or overly skeptical denials, by considering the power or faculty of reason itself, namely, an awareness of its operation, the conditions thereof, and its limitations (Kant 1996). Correspondingly, in ethics, Kant sought to respond to David Hume's claim that reason does—and should—serve as a slave of the passions, i.e., reason has merely an operational and justificatory role in human action (Hume 2007, II.III.3). For Kant, the role of reason in ethics is to determine the universal moral law that reason can discover and to defend the prominence of reason over instinct, emotion, and self-interest (Kant 1993).

Kant's work in metaphysics and epistemology was also provoked by Hume's recognition that cause-and-effect can never be experienced: lacking innate knowledge, all that one can say about previous events is that they happened in a certain way, perhaps according to a certain pattern, in the past, but there is no way to show that the laws of nature *had* to be followed in the past or could be said to be universalizable (Hume 1993 [original 1748], §V Pt I and §VII, Pt II). Hume identified what is called the problem of induction—one can draw either demonstrative conclusions, which cannot extend beyond experiences and what is already known, or probable conclusions, which are provisional. If this problem is real, then the natural sciences, including Newtonian physics, on which much of engineering rests, stand on unsteady ground. Kant argued that the problem is resolved by recognizing the existence of another kind of judgments, which he called synthetic *a priori* judgments, in which the conclusion extends beyond what is already known and what has been experienced. These judgments, which make it possible for mathematical and scientific knowledge to be universalized, can be recognized by reflecting on the necessary conditions for all knowledge.

Kant's systematic explanation provides a grounding for the mathematical, natural, and engineering sciences that distinguish engineering from related acts of craft or problem-solving. It provides a justification for universalizing the results of experimental science. Moreover, his designation of space and time as *a priori* forms of human perception aligns with the Euclidean spatial grid and undergirds Newtonian physics, which are still fundamental for engineering practice in many disciplines. In fact, statics, the study of loads or forces on an object, is the engineering course that nearly all majors have in common. The emphasis on causality, three-dimensional space, and temporal transformations in engineering education and practice today largely correspond with a Kantian view of the world.

The emphasis on the unifying and organizing cognitive faculty of reason also powers Kant's ethical work. Kant is somewhat dismissive of what can be gleaned from empirical studies of free human action, aiming instead for a universal moral law, that, by definition, does not depend on circumstances, contingencies, or uncontrollable consequences of one's action (Kant 2002). Accordingly, the only thing that Kant thinks can be considered good without qualification is a good will, without which no action has moral worth and for which a trade for any other good is mistaken. The will is good when it is aligned with duty. Two intermediaries are necessary for this move from the subjective will to universal duty. First, the will, which is subjective, must express itself in a maxim, a first-person principle of action, such as "I shall make a promise without intending to keep it." This maxim is then tested against the categorical imperative to determine whether the maxim has been crafted by an impure will. The maxim, and so the will, can be seen to be flawed in two ways. One, when it suffers from an internal contradiction, called a violation of strict or perfect duty, which occurs in the example listed earlier, as the concept of promising requires the intent to fulfill the promise. Two, absent internal contradiction, when it cannot be instilled as part of human nature without the will contradicting itself, a violation of what Kant called broad or imperfect duty. For example, Kant offers the cultivation of one's talents: no rational being could wish that his or her faculties remain undeveloped or atrophy.

Kant offers three formulations of the categorical imperative, which are different expressions of one law that applies to all humans on account of their rational natures. The first is universalization, wherein one may act only according to a maxim that can be made a universal law. Kant thought people are tempted to carve out exceptions for themselves, such as making a promise to repay borrowed money without intending to keep the promise. The first formulation shows that this is inconsistent—a promissory note depends on the intent to fulfill the promise—and thus immoral. The second is the humanity formula, whereby a human can never be used as a mere means. The third is the formula of autonomy, whereby one acts according to maxims that could legislate universal law for the community, or, expressed from a slightly different perspective, as the formula of the kingdom of ends, i.e., the recognition that one is united with other rational beings—each of which deserves dignity, an intrinsic value that surpasses his or her instrumental value—through shared and related goals, and that each person must be both sovereign and subject of the universal moral law that reason reveals.

Despite the fact that Kant's moral reasoning ignores contingencies important in ethics—expressly separated from technical rules of skill and counsels of prudence that yield welfare (Kant 1993: Ak. 416–417)—its rational basis and cosmopolitan characteristics have made it important in ethical considerations of engineering and technology. The principle of universalization grounds the ideal of equality, impartiality, and honesty emphasized in many engineering codes of ethics; it also explains why conflicts of interest, situations in which personal interest could reasonably be understood to affect professional decisions, are problematic. The humanity formula provides a grounding for what is called "respect for persons," concern for each individual based on the person's dignity, over and above any instrumental value the person has, which at times can run contrary to societal welfare. While engineering risk is often calculated as the sum of the products of the probability of an event and its consequences, the humanity formula adds another perspective on what makes risk acceptable, which depends on voluntary acceptance, informed consent, and, if appropriate, negotiated compensation. Extending the idea behind the third formulation supports the inclusion of more non-experts in technology assessments and public involvement in engineering projects that will affect them. Lastly, Kant's emphasis on the interaction between the will and reason results in a far simpler moral analysis than one that demands assessing whether an act led to an outcome that was at least as good as any other possible one, which is difficult (or even impossible) to determine, much less to have achieved, in many complex situations encountered in engineering projects.

While Kant's version of duty ethics is the most popular, it is not the only one. Whereas Kant's version derives a limited set of universal moral rules based on the nature of the power of reason, W.D. Ross (2002) understands ethics to be a task of identifying the moral principles that undergird thoughtful moral beliefs, which themselves have been tested over time, and through reflection. His approach, often called intuitionist, yields six *prima facie* duties, i.e., duties that may have moral salience in a particular situation. They are duties to:

- fidelity, to keep promises that one has made, and reparation, to make amends for acts that have caused harm;
- gratitude, based on what others have done for the agent;
- justice, the distribution of happiness based on desert;
- beneficence, to do what is possible to improve the lot of others;
- self-improvement, especially in terms of knowledge and virtue; and
- non-maleficence, i.e., not to cause harm.

Unlike Kantian duties, which are absolute and universal,[6] Ross's duties may or may not hold in a particular situation: it is the responsibility of the moral agent to adjudicate which duties are relevant in a particular situation and to determine the proper course of action, keeping in mind that *prima facie* duties may be in conflict and it might not be possible for an agent to perfectly satisfy all relevant duties. In some cases, imperfect satisfaction of such duties may lead to a moral remainder, i.e., a moral obligation that the agent may need to or try to satisfy in the future.

These duties connect closely with the ethical principles widely adopted by engineers and enshrined into their societal and professional codes of ethics (see Chapter 43, "Professional Codes of Ethics'" by Michael Davis, this volume). To give but a few examples, the United States National Society of Professional Engineers Code of Ethics requires engineers to "hold paramount the health, safety, and welfare of the public," which aligns with non-maleficence, the duty to which Ross gives priority, and secondarily, to beneficence; the duty of fidelity aligns with requirements to be "objective and truthful" in public statements and to avoid deceptive acts. ABET, a U.S.-based accreditor of engineering programs, and the European Network for Accreditation of Engineering Education (ENAEE) both emphasize the importance of lifelong learning, which aligns closely with self-improvement. The duty of gratitude offers a justification for commitment to one's profession, which has developed the public reputation, educational system, and tools that new engineers inherit.

While Ross's approach has been criticized for its unsystematic nature, a lack of a central principle, and its docility toward commonly accepted moral belief, its flexibility and responsiveness to varying material conditions and cultural influences make it, at least at first glance, a reasonable ground for assessing engineering activities and professionalism. Moreover, there are parallels between the Rossian moral agent, who must assess the relevant moral features of a situation and determine a good way forward, and an engineer, who is accustomed to assessing the relevant technical considerations for a particular project and determining which trade-offs to make.

Ross's duty ethics blends a weaker concept of duty developed by Kant with concern about the amount of virtue, knowledge, and pleasure (and their fair distribution, justice) in the world. It stands as a midway point between duty ethics and consequentialist concerns.

3.3. Consequentialism and Pragmatism

Consequentialism refers to a related set of moral theories that all hold that the moral assessment of an act should be based solely on its consequences, not on motives of the actor or any intrinsic features of the act.[7] Consequentialists believe that an act is morally right if and only if there is no other act that would bring about better consequences. Generally, the consequences taken into account include those experienced by anyone directly or indirectly affected by the act at the present and in the future.

According to all consequentialist theories, positive consequences are to be sought, and negative consequences avoided. How consequences are understood, however, varies among theories. Hedonistic consequentialists aim to maximize pleasure and minimize pain; preference-fulfillment consequentialists, as the name suggests, understand an act to be morally right if no other act would result in a better overall state of individual preferences satisfied and fewer left unsatisfied; objective-list consequentialists believe there is a set, common to all individuals, of positive consequences that should be sought and negative consequences that should be avoided. Some consequentialists believe that an act is right if it is expected to bring about the best possible consequences; others believe the assessment should be based on actual consequences brought about. Some believe that the assessment of consequences should be done based on what is expected for each specific act; others believe that the rightness of an act depends on whether it adheres to a set of moral rules that, if widely adopted, would lead to the best overall consequences for everyone.

In most consequentialist theories, the assessment is agent-neutral, i.e., no special concern or priority is given to the agent or his or her close relations. These theories are usually considered utilitarian. Among the others, ethical egoists consider only the consequences experienced by the agent, and some varieties give preferential treatment to the consequences experienced by the agent and those close to him or her.

Consequentialist principles attempt to neutralize the force of custom or dogmatic beliefs in determining right and wrong. Utilitarian principles work against exaggerated self-concern and encourage consideration of the broader impacts of what one does; they have often been cited by social reformers as a justification for considering the welfare of groups lacking political or economic power. Out of the various ethical theories discussed in this chapter, utilitarianism is perhaps easiest to extend to animal ethics, especially in its hedonist form, for it is clear that many animals experience pleasure and pain, and to environmental concerns more broadly, even if only for anthropocentric reasons.

A focus on bringing about the most positive consequences and fewest negative consequences can lead to acts being determined to be right even when positive impacts are small and spread widely and negative effects are experienced by a select group or individual. The need to consider the short- and long-term consequences experienced by all of those affected by the set of possible acts puts a significant, if not impossible, cognitive burden on an agent who is trying to identify the right act, a problem diminished but not solved by applying consequentialist analysis to assess an act after the fact.

A great deal of reasoning in engineering projects and about technological development in general aligns with consequentialist reasoning. The gap between expected outcomes and what actually comes to pass in the adoption of technological innovations mirrors the problem faced by consequentialists. At a societal level, technological adoption is often justified by the promised improvement to overall welfare, even though such innovation may harm certain individuals, especially in the short term. The determination of government regulations for technologies often includes an analysis of utility, especially in the form of a cost-benefit analysis, that takes into account the interests of everyone in society. Cost-benefit analysis has parallels with utilitarian reasoning, although consequentialists do not reduce happiness strictly to economic terms. Cost-benefit analysis is frequently used by corporations to determine which option to pursue based on an assessment of the overall consequences to the company (Hansson 2007), which, considering the corporation as a person and excluding external impacts, is an egoistic approach. Cost-benefit analysis and utility calculations have been connected to engineering since their origins around the 1840s: the practicing French engineer Jules Dupuit is considered the founder of the cost-benefit approach and has been called "the progenitor of public economics, welfare economics, demand analysis, price discrimination, the compensation principle and the theory of economic surplus" (Maneschi 1996: 411). Risk analysis, especially in probabilistic risk assessment, often uses a utilitarian form of reasoning to consider overall outcomes with little or no consideration given to what will happen to a select group of unfortunate individuals and without concern over who those individuals might be, which can result in issues of justice, especially environmental justice (see, e.g., Shrader-Frechette 2002).

More generally, Thomas Tredgold's (1828) early definition of engineering as "the art of directing the great sources of power in nature for the use and convenience of man" has largely been accepted through to the present. The term "use and convenience" is also found under the heading of utility in the writing of Hume, who accepts it as an unproblematic principle for morals (Mitcham and Briggle 2012).[8] Tredgold's definition concludes with a listing of infrastructure improvements for transportation, production, and commerce, aims that align closely with civil and mechanical engineering today. According to a definition from British engineer G.F.C. Rogers, adopted and modified slightly by Walter Vincenti, engineering is "the practice of organizing the design and construction [and operation] of any artifice which transforms the world around us to meet some recognized need" (Vincenti 1990: 6). This definition encompasses the labor of more disciplines of engineering and also includes consumer goods; the focus on "recognized needs" could be understood in consequentialist terms as a satisfaction of desire, an element from an objective list, or pleasure. To the degree that engineering focuses on efficiency as the consequence to be sought and inefficiency an evil to be avoided, it also aligns with consequentialist reasoning. Moreover, Hume's attempt to apply the scientific method to a study of human nature aligns with the scientific approach that engineers apply to technoscientific problems.

The idea of utility is also central in pragmatism. In *Pragmatism*, a book dedicated to John Stuart Mill, William James defined the pragmatic method as "the attitude of looking away from first things, principles, 'categories,' supposed necessities; and of looking towards last things, fruits, consequences, facts" (James 1965: 47). Whereas consequentialist theories focus almost exclusively on ethics, pragmatism gives greater emphasis to epistemic concerns, but not for their own sake. Whereas previous epistemologies emphasized correspondence between one's understanding and what existed, and coherence between the objects of one's understanding, pragmatists added another criterion, namely the usefulness of what one knows. James said, "any idea that will carry us prosperously from any one part of our experience to any other part, linking things satisfactorily, working securely, saving labor; is true for just so much, true in so far forth, true *instrumentally*" (James 1965: 49, italics in original). Importantly, these truths prioritize an experiential account of knowledge, with priority given to scientific knowledge. This account marries the trial-by-error origins of engineering, an approach which still is the foundation of engineering and manifests itself through the evolutionary development of our technology and the engineering practices that create it, with its emphasis on natural and engineering sciences; it shares the orientation toward usefulness and instrumentality found from Tredgold forward; moreover, a pragmatist's acquisition of knowledge is experiential and active, not primarily observational and receptive, i.e., it is a process of learning, not mere description of what is observed.

Similarly, compare fellow pragmatist John Dewey's process of inquiry to the engineering process. Dewey defines inquiry as "the controlled or direct transformation of an indeterminate situation into one that is so determinate in its constituent distinctions and relations as to convert the elements of the original situation into a unified whole" (Dewey 1982: 319–320, italics removed). Inquiry, which holds for the bodily, cultural, and natural realms as well as the mental realms, progresses through five stages. First, one recognizes something problematic (unease, uncertainty, incoherence, etc.) in the current state of one of the realms or between realms. Second, one needs to conceive and express the problem. Third, based on observable facts, relevant solutions are proposed, which is analogous to forming a scientific hypothesis. Fourth, one considers how well the possible proposed solutions work in the four realms, often iteratively moving the reasoner closer to the most relevant solution, which is analogous to experimental tests of a scientific hypothesis. Fifth, the solution is operationalized, i.e., this new state of affairs is tested against other observed facts, especially those that come to light as the agent acts (Dewey 1982: 320–329). This process parallels the engineering design process, where the steps would be gap analysis or opportunity identification; requirements definition; design; modeling and testing; and project execution. Billy Vaughn Koen (2003: 28) describes the engineering method in terms almost identical to Deweyan inquiry: "the engineering method (often called design) is the use of heuristics to cause the best change in an uncertain situation within the available resources." Heuristics, often called rules of thumb, are plausible and useful (but not necessarily true) aids that yield a solution to a problem. Henry Petroski (1985) notes the trial-and-error method of engineering, which aligns with the experimental sense of pragmatism. More recently, William M. Bulleit (2016) has also drawn connections between pragmatism and engineering.

Perhaps the closest connection between pragmatism and engineering comes by way of technology. Dewey came to see all inquiry as technology. In Larry Hickman's words, for Dewey *"every reflective experience is instrumental to further production of meanings, that is, it is technological"* (Hickman 1990: 40–41, italics in original). Mario Bunge (1967) similarly used a broad definition of technology, which includes material, social, conceptual, and systems realms, combined with logical positivism, to develop the pragmatism-technology-engineering connection. Bunge's attempt to find a technoscientific ground for philosophy has been called "perhaps the most comprehensive vision of engineering philosophy of technology" (Mitcham 1994: 38). Economist Herbert Simon uses utility and statistical decision theory, combined with heuristics, computational methods, and context appropriateness, to develop an engineering design methodology that includes engineering, operations, and management, one that can even explain most complex human behaviors as "the search for good designs" (Simon 1996: 138). More recently and from a Chinese perspective, Wang Dazhou has used Dewey to develop an experimental philosophy of engineering that recognizes the importance of human experimentation in shaping and responding to "the co-evolution of humans, artifacts, and engineering practices" (2018: 49).

3.4. Conclusion

Philosophy of engineering is still embryonic, and few investigations into engineering explicitly situate themselves in a philosophical tradition. Analyses of practices, methods, and artifacts often ignore their historico-philosophical context, and they often implicitly adopt pragmatism, the tradition most compatible with an engineering way of thinking. On the one hand, these common approaches avoid errors of dogmatism and the risk of importing obsolete ideas by focusing on the ethical and epistemic processes themselves. On the other hand, they ignore many resources from other philosophical traditions that have been developed as craft knowledge has evolved into contemporary engineering. These resources can be useful in thinking through engineering practice and its social institutions, and they counteract tendencies to think about engineering as a quasi-sufficient way of life, one in which in which the world and the people in it are merely problems to be solved or resources that need to be optimized.

The three traditions investigated in this chapter provide a reasonable composite view of the relationship between philosophical theory and engineering, but one that is of course nowhere near complete. It ignores feminist approaches to engineering (see Chapter 48, "Feminist Engineering and Gender" by Donna Riley, this volume) and Eastern philosophy (see Chapter 4, "Eastern Philosophical Approaches and Engineering" by Glen Miller et al., this volume), and it offers just a cursory mention of religiously influenced philosophies. These other intellectual resources can contribute to a more comprehensive and insightful understanding of engineering, especially in their critiques of engineering practice, its social institutions, and its way of thinking, and in their imaginative potential to address the problems, excesses, and deficiencies in engineering and the technology it produces. Furthermore, recontextualizing engineering and various philosophical traditions offers the advantage of stimulating the development of these traditions through serious consideration of engineering practice, its social organizations, and the sociotechnical systems that it produces and that surround us in what has been called the Anthropocene.

Related Chapters

Chapter 1: What Is Engineering? (Carl Mitcham)
Chapter 4: Eastern Philosophical Approaches to Engineering (Glen Miller, Xiaowei (Tom) Wang, Satya Sundar Sethy, Fujiki Atsushi)

Chapter 5: What Is Engineering Science? (Sven Ove Hansson)

Chapter 6: Scientific Methodology in the Engineering Sciences (Mieke Boon)

Chapter 48: Feminist Engineering and Gender (Donna Riley)

Chapter 53: Engineering and Contemporary Continental Philosophy of Technology (Diane P. Michelfelder)

Chapter 54: Engineering Practice From the Perspective of Methodical Constructivism and Culturalism (Michael Funk and Albrecht Fritzsche)

Further Reading

Michelfelder, D., Newberry, B. and Zhu, Q. (eds.) (2017). *Philosophy and Engineering: Exploring Boundaries, Expanding Connections*. Cham, Switzerland: Springer. (Develops philosophy of engineering as a joint effort of philosophers and engineers.)

Mitcham, C. (1994). *Thinking Through Technology: The Path Between Engineering and Philosophy*. Chicago: University of Chicago Press. (The classic and extensive historico-philosophical investigation into both engineering and philosophy.)

Mitcham, C., Li, B., Newberry, B. and Baicun, Zhang (eds.) (2018). *Philosophy of Engineering, East and West*. Cham, Switzerland: Springer. (An interdisciplinary and international collection.)

Notes

1. G.E.M. (Elizabeth) Anscombe and Etienne Gilson were two of the most prominent virtue ethicists during the early renewal; more recently, Alasdair MacIntyre, Rosalind Hursthouse, and Philippa Foot have been influential in developing the tradition.
2. In Thomistic thought, the political is subordinated to the supernatural, most importantly knowing God and acting according to his will, i.e., being related to Him, and other created beings, in charity or love.
3. The translations of the moral virtues are more adequate; the Greek for courage is *andreia*, temperance *sophrosune*, and justice *dikaiosune*.
4. For an extended discussion on *techne* and *episteme*, see Mitcham (1994: 119–125) and Mitcham and Schatzberg (2009).
5. The relationship between *episteme* and *nous*, described at length in Aydede (1998), is less clear than that between *episteme* and *techne*. For the purpose of this chapter, I treat *nous* as primarily grasping foundational truths without considering how that happens.
6. Self-improvement and beneficence (and non-maleficence) are imperfect Kantian duties, which are always in force but must not always be fulfilled.
7. Jeremy Bentham, John Stuart Mill, and Henry Sidgwick are the most important early utilitarians; more recent expositors include R. M. Hare, Derek Parfit, and Peter Singer.
8. I have developed this point to at some length in "What Ethics Owes Engineering" (Miller 2018).

References

Aristotle (1999). *Metaphysics*. Trans. J. Sachs. Santa Fe, NM: Green Lion Press.

Aristotle (2002). *Nicomachean Ethics*. Trans. J. Sachs. Newburyport, MA: Focus.

Aydede, M. (1998). Aristotle on *Episteme* and *Nous* in the *Posterior Analytics*. *Southern Journal of Philosophy*, 36(1), 15–46.

Bacon, F. (2000). *The New Organon*. Eds. L. Jardine and M. Silverthorne. Cambridge: Cambridge University Press.

Bulleit, W.M. (2016). Pragmatism and Engineering. In D. Michelfelder, B. Newberry, and Q. Zhu (eds.), *Philosophy and Engineering: Exploring Boundaries, Expanding Connections*. Cham, Switzerland: Springer.

Bunge, M. (1967). Toward a Philosophy of Technology. In C. Mitcham and R. Mackey (eds.), *Philosophy and Technology: Readings in the Philosophical Problems of Technology*. New York: Free Press.

Descartes, R. (2007). *Discourse on Method*. Trans. R. Kennington, Newburyport, MA: Focus.

Dewey, J. (1982). *Logic: The Theory of Inquiry*. In H.S. Thayer (ed.), *Pragmatism: The Classic Writings*. Indianapolis, IN: Hackett.

Hansson, S.O. (2007). Philosophical Problems in Cost-Benefit Analysis. *Economics and Philosophy*, 23, 163–183.

Harris, C.E. (2008). The Good Engineer: Giving Virtue Its Due in Engineering Ethics. *Science and Engineering Ethics*, 14(2), 153–164.

Heidegger, M. (1993). The Question Concerning Technology. In D.F. Krell (ed.), *Martin Heidegger: Basic Writings*. London: HarperCollins.

Hickman, Larry (1990). *John Dewey's Pragmatic Technology*. Bloomington, IN: Indiana University Press.

Hume, D. (1993). *An Enquiry Concerning Human Understanding,* 2nd ed. Ed. E. Steinberg. Indianapolis, IN: Hackett.

Hume, D. (2007). *A Treatise of Human Nature*. Eds. D.F. Norton and M.J. Norton. Oxford: Clarendon Press.

James, W. (1965). *Pragmatism: A New Name for Some Old Ways of Thinking*. Cleveland: Meridian Books.

Jonas, H. (1982). Technology as a Subject for Ethics. *Social Research,* 49(4), 891–898.

Kant, I. (1993). *Grounding for the Metaphysics of Morals,* 3rd ed. Trans. J.W. Ellington. Indianapolis, IN: Hackett.

Kant, I. (1996). *Critique of Pure Reason*. Trans. W. S. Pluhar. Indianapolis, IN: Hackett.

Kant, I. (2002). *Critique of Practical Reason*. Trans. W.S. Pluhar. Indianapolis, IN: Hackett.

Koen, B.V. (2003). *Discussion of the Method: Conducting the Engineer's Approach to Problem Solving*. New York: Oxford University Press. As cited by B.V. Koen in The Engineering Method and Its Implications for Scientific, Philosophical, and Universal Methods. *Monist 2009,* 92(3), 357–386.

Maneschi, A. (1996). Jules Dupuit: A Sesquicentennial Tribute to the Founder of Benefit-Cost Analysis. *The European Journal of the History of Economic Thought,* 3(3), 411–432.

Miller, G. (2018). What Ethics Owes Engineering. In Albrecht Fritzsche and Sascha Julian Oks (eds.), *The Future of Engineering: Philosophical Foundations, Ethical Problems and Application Cases*. Dordrecht: Springer.

Mitcham, C. (1994). *Thinking Through Technology: The Path Between Engineering and Philosophy*. Chicago: University of Chicago Press.

Mitcham, C. and Briggle, A. (2012). Theorizing Technology. In P. Brey, A. Briggle, and E. Spence (eds.), *The Good Life in a Technological Age*. New York: Routledge.

Mitcham, C. and Mackey, R. (2009). Comparing Approaches to the Philosophy of Engineering: Including the Linguistic Philosophical Approach. In Ibo Van de Poel and David Goldberg (eds.), *Philosophy and Engineering:. Philosophy of Engineering and Technology,* Vol. 2. Dordrecht: Springer.

Mitcham, C. and Schatzberg, E. (2009). Defining Technology and the Engineering Sciences. In A. Meijers (ed.), *Philosophy of Technology and the Engineering Sciences*. Amsterdam: Elsevier.

Petroski, H. (1985). *To Engineer Is Human: The Role of Failure in Successful Design*. New York: St. Martin's.

Ross, W.D. (2002). *The Right and the Good*. Ed. Philip Stratton-Lake. Oxford: Clarendon Press.

Schmidt, J.A. (2014). Changing the Paradigm for Engineering Ethics. *Science and Engineering Ethics,* 20(4), 985–1010.

Shrader-Frechette, K. (2002). *Environmental Justice: Creating Equality, Reclaiming Democracy*. Oxford: Oxford University Press.

Simon, H.A. (1996). *The Sciences of the Artificial,* 3rd ed. Cambridge: MIT Press.

Tredgold, T. (1828). Description of a Civil Engineer. *Minutes of the Proceedings of the Institution of Civil Engineers,* 2(1827–1835), 20–23, January 4, 1828. Quoted in C. Mitcham and D. Muñoz's *Humanitarian Engineering*. San Rafael, CA: Morgan & Claypool, 2010, p. 3.

Vincenti, W. (1990). *What Engineers Know and How They Know It: Analytical Studies From Aeronautical History*. Baltimore, MD: Johns Hopkins University Press.

Wang, D. (2018). Toward an Experimental Philosophy of Engineering. In C. Mitcham, B. Li, B. Newberry and B. Zhang (eds.), *Philosophy of Engineering, East and West*. Cham, Switzerland: Springer.

Zhu, Q. (2010). Engineering Ethics Studies in China: Dialogue Between Traditionalism and Modernity. *Engineering Studies,* 2(2), 85–107.

4

EASTERN PHILOSOPHICAL APPROACHES AND ENGINEERING

Glen Miller, Xiaowei (Tom) Wang, Satya Sundar Sethy, Fujiki Atsushi

While engineering practice and its related sciences developed organically within the Western philosophical and historical tradition (see also Chapter 3, "Western Philosophical Approaches and Engineering" by Glen Miller, this volume), it arrived somewhat suddenly in the East in spite of its long history of pre-engineering technical achievements and rich cultural histories. To date, connections between the Chinese, Indian, and Japanese philosophical traditions and engineering have developed to varying degrees, with Western philosophical ideas often imported along with technological expertise. In these countries, the development of engineering, closely associated with modern experimental science and technological development, has been closely yoked to political, military, and economic concerns. In China, the development occurred through their alliance with the Soviet Union and their shared Marxist ideals starting after World War II; in India, it arrived through British colonial activities beginning in the 1830s; in Japan, engineering was invited as part of a conscientious opening of society to Western ideals in the late nineteenth century. In these cultures, the introduction of engineering represents a discontinuity with traditional cultural forces and ethical ideals, one that has been addressed differently in each tradition. This chapter provides a philosophical and historical overview of the development of engineering and explores the cultural resources that have been connected to engineering in each country.

4.1. Chinese Philosophical Approaches and Engineering in China

Joseph Needham (Needham and Wang 1954) once asked why modern science did not originate in China, despite its great engineering achievements. "Needham's question" drew a great deal of attention from scholars in China and the West. A second similar question can be asked about why a philosophy or ethics of engineering did not develop alongside impressive Chinese engineering activities such as the Great Wall and the Jing-Hang Great Canal. It was not until the 1980s that several Chinese professors started to study philosophy of engineering, approaching the question from three different perspectives: traditional Chinese cultural resources, Marxism, and Anglo-American philosophy. To address this second question, a brief overview is given of (i) the tradition of engineering in China; (ii) engineering in practice and ethics in contemporary Marxist China; and (iii) the trajectory anticipated for Chinese engineering ethics in the future.

4.1.1 Revival of Chinese Traditions

Whereas in English, engineering is derived from ingenious (*ingeniare*), which means a device or skill to devise, the Chinese word 工程 (工Gong 程Cheng) has a different origin. It is a combination of two distinct characters. Gong originally represented a multifunctional tool, while Cheng meant to transport grain to its destination, which then also was used as a unit of length. Gong Cheng literally means a task that demands careful gauging and calculation.

Beyond the etymological differences, Gong Cheng and engineering have somewhat different foci: the former focuses on the harmony, the later on creating. Even in ancient Chinese geometry, measuring does not merely entail understanding the proportion of parts, but it also has a moral or normative dimension, aiming to give the order and harmony to the artifact, bringing the sacred unity—the *Dao*—into being through technical activities. How this ancient thought helps to render a different conception of engineering needs to be further articulated by contemporary philosophers.

Li Bocong (2002), often regarded as the founding father of the philosophy of engineering in China, made the first comprehensive effort to study philosophy of engineering. In *The Introduction to the Philosophy of Engineering*, he argued for the revival of Chinese conception of the *Dao-Qi* relation to better understand engineering. *Dao* literally means the way, representing the heavenly pattern, translated also as the law of nature (天理, tian li); while the *Qi* (气) means the container, representing the material. Oversimplifying considerably, *Dao* is the heavenly pattern that is imbued in every *Qi*, the artifacts. *Qi* is the instantiation of *Dao*, carries *Dao*, and actualizes it in the real world. *Dao* is constant, dynamic, transcendental, and omnipresent, while *Qi* is instantaneous, inert, empirical, and contained. The idea of *Dao* can be further elaborated as rendering an organic relationship between heaven, the divinity; earth, the carrier; and the human, the observer and the agent. The highest moral end of Confucianism is to achieve the sacred unity of these three beings, to embody *Dao*. According to Li Bocong, the interplay between *Dao* and *Qi* yields a tripartite philosophical system in which science is for knowing, technology is for creation, and engineering is for application. Li Bocong's effort was hailed by another pioneering scholar, Chen Changshu (Changshu and Chenfu 2002), as the most original and ingenious Chinese philosophy of engineering. It also has been developed by many other scholars (Wang Qian 2009; Hui 2016).

Several other scholars have also invoked traditional resources to articulate a Chinese philosophy of technology and engineering. Tom (Xiaowei) Wang (2016) adopts the Confucian notion of conscience (良知, *Liang Zhi*) to propose a reengineering of the way emotions are conveyed in social networks. He notes the influence of liberal values in the way emotions are expressed in mainstream social networks, where they are reduced to "likes" and "dislikes," consequently impoverishing communication. A Confucian approach identifies the presupposed values and introduces needed complements, such as compassion, an other-regarding emotion, which are more important than individual judgments of "like" and "dislike." For Confucians, humans, due to their unique moral status, deserve compassion, an entitlement paired with an obligation to act in a compassionate way. Thus, the presentation of emotions in social networks should be engineered to emphasize compassion rather than judgment of another.

Pak-Hang Wong (2013) suggested a Chinese philosophy of technology can be developed from three notions, the reconstruction of *Dao*, harmony, and personhood. These three notions help to (i) close the gap between what is right and what is good, a distinction commonly emphasized by Western scholars; (ii) shift the central goal of ethical discourse from finding ultimate moral principles to attaining harmonization between technology and man; and (iii) give more weight to the social dimension vis-à-vis understanding humans as atomistic individuals.

Zhang Wei (2018) investigated and reinvigorated the Chinese notion of *Cang Li Yu Qi* (藏礼于器), an ancient engineering idea that seeks to embed social norms into artifacts. To give an example using architecture, the number of houses, size of the rooms, and color of the roof one could have would be strictly regulated according to one's social status.

Yuk Hui (2016) discussed the concept of cosmotechnics in the context of *Dao-Qi* relationship. While technology was viewed by ancient Greek philosophers as something that stands against nature, Hui offers the concept of cosmotechnics, technical activities that helps to unify cosmic and moral orders, to overcome this opposition.

Most recently, Tom Wang (2018) contrasted the biological view of life with the Confucian conception of a sacred unity of humans, heaven, and earth, which results in an obligation to pursue this unity as part of an authentic existence. The unity promises a novel environmental ethic that draws upon engineering ethics.

These attempts all start from a reconsideration of the *Dao-Qi* relationship, the grounding notion of Chinese philosophy whose reconsideration and articulation remains the central task of contemporary Chinese philosophy of engineering. It should be noted that the narrative developed here is one that overlooks the diversity that exists between different schools of Confucianism and the various interpretations of the *Dao-Qi* relation. Moreover, while Confucianism has been considered the most relevant foundation for a Chinese philosophy of technology, Daoism and Buddhism are also among the important influences in Chinese culture.

4.1.2 Marxist Influences

Marxist philosophy has significantly influenced the focus and methodology of philosophy of engineering in China. Marx argued that the central task of philosophy is to change the world, not merely interpret it, and Li Bocong, inheriting this spirit, argued that philosophers in China should shift their focus from ontology and epistemology to philosophy of engineering, which is a philosophy of making and change.

While the Chinese philosophical landscape has not shifted dramatically, several major contributions have been made to investigate the philosophy of technology and engineering present in Marx and Engels's work. In their systematic studies of Marx and Engels, Liu Dachun (2018) and Wang Bolu (2009) discuss at length the concepts of the machine and technology and how they relate to social changes. Liu Dachun and Wang Bolu held that Marx adopted a dialectical view on technology: the machine in particular helps to boost productivity, emancipating people from the toils of labor, while at the same time, in alliance with capital, alienating them from each other. On the one hand, Marxism encourages people to engineer nature to make it more appealing to inhabit, as it is a proper object of human agency; on the other hand, relying too much on machine-based engineering deprives humans of their skills, making them appendages, as it were, to the machines.

Marx-inspired social critique of technology and engineering, such as the work of Herbert Marcuse (2013) and Andrew Feenberg (1991), has been widely discussed in China. Nevertheless, Chinese Marxists, situated in a developing country, have been less critical of engineering than many Western intellectuals, especially the Frankfurt School. Concerns about alienation and technocracy have had limited purchase as China desperately sought to acquire the capacity to survive and thrive by deploying engineering on a grand scale.

The Marxist influence can also be seen in practice. Since 1953, the People's Republic of China, largely influenced by the Chinese Communist Party (CCP), has adopted several five-years plans to promote engineering. The first National People's Congress held in 1954 proposed the first modernization of Chinese industry, agriculture, transportation, and military, commonly called the "Four Modernizations." In 1956, this task was again included in the party constitution enacted by the Eighth Congress of the CCP. Early on, engineering projects were politically oriented and were expected to show the advantages of socialism over capitalism. Later, Deng Xiaoping deemphasized ideological and philosophical discussions about socialism and capitalism, prioritizing practical concerns. Deng held that the central goal of socialism is to advance productivity so that Chinese have happy lives (Liu Dachun 1998, 2001), and science and technology—the fundamentals on which engineering develops—are the primary driving forces for productivity.

The early five-years plan emphasized mining and the construction of power plants and factories, completed with substantial help from the Soviet Union. At that time, China was committed to a self-transformation from an agricultural state to an industrial state so that it would be able to defend itself in potential wars against Western powers. Plans in the next stages aimed toward developing a basic transportation and telecommunication infrastructure in order to make China a unified state. China's vast territory and authoritarian power regime make it an ideal place to practice grand engineering, engineering on a massive scale. Reciprocally, the completion of grand engineering projects and their material impressiveness imbued citizens with a sense of pride, and their function promotes a homogenous Chinese identity.

Even though the goals of engineering have shifted, what underlies the CCP's understanding of engineering has been consistent: it is a technique that can lead to political and economic power. While CCP leaders have traditionally employed a paternalistic approach to engineering projects, more recently they have begun to pay more attention to the accompanying risks, environmental concerns, and the need for transparency and inclusiveness in engineering progress to respect human dignity. With the development of civil society, Chinese engineering practices are facing increasing resistance from its citizens. The not-in-my-backyard (NIMBY) movement found in Western countries has an echo in China. In the coming years, China will also face the challenge of maintaining and renovating its current infrastructure. Against this altered background, the Marxist idea of alienation may attract more attention.

4.1.3 Concluding Thoughts on Philosophy and Engineering in China

Contemporary Western discussions on philosophy of engineering have also caught the attention of Chinese scholars, particularly young intellectuals. Carl Mitcham, who now serves as a distinguished professor in Renmin University of China, has facilitated much of the communication between the East and West. The systematic work of Dutch philosophers, referred to in China as the Dutch School, is highly visible due to the effort made by Chinese scholars trained in the West. In particular, the works of Peter-Paul Verbeek, Peter Kroes, and Pieter Vermaas have been widely introduced in China, with considerable focus on the idea of moralizing technology.

In general, Chinese scholars are less critical than those in the West about concerns related to embedding values in technology and the nudges that result, not because they necessarily trust government and experts, but because they believe that engineering is a force for good. However, some young scholars (Yan Ping et al. 2014; Wang Xiaowei and Yao Yu 2017) argue that the acceptability of morally freighted technology depends on transparent and inclusive processes of engineering and technological development. They argue that a diverse group of participants, including engineering, ethicists, and even citizen representatives, should all have a say in the design stage. Along these lines, the idea of Responsible Research and Innovation (RRI) has become a catchword within the Chinese academy (e.g., Liu Zhanxiong 2015; Mei Liang 2018) (see also Chapter 20, "Towards an Ontology of Innovation" by Vincent Blok, this volume).

In addition to efforts directed at addressing philosophical concerns, China is also devoting resources to integrate ethics into engineering practice through education. More than 90% of Chinese universities offer bachelor's degrees in engineering, and 10.4 million students, 38% of all university students, are studying engineering (Yang Bin et al. 2017). In 2007, the first national conference on engineering ethics was held, led by Tsinghua University and Zhejiang University. Since 2010, engineering ethics has gradually become the red-hot topic. Many textbooks on engineering ethics have been published, including a recent influential one written by faculty at Tsinghua University (Li Zhengfeng et al. 2016), and it is likely that courses in engineering ethics will be offered more widely and be required at more universities. While the textbooks are mostly translations or reassemblages of Western books, attempts to incorporate Confucian ideas are underway.

As more and more engineering is done in China, the number of ethical controversies will likely grow, and non-governmental organizations (NGOs) with expertise in engineering ethics, especially those from countries who have experience with similar development trajectories, will likely be able to provide valuable insights.

4.2. Indian Philosophical Approaches and Engineering

Indian philosophy consists of many traditions that originated in the Indian subcontinent, including Hinduism, Jainism, and Buddhism. The Vedas (*Rig Veda, Sama Veda, Yajur Veda*, and *Atharva Veda*) are believed to be the original source of Indian philosophy. They are the repository of Indian philosophical wisdom and cultural history, including depictions of the oldest monuments of humanity and descriptions of many technological innovations that are predecessors to engineering.

Some of the technological innovations described in the Vedas include the construction of wells (*avata*) and canals (*kulya*) described in the *Rig Veda*; canals and dams (*sarasi*) in the *Yajur Veda* (Bhattacharyya 2011: 5); canals from rivers in the *Atharva Veda* (Sarava 1954); and deep wells in the *Sama Veda*. In *Brihaspati* and *Vishnu-Puran*, construction and repair of tanks and dams are considered pious works (Sharma 2017: 13). The Vedas were also attuned to social and military dimensions of technology, as the following examples show. In the *Manusmruti*, the sage Manu wrote that a person causing a breach in the dam deserved punishment of death. In *Arthasastra*, Kautilya mentioned that during a war, an enemy's dam should be broken to flood their land and disrupt their transport system (Bhattacharyya 1998). He argued that the agricultural system should be engineered to generate state revenue by preserving dam water the year round for varieties of crop cultivation, and prescribed fines for misuse of water from dams and failure to renovate and maintain water reservoirs (Agarwal and Narain 1997). According to Chakroborty and Das (2003), there is evidence of a precursor to engineering in irrigation, transport, architecture, and other areas in the Vedic period.

Temples, including the Minakshi, Jagannath, Kashi Visveshwara, and Kamakhya temples built in the Vedic period, are of both religious and technical significance. Both aspects were taken into account during a holistic design process that considered planning, design, construction, and maintenance; that so many are still used today make these *vastus* "structural engineering marvels" (Abhyankar 2017). *Vastu* means a dwelling and its surrounding land where the functions of various parts of the structure are integrated and harmony with nature maintained, often utilizing geometric patterns (*yantra*), symmetry, and directional alignments (Kramrisch 1976). During construction of the temples, architectural designers took local cultural traditions, accessibility, and safety concerns into account, while keeping the deity to whom the temple was dedicated the focal point of entire construction.

In India, what we know as modern engineering developed in three phases. During the pre-independence era (1617–1947), engineering was associated with completing civilian projects (e.g., road construction, electrical lines, and surveying lands) according to instructions from authorities. In this period, engineering work is closely linked to the satisfaction of duty (*karma*). During the post-independence era (1948–1990), engineers gave increased importance to protecting public health and protecting the environment while thinking about themselves and their work primarily in terms of virtue, most closely linked with *dharma*. In the contemporary phase, beginning in the early 1990s, engineers increasingly were treated as professionals because they were required to (i) integrate principles in engineering codes of ethics into their professional work to resolve ethical problems and dilemmas and (ii) direct their work toward the public good. They were also expected to achieve the highest standards in their design and production of artifacts that were meant for public use and welfare. The characteristics of engineering in this period have much in common with Jain ideas. The following sections explore each of the three phases in more detail.

4.2.1 Engineering in India in the Pre-Independence Era (1617–1947)

In the pre-independence era, technical works were directed toward attaining societal self-sufficiency. For example, tools were designed for cultivation, pots for collecting and preserving drinking water, and swords for protection from enemies and wild animals. Pre-modern engineers who worked in factories (*karkhanas*), supported by the Mughals and Sultans, received vocational training so that they could produce swords, scissors, fabrics of cotton and silk, etc., for public use and welfare.

After Mughals and Sultans, the British ruled India from the mid-nineteenth century until the mid-twentieth century. The British government sought to survey India for two main reasons: to earn land revenues by levying agrarian property tax on landowners and farmers (Travers 2004), and to determine the internal divisions and boundaries of the territories under their control (Michael 2007). A large number of civil engineers were required to complete the surveys, which led to the establishment of the first civil engineering college in Roorkee (1847), followed by the College of Engineering, Pune (1854); Calcutta Civil Engineering College, Calcutta (1857); and College of Engineering, Guindy, Madras (1858). Engineers were also asked to complete other public works, such as road and canal construction and maintenance.

In the period of British rule, engineers were primarily involved in these civil works and others such as railways, electricity, telecommunication, and irrigation, though they usually did not know why they did what they did in the projects. The measure of their work was how closely they followed the duties assigned by their authorities or higher officials, and they had no responsibility for the outcomes of their works. This mindset aligns with an important precedent in Chapter 18 of the *Bhagavad Gita* in which Krishna told Arjuna that an obligation, in this case to fight a war against one's enemies, must be fulfilled regardless of the consequences.[1]

Toward the end of this phase, Gandhi (1869–1948) made several influential assessments about modern engineering, science, and technology, which have been subject to varied interpretations. As recounted by Baldev Singh (1988), Aldous Huxley portrayed Gandhi as critical of modern technology. It is true that Gandhi had stated that the use of machinery at the time tended to concentrate wealth in the hands of a few, with total disregard for millions of men and women whose bread is snatched out of their mouths by it (1999, 61: 416). Yet shortly after Huxley's comments, Jawaharlal Nehru, the first prime minister of India, stated that perhaps Huxley did not understand the true spirit behind Gandhi's views on science and technology (Singh 1988). In *Hind Swaraj*, Gandhi wrote, "My opposition to machinery is much misunderstood. I am not opposed to machinery as such. I am opposed to machinery which displaces labour and leaves it idle" (1999, 85: 239–340), and that he has nothing to say in opposition to the development of industry in India by means of machinery (1999, 22: 401–402).[2] Thus, an understanding of Gandhi as antagonistic to machinery and technology is mistaken (Parel 2009).

4.2.2 Engineering and India's Post-Independence Era (1948–1990)

In the post-independence era, people in India were suffering from hunger, poverty, illiteracy, and belief in superstitions. To address these problems, scientists and engineers worked together to bring about three revolutions (Gupta 2015). The White Revolution transformed India from a milk-deficient nation in 1970 to the world's largest milk producer and distributor in 1998. The Green Revolution refers to the growth in grains production due to the adoption of modern agricultural technologies. The ongoing information or knowledge revolution refers to the development of engineering technologies that disseminate and analyze information in ways that transform human action, social relations, and economics.

In this period, the emphasis in Indian engineering shifted from obligatory acts to character development, toward ideas of virtue ethics. Virtue ethics is "agent-centered," not "act-centered" (Fink 2013), and as it relates to engineering, places emphasis on what sort of people engineers should be. While contemporary Western virtue ethics can often be traced to an Aristotelian root, Indian formulations usually have Buddhist origins. Buddhists believe that a virtuous person must be freed from lust and greed (*lobha*), hatred (*dosa*), and ignorance (*moha*), and possess the positive character traits of renunciation and generosity, kindness and compassion, and wisdom (Fink 2013: 673–674).

A Buddhist should act in conformance with these virtuous traits, and their acts should be altruistic. These character traits and altruistic intent can be seen in each of the three revolutions that occurred during this period.

In addition to the revolutions, two impressive engineering successes significantly contributed to the sense of national identity. One of India's great engineering achievements during this period is its development of its nuclear program, which has played a significant role in the development of the country. India started its nuclear program in 1947 aiming for economic, technological, and scientific benefits. Its weapons program began in 1964 in response to security concerns involving first China and later Pakistan (Ganguly 1999; Chakma 2004), and it was designed as a deterrent to maintain peace (Singh 2016: 5). Its technical accomplishments included successful nuclear bomb tests of Pokhran-I in 1974 and Pokhran-II in 1998. While domestic politics may have provided some impetus for the program—especially at its origin with the influence of Indian Atomic Energy Commission chairman Homi Bhabha, and later, some argue, as the Bharatitya Janata Party's (BJP) ideological inclination toward Hindu nationalism may have led Prime Minister Atal Bihari Vajpayee to order the Pokhran-II test (Bajpai 2009)—others have assessed this effect as weak (Chakma 2005). And while Bhumitra Chakma (2005: 234–235) argued geopolitical prestige was just a secondary motivation for nuclear development, Baldev Raj Nayar and T.V. Paul (2003) argue that India's international ascent can be linked to the successive successful nuclear tests. India became seen as a responsible stakeholder in nuclear power, which led to benefits in other spheres, such as diplomatic and economic connections. While nuclearization is often connected with ethical concerns, Pokhran-I and Pokhran-II were tested in such a manner that neither human life nor the environment were negatively affected, for altruistic reasons (for regional peace and stability), and done competently, aligning with the characteristics of virtue ethics mentioned earlier.

Another notable engineering success is the work of the Space Research Organisation (ISRO) of India, which, among many initiatives, has developed its own satellites since 1957. India's satellite ventures improved weather forecasting, which has also facilitated better disaster warnings and search and rescue services. In this context, scientists and engineers have collaborated for national progress and for the benefit of their fellow citizens.

In the midst of these successes, a well-known engineering disaster took place. In 1984, as the importance of virtue in engineering was increasing, India suffered the worst industrial disaster in history, the Bhopal gas tragedy. Methyl isocyanate gas leaked from Union Carbide Corporation's pesticide chemical plant in Bhopal, a city of Madhya Pradesh. Around 3,800 people died in the immediate aftermath, and around 102,000 people suffered permanent disabilities (Kumar 2004), some of whom also lost their lives prematurely due to this engineering disaster.

The incident investigation report stated that had engineers followed appropriate safety measures and acted virtuously rather than simply pursuing company profit, they would have saved human lives and prevented environmental damage caused by the accident (Shrivastava 1987; Fortun 2001). Further, the report mentioned that the company lacked adequate accident prevention measures to prevent or control the disaster. The disaster highlighted the need for enforceable international standards for environmental protection, safety, and industrial disaster prevention and preparedness. After this incident, India has enforced international standards for environment safety, incident prevention strategies, and industrial disaster preparedness to avoid a similar accident of this sort in the future while allowing companies to establish their plants in India (Broughton 2005). Engineers and politicians have also learned that they are expected to give utmost importance to protecting public health and limiting environment pollution. For example, the chemical company DuPont was planning to establish its plastics plant in Chennai, a city of Tamil Nadu, but the state government demanded more concessions for public health and environment protection. Eventually, the company stopped its plan to establish the plant in Chennai, citing "finance concerns" (Broughton 2005: 5).

In the post-independence period, scientific and engineering discoveries propelled the development of India. In addition to the events addressed here, engineers were expected to participate in production processes that add value to products, follow international standards for engineering projects, and contribute research that results in the growth of national wealth, among other aims. These pressures led to the third phase of Indian engineering, which is marked by shift toward professionalism.

4.2.3 Engineering and Contemporary India (1990–Present)

Since 1990, Indian engineering has been shaped by rapid industrial growth, which has coincided with economic improvement and social development. The stresses mentioned previously have resulted in a shift in engineering toward professionalism: engineers now see themselves and are seen as professionals who contribute to the public good and adhere to engineering codes of ethics. An especially important code of ethics is the one formulated by the Engineering Council of India, an umbrella institution that includes many engineering institutions and professional organizations established in 2002. Its code of ethics reminds engineers of their autonomy, roles, and responsibilities toward the profession and the society.[3]

The shift toward professionalism is aligned with a number of ideas from Jainism. For Jains, performing a task without right knowledge is visionless, while having knowledge about something and not acting with right conduct is futile (Jaini 1998). Self-reliance is achieved through the three jewels of right belief, right knowledge, and right conduct (Damodaran 1967), which are necessary for a person to act as a professional. Taken together, the jewels result in the improvement of civilization. Jainism believes that professionals can make ethical decisions and set ethical goals in their works (Radhakrishnan and Raju 1966).

In addition to technological ability and problem-solving skills, Indian engineers are expected to develop skills in time management and written, oral, and visual communication, as well as those needed for interpersonal relationships and working as a team. In addition, they need to understand the significance of commitment to safety, reliability, and sustainability in all engineering activities. In this context, Chitra Lakshimi (2009) writes that a strong sense of moral values in engineers helps them to arrive at ethical judgments that would lead to appropriate decision-making in their professional tasks.

4.2.4 Concluding Thoughts on Philosophy and Engineering in India

Through the processes of designing and manufacturing artifacts for public use and benefit, engineers shape people's views of various technologies and, given the ubiquity and influence of technology in the contemporary period, their approaches to life. Today, Indian engineering innovations not only provide comfort to human beings but also assist the nation in its rapid economic, educational, and healthcare development. Engineering is broadly supported as it aligns with the nation's desires and interests and allows India to participate in the global economy.

A recent project shows the multifaceted importance of engineering in India. In October 2018, 250 Indian engineers and 3,000 workers completed the construction of the world's tallest statue, a 182-meter-tall bronze figure of Sardar Vallabhbhai Patel, India's first deputy prime minister, on the bank of Narmada River in Gujurat state. The statue is called the "Statue of Unity" for Patel's work in creating a unified India in the post-independence era. It is a symbol of India's engineering skills, ability, growth, status, and reputation, as well as a reminder of the political importance of engineering in the country.

4.3. Japanese Philosophical Approaches and Engineering

The phrase "philosophy of engineering" is rarely used in Japan, where it is generally regarded as part of philosophy of technology. The most promising research in philosophy of technology comes from the Kyoto School, especially the work of Kitaro Nishida, Kiyoshi Miki, and Jun Tosaka, starting from the 1930s (Nakamura 1995), although it stopped for a while after World War II (Murata 2009). Their philosophical speculation on the relationship between technological artifacts and human society might be able to be a resource for resolving disputes over issues that arise with modern and emerging technology, but little progress has been made so far, perhaps because their discourse developed within the confined theoretical framework of Marxism, which limits its applicability to contemporary Japan (Murata 2009).

Along the same lines, the concept of engineering is quite often subsumed into the concept of technology (see also Chapter 53, "Engineering and Contemporary Continental Philosophy of Technology" by Diane P. Michelfelder, this volume). Work on technology is just a small of philosophical work in Japan today, and so what would be considered philosophy of engineering is but a fraction of contemporary philosophical inquiry. There have been a few scattered investigations into philosophy of engineering in recent years (e.g., Saito 2003), but there is no established line of inquiry independent from philosophy of technology.

This state of affairs means not that philosophy of engineering is less valuable, but rather that one must ask the more fundamental question of "What is engineering?" (see also Chapter 1, "What Is Engineering?" by Carl Mitcham, this volume) A sociohistorical approach shows how the process of accepting the concept and practice of "engineering" from Western countries affected the Japanese view of engineering and consequently the development of philosophy of engineering. Using this approach, the development of technology and engineering can be divided into four periods, which are used to structure this section.

4.3.1 Pre-Engineering: Era of Artisans and Craftsmen

People living in the Japanese archipelago have made various tools in order to survive the severe natural environment long before modern engineering came into existence. From ancient times, they have made and utilized chipped and polished stone tools, Jōmon (cord-marked) pottery, Yayoi earthenware, and the like. After the Jōmon (ca. 10,500–ca. 300 BCE) and Yayoi (ca. 300 BCE–250 CE) periods, people continued to make commodities and to construct structures ranging from buildings to keyhole-shaped burial mounds. Artisans and craftsmen emerged several hundred years later as divisions of labor arose. While the exact time is not easy to determine, it is certain that communities of artisans and their technical skills have acted as the foundation for Japanese society at least since the Middle Ages.

The production of firearms is an example that illustrates the role of artisanship in this period. Noel Perrin's (1979) history of rise and fall of guns in Japan from 1543 to 1879 explains that the skilled artisans, especially the swordsmiths, deserved credit for the mass production of firearms. The Japanese sword was a weapon of outstanding quality, he noted, and "people who could make weapons of this quality were not going to have much trouble adapting their technology to firearms" (Perrin 1979: 12). The activities of highly skilled artisans since at least the sixteenth century made the mass production of guns a success that changed Japanese history. The Japanese name for the period from 1490 to 1600 is *Sengoku Jidai*, or Age of the Country at War. The firearms completely changed conventional military strategies accompanied with swords, lances, and bows and arrows. These artisans, together with the members of the samurai class, also played an important role in the process of accepting "engineering" from Western countries in the Meiji period.

4.3.2 Meiji Period: The Acceptance of Engineering From Western Countries

The concept and practice of "engineering" was imported from Western countries during the Meiji period from 1868 to 1912 (Murakami 2006). During this period, also called the Meiji Restoration, political and economic threats from Western colonial and imperial powers led Japan to pursue rapid industrialization. To do so, the Japanese intentionally adopted Western practices, including those used in science, engineering, and military affairs, as well as the higher education system, selectively westernizing themselves as a protection against those same countries. This process is historically significant.

> Japan is the first non-Western nation to industrialize through a self-determined strategy. Through industrialization Japan was able to profoundly influence the nation's social, economic and strategic prospects. This established Japan's position irrevocably in its own geopolitical orbit and ultimately its place on the world stage. In a little over half a century, Japan was universally recognized as an industrial nation. The extraordinary achievement was characterized by a forthright process of adopting and adapting Western industrial technologies, especially in the heavy industrial sectors of iron and steel, shipbuilding and coal mining.
> *(World Heritage Council for the Sites of Japan's Meiji Industrial Revolution 2015)*

Engineering in its modern sense gradually arose with the acceptance and adaptation of these processes and systems.

Engineers who promoted and maintained this industrialization can be divided into two kinds, craftsman and samurai. Hisashige Tanaka (1799–1881), the founder of Toshiba, is of the first kind. Takehiko Hashimoto, Professor of History and Philosophy of Science at the University of Tokyo, explains that Tanaka bridged two worlds: the world of traditional craftsmen during the Edo period (1603–1868) and modern Western engineers during the Meiji period (Hashimoto 2009: 31). Tanaka was seriously engaged in importing Western technologies before and after the Meiji Restoration.

In 1852, the Saga domain that was responsible for garrisoning Nagasaki, the only port open to foreign trade at that time, established a physics and chemistry research institute called *Seirenkata* that investigated technologies such as the steam engine and the reverberatory furnace. Various craftsmen and young men in the samurai class participated in activities there and at other educational institutions such as the Nagasaki Naval Training Center. Naomasa Nabeshima, feudal lord of Saga domain, and his trusted adviser Tsunetami Sano, an expert in Western science who later founded the Japanese Red Cross Society, invited Tanaka to the Saga domain to work on further technological advances (Toshiba Science Museum). Of course, not all of the efforts at *Seirenkata* were successful, but those trials gave the participants a realistic sense of the severity of the Japanese situation and the limitations of their own industrial capacities. In addition to the craftsmen and samurai, neo-Confucianists and scholars of Western learning using the Dutch language played an important role in these trials (Hirota 1980: 180–182; Inkster 1991).

In 1863, five young men in samurai class from Choshu feudal domain, called the "Choshu Five," were smuggled out of Japan to travel to the United Kingdom in order to study at University College London. One, Yozo Yamao, is well known as "the father of Japanese engineering" for his achievements and his efforts toward the establishment of the Tokyo Imperial College of Engineering (*Kobu-Daigakko*) in 1873, which later became the faculty of engineering at the University of Tokyo (Embassy of Japan in the UK). All members of the Choshu Five later assumed important assignments in the Meiji government and thrived as civil servants and technocrats.[4]

During the Meiji Restoration, the samurai class was also the intellectual class, so they played an especially important role. The samurai class had their own codes of conduct and a spirit of realizing and fulfilling their obligations called "Bushi-do" ("The Way of the Warrior") (Shimada 1990; cf. Nitobe 1905). Their ideals, which included courage, selflessness, self-control, and benevolence, drew from Buddhism, Confucianism, and neo-Confucianism (Shimada 1990).

Higher education institutions that can cultivate engineers are a necessity for the development of modern technology. The first of these institutions in Japan is *Kobu-Daigakko*, which opened as the Ministry of Public Works (*Kobu-sho*) in 1873, with Henry Dyer its first principal (Nakaoka 2009:

437). Dyer and many other "hired foreigners" (*oyatoi gaikoku-jin*) who were employed as teachers by the Meiji government since around the early 1870s contributed to the establishment of the higher engineering education systems, and most of the students were from the samurai class. Dyer emphasized the importance of the humanities, including philosophy, as part of an engineering education (Miyoshi 1983: 40). He had planned that *Kobu-Daigakko* would provide a liberal education during the first and second years, but unfortunately, the priority given to classes on the engineering fundamentals displaced these courses (Miyoshi 1983: 40).

After the expiration of his contract in 1882, Dyer returned to his homeland and wrote and published two voluminous books, *Dai-nippon* (1904) and *Japan in World Politics* (1909). In the preface of the first book, Dyer wrote that "I believe that in material, intellectual, and moral influence Japan will fully justify her claim to be called the Britain of the East" (Dyer 1904: ix). In the second, he noted that

> it is now being recognised by statesmen and by students of sociology in the West that the changes which have taken place in Japan not only affect that country, but that they will have great influence on social and political conditions in the Far East generally, and directly or indirectly on all the commercial and industrial countries in the world, and especially on those in the Pacific area.
>
> *(Dyer 1909: vi)*

While Dyer was proud of the success of *Kobu-Daigakko* and of his own achievements of industrialization, philosophy and engineering never came together in his educational system and its successors over the long term. Although both disciplines were still isolated from one another, the fact remains that he is "the father of Japanese engineering education system" (Katoh 2004: 57).

4.3.3 Wars and After Defeat in World War II

Japan was involved in two wars in the late Meiji period, the Sino-Japanese War (1894–1895) and the Russo-Japanese War (1904–1905). Around this time, Japanese engineering was rapidly drawn into warfare against other countries. Hoshimi Uchida summarized the period between 1910 and 1935 as

> the emphasis of the government's technology policy was on the practical application of technological information by already highly trained engineers, technicians, and mechanics. The goal of the army and the navy was to help Japan become a military power, which coincided with the aspirations of the party politicians to make Japan a first-rate industrial nation.
>
> *(Uchida 1984: 79)*[5]

Even though there are several examples of engineering products made for scientific purposes during those days, such as the Kōken long-range mono-plane (Tomizuka 1998), the overwhelming majority of engineering resources and engineers were devoted to military concerns.

Japan was a combatant in both World Wars, and almost all Japanese, including engineers, had to engage in them. After the defeat in the Second World War, the circumstances of engineers changed completely, and a large number lost their jobs because their companies and factories were broken up on the order of General Headquarters (GHQ).

The broad-scale unemployment resulting from diminished military efforts ended up aiding postwar reconstruction. Engineers moved from satisfying military objectives to manufacturing consumer products. For example, Nakajima Aircraft Company, renamed New Fuji Industries in late 1945 and then Fuji Heavy Industries in 1953, manufactured the Fuji Rabbit scooter in 1946, the same year that Mitsubishi Heavy Industries, which had made submarines, aircraft carriers, and battleships, first produced the Silver Pigeon scooter (Alexander 2008: 67–71). Jiro Horikoshi, a chief designer of the Zero fighter and an aeronautical engineer in wartime, turned his attention to the design of the

YS-11, the first Japanese-made commercial aircraft put into service after World War II (cf. Samuels and Whipple 1989). Hideo Shima, who designed and provided the impetus for the building of the first bullet train, whose line was called the Shinkansen, had been a railway engineer of Japanese Government Railways during a period of war (Salpukas 1998).

According to Shigeru Hirota (1980), the foundation of Japan's rapid modernization and its process of transitioning from absorbing Western science and engineering to making its own technological developments was the pluralism derived from Shintoism, the indigenous religion of Japan, which recognizes the coexistence of myriads of deities. Yet while influences of Eastern philosophical traditions and their religious backgrounds, such as Shintoism, Buddhism, and Confucianism, as well as Western philosophy, are apparent in Japanese history, there is very little evidence of interplay between them and the development of engineering and technology in Japan.

Similarly, in spite of the importance and many successes of the engineers mentioned here, engineering has always been seen as its own realm, to a large degree distinct from philosophical and cultural discourse. Concerning this phenomenon, Takanori Maema (2004) wondered why the ideas of intellectuals, philosophers, literati, military personnel, activists, and politicians are regarded worthy of attention, but not those of engineers. Hideo Ohashi, a former chairman of the Japan Accreditation Board for Engineering Education, pointed out that the social invisibleness of engineers has been a serious problem in Japan (Ohashi 2008).

This problematic situation may ultimately be improved by the second importation of engineering, this time in the form of engineering education and engineering ethics, which took place in the mid-1990s. Engineering education reform was a response to increasing pressure associated with the globalization of the economy, imparted by foreigners. Japanese engineering leadership decided to professionalize and to import accreditation systems for engineering education, including engineering ethics as an accreditation criterion, from the United States (Fujiki 2012: 54–55). Many faculty of engineering of universities and engineering colleges responded to this decision by offering engineering ethics courses in the early 2000s (Kanemitsu 2018: 244). As a result, as of 2010, over eighty textbooks of engineering ethics were quickly published (Fujiki and Sugihara 2010). Some cooperation between engineers and philosophers arose through the process of creating these textbooks. Yet while some researchers of engineering ethics in Japan focus on the philosophical element of engineering practice, the importation on the whole has not yet bore much fruit.

4.3.4 Concluding Thoughts on Philosophy and Engineering in Japan

Other than recently, at no time in Japanese history have philosophy and engineering come into contact, whether in higher education or in engineering practice. The artisans and samurai who became the engineers in the early processes of industrialization did not have the chance to formally learn philosophy. Similarly, philosophers and ethicists have not yet shown a profound interest in engineering, beyond what is considered philosophy of technology. Finally, for reasons unknown, after the Second World War, engineers became socially invisible.

In addition to the work on engineering ethics and education mentioned earlier, a few philosophers and engineers in Japan have recently started to think about engineering practice and engineering (e.g., Saito 2003; Naoe 2008; Ohishi 2013; Hiyagon 2017). Such activities will hopefully result in more insights and the establishment of the philosophy of engineering in Japan.

4.4. Conclusion

As the world becomes increasingly engineered in what has been called the Anthropocene period and issues related to technological development arise on larger and larger scales, the search for cultural resources needed to direct and guide technological development becomes increasingly important. The rich historical traditions of Eastern cultures have yet to be integrated into the global conversation about how engineering should be institutionalized and practiced. Doing so may contribute to the development of engineering practices that align with cultural values in those cultures; it also may offer alternative paths for other peoples to consider when designing the engineering and technology of the twenty-first century. To give but a few examples, what does engineering and technology look like when its development and use embrace Chinese concepts such as the *dao* and *qi*; Indian concepts such as *karma*, understood as interdependence and interrelations between beings, and *ahimsa*, non-violence toward all other beings (Coward 2003); or Japanese insights from the Kyoto School on how to rethink modernity rather than continuing to follow its present inertia? Given the extensive connections between engineering, technology, economics, and military power, progress in philosophy of engineering is critical for geopolitical affairs. The task of developing national identities that incorporate engineering without resorting to hostile nationalism seems essential in creating a durable, sustainable, and just world order; the development of socioeconomic systems, again, also dependent on engineering, that retain human creativity, meaningful interaction, and uniqueness is necessary to make these systems habitable. The project of developing a philosophy of engineering, or, better, perhaps, philosophies of engineering, is an especially important task for the next hundred years, for both the East and the West.

Related Chapters

Chapter 1: What Is Engineering? (Carl Mitcham)
Chapter 2: A Brief History of Engineering (Jennifer Karns Alexander)
Chapter 3: Western Philosophical Approaches and Engineering (Glen Miller)
Chapter 20: Towards an Ontology of Innovation (Vincent Blok)
Chapter 53: Engineering and Contemporary Continental Philosophy of Technology (Diane P. Michelfelder)

Further Reading

Brajendranath, Seal. (1915). *The Positive Sciences of the Ancient Hindus*. London: Longmans, Green and Co. (An outline of engineering and scientific works traced from Indian philosophical systems.)

Christensen, Steen Hyldgaard, Mitcham, Carl, Li, Bocong and An, Yanming. (2012). *Engineering, Development and Philosophy: American, Chinese and European Perspectives*. Dordrecht: Springer. (An introduction to Chinese philosophy of engineering from a comparative perspective.)

Heisig, James W., Kasulis, Thomas P. and Maraldo, John C. (eds.) (2018). *Japanese Philosophy: A Source Book*. Honolulu: The University of Hawai'i Press. (A sketch of the history of Japanese philosophy, ethics, and religion in broad strokes.)

Hui, Yuk. (2016). *The Question Concerning Technology in China: An Essay in Cosmotechnics*. Falmouth, UK: Urbanomic Media. (The first systematic philosophical attempt to articulate a Chinese philosophy of technology.)

Mitcham, Carl, Bocong, Li, Newberry, Byron and Baichun, Zhang. (2018). *Philosophy of Engineering, East and West*. Cham, Switzerland: Springer. (Interdisciplinary edited volume on Chinese and Western reflections on engineering, technology, and development.)

Nakayama, Shigeru, Gotō, Kunio and Yoshioka, Hitoshi. (2001). *A Social History of Science and Technology in Contemporary Japan*, 4 vols. Melbourne: Trans Pacific Press. (English translation of respected history of science and technology in Japan.)

Ramnath, Aparajith. (2017). *The Birth of an Indian Profession: Engineers, Industry, and the State, 1900–47*. New Delhi: Oxford University Press. (An examination of engineers' identity, culture, and status in society, and their role in building modern India.)

Notes

1. As an example of its importance, the story is used by Amartya Sen (2000) to explain consequence-independent deontology in his "Consequential Evaluation and Practical Reason" article in *The Journal of Philosophy*.

2. Along the same lines, in *Hind Swaraj* Gandhi also writes that "What I object to, is the craze for machinery, not machinery as such. I am aiming, not at eradication of all machinery, but limitations . . . I want the concentration of wealth not in the hands of a few, but in the hands of all. The labourer must be assured not only daily living wage but also a daily task and that shall not be a drudgery" (1999, 25: 251–252).
3. At the time of writing, the code of ethics can be found at www.cidc.in/new/support/PE/ECI-Code%20 of%20%20Ethics.pdf.
4. The sense of "technocrat" meant here is a technically skilled person with government managerial power.
5. Minor syntactical corrections were made.

References

Abhyankar, Vivik G. (2017). *Indian Temples—A Structural Engineering Marvel: A Brief Engineering Report on Indian Temples*. www.scribd.com/document/126770777/Indian-Temples-A-Structural-Marvel-Vivek.

Agarwal, Anil and Narain, Sunita (eds.) (1997). *Dying Wisdom: Rise, Fall and Potential of India's Traditional Water Harvesting Systems*. New Delhi: Centre for Science and Environment.

Alexander, Jeffrey W. (2008). *Japan's Motorcycle Wars: An Industry History*. Honolulu: University of Hawai'i Press.

Bajpai, Kanti (2009). The BJP and the Bomb. In Scott D. Sagan (ed.), *Inside Nuclear South Asia*. Stanford: Stanford University Press.

Bhattacharyya, Kumkum (1998). *Applied Geomorphological Study in a Controlled Tropical River: The Case of the Damodar between Panchet Reservoir and Falta*. PhD dissertation, The University of Burdwan, West Bengal, India.

Bhattacharyya, Kumkum (2011). *The Lower Damodar River, India: Understanding the Human Role in Changing Fluvial Environment*. Heidelberg, Germany: Springer.

Broughton, Edward (2005). The Bhopal Disaster and Its Aftermath: A Review. *Environmental Health: A Global Access Science Source,* 4(6), 1–6.

Chakma, Bhumitra (2004). *Strategic Dynamics and Nuclear Weapons Proliferation in South Asia*. Bern: European Academic Publishers.

Chakma, Bhumitra (2005). Towards Pokhran II: Explaining India's Nuclearisation Process. *Modern Asian Studies,* 39(1), 189–236.

Chakroborty, Partha and Das, Animesh (2003). *Principles of Transportation Engineering*. New Delhi: PHI Learning.

Chen Changshu and Bei Chenfu (2002). 开创哲学研究的新边疆—评《工程哲学引论 [Initiating a New Frontier of Philosophical Research—Comment on 'Introduction to Engineering Philosophy']. 哲学研究 [*Philosophical Research*] 2002(10), 71–72.

Coward, Harold (2003). Ethics and Genetic Engineering in Indian Philosophy, and Some Comparisons with Modern Western. *Journal of Hindu-Christian Studies,* 16, 38–47.

Damodaran, K. (1967). *Indian Thought: A Critical Survey*. New York: Asia Publishing House.

Dyer, Henry (1904). *Dai Nippon: The Britain of the East: A Study in National Evolution*. London: Blackie & Son.

Dyer, Henry (1909). *Japan in World Politics: A Study in International Dynamics*. London: Blackie & Son.

Embassy of Japan in the UK. The Father of Japanese Engineering: Yozo Yamao. www.uk.emb-japan.go.jp/en/event/2013/choshu/Yamao.html. Accessed March 11, 2019.

Feenberg, Andrew (1991). *Critical Theory of Technology*. Oxford: Oxford University Press.

Fink, Charles K. (2013). The Cultivation of Virtue in Buddhist Ethics. *Journal of Buddhist Ethics,* 20, 668–701.

Fortun, Kim (2001). *Advocacy After Bhopal*. Chicago: University of Chicago Press.

Fujiki, Atsushi (2012). How Should We Accept the American-Style Engineering Ethics? *Proceedings of the 2nd International Conference on Applied Ethics and Applied Philosophy in East Asia*, 47–60.

Fujiki, Atsushi and Sugihara, Keita (2010). Kougaku rinri no kyouka-sho no hensen [On the Transition of the Textbooks of Engineering Ethics from 1998 to 2010 in Japan]. *Gijutsu rinri kenkyu [Journal of Engineering Ethics]*, 7, 23–71.

Gandhi (1999). *The Collected Works of Mahatma Gandhi*, 98 vols. New Delhi: Publications Division Government of India. www.gandhiashramsevagram.org/gandhi-literature/collected-works-of-mahatma-gandhi-volume-1-to-98.php.

Ganguly, Sumit (1999). India's Pathway to Pokhran II: The Prospects and Sources of New Delhi's Nuclear Weapons Program. *International Security,* 23(4), 148–177.

Gupta, Amitba (2015). Foundations for Value Education in Engineering: The Indian Experience. *Science and Engineering Ethics,* 21(2), 479–504.

Hashimoto, Takehiko (2009). *Historical Essays on Japanese Technology*. Collection UTCP 6. Tokyo: The University of Tokyo Center for Philosophy.

Hirota, Shigeru (1980). Nihon no shisou to kagaku-gijutsu [Japanese Thoughts and Science and Technology]. In *Gijutsu-sha no tame no tetsugaku [Philosophy for Engineers]*. Tokyo: Gakubun-sha.

Hiyagon, Hitoshi (2017). *Gijutsu no chi to rinri* [*Wisdom of Technology and Ethics*]. Tokyo: Rikou tosho.

Hui, Yuk (2016). *The Question Concerning Technology in China: An Essay in Cosmotechnics*. Falmouth, UK: Urbanomic Media.

Inkster, Ian (1991). Science, Technology and Economic Development—Japanese Historical Experience in Context. *Annals of Science,* 48(6), 545–563.

Jaini, Padmanabh S. (1998). *The Jaina Path of Purification*. New Delhi: Motilal Banarsidass Publishers.

Kanemitsu, Hidekazu (2018). New Trends in Engineering Ethics—A Japanese Perspective. In Albert Fritzsche and Sascha Julian Oks (eds.), *The Future of Engineering: Philosophical Foundations, Ethical Problems and Application Cases*. Dordrecht: Springer.

Katoh, Shoji (2004). Nichi-ei Kohryu no suishin-sha Henry Dyer no bohi-mei [Epitaph of Henry Dyer: Pioneer of International Interchange between Britain and Japan]. *Historical English Studies in Japan,* 2004(36), 57–72.

Kramrisch, Stella (1976). *The Hindu Temple*, Vol. 1. New Delhi: Motilal Banarsidass Publishers.

Kumar, Sanjay (2004). Victims of Gas Leak in Bhopal Seek Redress on Compensation. *BMJ: British Medical Journal,* 329(7462), 366.

Lakshimi, Chitra (2009). Value Education: An Indian Perspective on the Need for Moral Education in a Time of Rapid Social Change. *Journal of College and Character,* 10(3), 1–7.

Li Bocong (2002). [*Introduction to Engineering Philosophy*]. 工程哲学引论 [Zhengzhou: Daxiang Publishing Press].

Li Zhengfeng, Cong Hangqing and Wang Qian (2016). 工程伦理 [*Engineering Ethics*]. 清华大学出版社 [Beijing: Tsinghua University Press].

Liu Dachun (1998). 科技革命与当代社会主义运动 [Science and Technology Revolution and Contemporary Socialist Movement]. 哲学研究 [*Philosophical Research*], 12, 3–11.

Liu Dachun (2001). 现代科技对马克思主义哲学的挑战 [The Challenge of Modern Science and Technology to Marxist Philosophy]. 山东社会科学 [*Shandong Social Sciences*], 2, 2–5.

Liu Dachun (2018). 审度：马克思科学技术观与当代科学技术论研究 [*Reconsideration: Between Marx's View of Science and Technology and Contemporary Science and Technology*]. 中国人民大学学报 [Beijing: Renmin University Press].

Liu Zhanxiong (2015). 负责任创新研究综述：背景、现状与趋势 [Introduction to Responsible Research and Innovation]. 科学进步与对策 [*Journal of Scientific Development and Policy*], 32(11), 155–160.

Maema, Takanori (2004). *Gijutsu-sha tachi no haisen* [*The Defeat for Engineers*]. Sugano: Soshi-sha.

Marcuse, Herbert (2013). *One-Dimensional Man: Studies in the Ideology of Advanced Industrial Society*. London: Routledge.

Mei Liang (2018). 责任式创新："内涵-理论-方法"的整合框架 [A Synthetic Framework for Responsible Research and Innovation]. 科学学研究 [*Studies in Science of Science*], 33(2), 49–53.

Michael, Bernardo A. (2007). Making Territory Visible: The Revenue Surveys of Colonial South Asia. *Imago Mundi,* 59(1), 78–95.

Miyoshi, Nobuhiro (1983). *Meiji no engineer kyouiku: nihon to igiris no chigai* [*Engineering Education in Meiji Period: Difference Between Japan and the United Kingdom*]. Tokyo: Chuokoron-sha.

Murakami, Yoichiro (2006). *Ko-gaku no rekishi to gijutsu no rinri* [*History of Engineering and Ethics of Technology*]. Tokyo: Iwanami Shoten.

Murata, Jun-ichi (2009). *Gijutsu no tetsugaku* [*Philosophy of Technology*]. Tokyo: Iwanami Shoten.

Nakamura, Seiji (1995). *Shinban gijutsu-ron ronsou-shi* [*New Edition of a History of Controversy on the Theory of Technology*]. Tokyo: Sofu-sha.

Nakaoka, Tetsuro (2009). *Nihon kindai gijutsu no keisei: <dentou> to <kindai> no dynamics* [*The Formation of Modern Technology in Japan: Dynamics of Tradition and Modern*]. Osaka: Asahi Shinbun-sha.

Naoe, Kiyotaka. (2008). Design Culture and Acceptable Risk. In Pieter E. Vermaas, Peter Kroes, Andrew Light and Steven A. Moore (eds.), *Philosophy and Design: From Engineering to Architecture*. Dordrecht: Springer, pp. 119–130.

Nayar, Baldev Raj and Paul, T.V. (2003). *India in the World Order: Searching for Major Power Status*. New York: Cambridge University Press.

Needham, Joseph and Ling, Wang (1954). *Science and Civilization in China*. Cambridge: Cambridge University Press.

Nitobe, Inazō (1905). *Bushido, the Soul of Japan: An Exposition of Japanese Thought*. New York: GP Putnams & Sons.

Ohashi, Hideo (2008). Gijutsu-sha ga mie-nai [Invisible Engineers]. *Journal of JSEE,* 56(4), 10–14.

Ohishi, Toshihiro (2013). Ouyou rinri-gaku no houhou-ron to shiteno sekkeiteki shikou [Design Thinking as a Method of Applied Ethics]. *Kagaku tetsugaku* [*Philosophy of Science*], 46(2), 31–47.

Parel, Anthony J. (2009). Introduction to *Hind Swaraj and Other Writings,* Anthony J. Parel (ed.) . Cambridge: Cambridge University Press.

Perrin, Noel (1979). *Giving Up the Gun: Japan's Reversion to the Sword, 1543–1879*. Jaffrey, New Hampshire: David R. Godine Publisher.

Radhakrishnan, Sarvepalli and Raju, Poolla Tirupati (eds.) (1966). *The Concept of Man: A Study in Comparative Philosophy*, 2nd ed. London: Allen & Unwin.

Saito, Norifumi (2003). Kougaku no tetsugaku to rinri [Philosophy and Ethics of Engineering]. *Kagaku kougaku* [*Chemical Engineering of Japan*], 67, 194–196.

Salpukas, A. (1998). Hideo Shima, a Designer of Japan's Bullet Train, Is Dead at 96. *The New York Times,* March 20. www.nytimes.com/1998/03/20/business/hideo-shima-a-designer-of-japan-s-bullet-train-is-dead-at-96.html. Accessed March 11, 2019.

Samuels, Richard J. and Whipple, Benjamin C. (1989). Defense Production and Industrial Development: The Case of Japanese Aircraft. In Chalmers Johnson, Laura Tyson, and John Zysman (eds.), *Politics and Productivity: The Real Story of Why Japan Works*. Cambridge, MA: Ballinger.

Sarava, M.S.S. (1954). *Irrigation in India Through Ages*, 2nd ed. New Delhi: Central Board of Irrigation and Power.

Sen, Amartya (2000). Consequential Evaluation and Practical Reason. *The Journal of Philosophy,* 97(9), 477–502.

Sharma, S.K. (2017). *Treatise on Irrigation Engineering and Hydraulic Structures Including Engineering Hydrology, Dams, and Water Power Engineering*. New Delhi: S. Chand and Company.

Shimada, Akiko (1990). *Nihon-jin no syoku-gyo rinri* [*Professional Ethics of Japanese*]. Tokyo: Yuhikaku.

Shrivastava, Paul (1987). *Managing Industrial Crisis: Lessons of Bhopal*. New Delhi: Vision Books.

Singh, Baldev (ed.) (1988). *Nehru on Science and Society*. New Delhi: Nehru Museum and Memorial Library.

Singh, Smita (2016). Dynamics of India's Nuclear Identity. *World Affairs: The Journal of International Issues,* 20(1), 94–109.

Tomizuka, Rei (1998). *Kōken-ki: sekai kiroku juritsu heno kiseki* [*Kōken Long Range Mono-Plane: The Path to Establish the World Record*]. Tokyo: Miki Press.

Toshiba Science Museum. *Saga-Kurume Period—The Third Turning Point*. http://toshiba-mirai-kagakukan.jp/en/learn/history/toshiba_history/spirit/hisashige_tanaka/p01_4.htm. Accessed March 11, 2019.

Travers, T.R. (2004). The Real Value of Lands: The Nawabs, the British and the Land Tax in Eighteenth Century Bengal. *Modern Asian Studies,* 38(3), 517–558.

Uchida, Hoshimi (1984). A History of Technological Policy. In *Project on Technology Transfer, Transformation, and Development: The Japanese Experience—Final Report*. The United Nation University Press, Chapter 13. https://d-arch.ide.go.jp/je_archive/pdf/finalreport/x2_d16.pdf02. Accessed March 11, 2019.

Wang Bolu (2009). 马克思技术哲学纲要 [*Marx's Thought on Philosophy of Technology*]. 科学出版社 [Beijing: Science Press].

Wang Qian (2009). 道技之间—中国文化背景的技术哲学 [*Between Dao and Technique—Chinese Philosophy of Technology*]. 人民出版社 [Beijing: People's Press].

Wang Xiaowei (Tom) (2016). Designing Confucian Conscience Into Social Networks. *Zygon: Journal of Religion and Science,* 51(2), 239–256.

Wang Xiaowei (Tom) (2018). Confucian Cosmological Life and its Eco-Philosophical Implications. *Environmental Ethics,* 40(1), 41–56.

Wang Xiaowei (Tom) and Yao Yu (2017). 负责任地反思负责任创新 [Reconsidering 'Responsible Innovation']. 自然辩证法研究 [*Studies in Dialectics of Nature*], 39(6), 37–43.

Wong, Pak-Hang (2013). Confucian Social Media: An Oxymoron? *Dao,* 12(3), 283–296.

World Heritage Council for the Sites of Japan's Meiji Industrial Revolution (2015). *Outstanding Universal Value*. www.japansmeijiindustrialrevolution.com/en/site/ouv/index.html. Accessed March 11, 2019.

Yan Ping, Zhang Wei and Wang Qian (2014). 负责任创新理论与实践述评 [Reviewing the Approach of Responsible Innovation]. *Science and Technology Innovation Report,* 27, 84–90.

Yang Bin, Zhang Man and Shen Yan (2017). 推动面向未来发展的工程伦理教育 [Promoting China's Engineering Ethics Education for Future Development]. 清华大学教育研究[*Tsinghua University Education Research*], 38(4), 1–8.

Zhang Wei (2018). 藏礼于器，内在主义技术伦理的中国路径 [Embedding Values Into Artifacts: The Chinese Path to Ethics of Technology], 大连理工大学学报 [*Journal of Dalian University of Technology, Social Science Edition*], 2018(3), 116–121.

5

WHAT IS ENGINEERING SCIENCE?

Sven Ove Hansson

What is engineering science? And how can it be distinguished from other forms of science? This chapter begins with an outline of its historical background and origins. This is followed by discussions of how it can be defined or delineated, its distinguishing metaphysical and epistemological features, and its idealizations.

5.1. From Practical Arts to Engineering Science

The engineering sciences are just a couple of centuries old, but they have a background in a much older historical tradition.

5.1.1 Practical Arts

The ancients had no concept of engineering, nor, for that matter, technology. The closest equivalent was the concept of practical arts, which was much wider. By practical arts they meant all the knowledge and abilities needed to achieve the goals of practical life, including farming, the various crafts, sailing, tradesmanship, etc. These arts were all learnt by apprenticeship. They were largely performed by slaves and by others with a low social status, in contrast to the liberal arts, such as reading and writing, which were considered to be suitable for the higher strata in society. In ancient Greece and Rome, the practical arts were referred to with various pejorative terms such as *artes illiberales*, *artes vulgares*, and *artes sordidae* (Van Den Hoven 1996: 90–91; Ovitt 1983; Tatarkiewicz 1963; Whitney 1990).

In the Middle Ages, the practical arts were often referred to as the mechanical arts (*artes mechanicae*), a term that was probably introduced by Johannes Scotus Eriugena (c. 815–c. 877) (Noble 1997). Contrary to a common misconception, this term did not specifically signify what we today call technology. Although the word is derived from a Greek word meaning machine, in the Middle Ages it referred to manual work and craftsmanship in general, not exclusively to the trades and occupations whose work was mechanical in the modern sense of the word.

Medieval scholars produced a large number of classification schemes for human knowledge, usually with a tree-like structure organizing disciplines in groups and subgroups (Ovitt 1983; Dyer 2007; Hoppe 2011; Freedman 1994; James 1995). Early on, such classifications made no mention of the practical arts, but beginning with work by Hugh of Saint Victor in the late 1120s, they were mentioned increasingly often. Hugh had a more positive view of the practical arts than most scholars and churchmen of his time. In his knowledge classification he included seven practical arts, or rather, groups of practical arts:

1. lanificium: weaving, tailoring
2. armatura: masonry, architecture, warfare
3. navigatio: trade on water and land
4. agricultura: agriculture, horticulture, cooking
5. venatio: hunting, food production
6. medicina: medicine and pharmacy
7. theatrica: knights' tournaments and games, theatre.

(Hoppe 2011: 40–41)

The distinction between liberal and practical arts continued to be used in the early modern era. It had an important role in the great French *Encyclopédie*, published from 1751 to 1772, which was the most influential literary output of the Enlightenment. In its preface, Jean Le Rond d'Alembert (1717–1783) emphasized the high value for humanity of the practical arts. The encyclopedia contains a wealth of information about the practical trades and what we would today call technology.

5.1.2 Engineering

The word "engineer" derives from the Latin *ingenium*. In the classical period, *ingenium* usually meant talent or inventiveness, and it is the origin of our words "genius" and "ingenious". But already in classical times, *ingenium* could also refer to a clever construction (and it gave rise to our word "engine"). In the Middle Ages, *ingenium* was primarily used as a general term for catapults and other machines for siege war. A master builder of such devices was called an *ingeniarius* or *ingeniator* (Bachrach 2006; Langins 2004). As late as in the 18th century, "engineer" was still a military category, but due to changes in warfare, the tasks of engineering officers had changed. They were charged with cartography and the construction of fortifications, roads, and bridges. Several European countries had schools for engineering officers, in which they were taught cartography, building construction, machine design, and the mathematics needed for their profession (Langins 2004). But outside of the military, advanced technological tasks were still performed by master craftsmen without any theoretical education.

In the 18th century, increasingly advanced mathematics was employed in the more advanced practical crafts, in particular for construction purposes (Klemm 1966). The French military engineer Bernard Forest de Bélidor (1698–1761) published a famous four-volume book, *L'architecture hydraulique* (1737, 1739, 1750, and 1753) that represents the first extensive use of integral calculus to solve engineering problems. In 1773, the physicist Charles-Augustin de Coulomb (1736–1806), now best known for his work on electricity, published his *Essai sur une application des règles de maximis et de minimis à quelques problèmes de Statique relatifs à l'Architecture*, in which he applied mathematical analysis in innovative ways to what is now called structural mechanics. In 1775, the Swedish shipbuilder Fredrik Henrik af Chapman published a treatise on naval architecture that made use of Thomas Simpson's method for the approximation of integrals (Harris 2001).

Largely in response to this development, the first civilian school for engineering was founded in Paris in 1794. It is commonly known under the name École Polytechnique (Grattan-Guinness 2005). It was led by Gaspard Monge (1746–1818), who was an able mathematician and also a Jacobin politician. He considered mathematics and the natural sciences, including mechanics, to be the most important subjects (Hensel 1989a: 7). About a third of the curriculum hours were devoted to mathematics (Purkert and Hensel 1986: 27, 30–35). He also developed a new discipline, called descriptive geometry, which is a mathematically precise form of technical drawing that is particularly useful as a tool for machine construction (Lawrence 2003; Klemm 1966).

In addition to being practically useful, mathematical education contributed to creating a merito-cratic structure that was well in line with the anti-aristocratic ideology of the young republic. Math-ematical proficiency was an objectively verifiable standard that could replace previous aristocratic criteria for selection and promotion. This use of mathematics was largely modeled from the educa-tion of artillery engineers (Alder 1999).

The École Polytechnique became the paragon of polytechnic schools in other countries in Europe, and also in the United States. A sizable number of polytechnic schools were founded in the 1820s and 1830s in the German-speaking countries, and a similar development took place in other parts of Europe as well as in North America (Purkert 1990: 180; Schubring 1990: 273; Scharlau 1990). The new schools all followed the example of the École Polytechnique in providing their students with a high level of mathematical and natural science education (Hensel 1989a: 6–7; Grayson 1993). Gradually, their education became more and more theoretical. An increasing number of treatises and textbooks were published on the application of mathematics and mathematical physics to engineer-ing tasks, such as the construction of buildings, machines, and ships (Klemm 1966). The polytechnic schools increased their education in mathematics and hired prominent mathematicians as teachers (Schubring 1990: 273; Scharlau 1990: 264–279; Hensel 1989b; Purkert 1990: 188). The emphasis on mathematics was not always uncontroversial. In the 1890s, a zealous "anti-mathematical" movement emerged among teachers in technological disciplines at German university colleges. They distrusted modern, rigorous mathematics and demanded reductions in the mathematics curriculum. However, this was a short-lived movement that hardly survived into the next century (Hensel 1989a; Hansson 2018).

It was essentially in the practice of the polytechnic schools that the modern (civil) profession of engineering was defined and delimited. Their original purpose was to provide a thorough and sufficiently theoretical education for the advanced craftspeople of the towns and cities, who had previously learned everything they needed by apprenticeship. This excluded farm work and most other work outside of the towns. It also excluded work predominantly performed by women, such as cooking. Furthermore, the occupations for which there was already a theoretical education, such as medicine and pharmacy, were excluded. This initial demarcation still has a strong influence on our conception of engineering. We still do not consider farming, pharmacy, cooking, or surgery as engineering, although they involve equally extensive and sophisticated use of tools and machines as many of the tasks that we consider to be engineering.

5.1.3 *Engineering Becomes Science*

Thomas Hobbes's *Leviathan* contains a diagrammatic knowledge classification with a tree structure, dividing the various forms of knowledge into categories and subcategories. Its most general category was "Science, that is Knowledge of Consequences, which is also called Philosophy". On the next lowest level we find "Mechaniques", which is subdivided into three subcategories on the lowest level, namely "Science of Engineers", "Architecture", and "Navigation" (Hobbes [1651] 1909: 65, Ch. I: ix). This is the first reference to engineering science that I have been able to find in the litera-ture. However, Hobbes does not provide any indication of what topics were covered by the science of engineers.

The first, more extensive treatment of engineering science may have been a book from 1729 by the aforementioned French military engineer Bernard Forest de Bélidor, *La Science des ingénieurs dans la conduite des travaux de fortification et d'architecture civile* ("The Science of Engineers in the Manage-ment of Works of Fortification and Civil Architecture"). The translation of this book into German in 1757 seems to have introduced the word "Ingenieurwissenschaft" (engineering science) into the German language. I have not found any use of the term "engineering science" in English in the 18th century. An obituary for the prominent American engineer Robert Fulton (1765–1815) noted that he "devoted himself to the science of civil engineering" (Anon. 1815: 141). But the first known usage of the exact English phrase "engineering science" seems to be from a committee hearing in the British Parliament in 1826, in which a Mr. Rolfe asked a witness whether the obstacles to improving Yarmouth harbor (in East Anglia) might "be either palliated or removed by an engineering science" (British Parliament 1826).

In the decades that followed, the term continued to be used predominantly about large building and construction projects. In an address to the Royal Geographical Society in London in 1844, the Scottish geologist Roderick Murchison (1792–1871) expressed his support for the plans to build a canal between the Mediterranean and the Red Sea, but he refrained from expressing any opinion on a suitable line for the canal, "for these are questions of engineering science" (Murchison 1844: cvii). In the *Scientific American*, which was founded in 1845, a patent application for portable dams was described as "a great acquisition to the engineering science of our country" (Anon 1850). In 1869, the German-British engineer C.W. Siemens made a speech at the yearly meeting of the British Association for the Advancement of Science, in which he focused on two major examples of "the latest achievements in engineering science", namely the Pacific Railway and the Suez Canal (Siemens 1870: 202). This part of his speech was also printed in the *Scientific American* (Siemens 1869). (In the same speech, Siemens expressed his confidence in a "speedy and satisfactory solution" to the crossing of the British Channel. The Channel Tunnel opened 125 years later.)

In the 19th century, teachers at the polytechnic schools made considerable efforts to increase the status of their disciplines. They wanted advanced engineering to obtain the same high status as the natural sciences, not only in the popular press but also in the academic system. In particular, they craved for the right to award a doctoral degree. Two strategies were employed to achieve this. One was to use results from the natural sciences to investigate the workings of machines and other technological constructions. Formulas from mechanics were used to characterize the movements of machine parts, and the theory of electromagnetism was employed in the construction of electric machines and appliances. New disciplines, such as structural mechanics, were developed that broadened the basis of this type of calculations.

The other strategy was to apply scientific method directly to technological constructions. Machines and machine parts were built, and measurements were performed in order to optimize their construction (Faulkner 1994; Kaiser 1995). In many cases, this was the only way to solve practical technological problems, since the theories were not reliable enough or required complex computations that were far beyond the available capacity (Hendricks et al. 2000).

Engineering science as we know it today is still based on a combination of these two strategies: applications of natural science, and direct investigations of technological constructions. Official recognition as science came in the first half of the 20th century, when technological colleges in many countries were given the status and names of universities and granted the right to award a doctoral degree.

5.2. Delineating Engineering Science

Previously, engineering science was sometimes defined as applied natural science (Bunge 1966), but today there is consensus that this is a grossly misleading definition. Engineering scientists (and engineers in general) operate on a daily basis with concepts, constructions, and work methods that cannot be derived from natural science (Mitcham and Schatzberg 2009; Bunge 1988). Some authors have even claimed that it is more accurate to describe natural science as applied engineering than the other way around (Lelas 1993). However, the science–technology relationship is much too complex to be captured with reductive notions such as "application" in either direction. It is much more constructive to treat engineering science as a major branch of science on a par for instance with natural, social, behavioral, and medical science (Hansson 2007).

5.2.1 Methodology or Subject Matter?

There are two common ways to delineate a scholarly or scientific discipline or group of disciplines, namely according to its subject matter or its methodology. For instance, biology is usually defined as the study of living organisms, whereas economics is often demarcated from other social sciences as the science that focuses on incentives, incentives-driven behaviors, and their combined effects.

Engineering science is methodologically so broad that a delineation based on its methods would run the risk of covering so much that it would become useless. It seems more adequate to define engineering in terms of its subject matter. Contrary to most other sciences, engineering science is devoted to the study of objects and processes constructed by humans in order to be used for some practical purpose, such as combustion engines, airplanes, light bulbs, artificial heart valves, and computer programs. In contrast, the natural sciences are devoted to the study of objects and processes from nature, such as atoms, animals, volcanoes, chemical reactions, and the evolution of species. The distinction between natural and human-made objects is simple, and it is quite useful for distinguishing between the natural and the engineering sciences. However, some nuances have to be added in order to capture the demarcation between the two branches of science (Hansson 2007).

Importantly, the distinction refers to the *ultimate* study objects of the respective disciplines. Natural scientists often study objects that have been modified for the purpose of measurement or experiment, but they do this in order to understand objects or phenomena that occur naturally. For instance, in order to determine the structure of a protein it is often useful to produce a crystallized form of it. But although it is the crystallized protein that is analyzed in the laboratory, it is studied only in order to understand the structure and the workings of the naturally occurring protein, which is therefore the ultimate study object. In contrast, studies in the engineering sciences have technological objects as their ultimate study objects.

Furthermore, only some human constructions belong to the domains of the engineering sciences. Others belong to the social sciences, the humanities, or mathematics. For instance, money is a human construction, but most aspects of this construction belong to economics rather than to the engineering sciences. Language is a human invention studied in linguistics. Furthermore, molecules that are human-made, i.e. the outcomes of chemical synthesis, are study objects in chemistry (Schummer 1997; Hansson 2014). The central concern of the technological sciences is the construction and use of material objects (and of immaterial objects such as computer programs that are needed to employ some of these material objects in the intended ways).

5.2.2 The Science of a Profession?

But arguably, engineering can be defined in a third way, namely as the science of a profession. We can define it as consisting of those scientific studies that support the successful performance of the profession of an engineer. Several other branches of science may be defined analogously, such as medical, veterinary, dental, agricultural, and possibly legal science. Interestingly, engineering science is the only among these whose English denomination contains the name of the profession. That might be taken as an indication of a particularly strong association between the profession and its branch of science. (However, the Scandinavian languages have a word for medical science that literally means "physician's science": "läkarvetenskap" in Swedish, "lægevidenskab" in Danish, and "legevitenskap" in Norwegian.)

Much more could be said about this. Engineering science is so entangled in complex social processes that it may be impossible to achieve a definition that is both reasonably simple and well aligned with linguistic practice. Even if that is so, attempts at a precise definition will help us to better understand what we mean by the concept.

5.3. Metaphysics

Technology and the engineering sciences give rise to many metaphysical topics. Here, the focus will be on a particularly important such issue, namely the metaphysics of the objects produced in engineering.

5.3.1 *The Dual Nature Theory*

Seminal work by Peter Kroes and Anthonie Meijers has made it clear that technological objects have a "dual nature": They can be described either in terms of their physical–structural characteristics or in terms of their functions ("wooden cylinder, 3 mm thick and 94 mm in diameter"—"cup coaster"). Consequently, technological artefacts can be understood both as physical objects and as objects with certain functions. Whereas the physical properties of a technological object can be described without any reference to human intentions, its functional properties are closely related to the intentions of its designers and users (Kroes and Meijers 2006; Kroes 2006; Vermaas and Houkes 2006; Hansson 2006). Understanding the function of an object is essential for understanding its technological nature. For instance, you have not understood what a set of chimes is if you know only how it is constructed (with hanging metal tubes of different lengths). A set of chimes has to be understood in relation to its function as a musical instrument.

The dual nature theory is also helpful in the analysis of engineering design. A design process can be described as a process that translates functions into structures. For instance, to design a rice cooker means to construct a physical device that realizes the function of cooking rice. For some types of technological artefacts, the criteria of functionality can be very complex. It is no easy task to make a full list of our functional requirements on a car. Arguably, this task is impossible, since the list is unfinished and always open to additions due to technological or social developments. Our appraisal of a new car model may very well be influenced by features or properties that we had never thought of before in connection with motor vehicles.

5.3.2 *The Classification of Technological Objects*

The dual nature theory can be used to clarify the metaphysical nature of the concepts we use to classify the objects that are constructed and used by engineers. The emerging picture is surprisingly complex, since our terminology combines physical and functional concepts, sometimes in rather intricate ways (Hansson 2014).

Some of the categories that we use to classify technological objects are purely functional. In order to determine whether an object belongs to one of these categories, it is sufficient to ascertain its function. Hence, in order to determine whether an object is a plough, we have to find out whether or not its function is to turn over the upper layer of the soil. In order to determine whether a computer program is a search machine, we have to find out if it serves to find digitally stored information. We do not need to find out what its components are or how they have been combined. The vast majority of the categories we use to classify technological objects are specified according to functional characteristics. Screwdrivers, nutcrackers, calculators, pens, aircrafts, chairs, diodes, ladders, lamps, refrigerators, and particle accelerators are all functionally defined categories.

Other technological categories are predominantly structural (physical). This applies for instance to the notions of a plank, a steel wire, a rope, and a fiberboard. As these examples indicate, technological categories defined in structural terms tend to be raw materials or multipurpose components. In order to determine if an object belongs to one of these categories, it is sufficient to know its structure, i.e. what its components are and how they are put together.

Some categories of technological artefacts are delimited in both physical and functional terms. One example is cogwheels. We would probably not use that designation about a toothed wheel that was not constructed to be connected with another toothed component. Nor would we use it about a wheel that connected with another wheel through some mechanism other than interlinking teeth. Another example is helicopters. A helicopter is a hovering aircraft employing revolving rotor blades. We do not use the term "helicopter" about a hovering aircraft not operating with revolving rotor blades (such as a jet pack or a hovering jet plane). Nor would we use it about a device with revolving rotor blades that are mounted for some other purpose than flying (such as a hypothetical submarine constructed in that way). Yet another example is scissors. Saws and knives have essentially the same function as scissors. The reason why we do not call them scissors is their physical properties. We call a cutting instrument a pair of scissors only if its cutting function relies on two edges that slide against each other. In order to determine whether an object belongs to such a (mixed) category, it is necessary to have information about both its function and some of its structural (physical) properties.

The categorization of technological objects is often performed in several steps, in categories and subcategories. In the creation of subcategories, structural or functional categories are often subdivided into subcategories of the mixed type. Hence, "engine" is a functional category, but "two stroke engine" a mixed one. "Clock" is a functional category, but "pendulum clock" is a mixed one. "Plank" can be defined as a structural category ("a piece of sawn timber at least 50 mm thick and 225 mm wide" according to the *Oxford English Dictionary*), but "floor plank" is a mixed one. Some categories can be divided into subcategories in terms of either structure or function ("pipe", "copper pipe", "sewage pipe").

In summary, the terminology of engineering and engineering science is based on complex mixtures of functional and structural specifications. In this respect, the "dual natures" of technological objects are intertwined rather than juxtaposed, thereby providing us with a rich and varied set of concepts for the description of human-made objects.

5.4. Epistemology

Many attempts have been made to classify the knowledge of engineers. (For an overview, see Houkes 2009: 321–327.) The following account of major categories of technological knowledge is centered on learning, i.e. on the various processes through which we acquire knowledge in engineering and technology (Hansson 2013).

5.4.1 Tacit Knowledge

Tacit knowledge is knowledge that the knower cannot explain or justify. This type of knowledge has an important role in many forms of craftsmanship. For instance, painters can seldom explain the hand movements by which they even out a surface much faster, and with much less spackling paste, than an amateur. Tacit knowledge also has a place, although a less prominent one, in engineering. An experienced engineer can often "feel" that a construction will not work, or that one construction will be more reliable than another, without being able to provide a satisfactory argument why this is so.

The expression "tacit knowledge" was introduced by the Hungarian-British chemist and philosopher of science Michael Polanyi (1891–1976) (Polanyi 1966). However, his main interest was natural science rather than engineering or engineering science. He wanted to show that a strictly rule-bound road to scientific knowledge does not work, since there is always an element of intuitive human judgement.

Tacit knowledge is at the center of the discipline of knowledge management, which was established at the beginning of the 1990s. The Japanese researcher Ikujiro Nonaka (born in 1935) has had a leading role in applying Polanyi's concept to practically oriented management and organization research. His main focus is on how tacit knowledge can be transferred from one person to another. This is a problem with important engineering applications (Nonaka and Takeuchi 1995; Nonaka and von Krogh 2009). One of the most famous examples is the development of the first bread-making machine for household use, which was launched in 1987 (Nonaka and Takeuchi 1995). The early prototypes did not produce bread of sufficient quality. In order to improve the machine, the designers had to find out how to knead a dough. A member of the design team apprenticed with a master baker at a luxury hotel. The baker was unable to tell her in words what to do, but the design team member tried to imitate him, and in this way she gradually learnt the right movements. Finally she managed to express what she had learnt in words: a "twisting stretch" was required. She and her colleagues in the design team managed to construct a machine that performed a twisting stretch and baked (sufficiently) good-tasting bread.

What she achieved was an example of the *articulation* of tacit knowledge, by which is meant its expression in verbal instructions. Articulations of tacit knowledge have been performed with at least three different purposes. First, as in this example, articulation can facilitate the mechanization and automatization of a work process. Since the Industrial Revolution, engineers have mechanized work tasks previously performed by craftspeople. Articulation of tacit knowledge has been an important part of that process. Today, such articulation of tacit knowledge often takes the form of computer programming.

The second purpose of articulation is to facilitate teaching and learning. Learning a craft or profession would be incredibly inefficient and time-consuming if every learner had to repeat the mistakes of her predecessors. The articulation of tacit knowledge makes it possible to learn without long apprenticeships.

The third purpose of articulating tacit knowledge is to control other people's work. Ever since the Industrial Revolution, employers have systematically divided qualified tasks into simpler subtasks, most of which can be performed by cheaper labor. The extensive (tacit and explicit) knowledge of highly qualified workers has been codified and divided into small tasks that can more easily be taught and learnt. The assembly line is the most well-known example of this process. The American engineer Frederick Taylor (1856–1915), the pioneer of so-called scientific management, saw this dequalification of labor as a major purpose of new management practices. "All possible brain work should be removed from the shop and centered in the planning or laying-out department" (Taylor [1911] 2008: 50). But ever since the early days of the Industrial Revolution, critics, including Adam Smith (1776: V:i:ii) and Karl Marx (1867: I:12.5), have warned against the resulting deterioration of the quality of working life (cf.: Braverman 1974; Campbell 1989: 226; Wood 1982; Wood 1987).

Tacit knowledge has been romanticized and described as a superior type of knowledge, which should preferably not be articulated. From the viewpoint of engineering science, that is not a tenable approach to technological knowledge. We humans have developed language in order to convey insights, instructions, and other messages to each other. If someone knows how to use a tool or run a machine but cannot explain it to others, then it is certainly an improvement if that person's knowledge can be conveyed verbally so that others can learn how to operate the tool or machine. Articulation can also facilitate endeavors by engineers to introduce improvements. Both tacit and explicit knowledge can be in error, but explicit knowledge is more easily corrigible since it is more accessible to critical discussion and evaluation.

5.4.2 Practical Rule Knowledge

When designing certain load-bearing parts, engineers make them strong enough to carry twice the intended load. This is a practical and reasonably simple way to ensure that the constructions do not break, but there is no theoretical ground for choosing 2 as a safety factor (Doorn and Hansson 2011). This is an example of what we can call practical rule knowledge. Another example is the set of rules that electricians apply when connecting wire. Some of these rules have a theoretical justification, such as the rules on what types of cable to use at various tensions. Others are just convenient conventions, such as the choice of green and yellow for earth wires. These are "rules of thumb", rules that are easy to memorize and apply (Norström 2011). The term "rule of thumb" has been in use at least since the latter part of the 17th century. (It probably derives from the use of the thumb as a unit for length measurement.)

Rule-of-thumb knowledge has many advantages over tacit knowledge, but it also has the disad-vantage of requiring attention. It is therefore often an advantage to *routinize* it, i.e. make it possible to perform the task with at most sporadic attention. Routinization can be described as the opposite process of articulation. One typical example is when a novice driver picks up how to gear up or down without thinking of it. Another example of routinization is learning to play a musical instrument. In these and many other cases, the routinization of practical rule knowledge is necessary to make us able to perform various tasks without too much delay or effort. Routinization is also an important part of learning a practical craft. However, it has a less prominent role in engineering and in particular engineering science, since these activities tend to require a high degree of explicitness and articulability.

5.4.3 *Systematic Engineering Knowledge*

In addition to tacit knowledge and practical rule knowledge, there is an additional form of technological knowledge, namely systematic knowledge based on well-organized investigations of technological constructions. Today, such knowledge is obtained on a large scale through studies performed by engineering scientists in universities and industrial laboratories. We can use the term "technological science" to denote the scientific investigation of technological objects, constructions, and processes. However, it is important to realize that this type of technological knowledge has a much longer history than engineering (or natural science) (Hansson 2015). Through the centuries, technological development has largely been driven by craftspeople systematically trying out different constructions and methods. For instance, in order to find the right composition of mixtures such as mortar, cement, glue, and various alloys, it was necessary to try different proportions and compare the results. An interesting example of such experiments has been recorded in excavations in Raqqa, Syria, which was a world-leading center in glassmaking in the eighth to eleventh centuries. Analysis of debris from the workshops has revealed the use of a so-called dilution line in which different proportions of two components were tested in order to find the best one (Henderson et al. 2004, 2005).

The experimental tradition among craftspeople was one of the major sources of modern scientific methodology. Galileo Galilei (1564–1642) and other scientific pioneers learned much from skilled workers on the art of extracting information from nature by manipulating it, i.e. making experiments (Drake 1978; Zilsel 2000). But from the very beginning, experiments in the natural sciences had another goal than technological experiments. Craftspeople made experiments in order to solve practical problems, natural scientists in order to find out the workings of nature.

The experimental tradition in the crafts continued to develop in parallel with that of natural science. In the 18th and 19th centuries, millwrights performed advanced experiments and measurements, but they had little or no contact with the academic science of their times (Layton 1978). The experimental tradition in engineering science is largely a continuation of such traditions from advanced work by craftspeople. For instance, well into the 20th century, new metallurgic methods were tried out and tested in ironworks, based on experience rather than on principles and ideas from the natural sciences (Knoedler 1993). Today, there is no sharp distinction between experiments in engineering and in engineering science.

5.4.4 *Applied Natural Science*

As we saw earlier, engineering students in the early 19th century received a considerable grounding not only in mathematics but also in the natural sciences, in particular physics. But at that time, natural science still had a fairly small role in practical engineering. It was not until the latter half of that century that natural science began to be employed on a large scale to develop new technology. The chemical and electrotechnical industries were pioneers in this new development. Important inventions such as the telegraph were the outcomes of discoveries in university laboratories (Böhme 1978; Kaiser 1995).

Since then, the use of natural science in engineering, particularly in engineering science, has increased substantially. Knowledge from a wide range of areas in the natural sciences, including most branches of physics and chemistry and substantial parts of biology and the earth sciences, are now being incorporated in the development of new technology. We can therefore add incorporated natural science as a fourth type of technological knowledge, in addition to tacit knowledge, rule knowledge, and knowledge based on systematic investigations of technological constructions. Engineering scientists have to combine information from technological and natural science. As we saw at the beginning of Section 2, the relationship between these two forms of science is complex and goes far beyond the simplistic notion of application. Often, engineers also have to assimilate information from other branches of science, for instance behavioral science, which provides crucial information for the construction of human–machine interfaces.

5.5. Idealization in the Engineering Sciences

Idealization has a central role in all scientific endeavors. To idealize means in this context to restrict one's attention to certain properties of the study object that are presumed to be particularly important. Doing so is often necessary in order not to get lost in details. Idealization "may involve a distortion of the original or it can simply mean a leaving aside of some components in a complex in order to focus the better on the remaining ones" (McMullin 1985: 248). Studying the idealizations employed in a branch of science is often a good way to understand the perspective it applies to its study objects. The idealizations typically performed in the engineering sciences differ from those of the natural sciences in at least three ways (Hansson 2007).

First, the engineering sciences tend to stay at lower levels of abstraction than the natural sciences. High levels of abstraction are often conducive to the understanding of very general patterns. For instance, physicists working with string theory, or trying to understand black holes or the Big Bang, work with models that are more abstract, and at larger distance from the features of the material world that we can observe directly, than the models used in engineering work. Engineering science is concerned with the material objects that we have access to and can modify and transform, and for this purpose, such high-level abstractions are usually not helpful. However, it should be noted that over time, increasingly abstract and advanced physical theories have become practically useful. The use of quantum theory in the construction of electronic components is an example of this.

Secondly, in order to explain the underlying workings of the natural world, natural science investigates natural phenomena in isolation from each other. A physicist who studies electromagnetism uses models of electromagnetic phenomena in which gravitation is absent. Often experiments are performed under specially constructed conditions that are tailored to suit such simplified models. Hence, physical experiments are often performed in a vacuum in order to correspond to theoretical models in which the impact of atmospheric pressure has been excluded. Chemical experiments are conducted in gas phase in order to ensure that each pair of reacting molecules has no interaction with other molecules. This brings these experiments in closer correspondence to theoretical models of reaction mechanisms. Such idealizations are fruitful in the natural sciences, but they tend not to be workable in engineering or engineering science. The engineer who builds a machine based on electromagnetic principles cannot disregard the effects of gravitation, unless the machine is intended for use in outer space. Similarly, whereas investigations in theoretical mechanics can disregard weather conditions, an engineering scientist working with the construction of suspension bridges has to make sure that the construction resists severe storms and other extreme weather events (Layman 1989).

Thirdly, there is a difference between the ways in which mathematical models are used in the natural and the engineering sciences. Both branches of science use such models extensively, and both strive for a high degree of mathematical precision. However, their ideals of precision are not the same. The precision needed in the engineering sciences is usually of the same type as in practical engineering, which means that sufficient precision is always obtainable by means of a sufficiently good approximation. Hence, if the choice of wire dimensions for a suspension bridge depends on the solution of a complex system of equations, then the engineer does not need an analytical solution of the system. If a solution has been obtained with a sufficient number of decimals, then the engineering problem has been solved. In this, the engineer differs, for instance, from an astrophysicist or a population geneticist who needs to solve an equally complex system of equations. For the purposes of the natural sciences, an analytical solution is always preferred (although there are areas, such as quantum chemistry, in which it is seldom obtainable). This may partly be a matter of the aesthetic qualities of the solutions, but it is also a matter of their explanatory qualities. There is a good chance that the insights obtainable from an exact solution can contribute to the solution of similar problems in the future.

5.6. Conclusion

To sum up, engineering science has several important metaphysical and epistemological features that distinguish it from the natural sciences:

- It serves the needs of a practical profession, namely that of engineers.
- Its ultimate objects of study are human-made, rather than natural, objects.
- Its study objects have to be understood in terms of complex combinations of physical and functional characteristics.
- Its epistemology incorporates various forms of systematized action-guiding knowledge that is not based on natural science.
- It usually operates on a less abstract level than the natural sciences, and it refrains from many of the idealizations employed in these sciences.

These differences confirm that engineering science forms a major branch of science, on a par, for instance, with natural, social, behavioral, and medical science.

Related Chapters

Chapter 2: A Brief History of Engineering (Jennifer Karns Alexander)
Chapter 8: Prescriptive Engineering Knowledge (Sjoerd Zwart)
Chapter 9: The Art of Engineering and Engineering as Art (Lara Schrijver)
Chapter 15: Artifacts (Beth Preston)

Further Reading

Hansson, Sven Ove (ed.) (2015). *The Role of Technology in Science: Philosophical Perspectives*. Dordrecht: Springer. (A collection of studies of the relationship between technology and science.)
Hansson, Sven Ove (ed.) (2018). *Technology and Mathematics: Philosophical and Historical Investigations*. Cham: Springer. (A collection of studies of the relationship between technology and mathematics.)
Kroes, Peter (2012). *Technical Artefacts: Creations of Mind and Matter*. Springer 2012. (A thorough investigation of the ontology of technological artefacts. Contains important analyses of the notion of a technological function.)
Kroes, Peter und Meijers, Anthonie (eds.) (2000). *The Empirical Turn in the Philosophy of Technology*, Vol. 20 in Carl Mitcham (ed.), *Research in Philosophy and Technology*. Amsterdam: Elsevier. (This book was the

starting point for new developments in the philosophy of technology that put much greater emphasis on engineering.)

Meijers, Anthonie (ed.) (2009). *Handbook of the Philosophy of Science*, Vol. 9: *Philosophy of Technology and Engineering Sciences*. Burlington, MA: Elsevier. (With 48 chapters and 1,453 pages, this is an indispensable reference work for the philosophy of technology and engineering.)

Mitcham, Carl (1994). *Thinking Through Technology: The Path Between Engineering and Philosophy*. Chicago: University of Chicago Press. (An excellent introduction to philosophical studies of technology and engineering, with much historical material.)

Vincenti, Walter G. (1990). *What Engineers Know and How They Know It: Analytical Studies From Aeronautical History*. Baltimore: Johns Hopkins University Press. (Probably the most influential empirical study of engineering and its relationship with science.)

References

Alder, Ken (1999). French Engineers Become Professionals; or, How Meritocracy Made Knowledge Objective. In William Clark, Jan Golinski, and Simon Schaffer (eds.), *The Sciences in Enlightened Europe*. Chicago: University of Chicago Press, pp. 94–125.

Anon (1815). Obituary. *North-American Review and Miscellaneous Journal*, 1(1), 140–144.

Anon (1850). Improvement in Dams. *Scientific American*, 6(10), 76–76.

Bachrach, David Stewart (2006). English Artillery 1189–1307: The Implications of Terminology. *English Historical Review*, 121(494), 1408–1430.

Böhme, Gernot, van den Daele, Wolfgang and Krohn, Wolfgang (1978). The 'Scientification' of Technology. In Wolfgang Krohn, Edwin T. Layton Jr. and Paul Weingart (eds.), *The Dynamics of Science and Technology: Social Values, Technical Norms and Scientific Criteria in the Development of Knowledge*, Vol. II. Dordrecht: Riedel, pp. 219–250.

Braverman, Harry (1974). *Labor and Monopoly Capital: The Degradation of Work in the Twentieth Century*. New York: Monthly Review.

British Parliament (1826). *Norwich and Lowestoft Navigation Bill. Minutes of Evidence Taken Before the Committee on the Bill. For Making and Maintaining a Navigable Communication for Ships and Other Vessels, Between the City of Norwich and the Sea at or Near Lowestoft, in the County of Suffolk. Reports From Committees*, 2 February to 31 May, 1826, Vol. IV.

Bunge, Mario (1966). Technology as Applied Science. *Technology and Culture*, 7, 329–347.

Bunge, Mario (1988). The Nature of Applied Science and Technology. *Philosophy and Culture, Proceedings of the XVIIth Congress of Philosophy*, II, 599–604.

Campbell, Robert A. (1989). Work, Workers and Working-Class History. *Le Travail*, 23, 221–234.

Doorn, Neelke and Ove Hansson, Sven (2011). Should Probabilistic Design Replace Safety Factors? *Philosophy and Technology*, 24, 151–168.

Drake, Stillman (1978). *Galileo at Work: His Scientific Biography*. Chicago: University of Chicago Press.

Dyer, Joseph (2007). The Place of *Musica* in Medieval Classifications of Knowledge. *Journal of Musicology*, 24, 3–71.

Faulkner, Wendy (1994). Conceptualizing Knowledge Used in Innovation: A Second Look at the Science-Technology Distinction and Industrial Innovation. *Science, Technology and Human Values*, 19, 425–458.

Freedman, Joseph S. (1994). Classifications of Philosophy, the Sciences, and the Arts in Sixteenth- and Seventeenth-Century Europe. *Modern Schoolman*, 72, 37–65.

Grattan-Guinness, Ivor (2005). The Ecole Polytechnique, 1794–1850: Differences Over Educational Purpose and Teaching Practice. *American Mathematical Monthly*, 112, 233–250.

Grayson, Lawrence P. (1993). *The Making of an Engineer: An Illustrated History of Engineering Education in the United States and Canada*. New York: Wiley.

Hansson, Sven Ove (2006). Defining Technical Function. *Studies in History and Philosophy of Science*, 37, 19–22.

Hansson, Sven Ove (2007). Hansson "What Is Technological Science?". *Studies in History and Philosophy of Science*, 38, 523–527.

Hansson, Sven Ove (2013). What Is Technological Knowledge? In Inga-Britt Skogh and Marc J. De Vries (eds.), *Technology Teachers as Researchers*. Rotterdam: Sense Publishers, pp. 17–31.

Hansson, Sven Ove (2014). Values in Chemistry and Engineering. In Peter Kroes and Peter-Paul Verbeek (ed.), *The Moral Status of Technical Artefacts*. Dordrecht: Springer, pp. 235–248.

Hansson, Sven Ove (2015). Experiments Before Science. What Science Learned From Technological Experiments. In Sven Ove Hansson (ed.), *The Role of Technology in Science. Philosophical Perspectives*. Dordrecht: Springer, pp. 81–110.

Hansson, Sven Ove (2018). The Rise and Fall of the Anti-mathematical Movement. In Sven Ove Hansson (ed.), *Technology and Mathematics: Philosophical and Historical Investigations*. Cham: Springer, pp. 305–324.

Harris, Daniel G. (2001). *Fredrik Henrik af Chapman: The First Naval Architect and His Work,* revised ed. Stockholm: Literatim.

Henderson, J., Challis, K., O Hara, S., McLoughlin, S., Gardner, A. and Priestnall, G. (2005). Experiment and Innovation: Early Islamic Industry at al-Raqqa, Syria. *Antiquity,* 79(303), 130–145.

Henderson, J., McLoughlin, S.D. and McPhail, D.S. (2004). Radical Changes in Islamic Glass Technology: Evidence for Conservatism and Experimentation with New Glass Recipes from Early and Middle Islamic Raqqa, Syria. *Archaeometry,* 46(3), 439–468.

Hendricks, Vincent Fella, Arne Jakobsen, and Stig Andur Pedersen (2000). Identification of Matrices in Science and Engineering. *Journal for General Philosophy of Science,* 31, 277–305.

Hensel, Susann (1989a). Die Auseinandersetzungen um die mathematische Ausbildung der Ingenieure an den Technischen Hochschulen in Deutschland Ende des 19. Jahrhundert. In Susann Hensel, Karl-Norbert Ihmig, and Michael Otte, *Mathematik und Technik im 19. Jahrhundert in Deutschland. Soziale Auseinandersetzung und Philosophische Problematik.* Göttingen: Vandenhoeck and Ruprecht, pp. 1–111, 215–305.

Hensel, Susann (1989b). Zu einigen Aspekten der Berufung von Mathematikern an die Technischen Hochschulen Deutschlands im letzten Drittel des 19. Jahrhunderts. *Annals of Science,* 46(4), 387–416.

Hobbes, Thomas ([1651] 1909). *Leviathan,* with an Essay by W. G. Pogson Smith. Oxford: Clarendon Press.

Hoppe, Brigitte (2011). The Latin Artes and the Origin of Modern Arts. In Maria Burguete and Lui Lam, *Arts: A Science Matter.* Singapore: World Scientific, pp. 2, 35–68.

Houkes, W. (2009). The Nature of Technological Knowledge. In A. Meijers (ed.), *Handbook of the Philosophy of Science,* Vol. 9, *Philosophy of Technology and Engineering Sciences.* Burlington, MA: Elsevier, pp. 310–350.

James, Jamie (1995). *The Music of the Spheres.* London: Abacus.

Kaiser, Walter (1995). Die Entwicklung der Elektrotechnik in ihrer Wechselwirkung mit der Physik. In Lothar Schäfer and Elisabeth Ströker (eds.), *Naturauffassungen in Philosophie, Wissenschaft, Technik,* Vol. III: *Aufklärung und späte Neuzeit.* Freiburg and Munich: Verlag Karl Alber, pp. 71–120.

Klemm, Friedrich (1966). Die Rolle der Mathematik in der Technik des 19. Jahrhunderts. *Technikgeschichte,* 33, 72–90.

Knoedler, J.T. (1993). Market Structure, Industrial Research, and Consumers of Innovation: Forging Backward Linkages to Research in the Turn-of-the-century US Steel Industry. *Business History Review,* 67, 98–139.

Kroes, Peter (2006). Coherence of Structural and Functional Descriptions of Technical Artefacts. *Studies in History and Philosophy of Science,* 37, 137–151.

Kroes, Peter and Meijers, Anthonie (eds.) (2006). The Dual Nature of Technical Artefacts (special issue). *Studies in History and Philosophy of Science,* 37, 1–158.

Langins, Janis (2004). *Conserving the Enlightenment: French Military Engineering From Vauban to the Revolution.* Cambridge, MA: MIT Press.

Lawrence, Snezana (2003). History of Descriptive Geometry in England. In S. Huerta (ed.), *Proceedings of the First International Congress on Construction History, Madrid, 20th–24th January 2003.* Madrid: Instituto Juan de Herrera, pp. 1269–1281.

Layman, Ronald (1989). Applying Idealizing Scientific Theories to Engineering. *Synthese,* 81, 353–371.

Layton, Edwin (1978). Millwrights and Engineers, Science, Social Roles, and the Evolution of the Turbine in America. In Wolfgang Krohn, Edwin T. Layton Jr. and Paul Weingart (eds.), *The Dynamics of Science and Technology: Social Values, Technical Norms and Scientific Criteria in the Development of Knowledge,* Vol. II. Dordrecht: Riedel, pp. 61–87.

Lelas, Srdjan (1993). Science as Technology. *British Journal for the Philosophy of Science,* 44, 423–442.

Marx, Karl (1867). *Das Kapital. Kritik der politischen Oekonomie.* Erster Band. Hamburg: Otto Meissner.

McMullin, Ernan (1985). Galilean Idealization. *Studies in History and Philosophy of Science,* 16, 247–273.

Mitcham, Carl and Schatzberg, Eric (2009). Defining Technology and The Engineering Sciences. In Anthonie Meijers (ed.), *Handbook of the Philosophy of Science. Volume 9: Philosophy of Technology and Engineering Sciences.* Amsterdam: Elsevier, pp. 27–63.

Murchison, Roderick Impey (1844). Address to the Royal Geographical Society of London. *Journal of the Royal Geographical Society of London,* 14, xlv–cxxviii+336.

Noble, David F. (1997). *The Religion of Technology: The Divinity of Man and the Spirit of Invention.* New York: Alfred A. Knopf.

Nonaka, Ikujiro and Takeuchi, Hirotaka (1995). *The Knowledge-Creating Company: How Japanese Companies Create the Dynamics of Innovation.* New York: Oxford University Press.

Nonaka, Ikujiro and von Krogh, Georg (2009). Tacit Knowledge and Knowledge Conversion: Controversy and Advancement in Organizational Knowledge Creation Theory. *Organization Science,* 20, 635–652.

Norström, Per (2011). Technological Know-How From Rules of Thumb. *Techné. Research in Philosophy and Technology,* 15, 96–109.

Ovitt Jr., George (1983). The Status of the Mechanical Arts in Medieval Classifications of Learning. *Viator,* 14, 89–105.

Polanyi, Michael (1966). *The Tacit Dimension*. London: Routledge.

Purkert, Walter (1990). Infinitesimalrechnung für Ingenieure—Kontroversen im 19. Jahrhundert. In Detlef D. Spalt (ed.), *Rechnen mit dem Unendlichen. Beiträge zur Entwicklung eines kontroversen Gegenstandes*. Basel: Birkhäuser, pp. 179–192.

Purkert, Walter and Hensel, Susann (1986). Zur Rolle der Mathematik bei der Entwicklung der Technikwissenschaften. *Dresdener Beiträge zur Geschichte der Technikwissenschaften*, 11, 3–53.

Scharlau, Winfried (1990). *Mathematische Institute in Deutschland 1800–1945*. Braunschweig: Friedr. Vieweg & Sohn.

Schubring, Gert (1990). Zur strukturellen Entwicklung der Mathematik an den deutschen Hochschulen 1800–1945. In Winfried Scharlau (ed.), *Mathematische Institute in Deutschland 1800–1945*. Braunschweig: Friedr. Vieweg & Sohn, pp. 264–279.

Schummer, Joachim (1997). Challenging Standard Distinctions Between Science and Technology: The Case of Preparative Chemistry. *Hyle*, 3, 81–94.

Siemens, C. William (1869). The Latest Achievements in Engineering Science. *Scientific American*, 21(13), 193–194.

Siemens, C. William (1870). Mechanical Science. In *Report of the Thirty-ninth Meeting of the British Association for the Advancement of Science, Held at Exeter in August 1869. Notices and Abstracts of Miscellaneous Communications to the Sections*. London: John Murray, pp. 200–206.

Smith, Adam (1776). *An Inquiry Into the Nature and Causes of the Wealth of Nations*. London: W. Strahan and T. Cadell.

Tatarkiewicz, Wladyslaw (1963). Classification of Arts in Antiquity. *Journal of the History of Ideas*, 24, 231–240.

Taylor, Frederick Winslow (2008 [1911]). *Shop Management*. Sioux Falls: NuVision Publications.

Van Den Hoven, Birgit (1996). *Work in Ancient and Medieval Thought: Ancient Philosophers, Medieval Monks and Theologians and Their Concept of Work, Occupations and Technology. Dutch Monographs on Ancient History and Archaeology* 14. Amsterdam: Gieben.

Vermaas, Pieter E. and Wybo Houkes (2006). Technical Functions: A Drawbridge Between the Intentional and Structural Natures of Technical Artefacts. *Studies in History and Philosophy of Science Part A*, 37, 5–18.

Whitney, Elspeth (1990). Paradise Restored: The Mechanical Arts From Antiquity Through the Thirteenth Century. *Transactions of the American Philosophical Society*, 80, 1–169.

Wood, Stephen (ed.) (1982). *The Degradation of Work? Skill, Deskilling and the Labour Process*. London: Hutchinson.

Wood, Stephen (1987). The Deskilling Debate, New Technology and Work Organization. *Acta Sociologica*, 30, 3–24.

Zilsel, Edgar (2000). *The Social Origins of Modern Science*. Ed. Diederick Raven, Wolfgang Krohn, and Robert S. Cohen. *Boston Studies in the Philosophy of Science*, Vol. 200. Dordrecht: Kluwer.

6

SCIENTIFIC METHODOLOGY IN THE ENGINEERING SCIENCES

Mieke Boon

6.1. Introduction

Scientific research in the engineering sciences is ubiquitous and nowadays accounts for more research effort than the so-called *fundamental* or *basic* natural sciences. Nevertheless, within mainstream philosophy of science, relatively little attention has been paid to this type of *application oriented* scientific research.

The term *engineering sciences*, however, can be ambiguous and can elicit different understandings. Commonly, the emphasis is on the *engineering* part of the term, although in this chapter the focus will be on the *science* part—that is, on *scientific research in the context of technological applications*.

This chapter aims to address some salient aspects of the engineering sciences and their methodology, cumulating in a *methodology of scientific modeling* in the engineering sciences. Section 2 aims to characterize the engineering sciences by comparison with scientific research in the basic sciences. A noticeable difference is the role and character of phenomena (see Box 6.1), which in the basic sciences serve as aids in discovering and testing theories, while the engineering sciences analyze (physical-technological) phenomena in view of technological functioning or malfunctioning. Scientific research on technological problem-solving and innovation, therefore, is better cast in terms of design-concepts that are based on functional interpretations of phenomena, which is the topic of Section 3. Section 4 elaborates on the ways in which (physical-technological) phenomena are investigated and on the specific character of scientific knowledge for creating or controlling them by means of *physical-technological circumstances* (see Box 6.1). It will appear that scientific modeling of technological systems is central to the engineering sciences, encompassing both the modeling of physical-technological phenomena in specific physical-technological contexts as well as the modeling of technological artifacts producing specific phenomena. Section 5 explains how scientific models of (physical or physical-technological) phenomena are constructed, and proposes a methodology for their scientific modeling (the B&K method, Figure 6.2), which is on par with the well-known hypothetical-deductive methodology (the HD method, also called the empirical cycle, Figure 6.1). Each section concludes with some remarks about concepts (Boxes 6.1 and 6.2) and methodology (Boxes 6.3, 6.4, 6.5, and 6.6) that prepare for the B&K method of constructing scientific models.

6.2. What Is Engineering Science?

When it comes to the term *engineering science*, philosophers, as well as laypeople, often focus on the *engineering* part of the term, according to which it is synonymous with engineering and technology

---

Box 6.1 Clarification of Concepts: Phenomena and Target Systems

The term (physical) *phenomena* refers to existing or non-existent, observable or non-observable, often functionally interpreted physical *objects*, *properties* and *processes* (including *regularities* and *mechanisms*). The adjective 'physical' aims to indicate that phenomena *exist* or *potentially exist* in the real world (both nature and technology). The term *phenomenon* is often used in research-contexts to indicate the 'object of study.' Additionally, the term is used to indicate objects, properties and processes that (supposedly) cause or affect the investigated phenomenon or target-system (see question Q2, Figure 6.2).

The engineering sciences investigate *physical-technological* phenomena, which can be *technological* objects, properties and processes, but also *technologically produced* physical phenomena. For example, a membrane, an electromagnetic coil and a prosthesis are technological *objects*, which have specific *properties*, or which function by means of specific *processes*. Examples of technologically produced physical phenomena are light, sound, electricity, chemical compounds and all kinds of material properties. Additionally, the term *target-system* will be used, which is better suited when studying a technological artifact (see Figure 6.2, Q2).

(e.g., Vincenti 1990; Meijers, ed. 2009). With this interpretation, engineering science becomes synonymous with technological design (with bridges, cars and tools as paradigm examples), the development of technological artifacts (such as materials, instruments, devices, industrial processes and other systems), and research towards rational methods of engineering design and development. In this chapter, the emphasis is on the *science* part, taking the engineering sciences as scientific fields that in many respects resemble basic natural sciences, but with a number of salient differences.[1]

An example of an engineering science is scientific research in the context of chemical engineering. Chemical engineering as an *engineering* discipline aims at designing and building industrial processes for converting chemicals or raw materials into specific chemical compounds or materials that meet specific functions. Scientific research in the field of chemical engineering typically concerns *how to* produce a specific chemical compound or material that meets the functional and quality requirements in an effective, reliable, safe and economically feasible manner.

That is why the engineering sciences qua science deal both with scientific knowledge concerning questions such as how a product or technological device can be made that meets specific functional and quality requirements, and how industrial processes for the production of a product or device must be designed and built. These *how to* questions are usually analyzed in terms of a whole series of questions about mutually coordinated aspects, for example, parallel and serial sub-processes that together form the intended technological configuration or system. Examples of *how to* questions in chemical engineering concern sub-processes such as: *how to* produce a specific chemical compound (e.g., a medicine or agricultural chemical); *how to* achieve the desired chemical conversion or material property and to prevent the undesirable ones (e.g., by using a catalyst or by control of physical-chemical circumstances); *how to* separate the desired from the undesired compounds (e.g., by distillation or membrane filtration processes) and *how to* optimize these sub-processes to reduce costs related to chemicals, energy and waste. Similar types of questions are asked in the context of developing technological devices in material sciences, electrical engineering, mechanical engineering, biomedical engineering and so forth.

In engineering scientific research, these questions are phrased as being about *phenomena* (Box 6.1), which are physical and/or technological *objects* (e.g., a protein, a membrane, an electrical circuit, a prosthesis), *properties* (e.g., hydrophobic, magnetic, electrical resistant, elastic) and *processes* (e.g., chemical production processes, conduction of heat, electromagnetic processes, transfer of momentum, conversion of energy). Furthermore, a phenomenon often consists of sub-phenomena, which will be explained in greater depth in Section 4.

Scientific research is crucial to design and development in the sense that scientific research aims at *knowledge for* the design and development of technology. Here it is important to keep in mind that scientific research aimed at *knowledge for* is basically motivated by *how to* questions in technological contexts (see earlier in this section and Box 6.2).

Box 6.2 Clarification of Concepts: Knowledge of Phenomena

Similar to the distinction between the *real world* and *knowledge of that world* stressed by Giere et al. (2006), a conceptual distinction is needed between *phenomena* and *knowledge of phenomena*. The term *knowledge of phenomena* will cover different types of *epistemic artifacts*, such as: *descriptions* of phenomena (Figure 6.2, Q2 and Q5); *concepts* of phenomena (e.g., 'elasticity,' 'conductivity,' 'sound-waves,' 'prosthesis,' 'energy,' 'motor,' 'amplifier'; Figure 6.2, Q2); *explanations* of phenomena (e.g., Figure 6.2, Q8 and Q9); *measurements* of (aspects of) phenomena (Figure 6.2, Q6); and, *scientific models* of phenomena (centre in Figure 6.2).

Additionally, *phenomenological laws* are the verbal or mathematical representations of reproducible phenomena (Boon 2011b, 2017a). For example, technological devices in an experimental set-up (e.g., Boyle's, Hooke's and Ohm's experiments) produce reproducible *physical-technological phenomena* that are represented by sentences (e.g., 'there exists an inversely proportional relationship between the pressure and the volume of a gas,' or 'increasing the pressure causes a decrease of the volume') and/ or by mathematical equations that relate measured quantities (e.g., $P.V = c$, $F = k.x$, and $V = I.R$).

Engineering sciences aim at *knowledge of phenomena* that can be used to answer 'how to' questions. As this knowledge is meant to be used in *epistemic activities* by humans or computers (e.g., regarding questions 'how to' create, design, optimize, etc.) the types of knowledge just mentioned are also called *epistemic artifacts*. Hence, research aims at *knowledge for* specific epistemic purposes (Figure 6.2, Q3)—for example, a mathematical model of the phenomenon (i.e., the model in Figure 6.2) can be used to make predictive calculations.

Physical-technological phenomena can be created, controlled, optimized, etc., by means of *physical-technological circumstances*, (e.g., temperature, pressure, light, field-strength, chemical composition, material properties, and features of the technological device). Therefore, qualitative or quantitative knowledge of a physical-technological phenomenon must include the role of relevant physical-technological circumstances (see questions Q5 and Q6, Figure 6.2).

An example is scientific research into membranes. This concerns, for example, the question of *how to* make a material with which dissolved molecules such as sea salt or proteins can be separated from a fluid, for technological applications such as the production of drinking water from seawater, and medical applications such as kidney dialysis. Another example is research into technological artifacts such as prostheses.

Research topics in the engineering sciences can be motivated in various ways, ranging from *pragmatic* problem-solving to more *fundamental innovative* approaches, which are the subject of Section 3. In a pragmatic approach, research is usually motivated by specific problems of a technology, and it often starts off with a systematic 'trial-and-error' approach in an experimental setup to investigate which physical-technological factors have an effect (see Box 6.2). This pragmatic approach could be characterized as the transition between engineering and science. While a researcher in an *engineering* context who follows this approach is satisfied with finding a workable solution, *scientific* research in the engineering sciences usually aims at a *more fundamental understanding* of the phenomena involved. For example, if the technological problem is to improve the selectivity of a specific membrane, experiments will be set up to examine relevant interrelated physical-technological factors (e.g., the chemical structure of the membrane, selectivity, efficiency, permeability, fouling, strength, membrane thickness, pore size and size distribution, osmotic pressure and shear stress), and to investigate how changes of the material properties affect these factors and vice versa. Also, existing scientific understanding of these factors will be used to design the experimental setup and to interpret the measurements, which results in increasing scientific understanding of the specific *physical-technological phenomena* (i.e., materials, properties and processes) involved.

Summing up, similarity between the engineering sciences and the basic natural sciences consists in aiming to understand phenomena scientifically. In both cases, this can be done in a systematic 'trial and error' manner and in a more fundamental way, which often requires more innovative scientific approaches (Section 3). Both develop and use technological instruments for experimentally generating and investigating the phenomena. Also, both adopt systematic approaches and scientific methodology to check the results. Additionally, both aim at scientific modeling, the results of which are published in scientific articles. A salient difference, however, is due to the application-context of designing and developing technological artifacts. Therefore, in the engineering sciences, the *epistemic aim* of investigating phenomena and of developing technological instruments is not firstly to discover or test scientific theories, but rather, to obtain *knowledge for* how a functionally relevant phenomenon is created, produced, improved, controlled, manipulated, prevented or optimized through physical-technological circumstances (Box 6.2).

This application context of aiming at *knowledge for* reasoning about phenomena in specific physical-technological circumstances, i.e., knowledge that enables and guides epistemic agents' reasoning about a phenomenon in physical-technological circumstances actually implies that the term 'phenomenon' has a slightly different meaning in the engineering sciences (Box 6.1)—and thus also differs from how the philosophy of science usually thinks about phenomena (e.g. Hacking 1983; Bogen and Woodward 1988; Bogen 2011; Woodward 2011; Bailer-Jones 2009). As already said,

---

Box 6.3 *Methodology: Towards the B&K Method of (Re-)Constructing Scientific Models—From Technological Function to Phenomena*

How do the indicated similarities and differences between the engineering sciences and the basic natural sciences affect methodology? A few aspects can already be pointed out. To begin with, scientific research in the engineering sciences is motivated by a technological application context, which requires a specific methodological approach when specifying the research question (Figure 6.2, Q1). The research question relates to, say, the proper (or improper) functioning of technological artifacts, which is understood in terms of physical-technological phenomena (Figure 6.2, Q2). Accordingly, scientific research aims at practical and scientific understanding of how these phenomena are technologically produced (or created, prevented, controlled, improved, manipulated, optimized, affected, etc.), that is, at the construction of the scientific model of the phenomenon (center of Figure 6.2), which usually goes hand in hand with the development of technological devices and procedures (for their actual production, or their creation, prevention, control, improvement, etc.), and often also with the development of computer programs for simulating them in the development and design phase.

---

in basic natural sciences, phenomena are aids to research. Conversely, in engineering sciences, phenomena are part of the goals of research. Furthermore, phenomena are interpreted in terms of their physical features *and* in terms of their technologically relevant (dys)function. They have a dual nature, so to speak (Kroes 1998, 2010; Weber et al. 2013). It is also important to recognize that in the basic natural sciences, scientific understanding of phenomena is usually detached from technology—phenomena are often presented as somehow 'free-floating' in nowhere. Contrari-wise, in the engineering sciences, phenomena are firstly understood as 'embedded' in physical and physical-technological circumstances.

6.3. Design-Concepts in the Engineering Sciences

Innovative approaches often start with a new *design-concept* for a specific technological function. Whereas a more pragmatic approach aims to improve the performance of an already existing technol-ogy, not fundamentally changing the physical-technological way in which the specific technological function is achieved, a more innovative approach starts from asking whether the same technological function can perhaps be achieved in a different way. A well-known strategy is biomimicry or bio-inspired approaches, where scientists learn from nature by looking how specific functions are gener-ated in biological systems (e.g., how natural membranes work, how photosynthesis produces energy, or how musculoskeletal systems work). An innovative strategy can aim at solving existing problems, but also at new innovative opportunities that no one has ever thought of. Creative minds look at salient phenomena from a *functional perspective*, thereby inventing new design-concepts for new ways of achieving technological functions. Let us look at a few examples that elaborate on technologies for the production of sugar.

When one learns chemistry in high school and sees a stoichiometric equation of how glucose (sugar) is produced from carbon dioxide and water and vice versa—$6\,CO_2\,(g) + 6\,H_2O\,(l) = C_6H_{12}O_6$ (aq) $+ 6\,O_2\,(g)$—one may tend to believe that this is how glucose can be produced industrially. This thought is a rudimentary design-concept. However, this chemical conversion is not effective, nor economically feasible, and in any case impossible to do in a *simple* way.

What seems feasible is another design-concept in which carbon dioxide is converted by means of sunlight into glucose or other 'energy-rich' compounds such as hydrogen gas or ethanol, by using natural catalysts. This design-concept is attractive because it can help to solve the problem of storing solar energy (e.g., as produced by solar panels on roofs). 'Clean' energy-rich compounds thus pro-duced can then be supplied to already existing technological systems that produce electrical energy or heat by converting these compounds back to CO_2 (carbon dioxide) and H_2O (water). Hence, the general, 'to be realized' design-concept is *artificial photosynthesis to harness solar energy* (e.g., Pandit et al. 2006; Ong et al. 2018). Yet, this *general design-concept* needs to be made more concrete—it needs to be translated into design-concepts of realizable technological systems for fulfilling the technological function of artificial photosynthesis.

The development of an artificial photosynthesis technology faces major challenges. For instance, how to create an effective, robust, etc. (bio-inspired), catalyst for CO_2 reduction by sunlight; how to create a carrier for the catalyst; how to create a large surface area at which the sun shines and artificial photosynthesis takes place. One design-concept to meet these challenges is to create a high surface area on which sunlight can shine and artificial photosynthesis can take place, by means of a reactor consisting of shallow micro-channels. In this so-called *opto-fluidic micro-reactor*, a membrane separates the CO_2-containing gas phase from the water phase flowing through the micro-channels. The mem-brane is also the carrier for the catalyst that promotes photosynthesis. Hence, the membrane has three functions: it is the surface at which photosynthesis takes place; it selectively allows transport of specific molecules across the membrane; and it separates the water flow that transports the energy-rich products from the gas flow that transports CO_2 (Huang et.al. 2018).

Design-concepts also concern the bio-inspired catalyst, which are inorganic molecules called *artificial light-harvesting antennas*. Scientific research firstly aims at finding molecules that transfer charge (electrons), and next, at understanding the properties of these light-harvesting antennas—for instance, in order to design molecular structures that have the highest possible electron-transfer efficiency (e.g. Uosaki et al. 1997; Gust et al. 2001; Dubey et al. 2016).

Summing up, the engineering sciences typically start from problems regarding specific functions or dysfunctions. Crucially, scientific researchers aim to understand these problems in terms of *physical-technological phenomena* (i.e., objects, properties and processes). From an engineering science perspective, physical phenomena can be naturally or technologically generated—in this latter case, we call them physical-technological phenomena. Also, phenomena can cause the functioning but also the malfunctioning of a technology. The engineering sciences, therefore, can be understood as aimed at the creation and control of physical-technological phenomena through technological means. The epistemic task is scientific *knowledge for how to* do this. This knowledge concerns the physical-technological phenomena, including the technological devices that can be understood in terms of desired functional and undesired dysfunctional phenomena.

Box 6.4 *Methodology: Toward the B&K Method of (Re-)Constructing Scientific Models—From Phenomena to Design-Concepts*

The examples in this section show how scientific researchers in the engineering sciences are innovative by inventing new design-concepts for specific functions, which often start at a very general level. The challenge is to translate these into more concrete design-concepts; for instance, a concept of a technological system that actually does the job. This is where scientific research starts, and also, where many brilliant ideas ultimately fail (Figure 6.2, Q1 and Q2 cover this dynamics). The presented examples also show that although a general design-concept is often inspired by functional aspects elsewhere (e.g., bio-mimicking or bio-inspired technology), the functional parts (e.g., natural membranes, micro-channels and catalysts in natural photosynthesis and musculoskeletal systems) usually are not literally reproduced in the technology. Instead, the task of the engineering sciences is to find out how functional objects, properties and processes that are not yet existent, but that are presented as design-concepts, can be created by 'artificial' technological means. Importantly, this also implies that the engineering sciences aim not only at utilizing existing phenomena, but also at *creating* physical-technological phenomena that do not exist as yet (Boon 2012, 2017a).

6.4. Scientific Knowledge for Creating or Controlling Physical-Technological Phenomena

In scientific research that aims at (technologically) creating or developing the function indicated by the design-concept, *knowledge of physical-technological phenomena* can play several roles. Researchers may aim to technologically mimic the physical phenomena, which requires scientific research that gets them from a mere *description* of the phenomenon to a *scientific model* of it that allows for its technological reproduction in a more or a less literal fashion (Figure 6.2). For instance, researchers may try to technologically create chlorophyll molecules in order to develop artificial photosynthesis technology, but often these kinds of approaches are not feasible. Nevertheless, in-depth scientific understanding of the phenomenon (e.g., knowledge of the chemical structure of the chlorophyll molecule, its bio-chemical mechanism and its energetic efficiency) is of crucial importance (Pandit et al. 2006; Dau et al. 2017). As was illustrated earlier, rather than literally mimicking, researchers often aim only at utilizing the underlying 'physical principles,' which involves the interpretation of the functioning of the phenomenon at a more abstract conceptual level. For example, scientific research that aims at developing 'artificial photosynthesis' builds on knowledge of the phenomenon 'electron-transfer by chlorophyll molecules' as a principle by which solar energy is harnessed. This knowledge is used to develop artificial inorganic molecules that can do the electron transfer in a 'similar' fashion (e.g., Uosaki et al. 1997; Dubey et al. 2016).

Another role played by *knowledge of physical phenomena* in the (technological) creation, control or development of the function indicated by the design-concept is based on a modular approach that is made possible by thinking in terms of phenomena. In the engineering sciences—technologically produced—physical-technological phenomena are conceived as *physical building blocks* for physically creating, managing and developing technological functions. Scientific knowledge of physical-technological phenomena, therefore, consists of *epistemic building blocks* for the design and development of a specific technological function. Hence, modular knowledge of phenomena enables and guides the understanding of a design-concept in terms of mutually interacting phenomena—also indicated by higher-level/lower-level, horizontal/vertical and parallel/sequential phenomena as well as networks thereof. Knowledge of phenomena is usually represented through scientific models (center of Figure 6.2), and the modular approach allows for the construction of so-called multi-scale models. Hence, this scientific approach prompts scientific researchers to analyze a specific functional phenomenon in terms of a combination of interacting phenomena.

In chemical engineering, for example, the full chemical process for, say, producing a chemical compound is analyzed in terms of more basic phenomena, such as desired and undesirable chemical reactions in producing the intended chemical compound; the transport of liquids, gases and solids within the device; the transport of chemical compounds by means of fluid flow and diffusion processes within the fluid; the transport of heat by convection and conduction, and other physical processes such as absorption, dissolution, ionization, precipitation, vaporization and crystallization.[2]

A crucial strategy in the engineering sciences is to analyze physical phenomena in nature and as produced by technological devices in terms of more basic physical phenomena. Importantly, although physical phenomena are referred to as individual physical entities, they should not be understood as entities that exist *independently*. Many physical phenomena are technologically generated and are in that sense *dependent* on physical and technological circumstances. Examples of technologically produced phenomena—i.e., physical-technological phenomena—are electrical currents, electromagnetic waves, piezo-electricity, super-conductivity and selective permeability of membranes. Other 'naturally occurring' phenomena manifest only at specific, controlled physical-technological conditions (e.g., catalyzed chemical reactions, boiling point of a substance, standing sound waves, electron beams, and Röntgen rays). Nevertheless, in order to be recognized as a phenomenon at all, it must manifest itself in a reproducible manner (Bogen and Woodward 1988). From a technological point of view, it is precisely these characteristics of phenomena that make them technologically interesting and, moreover, manageable for scientific research.

Therefore, different from the more naïve view of independently existing phenomena as we often see in textbooks, it is stressed here that technologically relevant phenomena need to be understood as in a relationship with relevant physical and technological circumstances. The task of scientific research in the engineering sciences is then to generate scientific knowledge of physical or physical-technological phenomena significant to specific technological problems or functions that take into account the role of relevant physical-technological circumstances (see Box 6.2).

Furthermore, in the engineering sciences, descriptions, conceptions, explanations, measurements and scientific models of physical or physical-technological phenomena are closely related to *design-concepts*, but this relationship is multi-layered. As has been illustrated earlier, knowledge of salient physical or physical-technological phenomena often inspires a design-concept when the observed or purported phenomenon is interpreted in terms of a specific function. Photosynthesis (i.e., a purported, non-observable phenomenon that causes observed physical or physical-technological phenomena) is interpreted in terms of the function 'energy-production'; while the human musculoskeletal system can be interpreted in terms of 'mechanical movement and support of the body,' and next, as a design-concept for the exoskeleton. Making these creative leaps between 'knowledge of a phenomenon' to 'function of the phenomenon' involves an act of 'seeing as' by an engineering mind that asks how the phenomenon can be practically and technologically utilized.

Box 6.5 *Methodology: Toward the B&K Method of (Re-)Constructing Scientific Models—The Production of Scientific Knowledge*

Scientific research can produce different types of scientific knowledge about a phenomenon under study. An elementary form of scientific knowledge of a physical phenomenon is observed regularities in measurements, which are represented by means of phenomenological laws or mathematical equations that describe causal relations, correlations or statistical relationships between observed and measured variables (also see Box 6.2). Scientific models of the phenomenon are a more sophisticated form of scientific knowledge. These models can be mathematical, causal-mechanistic or diagrammatic models, or a combination of these, which includes multi-scale and network models (Figure 6.2, Q4). In the engineering sciences, scientific knowledge production is also guided by *epistemic purposes* in specific technological application contexts (Figure 6.2, Q1), while it is confined by available knowledge (Figure 6.2, Q8) and (experimentally or computationally generated) data about the phenomenon (Figure 6.2, Q5).

6.5. Constructing and Reconstructing Scientific Models: The B&K Method Next to the HD Method

In the engineering sciences, education in *mathematical* modeling of phenomena and technological systems is strongly developed. Several mathematical approaches can be distinguished (e.g., Dym 1980/2004). The most rudimentary approach starts from reproducibly measured, quantitative datasets and aims to find mathematical patterns or structures in them, represented by an algorithm, such as linear or exponential equations (considered *phenomenological* laws such as Boyle's law or Hooke's law) or a set of equations that forms a mathematical model for a specific phenomenon or system. Another approach is to translate a conceived mechanism into a mathematical model. Examples are the Lotka-Volterra model of predator-prey dynamics and the Maxwell-Boltzmann model for describing the mean distribution of non-interacting particles at random velocity in statistical mechanics. Yet another well-known approach is to interpret a system or phenomenon as being a specific type (e.g., a harmonic oscillator as being a Newtonian system), and then construct a mathematical model by derivation from the fundamental (axiomatic) theory (i.e., fundamental theories such as Newtonian mechanics, quantum mechanics, fluid mechanics, thermodynamics and electromagnetism) for this (idealized) type of system. In scientific practices, mathematical models are often constructed through a combination of these approaches, including the construction of diagrammatic models (Boon 2008).

Surprisingly, however, engineering science education often pays little attention to non-mathematical ways of scientific modeling. Rather, the *hypothetical-deductive* method (e.g., as in Hempel 1966) is usually put forward as a methodology to *test* explanations of observed phenomena (i.e., the *hypothesis* in the HD method; Figure 6.1), while assuming that the way in which hypotheses come about is either a highly creative process or rather superficially by inductive inference, as in 'all swans are white.' Hence, although HD reasoning definitely plays a role in testing hypotheses, it hardly provides

HD method in natural sciences

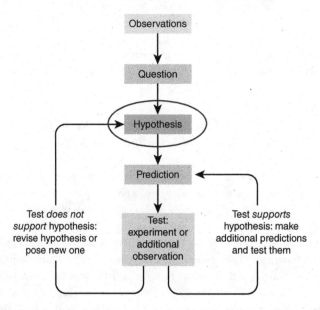

Figure 6.1 Hypothetical Deductive method in natural sciences. Schema based on Hempel (1966)

Source: Redrawn from Redrawn from Google Image: Scientific methodology. Accessed: November 30, 2005.

Box 6.6 *Methodology: Toward the B&K Method of (Re-)Constructing Scientific Models*

The B&K method for (re-)constructing scientific models considers scientific models for physical or physical-technological phenomena. It suggests that a model can be analyzed in terms of several elements. The B&K method consists of ten questions to systematically analyze these elements (see caption Figure 6.2, questions Q1–10). Thereby, it allows the analysis of the elements that play a role in constructing a scientific model or in reconstructing how a model is put together. Importantly, this list is not intended to present an algorithm by means of which the scientific model can be derived—as seems to be the case in mathematical modeling based on fundamental theories. Instead, the B&K method guides the process of gathering relevant information—including different types of scientific knowledge—about the phenomenon, which researchers need to fit together into a scientific model that suits their epistemic purpose(s). Vice versa, the B&K method guides the reconstruction of scientific models (e.g., models presented in scientific articles or research projects) by systematically retrieving the elements that went into them.

---

guidance to how scientific models that explain physical-technological phenomena (e.g., in terms of non-observable phenomena and physical-technological circumstances) are put together.

This defective representation of the construction of scientific models is a serious shortcoming of textbooks in the engineering sciences. This shortage is also present in the philosophy of science and in the philosophy of engineering.[3] Another deficiency due to not teaching students

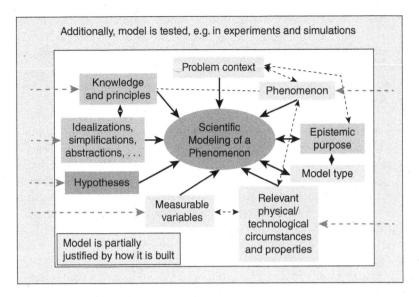

Figure 6.2   The B&K method

The B&K method consists of ten 'what is/are?' questions (Q1–10) to systematically determine the concrete elements that are 'built-in' to the scientific model:

(Q1) Problem context?
(Q2) Target-system or physical-technological phenomenon (P) for which the model is constructed?
(Q3) Intended epistemic function(s) of the model?
(Q4) Model type?
(Q5) Relevant (physical and/or technical) circumstances and properties (i.e., those that affect the phenomenon)?
(Q6) Measurable (physical-technological) variables (i.e., by which variables is a non-observable phenomenon connected to the tangible world, or, by which variables is the phenomenon or target-system affected)?
(Q7) Idealizations, simplifications and abstractions?
(Q8) Knowledge (theoretical knowledge, knowledge of sub-phenomena, phenomenological laws, empirical knowledge) and theoretical principles used in the construction of the model?
(Q9) Hypotheses (e.g., new concepts and explanations) 'built-in' to the model?
(Q10) Justification and testing of the model? The red arrows indicate elements that can be modified when testing and improving the model. The yellow square surrounding the modeling process indicates the testing phase in Q10.

how scientific models for phenomena are constructed is the way in which explanatory or causal-mechanistic models (e.g., the model of DNA) are usually presented in textbooks. These models suggest, first, that the phenomenon has been *observed* in science; second, that the phenomenon exists independent of physical or physical-technological conditions; and, thirdly, related to the above, it is usually not reported how the phenomenon is actually detected, nor how it is produced or manipulated by means of technological devices—that is, it is often not made clear how the purported (non-observable) phenomenon is connected with the tangible world. In short, explaining scientific research in terms of the HD method *only* does not serve to understand how scientific *knowledge for* specific epistemic tasks (such as those described in this chapter) is produced and utilized.

Since Morgan and Morrison's (1999) collection *Models as Mediators* it is generally accepted that scientific models often are autonomous epistemic artifacts, rather than merely derived from fundamental theories, as was assumed in the semantic view of theories.[4] Accordingly, Boumans (1999) argues that scientific models consist of several 'ingredients,' and that the way in which these ingredients are put together already provides quite a bit of the justification of the model. Knuuttila and I have elaborated on this idea, resulting in a more specific list of elements that usually play a role in building scientific models. Moreover, we argue that researchers can actually use scientific models for epistemic tasks *because* of how models are built, rather than models being first and foremost objective, human-independent representations of their real-world target (Boon and Knuuttila 2009; Knuuttila and Boon 2011).[5]

Basically, the collection of questions (Q1–10) that form the heart of the B&K method account for several important characteristics of scientific modeling in the engineering sciences. The first characteristic is that the *target-system or physical-technological phenomenon of interest* (Q2) is often investigated in a broader, more complex *problem-context* (Q1), and that the scientific model of the phenomenon must be such that it allows for *epistemic uses* in that problem-context (Q3). Thus, the intended epistemic uses may call for a specific *type of model* (Q4). For instance, if researchers aim at investigating the dynamic behavior of the system in computer simulations, one needs a mathematical model, but if researchers aim at knowledge of how to create a specific function, a (causal-)mechanistic or network model is needed. Hence, what the scientific model of the phenomenon 'looks like' is partially determined by this context.

A second characteristic is that researchers usually aim to investigate and to scientifically model a functional or dysfunctional phenomenon—or target-system—in isolation. Hence, whereas the *problem-context* (Q1) refers to the function of a technological artifact, the specific research usually focuses on *specific (sub-)phenomena or parts of the target-system* (Q2), for instance, (sub-)phenomena or parts that are held causally responsible for the (mal-)functioning of the technological artifact, or on discovered (sub-)phenomena found in measured data (Boon 2012).

A third characteristic is that the construction of scientific models builds on *existing knowledge* (Q5 and Q8). To begin with, scientific models are therefore considered hubs for bringing together and integrating relevant *information and scientific knowledge* about the phenomenon. Gathering knowledge about the phenomenon is not only based on textbook knowledge and scientific literature, but it is also often produced in an interaction with developing and using experimental (and computer) models (Nersessian 2009b; Nersessian and Patton 2009). The B&K method distinguishes two types of knowledge, although there may be overlap. Q5 focuses on *knowledge about the physical-technological circumstances* affecting the phenomenon, while Q8 focuses on all kinds of *scientific knowledge* about the phenomenon or target-system P—including knowledge of 'lower-level' phenomena that are held causally responsible for (properties and functioning of) the phenomenon or target-system P. Clearly, the construction of the model depends on available knowledge, i.e., the sophistication of a scientific field. Yet, the phenomenon or target-system do not provide information about what should be included in the model. Instead, the expertise of scientific researchers is crucial for deciding which information and scientific knowledge about the phenomenon is relevant to the model (and in view of the intended *epistemic aim* of the model, Q3).

A fourth characteristic is that the scientific model must account for how (aspects of) the phenomenon or target-system are latched to the tangible world—how are these aspects *detected in measurements* (Q6), and how is the phenomenon or target-system affected by *physical-technological conditions* (Q5)? It is important to notice that not all *relevant conditions* (Q5) are actually *measurable* (Q6). Pointing out what is measurable indicates the confines within which a scientific model is built. Additionally, the measurable variables and parameters determine the *testability* of the model (Q10). Aspects in the model that cannot somehow be latched to (imaginable) measurable or observable data do not make sense, which is why in scientific research scientific modeling often goes hand in hand with the development of experimental and measurement procedures (Boon 2017a), and also, why the construction of scientific models is confined by measurement procedures (even if only conceivable and not yet actual) and available data.

Fifth, the B&K method points out several elements common to *model-building*. Usually, modeling involves *idealizations, abstractions and simplifications* (Q7). Different from the naïve idea that models are an idealization of something else—implying that idealizations, etc., are made on the basis of some kind of unpolished picture of the non-observable (!) phenomenon or aspects of the target-system—idealizations, abstractions and simplifications rather concern the heterogeneous bits of information and scientific knowledge that go into the model (Q5 and Q8). Making the decisions on 'what goes in and what not,' involves, for instance, a trade-off between simplicity and exactness, which needs to be assessed in view of the *epistemic purpose* of the model (Q3). Another important aspect of model-building is that *hypotheses* of all kinds may be built into the model (Q9). These hypotheses are sometimes 'real discoveries' concerning explanations of the phenomena, but they are often just 'small insights' concerning assumptions on the relevance of *physical and technological circumstances* (Q5) or *sub-processes* (Q8), or the assumptions made in *simplifications* and *idealizations* (Q7).

Finally, a crucial characteristic of modeling is that the model needs to be justified and tested (Q10). As previously said, part of the justification is already done through justified decisions by researchers along the lines of the questions in the B&K method. Additionally, the model can be tested by comparison between model-predictions and outcomes of experiments or computer simulations—which is the type of justification basically assumed in the semantic view (e.g., Suppe 1989; Giere 2010). Negative test-results will require reconsidering justification of the elements built into the model.

It is not claimed that every scientific model necessarily entails these ten elements. Rather, the B&K method aims at providing a schema that assists the (re)construction of scientific models of phenomena, similar to the way in which the schema of the hypothetical-deductive method assists in recognizing elements playing a role in the dynamics of scientific research that aims at theories.

To conclude, the B&K method provides important insights into the general character of scientific models in the engineering sciences. In the philosophy of science, it is widely held that models *represent* their target (the physical or physical-technological phenomenon), and debate concerns what constitutes this representational relationship. As said, scientific models presented in the basic sciences seem to suggest that there is a *similarity relationship* as in a drawing or a photo. Yet, this suggestion is epistemologically unsound. Furthermore, in the philosophy of science it is often assumed that models are mere non-linguistic entities (mathematical structures, pictures, diagrams), whereas the B&K method emphasizes that scientific models also entail descriptive, numerical and theoretical, empirical and practical 'background' information relevant to the phenomenon or target-system, and also that this content is critical to epistemic uses of scientific models. Constructing and understanding scientific models in the engineering sciences, therefore, crucially involves the elements pointed out in the B&K method.

Acknowledgements

This work is financed by a Vidi grant (2003–2009) and an Aspasia grant (2012–2017, 409.40216) from the Dutch National Science Foundation (NWO) for the project 'Philosophy of Science for the Engineering Sciences.' A draft of this article was presented at the *Models and Simulations* conference (MS8) in 2018 at the University of South Carolina, and I wish to thank the organizers and the audience for helpful suggestions. I also wish to thank Henk Procee, Neelke Doorn, Diane P. Michelfelder and Thendral Govindaraj for their constructive suggestions. Thanks to Claire Neesham for reading the paper as native English speaker.

Related Chapters

Chapter 1: What Is Engineering? (Carl Mitcham)
Chapter 2: A Brief History of Engineering (Jennifer Karns Alexander)

Chapter 3: Western Philosophical Approaches and Engineering (Glen Miller)
Chapter 5: What Is Engineering Science? (Sven Ove Hansson)
Chapter 9: The Art of Engineering and Engineering as Art (Lara Schrijver)
Chapter 10: Creativity and Discovery in Engineering (David H. Cropley)
Chapter 13: Systems Engineering as Engineering Philosophy (Usman Akeel and Sarah Bell)
Chapter 15: Artifacts (Beth Preston)
Chapter 16: Engineering Objects (Wybo Houkes)
Chapter 18: Function in Engineering (Boris Eisenbart and Kilian Gericke)
Chapter 19: Emergence in Engineering (Peter Simons)
Chapter 20: Towards an Ontology of Innovation (Vincent Blok)
Chapter 21: Engineering Design (Peter Kroes)
Chapter 25: Sustainable Design (Steven A. Moore)
Chapter 27: Measurement (Lara Huber)
Chapter 28: Models in Engineering and Design: Modeling Relations and Directions of Fit (Michael Poznic)
Chapter 29: Scale Modeling (Susan G. Sterrett)
Chapter 30: Computer Simulations (Hildrun Lampe)
Chapter 31: Experimentation (Viola Schiaffonati)
Chapter 32: On Verification and Validation in Engineering (Francien Dechesne and Tijn Borghuis)
Chapter 38: Health (Marianne Boenink)

Further Reading

Hans Radder (2003) edited a collection, *The Philosophy of Scientific Experimentation*, which presents state-of-the-art philosophical studies of the role of technological instruments in scientific experimentation. In a review article about this collection, I distinguish between three different roles of technological instruments: to *measure*, to *model* and to *manufacture* (Boon 2004).

Sven Ove Hansson (2015), in a similar fashion, has edited a collection, *The Role of Technology in Science*, in which already more emphasis is put on scientific research aiming at the development of the technology, in addition to the role of technology in testing scientific theories.

Wybo Houkes and Pieter Vermaas (2010), in *Technical Functions: On the Use and Design of Artifacts*, addressed the issue of technological functions from the perspective of the design of technological artifacts, in which—different from the methodology presented in this chapter—the role of the users of artifacts is pivotal. This work is discussed by the authors and reviewers in Weber et al. (2013).

Anthonie Meijers (2009) has edited a handbook in the philosophy of science: *Philosophy of Technology and Engineering Sciences*. This handbook touches on many important issues in engineering and engineering sciences. One of the ideas put forward is the notion of technological knowledge as an autonomous kind of knowledge, which is critically discussed in Boon (2011a).

Mary Morgan and Margaret Morrison (1999) have edited the collection *Models as Mediators*, which has played a pivotal role in attention to the role of models and modeling in scientific research.

Daniela Bailer-Jones's (2009) monograph presents a comprehensive account of *Scientific Models in Philosophy of Science*.

Notes

1. See Boon (2011a). In a recent article, I argue that a so-called physics paradigm of science prevents us from recognizing engineering science as a scientific practice. As an alternative, I propose an engineering paradigm of science (Boon 2017b).
2. For example, Westerterp et al. (1984). See Schneider et al. (2003) for an accessible example on modeling interacting phenomena, and see Boon (2008) for an explanation of these types of diagrammatic models.
3. Some notable exceptions are: Magnani and Bertolotti (eds. 2017); Nersessian (2009a, 2009b, 2009c, 2012); Sterrett (2002, 2006, 2009, 2014, 2017).

4. Clear accounts of the semantic view can be found in Suppe (1989) and Giere (2010). Cartwright (1983, 1989, 1999) and Morgan and Morrison (1999) have convincingly argued that this view is very restricted on the role of models in science.

5. An elaborate example to introduce and illustrate this approach by reconstructing Sadi Carnot's model of the ideal heat-engine has been presented in Knuuttila and Boon (2011). This chapter presents only an outline. Nevertheless, based on teaching the B&K method in analyzing scientific articles and research projects in engineering science education, the original list of eight elements in Knuuttila and Boon (2011) has been expanded with 'problem-context' (Q1) and 'hypotheses' (Q9), resulting in the ten questions listed in Figure 6.2. Also, in teaching, I have started to call it the *B&K Method for (Re-)Constructing Scientific Models* (Figure 6.2 and caption), as an alternative to the traditional HD method (Figure 6.1).

References

Bailer-Jones, D.M. (2009). *Scientific Models in Philosophy of Science*. Pittsburgh: University of Pittsburgh Press.

Bogen, J. (2011). 'Saving the Phenomena' and Saving the Phenomena. *Synthese*, 182(1), 7–22. doi:10.1007/s11229-009-9619-4

Bogen, J. and Woodward, J. (1988). Saving the Phenomena. *The Philosophical Review*, 97(3), 303–352. doi:10.2307/2185445

Boon, M. (2004). Technological Instruments in Scientific Experimentation. *International Studies in the Philosophy of Science*, 18(2&3), 221–230.

Boon, M. (2008). Diagrammatic Models in the Engineering Sciences. *Foundations of Science*, 13(2), 127–142. doi:10.1007/s10699-008-9122-2.

Boon, M. (2011a). In Defense of Engineering Sciences: On the Epistemological Relations Between Science and Technology. *Techné: Research in Philosophy and Technology*, 15(1), 49–71.

Boon, M. (2011b). Two Styles of Reasoning in Scientific Practices: Experimental and Mathematical Traditions. *International Studies in the Philosophy of Science*, 25(3), 255–278.

Boon, M. (2012). Scientific Concepts in the Engineering Sciences: Epistemic Tools for Creating and Intervening With Phenomena. In U. Feest and F. Steinle (eds.), *Scientific Concepts and Investigative Practice*. Berlin: De Gruyter, pp. 219–243.

Boon, M. (2017a). Measurements in the Engineering Sciences: An Epistemology of Producing Knowledge of Physical Phenomena. In N. Mößner and A. Nordmann (eds.), *Reasoning in Measurement*. London and New York: Routledge, pp. 203–219.

Boon, M. (2017b). An Engineering Paradigm in the Biomedical Sciences: Knowledge as Epistemic Tool. *Progress in Biophysics and Molecular Biology*, 129, 25–39. doi:10.1016/j.pbiomolbio.2017.04.001

Boon, M. and Knuuttila, T. (2009). Models as Epistemic Tools in Engineering Sciences: A Pragmatic Approach. In A. Meijers (ed.), *Philosophy of Technology and Engineering Sciences*. Handbook of the Philosophy of Science, Vol. 9. Amsterdam: Elsevier, pp. 687–720.

Boumans, M. (1999). Built-in Justification. In M.S. Morgan and M. Morrison (eds.), *Models as Mediators—Perspectives on Natural and Social Science*. Cambridge: Cambridge University Press, pp. 66–96.

Cartwright, N. (1983). *How the Laws of Physics Lie*. Oxford: Clarendon Press; Oxford University Press.

Cartwright, N. (1989). *Nature's Capacities and Their Measurement*. Oxford: Clarendon Press; Oxford University Press.

Cartwright, N. (1999). *The Dappled World. A Study of the Boundaries of Science*. Cambridge: Cambridge University Press.

Dau, H., Fujita, E. and Sun, L.C. (2017). Artificial Photosynthesis: Beyond Mimicking Nature. *Chemsuschem*, 10(22), 4228–4235. doi:10.1002/cssc.201702106

Dubey, R.K., Inan, D., Sengupta, S., Sudholter, E.J.R., Grozema, F.C. and Jager, W.F. (2016). Tunable and Highly Efficient Light-harvesting Antenna Systems Based on 1,7-perylene-3,4,9,10-tetracarboxylic Acid Derivatives. *Chemical Science*, 7(6), 3517–3532. doi:10.1039/c6sc00386a

Dym, C.L. (2004 [1980]). *Principles of Mathematical Modeling*, 2nd ed. Amsterdam: Elsevier Academic Press.

Giere, R.N. (2010). An Agent-based Conception of Models and Scientific Representation. *Synthese*, 172(2), 269–281. doi:10.1007/s11229-009-9506-z

Giere, R.N., Bickle, J. and Mauldin, R. (2006/1979). *Understanding Scientific Reasoning*, 5th ed. Belmont, CA: Thomson/Wadsworth.

Gust, D., Moore, T.A. and Moore, A.L. (2001). Mimicking Photosynthetic Solar Energy Transduction. *Accounts of Chemical Research*, 34(1), 40–48. doi:10.1021/ar9801301

Hacking, I. (1983). *Representing and Intervening*. Cambridge: Cambridge University Press.

Hansson, S.O. (ed.) (2015). *The Role of Technology in Science: Philosophical Perspectives* (18 ed.). Dordrecht/Heidelberg: Springer.

Hempel, C.G. (1966). *Philosophy of Natural Science*: Englewood Cliffs, NJ: Prentice-Hall. www.nature.com/articles/nmeth.2651#supplementary-information

Huang, X.W., Wang, J.C., Li, T.H., Wang, J.M., Xu, M., Yu, W.X., . . . Zhang, X.M. (2018). Review on Optofluidic Microreactors for Artificial Photosynthesis. *Beilstein Journal of Nanotechnology*, 9(1), 30–41. doi:10.3762/bjnano.9.5

Knuuttila, T.T. and Boon, M. (2011). How Do Models Give Us Knowledge? The Case of Carnot's Ideal Heat Engine. *European Journal for Philosophy of Science*, 1(3), 309–334. doi:10.1007/s13194-011-0029-3

Kroes, P. (1998). Technological Explanations: The Relation Between Structure and Function of Technological Objects. *Techné: Research in Philosophy and Technology*, 3(3), 124–134. doi:10.5840/techne19983325

Kroes, P. (2010). Engineering and the Dual Nature of Technical Artifacts. *Cambridge Journal of Economics*, 34(1), 51–62. doi:10.1093/cje/bep019

Magnani, L. and Bertolotti, T. (eds.) (2017). *Springer Handbook of Model-Based-Science*. Dordrecht: Springer.

Meijers, A. (ed.) (2009). *Philosophy of Technology and Engineering Sciences. Handbook of the Philosophy of Science*, Vol. 9. Amsterdam: Elsevier.

Morgan, M.S. and Morrison, M. (eds.) (1999). *Models as Mediators—Perspectives on Natural and Social Science*. Cambridge: Cambridge University Press.

Nersessian, N.J. (2009a). *Creating Scientific Concepts*. Cambridge, MA: MIT Press.

Nersessian, N.J. (2009b). How Do Engineering Scientists Think? Model-Based Simulation in Biomedical Engineering Research Laboratories. *Topics in Cognitive Science*, 1(4), 730–757. doi:10.1111/j.1756-8765.2009.01032.x

Nersessian, N.J. (2012). Engineering Concepts: The Interplay Between Concept Formation and Modeling Practices in Bioengineering Sciences. *Mind, Culture, and Activity*, 19(3), 222–239. doi:10.1080/10749039.2012.688232

Nersessian, N.J. and Patton, C. (2009). Model-based Reasoning in Interdisciplinary Engineering. In A.W.M. Meijers (ed.), *Handbook of the Philosophy of Technology and Engineering Sciences* (pp. 687–718).

Ong, W.J., Lin, Z.Q. and Domen, K. (2018). Artificial Photosynthesis: Taking a Big Leap for Powering the Earth by Harnessing Solar Energy. *Particle & Particle Systems Characterization*, 35(1), 4. doi:10.1002/ppsc.201700451

Pandit, A., de Groot, H. and Holzwarth, A. (eds.) (2006). Harnessing Solar Energy for the Production of Clean Fuel. *White Paper by an International Task Force Under the Auspices of the European Science Foundation*. ISBN 978-90-9023907-1

Radder, H. (2003). *The Philosophy of Scientific Experimentation*. Pittsburgh, Penn.: University of Pittsburgh Press.

Schneider, R., F. Sander, A. Górak (2003). Dynamic Simulation of Industrial Reactive Absorption Processes. *Chemical Engineering and Processing*, 42, 955–964.

Sterrett, S.G. (2002). Physical Models and Fundamental Laws: Using One Piece of the World to Tell About Another. *Mind & Society*, 3(1), 51–66. doi:10.1007/bf02511866

Sterrett, S.G. (2006). Models of Machines and Models of Phenomena. *International Studies in the Philosophy of Science*, 20(1), 69–80. doi:10.1080/02698590600641024

Sterrett, S.G. (2009). Similarity and Dimensional Analysis. In A. Meijers (ed.), *Philosophy of Technology and Engineering Sciences. Handbook of the Philosophy of Science*, Vol. 9. Amsterdam: Elsevier, pp. 799–823.

Sterrett, S.G. (2014). The Morals of Model-making. *Studies in History and Philosophy of Science Part A*, 46, 31–45. doi:10.1016/j.shpsa.2013.11.006

Sterrett, S.G. (2017). Experimentation on Analogue Models. In L. Magnani and T. Bertolotti (eds.), *Springer Handbook of Model-Based Science*. Cham, Switzerland: Springer, pp. 857–878.

Suppe, F. (1989). *The Semantic Conception of Scientific Theories and Scientific Realism*, 1989 ed. Chicago and Urbana: University of Illinois Press.

Uosaki, K., Kondo, T., Zhang, X.-Q. and Yanagida, M. (1997). Very Efficient Visible-Light-Induced Uphill Electron Transfer at a Self-Assembled Monolayer With a Porphyrin−Ferrocene−Thiol Linked Molecule. *Journal of the American Chemical Society*, 119(35), 8367–8368. doi:10.1021/ja970945p

Vincenti, W.G. (1990). *What Engineers Know and How They Know It: Analytical Studies From Aeronautical History*. Baltimore and London: Johns Hopkins University Press.

Weber, E., Reydon, T.A.C., Boon, M., Houkes, W. and Vermaas, P.E. (2013). The ICE-Theory of Technical Functions. *Metascience*, 22(1), 23–44.

Westerterp, K.R., van Swaaij, W.P.M. and Beenackers, A.A.C.M. (1984). *Chemical Reactor Design and Operation*, 2nd ed. Chichester: John Wiley.

Woodward, J.F. (2011). Data and Phenomena: A Restatement and Defense. *Synthese*, 182(1), 165–179. doi:10.1007/s11229-009-9618-5.

PART II

Engineering Reasoning

7

ENGINEERING DESIGN AND THE QUEST FOR OPTIMALITY

Maarten Franssen

7.1. Introduction

'Let's make things better' used to be an advertisement slogan of a major multinational engineering firm. It was criticized for not being distinctive enough: Is this not what all engineering firms are doing? Is not even what an engineering firm comes up with always the best it can do, within the constraints set by time and money? It seems that it is widely accepted that optimality is what engineering deals in. Take, for example, the very first sentence of Pahl and Beitz's well-known textbook on engineering design (1996): "The main task of engineers is to apply their scientific and engineering knowledge to the solution of technical problems, and then to optimise those solutions within the requirements and constraints set by material, technological, economic, legal, environmental and human-related considerations". This raises the question whether engineering can live up to it, and this is a philosophical, because ultimately a normative, question. Can we assess to what extent engineers can ever be justified in claiming that what they deliver is the best that can be expected, given the objectives and the constraints?

In this chapter I look into how, for engineering design, optimization enters as a possible objective in the first place and what the prospects are for achieving it, depending on the nature of the design problem and the circumstances. In Section 2 I discuss how a design task comes to be fixed and that it can be considered determined by two sorts of specifications, demands and wishes, which are nominal and ordinal in character, respectively. In Section 3 I discuss how satisfying the nominal requirements ends in a design being good enough. In Section 4 I discuss how satisfying the ordinal requirements introduces the notion of being better, and how this can lead to a minimal notion of being best: Pareto optimality. In Sections 5 and 6 I argue that there are severe limits as to how far we can hope to establish certain designs as best in a full sense of best, first addressing the qualitative case and then the quantitative one. In the closing sections 7, 8 and 9, I first discuss an alternative to the valuation framework underlying the approach of Sections 5 and 6. I then argue that because design typically is a team activity in a social environment the noted problems are multiplied, and I finally draw some conclusions.

7.2. Fixating the Design Task

Engineers work to deliver devices that serve concrete, practical purposes or, in a broader interpretation, solutions to concrete, practical problems. Whether the purposes are indeed served or the problems indeed solved, and to what extent, is determined by a range of criteria that specify what is meant by a 'working' device or a solution. This goes from a minimal conception of 'mere working' to optimal working or complete resolution of the problem. To settle these criteria is not exclusively up to engineering. The ideal picture is that engineering, unlike science, is a practice that is thoroughly embedded in society. It is from society that engineering receives the purposes for which it designs devices and the problems for which it creates solutions. Still, textbooks of engineering design generally do not take things much further than acknowledging that at the origin of the design process are the 'needs' of 'customers' or 'clients' (e.g. French 1992: 1–3, Suh 1990: 27, 2001: 10–13, Cross 2008: 29–31). This, however, may include a wide variety of sources: single clients, typically incorporated business firms, or private customers whose needs are in some way perceived or even just anticipated, or the political representatives of the community of citizens that make up society.

Due to this wide variety of sources, the social origin of engineering problems is an idealization. Especially when designing mass-market consumer products, designers must construct the wants and needs of prospective customers. In doing so, they may well be tempted to project upon these customers what suits or fascinates them purely from an engineering perspective. In larger firms, it is exactly the task of marketing and sales departments to accurately represent the firm's potential customers and their needs and wishes, and to force a consumer's perspective if not on the firm's technology departments, then at least on its top managers. The 'purposes and problems received from society' therefore rarely if ever in themselves suffice to define a concrete design problem. But even when design starts with a wish list supplied by a specific client, these wishes will typically not be articulated to a sufficient level of detail to allow one to see how they could be met right away. The first phase of the design process, therefore, typically consists of a *translation* phase, which has a client's or customer's supplied or anticipated wish list as input and a comprehensive list of *design specifications* as output. These specifications are considered to become fixed in a dialogue with customers or clients: either directly, when the client is a single agent, typically a company or a government agency; or indirectly, when engineers are designing for the market and they have to anticipate the needs and wishes of their customers. In all cases, the issue which specifications *best* translate these needs and wishes already introduces the problem of optimization in all of its complexity, since many different stakeholders may be involved in this process, even when dealing with a single company as client. This issue forms part of the social dimension of engineering, to which I briefly return at the end of this chapter.

For now let us assume that the initial phase proceeds by a translation process on the basis of a clear wish or need: a purpose for which a customer is seeking a device, or a problem for which a solution is sought. However clearly felt the need or wish may be, it is usually not stated to the level of detail that is required to define a design problem. The translation process involves not just rephrasing what is on the client's wish list into engineering terms—which typically means making the requirements much more precise and if at all possible quantitative—but also *completing* the wish list in the light of the client's needs or problem. When, for example, a client asks for a car that is fast and cheap, then if it is pointed out that a car that is really fast and really cheap is likely to have safety issues, the client will realize that they forgot to mention that the requested car 'of course' should also be safe. It is a well-known problem in design, especially when a client has little knowledge of the technicalities of a product or system, that clients 'forget' to mention many of the requirements that for engineers would be crucial ingredients in the list of design specifications (see e.g. Middendorf and Engelmann 1998, Ch. 3; and for software engineering, where the problem is particularly notorious, Ewusi-Mensah 2003).

Moreover, not everything that goes on in this first phase of design is translation of the client's or customer's wish list into terms that engineers know how to handle. The design specifications will also come to include constraints that have their origin elsewhere, ranging from the laws of nature and the properties of materials at one extreme to government regulations and industry standards at the other extreme. The final outcome of this first phase, then, can be characterized as consisting of four types of specifications, determined by two distinctions. The first distinction is whether the specification is one that must be fulfilled, with the fulfilment being of a yes/no type; or whether it is one that is

Table 7.1 Different types of specifications generated by two basic distinctions

	Nominal (Yes/No)	Ordinal (Gradual)
Internal	Demands/requirements/constraints	Wishes/requirements
External	Constraints	(?)

going to be met to a certain extent, on a gliding scale in the direction of an ideal. The former can be characterized as a *nominal* type of specification, since any design concept can have only one of two possible values along the dimension corresponding to this distinction: it either meets or fails to meet the requirement or constraint. The latter can be characterized as *ordinal*, since a design concept will have a value only relative to other concepts: it meets the requirement better or equally well or worse, and which of these is the case has to be established in relation to every design alternative separately. The second distinction, next, is whether the specification originates in the client's wish list or whether it originates elsewhere, as a limitation imposed by the external world, be it nature or society.

There is no agreed, fixed terminology to indicate the four resulting types of specifications. Additionally, the two distinctions are not entirely independent. External constraints are rarely such that they can be met to some extent, or relatively better or relatively worse compared to alternatives. Safety regulations, for instance, may require that some voltage does not exceed 15V but do not further discriminate among all the different voltages that satisfy this. What seems relatively constant in terminology is that the term *requirements* is preferred for those specifications that originate with the client or customer, and that the term *constraints* is reserved for specifications that have a must-be-fulfilled character. Table 7.1 summarizes the four kinds of specifications distinguished here.

Once a full list of specifications has been determined, we have arrived at a well-defined engineering problem or engineering task. It should be noted, however, that there is no unique point in time that corresponds to this. The full list of design specifications that marks the end of the initial translation or clarification phase can very often not be considered stable or definitive. Design is an iterative process: investigating the options or design concepts that are promising to satisfy the design specifications may lead to a reconsideration of the specifications, which may—and arguably should—require a new involvement of the client or customer. Additionally, the choice of materials or the properties of available materials or components can introduce further constraints or can require the modification of existing constraints. If external constraints of either a material or social character turn out to be too demanding, modifications to the internal requirements may be necessary.

7.3. Meeting Nominal Requirements and Constraints

Once the design specifications or design objectives are fixed at least for the time being, engineering expertise takes over. It is now an engineer's task—or an engineering team's task—to come up with one or several designs that meet the specifications. The form of the specifications dictates what manoeuvring space there is for meeting them, and accordingly to what extent the issue of optimizing poses itself.

The foremost requirement that must be met is that a device must achieve the function required by the client. This will typically correspond to the device meeting a list of behavioural requirements. If the device under design is a stapling or punching machine, then it must actually have the capacity to drive staples through whatever needs to be stapled or punch holes in whatever needs to receive a hole. This is referred to as the requirement that the design must be *effective*: it must be able to do what it is supposed to do. The example serves to illustrate that the requirement of being fit for a particular purpose, or achieving a particular function, will correspond to a list of specifications that must include specification of the material or object on which the function is to be applied and on the operating conditions and environmental conditions. The design of a punching device, for example, cannot proceed, at least not with a guarantee of customer satisfaction, if it is not clear what materials are to be punched and, in the case of a hand-operated punching device, what force the person who operates the device will be able to apply. And there are further factors that will have to be taken into account, depending on the device: temperature, humidity, light conditions, and so forth, may all have an influence on what physical behaviour a device is capable of, and therefore have a say in whether the device is judged effective. Accordingly they will have to be present in the list of specifications, and in the right way.

The nominal specifications—those that must be met—define a conceptual space in which any device that meets the specifications must be located. There is no guarantee that the laws of nature, or rather our knowledge of the laws of nature, allow such a device to exist at all. If the specifications are equivalent to the definition of a perpetual-motion machine, then the design task is doomed from the start. But if it is possible to meet the specifications at all, then insofar as one can conceive of several different devices that meet them, the list of demands does not serve to discriminate among them. They all do what they must be able to do, in the environmental conditions in which they are supposed to function, and that's it. They are all 'good'. We may suspect, however, that they are not all 'equally good'. To further narrow down the set of design concepts and to arrive at a reasoned decision as to what device to build or offer to the client or the customer, of all the ones that could be built or offered, the list of ordinal specifications must be brought in as well.

This is strictly true, however, only if the design problem or design task is approached purely on the basis of its definition in terms of objectives. But if a design task is defined by a list of specifications that all have the form of constraints, that is, yes/no requirements, then this does not preclude the possibility of reasoned choice among all alternatives that satisfy the constraints. It may be that from the perspective of the client, all alternatives satisfying the constraints indeed can be considered on a par, but this leaves room for the engineer's perspective to be added to this as long as this does not lead to conflict. Hardly ever do the specifications define a space in which just one design concept exists. The choice of materials, of components and other off-the-shelf products to be made use of, of power sources, of manufacturing methods perhaps, will generally not be exhaustively determined by the specifications. It is here that the designing engineer can apply additional criteria, representing considerations that are typical of engineering. These considerations can be summarized as considerations of *efficiency*, although again there is no consensus on terminology. The striving for efficiency typically amounts to the avoidance of waste—waste of materials in manufacturing and waste of energy in operation. This does not mean, of course, that efficiency considerations are not among those that a client will formulate. Certainly avoidance of waste of energy in operation will often be covered, typically by a constraint of the type that operation costs should not exceed such-and-such an amount. Likewise, avoidance of waste of materials in manufacture will at least partly be covered by a constraint of the type that the (purchase) price of a device must not exceed such-and-such an amount. Insofar as there are constraints of this type among the design specifications, however, they will typically reflect the client's or customer's direct concern with costs, not a direct concern with waste. Engineers typically take this concern much further, beyond the constraints set by the client. This does not mean that it is not ultimately in the service of other design requirements, such as constraints on weight and size. Nor does it mean that it does not eventually boil down to money as well on the engineering side. Being efficient beyond the specified constraints can be used either to increase the designing firm's profit once a contract has been signed and a price agreed, or to reduce the price of a mass-market product and thereby gain a competitive advantage. Nevertheless, a concern with efficiency may also be inherently motivated as part of the engineering perspective, and particularly so in the last decades, with the arrival of sustainability as a general concern. Market conditions matter, though. Engineering's own conceptions of elegance and beauty in design may be strongly related to the notion of efficiency, but occasionally, when market conditions pulled in the direction of wasteful ornamentations, engineering products have also shown preciously little inclination to avoid a waste of material or energy.

7.4. Meeting Ordinal Requirements, Pareto Optimality

It is with the ordinal wishes rather than the nominal demands that optimality enters as a problem for engineering design. Depending on the amount of work that has gone into the initial clarification phase, such specifications may either explicitly introduce a quest for optimizing—e.g. when a specification is phrased as 'as light as possible' or 'as small as possible'—or they may do so implicitly—e.g. when a specification is phrased as 'easily maintainable', where it is implicit that the easier the device is to maintain, the better, or 'low in fuel consumption', where it is implicit that the less fuel is consumed, the better. Note that specifications may occasionally, if not often, combine demands and wishes, or constraints and requirements, e.g. 'as light as possible but definitely not heavier than 20 kg'.

The design task is now governed by the question of which of all the potential solutions in the space delineated by the nominal demands is the best solution, the design that best meets the totality of design constraints and requirements. They will have to be compared. This raises the crucial methodological question for engineering design whether there is a general method for finding the design concept that meets the totality of the specifications best. A little reflection will make clear that this is a 'deep' problem. Generally, there are many specifications to meet, and of any two design concepts—let's call them A and B—A may meet some specifications better than B does, but B may meet other specifications better than A does. These differences need somehow to be balanced to arrive at an overall comparison between A and B, and it can be shown that in many situations, every conceivable way of balancing runs into major, even insurmountable difficulties.

Before getting to an explanation of why this is so, let us first look into the possibilities of avoiding this balancing act altogether. Suppose that there are two options on the table between which a choice has to be made—let's refer to them again as A and B—and suppose there are three specifications that we wish to see satisfied—let's call these P, Q and R. Suppose that P, Q and R are all ordinal requirements, that is, of the 'as Φ as possible' type or 'the Φ-er the better' type. Then it may turn out that, when A's and B's performances are carefully measured, A satisfies all three of P, Q and R better than B—it is both Φ-er than B, less Ψ than B and more Ξ than B, and all of these is what we aim our design to be. In this case, there can be no question that A is the better design of the two. If this is satisfied for a pair of options A and B, A is termed *strictly better* than B. If A outperforms B on several specifications and performs equal to B on the remaining specifications, A is termed *weakly better* than B.

In design problems where a choice has to be made among an unspecified number of design options on the basis of an unspecified number of specifications, one can start to look for all options that are strictly or weakly worse than one or several other options. It can never be justified to present one of these as the final design, since it was possible to come up with another design that is strictly or weakly better, and that one should therefore have been presented instead if anything was. If all strictly or weakly worse options are eliminated from the set of options, then the remaining are in the following weak sense optimal: for each of them, no option exists (among the set of options that are considered at all) which outperforms it on all specifications. This weak sense of optimality is called *Pareto optimality*—after the Italian economist Vilfredo Pareto who first emphasized its importance (Pareto 1900)—and the set of options that are Pareto-optimal is called the *Pareto set*. Clearly, a rational solution to a design task must be a member of the task's Pareto set if that set is known. But charting it is a problem in itself.

Many design tasks can be looked upon as continuous in that by continuously varying certain design parameters, an infinity of options can be generated. The Pareto set then can be represented as a manifold in the parameter space. Let's look at a simple example. Eggert (2005, Ch. 8.3) discusses a design where an electric motor will be used to drive a grinding wheel by a belt-and-pulley drive system. The rotational speed of the motor and the required rotational speed of the grinding wheel are given and fixed, as well as the diameter of the motor pulley—these are nominal requirements or demands, as are the maximally allowed belt tension and the maximally allowed size of the whole setup. Figure 7.1 offers an elementary representation. Maximal power transfer from the motor to the grinding wheel and absence of belt slip are also demands, whereas compactness—the smaller the whole set-up the better—and small maximum belt tension—the smaller the maximum tension the better—are ordinal requirements. The laws of physics then dictate the diameter of the wheel pulley, given the demands of maximal power transfer and absence of slip. If the distance separating the two pulleys and the tension in the belt could be chosen independently, then to come up with a design that is optimal given the requirements would be easy: it would consist of the configuration in which the two pulleys would be as close to one another as their dimensions allowed and in which the maximum tension in the belt would be relatively smallest. However, the laws of physics imply that a law-like relation exists between pulley distance and maximum belt tension: the closer the two pulleys are together, the greater the angle of wrap the belt has to the pulley, and the greater the maximum tension in the belt. As a consequence, the complete set of design options is identical to the Pareto set, since every design can be characterized by a parameter pair (d, t), where d is the distance between the pulleys and t the maximum belt tension, and for each pair of designs (d_1, t_1) and (d_2, t_2) we have $d_1 < d_2$ if and only if $t_1 > t_2$, meaning that if (d_1, t_1) is preferred to (d_2, t_2) on the requirement of compactness, then (d_2, t_2) is preferred to (d_1, t_1) on the requirement of small maximum belt tension. The Pareto set therefore corresponded to a curve in (d, t) space, with the end points determined by the maximally allowed maximum belt tension and the maximally allowed pulley distance.

Eggert's example of the belt-and-pulley design clearly shows the limited relevance of the notion of Pareto optimality. If Pareto optimality were all the optimality we had, every design satisfying the demands alone would be optimal and the two requirements of compactness and small maximum belt tension would not have much of a bite. Not every design problem will be like this, of course. Take the problem where a metal object of a specified size is wanted, presumably to serve as a component in a larger device, which has to have as low weight as possible, as high electrical conductivity as possible and as low heat conductivity as possible. Since an infinity of different alloys could be produced, each with its specific values for density d, electrical conductivity e and heat conductivity h, possible designs will correspond to a subspace of (d, e, h) space, and the Pareto set will correspond to, or at least be included in, a manifold that forms part of the boundary of this subspace.

For design problems that are of this type, the task is first of all to determine the Pareto set, that is, to identify design options that are included in it or close to it. This offers a new perspective on what makes a design task well-defined. Should designers now start to compile a complete overview

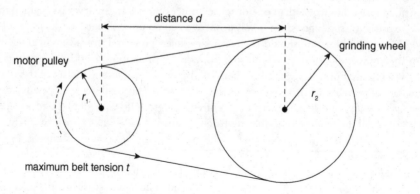

Figure 7.1   Eggert's belt-and-pulley example

of alloys for which density, electrical conductivity and heat conductivity have all been measured? On the assumption that this overview, even when complete, does not contain all possible alloys, should the designers first complete it by starting to manufacture samples of new alloys and measure their density and conductivity values? Clearly this will hugely inflate both the design task's time frame and its cost frame.

It is partly with a view to situations like this that Herbert Simon introduced the notion of *satisficing* (Simon 1957, 1982). In many cases, rational agents should not want to optimize, because this will involve an explosion of the costs involved, or even a complete loss of cost control because feeding cost considerations into the design task will introduce loops. The quest for optimality can be self-defeating. Instead, such problems should be solved by settling on a solution that is good enough. In the terminology applied here, this means looking at a design task as determined exclusively by demands: specifications that have to be met, and anything that meets them is an acceptable design option.

But let us take the discussion back to cases where this quest is not, at least not obviously so, self-defeating. This could either be a case where the set of design options is restricted, for whatever reasons, to a small number of alternatives all satisfying the demands, or a case where the set of design options consists of options that differ only in the choice of a small number of parameters, with the designer controlling which values these parameters will have in the final design—as is the case in Eggert's belt-and-pulley design problem. Suppose that all options in this set are Pareto-optimal: for any option, there is another one that outperforms it on at least one of the design requirements, but no option is outperformed by another option on all the design requirements—if there were such options, they have all been rejected by now. What room do we have for optimizing still further, beyond the notion of Pareto optimality? Clearly one would want to resist the claim that with narrowing down the set of options to the Pareto set, all room for optimizing has been spent. Nothing compels us to accept that all remaining options are equally good, and that the choice of one of them as the final design could be settled by a coin flip. To claim, however, that one of the remaining options *is* the best one and the designer's task is to *find* it would also misrepresent the character of the problem. The problem is rather to define the notion of 'optimal', to operationalize it in a sense, so that, on the one hand, justice is done to our intuitive but vague conception of 'optimal' and, on the other hand, a single option satisfies this definition. Or perhaps a few options jointly, which can then justifiably be termed equally good.

7.5. A Fundamental Barrier to Optimization: Arrow's Theorem

The ordinal design requirements are typically of the form that the object under design should be as Φ as possible, or that the Φ-er it is, the better, where Φ is some measurable characteristic. Once the performance of the various design concepts is known—either in the form of an outcome of an accurate test or in the form of an estimate—then we can, for each requirement, rank-order the options in the set under consideration from better to worse from the perspective of this requirement. (Of course, two options may also perform equally well with respect to a requirement.) It is when the resulting rank-orders, once we have done this for all requirements, do not match that we have a problem. Of two options A and B, A may be preferable to B as far as one requirement is concerned but B preferable to A as far as another requirement is concerned. The designer's task, however, is to judge whether A is preferable to B, or the reverse, from the perspective of *all requirements jointly*. If an option can be found that, from this perspective, is preferable to all other options, that option should definitely be identified as the optimal design or the optimal solution to the design task. So, one way or another, the conflicts between the rank-orders have to be compared and balanced in some sort of accounting system.

The intensive study of this type of problem dates back to 1950, when a formal proof was published (Arrow 1950) that no accounting system is possible that will do this once the number of options is at least three and the number of requirements is at least two—hardly a significant restriction—given some features that one would want any such accounting system to satisfy. These desired features are the following:

1. The accounting system matches to every logically possible profile of rank-orders of the options—one rank-order per requirement—an overall rank-order of all options;
2. If an option is on top of all rank-orders, then it will also be on top of the overall rank-order (in other words, the accounting system honours Pareto optimality);
3. Since the input consists entirely of rankings, the overall ranking between any two options is determined entirely by the relative ranking of these two options on all requirements separately;
4. None of the requirements determines the overall ranking on its own, or, every requirement has some say in the overall ranking, in that, if overall A comes out as better than B, then by sufficiently raising the performance of B on anyone (or several) of the requirements, B can be made to come out better than A overall.

This result is known as Arrow's impossibility theorem (see Moreau 2014 for an extensive overview). It was initially proved for a situation where voters in a population have preferences over options and an overall ranking of these options has to be determined by some form of vote-counting system. Since then, however, it has been recognized that the result applies to any situation where several rankings, which all rank-order the same set of items, are to be aggregated into a single ranking that can count as the 'best representation' of the profile of separate rankings. After all, what goes into the notion of 'best representation' at issue here is quite independent of what is being ranked, just as the validity of a deductive argument is independent of what the statements that make up the argument are about. Once a situation gives rise to several rankings, which all need to be taken into account in some way or other, we enter the domain where Arrow's theorem holds sway. Clearly such situations abound in engineering, where design tasks are typically defined by lists of requirements which all have to be met as best as possible (see Franssen 2005 for an extensive discussion). Both Eggert's belt-and-pulley example, where a design is better the more compact it is *as well as* the smaller its maximum belt tension is, and my toy example of a component which is better the smaller its weight is *and* the larger its electrical conductivity *and* the smaller its heat conductivity is, are situations of this kind.

The theorem's implications are that whatever the method is for determining the 'overall best' option, it will *occasionally* (and not inevitably) lead to undesirable outcomes. What these are depends on the precise method used. It may be that sometimes it fails to rank options at all, or that cycles between options emerge where one option is ranked above another one, which is ranked above a third one, which is again ranked above the first option, or that adding or removing options that are low in the ranking results in a reversal of the ranking at the top end of the scale.

Insofar as the impossibility theorem is taken to apply to design tasks where various options end up in a rank order based on their relative performance with respect to several requirements, the conclusion would have to be that the notion of the optimal design is simply not, in general, well-defined. We must take the non-existence of an algorithm to mean that there is no rank-order of all options 'implied' by the rankings belonging to the individual requirements, and therefore no rank order that we can consider to be the one 'best representing' them.

Whether the theorem straightforwardly applies to design can be questioned, however. To start with, requiring an algorithm to produce a complete rank-order of the options under consideration may be considered too strong; it would be sufficient if the process results in the identification of a top-ranking option. For this weaker case a modified impossibility has been proved (Mas-Colell and Sonnenschein 1972) that is equally embarrassing and just as much implies that any algorithm that is used to determine a winner will occasionally behave in an undesirable way.

Secondly, it has been argued (Scott and Antonsson 1999) that it is unreasonable to require that an aggregation algorithm should accept every possible combination of rankings as input: in design one may expect causal dependence to exist among requirements and therefore correlations between rankings. It is hard to deny that such correlations occur—Eggert's belt-and-pulley problem is a case in point, as there the ranking of options on the requirement of compactness is the reverse of the ranking of options on low maximum-belt tension. However, one cannot rely on this always being the case. It is sufficient for the theorem to apply that there are three requirements among which no correlations exist. Obviously a designer must expect many design problems to satisfy this—certainly if we accept the picture that engineering itself promotes, namely that engineering does not itself define design tasks in terms of their demands and requirements but that these are passed by its clients or customers.

The only promising objection against the theorem's relevance is that it relies on the design options' performances on the requirements being assessed only qualitatively, meaning that the relations engineers can establish to exist between two design options A and B are exclusively of the type 'A is a better option than B as far as requirement P is concerned' or occasionally 'A and B are equally good options as far as requirement Q is concerned'. Engineers typically have quantitative access to the performance of options. Even if a requirement is stated qualitatively—the device's weight should be as low as possible—every design option has a numerical weight in kilograms (or micrograms) and one option can be said to be twice as light as another, or to weigh 30 percent more than another. Indeed, as is argued in the next section, the only escape that engineering design has from Arrow's impossibility theorem is through quantitative measures of performance. However, while it is a necessary way to go, it is not sufficient.

7.6. Optimizing With Quantitative Assessments: Trade-Offs

There are two reasons why quantitative measures for requirement performance are necessary to find a way around Arrow's theorem but not sufficient. First, it is not a design's physical performance that matters here, but a designer's *assessment* of that performance, so it is this assessment which needs to be quantitative. Second, merely assessing requirement performance quantitatively is not enough; all assessments must also use a common unit.

The first of these reasons is not controversial. Engineers understand very well that the quantitative measures required are ultimately measures of an *assessment* of performance, not the actual physical performances. The mere value of 2 kg for an option's weight bears no intrinsic relation to the requirement that the device under design should be as light as possible. What matters is an assessment of where this weight will make an option end up among all possible solutions to the design problem under consideration. It is not obvious who is responsible for this assessment: the designing engineer(s) or the client or customer. Eggert, for instance, in his belt-and-pulley example, assumes that the designer will base the final selection of an optimal configuration on the *client's* assessment that a concept's worth or value decreases linearly with increasing distance between the two pulleys and also decreases linearly with increasing maximum-belt tension. It must be assumed, then, either that the client furnishes these assessments jointly with the list of design criteria at the start of the process, or that the designers solicit them from the client once they arrive at the stage when a single best design has to be selected. Clearly this is not realistic in most design tasks. Typically, the client is unable to oversee the relative contributions of each requirement to the overall device and will not be able to come up with such detailed assessments at the start. At a later stage, the client is often unavailable for such assessments, or may declare that the problem has been handed over to the designers and it is now their job to bring it to closure.

Additionally, of course, many design tasks are of such complexity that the client's wishes serve as constraints for a huge number of sub-tasks for which a client's input is out of the question. In designing a photocopying machine, the client's requirements concern image quality, speed, and the like, but the design of the machine's separate modules for image capture, image fixation, paper transportation, and so forth, is up to the design team. Questions of what the maximum or optimal size of a module is, or where best to place it, are beyond the client's involvement. The worth or value assessment of a particular concept's numerical state or performance indicators—such as weight, operational speed, ease of access, or whatever is considered relevant to the final reckoning—is therefore typically up to the designing engineers themselves. But no method exists for doing this: in the end a *decision* has to be made to adopt a particular assessment of how much better or worse one concept is *judged* to be with respect to another with respect to any of the relevant requirements.

Even having quantitative values for relative worth of options is not sufficient for an algorithmic solution. Take the resulting schedule for the belt-and-pulley case. The number of design concepts is in principle infinite, since the distance between the pulleys can be varied continuously between two extremes: 5 in., the distance at which the maximum belt tension reaches the safety limit of 35 lbs. for the type of belt, and 20 in., which is the maximal value the client will accept. With value decreasing linearly with increasing distance and increasing maximum belt tension, a table can be drawn (Table 7.2) that gives the values for a limited number of distances.

This is the simplest example of what any design problem with a number of concepts that all score differently with respect to a number of requirements will look like. Typically engineering methodology textbooks recommend that to determine what the optimal configuration is, the average weighted value of each option should be calculated and the option with the highest weighted average value must count as the best one (Dym and Brown 2012 is an exception). No textbook that I know of seems interested in justifying this approach (see Pahl and Beitz 1996, Ch. 6; Eggert 2005, Ch. 8; Cross 2008, Ch. 11). A justification can be given, however. It can be proved that the only way to arrive at a final quantitative, interval-scale measure of the overall worth or value of the options—and such a measure will also imply a final rank order of options and a best option—which satisfies the desirable features introduced earlier that any order-aggregating algorithm should satisfy, is weighted-averaging. In order to prove this, however, the third of the desirable features that gave rise to the original impossibility theorem must be modified, since now the input information that is to be aggregated is quantitative and stronger. The feature that takes this quantitative aspect adequately into account is:

3′.   Since the input consists entirely of relative quantitative distances separating the options, the overall ranking between options must depend only on these distances and must therefore be invariant under all transformations of value numbers that leave their relative distances unaffected.

These relative distances, however, are defined per requirement. From the above matrix for the belt-and-pulley case, we can derive that with regard to the requirement of compactness, a 6-in. setup is just as much better than an 8-in. setup as an 8-in. setup is better than a 10-in. setup, but with regard to the requirement of small maximum belt tension, a 10-in. setup is only half as much better than an

Table 7.2 Overview of acceptable physical values for the design parameters and their assessment values

Pulley Distance	Value Score on Compactness	Maximum Belt Tension	Value Score on Small Belt Tension
5 in.	1.00	35.0 lbs.	0.00
6 in.	0.93	33.5 lbs	0.30
8 in.	0.80	32.0 lbs.	0.60
10 in.	0.67	31.3 lbs.	0.76
14 in.	0.40	30.4 lbs.	0.92
20 in.	0.00	29.9 lbs.	1.00

8-in. setup as an 8-in. setup is better than a 6-in. setup. It is not clear, however, what, if anything, connects these two assessments to one another. If they are not connected, then we can replace the value scores of the various setups for small belt tension by 0, 6, 12, 15.2, 18.4 and 20, respectively, and all the information will still be there. These numbers just as well convey that with regard to the requirement of small maximum belt tension, a 10-in. setup is only half as much better than an 8-in. setup as an 8-in. setup is better than a 6-in. setup. But no aggregation method can deliver results that are invariant to that much freedom—no constraints on how to score the assessments per requirement as long as the relative distances of options remain intact. It can be proved, then, that weighted-averaging is the unique method that satisfies the four desirable features—with 3 replaced by 3'—exclusively if the choice of the scoring values is made for all requirements jointly; in other words, if the valuations of how the options perform with respect to the requirements all have the same unit. (See for the formal details Aczél and Roberts 1989; Bossert and Weymark 2004, grounded in measurement theory and welfare economics respectively, but with matching results.)

A common unit implies that the decrease in the overall value of the design concept due to an increase of the distance between the pulleys by 2 in. must be *numerically equivalent* to the increase in the overall value due to the maximum belt tension decreasing from 32.0 lbs. to 31.5 lbs. Such judgements, that making a design worse through increasing its size from, say, 7 in. to 9 in., can be exactly compensated by making it better through decreasing the maximum belt tension from 32.8 lbs. to 31.5 lbs., are called *trade-offs*. Trade-offs are ubiquitous in engineering, and that engineers should be capable of making them is, then, a presupposition of weighted-averaging being a general method for solving optimization problems in engineering design. So it is for the notion of a particular design concept being the optimal one to be well-defined in the first place. Again, however, the question emerges where these judgements come from. In economics and decision theory, methods have been developed for measuring perceived quantitative value. To determine how a person judges the value distance between options A, B and C, the person is asked to specify the probability p such that a lottery giving A with probability p and C with probability $1-p$ is judged equivalent to receiving B with certainty. Such methods become too complicated when applied to trade-offs—that is, people no longer grasp what they are asked to judge. As a consequence, in the current state of our understanding, the judgements of engineers when making trade-offs remain hidden behind a 'veil of intuition': it seems that engineers cannot point to anything but their expertise when they are asked to justify trade-offs made in practice. It has been argued that for the purpose of justification, this is sufficient (e.g. Scott and Antonsson 1999). However, rather than accepting that part of engineering expertise cannot be made explicit—which could be considered equivalent to accepting that engineering is, to some extent at least, an art—one should perhaps rather accept that engineering expertise consists in understanding clearly which engineering claims require strong justification, and what that justification looks like, and which require weaker forms of justification, or even, occasionally, none at all. In my conclusion I argue for the latter view.

Note, finally, that the method of weighted averaging does not in any way prescribe how the weights must be chosen or what they represent. Customarily, engineers use weights to express the relative importance of requirements. As with the valuations themselves, there is the question of who is responsible for choosing the weights, and the considerations discussed there apply here as well.

7.7. Alternative Forms of Assessment

Ordinal and interval value scales, or cardinal value scales as the latter are called in economics, are not the only ones available. A scale that does not fit into the standard range of scales of measurement of increasing strength is the one underlying any form of *grading*—a scale that has several categories of increasing value to which items are assigned, as in the A-to-F scale of school and college grades in English-speaking countries. Grading implies ranking, since an A grade has higher 'quality value' than a C grade, but the ranking is only partial, as grades are not equivalence classes and two items both receiving an A grade are not considered therefore to be of equal value. The grades themselves, however, are taken to be absolute categories: what is required for receiving an A grade is supposed to be recognizable by everyone on the basis of a list of criteria that define an A grade.

Grading would escape from Arrow's impossibility theorem if we accept that grades have absolute meaning, which would entail that no required feature concerning invariance corresponding to Arrow's 3rd one can be formulated. Exactly this has recently been claimed to be what characterizes grading (Balinski and Laraki 2010). However, no way has yet been proposed for grounding this property of grading scales in human behaviour that can be empirically tested, similar to the way in which this is done for the ordinal and cardinal scales, which allows for their axiomatization. No justification for any way of aggregating grades seems available, therefore. Nevertheless, it has recently been shown that a particular method which arrives at an aggregate ranking of options using the median grade in a profile of grades uniquely satisfies a number of desirable features similar to the features that figure in Arrowian aggregation (Balinski and Laraki 2007, 2010). What the relevance of this development is for engineering design is difficult to say, however, since it is not clear—at least not to me—how often grading is used. EU Directive 92/75/EC, which prescribes a grading scale consisting of seven grades to indicate the energy efficiency of a certain class of consumer products (white goods, light bulbs, cars) is an example that is at least close to engineering.

7.8. The Social Dimension of Design and the Multiplication of Obstacles to Optimization

As was mentioned earlier, design is a social process in which typically various stakeholders are involved in defining the design problem in the first place. Additionally, design is done in teams of engineers who represent different disciplines and who are organized through companies that represent further interests—marketing and finance next to design, for instance. Every stakeholder or designer represents a particular perspective to which belong particular valuations of every aspect involved. Clearly this complicates enormously the task of aggregating or unifying all these perspectives so that a particular design emerges as 'the best solution' to the design problem. All problems discussed earlier reoccur, and to some extent in a more inscrutable form. Where the expertise of a single engineer, or the authority of a single client, can perhaps be accepted as a justification of the trade-offs between the performances of options on the requirements that are necessary for singling out an optimal design, no such justification is available once multiple stakeholders, including multiple designers, make up the scene. There is no vantage point from which it can be assessed that client X's preference of option A over B regarding weight equals engineer Y's preference of C over D regarding stability. And any algorithmic approach that is to pass the Arrowian criteria requires such equivalences.

There is more than enough reason to suppose that such formal approaches could at most be extremely idealizing rational reconstructions anyway. It has been argued (Bucciarelli 1994) that even within a firm the dealings between engineers in a design team with different disciplinary backgrounds is a messy process of ongoing negotiation and exchange, rather than an exercise in applying algorithms for decision-making.

This is not to say that due to its social nature, design is bound to be suboptimal at best. Against tall claims that, because of the Arrowian structure of the problem they face, designers cannot optimize and that therefore design must occasionally be irrational (Hazelrigg 1996), it has been argued that as long as communication is a possibility, so is an outcome that is at least Pareto-optimal (Franssen and Bucciarelli 2004). To achieve this, however, communication will typically have to be intensive and continuous—precisely what the context of a firm can offer to design teams, as Bucciarelli (1994) has documented. between several stakeholders who all have a say in what the design should achieve or what the problem is for which a design solution is sought, an equivalent form of communication can be offered by the context of political deliberation, broadly conceived. This can also often be represented as a negotiation process, although one that is more ambiguous and likely to be settled by extrapolation—enlarging the scope and adding complexity—rather than interpolation. It is an open question, however, to what depth the communication can and must go between the external stakeholders who own a problem and the designers who are invited to solve it. This would involve a mixing of contexts that are worlds apart, and indeed, in the modelling of design processes, scarce attention seems to have been paid to the translation phase.

7.9. Conclusion

If there is a main conclusion to be drawn from this overview, it is, I suggest, that optimality in engineering design is overrated. In many cases, the claim that a particular design is the best answer to the problem posed in the design task, the option that best satisfies the design specifications, can be shown to rest on very thin ice. But engineering can get away with this, because in the shadow of the optimal, several less enticing notions are doing the real work. Engineering first of all secures effectivity and avoids failure by yes/no demands that can be relied upon to deliver a design that is good enough—to be sure, within the bounds put by the recognition that all worldly knowledge is fallible. Additionally, the ordinal requirements can be used to avoid designs that can be improved upon in an obvious and uncontroversial way. What reason do we have to ask for more? The justification for the more that we seem to be getting, or are told we are getting, is not to be found in the strict methodology of engineering design. But this is not a problem, because at this stage engineering can afford to do without.

Related Chapters

Chapter 17: Use Plans (Auke Pols)
Chapter 18: Function in Engineering (Boris Eisenbart and Kilian Gericke)
Chapter 34: Engineering and Sustainability: Control and Care in Unfoldings of Modernity (Andy Stirling)

Further Reading

Jacobs, J.F., Van de Poel, I. and Osseweijer, P. (2014). Clarifying the Debate on Selection Methods for Engineering: Arrow's Impossibility Theorem, Design Performances, and Information Basis. *Research in Engineering Design,* 25, 3–10. (Gives a reasonably short overview of the various ways in which Arrow's theorem is relevant to the methodology of engineering design.)

Kroes, P., Franssen, M. and Bucciarelli, L.L. (2009). Rationality in design. In A. Meijers (ed.), *Philosophy of Technology and Engineering Sciences* (Handbook of the Philosophy of Science, Vol. 9). Amsterdam etc.: North Holland, pp. 565–600. (Discusses to what extent the engineering design process can be seen as a structured decision-making process subject to rational constraints.)

References

Aczél, J. and Roberts, F.S. (1989). On the Possible Merging Functions. *Mathematical Social Sciences,* 17, 205–243.

Arrow, K.J. (1950). A Difficulty in the Concept of Social Welfare. *Journal of Political Economy,* 58, 328–346.

Balinski, M. and Laraki, R. (2007). A Theory of Measuring, Electing, and Ranking. *Proceedings of the National Academy of Sciences,* 104(21), 8720–8725.

Balinski, M. and Laraki, R. (2010). *Majority Judgment: Measuring, Ranking, and Electing.* Cambridge, MA and London: MIT Press.

Bossert, W. and Weymark, J.A. (2004). Utility in Social Choice. In S. Barberà, P.J. Hammond and C. Seidl (eds.), *Handbook of Utility Theory.* Boston, Dordrecht and London: Kluwer, pp. 1099–1177.

Bucciarelli, L.L. (1994). *Designing Engineers.* Cambridge, MA and London: MIT Press.

Cross, N. (2008). *Engineering Design Methods: Strategies for Product Design,* 4th ed. Chichester: John Wiley.

Dym, C.L. and Brown, D.C. (2012). *Engineering Design: Representation and Reasoning,* 2nd ed. Cambridge: Cambridge University Press.

Eggert, R.J. (2005). *Engineering Design.* Upper Saddle River, NJ: Pearson Prentice Hall.

Ewusi-Mensah, K. (2003). *Software Development Failures.* Cambridge, MA and London: MIT Press.

Franssen, M. (2005). Arrow's Theorem, Multi-criteria Decision Problems and Multi-attribute Preferences in Engineering Design. *Research in Engineering Design,* 16, 42–56.

Franssen, M. and Bucciarelli, L.L. (2004). On Rationality in Engineering Design. *Journal of Mechanical Design,* 126, 945–949.

French, M. (1992). *Form, Structure and Mechanism.* Houndmills and London: Palgrave Macmillan.

Hazelrigg, G.A. (1996). The Implications of Arrow's Impossibility Theorem on Approaches to Optimal Engineering Design. *Journal of Mechanical Design,* 118, 161–164.

Mas-Colell, A. and Sonnenschein, H.A. (1972). General Possibility Theorem for Group Decision. *Review of Economic Studies,* 39, 185–192.

Middendorf, W.H. and Engelmann, R.H. (1998). *Design of Devices and Systems,* 3rd revised and expanded ed. New York, Basel and Hong Kong: Marcel Dekker.

Moreau, M. (2014). Arrow's Theorem. *Stanford Encyclopedia of Philosophy.* https://plato.stanford.edu/entries/arrows-theorem/

Pahl, G. and Beitz, W. (1996). *Engineering Design: A Systematic Approach,* 2nd ed. London: Springer-Verlag.

Pareto, V.F. (1900). Sunto di alcuni capitoli di un nuovo trattato di economia pura del Prof. Pareto. *Giornale degli Economisti,* 20, 216–235, 511–549.

Scott, M.J. and Antonsson, E.K. (1999). Arrow's Theorem and Engineering Design Decision Making. *Research in Engineering Design,* 11, 218–228.

Simon, H.A. (1957). *Models of Man, Social and Rational: Mathematical Essays on Rational Human Behavior in a Social Setting.* New York: John Wiley.

Simon, H.A. (1982). *Models of Bounded Rationality.* Cambridge, MA and London: MIT Press.

Suh, N.P. (1990). *The Principles of Design.* New York and Oxford: Oxford University Press.

Suh, N.P. (2001). *Axiomatic Design: Advances and Applications.* New York and Oxford: Oxford University Press.

8

PRESCRIPTIVE ENGINEERING KNOWLEDGE

Sjoerd Zwart

8.1. Introduction

Mario Bunge, the remarkable Argentinean philosopher of science and technology, distinguishes *substantive* and *operative* theories. The first are "essentially applications, to nearly real situations, of scientific theories" whereas "[o]perative technological theories . . . are from the start concerned with the operations of men and man–machine complexes in nearly real situations" (1967: 122). According to Bunge, action-oriented research therefore focuses on establishing *technological rules*. Such a rule "*prescribes* a course of action: it indicates how one should proceed in order to achieve a predetermined goal" (1967: 132, original emphasis). From a more historical and empirical perspective, aeronautic engineer and historian of engineering Walter Vincenti draws on a similar contrast. In his ground-breaking *What Engineers Know and How They Know It*, he distinguishes between *descriptive* and *prescriptive* engineering knowledge: "Descriptive knowledge is . . . knowledge of truth or fact; it is judged in terms of veracity or correctness. Prescriptive knowledge is knowledge of procedure or operation; it is judged in terms of effectiveness, of degree of success or failure" (1990: 197). Somewhat further, Vincenti adds: "[F]or epistemological discussion, classification of technological knowledge according to its nature as descriptive or prescriptive may be more fundamental than according to its purpose for production or design."[1]

This chapter is about prescriptive engineering knowledge (which will be referred to in what follows as simply "prescriptive knowledge") as distinguished by Bunge and Vincenti. Its purpose is to raise awareness of the special character of prescriptive knowledge and its importance for engineers as problem-solvers. To that end we will first present our systematic treatment of prescriptive knowledge and then will discuss important engineering initiatives to cope with its specifics. Following that, we will ask about the specifics of prescriptive knowledge, and how the engineering literature deals with them. Broadly conceived, prescriptive knowledge is about *physical engineering actions*. More specifically, it consists of recommendations or rules for engineers on *how to act* or *intervene* within some engineering context to achieve a technical goal specified in advance.[2] These recommendations are *normative*, they have empirical content and can be written out, i.e., they are not *tacit*. Let us consider some typical examples.[3]

Between the first and second Industrial Revolution, Sadi Carnot formulated his prescription of how to increase the efficiency of heat (steam) engines: (1) "The temperature of the fluid should be made as high as possible"; (2) "the cooling should be carried as far as possible"; (3) "it should be so arranged that the cooling of the gas should occur spontaneously as the effect of rarefaction" (Carnot 1824: 96). About fifty years later William Froude developed his extrapolation method, which prescribes how to predict the resistance of a prototype ship hull. He found that to predict this resistance using a scaled-down, geometrically similar model, one must separate the model's skin friction due to viscosity from its wave-making friction and extrapolate them independently (Zwart 2009). In any event, scaling is an old engineering method. As early as the third century BC, Philon of Byzantium wrote in his *Belopoiika* how to scale up and down missile-throwing machines (Philon of Byzantium 1918). His main prescriptive knowledge indicates how to calculate the required diameter of a bore hole (for the twisted skein) to throw stones ranging between 10 and 180 minas (one mina being approximately half a kilogram). Philon indicates that his formula mainly summarizes systematic "experiment and investigation"; and regarding the validations he remarks that the outcomes were in accordance with the machines of the master builders of Rhodes (Cohen and Drabkin 1948: 319).

Another example is Reginald Fessenden's prescription to use *continuous waves* to transmit voice and music via the "ether" and to avoid Marconi's original spark transmitters.[4] The Wright brothers produced much know-how about flying and manipulating heavier-than-air flying machines (Newberry 2013). And again, Vincenti showed how much prescriptive knowledge is implied by the developments of flush riveting for aircraft (Vincenti 1990: Chapter 6). All working principles for artifacts harbor prescriptive knowledge: e.g., about how to build a refrigerator that uses a flame to drive the cooling process; or, about how to generate and maintain aerobic granular sludge in a sequencing batch reactor for a municipal wastewater treatment plant (Nereda®). All these examples[5] deal with how to achieve a complex technical goal in a specific context, and none of them reduces to just the application of science. It follows from the vast conceptual gap separating descriptive and prescriptive knowledge explored in this chapter that for the latter to be trustworthy, engineers explore many more sources of knowledge than only scientific descriptive knowledge.

Developers of prescriptive knowledge undeniably sometimes use scientific results, true or false. Carnot, for instance, mentions caloric and what he called Mariotte's law. This phenomenon, however, hardly justifies the well-established custom in the Anglo-Saxon world to equate engineering science with nothing more than the application of science ("applied science"). Engineering science requires too much creativity and problem-solving ingenuity to be characterized as just the application of science (Vincenti 1990). Moreover, scientific results are applied for many other purposes than only engineering ones.[6]

Prescriptive knowledge often precipitates in patents. For instance, take the question of how to make an incandescent filament. Thomas Edison states in his patent on this subject:

> The invention further consists in the method of manufacturing carbon conductors of high resistance, so as to be suitable for giving light by incandescence, and in the manner of securing perfect contact between the metallic conductors or leading-wires and the carbon conductor.
>
> *(Meijers and Kroes 2013)*

Or consider the principal Nereda® technology patent (EP 1542932), which is called *Method for the Treatment of Waste Water With Sludge Granules* and covers ten methodical how-to claims.

Although prescriptive knowledge plays a prominent role in engineering and technology, its particularities are scarcely recognized and discussed in the literature; however, it deserves careful scrutiny and "warrant[s] further attention" (Houkes 2009: 342). Despite its profound epistemological consequences, here we have the opportunity to discuss prescriptive knowledge only from the perspective of engineering practices, where examples abound.[7] Thus we discuss questions regarding the identification, formulation and reliability of prescriptive knowledge. However, the question whether prescriptive knowledge can *in principle* be reduced to descriptive knowledge (Stanley and Willamson 2001) is considered less relevant for this chapter because regarding engineering practices, it definitely cannot. We take sides unreservedly with Per Norström when he contends: "there are strong reasons to keep the *knowing how–knowing that* dichotomy in technological contexts" (2015: 553, emphasis in original), and we will discuss these reasons (conceptual differences) in this chapter. For engineering practices the dichotomy is pivotal because the two types of knowledge require distinctive methodologies.

This chapter is organized as follows. Section 2 begins by comparing prescriptive knowledge with other related teleological knowledge notions and discussing their similarities and differences. It continues by focusing on the characteristic properties of prescriptive knowledge, comparing them with those of descriptive knowledge. After these conceptual analyses, in Section 3 we sketch the ways engineers *themselves* conceptualize prescriptive knowledge; remarkably, nearly all relevant literature was found in the information and management sciences. Section 4 discusses methodological strategies to increase the reliability of prescriptive engineering knowledge. In Section 5, we contrast the emancipatory and the eliminativist perspectives on prescriptive knowledge; and Section 6 finishes with conclusions and a call for the emancipation of prescriptive knowledge, worthy of being investigated in its own right.

8.2. Prescriptive vs Descriptive Knowledge

We have seen that prescriptive knowledge consists of advice or prescriptions for which engineering *actions* to take to achieve some technical goal. Being *prescriptive*, it is normative means-end action knowledge, it has empirical content, and it can be explicated in words. More precisely, in this chapter prescriptive engineering knowledge is identified with all concrete engineering action-knowledge that can be paraphrased as *technical norms* of this (canonical) form:

(★) *"If you want to achieve technical goal G and you are in context C, you should perform engineering action A."*

This comes close to the form of technical norms as introduced by von Wright (1963) (elaborated by Niiniluoto (1993)), and Bunge, which he poses as "an instruction to perform a finite number of acts in a given order and with a given aim" (1967: 132).

Let us introduce some terms of assessment. We call a (★) rule *validated* if the action prescribed *effectively* achieves the technical goal under the intended circumstances mentioned; it is *justified* if the action is as least as *efficient* as all other reasonable alternatives. A technical rule (★) is *reliable* if it is validated and justified. Moreover, if the action and the goal of a technical rule are morally and juridically allowed and avoid unwanted side effects, we call this rule *feasible*.[8] Strictly speaking, although effectiveness and efficiency apply to the action prescribed in the rule, in a *pars pro toto* manner, we apply them also to the (★) rules that prescribe these actions; moreover, the reliability of prescriptive knowledge reduces to that of its rules.[9]

Many other teleological knowledge terms relate to prescriptive knowledge. We do not distinguish it sharply from "know(ing)-how" or "how-to knowledge," although the latter might, but need not, be more personalized than prescriptive knowledge. The terms "engineering means-end knowledge" and "functional ascription" come close to prescriptive knowledge as well, although they lack the necessary normative aspect; the same holds for "rules for technical interventions"; and "functional knowledge" and "engineering science" need not even concern actions. Although applying rules requires skills, we do not identify prescriptive knowledge with "skills," which are more personalized and most of the time tacit, defying the challenge of being articulable; the same holds for "knowledge-as-ability" and "tacit knowledge," which is inexpressible by definition. Neither is prescriptive knowledge, rather, just a set of general "heuristics" (Koen 2003). Although close in intention, heuristics are often less committal and have less empirical content than prescriptive knowledge. Finally, "operational or procedural knowledge" seems to come close but is more recipe-like, whereas prescriptive engineering knowledge does not come as a recipe. Although the former is strongly context-related, it is not "ad hoc." It is general to some extent within (some type of) context, whereas recipes are far less related to contexts—following the prescribed steps exactly does result in a cake with precise characteristics—whereas Carnot's guidelines are not concrete steps to design an actual heat engine.

Research engineers are reluctant to formulate their prescriptive knowledge in the (★) form. Instead they often use weaker modal formulations of the form:

(†) "With *A* one can achieve *B*."

Many engineers use (†) to formulate the outcomes of their project, which they consider a "hypothesis." As these formulations cannot be falsified, (†) claims make bad hypotheses (Zwart and Vries 2016: 231). Some engineers are even unaware of this weakness of (†) propositions; they may even refuse to accept the non-falsifiability as a drawback of this form.

Let us consider some reasons why engineers are attracted to (†). First, many of their research projects are *opportunity-driven*. Such projects typically begin with the study of a new material, process, or technique, and then the researchers try to find new applications for them. The results then naturally take the (†) form. A second reason for proposing a (†) "hypothesis" is the lack of a handbook on *engineering methodology* or epistemology. Consequently, engineers refer to standard methodology literature, which advocates the formulation of hypotheses, without considering the specific characteristics of prescriptive knowledge. The final reason for engineers to embrace the (†) form is their reluctance to accept *accountability* for the contents a claim in the (★) form bestows on them. They are often aware of the difference in accountability between claims of the (★) form and the (†) form.[10]

To show that the prescriptive part of engineering knowledge may be studied as an epistemological category in its own right, we contrast descriptive and prescriptive knowledge regarding their main purpose and four significant characteristics.

To begin with the first, the goal of developing descriptive knowledge is to come to *approximately true description*s of some subjective or objective aspect of the natural or artificial world. Scientific laws and theories, hypotheses, and descriptive models serve as examples—despite empiricist scrutiny, most engineers consider causal explanations also to be descriptive knowledge. In contrast, the goal of prescriptive knowledge is to prescribe an *effective engineering action* that, within the indicated context, is efficient to achieve the preconceived technical goal. Kirchhoff's circuit laws are examples of descriptive knowledge in engineering, and prescriptions about how to design an electrical amplifier with a high input impedance are examples of prescriptive knowledge. Descriptive and prescriptive knowledge have drastically different properties because their goals are different. Let us consider some of them.

First, the *syntactic form* of descriptive knowledge differs from that of prescriptive knowledge because the former are descriptions and the latter prescriptions. Consequently, their logics are different as well. Whereas the logic of descriptive knowledge is *first* (or perhaps higher) *order* with static truth-value distributions, the logic of prescriptive knowledge must be captured in some form of *dynamic logic* where actions may change the truth-value of a sentence from one state of the world to another (Fervari et al. 2017; Wang 2017; Zwart et al. 2018). In these logics, the fact-value gap is unbridgeable.

Second, the *mode* of descriptive knowledge is structural. It does not imply any functionality or intentionality. The statement "the yield strength of stainless steel is 520 MPa" reveals no directedness or "is-for" quality. Prescriptive knowledge, however, formulated as (★) rules, is teleological by definition, such as, e.g., "use quenching to strengthen steel." As a result, alleged descriptive knowledge is true or false. In contrast, prescriptive knowledge is valued for its reliability or feasibility but it is not true or false. The judgement how to accomplish a goal always concerns a multiple criteria problem about advantages and disadvantages implied by the actions considered. Descriptive knowledge does not involve these multiple criteria problems.

Third, the *value-relatedness* of prescriptive knowledge differs from that of descriptive knowledge; whereas prescriptive knowledge is inherently value-laden on the object level—the prescription imposes what should be done—descriptive knowledge is often considered value-free, at least on the object level. Even if the meta-level normativity reflects on the object level (e.g., as in a hypothesized correlation between aggressiveness and eating meat) one could consider the object-level description to be value-free.

Fourth, and finally, descriptive knowledge is valued for being *as little dependent on context* as possible. Newton's mechanics produce concrete results only if provided with precise initial conditions. Prescriptive knowledge, by contrast, is valued if geared to an exactly circumscribed type of context. Decontextualized prescriptive knowledge is often impaired by becoming increasingly vague. Table 8.1 lists the differences just discussed.

Clearly, our claims about the difference between descriptive and prescriptive knowledge deserve much more philosophical scrutiny. Philosophical depth notwithstanding, the profound difference in goals and the ensuing contrasts of characteristics emphasize the sheer disparity between descriptive and prescriptive knowledge. Considering this disparity, one may wonder why engineering prescriptive knowledge has hardly been studied in isolation. One reason is probably the mutual interdependence of descriptive and prescriptive knowledge. Any production of substantial prescriptive knowledge does depend on a good deal of true description and vice versa: the development of non-trivial descriptive knowledge always requires prescriptions about experimental setup, statistics, finding literature, etc. Another approach to the contrast between descriptive and prescriptive knowledge is simply *to deny* that any prescriptive knowledge can be scientific: this eliminativist position will be discussed at the end of this chapter.

A final remark concerns the difference between prescriptive knowledge and *design methodology* as pursued in industrial design and in architecture and urban design. Although not unchallenged, standard design methodology often differentiates phases of the design process: setting up problem criteria, synthesizing, prototype construction, testing and evaluation (ABET 1998).[11] On the object level, this methodology hardly considers the empirical content of the design at all. Design methodology is about how to come to design decisions in general. In contrast, prescriptive knowledge is mainly concerned with empirical content: the material intervention prescribed should achieve the specified goal within the context described. Thus, whereas the former concerns the methodological steps and is largely void of empirical content, the latter is about the material connection between the prescribed action and the technical goal to be achieved and is based on underlying causal mechanisms.

Table 8.1 Differences between descriptive and perspective knowledge

	Descriptive Knowledge	*Prescriptive Knowledge*
Goal of development	Approximately true hypotheses describing some aspect of the world	Prescriptions about how to act to achieve some (technical) goal in a certain engineering context
1. *Form*	Descriptive (first- or higher-order logic with static truth-values)	Normative means-end (dynamic logic with changing truth-values)
2. *Mode*	Structural; non-intentional; non-functional (approximately true or false)	Intentional and functional (reliability and feasibility)
3. *Values*	As value-free as possible on the object level	Inherently necessarily value-laden on the object level
4. *Context*	As decontextualized (general) as possible	Strongly, and inherently, context dependent

8.3. Prescriptive Knowledge as Characterized by Engineers

Prescriptive engineering knowledge has hardly been the subject of profound epistemological study. Despite this lack of philosophical encouragement, engineers recognized themselves the importance of the distinction between prescriptive and descriptive knowledge. Nearly all influential research about prescriptive knowledge has been carried out by *information* and *management engineers*, and in this section we discuss their most noticeable publications. The field of information systems (IS) features a well-established research area called *design science*, whose aim is to offer prescriptions for information engineering projects. Less predominantly, but still successfully, Joan van Aken also has taken initiatives to incorporate action knowledge in *management science* because its main goal is to develop reliable and feasible managerial interventions.

Many prescriptive knowledge studies in engineering recognize *Sciences of the Artificial* (1996) by Nobel laureate Herbert Simon to be an important source of inspiration. This comes as no surprise. Whereas Simon's teacher, logical empiricist Rudolf Carnap, was mainly concerned with the rational reconstruction of mature descriptive science, Simon entered the world of engineering (1996: 4). He showed much more interest in knowledge about the artificial, which is normally determined by its goals and functions, and often is formulated using descriptions *and imperatives* (1996: 5). About the difference between science and engineering, Simon observes: "Historically and traditionally, it has been the task of the science disciplines to teach about natural things . . . It has been the task of engineering schools to teach . . . how to make artifacts that have desired properties and how to design" (1996: 111). Simon sustained an interest in the action side of science. For instance, in "Rationality as Process and as Product of Thought," he writes:

> we must give an account not only of substantive rationality—the extent to which appropriate courses of action are chosen—but also procedural rationality—the effectiveness, in light of human cognitive powers and limitations, of the procedures used to choose actions.
>
> *(1978: 9)*

Simon identifies two "branches of applied science" concerned with procedural rationality, viz., *artificial intelligence* (as part of information science) and *operation research* (as part of management science; 1996: 27). The next two subsections discuss the most influential engineering publications about prescriptive knowledge.

8.3.1 Prescriptive Knowledge in Design Science as Described by Information Engineers

In the first seminal paper, Joseph Walls et al. (1992) take up Simon's challenge to formulate sciences of the artificial; the paper starts with the motto: "The professional schools will reassume their . . . responsibilities to the degree that they can discover a science of design, a body of intellectually tough, analytic, partly formalizable, partly empirical teachable doctrine about the design process" (Simon 1981). They claim that IS needs *design theories*, which are prescriptive, and are "based on theoretical underpinnings . . . [to] say how a design process can be carried out in a way which is both effective and feasible" (1992: 37). According to the authors, *goals* should be taken as an integral part of design theories, which are *prescriptive*, in contrast to descriptive theories. Moreover, design theories should say *how to achieve a goal*, and, finally, they claim design theories to be theories of Simon's *procedural rationality* (1992: 40–41). Clearly design theories, as characterized by Walls and his colleagues, are at the heart of prescriptive knowledge as formulated earlier. In the second part of the paper, the authors develop a substantive design theory for *vigilant information systems*, which helps managers to remain alert to threats and opportunities for their businesses; it offers a clear example of engineering know-how production.

The authors did so because they thought in 1992, time was ripe. IS had matured sufficiently by then to develop endogenous IS theories, while before it had to be based mainly on approaches of other disciplines. By showing how to develop a design theory, they intended to encourage colleagues to follow their example and also to develop internal IS design theories.

In the second paper, Salvatore March and Gerald Smith (1995) aim to stimulate progress in information technology (IT) by improving the answers to *why* and *how* some IT systems succeed and others fail. For that purpose they state that we should distinguish between "design science," for which they refer to Simon (1996), and the natural science aspects of IT research.[12] About the difference of intention, they write: "Whereas natural science tries to understand reality, design science attempts to create things that serve human purposes" (1995: 253). They distinguish the two as follows: "Natural science is descriptive and explanatory in intent. Design science offers prescriptions and creates artifacts that embody those prescriptions" (1995: 254). Their examples of prescriptive sciences are operations research and management science (1995: 253). March and Smith's distinction between design science and natural science is clearly similar to the contrast between descriptive and prescriptive knowledge; however, an important difference is that March and Smith's design science harbors prescriptive knowledge *and* design, and therefore comprises two distinct and quite different methodologies: one for producing successful artifacts, and another for developing reliable prescriptive knowledge for design.[13]

Interestingly, according to March and Smith, both natural and design science should be part of research in IT. They distinguish between *products* (constructs; models; methods; instantiations) and *activities* (building; evaluating; theorizing and justifying), and consider building and evaluating to be design activities, viewing theorizing and justifying as part of natural science. Thus in their view of IT, design science and natural sciences intertwine intricately. Artifacts and their workings provide new phenomena to be explained; the building of artifacts is often helped by thorough understanding of the natural phenomena; finally, artifacts may provide substantive tests for claims in natural science. In the meantime, the term "design science" covering both design (of artifacts) and prescriptive design knowledge has become a more-or-less standard IS concept.

Salvatore March is also involved in the third seminal paper. With Alan Hevner, Jinsoo Park and Sudha Ram, he wrote the famous "Design Science in IS Research" (2004). It is the highest recognized and most-cited publication of its sort. Contrasting the behavioral science with the design science paradigm, the authors set up instructions to understand, carry out, and evaluate IS design-science research, which they acknowledge to be rooted in Simon (1996). Like March and Smith (1995), Hevner and his colleagues consider solutions in design science to be achieved by constructing an artifact and putting it to work. Consequently, they assume truth and usability to be inseparable: "Truth informs design and utility informs theory" (2004: 80). Within this pragmatist point of view, not distinguishing between design and design knowledge is coherent with closely relating truth and usability.

Hevner et al. formulate seven guidelines of design science, which display the dualistic nature of design science outcomes (2004: 83). On the one hand, according to Guideline 1, "Design-science research must produce a viable artifact in the form of a construct, a model, a method, or an instantiation." On the other hand, Guideline 4 describes the prescriptive knowledge part of IS design science: "Effective design-science research must provide clear and verifiable contributions in the areas of the design artifact, design foundations, and/or design methodologies." This guideline is also about means-end knowledge, where the authors add: "Effective design-science research must provide clear contributions in the areas of the design artifact, design construction knowledge (i.e., foundations), and/or design evaluation knowledge" (p. 87).[14]

In the fourth paper that we discuss, Juhani Iivari (2007) very explicitly defends prescriptive knowledge as a separate goal of IS design research. Partly in reaction to Hevner et al. (2004), Iivari "emphasizes the prescriptive knowledge of IT artifacts *as a distinct knowledge area* that cannot be reduced to descriptive knowledge" (Iivari 2007: 41, my emphasis). Thus, Iivari explicitly differentiates between prescriptive and descriptive knowledge in IS research design. He characterizes design process knowledge, which is almost identical to prescriptive knowledge, using Bunge's (1967) notion of technical rules and Niiniluoto's technical norms. In contrast to Niiniluoto, however, according to Iivari, technical norms lack truth-values. Besides recognizing prescriptive knowledge as a distinct knowledge area, and referring to technical rules or norms to characterize prescriptive knowledge, Iivari also observes explicitly that design science is value-laden. He summarizes his paper in twelve theses, of which the eleventh reads: "design science cannot be value-free, but it may reflect means-end, interpretive or critical orientation." Iivari's last thesis prescribes that design science researchers should make the values of their products as explicit as possible, which closely fits the value-ladenness characteristic of prescriptive knowledge.

Although Alan Hevner (2007) claims to agree with Iivari's twelve theses, he does not view prescriptive knowledge as a self-contained ingredient in his triple-cycle model, which is supposed to guarantee high quality in IS design science research. It comprises a *design cycle* to reach an appropriate design, a *relevance cycle* for field tests and requirements, and a *rigor cycle* to apply descriptive scientific knowledge to design. Refraining from identifying prescriptive knowledge as a separate knowledge category, the output of Hevner's design science research is still artifacts (or processes) *and* knowledge. Hevner's systematization is not false, but it fails to attribute explicitly different methods for the design of artifacts and the development of prescriptive knowledge; this renders the model less appropriate for methodological purposes.[15]

8.3.2 *Prescriptive Knowledge as Described in Management Science*

In characterizing prescriptive knowledge, the publication that comes closest to Iivari (2007) and this chapter is in the management science literature: van Aken (2004). That paper's purpose is to mitigate the relevance problems of management research by complementing descriptive and explanatory organization theory with a prescription-driven management theory (MT). The latter typically consists of field-tested and grounded technological rules and is developed within the design sciences paradigm. Van Aken acknowledges Simon (1996) to be an important source for the distinction between explanatory and design sciences for which he mentions medicine and engineering as important examples. Unlike Walls, March and Hevner, van Aken takes design science research to be only about prescriptive knowledge and not about artifact production. It should "support the design of interventions or artefacts . . . a design-science is not concerned with action itself, but with knowledge to be used in designing solutions, to be followed by design-based action" (2004: 226). This knowledge comes in the form of technological rules (Bunge 1967), which van Aken defines as "a chunk of general knowledge, linking an intervention or artefact with a desired outcome or performance in a certain field of application."

In accordance with (★) in Section 2, van Aken describes "The logic of a prescription" of the technical rules this way: "if you want to achieve Y in situation Z, then perform action X" (2004: 227). Before accepting them as part of design science, he requires these rules to be *field-tested*, i.e., being "systematically tested within the context of its intended use," and *grounded*, i.e., based "on scientific knowledge . . . including law-like relationships from the natural sciences" (2004: 228). Niiniluoto (1993: 13) refers to this distinction as support "from below" and support "from above."[16] He and van Aken both take Bunge's groundedness as an example according to which a "rule is grounded if and only if it is based on a set of law formulas capable of accounting for its effectiveness" (Bunge 1967: 132–133). Van Aken, however, must weaken his notion because of the lack of law formulas in the management sciences.

Interestingly, van Aken also develops a framework to produce technological rules, which is typically based on solving different but similar professional problems. Regarding this problem-solving, he refers to van Strien's regulative cycle (1997: 689), which develops and assesses an *intervention* or *plan of action* rather than establishing a hypothesis as in the empirical cycle. Van Aken designs a *reflective* cycle (2004: 229), which consists of "choosing a case, planning and implementing interventions," then "reflecting on the results," and finally results in developing design knowledge that should be "tested and refined" in future cases. In this way, repeated problem-solving based on scientific descriptive knowledge produces prescriptive knowledge in the form of field-tested and grounded technological rules.

Although cycles as those of van Strien and van Aken are definitely part of the development of prescriptive knowledge, we mentioned that it might be *opportunity* driven as well. As observed by Iivari (2007) and Hevner (2007), it also comes through curiosity about how new technological opportunities may be explored. With some newly developed material, mechanism or process *A*, we can bring about state-in-the-world *B* in a new, more efficient or cheaper way than before. Let us consider an example.

The beginning of the revolutionary Nereda® wastewater treatment technology provides a typical example of opportunity-driven research. At the end of the 1990s, Mark van Loosdrecht and his colleagues at the Delft University of Technology succeeded in undermining the biological explanation of granulation in terms of mutual exchange of indispensable nutrients by producing aerobic granules of fast-growing organisms. They showed that fast-growing organisms granulate in an airlift, with its many shear forces, but resist granulation in a bubble column, with fewer shear forces. This outcome enabled them to develop granules with an aerobic autotrophic outside layer where nitrification can take place, and an anaerobic heterotrophic core for denitrification. This phenomenon enabled the "plant-in-the-model" granule technique, which reduced the two stages of traditional wastewater treatment into one granular-sludge sequencing batch reactor (GSBR). The opportunity of layered granules initiated the now-famous Nereda® wastewater treatment technology.[17]

Let us summarize the findings of this section. Acknowledging Simon's *Sciences of the Artificial* (1996) as an important source of inspiration, *engineers* in IS and management have recognized themselves the importance of discerning prescriptive knowledge, and they contrast it with descriptive knowledge. Some, but not all, of them distinguish explicitly between design and prescriptive knowledge. Moreover, they recognize that prescriptive knowledge may be the product of problem- and of opportunity-driven research. Finally, one of these engineers refers to the regulative cycle instead of the empirical one to found explicit engineering know-how.

8.4. Reliability of Prescriptive Knowledge

Analogous to the logical empiricists' distinction between the discovery and justification of descriptive knowledge, we may wonder how prescriptive knowledge comes into being, and how its reliability is established. Because of the means-end character of prescriptive knowledge, its context dependency and normativity, and the assessment of its reliability and feasibility, differ substantially from the verification or confirmation of descriptive knowledge. If van Strien and van Aken are right, and prescriptive knowledge is developed in repeated cycles where the effectiveness of a technical rule repeatedly has to prove its mettle in practice, the formulation-evaluation distinction is even harder to maintain for prescriptive than for descriptive knowledge. Remarkably, Simon (1996) does not discuss the reliability of prescriptive knowledge, and for Hevner et al. (2004), evaluating design science relates mainly to the assessment of designed artifacts.[18] Engineers use various methods to increase the reliability of their prescriptive knowledge. Since no handbook of epistemology or general engineering attempts to assess the reliability of prescriptive knowledge, we must take up this task ourselves. We will discuss three general strategies to increase the reliability of prescriptive knowledge: support "from below," support "from above" and computational modeling.

Systematic Parameter Variation (PV) is the first form of support from below, or field testing. Vincenti's chapter on the Durand-Lesley Propeller Tests (1990: 142) provides a thorough introduction to the phenomenon. According to Vincenti, PV is a "procedure of repeatedly determining the performance of some material, process, or device while systematically varying the parameters that define the object of interest or its conditions of operation" (1990: 139). Systematically varying the geometric possibilities of then-current reasonable propeller forms, Durand came to forty-eight propellers. To choose the most appropriate form, the performance of each propeller was tested and compared under different speeds, rates of rotation, and diameters. PV is a commonsense way for engineers to proceed when science fails to help them. In his chapter on the history of flush riveting in the airline industry, Vincenti observes: "The development of production techniques [of flush riveting] took place entirely empirically by trial and error or parameter variation of a commonsense engineering kind" (1990: 193). PV first identifies the relevant variables, then systematically varies them, and finally determines which combination is the most effective for the preconceived goal in given contexts. It is an old method to find out how to scale some artifact (cf. Philon of Byzantium), or to increase the effectiveness of artifacts or yields in agriculture. Another famous and more recent example of PV is the Wageningen B-screw series (1930s–1970s). In this series, the torque and thrust of 120 propeller models were determined (with from two to seven blades), at different speeds, and for blade-propeller area ratios between 0.3 and 1.05. The results are still being discussed in modern handbooks on marine propulsion (e.g., Carlton 2012: 93).

As a second form of support from below, engineers (and other researchers in the practical sciences) use *Practical Experiments* (PE). These are experiments in which the dependent or independent variables are not measurable on an ordinal or quantitative scale.[19] Determining the effect of an intervention or population treatment, and comparing different types of materials for some purpose are examples. In these cases, the *independent variable* ranges only over a nominal scale, because differences in interventions or treatments, and the choice of material, often evade measurement; consider, for instance, the different ways of preparing surfaces for welding or gluing them together. In other examples the *dependent variables* may be impossible to measure because no scales exist, such as for the sharpness of cutting tools. PEs may be viewed as precursors of controlled descriptive experiments, where independent and dependent variables are measured on qualitative or quantitative scales and are construed to test hypotheses that map independent variable values on the dependent ones. These experiments are appropriate for testing descriptive knowledge, whereas PEs often serve prescriptive knowledge.[20] The underlying causal mechanisms of PEs are typically too complex and too fuzzy to be described in terms of the (in)dependent variables, such as the forms of sand precipitation in estuaries, or the behavior of large crowds in social technical systems. When engineers must achieve a goal in situations where ignorance of the underlying mechanisms blocks testing of quantitative relations between (in)dependent variables, they are typically forced to perform PEs.

The third way in which prescriptive knowledge is supported from below is by *Established Technology* (ET) or "best practices."[21] Prescriptive knowledge is most reliable if it has proven its validity and efficiency over many years. When a design has been used successfully in commercial applications for a long time, then the rules by which this design came into existence have withstood the strongest possible test of reliability; they are valid because the design is successful and efficient because competitors have failed to find alternatives. This applies to all prescriptive knowledge whatever it concerns: the building of ships, dikes, air- and spacecraft, computers, internal combustion engines, and even software.[22]

Support from below is the oldest and most respected way to create reliable technical rules. Engineers have built amazing edifices, ships and contrivances based on practical experiments or other empirical experience millennia before the Scientific Revolution (e.g., Philon of Byzantium). The development of fluid- and thermodynamics, chemistry, electromagnetism, or even quantum mechanics, partially thanks to engineers, increased technological possibilities considerably and led to the "scientification" of engineering and technology. Besides support from below, many technological rules are therefore supported from above, i.e., they are *grounded* in scientific theories, laws or models able to explain the effectiveness of the rules (Bunge 1967: 132).[23] As an example, Bunge shows how the rule "In order to demagnetize a body heat it above its Curie point" follows from the statement "Magnetism disappears above the Curie temperature" (1967: 133–134).

Grounding technological rules in scientific theories, laws and models increases their reliability and prestige, and backing up experimental rules with scientific knowledge increases their reliability. Grounding protects engineers from pseudo-prescriptive knowledge such as John Scott Russell's "wave line" rule for the lowest ship resistance, which was falsified by Froude's famous Swan-Raven experiments. Many technical rules, however, are still without scientific ground: witness the need for practical experiments and parameter variation in the engineering sciences. Scientific theories and laws enable engineers to adjust undesired outcomes to desired ones by intervening effectively in the underlying causal mechanism. This applies both to problem- and opportunity-driven projects. Regarding the former, the examples of Froude (application of Newtonian Mechanics via the Reech formula) and Carnot (application of Mariotte's law to define a reversible heat engine) show how theories and laws may be used to develop rules for solving technical problems; and regarding opportunity-driven research, quantum mechanical developments have established rules to build lasers, MRI devices, scanning tunneling microscopes, quantum computers, etc.

By the standards of these cases, Bunge's demagnetization example is atypically simple. Technical rules are seldom action-paraphrased laws such as "to increase the pressure of this gas, you should decrease its volume." The gap between abstract laws and highly context-dependent engineering problems is unimaginably larger in practice, and a lot of expertise and clever decision-making is needed to apply laws to the advantage of the problem-solvers. This gap is also nicely illustrated by alleged technical "impossibilities" that have been "deduced" from scientific knowledge such as the impossibility of heavier-than-air flight or transatlantic communication cables.[24] In brief, grounding technical rules in science does contribute to their reliability and may protect us from relying on pseudo-rules, but it should not make us think that engineering is nothing more than simply the application of science. The relation between normative technical rules and descriptive laws and theories is overly complex and still warrants close attention of epistemologists.

In recent decades, because of a revolutionary increase in computational power and the development of dedicated software, modeling and simulation (M&S) has become an everyday tool for design engineers. M&S is a fascinating computational parallel combination of support from above and below. Insofar as computational models are based on scientific theories such as classical mechanics, fluid- and thermodynamics, and electromagnetism, they instantiate support from above. But to the extent to which contextual contingencies of the problem situation are implemented and display the design's behavior, these models are a form of support from below. M&S allows for "computational experiments," possibly supplemented with real data; or for automatic parameter variation. In engineering practice, the reliability of technical rules hinges largely on computational experimentation. Finite element models (e.g., as in ANSYS) are used to find stresses and strains in complex designs. Other M&S packages are applied to more dynamic modeling (e.g. MATLAB/Simulink and LabVIEW; note the abbreviation of "laboratory" in these names). COMSOL is used for computational fluid-dynamics and even for multi-physics applications (combinations of different physical processes such as, e.g., modeling plasma reactors). And using computational mock-up models in AUTOCAD, designers show how their envisaged designs will appear in reality. Although some major accidents can be contributed to overconfidence in computational modeling (Collins et al. 1997), in practice M&S support establishes substantial reliability for prescriptive knowledge.

8.5. Emancipation vs Elimination of Prescriptive Knowledge

Referring to the normativity and goal-directedness of prescriptive knowledge, many academics doubt that this type of knowledge deserves the epithet "scientific." According to Simon, the genuine difficulty for prescriptive knowledge is to produce empirical claims at all "about systems that, given different circumstances, might be quite other than they are" (1996: xi). Without truth-values, prescriptive knowledge cannot be justified true belief, and the mere act of following recipes can hardly be called academic.[25] For a technical university to be a real *university*, and not retreat to being an engineering school or a professional training facility, requires the production of real descriptive knowledge. Additionally, prescriptive knowledge is intersubjective and is indelibly context-dependent. This intersubjectivity emerges in the decision matrix required to choose the "optimal" means for achieving the intended technical goal in the given context. In this matrix the scores of the various options regarding the criteria are weighed and balanced against each other, something that almost never can be done objectively.[26] Moreover, context-dependency plays tricks on prescriptive knowledge as well. First, the state of the technology strongly influences the choice of the "optimal" technical means which makes this optimum time- and context-dependent. Second, the applicability of this knowledge to other context-dependent problems remains uncertain, as the description of contexts always is partial. These reservations about prescriptive knowledge prompt us to discuss the eliminativist position, which claims to be able to eliminate prescriptive knowledge altogether from (academic) design.

Roel Wieringa (2009) elaborates an interesting attempt to purge design science completely of all prescriptive knowledge because it is "philosophically suspect," since "prescriptions do not follow from observations" and involve human decisions. Thus, Wieringa uses a dichotomy between descriptive knowledge and design as follows. Information-system engineering should be considered to address only *descriptive conceptual and empirical knowledge questions* on the one hand, and *practical problems* concerning design and implementation on the other. The descriptive questions are answered using traditional scientific methodology such as testing hypotheses in the empirical cycle; and for the practical problems Wieringa refers to van Strien's regulative cycle, which van Aken used as well (see Section 3). Designing an artifact is considered a practical problem that requires only the regulative cycle and descriptive (and conceptual) knowledge to be resolved and can do without prescriptive knowledge. In his dual perspective Wieringa distinguishes four types of practice-oriented theories in design science research: (1) *diagnostic theories* that formulate explanations for a class of problems; (2) *solution theories* that predict the effects of a class of design solutions; (3) *goal theories* that provide operationalizations for a class of goals and (4) *impact theories* that describe and explain the impact of a class of implemented solutions once the design has been implemented. According to Wieringa, all of these are purely descriptive except (3), which is conceptual. Even evaluative questions, such as how the criteria compare or relate to the facts, causes and impacts, are considered only empirical. In this way Wieringa maintains that (information system) engineering can do without a methodology for know-how or prescriptive knowledge.

Wieringa's eliminativist perspective, in which all technical rules are cunningly reduced to causal claims, has important advantages. It evades all problems with methodological prescriptive knowledge regarding form, status and reliability, and makes engineering science descriptive and more worthy of respect, because all normativity and intersubjectivity are completely relegated to problem-solving. It fails, however, to avoid certain disadvantages. For instance, it falls prey to van Strien's problem of self-fulfilling prophecy. To an important extent, practical scientists validate their own "hypothesis" about what to do in a specific case, and thus confirm solution theories by mechanisms of self-fulfilling prophecy. This applies to social scientists as well as, for example, to (software) design engineers. "This contamination is another argument by academic science that denies scientific status to practice" (Strien 1997: 688). More importantly, however, historical studies clearly exhibit that the eliminativist picture is astoundingly empirically inadequate.[27] Most authors discussed earlier acknowledge prescriptive knowledge because know-how development does in fact play a large role in engineering research projects. To dissect Froude's extrapolation method or Nereda® GSBR know-how into nothing more than descriptive knowledge and problem-solving requires too much *ex post facto* reconstruction to be historically adequate and methodologically feasible. Even according to van Strien, treatment theories, which Wieringa calls solution theories, are *action* theories and are not purely causal.

8.6. Conclusions and Outlook

In this chapter we considered prescriptive engineering knowledge as explicit normative recommendations of physical engineering actions intended to achieve a specific technical goal within some engineering context. Our examples showed that prescriptive knowledge plays a prominent and irreducible role in engineering practices. A brief analysis illustrated the distinct conceptual gap between descriptive and prescriptive engineering knowledge. Skimming the relevant engineering literature, we saw how information and management engineers make the effort to model the way that both types of knowledge intertwine in the design and production of successful artifacts or interventions. Next, we concluded that prescriptive knowledge is validated in various empirical and theoretical ways and by using models. Finally the eliminativist picture of prescriptive knowledge illustrated a tendency to be reluctant to call prescriptive knowledge scientific. These observations all emphasize the urgent need to *emancipate prescriptive engineering knowledge* (and not technological knowledge in general, since that covers also descriptive knowledge; Houkes 2009). Far too often its intricacies and complexity have been mocked, its importance to engineering practice grossly underestimated and its right to be investigated for its own sake curtailed or even completely denied.

In this chapter, the relevant issues could be touched upon only briefly; the chapter may therefore be taken to set out a research agenda. We require more insight into: the definition of prescriptive knowledge and its relations with adjacent knowledge forms; empirical case studies; its relation with descriptive scientific knowledge; its logic and its methods of validation; and its relation to models and modeling.

As long as engineering and technology are developed and taught within and outside our universities, we must acknowledge that in fact it concerns the development of something—let us call it prescriptive knowledge—which engineers apply while solving their technical problems, and that has the characteristics discussed in the previous sections. This prescriptive knowledge also results in patents and has provided names for the major eras in our cultural history, such as the Stone, Bronze and Iron Ages. The differences between the properties of prescriptive knowledge and what has been called (descriptive) knowledge make some academics to refuse it access to the realms of scientific knowledge. For engineering, however, the major issue is not whether this refusal is justified or not; it is whether the research agenda described earlier is carried out, so it may finally be acknowledged that universities produce, besides justified and true, also *justified and effective beliefs*.[28]

Acknowledgements

I thank the members of the TU/e philosophy section, Peter Kroes, the editors, and especially Peter Kaiser for their kind suggestions and constructive ideas about how to improve the contents and readability of this chapter. Of course, all errors in form and content remain completely mine.

Related Chapters

Chapter 5: What Is Engineering Science? (Sven Ove Hansson)
Chapter 6: Scientific Methodology in the Engineering Sciences (Mieke Boon)
Chapter 21: Engineering Design (Peter Kroes)
Chapter 23: Design Methods and Validation (Sabine Ammon)
Chapter 29: Scale Modeling (Susan G. Sterrett)

Chapter 30: Computer Simulations (Hildrun Lampe)
Chapter 31: Experimentation (Viola Schiaffonati)

Further Reading

Aken, J.E. van (2004). Management Research Based on the Paradigm of the Design Sciences: The Quest for Field-tested and Grounded Technological Rules. *Journal of Management Studies,* 41, 219–246. (Classic application of van Strien's regulative cycle to management science eventuating in field-tested and scientifically grounded technological rules.)

Iivari, J. (2007). A Paradigmatic Analysis of Information Systems as a Design Science. *Scandinavian Journal of Information Systems,* 19(2), 5. (Most clearly distinguishes between the prescriptive knowledge about artifacts and theoretical descriptive knowledge in information science.)

Niiniluoto, I. (1993). The Aim and Structure of Applied Research. *Erkenntnis,* 38(1), 1–21. https://doi.org/10.1007/BF01129020 (Modern philosophical discussion of von Wright's notion of "technical norm" within the context of the basic and applied research distinction.)

Strien, P. J. van (1997). Towards a Methodology of Psychological Practice: The Regulative Cycle. *Theory & Psychology,* 7(5), 683–700. (Prominent introduction of the "regulative cycle" for professional interventions, contrasting it to the empirical cycle of "nomological science".)

Vincenti, W. (1990). *What Engineers Know and How They Know It: Analytical Studies From Aeronautical History.* Baltimore: Johns Hopkins University Press. (The great classic in modern philosophy of technology. Chapter 6, section "Observation and Reflections" addresses prescriptive engineering knowledge.)

Notes

1. In the same vein, Joel Mokyr distinguishes between propositional and prescriptive knowledge (2002: 4). Other authoritative taxonomies of engineering knowledge cover similar distinctions. Ropohl (1997) considers functional rules, "which specify . . . what to do, if a certain result is to be attained under given circumstances." De Vries (2005) defines functional nature knowledge: "X knows that carrying out action Ac with artifact A will result in a change in state of affairs." Sven Ove Hansson describes technical rule knowledge, which consists of technical "rules of thumb" (2013: 21). Finally, Meijers and Kroes (2013) also identify prescriptive knowledge: "It consists . . . of procedures, which are "series of prescribed actions that will lead to a desired result: the produced artefact." For further philosophical references cf. Hendricks et al. (2000) and Houkes (2009).
2. Even *mental* actions may define cognitive prescriptive knowledge, which we do not consider here.
3. Engineering is taken here to be problem-solving in technological developments.
4. In the same context, one may wonder if the recommendation to use tubes to design high-end audio amplifiers is prescriptive *knowledge* or just superstition.
5. Obviously, I do not pretend the list of examples to be complete, nor does it exhaust all types of prescriptive knowledge. It is just a contingent list of examples I have encountered over time.
6. Applied science is the result of successfully applying abstract knowledge (as abstract as possible) to practical problems within a concrete context. This application remains descriptive; from descriptive science it never follows how one *should act.*
7. See Zwart and Vries (2016).
8. Here the reliability of prescriptive knowledge is assumed to be an internal technical affair whereas feasibility implies societal values and considerations.
9. Even if applying the rule were always to be accompanied with an explicit description of the rule, this would not render the rule itself descriptive knowledge. The rule's description may be more or less accurate (true or false) but the rule's content can only be reliable (i.e., validated and justified) or not.
10. These three reasons stem from personal experience and should be tested by serious social scientific research.
11. See also the Generator-Test Cycle of Simon (1996).
12. Remarkably, the term "design science" does not occur in Simon (1996).
13. See Zwart and De Vries (2016).
14. Hevner et al. do not discuss different methods for these disparate goals.
15. "Action research" in social sciences and "Research through Design" suffer from the same shortcomings (i.e., confusing different types of goals and insufficiently distinguishing between different kinds of methods). According to Denscombe (2014: Table P1.1) the purpose of the first is to "solve a practical problem" and to "produce guidelines for best practice," whereas for the second, new knowledge is gained from carrying out design process.

16. Hevner (2007) also grounds design science research in scientific theories and methods (Figure 1).
17. See Zwart and Kroes (2015: 387–394) for a more detailed history of the development of the Nereda® technology.
18. The same holds for Venable et al. (2012), one of the very few papers dedicated to design science research evaluation but also concentrating on the resulting artifacts and not on the know-how produced.
19. PEs are similar to Hansson's (2015) "action-guided experiments." The discussion about their exact characterization must be left to another occasion. See also Kroes (2018).
20. PEs are also applied in the explorative phase in descriptive science and controlled experiments for subprojects in establishing prescriptive knowledge. (Froude validated his scaling method with the *HMS Greyhound* experiment.)
21. Koen (2003: p. 32, emphasis in original): "The final signature of a heuristic is that its acceptance or validity is based on the pragmatic standard *it works or is useful in a specific context* instead of on the scientific standard *it is true or is consistent with an assumed, absolute reality.*"
22. Other forms of "support from below" not discussed here are definition of the working principles, proof of concept/principles, feasibility studies and scaling, although these may also apply to the reliability of *a design.*
23. Van Aken (2004) also elaborates on Bunge's notion of "groundedness."
24. Grounding and the underestimation of the gap between theories and laws, and technical rules are also sources of the shortsighted *linear model,* which sees engineering as nothing more than the application of science.
25. See also van Aken (2004: 220).
26. Differences in antibiotics prescriptions provide a telling example. Until 2001 French physicians prescribed antibiotics four times as often as did their Dutch colleagues (Cars et al. 2001), so they clearly gave different weights to the values "patient suffering" and "the risk of large percentages of resistant bacteria."
27. See e.g. Vincenti (1990).
28. Where "beliefs" is to be taken in a broader way than in the traditional representationalist one.

References

ABET (1998). *Annual Report for the Year Ending September 30.* New York: Accreditation Board for Engineering and Technology, Inc.

Aken, J.E. van (2004). Management Research Based on the Paradigm of the Design Sciences: The Quest for Field-tested and Grounded Technological Rules. *Journal of Management Studies,* 41, 219–246.

Bunge, M. (1967). *Scientific Research II: The Search for Truth.* Berlin, Heidelberg: Springer-Verlag.

Carlton, J. (2012). *Marine Propellers and Propulsion,* 3rd ed. Butterworth-Heinemann.

Carnot, S. (1824). *Reflections on the Motive Power of Fire, and on Machines Fitted to Develop that Power.* Trans. R.H. Thurston, 1943. New York: The American Society Of Mechanical Engineers.

Cars, O., Mölstad, S. and Melander, A. (2001). Variation in Antibiotic Use in the European Union. *The Lancet,* 357, 1851–1853. doi:10.1016/S0140-6736(00)04972-2

Cohen, M.R. and Drabkin, I.E. (1948). *A Source Book in Greek Science.* New York: McGraw-Hill Book Company, Inc.

Collins, M.R., Vecchio, F.J., Selby, R.G., Gupta, P.R. (1997). The Failure of an Offshore Platform. *Concrete International,* 19, 28–36.

Denscombe, M. (2014). *The Good Research Guide: For Small-scale Social Research Projects.* McGraw-Hill Education.

De Vries, M.J. (2005). The Nature of Technological Knowledge: Philosophical Reflections and Educational Consequences. *International Journal of Technology Design & Education,* 15, 149–154. doi:10.1007/s10798-005-8276-2

Fervari, R., Velázquez-Quesada, F.R. and Wang, Y. (2017). Bisimulations for Knowing How Logics. *Presented at the 5th International Workshop on Strategic Reasoning, SR 2017.*

Hansson, S.O. (2013). What Is Technological Knowledge? In *Technology Teachers as Researchers, International Technology Education Studies.* Rotterdam: Sense Publishers, pp. 17–31. doi:10.1007/978-94-6209-443-7_2

Hansson, S.O. (2015). Experiments Before Science. What Science Learned from Technological Experiments. In S.O. Hansson (ed.), *The Role of Technology in Science: Philosophical Perspectives, Philosophy of Engineering and Technology.* Dordrecht: Springer, pp. 81–110. doi:10.1007/978-94-017-9762-7_5

Hendricks, V.F., Jakobsen, A., Pedersen, S.A. (2000). Identification of Matrices in Science and Engineering. *Journal for General Philosophy of Science,* 31, 277–305. doi:10.1023/A:1026512011115

Hevner, A.R. (2007). A Three Cycle View of Design Science Research. *Scandinavian Journal of Information Systems,* 19(2), 4.

Hevner, A.R., March, S.T., Park, J. and Ram, S. (2004). Design Science in Information Systems Research. *MIS Quarterly,* 28(1), 75–105.

Houkes, W. (2009). The Nature of Technological Knowledge. In A. Meijers (ed.), *Philosophy of Technology and Engineering Sciences, Handbook of the Philosophy of Science*. Amsterdam, Boston and Oxford: Elsevier, pp. 309–350.

Iivari, J. (2007). A Paradigmatic Analysis of Information Systems as a Design Science. *Scandinavian Journal of Information Systems*, 19(2), 5.

Koen, B.V. (2003). *Discussion of the Method*. Oxford and New York: Oxford University Press.

Kroes, P. (2018). Control in Scientific and Practical Experiments. In I.R. Van de Poel, L. Asveld and D.C. Mehos (eds.), *New Perspectives on Technology in Society* (pp. 16–35). London ; New York: Routledge. doi:10.4324/9781315468259-2

March, S.T. and Smith, G.F. (1995). Design and Natural Science Research on Information Technology. *Decision Support Systems*, 15(4), 251–266.

Meijers, A.W.M., Kroes, P.A. (2013). Extending the Scope of the Theory of Knowledge. In M.J. De Vries, S.O. Hansson and A.W.M. Meijers (eds.), *Norms in Technology, Philosophy of Engineering and Technology*. Dordrecht: Springer, pp. 15–34. doi:10.1007/978-94-007-5243-6_2

Mokyr, J. (2002). *The Gifts of Athena: Historical Origins of the Knowledge Economy*. Princeton: Princeton University Press.

Newberry, B. (2013). Engineered Artifacts. In D.P. Michelfelder, N. McCarthy and D.E. Goldberg (eds.), *Philosophy and Engineering: Reflections on Practice, Principles and Process* (pp. 165–176). Dordrecht: Springer. doi:10.1007/978-94-007-7762-0_13

Niiniluoto, I. (1993). The Aim and Structure of Applied Research. *Erkenntnis*, 38(1), 1–21. doi:10.1007/BF01129020

Norström, P. (2015). Knowing How, Knowing That, Knowing Technology. *Philosophy & Technology*, 28(4), 553–565. doi:10.1007/s13347-014-0178-3

Philon of Byzantium (1918). *Belopoiika*. Abband. Der Preussischen Akad. *Der Wissenschaften, Phil.-Hist. Klasse*, (16).

Ropohl, G. (1997). Knowledge Types in Technology. In *Shaping Concepts of Technology*. Dordrecht: Springer, pp. 65–72. doi:10.1007/978-94-011-5598-4_6

Simon, H.A. (1978). Rationality as Process and as Product of Thought. *The American Economic Review*, 68(2), 1–16.

Simon, H.A. (1981). *Sciences of the Artificial*, 2nd ed. Cambridge, MA: MIT Press.

Simon, H.A. (1996). *The Sciences of the Artificial*, 3rd ed. Cambridge, MA: MIT Press.

Stanley, J. and Willamson, T. (2001). Knowing How. *The Journal of Philosophy*, 98(8), 411–444.

Strien, P.J. van (1997). Towards a Methodology of Psychological Practice: The Regulative Cycle. *Theory & Psychology*, 7(5), 683–700.

Venable, J., Pries-Heje, J. and Baskerville, R. (2012). A Comprehensive Framework for Evaluation in Design Science Research. In *International Conference on Design Science Research in Information Systems*. Berlin, Heidelberg: Springer, pp. 423–438.

Vincenti, W. (1990). *What Engineers Know and How They Know It: Analytical Studies From Aeronautical History*. Baltimore: Johns Hopkins University Press.

von Wright, G.H. (1963). *Norm and Action: A Logical Enquiry*, 1st edition ed. London: Routledge & Kegan Paul PLC.

Walls, J.G., Widmeyer, G.R. and El Sawy, O.A. (1992). Building an Information System Design Theory for Vigilant EIS. *Information Systems Research*, 3(1), 36–59.

Wang, Y. (2017). A Logic of Goal-directed Knowing How. *Synthese*, 1–21.

Wieringa, R. (2009). Design Science as Nested Problem Solving. In *Proceedings of the 4th International Conference on Design Science Research in Information Systems and Technology*. ACM, New York, NY, USA, p. 8.

Zwart, S.D. (2009). Scale Modelling in Engineering: Froude's Case. In A. Meijers (ed.), *Philosophy of Technology and Engineering Sciences*. Amsterdam: Elsevier, pp. 759–798.

Zwart, S.D., Franssen, M. and Kroes, P. (2018). Practical Inference—A Formal Analysis. In A. Fritzsche and S.J. Oks (eds.), *The Future of Engineering: Philosophical Foundations, Ethical Problems and Application Cases* (pp. 33–52). Cham, Switzerland: Springer. doi:10.1007/978-3-319-91029-1_3

Zwart, S.D. and Kroes, P. (2015). Substantive and Procedural Contexts of Engineering Design. In *Engineering Identities, Epistemologies and Values, Philosophy of Engineering and Technology*. Cham, Switzerland: Springer, pp. 381–400. doi:10.1007/978-3-319-16172-3_22

Zwart, S.D. and De Vries, M.J. (2016). Methodological Classification of Innovative Engineering Projects. In M. Franssen, P.E. Vermaas, P. Kroes and A.W.M. Meijers (eds.), *Philosophy of Technology After the Empirical Turn*. Springer, pp. 219–248. doi:10.1007/978-3-319-33717-3_13

9

ENGINEERING AS ART AND THE ART OF ENGINEERING

Lara Schrijver

9.1. Art and Engineering, or, Does Form Follow Function?

The 'art of engineering' is often seen as fundamentally distinct from the concerns of 'art' as a discipline. Yet this simplified distinction fails to acknowledge the common foundations of the two. In essence, by polarizing the domains of engineering and art, a continuation is provided of the conceptual distinction between 'creative' and 'rational', a long-standing trope in public debates on science and art, which is nevertheless difficult to find evidence for.

This perceived discrepancy between the broadly defined fields of art and engineering sits at the center of debates on the future of engineering, and merits further examination (see Chapter 5, "What Is Engineering Science?" by Sven Ove Hansson, this volume). In order to do so, however, a number of aspects should be distinguished. It is vital to separate the internal mechanisms of engineering, which is understood as focusing primarily on functionality—'Will it work?'—from the concerns of art, seen as primarily concerned with perception—'How does it look?' Moreover, there are external influences that affect how the profession sits in society. The understanding of what constitutes engineering has changed over time, from the historical focus on military and infrastructure concerns to the contemporary domains of chemical, electrical and computer engineering. These changes are at times due to innovations or developments in technology, and at times driven by societal needs or transformations. As such, the historical context includes both external transformations in society and cultural perceptions, as well as internal developments in the field and technological progress within science and engineering.

There is currently a good reason to bring the relationship between art and engineering into focus: recent discussions in higher education proclaim the need for more students in the so-called 'STEM' fields: science, technology, engineering and mathematics (President's Council 2012; Gates and Mirkin 2012). This appeal is related to the increasing presence of technology in all domains of life, from banking and healthcare to education and food provision. As the systems that our increasingly urbanized society are founded on become more and more complex, there is a growing need for professionals who can handle complex logistical systems and guide innovations that aid the efficient organization of many aspects of daily life. At the same time, a 'pure' engineering approach is seen as too rationalized in today's world: an increasing number of scholars and policymakers support the need to add 'art' to the mix (leading to the so-called 'STEAM' disciplines), signaling the importance of interfaces, user experience and aesthetics and perception even in highly technological fields (Lachman 2018). The true Renaissance designer is perhaps more needed than ever, and a historical understanding of engineering as art may contribute to this transition. As such, this chapter sets out to examine the central connections between art and engineering, based on a historical overview of the position and nature of engineering in relation to both science and the arts (Black 1961; Faber 1945).

Indeed, the very foundations of 'art' and 'science' may be etymologically traced through their Latin roots: *ars* meaning a (practical) art of application, and *scientia* denoting knowledge (i.e. facts). Yet these simplified distinctions are inadequate for contemporary understanding, and indeed may even be approached from the perspective of shared foundations in the ancient world. In the earliest beginnings of philosophy, architecture, arts and science, there may be certain words that denote areas we might now consider to belong more to the one or the other—such as the Greek *techne*, typically considered to be the (craftsman-like) technique of making things, eminently human, while *poiesis* implies a more natural 'growth' of something. According to Vitruvius the architect was tasked with designing houses and buildings, but also with military works. In fact, the original Latin term 'ars' included sciences along with what we would now call arts and humanities, and a plea for the reintegration of art and technology is increasingly visible (Faste 1995; Jeon and Fishwick 2017; Lachman 2018). Could it be that we are now beginning to see a turning point in the increasing distinction into separate disciplinary fields with their own approaches, methodologies, vocabularies and habits?

In order to explore this perspective, this chapter traces back to the ancient world, to the shared common stem of the now distinct and delimited domains of engineering, art and science in order to bring to the surface the shared origins, and by tracing them through a number of fundamental shifts that have increased the differences, brings it to today, where the need for an integrated and multidisciplinary approach is gaining traction throughout many fields (Ollis et al. 2004).

In this chapter, a historical perspective helps to show a number of moments in which the manner of thinking about engineering, architecture and art is transformed, either through internal concerns of the domains, or through external transformations in society. For example, the Industrial Revolution caused large-scale transformations in technologies of transportation and production, which in turn influenced the manner technology was perceived and utilized. Today, the transformations both in materials science (particularly on the scale of nanotechnology) and in digital networks potentially are equally influential to many domains of engineering.

Engineering as such is a very broad domain including many different subdomains, or as Antoine Picon notes, it "looks more like a continent marked by striking contrasts than like a unified field" (Picon 2004: 422). Picon identifies civil, mechanical and electrical engineering as the primary domains. In this chapter, I depart from these three central domains, the first two of which are already included in the first known treatise on architecture by Vitruvius. The primary focus of this chapter will remain civil engineering as the area historically most aligned with architecture and concerned with aesthetics. Since the Industrial Revolution, civil engineering has played a central role in constructing infrastructure and large-scale objects that are now part and parcel of the everyday environment. In these eminently visible contributions to the built environment (bridges, roads, hydraulic infrastructures), civil engineering provides a prime example of how the theoretical foundations and values of engineering take shape and are perceived. Moreover, the domain of civil engineering has a historically well-developed debate on the distinctions between art (or architecture) and engineering, which is particularly tangible in the distinction between the French École Polytechnique and École des Beaux-Arts (Kruft 1994: 272–275; Picon 2004). These two schools operated on different views: where the Beaux-Arts curriculum was founded primarily on the view that architecture was an art of formal and classical design principles, the École Polytechnique addressed issues of engineering, structure and stability as the primary concerns of architectural design. As such, the privileged position of civil engineering (and to a lesser extent, mechanical engineering) provides a good foundation for clarifying a number of the central questions on aesthetics, art and engineering.

What might the art of engineering be? Is engineering not aimed at utility first and foremost, with other dimensions subsidiary to functionality? Certainly a commonly held conception of the engineer is that of a 'problem-solver' aimed at function over form. As such, even speaking of the art of engineering raises questions from the start. Yet, the literature on engineering as providing its own art form is equally unmistakable (Billington 1985; Committee on General Structures, Subcommittee on Bridge Aesthetics 1991). Eminent scholars have provided convincing arguments that demonstrate the unique and often aesthetically striking solutions that derive from a function-oriented yet creative approach to new questions raised by technological innovations and social transformation. For example, Robert Mark has studied the structures and construction of Gothic cathedrals, and as such begun to understand more of the ingenuity of the craftsman tasked with building these structures (Mark 1990). David Billington demonstrates how naturally intertwined the requirements for engineering the concrete bridges of Maillart are with a sense of elegance and beauty (Billington 1985). And David Gelernter opens up a world of beauty through mechanical and technological objects, exploring how we begin to understand beauty in more than a traditional sense (Gelernter 1998).

Moreover, the question of engineering as an art is not only tied to issues of aesthetics but is also founded on a number of salient features such as creativity and subjective judgment (Bronowski 1961: 1–24; Langer 1953). In the recent debates, two dominant considerations stand out: the relation between knowledge and practice, and the relation between aesthetics and functionality (Parsons and Carlson 2008; Abrams 2015; Devaney 2016). This last feature may be approached from the perspective of the products of engineering, or from the perspective of the design process (Black 1976). When discussing the objects, issues revolve around evaluation criteria typically considered subjective, versus the criteria typically considered objective, such as functionality and efficiency. When approached from the view of the design process, debates typically center on the relation between rationality (as a problem-solving approach) and creativity (as a problem-finding approach). (Picon 2004; Sennett 2008). An enhanced understanding of these varieties of problem-solving and evaluation criteria is seen as a potential added value of STEAM curricula over the earlier STEM debates (Connor et al. 2015).

9.2. Shared Foundations: Architecture and Engineering in the Ancient World

Some of the most awe-inspiring constructions in the world were built in a time when there were immense limitations to the resources available. Indeed, this might be considered one of the fundamental drives of engineering: how to exceed the limitations of human or animal power (Landels 1978: 9; see also Chapter 2, "A Brief History of Engineering" by Jennifer Karns Alexander, this volume). In other words, harnessing force is one of the central features of early engineering works. While this continues in an increasingly sophisticated form today, many early machine technologies such as levers, pulley systems, and hydraulics serve primarily to harness natural forces or extend the capacity of muscle power. Examples of these kinds of machines served for military purposes (catapults), constructions (cranes) or facilitating human settlements (transport and water supply). Even now, the pyramids, the Chinese wall and the Roman aqueducts stand for the innovative capacity of humans to intervene in the natural environment in order to provide for their needs, be they symbolic or functional. As such, they remain a powerful example of human imagination and skill. Some consider the first engineer to be Imhotep, who built the pyramid of Djoser (2650–2600 BCE), first of the great pyramids in Egypt (Alba 2017).

In the ancient world, architecture and engineering were largely considered one domain of activity and design. The structures we might now more easily identify as belonging to the domain of engineering tend to be infrastructural, such as the aqueducts, roads and sewage systems. At the same time, there are also the exemplars of architecture as a feat of engineering: structures that strike the imagination by their sheer scale, such as the pyramids; the remarkable innovations in techniques for challenging locations such as Machu Picchu high in the Andes; or structures that exceed what we would expect for their time: the dome-span of the Pantheon (ca. 118 AD) remained the largest free span for nearly a century and a half, until the Renaissance.

Although there is relatively little written documentation available on architecture and engineering approaches, a few notable works survived from ancient Greece and Rome. Pliny the Elder provides an overview of engineering solutions throughout the ancient world that was still used throughout the Middle Ages and the Renaissance (Landels 1978: 215–217).

In the earliest known treatise on architecture, *The Ten Books on Architecture* (ca. 27 BCE), Vitruvius distinguishes three 'departments' of architecture: "the art of building, the making of time-pieces, and the construction of machinery" (book I, Ch. III, section 1.) This document, rediscovered in the mid-15th century and published by Leon Battista Alberti, positions architecture as the overall discipline of designing mechanical devices as well as public and private environments. The book itself discusses various subdomains and contributing skills such as drawing, geometry, music and astronomy, but also what we might now consider materials engineering, treating at length the different mixtures required for different types of bricks, lime and pozzolano. It provides instructions on types of cranes, catapults and defensive structures, showing how broad at that point the domain of 'architecture' was.

Crucial in the work of Vitruvius is what he is best known for in architecture: the triad 'firmitas, utilitas, venustas', or firmness, commodity and delight as it is usually translated. In other words, all built structures (including machines) should answer to the requirement of being sturdy and appropriate to their use, but also answer to the aesthetic sense. The shared roots of engineering and architecture thus support an understanding of engineering as art.

Books 8 and 10 treat infrastructural works such as aqueducts and cisterns, and the construction of machines to aid in transport and building. While the Greeks provided many of the theoretical foundations for knowledge, it is particularly in the Roman period that many innovations were provided, expanding the practical applications available. This distinction between knowledge and application shows the initial demarcations between science on the one hand and engineering on the other. A tendency to accord greater value to theoretical knowledge than to practical application begins to arise. Some attribute this to Plato's writings, and in particular an 'anti-physical' dimension manifest in the notion of 'Forms' or 'Ideas', "eternal and unchanging objects of true knowledge, by relation to which material things could be studied and reasonably interpreted, though never truly 'known'" (Landels 1978: 187). By referring to the metaphysical notion of 'Forms' as the only source of solid knowledge, Plato not only relegates application and practice to a lower position in the hierarchy of knowledge, but he also casts suspicion upon observation and empirical evidence as the basis for knowledge. His student Aristotle, on the contrary, takes on empirical observation as a solid foundation for knowledge of how the world works. This valuation of abstract knowledge and general principles still at times colors the discussion on engineering and science.

It is useful to take note of these respective positions, as debates on knowledge and scientific method have often continued to grant priority to abstract knowledge, even as the interest in knowledge-in-the-making has increased in recent years (Sennett 2008; Bertucci 2017). While the full extent of engineering in the ancient world is beyond the scope of this chapter, a number of insights from Landels provide a useful foundation for a consideration of the role of aesthetics in engineering, and the status of knowledge in engineering versus 'general' science. Some of the concerns can be traced back not only to a hierarchical treatment of abstract knowledge, but also to a lack of vocabulary for approaching the type of knowledge that sits within domains of application, treated variously as 'tacit knowledge', 'artisanal knowledge' or 'knowing how' (Polanyi 1966; Bertucci 2017; Ryle 1946; see also Chapter 2, "A Brief History of Engineering" by Jennifer Karns Alexander, and Chapter 5, "What Is Engineering Science?" by Sven Ove Hansson, this volume).

By its very etymology, engineering is intimately bound to technology. With ties to the medieval Latin *ingenium* (device or machine) or *ingeniare* (contrive), it is from its origins related to the design and creation of (mechanical) devices. In the medieval period and the Renaissance, a continued development of the (mechanical) art of engineering may be traced as relatively integrated with the concerns of architecture and design. Innovations of the Gothic cathedrals, for example, such as fly-ing buttresses, contributed to their structural integrity while also achieving a more slender silhouette even with their greater height (Mark 1990: 91–135). The dome span of the Pantheon was first exceeded by Brunelleschi's double-shell dome in Florence (1436), which equally provided a creative approach to the problem of an octagonal dome (Corazzi 2017). In this period, the Turkish scholar/engineer/inventor Ismail Al-Jazari also published his *Book of Knowledge of Ingenious Mechanical Devices* (1206), which provides designs for various automata and clocks, among others (Alba 2017).

9.3. Embracing Rationality: Technology and the Rise of Engineering in the 18th–20th Centuries

While the Renaissance has its own historical lessons that are valuable, and even may lay claim to the exemplar of the eminent 'artist-engineer' in the work of Leonardo Da Vinci, it is in the 18th and 19th centuries that we find a new turning point. In this period, a number of developments come together. The invention and improvement of the steam engine towards the end of the 18th century transformed the scale of technological impact and was a key element in the Industrial Revolution, typically seen to have begun around 1760, and which continues to have a broad impact in society. The leaps in technological development, including the steam engine, electricity and mechanical production, combined with the rapidly growing insights into the laws of physics, mechanics and other areas of natural science, form the foundation of the contemporary age (see also Chapters 2, "A Brief History of Engineering" by Jennifer Karns Alexander, and Chapter 5, "What Is Engineering Science?" by Sven Ove Hansson, this volume). The belief in the power of rational and scientific thought took precedence, while automotive developments had an impact throughout many modes of production, from farming to milling and industrial production.

It is arguably in this period that the primary distinction arises between engineering and architecture that we continue to see today, and that the 'art of engineering' fades to the background in favor of its technical and scientific characteristics (Weber and Sigrist 2009). The historical aim of harnessing power that we find throughout the ancient world begins to turn to the production of power, and as such transforms the relation with nature. It is little wonder, then, that the 18th and 19th centuries may be considered as the precursors to modernity as we know it today.

In the same period, the common base of civil engineering and architecture split off into distinct trajectories of education in France. The clichéd perception of engineering as a functional art, and architecture aimed more at aesthetics and style, is visible in the split between the École Polytech-nique in Paris, founded in 1794, and the École des Beaux-Arts, the architecture academy, originat-ing in the 17th century (Kruft 1994: 275–277). The founding of the École Polytechnique divided the common ground of architectural design and building construction into two distinct trajectories: the Beaux-Arts-educated architect, well-versed in art history and styles, and the engineer-architect, educated primarily in the topics of functionality and 'commodité', or appropriateness to purpose. It is in the École Polytechnique that a number of teachers originally affiliated with the École des Beaux-Arts proposed that transformations in society and technology required a different focus in building approaches and construction, premised on arguments of structure, durability and suitability over considerations of form, style and composition. Moreover, it is precisely in the arguments of durability and structure that beauty and elegance is to be found, according to the early teachers of the École Polytechnique (Durand 1805: 5–15; Kruft 1994: 273–275).

This leads to a particular view of beauty in engineering, as naturally arising from the most suitable solution to a series of functional requirements. In the modern era, or perhaps with the rise of indus-trialization, this beauty has become ever more central to the understanding of engineering—not as a goal in itself, but as a natural side effect of something perfectly suited to its purpose. Notwithstanding the instances in which functionality and aesthetics struggle to find the right balance, in many other examples, the key characteristics of a well-designed object appeal to similar properties: elegance, coherence, purpose. Or, as suggested in one of the handbooks on engineering and design:

> the art will be the natural byproducts of our attempts to produce an efficient structure, one based on minimizing cost or materials . . . there is a unique and undeniable beauty to structures that have been engineered with functionality and efficiency as the main priorities.
>
> *(Kosky et al. 2010: 237)*

This position is to be found throughout the 19th century: here, the notion of function is still broader, and fulfilling a purpose well is more than just functionality. It stands in for the type of beauty found in nature, where there is no more or no less than needed to fulfill its purpose (Greenough 1947; Ruskin 1849). In other words, the assumption is that if it works well, it will also be beautiful. 'Machine beauty' might denote this particular aspect of aesthetics, though it does not quite do justice to the scope of civil engineering. However, we might consider the early modern age as one in which the breathtaking constructions of engineers began to create a new sensibility. Examples of the type of beauty that are seen to naturally result from the basic principles of engineering include such works as the Eiffel Tower in Paris (1889), for which Gustav Eiffel made use of the structural moment diagram to minimize wind loads towards the top of the tower. Earlier, he had already gained experience with minimizing loads in his design for the Garabit viaduct in the south of France (1882–1884), which bridges a deep valley with a cast-iron truss bridge reminiscent of the structure for the Eiffel Tower. The Swiss engineer Maillart is best known for his concrete bridges where the minimal material possible is used in accordance with structural diagrams, creating a very thin, elegant structure. This type of beauty is not limited to appearance but also includes the experience of awe in the ability for such an apparently light structure to carry substantial loads (Billington 1985).

It is here that we might begin to speak of the 'industrial sublime'. The discourse on the sublime began in the 18th century with Edmund Burke's famous *Philosophical Enquiry Into the Origin of Our Ideas of the Sublime and Beautiful* (1757). Burke distinguishes the experience of the sublime from the experience of beauty particularly through the sensations of awe, terror and pain, while in the knowledge of being safe. This sense of the sublime underpins much of the Romantic era (such as Shelley's Frankenstein and the paintings of Caspar David Friedrich). The sense of awe, and of the overpowering presence of—in the case of Burke—the power of nature, is increasingly transposed to the products of human ingenuity in the industrial era, leading increasingly to a fascination with great works of infrastructure, which will continue throughout the modern era (Jensen 2013). Structural innovations and indeed challenging structures, such as the thin-shell concrete works of Nervi or the delicately balanced bridges of Calatrava, challenge our perceptions of gravity and stability, and as such appeal to the sense of art and beauty from traditional aesthetic discourse.

The advent of the industrial era, beginning with the steam engine and expanding into many areas of infrastructure and machine production, has positioned engineering in general in a different place than the ages before. The issues raised by the Industrial Revolution extended into what we now tend to call 'modernity'. It is in this time that the scale and impact of infrastructure and built works begins to transform everyday life. If architecture was initially to service humankind, to facilitate social structures and to aid in the presentation of institutional structures, for example, engineering in its purest form tends towards the ambition to tame nature: it provides awe-inspiring structures that transcend the power of the individual and may even compete with the natural environment in its overpowering greatness. Bridges and tunnels, railways and airports, windmills and hydraulic pumping stations are perhaps some of the most visible of these tangible products of the engineering mindset. While these structures tend to be evaluated by their functionality—the Hoover Dam, the Göteborg Bridge—they may equally be understood in terms of their symbolic and aesthetic value.

9.4. Reintroducing Human Sensibilities: From the Art of Engineering to Sciences of the Artificial

It is in this idea of the 'industrial sublime' that the main focus of the modern era may be found. From Le Corbusier's championing of grain silos and ocean liners in his 1923 manifesto *Vers une architecture*, to Dziga Vertov's 1929 film *Man With a Movie Camera*, the awe-inspiring structures of engineers take the lead in determining a new aesthetic in the early 20th century (Weber and Sigrist 2009).

The constructions that stand as symbols for the industrial era increasingly emphasize the distinction between architecture and engineering. One may, from the perspective of today, identify traditional engineering works as those that are focused primarily on use and function, but also as those that serve other functions than human occupation. While architects (save a few exceptions) focus mainly on structures meant to be occupied by humans, the large-scale structures of the 19th century needed engineers to envision new machinery and technologies in order to allow for the many consequences of the rising mass society, such as infrastructure (roads, dams, railways) and industries.

One of the striking distinctions between the art of engineering and the art of architecture is that the element of human inhabitation is typically somehow visible in architectural constructions—whether it is the distinct, repetitive punched windows in a facade, or clearly human-scale stairs and escalators that show us the promise of human occupation, the structures of engineering include no such constraints: they may have fully blank facades, oversized towers or incomprehensible infrastructural elements—they are made for machines, for utility, and they speak to the awesome power of the engineer's imagination. As such, these objects are not to human scale, but they speak to human capacity: to conceive of as yet unthinkable objects that make their mark on the natural environment. With the rise of the 19th century, these elements of engineering prowess became particularly visible. The power of these objects unto themselves marked the first half of the 20th century, extending the notion of the industrial sublime throughout all domains of life.

Around the middle of the 20th century, however, something changed. A key polemic in this issue is the problem of 'disenchantment' seen to derive from the excessive focus on functionality, rationalization and quantification. While it may seem easy to place responsibility for this excess in the fields of natural science and engineering, this polemic becomes too simplistic a rendering of an increasingly complex condition. It therefore becomes crucial to examine the issue more closely: how much rationalization helps us control our environment, ensuring that departing and arriving airplanes pass each other smoothly and efficiently. When is it too much? When do we lose the capacity to step outside of the rational systems in order to conceive of an unheard-of, radically different solution, that changes not only the approach but even our relation to the problem?

This discomfort with excessive rationalization and technocratic focus underpins contemporary pleas to introduce aesthetics in engineering programs (even since the 1960s) which have begun to test the simplified boundaries between these areas that distinguish between rational and creative approaches, and instead suggest that the dominant modes of thinking about these disciplines is ripe for revision. Indeed, perhaps one of the most notable revisions of typical delineations was Herbert Simon's introduction of a 'sciences of the artificial'. In his 1969 book of the same title, he challenged traditional divisions in science and humanities by adding 'objects', in the sense that the sciences of the artificial actually transform reality, not simply observe or understand it (Simon 1969: 3–5). In his model, the natural sciences seek to understand the laws of nature by careful observation and by modeling underlying forces that shape and determine the natural environment. In contrast, the sciences of the artificial not only observe—they model possible solutions and actively transform or adapt reality by introducing new structures, systems or other interventions. Although Simon primarily discusses fields that are seen as having some scientific or technical elements (such as management, computer science, engineering and planning, to name a few), this model does open up the possibility of moving beyond the typical clichés of rationality and technicality as the prime concerns of engineering.

Instead, the shared foundations of art and engineering then may lie in problem-solving (and its partner 'problem-finding'), in material sensibility, and in creatively adapting to changing conditions.

Questions such as these led to the restructuring of the Stanford program in mechanical engineering in order to bring together art and science now nearly 60 years ago (Faste 1995). In its introduction of aesthetics as an integral part of the curriculum, it suggests that a human connection with technology can be aided by paying attention to the kind of interest that beauty can trigger. Moreover, the program argues that aesthetics is part of the creative approach, which allows engineers to transcend solely functional or quantitative requirements. These ideas are formulated earlier by Eugene Ferguson, who argues for the crucial role of aesthetics in engineering education, but the sustained incorporation of the arts in engineering programs is only now beginning to become more common (Ferguson 1977, 1994).

The artist Alexander Calder may stand as an example of this transformed relation between technology and art. Trained as an engineer, Calder became an artist by profession, and his works testify to the beauty inherent in bridging the two domains (and a little-known anecdotal moment: in the *Joy of Cooking*, reference is made to a spectacular birthday cake, that unfortunately is not included as an image) (Rombauer and Rombauer Becker 1995: 724).

As many domains of engineering and design are becoming increasingly integrated in everyday life, the ability of specific professions to integrate and coherently solve a multitude of (wicked) problems is increasingly put to the test. In other words, a bridge is no longer a manner to get from here to there, but also a branding item for local government. At the same time, vast areas of our globe are becoming increasingly dependent on the creative mindset of engineering solutions in order to maintain the level of comfort and innovations we have grown comfortable with over the course of modernity.

9.5. Engineering the Future: STEAM and the Expansion of Art and Engineering

In some sense, the provocative position put forward by C.P. Snow in *The Two Cultures* (1959) is an early appeal to the need to take scientific knowledge and technological insight seriously as products of culture. His appeal was particularly aimed at ensuring a clear knowledge of basic principles of science in a society he deemed overly focused on the products of culture (i.e., literature, the arts). In other words, he suggested that the laws of thermodynamics should receive equal attention to Shakespeare in early education. While at the time his position met with as much resistance as support, the underlying valuations of the domains of scientific knowledge and humanities remain present in many areas of study and policymaking. Currently, many elements of technological skill are becoming more embedded in secondary school programs, which reflects not only the societal need but also the development of these formerly often specialized fields into a more intuitive approach, possible to convey even to elementary school children. As an example, Scratch was developed as a visual programming language in order to help children to intuitively grasp the structures of computer programming, and is increasingly used not only in high schools, but also in elementary and middle schools.

Even as we continue to maintain a simplified version of the 'two cultures' of art and science, or continue to see engineering as the domain of 'nerds', it is rapidly becoming difficult to maintain these simplifications. Fundamental aspects of engineering include not only technology but also creativity in its design process, and now, more often than not, some sense of aesthetics or understanding of the (future) user. As technologies become increasingly dominant in all spheres of activity, they also require a more thorough consideration of their position as social and aesthetic objects, including not only ergonomics but also aesthetic preferences and cultural codes, for example (Ollis et al. 2004). As the fields of engineering expand beyond the traditional triad of civil, mechanical and electrical engineering, their impact becomes a more integral part of everyday life, and aesthetics thereby becomes more integral to our understanding of these technologies. Should we address the technology and impact of Facebook in a restrictive manner, as a primarily technical concern, or should we take into account the psychological dimension of 'clickbait' as part and parcel of its mechanisms? If the latter, should we then analyze it from the user's perspective of satisfying a need for attention and rewards, or should we equally examine the forms in which these triggers are presented? (Lanier 2018; Wadhwa and Salkever 2018). Surely many of these questions can no longer be seen as limited to only a single domain of knowledge or application.

Many of these questions may well become central in the immediate future. How indeed should we examine the newer aspects of engineering—the smaller-scale, the ephemeral and the invisible? They are of a different order in terms of their perception: less tangible, less visible, yet they fundamentally impact our experience of the world. Whether it concerns the chemical engineering of nano-materials and self-healing materials, or the hidden structures of digital networks, transformative elements are being added to our formerly natural environment that support the identification of our current time as the 'anthropocene'. This term, although it has been around for some time, has been popularized in recent decades to denote that we have entered an era in which the human impact on the environment had expanded to a degree that we can no longer distinguish a 'natural' world from the human interventions that have become part of its fabric. In light of this transition, it is only sensible that many engineering and science programs now include courses in ethics, debates on values and aesthetics, and other approaches from within the 'fuzzy' fields of the humanities.

It is this concern with the complex systems of a time in which technologies and engineering are thoroughly integrated with all domains of life that drives the STEM-to-STEAM discussions. Given current developments, rather than expecting our engineers to all sit in a laboratory or be engaged with particular finesses of further rationalization, there is a growing need for many kinds of engineers to understand the consequences of their designs and their systems (Connor Karmokar and Whittington 2015). While the core skills of engineering are still needed, the issues at stake are not only technological and functional. As Richard Lachman notes: "They contain technological issues but they are not fundamentally technological issues. They are ethical ones" (Lachman 2018). As he continues his argument, he argues the need for experts that can debate the consequences for society, and even "to plan what, and if, we need to create" (Lachman 2018). In other words, the questions 'Will it work?' and 'How does it look?' are no longer easily separated. Instead, both are intimately related to our values and how people wish to shape the societies to which they belong. As such, introducing a more thorough study of the arts into the traditional fields of engineering is building a new sensibility that is meant to aid in navigating the increased complexity and above all interdependency of the mechanical and technological domains with the human and cultural.

Related Chapters

Chapter 2: A Brief History of Engineering (Jennifer Karns Alexander)
Chapter 5: What Is Engineering Science? (Sven Ove Hansson)
Chapter 10: Creativity and Discovery in Engineering (David H. Cropley)

Further Reading

Billington, D.P. (2003). *The Art of Structural Design: A Swiss Legacy*. New Haven: Yale University Press. (A study of the artistic approach to bridge design through the work of four Swiss engineers, accompaniment to an exhibition.)

Henderson, K. (1999). *On Line and on Paper: Visual Representations, Visual Culture, and Computer Graphics in Design Engineering*. Cambridge, MA: MIT Press. (A study of how designers use drawings to organize work and knowledge, and of the impact of computer drawing.)

Hindle, B. (1982). *Emulation and Invention*. New York: New York University Press. (Study of how technologists think similarly to artists: visually, spatially, non-verbally, explored through two cases.)

Peters, T.F. (1996). *Building the Nineteenth Century*. Cambridge, MA: MIT Press. (A study of tunnels, bridges and canals to investigate what a 'culture of construction' might be.)

Zwijnenberg, R. (2013). On the Need for Cooperation Between Art and Science. In Judith Thissen, Robert Zwijnenberg and Kitty Zijlmans (eds.), *Contemporary Culture: New Directions in Art and Humanities Research*. Amsterdam: Amsterdam University Press, pp. 169–174. (A statement on hybrid practices between art, technology and science, based on experimental artistic practices.)

References

Abrams, M. (2015). The Beauty of Movement. *American Society of Mechanical Engineers* (ASME). www.asme.org/engineering-topics/articles/technology-and-society/the-beauty-of-movement. Accessed June 20, 2018.

Alba, M. (2017). Who Was the First Engineer? *Designer Edge*, Engineering.com. www.engineering.com/DesignerEdge/DesignerEdgeArticles/ArticleID/14451/Who-was-the-First-Engineer.aspx. Accessed June 20, 2018.

Bertucci, P. (2017). *Artisanal Enlightenment: Science and the Mechanical Arts in Old Regime France*. New Haven: Yale University Press.

Billington, D.P. (1985). *The Tower and the Bridge: The New Art of Structural Engineering*. Princeton: Princeton University Press.

Black, M. (1961). Art and the Engineer. *Nature,* 191(4792), 949–953.

Black, M. (1976). The Aesthetics of Engineering: The Interaction of Art and Engineering. *Interdisciplinary Science Reviews,* 1(1), 31–42.

Bronowski, J. (1961). *Science and Human Values*. London: Hutchinson.

Committee on General Structures, Subcommittee on Bridge Aesthetics (1991). *Bridge Aesthetics Around the World*. Washington, DC: National Research Council.

Connor, A., Karmokar, S. and Whittington, C. (2015). From STEM to STEAM: Strategies for Enhancing Engineering & Technology Education. *International Journal of Engineering Pedagogy*, 5(2), 37–47.

Corazzi, R. (2017). *Brunelleschi's Dome: The Secret of Its Construction*. Florence: Angelo Pontecorboli.

Devaney, K. (2016). Beauty as a Guiding Principle for Systems Engineering. *INCOSE* International Symposium, 26(1), 1448–1462.

Durand, J. (1805). *Précis de Leçons d'Architecture*, Paris: Bernard, Libraire de l'Ecole Polytechnique.

Faber, O. (1945). *The Aesthetic Aspect of Civil Engineering Design*. Institution of Civil Engineers, issued as a separate publication.

Faste, R. (1995). The Role of Aesthetics in Engineering. *Japan Society of Mechanical Engineers* (JSME) Journal, Winter.

Ferguson, E. (1977). The Mind's Eye: Nonverbal Thought in Technology. *Science,* 197(4306), 827–836.

Ferguson, E. (1994). *Engineering and the Mind's Eye*, Cambridge, MA: MIT Press.

Gates Jr., S. and Mirkin, C. (2012). Encouraging STEM Students is in the National Interest. *Chronicle of Higher Education Commentary*. www.chronicle.com/article/Encouraging-STEM-Students-Is/132425. Accessed October 12, 2018.

Gelernter, D. (1998). *Machine Beauty: Elegance and the Heart of Technology*. New York: Basic Books.

Greenough, H. (1947). *Form and Function: Remarks on Art, Design and Architecture*. Ed. Harold A. Small. Berkeley, CA: University of California Press.

Jensen, K. (2013). *Industrial Sublime: Modernism and the Transformation of New York's Rivers, 1900–1940*. New York: Fordham University Press.

Jeon, M. and Fishwick, P. (2017). Homo Ludens in Virtual Environments. *Presence: Teleoperators and the Virtual Environment*, 26(2), iii–vii.

Kosky, P., Balmer, R., Keat, W. and Wise, G. (2010). *Exploring Engineering: An Introduction to Engineering and Design*. New York: Elsevier.

Kruft, H.W. (1994). *A History of Architectural Theory From Vitruvius to the Present*. New York: Princeton Architectural Press.

Lachman, R. (2018). STEAM not STEM: Why Scientists Need Arts Training. *The Chronicle of Education*, January 19. http://thechronicleofeducation.com/2018/01/19/steam-not-stem-scientists-need-arts-training/. Accessed October 12, 2018.

Landels, J.G. (1978). *Engineering in the Ancient World*. Berkeley: University of California Press.

Langer, S.K. (1953). *Feeling and Form*. London: Routledge and Kegan Paul.

Lanier, J. (2018). *Ten Arguments for Deleting Your Social Media Accounts Right Now*. New York: Henry Holt and Co.

Mark, R.E. (1990). *Light, Wind and Structure: The Mystery of the Master Builders*. Cambridge, MA: MIT Press.

Ollis, D.F., Neeley, K.A. and Luegenbiehl, H.C. (eds.) (2004). *Liberal Education in Twenty-first Century Engineering*. New York: Peter Lang.

Parsons, G. and Carlson, A. (2008). *Functional Beauty*. Oxford: Oxford University Press.

Picon, A. (2004). Engineers and Engineering History: Problems and Perspectives. *History and Technology*, 20(4), 421–436.

Polanyi, M. (1966). *The Tacit Dimension*. Chicago, IL: University of Chicago Press.

President's Council of Advisors on Science and Technology (2012). *Engage to Excel: Producing One Million Additional College Graduates With Degrees in Science, Technology, Engineering, and Mathematics*, Report to the President, February. https://obamawhitehouse.archives.gov/sites/default/files/microsites/ostp/pcast-engage-to-excel-final_2-25-12.pdf. Accessed October 12, 2018.

Rombauer, I. and Rombauer Becker, M. (1995). *Joy of Cooking*. New York: Scribner.

Ruskin, J. (1849). *The Seven Lamps of Architecture*. New York: John Wiley.

Ryle, G. (1946). Knowing How and Knowing That: The Presidential Address. *Proceedings of Aristotelian Society*, 46, 1–16.

Sennett, R. (2008). *The Craftsman*. New Haven: Yale University Press.

Simon, H.A. (1969). *The Sciences of the Artificial*. Cambridge, MA: MIT Press.

Snow, C.P. (1959). *The Two Cultures and the Scientific Revolution*. London: Cambridge University Press.

Wadhwa, V. and Salkever, A. (2018). *Your Happiness Was Hacked: Why Tech Is Winning the Battle to Control Your Brain—and How to Fight Back*. Oakland, CA: Berrett-Koehler Publishers.

Weber, J. and Sigrist, V. (2009). The Engineer's Aesthetics—Interrelations Between Structural Engineering, Architecture and Art. Conference Proceedings, *Third International Congress on Construction History*, Cottbus.

10

CREATIVITY AND DISCOVERY IN ENGINEERING

David H. Cropley

10.1. Introduction

What is the core purpose of engineering, and the primary function that engineers undertake? Leaving aside obvious and specific things such as "engineers build bridges" or "engineers design computers", what is a single, shared and underpinning *purpose* behind all specializations in engineering, whether mechanical, aeronautical, civil or any other sub-branch? In the engineering literature, there has been a consistent answer to this question for many years. Jensen (2006), for example, stated that "Engineers solve problems" (p. 17). He noted, furthermore, that problem-solving requires the ability to understand and define the problem, to apply standard approaches to solving the problem, and to "supplement the standard solution methods with creativity and insight" (p. 18). Burghardt (1995) described the engineering profession as one "devoted to the creative solution of problems" (p. 2), while Voland (2004) described a process of "solving technical problems in creative yet practical ways" (p. xix).

The term *design* is also frequently used to describe this technological problem-solving process. Buhl (1960) stated that "a designer is one who satisfies mankind's needs through new answers to old problems" (p. 9). He went on to say, "The designer must deliberately create new products and processes which will fulfill mankind's needs. He must be creative in all stages of problem solution" (pp. 9–10). Horenstein (2002) also explained that "design" is what engineers do (p. 22), and that "design can be defined as any activity that results in the synthesis of something that meets a need" (p. 22). Dieter and Schmidt (2012), in similar fashion, noted that design involves synthesizing "something new or to arrange existing things in new ways to satisfy a recognized need of society" (p. 1). Brockman (2009) also linked engineering to needs-driven problem-solving, reminding us that these problems arise from a desire to "satisfy mankind's complex needs and desires" (p. 3).

The needs of society are the sources of problems, and engineers *design* technological solutions to these problems. We have seen some clues as to the kinds of reasoning required for this process—creativity, for example. However, to understand fully the forms of engineering reasoning that underpin engineering design—i.e. technological problem-solving—we must first go back a step, and discuss the driving force behind *society's complex needs and desires*.

10.2. Change

Since the time of the earliest modern humans, technology—for example, spears, stone tools and oars—has been critical to progress. Each of these artefacts represents the synthesis of something that meets a need—for hunting, for cutting, for transport. However, the driving force behind those needs has always been *change*. Burghardt (1995) described three "major revolutions in the way society has been organized" (p. 22) and that these came about as a result of major *changes* that stimulated human-kind's needs and desires, resulting in technological problems that could be solved through design, by engineers. The first major change, which saw a shift from hunter-gatherer to agrarian societies, was environmental—the end of the Ice Age. The dramatic change—a warming of the climate about 12,000 years ago—meant that old methods of subsistence were no longer able to meet the needs of a growing human population. This change created new needs—for better food supply—and therefore resulted in problems that could be solved, potentially at least, by technology.

The second major change, which stimulated the shift from agrarian to industrial societies, resulted from a combination of famine and disease. As European societies had grown under the agrarian model, population growth outstripped food supply. Famine resulted at the same time that more and more land was cleared, resulting in a shortage of the main fuel of the time, wood. The addition of the bubonic plague epidemic—the Black Death—that ravaged Europe in the 14th century left a short-age of people to work the land. These changes together provided the stimulus for new needs—for alternative sources of fuel, for less labor-intensive ways to work the land—that again were the source of problems with potential technological solutions.

The third major change, which has seen the shift to the modern, post-industrial, information age, is underway, and therefore a little harder to characterize. However, we can say that changes to the earth's population—both its sheer size, and its age profile—along with changes to the climate and changes in the interconnectedness of people around the planet, i.e. globalization, are all generating needs that are the basis of problems ripe for technological solutions.

Even on a smaller scale, we see numerous examples from the history of humankind that illustrate *change* as a driver of society's needs and desires, and therefore as a source of engineering problems. It is also at this point that we can speculate that the effects of change manifested themselves in two different but related ways. The examples we have considered so far represent new needs generated by external changes. Thus, climate change generates a need for better food supply, and therefore problems like "How can we *grow* food to supplement our supplies?" However, it is apparent that a second pathway also occurs in parallel. External changes also generated *new solutions*. We will never know who first hit upon the idea that a stick used as a spear thrower (a *Woomera* in Australian Aboriginal culture) could improve the range and effectiveness of that weapon; however, it is plausible that the spear thrower was *not* developed *in response to* an identified need—"How can I throw my spear further?"—but was simply a lucky discovery—"Look what happens when I do this?!" Whether driven by the need (i.e. a process of *invention*), or driven by the solution (i.e. a process of *discovery*), the result is the same.[1] New technologies—solutions—and new problems are linked together in a mutually beneficial way, with this process *driven by change*.

A modern example serves to illustrate the two pathways stemming from change. Human-made climate change resulting from factors such as excessive atmospheric carbon has two effects. On the one hand, this change leads to new needs—for example, the *market pull* (i.e. the demand) for cleaner air travel. This defines, in effect, a new problem: How can we reduce carbon emissions from commercial aircraft? On the other hand, climate change produces new technologies—for example, the *technology push* of low-drag surfaces in aircraft. This can be said to generate a new solution: for example, the *winglets* that are now standard on most commercial passenger aircraft. Engineer-ing problem-solving—design—is the process of connecting these new problems and new solutions together (Figure 10.1).

In the early days of agrarian society—in the Nile valley, for example—we see more examples of change driving problems, resulting in technological solutions—in other words, *engineering design*. Growing populations created new needs—for clothing, shelter and food—that could not be met by existing means. These new needs therefore drove technological solutions that could replace animal

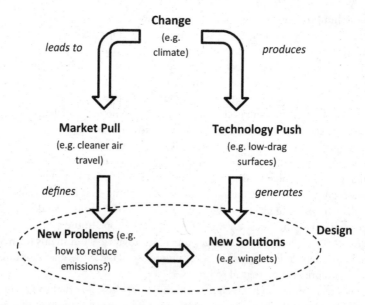

Figure 10.1   Change as a driver of problems and solutions

skins with woven fabrics and hand tilling of soil with ox-drawn ploughs. It is also likely that individuals developed new technologies—new solutions—either serendipitously or deliberately, that could then be matched to a need. In other words, once a society develops (discovers) the wheel, they suddenly realize how many needs they can satisfy with it. This *chicken and egg* situation does not, however, present us with any paradox because the end result is the same. Whether deliberately invented or serendipitously discovered, change generates both new problems and new solutions, and joining these two together, in a technological sense, is engineering design (Figure 10.1).

10.3. The Need for Creativity

To understand the role that creativity plays in the engineering problem-solving process (and to understand the forms of reasoning required), I now draw attention to a crucial feature of the process outlined in Figure 10.1. The examples we have considered, and the model of the process, have both identified the fact that change stimulates *new* problems and *new* solutions. The changes that occurred to our ancestors were all *unprecedented*—there was no previous case or example to follow, and no prior solution that could be applied. Humans had never tackled the after-effects of an ice age. They had never before dealt with the problem of feeding large numbers of people, and they had never needed anything other than animal skins for warmth. New problems may look similar to past cases, but they are new precisely because they have unique characteristics and constraints that have never been encountered before. New solutions are new because they are identified as unique, original and *never seen before*. It is the characteristic, indeed the *requirement*, of newness—in the problems and in the solutions—that is why creativity is central to engineering problem-solving. As engineers, however, we should note that there is also an important place for old solutions satisfying old needs, which I will discuss shortly.

It appears, therefore, that there are, in effect, *five different pathways* through our problem-solving (design) process. Most are driven by change, as I have already discussed, but some are driven by what I will call *stasis*—the maintenance of the status quo. It is important to note that there is nothing bad about stasis in the engineering context. Not every problem needs a creative solution, if no change is introduced into the system. Buhl (1960) noted that if "we consciously try to solve each and every problem which arises as we drive from here to there we should soon lose our sanity" (p. 16). Buhl's point was that *precedented* problems, in other words, *old* or *existing* problems, can be addressed perfectly well with old or existing solutions. This is also important economically because if there is no requirement for creativity—because the need/problem and/or solution are routine—then costs can be lowered, risks reduced and time saved. The urban myth of the NASA Space Pen illustrates this well. The popular story tells us that NASA spent 10 years and $12 million developing a ballpoint pen that would work in zero gravity and extremes of temperature. Meanwhile, the Soviet cosmonauts simply used pencils! In other words, the myth is making the point that sometimes the *old*, routine solution is perfectly satisfactory. However, when change is injected, as is usually and eventually the case, then stasis can only ever be a temporary state of affairs in engineering problem-solving. Figure 10.2 illustrates the five pathways through our engineering problem-solving process.

With *change* as the primary driver, the following pathways can be traced (Figure 10.2):

1. A new solution, i.e. a discovery, satisfies a new, previously unrecognized, problem. This is sometimes referred to as *technology-push*, or a *supply-side problem*. Another label for this pathway is *redirection*[2] (in the sense that thinking is propelled in a new direction).
2. A new problem is satisfied by a new solution. This is often referred to as *market-pull*, or a *demand-side problem*. We can also call this *reinitiation* (in the sense that a paradigm shift restarts thinking along wholly new lines).
3. A new solution improves an old problem. The wheel, for example, satisfied the problem of moving large loads easily and quickly, for example for trade or for food production. In engineering, this is quite common and is described by the mantra "better, faster, cheaper". We can also call this *incrementation*.

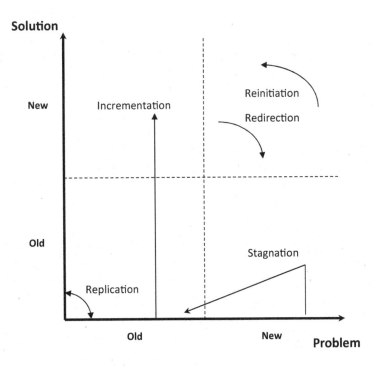

Figure 10.2 Five pathways in engineering problem-solving

4. A new problem is met with an old solution, which, unable to solve the new problem, effectively drags the need back to what *can* be satisfied—an old problem. This pathway has the effect of pulling change back to stasis and can be characterized as *stagnation*. This is also what Albert Einstein was referring to when he said, "We can't solve problems by using the same kind of thinking we used when we created them". In the end, the new problem remains unsolved and the progress implicit in change is halted. The reader will also notice that creativity is absent; highlighting the fact that change can only be met successfully with the creativity that is implicit in novelty.

If the primary driver, by contrast, is *stasis*, then a final, albeit temporary, pathway exists:

5. An old, or existing, problem is satisfied by an old solution. This case is common to many engineering scenarios, and is not inherently inferior or deficient. If I need to hammer in a nail, then my trusty old hammer is a perfectly good solution. If I need to get traffic across a river, then a standard bridge design will almost certainly be satisfactory. This "more of the same" approach, or *replication*, is the bread and butter of engineering. This is how engineers provide solutions to many of society's problems—and satisfy needs—in a cost-effective, timely, safe and low-risk manner. While it involves no creativity, at least at the macro-level, it should not be dismissed for at least two reasons. First, it comprises a large percentage of what engineers do, and second, because there may be a requirement for creativity buried deeper within it. For our current purposes, however, and for the case where old problems are satisfied by old solutions, we will set this aside as outside of the scope of our discussion of creativity in engineering. *Provided nothing changes*, Replication is a valid pathway in engineering problem-solving.

The need for creativity as a core component of engineering is now clear. Change—in the form of ageing populations, carbon taxes, terrorism, globalization, food security, scientific discoveries and more—drives new needs and new technologies. Engineering is a problem-solving process that connects new needs to new technologies. Creativity is concerned with the generation of effective, novel solutions, and therefore creativity and engineering are, in essence, two sides of the same coin. In fact, we can characterize engineering as a special case of the more general process of generating effective, novel solutions to problems—i.e. creativity (Cropley, 2015a).

Gertner (2012) further illustrates the value of creativity in engineering, describing the driving force behind the activities of the famous Bell Labs: "that the growth of the system [the US telephone system run by AT&T] produced an unceasing stream of operational problems meant it had an unceasing need for inventive solutions" (p. 45). The impetus for creativity, moreover, was not simply technological in nature: "the engineers weren't merely trying to improve the system functionally; their agreements with state and federal governments obliged them to improve it economically, too" (p. 45).

10.4. Diminishing Returns

Adding to the complexity of the engineering problem-solving process is its dynamic nature. Certainly in the case of replication, and even after a new problem is tackled with a new solution, the pressure of change continues to exert an influence. Take the case of the problem of feeding a growing population. Farmers have long known that the yield of a crop can be increased through the use of fertilizer. The existing problem (growing a crop) is incrementally improved through the application of a new solution (fertilizer). However, like many, if not all, natural systems, this scenario is subject to the phenomenon of *diminishing returns*—i.e. a progressive decline in the relative increase in output, in this case resulting from the application of fertilizer. Indeed, many systems are also subject to *negative returns*, whereby the output will not merely level off but will go into decline. For this reason,

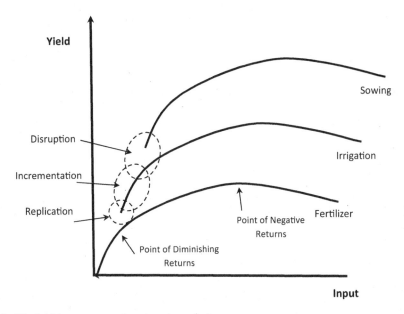

Figure 10.3 Diminishing returns and engineering solutions

engineers cannot simply rely on an incremental solution to keep pace with change. Once the point of diminishing returns has been passed, the cost of an incremental improvement will become prohibitive, meaning that some other approach is required as further change places demands on the system.

The solution to the effect of this constant change is a combination of incrementation, to the point of diminishing returns, and periodic redirection or reinitiation (collectively, I will call these *disruption*), to move the problem-solving situation to a new curve, itself eventually subject to diminishing returns. Figure 10.3 illustrates this.

Thus, fertilizer is applied only sufficiently to take the crop yield to the point of diminishing returns. At that point, a farmer might then add irrigation as a means of achieving further increases in yield. Once the irrigation solution has hit its point of diminishing returns, novel sowing technology may be added, placing each seed at the optimal depth to ensure maximum germination and growth. Changes drive a need for constant creativity, both incremental and disruptive, and creativity therefore remains at the core of problem-solving process.

10.5. Creative Engineering Problem-Solving

Simplifying the previous discussion of needs, solutions and driving factors, we can now separate out three generic types of creative engineering problem-solving, and one type of routine engineering problem-solving. Where problem and solution are *old* and the engineering process is one of matching these together, we are dealing with *routine*, but nevertheless important, engineering *replication*. Where *new* problems and *new* solutions are combined in some way, we are dealing with creative engineering problem-solving that can be further characterized as *incrementation* (where new solution satisfies an old problem) and *disruption* (comprising *redirection* and *reinitiation*), shown in Figure 10.4.

The link between creativity and engineering is clear. Where new customer demands can be met by new technological solutions, we need engineers who are equipped—both technically and creatively—to generate those solutions. Where new technologies become available through scientific discovery, research and development, we need engineers who are able to find new possibilities and new markets to exploit these possibilities. There will always remain a place for the application of engineering

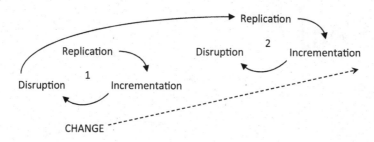

Figure 10.4   Change and the cycle of replication–incrementation–disruption

knowledge to the solution of routine—well-understood, straightforward—problems. However, as the pace of change accelerates in the 21st century, we will see a growth in new problems that require creative—in other words, effective *and* novel—technological solutions. To meet the challenge of these changes, we need engineers equipped with the blend of technical knowledge and *creativity*.

Another way to understand the need for creativity in technological problem-solving is to imagine a world *without* novelty. Imagine what our world would look like, and how our society would function, if bridges were still made only of tree trunks, or if the only form of heating was from individuals burning wood. Imagine if the only form of augmented transport (i.e. something other than *by foot*) was the horse, or if the locomotive, to paraphrase Buhl (1960), had *not* been displaced by the automobile?

The value and importance of creativity—the *new* solutions—emerges in Buhl's discussion where he points out (p. 10) that "we expend a great deal of effort in modifying modification rather than attacking the problems at their core" and makes the vital point that "Industries are continually being supplanted, not by modifications but by innovations." He gives a specific example that illustrates this point very well: "Locomotives were not displaced by modified locomotives but by a *new approach* to transportation needs—the car" (p. 10, emphasis added).

At a very general level, Sternberg (2007) expresses a sentiment common in discussions of creativity and innovation, and the value that they bring to society: "The problems we confront, whether in our families, communities, or nations, are novel and difficult, and we need to think creatively and divergently to solve these problems" (p. 7).

10.6. Engineering Reasoning: Critical and Creative Thinking

What sort of reasoning skills—in other words, *thinking* skills—are required to support the incrementation and disruption that engineering problem-solving demands? Notwithstanding the emphasis I have placed on creativity in this chapter, the defining characteristic of engineering reasoning, in fact, is the necessity of switching back and forth between critical thinking—i.e. reasoning, logic, evaluation, analysis—and creative thinking. The former can be summarized as convergent thinking, while the latter can be characterized best as divergent thinking.

Convergent (or critical) thinking involves deriving the single best (or one and only correct) answer to a clearly defined question. It emphasizes speed, accuracy, logic, and the like, and focuses on accumulating information, recognizing the familiar aspects of new situations, reapplying already-proven techniques and similar processes. It is thus most effective in situations where the required answer can be worked out from what is already known by applying conventional and logical search, recognition, decision-making and evaluation strategies. One of the most important aspects of convergent thinking is that answers are either right or wrong. Convergent thinking is also intimately linked to knowledge, because it involves manipulation of existing knowledge by means of standard procedures, which are themselves part of existing knowledge, and its typical result is improvement or extension of the already known along known lines.

Table 10.1 Convergent (critical) thinking versus divergent thinking

Convergent Thinking	Divergent Thinking
• thinking logically	• thinking unconventionally
• homing in on the best answer	• branching out to generate multiple answers
• recognizing the familiar	• seeing the known in a new light
• grasping the facts accurately	• finding a new perspective on the facts
• retrieving relevant information	• retrieving a broad range of information
• staying within the limits	• going beyond the limits
• combining what belongs together	• combining the apparently disparate
• making associations from adjacent fields	• making remote associations
• reapplying set techniques	• devising new techniques
• playing it safe	• taking a risk
• preserving the already known	• challenging or transforming the already known
• ensuring feasibility	• ensuring novelty

Source: Based on Cropley and Cropley (2009: 48).

Divergent thinking, by contrast, involves producing multiple or alternative answers from available information by, for instance, making unusual combinations, recognizing remote links among pieces of existing information, transforming existing information into unexpected forms, or seeing obscure implications. Divergent thinking may lead to a variety of conclusions based on the same starting point, and the ideas it yields may vary substantially from person to person. The results of divergent thinking may never have existed before: Sometimes this is true merely in the experience of the person producing the variability in question, or for the particular setting, but it may also be true in a broader sense. The key characteristics of convergent and divergent thinking are set out in Table 10.1.

In the engineering design process, these two forms of reasoning come together in the way the design process is executed—i.e. in the stages of problem-solving.

10.7. The Engineering Design *Process*

There are many models of the engineering design process. Dieter and Schmidt (2012) serves as a representative example: *problem definition* and *information gathering* set the scene for *concept generation*. This is followed by *concept evaluation and selection*, and then a series of progressively more detailed steps that define *product architecture, configuration design, parametric design* and *detailed design*. In many models this is followed by a stage of implementation and/or validation of the design.

Problem-solving in a more general sense has long recognized not only that distinct stages do exist, but that each stage has certain characteristics with respect to the *kind of reasoning* required. Guilford (1959), exploring the psychology of creative thinking, identified four stages in problem-solving, along with the key form of reasoning that supports each stage (Table 10.2).

Similar stages have been identified by many authors, both from a psychological and a cognitive perspective (e.g. Bartlett 1932; Wallach and Kogan 1965; Cropley 1999) and from a business or technical/engineering perspective (e.g. Prindle 1906; Wallas 1926; Pugh 1991; Higgins 1994; Cropley 2017). Indeed, many modern *creative* problem-solving processes (e.g. *Design Thinking* and *Appreciative Enquiry*—see Puccio and Cabra 2010) follow a similar pattern.

In my own previous work exploring creativity in the context of engineering, I have made extensive use of an extended phase model of creativity (see Cropley and Cropley 2000; Cropley and Cropley 2008; Cropley 2015b). Based on Wallas (1926), the extended phase model corresponds closely to the key stages of engineering design, reinforcing the importance of both convergent and divergent thinking as components of engineering reasoning. Table 10.3 shows the correspondence between different engineering design models, and the forms of engineering reasoning that support these.

Table 10.2 The stages of creative problem-solving

Stage	1	2	3	4
Description	Recognition that a problem exists	Production of a variety of relevant ideas	Evaluation of the various possibilities produced	Drawing of appropriate conclusions that lead to the solution of the problem
Summary	Problem Recognition	Idea Generation	Idea Evaluation	Solution Validation
Characteristic Reasoning	Convergent	Divergent	Convergent	Convergent

Source: Based on Cropley (2015b: 43).

Table 10.3 Mapping phases to engineering design models

Generic Phases (EPM)	Preparation	Activation	Generation	Illumination	Verification	Communication	Validation
Super-phases	*Invention*					*Exploitation*	
Character	Convergent	Divergent	Divergent	Convergent	Convergent	Mixed	Convergent
Pugh (1991)	Market/user needs and demands	Product design specification	Generation of solutions (conceptual design)	Evaluation of solution (conceptual design)		Conceptual Phase outputs	
Cropley and Cropley (2000)	List system elements	Identify element option descriptors for system elements	Define the design space envelope	Define a process to generate a range of element options	Select element options to populate the design space	Describe element choices in greater detail	
Dieter and Schmidt (2012)—	Define problem	Gather information	Concept generation	Evaluate and select concept			
Buhl (1960)	Recognition, analysis	Definition, preparation	Synthesis	Evaluation		Presentation	

Source: Based on Cropley (2015b: 60).

10.8. Summary and Conclusions

Engineering reasoning—the question of how engineers think—is characterized by two contrasting, but intimately intertwined forms of thinking. Convergent (critical) thinking, associated with analysis, logic, evaluation and reasoning, is central to the process of solving technological problems. It focuses on accuracy and correctness, reinforced by specialist knowledge, to deliver safe, functional solutions. Yet, without divergent thinking, characterized by novelty, unconventionality and risk-taking, engineers would be trapped in a world of *replication*, unable to respond to the new problems generated by change.

Despite extensive lip service, the fact remains that engineering education has tended to focus heavily on convergent thinking, frequently at the expense of divergent thinking (see, for example, Cropley (2015b, 2015c). Driven by a growing understanding of the importance of creativity, problem-solving and associated *soft* skills—stemming as much from the emergence of the paradigm of Industry 4.0, as anything else—there is an urgent need to reexamine how engineers are educated, and how they are equipped for the future work environment.

The core purpose of engineering is solving technological problems, i.e. design. This is driven by change, and change brings with it an ever-present need for creativity (in the form of new problems and new solutions). Therefore, creativity becomes an inherent part of engineering design. What sort of thinking does this entail? Although creativity is often associated only with divergent production, the wider creative problem-solving process in fact oscillates back and forth between divergent thinking and convergent thinking. Engineering reasoning—the kinds of thinking skills that engineers need in order to solve problems—therefore requires skills in both convergent and divergent thinking. Education typically does a good job of supporting the former, and not such a good job at supporting the latter. Engineering education, therefore, must ensure that students develop the capacity for both, in order to be effective problem-solvers.

Related Chapters

Chapter 1: What is Engineering? (Carl Mitcham)
Chapter 20: Towards an ontology of innovation (Vincent Blok)
Chapter 21: Engineering Design (Peter Kroes)

Further Reading

For a deeper analysis of creativity, innovation design and discovery in engineering, readers are directed to:

Buhl, H. R. (1960). *Creative Engineering Design*. Ames, IA: Iowa State University Press (an insightful discussion linking creativity and engineering from early in the modern creativity era)

Cropley, D. H. (2015). *Creativity in Engineering: Novel Solutions to Complex Problems*. San Diego: Academic Press (which links creativity in engineering to the psychological foundations of creativity research)

Cropley, D. H. (2017). Nurturing Creativity in the Engineering Classroom. In R. Beghetto and J. C. Kaufman (eds.), *Nurturing Creativity in the Classroom*. New York: Cambridge University Press (a discussion of changes needed in engineering education to support the development of creativity)

Sternberg, R. J. (2007). Creativity as a Habit. In A.-G. Tan (ed.), *Creativity: A Handbook for Teachers*. Singapore: World Scientific, Chapter 1, pp. 3–25 (a more general, psychological discussion of the development of creativity in education).

Notes

1. Questions about the related processes of discovery and invention, and their relationship to creativity, have been addressed for more than a century. Ernst Mach (1896) tackled these in his inaugural lecture as Professor

of History and Theory of Inductive Science at the University of Vienna. I have also reflected on Mach's paper, and what it tells us in the framework of modern creativity research in Cropley (2019).
2. I have drawn on the Propulsion Model of Kinds of Creative Contributions that is described by Sternberg et al. (2002) for labels such as *redirection*. However, I have not used all of the eight that they describe, and not necessarily with exactly the same meaning. The general spirit of their model, however, is at the core of my use of the terms.

References

Bartlett, F.C. (1932). *Remembering*. Cambridge, UK: Cambridge University Press.

Brockman, J.B. (2009). *Introduction to Engineering: Modeling and Problem Solving*. Hoboken, NJ: John Wiley & Sons Inc.

Buhl, H.R. (1960). *Creative Engineering Design*. Ames, IA: Iowa State University Press.

Burghardt, M.D. (1995). *Introduction to the Engineering Profession,* 2nd ed. New York: HarperCollins College Publishers.

Cropley, A.J. (1999). Creativity and Cognition: Producing Effective Novelty. *Roeper Review*, 21(4), 253–260.

Cropley, A.J. and Cropley, D.H. (2008). Resolving the Paradoxes of Creativity: An Extended Phase Model. *Cambridge Journal of Education*, 38(3), 355–373.

Cropley, A.J. and Cropley, D.H. (2009). *Fostering Creativity: A Diagnostic Approach for Education and Organizations*. Cresskill, NJ: Hampton Press.

Cropley, D.H. (2015a). *Creativity in Engineering: Novel Solutions to Complex Problems*. San Diego: Academic Press.

Cropley, D.H. (2015b). Promoting Creativity and Innovation in Engineering Education. *Psychology of Aesthetics, Creativity, and the Arts*, 9(2), 161–171.

Cropley, D.H. (2015c). Teaching Engineers to Think Creatively: Barriers and Challenges in STEM Disciplines. In R. Wegerif, L. Li and J.C. Kaufman (eds.), *The Routledge International Handbook of Research on Teaching Thinking*. New York: Routledge, Chapter 33, pp. 402–410.

Cropley, D.H. (2017). Nurturing Creativity in the Engineering Classroom. In R. Beghetto and J.C. Kaufman (eds.), *Nurturing Creativity in the Classroom*. New York: Cambridge University Press, Chapter 33, pp. 212–226.

Cropley, D.H. (2019). Do We Make Our Own Luck? Reflections on Ernst Mach's Analysis of Invention and Discovery. In Vlad Petre Glaveanu (ed.), *The Creativity Reader*. Oxford: Oxford University Press.

Cropley, D.H. and Cropley, A.J. (2000). Creativity and Innovation in the Systems Engineering Process. Paper presented at the *Proceedings of the Tenth Annual International Symposium on Systems Engineering*.

Dieter, G.E. and Schmidt, L.C. (2012). *Engineering Design,* 5th ed. New York: McGraw-Hill Higher Education.

Gertner, J. (2012). *The Idea Factory: Bell Labs and the Great Age of American Innovation*. London, UK: Penguin Press.

Guilford, J.P. (1959). Traits of Creativity. In H.H. Anderson (ed.), *Creativity and its Cultivation*. New York: Harper, pp. 142–161.

Higgins, J.M. (1994). *101 Creative Problem Solving Techniques: The Handbook of New Ideas for Business*. Winter Park, FL: The New Management Publishing Company.

Horenstein, M.N. (2002). *Design Concepts for Engineers,* 2nd ed. Upper Saddle River, NJ: Prentice-Hall, Inc.

Jensen, J.N. (2006). *A User's Guide to Engineering*. Upper Saddle River, NJ: Pearson: Prentice Hall.

Mach, E. (1896). On the Part Played by Accident in Invention and Discovery. *The Monist*, VI(2), 161–175.

Prindle, E.J. (1906). The Art of Inventing. *Transactions of the American Institute for Engineering Education*, 25, 519–547.

Puccio, G.J. and Cabra, J.F. (2010). Organizational Creativity: A Systems Approach. In J.C. Kaufman and R.J. Sternberg (eds.), *The Cambridge Handbook of Creativity*. New York: Cambridge University Press, pp. 145–173.

Pugh, S. (1991). *Total Design: Integrated Methods for Successful Product Engineering*. Wokingham, UK: Addison-Wesley.

Sternberg, R.J. (2007). Creativity as a Habit. In A.-G. Tan (ed.), *Creativity: A Handbook for Teachers*. Singapore: World Scientific, Chapter 1, pp. 3–25.

Sternberg, R.J., Kaufman, J.C. and Pretz, J.E. (2002). *The Creativity Conundrum: A Propulsion Model of Kinds of Creative Contributions*. New York: Psychology Press.

Voland, G. (2004). *Engineering by Design,* 2nd ed. Upper Saddle River, NJ: Pearson Prentice Hall.

Wallach, M.A. and Kogan, N. (1965). *Modes of Thinking in Young Children*. New York: Holt: Rinehart and Winston.

Wallas, G. (1926). *The Art of Thought*. New York: Harcourt Brace.

11

UNCERTAINTY

William M. Bulleit

11.1. Introduction

There is no certainty, only less uncertainty. This statement is, of course, true, but there are many events that for all intents and purposes are certain; for instance, the sun will rise tomorrow, and each of us will eventually die. These types of events have been referred to by Habermas as *behavioral certainties* (Habermas 2000: 49); we behave as if our knowledge is certain. They could also be referred to as *practical certainties* (Bernstein 2010: 184). In this chapter, we will be discussing the ways that engineers are able to reach a point where they feel like the effects of their engineering decisions are practically certain to occur, or at least they behave as if their decisions produce effects that are certain to occur. But, obviously, engineers are sometimes wrong about an action being a practical certainty, since failures do occur on a more frequent basis than we might think reasonable. Dewey (1958: 21) would say, "Strain thought as far as we may and not all consequences can be foreseen or made an express or known part of reflection and decision." Thus, failures will need to be discussed in this chapter as well.

To begin, consider two broad categories of uncertainties that are important from the standpoint of deciding how to deal with an uncertain future. They are *aleatory* and *epistemic* uncertainties. Aleatory means related to luck or chance. Epistemic means related to knowledge. Truly aleatory uncertainties can generally be dealt with only by using statistical and probabilistic techniques. Epistemic uncertainties can be dealt with by gaining better knowledge of the phenomenon under consideration but still may require statistical methods for that portion of the uncertainty where no additional knowledge can be gained (Der Kiureghian and Ditlevsen 2009). It may not always be clear which type of uncertainty you are dealing with, particularly in a system with strong interconnectedness and interactions among its components. Consider as an example the flipping of a fair coin (e.g., Bulleit 2008). The result of a coin flip is usually thought of as an aleatory uncertainty. But what if we could model the coin flipping event, including say, the air flow around the coin, the strength of the flip, and the coin characteristics that affect the result of the flip. If this model, probably a physics-based, mathematical model, allowed us to predict whether we get heads or tails with some level of accuracy better than 50–50, but not perfect, then we would know that some portion of the uncertainty in flipping a coin is epistemic, not aleatory. In large-scale systems, the distinction between aleatory and epistemic uncertainty becomes even more problematic.

11.2. Systems

Four types of systems will be considered because the level of uncertainty, both aleatory and epistemic, is different for each. These four system types encompass the vast majority of systems that engineers will encounter. The four types are (1) simple, (2) complicated, (3) complex, and (4) complex adaptive. Keep in mind that the boundary between each of these system types is fuzzy. There will be situations where it is difficult to discern if, say, the system is complicated or complex. We will consider each one and the basic sources of uncertainty for it. Later in the chapter, we will examine how the basic uncertainties can be mitigated. The differences between the systems will be primarily related to the interconnectedness and interaction among the components of the system. We will not consider atomic- or sub-atomic-level systems in this chapter, so we will not need to consider, for example, quantum effects.

A *simple* system is a system with one component or maybe a few components exhibiting little or no interconnectedness or interactions between them. In many cases, a simple system will be a small portion of a larger system, e.g., a complicated or complex system, separated off for design purposes, referred to as a *control volume* (Vincenti 1990). The effect of control volumes will be examined later in the chapter. An example of a simple system is a beam in a girder-deck highway bridge. These bridges consist of a number of beams, with a concrete deck on top of them, spanning over, for instance, a road or a river. We can, for all practical purposes, design a typical beam as if it is a single entity. Another example is a weir in a stream. A weir is a structure that alters the flow of the stream. We can design the weir considering only the input flow and the desired output flow. Both of these examples are control volumes, a portion of a larger system separated out for design, and are simple systems.

A *complicated* system is a combination of elements with some interaction between those elements. Some examples of complicated systems are aircraft, automobiles, buildings, and large bridges. There will be interactions among the elements, such as between the beams and columns in a building or the parts of an automobile engine, but these interactions will not lead to *emergence* (Holland 1998) of non-intuitive behaviors such as self-organized criticality (Bak 1996) or "black swan" events (Taleb 2007). Emergence can be thought of as the occurrence of a global pattern that arises from local actions and interactions. Often the global pattern is an unexpected result of the local behavior. (See also Chapter 19, "Emergence in Engineering" by Peter Simons, this volume.) Some complicated systems can exhibit "domino behavior" (Harford 2011) where a single portion of a system fails and leads to collapse of the entire system or a portion of it. An example of domino behavior is the partial collapse of the Ronan Point apartment building in London (Delatte 2009). In this failure, a gas explosion occurred in a corner apartment near the top of the building, causing all the apartments above and below it to collapse. This is a specific type of domino behavior and is referred to by structural engineers as progressive collapse.

Complex systems, on the other hand, have strong interaction and interconnectedness among elements in the system. These conditions can lead to emergence of non-intuitive system behaviors (Holland 1998). An example of a complex system, which will be considered later in this chapter, is a sand pile being built by dropping single grains of sand on top of the pile. At some point, a portion, possibly a large portion, of the pile will collapse from the dropping of a single grain of sand (Bak 1996). *Complex adaptive* systems also have strong interaction and interconnectedness among the elements in the system, but many of the elements are agents, such as humans, who can adapt their behavior to changes in the system, be it changes in the environment or the behavior of other agents (Miller and Page 2007). Human society is an example of a complex adaptive system. The range of behaviors that can be exhibited in either complex or complex adaptive systems is larger than the other systems and presents types of uncertainty that can be vastly different than those typically dealt with in engineering of simple and complicated systems (e.g., Bak 1996; Miller and Page 2007; Taleb 2007).

11.3. Sources of Uncertainty

There are a range of uncertainties that occur in each of the system types. In this section we will discuss the basic sources of uncertainty that occur in each of the four systems. Techniques to mitigate these uncertainties will be discussed in the next section. In this chapter, the sources of uncertainty will be separated into those that are inherent to the system, referred to as basic sources, and those that arise from attempts to mitigate the basic uncertainties. In previous discussions of sources of uncertainty, Bulleit (2008, 2013, 2018) combined these types of sources into one set.

Considering simple systems, there are four basic sources of uncertainty: *time, randomness, contingency*, and *human influences*, primarily *human error*. Consider *time* first. Properties of a system and environmental effects on a system are functions of time. A few examples include: (1) the wind speed varies over time; (2) the properties of the system change over time, e.g., decay and wear out; (3) rainfall varies over time; and (4) traffic volume varies over time. Furthermore, some effects not only vary over time but change in average magnitude. Cars and trucks get larger and heavier, climate change may affect rainfall and snowfall, and more options on a car may draw more power from the battery. Last, a system built many years ago might, even though it is still available for use, not be useful anymore. For example, bridges that are still in existence may no longer be fully functional. Old wood-covered bridges, even though still used for cars, are functionally obsolete since they cannot support larger vehicles like trucks. The loads and environmental effects as well as the capacity of the system vary due to *randomness*. For instance, the strength of a beam is variable, the amount of rain or snowfall varies over the years even if the average doesn't change, and the magnitude of floods and earthquakes vary from event to event. Engineers must visualize and then design a system that doesn't exist yet. This produces *contingency* in the final built system. The final built system will not be exactly like the visualized system, due to both limits on visualization and limits about what can be known about the final system. Simon (1996) has suggested that contingency is one of the major differences between engineers and scientists. Science is primarily about knowing, and there is always more to know. Engineering is about doing, so decisions must be made with limited information, which introduces contingency. Simon (1996: 119) refers to this approach as satisficing. "We satisfice by looking for alternatives in such a way that we can generally find an acceptable one after only moderate search" (Simon 1996: 120). Last, engineers make mistakes during the design process and fabricators make errors during construction, *human errors*.

Complicated systems are affected by the same basic sources of uncertainty as simple systems, but the magnitude of the uncertainties may be increased due to the interaction of the components in the system. As an example, consider an automobile engine. A single component of the engine, say the battery, has variable life. If the battery fails, the engine won't start. But, there are a lot of other components that can fail to prevent the engine from starting. The alternator might fail to charge the battery. Thus, the failure is in the alternator, but the battery doesn't get charged, so the engine doesn't start. So, the variability of engine performance is some composite of all the components, which means that the variability is greater than that of a single component. Another example is a building frame. The variability of the behavior of the entire frame is greater than the variability of a single beam in the building due to the interactions among all the components in the frame. It should be apparent that the system in each of these examples is also affected by time and randomness in a similar manner as simple systems. Furthermore, each system will vary somewhat from the original design due to contingency, and human error can occur in a number of ways, all of them uncertain. So, in general, the variability of a complicated system of components, each of which are simple systems, is greater than the variability of a single component due to the number of components and their interactions, even though the basic sources of uncertainty are the same as simple systems.

Complex and complex adaptive systems are affected by all the sources of uncertainty that affect simple and complicated systems, but they are also affected by events that emerge from the interactions and interconnectedness among the components (Holland 1998). Thus, we will add *emergence* as another basic source of uncertainty. The first aspect of complex systems that may cause emergence of unexpected events is extreme sensitivity to initial conditions (Dekker 2011). The canonical example of this behavior is the so-called "butterfly effect." A butterfly flapping its wings in Singapore could cause, at some later time, a storm to emerge in New York City. Although the butterfly effect is seemingly far-fetched, mathematical complex systems can be affected by initial conditions that are altered only in apparently insignificant ways, say, a digit changed in the tenth decimal place. The second example is the near meltdown of a nuclear reactor at Three Mile Island in 1979. This is an example of strong interaction and interconnectedness between components, referred to as tight-coupling by Perrow (1999), and cascading of events, i.e., a domino effect. The failure began with a moisture leak into an instrumentation line. This leak caused the turbines to be unnecessarily shut down, which in turn caused the reactor to begin to heat up. From here a cascade of component failures, including safety devices, led to a near loss-of-coolant accident, which would have caused a meltdown of the reactor. The cascade of events occurred in 13 seconds and, yet, took the operators over 2 hours to remedy. It should be noted here that safety devices increase the uncertainty in system response, particularly in complex systems. (See also Chapter 33, "Values in Risk and Safety Assessment" by Niklas Möller, this volume.) The third example is the September 11, 2001, attack on the World Trade Center towers. This is an example of a "black swan" event (Taleb 2007). The name comes from the belief that all swans were white until the surprising discovery of black swans in Australia. A black swan event is an event that is, for all practical purposes, not predictable because it is so far beyond anyone's experience and, generally, is also a high consequence event. Knight (1948), as far back as 1921, considered what he called "measurable uncertainties" and "unmeasurable uncertainties." Measurable uncertainties have probabilities associated with them, so that they can be dealt with using tools such as insurance. Unmeasurable uncertainties are not predictable, and may be black swan events. The Twin Towers disaster emerged from the complex adaptive system that is human society and shows the effect of human influences, beyond human error, on the uncertainty of complex adaptive systems.

Before we move on to consider ways to mitigate uncertainty, we need to examine whether the five basic sources produce aleatory and/or epistemic uncertainties. Time produces primarily aleatory uncertainties. We can learn more about past occurrences (e.g., keeping better records of rainfall), but no matter how good our past data, we still will be limited as to how well we can predict the future. True randomness is aleatory, but some events that we presently consider random may be somewhat predictable as we learn more. For instance, mathematical models of wind forces on buildings combined with data from wind tunnel tests have enhanced our ability to predict wind forces that were once considered random. So, some portion of the uncertainty in wind forces is epistemic and the rest, at least for now, is considered aleatory. Contingency produces both aleatory and epistemic uncertainties. As we develop better ways to represent designs, e.g., building information models, the closer we can make the final building to the design. Of course, uncertainty of the final nature of the building will still exist. Human error is primarily aleatory because errors in design and fabrication are essentially unpredictable. They arise from, for example, communication errors between the designer and the fabricator, conceptual errors in modeling the system, misinterpretation of codes of practice, and even something as mundane as placing a value in the wrong system of units. Emergence presents some additional problems separating the uncertainty. It produces both types of uncertainties. The question of prediction of events in complex and complex adaptive systems boils down to determining the point at which, for a given system, emergent events are absolutely unpredictable. We likely will never know for sure where that point will occur. Bak (1996) used the example of building a sand pile by dropping single grains of sand one at a time on the top. At some point a single grain will cause a significant collapse of some portion of the pile. This is an emergent event, and predicting which grain of sand will cause the collapse is not possible with present-day techniques and will likely never be predictable. We may someday be able to develop a computer model to narrow the answer, but narrowing the answer will be the best that we can do. The sand pile example is a complex system; complex adaptive systems, e.g., human society, present even more difficulties in predicting events, and human influences will go beyond just human errors.

11.4. Mitigation of Uncertainty

The discussion in this section will be about approaches that are used to mitigate the uncertainty caused by the basic sources of uncertainty. These methods reduce some uncertainty but then introduce their own uncertainties, which will also be discussed. Due to the nature of engineering design, the techniques used can be referred to as *heuristics* (Koen 2003). From an engineering standpoint, heuristics are anything that engineers use to help solve problems and perform designs that would otherwise be intractable or too expensive. "A *heuristic* is anything that provides a plausible aid or direction in the solution of a problem but is in the final analysis unjustified, incapable of justification, and potentially fallible" (Koen 2003: 28, italics in original). These heuristics may appear in codes of practice, may be widely used by engineers but not appear in codes of practice, or could be techniques for use in the future for design of systems like complex and complex adaptive systems. The techniques described in this section and the section on living with uncertainty are all heuristics, or more specifically engineering heuristics.

The uncertainties due to the effects of *time* are mitigated in a number of ways. First, consider uncertainties related to the magnitude of environmental effects, such as snow loads and floods. These kinds of effects are mitigated using probability concepts, which will be discussed in some detail, but the basic heuristic is that the future can be represented by the past. We are using the past to estimate the future, and we use some sort of model, generally a mathematical model, that is based on the past to predict the future. This approach means that not only do we have uncertainty about what the environmental effects will be if the future is exactly like the past, but we have uncertainty about whether the past is representative of the future. Climate change will likely alter some environmental effects (e.g., hurricane wind speeds), but how to incorporate that possible change into design is not clear (uncertain) at this point. Thus, most in-time variation of environmental effects is modeled using probabilistic methods to mitigate the uncertainty in the design, and we assume that the past is representative of the future. Second, loss of strength over time due to factors such as decay or rust is generally handled using inspection and maintenance, but prediction of the amount of loss can be done using mathematical models of the loss of strength for the material and, possibly, historical data about how that material has behaved in the past under conditions similar to what is expected for the system being designed. Models of component behavior will be discussed next.

Models allow mitigation of the uncertainties that arise from all sources, and as we continue our discussion of that mitigation, it will become apparent that models are one of the primary ways to mitigate uncertainties from whatever source they arise. All models are imperfect representations of reality, so the use of models also introduces uncertainty, referred to as *model uncertainty*. Further details on model uncertainty can be found in Bulleit (2008: 28, 2013: 319, 2018: 34).

The next basic source of uncertainty that needs to be mitigated is *randomness*. Consider first how randomness in external stressors, such as environmental effects, can be mitigated. For instance, snow loads are an external stressor, and past maximum snow loads can be used to estimate future maximum snow loads (Bulleit 2008). The same can be done for flood magnitudes. A 100-year flood, a term most readers have heard, is a flood magnitude that on average will occur with probability 1/100 in any given year. Thus, the return period for this flood magnitude is 100 years. Engineers can use the return period flood or return period snow load for design, but to do that they must choose a design life. Many structures in the United States are designed using a design life of 50 years. This doesn't mean that the structure is no longer usable after 50 years or even that it is guaranteed to last 50 years. It simply means that a load will be used in design that has a design return period combined with a defined design life. In some sense, both of these are arbitrary but, once chosen, are built into design specifications where the required capacity of the system is determined using these chosen design values (Bulleit 2008).

Using return periods is an implicit way to deal with the effect of time. It would also be possible to use stochastic processes (Bhat 1972) to model the variation in time, but that approach is unusual in design. Design loads have been developed using stochastic processes, e.g., Corotis and Tsay (1983) for duration of building occupancy live loads, but stochastic processes typically do not appear explicitly in a design specification.

Engineers also must deal with the variability of the properties of components in the system. For instance, the strength of a component, such as the drive shaft of an automobile, a beam or column in a building, the soil or rock that a building rests on, and the concrete strength in a highway pavement is a part of a design. All of these strengths are variable with a probability distribution describing their variation. In some cases, say a steel beam in a building, the variation depends on the type of steel, and its probability distribution is, for all practical purposes, known since we have a lot of data about the material. But some portions of a system are highly location dependent, say soil strength at a building site, and we must use *statistical methods* to estimate the probability distribution, usually in terms of parameters defining the distribution, e.g., mean and standard deviation. Note that all estimates of the parameters of a probability distribution require statistical methods, but the effects on uncertainty are most apparent where the methods must be used for estimations using small data sets. If we think about determining the soil strength on a building site, then we should realize that tests to estimate the strength cost money (and time), so only a limited number of tests can be performed, thus a small data set. That immediately means that our estimates of, say the mean and standard deviation, will have a lot of variation, so even though we are reducing uncertainty compared to no testing, we still have uncertainty remaining. For instance, we may want to do some drilling to bring up long samples (cores) of the ground under the site. In a site that is an acre or more, we may only be able to get a few cores. Yes, we have reduced uncertainty some, but we still have to live with a significant amount of remaining uncertainty.

We have similar limitations for snow and flood data. Stations that gather that data are relatively few and far between so areas between them must be estimated, but worse, even at the site where the data is gathered there is only one maximum value per year, and we are mostly interested in maximum values. Yes, there are statistical methods to make estimates of values that we cannot collect, but, as described earlier, the methods leave some remaining uncertainty. There would be statistical uncertainty in estimating the values between stations and additional uncertainty from determining maximum values, say the 100-year flood, from annual maximum data. In design, engineers have techniques, e.g., factors of safety, to help them live with this uncertainty. We'll examine these approaches in the section on living with uncertainty.

The next source to consider is *contingency*. The problem here is that no matter how well you represent the artifact that is being designed, there are still aspects of the final version that will be different from the design. This uncertainty is unavoidable. The general systems theorist Alfred Korzybski supposedly said: "The map is not the territory." That is probably the best way to think about contingency in engineering: no matter how good your plans and specifications (the maps of the territory), they are not the built system. Mitigation of contingency is done by developing better models of the system (the territory). Here we mean a range of model types including numerical models, computer visualization/graphics models such as building information models that include as much of the system as possible, and possibly even scale models. The goal of the models is to understand as well as possible how the system will behave, what it will look like, and how it can be built. But of course, no matter how good the representation, reality will be different. Another way to mitigate some portions of contingency is quality control measures such as construction inspection. Inspection helps make sure that what appears in the design is put in the system. For certain types of systems, e.g., automobile engines, it is possible to build a prototype. Considering an automobile engine, it is designed using mathematical models, large 3-D computer models, and other design techniques. Then, that design is built and run to see what kinds of problems arise. The engine may first be run on a test stand, modified based on those tests, then put in a car and run on a test track to see how it behaves in a car, modified again, and then put back in the car and run on city and county streets. The tests still will be relatively short term, so there will be some remaining uncertainty about how the engine will behave over many years, but the prototype testing has significantly reduced the uncertainty in the original design. Systems like buildings and bridges are non-prototypical systems because it is impossible to build the system and test it. Non-prototypical systems will have greater uncertainty than prototypical systems (Bulleit 2013).

Human error is primarily dealt with using quality control methods. For example, in structural design offices, all design calculations are checked. The check is a complete examination of assumptions, interpretation of design code provisions, use of design code provisions, correctness of analysis procedures, and details of the final design. As we move into complex and complex adaptive systems, *human influences* will contribute to the uncertainty and will be more difficult to mitigate because the influences, likely more than errors, will occur during the operation of the system, or will be a part of the system in complex adaptive systems like human society. In recent years, there have been significant advances in using agent-based modeling to examine the behavior of complex adaptive systems (e.g., Holland 1998; Miller and Page 2007). See Bulleit and Drewek (2011) for an example of using a specific type of agent-based modeling approach to examine a complex adaptive system.

Mitigation of the uncertainty produced by *emergence* requires changing the way we think about system design (Bulleit 2018; Dekker 2011; Popper 1985). Emergence means that small changes to complex and complex adaptive systems can lead to failures in a major part of the system or the entire system. In 2003, a tree brushed against a power line in Ohio and caused a power blackout over a large portion of the eastern United States (Minkel 2008). This is an extreme example of the emergence of a domino-effect failure. A single local or global event is not the only way that complex and complex adaptive systems can fail. Consider what Dekker (2011) refers to as "drifting into failure." As Dekker (2011: xii) says: "Drifting into failure is a gradual, incremental decline into disaster driven by environmental pressure, unruly technology, and social pressures that normalize risk." This means that complex and complex adaptive systems can fail without the causes of the failure being apparent as they occur. An example of drifting into failure is the loss of the space shuttle *Columbia* (Dekker 2011: 163). The ultimate cause of the failure was foam damage to the heat tiles. An original design criterion stated that the external fuel tank was to shed no debris. From early in the life of the shuttle *Columbia*, the external tank lost foam pieces that hit the shuttle. The foam loss became so commonplace that it was not even mentioned in flight readiness reports. Foam loss became a part of the flights. Then on the 113th mission, the failure occurred. The system had drifted into failure. Of course, it is also possible to cause a failure by perturbing the system, in a small local way or in a large global way. The primary mitigation approach is to think of design of complex and complex adaptive systems more as maintenance than design, in the common usage of the word (Bulleit 2018). This approach makes design of complex and complex adaptive systems more a form of adaptation. Harford (2011: 243) suggests three principles of adaptation: first, "try new things, expecting that some will fail," second, "make failures survivable: create safe places for failure or move forward in small steps," and third, "make sure you know when you have failed, or you will never learn." This approach where we perturb the system in a small local way, possibly modeling the effect of the perturbation prior to making it (maybe with agent-based models), and then observing the effects of the perturbation helps mitigate the uncertainty of the change. Consider Dekker (2011: 180–181) again:

> The feature of drift discussed above, small steps, is actually a wonderful property here. Small steps can mean small experiments. Many small steps mean many small experiments and opportunities for discovery. These are small experiments without necessarily devastating consequences, but at the same time small experiments with the potential to create hugely important results and insights.

The idea of using information gained from experiences in the system to alter the system has been referred to as "design without final goals" by Simon (1996: 162). Karl Popper (1985: 309) also suggested this approach for the complex adaptive system that is society.

> The characteristic approach of the piecemeal engineer is this. Even though he may perhaps cherish some ideals which concern society "as a whole"—its general welfare, perhaps—he does not believe in the method of redesigning it as a whole. Whatever his ends, he tries to achieve them by small adjustments and re-adjustments which can be continually improved upon.

In complex and complex adaptive systems, this maintenance-as-design approach is truly the only safe way to alter the system. (See also Chapter 35, "The Role of Resilience in Engineering" by Neelke Doorn, this volume.) It does not guarantee that no failures will occur, but it works as much as possible to make failures survivable so that we can learn from them.

11.5. Living With Uncertainty

Clearly the efforts to mitigate uncertainty described in Section 4 help engineers to live with uncertainty, but living with uncertainty will require further heuristics. Consider one heuristic that was mentioned earlier, use control volumes. A control volume is a small portion of a system that is separated from the remainder of the system to aid in design (Bulleit 2018; Vincenti 1990). The control volume is often small enough that it can be considered a simple system for design purposes. Since it is a small part of the system, the uncertainty about its behavior is reduced. It should be clear that this approach is necessary and is reductionist. The danger is that the control volume chosen does not meet the assumption that it merely converts an input to an output in its local region and that is its only effect on the overall system. The more complicated or complex systems become, the more likely that the assumption of local-effect-only is not met. Certainly, control volumes will be required for complex and complex adaptive systems, but the possible global effect of alterations to the control volume must be considered. If we think about complex and complex adaptive systems, then a control volume could not only be a small portion of the system in a spatial sense, but also in a temporal sense. In this case, we need to also think about a *prediction horizon*. This is a length of time over which we can model the behavior of the control volume with enough confidence to say that our prediction of behavior has a better than, at least, 50–50 chance of occurring (Bulleit 2018). From a design standpoint, we may want the probability of a correct prediction to be much better than 50–50.

The prediction horizon is one way to determine if we are moving in small steps in our perturbation of the system. The second criterion for a small step is that predicted failures that might occur in the overall system, if the control volume modification does not work as planned, are small enough that complete system collapse does not occur. Of course, altering complex and complex adaptive systems is a highly uncertain undertaking and black swan events may occur. This problem was addressed by many of the writers in Vitek and Jackson (2008) and was referred to as *ignorance*. To quote from one of the authors (Talbot 2008: 103):

> We can never perfectly know the consequences of our actions because we are not dealing with machines. We are called to live between knowledge and ignorance, and it is as dangerous to make ignorance the excuse for radical inaction as it is to found action on the boast of perfect knowledge.

I would argue that even with machines, we can never perfectly know the consequences of our actions, but the sentiment of the quote fits well with the direction of this chapter. Ignorance, lack of knowledge, is what the heuristics described earlier are attempting to account for. But, decisions must be made and should be made cautiously and without hubris. The question is, how do we find heuristics that in the long run allow us to deal with the uncertainty in the results of those decisions? One approach to dealing with complex and complex adaptive systems and their effect on engineered systems is to use what Olsen (2015) refers to as the *observational method*. This method, basically adaptation, has been suggested as a way for civil engineers to deal with the effects of climate change on large projects. "When it is not possible to fully define and estimate the risks and potential costs of a project and reduce the uncertainty in the time frame in which action should be taken, engineers should use low-regret, adaptive strategies, such as the observational method to make a project more resilient to future climate and weather extremes" (Olsen 2015: 62) Whether the method is referred to as the observational method or other names such as adaptive management or robust decision-making, it is a form of adaptation and requires a wide range of heuristics, especially a mindset that considers design to be a form of maintenance (Bulleit 2018).

Although we are not able to have perfect knowledge about the consequences of our actions, even with machines, we may be able to extend heuristics for dealing with machines into more complex problems. One heuristic that is used both in the design of machines, many of which are prototypical, and in design of non-prototypical systems like buildings are *safety factors* (Bulleit 2008, 2018). One type of safety factor separates the required capacity of the system from the maximum design demand by a multiplicative constant. So, for instance, the load that a roof beam should support must be two times the 50-year return period snow load. More complicated versions of safety factors used in structural engineering can be found in Bulleit (2008). Safety factors are useful in situations where it is not necessary, or even possible, to predict exactly what will occur in the future, just design for a safe value. Another approach used in process control and in machine design are *tolerance limits*. A simple example is to say that a circular part that will go into a machine must be within a certain diameter range to be used, e.g., it must be between 9.98 and 10.02 mm. If it falls outside this range, then it must be discarded.

Statistical techniques are required to help us live with small data sets. A small data set, say only a few values of the soil strength at a building site, as described earlier, is unlikely to produce good estimates of probability distribution information such as the mean and standard deviation. Statistical techniques are available to give *confidence intervals* on the parameters, and the intervals get smaller as we get more data and larger for small amounts of data. So, in the case of the soil strength, we might to be able say that we are confident that the mean (average) soil strength will not fall below a certain value some relatively large percentage of the time. Then we could use that low value of the mean in the design. Similar techniques would be required if the standard deviation was also needed (Ayyub and McCuen 2003).

As suggested, engineering of complex and complex adaptive systems will likely require a mindset that is different from how we design simple and complicated systems, particularly in the ways that uncertainty is handled. Pool (1997) has argued that the incident at the Three Mile Island nuclear power plant was primarily caused by an inappropriate mindset. "Most of the causes of the accident could be traced back, directly or indirectly, to a mindset and an organizational structure inherited from a time when the utility operated only fossil-fuel plants" (Pool 1997: 271). One change in mindset is to think of large-scale systems as things that evolve so that changes to the system must be small and appropriately infrequent. They must be small enough to minimize the possibility of system-wide damage or collapse and infrequent enough to allow small failures that do occur to be learning experiences. But small is a relative term, and the possibility exists that a small change might cause a large effect, or the system might be acted upon by an external stressor that produces failure, possibly a black swan event. Thus, a second mindset is to try to make the overall system *resilient* with respect to both internal perturbations and external stressors. Resilience is often thought of as having two parts: robustness and rapidity. Robustness is the ability of the system to respond to events that are beyond what it was designed for in a way that minimizes damage, and rapidity is the ability of the system to recover from failures. For example, the Twin Towers responded in a robust way to the attacks since they did not collapse immediately, but stood long enough to allow the majority of the occupants to escape. Determining the resilience of complex and complex adaptive systems is a hard problem, as is evident by the effect of storms and other events on communities. (See also Chapter 35, "The Role of Resilience in Engineering" by Neelke Doorn, this volume.)

Heuristics used in engineering are context dependent. Some heuristics, such as computer models of machines, e.g., finite element models, combined with the use of prototypes, are refined enough that it is readily apparent that we have practical certainty. That is not so much the case with large buildings subjected to seismic events. Yes, we have sophisticated models of these structures, and also finite element models, but the uncertainty about the complicated system behavior, the earthquake ground motion, and the lack of prototypes makes it clear that we may behave as if we are certain, but we know we have less certainty than the machine designer. The machine designer might use analysis procedures that attempt to account for the action and interaction of all the parts in the machine in a precise way, whereas the structural designer will not be able to be nearly as precise since the uncertainty in the way the system will be built and the highly uncertain ground motion make high levels of precision much less meaningful. This limit on preciseness will get even worse as we begin to seriously attempt to engineer complex and complex adaptive systems. As K. T. Fann (1971: 59) said: "There is no point in criticizing the cook for failing to match the druggist's standards of exactness."

11.6. Final Remarks

The physicist William Thomson, Lord Kelvin, reputedly said: "It's no trick to get the answers when you have all the data. The trick is to get the answers when you only have half the data, and half that is wrong, and you don't know which half." He was referring to science, but his statement is applicable to the problem of reaching practical certainty in engineering.

Henry Petroski (2006) has argued for years that engineering advances by failures. The need to learn from failures arises from what he calls the paradox of design: "Things that succeed teach us little beyond the fact that they have been successful; things that fail provide incontrovertible evidence that the limitations of design have been exceeded" (Petroski 2006: 114). This paradox is unavoidable since design and fabrication of any artifact is an alteration of reality, and any alteration of reality is an uncertain proposition, maybe highly uncertain. Even with techniques to mitigate uncertainty, failures will occur. Engineers have learned how to live with that uncertainty for systems like machines and buildings. But, those techniques will likely not be adequate as we attempt to design complex and complex adaptive systems.

Those types of systems will require a mindset that approaches design as a form of adaptation. Yes, engineering has evolved by learning from failures, but that occurred over decades and even centuries. Design of complex and complex adaptive systems will require that the learning occur during the life of the system. The heuristics for this type of design are only now beginning to be developed, and living with the uncertainty in the behavior of these types of systems will be one of the main challenges in designing and maintaining systems such as human society and the global climate.

Related Chapters

Chapter 19: Emergence in Engineering (Peter Simons)
Chapter 33: Values in Risk and Safety Assessment (Niklas Möller)
Chapter 35: The Role of Resilience in Engineering (Neelke Doorn)

References

Ayyub, B. and McCuen, R. (2003). *Probability, Statistics, and Reliability for Engineers and Scientists*, 2nd ed. Boca Raton: Chapman & Hall/CRC.

Bak, P. (1996). *How Nature Works: The Science of Self-Organized Criticality*. New York: Copernicus, Springer-Verlag.

Bernstein, R. (2010). *The Pragmatic Turn*. Malden: Polity Press.

Bhat, U. (1972). *Elements of Applied Stochastic Processes*. New York: John Wiley & Sons.

Bulleit, W. (2008). Uncertainty in Structural Engineering. *Practice Periodical on Structural Design and Construction*, 13(1), 24–30.

Bulleit, W. (2013). Uncertainty in the Design of Non-Prototypical Engineered Systems. In D. Michelfelder, N. McCarthy, D. Goldberg. Dordrecht (eds.), *Philosophy and Engineering: Reflections on Practice, Principles, and Process*. Dordrecht: Springer, pp. 317–327.

Bulleit, W. (2018). Uncertainty in the Design and Maintenance of Social Systems. In C. Garcia-Diaz and C. Olaya (eds.), *Social Systems Engineering: The Design of Complexity* (31–43), Chichester: John Wiley.

Bulleit, W. and Drewek, M. (2011). Agent-Based Simulation for Human-Induced Hazard Analysis. *Risk Analysis*, 31(2), 205–217.

Corotis, R. and Tsay, W. (1983). Probabilistic Load-Duration Model for Live Loads. *Journal of Structural Engineering*, 109(9), 859–874.

Dekker, S. (2011). *Drift Into Failure: From Hunting Broken Components to Understanding Complex Systems*. Burlington: Ashgate.

Delatte, Jr., N. (2009). *Beyond Failure: Forensic Case Studies for Civil Engineers*. Reston: American Society of Civil Engineers.

Der Kiureghian, A. and Ditlevsen, O. (2009). Aleatory or Epistemic? Does It Matter? *Structural Safety*, 31, 105–112.

Dewey, J. (1958). *Experience and Nature*, New York: Dover Publications, Inc.

Fann, K. (1971). *Wittgenstein's Conception of Philosophy*, Berkeley: University of California Press.

Habermas, J. (2000). Richard Rorty's Pragmatic Turn. In Brandom (ed.), *Rorty and His Critics (31–55)*, Baton Rouge: Louisiana State University Press.

Harford, T. (2011). *Adapt: Why Success Always Starts With Failure*. New York: Picador.

Holland, J. (1998). *Emergence: From Chaos to Order*. Cambridge: Perseus Books.

Knight, F. (1948). *Risk, Uncertainty, and Profit*. Boston: Houghton Mifflin, Co. Originally published in 1921.

Koen, B.V. (2003). *Discussion of the Method*. Oxford: Oxford University Press.

Miller, J. and Page, S. (2007). *Complex Adaptive Systems: An Introduction to Computational Models of Social Life*. Princeton: Princeton University Press.

Minkel, J. (2008). The 2003 Northeast Blackout—Five Years Later. *Scientific American*. www.scientificamerican.com/article/2003-blackout-five-years-later/. Accessed February 9, 2018.

Olsen, J. (2015). *Adapting Infrastructure and Civil Engineering Practice to a Changing Climate*. Reston: American Society of Civil Engineers. http://ascelibrary.org/doi/pdfplus/10.1061/9780784479193. Accessed March 12, 2018.

Perrow, C. (1999). *Normal Accidents: Living With High Risk Technologies*, 2nd ed. Princeton: Princeton University Press.

Petroski, H. (2006). *Success Through Failure: The Paradox of Design*. Princeton: Princeton University Press.

Pool, R. (1997). *Beyond Engineering: How Society Shapes Technology*. Oxford: Oxford University Press.

Popper, K. (1985). Piecemeal Social Engineering (1944). In David Miller (ed.), *Popper Selections*. Princeton: Princeton University Press, pp. 304–318.

Simon, H. (1996). *The Sciences of the Artificial*, 3rd ed. Cambridge: MIT Press.

Talbot, S. (2008). Toward an Ecological Conversation. In B. Vitek and W. Jackson (eds.), *The Virtues of Ignorance: Complexity, Sustainability, and the Limits of Knowledge* (101–118), Lexington: The University Press of Kentucky.

Taleb, N. (2007). *The Black Swan: The Impact of the Highly Improbable*. New York: Random House.

Vincenti, W. (1990). *What Engineers Know and How They Know It*. Baltimore: The Johns Hopkins University Press.

Vitek, B. and Jackson, W. (2008). *The Virtues of Ignorance: Complexity, Sustainability, and the Limits of Knowledge*, Lexington: The University Press of Kentucky.

12
SCENARIOS

Christian Dieckhoff and Armin Grunwald

12.1. Introduction

Scenarios describe possible future developments in certain parts of the world. These could be in a nation's energy system, a traffic situation which an autonomously driving car has to cope with, or any other aspect of interest. Future possibilities can be both positive and negative—they can pose dangers or opportunities, and often they are contested. Scenario methods are used both as a means to develop actions to shape the future and as a means of precaution to cope with the uncertainty connected to the future.

Technologies in particular strongly shape modern societies and their futures—what life will look like for us, and in what environment this life will take place. Technologies most often arrive as both a means to open up possibilities by enabling actions and as a danger due to potential negative side effects or misuse. Therefore, it is not surprising that scenarios are most prominently used in the context of technology development, which is the focus of this contribution.

As will be discussed in detail in this chapter, there is no single understanding of what a scenario is, or a standard method to generate one. Instead, there is a spectrum of understandings and methods, which are in turn applied for a range of different purposes. This chapter's task is therefore to provide a structured overview of these variants. But there is a danger in this approach: the danger of getting lost in rather abstract distinctions. Therefore, we decided to root this chapter as strongly as possible in actual practice, and we refer to exemplary scenario studies wherever possible. Due to its relevance (and our own expertise), we selected mobility and transport as the field from which to draw our examples.

We start out in Section 2 with a general discussion of the role scenarios can play in technological development, illustrated for the field of mobility and transport. In Section 3 we give a short introduction to the historical background of scenario thinking. From there we dive into the topic, by reflecting on how to understand scenarios (Section 4), explaining how to create them with different methods (Section 5), and discussing the purposes they are used for (Section 6). We end with some concluding remarks, in Section 7.

12.2. Scenarios in Technological Development—the Example of Mobility and Transport

As will be clarified throughout the chapter, envisioning possible futures is a major purpose of scenarios. To some degrees, all scenarios are therefore creative and visionary products. But at the same time, scenarios serve quite serious purposes, namely to support orientation and decision-making on far-reaching technological changes. Today, we find scenarios on at least three levels:

1. On the level of *national and international politics and public discourse*—scenarios inform decisions about the appropriate regulatory framework for technological development.
2. In the *decision-making of organizations*—especially in corporate strategy and management, scenarios support decisions about which technology to pursue.
3. In *engineering and design*—scenarios of a technology's future application guide its construction and design.

Besides expressing relevant possible futures at the respective levels, scenarios also contribute to the coordination of decisions and the actions taken between these levels, and thereby have a major role in directing technological development as a whole (cp. Grunwald 2011 for the case of energy scenarios).

An excellent case to illustrate this is the field of mobility and transport. Due to its strong reliance on technologies, its crucial role for modern societies and its high contribution to global warming and air pollution, this field will serve as a guiding example throughout this chapter. Since mobility and transportation is part of the energy system and is often analyzed and regulated as such, we also draw on energy scenarios in general.

Starting at the first of the aforementioned levels, national and international politics and public discourse, the scenario analysis in the Assessment Reports of the Intergovernmental Panel on Climate Change (IPCC) is the most influential today on a global scale, since these reports are a central input to the negotiation of international climate policy agreements. In order to estimate future global greenhouse gas emissions, the reports integrate scenarios for various emission sources, including the transport sector (Edenhofer et al. 2014, Ch. 6). For example, the IPCC draws upon the International Energy Agency's annual *World Energy Outlook* (IEA 2017b)—a study often called the world's "bible of the energy industry" (van de Graaf and Zelli 2016: 48). By informing the 2015 Paris Agreement to restrict global warming to at most 2 but preferably not more than 1.5 degrees Celsius, the IPCC's scenario analysis is directly linked to the ratifying states' energy and transport policies. For example, the European Union is currently preparing CO_2 restrictions for new passenger cars and light commercial vehicles for the period after 2020, stimulating car manufacturers to develop and deploy zero- and low-emission vehicles. So, via the climate targets, the scenarios of possible future emission pathways shape the scenarios of the business environment which car manufacturers have to consider today. These business scenarios in turn translate into requirements for the design and engineering of future cars, and therefore shape the technical scenarios which car designers and engineers investigate.

Focusing on the second and third of the aforementioned levels (the decision-making of organizations, and engineering and design), the rapidly advancing development of autonomously driving vehicles is a case in point. On the third level of engineering and design, development is driven by the advances in sensor and data interpretation technology. Here we find scenarios in a very specific form: the Advanced Driver Assistance Systems (ADAS) already implemented are not yet able to realize fully autonomous cars. Yet they are seen as a necessary step towards them. The most important norm regulating the design of ADAS is ISO 26262–3 (International Organization for Standardization 2011). To ensure the system's robust operation in a variety of traffic situations, even in the event of a malfunction, the norm outlines a scenario approach to test this requirement. Here, a scenario is understood as a possible driving situation (ibid., Annex B). On the level of decision-making in organizations, even the established automotive companies now take autonomously driving cars seriously as an option or even as a necessity to maintain their business (Welch and Behrmann 2018). In turn, this advancing technology has already been adopted in scenarios for the future development of mobility as a whole. For example, Fulton et al. (2017) analyze three scenarios up to 2050, and show that especially in combination with electric and shared mobility, autonomous driving can dramatically reduce CO_2 emissions and energy usage in urban transport. At the same time this development has currently been taken up on the level of policymaking. For example, the European Commission has published a strategy on intelligent transport systems, including automated mobility (European Commission 30.11.2016), and in parallel has established an official high-level stakeholder group to conduct a scenario analysis of the competitive position of the EU automotive industry in view of autonomous vehicles (Asselin-Miller et al. 2017).

12.3. Historical Scenario Schools

As Bradfield et al. (2005) analyze in detail, modern-day scenario methods emerged in parallel in two geographical centers during the 1950s and 1960s. They identify corresponding "scenario schools", which are, to different degrees, present today but which have been complemented by approaches developed without a clear connection to these "schools".

The "intuitive logics school" has its origins in Cold War military strategizing in the USA. To force military strategists to take into account unconventional possibilities, such as a thermonuclear war (Kahn 1960), Herman Kahn at the RAND Corporation notably introduced scenarios into strategic thinking (Ghamari-Tabrizi 2005). His approach was rapidly developed by others and taken up outside the realm of national defense. Most importantly, it entered corporate strategy, most prominently at Royal Dutch Shell in the early 1970s (Wack 1985; Chermack 2017). Today, the label "scenario planning" is often, yet not exclusively, used for approaches in the line of the intuitive logics school (e.g. Schoemaker 1995; Ringland 1998).

At the same time as Kahn, Gaston Berger started to develop a scenario approach in France under the label "la prospective" or "prospective thinking", leading to the "French scenario school". Central to this approach is the premise that the future is not predetermined by the past, but can be actively shaped by society. Therefore, its scenarios serve as positive visions for policymakers and the public. Berger died in 1960, and it was his successors Pierre Masse and Bertrand de Jouvenel who succeeded in introducing the approach into the political process in France in the 1960s and 1970s. Yet, since it was applied mostly in France, the approach received much less international attention than the intuitive logics approach (de Jouvenel 1967; Bradfield et al. 2005).

With respect to today's application of scenarios on the three levels of technological development (see Section 2), a strong connection to the schools is not always visible. As will be pointed out in Section 5, the intuitive logics approach is still a prominent reference in organizational decision-making and in some cases in the context of engineering and design. But especially for scenarios addressing the level of politics and public discourse, such as the *World Energy Outlook*, no such connection is visible, at least in the realm of transport, energy and climate politics.

12.4. Understanding Scenarios

12.4.1 Scenarios as Constructs

In the standard interpretation, thinking in scenarios refrains from making strong prognostic statements about the future. Instead, it aims at a broader and more explorative view of the future, acknowledging its openness as well as its uncertainty, and aiming to inform opinion-formation, orientation and decision-making. From a broad perspective, scenarios can be understood as technological futures (Grunwald 2016, Ch. 5) or sociotechnical imaginaries (Jasanoff and Kim 2015), analytical terms that emphasize the openness of the future and the resulting necessary multiplicity of individual and social attempts to picture it.

Our investigation of what a scenario is begins with the observation that scenarios are *constructs* (Brown et al. 2000; Grunwald 2011). They are not discovered but are created by authors, individuals or teams, often commissioned by organizations, and therefore are constructed with specific intentions and methods and in specific social constellations. In short, scenarios are means to certain ends.

As will be discussed in Section 5, different methods for the generation of scenarios have been developed, ranging from more creative approaches to quantitative and formal ones. This variety of methods reflects the diversity of intentions, targets, and contexts. However, with this variety, there also comes the challenge that scenarios can incorporate quite diverse ingredients, amongst others:

- *Present knowledge*, which is accepted as knowledge according to the prevailing criteria in the scientific disciplines or groups involved in the scenario generation (e.g. statistical validity or argumentative soundness).
- *Assumptions* of future developments, based on present knowledge and theories of how the future might unfold, including mere extrapolation (e.g. trends in demographic change or energy needs).
- *Ceteris paribus assumptions*, which implicitly assume developments and conditions that will not deviate from present states or trends, thereby providing a continuity framework for prospective thinking (e.g. the exclusion of the prospect of the European Union's collapse in standard transport scenarios).
- *Values and norms* expected to be effective in the future society, for example in the relation between humans and autonomous technologies, or regarding fair access to energy.
- *Ad hoc suppositions*, which are not substantiated by knowledge yet are necessary to fill gaps in the picture of the future; in a model these include numerical values for variables which must be defined, but which are not empirically validated.
- *Imaginations* of worlds where everything could be different in the future, possibly of utopian or dystopian nature.

Scenarios also come in quite different forms: in principle, any means to express a particular future state can be used. A common distinction is between *qualitative* or *narrative* in contrast to *quantitative scenarios*. The first type mostly describes a future in a text, sometimes enhanced or substituted by images or other artistic means, whereas the second uses quantitative information, such as timelines for certain variables, typically presented in charts or tables. Yet, in many scenarios, both elements are combined. Here, *storylines* are an important tool: they are used as the narrative backbone of a scenario which is then elaborated qualitatively in subsequent steps or, as in the Storyline and Simulation (SAS) approach (Alcamo 2008), translated into quantifications.

12.4.2 Scenarios as Possibilities

When scenario-building, one does not want to explicate just any arbitrary picture of the future that comes to mind when letting fantasy run wild. Instead, scenarios are meant to fulfill a purpose for some addressee, for example, to support an investment decision in a company. Therefore, the typical scenario comes with a claim that the future depicted in it is *possible*. But what does that actually mean?

First of all, it is important to note that even if most scenarios claim to (merely) describe possibilities, there is no consensus that this is the only reasonable claim one can raise about them, and indeed, the situation is frequently described as confusing or chaotic (Bradfield et al. 2005: 796). For example, Börjeson et al. (2006) propose a taxonomy of scenarios which includes "predictive forecasts"; in our understanding this goes far beyond the claim to possibility, as will be explained in this chapter. Scenarios are also sometimes classified as "imaginable", "plausible" or "probable", and often it remains unclear whether authors mean something beyond mere possibility by those terms.

One interpretation provides a much clearer picture. Here, any statement about the future is called a "prediction". Following the classical distinction of Knight (1921), one then distinguishes *possibilistic* from *probabilistic* and *deterministic predictions*, which respectively state what *can be*, what *will be with a certain probability* or what *certainly will be the case*. Which of the three types of statements can be justified in a given situation depends on the type of knowledge at hand and the prevailing uncertainties. Hansson (2016b) provides a more detailed account of this typology. We propose to use the term *scenario* exclusively to denominate possibilistic (*what can be*) predictions.

The second part of this interpretation of scenarios gives an answer to the question of what "possible" actually means. Betz (2010, 2016a) introduced the *epistemic account of possibility* to the interpretation of scenarios. Here, "possible" simply means *logically consistent with the relevant knowledge*. As Betz (ibid.) elaborates in greater detail, this interpretation has far-reaching consequences. For example, we certainly can *imagine* many future developments. But given the epistemic account, not all imaginable scenarios are also possible. For example, we can easily imagine a situation in the year 2025 where cars are powered merely by the power of thought—a vision which is obviously inconsistent with our knowledge about the laws of nature. Or, we can imagine that in the year 2025, all cars in the United States are electric ones, yet we have to concede that even if this vision is consistent with the laws of nature, it is inconsistent with respect to our knowledge about the speed at which such a technological transformation can take place. Since the claim of something being possible is much stronger than the claim of it being imaginable, we can see why the sophisticated methods discussed in Section 5 are needed, to separate the possible from the fantastic.

12.4.3 Scenarios as Arguments

The composition of the aforementioned ingredients of scenarios and thereby their internal structure is still widely opaque. Yet, in recent years philosophical analyses have begun to shed light on their logical and epistemological structure by analyzing their reliance on knowledge, assumptions and models. In part, this work has recently been summarized as an *argumentative turn* in policy analysis (Hansson and Hirsch Hadorn 2016). Its central idea is to apply argumentation analysis and logic to better understand political debates, decision-making and scientific policy advice under conditions of uncertainty. In line with this approach, scenarios can be understood in two respects. On the one hand, analyzing *scenarios as arguments* and therefore by identifying central premises and conclusions *in scenario studies* allows for an analysis of their internal logical structure. On the other hand, by analyzing *scenarios as premises in larger arguments or even debates*, one can also investigate how conclusions are drawn from the scenarios and what role they play in debates and decision-making (Brun and Betz 2016).

Betz (2016a, 2016b) as well as Hansson (2016a) pay attention to the limitations that scenarios have in practical argumentation. Due to their political relevance, energy scenarios have recently received much attention. The contributions in Dieckhoff and Leuschner (2016) shed critical light on their role in policy advice, and Dieckhoff (2015) provides a detailed analysis, by reconstructing the fundamental argumentative structure of model-based energy scenario studies. We assume that this focus can be applied to scenarios on other subjects and, as outlined at the end of this subsection, also to qualitative scenarios.

So how is the claim for possibility justified for a model-based scenario? Dieckhoff draws on the epistemic interpretation of possibility (ibid., Ch. 6.2.1). In a simplified form this can be reconstructed as:

Premise 1	The set of assumptions is possible (consistent with the relevant knowledge).
Premise 2	*If* the set of assumptions is to be found in reality, *then* one would also find the set of results calculated by the model in reality.
Conclusion	Hence, the calculated set of results is possible (consistent with the relevant knowledge).

Figure 12.1 Simplified argumentative structure of a model-based scenario

As will become clearer in Section 5.3, this argument reflects the generation process of model-based scenarios. In a first step, the modeler needs to define assumptions for the variables the model is not calculating itself. These are also called "exogenous" variables. Based on these, the model then calculates the results for the remaining—also called "endogenous"—variables. Put differently, in the first step the modeler is painting one half of the picture of a future, which is then completed by the model.

Two aspects are important to note. First, for the conclusion to follow, it is not sufficient for each single assumption in a scenario to be consistent with the relevant knowledge. Instead, the whole set needs to be consistent with the relevant knowledge, which implies that the assumptions should also be consistent with each other. This also holds for a scenario which has not been generated with an explicit model.

Since many models have hundreds of exogenous variables, the justification of the first premise is not trivial, which is why additional methods such as the CIB approach (Cross-Impact-Balancing Analysis) are sometimes applied (see Section 5.2). The approach of writing storylines in scenario generation is also an attempt to ensure internal consistency.

The second premise is a conditional deterministic prediction and therefore quite demanding. In order to be justified, the model needs to be at least empirically adequate (Van Fraassen 1980) or, put more simply, the model has to be sufficiently correct in grasping the interrelation of exogenous and endogenous variables. This is at least dubious for models failing to (sufficiently accurately) reproduce the historical data for the endogenous variables when given the corresponding historical data for the exogenous variables, which is a notorious problem in energy modeling (Dieckhoff 2015: 220 ff.; Trutnevyte 2016). Additionally, as Parker (2006) points out for the case of climate modeling, when systems that are highly afflicted by uncertainties are analyzed, there are often a number of models available, and for practical or fundamental reasons it is not possible to identify the single best model. One methodological answer to this "model pluralism" is the performance of ensemble and model-intercomparison studies. Here, a multitude of scenarios is calculated with the multiple models and then jointly analyzed; this approach is at the heart of the IPCC's scenario analysis.

The outlined justification of a model-based scenario can also serve as a template for the analysis of qualitative scenarios. Let us use the scenario of a hydrogen economy for illustration, inspired by the Hydrogen Council (2017). For example, one could put the following aspect of such a scenario forward (and elaborate it with pictures, charts and texts): "It is possible that the production of green hydrogen will rise dramatically until 2050 due to the rising demand in the industrial sector". A different aspect of the same scenario could be summarized in a second statement: "It is possible that the worldwide sales of fuel cell electric vehicles will rise dramatically until 2050". If these two aspects of a scenario are put forward, we can (logically) connect them with "and" into a single possibilistic prediction and then scrutinize it with regards to its consistency with our knowledge—this is equivalent to the first premise in the earlier scheme. But we could also assume some kind of principle, correlation or causation connecting the two aspects. We could then try to reconstruct this as an "if-then statement", analogous to the second premise, and scrutinize it independently. For example, the following could express such a connection: "If the production of green hydrogen rises, the price for hydrogen will drop. Then fuel cell vehicles will be economically more attractive and in turn bought more frequently". This statement would then become the second premise in the scheme, used to justify the second part of the scenario as the conclusion. Whether such a connection is really made in a qualitative scenario, and how to reconstruct it in an analysis, is often a matter of interpretation.

12.5. How Are Scenarios Created?

Put in general terms, to "make" a scenario involves two stages: (i) to express or even create a vision of one or more future states—be it in a text, a picture or a row of numbers—and (ii) to—at least roughly—chart the space of possible developments relevant to the question at hand and select the ones deemed worthy of a closer look. In practice these two stages are inseparable from one another. But, as we will see throughout this section, scenario projects and in turn the methods applied to them put different emphasis on these two aspects—for example if a model is at hand from the inception of the project, the emphasis typically lies on the selection of scenarios.

Nevertheless, all scenario projects are naturally confronted with a large space of the possible developments which are potentially relevant for the question. Therefore, a number of strategies have been developed to help select a reasonable number of scenarios.

One is rooted in the distinction between *backcasting scenarios* and *explorative scenarios* (Greeuw et al. 2000; Van Notten et al. 2003): the idea behind the former, also called *normative scenarios*, is to determine the desired future states of the system under consideration (e.g. a CO_2-emission target) and then to identify possible pathways to reach these by alternative sets of actions and decisions. *Explorative scenarios*, also called *descriptive scenarios*, depart from a system's status quo to explore the space of possible developments as broadly as possible without defining normative targets.

Another strategy is the identification of *worst-case* and *best-case scenarios*, describing developments where extreme values are assumed for some aspects under consideration in order to somehow cover the possibility space in between. Voigt (2016) provides a critical account of this strategy. An approach that goes beyond the mere selection of scenarios and instead embraces the idea of analyzing all relevant scenarios for a given question was proposed by Lempert et al. (2003) under the label Robust Decision Making.

In addition to all these strategies, in most studies, one scenario is often called the *business-as-usual scenario* and describes a development where the system under consideration runs its due course without interference by any new (for example, political) measures.

As there is a multitude of interpretations of scenarios, there is also a wide variety of techniques and approaches ("methods" in short) to generate them. For example Mietzner and Reger (2005) compare four methods for application in corporate planning. Bishop et al. (2007: 6) identify 23 scenario methods and, guided by their starting points, processes and products, aggregate them into eight types. Börjeson et al. (2006) distinguish methods with respect to their approach to generating, integrating and ensuring consistency of scenarios. Amer et al. (2013) review qualitative and quantitative methods.

In this chapter we aim for a simple systematization of scenario methods. First, we order important types of methods on a spectrum ranging from creative, through those which emphasize cooperative development, to formal methods. Secondly, we distinguish four phases in which the development of scenarios can typically take place in a project:

1. *Scoping*: clarification of the project's aim and the question to be answered.
2. *Modeling*: identification of the aspects of the world which are relevant to answering the question, and the (causal) relationship of these aspects to one another. Modeling is understood in a broad sense here, including formal and non-formal approaches.
3. *Development of scenarios*: apply a method to construct one or more scenarios, including rigid methods as well as creative thinking.
4. *Interpretation and use of scenarios*: take the scenarios as a given, and examine what one can learn and conclude from them.

As will be explained in the following three subsections, some methods separate these phases clearly from one another, while others integrate them or connect them in iterative loops (Rowland and Spaniol 2017). Our main aim here is to explain the individual foci of the methods, including a discussion of the claims typically raised with their scenarios, and to introduce characteristic tools applied within them. Therefore, we concentrate on the second and third phases introduced earlier (scenario modeling, and development), while the fourth is addressed in Section 6.

12.5.1 Creative Methods

In the extreme, one might aim for depicting visionary future worlds, for example by writing science-fiction stories or painting futuristic pictures. Whether or not there are methods applied here in a narrower sense might be a point of dispute, since one might state that the very point of visual and literary arts is not to be restricted by methods. But in any case, we can reasonably treat the resulting stories or pictures as scenarios, and clearly the very explication of these futures is what they are concerned with. Interestingly, multiple authors describe the scenario method itself as an art (e.g. de Jouvenel 1967; Schwartz 1996; van der Heijden 1996).

At first sight one might also say that such products of the arts do not come with a strong claim such as consistency with our current scientific knowledge. But, for example, the science-fiction universe of *Star Trek* is famous for its strong adherence to physics, and as Krauss (2007) discusses, some of the seemingly impossible technologies presented are at least consistent with our scientific knowledge.

Leaving the field where scenarios are primarily products of the arts, we enter a realm where creativity is still of high importance, yet scenarios are required for a distinct purpose. One typical situation is a company confronted with the decision of what new product to develop. Keinonen and Takala (2006) provide a review of concept design which is the process of developing rather abstract concepts of new products in a first step. As they explain, this is a recent methodological development addressing uncertainties in dynamic markets (ibid., v ff.). Inspired by the intuitive logics school, Saaskilahti and Takala (2006) outline how scenarios can be used here to develop visionary product concepts, such as a light step-in car designed to safely transport persons in times of global conflict, terrorism and natural disaster.

MacKay and McKiernan (2010) provide an analysis of the benefits and drawbacks of creativity in scenario planning. For example, they point out that a scenario process in a company benefits from the multiplicity of ideas and perspectives yielded by a diverse set of participating employees. But at the same time, systematic engagement in the analysis of future uncertainties can cause personal uncertainty, for example about the question itself, and whether one's job is still needed in the scenarios discussed, which can in turn undermine the company's morale (ibid., 277).

12.5.2 Cooperative Methods

Scenarios are meant to inform us about possible future developments in a specific area of society, for example an economic branch, a region, an infrastructure or a line of innovation. Therefore, a necessary step when generating any informative scenario is to identify this area and decide what is relevant in order to answer a given question. What seems to be trivial at first sight is often rather challenging in situations where we typically use scenarios—situations created by a need for orientation, where we often start without a clear question and do not yet know what the relevant aspects of the world are.

In such situations cooperative methods are applied, often in the line of the intuitive logics school. These emphasize the aim of explicating future developments and are strong in providing creative stimuli while at the same time providing the structure and guidance needed for efficient decision-making. Numerous adaptations and extensions have been made to the original proposals by Wack (1985), Schoemaker (1995), Schwartz (1996), van der Heijden (1996) and Ringland (1998), many of them documented in two recent special journal issues (Wright et al. 2013; Wright et al. 2017).

Wright et al. (2013: 634) identify eight basic stages for a cooperative scenario approach in the line of the intuitive logics school. We present the stages as follows, adding our own comments and explanations to them:

1. *Setting the agenda*: defining the issue to be analyzed.
2. *Determining driving forces* (sometimes also called "drivers"): these are the factors that have a high impact on the issue at hand. They can be of political, economic, social, technological, ecological, legal or other natures. For example, the growth of the population and the resulting demand for energy and mobility is such a driver in most energy and transport scenarios.
3. *Clustering the driving forces*: this stage aims at reducing the multitude of factors which the second stage typically delivers.
4. *Defining cluster outcomes*: the ranges in which the clustered drivers can develop in the future are determined by identifying the two extreme outcomes for each cluster.

The Delphi method is frequently applied in stage 4. It addresses the challenge that knowledge about the far future of complex systems is notoriously hard to come by, and it approaches this problem by eliciting expert opinions in a sequence of surveys (Nowack et al. 2011). Melander (2018) provides a review of scenario studies on the future of transport using the Delphi method.

5. Via an *impact and uncertainty matrix*: the causal relationships among the clustered drivers are identified. At the same time the two most influential but uncertain factors are identified.
6. *Framing the scenarios*: relevant combinations of the extreme values for these key factors are selected. These combinations are then interpreted as scenarios for the most relevant factors external to the issue (for example, the business decision) at hand.

The Cross-Impact-Balancing (CIB) method is frequently applied in stages 5 and 6. This method is based on expert estimates but concentrates on the elicitation of their opinions on causal relations or correlations between factors (also called "descriptors" in the standard terminology), deemed to be relevant for an issue at hand (Weimer-Jehle 2006). For example, Ecola et al. (2015) applied the method to generate scenarios for China's mobility in 2030.

7. *Scoping the scenarios*: for the resulting four scenarios, broad descriptors are outlined.
8. *Developing the scenarios*: the resulting, rather abstract scenarios are "fleshed out" by developing storylines or narratives for them.

The typical result of scenario projects in the line of the intuitive logics approach is a set of differentiated, qualitative or semi-quantitative scenarios. In practice, this process is often adjusted to the project's aims. The study by Ecola et al. (ibid., pp. 4) is a good illustration, where the process leads to two main scenarios for China's future mobility. They are illustrated by the narratives "The Great Reset" and "Slowing but Growing", indicating differences in economic growth as the main drivers identified for this subject.

While stages 2 to 5 correspond to the second phase ("Modeling") in in our earlier schematic, and stages 6 to 8 to our third phase ("Scenario development"), our last phase ("Interpretation and use of scenarios") is not included in the proposal by Wright et al. (2013). Indeed, they come to the conclusion that the contribution of scenarios to actual decision-making remains rather unclear (ibid., p. 640). Again here, Ecola et al. (2015: 67) provide an example of how this phase can be realized in practice.

The reason we characterize these methods as "cooperative" is that they aim at developing scenarios through structured cooperation by a group of people. In many such projects this group, for example a company's management board, is the primary addressee of the scenarios. The aim is to integrate the individuals' experience, knowledge, normative goals and uncertainty estimates in these scenarios. The scenarios are not only a product at the end of the process; they also (or only) serve as a tool used along the way to clarify and structure the decision situation. Hence, much effort is put into the design of the development process, for example, by carefully selecting the participants (Franco et al. 2013; Ernst et al. 2018).

12.5.3 Model-Based Methods

In cooperative methods the identification of the relevant aspects of the world and their interdependencies or causal relations, the modeling phase in our terminology, has typically already been performed in a structured manner. And all scenario methods imply some sort of modeling, if only a mental one.

But model-based methods take this a significant step further. Here, models are at the heart of the method. The relevant aspects of the world, also called the model's target system in model theory, are represented by quantitative variables and mathematical functions, often assembled to large equation systems, which can also be called "formal" models. They are typically solved with the aid of computers, which is why the term "computer model" or "simulation model" is also used. In the following, we will use the term "models" and for the scenarios generated by them "model-based scenarios".

The variety of models applied for generating scenarios in the field of technological development is too large to systematize in this contribution. A model-type broadly used in energy systems analysis is the optimization model. Here, a geographical region's energy system is represented by its present and prospectively available technologies for the production and use of energy, such as wind turbines, and domestic heating systems or cars. Characteristic of optimization models is that an algorithm identifies cost-minimal scenarios for the future amount of utilization of the technologies; for example, the share of electrical, fuel cell and conventional cars in a fleet up to 2050. Such a model is used in the *World Energy Outlook* (IEA 2017b). Mai et al. (2013) provide a good introduction for this and other model types used in energy systems analysis.

With the application of computer models, there comes a major practical difference between model-based and other scenario methods: the modeling phase, here the construction and maintenance of the mathematical models and their algorithmic equivalents, is often temporarily separated from the phase of scenario development (here the calculation of scenarios). The reason is that the models are typically large and complicated mathematical and algorithmic structures, incorporating huge amounts of empirical data. Therefore, these models are not developed for a single scenario study but are applied in many of them. So, while in cooperative methods the modeling is typically part of the individual project, here there is normally already a model to be applied. Dieckhoff (2015) has analyzed this in detail for energy scenarios. To overcome the resulting opaqueness of the modeling process to external groups such as stakeholders or decision-makers, participatory model-based approaches have recently been developed (e.g. Ulli-Beer et al. 2017).

But what does it actually mean to "calculate" a scenario with such a model? We apply scenario analyses in situations of uncertainty. We know that the future development of a target system is strongly influenced by a number of exogenous factors, and for some of them we may have good deterministic or probabilistic predictions. But for many exogenous factors we can typically only determine possible future developments, represented by more or less well-delimitable numerical value spans. For example, in transportation we cannot reliably predict the future cost reduction of new technologies such as fuel cells or batteries up to the year 2050, but we know that this has a massive impact on the technologies' deployment.

Therefore, the idea of model-based scenario analysis is to vary the assumptions for these uncertain exogenous variables in different model-runs. Each set of assumptions produces a corresponding set of numerical results for the endogenous variables, and together they form a scenario.

So, the calculation is actually the second step of the model-based scenario generation, while the first is the selection of the numerical assumptions for the exogenous variables to be used in the model-run. In a typical model-based scenario project, this first step poses a major challenge, as a typical model has hundreds of exogenous variables, and for many of them we can determine only rather large spans of possible values. So in order to analyze all possible scenarios for the corresponding target system, one would have to vary all these variables in their complete spans in all possible combinations. Additionally, many models, for example the optimization models used in energy systems analysis, take multiple hours or even days to calculate the solution for one set of assumptions.

Since this complete analysis of all scenarios is often practically impossible, the standard approach is presently to select from one to fifteen sets of assumptions characterizing central uncertainties and calculating only these scenarios, by applying the aforementioned selection strategies. Additionally, in most studies, some uncertain variables, believed to have a large impact on the results, are varied in *sensitivity analyses*, resulting in additional scenarios.

12.6. What Can Scenarios Be Used For?

The strength of scenario thinking and methods lies in its embrace of the openness and uncertainty of the future. Therefore, in most contexts the value of this approach lies in the development and analysis of *multiple* scenarios for the issue at hand, even if the generation of each scenario often comes with substantial effort. As we have seen in the previous sections, scenarios can be used for a wide variety of purposes. In the following we systematically discuss the most prominent ones, including the purposes mentioned before.

12.6.1 Communication

As discussed in Section 1, scenarios are a means to present and discuss imaginations of possible futures on different levels of discourse, ranging from public and political debate to communication in companies or smaller groups of people. This communication can be of an external nature, for example when an organization is expressing its mission, corporate identity or aims via scenarios, as is a major purpose of the Energy [R]evolution scenarios frequently published by Greenpeace (e.g. Greenpeace, GWEC and SolarPowerEurope 2015). Or it can be of an internal nature, when scenarios are a means of decision support inside an organization.

When it comes to the society as a whole, scenarios are also a means for the discussion of normative questions, most of all the question of how we want to live in the future. Therefore, they are (and should be) contested (Brown et al. 2000). But for this purpose, the transparency of the normative ingredients discussed in Section 4.1, such as implicit political aims, is of utmost importance. Especially in cases where scientific expert knowledge is a major resource for decision-making under uncertainty, such as in climate, energy and transport politics, many scenario studies fail to meet the communicative requirements for reception in the political, public and often scientific discourses (Leopoldina et al. 2016). For example, Pissarskoi (2016) shows how energy scenario studies fail to adequately represent the uncertainties relevant to their results by using ambiguous or even misleading graphs and formulations.

12.6.2 Integration

Thinking about the future necessarily requires a great deal of fantasy and creativity. But when we want to develop shared visions of the future, it is necessary to express individual estimates of what is relevant, possible and desirable and to integrate these individual positions. One important purpose of scenario methods is therefore to support this, even if the approaches differ in the means to do so. The aim is always to come up with a set of scenarios representing (or being consistent with) what is the accepted relevant knowledge and the normative premises and goals one wants to address or presuppose in the context at hand. For example, the annual Global EV Outlook integrates the knowledge and estimates of experts and members of a number of governments on the future of electric mobility (IEA 2017a).

12.6.3 Insight

As with any foresight method, a main purpose of scenarios is of an epistemic nature: we want to gain insights about the future, even if under conditions of uncertainty that means to gain insights about multiple possible futures.

Following the analysis of Betz (2010), an important first step is therefore to formulate *hypotheses* – in other words, images or visions – about what might be possible. Here, there is no limit to creativity, since in this step a proposed scenario does not have to be consistent with our knowledge. This was one major aim of Kahn's when he developed his scenario approach: he wanted to shake the public and politicians awake into accepting that there are possibilities such as a thermonuclear war, which one normally refrains from thinking about (see Section 3). Yet, when scenarios are meant to support decisions, they need to be possible in a stricter sense, e.g. consistent with the relevant knowledge. Betz introduces two principal methods to *verify* or *falsify a possibility*. Here, even a single scenario can be of great relevance, namely, when it shows the possibility of a development previously neglected, and catastrophic global climate change due to greenhouse gas emissions is surely a case in point.

But, as mentioned before, the strength of scenario methods lies in the development and analysis of multiple scenarios. The underlying aim is to investigate the possibility space one faces in a certain situation to either *assess the uncertainty range* or *to identify possible actions*. So while each single scenario can be of interest, in these cases it is the insights one gains from their joint analysis that matters.

The most important aim is the identification of commonalities of a set of scenarios, which are not only pursued in scenario studies on their own, but also in meta-analyses which include the scenarios from different studies (Dieckhoff 2016). Commonalities are then often interpreted as *robust developments*, or used to identify *robust* or *no-regret actions*. The idea is that such a common development will occur under any circumstances; indeed, a conclusion to a deterministic prediction about this development is then made, or it is concluded that the action will have the intended effect under any circumstance.

But as Betz (2016a) analyzes in detail, these are quite demanding conclusions, since a development is robust only when it is present in *all* relevant scenarios. Yet, the requirement of incorporating all relevant scenarios in an analysis is hard to meet when facing a large and complex possibility space such as we encounter for the development of future energy, information or transportation systems. Therefore, conclusions on robustness must always be scrutinized. Additionally, the identification of robust actions presupposes normative premises such as targets to be reached under any circumstance which, therefore, is not solely a scientific matter.

Especially in scientific contexts, scenarios are also used as a vehicle to understand causal relations in a modeled target system, which is expressed in the famous catch-phrase "modeling for insights, not numbers" (e.g. Huntington et al. 1982). Here, the idea is also to calculate multiple scenarios with a model, but then to identify effects in the results that occur under certain common assumptions.

12.6.4 Decision-Making

The most demanding purpose of scenarios is their use in decision-making. Until now there has been a lack of systematic analysis on how scenarios are actually used in different decision contexts, for example, in strategic decision-making inside an organization, or in the political process. Yet, what is visible is a spectrum ranging from merely informing the decision-makers to the actual use of scenarios as premises in the justification of actions, hence in practical argumentation. It is especially in political decision-making where scenarios have to meet quality standards such as transparency, unbiasedness, and scientific validity, which implies a stronger claim than merely the possibility raised by them (Leopoldina et al. 2016).

When it comes to the use of scenarios in practical argumentation, we face a number of pitfalls, as discussed in detail by Betz (2016a, 2016b) and Hansson (2016a). One of these results from the aforementioned fact that most scenario studies, and even meta-analyses, will fail to grasp all relevant scenarios. In such a case, a development or action identified as robust might not be robust at all and could lead to wrong decisions. Hirsch Hadorn (2016) provides an overview of strategies for decisions under uncertainty which also addresses the problem of incomplete knowledge.

12.7. Concluding Remarks

Scenarios and the methods to develop them come in a wide variety, ranging from complex modeling, through cooperative and participatory methods, to more explorative approaches based on creativity. Besides their explicit treatment of the openness of the future and the resulting uncertainty by analyzing alternative possible futures, it is hard to identify a single common core. It seems more appropriate to treat scenario development as a methodological idea around which a variety of practices and methods has evolved.

Precisely because of this variety, scenarios are a flexible medium of reasoning, deliberation, creating insight, organizing communication and supporting decisions in a number of social fields, including the economy, politics and public discourse. But even if they seem to come with the rather humble claim of describing "only" possible future developments, it is important to note that this claim itself, and more so the conclusions based on it, can be quite hard to justify and often require substantial scientific work and careful reasoning.

Related Chapters

Chapter 11: Uncertainty (William M. Bulleit)
Chapter 21: Engineering Design (Peter Kroes)
Chapter 28: Models in Engineering and Design (Michael Poznic)
Chapter 30: Computer Simulations (Hildrun Lampe)
Chapter 45: Engineering as a Political Practice (Govert Valkenburg)

Further Reading

Betz, G. (2010). What's the Worst Case: The Methodology of Possibilistic Predictions. *Analyse & Kritik*, 32(1), 87–106. (An accessible introduction to the epistemic interpretation of scenarios, giving an answer to what "possible" means.)

Börjeson, L., Höjer, M., Dreborg, K.-H., Ekvall, T. and Finnveden, G. (2006). Scenario Types and Techniques. Towards a User's Guide. *Futures*, 38(7), 723–739. (A systemization of scenario methods, ranging from surveys to computer modeling.)

Hansson, S.O. and Hirsch Hadorn, G. (eds.) (2016). *The Argumentative Turn in Policy Analysis*. Cham, Switzerland: Springer. (An overview of recent contributions on decision-making under uncertainty.)

Wack, P. (1985). Scenarios: Uncharted Waters Ahead. *Harvard Business Review*, 73–89. (A classical contribution to scenario thinking in organizations.)

Wright, G., Meadows, M., Tapinos, S., O'Brien, F. and Pyper, N. (2017). Improving Scenario Methodology. Theory and Practice, introduction to the special issue. *Technological Forecasting and Social Change*, 124, 1–5 (An overview of recent developments in cooperative scenario approaches.)

References

Alcamo, J. (2008). The SAS Approach: Combining Qualitative and Quantitative Knowledge in Environmental Scenarios. In J. Alcamo (ed.), *Environmental Futures. The Practice of Environmental Scenario Analysis*. Amsterdam and Boston: Elsevier, pp. 123–150.

Amer, M., Daim, T.U. and Jetter, A. (2013). A Review of Scenario Planning. *Futures*, 46, 23–40.

Asselin-Miller, N., Horton, G., Amaral, S., Figg, H. and Sheldon, D. et al. (2017). *GEAR 2030 Strategy 2015–2017. Comparative Analysis of the Competitive Position of the EU Automotive Industry and the Impact of the Introduction of Autonomous Vehicles*.

Betz, G. (2010). What's the Worst Case: The Methodology of Possibilistic Predictions. *Analyse & Kritik*, 32(1), 87–106.

Betz, G. (2016a). Accounting for Possibilities in Decision Making. In S.O. Hansson and G. Hirsch Hadorn (eds.), *The Argumentative Turn in Policy Analysis*. Cham, Switzerland: Springer, pp. 135–169.

Betz, G. (2016b). Fehlschlüsse beim Argumentieren mit Szenarien. In C. Dieckhoff and A. Leuschner (eds.), *Die Energiewende und ihre Modelle. Was uns Energieszenarien sagen können—und was nicht*. Bielefeld: transcript, pp. 117–136.

Bishop, P., Hines, A. and Collins, T. (2007). The Current State of Scenario Development: An Overview of Techniques. *Foresight*, 9(1), 5–25.

Börjeson, L., Höjer, M., Dreborg, K.-H., Ekvall, T. and Finnveden, G. (2006). Scenario Types and Techniques. Towards a User's Guide. *Futures*, 38(7), 723–739.

Bradfield, R., Wright, G., Burt, G., Cairns, G. and van der Heijden, K. (2005). The Origins and Evolution of Scenario Techniques in Long Range Business Planning. *Futures*, 37(8), 795–812.

Brown, N., Rappert, B. and Webster, A. (2000). *Contested Futures. A Sociology of Prospective Techno-Science*. Aldershot et al.: Ashgate.

Brun, G. and Betz, G. (2016). Analysing Practical Argumentation. In S.O. Hansson and G. Hirsch Hadorn (eds.), *The Argumentative Turn in Policy Analysis*. Cham, Switzerland: Springer, pp. 39–77.

Chermack, T.J. (2017). *Foundations of Scenario Planning. The Story of Pierre Wack*. Milton: Routledge.

de Jouvenel, B. (1967). *The Art of Conjecture*. New York: Basic Books.

Dieckhoff, C. (2015). *Modellierte Zukunft. Energieszenarien in der wissenschaftlichen Politikberatung*, 1st ed. Bielefeld: transcript.

Dieckhoff, C. (2016). Epistemische Meta-Analyse. Ein konzeptioneller Vorschlag für die Analyse und den Vergleich von Szenarien. In C. Dieckhoff and A. Leuschner (eds.), *Die Energiewende und ihre Modelle. Was uns Energieszenarien sagen können—und was nicht*. Bielefeld: transcript, pp. 137–166.

Dieckhoff, C. and Leuschner, A. (eds.) (2016). *Die Energiewende und ihre Modelle. Was uns Energieszenarien sagen können—und was nicht*. Bielefeld: transcript.

Ecola, L., Zmud, J., Gu, K., Phleps, P. and Feige, I. (2015). *The Future of Mobility. Scenarios for China in 2030*. Santa Monica, CA: RAND.

Edenhofer, O. et al. (eds.) (2014). *Climate Change 2014: Mitigation of Climate Change. Working Group III Contribution to the Fifth Assessment Report of the Intergovernmental Panel on Climate Change*. Cambridge et al.: Cambridge University Press.

Ernst, A., Biß, K.H., Shamon, H., Schumann, D. and Heinrichs, H.U. (2018). Benefits and Challenges of Participatory Methods in Qualitative Energy Scenario Development. *Technological Forecasting and Social Change*, 127, 245–257.

European Commission (2016). *A European Strategy on Cooperative Intelligent Transport Systems, a Milestone Towards Cooperative, Connected and Automated Mobility*. Communication from the Commission, November 30.

Franco, L.A., Meadows, M. and Armstrong, S.J. (2013). Exploring Individual Differences in Scenario Planning Workshops. A Cognitive Style Framework. *Technological Forecasting and Social Change*, 80(4), 723–734.

Fulton, L., Mason, J. and Meroux, D. (2017). *Three Revolutions in Urban Transportation: How to Achieve the Full Potential of Vehicle Electrification, Automation, and Shared Mobility in Urban Transportation Systems Around the World by 2050*.

Ghamari-Tabrizi, S. (2005). *The Worlds of Herman Kahn. The Intuitive Science of Thermonuclear War*. Cambridge, MA [u.a.]: Harvard University Press.

Greenpeace, Global Wind Energy Council (GWEC), and SolarPowerEurope (2015). *Energy [R]evolution. A Sustainable World Energy Outlook*: Greenpeace, EREC, GWEC.

Greeuw, S.C.H., Van Asselt, M.B.A., Grosskurth, J., Storms, C.A.M.H. and Rijkens-Klomp, N. et al. (2000). *Cloudy Crystal Balls: An Assessment of Recent European and Global Scenario Studies and Models*. Luxembourg: Off. for Off. Publ. of the Europ. Communities.

Grunwald, A. (2011). Energy Futures: Diversity and the Need for Assessment. *Futures*, 43(8), 820–830.

Grunwald, A. (2016). *The Hermeneutical Side of Responsible Research and Innovation*. London: John Wiley & Sons Incorporated.

Hansson, S.O. (2016a). Evaluating the Uncertainties. In S.O. Hansson and G. Hirsch Hadorn (eds.), *The Argumentative Turn in Policy Analysis*. Cham, Switzerland: Springer, 79–104.

Hansson, S.O. (2016b). Introducing the Argumentative Turn in Policy Analysis. In S.O. Hansson and G. Hirsch Hadorn (eds.), *The Argumentative Turn in Policy Analysis*. Cham, Switzerland: Springer, pp. 11–35.

Hansson, S.O. and Hirsch Hadorn, G. (eds.) (2016). *The Argumentative Turn in Policy Analysis*. Cham, Switzerland: Springer.

Hirsch Hadorn, G. (2016). Temporal Strategies for Decision-making. In S.O. Hansson and G. Hirsch Hadorn (eds.), *The Argumentative Turn in Policy Analysis: Reasoning About Uncertainty*. Cham, Switzerland: Springer, pp. 217–242.

Huntington, H.G., Weyant, J.P. and Sweeney, J.L. (1982). Modeling for Insights, Not Numbers. The Experiences of the Energy Modeling Forum. *Omega*, 10(5), 449–462.

Hydrogen Council (2017). *Hydrogen, Scaling Up. A Sustainable Pathway for the Global Energy Transition*. http://hydrogencouncil.com/wp-content/uploads/2017/11/Hydrogen-Scaling-up_Hydrogen-Council_2017.compressed.pdf.

International Energy Agency (IEA) (2017a). *Global EV Outlook 2017. Two Million and Counting*. Paris: OECD Publishing.

International Energy Agency (IEA) (2017b). *World Energy Outlook 2017*. Paris: OECD Publishing.

International Organization for Standardization (2011). *ISO 26262–3: Road Vehicles—Functional Safety—Part 3: Concept Phase, Review*, 1.

Jasanoff, S. and Kim, S.-H. (eds.) (2015). *Dreamscapes of Modernity. Sociotechnical Imaginaries and the Fabrication of Power*. Chicago et al.: University of Chicago Press.

Kahn, H. (1960). *On Thermonuclear War*. Princeton: Princeton University Press.

Keinonen, T. and Takala, R. (eds.) (2006). *Product Concept Design. A Review of the Conceptual Design of Products in Industry*. Berlin and London: Springer.

Knight, F.H. (1921). *Risk, Uncertainty, and Profit*. Boston: Hart, Schaffner & Marx; Houghton Mifflin Co.

Krauss, L.M. (2007). *The Physics of "Star Trek"*. New York: Basic Books.

Lempert, R.J., Popper, S.W. and Bankes, S.C. (2003). *Shaping the Next One Hundred Years. New Methods for Quantitative, Long-term Policy Analysis*. Santa Monica: RAND.

Leopoldina (German National Academy of Sciences), acatech (National Academy of Science and Engineering), and Union (Union of the German Academies of Sciences and Humanities) (eds.) (2016). *Consulting with Energy Scenarios. Requirements for Scientific Policy Advice*. München: National Academy of Science and Engineering.

MacKay, B. and McKiernan, P. (2010). Creativity and Dysfunction in Strategic Processes. The Case of Scenario Planning. *Futures*, 42(4), 271–281.

Mai, T., Logan, J., Nate, B., Sullivan, P. and Bazilian, M. (2013). *Re-Assume: A Decision Maker's Guide to Evaluating Energy Scenarios, Modeling, and Assumptions*.

Melander, L. (2018). Scenario Development in Transport Studies: Methodological Considerations and Reflections on Delphi Studies. *Futures*, 96, 68–78.

Mietzner, D. and Reger, G. (2005). Advantages and Disadvantages of Scenario Approaches for Strategic Foresight. *International Journal of Technology Intelligence and Planning*, 1(2), 220–239.

Nowack, M., Endrikat, J. and Guenther, E. (2011). Review of Delphi-based Scenario Studies: Quality and Design Considerations. *Technological Forecasting and Social Change*, 78(9), 1603–1615.

Parker, W. (2006). Understanding Pluralism in Climate Modeling. *Foundations of Science*, 11(4), 349–368.

Pissarskoi, E. (2016). Die Bürde des Möglichen. Zum verantwortlichen Umgang mit Unsicherheiten in Energieszenarien. In C. Dieckhoff and A. Leuschner (eds.), *Die Energiewende und ihre Modelle. Was uns Energieszenarien sagen können—und was nicht*. Bielefeld: transcript, pp. 89–115.

Ringland, G. (1998). *Scenario Planning. Managing for the Future*. Chichester et al.: John Wiley.

Rowland, N.J. and Spaniol, M.J. (2017). Social Foundation of Scenario Planning. *Technological Forecasting and Social Change*, 124, 6–15.

Saaskilahti, M. and Takala, R. (2006). Vision Concepts. In T. Keinonen and R. Takala (eds.), *Product Concept Design: A Review of the Conceptual Design of Products in Industry*. Berlin and London: Springer, pp. 178–195.

Schoemaker, P.J.H. (1995). Scenario Planning: A Tool for Strategic Thinking. *Sloan Management Review*, 36(2), 25–40.

Schwartz, P. (1996). *The Art of the Long View. Planning for the Future in an Uncertain World*. New York: Doubleday.

Trutnevyte, E. (2016). Does Cost Optimization Approximate the Real-world Energy Transition? *Energy*, 106, 182–193.

Ulli-Beer, S., Kubli, M., Zapata, J., Wurzinger, M. and Musiolik, J. et al. (2017). Participative Modelling of Socio-Technical Transitions: Why and How Should We Look Beyond the Case-Specific Energy Transition Challenge? *Systems Research and Behavioral Science*, 34(4), 469–488.

van de Graaf, T. and Zelli, F. (2016). Actors, Institutions and Frames in Global Energy Politics. In T. Graaf (ed.), *The Palgrave Handbook of the International Political Economy of Energy*. London: Palgrave Macmillan, pp. 47–71.

van der Heijden, K. (1996). *Scenarios: The Art of Strategic Conversation*. Chichester et al.: John Wiley.

Van Fraassen, B.C. (1980). *The Scientific Image*. Oxford: Oxford University Press.

Van Notten, P.W.F., Rotmans, J., Van Asselt, M.B.A. and Rothman, D.S. (2003). An Updated Scenario Typology. *Futures*, 35(5), 423–443.

Voigt, C. (2016). Wie viele Szenarien brauchen wir? Eine wissenschaftstheoretische Kritik des Netzentwicklungsplans. In C. Dieckhoff and A. Leuschner (eds.), *Die Energiewende und ihre Modelle. Was uns Energieszenarien sagen können—und was nicht*. Bielefeld: transcript, pp. 13–43.

Wack, P. (1985). Scenarios: Uncharted Waters Ahead. *Harvard Business Review*, 73–89.

Weimer-Jehle, W. (2006). Cross-impact Balances: A System-theoretical Approach to Cross-impact Analysis. *Technological Forecasting and Social Change*, 73(4), 334–361.

Welch, D. and Behrmann, E. (2018). *Who's Winning the Self-Driving Car Race?* [online]. Bloomberg. www.bloomberg.com/news/features/2018-05-07/who-s-winning-the-self-driving-car-race

Wright, G., Bradfield, R. and Cairns, G. (2013). Does the Intuitive Logics Method—and its Recent Enhancements—Produce "Effective" Scenarios? *Technological Forecasting and Social Change*, 80(4), 631–642.

Wright, G., Cairns, G. and Bradfield, R. (2013). Scenario Methodology. New Developments in Theory and Practice—Introduction to the Special Issue. *Technological Forecasting and Social Change*, 80(4), 561–565.

Wright, G., Meadows, M., Tapinos, S., O'Brien, F. and Pyper, N. (2017). Improving Scenario Methodology: Theory and Practice, Introduction to the Special Issue. *Technological Forecasting and Social Change*, 124, 1–5.

13

SYSTEMS ENGINEERING AS ENGINEERING PHILOSOPHY

Usman Akeel and Sarah Bell

13.1. Introduction

Systems engineering emerged in the second half of the twentieth century as a powerful set of methods for delivering complex engineered projects, products and infrastructure. It addresses the integration of components to achieve a specific purpose, including attention to interactions between elements that lead to overall system properties and performance. Systems engineering provides generic procedures and techniques for engineering design, management and operations, which are applied across conventional engineering disciplines. Systems engineers can be found at work in electricity networks, water resources, industrial processes, aerospace, software development, railways, manufacturing, telecommunications and other sectors where overall performance depends upon the complex interactions of many constituent parts and their surrounding environment.

Systems engineering methods have been formalised through handbooks, university curricula and professional bodies. Its claims to span the breadth of engineering disciplines to provide generic procedures for solving complex problems indicate concern with higher methodological, epistemological and ontological questions than are traditionally featured in monodisciplinary engineering fields. Systems engineering might therefore be characterised as engineering philosophy, an intellectual tradition within engineering, with formalised, abstract conceptions of the nature of reality and the means for intervening to achieve desired outcomes.

This chapter makes the case for systems engineering as engineering philosophy. It begins with a review of the concept of engineering philosophy, as distinct from the philosophy of engineering. Systems engineering is then characterised through a short history of the discipline, core concepts in systems theories and the nature of systems thinking. Within systems engineering three distinct discourses are identified—*ex novo*, good engineering and meta-methodology—each with their own conceptions of the nature of the discipline and its relationship to engineering and problem-solving more broadly. Explicit attention to ontological and methodological foundations and diverse conceptions of the position of the discipline within engineering and the wider world show the dynamic, yet largely internalised, reflections that constitute systems engineering as engineering philosophy.

13.2. Engineering Philosophy

In *Thinking Through Technology*, Mitcham (1994) distinguishes engineering philosophy of technology from humanities philosophy of technology based on the approach to thinking about technology and its relation to human life. Mitcham maintains that the two philosophies differ primarily in their ways of viewing, understanding and interpreting technology, but also in their historical origins and basic orientations.

Engineering philosophy of technology, according to Mitcham, is an attempt by technologists or engineers to expound a technological philosophy. It is a way of thinking about technology which sees it as the core of being human, and thus seeks to expand this technological view to all facets of human life. The approach is pro-technology and analytic, as the following quote suggests: "analyses of technology from within, and oriented toward an understanding of the technological way of being-in-the-world as paradigmatic for other kinds of thought and action" (Mitcham 1994: 39). Furthermore, engineering philosophy of technology can be seen as a self-conscious activity by engineers or technologists to reflect on their discipline and defend it against hostile criticisms.

Humanities philosophy of technology, on the other hand, is said to be "an effort by scholars from the humanities, especially philosophers, to take technology seriously as a theme for disciplined reflection" (Mitcham 1994: 17). This approach to thinking about technology proceeds by dismissing the idea that technology is at the core of human life. Rather, it sees technology as simply belonging to an expansive framework of human thought and action. The approach is more critical and uses non-technological criteria to investigate and interpret technology. It argues that there are other ways of being in the world which are not necessarily technological, but as technology obscures these ways, it portends danger for contemporary life (Mitcham 1994).

These two versions of philosophy of technology are seen to permeate the emerging philosophy of engineering. While the philosophy of science is well established and the philosophy of technology still maturing, the philosophy of engineering is just emerging (Van de Poel and Goldberg 2010). However, as it evolves into a distinct discipline, the philosophy of engineering seems to take on some aspects of both philosophies of science and technology. There is a kind of concept borrowing in which epistemology, Weltanschauung, ontology and ethics are employed to question what constitutes engineering knowledge, the worldview that guides the engineering approach to problem-solving, the existence and essence of engineering projects and the rightness or wrongness of engineering activities, respectively. Nonetheless, Saarinen et al. (2008) stress the distinction between philosophy of engineering and engineering philosophy. They propose that philosophy of engineering assesses engineering from the outside using the concepts and methodologies of the discipline of philosophy, while engineering philosophy involves a concerted philosophical effort at analytical thinking of engineering from within.

Similar to Saarinen et al., Hillier (2010) uses the terms engineering philosophy and philosophy of engineering, arguing that the former is defensive and the latter is inclusive. Discarding the 'of technology' component of Mitcham's phrase, Saarinen et al. and Hillier suggest that engineering philosophy implies engineers thinking philosophically about engineering, while philosophy of engineering denotes humanities philosophers reflecting on engineering. The concept of engineering philosophy as initially expounded by Mitcham, adopted by Saarinen et al. and maintained by Hillier is used in this chapter to characterise systems engineering. Engineering philosophy will be taken *as an introspection and utilisation of engineering concepts, methods, cognitive structures, objective manifestations and limits to explain the natural and social world.* Thus, introspection *per se* is not the differentiator of engineering philosophy, as it is common for engineers, for example, to reflect on concepts, procedures and methods in the course of a design activity. However, engineering philosophy has an endogenous point of departure that extends outward to facilitate an understanding of the world using engineering criteria (Mitcham 1994). It is a fundamentally optimistic philosophy that justifies engineering activities and looks for how best to effect changes in a complex situation.

13.3. Systems Engineering

13.3.1 Brief History

Systems engineering evolved from the management of heightened complexity in engineering processes, products and projects. Systems engineering would eventually become associated with the space, defence, telecommunications and aerospace industries. The various accounts of the emergence of systems engineering in the literature stress the need to deal with complexity that was constraining successful engineering outcomes. A version of systems engineering history suggests that the principle of systems engineering might have guided the construction of the Egyptian pyramids some 4600 years ago (Sage 1992). Informing this idea is obviously the enormity and complexity of activities required to build the pyramids. According to this account, ideas and principles that constitute systems engineering today have guided engineering since antiquity. Notwithstanding this variant of systems engineering roots, the early twentieth century witnessed an ascendance of systems engineering within the engineering community.

The 1920s and 1930s were particularly periods of grand ideas about general problem-solving. For example, John Dewey, the American philosopher, proposed in the 1930s a six-phase activity towards problem resolution. Dewey's proposal involves a problem identification stage, solution suggestion phase, intellectualisation of the problem, evaluation of the suggested solution, idea rationalisation and hypothesis testing (Rodgers 2002). These ideas proved useful in later developments of problem-solving (pro-sol) techniques during the Second World War. At some point during the conflict, the war efforts necessitated the involvement of the scientific and engineering community, especially in the UK, where a robust air defence system was needed. This convocation of scientists and engineers eventually resulted in the problem-solving strategy known as operations analysis which subsequently provided the impetus for operations research, systems analysis, management science and systems engineering (Hitchins 2007).

The years following the Second World War witnessed a widespread adoption of the problem-solving ideas emerging from the military-industrial complex. In the 1940s, Bell Telephone Laboratories, a subsidiary of American Telephone & Telegraph (AT&T), had to deal with a large, complex telephonic network that had resulted from market expansion and adoption of new technology (Züst and Troxler 2006). Furthermore, AT&T was experiencing an increasing complexity with product management and was also involved in military radar development (Adams and Mun 2005). Bell Labs is said to have been the first entity to use the term 'systems engineering' to describe the ambitious efforts of managing complexity (Weigel 2000). Hall (1962) opines that possible triggers of the systems engineering approach at Bell Labs include increasing complexity of products, rapid expansion of consumer needs, expansion of business environment and acute shortage of experts to manage such challenges.

Perhaps owing to the achievements of Bell Labs, which included the invention of the transistor in 1947, development of solar cell in 1954 and launching of Telstar I communication satellite in 1962 (Alcatel-Lucent 2010), other industrial organisations and institutions of learning embraced systems engineering. The Director of Bell Labs, G.W. Gilman, pioneered a systems engineering course at Massachusetts Institute of Technology (MIT) in 1950. Systems engineering was adopted by the National Aeronautics and Space Administration (NASA) to manage its space projects. NASA officials announced the Apollo Project in 1960 with the goal of transporting humans to the moon and returning them safely to Earth (Brill 1998). The success of Apollo 11 in July 1969, which saw two humans walk the lunar surface and returning safely to Earth, is well documented. Systems engineering was largely credited with the success of this undertaking.

Whether this great success actually resulted from the application of systems engineering remains a topic of debate. Nonetheless, some systems engineers contend that the Apollo 11 project would not have been successful without the use of systems engineering principles and methods (Hitchins 2007). This claim is fairly substantiated by the eventual codification of systems engineering practices in NASA in the succeeding years. The space agency published a document entitled *NASA Systems Engineering Handbook* in 1995 (Shishko 1995) which is revised periodically with the latest revision issued in February 2017 (NASA 2017). The handbook details the fundamentals of systems engineering as conceived and applied in the NASA environment. The overall aim of the NASA handbook is to provide top-level guidelines for systems engineering practices (NASA 2017).

With the successes of Bell Labs and NASA projects attributed to systems engineering, the adoption of the discipline by more industrial institutions was widespread. Separate departments dedicated to systems engineering were created and supervised at the top hierarchy of various institutions. Systems Engineering Company (SEC) and Systems Engineering Division (SED) of the National Defence Industrial Association (NDIA) in the United States were established in 1986 and 1997 respectively. In the UK, the notion of sociotechnical systems emerged around the same time as these systems ideas. The Tavistock Institute employed the term 'sociotechnical system' to emphasise the interaction of humans with technology; an idea which permeated several corporations including British Aerospace and the Lucas Group amongst others.

In 1990, in the United States, the National Council on Systems Engineering (NCOSE) was established following a meeting of a group of professionals interested in the systems perspective. Subsequently, NCOSE metamorphosed into the International Council on Systems Engineering (INCOSE), a not-for-profit organisation intended to develop and disseminate systems engineering knowledge and ideals. Since its formation in 1990, INCOSE has been populated by a broad spectrum of members including scientists, engineers and business experts. As of December 2015, INCOSE boasted having over 11,000 members (INCOSE 2015). The creation of INCOSE significantly advanced the practice of systems engineering and contributed to enriching its body of knowledge. Systems engineering is currently considered a professional discipline whose application stretches from combating terrorism (Mackey et al. 2003) to improving healthcare (Kopach-Konrad et al. 2007).

13.3.2 Objectives and Methods

The question of what activities or roles systems engineers perform is somewhat tricky. The literature is replete with different examples of systems engineering activities, ranging from technical and sociotechnical to generic problem-solving roles. To appreciate these issues, it is pertinent to consider the various objectives of systems engineering. Part of these objectives is to scope the problem space (Hitchins 2007) but also to assist in developing a clear rational thought process (Sage 1992). Systems engineering also aims to explore a problem space through sceptical positive thinking (Kossiakoff et al. 2011) as well as encourage rational conflict resolution (Sage 1992). Another goal of systems engineering is to promote the incorporation of risk and uncertainty considerations into planning and decision-making (Hipel et al. 2015). In addition, systems engineering seeks to characterise the whole problem by establishing and managing requirements (Calvano and John 2004) and to conceive potential remedies to problems (Bahill and Gissing 1998). Two other objectives of systems engineering are to formulate and manifest the optimum solution and to solve, resolve or dissolve the whole problem (Hitchins 2007).

An interesting observation is made by Hitchins regarding the last objective. Referring to Professor Russell L. Ackoff (1919–2009), Hitchins explains that problems generally can be addressed in three ways—they can be resolved, dissolved or solved. To resolve a problem is to provide an answer or response that is simply 'good enough' or satisfices. It involves dealing more with symptoms than with the roots of a problem to offer a pragmatic resolution of the problem. Although it sometimes leads to decisions taken without fully comprehending the problem, satisficing results in greater knowledge about the problem. To dissolve a problem, however, is to alter the situation in some way so as to make

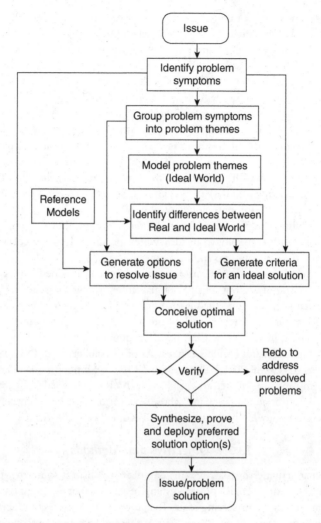

Figure 13.1 Generic systems engineering methodology

Source: Hitchins (2007).

the problem disappear. The deployed intervention seeks to create a condition where the problem no longer exists. An example of problem dissolution is seen in politics, where an adversary is compensated with a political appointment to soften opposition. The third approach to addressing a problem is to solve it. This simply means finding the right answer or best solution to the problem. Finding the best solution to a problem is recognised as quite unattainable; nonetheless, the existence of this notion provides a benchmark for judging potential remedies. Figure 13.1 summarises the generic approach to attaining systems engineering objectives.

13.4. Concepts in Systems Engineering

Some concepts have been derived variously from a range of disciplines to underpin systems engineering. What follows is a brief description of these precepts beginning with the concept of a system and ending with the notion of emergence.

13.4.1 System

'System' is a ubiquitous word in the English language. Almost anything could be couched with the term 'system' as either a qualifier or an affix. Expressions such as belief system, educational system, gambling system and solar system are quite common. However, these are relatively modern constructs and usages of the word. System has an interesting etymology that is traceable to Latin and Greek roots. In Latin and Greek, *Systéma* is defined as 'an arrangement' and 'an organised whole' respectively (Harper 2017). According to available records, the word 'system' began to appear in English writings in the early seventeenth century (OED 2010). 'System' gradually gained literary ubiquity occurring, for example, nearly 100 times in *The Age of Reason* (1794)—the seminal work of the iconic author Thomas Paine. Other influential works such as Littel's *The Living Age* (1846) employed the word in several phrases and contexts. The 1600s and 1700s denotation of 'system' has been somewhat retained in modern usage. A typical modern English dictionary contains multiple lexical entries for 'system'. Such multiplicity of entries is not unexpected of a word that has become useful to several fields of knowledge. However, the recent spread of 'system' in the intellectual community did not occur in a vacuum. It resulted partly from the vast scientific and technological activities that ensued following the First and Second World Wars. The idea of a system existed in activities ranging from the production of the smallest widget to the construction of the largest aircraft carrier.

In systems engineering—a discipline that is basically about creating systems—the concept of system has been repeatedly discussed since the field began to evolve. Leveraging the eclecticism of systems theory with its array of principles and terms such as complexity, emergence and wholeness (or holism) (described in the succeeding subsections), systems engineers either singly or multiply advanced several definitions of a system. Writing in 1970, Kast and Rosenzweig define system "as an assemblage or combination of things or parts forming a complex unitary whole" (p. 110). INCOSE states that a system is "a construct or collection of different elements that together produce results not obtainable by the elements alone" (INCOSE 2015). Another meaning of system is "an open set of complementary, interacting parts with properties, capabilities and behaviours emerging both from the parts and from their interactions to synthesise a unified whole" (Hitchins 2007: 76). Systems engineers further explicate that the delineation of system is subjective and dependent on the perspective of the person offering the definition. What counts as a system to one person may not be considered a system by another. Thus, the idea of subsystem and supersystem is expounded, which can be construed as system hierarchy. The notion of system boundary aids in delineating a system of interest (whether subsystem or supersystem) in any given circumstance.

Systems engineers also explain that a system may be tangible wholes or ideas, and that system elements or parts can include people, hardware, facilities, and even policies. Systems engineers equally point out that the value added by the system as a whole, which transcends the independent contribution of its parts, is a consequence of the relationship among the parts. The trio of constraints, choice and demands constitutes powerful notions that aid systems engineers in system realisation. A vital notion of the system concept is the idea of open and closed systems (elaborated in the succeeding section)—the former receives input and produces output from the environment, whilst the latter does not. As will be discussed later in the chapter, these concepts are not unanimously shared within the systems engineering community. Figure 13.2 represents the concept of a typical (open) system in systems engineering.

13.4.2 Closed and Open Systems

Closed and open systems are indispensable concepts in systems engineering. The criterion for establishing whether a system is closed or open is the relationship it has with its environment. Within a systems environment, energy and matter, which can either be inputs or outputs, subsist. A system

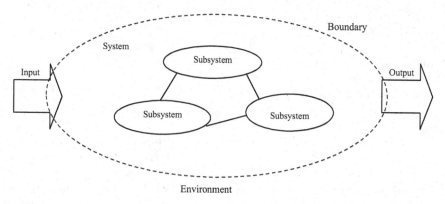

Figure 13.2 An open system

is considered closed if it does not allow the exchange of inputs and outputs with its environment. An open system, by contrast, allows such interchange. The understanding of closed systems differs slightly across scientific disciplines which deal with systems. In the physical and chemical sciences, for example, a closed system is thought of as a system that allows energy transfer without exchange of matter. Hence, a covered cooking pot is considered a closed system since vapour is trapped but heat energy gets released from the sides of the container.

Although knowledge of closed systems is important, systems principles and ideas are primarily premised on the concept of open systems. The systems notions and features that will be discussed shortly, such as complexity and emergence, can be comprehensible only in the context of an open system. These wide-ranging characteristics require an exchange of inputs and outputs to materialise. Therefore, it could be posited that the systems of interest in systems engineering and most systems disciplines are basically open systems. Examples of open systems thus abound including Earth, human activity systems (such as urban traffic and national economies) and large engineering projects (such as Solar Energy Generating Systems in California and the International Space Station).

13.4.3 Complexity

Another important concept in systems engineering is complexity. Complexity is frequently stated as one of the triggers of systems engineering, and how to address complexity informs systems engineering methods and techniques. Thus, at the heart of systems engineering is the requirement to manage complexity. Various definitions of complexity have largely distinguished it from the term 'complicated'. Regarded as highly relevant for the 21st century, complexity is considered a characteristic of such phenomena as insect colonies, national economies, urban traffic, the Internet, the human brain and so on. Complexity is framed around the trio of variety, connectedness and disorder (Hitchins 2007). Complexity has also been explained in terms of logical depth, thermodynamic depth and fractal dimension (Mitchell 2009). Logical depth concerns difficulty in constructing an object, while thermodynamic depth entails the interplay of thermodynamic and informational resources in the production process. Fractal dimension is a measure of self-similar objects replicable at every magnification level of an object (Mitchell 2009). These efforts to pin down complexity reinforce an all-important rule that no complex state of affairs can be understood fully. Hence, complexity shares overlapping similarities with the ideas of diversity, unpredictability, nonlinearity, adaptability and connectivity in the Wittgensteinian family resemblance sense. The distinction between complicated and complex is usually illustrated by a continuum of three elements as depicted in Figure 13.3. System engineering is greatly enriched by these insights.

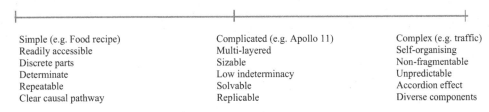

Simple (e.g. Food recipe)	Complicated (e.g. Apollo 11)	Complex (e.g. traffic)
Readily accessible	Multi-layered	Self-organising
Discrete parts	Sizable	Non-fragmentable
Determinate	Low indeterminacy	Unpredictable
Repeatable	Solvable	Accordion effect
Clear causal pathway	Replicable	Diverse components

Figure 13.3   Continuum of system complexity

13.4.4 Emergence

Emergence is another useful notion arising from complexity in systems engineering, inspired by theories from the biological, physical and chemical sciences. In the biological sciences, an example of emergence is life—the very essence that separates living things from inanimate ones.[1] When a live frog is dissected, its life is not restored by merely reassembling the dismembered parts. Failure to resuscitate the anatomised amphibian by reintegrating its parts explains the emergent nature of life. Neither the individual organs, genes and cells nor even the collection of these constituents exclusively accounts for the spark of life. Wetness and consciousness are further examples of emergent properties. Their manifestations cannot be traced to either hydrogen or oxygen molecules (in the case of wetness) or to either neurons or synapses (in the case of consciousness). These insights have contributed immensely to systems engineering. Emergence is a product of interrelationship between objects forming a whole. Simply bringing items together devoid of any meaningful interaction will not yield emergence. To imagine any outcome from a collection of lamp, laptop, mobile phone, pen, books and speakers is impossible. In systems engineering, a suggested application of emergence is to assemble carefully selected parts and cause among them an interaction that will yield desirable emergent properties. The manipulation of interrelationship between constituent parts of a system is a goal that continues to drive systems design. Honda engineers recently designed cars that can resist the accordion effect (Quick 2012)—an emergent phenomenon in vehicular traffic that ensues from speed fluctuations in vehicles mushrooming into a gridlock as the effect propagates down the line.

13.5. Reductionism, Holism and Systems Thinking

Reductionism, holism and systems thinking are essential to systems engineering. These ideas have contributed to the study and understanding of the mechanisms of complex reality in the world as well as how to approach problem-solving involving such complexities.

13.5.1 Reductionism and Holism

The narrative of reductionism and holism features both classical periods in which natural philosophers espoused holistic principles to explain reality and the Enlightenment age in which reductionism was the default approach to investigation. The Aristotelian quote about the whole being greater than the sum of its parts has already been mentioned, which indicates traces of holistic ideals at the time of the classical philosophers. René Descartes (1596–1650) popularised the reductionist method so intensely that his name became an eponym of its full expression: Cartesian reductionism. Reductionism and holism diverge on epistemology, ontology and methodology (Fang and Casadevall 2011).

As an epistemology, reductionism considers relationships among scientific disciplines with the object of reducing one to the other (e.g. explaining biology in terms of physics and chemistry). Reductionism as ontology is concerned with conceiving reality as essentially constituted by discrete elements and/or individual components. Methodological reductionism is an approach to understanding and knowledge based on simplistic analysis. This is perhaps the commonest rendition of reductionism; an approach to problem-solving that proceeds by breaking down a problem or system into its constituent parts. It is based on the assumption that understanding ensues from disaggregating a problem. For most of modern science, reductionism informed intellectual inquiry that led to physical bodies being reduced to atoms, electricity to electrons, etc. The goal of reductionism is to explain and possibly predict the behaviour of a system or problem through separation and simplification of parts. Given its successful application in problem-solving since the Enlightenment, reductionism is still relevant in systems engineering. The fact that systems engineering is considered "the art and science of creating whole solutions to complex problems" (Hitchins 2007: 91) does not nullify reductionist thinking. Indeed, focusing on a subsystem before considering its containing system can be regarded as a form of reductionism, especially from the perspective of the system.

With respect to holism, knowledge, problems and systems are better acquired, explained and understood as wholes. Instead of attempting to reduce one scientific field to another, holistic epistemology argues for metadisciplinarity in which all disciplines are embodied by the universe of knowledge. Such stance results from a holistic ontology which stresses the existence of reality in terms of irreducible entities or ensembles. The goal of the holistic method is to ensure that interaction, which is ineluctably lost in reductionism, is eventually accounted for. The philosophies of reductionism and holism are methodologically differentiated by analysis and synthesis respectively. The holistic viewpoint is prominently upheld in systems engineering as it guides the delivery of complex systems. Holism is typically invoked in the comprehensive consideration of the pre- and post-implementation behaviour of a system. Since holism defines a system's property not manifested by individual system elements, systems engineering leverages this strength to anticipate the performance of an engineered system. For example, the air defence capability of a nation is a complex system that involves economic, political and technological issues. To achieve a robust air defence system, those myriad issues and components have to be holistically captured and synthesised through detailed systems engineering processes.

13.5.2 Systems Thinking

Applying systems ideas to problem-solving endeavours is called the systems approach, and system thinking is its key cognitive tool. As the term suggests, systems thinking is a way of thinking or engaging with the world based on the systemic perspective. Such thinking expectedly harnesses the concepts of complexity, holism, emergence and open system amongst several others. Systems thinking has been characterised variously, albeit within a framework of system principles and philosophy. Accordingly, definitions of systems thinking advanced in the literature have been framed around the need to consider wholes as opposed to individual parts. Peter M. Senge defines systems thinking as a "framework for seeing interrelationships rather than things [and] for seeing patterns of change rather than static snapshots" (Senge 2006: 69). Another characterisation of systems thinking from the literature is "analysis, synthesis and understanding of interconnections, interactions, and interdependencies that are technical, social, temporal and multilevel" (Davidz 2006: 44) Similarly, systems thinking is "a vantage point from which you see a whole, a web of relationship [with] events seen in the larger context of a pattern that is unfolding over time" (WF 2008: 1).

Systems thinking essentially calls for a conceptual departure from the conventional approach to problem-solving, which has been reductionist or mechanistic, with systems typically reduced to their constituent parts to aid analysis. Early systems thinking might have unwaveringly challenged reductionist thinking, insisting on holistic thinking at all times. However, a more nuanced position with pluralistic undertones is favoured by contemporary systems thinking scholars (Cabrera 2006). Increasingly, system thinking is represented as "balanced thinking" characterised by both holistic and reductionist approaches. The idea is that both approaches can prove useful depending on the required level of abstraction in a problem-solving activity. A situation might arise that will require first the use of a reductionist method followed by a holistic approach or vice versa. A typical example is an exploratory research into the influence of socio-religious factors on sustainability acceptance in a society. The research could begin with a qualitative investigation of a small sample size (reductionism) seeking stakeholders' perception on the idea of sustainability vis-à-vis their social and religious values. A quantitative study could then be undertaken to ascertain the spread of the qualitative findings across the whole community (holism). Hence, portraying systems thinking as an essentially anti-reductionist framework not only fails to appreciate its inherent pragmatism but also constrains its utility. A complementary relation between holistic and reductionist perspectives is therefore the preferred view in systems thinking.

The pluralistic outlook of systems thinking further recommends thinking creatively and broadly, having multiple perspectives and condoning ambiguity and uncertainty (Behl and Ferreira 2014). Part of the motivation for systems thinking is to facilitate the management of complexity. As such, system thinking is useful in complex decision-making, identifying causality, understanding dynamic behaviour, identifying feedback and recognising interconnections within systems. Tools which aid in systems thinking have been developed within four broad categories delineated by Daniel Kim (2016): brainstorming tools, dynamic thinking tools, structural thinking tools and computer-based tools. Systems mapping and cause-and-effect diagrams are examples of brainstorming tools. Examples of dynamic thinking tools include graphical analysis aids such as frequency plot and time series plots in the form of histograms, dot plots, etc. Tools in the structural thinking category include graphical function diagrams and conceptual models.

System thinking is typically divided into hard systems thinking and soft systems thinking. This dichotomy is based on the nature of the problem or system of interest and the suggested problem-solving method. Hard systems thinking describes thinking intended to address structured problems, i.e. problems for which the existing state and desired state are defined. The traditional engineering approach is often associated with hard systems thinking, although the principle of holism is an important refinement. Emphasising the distinction between hard and soft systems thinking, Checkland (2000) asserts that the former is similar to thinking that is natural to design engineers. Such thinking, according to Checkland, occurs when there is a desired state, S_1, and a present state, S_0, with alternative ways of moving from S_0 to S_1. The mental efforts exerted in arriving at an appropriate solution to such a defined problem constitute hard systems thinking. The hallmark of hard systems thinking is therefore reduced ambiguity between the problem situation and problem solution. For example, remotely piloted vehicles (RPVs) such as the Raptor, which is shaped like a bird of prey indigenous to the area of action, or the Dragonfly, which resembles a real dragonfly with its wings beating rapidly to maintain hovering flight, can be developed using hard systems thinking. The targeted RPV functions of remote identification and engagement as well as minimal interference with delicate ecosystems and prevention of casualties to own and friendly forces are the desired state, S_1. The various methods and activities undertaken to produce these vehicles are rooted in hard systems thinking.

Soft systems thinking is the opposite of hard systems thinking. It denotes an approach to addressing an unstructured or ill-defined problem or system. A system or problem is termed 'ill-defined' or 'unstructured' if it cannot be explicitly delineated (Checkland 2000). The term 'soft' in soft systems thinking suggests human activity which contributes to the fuzziness of a problem situation. Soft system thinking thus deals mostly with sociotechnical systems involving the interaction of people, the environment and technology. Examples of soft problems are congested urban transportation systems and inefficient urban drainage systems. The evident role of human activity in aggravating these problems must be factored into any problem-solving attempt, which can be facilitated by a soft systems thinking approach. Systems engineering is therefore undergirded and enriched by the systemic ideals inherent in systems thinking. The tools of systems thinking mentioned earlier are extensively employed in systems engineering.

13.6. Representations of Systems Engineering

As early as 1969, criticisms of systems engineering began to emerge mostly focusing on its claims to versatility (Frosch 1993). Conflicting remarks were made about the origins, ambitions and limits of systems engineering. Similarly, a discourse analysis of systems engineering literature and qualitative data obtained from sampling a UK-based systems engineering community found three representations of systems engineering (Akeel and Bell 2013). These were systems engineering as *ex-novo*, as good engineering and as a metamethodology. The investigation further revealed that the three discourses share the systemic perspective but diverge on boundaries of engineering knowledge and limits to the engineering method.

Systems engineering as *ex novo* discourse conceives systems engineering as a new and distinct discipline but also as fundamentally a technical field. This view of systems engineering occurs variously in the narration of the concept, origin and purview of systems engineering. The origin of systems engineering is situated in the first half of the twentieth century following the Second World War, while the purview of the discipline is confined to the technical community. The technical nature of systems engineering is emphasised by reference to its espousal in military missile and radar developments, and also in the NASA projects. Furthermore, the view of systems engineering as something new is maintained in the description of the role and training requirement of a systems engineer. Systems engineers are shown to perform essentially technical roles and to acquire their training formally. In addition, the epistemological position and worldview of systems engineering as *ex novo* are empiricism and hard systems thinking respectively. The *ex-novo* discourse is basically intended to depict systems engineering discipline as an unprecedented field of knowledge confined to hard engineered systems (Akeel and Bell 2013).

Systems engineering as good engineering presents the field of systems engineering as simply conventional engineering done properly. This view is maintained in the narration of the concept, origin and purview of systems engineering. The origin of systems engineering is traced to as far back as the beginning of engineering practice with references to the Egyptian pyramids and Noah's Ark. Similarly, the purview of systems engineering in the good engineering discourse extends from the technical domain to the social issues of engineered systems. This discourse views engineering through the prism of sociotechnical systems stressing the systems engineer's role as involving both technical and social aspects of a project. The requirement for acquisition of special academic training is downplayed to promote post-academic experience. In addition, the epistemology and worldview espoused in the systems engineering as good engineering discourse are empiricism and rationalism, and hard and soft systems thinking respectively (Akeel and Bell 2013).

In the systems engineering as metamethodology discourse, systems engineering is conceived as an all-encompassing problem-solving framework applicable to any problem situation. The origin of systems engineering is depicted as a consequence of eclecticism with ideas contributed from various disciplines and sources. This discourse of systems engineering envisions systems engineering as accommodating all kinds of problems and transcending familiar spheres of thinking—technical, social, sociotechnical, and human activity. Also, the metamethodology discourse prescribes the role, training, epistemology and worldview of systems engineering in distinguishable terms. Systems engineers are professional 'problem-solvers' who use miscellaneous tools in a generic paradigm to create tangible solutions to complex world problems. In the same vein, the discourse accentuates a multidisciplinary training requirement for the systems engineer. Three epistemological positions are espoused in the metamethodology representation, namely empiricism, rationalism and pragmatism, while the worldview amenable to the discourse is a combination of hard and soft systems thinking—a metamethodological worldview (Akeel and Bell 2013).

These conceptions of systems engineering are evidently grounded in systems engineering literature. Each of the three discourses derives its justification from a shared language of systems and systems engineering philosophy. The discourses also align with the history of the systems engineering community represented by INCOSE, which, at inception, allowed multiple perspectives on systems engineering amongst attendees with a view to benefiting from their enriching potentials (Lake 1996). Anticipating conceptual homogeneity over time, the organisers of the first INCOSE meeting did not attempt to define systems engineering, but rather tasked participants to submit individual definitions of systems engineering, which were highly varied. The three discourses emerging from the discourse analysis could therefore point to these multiple perspectives with which the discipline started. Thus, the initial differing views of systems engineers may have survived the almost-three-decade history of the systems engineering community.

13.7. Systems Engineering as Engineering Philosophy

What qualifies systems engineering as engineering philosophy is a central question in this chapter. The derived definition of engineering philosophy is *an introspection and utilisation of engineering concepts, methods, cognitive structures, objective manifestations and limits to explain the natural and social world.* The hallmark of engineering philosophy based on this working definition is its endogenous point of departure that extends outward to facilitate an understanding of the world. Systems engineering has been shown to embody myriad subtleties through the espousal of several philosophies and especially systems thinking. Accordingly, a nexus between systems engineering and engineering philosophy can be established.

Systems engineering can be considered an engineering philosophy in the light of its provenance. As previously highlighted, systems engineering developed through a deliberate introspection of engineering concepts, processes and methods in the face of increasing complexity. The three discourses of systems engineering all agree on the triggering influence of complexity[2] and the need for engineering reflection in their narratives of systems engineering origins. Introspection of engineering coupled with the paradigmatic role of explaining the world using engineering criteria is a key feature of engineering philosophy, and on this basis, systems engineering qualifies as engineering philosophy. In the same vein, systems engineering addresses cognitive structures in engineering epistemology through the application of systems thinking. The use of both hard and soft systems thinking is fundamental to the three representations of systems engineering. The elastic concept of system which the metamethodology discourse of systems engineering adopts is in tandem with the depiction of engineering philosophy as a mindset and an orientation. The utility of systems thinking as an essentially conceptual or cognitive framework to manage complexity qualifies systems engineering as engineering philosophy.

In the metamethodology construct of systems engineering, the engineering view of the world is expanded to all facets of human life. The assertion that systems engineering is an art evidenced by creative thinking of possible solutions to problems presents a paradigm for understanding the world. This is in congruence with the *homo faber* view of humans as essentially creators or makers who make sense of the world through artefact-making. The universal problem-solving framework of systems engineering suggests an engineering way of being-in-the-world that can effectively address all global challenges. All problems are resolvable by engineering rightly the right systems. Another compelling pointer to the qualification of systems engineering as engineering philosophy are the underpinning notions of emergence, reductionism and holism. These concepts require cognitive efforts for their application in systems engineering. Emergence, for example, has been a particularly useful concept in systems engineering employed to anticipate consequences at the level of a system. To predict a system's behaviour, however, is not wholly achievable given the ideal of system darkness, which states that no system can be completely known (Skyttner 2001). This principle is rooted in the idea that the best representation of a complex system is the system itself, and that any representation other than the system itself is inherently flawed. Thus, until a system is created, it is impossible to anticipate its behaviour. This is a familiar theme in epistemology reminiscent of Socrates's belief that "the only true wisdom is in knowing you know nothing" (Lavine, 1984: 15). Such a philosophical undertone justifies the framing of systems engineering as engineering philosophy.

The other systems engineering idea suggestive of engineering philosophy is the concept of a system boundary. The boundary of a system is not always defined especially in the case of sociotechnical systems. Arriving at a decision to establish a system boundary has been recognised to involve subjective constructions of reality. Such activity is akin to metaphysical inquiry of depicting the nature of reality. To determine the boundary of a sociotechnical system, for example, a systems engineer has to decide on the system of interest, its enabling system, degree of influence from surrounding systems and so on. The fact that the systems engineering discipline admittedly deals with such metaphysical issues, especially with abstract systems, qualifies it as engineering philosophy. Additionally, the terms employed to describe a system such as parts, wholes, layers, connections and processes often lend themselves to human perceptual skills which are significant in engineering philosophy.

Through the prism of engineering philosophy, we can appreciate that systems engineering generates a spectrum of views because systems engineers construct the nature of their discipline and its broader contribution to humanity differently. The three narratives of systems engineering all demonstrate introspection on the nature of engineering. Since the increased complexity of engineered systems necessitates reflection within engineering practice, the three systems engineering discourses can be rationalised as addressing various manifestations of complexity. Accordingly, the *ex-novo* construct addresses technical complexity, the good engineering discourse deals with complexity at the level of sociotechnical systems, whilst the metamethodology representation addresses generic complexity.

13.8. Conclusion

Systems engineering has been involved in some of the most significant engineering projects of the last century. Its power as a set of methods and procedures for delivering complex projects is based on specific ontological and methodological frameworks, and it has provided the basis for broader reflection on its potential as the foundation for generic problem-solving. In addition to complexity being suggested as the catalyst for the rise of systems engineering, the centrality of introspection as a strategy for problem-solving is evident in the three constructs of systems engineering. The ambitious goal of generic problem-solving subjects the discipline to a plethora of intellectual queries of which justification for the universality claim remains pertinent. How systems engineering serves a universal problem-solving role is not readily intelligible. However, viewing systems engineering as engineering philosophy—introspection within engineering about the nature of its methods and relationship to the wider society and humanity—has proved useful in making sense of the potential generic utility of systems engineering.

Characterising systems engineering as engineering philosophy reveals a particular style of intellectual work within the discipline, beyond the instrumental development of a set of methods for delivering engineering projects and programmes. As an engineering philosophy, systems engineering may well facilitate the creation of whole systems, but it also aids the human ability to make sense of complex modern societies. Furthermore, the discourses within systems engineering are better appreciated through the lens of engineering philosophy. Systems thinking and core concepts such as complexity, emergence and holism show efforts to learn from other disciplines. Such dynamism and syncretism are characteristic of a discipline transcending conventional monodisciplinary boundaries. Systems engineering as engineering philosophy leverages the critical perspectives of the social sciences and humanities fields to capacitate systems engineers for deeper intellectual engagement and reflection on engineering.

Related Chapters

Chapter 1: What Is Engineering? (Carl Mitcham)

Chapter 53: Engineering and Contemporary Continental Philosophy of Technology (Diane P. Michelfelder)

Chapter 54: Engineering Practice From the Perspective of Methodical Constructivism and Culturalism (Michael Funk and Albrecht Fritzsche)

Notes

1. Life is a puzzling phenomenon that defies universal definition, and the age-old question of "What is life?" remains profoundly unanswered.
2. What may be in contention is the type of complexity, as *ex-novo* stresses technical complexity and good engineering emphasises sociotechnical, while metamethodology complexity is uncategorised.

References

Adams, K.M. and Mun, J.H. (2005). The Application of Systems Thinking and Systems Theory to Systems Engineering. In *Proceedings of the 26th National ASEM Conference: Organizational Transformation: Opportunities and Challenges*. Rolla, MO: American Society for Engineering Management, pp. 493–500.

Akeel, U. and Bell, S. (2013). Discourses of Systems Engineering. Engineering Studies, 5(2), 160–173. doi:10.1080/19378629.2013.795575.

Alcatel-Lucent (2010). *About Bell Laboratories—History and Top Ten Innovations*. www.alcatel-lucent.com/wps/portal/bell-labs. Accessed June 15, 2010.

Bahill, A.T. and Gissing, B. (1998). Re-evaluating Systems Engineering Concepts Using Systems Thinking. *IEEE Transactions on Systems, Man and Cybernetics, Part C (Applications and Reviews)*, 28, 516–527. doi:10.1109/5326.725338.

Behl, D.V. and Ferreira, S. (2014). Systems Thinking: An Analysis of Key Factors and Relationships. *Procedia Computer Science*, 36, 104–109. doi:10.1016/j.procs.2014.09.045.

Brill, J.H. (1998). Systems Engineering—A Retrospective View. *Systems Engineering*. doi:10.1002/(SICI)1520–6858(1998)1:4%3C258::AID-SYS2%3E3.0.CO;2-E.

Cabrera, D. (2006). *Systems Thinking*. Ithaca, NY: Cornell University Press.

Calvano, C.N. and John, P. (2004). Systems Engineering in an Age of Complexity. *Systems Engineering*, 7, 25–34. doi:10.1002/sys.10054

Checkland, P. (2000). Soft Systems Methodology: A 30-year Retrospective. *Systems Research and Behavioral Science*, 17, S11–S58.

Davidz, H.L. (2006). *Enabling Systems Thinking to Accelerate the Development of Senior Systems Engineers*. Massachusetts Institute of Technology. http://esd.mit.edu/staging/HeadLine/pdfs/davidz1_incose_092004.pdf.

Fang, F.C. and Casadevall, A. (2011). Reductionistic and Holistic Science. *Infection and Immunity*, 79(4), 1401–1404. doi:10.1128/IAI.01343-10

Frosch, R.A. (1993). A Classic Look at Systems Engineering. In F.T. Hoban and W.M. Lawbaugh (eds.), *Readings in Systems Engineering*. Washington DC: NASA, pp. 1–217.

Hall, A.D. (1962). *A Methdology for Systems Engineering*. London: van Nostrand.

Harper, D. (2017). *Etymonline.com*. www.etymonline.com/index.php?term=system. Accessed January 10, 2017.

Hillier, J. (2010). The Philosophy of Engineering: A Critical Summary. *Proceedings of the Institution of Civil Engineerscivil Engineering*, 163(2), 91–95. doi:10.1680/cien.2010.163.2.91.

Hipel, K.W., Fang, L. and Bristow, M. (2015). System of Systems Thinking in Policy Development: Challenges and Opportunities. *Contemporary Issues in Systems Science and Engineering*, pp. 21–70. doi:10.1002/9781119036821.ch2.

Hitchins, D.K. (2007). Systems Engineering Philosophy. *Systems Engineering: A 21st Century Systems Methodology*, (1986), 83–101. doi:10.1002/9780470518762.ch4.

INCOSE (2015). *International Council on Systems Engineering*. www.incose.org/. Accessed December 15, 2015.

Kast, F.E. and Rosenzweig, J. (1970). *Organization and Management, A Systems Approach*. New York: McGraw and Hill.

Kim, D. (2016). *The Systems Thinker—Palette of Systems Thinking Tools—The Systems Thinker*. https://thesystemsthinker.com/palette-of-systems-thinking-tools/. Accessed February 17, 2017.

Kopach-Konrad, R., Lawley, M., Criswell, M., Hasan, I., Chakraborty, S., Pekny, J. and Doebbeling, B.N. (2007). Applying Systems Engineering Principles in Improving Health Care Delivery. *Journal of General Internal Medicine*, 431–437. doi:10.1007/s11606-007-0292-3.

Kossiakoff, A., Sweet, W.N., Seymour, S.J. and Biemer, S.M. (2011). *Systems Engineering Principles and Practice, Computer Methods and Programs in Biomedicine*. doi:10.1016/j.cmpb.2010.05.002.

Lake, J. (Jerry) G. (1996). Unraveling the Systems Engineering Lexicon. *INCOSE International Symposium*, 6(1), 506–515. doi:10.1002/j.2334-5837.1996.tb02046.x.

Lavine, T. (1984). *From Socrates to Sartre: The Philosophic Quest*. London: Bentham Publishers.

Mackey, W.F., Crisp, H., Cropley, D., LongPresident, J., Mayian, S. and Raza, S. (2003). 1.2.1 The Role of Systems Engineering in Combating Terrorism. *INCOSE International Symposium*, 13(1), 146–164. doi:10.1002/j.2334-5837.2003.tb02607.x.

Mitcham, C. (1994). *Thinking Through Technology: The Path Between Engineering and Philosophy*. London: The University of Chicago Press, Ltd.

Mitchell, M. (2009). *Complexity: A Guided Tour*. Oxford: Oxford University Press. https://read.amazon.co.uk/.

NASA (2017). *NASA Systems Engineering Handbook—NASA Sp-2016–6105*. 2nd ed. Washington, DC: CreateSpace Independent Publishing Platform.

OED (2010). *'System' Etymology, Online*. www.etymonline.com/index.php?search=suffering&searchmode=none. Accessed January 18, 2016.

Quick, D. (2012). *Honda Develops Technology Designed to Prevent Traffic Jams*. http://newatlas.com/honda-traffic-congestion-detector/22327/. Accessed January 23, 2017.

Rodgers, C. (2002). Defining Reflection: Another Look at John Dewey and Reflective Thinking. *Teachers College Record*, 104(4), 842–866. doi:10.1111/1467-9620.00181.

Saarinen, M.M.E., Hamalainen, R.P., Martela, M. and Luoma, J. (2008). Systems Intelligence Thinking as Engineering Philosophy. *Workshop on Philosophy & Engineering (WPE2008)*.

Sage, A.P. (1992). *Systems Engineering*. John Wiley & Sons, Inc.

Senge, P.M. (2006). The Fifth Discipline: The Art and Practice of the Learning Organization. *5Th Discipline*, 445. doi:10.1016/0024-6301(95)90931-1.

Shishko, R. (1995). *NASA Systems Engineering Handbook*. Washington, DC: National Aeronautics and Space Administration.

Skyttner, L. (2001). *General Systems Theory*. New York. doi:10.1142/9789812384850.

Van de Poel, I. and Goldberg, D. E. (2010). *Philosophy and Engineering: An Emerging Agenda*. Eds. I. van De Poel and David E. Goldberg. London: Springer Science & Business Media.

Weigel, A.L. (2000). An Overview of the Systems Engineering Knowledge Domain. *Research Seminar in Engineering Systems*, Fall, 8, October.

WF (2008). Systems Thinking in Schools—A Waters Foundation Project. *Waters Foundation*, p. 1. www.waters-foundation.org/.

Züst, R. and Troxler, P. (2006). *No More Muddling Through: Mastering Complex Projects in Engineering and Management*. doi:10.1007/978-1-4020-5018-3.

14

ASSESSING PROVENANCE AND BIAS IN BIG DATA

Brent Mittelstadt and Jan Kwakkel

14.1. Introduction

Data can now be generated by ever more sources, aggregated and assessed at an unprecedented rate and scale, heralding what is often referred to as the era of 'Big Data' (Bail 2014).[1] At the heart of many of these datasets are people, whose behaviours and preferences are increasingly recorded in meticulous detail, and analysed to provide personalized, linked-up experiences. Analysis of Big Data is said to provide unprecedented opportunities for business, policy, and research, allowing for products, services, policies, and decision-making strategies that can be better targeted, more effective, and insightful (IBM 2014).

For some, the mere proliferation of data and analytics inevitably leads to more accurate, objective, and efficient knowledge discovery and decision-making (Crawford 2013; Crawford et al. 2014; Fairfield and Shtein 2014; Puschmann and Burgess 2014). This attitude suggests that the epistemological challenges traditionally facing analytics and data-intensive sciences disappear as our ability to collect and analyse data grows. This view is, however, misguided (Mittelstadt and Floridi 2016). Problems associated with the techniques used to derive meaningful information from data (e.g. statistics; Huff 1954) or data visualization (Tufte 1983) inevitably accompany the proliferation of Big Data, and arguably become even more pressing due to the scale and variety of data involved.

Unfortunately, these type of well-known epistemological challenges are not the only difficulty facing Big Data. Historical and real-time behavioural data now drive decision-making at scale, underlying the ubiquitous platforms and services characteristic of modern life in mature information societies (Floridi 2016). These datasets necessarily reflect reality and society as it exists, both for good and bad. Biases and usage gaps observed in society and social institutions are inevitably reflected in datasets describing them (Crawford 2013). When applied at scale, for example when used to train machine learning decision-making models, these biases and gaps can easily and inadvertently be reinforced and deployed at scale, causing harm to new individuals and groups at scale.

Given these epistemological challenges, understanding the history, strengths and weaknesses, and blind spots of datasets and analytics takes on new importance in the age of Big Data. Thankfully, work in the areas of data and process provenance, fairness- and discrimination-aware data mining aim to achieve precisely this epistemologically rich and socially aware view of Big Data.

In this chapter we connect the established field of data and process provenance, which address the history of a dataset and analytic processes to establish the reliability of outputs, to the burgeoning fields of fairness- and discrimination-aware data mining, which similarly assess the history and completeness of input data and learned models to address biased or unfair outputs. We start by introducing Big Data analytics through the idea of a 'data chain', which maps the collection and analytic timeline from the initial recording of observations to the eventual use of the derived knowledge. We then review major themes and recent work in the fields of data and process provenance, before connecting them to specific methods for assessing bias and fairness in datasets and machine learning models. We conclude by discussing evidence and remaining barriers to greater uptake of methods to assess data and process provenance in Big Data.

14.2. General Introduction to Big Data: the Data Chain

To clarify where epistemological challenges can arise in Big Data analytics, we look at the data chain from raw data to meaningful and useful information (Figure 14.1).

In each step of this data chain, assumptions have to be made by analysts. Other analysts involved elsewhere in the data chain do not know many of these assumptions. It thus remains unknown to what extent conclusions made at the end are due to the data itself, or due to choices made somewhere in the data chain.

14.2.1 Stage 1: Observation

In the observation stage, important choices are made with respect to the method or tools used. What are the methods or sensors that are used able to measure, and what do they miss? What is the quality of the sensors: what is the signal-to-noise ratio? What was the context in which measurements took place? That is, what external conditions might have influenced the data, or have caused or can explain specific patterns in the data?

An example pertaining to data on the flow of cars on a road system can clarify the relevance and importance of these questions. Are the GPS sensors precise enough to identify the lane being used by a car? Is the prevalence of sensors properly distributed over different types of vehicles? Is the time interval between measurements short enough to observe shifts in lanes in both time and place? Do we keep track of weather and accidents, both of which might influence observed driver behaviour? Are the clocks in the different sensors set to the same time? If any of these questions are answered negatively, there are data quality problems already in the first stage of the data chain.

14.2.2 Stage 2: Transformation

Often, it is not possible to send all the raw data gathered by the sensors directly. Instead, the data is often processed before transmission. For example, GPS sensors typically send data every few seconds. This can cause an implicit bias in the data and potentially relevant phenomena are lost because of it. Because of bandwidth concerns, data might also have to be compressed before it can be transmitted. Compression can be lossless, meaning that no data is lost. However, compression can also be lossy. This means that some of the data is being lost. A simple example is in image processing. The raw data coming of a high-end camera for a single photo might be roughly 30–50 Mb. A lossy compression of this data in the form of a jpg file might be one or two orders of magnitude smaller. However, colour information and image details are lost because of the compression.

Observation → Transformation → Transmission → Aggregation and combination → Use

Figure 14.1   The data chain

14.2.3 Stage 3: Transmission

When transmitting the data, often there are bandwidth problems. This is particularly the case if something interesting or unusual happens that creates additional data. This can result in delays in the transmission, or the accidental loss of data. An example is river runoff data, where during floods, the sensors might be washed away or not be functioning properly. In turn, data will be missing in the river runoff time series. This missing data is not uniformly distributed over time but concentrated in periods of extreme runoff, which are often the most relevant parts of the time series for decision-making.

14.2.4 Stage 4: Combination and Aggregation

Combination and aggregation typically take place in the most complex phase in the data chain. For most end users of the data, however, this phase is a black box where analysts apply their magic sauce. Combining data is often very hard because relevant metadata required to adequately link disparate data sources are missing. For example, in a recent project on container transport by inland barges, data was gathered from a multitude of public and private data sources. However, several key inconsistencies across the data sources made combining them difficult: (1) the timestamps of the data differed from one source to the next; (2) some inland terminals were missing from some data sources; (3) maps from the national government were not consistent with GIS data from other sources; and (4) the naming of bridges and sluices differed across data sources.

To address each of these inconsistencies, analysts are forced to make a judgement call. In some cases, the only choice is to leave the data out, but this might unintentionally bias the results. In others, it might be possible to reconcile the discrepancies between data sources on the basis of assumptions or expert judgment. The magnitude of the data in this project—over 20 data sources, and hundreds of thousands of records—meant manual inspection of the data was impossible. Missing data in a time series could be filled in using interpolation, but this approach will not reveal potentially relevant causes for the missing data. For example, if the missing data was due to a strike, the number of containers actually handled was 0, while the next day there would be a much higher number of handled containers. Simply leaving out the missing data, or filling it in through interpolation, in such cases is not justifiable.

Aggregation involves a similarly normative set of choices. Aggregation involves transforming data into something useful. Individual observations are transformed into averages, the number of observations is counted, correlations are calculated. In this, the choice of how to aggregate is critical. The hourly number of vehicles on a road over a long time period is something quite different from the hourly number of vehicles on the road between 07:00 and 20:00 over 20 Mondays outside holiday periods. The number of car accidents per year is something quite different from the average number of car accidents per transport kilometre or per movement. The meaning of these statistics changes again depending on whether we include accidents on the highways only, or also those taking place on other roads. When a newspaper headline states, 'the number of car accidents increases again after years of decline', it is completely obscure what this is based on. In fact, a slightly different aggregation of the data might have resulted in the opposite headline.

14.2.5 Stage 5: Use

Use is the final stage of the data chain. It is typically the phase that attracts most attention. However, there is a substantial risk that any insights produced might be due to choices made in the chain rather than being an accurate reflection of the data. This risk is being stimulated by the fact that increasingly, the combination and aggregation of data sources is left to automated algorithms, rendering the subjectivity of the process, or the ways in which inconsistencies are resolved and data points transformed, opaque when using the data. Moreover, it is not uncommon that data gathered for other purposes in unknown ways is used in Big Data analytics. Again, rendering opaque how data were gathered, transformed, and aggregated means that any conclusions rest on unclear epistemic grounds. Because of this, next we discuss data provenance approaches for addressing this risk.

14.3. Data and Process Provenance

According to the *Oxford English Dictionary*, provenance in general means the place of origin or earliest history of an object. Provenance is a well-established concept in the art world, referring to the record of ownership which is used as a guide to establish the authenticity and quality of a work of art. Increasingly, there is also an interest in provenance in data science (Freire et al. 2008; Moreau et al. 2011). The dominant motivation for considering provenance in data science is reproducibility, although validation and fostering trust are sometimes mentioned as well (Carata et al. 2014; see also Chapter 32, 'On Verification and Validation in Engineering' by Francien Dechesne and Tijn Borghuis, and Chapter 36, 'Trust in Engineering' by Philip J. Nickel, this volume).

In data science, provenance can focus on either the data itself or the analytic processes. Data provenance focuses on the inputs, entities, systems, and processes that influence data of interest. That is, it focuses on describing the data chain from initial observation to immediately before actual use. Data provenance should provide answers to questions such as where the data come from, who collected it and when, from which other data source(s) it was derived, and so on. Capturing data provenance ensures that analysts are able to assess the extent to which results are grounded in the data, or could be explained by choices made in the data chain.

In contrast, process provenance focuses on the causality chain or network which describes the workflow of how a data product (e.g. a chart, table or figure) was created: which input data were used, which sequence of transformations took place, how these were parameterized, which other resources (e.g., software, scripts) were used, and where in the data chain, when was this done and which version of mechanisms (e.g. a script) were used.

The provenance of the entire data chain is important to establish the validity of the process and outputs. In experimental sciences, this often takes the form of lab notebooks. In data science, provenance often takes the form of a more or less ad hoc practice or breaking down a computational task into a sequence of subtasks, each of which takes the form of a separate script or function.

There is a growing interest in formalizing or automating this ad hoc practice under the label of workflow (Reichman et al. 2011). A workflow is a formal representation of the sequence of steps taken to achieving a predefined computational task. Workflows can be seen as a directed graph, with processes as nodes, and directed edges describing the flow of data between processes (Freire et al. 2008; Moreau et al. 2011).

Data and process provenance are necessarily intertwined. Only by capturing both does it become possible to properly evaluate the final product, enable reproduction, and thus foster trust. Moreover, knowing the various data and process dependencies also means that it becomes easier to recreate a product if the underlying data is changed or an error in one of the components is discovered.

Presently, a variety of data provenance management and process provenance (i.e. workflow) management systems exist, such as VisTrails and Kepler. For a comprehensive overview, see Pérez et al. (2018). These systems all involve the automated capturing of provenance-relevant details including data processing steps, execution information, and any additional user specified annotation that provide a motivation for specific choices or an interpretation of intermediate results. These systems also offer some way of representing provenance information. The representational model is typically domain-specific and layered. A layered representation enables presenting provenance at different levels of abstraction. For example, one layer might capture the overall workflow, while another layer captures individual traces of the execution of individual steps of this workflow. Storing and querying is the third component of these management systems. Here different researchers are experimenting with domain-specific querying approaches, as well as querying mechanisms dictated by the storage format (e.g. SQL). That is, can a query of the provenance data be expressed in a language familiar to the user, or must it be expressed in a language that is related to the database in which the provenance data is being stored? The advantage of the former is that it enables the reuse of domain language and abstracts the storage system away from the user.

14.4. Assessing Data and Process Provenance in Machine Learning: Bias and Fairness

As described in Section 3, provenance is closely linked to the validity of data analysis. This is not, however, the only potential aim of data and process provenance. By identifying the history, context, assumptions, and gaps in a dataset, information about provenance can clarify the biases and fairness of a dataset, and how these characteristics can be adopted, reinforced, and evolved as the data is processed and used for different purposes.

In this context, it is worth noting that Big Data often involves machine learning. At a minimum, the description of the data chain, as well as data and process provenance, applies to the training and deployment of machine learning models. Given this link between Big Data and machine learning, in this section we examine a specific type of provenance analysis increasingly popular in machine learning, which examines bias and fairness in training data, learned models, and their outputs (e.g. classifications).

With regards to data provenance, biased and discriminatory outputs can emerge from many sources in the data chain. Statistical and subjective (i.e. taste-based) biases and prejudices can be implicit in datasets and inherited by machine learning models (Barocas 2014; Datta et al. 2017). According to Lepri et al. (2017: 614), input data may be biased, non-representative, or "poorly weighted", resulting in disparate impact or indirect discrimination. Romei and Ruggieri (2014) similarly identify three potential sources of bias in datasets: prejudice, which stems from a lack of information about both individuals and the groups to which they belong; statistical thinking, which is based on information about the behaviour or performance a group but not its individual members; and unintentionality, where information is available about individuals in the dataset but not about the impact of decisions upon them. Prejudice consists of the formation of beliefs regarding an individual or group prior to any interactions. Statistical thinking consists of the usage of pre-existing beliefs about a group in place of actual observations of an individual member of the group according to a targeted attribute. Unintentional discrimination consists of a lack of awareness of implicit bias and the effects of decisions.

Concerning process provenance, algorithms can produce feedback loops in which "biased training data can be used both as evidence for the use of algorithms and as proof of their effectiveness" (Lepri et al. 2017: 615). Profiling has, for example, been criticized for creating an evidence base for discriminatory treatment (Vries 2010). Further, the application of models in contexts beyond where they were originally designed can produce discrimination due to a poor fit between the model and ground truth (Lepri et al. 2017: 615). Discriminatory outputs can also unexpectedly emerge in the 'use' stage. A machine learning model can, for example, unexpectedly learn and use discriminatory decision-making rules, for example, by basing lending decisions on post codes which unintentionally act as a proxy for ethnicity or socioeconomic status. Discriminatory correlations can be discovered by the algorithm during training, resulting in a discriminatory model (Datta et al. 2017).

14.4.1 Pre-, In-, and Post-Processing Methods to Assess Bias, Fairness, and Discrimination

Fairness- and discrimination-aware data mining seek to prevent or detect unfair, discriminatory, or invasive outputs. Romei and Ruggieri (2014) offer an in-depth review of relevant techniques under the moniker of 'discrimination analysis'. Generally speaking, discrimination can be defined as an "unjustified distinction of individuals based on their membership, or perceived membership, in a certain group or category" (Romei and Ruggieri 2014: 583). Discrimination analysis consists of methods of 'discrimination discovery' and 'discrimination prevention'.[2] Concerning the former, discrimination discovery generally aims to uncover instances of discrimination in prior decisions, whereas discrimination prevention seeks to design, train, and correct models and their predictions to be non-discriminatory.

Concerning the latter, methods of discrimination prevention seek to identify and prevent discriminatory algorithmic decision-making. Romei and Ruggieri (2014) observe four overlapping strategies for discrimination prevention which reflect implementation at different stages of the data chain: (1) controlled distortion of training data (Stage 2, Transformation); (2) integration of anti-discrimination criteria into the classifier algorithm (Stage 2, Transformation, and Stage 4, Aggregation and Combination); (3) post-processing of classification models (Stage 2, Transformation, and Stage 5, Use); (4) modification of predictions and decisions to maintain a fair proportion of effects between protected and unprotected groups (Stage 5, Use).

With these general distinctions and strategies in mind, a further distinction can be drawn based on the stage of data processing at which the assessment occurs. Methods are often divided between pre-processing, in-processing, and post-processing approaches to identify biases in datasets, measure fairness in data processing and outputs, and detect and prevent discriminatory outcomes.

'Pre-processing' methods seek to prevent bias from ever entering the data chain by identifying bias in datasets and filling in identified gaps or otherwise modifying the data to prevent bias from being passed further down the chain. Training data can, for example, be distorted to prevent a model from being trained on biased data and thus learning unfair or discriminatory decision-making rules (d'Alessandro et al. 2017). Examples include modification of class labels for discriminated individuals, for example by changing the 'ethnicity' of particular cases in a dataset (Luong et al. 2011), reduction of indirect discrimination and proxy rules through perturbation of the training set (Hajian and Domingo-Ferrer 2012), and preservation of explanatory variables correlated with justified grounds for discrimination (e.g. job requirements) to ensure post-hoc analysis can assess the basis and justification of discriminatory outcomes (Zliobaite 2015).

Sources of discrimination and proxy variables can also be identified via patterns contained in sets of historical decisions. Pedreschi et al. (2008) and Ruggieri et al. (2010) describe how 'potentially discriminatory' (PD) classification rules can be extracted from historical datasets containing decisions or classifications that have themselves been discriminatory. Similarly, Feldman et al. (2015) describe a method to modify input data to preclude prediction of protected attributes, without a significant loss of signal in the modified attributes. An overview of sanitization techniques is offered by Kamiran and Calders (2012).

'In-processing' methods aim to modify the analytic process; for example, a trained model prior to deployment 'in the wild', to prevent biased or discriminatory outputs. Approaches often involve injection of anti-discriminatory or corrective criteria into the analytic process. In the case of machine learning, these criteria would be added to the classifier or learning algorithm (d'Alessandro et al. 2017). Such methods include training separate models for each protected group (e.g. a different classifier for different ethnic groups), which requires information on the protected attributes to be available; adding a latent variable to allow class value to be modelled without discrimination (Calders and Verwer 2010), modifying criteria for branching in a decision tree to ignore or account for protected attributes such as gender (Kamiran et al. 2010), and "regularization (i.e. a change in the objective minimization function) to probabilistic discriminative models, such as logistic regression" (Kamishima et al. 2012a; quote from Romei and Ruggieri 2014: 623). Automated systems that detect discriminatory criteria in a trained model and introduce appropriate modifications are also feasible, for instance to detect and correct disparate impact (Calders and Verwer 2010; Feldman et al. 2015; Kamishima et al. 2011; Tramer et al. 2015; Zemel et al. 2013). It is worth noting that discrimination based on proxy variables adds additional complexity to in-processing approaches; detection of proxies for sensitive attributes is therefore critical to the success of in-processing approaches (see Section 4.2).[3]

'Post-processing' methods occur at the very end of the data chain, once the outputs of the analytic process are known. Post-processing approaches are particularly appropriate when elements of the data chain are inaccessible or poorly interpretable, for instance to protect commercial interests (Burrell 2016; Mittelstadt et al. 2016). Such approaches involve post hoc modification of the learned model to remove discriminatory criteria, or correction of predictions to meet an external standard of outcome distribution across the affected population (d'Alessandro et al. 2017). Concerning the former, approaches to model modification include alteration of confidence intervals, weights (Pedreschi et al. 2009), or probabilities (Calders and Verwer 2010). Kamiran et al. (2012) suggest changing predictions falling close to a decision boundary, or those where a particular decision lacks strong support, to correct for statistical discrimination.

Concerning the latter, outputs can be examined for specific harms, for example discrimination against legally protected groups (Barocas and Selbst 2016; Kroll et al. 2016; Sandvig et al. 2014), and in some cases modified to redistribute outputs. Disparate impact can be detected with audit studies involving surveys, comparative testing of different decision-making inputs, models or classifiers (i.e. A/B testing), non-invasive data scraping, sock puppet audits, and crowdsourced audits involving multiple users of a system testing different inputs and reporting on their results (Sandvig et al. 2014). 'Code audits' are also feasible in some cases, although systems designed by large teams over time (which includes most major online personalization systems) are often prohibitive (Sandvig et al. 2014) due to the required human effort and expertise to "untangle the logic of the code" (Burrell 2016: 4–5). Further, audits often do not engage with the system's development history and design culture (Ananny 2016), which can undermine detection of latent bias (Friedman and Nissenbaum 1996; Hildebrandt 2011; Morek 2006). Ananny (2016) recommends complementary ethnographic studies to document the process and culture that created the system, as well as potential biases that may result which can then be targeted by detection methods.

The effectiveness of such post-processing 'black box audits' has recently been called into question (Datta et al. 2015; Lepri et al. 2017) in relation to Big Data analytics involving highly complex machine learning models. Many varieties of machine learning prohibit reconstructing the rationale of a decision via replication. Approaches to machine learning based on trial and error or estimates with random number generation will, for instance, produce similar but not identical results when given identical training data. Algorithms that use such probabilistic techniques will perform unpredictably even if the code, inputs, outputs, operating environment, and other relevant information is fully transparent. Finally, bias and discrimination often emerge only when an algorithm processes particular data, meaning a code audit alone would detect a limited range of problems (Sandvig et al. 2014). 'Active learning' or models that are updated over time therefore require iterative scrutiny.

No single method for detecting and correcting discrimination is sufficient by itself. As discrimination can emerge due to historical biases as well as unanticipated pattern recognition or the discovery of proxies for protected attributes, both pre-processing of data and sets of historical decisions as well as in-processing correction of identified discriminatory classifications are essential. In many cases the systems in question will effectively be 'black boxes', owing either to technical or intentional opacity, necessitating a 'black box' approach involving systematic examination of the system's inputs and outputs where in-processing access is not available. Each approach has its limits, but together, pre-, in-, and post-processing methods can provide a holistic assessment to identify when automated classification and decision-making systems result in unjustified discrimination.

14.4.2 Fairness Methods and Metrics

In the assessment of fairness in data analytics and machine learning, numerous quantitative metrics are available to measure fairness in datasets and decision-making records (Kamishima et al. 2012b). Choosing between metrics in effect determines which inputs and outcomes in a given set will be seen as normatively problematic, or unfair (Pedreschi et al. 2012). As data inevitably reflects the inequalities and biases present in society, a key challenge for fairness-aware data mining is to employ methods that consistently detect subtle biases inherited by algorithms from training and inputs (Barocas and Selbst 2016; Romei and Ruggieri 2014). This is not a simple task, as unfairness in algorithmic decision-making can arise from many sources, owing, for example, to historically biased distributions, selection bias (non-representative training data), and prediction bias (predictions based on protected attributes or proxies) (Kusner et al. 2017). Each potential source imposes different requirements on fairness detection and prevention methods.

Diversity in fairness metrics should be both expected and encouraged. Many disciplines have a rich history of discussion of fairness and equality. Ideally, lessons from philosophy, political science and other social sciences, and other fields can be drawn upon by the data mining and machine learning communities to ensure datasets, models, and decision sets can be examined from multiple angles, representing different fairness norms and values. Such an approach is essential given the broad scope of Big Data, with possible applications in a myriad of sectors, each with a unique professional, organizational, and social norms and history, and related requirements. Consensus should not be expected, as perspectives of "what constitutes fairness changes according to different worldviews" (Lepri et al. 2017: 618). Put simply, fairness metrics appropriate in one sector may be both legally and ethically inappropriate in others. Critical investigation of sector-specific problems which produce unfair outcomes, and the local requirements that arise from them, is essential to determine sector-appropriate fairness metrics.

Existing fairness metrics can be broadly and roughly classified into four groups: group fairness, individual fairness, fairness through awareness, and counterfactual fairness.

'Group fairness', or statistical parity, measures the distribution of outcomes across groups in a population. In a population consisting of multiple groups, fairness as statistical parity requires outcomes to be equally distributed across the groups (Calders and Verwer 2010; Kamishima et al. 2012b).[4] Variations incorporating tolerances not requiring strict equality across groups are also common. 'Disparate impact' is an approach to statistical parity where a buffer is introduced for the distribution of outcomes. In American anti-discrimination law, disparate impact is generally measured by the '80% rule', which requires the ratio between groups of all possible outcomes to be at least 4:5, or 80%. The notion applies primarily to decision-making in housing and employment but can be used in other sectors as well where discrimination against legally protected groups (e.g. ethnic groups) may arise (Barocas and Selbst 2016; Feldman et al. 2015; Zafar et al. 2015).

A primary shortcoming of group fairness metrics is that they produce inaccurate results when actual differences exist between groups in a target variable or prediction (e.g. ability to pay back a loan). Individuals from different groups are not guaranteed fair treatment based on their respective merit: due to balancing between groups, two individuals with equal merit in different groups can receive different outcomes (e.g. Dwork et al. 2011). As a result, group fairness can be perceived as unfair to individuals who, according to individual merit, deserve an outcome they do not receive.

'Individual fairness' addresses precisely this problem. Individual fairness requires that individuals similar with respect to a target variable receive similar treatment (Dwork et al. 2011),[5] meaning individuals are judged based on merit as opposed to membership in a target group. Sensitive attributes are therefore not strictly necessary to achieve individual fairness; rather, individual fairness requires at a minimum that one or more target attributes be defined which reflect merit or some other condition. Analysis and decision-making can then be optimized towards these attributes.

A prominent variation of individual fairness is 'strict equality of opportunity'. According to this approach, fairness is defined by offering the "same ability to achieve advantage through free choices" to individuals, irrespective of aspects of their lives beyond their control (Lepri et al. 2017: 617).[6] Strict equality of opportunity is distinct from individual fairness in that equal accuracy is required for a defined attribute (Kusner et al. 2017). Specific sensitive attributes can thus be targeted for decision-making to achieve equal treatment according to one or more aspects of the individual. A closely related approach uses a Rawlsian definition of fairness describing fair equality of opportunity, according to which individuals "who are at the same level of talent and have the same willingness of using it, should have the same perspectives of success regardless [of] their initial place in the social system" (Rawls 1971: 73). Joseph et al. (2016) describe an implementation of Rawlsian fairness into sequential algorithmic decision-making. At each step in the sequence, individuals with a lower score or performance on a targeted attribute should never be preferred.

A potential shortcoming of strict equality approaches is that sensitive attributes (e.g. ethnicity, religious belief) must be recorded in datasets for fairness to be evaluated. This requirement cannot be taken for granted. Alternative approaches are, however, available which explicitly exclude sensitive attributes from consideration.

'Fairness through unawareness' requires sensitive attributes that define protected groups to be excluded from a dataset prior to decision-making (e.g. Calders and Verwer 2010; Kamiran and Calders 2010; Schermer 2011; Grgic-Hlaca et al. 2016; Calders et al. 2009).[7] 'Blindness approaches' thus achieve fairness by purposefully ignoring attributes against which discrimination is defined.

Such approaches are vulnerable to proxy variables that implicitly contain or correlate with sensitive attributes (d'Alessandro et al. 2017). Many proxy variables are well established; postal codes are, for example, known to be a strong indicator of ethnicity (Schermer 2011). Proxies are particularly a risk when algorithms access linked datasets (Barocas and Selbst 2016; Macnish 2012; Schermer 2011). Proxies for protected attributes are also not easy to predict or detect (Romei and Ruggieri 2014; Zarsky 2016), often emerging only through training a model and subsequent analysis. Each of these challenges shows that merely removing sensitive attributes from a dataset at the pre-processing stage will not reliably prevent group unfairness or discrimination. Rather, subsequent investigation of the trained model or decision set is also required.[8]

Finally, 'counterfactual fairness' involves modelling fairness in a dataset using counterfactual inferences. Counterfactuals are calculated to determine changes in outputs caused by a targeted variable. A fair decision is defined as one which "gives the same predictions in (a) the observed world and (b) a world where the individual has always belonged to a different demographic group, other background causes of the outcome being equal" (Kusner et al. 2017: 1). The approach is notable for directly engaging with protected sensitive attributes. Counterfactual inference can produce causal models that reveal explicit biases towards protected attributes, and thus map "social biases . . . and the implicit trade-off between prediction accuracy and fairness in an unfair world" (Kusner et al. 2017: 9). Historical biases and prejudices in a dataset can be identified that would be ignored with other metrics.

14.4.2.1 The Impossibility of Universal Fairness

As argued earlier, different fairness metrics reflect different ethical values and norms of equality. It is therefore unsurprising that different fairness metrics are fundamentally incompatible, meaning an objectively or universally 'fair' dataset, model, or decision is impossible. Friedler et al. (2016: 1) note that individual and group fairness are mutually incompatible due to differing assumptions about the "relationship between constructs and observations". Similarly, Kleinberg et al. (2016) argue that statistical parity and strict equality of opportunity are fundamentally incompatible, due to three conditions which cannot be met simultaneously in most cases:

- *Calibration within groups*—probability estimates or predictions by an algorithm should match ground truth. So, if an algorithm predicts an 80% probability that a set of people have a predicted attribute, 80% of the people should actually have the attribute. The condition should be satisfied across all groups targeted.
- *Balance for the positive class*—for all positive cases in each group, the average score or probability received by those people should be equal across the groups. False positive rates should be equal across all groups.
- *Balance for the negative class*—an inversion of the positive class condition. False negative rates should be equal across all groups.

Given these conditions, internally consistent but mutually incompatible metrics can justifiably be preferred across different sectors and stakeholders. False positives and false negatives impose different costs which must be considered when determining which metric of fairness is most appropriate in a given context. The specific costs in a use case will require prioritisation of one of the three conditions over others (Chouldechova 2017; Kleinberg et al. 2016). As suggested by the broader discourse on fairness in Big Data in policy and academic circles, deliberation and consensus between stakeholders in a given use case is essential.

14.5. Discussion and Conclusions

With methods for establishing and assessing data and process provenance, bias and fairness in Big Data now explored, the question remains as to the adoption of such techniques in academic and commercial projects and applications. Despite the ongoing discussion of data and process provenance, the development of systems for it, and its importance for fairness and validity in the use of big data, the uptake of formal systems appears to be confined primarily to scientific applications (Deelman et al. 2015). Here the key concern motivating the use of standardized ways of capturing provenance is reproducibility, and a concern with provenance is a natural extension of existing practices in, e.g., experimental sciences. Increasingly, journals are now updating their publication guidelines to highlight the importance of computational reproducibility, thus fostering the adoption of provenance systems and practices, although the effectiveness of this in achieving computational reproducibility is being questioned so far (Stodden et al. 2018).

Various reasons can be offered that explain the limited uptake of formal provenance systems. A key concern is their overhead. They can slow down applications, impose additional overhead on available storage, as well as introduce additional steps for the analyst (Carata et al. 2014). A second issue with formal provenance systems is security. A fine-grained provenance system, if breached, can reveal proprietary and privacy sensitive information. Moreover, if breached, it might also make it possible to change the data in a way that is difficult to detect. A third issue is familiarity and inclusion of formal provenance systems in established software systems.

Here, the emerging ecosystem surrounding Jupyter notebooks (Kluyver et al. 2016) is an interesting example. A notebook contains in a single document the code, the output of the code (e.g. figures) and text explaining the code, and the interpretation of the output. Jupyter notebooks are increasingly popular and being shared through subversion systems. Moreover, their use is not limited to the sciences. Data-driven journalism, for example, is another space where they are currently used. At the same time, people make tools available for automatic workflows where parts of the processing chain are automatically rerun if either the code or the input data has changed. However, the fact that these tools are not part of the core Jupyter ecosystem, require specialized knowledge, and impose specific constraints on the way in which one has to work limits their broader uptake.

The adoption of fairness and bias assessment systems is in a similarly early but encouraging stage. Growth in academic interest is evident from the rapid expansion of the Fairness, Accountability and Transparency in Machine Learning (FAT-ML) research network since 2014, including the introduction of an annual conference. Similar issues are increasingly a focal point at major machine learning conferences, including the ACM SIGKDD Conference on Knowledge Discovery and Data Mining (KDD), the Annual Conference on Neural Information Processing Systems (NIPS), and the ACM CHI Conference on Human Factors in Computing Systems (CHI). Commercial interest is also evident, given the involvement in major technology companies in global initiatives around ethics and accountability for machine learning and AI, such as the 'Partnership on AI to Benefit People and Society'. As of September 2018 Google has also introduced an interpretability and fairness 'What-if' interface into TensorFlow, an open-source machine learning software package developed by Google, further signalling their interest in this area.

Finally, numerous calls have been made in 2018 to establish standardized documentation and interfaces to establish and assess data provenance in Big Data analytics and machine learning (see also Chapter 42, 'Standards in Engineering' by Paul B. Thompson, this volume). These calls are largely motivated by the usage, sharing, and aggregation of diverse datasets in these practices, which runs a risk of introducing biases and context-ignorant analysis (Bender and Friedman 2018; Gebru et al. 2018; Holland et al. 2018; Yang et al. 2018). Currently, no universally standardized form of documentation for training datasets is required that describes their creation, contents, intended uses, and relevant ethical and legal concerns (Gebru et al. 2018). Manual approaches involve labelling a dataset with relevant information to accompany it as it is passed to other data controllers and reused. This information is explicitly intended to interrogate datasets and identify potential biases in datasets and models prior to and during processing (Holland et al. 2018), and to inform future users of biases and gaps in the dataset to avoid analytics or machine learning models which inherit and reinforce these biases, unknown to the analyst or modeller (Gebru et al. 2018; Holland et al. 2018). As a secondary effect, standardization of provenance documentation procedures may also drive better data collection practices (Holland et al. 2018) as well as consideration of contextual and methodological biases more generally.

Major commercial and research bodies are pursuing projects to establish documentation standards. Gebru et al. (2018: 1), in a project sponsored by the AI Now initiative and with involvement from Microsoft Research, propose a type of 'datasheet' to accompany datasets, comparable to information sheets required in the electronics industry that accompany all components describing their "operating characteristics, test results, recommended usage, and other information". In comparison, a datasheet for public datasets, APIs, and pretrained models would describe "dataset creation; dataset composition; data collection process; data preprocessing; dataset distribution; dataset maintenance; and legal and ethical considerations". Each of these pieces of information is intended to help potential users decide "how appropriate the corresponding dataset is for a task, what its strengths and limitations are, and how it fits into the broader ecosystem" (Gebru et al. 2018: 3).

A similar approach is taken by Holland et al. (2018), who propose a 'dataset nutrition label'. Following labelling requirements in product safety, pharmaceuticals, energy, and material safety, the approach is comparable to the aforementioned 'datasheets for datasets', albeit with greater depth. In addition to manual analysis and labelling of datasets, the creation of a 'nutrition label' is aided by a modular infrastructure featuring "probabilistic computing tools to surface potential corollaries, anomalies, and proxies" (Holland et al. 2018: 5). Specifically, seven modules are described, each of which generates information about the characteristics of the dataset: Metadata, Provenance, Variables, Statistics, Pair Plots, Probabilistic Model, and Ground Truth Correlations. The infrastructure seeks to minimize the time between data acquisition and model development and deployment by providing essential information and technical features necessary for the 'exploratory' phase immediately following data collection. Several of these modules address similar areas as the datasheet approach described earlier, albeit supported by built-in statistical tools.

Such approaches seek to create a standardized set of information to accompany datasets as they are shared and reused. In other words, they seek to establish a baseline set of information describing the provenance of datasets (broadly understood), with particular focus paid to potential biases, gaps, proxies, and correlations which could be inadvertently picked up and reinforced by machine learning systems making use of the data.

Big Data faces both well-understood and novel epistemological challenges, which grow in step with the scale and variety of data being assessed. To address these challenges, the provenance, blind spots, and biases of relevant datasets and analytic processes must be understood. Uptake of systems to establish and assess these features of Big Data is far from universal, but progress is being made. To ensure Big Data delivers the greater accuracy and efficiency promised without reinforcing existing biases and marginalising segments of society, contextual awareness of the history, gaps, and biases of the data we use is essential.

Related Chapters

Chapter 7: Engineering Design and the Quest for Optimality (Maarten Franssen)
Chapter 11: Uncertainty (William M. Bulleit)
Chapter 22: Values and Design (Ibo Van de Poel)
Chapter 24: Human-Centred Design and Its Inherent Ethical Properties (Marc Steen)
Chapter 28: Models in Engineering and Design: Modeling Relations and Directions of Fit (Michael Poznic)
Chapter 32: On Verification and Validation in Engineering (Francien Dechesne and Tijn Borghuis)
Chapter 36: Trust in Engineering (Philip J. Nickel)
Chapter 40: Ethical Considerations in Engineering (Wade L. Robison)
Chapter 42: Standards in Engineering (Paul B. Thompson)

Further Reading

Kleinberg, J., Mullainathan, S. and Raghavan, M. (2016). *Inherent Trade-offs in the Fair Determination of Risk Scores.* ArXiv Prepr. ArXiv160905807. (Offers proof for inherent trade-offs between different fairness metrics.)

Lepri, B., Oliver, N., Letouzé, E., Pentland, A. and Vinck, P. (2017). Fair, Transparent, and Accountable Algorithmic Decision-making Processes: The Premise, the Proposed Solutions, and the Open Challenges. *Philosophy & Technology.* doi:10.1007/s13347-017-0279-x. (An informative overview of transparency methods and fairness metrics.)

Mittelstadt, B., Allo, P., Taddeo, M., Wachter, S. and Floridi, L. (2016). The Ethics of Algorithms: Mapping the Debate. *Big Data Society*, 3. doi:10.1177/2053951716679679. (A systematic review and conceptual framework for ethical issues with algorithms and AI.)

Pérez, B., Rubio, J. and Sáenz-Adán, C. (2018). A Systematic Review of Provenance Systems. *Knowledge Information System*, 1–49. (A comprehensive overview of provenance systems.)

Notes

1. The term Big Data is used here to refer to a combination of large, typically aggregated datasets and the processes through which they are analysed.
2. Custers et al. (2013) also provide an in-depth overview of methods of discriminatory prevention and discovery.
3. Datta et al. (2017: 1) define proxy discrimination as "the presence of protected class correlates that have causal influence on the system's output".
4. Other closely related, but distinct formal definitions have also been advanced. Lepri et al. (2017: 616) see statistical parity to require that "an equal fraction of each group should receive each possible outcome".

According to Kusner et al. (2017: 2), "an algorithm is fair if its predictions are independent of the sensitive attributes A across the population", where sensitive attributes define the groups to be protected.

5. Similarly, Kusner et al. (2017: 2) offer the following formal definition for individual fairness: "an algorithm is fair if it gives similar predictions to similar individuals".

6. Hardt et al. (2016), for example, describe an implementation of strict equality to assess discrimination in a supervised learning system seeking to predict another attribute. In this case, fairness is defined against one sensitive attribute of interest.

7. Kusner et al. (2017: 2) offer the following formal definition: "an algorithm is fair so long as any sensitive attributes A are not explicitly used in the decision-making process".

8. 'Blindness approaches' have also been shown in some cases to produce highly inaccurate models. Dwork et al. (2011) argue that ignorance of protected attributes can lead to decision rules selecting for the opposite of the intended result, due to undiscovered correlations between hidden attributes and the target variable.

References

Ananny, M. (2016). Toward an Ethics of Algorithms Convening, Observation, Probability, and Timeliness. *Science Technology & Human Values*, 41, 93–117. doi:10.1177/0162243915606523

Bail, C.A. (2014). The Cultural Environment: Measuring Culture With Big Data. *Theory Soc.*, 43, 465–482. doi:10.1007/s11186-014-9216-5

Barocas, S. (2014). Data Mining and the Discourse on Discrimination. In *Data Ethics Workshop, Conference on Knowledge Discovery and Data Mining, KDD'14*, pp. 1–4.

Barocas, S. and Selbst, A.D. (2016). Big Data's Disparate Impact. *California Law Review*, 104. doi:10.15779/Z38BG31

Bender, E.M. and Friedman, B. (2018). Data Statements for Natural Language Processing: Toward Mitigating System Bias and Enabling Better Science. *Transactions of the Association for Computational Linguistics*, 6, 587–604.

Burrell, J. (2016). How the Machine "Thinks:" Understanding Opacity in Machine Learning Algorithms. *Big Data Society*. doi:10.1177/2053951715622512

Calders, T., Kamiran, F. and Pechenizkiy, M. (2009). Building Classifiers With Independency Constraints. In *Data Mining Workshops, 2009. ICDMW'09. IEEE International Conference On*. IEEE, Stanford, CA, USA, pp. 13–18.

Calders, T. and Verwer, S. (2010). Three Naive Bayes Approaches for Discrimination-free Classification. *Data Min. Knowl. Discov.*, 21, 277–292. doi:10.1007/s10618-010-0190-x

Carata, L., Akoush, S., Balakrishnan, N., Bytheway, T., Sohan, R., Selter, M. and Hopper, A. (2014). A primer on provenance. *Communication ACM*, 57, 52–60.

Chouldechova, A. (2017). *Fair Prediction With Disparate Impact: A Study of Bias in Recidivism Prediction Instruments*. ArXiv170300056 Cs Stat.

Crawford, K. (2013). The Hidden Biases in Big Data. *Harvard Business Review*.

Crawford, K., Gray, M.L. and Miltner, K. (2014). Critiquing Big Data: Politics, Ethics, Epistemology| Special Section Introduction. *International Journal of Communication*, 8, 10.

Custers, B., Calders, T., Schermer, B. and Zarsky, T. (eds.) (2013). *Discrimination and Privacy in the Information Society, Studies in Applied Philosophy, Epistemology and Rational Ethics*. Berlin, Heidelberg: Springer. doi:10.1007/978-3-642-30487-3

d'Alessandro, B., O'Neil, C. and LaGatta, T. (2017). Conscientious Classification: A Data Scientist's Guide to Discrimination-Aware Classification. *Big Data*, 5, 120–134. doi:10.1089/big.2016.0048

Datta, A., Fredrikson, M., Ko, G., Mardziel, P. and Sen, S. (2017). *Proxy Non-Discrimination in Data-Driven Systems*. ArXiv Prepr. ArXiv170708120.

Datta, A., Tschantz, M.C. and Datta, Anupam (2015). Automated Experiments on Ad Privacy Settings. *Proc. Priv. Enhancing Technol.* doi:10.1515/popets-2015-0007

Deelman, E., Vahi, K., Juve, G., Rynge, M., Callaghan, S., Maechling, P.J., Mayani, R., Chen, W., Ferreira Da Silva, R., Livny, M. and Wenger, K. (2015). Pegasus, a Workflow Management System for Science Automation. *Future Generation Computing System*, 46, 17–35.

De Vries, K. (2010). Identity, Profiling Algorithms and a World of Ambient Intelligence. *Ethics Information Technology*, 12, 71–85. doi:10.1007/s10676-009-9215-9

Dwork, C., Hardt, M., Pitassi, T., Reingold, O. and Zemel, R. (2011). *Fairness Through Awareness*. ArXiv11043913 Cs.

Fairfield, J. and Shtein, H. (2014). Big Data, Big Problems: Emerging Issues in the Ethics of Data Science and Journalism. *Journal of Mass Media Ethics*, 29, 38–51. doi:10.1080/08900523.2014.863126

Feldman, M., Friedler, S.A., Moeller, J., Scheidegger, C. and Venkatasubramanian, S. (2015). Certifying and Removing Disparate Impact. In *Proceedings of the 21th ACM SIGKDD International Conference on Knowledge*

Discovery and Data Mining, KDD '15. ACM, New York, NY, USA, pp. 259–268. doi:10.1145/2783258. 2783311

Floridi, L. (2016). Mature Information Societies—a Matter of Expectations. *Philos. Technology,* 29, 1–4. doi:10.1007/s13347-016-0214-6

Freire, J., Koop, D., Santos, E. and Silva, C.T. (2008). Provenance for Computational Tasks: A Survey. *Computer Science & Engineering,* 10, 11–21. doi:10.1109/MCSE.2008.79

Friedler, S.A., Scheidegger, C. And Venkatasubramanian, S. (2016). *On the (Im) Possibility of Fairness.* ArXiv Prepr. ArXiv160907236.

Friedman, B. and Nissenbaum, H. (1996). Bias in Computer Systems. *ACM Transactions on Information Systems (TOIS),* 14(3), 330–347.

Gebru, T., Morgenstern, J., Vecchione, B., Vaughan, J.W., Wallach, H., Daumeé III, H. and Crawford, K. (2018). *Datasheets for Datasets.* ArXiv Prepr. ArXiv180309010.

Grgic-Hlaca, N., Zafar, M.B., Gummadi, K.P. and Weller, A. (2016). The Case for Process Fairness in Learning: Feature Selection for Fair Decision Making. In *NIPS Symposium on Machine Learning and the Law,* NIPS 2016, Barcelona, Spain.

Hajian, S. and Domingo-Ferrer, J. (2013). A Methodology for Direct and Indirect Discrimination Prevention in Data Mining. *IEEE Transactions on Knowledge and Data Engineering,* 25(7), 1445–1459. doi.org/10.1109/TKDE.2012.72

Hardt, M., Price, E. and Srebro, N. (2016). Equality of Opportunity in Supervised Learning. In *Advances in Neural Information Processing Systems,* NIPS 2016, Barcelona, Spain, pp. 3315–3323.

Hildebrandt, M. (2011). Who Needs Stories If You Can Get the Data? ISPs in the Era of Big Number Crunching. *Philosophy & Technology,* 24(4), 371–390. doi.org/10.1007/s13347-011-0041-8

Holland, S., Hosny, A., Newman, S., Joseph, J. and Chmielinski, K. (2018). *The Dataset Nutrition Label: A Framework to Drive Higher Data Quality Standards.* ArXiv180503677 Cs.

Huff, D. (1954). *How to Lie With Statistics.* New York: W.W. Norton.

IBM (2014). The Four V's of Big Data [WWW Document]. www.ibmbigdatahub.com/infographic/four-vs-big-data. Accessed October 23, 2014.

Joseph, M., Kearns, M., Morgenstern, J., Neel, S. and Roth, A. (2016). *Fair Algorithms for Infinite and Contextual Bandits.* ArXiv161009559 Cs.

Kamiran, F. and Calders, T. (2010). Classification With No Discrimination by Preferential Sampling. In *Proceedings of the 19th Machine Learning Conference of Belgium and The Netherlands.* Benelearn'10, Leuven, Belgium, pp. 1–6.

Kamiran, F. and Calders, T. (2012). Data Preprocessing Techniques for Classification Without Discrimination. *Knowledge Information System,* 33, 1–33. doi:10.1007/s10115-011-0463-8

Kamiran, F., Calders, T. and Pechenizkiy, M. (2010). Discrimination Aware Decision Tree Learning. In *Data Mining (ICDM), 2010 IEEE 10th International Conference On.* IEEE, Stanford, CA, USA, pp. 869–874.

Kamiran, F., Karim, A. and Zhang, X. (2012). Decision Theory for Discrimination-aware Classification. In *Data Mining (ICDM), 2012 IEEE 12th International Conference On.* IEEE, Stanford, CA, USA, pp. 924–929.

Kamishima, T., Akaho, S., Asoh, H. and Sakuma, J. (2012a). Fairness-aware Classifier with Prejudice Remover Regularizer. In *Joint European Conference on Machine Learning and Knowledge Discovery in Databases, ECMLP-KDD'12.* Springer, Berlin, Heidelberg, Germany, pp. 35–50.

Kamishima, T., Akaho, S., Asoh, H. and Sakuma, J. (2012b). *Considerations on Fairness-Aware Data Mining.* IEEE, Stanford, CA, USA, pp. 378–385. doi:10.1109/ICDMW.2012.101

Kamishima, T., Akaho, S. and Sakuma, J. (2011). *Fairness-aware Learning through Regularization Approach.* IEEE, Stanford, CA, USA, pp. 643–650. doi:10.1109/ICDMW.2011.83

Kleinberg, J., Mullainathan, S. and Raghavan, M. (2016). *Inherent Trade-offs in the Fair Determination of Risk Scores.* ArXiv Prepr. ArXiv160905807.

Kluyver, T., Ragan-Kelley, B., Pérez, F., Granger, B., Bussonnier, M., Frederic, J., Kelley, K., Hamrick, J., Grout, J., Corlay, S., Ivanov, P., Avila, D., Abdalla, S. and Willing, C. (2016). Jupyter Notebooks—A Publishing Format for Reproducible Computational Workflows. In F. Loizides and B. Schmidt (eds.), *Positioning and Power in Academic Publishing: Players, Agents and Agendas.* Amsterdam: IOS Press, pp. 87–90.

Kroll, J.A., Huey, J., Barocas, S., Felten, E.W., Reidenberg, J.R., Robinson, D.G. and Yu, H. (2016). *Accountable Algorithms* (SSRN Scholarly Paper No. ID 2765268). Social Science Research Network, Rochester, NY.

Kusner, M.J., Loftus, J.R., Russell, C. and Silva, R. (2017). *Counterfactual Fairness.* ArXiv Prepr. ArXiv170306856.

Lepri, B., Oliver, N., Letouzé, E., Pentland, A. and Vinck, P. (2017). Fair, Transparent, and Accountable Algorithmic Decision-making Processes: The Premise, the Proposed Solutions, and the Open Challenges. *Philosophy & Technology.* doi:10.1007/s13347-017-0279-x

Luong, B.T., Ruggieri, S. and Turini, F. (2011). K-NN as an Implementation of Situation Testing for Discrimination Discovery and Prevention. In *Proceedings of the 17th ACM SIGKDD International Conference on*

Knowledge Discovery and Data Mining, KDD '11. Association for Computing Machine, New York, NY, USA, pp. 502–510. doi.org/10.1145/2020408.2020488

Macnish, K. (2012). Unblinking Eyes: The Ethics of Automating Surveillance. *Ethics Information Technology,* 14, 151–167. doi:10.1007/s10676-012-9291-0

Mittelstadt, B., Allo, P., Taddeo, M., Wachter, S. and Floridi, L. (2016). The Ethics of Algorithms: Mapping the Debate. *Big Data Society,* 3. doi:10.1177/2053951716679679

Mittelstadt, B. and Floridi, L. (2016). The Ethics of Big Data: Current and Foreseeable Issues in Biomedical Contexts. *Science, Engineering & Ethics,* 22, 303–341. doi:10.1007/s11948-015-9652-2

Moreau, L., Clifford, B., Freire, J., Futrelle, J., Gil, Y., Groth, P., Kwasnikowska, N., Miles, S., Missier, P., Myers, J., Plale, B., Simmhan, Y., Stephan, E., Den Bussche, J.V. (2011). The Open Provenance Model Core Specification (v1.1). *Future Generation Computer Systems,* 27, 743–756.

Morek, R. (2006). Regulatory Framework for Online Dispute Resolution: A Critical View. *University of Toledo Law Review,* 38, 163.

Pedreshi, D., Ruggieri, S. and Turini, F. (2008). Discrimination-aware Data Mining. In *Proceedings of the 14th ACM SIGKDD International Conference on Knowledge Discovery and Data Mining, KDD '08.* ACM, New York, NY, USA, pp. 560–568. doi:10.1145/1401890.1401959

Pedreschi, D., Ruggieri, S. and Turini, F. (2009). Measuring Discrimination in Socially-Sensitive Decision Records. In *Proceedings of the 2009 SIAM International Conference on Data Mining, Proceedings.* Society for Industrial and Applied Mathematics, pp. 581–592. doi:10.1137/1.9781611972795.50

Pedreschi, D., Ruggieri, S. and Turini, F. (2012). A Study of Top-k Measures for Discrimination Discovery. In *Proceedings of the 27th Annual ACM Symposium on Applied Computing.* ACM, New York, NY, USA, pp. 126–131.

Pérez, B., Rubio, J., Sáenz-Adán, C. (2018). A Systematic Review of Provenance Systems. *Knowledge Information Systems,* 1–49.

Puschmann, C. and Burgess, J. (2014). Big Data, Big Questions| Metaphors of Big Data. *International Journal of Communication,* 8, 20.

Rawls, J. (1971). *A Theory of Justice.* na.

Reichman, O.J., Jones, M.B. and Schildhauer, M.P. (2011). Challenges and Opportunities of Open Data in Ecology. *Science,* 331, 703–705. doi:10.1126/science.1197962

Romei, A. and Ruggieri, S. (2014). A Multidisciplinary Survey on Discrimination Analysis. *Knowledge Engineering Review,* 29, 582–638. doi:10.1017/S0269888913000039

Ruggieri, S., Pedreschi, D. and Turini, F. (2010). DCUBE: Discrimination Discovery in Databases. In *Proceedings of the 2010 ACM SIGMOD International Conference on Management of Data.* ACM, New York, NY, USA, pp. 1127–1130.

Sandvig, C., Hamilton, K., Karahalios, K. and Langbort, C. (2014). *Auditing Algorithms: Research Methods for Detecting Discrimination on Internet Platforms.* Data and Discrimination: Converting Critical Concerns Product. Inq.

Schermer, B.W. (2011). The Limits of Privacy in Automated Profiling and Data Mining. *Computing Law Security Review,* 27, 45–52. doi:10.1016/j.clsr.2010.11.009

Stodden, V., Seiler, J. and Ma, Z. (2018). An Empirical Analysis of Journal Policy Effectiveness for Computational Reproducibility. *Proceedings of National Academy of Science USA,* 115, 2584–2589.

Tramer, F., Atlidakis, V., Geambasu, R., Hsu, D.J., Hubaux, J.-P., Humbert, M., Juels, A. and Lin, H. (2015). *Discovering Unwarranted Associations in Data-driven Applications With the Fairest Testing Toolkit.* ArXiv Prepr. ArXiv151002377.

Tufte, E.R. (1983). *The Visual Display of Quantitative Information,* Vol. 2. Cheshire, CT: Graphics Press.

Yang, K., Stoyanovich, J., Asudeh, A., Howe, B., Jagadish, H.V. and Miklau, G. (2018). A Nutritional Label for Rankings. *Proc. 2018 Int. Conf. Managing Data—SIGMOD,* 18, 1773–1776. doi:10.1145/3183713.3193568

Zafar, M.B., Valera, I., Rodriguez, M.G., Gummadi, K.P. (2015). *Fairness Constraints: Mechanisms for Fair Classification.* ArXiv150705259 Cs Stat.

Zarsky, T. (2016). The Trouble with Algorithmic Decisions: An Analytic Road Map to Examine Efficiency and Fairness in Automated and Opaque Decision Making. *Science, Technology & Human Values,* 41, 118–132. doi:10.1177/0162243915605575

Zemel, R., Wu, Y., Swersky, K., Pitassi, T. and Dwork, C. (2013). Learning Fair Representations. In *Proceedings of the 30th International Conference on Machine Learning (ICML-13), PMLR,* 28(3). Atlanta, GA, USA, pp. 325–333.

Zliobaite, I. (2015). *A Survey on Measuring Indirect Discrimination in Machine Learning.* ArXiv151100148 Cs.

PART III

Ontology

15

ARTIFACTS

Beth Preston

15.1. Introduction

It is clear that engineers have an interest in artifacts. The stereotypical engineer is a person who designs and makes things. And the stereotypical artifact, as the etymology of the word indicates, is something made by art or skill. So the products of engineering are artifacts. Moreover, engineers make artifacts using other artifacts—you can't design a bridge without pencil and paper (or the computer-based equivalent); and you can't build one without pile drivers, wire cutters, shovels, welding gear, and the like. So the activity of engineering is dependent on artifacts as well. However, the clarity of this intuition about the engineer's deep and abiding interest in artifacts is offset by the murkiness of any inquiry that follows up by asking exactly what an artifact is, or even more alarmingly, exactly who the engineers are. These murky depths are inhabited by philosophers, working away busily on problems with alluring titles, such as Special Composition and Arbitrary Fusion. Some say that it is the philosophers themselves who generate the murk. Others say that, on the contrary, it is the nature of the artifacts and other objects about which the philosophers inquire that is murky. The truth, perhaps, lies somewhere in between. We will attempt to discern it by reviewing the philosophers' results in four areas of inquiry, and asking about their usefulness and importance to the engineers. In the next section, we will focus on the existence of artifacts. Section 3 takes up matters of definition. Section 4 addresses questions concerning artifact kinds, and Section 5 focuses on artifact parts.

15.2. Do Artifacts Exist?

The leading philosophical concern about artifacts has traditionally been the simple question: Do they truly exist? This concern goes back to Aristotle, who argued that only natural things, which have their own principle of movement and growth within them, truly exist (Aristotle 1984: *Physics* 192b 8–39, *Metaphysics* 1043b 15–25). He called them substances, thus establishing the definition of substance as that which exists independently and undergirds everything else. Artifacts, which depend on humans for their construction, are not substances in this sense, and thus do not truly exist. In short, for Aristotle, the paradigm existent is a living thing—a goat or a dandelion, for instance. Everything else is suspect.

This ontological downgrading of artifacts has been revived and extended by contemporary metaphysicians, some of whom have denied the existence not only of artifacts but also of other ordinary objects, such as rivers and stones, and even sometimes living things. Their underlying intuition is that while the ultimate, simple building blocks of the universe—be they particles or strings or whatever—must exist independently and thus be true substances, the things allegedly composed of them do not necessarily have a similar claim. This intuition is bolstered on one side by what Daniel Korman (2015: 4–7) calls debunking arguments. The ordinary objects that we unwarily take to exist in our everyday world are merely the composites that happen to correspond to our needs and interests as human beings. These needs and interests, in turn, are merely accidents of our biology or of our culture, both of which are in a state of continual evolutionary flux. So there is no reason to think that they give us any privileged handle on what composites truly exist.

The intuition of the metaphysicians is bolstered from another direction by problems that arise when they try to answer what Peter van Inwagen (1990: 20) calls the Special Composition Question—under what conditions *do* truly unified wholes arise out of parts? An answer to this question would give us a way of rehabilitating at least some ordinary objects by explaining what the principles of composition are, such that there are true wholes composed of proper parts.[1] But there is little agreement about what these principles might be. At the nihilistic end of the spectrum is the position that there are no such principles, and thus no truly existing composites. At the other end is the position that there is a plethora of such principles, and that therefore any arbitrary fusion of things—your left eye, Alpha Centauri, and the third daffodil from the right in the clump next to my mailbox, for instance—constitutes a fourth thing. Most occupied positions on the spectrum fall somewhere in between these two extremes.[2] Van Inwagen himself, echoing Aristotle, claims that only simples whose combined activity constitutes a life are proper parts of a new whole, which their activity brings into existence. So although we do move simples around when we make artifacts, we do not bring anything new into existence by so doing. Thus, like nihilism, van Inwagen's view does not rehabilitate artifacts. Arbitrary fusion does rehabilitate them, but at a steep price, since all sorts of unlikely objects must be countenanced as well.

In response to the animus against ordinary objects represented by this spectrum, a new conservatism has arisen, often with a particular focus on artifacts. Lynne Rudder Baker (2007) argues for a constitution view on which some things are (non-reductively) made up of other things under certain conditions. In the case of artifacts, these conditions include intelligent agents and materials. This allows artifacts to be duly constituted out of existing materials by an agent with the intention to make something out of them and the knowledge of how to do this. Similarly, Simon Evnine (2016) espouses a version of hylomorphism, in which artifacts are the paradigmatic existents. For him, the crucial feature is the origin of the thing, the act or event that forms it. Artifacts are paradigmatic because their origin is clearly a process involving labor on materials that is guided by the intentions of the laborer. Amie Thomasson (2007a) takes a different tack, arguing that our concepts incorporate a specification of the conditions for their application. If these conditions are satisfied, then we should say that the kind of thing specified exists. Thus artifacts such as cooking pots exist, because we can verify in any kitchen that the conditions specified in the concept of a cooking pot are met. Daniel Korman (2015) does not focus on artifacts but does provide a compendium of ways to defuse the problematic arguments against ordinary objects in general. We may thus eventually get our ordinary object ontology back.

But do engineers need to pay attention to this debate? Arguably, they do not. The practice of engineering is unlikely to be affected in any way, because it operates at the scale of human interests and needs, where ordinary objects are the taken-for-granted ontology. Or, to put it more simply, regardless of what the philosophers think about the ontological status of artifacts in the grander metaphysical scheme of things, the engineers still have to deal with them in the humbler ordinary scheme of things, where their effects are real enough. When a pedestrian bridge falls onto a busy highway and crushes out the lives of six people, as happened recently in the United States, it is the engineers who will be blamed, required to investigate, and tasked with making sure it does not happen again. One can imagine scenarios in which a hotline to a metaphysician might prove useful, but no one in their right mind would call a metaphysician to deal with events of this type. (An ethicist, perhaps. But not a metaphysician.) So while individual engineers may find the metaphysical debate about whether artifacts exist of interest for general philosophical reasons, the importance of the philosophy of artifacts for the philosophy of engineering lies elsewhere.

15.3. What Is an Artifact?

The term 'artifact' in everyday English is used in a quite narrow sense to mean something made by humans from an earlier historical period—something that an archaeologist would unearth, or that you would find in a museum. Thus, we speak colloquially of artifacts from the Atlantic slave trade, for instance, or the artifacts, now in the British Museum, from the 7th-century ship burial at Sutton Hoo.[3] But in philosophy, 'artifact' is a term of art, used much more broadly to distinguish artificial things from natural things. So philosophers speak of an artifact as something made by humans or other intelligent beings, including non-human animals. 'Artifact' in this sense is what philosophers call a 'category.' Categories are classifications of things at the most abstract levels. We do, of course, also classify things at more specific levels, and these classifications are usually called 'kinds.' So under the abstract 'artifact' category we find specific kinds of artifacts such as baskets, arrows, cooking pots, shoes, and the like. We will investigate questions about artifact kinds in the next section. Here we deal with 'artifact' as an upper-level category.

Categories were originally proposed by Aristotle in a realist spirit, as the broadest classifications into which things naturally fall, delineating the main joints in the world itself. He offered a comprehensive system of such categories (*Categories*; see also Studtman 2017). Aristotle's realism about categories fell out of favor by the time of Immanuel Kant, for whom categories were classifications in our conceptual system—the main joints in the way we think about the world rather than in the way the world itself is. Thus Kant's (1781) comprehensive system of categories does classify the objects of our experience, but only *as* experienced, not as they are in themselves. Both realist and non-realist systems of categories have been proposed more recently by 20th-century philosophers such as Edmund Husserl, Roderick Chisholm, and E.J. Lowe (Thomasson 2018).

An important 20th-century development is skepticism about the project of producing a unique and universally accepted system of categories. Systems produced over the last couple of millennia are all different and were constructed using different methods, thus casting doubt on the plausibility of an ultimate consensus. Philosophers have not given up on categories, but they have wielded them in a much more limited way, focusing on individual categorial distinctions that have a bearing on specific problems. The classic case is Gilbert Ryle (1949), who argued that Cartesian dualists make a category mistake when they characterize mental processes as invisible *things*—processes in some ghostly substance—rather than as dispositions. The telltale signs of this category mistake, Ryle thought, are the absurdities generated by Cartesian dualism, such as the apparently insoluble problem of how body and mind could possibly interact on this view. This piecemeal approach to categories allows us to identify categorial fault lines on a case-by-case basis, when and as they are needed to regularize how we talk about the world, rather than insisting on *a priori* system building. Adopting this approach, and generalizing a bit, we may then ask what epistemically useful work, if any, the categorial distinction between artifacts and naturally occurring objects does for us.

We may begin with the standard philosophical definition of 'artifact,' which purports to give the identity conditions for belonging to this category. It can be traced back ultimately to Aristotle's distinction mentioned earlier, between things that exist by nature and have their own principle of movement and change in themselves, and things that exist by craft, whose principle of movement and change depends on the intentions and activities of their makers (*Metaphysics*, 1033a ff., *Nicomachean Ethics*, 1140a ff., *Physics*, 192b ff.). This distinction grounds the standard philosophical definition of artifacts as *objects made intentionally in order to accomplish a specific purpose* (Hilpinen 1992, 2011).

On this view, artifacts must satisfy three conditions. First, they must be *intentionally* made. This condition distinguishes artifacts not only from naturally occurring objects such as sticks, but also from unintended by-products of intentional actions, such as the sawdust that results from sawing boards. Second, they must involve *modification* of materials. This rules out naturally occurring objects even when used intentionally for a purpose, such as the stick thrown to amuse your dog. Finally, they must be made for a specific *purpose*. This is supposed to rule out naturally occurring objects on the assumption that even when they appear to have functions, such as the food processing function of stomachs, these functions are the product of mechanical evolutionary processes and thus not the designed purpose of the thing in the way that food processing is the designed purpose of your Cuisinart blender (but see Evnine 2016). This condition also rules out intentionally modified objects that are not made for a purpose—for example, when you intentionally produce one, long spiral apple peel only to drop it in the garbage. These three conditions aim to provide individually necessary and jointly sufficient conditions that distinguish artifacts from naturally occurring objects.

But there is a problem. Along all three definitional dimensions we encounter a continuum of cases (Grandy 2007; Sperber 2007). Take the *intention* condition. Paths are often created unintentionally by people repeatedly walking back and forth between two points. But it seems odd to insist that such a path is not an artifact, whereas an exactly similar one that was created intentionally by exactly the same process is. Furthermore, there are questions about what would change the status of the unintentionally created path from natural landscape feature to artifact. It might be sufficient to notice it, approve, and intend from then on that there be a path there. But then you might equally well notice, approve, and intend the growing pile of sawdust as your logs are sawn into boards, perhaps because you want it to mulch your strawberries.

The *modification* condition is even more clearly a matter of degree. Suppose I collect a particularly nice white stone from the woods behind my house to use as a grave marker for my late, lamented cat. The question here is whether the transport alone counts as a modification, or whether I would also have to clean it up, or whether I would actually have to chip a piece or two off. Complicating this issue is the fact that use of naturally occurring objects can itself cause modification. A stone of any shape picked up and used as a hammer rapidly acquires a spherical shape, for instance (Schick and Toth 1993: 130 ff.). Since the very first blow undoubtedly struck off some fragments, it seems the stone should count as modified from that point on. The modification condition is thus just as slippery as the intention condition.

The *specific purpose* condition might seem immune to this problem. But Dan Sperber (2007) has argued that even function is continuous between nature and culture. What he calls biological artifacts—domesticated plants and animals—have both biological and cultural functions. Moreover, they perform their cultural functions by performing their biological functions, and vice versa. Seedless grapes, for example, retain the biological function of all fruit—attracting animals to eat the fruit and thus propagate the plant. But in this case, we do that by vegetatively propagating grapevines rather than by depositing the seeds elsewhere with a dollop of fertilizer. This entanglement of biological and cultural functions in domesticates thus exhibits the imperceptible merger of nature and culture in a particularly vivid form.

Philosophers have typically reacted to this 'continuum problem' by holding tight to 'artifact' as a basic category. Risto Hilpinen (2011), for instance, acknowledges that there are phenomena that 'bridge' the categorial divide between clear cases of artifacts and clear cases of naturally occurring objects. He proposes several auxiliary classifications, such as 'residue' (sawdust, for example), to accommodate them. Similarly, Randall Dipert (1993: 17) proposes a more refined classification.

- Instruments—naturally occurring objects that have been intentionally used for a purpose, but not modified, such as stones used as hammers.

- Tools—intentionally modified instruments, such as stones shaped and polished for use as hammers.
- Artifacts—tools intended to be recognized as tools, such as claw hammers made from standard materials and with a standardized, easily recognizable shape.

These efforts are clearly aimed at taming the continua and thus preserving the traditional divide between artifacts and naturally occurring objects.

Sperber, on the other hand, takes the opposite tack. He concludes that 'artifact' as a theoretical term of art cannot be usefully defined—any attempt to do so will be frustrated by the continuum problem detailed earlier—and that this shows that at best naturalistic social sciences simply do not need 'artifact' or its correlative 'natural object' as basic categories. At worst, he continues, by insisting on this categorial distinction we disadvantage ourselves epistemically by buying into "a doubly obsolete industrial-age revival of a Paleolithic categorization" (2007: 136). It is obsolete, first, because before the advent of agriculture, the technologies available to people were paradigmatic artifacts—stone tools, baskets, beads—and we evolved a psychological disposition to classify in accordance with the clear distinction between such objects and naturally occurring ones. We have retained this disposition, even though the Neolithic transition to agriculture 12,000 years ago made biological artifacts—that is, domesticated plants and animals—the predominant type of artifact in human experience. Second, it is obsolete because information technology makes increasingly available artifacts that act on their own, beyond any intention their creators may have; and biotechnology simultaneously makes impressing our intentions on our biological artifacts increasingly effective. In short, our non-biological artifacts are increasingly lively and our biological artifacts increasingly controlled, thus erasing any erstwhile gap between nature and artifact. These differences between our situation and the Paleolithic situation mean that Paleolithic classification practices and the psychological dispositions that drive them simply do not serve us well in understanding ourselves and the world we now inhabit.[4] Thus, there is reason to doubt that a categorial distinction between artifacts and naturally occurring objects is really legitimate or can be applied unproblematically in any domain.

But is this emerging philosophical debate any more important for engineers than the traditional debate as to whether artifacts really exist? The answer is emphatically 'yes.' In part, this is just because classification systems are essential for engineering, as for the sciences generally, for the representation and management of knowledge in such a way that it is easily communicable—ideally across disciplines—and easily manipulable—ideally by information technologies (Smith 2003; Munn and Smith 2008; Simons 2011). The study of classification in this context is usually called formal or applied ontology. It has inherited the major issues and debates of traditional philosophical ontology mentioned earlier—for example, the face-off between the (Aristotelian) realists, who take themselves to be categorizing things as they are in themselves, and the (Kantian) conceptualists, who take themselves to be categorizing the ways we think (and/or talk) about things (Munn and Smith 2008). Similarly, the applied ontologists have added to the already burgeoning list of competing category systems, still with no consensus candidate in sight.[5] This has prompted some to limit their ambitions to a search for a unique upper-level category system, allowing lower-level systems to proliferate as they may. But applied ontologists have proven more immune than philosophers to the skepticism that prompted the latter to turn away from system building in favor of a focus on individual categorial distinctions. In applied ontology, the proliferation of different category systems has more frequently been met with efforts to integrate ontologies, or to reuse them in different domains.[6] This may well be due to the overwhelming practical advantages for the engineering and information sciences of a shared basis for knowledge management, at least within domains or general project areas.

But what of the 'artifact' category, specifically? As Borgo and Vieu (2009: 282) report, despite the central role of artifacts in human life, very few of the upper-level ontologies developed so far include 'artifact' as a basic category. Moreover, they argue, the characterization of artifacts in these systems is far from adequate. They attempt to remedy this by developing their own 'artifact' category, characterizing artifacts as "created entities in which 'created' refers to a mental event, not to a physical modification" (292). Thus, their definition is even weaker than Dipert's definition of mere instruments, since a creator need only intend that a stone be a paperweight, for instance, to turn it into an artifact. As Borgo and Vieu themselves note, this choice will not appeal to everyone, so it is no surprise that several alternative definitions have since appeared, all aimed at integrating 'artifact' into ontologies suitable for use by engineers and scientists (Kassel 2010; Kitamura and Mizoguchi 2010; Houkes and Vermaas 2014; see also the comparative overview in Borgo et al. 2014). In addition, Paweł Garbacz (2013) argues that 'artifact' cannot be captured in an ordinary definition focused on necessary and sufficient conditions, but requires a Wittgensteinian family resemblance analysis. Thus, in applied ontology as in philosophical ontology, we are faced with an embarrassment of riches with regard to the definition of 'artifact.' This suggests that engineers, like philosophers, would do well to take seriously the questions raised by Sperber and others about the ultimate viability of 'artifact' as a fundamental category. If it is not viable, then any insistence on incorporating it into applied ontologies intended for use in engineering disciplines will ultimately end in failure. Moreover, this will be a failure in practice, not just in theory. Applied ontology focuses on the management of knowledge systems, especially by means of sophisticated information technologies. Information technologies are only as useful as their programming makes them. So programming into them an ontology with a fundamental category that is not coherent and cannot be made so may well result in degraded output. "Garbage in, garbage out," as the saying goes. From this point of view, the observation of Borgo and Vieu (2009) that most current upper-level ontologies do not include 'artifact' as a fundamental category may well signal a strength of these ontologies rather than a weakness requiring remediation.

15.4. Artifact Kinds

Kinds are the classifications of things at more specific levels than categories. Traditionally, it is thought that naturally occurring things—organisms, in particular—also fall naturally into different 'natural kinds' that are independent of human intentions and interests, and thus reflect real cleavages in the world. On the other hand, it is thought, artifacts are created by us, in accordance with our shifting intentions and local interests. So they are dependent on us, and the kinds into which we classify them are therefore not natural or real. To put it another way, natural kinds reflect mind-independent essences shared by the individuals making up the kind, whereas artifact kinds reflect only mind-dependent human preferences for grouping artifacts in various ways. On this traditional view, no unified account of natural kinds and artifact kinds is possible. The basis of the classification in the two cases is entirely different, and it results in the identification of real, natural kinds only in the case of naturally occurring objects.

This tidy scheme was rudely interrupted by the Darwinian theory of evolution, which showed that species—up to that point considered the very paradigm of natural kinds—are historically conditioned and continuous with each other. Similar phenomena have now come to light even with regard to the kinds of chemistry and physics (Khalidi 2013). Thus we find the continuum problem that infected the erstwhile sharp distinction between artifacts and naturally occurring objects returning in spades at the lower levels of classification. On the other hand, it is clear that grouping natural objects into kinds does promote useful inferences and successful explanations. This observation has led to what Thomas Reydon (2014) calls an 'epistemological turn' in accounts of classification for naturally occurring objects. Whereas traditionally the criteria were taken to be metaphysical—that is, independent of human interests and practices—now the criteria are increasingly taken to be epistemological—that is, dependent to some extent on human epistemological practices. The practical and scientific success of the best classification practices may be taken as evidence that the classifications do correspond in some, perhaps inexact, way to real differences in the things classified.

However, on this view, natural kinds are no more or less mind-dependent than artifact kinds, so there is in principle no barrier to a unified account, and no reason to deny that artifact kinds are real kinds on a par with natural kinds.

Crawford Elder (2004, 2007) provides a good example of a unified account. In his view, artifact kinds and natural kinds are both copied kinds. A copied kind is characterized by three linked properties—a distinctive 'shape,' a proper function established by a copying mechanism that reproduces successfully performing things of that shape, and a historically proper placement. For example, cats' whiskers are distinctively shaped organs that are copied by biological reproduction because they successfully enable cats to navigate in the dark, and are historically located on the cat's body in the proper place to perform this function. Similarly, flashlights are distinctively shaped artifacts that are copied by industrial reproduction because they successfully enable humans to navigate in the dark and are historically located in households, cars, and the like where they are readily available to perform this function. One acknowledged problem with Elder's account is that many erstwhile artifact kinds turn out not to be copied kinds. Neckties, for example, do not qualify because they do not appear to have a proper function (Elder 2004: 158–159). But Elder's account does have the virtue of drawing out useful analogies between natural kinds and artifact kinds, thus suggesting that if the former are real kinds, so too are the latter.

Amie Thomasson (2003, 2007b) takes a more direct route to this conclusion. Realists about kinds, she points out, are not forced to choose between showing that artifact kinds are mind-independent just like natural kinds, on the one hand, or conceding that artifact kinds are not real, on the other. Instead, they can just deny that mind-independence is the touchstone of reality. Thomasson then proceeds to build mind-dependence right into her account of artifact kinds.

> Necessarily, for all x and all artifactual kinds K, x is a K only if x is the product of a largely successful intention that (Kx), where one intends (Kx) only if one has a substantive concept of the nature of Ks that largely matches that of some group of prior makers of Ks (if there are any) and intends to realize that concept by imposing K-relevant features on the object.
> *(Thomasson 2003: 600)*

On this account, human intentions and concepts establish artifact kinds and determine their membership. Thomasson (2014) also argues that concepts of artifact kinds do not revolve exclusively around intended function. We do often label artifact kinds in accordance with function—flashlight, toothpick, windshield, etc.—but artifacts actually have an array of features captured in their concepts. These include structural or perceptible features that are also sometimes reflected in our terms—fork, red light, etc.—as well as normative features specifying how that kind of artifact is to be treated or regarded. You can blow your nose just as well with a cloth handkerchief as with a paper facial tissue, but you are expected to wash and reuse the former, as opposed to disposing of it in the trash.

The main issue here is similar to the issue we discussed in Section 2. There, it was a question of the reality of artifacts; here, it is a question of the reality of artifact kinds. And if, as we said, there are no pressing reasons for engineers to worry about the reality of artifacts, it might seem that there are no pressing reasons for them to worry about the reality of artifact kinds. But our topic in the present section has even stronger connections with Section 3. We said there that whether or not the *category* 'artifact' is epistemically useful and, if so, how it should be defined, is of concern to anyone working in applied ontology. Similarly, then, how the lower-level classification of artifacts into kinds should be done, and which artifact kinds turn out to be epistemically useful, is of even greater interest to applied ontologists—and more broadly, to the knowledge engineers who construct the knowledge bases used in science, engineering, and other technical fields—than the overarching 'artifact' category itself. A civil engineering knowledge base might well be able to get along without 'artifact' in its upper-level ontology, but it cannot get along very well without some specification in its lower-level ontologies of the artifact kinds in its domain, such as 'bridge,' 'cable,' 'pile driver,' 'steel,' and the like.

The most pressing questions concerning these lower-level ontologies are methodological. How should knowledge engineers and applied ontologists go about identifying artifact kinds? What sorts of evidence for the proper classification of artifacts are most reliable? These methodological questions, oddly enough, initially land us right in the thick of an ancient philosophical debate mentioned several times already. Does human knowledge give us access to the world as it is (realism), or does it give us access only to the world as we conceive or represent it (conceptualism)? Several well-known applied ontologists have recently argued strenuously in favor of realism as the proper basis for knowledge engineering and applied ontology in general (Munn and Smith 2008).[7] One consideration is that the conceptualist view unnecessarily substitutes a world of concepts for the real world of things (Munn 2008). This objection is familiar from decades of work in epistemology that tries to counter the baleful influence of Cartesian- and Kantian-inspired representationalism—a family of related views that boil down to the thesis that what we know is the world as we represent it to ourselves, not the world as it is in itself. But on the opposing, realist family of views, a better understanding of our epistemic situation is that we use representations and concepts as tools through which we gain access the world, not as substitutes for it. The interesting question here for philosophy of engineering, though, is whether there is any methodological fallout from this esoteric dispute. Does adopting a realist as opposed to a conceptualist stance make any difference to how a knowledge engineer or applied ontologist does her job, or to the quality of her results?

Barry Smith (2008) argues for an affirmative answer to these questions. He claims, for example, that the conceptualist knowledge engineer is motivated to define a concept for any term that shows up in reports of beliefs or experiences, such as when a medical patient reports experiences of unicorns. Since the objects of our knowledge are concepts, on the conceptualist view, the unicorn concept must be defined, and unicorns must be admitted as denizens of our ontology, along with horses, narwhals, and the like. But this practice tends to obscure the distinction between unicorns, which are not a natural kind, and horses or narwhals, which are. In turn, this threatens the integrity of the knowledge base, which may generate results that are internally inconsistent, and that certainly will not be externally consistent with the deliverances of science. Furthermore, as Pierre Grenon (2008) points out, the conceptualist approach may yield knowledge bases that, even if internally consistent, cannot be easily connected with each other. If I have unicorns in my conceptualist ontology and you do not, how do we settle the issue in favor of a general ontology that would make our systems interoperable? These difficulties are exacerbated by the need for machine processing of large amounts of data in technical fields, which requires a much higher degree of formalization and standardization than processing in the old-fashioned way by humans ever did. A realist approach, Smith and Grenon both argue, is focused not on what concepts we have, but on what natural kinds really exist, so it can avoid these difficulties by taking the world as the arbiter of ontologies. But it can still handle reports of unicorn experiences, which are, after all, facts about what some people believe. Thus, methodological considerations alone dictate the adoption of realism over conceptualism by knowledge engineers and applied ontologists.

However, this leaves us in a bit of a quandary with regard to artifact kinds. There is no science of artifacts that would tell us what the 'natural' artifact kinds are, the way the science of biology tells us what the natural animal or plant kinds are. As we saw, Crawford Elder does propose a way of unifying our accounts of artifact and natural kinds, but there is no consensus around his or any other unified account. Worse yet, if Amie Thomasson is right, we cannot avoid appeal to concepts in identifying artifact kinds, for it is the history of human concepts and human intentions to realize them in specific artifacts that establishes what kind of artifact we are dealing with in the first instance. On this view, human concepts, not the world, are the arbiters of artifact ontologies. So the debate between conceptualism and realism is hardly settled, and the question of its methodological implications for the work of knowledge engineers and applied ontologists remains open. This is a particularly important issue, because for most branches of engineering, artifacts, not naturally occurring objects, predominate in their domain of activity, and thus in their ontologies.

15.5. Artifact Parts

Our final topic is mereology—the study of parts and wholes—as applied to artifacts. In philosophy, mereology is an established research area, with a history going back to Aristotle, and an impressive set of formal methods developed more recently (Varzi 2016). We already encountered it in Section 2, where it figured in some arguments against the very existence of artifacts. Peter van Inwagen (1990), for example, argues that only living organisms—who can, in effect, assemble their own parts into integrated wholes—really exist. Artifacts, he claims, are not true wholes, but merely elementary particles arranged in a certain way—a potholder is merely a set of particles arranged 'potholderwise,' a teapot a set arranged 'teapotwise,' and so on. If van Inwagen is right, there are no philosophical problems about artifact parts, because there are no artifacts in the first place. However, we concluded earlier that whether or not artifacts exist in this metaphysical sense, engineers still have to deal with erstwhile artifacts in practice. Similarly, we must conclude that even if van Inwagen is right about the metaphysics of parts and wholes, engineers must still take artifacts apart and put them together on a daily basis. So we must ask a further question—do mereological methods for analyzing the relation of parts to wholes have anything to offer the engineer, or at least the philosopher of engineering?

Peter Simons gives a resoundingly negative answer to this question.

> [A] large number of the mereological problems which preoccupy metaphysicians have little or no relevance to engineering practice or theory. Despite this, the concept of part–whole in engineering is not a mere simple application, to be indicated in passing while sticking to the theoretical high road. On the contrary, the mereology of artefacts is rife with problems, for which the philosophical ontologist's mereology is of little or no use.
>
> *(Simons 2013: 152–153)*

To understand this criticism, we must look more closely at the philosophical problems on which mereologists traditionally focus. A good starting point is the straightforward view that the world is made up of simple things—we may call them 'atoms'—which cannot be divided any further into smaller parts, and which are themselves the parts of more complex wholes, such as trees, teapots, stars, and bridges. This commonsense atomism has been problematized in several different ways, depending on what sort of alternate view is contrasted with it. The first alternative which has mightily exercised philosophical mereologists is the possibility that the world is actually made up of 'gunk,' that is, of stuff that can be divided into smaller and smaller parts *ad infinitum*. Another alternative, usually called 'monism,' is that there really is only one thing, viz., the entire universe, and it has no proper parts at all. On these two alternatives there is no plurality of commonsense atoms, either because everything is infinitely divisible or because nothing is divisible. But let us suppose there *are* atoms. What are the principles of composition according to which they form wholes? You may have thought that van Inwagen's answer to this question—that only living things are true wholes composed of proper parts—is unhelpfully radical. But an even more radical alternative favored by some mereologists is that there are no principles of composition at all, so no atoms ever truly combine into wholes. On this view—variously termed 'nihilism' or 'eliminativism'—not only are there no artifacts, there are no organisms or other natural objects, either. At the other end of the spectrum is the equally radical thesis of 'unrestricted'—or 'universal'—composition. On this view, any atom and any other atom (or atoms) form a whole. So there are an infinite number of composite objects, and the world is a vast collection of 'junk,' as some mereologists like to say. You can understand Simons' despair about the relevance of contemporary philosophical mereology for engineering, if these are indeed the leading issues of the day.

Philosophers should at least have been able to help the engineer with definitions of 'part' and/or taxonomies of different kinds of parthood. This is important, because in English—and presumably in other languages with corresponding terms as well—'part' is used in a myriad of ways. Instead, mereologists have defined 'part' in a restrictive way, largely, perhaps, in order to have a concept that lends itself to the formal methods of analysis they favor for discussing the issues we have just reviewed (Varzi 2016; Simons 2013; Simons and Dement 1996). Worse yet, this restricted concept will not get the engineer very far in understanding the parts of her bridge, because it rules out of court some varieties of parthood that are of pressing concern to anyone working with bridges as opposed to, say, trogs (a trog is a thought-experimental object, devised to illustrate unrestricted composition, and consisting in each case of a dog and the trunk of a tree; Korman 2015). For example, as Peter Simons (2013: 156) points out, we do speak of the materials or ingredients of an artifact as parts of it, and these varieties of parthood are clearly of great importance for engineering. But they are explicitly excluded from the standard concept of 'part' favored by philosophical mereology (Varzi 2016: §1).

Simons (2013) goes on to list several additional concepts of 'part' that would be important for a mereology of artifacts suited to engineering. These include:

- Physical part—a part that can stand on its own as a physical object if separated from the whole artifact; for example, the left half of a birthday cake.
- Salient part—a part that stands out, either by design or relative to the interests of some observer(s); for example, the frosting on the birthday cake.
- Engineering part—a part that has a unitary role to play in the engineering processes of design, assembly, repair, or retirement of an artifact. For example, the frosting plays a unitary role in assembling the birthday cake, but does not have a unitary role in the retirement (that is, eating) of the cake.
- Functional part—a part that has a specific function in the operation of the whole artifact; for example, the eggs in a cake function to bind the other ingredients together.

Simons also mentions that a mereological account of processes would be useful for engineers, whose activities are as likely to involve management of the temporal parts of physical, chemical, and biological processes as management of the spatial parts of artifacts. Complicating this picture, a number of philosophers have argued that processes such as performances (Dipert 1993) or actions (Evnine 2016) are artifacts, so a mereology of artifacts may ultimately require a mereology of processes as one of its components. Be that as it may, a mereological framework for analyzing processes into their component temporal parts and describing the relationships between them—a framework that would utilize the sophisticated logical and mathematical tools already developed by mereologists for enduring objects—would certainly be a boon to process engineers. But, as Simons notes, this is a project the mereologists have so far not taken up.

15.6. Conclusion

The study of artifacts in philosophy is a growth area, but it is still in its infancy. Moreover, its results so far are often too remote from the actual work and professional interests of engineers to be very helpful for the study of artifacts by philosophers of engineering, or by even the most theoretically minded of engineers. Nevertheless, we have identified some important points of contact that have much potential for further development. These include a common interest in the status of 'artifact' as a basic category, the nature and status of artifact kinds, and the mereology of artifacts. The open questions in these areas are substantial, and merit the attention of philosophers and engineers alike.

Related Chapters

Chapter 16: Engineering Objects (Wybo Houkes)

Chapter 21: Engineering Design (Peter Kroes)

Chapter 53: Engineering and Contemporary Continental Philosophy of Technology (Diane P. Michelfelder)

Chapter 54: Engineering Practice From the Perspective of Methodical Constructivism and Culturalism (Michael Funk and Albrecht Fritzsche)

Further Reading

Applied Ontology (Journal specializing in ontological analysis and the conceptual modeling of the resulting ontologies in applied contexts, including engineering.)

Dipert, R. (2014). Artifact. In M. Kelly (ed.), *Encyclopedia of Aesthetics*, 2nd ed. New York: Oxford University Press. (State-of-the-art encyclopedia article on the philosophical study of artifacts, with particular reference to aesthetic issues.)

Margolis, E. and Laurence, S. (eds.) (2007). *Creations of the Mind: Theories of Artifacts and Their Representation.* Oxford: Oxford University Press (This interdisciplinary anthology deals with the metaphysics and epistemology of artifacts.)

Preston, B. (2018). Artifact. In E. Zalta (ed.), *Stanford Encyclopedia of Philosophy* (State-of-the-art encyclopedia article on the philosophical study of artifacts, with particular reference to metaphysical and epistemological issues.)

Notes

1. The study of how wholes are composed out of proper parts is called 'mereology.' We will return to it in Section 5 of this chapter.
2. For a complete catalog of these positions, see Korman (2015), Chapter 3. See also Simons (2013) for a somewhat more pessimistic view of the baleful influence of the extremes on contemporary philosophical mereology.
3. There is another technical use of 'artifact' in the sciences, to indicate something in the data or results that would not occur naturally but was produced as a side effect of the investigative procedures employed.
4. Steven Vogel (2015) and Beth Preston (2013) have also recently questioned the integrity and usefulness of the categorial distinction between artifacts and naturally occurring objects, although their concerns and arguments are different, both from each other and from Sperber.
5. See Ludger Jansen (2008) for a review of the current top competitors in the applied ontology arena.
6. See, for example, the recent *Applied Ontology* special issue edited by Mike Bennett and Kenneth Baclawski (Vol. 12, nos. 3–4, 2017).
7. See especially the introduction by Katherine Munn, and the articles by Pierre Grenon (Chapter 3), Barry Smith (Chapter 4), and Ludger Jansen (Chapter 7).

References

Aristotle (1984). Categories, Nichomachean Ethics, Metaphysics, Physics. In J. Barnes (ed.), *The Complete Works of Aristotle*, Vols. I and II. Princeton: Princeton University Press.

Baker, L. (2007). *The Metaphysics of Everyday Life: An Essay in Practical Realism.* Cambridge: Cambridge University Press.

Borgo, S., Franssen, M., Garbacz, P., Kitamura, Y., Mizoguchi, R. and Vermaas, P. (2014). Technical Artifacts: An Integrated Perspective. *Applied Ontology*, 9, 217–235. doi:10.3233/AO-140137

Borgo, S. and Vieu, L. (2009). Artefacts in Formal Ontology. In A. Meijers (ed.), *Philosophy of Technology and Engineering Sciences (Handbook of the Philosophy of Science*, Vol. 9). Amsterdam: Elsevier, pp. 273–307.

Dipert, R. (1993). *Artifacts, Art Works, and Agency*. Philadelphia: Temple University Press.

Elder, C. (2004). *Real Natures and Familiar Objects*. Cambridge: MIT Press.

Elder, C. (2007). The Place of Artifacts in Ontology. In E. Margolis and S. Laurence (eds.), *Creations of the Mind: Theories of Artifacts and Their Representation*. Oxford: Oxford University Press, pp. 33–51.

Evnine, S. (2016). *Making Objects and Events: A Hylomorphic Theory of Artifacts, Actions, and Organisms*. Oxford: Oxford University Press.

Garbacz, P. (2013). Artifacts and Family Resemblance. *Review of Philosophy and Psychology,* 4, 419–447. doi:10.1007/s13164-013-0145-4

Grandy, R. (2007). Artifacts: Parts and Principles. In E. Margolis and S. Laurence (eds.), *Creations of the Mind: Theories of Artifacts and Their Representation*. Oxford: Oxford University Press, pp. 18–32.

Grenon, P. (2008). A Primer on Knowledge Representation and Ontological Engineering. In K. Munn and B. Smith (eds.), *Applied Ontology: An Introduction*. Frankfurt: Ontos Verlag, pp. 57–81.

Hilpinen, R. (1992). Artifacts and Works of Art. *Theoria,* 58, 58–82.

Hilpinen, R. (2011). Artifact. In Edward N. Zalta (ed.), *The Stanford Encyclopedia of Philosophy,* Winter 2011 ed. https://plato.stanford.edu/archives/win2011/entries/artifact/

Houkes, W. and Vermaas, P. (2014). On What Is Made: Instruments, Products and Natural Kinds of Artefacts. In M. Franssen, P. Kroes, T. Reydon and P. Vermaas (eds.), *Artefact Kinds: Ontology and the Human-Made World*. Cham, Switzerland: Springer, pp. 167–190.

Jansen, L. (2008). Categories: The Top-Level Ontology. In K. Munn and B. Smith (eds.), *Applied Ontology: An Introduction*. Frankfurt: Ontos Verlag, pp. 173–196.

Kant, I. (1781/1958). *Critique of Pure Reason*. Trans. Norman Kemp Smith. London: Palgrave Macmillan.

Kassel, G. (2010). A Formal Ontology of Artefacts. *Applied Ontology,* 5, 223–246. doi:10.3233/AO-2010-0078

Khalidi, M. (2013). *Natural Categories and Human Kinds: Classification in the Natural and Social Sciences*. Cambridge: Cambridge University Press.

Kitamura, Y. and Mizoguchi, R. (2010). Characterizing Functions Based on Ontological Models From an Engineering Point of View. In A. Galton and R. Mizoguchi (eds.), *Formal Ontology in Information Systems*. Amsterdam: IOS Press, pp. 301–314.

Korman, D. (2015). *Objects: Nothing Out of the Ordinary*, Oxford: Oxford University Press.

Munn, K. (2008). Introduction: What Is Ontology For? In K. Munn and B. Smith (eds.), *Applied Ontology: An Introduction*. Frankfurt: Ontos Verlag, pp. 7–19.

Munn, K. and Smith, B. (eds.) (2008). *Applied Ontology: An Introduction*. Frankfurt: Ontos Verlag.

Preston, B. (2013). *A Philosophy of Material Culture: Action, Function, and Mind*. New York: Routledge.

Reydon, T. (2014). Metaphysical and Epistemological Approaches to Developing a Theory of Artifact Kinds. In M. Franssen, P. Kroes, T. Reydon, and P. Vermaas (eds.), *Artifact Kinds: Ontology and the Human-Made World*. Cham, Switzerland: Springer, pp. 125–144.

Ryle, G. (1949). *The Concept of Mind*, Chicago: University of Chicago Press.

Schick, K. and Toth, N. (1993). *Making Silent Stones Speak: Human Evolution and the Dawn of Technology*, New York: Simon & Schuster.

Simons, P. (2011). Ontology in Engineering. In K. Guy (ed.), *Philosophy of Engineering*, Volume II, London: Royal Academy of Engineering. www.raeng.org.uk/philosophyofengineering, pp. 21–25.

Simons, P. (2013). Varieties of Parthood: Ontology Learns From Engineering. In D. Michelfelder, N. McCarthy and D. Goldberg (eds.), *Philosophy and Engineering: Reflections on Practice, Principles and Process*, Dordrecht: Springer, pp. 151–163.

Simons, P. and Dement, C. (1996). Aspects of the Mereology of Artifacts. In R. Poli and P. Simons (eds.), *Formal Ontology*. Dordrecht: Kluwer, pp. 255–276.

Smith, B. (2003). Ontology. In L. Floridi (ed.), *Blackwell Guide to the Philosophy of Computing and Information*, Oxford: Blackwell, pp. 155–166.

Smith, B. (2008). New Desiderata for Biomedical Terminologies. In K. Munn and B. Smith (eds.), *Applied Ontology: An Introduction*. Frankfurt: Ontos Verlag, pp. 83–107.

Sperber, D. (2007). Seedless Grapes: Nature and Culture. In E. Margolis and S. Laurence (eds.), *Creations of the Mind: Theories of Artifacts and Their Representation*. Oxford: Oxford University Press, pp. 124–137.

Studtmann, P. (2017). Aristotle's Categories. In Edward N. Zalta (ed.), *The Stanford Encyclopedia of Philosophy,* Fall 2017 ed. https://plato.stanford.edu/archives/fall2017/entries/aristotle-categories/

Thomasson, A. (2003). Realism and Human Kinds. *Philosophy and Phenomenological Research,* 67, 580–609.

Thomasson, A. (2007a). *Ordinary Objects*. Oxford: Oxford University Press.

Thomasson, A. (2007b). Artifacts and Human Concepts. In E. Margolis and S. Laurence (eds.), *Creations of the Mind: Theories of Artifacts and Their Representation*. Oxford: Oxford University Press, pp. 52–73.

Thomasson, A. (2014). Public Artifacts, Intentions, and Norms. In Maarten Franssen, Peter Kroes, T.A.C. Reydon, and Pieter E. Vermaas (eds.), *Artefact Kinds: Ontology and the Human-Made World*. Cham, Switzerland: Springer, pp. 45–62.

Thomasson, A. (2018). Categories. In Edward N. Zalta (ed.), *The Stanford Encyclopedia of Philosophy,* Spring 2018 ed. https://plato.stanford.edu/archives/spr2018/entries/categories/

van Inwagen, P. (1990). *Material Beings*. Ithaca: Cornell University Press.

Varzi, A. (2016). Mereology. In Edward N. Zalta (ed.), *The Stanford Encyclopedia of Philosophy,* Winter 2016 ed. https://plato.stanford.edu/archives/win2016/entries/mereology/

Vogel, S. (2015). *Thinking Like a Mall: Environmental Philosophy After the End of Nature*, Cambridge: MIT Press.

16

ENGINEERING OBJECTS

Wybo Houkes

16.1. Introduction

Engineering objects—the products of engineering activities—are everywhere. Perceptually, high-rise offices, the smell of kerosene or notification sounds from social media command our attention. Practically, processed food and drinks, public transportation systems and smartphones enable us to fulfil our basic needs, earn an income and maintain social relations. More personally, we use prosthetic devices and medication—and perhaps, someday, neuro-enhancement—to engineer our bodies. On a large scale, transnational supply chains and communication systems—and perhaps, someday, geo-engineering projects—span the entire planet.

There is even more than meets the eye, palate or user interface. Like all experts, engineers distinguish many more objects than non-experts in their fields. Electrical engineering features metal-oxide-semiconductor field-effect transistors (MOSFETs) and active terminators; software engineers may be familiar with decorators and cloned repositories; mechanical engineering results in such objects as aeolian anemometers and spiral concentrators; and textbooks in chemical engineering mention IsoSiv and Weldon processes—to give only a handful of examples. Some of these engineering objects are what ordinary-language philosopher J.L. Austin called 'medium-sized dry goods', such as "chairs, tables, pictures, books, flowers, pens, cigarettes" (1962: 8), which are familiar enough in everyday life. Other engineering objects occupy different levels of organization and are only perceptually accessible and intelligible—if at all—to specialists; they range from micro-components (e.g., the circulators of quantum computers) to supersystems (e.g., international civil aviation), and they include products as well as processes.

This chapter examines engineering objects through the lens of ontology, i.e., the philosophical sub-discipline concerned with questions regarding what exists and what exists most fundamentally. In Section 2, I give a brief introduction into this sub-discipline, emphasizing one currently influential perspective—that of Willard Van Orman Quine's meta-ontology. The focus in this perspective, and in the resulting tradition of analytic ontology, lies on making an inventory of real entities through logical analysis of statements, revealing expressions of ontological commitment. After explaining this focus, I outline two questions and multiple answers phrased within analytic ontology regarding engineering objects, namely:

- How, if at all, are we committed to the existence of engineering objects? (§3)
- Are there real kinds of engineering objects? (§4)

As shall become clear, the ontology of engineering objects is still at best an emerging topic or side issue. Tackling the issues posed by the products of engineering activities requires in-depth, specific analysis and a broader perspective than that of the dominant tradition in analytic ontology. To analyse engineering objects in appropriate terms, philosophers need to go beyond their typical range of examples, which mostly involve medium-sized dry goods that may be products of artisanship rather than engineering and that are considered from a user perspective. There is, to put it bluntly, more to engineering than sitting on chairs and using stones as paperweights. A broadening of perspective and deepening of analysis is required in at least the following respects: (a) reflecting on the full variety of engineering objects, including processes and systems; (b) finding appropriate criteria for ontological commitment, grounded in engineering practice and non-linguistic modes of involvement with engineering objects; (c) explicating similarities and respecting differences between biological organisms, art works, social institutions, scientific entities, and engineering objects.

16.2. What Is Ontology?

The word 'ontology' goes back some 350 years. As a philosophical sub-discipline, i.e., a set of questions and basic approaches in answering them, ontology is much older, going back to the discussion of a 'science of being qua being', or of 'being as such' in Aristotle's *Metaphysics*. Prototypical artefacts—mid-sized, artisanally produced dry goods such as tables and statues—play an important role in this work, for instance in the idea of substances as 'hylomorphic' compounds (*Metaphysics IV*) and the distinction between four types of causes (*Metaphysics V*). The influence of these ideas in Western philosophical thought can hardly be overestimated and has not ended. One contemporary tradition in ontology consists of the ideas of 'neo-Aristotelian' metaphysicians such as E.J. Lowe and Tim Crane. Another stems from Heidegger's specific take in *Being and Time* on Aristotle's questions and answers, and from Merleau-Ponty's phenomenology of perception. These reflections are again largely grounded in artisanal practice and user involvement with tools and utensils. This tradition continues, with extensions to scientific instrumentation and engineering practice, in 'continental' and post-phenomenological approaches such as Ihde's (1990) and Verbeek's (2005).

This chapter focuses largely on a perspective on ontology that is more circumscribed both in its method and in its reception of the subject's long history. This perspective originated with Quine's essay "On What There Is" (1948) and has become so dominant in the 'analytic' or 'Anglo-American' philosophical outlook that a recent textbook can present it as 'The Standard View' (Berto and Plebani 2015, especially Chs. 2 and 3). One reason for its dominance is its seemingly straightforward and explicit methodology for posing and answering ontological questions, i.e., its meta-ontology. As a main objective for ontology, Quine proposes—as can be gathered from the title of his essay—the following:

1. Inventory Objective: ontology is first and foremost concerned with determining what there is, i.e., making a complete inventory of entities.

At first glance, this objective seems clear and manageable. Inventorying entities is a familiar enough task in many practical contexts—think of sorting your clothes or describing the complete collection of the Louvre—and it is often combined with characterizations or distinctive labels for the entities— think of the specification of various terms in a contract (e.g., "beneficiary" and "premium" in a life-insurance contract) or a mathematical theory (e.g., "polygonal knot" and "wild knot" in knot theory), or a list of names for the staff members in a philosophy department.

What distinguishes the proposed task from these more mundane activities is that it does not ask for a list of things in some circumscribed context or section of reality, but rather for a complete list of *everything*. This is not only a task of mind-blowing scope but, in the absence of a clear practical or theoretical context, also one that creates difficulties in making inventorying decisions. Suppose, for instance, that you would start the grand project at home, by making a list of your clothes. Do you put each sock on the list as a separate entry, or only pairs? What about the loose button in the drawer that was once on some long-lost shirt? Typically, we get our guidance in answering such questions from the practical context: if you try to find out what to wear, pairs of socks belong on the list and loose buttons do not. Yet free of context, any choice seems equally defensible. The same goes for the Louvre collection: given the practical context, it makes sense to put some things in the building— the Mona Lisa, Venus de Milo—on the list and not other things—Mona Lisa's smile, Venus de Milo postcards in the museum shop. Likewise, it would be wise to put on the list some things that are not in the building at all—such as items loaned to the Rijksmuseum. By abstracting from such contexts, Quine's Inventory Objective creates the need for a criterion that aids in sorting and counting entities, and in telling what belongs on the list from what does not.

Here, Quine offers a proposal that is largely in line with the 'linguistic turn' taken by (analytic) philosophy in the late 19th and early 20th centuries:

2.   Criterion of Ontological Commitment ("To be is to be the value of a bound variable"): to determine what there is, one should paraphrase sentences into canonical notation and ascertain their domains of quantification.

This criterion—which might seem arcane at first glance—contains several distinct restrictions and recommendations for doing ontology. Some of these are specific to Quine's approach and some are shared much more broadly in analytic ontology.

To start with one shared element: the criterion focuses exclusively on *language use* as a guide to ontology. Putting it crudely: for your clothes-inventorying project, it does not matter what you wear, but only what you say you wear. Although broader conceptions of ontological commitment than Quine's have been offered in the past decades (e.g., Rayo 2007), they retain the focus on language use. A second broadly shared element of the criterion is that language use should not be taken at face value. You might for instance claim—probably correctly—that you do not possess any four-armed jackets. Do we now need to take you at your word and maintain that you, by using the term, claim that there are four-armed jackets? Fortunately, this statement can be suitably paraphrased such that they do not 'express ontological commitment' to four-armed jackets; it does not provide grounds for adding these items to your list. You are claiming of all things in your wardrobe that they are not four-armed jackets—rather than claiming of something that is not in your wardrobe that it is. Thus, terms must be studied in their *linguistic* context (minimally, full sentences) to find out whether they denote entities or play some other role. Careful reformulation may be needed to find out how statements may be true without requiring belief in all manner of strange objects. For this reformulation, criterion 2 suggests a preferred technique: paraphrasing into 'canonical notation'. For most ontologists, this means some formal language; for Quine, it specifically means first-order predicate logic, in which quantifiers ('all', 'some') range over or bind individual variables in expressions such as:

$$\forall x \, (Wx \rightarrow \neg Fx)$$

where $\forall$ is the all-quantifier, W means 'is a wardrobe item', and F means 'is a four-armed jacket'. Here, the domain of quantification are wardrobe items rather than four-armed jackets, so only the former (whatever they are) go on the inventory list. Quine's implementation of the criterion does not allow richer formal language, such as modal logic, nor second-order logic in which quantifiers can bind predicate variables (potentially committing us to entities like four-armed-jackethood)—but this is an element, implicit in the criterion, that is not adopted by everyone in analytic ontology.

The example shows that Quine's criterion does not involve 'reading off' an ontology from what people are saying. Paraphrasing into canonical notation can have the effect of dismissing many objects to which ordinary-language statements appear to refer. To some extent, this amounts to a sanity check—witness the four-armed jackets—but it risks turning into a sanitization program. "On What There Is" starts by stating that, simply, everything is there, but the rest of the essay shows a strong concern with tidiness, for instance in referring to Plato's tangled beard, allergies, a taste for desert landscapes and—perhaps most colourfully—a "slum . . . [that] is a breeding ground for disorderly elements" (1948: 4).

Here, Quine's project becomes a kind of philosophical Lean Inventory Management: the objective is not just to make an inventory, but to make it as small as possible. This turns paraphrasing into an economizing technique: it can show how apparent commitments to entities are "avoidable manners of speaking" (1948: 13). Neo-Quineans such as Peter van Inwagen (1990, 1998) and Ted Sider (2011) have developed Quine's rather sloganesque guidelines and presented arguments in favour of drastically reducing the list of real entities or otherwise revising our intuitions about what is there. In the next section, I present some of their arguments and underlying principles, with a specific eye to engineering objects—which, as we shall see, are all 'disorderly elements' in the neo-Quineans' book.

Not everyone working in the tradition of analytic ontology shares the neo-Quinean quest for austerity, with its strongly revisionary analysis of ordinary language. Other ontologies are primarily descriptive: in line with the Inventory Objective, they do not go beyond identification of all alleged constituents of reality and use analysis only to weed out inconsistencies and strongly counterintuitive consequences. For some, such a descriptive ontology looks for fundamental concepts in a scheme shared by all human beings: what, as humans, we cannot help believing that there is (Strawson 1959). Other, 'deflationist' or 'easy' ontologies (e.g., Chalmers 2009; Thomasson 2015) have dropped this aspiration. They take existence claims in ordinary discourse more or less at face value, without insisting on paraphrases into canonical notation as required by Quine's criterion of ontological commitment. In the next section, I briefly discuss how this offers a route to save a commitment to engineering objects, albeit one that is still in need of further specification and extension.

16.3. How, if at All, Are We Committed to the Existence of Engineering Objects?

As indicated earlier, Quine's meta-ontology has led to a radical austerity program, to which many alleged entities fall prey. Although Quine himself did not consider the products of engineering or artisanship, others did. They concluded, after some reasoning, that we had best maintain that there are in fact no tables or keyboards, but only 'simples' arranged table- or keyboard-wise. In this section, I review some central principles and arguments that have been used in support of this eliminativist or nihilist view regarding engineering objects, as well as some actual and potential responses.

Crucially, in developing his meta-ontology, Quine does not appeal only to the broad Inventory Objective (1) and a criterion of ontological commitment that focuses on quantifiers in canonical notation (2). He endorses a strong form of *scientism*, which determines both the contents of the desired inventory and the method for making it. In terms of content, our best current scientific theories provide the statements that should be translated into canonical notation: these theories, rather than, say, commentaries on a soccer match or the lexicon of engineering textbooks, express our ontological commitment. In terms of method, ontology should resemble the natural sciences: it should be guided by epistemic virtues such as explanatory power, empirical adequacy, and—as discussed earlier—simplicity. Furthermore, Quine insists on the importance of *identity criteria*—through the slogan "No entity without identity": presumably to prevent double counting and other forms of disorderliness, types of entities should have clear conditions under which they are identical and non-identical.

These additional principles set high standards. Many alleged entities, such as light sabres and hor-cruxes, fail to feature in our best scientific theories. Others do not feature 'indispensably', as Quine and others have argued by offering set-theoretical paraphrases of mathematical concepts such as numbers. Finally, the focus on clear identity criteria discredits entire types of entities, such as properties (e.g., the colour red), propositions (e.g., that there are no horcruxes) and possible entities (e.g., an Olympic-medallist Wybo Houkes).

None of this may seem especially worrying to engineers. Yet Quinean standards also create scepticism about the existence of the products of engineering activities. The reason is that engineering, or any artisanal activity, is not *creatio ex nihilo*. Engineering objects are, perhaps with the exception of synthetic elements, composites of more basic components, materials, or processes. Smartphones are composed of silicon and polymers; the Weldon process is composed of basic reactions involving calcium manganite; and the civil-aviation system is composed of, among others, airplanes, air traffic controllers, and bodies of legislation. As composites, all these engineering objects fall under the scope of one of the thorniest issues in analytic ontology.

This *composition problem* may be set up as follows. Suppose that we accept that physical objects are identical if and only if they have the same spatiotemporal location. Then, since every composite object, i.e., everything that has parts, coincides with its parts, it is identical with the sum of its parts. It has no clear identity criteria over and above those of this mereological sum, and thus it can and should be scratched from the inventory list. Moreover, any causal effect of the composite object can likely be understood as an effect of its composing parts and their arrangement. So, what you think of as your smartphone does not, strictly speaking, exist or do anything: there are only suitably arranged chunks of various materials, ultimately of whatever physics currently identifies as smallest, indivisible particles or 'simples'. Hence, Neo-Quineans draw their conclusion about there being only simples being arranged everywhich-engineering-object-wise (see, e.g., van Inwagen 1990; Sider 2001; Merricks 2001; Renz 2016).

The composition problem is not exclusive to the Quinean meta-ontology, and the approach is compatible with other positions in analytic ontology than the 'mereological' or 'compositional' nihilism described earlier. It does, however, put in sharp relief both the problem and some rather counterintuitive solutions. Part of what makes the problem so salient is its broad sweep: seahorses and comets are just as much composite objects as smartphones and cars. Discussions cluster around possible exceptions—van Inwagen (1990), for instance, recognizes organisms as simples—and specific puzzle cases, some of which are artisanal, e.g., the Ship of Theseus and a statue of the biblical king David made from a single lump of clay (see, e.g., Rea 1997; Paul 2010).

Mereological nihilism does have some unattractive consequences specific to engineering, however. If we accept it, we need to paraphrase not only our entire discourse about tables or smartphones— but also about engineering activities themselves. Engineers often describe their core activity of designing as description of a physical object (say, a smartphone) that can fulfil a function (Vermaas et al. 2011: 21). If there would, strictly speaking, be no such objects to be described, engineers would do better to characterize their core activity as describing a new arrangement of simples, say, suspension-bridge-wise.

Engineers might find it annoying, or just unnecessary, to have their self-understanding corrected by philosophers, only on the basis of a preference for ontological austerity. Then, analytic ontology offers several alternatives. Many of those distinguish composite objects from their constitutive elements in terms of *persistence* conditions rather than mere identity conditions (under the slogan "Constitution is not identity"). Your smartphone, intuitively, does not cease to exist if you replace its battery; nor do you bring into existence a new object by leaving fingerprints on the screen; and few people would want to call either type of activity 'engineering'. Still, the smartphone *depends* on the arrangement of simples; it cannot change without a change in the simples, and its properties are intimately but intricately related to those of the simples. Despite (or perhaps because of) the best efforts of ontologists, there is no widely accepted, uncontroversial way to spelling out the dependence relation (see, e.g., Correia 2008) and (the role of) persistence conditions in adequate detail (e.g., Haslanger and Kurtz 2006).

A shared problem for alternatives to nihilism is that of arbitrariness. If one accepts that there is a smartphone in addition to—but depending on—the simples composing it, why restrict oneself to the phone? There are also its battery, its screen, the composite of battery and screen, and the composite of the upper right-hand corner of the screen and the midsection of its battery. Conditions on what counts as a 'proper' or 'real' composite seem called for, to prevent an unlimited proliferation of (partly or wholly coincident) composites. Such conditions need to be non-circular and non-arbitrary. Many also exclude reference to social conventions, ordinary language, or even any intentional activity in selecting which composites to save from elimination.

At this point, an advocate of engineering objects might seek support in Quine's scientism rather than a specification of identity or persistence conditions. After all, many theories in the engineering sciences feature terms that denote entities. For example, discontinuous automatic control theory, as described in Irmgard Flügge-Lotz's 1953 textbook *Discontinuous Automatic Control*, commits us to the existence of 'bang-bang controllers'—systems that switch abruptly from one state to another when a set point is reached, the way a simple thermostat responds to a temperature setting.

Neo-Quineans do not, however, accept just any theory as the raw material for paraphrasing and reading off ontological commitments: we are not committed to horcruxes through Albus Dumbledore's explanation of their construction and operation, as described in J.K. Rowling's 2005 novel *Harry Potter and the Half-Blood Prince*. Only our current *best* scientific theories qualify. Yet it is difficult to set workable standards of merit for theories that do not beg the ontological question and that are sufficiently welcoming to engineering objects. Requiring theories to be true, for instance, either sets a standard that is impossible to meet for any empirical theory in insisting that it be *de facto* true or merely repeats that we paraphrase those theories that we *accept* to be true. Likewise, focusing only on 'fundamental' theories assumes that we already know which theories feature the fundamental constituents of reality—which is a deliverable for an ontology, rather than its input.

One prominent standard proposed in the philosophy of science invokes explanatory power. We are committed to electrons, but not to their orbits, because the latter feature only in phenomenological models such as the Bohr atom and the former in genuinely explanatory theories like quantum field theory. Focusing on theories that are currently held to have the most explanatory power is not obviously circular, although it does require one to explicate what makes one explanation better than another. However, this standard favours (objects that feature in) theories offered in the natural sciences, such as electrons, and is unwelcoming to engineering objects such as bang-bang controllers. Proposed distinctions between the natural sciences and the engineering sciences (see, e.g., Radder 2009 for an overview) often emphasize that the former revolve around predicting and explaining reality, whereas the latter focus on changing it. It does not follow from the comparative lack of focus on explanation that engineering theories have *no* explanatory power. Yet it seems *prima facie* unlikely that some of our best explanations are produced by researchers whose focus lies elsewhere. One of the most important and general theoretical branches of engineering science—control theory—aids first and foremost in controlling dynamic systems rather than explaining or predicting their behaviour. Contributions to control theory, such as those discussed in the Flügge-Lotz textbook mentioned earlier, feature objects such as bang-bang controllers and open-loop systems for which clear and distinct control strategies, with practical advantages and drawbacks, may be formulated. Taking such theories as expressing ontological commitments requires a more diverse set of standards for appraising the merit of theories, or of any kind of knowledge produced in the engineering sciences.

Quine's meta-ontology, with its arsenal of paraphrases, its emphasis on simplicity and identity conditions, and its scientism, leaves no room for engineering objects: a desert landscape does not feature smartphones or suspension bridges. A more welcoming approach may be found in the deflationist ontologies that were mentioned at the end of the previous section. These, roughly put, maintain that issues regarding the existence of objects can typically be easily resolved after some conceptual analysis and empirical investigation, within certain rule-governed ways of speaking. Thus, the question "Is there a prime number greater than 131?" affords a clear, positive answer in number theory; and we can likewise affirm that bang-bang controllers exist by consulting control theory. This does not mean that the existence of bang-bang controllers is somehow relative to this theory—only that this theory, at least for now, governs use of the concept 'bang-bang controller'. Indeed, most deflationists regard questions such as "Do numbers exist?" outside the context of systems of rules that govern number concepts, as meaningless or, at best, as questions about the practical merits of the systems (Carnap 1950). We may well have several practically useful ways of speaking about the same spatiotemporal region, calling something a 'bang-bang controller' or 'thermostat', or describing it as an aggregate of plastic and metal. There is no single correct answer to the question what it is, unless we can do away with all ways of speaking but one.

Deflationism is, arguably, more welcoming to engineering objects, although not specifically so: it is tolerant towards all objects that feature in sufficiently well-specified ways of speaking. Yet—in line with Quine's original criterion—non-linguistic forms of practical involvement are still not regarded as ways of expressing ontological commitment. However, the more extensive and fine-grained awareness of engineering specialists does not seem grounded in *theories* or other ways of speaking alone. Also in terms of persistence, relational facts and causal effects—to mention a few concepts around which alternatives to nihilism revolve—our conception of engineering objects has deep roots in the role that these objects play in our lives and society. You may find that, after replacement of the battery or the latest software update, your smartphone is just not what it used to be; and any attempt at distinguishing stages of the civil-aviation system or generations of mobile telephony involves characterization of its societal impact and embedding. As contentious as such an 'anthropocentric' ontology might be for comets or seahorses, it seems fitting for engineering objects. Analysing the persistence of smartphones or aviation systems *without* considering their role in collective and individual use practices would seem more arbitrary (or outright pointless) than analysing them in close connection with such practices; and analysing these objects exclusively through the lens of language use appears—for engineering objects as well as for, say, musical works—to be curiously 'linguo-centric'.

16.4. Which Kinds of Engineering Objects Are Real?

The issues discussed in the previous section concern the ontology of engineering objects in general: they do little to differentiate these objects in terms of our ontological commitment to them. Some contrast is provided by recent work on artefactual kinds (often labelled 'artefact kind'; I prefer the broader term here to avoid an implicit focus on medium-sized dry goods).

Considerations of kind membership are central to analytical ontology, because kinds have strong connections to persistence and identity conditions. Things can, however, be classified in many ways: a bear may also be a zoo animal (like a penguin), a predator (like a puma), or a thing of beauty (like a poem). One concern has therefore been to give general conditions for kind terms that are more than conventional labels and refer to real kinds. Most of these conditions rule out artefactual kinds as real kinds, with few exceptions and together with most kinds of objects distinguished in the social and behavioural sciences.

David Wiggins (2001), for example, requires that there be laws of nature that govern the behaviour of objects *qua* members of a kind. This does not hold for artefactual kinds: screwdrivers come in all kinds of shapes and sizes, and operate on a variety of mechanisms (e.g., hand-held or electric). There are, therefore, no laws that govern screwdrivers as such. Rather, 'screwdriver' and other artefactual kind terms are closely related to ascriptions of functionality: screwdrivers are paradigmatic examples, but the same goes for the terminators, controllers, and repositories mentioned earlier. The functions of engineering objects are, moreover, often understood as whatever their designer intended them to do; this in contrast to other dispositions the objects might have, or any side effects of the realization of their functions. This appears to make artefactual-kind membership directly dependent on design activities, which are paradigmatically intentional—and ontologists tend to distrust reference to intentional activities in specifications of kind terms, because they introduce an element of conventionality: 'screwdriver' might be an arbitrary or merely instrumental label rather than a term for a real kind (see Baker 2007; Elder 2014 for a presentation of some of the reasons from different perspectives).

Few choose to accept that functions are essential to artefactual kinds, that they are essentially dependent on designer intentions, and that nevertheless virtually all artefactual kinds are real (but see Baker 2004). Most attempts to save artefactual kinds can be distinguished by how they resist this conclusion, and by the scope of their ambitions: they all select some but not all artefactual kind terms as corresponding to real kinds.

A first attempt agrees that artefactual-kind membership is (co-)determined by functions. However, in its analysis of these functions, it builds on analyses that do not necessarily refer to intentions, such as the etiological theory of functions (e.g., Millikan 1984; or, more specifically for artefacts and more explicitly non-intentionalist: Preston 2013, Part II). This theory associates an item's so-called *proper* functions with historically selected effects of the item's predecessors. In most variants of the theory, selective forces are those of natural selection; in some versions specific for artefacts, those of market economies. Thus, roughly, screwdrivers have their proper functions not because of anyone's intentions or decisions, but because they stand in long traditions of commercially successful items. One way of developing this 'non-intentionalist' ontology of artefactual kinds (Elder 2004, 2007) identifies 'copied artefact kinds' by a cluster of three properties: their proper function, their shape, and their 'historically proper placement' (i.e., realization of selected effects alongside members of other copied kinds, such as specific kinds of screws for specific kinds of screwdrivers). This only admits the reality of those artefactual kinds that are associated with a characteristic shape and a stable and well-circumscribed 'commercial history', such as specific types of chairs, tables and screwdrivers; one stock example is the 1957-design Eames desk chair. Broad artefactual kinds such as jet engines or capacitators, as well as kinds of processes, artificial materials, digital items, and systems, do not exist as such because they lack a characteristic shape (Thomasson 2009). The same goes for kinds that successfully adapted to changing markets; or that are successful because they fit a (potentially changing) variety of marginal niches. Elder (2014, Sect. 3.5) argues that his proposal does admit the reality of broad kinds such as corkscrews and paperclips; characteristically, for his function-oriented ontology, he does so by distinguishing between various proper functions. Still, this non-intentionalist, proper-function-based ontology saves little more than kinds of medium-sized dry goods: it would entail that desk chairs exist as such, but that the civil aviation system is just a collection of airplanes, landing strips, and other things that do have a characteristic shape, proper function, and proper placement.

A second attempt to save artefactual kinds acknowledges that such kinds are conventionally described in terms of some intentional activities or of their functionality, but maintains that they may nevertheless have characteristic principles of activity, or may be governed by kind-specific laws. Again, this entails that only some artefactual kind terms refer to real kinds. Functional terms typically do not: paperclips and corkscrews vary as much in principles of activity (i.e., specific operational principles or mechanisms) as they do in shape. One specific 'activity-based' proposal therefore rules out the existence of 'utensils' such as cooking pots and hammers and only admits the reality of members of so-called 'machine kinds', the members of which "do something by their own nature" and "embody a distinctive *unifying principle of activity* which is constrained by *sortal-specific . . . laws of engineering*" (Lowe 2014: 24, italics in original). Examples given include mechanical pendulum clocks and piston engines. This proposal puts (a selection of) artefactual kinds on a par with other real kinds, since it imposes similar conditions. Still, implementing it firstly requires non-question-begging ways of distinguishing *laws* of engineering from other sortal-specific descriptions (e.g., of hammers or paperclips). Secondly, the proposal seems curiously limited to *mechanical* engineering in its choice of examples. Do cellular automata or smart grids "do something by their own nature"? Human interventions often start or terminate the operation of engineering objects, or are—in the form of maintenance—required for their continued reliability. Still, all engineering objects do something (i.e., transform energy, matter, or information) through some internal principle of activity, on pain of being practically useless or redundant. Thus, without additional criteria, both rejection and acceptance of any or all engineering objects as members of real kinds seem compatible with the proposal.

A third type of proposal accepts that artefactual kinds are irredeemably dependent on intentional activities, but maintains that they can nonetheless be accepted as real. As Thomasson (2003) has argued, hesitations or refusals to admit the reality of artefacts and other mind-dependent objects are largely a consequence of the traditional focus of (analytic) ontology on the natural sciences. Furthermore, that functions are taken as essential to artefactual kinds and are defined exclusively in terms of designers' intentions has misled ontologists into concluding that artefactual kinds cannot be characterized by reliable clusters of properties. Thomasson (2003) instead offers a definition in terms of substantive and substantively correct intentions on the part of a maker or designer to realize a member of the kind. Similarly, Houkes and Vermaas (2009, 2014) argue that what they call 'instrumental kinds'—such as screwdrivers—can be characterized in terms of contributions to the realization of use plans. More recently, Thomasson has emphasized the importance of intended normative features in characterizing artefactual kinds: what makes something a table is that it is "recognizable as something that is to be treated, used, or regarded, in some ways rather than others (in some contexts, by some individuals, . . .)" (2014: 51–52). The underlying norms of use or, more broadly, comportment are not a matter of individual (designer or user) intentions, but rather of collective practices—which, for many engineering objects, may be restricted to specialists: the normative features of MOSFETs or cloned repositories are hardly recognizable to non-engineers.

Like the other two attempts, much about this third, 'intentionalist' or 'human-centred' ontology remains to be developed. Yet its emphasis on norm-guided practices allows it, in contrast to both other attempts, to save the reality of virtually all kinds of engineering objects without reducing to an 'anything-goes' ontology. The civil aviation system, digital repositories, and MOSFETs feature as prominently in some norm-guided practices as any hammer or desk chair. Perpetual motion machines, by contrast, do not exist as such: their designers' intentions may be substantive but fail to be substantively correct. Furthermore, the emphasis on norm-guided practices aligns the ontology of engineering objects with substantive bodies of work in social ontology, concerned with the reality of social and institutional kinds such as weddings and currencies. This alignment is still underexplored but may prove fruitful (see, for instance, Pearce 2016).

All three attempts potentially save some or all kinds of engineering objects. All do so by resisting, to some extent, a set of conditions for real kinds that are excessively focused on the natural sciences. Instead, they propose conditions that are heavily inspired by the human and social sciences (the third proposal) or that, in referring to long-term selection and internal principles of activity, adopt central concepts from the life sciences. Still, none of the proposals may convince advocates of Quinean austerity (e.g., Renz 2016). Moreover, all offer some implicit corrective of the self-understanding of engineers: ontologically, if perhaps not epistemically, their objects of expertise would not resemble most closely those studied in particle physics and organic chemistry, but those in the life sciences, or the social sciences, instead.

16.5. Conclusions and Outlook

As this brief overview made clear, few efforts in analytic ontology have focused on engineering objects as such. Some discussions prominently feature material artefacts as illustrations of more general problems and proposals, but these examples largely involve medium-sized dry goods only. Moreover, objects are mostly considered from an artisanal or user-centred perspective, rather than an engineering perspective. This makes it, in a way, unsurprising that analytic ontology offers few tools for developing an appropriate ontology of engineering objects. In particular, the emphasis on scientific theorizing or, more broadly, linguistic practices as expressions of ontological commitment seems an uneasy fit for objects with which we are primarily involved practically. The Quinean meta-ontology led to positions such as mereological nihilism that are especially unwelcoming to engineering and its products. Still, alternatives have difficulties adjusting to primarily practical modes of engagement with objects that have highly diffuse or unusual persistence conditions, such as repositories and large-scale systems. Promising options for further exploration, discussed earlier, hint at strong resemblances between engineering objects on the one hand, and organisms and social institutions on the other. Another route, not discussed here, would be to explore similarities between engineering objects and works of art. Some (e.g., Wollheim 1968; Levinson 1980) have argued that central intuitions regarding authorship and persistence are best captured by taking some or all art works as *abstract* structures or types that are instantiated in, but not identical to, the concrete individuals or tokens with which we interact. Saving similar intuitions for some or all engineering objects might thus bring us to the conclusion that the jumbo jet is, despite first impressions, not a concrete material item, but as abstract as a prime number.

It remains to be seen, of course, whether a thorough analysis of these resemblances can be given, while respecting distinct characteristics of engineering objects. Moreover, it is an open question whether a one-size-fits-all ontology of engineering objects is feasible, given their diversity. The persistence conditions of quantum circulators, anemometers, and the civil-aviation system may perhaps not be captured in the same terms; and our involvement with them, in the form of practical engagement but also of authorship, is not evidently of the same kind. Seventy years after Quine published "On What There Is", ontologists willing to abandon their lumps of clay and Eames desk chairs may be surprised at what else is still there to be investigated.

Related Chapters

Chapter 5: What Is Engineering Science? (Sven Ove Hansson)
Chapter 15: Artifacts (Beth Preston)
Chapter 18: Function in Engineering (Boris Eisenbart and Kilian Gericke)
Chapter 21: Engineering Design (Peter Kroes)

Further Reading

The *Stanford Encyclopedia of Philosophy* has excellent entries, with extensive bibliographies, on several topics that this chapter only touched upon or discussed briefly, such as: Aristotle's Metaphysics (S. Marc Cohen), Heidegger (Michael Wheeler), ontological commitment (Philip Bricker), ontological dependence (Tuomas E. Tahko and E. Jonathan Lowe), material constitution (Ryan Wasserman), the ontology of art (Paisley Livingston), and ordinary objects (Daniel Korman).

References

Austin, J.L. (1962). *Sense and Sensibilia.* Ed. G.J. Warnock. Oxford: Oxford University Press.
Baker, L.R. (2004). The Ontology of Artifacts. *Philosophical Explorations,* 7, 99–111.
Baker, L.R. (2007). *The Metaphysics of Everyday Life.* Cambridge: Cambridge University Press.
Berto, F. and Plebani, M. (2015). *Ontology and Metaontology.* London: Bloomsbury.
Carnap, R. (1950). Empiricism, Semantics, and Ontology. *Revue Internationale de Philosophie,* 4, 20–40.
Chalmers, D. (2009). Ontological Anti-Realism. In D. Chalmers, D. Manley and R. Wasserman (eds.), *Metametaphysics.* Oxford: Oxford University Press, pp. 77–129.

Correia, F. (2008). Ontological Dependence. *Philosophy Compass,* 3, 1013–1032.

Elder, C. (2004). *Real Natures and Familiar Objects.* Cambridge, MA: MIT Press.

Elder, C. (2007). On the Place of Artifacts in Ontology. In E. Margolis and S. Laurence (eds.), *Creations of the Mind.* Oxford: Oxford University Press, pp. 33–51.

Elder, C. (2014). Artifacts and Mind-Independence. In M. Franssen et al. (eds.), *Artefact Kinds: Ontology and the Human-Made World.* Cham: Springer, pp. 27–44.

Haslanger, S. and Kurtz, R. (eds.) (2006). *Persistence.* Cambridge, MA: MIT Press.

Houkes,. W. and Vermaas, P.E. (2009). Contemporary Engineering and the Metaphysics of Artefacts. *The Monist,* 92, 403–419.

Houkes, W. and Vermaas, P.E. (2014). On What Is Made. In M. Franssen et al. (eds.), *Artefact Kinds: Ontology and the Human-Made World.* Cham: Springer, pp. 167–190.

Ihde, D. (1990). *Technology and the Lifeworld.* Bloomington, IN: Indiana University Press.

Levinson, J. (1980). What a Musical Work Is. *The Journal of Philosophy,* 70, 5–28.

Lowe, E.J. (2014). How Real Are Artefacts and Artefact Kinds? In M. Franssen et al. (eds.) *Artefact Kinds: Ontology and the Human-Made World.* Cham: Springer, pp. 17–26.

Merricks, T. (2001). *Objects and Persons.* Oxford: Oxford University Press.

Millikan, R.G. (1984). *Language, Thought and Other Biological Categories.* Cambridge, MA: MIT Press.

Paul, L.A. (2010). The Puzzles of Material Constitution. *Philosophy Compass,* 5, 579–590.

Pearce, D. (2016). Collective Intentionality and the Social Status of Artifactual Kinds. *Design Science,* 2, e3. doi:10.1017/dsj.2016.3

Preston, B. (2013). *A Philosophy of Material Culture.* London: Routledge.

Quine, W.V.O. (1948). On What There Is. In *From a Logical Point of View.* Cambridge, MA: Harvard University Press, 1953, pp. 1–19.

Radder, H. (2009). Science, Technology, and the Science-Technology Relationship. In A. W. M. Meijers (ed.), *Handbook of the Philosophy of Science, Vol. 9: Philosophy of Technology and Engineering Sciences.* Amsterdam: North-Holland, pp. 65–91.

Rayo, A. (2007). Ontological Commitment. *Philosophy Compass,* 2, 428–444.

Rea, M. (ed.) (1997). *Material Constitution.* Oxford: Rowman & Littlefield.

Renz, G. (2016). It's All in Your Head. *Philosophia,* 44, 1387–1407.

Sider, T. (2001). *Four-Dimensionalism.* Oxford: Oxford University Press.

Sider, T. (2011). *Writing the Book of the World.* Oxford: Oxford University Press.

Strawson, P. (1959). *Individuals.* London: Methuen.

Thomasson, A.L. (2003). Realism and Human Kinds. *Philosophy and Phenomenological Research,* 67, 580–609.

Thomasson, A.L. (2009). Artifacts in Metaphysics. In A.W.M. Meijers (ed.), *Handbook of the Philosophy of Science, Vol. 9: Philosophy of Technology and Engineering Sciences.* Amsterdam: North-Holland, pp. 192–212.

Thomasson, A.L. (2014). Public Artifacts, Intentions, and Norms. In M. Franssen et al. (eds.), *Artefact Kinds: Ontology and the Human-Made World.* Cham: Springer, pp. 45–62.

Thomasson, A.L. (2015). *Ontology Made Easy.* Oxford: Oxford University Press.

Van Inwagen, P. (1990). *Material Beings.* Ithaca: Cornell University Press.

Van Inwagen, P. (1998). Meta-Ontology. *Erkenntnis,* 48, 233–250.

Verbeek, P.-P. (2005). *What Things Do.* University Park, PA: Penn State University Press.

Vermaas, P.E., Kroes, P., Van de Poel, I., Franssen, M. and Houkes, W. (2011). *A Philosophy of Technology.* San Rafael, CA: Morgan & Claypool.

Wiggins, D. (2001). *Sameness and Substance Renewed.* Cambridge: Cambridge University Press.

Wollheim, R. (1968). *Art and Its Objects,* 2nd revised ed. New York: Harper and Row.

17

USE PLANS

Auke Pols

17.1. Introduction

Designers do many things in the course of their working day: they sketch, tinker, consult with clients, drink coffee, etc. Some of these activities seem to belong to the core of actual design work, while others are more tangential to it. A major question in the philosophy of engineering design is thus: *What is engineering design?* This question cannot be answered without taking a stance on the ontology of design, or giving an answer to the question: *What is the object of engineering design?*

The use plan approach provides answers to both questions. Developed by Dutch philosophers of technology Wybo Houkes and Pieter Vermaas, most recently and comprehensively in their (2010) book *Technical Functions*, it argues that the object of design is a *use plan*, and thus, that design fundamentally is *the construction and communication of a use plan*. Characteristic of the use plan approach is that it is not so much focused on the *technical artefacts* that come out of design processes (see Chapter 15, "Artifacts" by Beth Preston, this volume) or their *functions* (see Chapter 18, "Function in Engineering" by Boris Eisenbart and Kilian Gericke, this volume). Instead, the focus is on *users* and what they do. Particularly, the use plan approach holds that we should see design fundamentally as enabling a series of user actions (with one or more technical artefacts) that allow that user to achieve a particular goal.

This chapter describes the use plan approach. First, it goes into what use plans are, how they are designed, communicated and executed, and how they can be evaluated. Second, it compares the use plan approach to two other philosophical approaches to the objects of design, namely, the function-based and affordance-based approaches. This serves to elucidate the workings of the use plan approach as well as some of its advantages and disadvantages. Finally, the chapter treats three criticisms of the use plan approach as well as replies to those criticisms, and reflects on some outstanding issues for the use plan approach in the conclusion.

17.2. What Are Use Plans?

The use plan approach assumes that design is best analysed according to the same framework by which actions in general are analysed in analytic philosophy: practical reasoning. Practical reasoning is the (rational) way by which an agent comes to answer the question: 'What should I do?' The use plan approach particularly follows Bratman's (1987) plan-based account of practical reasoning. In a nutshell, according to Bratman, agents address the question of what they should do by developing and prioritising *goals*, and then making a *plan* to achieve those goals. *Plans* are 'orderings of considered actions, undertaken for achieving a goal' (Houkes et al. 2002: 304). Use plans are then specific types of plans, namely, those plans that involve the use of one or more objects. The use plan approach has been inspired by works on artefacts and the importance of communicated intentions for their use (Dipert 1993), and earlier work on conceptualising design such as the Theory of Technical Systems (Hubka and Eder 1998).

By starting from practical reasoning, the use plan approach may seem overly rational and empirically not very accurate: when we act, we often do so from habit or impulse, without practical reasoning. Isn't design similarly messy? To answer this question, it is important to note that the use plan approach is an *ideal theory* of design. That is, it is not an attempt to generalise over a series of empirical observations on how actual design processes take place, nor is it a prescriptive model of the steps that designers should follow in their daily practice. Rather, it is a *rational reconstruction* of steps that need to figure in *any* design process in some way, whether in the field of mechanical engineering, architecture or anywhere else (Houkes et al. 2002). This means that those steps need not be taken explicitly, always in the same way, or only by one designer, but they have to be present in some way for a design process to count as such. Moreover, as a rational reconstruction, the use plan approach can be used as a normative tool to evaluate the rationality of design. It does this by investigating whether the ordering of considered actions that constitutes the designed plan can reasonably be expected to lead to goal achievement. For example: using a sieve to hold water is irrational—one cannot reasonably expect that filling a sieve with water will lead to its holding water. Similarly, if your goal is to design a container for holding water but you design a sieve as the artefact that should hold the water, you have designed an irrational use plan. The use plan approach thus is a theory of what design is and what the objects of design are, but it also offers us the tools to determine whether a given use plan is practically rational.

17.3. The Design of Use Plans

Houkes et al. (2002) work out the rational reconstruction of the design process in detail (see also Houkes and Vermaas 2010, Ch. 2). The first step in designing a use plan (in that paper still called 'user plan') is *goal-setting*. In practice, the goal will often be set by a client: it does not have to be a personal goal of the designer, though designer and user may be the same person. For example, a client may state that she prefers toast for breakfast over plain bread and ask a designer for help. Part of goal-setting is identifying the various *constraints* under which the designer has to work (time and money, but also skills and material available, etc.), as this determines which goals are achievable, or how close the 'ideal' goal can likely be approximated. Also, sometimes there may not be one designer, but several, or a design team to design the plan—how this collective plan design could work is analysed by Pearce (2016).

The second step is *designing the means to achieve the goal*. In this step the actual plan is designed. The designer investigates which actions should contribute to the realisation of the goal, and particularly, which artefacts should be designed and developed in order to enable those actions. To continue the example, the designer may consider various plans (toasting bread under the grill, baking it in a frying pan) before settling on a design for a toaster. The design follows the plan: a toaster designed for a single user will have different characteristics from one designed for a hotel breakfast buffet, not because the goal of toasting bread is different, but because the latter is used in a different context (many users, more space available for the toaster, etc.) In either case, it should be possible to undertake a series of considered actions to achieve that goal with the artefact, e.g. put the bread in the toaster, determine the browning level, turn the toaster on and take the toast out when ready.

In the third step, the *use plan* is *constructed*, the necessary *artefacts* are *designed and constructed* and the *use plan* is *communicated* to the relevant users. Houkes et al. explicitly list this as only one step because this step fully encompasses *artefact design*. They thus show how artefact design always follows plan design. Note, however, that plan design does not always have to be followed by artefact design. A designer may design a plan for toasting bread using a toaster, for example, but never design and construct the actual toaster, or leave it to another party.

The use plan approach considers artefact design to be 'product design' (Houkes et al. 2002), as it is but a part of the full use plan design. Artefact design consists itself of several substeps, namely, designing and executing a *make plan*: a series of considered actions, possibly (and typically) including the manipulation of one or more tools, aimed at producing an object with particular properties (cf. Houkes 2012). Houkes and Vermaas (2009) call the execution of this make plan '*manufacturing*': the physical construction of the technical artefact. Again, while manufacturing always has to be preceded by make plan design, a designer can design a make plan ('these are the specifications for making a toaster') without necessarily following it up with the manufacturing of the artefact. Note here that there is a difference between the use plan (a plan for achieving a goal, in this case, toasting bread) and the make plan (a plan for producing an object needed to execute the use plan, in this case, a toaster). However, the make plan can itself be a use plan for other artefacts; in this case, the tools needed to achieve the goal of producing a toaster).

In the fourth step, the outcome of the plan is *compared* to the goal or desired outcome. The result is then *evaluated* in the fifth step: determining whether the use plan is rational, that is, whether the designer can reasonably expect that users can achieve their goal by executing the designed plan. If so, the design process is complete. If not, the designer has to go through the steps again to check where the problem has originated, and remedy it. This can be done in various ways, from redesigning the artefact to modifying the original goal or constraints.

One characteristic of the use plan approach is that the threshold for what counts as 'design' is very low. Most humans can engage in practical reasoning, or think about ways to achieve their goals. Similarly, most humans will be able to appropriate artefacts for achieving novel goals, such as standing on chairs to change light bulbs or using crates as door-stoppers. Even hapless natural objects may be incorporated in impromptu use plans, such as when a hiker uses a log as a bridge to cross a stream, or a rock to hammer tent pegs into the soil. This already counts as use plan design, as all steps to create a new use plan are gone through (quickly).

Houkes (2008) holds that, while most humans can thus design use plans, there is a huge difference— in degree, not in kind—between impromptu plan design and the skilled and responsible activities of professional designers. Most people can pick up a rock and use it as a hammer, but designing an effective, efficient, safe and user-friendly toaster requires specialised skills and knowledge. This also gives designers a special, privileged position with regard to assigning functions to technical artefacts. Houkes and Vermaas (2010, Ch. 4; see also Vermaas and Houkes 2006a) show how this works by developing their ICE theory of function ascription on the basis of the use plan approach. Basically, the ICE theory justifies function ascriptions to artefacts on the basis of beliefs about their use plans, rather than on the basis of their physical properties. With regard to what needs to be believed in order for function ascriptions to be justified, the theory takes elements from three different kinds of theories of function ascription: the Intentionalist kind, where functions are ascribed on the basis of designer intentions; the Causal-role kind, where functions are ascribed on the basis of the causal role artefacts play in larger systems; and the Evolutionary kind, where functions are ascribed on the basis of the traits artefacts have been selected for over the long term. In this way, Houkes and Vermaas aim to use the strengths of each of these theories as well as to compensate for their individual weaknesses.

A different position on the issue of function ascription is taken by Scheele (2006) and Schyfter (2009). They argue that actual use is much more relevant for determining an artefact's function than the intentions of designers. In line with a rich literature on sociotechnical systems and social constructivist views of technology, they argue that theories of technical functions should take into account the social context and the social institutions in which artefacts are used. Scheele gives the example of a church that is now used/now functions as an event hall, while Schyfter points out that even for simple waiter's corkscrews, the 'right' way to use them is strongly determined by societal norms and institutions, not to mention that their function is constituted and sustained by a myriad of collective practices: particular ways of bottling wine, serving wine at restaurants, etc.

Houkes et al. (2011) respond that both frameworks, though they have a different focus, actually converge to a large degree. They argue that a social, 'collectivist' account of artefact use still needs a use plan-like account to properly explain what design is, and what the objects of design are. At the same time, they admit that the use plan account could use a collectivist account of actions to explain how social constraints affect plan design and execution, as well as how use practices ground and inspire design. Nevertheless, they also argue that a 'collectivist' account of artefact functions runs into several difficulties that the use plan account is better equipped to deal with. This holds particularly for properly analysing one-of-a-kind artefacts (e.g. a custom-made mould for casting a bell that has a function only for the craftsperson casting that bell) and defect types (the tokens of which only accidentally fulfil their intended functions), that are therefore no longer part of use practices.

17.4. Communicating Use Plans

In the previous section, the reader will have noted that communication of a use plan from the designer to the user is considered an inherent part of the design (step three) rather than a contingent activity. This may seem counterintuitive if one only considers enclosing a user manual with the artefact as a way of communicating (Houkes 2008). However, a use plan can also be communicated through advertisements, trainings or product demonstrations organised by or in close collaboration with the design team. Also, bear in mind that a large part of the communication of a use plan is done through artefact design itself: lights, buttons, handles, etc. Toasters, for example, often have big flat knobs that invite pressing down on them, to lower the toast into the machine. Indeed, the extensive literature on affordances, a concept from behavioural ecology introduced in the design literature by Norman (1988/2002; cf. Pols 2012, 2015), is an explicit recognition of the fact that design features do—and should—communicate to users how artefacts are to be used. In terms of the use plan approach, affordances are a very salient and immediate way in which designers can communicate (parts of) a use plan to users. Houkes and Vermaas (2006) distinguish three main ways in which users can (though not always do) come to rational beliefs about use plans: from the physical properties of artefacts (such as their affordances); from information about designer intentions (e.g. manuals, advertisements, training) and from behaviour and stories of fellow users.

An interesting difference between the practice-oriented affordance-based approach and the ideal theory of the use plan approach is how both approaches deal with misleading communication. Maier and Fadel (2009) give the example of a household ladder of which the horizontal brace looks like a step, but sports a sign stating: 'This is not a step'. For the use plan approach, this would count as rational design: the sign communicates part of a use plan that users violate at their peril. (Though if the sign is hidden or too small, a case can be made that the designer failed in communicating the use plan.) For Maier and Fadel, however, if the brace affords stepping, it is an undesired affordance and thus an example of bad design. This difference is not surprising, given that an ideal theory of design would assume an ideal, rational user, while Maier and Fadel aim for 'idiot-proof' design, or design for real users who at best exhibit bounded rationality.

Though proper communication of a use plan is necessary for ensuring that the user can properly use an artefact, it is not sufficient. Users may lack relevant skills to use the artefact, auxiliary items (such as batteries) or access to the physical context in which the artefact should be used (e.g. an electricity outlet to power the toaster). While it is not the responsibility of the designer to provide all these, communication of the use plan should include mention of necessary auxiliary items, needed skills or a certain context that might not immediately be obvious (Houkes and Vermaas 2010, Ch. 2).

A counterargument to the relevance of the designer for communicating use plans may be that many artefacts have been around for ages. The inventors of the wheel, the raft and the toothbrush have long been forgotten, yet their inventions are still used. Along the same lines, artefacts such as roads and paper have such obvious uses that explicit communication of a use plan by its designer seems unnecessary (Vermaas 2006).

Houkes (2008) argues that this counterargument has very little force for disproving the use plan approach. First, the use plans of these artefacts have simply been passed on through many different channels, including prior users, designers who adapted the original, etc. The use plan approach requires only that the use plan is communicated, not that there is always one-on-one communication between designer and user. Even in the case of artefacts for which the original use plan is lost, such as mysterious tools discovered in forgotten tombs, archaeologists often try to *reconstruct* use plans based on contextual information and secondary sources, showing the importance they assign to designer intentions (see also Vermaas and Houkes 2006a). Second, it is important to keep in mind the difference between kinds and types. While the toothbrush 'kind' has long been in existence, numerous 'types' have appeared on the market with flexible heads, extra-soft brushes, etc., and the specific use plans of those types are being communicated through packaging, advertisements, etc. (Houkes 2008).

17.5. Executing Use Plans

The use plan approach gives us not only an answer to the question of what *design* is, but also to the question of what *use* is. For the use plan approach, using an artefact is executing a use plan for that artefact (Houkes 2008).

If we spell out use in use plan terms in a similar way to design, the first step is for the user to *want/ desire* to bring about some goal, and to *believe* that the goal does not (yet) obtain. Second, the user has to *choose* a use plan from a set of possible alternatives for bringing about the goal. (Not all possible alternatives have to be use plans for artefacts, e.g. when I consider whether I should use my bike to get to the store or simply walk.) Third, the user needs to *believe or verify* that the use plan is an effective way to achieve the goal. Fourth, the user has to *believe or verify* that the physical circumstances and their set of skills support realising the use plan—considering making toast for breakfast with my electric toaster makes little sense if I believe that there is a power outage in my neighbourhood. All these beliefs, together with knowledge of the use plan and possession of the skills needed to execute it, form *use know-how*. This use know-how is what serves as justification for users' claims that they know that an artefact can be used for achieving a certain goal (Houkes 2008). Fifth, the user has to intentionally *execute* the use plan. The sixth step is that the user compares the achieved state with the desired goal state. Seventh, if the user is confident that the achieved state is sufficiently close to the desired goal state (if not identical to it), s/he is done. Otherwise, several options are open to the user, including retries, repairs to the artefact and abandoning the original goal (cf. Houkes and Vermaas 2010, Ch. 2).

17.6. Evaluating Design With the Use Plan Approach

Because the use plan approach is a rational reconstruction of the design process, it can be used to check whether both use plans and their execution are *practically rational*. This section will first examine how the use plan approach evaluates artefact use, followed by how it evaluates use plans themselves.

When it comes to artefact use, the use plan approach allows one to evaluate whether it is *rational*, that is, whether the user justifiably expects that the artefact can be used in that way to achieve the envisioned goal, and *proper*, that is, whether it is the kind of use for which the artefact has been designed (Houkes 2006). If an artefact is used according to its communicated use plan, we speak of *rational proper use*, e.g. using a screwdriver to twist a screw into a board. Use can also be *improper*, when users create a new use plan for an artefact. This can be rational, that is, based on reasonable expectations: an example of *rational improper* use would be using a screwdriver to open a tin of paint after seeing one's neighbour do so successfully. Improper use can be irrational if not based on reasonable expectations: an example of *irrational improper* use would be trying to use a screwdriver as a toaster because one is confused about its physical properties. The final possibility is *irrational proper* use, where the designer's use plan is followed but one cannot justifiably expect it to lead to goal achievement, e.g. when a screwdriver breaks as soon as the user tries to tighten a screw due to some material defect. One reason why this distinction is so important is the division of responsibility: while the user is responsible for improper use of an artefact and its consequences, the designer is in principle responsible for irrational proper use and its consequences (Pols 2010) and may even be held liable (Houkes 2006).

Note that a use plan's being rational does not guarantee that every use according to that plan will be *successful*. One might justifiably expect a screwdriver to be usable to tighten a screw (rational proper use), but fail due to an unexpected lack of strength or a material failure that could not reasonably have been foreseen. Conversely, executing an irrational use plan may be successful through blind luck or having expectations about a possible use that, though not justifiable, turned out to be right.

About rational use, Houkes and Vermaas (2004: 59) write:

> In a rational plan, the user believes that the selected objects are available for use—present and in working order—that the physical circumstances afford the use of the object, that auxiliary items are available for use, and that the user herself has the skills necessary for and is physically capable of using the object.

Of course, there may be cases where the user is not fully responsible for irrational artefact use, e.g. when under pressure from an employing organisation to use the artefact in that way. This, however, is a situation where one of the general conditions for taking up responsibility is not met (in this case, being able to act freely; see Pols 2010) rather than a consequence of the way the use plan approach is structured. Houkes and Pols (2013) develop an account of what makes acceptance of technology in an organisation rational, combining the use plan approach with the Unified Theory of Acceptance and Use of Technology (Venkatesh et al. 2003), a model that predicts the adoption of information systems in organisations.

The use plan approach enables evaluating use plans according to the standards of practical rationality. Basically, this means that it allows one to check whether executing the use plan will lead to the desired goal; in other words, that the use plan is *effective* and *efficient* (Houkes and Vermaas 2010, Ch. 2). Houkes and Vermaas, though, are quick to remark that assessing effectiveness is always relative to other use plans (some methods might be more effective or efficient than others), context (a bike is not an effective method of transportation in the desert), specification of the goal (preparing a microwave meal might be an effective plan for a quick dinner, but not for a family Christmas dinner), the availability of auxiliary items and the user possessing the right skills and being able to execute the plan.

Besides effectiveness and efficiency, the use plan approach brings in other evaluative standards for use plans. Houkes and Vermaas (2010, Ch. 2) mention the following: *Goal consistency*, where if a use plan is meant to serve multiple goals, the user must reasonably believe that the use plan will do so (e.g. a phone cannot be used to take a picture and send it to someone else if it doesn't have a camera). *Means-ends consistency*, where the user must reasonably believe that all auxiliary items are available to execute the use plan. And *belief consistency*, where the user must reasonably believe that the use plan can be correctly executed if all the user's beliefs are correct. Whether this is actually the case does not matter for plan rationality. If I have good reasons to believe that I will be able to tighten screws with my screwdriver, but unbeknownst to me, the metal is so brittle that it will break as soon as I start, my use is rational, though as it turns out, unsuccessful.

The use plan approach is very good in evaluating effectiveness and efficiency, but it is also important to know what it *cannot* do. It tells the designer nothing about which moral values should be instantiated in the artefact and how trade-offs between them have to be made (Houkes 2008), nor is it intended to do so. Nor does it tell the designer anything about how to incorporate more practical values such as marketability and ease of manufacturing (Vaesen 2011) beyond that they could be applied as constraints in goal-setting. The only exception would be where an artefact becomes so unsafe or risky to use that values of practical rationality such as effectiveness and means-ends consistency become compromised. A toaster that has a 50% chance of exploding every time it is turned on is not only unsafe, it is also not a good toaster. Similarly, the use plan approach does not say anything about how stakeholders should be involved in design (see Chapter 24, "Human-Centred Design and Its Inherent Ethical Qualities" by Marc Steen, this volume), nor does it prescribe any methodology to guide the assembly of the technical artefact itself. Finally, while the use plan approach describes how the use plan can be *communicated* properly, it does not analyse how artefacts or use plans could *prescribe* or *invite* particular actions or *mediate* our perception of the world. This has traditionally been the domain of science and technology studies (e.g. Latour 1992) and (post-)phenomenology (e.g. Verbeek 2005). However, Pols (2013) shows how these phenomena could in principle be analysed as changes in our reasons for action, and thus that they could be analysed under the umbrella of practical reasoning, just like the use plan approach.

17.7. Use Plans, Functions and Affordances

The use plan approach has been developed in response to the primacy of technical *functions* as the focus of the philosophy of design and technical artefacts. It is not the only alternative design approach that arose from dissatisfaction with the limits of function-based approaches: affordance-based design has sprung from that origin as well. Though functions are the subject of Chapter 18, this section will compare function-based, plan-based and affordance-based accounts, and show why Houkes and Vermaas consider a use-theoretic account more fundamental and more accurate than a function-theoretic account.

The use plan account has first been described in Houkes et al. (2002), who suggest that it best fits Cummins's (1975) account of functions as, basically, the causal contribution something makes to the capacities of systems that contain it. (For example, the function of a jet engine is to generate thrust, because that is the causal contribution it makes to an aircraft's capacity to fly.) However, the use plan account is only properly compared to function-theoretical accounts in Houkes and Vermaas (2004). Here, they claim that use plans are more fundamental than functions and thus, are the proper answer to the question of what the object of engineering design is. The reason for this is given in Section 3: the design of the use plan always precedes the artefact design phase during which the actual technical function is implemented. Indeed, for their ICE theory of function ascription, Houkes and Vermaas (2010, Ch. 4) define functions in terms of use plans.

The reason Houkes and Vermaas consider use plans more accurate is that theories of technical functions tend to focus on *proper* functions. Use plans similarly consider 'proper use' but are much better situated to also take *rationality* into account. This is because this factor is not dependent on artefactual properties, but on the properties of agents and their physical and social context. For instance, the proper function of my car might be to quickly transport me, but if I lack driving skills, and moreover, a hurricane has blown my car into a tree, I cannot properly use it, even if it is otherwise undamaged. Due to the physical circumstances and my lack of skills, I cannot execute a rational proper use plan for my car. (Alternatively, we could say that it is impossible for me to design a proper use plan for my car that is also rational: under the current circumstances, there is no way in which I could use my car to transport me quickly.)

Another advantage of the use plan approach is ontological parsimony. That is, if I use an artefact for another purpose than the designer intended it for (e.g. standing on a chair to change a light bulb), the artefact does not suddenly gain an extra property/(accidental) function. Rather, the artefact remains what it is; I just plan a new use for it.

It should be noted here that some proponents of function-theoretic accounts are aware of the relevance of contextual factors and have taken them into account as well (cf. Pols 2015; Vermaas and Houkes 2006a; Chandrasekaran and Josephson 2000). Houkes and Vermaas have done so by using the use plan approach as a basis for their ICE account of functions (Vermaas and Houkes 2006a; Houkes and Vermaas 2010). Thus, function-theoretic accounts have been made more accurate through the introduction of contextual factors. To the author's knowledge, however, no attempt has been made so far to argue that functions are more fundamental than use plans.

Houkes and Vermaas are not the only ones who have developed an alternative account of design and the objects of design in response to the (perceived) shortcomings of function-based accounts. Another alternative is the affordance-based account of Maier and Fadel (2009). Maier and Fadel consider functionalist accounts to be overly concerned with transformative aspects of design (e.g. how an artefact transforms electricity into motion, or vice versa) and neglect non-transformative considerations. For them, the fundamental objects of design are *affordances*, opportunities for behaviour and relational entities that depend on characteristics of both artefact and user. For example, a chair affords sitting because of its material characteristics as well as because of characteristics of human anatomy. If a chair is covered in barbed wire, it does not afford sitting (unless one is wearing armour). Neither, however, does a regular chair afford sitting to babies, who lack the capacity to remain upright by themselves. Design, for Maier and Fadel, is 'the specification of a system structure that does possess certain desired affordances in order to support certain desired behaviours, but does not possess certain undesired affordances in order to support certain undesired behaviours' (2009: 23).

Pols (2015), who has compared the notions of use plans, functions and affordances in greater detail, has argued that this definition entails that use plans are more fundamental than affordances as a description of the design process, for the notion of 'desired affordance' inevitably begs the question of what makes an affordance desired. This would be the use plan, or the combination of the goal that is to be achieved by using the artefact, and the series of actions by which it is to be achieved. Nevertheless, plan design according to the use plan approach would be quite compatible with artefact/product design according to affordance-based design: Pols (2012) explains in detail how both are related.

17.8. Criticism

The use plan account is not without its critics. In this section I discuss three general criticisms that have been levelled at it and Houkes and Vermaas's responses: some actual, some hypothetical. The first is that, contrary to most classical function accounts, it is not able to properly describe artefacts that require little or no user actions to operate, such as jet engines. The second is that it is too far removed from actual use practices to even count as a proper rational reconstruction. The third is that its method of rational reconstruction itself may be structurally biased against particular groups of designers.

The first criticism is that there is a whole class of artefacts that work with little or no user interactions, namely those artefacts that are components of other artefacts. For example, a jet engine consists of many individual artefacts, none of which are explicitly operated by any user. Function-based approaches can ascribe subfunctions to these artefacts through functional decomposition (e.g. van Eck 2011). Likewise, affordance-based approaches like those of Maier and Fadel (2009) have artefact-artefact affordances to deal with interacting components (e.g. 'this cog affords being rotated by another cog'). As components are not directly interacted with by the user, however, no use plan is available, the argument goes. Thus, they are a significant blind spot for the use plan approach.

The reply here is that this is not so much a problem for the use plan approach, as well as a salient difference between the rational reconstruction of the use plan approach and engineering practice. Added to that is the fact that this is an example that function-based approaches seem made for, for the relational/contextual factors for components are always clearly established in advance. The context for a jet engine component is always a jet engine. Vermaas (2006) has shown that for the design of such components, the use plan approach can be 'bracketed': component design is covered almost completely by the technical product design phase in that sense. However, even component functioning is always embedded in a use plan—if not for the jet engine, then for the jet itself. The description of components in function terms is thus not plan-less but plan-relative. Houkes and Vermaas (2010, Ch. 6.2) argue that for components that are complete artefacts, plan aspects become almost impossible to bracket. Thus, they argue that, while many components can be adequately described in terms of (sub-)functions, sooner or later, reference to plans is inevitable.

A second criticism can be found in Lucy Suchman's (1987/2007) book on plans and situated actions, in which she criticises a plan approach to human action in general. Her arguments form a challenge for the use plan approach's conceptualisation of use as the execution of a use plan as well. She argues that users *rarely plan their use in advance*. Instead, they mostly perform 'situated actions', where they check the opportunities for action an artefact offers at each given moment, and do what they think is likely to help them achieve their goal at that moment. For example, when using a copying machine, users tend to act on the information presented to them in subsequent menus, rather than planning every step in advance. Similar observations have been made by De Léon (2003) on the use of artefacts in cooking.

Vermaas and Houkes (2006b) agree with Suchman's empirical observations about users but argue that they do not threaten their account. Rather, they argue that designers should make sure that a rational use plan exists for the artefact, but that there is no reason why users cannot engage it with a 'light' or 'high-level' use plan that allows for a lot of situated actions and responses to the environment. Indeed, even detailed plans have to leave many specific actions open to the user: those that do not matter for goal achievement (e.g. where exactly in the kitchen to put your toaster) or because more specific instructions would make no sense (e.g. how exactly to move your finger to push the knob on the toaster down). Similarly, use plans have to be communicated to users, but there is no requirement that the full plan has to be communicated in advance rather than step by step. Thus, though the use plan account is at its core rationalistic, it is also very flexible and able to incorporate many different design and use practices, as long as they adhere to some basic criteria of practical rationality.

A critical reader might not be satisfied by this answer, noting that defining 'executing plans' so broadly as to include all kinds of non-planned, situated behaviour seems more like a conceptual slight-of-hand than a robust analysis. Hypothesising about the response, it seems that Houkes and Vermaas would likely address this concern by pointing out that their central concern is to develop a theory of engineering design and its objects, rather than a theory of their use. Looking at the division of labour between designers and users underscores this: the more users deviate from the rational ideal and exhibit bounded rationality (not to mention forgetfulness, irascibility, carelessness, etc.), the more important it becomes for designers to plan for this and ensure that this plan is presented to the user clearly and step by step. Thus, if users are successful in achieving their goals by situated actions, this might attest to the quality of the underlying plan and its communication rather than serve to disprove it.

A third criticism that could be brought against the use plan account, or more broadly, against the use of rational reconstructions and ideal theory in analytic philosophy, regards its validity as a method of analysis. Houkes and Vermaas stress that the use plan approach is intended not to reflect actual design practices, but rather as an ideal theory of design. This allows them to focus on rational rather than idiot-proof design (as Maier and Fadel 2009 do), but also forces them to translate Suchman's

(1987/2007) situated actions and Scheele's (2006) and Schyfter's (2009) social institutions into a framework in which humans are individualist, plan-based reasoners.

Ideal theory and its corresponding concept of humans as rational individuals, however, has been criticised, most notably by feminist philosophers. In a nutshell, the criticism is that such theories are not so much *ideal*, identifying how humans would 'ideally', or 'in perfect circumstances' behave, but *biased* in valuing particular aspects of humans over others, particularly their capacity for practical reasoning. While every model is a simplification of reality, this particular simplification has been accused of being problematic because it favours the thought and behaviour of particular kinds of humans, namely, Western higher-educated white males, over others. For example, Jaggar (1983) has criticised Rawlsian ideal political theory of over-valuing individual characteristics such as rationality over characteristics that humans share as family or community members, and incorrectly seeing rationality as value-neutral and detached. Code (1995) has similarly challenged analytic ideal epistemology for being overly concerned with propositional knowledge ('X knows that *p*') and hardly with the personal and social characteristics one must have before one can say, e.g., that one 'knows a person'.

To substantiate this criticism against the use plan approach specifically, one could go down several paths. One could be to analyse whether there are indeed cultural or gendered differences in conceptualising (the objects of) engineering design (e.g. Faulkner 2007; see also Chapter 48, "Feminist Engineering and Gender" by Donna Riley, this volume) and if so, whether the use plan approach can adequately incorporate them. Another, more radical one could be to question the prominence of analytic practical rationality itself in engineering design and build on a different conceptual foundation altogether, as Schyfter (2009) does.

Following either path lies outside the scope of this chapter. However, if we again hypothesise about the answer, Houkes and Vermaas would likely refer back to the introductory chapter of their (2010) book, in which they explicate not only the goals of the use plan approach and the ICE theory of function ascription, but also explain their 'design specifications' for a theory of technical artefacts. These are: being able to distinguish between stable, 'proper' functions and more transient 'accidental' functions; being able to accommodate the concept of malfunctioning; offering support for function ascriptions; and being able to accommodate functions of innovative artefacts (p. 5). Criticism that shows that the use plan approach (and thus, the ICE theory) is not actually able to meet its design specifications, such as Schyfter (2009) aims to give, is a real challenge for Houkes and Vermaas. More radical criticism that were to pose different design specifications for a theory of technical artefacts, however, might not necessarily be such a challenge. Rather, it would be an opportunity to engage in a dialogue about exactly what kind of answer we are looking for when we ask what engineering design or its object is.

17.9. Conclusion and Outstanding Issues

Though the use plan approach has been around for less time than function-based approaches to design, it has established itself as an elaborate and solid theory of what design and its object is. As such, it is not only a theory of the design process, but also of what designers need to communicate to users, and what makes use of an artefact rational and proper.

The claim of the use plan approach to be more fundamental than function-based approaches holds so far, though its claim that it is more accurate is being contested by function-based approaches that take rational or contextual and user aspects into consideration next to designer intentions. As a rational reconstruction or ideal theory of design, it has particular strengths and weaknesses. Among its weaknesses are that its empirical validity is hard to prove, that various other ideal theories in philosophy have been shown to be biased rather than value-free in their abstraction, and that its prescriptive value for (beginning) designers may be limited. Any prescriptive value of the use plan approach can be said to be incidental rather than intended, given its primary aims of providing a rational reconstruction of design and a conceptual basis for the ICE theory of function ascription. Among its strengths are that it sketches what an ideal design process should look like, and thereby provides us with a tool to evaluate actual design processes according to the standards of practical rationality.

The use plan approach is not a closed method, though the number of publications on it has dropped off sharply after publication of the Houkes and Vermaas (2010) *Technical Functions* book. Some open issues that remain are the following. It currently ignores the creative aspect of design— how designers get their ideas and how they 'play around' with different designs to come to new insights. It assumes one designer and one user, and it does not describe the interactions within a design team (but see Pearce 2016), or between multiple involved stakeholders such as product designers, testers, clients, prospective users, affected third parties, etc., who may all be different. This matters for issues of responsibility, such as whether a design team can be properly held responsible for an irrational proper use plan. It does not look at integration and coherence in artefact design (all from Houkes et al. 2002). More generally, the claim that the use plan approach can count as a rational reconstruction of the wide variety of actual design practices, including across cultures and genders, remains under-investigated. Remedying this would require interesting yet considerable empirical and conceptual work.

Related Chapters

Chapter 8: Prescriptive Engineering Knowledge (Sjoerd Zwart)
Chapter 15: Artifacts (Beth Preston)
Chapter 16: Engineering Objects (Wybo Houkes)
Chapter 18: Function in Engineering (Boris Eisenbart and Kilian Gericke)
Chapter 21: Engineering Design (Peter Kroes)
Chapter 48: Feminist Engineering and Gender (Donna Riley)

Further Reading

Brown, D.C. and Blessing, L. (2005). The Relationship Between Function and Affordance. In *ASME 2005 IDETC/CIE Conference, September 24–28, 2005, Long Beach, California, USA*. DECT2005–85017. (An earlier comparison of the relation between functions and affordances.)

Crilly, N. (2010). The Roles That Artefacts Play: Technical, Social and Aesthetic Functions. *Design Studies*, 31(4), 311–344. (A consideration of the non-technical functions of artefacts, including social, ideological and aesthetic ones.)

Kroes, P.A. and Meijers, A.W.M. (eds.) (2006). The Dual Nature of Technical Artefacts. Special issue of *Studies in History and Philosophy of Science*, 37(2). (Special issue of a research project in which Houkes and Vermaas further developed their use plan approach.)

Van de Poel, I. (2001). Investigating Ethical Issues in Engineering Design. *Science and Engineering Ethics*, 7, 429–446. (An introduction to the kinds of ethical choices that engineers face in the design phase).

References

Bratman, M. (1987). *Intention, Plans and Practical Reason*. Cambridge, MA: Harvard University Press.

Chandrasekaran, B. and Josephson, J.R. (2000). Function in Device Representation. *Engineering With Computers*, 16(3–4), 162–177.

Code, L. (1995). *Rhetorical Spaces: Essays on Gendered Locations*. New York: Routledge.

Cummins, R. (1975). Functional Analysis. *Journal of Philosophy*, 72, 741–765.

De Léon, D. (2003). Actions, Artefacts and Cognition: An Ethnography of Cooking. *Lund University Cognitive Studies*, 104.

Dipert, R.R. (1993). *Artefacts, Art Works and Agency*. Philadelphia: Temple University Press.

Faulkner, W. (2007). 'Nuts and Bolts and People': Gender-Troubled Engineering Identities. *Social Studies of Science*, 37(3), 331–356.

Houkes, W.N. (2006). Knowledge of Artefact Functions. *Studies in History and Philosophy of Science*, 37(2), 102–113.

Houkes, W.N. (2008). Designing Is the Construction of Use Plans. In P. Kroes, P.E. Vermaas, S.A. Moore and A. Light (eds.), *Philosophy and Design: From Engineering to Architecture*. Berlin: Springer, pp. 37–49.

Houkes, W.N., Kroes, P., Meijers, A. and Vermaas, P.E. (2011). Dual-nature and Collectivist Frameworks for Technical Artefacts: A Constructive Comparison. *Studies in History and Philosophy of Science*, 42, 198–205.

Houkes, W.N. (2012). Rules, Plans and the Normativity of Technological Knowledge. In M. De Vries, S. Hansson and A. Meijers (eds.), *Norms in Technology*. Dordrecht: Springer, pp. 35–54.

Houkes, W.N. and Pols, A.J.K. (2013). Plans for Modeling Rational Acceptance of Technology. In D.P. Michelfelder, N. McCarthy and D.E. Goldberg (eds.), *Philosophy and Engineering: Reflections on Practice, Principles and Process*, Dordrecht: Springer, pp. 291–303.

Houkes, W.N. and Vermaas, P.E. (2004). Actions Versus Functions: A Plea for an Alternative Metaphysics of Artefacts. *The Monist*, 87(1), 52–71.

Houkes, W.N. and Vermaas, P.E. (2006). Planning Behavior: Technical Design as Design of Use Plans. In P.-P. Verbeek and A. Slob (eds.), *User Behavior and Technology Development: Shaping Sustainable Relations Between Consumers and Technologies*. Dordrecht: Springer, pp. 203–210.

Houkes, W.N. and Vermaas, P.E. (2009). Contemporary Engineering and the Metaphysics of Artefacts. *The Monist*, 92, 403–419.

Houkes, W.N. and Vermaas, P.E. (2010). *Technical Functions. On the Use and Design of Artefacts*. Dordrecht: Springer.

Houkes, W.N., Vermaas, P.E., Dorst, C.H. and De Vries, M.J. (2002). Design and Use as Plans: An Action-theoretical Account. *Design Studies*, 23(3), 303–320.

Hubka, V. and Eder, W.E. (1998). *Theory of Technical Systems: A Total Concept Theory for Engineering Design*. Berlin: Springer.

Jaggar, A.M. (1983). *Feminist Politics and Human Nature*. Totowa, NJ: Rowman and Allanheld.

Latour, B. (1992). Where Are the Missing Masses? The Sociology of a Few Mundane Artifacts. In W.E. Bijker and J. Law (eds.), *Shaping Technology/Building Society: Studies in Sociotechnical Change*. Cambridge, MA: MIT Press, pp. 225–258.

Maier, J.R.A. and Fadel, G.M. (2009). Affordance Based Design: A Relational Theory for Design. *Research in Engineering Design*, 20(1), 13–27.

Norman, D.A. (1988/2002). *The Psychology of Everyday Things/The Design of Everyday Things*. New York: Basic Books.

Pearce, D. (2016). Collective Intentionality and the Social Status of Artefactual Kinds. *Design Science*, 2(3).

Pols, A.J.K. (2010). Transferring Responsibility Through Use Plans. In I. Van de Poel and D. Goldberg (eds.), *Philosophy and Engineering: An Emerging Agenda*. Dordrecht: Springer, pp. 189–203.

Pols, A.J.K. (2012). Characterising Affordances: The Descriptions-of-affordances-model. *Design Studies*, 33, 113–125.

Pols, A.J.K. (2013). How Artefacts Influence Our Actions. *Ethical Theory and Moral Practice*, 16(3), 575–587.

Pols, A.J.K. (2015). Affordances and Use Plans: An Analysis of Two Alternatives to Function-based Design. *Artificial Intelligence for Engineering Design, Analysis and Manufacturing*, 29(3), 239–247.

Scheele, M. (2006). Function and Use of Technical Artefacts: Social Conditions of Function Ascription. *Studies in History and Philosophy of Science*, 37, 23–36.

Schyfter, P. (2009). The Bootstrapped Artefact. *Studies in History and Philosophy of Science*, 40, 102–111.

Suchman, L. (2007). *Human-machine Reconfigurations: Plans and Situated Actions*, 2nd ed. Cambridge: Cambridge University Press. First published as (1987) *Plans and Situated Actions. The Problem of Human-machine Communication*.

Vaesen, K. (2011). The Functional Bias of the Dual Nature of Technical Artefacts Program. *Studies in History and Philosophy of Science*, 42, 190–197.

Van Eck, D. (2011). Supporting Design Knowledge Exchange by Converting Models of Functional Decomposition. *Journal of Engineering Design*, 22(11), 839–858.

Venkatesh, V., Morris, M.G., Davis, G.B. and Davis, F.D. (2003). User Acceptance of Information Technology. *MIS Quarterly*, 27, 425–478.

Verbeek, P.-P. (2005). *What Things Do. Philosophical Reflections on Technology, Agency, and Design*. University Park, PA: Penn State University Press.

Vermaas, P.E. (2006). The Physical Connection: Engineering Function Ascriptions to Technical Artefacts and Their Components. *Studies in History and Philosophy of Science*, 37, 62–75.

Vermaas, P.E. and Houkes, W.N. (2006a). Technical Function: A Drawbridge between the Intentional and Structural Natures of Technical Artefacts. *Studies in History and Philosophy of Science*, 37, 5–18.

Vermaas, P.E. and Houkes, W.N. (2006b). Use Plans and Artefact Functions: An Intentionalist Approach to Artefacts and Their Use. In A. Costall and O. Dreier (eds.), *Doing Things With Things: The Design and Use of Everyday Objects*. London: Ashgate, pp. 29–48.

18

FUNCTION IN ENGINEERING

Boris Eisenbart and Kilian Gericke

18.1. Introduction

'Function' is one of the most frequently used concepts in engineering theory and methodology. Practising engineers, similarly, think in a clearly function-oriented manner when developing new products and systems. From a philosophical viewpoint, it is very often argued that humans create technical systems, whether physical, digital, or procedural, to serve a purpose. These systems are, ultimately, designed to transform—or support humans in transforming—natural resources into something new. This purpose defines the link between a naturally existent resource and a human-made output in a deliberate manner that follows a fundamental principle of causality (Hubka and Eder 1988). This means that engineering products and systems are subject to means-to-end causality under which the central objective of the humans who design them is to produce the targeted output, goal or to serve a specific use plan. Eventually, this causality is the central premise for their creation. All the mechanisms, components, and technical processes incorporated into a system are thus intended to fulfil the necessary causal relation between the input and said output. When this system is correctly designed, following the paradigm of causality, a human user can employ it in a deliberate manner to deliver the desired outcomes at the desired time. At this abstract level, technical systems can be described through a black box that shows inputs and outputs (Figure 18.1).

Historically, particularly in the German-speaking engineering research community in the 1960s and 1970s, this black box approach has been used to manifest the first abstract representation of the system to be designed. The representation is then broken down into inherent actions or technical processes that transform the given inputs, both in combination and successively, until they reach the final output states (Rodenacker 1970; Pahl and Beitz 1977; Hubka 1980; Hubka and Eder 1988). Combining all relevant operations into a causal, logical chain, in which the output of one serves as input for the next, establishes the functional structure of the system. This structure then serves as a coherent representation of the inherent mechanisms, processes, transformations, and their relations of how the system *functions*, i.e. how it operates, to establish the causal relation between inputs and outputs. All these internal operations thus comprise the system's *functions*. As such, 'function' is an abstract concept that describes the distinct actions and effectual causes, and their relations, taking place in an engineering system in order to achieve a desired outcome.

Functions enable engineers to gradually build a thorough comprehension of the central operations that are necessary in the system under development. The fundamental, underlying premise is that engineers can then infer the technical solutions for implementing the individual functions and hence, logically, arrive at a system that can process the inputs to produce the originally intended

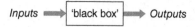

Figure 18.1 Black box that causally links an input to a (desired) output

outcome, following the same causal relationship. Functions hence make it easier for designers to perform the so-called 'creative leap' (Blessing and Upton 1997) from a problem to a solution.

Conceptually speaking, it is in most cases impossible to jump from a given engineering problem to its solution directly, particularly when the problem itself is not fully understood yet, which is typically the case at the start of an engineering project (Chakrabarti and Bligh 2001; Braha and Reich 2003). Establishing a system's functions helps to break down a complex problem into manageable portions that can more easily be matched to a suitable technical solution or working principle (Paetzold 2006; Erden et al. 2008).[1] At the same time, this gives engineers the necessary means to make a system's functioning explicit to themselves or to others. The former, i.e. decomposition of the problem, can be vital particularly in the early phases of product development, when there is as yet no clear mental model of what the final system will look like. Formal explication of what the functions are and how they are related can be vital in enabling engineering teams to build a shared mental model of what the system is supposed to do and achieve (Eisenbart and Kleinsmann 2017). This is essential for both dividing up work in a team (and the subsequent seamless integration of the developed partial solutions) and providing feedback on progress to project leaders and customers. Comprehensive descriptions of the functions assist the engineers in probing emerging solution concepts early and allow recurring verification and validation of the emerging or final design by comparing it with the original intentions. The longer engineers stay at the abstract level of functions, the less they are inclined to become fixated on a single potentially unsuitable solution too early (Jansson and Smith 1991).

Methodologically, in terms of its position in the overall development process of a product or system (see Figure 18.2), function is part of the early stages, when the engineering team is still engaged in clarifying the central tasks and has not as yet generated an initial concept. This is typically referred to as the *conceptual design* stage.[2] Conceptual design is based on the design state of *requirements specification* and essentially encompasses *system functionality* and the *principle solution* or *concept*. Conceptual design hence contains the central transition from a problem description towards an initial (mental) establishment of what the solution as a whole might entail. Both researchers (Bonnema and Van Houten 2006; Ehrlenspiel 2007) and practitioners (Gericke et al. 2013) consider conceptual design to be the stage that has the greatest influence on system design and thereby on all subsequent design activities. The overall concept and structure of the solution can be changed most easily at this stage. The decisions made on these aspects will eventually predetermine the manufacturing costs and applicability of the system in later phases of its life cycle.

Through this, function plays the key role in the essential transition from a description of the desired system or product to an emerging engineering solution for creating it. Function remains an abstract concept in this process. In the end, it is the form or behaviour of a product or system that is visible to humans, rather than its function as such.

When working towards a solution, almost every engineering designer forms some kind of abstract, conceptual model, even if it remains in their mind (Fowler 1998). Function is a way to consider, in explicit terms, what the proposed system is intended to achieve, even before it has been designed, thereby enabling discussions about these issues during joint solution synthesis. Thereby, function helps engineers to thoroughly explore the potential solution space and gradually determine a physical, procedural or virtual implementation. Research discussions have long centred on the way people reason in designing, but the views of scholars on this issue vary greatly. This has significant implications for how function is defined and theorised to link to a final design.

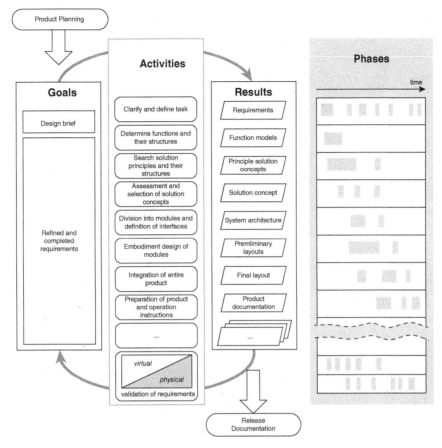

Figure 18.2   General process model for system development

Source: After VDI (2018).

18.2. Function as an Ontology in Conceptualising Solutions

An engineering design task is widely regarded to be an *ill-structured* problem, as often neither the problem nor the desired solution is sufficiently defined at the beginning of a project (Simon 1973). Therefore, conceptualising a solution is characterised by *co-evolution*: a stepwise accumulation of information about the addressed problem in parallel to the gradual emergence of a solution (Poon and Maher 1997). It is an iterative process, as an engineer synthesising and analysing the emerging solution will inevitably begin to comprehend the complexities inherent to the originally stated problem (see Figure 18.3).

As discussed earlier, function is a way of guiding the reasoning in the transition from a problem to a solution, i.e. it supports the iterative synthesis and evaluation steps. Over the years, researchers have produced a host of theoretical frameworks and approaches to describe—and prescribe an effective way of doing—this process (see Erden et al. 2008; Crilly 2010; Eisenbart et al. 2013 for an overview). On the one hand, these approaches enable the analysis of the intricacies of this process, and on the other hand they prescribe a way for engineers to effectively apply it themselves. Perhaps the most widespread approach has been proposed by Pahl et al. (2007, 1977). The model incorporates the discussed abstraction of the overall transformation of basic inputs to system outputs using a black box similar to Figure 18.1. Sometimes, only the desired outputs are known at the start of the process, which means that one must first establish appropriate inputs. As the next step, the black box is then 'opened' through

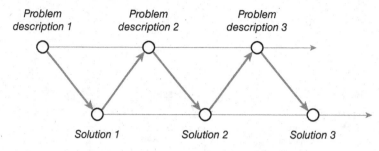

Figure 18.3 Iterative synthesis and evaluation steps during conceptual design

Source: After Poon and Maher (1997).

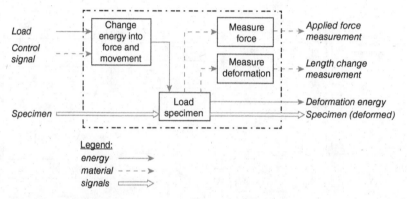

Figure 18.4 Example of a functional structure of a tensile testing machine

Source: After Pahl et al. (2007).

determining what the 'essence' or 'crux' is in terms of what the system as a whole is intended to do (see Pahl et al. 2007: 169). This can be captured in a short statement, such as "provide pulling load to deform solid test specimen and measure specimen deformation over force", which specifies the essence of what a tensile testing machine does. Based on this, the engineer can then determine that an object and some type of energy are required as inputs, and that a deformed object and measurement information are created as outputs. The overall function, represented by the black box, is then decomposed into a few main functions and their related inputs and outputs. These relations can be modelled as a functional structure (see Figure 18.4). Gradually, the initially established function(s) is/are refined by logically dissecting the causal relations and steps required to enable the overall state change to take place. Concurrently, any additionally required inputs and logically resulting outputs are added.

The derived combination of functions specifies the causal transformations carried out by the system in relation to flows of operands going through it. These represent the inputs and will eventually manifest the desired outputs of the system. Operands are traditionally divided into three different types: energy, materials, and signals.

Each of the decomposed subfunctions is eventually mapped to a function carrier. Methodologically, in order to avoid becoming fixated on a particular solution too early on, one should postpone this step until the functional structure has been sufficiently refined. Even when selecting potential solution principles, the engineer should collect as many alternatives as possible and evaluate them before making a final selection. Morphological charts, for instance, may be useful in this process (see e.g. Zwicky 1989). Figure 18.5 schematically shows the whole process from a main overall function to the mapping and selection of alternative solution principles.

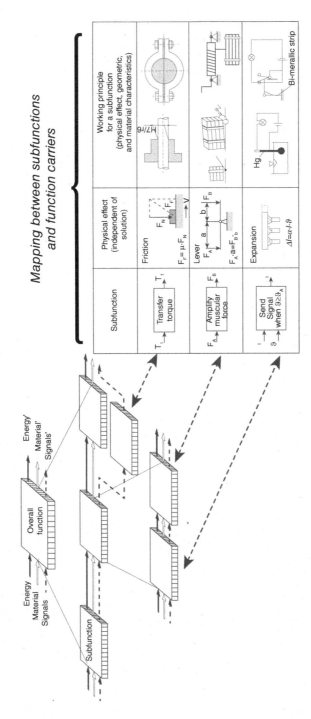

Figure 18.5 Function decomposition and mapping to function carriers

Source: After Pahl et al. (2007).

The underlying understanding of function is defined as the "intended input/output relationship of a system whose purpose is to perform a task" (Pahl et al. 2007: 31). As such, functions are established by the desired state change from required inputs to desired outputs, whether for the system as a whole (which means the overall function) or for the individual subfunctions, where inputs and outputs are determined by the flows of operands and their intermediate states between blocks.

Since their introduction in the early 1970s, these basic principles have been widely adopted in literature in the disciplines of mechanical engineering (e.g. Roozenburg and Eekels 1995; Ullman 2010; Ulrich and Eppinger 2008), electrical engineering (Rajan et al. 2003), mechatronics (e.g. VDI 2018, 2004; Eigner et al. 2010), and Product-Service System design (e.g. Spath and Demuss 2005; Schneider et al. 2005; Welp et al. 2007).

Complementary research has taken the Pahl et al. approach further to establish basic, fundamental function classes abstracted to simple verb/noun combinations with distinct inputs and outputs. These are typically referred to as function taxonomies and specify "a standard language of function" (Ahmed and Wallace 2003: 1), typically including a "standard set of functions and flows" (Stone and Wood 2000: 360). The functions are represented by verbs indicating the specific transformation of an operand, such as 'change' and 'connect'. Nouns represent the operands in the flows, i.e. instantiations of material (like 'shaft'), energy ('heat'), etc. One particularly prominent example of a function

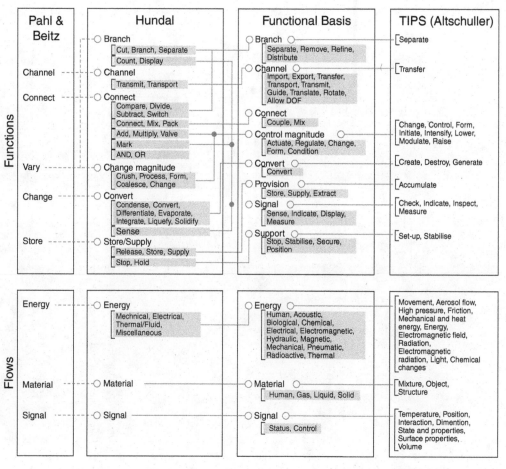

Figure 18.6 Comparison by Stone and Wood (2000) of their Functional Basis with the function taxonomies proposed by Pahl et al. (2007), Hundal (1990), and Altschuller (1999).

taxonomy is the "Functional Basis" developed by Stone and Wood (2000).[3] This and other examples are shown in Figure 18.6. A central aim behind these endeavours is to potentially advance function modelling to a state where it can be computed. Seminal work in this area (see Kurfman et al. 2003; Chakrabarti and Bligh 2001; Sen et al. 2013) suggests that there are huge opportunities in automated or 'smart' engineering, quality and change management, and many other areas.

However, the engineering body of research has produced a host of alternative theorisations of function to Pahl et al. over the years, with diverging views on how to position function. As will be discussed in the following section, scholars tend to reference function to the physical structure of a system under development, its intended purpose, task, or its intended/actual behaviour (compare Ericson and Larsson 2005), which then also affects how function is described and, by extension, how the system as a whole is perceived.

18.3. Function Archetypes

Merriam-Webster (2003) defines function as "the special purpose or activity for which a thing exists". In mathematics, a function is a clear relation between a set of inputs and admissible outputs, and terms such as *transformation* or *operator* are sometimes used synonymously (Halmos 1974). In the engineering literature, function can have very different meanings. Several scholars refer to function as the *ability* of a system to achieve a specific goal, e.g. by showing a certain *behaviour* (e.g. Roozenburg and Eekels 1995; Buur 1990). Others refer to an intended or required *transformation* or *conversion* of operands (e.g. Rodenacker 1970; Pahl et al. 2007; Fowler 1998; Cockburn 2000), which may be associated with the *input/output relations* discussed earlier. Others, yet again, refer to the *purpose* or *objective* of the system, e.g. to fulfil a goal or provide a specific value, as its function (US DoD 2001; Sakao and Shimomura 2007; Bucciarelli 2010; Ullman 2010). This is often discussed as a *teleological* notion of function (Hubka and Eder 1988). As one can see, all of these definitions of function revolve around similar themes and partially overlap. However, no *single* definition of function incorporates all these notions or has superseded the other definitions.

Based on comprehensive reviews Vermaas (2010) and Carrara et al. (2011) conclude that despite the centrality of function to system development, it is "a term that has a number of coexisting meanings, which are used side-by-side in engineering" (Vermaas 2011). However, there are several central themes related to describing function which can be considered ontologically archetypical to engineering, i.e. specific notions of function that represent the essence of the definitions in engineering literature. In general, three such function archetypes can be discerned (Carrara et al. 2011; Vermaas 2011):

1. *behaviour-related notion*: function as the intended behaviour of an entity.
2. *outcome-related notion*: function as the desired effects of the behaviour of an entity.
3. *task or goal-related notion*: function as the purpose for which an entity is designed.

In addition, Vermaas (2013) discusses the concept of the *capability* of a system or artefact—through its particular structure—to show a certain behaviour. Here, behaviour may serve not only the originally intended purpose, but also completely different use plans (Houkes and Vermaas 2010). Note that in all of these theories, function—as opposed to so-called 'affordances'—is considered as something deliberately designed into a system to fulfil a particular task. That makes function something *intended* for the system to possess, which is different from how people may intuitively discuss functions of a system, namely as something that the system 'does' or 'has' (compare Alink et al. 2010). Affordances (cf. Maier and Fadel 2001) cover the entirety of uses that a system can be put to due to the specific characteristics (after Weber 2007) it possesses, though they may not have been originally intended by the designers (Brown and Blessing 2005). This difference is best explained using a simple example. If one were to look at the functions of an everyday object like a shoe, things that come to mind immediately might be 'supporting foot', 'protecting foot from sharp obstacles', 'distributing forces equally', 'providing comfort', and many similar *functions*. All of these are things that the shoe is for to provide to the user, i.e. these are related to its purpose and were intended when the shoe was designed. They can, however, support many other, not originally intended applications, like 'holding a door open' when wedged between the door and its frame or 'holding a beach towel in place against the wind' when it is used as a weight. These are *affordances*; they were not originally intended when the shoe was designed and made, but because of its physical structure (with resulting solidity and weight), it can be used for more purposes than just its original intentions. Because purpose is often used synonymously with function, it is important to differentiate between function and affordance: function is what drives a design, affordances can permit misuse, but can also repurpose objects for alternative applications.

Finally, it should also be noted that some authors differentiate between different types of function, depending on what they aim to emphasise or describe, often at different levels of abstraction. For instance, Hubka and Eder (1988) differentiate between what they call *purpose functions* and *technical functions*. The former pertain to the overall purpose of a system and hence describe the main, overall function (which relates to the uppermost level in Figure 18.5), whilst the latter address subfunctions of the system and follow the classical input/output paradigm.

The specific definition of function has received such widespread theoretical attention because a change in how function is defined ultimately leads to a change in how systems are ontologically perceived. By extension, this has significant implications for the design process. A different conception of function underlying one's perception of a system invariably changes the way one reasons from a given problem to a potential solution, as will be discussed in the following sections in more detail.

18.4. Reasoning Based on Alternate Notions of Function

As discussed earlier, the transition from a problem to a solution is generally characterised by co-evolution (see Figure 18.3), essentially comprising recurring comparative analysis of which functions are required and to what extent they are fulfilled in the emerging solution. Theories on function proposed in the literature provide guidelines for reasoning in order to ensure nothing is forgotten and to arrive at a solution in an effective manner. In addition to the approach by Pahl et al. (2007) discussed earlier, three other fundamental approaches will be presented here: the Function-Behaviour-Structure framework, the Theory of Domains, and the Use Case–Based design. The chosen examples are considered particularly interesting, as they are prominent across disciplines.[4]

18.4.1 Function-Behaviour-Structure Framework

In the Function-Behaviour-Structure (FBS) framework proposed by Gero (1990), system development is described as serving the purpose of "transforming function **F** . . . into a design description **D** in such a way that the artefact being described is capable of producing these functions" (Gero 1990: 28). However, it is argued that the direct transition from **F** to **D** is hardly possible (particularly for complex systems as discussed earlier; see also Gero and Kannengiesser 2002).

In the FBS framework, function is defined as "the relation between the goal of a human user and the behaviour of a system" (Gero 1990: 28). Behaviour is differentiated into *expected behaviour* (**Be**), which is what the designer expects the artefact to behave like, and *behaviour of the structure* (**Bs**), which can be directly derived from the *structure* (**S**) as its concrete solution elements and the actual behaviour they show. The proposed framework (Figure 18.7) illustrates how designers are expected to move towards a solution (i.e. a structure **S**) that implements the required functions (**F**).

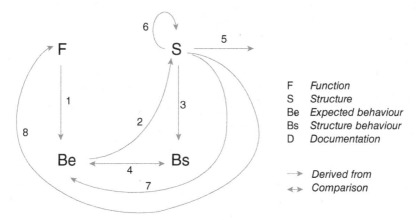

(1) formulation, (2) synthesis, (3) analysis, (4) evaluation, (5) documentation,
(6) reformulation type 1, (7) reformulation type 2, (8) reformulation type 3.

Figure 18.7   The FBS framework

Source: After Gero and Kannengiesser (2002).

In progressing to the engineering solution (**S**), one is to reason towards the final design of the system by deriving and formulating (*arrow 1* in Figure 18.7) an expected behaviour (**Be**) from the desired functions (**F**). From the determined expected behaviour (**Be**), a structure (**S**) is synthesised (*arrow 2*), which is intended to show the expected behaviour. The determined structure is documented (*arrow 5*). From the determined structure (**S**), the structural behaviour (**Bs**) can be derived through analysis (*arrow 3*). Structural behaviour (**Bs**) is compared with the formulated expected behaviour (**Be**) in an evaluation step (*arrow 4*). The comparison of **Be** and **Bs** then shows the current degree of function fulfilment provided by the determined structure (**S**). Based on the comparison, a new or adapted structure (**S**) may be determined, which shows an adapted behaviour (**Bs**) that is (hopefully) closer to the eventually expected behaviour (**Be**). This step is referred to as *reformulation type 1* (*arrow 6*). Other types of reformulation concern the specification of the expected behaviour (*arrow 7*) or of the functions themselves (*arrow 8*).

Gero and Kannengiesser (2002) discuss different alternatives for designers to move through the proposed framework, depending on the specific design situation at hand. The different steps in the framework are performed until the expected behaviour and the structural behaviour are sufficiently close. Thereby, the concept of co-evolution is very explicitly represented in this approach, whilst Pahl et al. (2007), as described in Section 2, postulate that engineers should stay on the abstract level until the functions have been sufficiently explored and only then advance to finding a solution.

18.4.2 Theory of Domains

The Theory of Domains was established by Andreasen (1980, 1992). It discerns four *domains* in which a system can be described and links these in the so-called Chromosome Model: *process domain, function domain, organ domain*, and *parts domain* (see Figure 18.8). Each domain represents the system under development on a different level of abstraction and is associated with a different representation of a system (using separate models) that describe the system from an alternative viewpoint. The approach borrows the associated models for each domain from the Theory of Technical Systems by Hubka and Eder (1988): the *transformation process structure, function structure, organ structure*, and *parts structure* (omitted in Figure 18.8). The chromosome model makes the causal links among the entities represented in these models explicit.

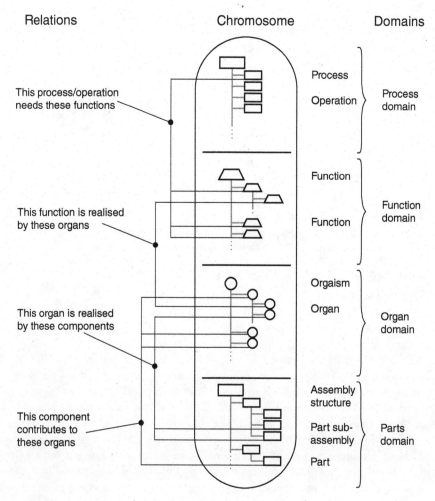

Figure 18.8 Chromosome model

Source: After Andreasen et al. (2014).

The specific relations between the different domains are defined by the capacity-related inter-pretation of function proposed in the Theory of Technical Systems. In contrast to Pahl et al. (2007), Hubka and Eder (1988) do not consider the transformation processes resulting in a state change of operands between inputs and outputs of a system to be the function of a system. Instead, function is defined in relation to the capability of a system to create a desired overall effect. In other words, function is the capacity to change operands from an input state to a desired output state, not the transformation process by which this is achieved.

The desired overall effect determines the purpose (or goal) of the system. The desired output state of operands is the result of a chain of transformation processes that are realised by operators through the functions they provide. Functions are hence defined as the *internal tasks* of the system that realise the transformation processes and thereby provide the capability to realise the desired overall effect. The different domains in the chromosome model can therefore be regarded as being linked from top to bottom through a "by-means-of" relation.

The *process domain* addresses the relation between transformation processes and operators that are being transformed. Operators can be humans, technical systems, and environmental influences, for instance. The desired overall effect is initially represented as a single box indicating the desired output states of operands in the transformation process structure.

The *function domain* focuses on the functions of the technical system that is to be developed and its functional interrelations with the surrounding operators that also contribute to the required transformation processes. The required inputs and main transformation processes are determined in such a way that the desired overall effect can be realised. The transformation processes are then decomposed and allocated to different operators that will realise them. In this step, therefore, engineers effectively allocate the specific transformation processes that must be realised by the technical system that is to be developed as opposed to those that will be realised by other means. Transformation processes are modelled in relation to a flow of operands and indicate the operators involved in their execution. For the transformation processes that have to be realised by the product under development, the required functions are subsequently determined and modelled in a function structure similar to Figure 18.4.

The *organ domain* describes the structure of organs required for realising the functions of the technical system to be developed. Organs are abstract formulations of the function carriers. They are defined as "function elements (or 'means') of a product, displaying a mode of action and a behaviour, which realise its function and carry its properties" (see Andreasen et al. 2014). Finally, the *parts domain* details how organs are realised by the system's components and individual parts.

The underlying reasoning approach focuses on determining the required transformation processes and functions, in order to realise the desired overall effect, as well as the function carriers (which may be specified as organs or parts) realising them. Hence, in the Theory of Domains, the engineers are first to consider which processes should be supported to achieve the overall goal. Then, they are to determine what the system to be developed must be able to do in order to realise those specific processes.

The reasoning between transformation processes, functions, and function carriers is highly iterative. This includes switching flexibly between the different domains and the related models (Eder 2008). The chromosome model adds context to the individual models proposed by Hubka and Eder by revealing the interdependencies between the different domains. It also highlights the central role functions have for understanding the overall goal of a system and for gradually detailing the means required for realising this goal.

18.4.3 Use Case-Based Function Reasoning

Use case-based design was originally proposed in the 1980s and 1990s in software engineering as one of the so-called object-oriented design approaches (Jacobson et al. 1992). The most prominent approach is the one proposed by Cockburn (2000, 2003). It essentially involves deriving the central goals of the prospective users of a system. Based on these goals, the use cases realising these goals are to be determined as well as the required functions and their carriers in the system realising the respective use cases.

A use case is a set of distinct activities or process steps that are intended to achieve a specific goal of a user. These processes typically include interaction processes between the users and different elements of the system under development, as well as processes provided by the system itself. The goal of the user is "the goal that the primary actor has in trying to get work done" or in order to receive some kind of value from the system (Cockburn 2000: 57). Functions are defined as the behavioural aspects (i.e. the processes) of the system or as "the services [that the] system offers" in fulfilling the user's goals (Cockburn 2000: 45).

In using this approach to derive solutions, one starts by determining the goals of prospective users through deriving the particular (change in the) situation that the user desires (see Figure 18.9). Then, the specific use cases are determined that will realise this desired new or changed situation. The use cases are initially described textually, also specifying the involved users and their behaviour

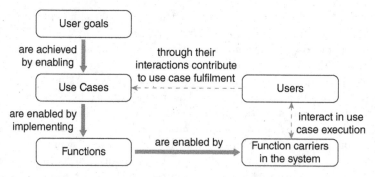

Figure 18.9   Schema of reasoning from user goals to required function carriers

Source: After Cockburn (2000).

in interacting with different sub-systems (i.e. the function carriers). At this stage, the designers are, hence, already expected to designate an initial set of function carriers. Function carriers can be concrete physical parts (such as, e.g., a touchpad or buttons) or software components (such as, e.g., data objects, classes, etc.). The use cases further include the specific processes (i.e. the functions or "services") that the system needs to provide in order to address and answer the different user requests. The modelled use cases and process flows are iteratively detailed and refined. In parallel, further function carriers are determined, so that the use cases (and eventually the user goals) can be fulfilled in an iterative manner.

These descriptions show that the concrete notion of function used or followed, respectively, directly affects the approach that authors prescribe for designers to take in conceiving and designing solutions to a given problem. The following section describes a more recent endeavour to help transition between different interpretations of function and the resulting variances in how a system is to be designed based on them.

18.5. A Consensus Model of Function-Based Design: 5-Key-Terms Ontology

The 5-key-terms approach, or ontology, can be regarded as an attempt to provide engineering research with some kind of consensus model on how to describe a system at an abstract level. It aims to provide a vocabulary relating relevant terms in a distinctive manner. The approach has been proposed by Vermaas (2011; see also Vermaas 2013) and is based on comprehensive reviews of engineering design approaches from literature. Five "key terms" are proposed that grasp the essence of what the previously discussed approaches and alternative definitions entail: *goal of the device, actions with the device, functions of the device, behaviour of the device*, and *structure of the device* (see Figure 18.10).
   Therein:

* *Goals of the device* are the "state of affairs the prospective users of the device are to achieve with the device".
* *Actions with the device* describe "a deliberate manipulation of the device by the user".
* *Functions of the device* describe the "physiochemical capacity of the device" enabling these actions (i.e. the actions with the device).
* *Behaviour of the device* "is the physiochemical evolution of the device including the evolution of its structure and the device's physicochemical interactions with its environment".
* Finally, *structure of the device* "describes the physiochemical configuration of the device" (Vermaas 2013: 196).

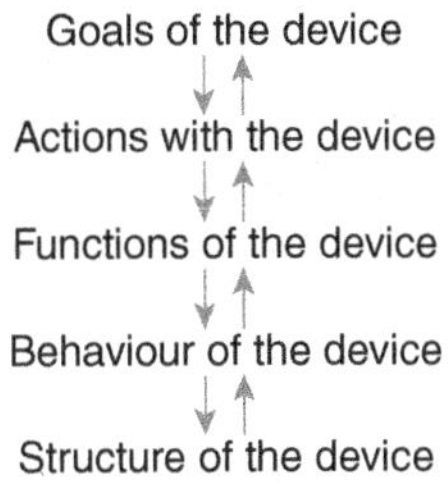

Goals of the device
↓ ↑
Actions with the device
↓ ↑
Functions of the device
↓ ↑
Behaviour of the device
↓ ↑
Structure of the device

Figure 18.10   The 5-key-term approach

Source: After Vermaas (2011, 2013).

Goals and related *actions with the device* depend on the specific "use plan" of a user, which refers to "a goal-directed series of actions, including manipulations of a [device] and its components" (see Houkes and Vermaas 2010: 19).

Vermaas describes the proposed key terms as different "conceptual layers" that one's reasoning passes through in an "ordered sequence of steps" towards a solution (Vermaas 2013: 197). The process starts by considering the goals of the prospective users and the associated actions they will take with the device, corresponding to its specific use plan. In order for the system to be capable of enabling this specific use plan, the designers have to determine the functions that the system needs to possess and the required behaviour of a system in correspondence with the use plan to be realised. Finally, the function carriers (i.e. the structure of the system) need to be determined in a way that the system can exhibit the expected behaviour and provide the desired functions. This process is expected to be iterative and span multiple layers at a time.

The 5-key-terms approach can be regarded as the most detailed in terms of explaining the different stages that function-based reasoning passes through during conceptualisation of a potential solution. When comparing it with the approaches highlighted earlier, it can be argued that these either combine or bypass one or more of the conceptual layers. This is referred to as *conceptual bypassing*, and it directly relates to the specific archetypical notion of function that the approaches are based on.[5] For instance, in Gero's FBS framework, the *actions with the device* are not addressed. Function is used synonymously with the *goals of the device* and directly relates the goals to the specific *behaviour of the device* that fulfils these goals. The inherent archetypical notion of function is thus *goal-related*. Gero explicitly addresses the *behaviour of the device*. This is either the expected (**Be**) or actual behaviour (**Bs**), established by the *structure of the device* (**S**). The classification of the FBS framework after Vermaas (2013) is illustrated in Figure 18.11a.

Pahl et al. (2007) consider the transformation of operands between the input and output of the system as the overall function of the system. With respect to the identified archetypical notions of function (see Section 3), this clearly conveys a *behaviour-related* notion of function. That means, the key-term of *behaviour of the device* is bypassed and integrated into *functions of the device* (see Figure 18.11b). *Actions with the device* are not explicitly addressed. Human activity impacting the system to be developed may inherently be considered, e.g. by including "hand force" as energy (i.e. an operand), as input to the system and/or affecting some of its subfunctions (compare Stone and Wood 2000: 363). As such, Pahl et al. take a decisively more product-focused view on the design process, and humans are only peripherally/implicitly considered. This is a stark contrast to user case-based design, where the human is considered a concrete actor that can be part of function fulfilment or at least is explicitly included as interacting with an (otherwise self-sufficient) system.

The Theory of Domains is based on a *goal-* or *task-related* notion of function. The function reasoning approach essentially starts by determining the central goal of the system, which is the desired overall effect. This corresponds to the *goals of the device*. Subsequently, the different transformation

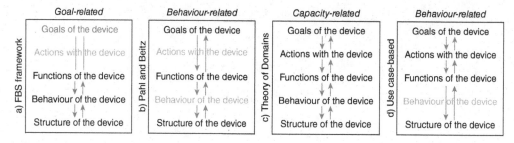

Figure 18.11   Classification of (a) FBS framework, (b) Pahl and Beitz approach, (c) Theory of Domains, and (d) Use case-based reasoning

processes that will bring about this desired overall effect are determined. Andreasen et al. (2014) have argued that transformation processes also incorporate the *actions with the device*, if a human operator is involved. This interpretation is apt, bearing in mind that in the Theory of Domains, humans interacting with the technical entity to be developed are viewed as integral to the wider system under consideration. Mainly though, transformation processes represent the *behaviour of the device* itself. Organs and parts, finally, establish the *structure of the device*. Hence, all five of the key-terms after Vermaas are addressed in the Theory of Domains (see Figure 18.11c).

Use case-based approaches start by determining the specific goals of the prospective users and subsequently derive the use cases that realise these goals. Use cases include the specific interaction processes between the user and the system to be developed. Hence, they explicitly address the *goals of the device* and *actions with the device*. Use cases further include the specific processes that the system requires to address the user requests. These processes are directly associated with the functions the system has to provide. Hence, use case-based approaches employ a *behaviour-related* notion of function. The *behaviour of the device* is incorporated into its function (see Figure 18.11d). Function carriers (i.e. *the structure of the device*) are determined in parallel to the determination and detailing of the use cases.

Note that there is no 'right' or 'wrong' in how to define function; by extension, the same applies to conceptually bypassing a notion that function may have. Whilst some authors differentiate in greater detail as to what exactly function is and how it relates to behaviour, aims, etc., others allow the engineers to take a wider perspective, potentially encompassing actions with a device into its function and behaviour (as it could be with services and product-service systems), which can be helpful in the early stages when the system is not yet clearly defined.

18.6. Discussion and Conclusion

Function is a central concept in theories about the emergence of a technical solution through engineering design work. From a philosophical viewpoint, it provides a vocabulary to describe and discern—and thereby cognitively grasp—an emerging system and its internal/external relations at an abstract level. The comparisons of various approaches in the previous section illustrate how varied the interpretation of function can be in such theories. Ontologically, following these approaches significantly changes the perception of a system. Humans using a technical entity may or may not be explicitly included in these interpretations. Views on the system to be developed may be purely focused on the internal causal relations or the outcomes one seeks to achieve (by themselves or, again, directed towards human activity and desires). Ultimately, all of these theorisations aim to assist the designers' reasoning from what a technical system must be able to do to the concrete (partial) solution elements that are able to achieve this. That said, there are also differences between them as to whether this should be inherently considered an iterative process, following the principles of co-evolution, or should be a deliberate jump at a later stage to avoid fixation.

Engineers in practice have been found to develop a much more thorough comprehension of a system when explicitly using a function-based approach during design (Alink et al. 2010; Caldwell et al. 2011; Eisenbart and Kleinsmann 2017). Yet, in doing so, practitioners very flexibly jump between viewing functions of a system using alternating notions of function, very similar to the archetypical notions discussed prior, and from frequently changing levels of abstraction (Eckert 2013; Tomiyama et al. 2013). This is very different from the extremely rigorous differentiations of what function means as put forth by different scholars. Scholars have argued that a more flexible way of cognitively construing what a system will need to do enables viewing—and eventually perceiving—the system from alternative angles and viewpoints. Thereby, engineers can understand the system to be developed more quickly and on a deeper level (Goel 2013). Yet, doing so reduces rigour in discerning function from related terms in engineering, which counters attempts to make function a computable entity for automated design, change management, and validation. To date, a suitable compromise has not been developed.

Eventually, function remains an abstract entity that allows engineering designers to build a mental model of the system to be developed, what is and what is not part of it, and the relation between how it is constructed, how it is built, and what it is supposed to do. This very interpretation is reflected in more recently developed function-based approaches in research. The IFM Framework (Eisenbart et al. 2016), the Function Analysis Diagram (Aurisicchio et al. 2012), and the SAPPhIRE model (Chakrabarti et al. 2005), among others, explicitly link procedural descriptions of a system, state changes, use cases, etc., with a system's structure and users (both human and other). This makes the described conceptual links tangible and provides a holistic, coherent model of the system and its internal and external relations.

Related Chapters

Chapter 17: Use Plans (Auke Pols)
Chapter 40: Ethical Considerations in Engineering (Wade L. Robison)

Further Reading

Chakrabarti, A. and Bligh, T.P. (2001). A Scheme for Functional Reasoning in Conceptual Design. *Design Studies*, 22(6), 493–517. (A very profound scholarly piece of different types of function in engineering.)
Crilly, N. (2010). The Role That Artefacts Play. Technical, Social and Aesthetical Functions. *Design Studies*, 31(4), 311–344. (A discussion of the differences in the perception of functions in technical and non-technical contexts.)
Hubka, V. and Eder, W. (1988). *Theory of Technical Systems: A Total Concept Theory for Engineering Design*. Berlin, Heidelberg, New York and Tokyo: Springer-Verlag. (One of the earliest and probably most comprehensive theoretical discussions of what constitutes a technical system and the role of functions in this.)
Pahl, G., Beitz, W., Feldhusen, J. and Grote, K.-H. (2007). *Engineering Design: A Systematic Approach*. Berlin, Heidelberg, New York and Tokyo: Springer-Verlag. (A profound piece on how functions determine the design concept of a technical product.)

Notes

1. It is to be noted here that whilst there is a belief of a causal relation between inputs and outputs, usually a multitude of alternative partial solutions may fulfil an individual function, and it is up to the engineering design to make the best selection.
2. For comprehensive discussions of the typical breakdown of a design process see Gericke and Blessing (2012), Eisenbart et al. (2011), Howard et al. (2008), Blanchard and Fabrycky (1998), or Blessing (1996).
3. Further examples and comparisons of function taxonomies can be found e.g. in Szykman et al. (1999), Hirtz et al. (2001), Kitamura et al. (2007), and Sen et al. (2010, 2013).

4. Other examples are C–K theory (Hatchuel and Weil 2003), Axiomatic design (Suh 2001), TRIZ (Altschuller 1999), and Property Driven Design (Weber 2007).
5. The following comparison builds on similar work by Vermaas (2011, 2013) and Howard and Andreasen (2013).

References

Ahmed, S. and Wallace, K. (2003). Evaluating a Functional Basis. In *Proceedings of the ASME Design Engineering Technical Conferences and Computers and Information in Engineering Conference IDEC/CIE.*

Alink, T., Eckert, C., Ruckpaul, A. and Albers, A. (2010). Different Function Breakdowns for One Existing Product: Experimental Results. In J. Gero (ed.), *Design Computing and Cognition—DCC,* Springer-Verlag, Dordrecht, pp. 405–425.

Altschuller, G. (1999). *The Innovation Algorithm: TRIZ, Systematic Innovation, and Technical Creativity.* Technical Innovation Center, Worcester (USA).

Andreasen, M.M. (1980). *Syntesemetoder på Systemgrundlag—Bidrag til en Konstruktionsteori.* Dissertation, Institute for Machine Design, Lund University, Lund.

Andreasen, M.M. (1992). The Theory of Domains. In D. Ullman and K. Wallace (eds.), *Understanding Function and Function-to-From Evolution: Workshop Report, CUED/C-EDC/TR 12,* Engineering Design Centre, Cambridge (UK), pp. 21–47.

Andreasen, M.M., Howard, T. and Bruun, H. (2014). Domain Theory, Its Models and Concepts. In A. Chakrabarti and L.T.M. Blessing (eds.), *An Anthology of Theories and Models of Design: Philosophy, Approaches and Empirical Explorations.* Berlin: Springer-Verlag.

Aurisicchio, M., Bracewell, R. and Armstrong, G. (2012). The Function Analysis Diagram: Intended Benefits and Coexistence With Other Functional Models. *Artificial Intelligence for Engineering Design, Analysis and Manufacturing—AIEDAM,* 27(3), 249–257.

Blanchard, B. and Fabrycky, W. (1998). *Systems Engineering and Analysis, International Series in Industrial & Systems Engineering,* Prentice Hall, Upper Saddle River, NJ, USA.

Blessing, L.T.M. (1996). Comparison of Design Models Proposed in Prescriptive Literature. In *Proceedings of the COST A3 / COST A4 International Workshop on "The Role of Design in the Shaping of Technology".*

Blessing, L.T.M. and Upton, N. (1997). A Methodology for Preliminary Design of Mechanical Aircraft Systems. *AIAA/SAE World Aviation Congress,* American Institute of Aeronautics and Astronautics, Reston (USA).

Bonnema, G.M. and van Houten, F.J.A.M. (2006). Use of Models in Conceptual Design. *Journal of Engineering Design,* 17(6), 549–562.

Braha, D. and Reich, Y. (2003). Topological Structures for Modeling Engineering Design Processes. *Research in Engineering Design,* 14(4), 185–199.

Brown, D.C. and Blessing, L.T.M. (2005). The Relationship Between Function and Affordance. In *Proceedings of the ASME International Design Engineering Technical Conferences & Computers and Information Engineering Conference IDEC/CIE.*

Bucciarelli, L.L. (2010). *From Function to Structure in Engineering Design.* Online publication. http://hdl.handle.net/1721.1/51789. Accessed May 30, 2013.

Buur, J. (1990). *A Theoretical Approach to Mechatronics Design.* Dissertation, Technical University of Denmark, Copenhagen.

Caldwell, B.W., Sen, C., Mocko, G.M. and Summers, J.D. (2011). An Empirical Study of the Expressiveness of the Functional Basis. *Artificial Intelligence for Engineering Design, Analysis and Manufacturing (AI EDAM)* 25(3), 273–287.

Carrara, M., Garbacz, P. and Vermaas, P. (2011). If Engineering Function Is a Family Resemblance Concept: Assessing Three Formalization Strategies. *Applied Ontology,* 6(2), 141–163.

Chakrabarti, A. and Bligh, T.P. (2001). A Scheme for Functional Reasoning in Conceptual Design. *Design Studies,* 22(6), 493–517.

Chakrabarti, A., Sarkar, P., Leelavathamma, B. and Nataruja, B. (2005). A Functional Representation Supporting Process and Product Knowledge in Biomimetic Design. *Artificial Intelligence for Engineering Design, Analysis and Manufacturing—AI EDAM,* 19(2), 113–132.

Cockburn, A. (2000). *Writing Effective Use Cases,* Vol. 1, Addison Wesley Professional, Indianapolis, IN, USA.

Cockburn, A. (2003). *Agile Software-Entwicklung: Die Prinzipien der Agilen Software-Entwicklung Dargestellt und Erläutert, Welche Methodik für Welches Software-Projekt?, Software-Entwicklung im Team.*Bonn: mitp.

Crilly, N. (2010). The Role That Artefacts Play. Technical, Social and Aesthetical Functions. *Design Studies,* 31(4), 311–344.

Eckert, C. (2013). That Which Is Not Form. The Practical Challenges in Using Functional Concepts in Design. *Artificial Intelligence for Engineering Design, Analysis and Manufacturing (AI EDAM),* 27(3), 217–231.

Eder, W. (2008). Aspects of Analysis and Synthesis in Design Engineering. In *Proceedings of the Canadian Engineering Education Association*.

Ehrlenspiel, K. (2007). *Integrierte Produktentwicklung: Denkabläufe, Methodeneinsatz, Zusammenarbeit*. München: Hanser-Verlag.

Eigner, M., Faisst, K., Hollerith, T. and Nem, F. (2010). A View-based Modelling Approach for Representing Multidisciplinary Functions in PDM Systems. *Proceedings of 11th International Design Conference—DESIGN*.

Eisenbart, B., Gericke, K. and Blessing, L.T.M. (2011). A Framework for Comparing Design Modelling Approaches Across Disciplines. *Proceedings of the 18th International Conference on Engineering Design—ICED*, 2, 344–355.

Eisenbart, B., Gericke, K. and Blessing, L.T.M. (2013). An Analysis of Functional Modell Approaches Across Disciplines. *Artificial Intelligence for Engineering Design, Analysis and Manufacturing (AI EDAM)*, 27(3), 281–289.

Eisenbart, B., Gericke, K. and Blessing, L.T.M. (2016). A DSM-based Framework for Integrated Function Modelling: Concept, Application and Evaluation. *Research in Engineering Design*, 28(1), 25–51.

Eisenbart, B. and Kleinsmann, M. (2017). Implementing Shared Function Modelling in Practice: Experiences in Six Companies Developing Mechatronic Products and PSS. *Journal of Engineering Design*, 28(10–12), 765–798.

Erden, M., Komoto, H., van Beek, T.J., D'Amelio, V., Echavarria, E. and Tomiyama, T. (2008). A Review of Function Modeling: Approaches and Applications. *Artificial Intelligence for Engineering Design, Analysis and Manufacturing (AI EDAM)*, 22(2), 147–169.

Ericson, A. and Larsson, T. (2005). A Service Perspective on Product Development: Towards Functional Products. *Proceedings of 12th International Product Development Management Conference—CBS*.

Fowler, M. (1998). *Analysis Patterns: Reusable Object Models*. Boston: Addison Wesley Longman Inc.

Gericke, K. and Blessing, L.T.M. (2012). An Analysis of Design Process Models Across Disciplines. *Proceedings of 12th International Design Conference—DESIGN*.

Gericke, K., Qureshi, A.J. and Blessing, L.T.M. (2013b). Analyzing Transdisciplinary Design Processs in Industry: An Overview. In *Proceedings of the ASME International Design Engineering Technical Conferences & Computer and Information in Engineering Conference IDEC/CIE*.

Gero, J.S. (1990). Design Prototypes: A Knowledge Representation Scheme for Design. *AI Magazine*, 11(4), 26–36.

Gero, J.S. and Kannengiesser, U. (2002). The Situated Function—Behaviour—Structure Framework. *Artificial Intelligence in Design*, 2, 89–104.

Goel, A. (2013). One Thirty Year Long Case Study. Fifteen Principles: Implications of the AI Methodology for Functional Modelling. *Artificial Intelligence for Engineering Design, Analysis and Manufacturing (AI EDAM)*, 27(3), pp. 203–215.

Halmos, P.R. (1974). *Naive Set Theory*. New York: Springer-Verlag.

Hatchuel, A. and Weil, B. (2003). A New Approach of Innovative Design. In *Proceedings of the 14th International Conference on Engineering Design—ICED*.

Hirtz, J., Stone, R.B., Szykman, S., McAdams, D. and Wood, K.L. (2001). Evolving a Functional Basis for Engineering Design. In *Proceedings of the ASME Design Engineering Technical Conference: DETC2001*.

Houkes, W. and Vermaas, P.E. (2010). *Technical Functions: On the Use and Design of Artefacts, Philosophy of Engineering and Technology*, Springer-Verlag, Dordrecht.

Howard, T. and Andreasen, M.M. (2013). Mindsets of Functional Reasoning in Engineering Design. *Artificial Intelligence for Engineering Design, Analysis and Manufacturing (AI EDAM)*, 27(3), 233–240.

Howard, T., Culley, S. and Dekoninck, E. (2008). Describing the Creative Design Process by the Integration of Engineering Design and Cognitive Psychology Literature. *Design Studies*, 29(2), 160–180.

Hubka, V. (1980). *Principles of Engineering Design*. Oxford: Butterworth-Heinemann Ltd.

Hubka, V. and Eder, W. (1988). *Theory of Technical Systems: A Total Concept Theory for Engineering Design*. Berlin, Heidelberg, New York and Tokyo: Springer-Verlag.

Hundal, M. (1990). A Systematic Method for Developing Function Structures, Solutions and Concept Variants. *Mechanism and Machine Theory*, 25(3), 243–256.

Jacobson, I., Christerson, M., Jonsson, P. and Övergaard, G. (1992). *Object-oriented Software Engineering: A Use Case Driven Approach*. New York, Wokingham and Reading, MA: ACM Press; Addison-Wesley Publications.

Jansson, D. and Smith, S. (1991). Design Fixation. *Design Studies*, 12(1), 3–11.

Kitamura, Y., Takafuji, S. and Mizogushi, R. (2007). Towards a Reference Ontology for Functional Knowledge Interoperability. In *Proceedings of the ASME International Design Engineering Technical Conferences & Computer and Information in Engineering Conference IDETC/CIE*.

Kurfman, M., Stock, M., Stone, R.B., Rajan, J. and Wood, K.L. (2003). Experimental Studies Assessing the Repeatability of a Functional Modeling Derivation Method. *Journal of Mechanical Design*, 125(4), 682–639.

Maier, J.R. and Fadel, G.M. (2001). Comparing Function and Affordance as Bases for Design. In *Proceedings of the ASME International Design Engineering Technical Conferences & Computer and Information in Engineering Conference IDEC/CIE*.

Merriam-Webster (2003). *Merriam-Webster's Collegiate Dictionary*, Merriam-Webster Incorporated, Springfield.

Paetzold, K. (2006). On the Importance of a Functional Description for the Development of Cognitive Technical Systems. *Proceedings of 9th International Design Conference—DESIGN*.

Pahl, G. and Beitz, W. (1977). *Konstruktionslehre*, Springer-Verlag, Berlin.

Pahl, G., Beitz, W., Feldhusen, J. and Grote, K.-H. (2007). *Engineering Design: A Systematic Approach*. Berlin, Heidelberg, New York and Tokyo: Springer-Verlag.

Poon, J. and Maher, M. (1997). Co-evolution and Emergence in Design. *Artificial Intelligence in Design*, 11(3), 319–327.

Rajan, J., Stone, R.B. and Wood, K.L. (2003). Functional Modelling of Control Systems. In *Proceedings of the 14th International Conference on Engineering Design—ICED*.

Rodenacker, W.G. (1970). *Methodisches Konstruieren*. Berlin, Heidelberg, New York and Tokyo: Springer-Verlag.

Roozenburg, N.F.M. and Eekels, J. (1995). *Product Design: Fundamentals and Methods, A Wiley Series in Product Development Planning, Designing, Engineering*. Chichester (a.o.): John Wiley.

Sakao, T. and Shimomura, Y. (2007). Service Engineering: A Novel Engineering Discipline for Producers to Increase Value Combining Service and Product. *Journal of Cleaner Production*, 15(6), 590–604.

Schneider, K., Daun, C. and Wagner, D. (2005). Vorgehensmodelle und Standards zur systematischen Entwicklung von Dienstleitungen. In H.-J. Bullinger, K. Schneider and A.-W. Scheer (eds.), *Service Engineering: Entwicklung und Gestaltung innovativer Dienstleistungen*. Berlin, Heidelberg: Springer-Verlag, pp. 113–138.

Sen, C., Summers, J.D. and Mocko Gregory, M. (2010). Topological Information Content and Expressiveness of Function Models in Mechanical Design. *Journal of Computing and Information Science in Engineering*, 10(3), 1–11.

Sen, C., Summers, J.D. and Mocko Gregory, M. (2013). A Formal Representation of Function Structure Graphs for Physics-based Reasoning. *Journal of Computational and Information Science in Engineering*, 13(2).

Simon, H. (1973). The Structure of Ill-Structured Problems. *Artificial Intelligence for Engineering Design, Analysis and Manufacturing (AI EDAM)*, 4(3–4), 181–201.

Spath, D. and Demuss, L. (2005). Entwicklung Hybrider Produkte. Gestaltung Materieller und Immaterieller Leistungsbündel. In H.-J. Bullinger, K. Schneider and A.-W. Scheer (eds.), *Service Engineering: Entwicklung und Gestaltung Innovativer Dienstleistungen*. Berlin, Heidelberg: Springer-Verlag, pp. 463–502.

Stone, R.B. and Wood, K.L. (2000). Development of a Functional Basis for Design. *Journal of Mechanical Design*, 122(4), 359–370.

Suh, N. (2001). *Axiomatic Design: Advances and Applications*. London: Oxford University Press.

Szykman, S., Racz, J. and Sriram, R. (1999). The Representation of Function in Computer-Based Design. In *Proceedings of the International Conference on Design Theory and Methodology—DTM*.

Tomiyama, T., van Beekm, T.J., Alvarez Cabrera, A.A., Komoto, H. and D'Amelio, V. (2013). Making Function Modeling Practically Usable. *Artificial Intelligence for Engineering Design, Analysis and Manufacturing (AI EDAM)* 27(8), 301–309.

Ullman, D. (2010). *The Mechanical Design Process, McGraw-Hill Series in Mechanical Engineering*, 4th ed. Boston: McGraw-Hill Higher Education.

Ulrich, K. and Eppinger, S.D. (2008). *Product Design and Development*. New York: McGraw-Hill Higher Education.

US DoD (2001). *Systems Engineering Fundamentals*. Fort Belvoir (USA): Defence Acquisition University Press.

VDI (2004). *VDI 2206—Design Methodology for Mechatronic Systems*. Berlin: Beuth Verlag.

VDI (2018). *VDI 2221—Systematic Approach for the Design of Technical Systems and Products*. Berlin: Beuth Verlag.

Vermaas, P. (2010). A Conceptual Ambiguous Future for Engineering Design. *Copenhagen Working Papers on Design*, 2, 1–6.

Vermaas, P. (2011). Accepting Ambiguity of Engineering Functional Descriptions. *Proceedings of 18th International Conference on Engineering Design—ICED*.

Vermaas, P. (2013). On the Co-Existence of Engineering Meanings of Function. Four Responses and Their Methodological Implications. *Artificial Intelligence for Engineering Design, Analysis and Manufacturing (AI EDAM)*, 27(3), 191–202.

Weber, C. (2007). Looking at 'DFX' and 'Product Maturity' from the Perspective of a New Approach to Modelling Product and Product Development Processes. In *Proceedings of the 17th CIRP Design Conference*.

Welp, E.G., Sadek, T., Müller, P. and Blessing, L.T.M. (2007). Integrated Modelling of Products and Services: The Conceptual Design Phase in an Integrated IPS2 Development Process. *Beiträge zum 18. Symposium "Design for X"*.

Zwicky, F. (1989). *Entdecken, Erfinden, Forschen im Morphologischen Weltbild. Mit Diagrammen*. Baeschlin-Verlag, Glarus (Germany).

19

EMERGENCE IN ENGINEERING

Peter Simons

19.1. Introduction

The idea of emergence was first coined in philosophy in the nineteenth century and came to play a prominent part in philosophical debates in the early and late twentieth centuries. From there it spread much more recently to engineering, particularly in discussions of the area of complex systems. Unfortunately, neither in philosophy nor in engineering theory is there a unique meaning of the term 'emergence', but an unruly host of closely related meanings, which are often not clearly distinguished, and which add to the general confusion surrounding uses of the term. This chapter outlines the origins and intended purpose of the notion of emergence and attempts to differentiate the various related notions and to indicate which ones may be of some use in engineering theory. The nature of the copious and often confusing (and confused) literature means that we cannot simply report on how 'the' concept of emergence applies to engineering, but we must first clarify what it has been and can be taken to mean.

19.2. The Emergence of Emergence

The general notion of emergence was formulated in the nineteenth century to provide a third or compromise position between dualism and monism, specifically in two areas: life and mind. Dualism in regard to such a subject matter is a view that there are two utterly distinct and equally basic principles governing the area, neither of which depends on or is reducible to the other. Psychophysical dualism, the version for minds, can be substance dualism, the view that there are mental substances (souls) distinct from physical substances (bodies); or it can be property dualism, the view that there are mental properties (such as consciousness and intentionality) distinct from physical properties (such as mass and electric charge). Dualism is the view that while these may co-occur, as in persons, neither is more basic than the other, or the foundation of the other: the duality must be taken as fundamental. Monism in this debate is often called *physicalism*, because it claims that minds are wholly constituted by physical entities and processes: the dualist position is usually known simply as *dualism*. In the case of life, dualism takes the form of claiming that in addition to inorganic things like rivers and properties like mass, there are irreducibly organic principles constituting life and living things. In this case the dualist position is known as *vitalism* and the monist position—denying any fundamental distinction—is called *mechanism*.

Both dualism and monism face difficult problems. In dualism the problem is to explain how two fundamentally different principles can interact if they have nothing basic in common. This problem for psychophysical dualism came to its notorious climax in the mind–body problem of Descartes, who accepted interaction but could not satisfactorily explain how it worked, leaving a legacy that divided philosophers for centuries and continues to do so. The problem for vitalism was to specify the nature of the essential vital ingredient that distinguished inorganic from organic bodies, while remaining connected to serious biological science. The problem for monism is that the nature and properties of the basis—the physical or the inorganic respectively—appear so different from those of the other sorts of things—mentality or life—that there is a huge explanatory gap, so that the monist position appears at best aspirational rather than scientifically grounded.

The compromise that is emergence is intended to find a way between the two difficult extremes. The general idea is that while emergent properties are *novel* with respect to the basis (physical or inorganic, according to the case)—hence the term 'emergent'—they do not represent a fundamentally different basic principle as in dualism but arise as a result of an increasingly complex configuration of the elements of the basis. Only when this configuration reaches a suitable level of complexity and is of the right kind do the novel properties 'emerge'. Emergence is often compared with supervenience, which is the idea that one set of properties of a thing depend on and co-vary with another set. Emergent properties are frequently viewed as supervenient on more basic properties. Indeed, sometimes the attempt has been made to *define* emergence in terms of supervenience (Van Cleve 1990, cf. McLaughlin 1997). While the two notions are clearly related, emergence stresses the utter *novelty* of the emergent properties (Kim 2003: 567). The chief problem confronting this notion is that the idea of novelty is unclear, and that different ways to spell out what this novelty consists in lead to different concepts of emergence. There is also a concern that, depending on how novelty is interpreted, emergence may collapse into one of the original extremes, and thus fail to be a 'third way' between them.

19.3. A Tiny History of Emergence

The term 'emergent' was first used in a modern philosophical sense by the English philosopher George Henry Lewes, in his multivolumed *Problems of Life and Mind* (Lewes 1875), though there had been hints of the notion in much older philosophers, such as Aristotle. Earlier, in his *System of Logic*, John Stuart Mill (Mill 1843, Bk. III, Ch. 6, §1) had distinguished between *homopathic* laws, in which the joint effect of several causes is the sum of the effects of each cause acting separately, as in the parallelogram of forces, and *heteropathic* laws, where the joint effect is not the sum of the several effects, as in the appearance of life based on physical constituents. Use of emergence reached its interim zenith with the so-called British emergentists of the early twentieth century, notable among them Samuel Alexander (Alexander 1920), C. Lloyd Morgan (Morgan 1923) and C. D. Broad (Broad 1925) . Here a number of subtly different notions of emergence were in play. After several decades in eclipse, the idea of emergence began to be revived in the late twentieth century. By this time, with the chemistry of life having been more thoroughly researched and the idea of a vital life-force having been widely discarded, the vitalism–mechanism controversy had lost force, and the focus of discussion was again on whether mental properties were in some way emergent from the physical properties of the brain (cf. O'Connor 1994; Kim 1999). The many meanings of 'emergent' resurfaced, and contemporary discussion in the philosophy of mind is ongoing—and inconclusive (for an overview, see Clayton and Davies 2008).

By analogy with the novel characteristics of consciousness, phenomenal content and intentionality that characterise mind, and which clearly arise only when a certain level of organic complexity has been achieved, complex engineered systems likewise sometimes exhibit behaviour that is in some way surprising or unexpected, and which may be hard or impossible to foresee, predict, understand, or explain. In such circumstances, a concept of emergence may appear apt to describe the phenomena in question, and this has led engineering theorists to reach for the notion of emergence. In so doing they have inherited the ambiguities and uncertainties of the philosophical discussion. So it is imperative, if some notion or notions of emergence are to be of use to engineering, that the different conceptual strands composing the various notions of emergence be separated as a prelude to assessing their applicability.

19.4. A Template for Emergence

In order to tame the varieties of emergence and not simply end up with an unstructured list of variants as proposed by this or that thinker, we set out a form or template to which we require candidate accounts of emergence to conform. The variety will come in the multiplicity of ways in which the variable parts of the template can be interpreted.

The elements of the template and the schematic letters used for them are as follows:

E: a candidate emergent entity, be it an object, property, process or mode of behaviour.

S: a bearer. This is an object or system whose parts, properties or behaviour include such a candidate emergent entity.

C: a context: an environment or conditions in which the candidate emergent entity occurs or appears.

B: a basis of supposedly more fundamental entities composing or constituting S. These are typically parts of S, whether parts at a next lower level of organisation, or its most basic parts.

O: characteristic properties, interrelations and modes of operation of the basis.

N: the respect or respects in which E is taken to be novel over against O.

The template, then, is as follows:

> E belonging to S occurring in context C is novel in respect N with regard to O of B

The principal differentiating factors concern N, the way in which E is novel. We first list the candidates, then comment on them, and finally assess their utility in and for engineering.

19.5. Modes of Novelty

The following list of ways in which a candidate emergent entity may be considered novel is distilled from the extremely copious literature on emergence (for an authoritative survey see O'Connor and Wong 2015). I have attempted to be as complete as practicable while attempting to keep the modes distinct as far as possible. In each definition, the word highlighted in small capitals will be used as a name for the mode of novelty in further discussion. We will avoid prematurely labelling any of them as 'emergent' until discussion and assessment is further advanced. For brevity we will call the candidate emergent entity or phenomenon E a *property* of S, but it should be remembered that it need not be a static characteristic but can be an object, process or mode of action or behaviour associated with S.

1. E is of a new kind: one that exists or is occurring for the FIRST TIME (has not existed or occurred earlier).
2. E is surprising or UNEXPECTED (given O of B).
3. E is unpredicted or UNFORESEEN (given O of B).
4. E cannot be predicted or is UNFORESEEABLE (on the basis of O of B).

5. *E* is a property of *S* as a WHOLE which is one that the members of *B* do not have.
6. *E* is a property of *S* which is NON-SUMMATIVE with respect to *O* of *B*.
7. *E* is a property of *S* which is not deduced or is UNDERIVED from *O* of *B*.
8. *E* is a property of *S* which cannot be deduced or is UNDERIVABLE from *O* of *B*.
9. *E* is a property of *S* which is UNEXPLAINED given those of *O* of *B*.
10. *E* is a property of *S* which is UNEXPLAINABLE given those of *O* of *B*.
11. *E* is a property of *S* which is UNGENERATED by *O* of *B*.

19.6. Commentary

FIRST TIME: It is to be expected than any emergent entity occurs for a first time. This is indeed one of the standard everyday meanings of the term 'emerge'. But many things occur for a first time that are not emergent: the first internal combustion engine, the first digital computer, the first moon landing and so on. Being something new falls far short of being emergent, so this idea is too weak and general to be of use.

UNEXPECTED: Surprise is relative to the beholder and context. Surprise of the ignorant or uninitiated cannot be counted. When John Wilkinson launched the first iron barge in 1787, most onlookers expected it to sink, but Wilkinson was well apprised of the principles of buoyancy and knew it would float. So only surprise of the highly qualified is relevant. Even they, however, may be surprised at unfamiliar phenomena. The anti-bacterial action of the penicillium mould was surprising to Fleming (Fleming 1929), but that stemmed from the then-unknown biochemistry of the mould itself, not from emergence. Despite the abundant evidence from birds, it was widely claimed in the nineteenth century, even by such notable scientists as Helmholtz, Kelvin and Edison, that manned heavier-than-air flight was impossible. Though other scientists, notably Boltzmann, were optimistic that it could be achieved (Boltzmann 1894), it took the experimental development of aerofoils by Cayley, Lilienthal and the Wright brothers to demonstrate to the contrary and to the surprise of many that heavier-than-air transport was possible (Anderson 2004). Thereafter, such flight became commonplace. As this shows, features may surprise initially but can soon become familiar, and this holds whether or not they are emergent. If consciousness is emergent, as many believe, it is hardly something that surprises us, being itself the condition for surprise to occur at all. So this mode of novelty is also too weak to be a good candidate definition for emergence. Nevertheless, if something is unexpected, it *may* be (weak) evidence for emergence in a stronger sense, but it may also simply be an indication that the phenomenon is insufficiently well understood.

UNFORESEEN: This is rather more useful. If some phenomenon is not predicted, especially if it is not predicted despite intensive efforts to foresee what will happen, then it has the beginnings of a claim to be emergent. Nevertheless, a phenomenon can be unforeseen or unpredicted for many reasons, such as lack of time, information or resources. The catastrophic decompression and in-flight break-up of early De Havilland Comet jetliners was unforeseen, but with more knowledge could have been foreseen, and it was certainly explained after the fact. A phenomenon may be unforeseen due simply to insufficiently well-developed theory as well as a lack of experience of the phenomenon, or, more mundanely, because extant scientific theory had not penetrated through to engineering design. Metal fatigue, the factor that contributed critically to the Comet accidents, was a phenomenon that had been under scientific investigation since a fatal railway accident in Meudon, France, in 1843, but even though there were warnings of the dangers of axle breakages due to fatigue cracks, the warnings were disregarded and accidents continued to occur due to fatigue.

UNFORESEEABLE: This is the first case of what we may call a *modal strengthening* of a previous kind of novelty. Being unforeseen is one thing: being unforesee*able* is quite another. Thinking in terms of engineering disasters, such as the aforementioned Comet accidents, the *Challenger* disaster of 1986 or the Tacoma Narrows Bridge collapse of 1940, these were not foreseen—or if they were, by few—but were foresee*able* with greater knowledge and/or diligence. It is a rather popular idea of emergence that emergent phenomena are those which cannot be predicted. However, if some phenomenon is indeed emergent, while it will typically not be foreseen at first, and may not at that time (or indeed, later) have a satisfactory explanation, successive occurrences, especially if occurring regularly in similar conditions, will build up a body of experience so that future occurrences may then be more or less reliably predicted. Even then, the phenomenon may simply remain unexplained, perhaps temporarily, perhaps not. For example, the early observed effects of transonic flight, such as buffet, flutter and Mach tuck, were soon reliably predicted in later flights before they were properly understood. To take the obvious example again, if consciousness is indeed emergent, it is highly predictable: all normal human infants and many animals can be reliably and accurately predicted to be conscious.

Unforeseeability is also not necessarily due to emergence. In chaotic systems, the state of the system at any given future time will by definition be unpredictable due to inherent limits of observational accuracy, even though, where it occurs, such unpredictability is due not to emergence but to the nature of the system, which may even be wholly deterministic. The weather is notoriously impossible to predict with any but short-term accuracy, but storms and other meteorological phenomena are neither mysterious nor ill-understood, just very complex. So unforeseeability is neither necessary nor sufficient for emergence. Nevertheless, where it occurs, it may be considered as *prima facie* evidence for emergence in some other sense.

Many accounts of emergence dwell on the nature of the relationship between a bearer and its parts or subsystems.

WHOLE: In popular and unscientific accounts of emergence, it is often claimed that emergence consists in the whole S having properties not possessed by its parts B. Taken strictly as read, this definition is hopelessly wide, since all or almost all non-atomic objects will have properties their parts do not have: none of the parts of a 1 kg body weighs 1 kg.

NON-SUMMATIVE: a more useful idea is the non-summativity of properties of a whole. This rules out the weight case, because any complete collection of disjoint parts of a 1 kg body add up to 1 kg. (Parts are disjoint which have no part in common: this condition is obviously required to avoid double counting.) Taking into account the modes of interaction and interrelation O among the parts B also rules out many properties that would otherwise count as emergence. An automotive machine such as a ship or a car has a certain power that it gets from the way in which its parts are interrelated and interact, the moving parts, their various shapes and the processes in which they are involved. The whole has properties the parts individually do not, but how it gets these is not mysterious.

UNDERIVED: a whole's properties may—as in the case of the last example—be derived or inferred from the properties of its parts and their interoperation. In some cases, however, truths about the whole are not inferred or deduced from those about the parts and their operation. This may result in the whole having properties or exhibiting behaviour that is unexpected or surprising, despite knowledge of O and B. This may again be *prima facie* evidence of some sort of emergence, but again, it may be due simply to the correct deduction not having been carried out. There are all sorts of reasons why deductions are not carried through: the derivation might be extremely complex, there may be temporal and resource limitations, human failings, or just a lack of motivation to think things through. Numerous engineering disasters such as those already mentioned came about because implications of knowledge present in perhaps a scattered way were not brought together or thought through. So merely factual lack of derivation is not sufficient to count as emergence.

UNDERIVABLE: this is a modal strengthening of the previous mode. Here the idea is that there is no way to deduce, even in principle, the emergent item E from the best knowledge about B and O. This is a very strong condition, and it is the opposite of derivability, which is often also called *reducibility*. David Chalmers calls this *strong emergence* (Chalmers 2006), reserving the term 'weak emergence' for what is merely *unexpected* (his term). It is a favoured candidate for 'the' concept of emergence. Once again, consciousness is touted as an example of underivability: no amount of knowledge about brains, their parts and operation is said to enable us to infer that the creatures possessed of such brains have conscious experience. Notice the difference from UNFORESEEABLE: there we can base predictions on past experience of cases even where the phenomenon is UNDERIVABLE. On the other hand, the very strength of the condition (something always signified by the slippery words 'in principle') means that the status of a candidate emergent phenomenon *E* as underivable is always provisional. At any time at which a phenomenon appears underivable, this may simply be because we have not yet hit upon the correct deduction, or discovered all the facts about *O* and *B* that would make such a deduction work. This tends to divide researchers in such an area into optimists, who are confident that a derivation will be forthcoming at some point in the future, and pessimists, who are sceptical or doubtful that this will happen. A glance at the state of contemporary philosophy of mind reveals just such a dichotomy regarding the mental. On the other hand, when a generally satisfactory derivation has been accomplished, either in detail or 'in principle', that counts as strong evidence against emergence. The once mooted emergence of chemical and biological properties is now generally denied, and reducibility to physical properties in principle is accepted, even if the characteristic phenomena of chemistry and life are of a complexity and specificity taking them beyond the focus or interest of physicists.

UNEXPLAINED: if a phenomenon is not understood and defies explanation by current theories, this is taken as evidence that it is a candidate emergent. As in all cases of a mere lack, this is insufficient in itself to count as genuine emergence, since it may be that not all the relevant facts are known, or that no one has happened on the right explanation, or current theory is deficient and needs augmentation or revision. Unexplained phenomena are a motor of scientific advance, and often the science has later caught up and furnished an explanation. Examples in the history of science are legion: from Lavoisier's discovery that oxides are heavier than their unoxidised counterparts, and hence to the discovery of oxygen's role in combustion, to Einstein's relativistic explanation for the puzzling perihelion of Mercury. Other paradigmatic advances greatly strengthening theories' explanatory powers were the Darwin–Wallace theory of natural selection and the theory of plate tectonics. However, if a phenomenon is unexplained, this is a challenge to find an explanation. The longer such a phenomenon resists explanation, the stronger its claim to be emergent.

UNEXPLAINABLE: strengthening 'unexplained' to 'unexplainable' again rises the bar, and a phenomenon that is unexplainable in principle by any theory about *O* and *B* does indeed have a strong claim to be emergent. This, alongside underivability, is a popular candidate notion of emergence. Once again, however, the modal strengthening places beyond verification an assertion that some phenomenon is unexplainable, since we are never in possession of all possible or as yet undiscovered future theories, and one can again expect there to be both optimists and pessimists on the question. Only when an actual explanation is forthcoming can it be safely stated that a candidate emergent phenomenon is in fact not unexplainable.

Expectation, prediction, derivation and explanation are all *epistemic* or cognitive matters. The last candidate for being a notion of emergence is by contrast wholly *ontological*.

UNGENERATED: a phenomenon *E* which does not arise naturally out of the properties and actions *O* of a basis *B*, one that is, in our terminology, not *generated* by them, but which constitutes a wholly new entity spontaneously appearing when they reach a certain complexity, is emergent in a strongly ontological sense. Its novelty is in no way hostage to human knowledge, theories, science, method or expectations. It is the world autonomously coming up with something wholly new. That this is a very strong notion can be seen from the fact that certain naturally occurring phenomena, such as crystal formation, are sometimes described as 'emergent', even though it is obvious that they arise naturally. Again, the very strength of the idea makes it a favoured candidate to be 'the' notion of emergence. As ever, consciousness and mentality have been tipped to be ontologically emergent in this sense. It is, however, subject to some familiar problems and one unfamiliar problem. The familiar problems have to do precisely with the fact that what the world may naturally bring forth is something that has to be probed by experience and theory, so that any statement that such and such a phenomenon is ungenerated is provisional unless and until proven false. Secondly, however, the spontaneous appearance of the ontically new in this strong sense looks uncomfortably like magic, and so akin to some kind of dualism. The new phenomenon E does not arise naturally, so its appearing is—not natural. What is this but magic? The main difference from classical dualisms is that ungenerated emergence is *dosed* magic: only when the preconditions of requisite complexity and suitable environment are satisfied does the magic kick in. So, while this was a popular notion of emergence among some British emergentists, particularly Broad, it is one whose very remoteness from accessibility renders suspect (Simons 2009). Claims that this or that phenomenon E is ungenerated again take on the status of professions of faith, a faith that can be undermined by subsequent derivation or explanation, but which need not be. Given the inherent limitations on knowledge, any claim that a phenomenon E is ungenerable, that is, is ontologically emergent, can never be conclusively verified, whereas it can be falsified if a derivation or explanation of the way the phenomenon is generated is forthcoming. For this reason, any claim that a phenomenon is ungenerable has no advantage—beyond metaphysical drama—*for engineering* over a claim that it is underivable or unexplainable. This is despite the fact that some claims for the usefulness and even engineering aimworthiness of emergence may be interpreted as claims of ontological emergence (e.g., Stepney et al. 2006).

19.7. Usefulness for Engineering

The focus of engineering is obviously very different from that of science or philosophy, so it is to be expected that not all of these extant notions will be of equal use or interest to engineers wondering about the possible emergence of this or that phenomenon. We leave aside those conceptions which are too broad or weak to be of interest, namely numbers 1, 5 and 6. The ontological conception 11, as argued earlier, is too metaphysical to be of use; that leaves conceptions 2–4 and 7–10. Of these, the modally strengthened versions 4, 8 and 10 are, perhaps paradoxically, while more generally favoured definitions of emergence, of less help for engineering purposes than they are for theoretical discussion, because the strength of the claims means that they are always provisional unless and until actually refuted. That leaves the following conceptions as of interest, where we repeat the formulations for ease of reference:

2. E is surprising or UNEXPECTED (given O of B).
3. E is unpredicted or UNFORESEEN (given O of B).
8. E is a property of S which is not deduced or is UNDERIVED from O of B.
9. E is a property of S which is UNEXPLAINED given those of O of B.

The function of UNEXPECTED or surprising phenomena is only weakly indicative of anything like emergence, but it is a useful marker nevertheless, since it indicates that something more needs to be done, either theoretically, or in design, or in production. Generally speaking, since engineering is about designing and implementing artefacts and systems with particular aims and requirements in view, unexpected phenomena are usually undesirable, as showing that the design and implementation are not as well understood or controlled as could be wished. To give a concrete example, the massive Northeast Blackout power outage of 14 August 2003, which affected millions of customers in the north-eastern United States and in Ontario, was (obviously) unexpected and unpredicted, even though the series of mishaps which caused a cascade of shutdowns was afterwards fully explained (U.S.-Canada Power System Outage Task Force 2004). Obviously, steps were taken to strengthen and improve systems as well as regulatory requirements in the light of the unexpected and unfortunate event. There can of course also be positive surprises in engineering, as when an artefact performs better than expected on some metric (such as efficiency, reliability or cost) or exhibits unexpected but welcome unintended behaviours. The invention of Post-It notes is often cited as a case, though this is more a question of the glue that was unsuccessful in its intended use turning out to have a beneficial different use (Post-It). Cardiovascular benefits of low-dosage aspirin, which was developed as an analgesic, are another candidate case (Ittaman et al. 2014). Another example is the fortuitously low radar cross-section at some (but not all) orientations to the radar of the British Avro Vulcan bomber, which resulted from its shape, long before stealth was an aircraft design consideration (Sweetman 1982, though the claims there are exaggerated).

In standard engineering, UNFORESEEN features of an artefact are generally unwelcome, or at best neutral, since the point of engineering is to attain goals through design. However, as complex systems become more common and the gap between detailed design and outcomes becomes less tractable, it may be that experimentation for the sake of giving serendipity a chance becomes more common. While it is unlikely that the idea of "Let's try it and see what happens" will ever be of widespread appeal in engineering, especially among critical systems where safety is paramount, there may be a place for it in less sensitive areas of development.

As systems become more and more complex, so that features that result from the implementation of local design decisions are increasingly UNDERIVED, engineers may well be content to rely on desired or fortuitous features that are regularly attainable (and so predictable) while waiting for scientific explanation to catch up. While there is no doubt that engineers will always prefer to exploit explanations of how and why the systems they design have the features they require, and also explanations as to why undesirable features may arise (and so how they can be engineered out), where such explanations are not—or not yet—forthcoming, practicality concerns will no doubt tend to encourage the production of designs with UNEXPLAINED features, provided the latter are beneficial and reliably attainable.

19.8. Areas of Application

While many engineering projects have produced features from our four distinguished types, it is complex systems, particularly complex digital systems, that are the most likely areas where features which are any, some or all of the shortlist of UNEXPECTED, UNFORESEEN, UNDERIVED and UNEXPLAINED occur with some frequency. The use of neural nets, artificial intelligence and artificial life allow algorithms to be developed, the outcomes of whose operation go beyond what their engineers consciously expect, intend, understand or can explain, and may produce results which are welcome and can be exploited. In the area of advanced digital engineering known as *system of systems*, where several previously designed systems are brought together in order to attain more ambitious goals, both the complexity and the paucity of prior detailed knowledge mean that outcomes tend to have features from our shortlist. This applies in particular to what are called (in the USA) *ultra-large-scale systems* or (in the UK) *large-scale complex IT systems*, with areas of application such as healthcare management, large enterprise management, and financial markets. Approaches which exploit or deliberately aim for such (preferentially favourable) outcomes are sometimes called 'emergence engineering'. While it might be more accurately termed 'engineering *for* emergence', as a generic label, this is useful. But I trust this chapter will have made it clear that there is no single well-defined thing that is emergence, so nor is there likely to be a single well-defined thing that is emergence engineering.

Related Chapters

Chapter 6: Scientific Methodology in the Engineering Sciences (Mieke Boon)
Chapter 10: Creativity and Discovery in Engineering (David H. Cropley)
Chapter 11: Uncertainty (William M. Bulleit)
Chapter 13: Systems Engineering as Engineering Philosophy (Usman Akeel and Sarah Bell)
Chapter 14: Assessing Provenance and Bias in Big Data (Brent Mittelstadt and Jan Kwakkel)

Chapter 20: Towards an Ontology of Innovation: On the New, the Political-Economic Dimension and the Intrinsic Risks Involved in Innovation Processes (Vincent Blok)

Chapter 35: The role of Resilience in Engineering (Neelke Doorn)

Further Reading

General Philosophy Overview

O'Connor, Timothy and Wong, Hong Yu. (2015). Emergent Properties. In Edward N. Zalta (ed.), *The Stanford Encyclopedia of Philosophy*, Summer 2015 ed. https://plato.stanford.edu/archives/sum2015/entries/properties-emergent/. (A highly readable general account of the history of the idea and the current state of the emergence issue in philosophy, together with a bibliography of relevant literature. We refer to this bibliography for the classic sources and individual papers, but recent useful anthologies are listed next.)

Anthologies

Bedau, M. and Humphreys, P. (2008). *Emergence: Contemporary Readings in Philosophy and Science*. Cambridge: MIT Press. (Collects classical as well as modern discussions.)

Clayton, P. and Davies, P. (eds.) (2008). *The Re-Emergence of Emergence*. Oxford: Oxford University Press. (Modern contributions to the philosophy.)

Corradini, Antonella, and O'Connor, Timothy (2010). *Emergence in Science and Philosophy*. New York: Routledge. (The title describes the contents.)

Kistler, Max (ed.) (2006). *New Perspectives on Reduction and Emergence in Physics, Biology and Psychology*, Special issue of *Synthese*, 151(3). (Likewise, a descriptive title.)

Macdonald, Cynthia and Macdonald, Graham (2010). *Emergence in Mind*. Oxford: Oxford University Press. (More restrictive in its scope.)

Rainey, Larry B. and Jamshidi, Mo. (eds.) (2018). *Engineering Emergence: A Modeling and Simulation Approach*. Boca Raton: CRC Press. (A first collection intended specifically for engineers, this examines both theoretical and practical aspects of emergence as applied to artefacts, with an emphasis on emergence (more particularly, its modelling and simulation) in system-of-systems.)

Emergence in Engineering

Chen, C. C., Nagl, S.B. and Clack, C.D. (2009) Complexity and Emergence in Engineering Systems. In A. Tolk and L.C. Jain (eds.), *Complex Systems in Knowledge-based Environments: Theory, Models and Applications. Studies in Computational Intelligence*, Vol. 168. Berlin, Heidelberg: Springer.

Hitchins, Derek (2007) *Systems Engineering: A 21st Century Systems Methodology*. Hoboken, NJ: Wiley. (An enthusiastic proponent of emergence in the systems engineering context.)

References

Alexander, S. (1920). *Space, Time, and Deity*. London: Palgrave Macmillan.

Anderson, J.D. (2004). *Inventing Flight. The Wright Brothers and Their Predecessors*. Baltimore: Johns Hopkins University Press.

Boltzmann, L. (1894). Ueber Luftschifffahrt. *Verhandlungen der Gesellschaft Deutscher Naturforscher und Ärzte*. Erster Theil. *Die allgemeinen Sitzungen* (1894), 89–96. Reprinted in *Populäre Schriften*. Leipzig: Barth, 1905, 81–91.

Broad, C.D. (1925). *The Mind and Its Place in Nature*. London: Routledge & Kegan Paul.

Chalmers, D. (2006). Strong and Weak Emergence. In P. Clayton and P. Davies (eds.), *The Re-Emergence of Emergence*, Oxford: Oxford University Press.

Clayton, P. and Davies, P. (eds.) (2008). *The Re-Emergence of Emergence*. Oxford: Oxford University Press. (Modern contributions to the philosophy.)

Fleming, A. (1929). On the Antibacterial Action of Cultures of a Penicillium, with Special Reference to Their Use in the Isolation of B. influenzæ. *British Journal of Experimental Pathology*, 10, 226–236.

Ittaman, S.V., Van Wormer, J.J. and Rezkalla, S.H. (2014). The Role of Aspirin in the Prevention of Cardiovascular Disease. *Clinical Medicine and Research,* 12, 147–154.

Kim, J. (1999). Making Sense of Emergence. *Philosophical Studies,* 95, 3–36.

Kim, J. (2003). Supervenience, Emergence, Realization, Reduction. In M.J. Loux and D.W. Zimmermann (eds.), *The Oxford Handbook of Metaphysics.* Oxford: Oxford University Press, pp. 556–586.

Lewes, G.H. (1875). *Problems of Life and Mind,* Vol. 2. London: Kegan Paul, Trench, Turbner, and Co.

McLaughlin, B.P. (1997). Emergence and Supervenience. *Intellectica,* 25, 25–43.

Mill, J.S. (1843). *A System of Logic.* London: Longmans.

Morgan, C.L. (1923). *Emergent Evolution.* London: Williams and Norgate.

O'Connor, T. (1994). Emergent Properties. *American Philosophical Quarterly,* 31, 91–104.

Post-It. www.post-it.com/3M/en_US/post-it/contact-us/about-us/. Accessed August 3, 2019.

Simons, P. (2009). Ontic Generation: Getting Everything From the Basics. In H. Leitgeb and A. Hieke (eds.), *Proceedings of the 2008 Wittgenstein Symposium.* Frankfurt am Main: Ontos, 2009, pp. 137–152.

Stepney, S., Polack, F. and Turner, H. (2006). Engineering Emergence, *CEC 2006: 11th IEEE International Conference on Engineering of Complex Computer Systems,* Stanford, CA, USA, August.

Sweetman, B. 1982 The Bomber that Radar Cannot See. *New Scientist,* 93, 565–568.

U.S.-Canada Power System Outage Task Force (2004). *Final Report on the August 14, 2003 Blackout in the United States and Canada: Causes and Recommendations.* United States Department of Energy. Energy.gov—Office of Electricity Delivery & Energy Reliability (Report), April.

Van Cleve, J. (1990). Emergence vs. Panpsychism: Magic or Mind Dust? In J.E. Tomberlin (ed.), *Philosophical Perspectives.* Vol.4. Atascadero, CA: Ridgeview Publishing Company, pp. 215–226.

20

TOWARDS AN ONTOLOGY
OF INNOVATION

On the New, the Political-Economic Dimension and the Intrinsic Risks Involved in Innovation Processes

Vincent Blok

20.1. Introduction

Innovation is often uncritically seen as a good thing (Rogers 1976) and considered as a panacea for all kinds of socioeconomic challenges (Godin 2015). For institutions like the Organization for Economic Co-operation and Development (OECD) and the European Union (EU), it is for instance self-evident that "most current social, economic and environmental challenges require creative solutions based on innovation and technological advance" (OECD 2010: 30; cf. European Commission 2010). At the same time, the notion of innovation itself remains undefined in these policy documents, while its meaning seems to be taken for granted in the scientific literature.

This is also the case in the context of philosophy of engineering and technology, where the concept of innovation is relatively new. Because it is often self-evidently associated with technology as *technological innovation*, it solely appears within the context of technology (Godin 2015; Bontems 2014). Although in recent years, the concept of innovation seems to attract particular attention with the emergence of the concept of 'Responsible Innovation' in the European policy context, the self-evident association with technology remains prevalent (Timmermans and Blok 2018).

What is more, innovation is associated not only with the exploration of new technologies, but also with the commercial exploitation of these new technologies (Blok and Lemmens 2015; Schomberg and Blok 2018). This becomes for instance clear in management and economics of innovation textbooks. In these textbooks, innovation is defined as "the first commercial application or production of a new process or product" (Freeman and Soete 1997: 1). And although the innovation management literature acknowledges that innovation can also take place in new services, it self-evidently associates innovation with a *technological* invention—the technology behind Facebook's or Amazon's services—which enables the company to provide new services like social media and online bookstores:

> Hence innovation embraces both a technological and a creative dimension, that we normally refer to as invention, together with a commercial dimension that involves the exploitation of the invention to turn it from a model or prototype into something that is available in the market for consumers to purchase. This latter aspect is much less heroic and less glamorous then invention, but it is crucial. Without it an invention is little more than a

great idea, and all too often this is an element of innovation that is neglected, with disappointed consumers the result. Only when both aspects have been effectively handled does one have an innovation.

(Smith 2006: 6)

Even if we accept the 'innovation imperative' that is dominant in engineering and business schools (Bessant and Tidd 2007), it remains unclear what the philosophical underpinnings of this notion of innovation are in contrast to related terms like 'invention' and 'imitation'. Why is innovation nowadays self-evidently associated with economic growth and the solution of societal challenges? What does it mean that the ideal of innovation is extended to all aspects of social life, ranging from innovation in healthcare to innovation in politics? (Blok 2018). While researchers in the domain of engineering ethics and science and technology studies (STS) primarily focus on the governance of the outcomes of technological innovations, for instance by engaging stakeholders during the innovation process, in this chapter we reflect on the nature of innovation itself.

Philosophical reflection on basic concepts like innovation is important, because they structure the way we understand the world around us. If we for instance understand innovation as technological innovation which is primarily executed by engineers in private R&D departments and laboratories, then we miss the whole potential of contemporary phenomena that can be associated with system innovation (for instance, agro-ecological innovations), social innovations (for instance, political innovations like online petition websites) or attitudinal innovations (for instance, prevention or lifestyle interventions), as well as the part of the innovation process that can be associated with the diffusion of innovations. Philosophical reflection on innovation can also help to assess whether phenomena fall under the concept or not; for instance, the new paradigm of technological developments and engineering practices which can be associated with *biomimicry*, i.e., with the imitation of natural processes in technological design (Blok and Gremmen 2016). Finally, philosophical reflection can help to develop a critical attitude towards the self-evident use of the concept of innovation, to highlight contradictions and tensions in its use, and to raise questions regarding the limitations of its use and the conditions of *responsible* innovation. Is innovation good *per se* (Rogers 1976), or should we reflect on its consequences in relation to the problems it intends to solve, the risks involved as well as the potential negative side effects?

One way to open up the concept of innovation for philosophical reflection is by tracing the different meanings it has in history. Historical analysis can help to question the self-evidence of the association between innovation and technology and commercialization, to deconstruct the presupposed concepts that always already structure our understanding of the world, and to explore the sedimentary conceptual structures which show themselves in the words and notions we self-evidently use in our dealings with the world (Blok 2020). In this, we are indebted to the valuable work by Benoit Godin, who wrote an intellectual history of the concept of innovation (2008, 2015). While his main contribution can be placed in the domain of the history of science and technology without the ambition to *theorize* about the concept (Godin 2015: 4), our objective in this chapter is precisely to philosophically reflect on the sources his studies brought forth.

In order to grasp the roots from which the self-evident conceptualization of innovation as technological and commercial innovation stems, in Section 2 we consult the work of the economist Joseph Schumpeter. In Section 3, based on findings in the history of innovation, we open up the concept by reflecting on two aspects of Schumpeter's conceptualization of innovation, namely its destructive (Xenophon and Plato) and its constructive aspect (Machiavelli and Bacon). In Section 4, we synthesize our findings and propose an ontic-ontological conceptualization of innovation as ontogenetic process and outcome with six dimensions—newness, political dimension, economic dimension, temporal dimension, human dimension and risk—that moves beyond its technological and commercial orientation.

20.2. Schumpeter as Founding Father of Innovation as Technological and Commercial Innovation

The founding father of our understanding of innovation and its intrinsic relation to technology and economy is the economist Joseph Schumpeter (1883–1950). According to Schumpeter, "capitalist enterprise" and "technological progress" are "essentially one and the same thing" (Schumpeter 1943: 110). The entrepreneur is always looking for new business opportunities. By doing things differently than others, i.e., by the introduction of innovative technologies, the entrepreneur enhances and secures his competitive advantage over competitors. Because large competitors will try to copy the entrepreneur's innovation to secure the market for themselves, and because large firms have an advantage over small firms according to Schumpeter, the entrepreneur has to make the difference anew and explore new innovative business ideas, etc. This cycle of entrepreneurs exploring and exploiting innovations to achieve a temporary monopoly, which are then copied by large firms and call for new innovations by the entrepreneur etc., is what is driving the economy, according to Schumpeter (Schumpeter 1943).

For Schumpeter, innovation not only concerns an invention at the product or service level but also relates to economic waves. Following the initial work by the Russian economist Nikolai Kondratieff (1892–1938), Schumpeter studied long economic waves that were driven by particular clusters of industries and can be associated with technological shifts; for instance, the wave starting around 1845 associated with steam power and innovations in the railway industry, or the wave starting around 1900 associated with electricity and innovations like the internal combustion engine (Schumpeter 1983). For Schumpeter, it is therefore *technological* innovation that plays a key role in economic development.[1]

So far, we recognize our contemporary notion of innovation as technological and commercial innovation in Schumpeter's conceptualization of the term. Yet, he already differs from our ordinary understanding when he talks about *waves* and not about an endless economic progress. Entrepreneurs disrupt the status quo or economic equilibrium with their innovations. These disruptive innovations will lead to economic growth (upswing), which will in the end decline because a new economic equilibrium is reached which is dominated by large firms, and in which there is no role for the entrepreneur anymore. The periodical economic decline or depression is explained by two factors: (1) the capitalist concentration of power and capital by large firms and corporate groups, in which no place is left for entrepreneurial behaviour; (2) by the intellectual class that on the one hand emerges as a result of economic growth but holds social democratic values that are hostile to capitalism on the other. Although we currently do not experience such an economic decline, we can recognize Schumpeter's ideas in our current society, where competition is crushed by technology giants like Google and Amazon, and where it becomes difficult for new entrants to enter the market, while social democratic movements against Transatlantic Trade and Investment Partnership (TTIP), for instance, are hostile to capitalism. This intrinsic tendency towards power concentration by large corporations is inherent in capitalism, according to Schumpeter—in this respect, he is pessimistic about

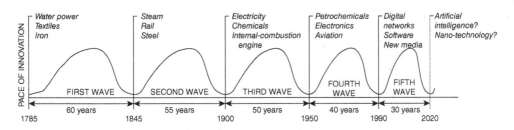

Figure 20.1 Schumpeter's business cycles

the abilities of capitalism to serve economic progress—and can only be disrupted by innovations that prevent the collapse of the capitalist system.

The role of innovation in the upswing of economic cycles becomes clear in Schumpeter's notion of creative destruction, which he borrowed from Marx. According to Schumpeter,

> Capitalism . . . is by nature a form or method of economic change and not only never is but never can be stationary. . . . The fundamental impulse that sets and keeps the capitalist engine in motion comes from the new consumers' goods, the new methods of production or transportation, the new markets, the new forms of industrial organization that capitalist enterprise creates. . . . The opening up of new markets, foreign or domestic, and the organizational development from the craft shop and factory to such concerns as U.S. Steel illustrate the same process of industrial mutation . . . that incessantly revolutionizes the economic structure *from within*, incessantly destroying the old one, incessantly creating a new one. This process of Creative Destruction is the essential fact about capitalism. It is what capitalism consists in and what every capitalist concern has got to live in.
>
> *(Schumpeter 1943: 82–83)*

The innovation of the diesel engine in locomotives, for instance, was not just the creation of a new technology but destructed at the same time the existing industry in steam engines, just like the innovation of the compact disc destructed the industry of cassette tapes and vinyl and is now replaced by MP3 and streaming services.

This brief consultation of the origin of our taken-for-granted understanding of innovation as technological and as commercial innovation shows that Schumpeter can on the one hand legitimately be seen as the founding father of our current understanding of innovation. At the same time, our initial reflections show a clear difference between Schumpeter's conceptualization of innovation and the contemporary taken-for-granted notion. While innovation is nowadays seen as contributing to economic growth *per se*, and as a *panacea* for all kinds of societal challenges, Schumpeter's notion of economic *waves* and creative *destruction* already enables us to question the unilateral progressive and constructive connotation of the concept.

Even if we do not agree completely with Schumpeter's diagnosis—the idea that large firms are better able to foster innovation, for instance, is challenged in the literature (cf. Deakins and Freel 2009)—the idea that economic decline follows every upswing of the economic cycle based on new technologies makes clear that innovation may be a necessary but not sufficient condition for economic growth; innovation may account for the upswing of the economic cycle but is in need of additional and maybe even non-economic interventions to prevent its decline. We may even argue that innovation, despite its contribution to the upswing of the economic cycle, is itself non-economical, to the extent that innovation *limits* the tendency to power and capital concentration in the capitalist economy, which would collapse without its temporary disruption by innovations. If innovation prevents the collapse of the capitalist system, then we can formally conclude that innovation itself doesn't belong to the capitalist economic system but constitutes its limit. Furthermore, the idea that every upswing of the economic cycle involves the construction of new and innovative solutions *and* the destruction of existing markets, industries and firms, i.e. the idea that the positive impact of innovation is accompanied by negative impacts elsewhere, makes clear that innovation may be a necessary but not sufficient condition for the solution of the societal challenges we face today; innovations that address societal challenges are accompanied with negative impacts elsewhere, and therefore raise new societal challenges. This intrinsic *Faustian* aspect of innovation is largely ignored in the policy documents dedicated to innovation (Blok and Lemmens 2015). In other words, our brief reflection on Schumpeter's notion of innovation puts the presupposed notion of innovation as economic progressive *per se* and as solution of societal challenges between brackets and raises questions about the extra-economical conditions that must be fulfilled to enable innovation to contribute to economic growth and to the solution of societal problems.

20.3. Philosophical Reflections on Creative Destruction as Characteristic of the Ontogenesis of Innovation

Schumpeter's thoughts can also provide a first grip for our philosophical reflection on the concept of innovation. It is clear, for instance, that innovation can both be seen as a process—i.e. the innovation process—and as the result of this process—i.e. the innovative product or service as an outcome of the process. Although according to Schumpeter it is ultimately the innovative product or service that provides competitive advantage, his reflections focus primarily on the process of innovation; creative destruction is not a characteristic of the innovative product or service, but of the innovation process.

If we consider the outcome of the innovation process as a concrete *individual* product or service, the innovation process itself can be conceived as the *pre*-individual. This reality of the innovation process *before* its individuation in a concrete innovation outcome can be conceived as the *ontogenesis* of this outcome. This ontogenetic process of innovation is missed in many typologies of innovation in the literature, like incremental versus radical innovation (Freeman and Soete 1997) and architectural versus modular innovations (Henderson and Clark 1990), which all take the outcome of the innovation process—concrete individual products or services, its components or the assemblage of these components—as point of departure. This focus on the innovation outcome may be explained by what is called the 'culture of things' or material culture: "The origin of this culture goes back to the Renaissance: due to commercial exchanges, exploration and travel, natural and artificial objects have been what is valued in arts, science, and real life" (Godin 2008: 21). But if innovation concerns both the process *and* the outcome of the process, a philosophical reflection on the innovation process as a distinct but integral part of the phenomenon of innovation can no longer be avoided. In this section, therefore, we focus on the innovation process of creation and destruction.

In order to explore the creative and destructive aspect of the innovation process, we consult the history of the innovation concept. In his history of the concept of innovation, Godin shows that while the study of innovation started in the late nineteenth century and the beginning of the twentieth century (Gabriel Tarde, Joseph Schumpeter) and accelerated in the 1960s with a variety of approaches, by the mid-1970s this variety of approaches was replaced by one dominant representation based on Schumpeter's work, namely as technological innovation and as commercial innovation (Godin 2015). While many scholars in the field of innovation studies uncritically adopt this progressive history of the origin of technological innovation, starting with Schumpeter, Godin points to the fact that they neglect the broader history of the concept, by for instance disregarding the literature on technological change and the religious and political connotations of the concept. Without neglecting the significant role of Schumpeter's conceptualization for our contemporary understanding of innovation, this broader history of the concept may provide a further grip for our philosophical reflections on the concept of the process of creative destruction as ontogenesis of innovation.

20.3.1 The Destructive Aspect of Innovation: The Ontological Level of the Ontogenesis of Innovation

In this subsection, we focus on the destructive aspect of innovation on the basis of mainly Plato's work. The concept of innovation originates from Ancient Greece, where it is named *kainotomia*. *Kainotomia* means change or the introduction of something new. It comes from *kainon* (new) and *tom* (cut, cutting) and originally meant 'cutting fresh into'. It was originally used in the context of the opening of new mines (Godin 2015: 19). Even though it is hard to argue that innovation is established as a theoretical concept in Greek philosophy, we can consider it as a proto-concept that can provide useful insights and form our understanding of the creative and destructive aspects of the innovation process.

It is in this context that Xenophon (430–355 BC) used the term in his *Ways and Means*. In a chapter on mines, Xenophon indicates a problem with the exploration of silver mines:

> Well then, it may be asked, why is it that there is not the same rush to makes new cuttings now as in former times? The answer is, because the people concerned with the mines are poorer nowadays. The attempt to restart operations, renew plant, etc., is of recent date, and any one who ventures to open up a new area runs a considerable risk. Supposing he hits upon a productive field, he becomes a rich man, but supposing he draws a blank, he loses the whole of his outlay; and that is a danger which people of the present time are shy of facing.
>
> (Xenophon 2014: IV 27–30)

His proposal is that the state of Athens should possess public slaves and make them available for hire to businesses that want to explore the mines. The state can cover the enormous investment costs needed to hire this labour force, and because of the spreading of the investment over various explorations of new galleries, the risks of failure and success are better balanced. For this reason, Xenophon calls for a *gradual* introduction of his innovation, instead of an introduction at once:

> If we proceed tentatively, as we find ourselves able, we can complete any well-devised attempt at our leisure, and, in case of any obvious failure, take warning and not repeat it. Again, if everything were to be carried out at once, it is we, sirs, who must make the whole provision at our expense. Whereas, if part were proceeded with and part stood over, the portion of revenue in hand will help to furnish what is necessary to go on with. But to come now to what every one probably will regard as a really grave danger, lest the state may become possessed of an over large number of slaves, with the result that the works will be overstocked. That again is an apprehension which we may escape if we are careful not to put into the works more hands from year to year than the works themselves demand.
>
> (Xenophon 2014: 36)

By taking over this responsibility, the state can reduce the risks of individual mining companies significantly and contribute to local business development. In return, the state can raise revenues.

It is interesting to see that 'innovation' in the literal sense of the word *kainotomia*, namely as the making of new cuttings and opening of new galleries in the silver mines, is a phenomenon that Xenophon not only describes in *Ways and Means*, but characterizes at the same time his own efforts in this book:

> And given that my proposal were carried into effect, the only *novelty* in it is that, just as the individual in acquiring the ownership of a gang of slaves finds himself at once provided with a permanent source of income, so the state, in like fashion, should possess herself of a body of public slaves, to the number, say, of three for every Athenian citizen.
>
> (Xenophon 2014: IV 17)

It is primarily his proposal for the state to own and possess slaves that is considered innovative.

Three characteristics of an ancient concept of innovation can be derived from Xenophon's notion of *kainotomia*: (1) innovation concerns something new in the literal sense of the word; (2) innovation is performed by the state; (3) innovation is a risky business (Godin 2015: 22). Although Godin stresses the political dimension of Xenophon's concept of innovation, we must acknowledge that the intrinsic economic orientation of innovation is at work as well already in case of Xenophon. This is confirmed by Aristotle, who argues that the introduction of innovations is often connected with private interests and therefore requires political authority to safeguard the political order (Aristotle 1944: 1308b20–25).[2] Nonetheless, despite this economic orientation, it is true that the juridico-political connotation of innovation becomes central in the ancient concept of innovation.

In his *Laws*, for instance, Plato introduces the concept of innovation in the context of the political order. He argues that novelty and innovation should be excluded from education:

> They fail to reflect that those children who innovate [*neotherizein*] in their games grow up into men different from their fathers; and being thus different themselves, they seek a different mode of life, and having sought this, they come to desire other institutions and laws; and none of them dreads the consequent approach of that result which we described just now as the greatest of all banes.
>
> *(Plato 1967: 798c)*

Children love innovations (new games, new toys, new devices), but that will lead them to despise old habits and traditions and embrace the new (Plato 1967: 797b), which will in the end lead to political instability. The same holds for Aristotle, who argues that the best possible political order is already discovered and that any change in it will make it worse (Aristotle 1944: 1264a20–25). Innovation concerns the introduction of change in the established political order, and the Greek philosophers are negative about innovation just because it can disrupt the political order and can lead to revolution. In light of this, we receive a first indication of the destructive aspect of the innovation process. While in current society, disruptive innovations like the Internet destruct the established *economic* order, in the ancient notion of innovation, it is primarily the *political* order which is destructed.

With regard to the question concerning what aspect of the political order is destructed by innovation according to Plato, a possible answer is found in his connection of innovation with the introduction of something new that threatens the established political order. It is important to consider that according to philosophers like Plato, the destructive aspect of innovation is not so much found in the destruction of *things* in the world, such as for instance the natural environment, but rather in the destruction of the political *order* of the world. The nature of this destruction becomes clear if we ask for the measure or unity of the order of the world. The philosophical tradition starting with Plato finds this measure or unity in the ontological characteristic of the being of beings, i.e. in the transcendental horizon of the Platonic *idea*. The *idea* is a fixed category or measure, within which the world appears as an ordered whole that makes sense. In light of the *idea* 'human being' for instance, various people appear *as* human being and we can understand this variety of humans *as* human beings. The *idea* human being is itself not a human being, but concerns a given measure, category or value within which the variety of people appears *as* unity. In the philosophical tradition there is a fundamental difference between the *idea*, category or value that establishes the order of beings in the world, and these beings themselves, which can be perceived and understood only in light of the *idea*. What is destructed by innovation, according to Plato, has to be sought at the ontological level of the *idea* as measure for the established political order, and not at the ontic level of things in the world.

This also becomes clear in Plato's *Republic*. Here, Plato argues that the state should be ruled by the philosopher king, who has the necessary training and education that enables him to intellectually grasp ethical notions such as the *idea* of justice, and who has the insights that are required to safeguard the political order. Here the problem with innovation becomes clear. If innovation transgresses the established political *order* of the world, it primarily intervenes at the level of the *ideai*, categories or values within which the world functions as *order*. Innovations are primarily disrupting the existing *ideai*, and consist in the human construction and introduction of new *ideai*. This means that the destructive aspect of innovation does not concern primarily the ontic level of things in the world, but the ontological level of the *ideai*, categories or values that establish and safeguard a world *order*. The idea that innovation primarily intervenes at the ontological level of the world *order* is also confirmed by a later writer on innovation, Francis Bacon, who argues that innovations "have altered the whole face and state of things right across the globe" (cited in Godin 2015: 182).

Although Plato rejects innovation because it intervenes at the ontological level and destructs the established order, he also has a second reason to reject innovation. The idea that innovation intervenes at the ontological level and destructs the established order testifies namely to the human ability to construct a new *idea*, category or value that establishes a new political-economic order. Two aspects are therefore important to consider in Plato's rejection of innovation. First, the human construction of a new *idea* should be rejected from the perspective that the transcendent world of the *ideai* is fixed and eternal according to Plato and can cannot be replaced by new ones. This is consistent with Aristotle's idea that the ideal political order is already established. Second, he would object to the idea that the role of the human being is to innovate and construct such a new *idea*. The philosopher king shouldn't be educated in innovation (we can associate this with the *vita activa*) but should be enabled to *grasp* the eternal idea of justice to safeguard the political order in light of this *idea* (we can associate this with the *vita contemplativa*). For Plato, innovation should be rejected therefore for two reasons: the innovator is primarily *guilty* of denying the eternal truth of the transcendental horizon of the *ideai*, categories or values that establish the world order by his effort to introduce a new one; but with this, secondly, he denies the nature of the human being who should primarily grasp and contemplate these pre-given ideas, categories or values in which he or she can live the good life as a political actor, instead of *producing* new ones.[3]

We leave this discussion of the human role in innovation for a moment and return to the difference between the ontic and ontological level of innovation, because it provides a second indication of the ontogenetic process of innovation, reflected in innovations such as the steam engine. To be sure, the steam engine does not only concern the ontogenesis of the engine at an ontic level. It also concerns the ontogenesis of the economic order of the world associated with steam at the ontological level, in which the steam engine can emerge, can be applied in various automated machines in factories like the spinning mill and is adopted and used by humans.[4] This distinction between the ontic and the ontological level of innovation provides a new perspective on Schumpeter's conceptualization of innovation as creative destruction. Innovations like the steam engine are definitely innovations at the ontic level of the creation of a new thing or artefact—the first engine, for instance—but their destructive character consists in the fact that they change 'the rules of the game'; they destruct the economic order of the world that is associated with water in which the water mill was embedded, applied and adopted by humans (in the textile industry, for instance), and give rise to a new world order associated with steam (see Figure 20.1). Likewise, the innovation of streaming services didn't destruct the CD in the literal sense of the word—there are still CDs in the world—but it changed the way value is created and captured via markets in the economic order associated with digital networks like the Internet.

What we have learned so far from our reflection on the history of innovation is that innovation primarily concerns the ontogenesis of innovation at an ontological level, namely the destruction of the *ideai*, categories or values that establish a world order. The other lesson we learned is that the economic orientation of the contemporary notion of innovation is not self-evident and could be extended to the political domain. Although it is clear that economists like Schumpeter focus on the impact of innovation on the *economic* order and assume that the articulation of a new world order is often established via markets,[5] the ancient notion of innovation makes us sensitive to the need to extend the ontogenesis of innovation to the political-economic domain, and maybe even to the domain of religion and art, as Godin shows in his history of the concept of innovation (Godin 2015).

In the next section, we continue our analysis by elaborating on the creative aspect of innovation and the creation of the *idea*, category or value. To this end, we focus on later writings regarding the concept of innovation.

20.3.2 *The Creative Aspect of Innovation: On the New*

The juridico-political connotation of innovation is continued in the Christian and Roman political writings up to the Renaissance. *In-novare* means the *introduction* of something *new*. Like in ancient Greece, in the Bible it is used in the political sense of the word, for instance, in the book of Samuel where Saul is called to renew the monarchy after his victory against the Ammonites (cf. Godin 2015).

Innovare also occurs in the work of Machiavelli (1469–1527) in the juridico-political context, namely in his reflections on *The Prince*. Here, Machiavelli discusses how the prince can break with established habits and customs and can take initiatives to renew them. According to Machiavelli, the prince can act either innovatively, by suddenly reforming customs and habits, or in a more cautious way, by acting safely without upsetting habits. Machiavelli shows that both attitudes have their advantages and disadvantages and are suitable in different situations. While the introduction of new rules and regulations may be needed in case of a political crisis, it becomes counterproductive in case the prince wants to stabilize the established new order. In such a case, the cautious way of acting would be more appropriate according to Machiavelli. For Machiavelli, just as for Plato, innovation—as introduction of new rules and regulations—serves the establishment of the political order on the ontological level, but it also involves the introduction of new institutions and new practices at the ontic level (cf. Godin 2015: 66). This more positive connotation of innovation in the case of Machiavelli may be explained by the fact that he experienced the world as continuously changing and even regressing, which then calls for the stabilization and founding of a new political order:

> One innovates because there is a changing situation that requires new ways of doing things or new things to do. One innovates when, in the face of changes, he himself changes things by introducing something new to stabilize a turbulent environment.
>
> *(Godin 2015: 65)*

For Machiavelli, the starting point of innovation is found in the absence of an *idea*, category or value that establishes the political order—a political crisis or chaos—and it is this absence of the political order that stimulates innovation as the introduction of a new *idea*, category or value to establish and safeguard the political order again.

In Machiavelli, the creative aspect of the innovation process is not necessarily found in the creation of something completely new to the world by the prince as creative actor. Machiavelli, for instance, argues for the return to the original political order that is corrupted over time (Godin 2015: 58). For this reason, we have to put 'newness' between brackets because in the first instance, and in line with the ancient idea that the *idea* or *morphe* (form) is eternal according to the Greek philosophers, the newness of *innovare* is understood as *renewal* or reformation of the original *idea*, category or value of beings; *re*-newal stresses newness as a *return* to or as a taking back into an original situation. This shows that innovation is in the first instance not something completely new to the world without any predecessor, as is sometimes said in case of disruptive innovations like the Internet and the combustion engine. Innovation is the product of a historical process of renewal, in which the new *idea*, category or value emerges out of a previous one and remains connected to it. This historical process of renewal can be associated with a return to an original starting position—such 'innovations' are associated with the renewal of the skin of the snake as renewal of their strength, or with the Phoenix which is reborn from its ashes in the ancient notion of innovation (Godin 2015: 49–50)—and therefore concerns nothing 'new' in the strict sense of the word. But this historical process can also result in something completely new, such as for instance the transf*orm*ation toward a new form (*idea*, category, value), which is the case with the transformation of a larva to a butterfly. Innovation may therefore consist in a repetition of an original state, or in the transformation of the current state (renovation) or in the renewal in a completely new state.

The history of innovation teaches us that the new does not have to be found in a unique characteristic of the product or service that did not exist before and cannot be associated with something that was originally already there. Seen from the perspective of our reflection on the innovation process, the new concerns the process of ontogenesis from pre-individual to individual, which remains embedded in the temporal dimension of past, present and future. On the one hand, innovation is a break with the past (discontinuity). On the other hand, it remains embedded in the history it emerges from. Innovation is therefore characterized by iterability (cf. Derrida 1982), i.e. by the paradoxical simultaneity of sameness and otherness. Framed in terms of the ontogenetic process: to the extent that innovations always remain embedded in the history they emerge from, the 'new' of innovation is always less than itself (pre-individual), and to the extent that the 'new' of innovation always breaks with this past and is on its way to a possible future, it is always more than itself (post-individual). The innovation process itself is primarily characterized by temporal renewal, renovation, etc., in which human existence is involved without being the subject of innovation, and this may explain why in the ancient and medieval notion of innovation, the human actor is not yet necessarily seen as the *origin* and *creator* of innovation (cf. Godin 2015: 66).

The creation of the new is also important to consider in light of the connection between innovation and risk (cf. §3.1). The tendency to conceive innovation as something good in itself becomes questionable if we consider the *Faustian* aspect of innovation (cf. §2). According to the ancient and medieval notion of the concept, innovation is intrinsically a risky business because it can undermine the established order. Machiavelli, for instance, argues:

> And it ought to be remembered that there is nothing more difficult to take in hand, more perilous to conduct, or more uncertain in its success, then to take the lead in the introduction of a new order of things. Because the innovator has for enemies all those who have done well under the old conditions, and lukewarm defenders in those who may do well under the new.
>
> *(Machiavelli 2017: IV)*

According to Francis Bacon, all innovation is ill-shapen:

> As the births of living creatures, at first are ill-shapen, so are all innovations, which are the births of time. Yet notwithstanding, as those that first bring honour into their family, are commonly more worthy than most that succeed, so the first precedent (if it be good) is seldom attained by imitation. For ill, to man's nature, as it stands perverted, hath a natural motion, strongest in continuance; but good, as a forced motion, strongest at first.
>
> *(cited in Bontems 2014: 43)*

While for Machiavelli this risk implies that innovation should be introduced radically because people will forget the innovation with time and get accustomed to the changes involved (Godin 2015: 68), Bacon argues that we should only engage in innovation in case we see a clear need or a clear advantage, and that we should implement innovations gradually and take time to reflect on their implementation in practice (Bontems 2014: 44).

With regard to the creative aspect of innovation, we may conclude that innovation is not ethically neutral and does not only become morally good or bad in the hands of men. Innovation is intrinsically a risky business that calls for ethical considerations. The intrinsicality of the ethical dimension of innovation can already be considered in the context of the original meaning of the term. If *kainotomia* is understood as the opening of new mines, the inherent risk involved is the risk that the new mine collapses. We could argue, with Xenophon, that the state should take this risk, but we could also argue that the opening of new mines (innovation in the literal sense of the word) should be accompanied with efforts to shore up the mine to prevent its collapse if we move on cutting. This idea, that innovation should be accompanied with supporting activities to prevent its collapse, can be associated with Bacon's call for radical gradualism, but also with more contemporary calls for radical incrementality in innovation to prevent lock-in effects (Collingridge 1981). In geoengineering, for instance, engineers make decisions under ignorance, which requires additional supportive actions.

20.4. Conclusions

In this chapter, we reflected philosophically on the concept of innovation presupposed as technological and commercial innovation in both the policy and scientific literature. As a first step, we consulted Schumpeter's conceptualization of innovation, who is often seen as the founding father of the contemporary notion of innovation. We saw that on the one hand, Schumpeter can legitimately be seen as the founding father of the techno-economic paradigm of innovation. On the other hand, it became clear that Schumpeter's conceptualization of innovation already differentiates from the contemporary notion in two important respects: (1) innovation doesn't contribute to economic growth *per se*; (2) innovation doesn't contribute to the solution of societal problems *per se*. As a second step, we consulted the history of the notion of innovation and found several grounds to open up the presupposed notion of technological and commercial innovation, and to reflect on the ontic and ontological dimension of innovation as an ontogenetic process and outcome. Our findings are summarized in Table 20.1.

Table 20.1 Differences between the contemporary, Schumpeterian and ontological concepts of innovation

Contemporary self-evident understanding of innovation	*Schumpeter's concept of Innovation*	*Proposed Dimensions of an Ontological Concept of Innovation*
1. Newness (product, process, marketing method, organizational method, workplace organization (OECD)), ranging from new to the firm to new to the world	1. New to the World (good, process, market, source of supply, industrial organization)	1. Newness (*idea*, category, value), ranging from re-newal and re-novation to new to the world.
2. *Technological* Innovation	2. *Technological* Innovation	2. *Political* dimension of Innovation
3. Serves economic progress *per se*	3. Serves economic cycles with temporary progression *and* depression	3. Serves the economy
4. Primacy of the innovation outcome (culture of things)	4. Primacy of the innovation process (creative destruction)	4. Primacy of the temporal dimension of the innovation process (ontic-ontological level)
5. Human actor (businessman) as subject of innovation	5. Human actor (entrepreneur) as subject of innovation	5. Human existence involved in the innovation process (but not as primary subject of innovation)
6. Conceived as good in itself and as solution for societal challenges	6. Faustian aspect of all Innovation acknowledged	6. Intrinsicality of Risk

We do not yet intend to draw conclusions about the nature of innovation based on our philo-sophical reflection on the history of the concept yet. Our primary aim is to open up the self-evident notion of innovation for future philosophical reflection on the concept (Blok 2018).

First, the distinction between the current presupposed concept of innovation and conceptualiza-tions that can be found in history enables us to question the *self-evidence* of the techno-economic paradigm of innovation. Second, we experienced several tensions between the current concept of innovation and this techno-economic paradigm. An example is that the dominance of large cor-porate actors hampers innovation instead of stimulating it. Another example is that the dominant political-economic system limits the development of more *responsible* innovations and calls for recon-sidering the currently dominant economic growth paradigm (Blok and Lemmens 2015). In the case of such tensions, philosophical reflection on historical conceptualizations of innovation can help to articulate these problems and find possible building blocks for its solution (Blok 2021). If the ancient concept of innovation highlights its establishment of a political-economic order of the world, it can inspire us to acknowledge the political dimension of all innovation and engineering practices. This is highly relevant in contemporary debates about innovation and engineering *for* society. If the ancient concept of innovation highlights the temporal process of innovation, it can inspire us to shift our perspective from the new at the product or service level to the process level of renewal and renova-tion. This is highly relevant in contemporary discussions about, for example, the bio-based or circular economy, that is currently mainly focusing on re-cycling practices, but should shift to re-pair and reuse of material as well to achieve its aspiration of a zero-waste economy (Zwier et al. 2015). If the ancient concept of innovation highlights the intrinsic risks involved in innovation, it can inspire us to acknowledge the necessity of the ethical dimension of innovation and engineering practices. This is highly relevant in contemporary policy debates about mainstreaming *responsible* research and innova-tion. Third, by reflecting on historical conceptualizations of innovation, we can find building blocks and dimensions of *another, possible* and *future* concept of innovation that is in fact able to address the grand challenges of our time.

Acknowledgements

I would like to thank Lucien von Schomberg for his comments on an earlier draft of this chapter, and Jilde Garst for the help with the development of Figure 20.1.

Related Chapters

Chapter 9: Engineering as Art and the Art of Engineering (Lara Schrijver)
Chapter 10: Creativity and Discovery (David H. Cropley)
Chapter 19: Emergence in Engineering (Peter Simons)

Notes

1. In this respect, we disagree with Godin's suggestion that "technological innovation" is a phrase that appeared only after World War II, and that Schumpeter uses the phrase only sporadically in his work (Godin 2015: 250). Even if this is the case, the idea of economic cycles that are associated with technology makes clear that for Schumpeter, innovation is primarily technological.
2. Plato and Aristotle use the word *neotherizein,* which has a comparable meaning as *kainotomia* (Godin 2015).
3. In this respect, we can argue that innovation operates here as an example of Nietzsche's revaluation of all values *avant la lettre*. This revaluation consists not only in the devaluation of the Platonic *idea* if innovation is heralded; the *nihil* of the Platonic *idea* in nihilism. These values (*ideai* or categories) are not only replaced by new ones in this revaluation, but the second aspect of the revaluation is, according to Nietzsche, that humans now become the source of newly constructed values (*ideai* or categories). Based on our findings in

this subsection, we can conclude that innovation concerns precisely these two aspects of revaluation, which provide the two reasons why it is rejected by Plato.

4. According to Simondon, the emergence of the steam engine changes the human-technology relation as a whole: "The factory uses true technical individuals, whereas, in the workshop, it is man who lends his individuality to the accomplishment of technical actions" (Simondon 2017: 131).

5. A similar idea can be found in Rogers's idea that "the adoption of a new idea almost always entails the sale of a new product" (Rogers 1962: 261).

References

Aristotle (1944). *Politics*. Cambridge, MA: Harvard University Press.

Bessant, J. and Tidd, J. (2007). *Innovation and Entrepreneurship*. West Sussex: Wiley.

Blok, V. (2018). Philosophy of Innovation: A Research Agenda. Guest Editorial. *Philosophy of Management.*, 17, 1–5.

Blok, V. (2020). *Heidegger's concept of philosophical method. Innovating philosophy in the age of global warming*. London: Routledge.

Blok, V. (2021). "What is innovation? Laying the ground for a philosophy of innovation". *Techne: Research in Philosophy and Technology*, DOI: 10.5840/techne2020109129.

Blok, V. and Gremmen, B. (2016). Ecological Innovation: Biomimicry as a New Way of Thinking and Acting Ecologically. *Journal of Agricultural and Environmental Ethics*, 29(2), 203–217. doi:10.1007/s10806-015-9596-1

Blok, V. and Lemmens, P. (2015). The Emerging Concepts of Responsible Innovation. Three Reasons Why It Is Questionable and Calls for a Radical Transformation of the Concept of Innovation. In B. Koops, I. Oosterlaken, J. van den Hoven, H. Romijn, and T. Swierstra (eds.), *Responsible Innovation 2: Concepts, Approaches, and Applications*. Dordrecht: Springer, pp. 19–35.

Bontems, V.K. (2014). What Does Innovation Stand For? Review of a Watchword in Research Policies. *Journal of Innovation Economics and Management,* 3, 39–57.

Collingridge, D. (1981). *The Social Control of Technology*. London: Palgrave Macmillan.

Deakins, D. and Freel, M. (2009). *Entrepreneurship and Small Firms*. Berkshire: Mc-Graw-Hill.

Derrida, J. (1982). *Margins of Philosophy*. Chicago: University of Chicago Press.

European Commission (2010). *Europe 2020. Flagship Innovative Innovation Union*. Brussels: EU. https://ec.europa.eu/research/innovation-union/pdf/innovation-union-communication-brochure_en.pdf. Accessed February 15, 2018.

Freeman, C. and Soete, L. (1997). *The Economics of Industrial Innovation*. London: Continuum.

Godin, B. (2008). Innovation: The History of a Category. *Working Paper No. 1.*

Godin, B. (2015). *Innovation Contested. The Idea of Innovation Over the Centuries*. New York: Routledge.

Henderson, R.M., Clark, K.B. (1990). Architectural Innovation: The Reconfiguration of Existing Product Technologies and the Failure of Established Firms. *Administrative Science Quarterly*, 35, 9–30.

Machiavelli, N. (2017). *Delphi Collected Works of Niccolò Machiavelli*. Delphi Publishers.

OECD (2010). *Innovation and the Development Agenda*. Paris, OECD.

Plato (1967). *Plato in Twelve Volumes*. Cambridge, MA: Harvard University Press.

Rogers, E.M. (1962). *The Diffusion of Innovations*. New York: Free Press.

Rogers, E.M. (1976). Where Are We in the Understanding of Diffusion of Innovations? In W. Schramm and D. Lerner (eds.), *Communication and Change: The Last Ten Years—and the Next*. Honolulu: University Press of Hawaii, pp. 204–222.

Schomberg, L. von, Blok, V. (2018). The Turbulent Age of Innovation. Questioning the Nature of Innovation in Responsible Research & Innovation. *Synthese* (forthcoming).

Schumpeter, J. (1943). *Capitalism, Socialism and Democracy*. London: Routledge

Schumpeter, J. (1983). *The Theory of Economic Development*. New Brunswick: Transaction Publishers.

Simondon, G. (2017). *On the Mode of Existence of Technical Objects*. Minneapolis, MN: Univocal Publishing.

Smith, D. (2006). *Exploring Innovation*. Berkshire: McGraw-Hill Higher Education.

Timmermans, J. and Blok, V. (2018). A Critical Hermeneutic Reflection on the Paradigm-Level Assumptions Underlying Responsible Innovation. *Synthese* (forthcoming).

Xenophon (2014). *Ways and Means*. eBooks@Adelaide https://ebooks.adelaide.edu.au/x/xenophon/x5wa. Accessed October 3, 2018.

Zwier, J., Blok, V., Lemmens, P., Geerts, R.J. (2015). The Ideal of a Zero-Waste Humanity: Philosophical Reflections on the Demand for a Bio-Based Economy. *Journal of Agricultural & Environmental Ethics*, 28(2), 353–374. doi:10.1007/s10806–015–9538-y

PART IV

Engineering Design Processes

21

ENGINEERING DESIGN

Peter Kroes

21.1. Introduction

We start with some general remarks about engineering design and its philosophy. First, there is the issue of where to locate the philosophy of engineering design in the philosophical landscape. Its most natural niche appears to be somewhere within the region of the philosophy of action, since engineering design is intimately related to *making* technical artifacts which is a form of intentional action. However, one will search in vain for a systematic discussion of making technical artifacts or making in general within the philosophy of action or in mainstream philosophy. Making technical artifacts, which according to Aristotle is a specific form of intentional action, namely productive action, appears not to have been given much consideration. The same applies also to *using* technical artifacts which is also a form of intentional action. The only subfield in philosophy that may provide some bearings for a discussion of engineering design is the philosophy of technology. In this field, however, engineering design has only recently attracted sustained attention (see Meijers 2009: part III). It is by itself a rather remarkable observation that in spite of the fact that making and using (technical) things plays such a dominant role in modern life, it does not do so in philosophy; a general philosophy of making is still missing.

Second, that engineering design is an integral aspect of making things becomes clear when we realize that modern engineering design is more or less the result of a division of labor between the mental and physical work in making technical artifacts; the process of making is, so to speak, split up into a phase of conceiving and of realizing. Engineering design is considered primarily an intellectual activity: in a nutshell, it is about proposing and working out new ideas and plans for technical artifacts. The actual realization of these ideas or execution of the plans then takes place in the production phase. This division of labor is a rather recent phenomenon that appears to run parallel to the emergence of modern engineering (sciences). It is absent in the early crafts. The craftsman usually performs the mental and physical work involved in making technical things without separating them into two distinct activities.

A further remark concerns the kind of things in the making of which engineering design plays a role; all kinds of things may be designed and made without falling in the province of engineering. Engineering is, as stated earlier, about making *technical* artifacts. But how to demarcate technical artifacts from other kinds of artifacts, such as pieces of art or social artifacts? This is an intricate demarcation problem that falls outside the scope of this chapter. The following observations will have to suffice here (for a more elaborate discussion, see Kroes 2012: ch. 1). Engineering design is primarily focused on making changes in our physical (material) world as opposed to changes in our social world. This does not mean that the things designed are necessarily physical objects; they may also be processes that, when implemented, produce some desired effect in our physical world. In recent times there has been a significant shift from the design of technical objects to the design of services (i.e. processes) that involve technical objects. Furthermore, technical artifacts are generally considered to be different from pieces of art because they serve 'practical' functions.[1] From here on, when speaking simply of design, we have in mind this characterization of engineering design.

Finally, the term '(engineering) design' may be used as a verb and as a noun. As a verb it denotes the activity of designing, that is, of solving design tasks. As a noun it refers to the outcome of this activity: a design of a technical artifact. The first thing to be done in this chapter is to discuss in more detail how to characterize these two senses of the notion of engineering design. Suffice it here to stress that both senses play a key role in making new kinds of technical artifacts, which presupposes a design activity (designing) that leads to a successful description of the design of a new kind of technical artifact.

In order to avoid confusion, any discussion of engineering design will have to distinguish carefully between the two uses of the term, as a verb and as a noun. So, in the following we will first focus on philosophical issues related to design as an activity (Section 2) and then on design as a feature of a technical artifact (Section 3). Thereafter we will turn to values in engineering design; we will see that values play an important, but different role in both senses of engineering design (Section 4). A value that is often considered of vital importance in engineering design is creativity; we will discuss two different ways of interpreting this notion that are relevant for design practice (Section 5). In the final section (6), we will have a brief look at how the social world enters the world of engineering design by looking at the design of techno-symbolic artifacts and of sociotechnical systems; the design of these kinds of objects involves taking into account not only technical but also social and symbolic functions.

21.2. Design as an Activity

Designing comes into play when the artifact to be made is expected to have some new features in comparison to already existing technical artifacts: simply making more copies of an already existing artifact does not involve designing. Schematically, these new features may concern the overall function of a technical artifact that may be directly relevant for its users (e.g., the design and introduction of the first color TV sets) or changes in the internal, technical operation of a technical artifact without a direct effect on its overall function (e.g., the design and introduction of solid state devices for replacing vacuum tubes in TV sets). In both cases the design task consists of how a function, whether overall function or internal subfunction, may be realized by a physical structure or process. Designing may therefore be characterized in its core as an activity that starts with a description of a set of desired or needed features (behaviors) of the artifact that is to be designed, that is, with a description of its function. The outcome of the design process is a detailed description of a physical structure[2] that is able to perform the function. In other words, in designing, a description of the function of a future technical artifact is transformed or translated into a description of its physical features.

For a better understanding of what is going on in this transformation, we will first have a closer look at both kinds of description and how they are related for an already existing technical artifact. A functional description of a given artifact states what features the artifact *should* have or what it *should* do (under certain conditions and circumstances). Given a certain input (including actions on the part of the user), the technical artifact should produce a certain output. A purely functional description does not describe how this is to be done, and therefore it does not include any of its physical features. A functional description is a black box description of the technical artifact as far as its physical properties are concerned. The physical description, by contrast, takes only its physical properties into account; it abstracts from all functional features of the technical artifact and in fact reduces it to simply a physical object. It states in detail all properties of the physical structure of the technical artifact.[3] It is the mirror image of the functional description: it black boxes all functional properties of a technical artifact.

These functional and physical descriptions play a key role not only in characterizing already existing technical artifacts but also in designing and making new ones. In brief, the functional description states what the technical artifact should do and the physical one, how this is to be achieved. The functional description is the starting point for designing a technical artifact, the physical one for making it. For actually making a technical artifact a purely physical description of all its components and how they fit together is necessary. As long as the description of the structure of a technical artifact contains components that are only described in terms of their functional properties, i.e., are described as a physical black box, it will not be possible to make it. In principle it should be possible to produce the physical structure that is able to perform the required function without knowing anything about this function. That is why part of the outcome of a design activity is a description of a physical structure that is able to perform the required function.

From a philosophical point of view, the relation between a functional and a physical description of a technical artifact and how a functional description is transformed into a physical one in designing raises a number of issues. The two descriptions refer to different kinds of properties, each of which plays an essential role in the characterization of technical artifacts (more about this in Section 3). The nature of the functional properties of technical artifacts has been and still is a matter of debate (see Krohs and Kroes 2009). However, there appears to be wide consensus that whatever theory of technical functions is adopted, it should be able to account for the fact that normative statements with regard to the performance of technical functions make sense (for instance, of the kind 'function x is performed well or badly'). So, to attribute functional properties to an object is to make normative claims about it. This is completely in line with our claim that the design activity starts with a statement of what the object to be designed should do. A physical description of a technical artifact refers only to its physical properties; it is factual in nature and therefore either true or false. It does not make sense to make normative claims with regard to the physical properties of a technical artifact (from a purely physical point of view, claims such as 'property y is good or bad' are meaningless).

So at first sight, the transformation of a functional into a physical description in designing appears to involve a problematic shift from a normative description to a factual one. If that is indeed the case, then this raises serious philosophical problems because of the alleged gap between both domains, that is, because of the infamous 'is-ought' problem. However, there is a significant difference between the physical description of the structure of an already existing technical artifact and the one that is the outcome of a design process. The latter refers to an object that still has to be made, and therefore it may be characterized as a physical description with normative force: it is a description of the physical properties that the object to be made *should* have. It defines the object in a stipulative and therefore normative (prescriptive) way. This description with normative force is part of what is called the design of a technical artifact. It is different from the physical description of an already existing technical artifact that was discussed earlier, which has no normative force. From this perspective, the transformation of a functional description into a physical one (with normative force) does not involve a problematic shift from the normative to the descriptive domain.

Nevertheless, designers appear to be able to bridge the alleged gap between the normative and the descriptive domains. This comes to the fore when we take a brief look at how they explain their designs. For instance, in explaining how a steam engine works, they start from its overall function and explain it in terms of the subfunctions of its parts (the cylinder, the piston, the valves, etc.) and how they are arranged. This step is repeated for the explanation of the subfunctions of the parts themselves; they in turn are explained in terms of the subfunctions of their parts. So far, nothing special is taking place; we are dealing with functional explanations, that is, the explanation of one function in terms of other functions. There is no crossing of the border of the functional (normative) and physical (descriptive) domains. But at the lowest level of this chain of explanations, when the part under consideration is itself not composed of components with subfunctions, another kind of explanation appears to enter the scene, one in which functions of parts are 'explained' in terms of their physical properties. Take a standardized technical component, for example a hexagon nut with flange, as described in ISO-norm 10663; its function is to fasten objects, and it is classified as such in this norm. Challenged to explain how this nut performs its function, engineers will point to its geometrical form, in particular of its 'thread', and its physical properties (in fact they will refer to many of the physical properties described in minute detail in the norm). It appears that at that stage in the chain of explanation, the only resource for explaining a function are physical properties. If so, this implies that they are able to move from the normative to the descriptive domain.

A similar kind of reasoning, that also appears to defy the gap between the normative and descriptive domain, occurs in reverse engineering, or in archeology in cases where, given a physical object with an unknown function, the challenge is to deduce its function from its physical properties (a famous example is the Antikythera mechanism). In these forms of reasoning the fact that the function and the physical structure of a technical artifact constrain each other plays an important role: not any physical structure can perform whatever physical function. The functional and physical properties of technical artifacts are intimately related to each other since the function has to be performed on the basis of its physical properties.[4] Thus, the physical properties of technical artifacts are relevant for reasoning about functional properties. That they are relevant is not what makes this reasoning from physical to functional properties remarkable; clearly facts matter also in reasoning about normative issues. What happens in explaining functional properties of a technical artifact in terms of its physical properties, however, appears to run very much parallel to what according to Hume (1969 [1739–40]: 521) often takes place in moral reasoning and what he considers to be very problematic: in the nut example, designers start to reason from claims about the physical properties of an object only (the 'is') and then finally end up with claims about its functional properties (the 'ought'). For Hume, this amounts to committing a logical fallacy. Whether the reasoning that designers use in the explanation of their designs indeed involves this problem remains an open issue.

From a different perspective on engineering reasoning, the same issue about the relation between normative and descriptive claims emerges. Reasoning in engineering in general and in engineering design in particular is often characterized as means–end reasoning or instrumental reasoning. The idea seems obvious: technology is all about designing, making and using the appropriate (physical) means to achieve certain ends. This instrumental nature of engineering thinking and reasoning is reflected in the prime criteria for evaluating means, namely efficacy and efficiency. Means–end reasoning is very closely akin to what von Wright (1963) calls practical reasoning. von Wright's standard example of practical reasoning is the following: '*A* wants to make the hut habitable. Unless *A* heats the hut it will not become habitable. Therefore, *A* must heat the hut'. The first premise states a factual claim about an end and the second a factual claim about a causal relation. The conclusion states an action that must be performed. Here, the conclusion states what a person must do, not what a technical artifact should do, so the means is not an object but an action. Certainly, the conclusion has a normative ring to it. If indeed it is a normative claim (for a discussion of this issue, see Black 1964) and if von Wright's example captures the core of what goes on in means–end or practical reasoning that is employed in engineering design, then we are back to the issue about how designers manage to bridge the gap between the descriptive and the normative domains.

We now turn to another, more pragmatic aspect of design as a process, namely design methodology. Design methodology has attracted a lot of attention, in particular from design engineers themselves (Dym 1994; Roozenburg and Eekels 1995; Hubka and Eder 1996). Design methodology studies the general, domain-independent methods by which to solve design problems. This has resulted in numerous flow diagrams that describe in more or less detail the different steps to be performed; most of these flow diagrams appear to be variations on the analyze, synthesize and evaluate scheme. On the one hand, these flow diagrams are intended to provide insight into the rational solution of design problems. They play a role in rational reconstructions of real design processes, that is, in analyzing the rationally prescribed steps and the rationally prescribed design decisions that an 'ideal designer' (one purely thinking on the basis of instrumental rationality) would have taken in order to reach the solution that was actually found. On the other hand, they have a normative role: they prescribe which steps to take in solving a design problem.

One of the interests of designers in design methodology is that they expect (or hope) that the implementation of the proposed methodologies (of their flow diagrams) will somehow improve actual design processes. Here we will not enter into a discussion whether there is empirical evidence to support this expectation. Instead we restrict ourselves to pointing out an ambiguity in the idea of improving actual design processes. This ambiguity is closely related to the double meaning of the term 'design'. As a process, two of the core criteria for evaluating design are efficacy and efficiency (see also the earlier remarks about means-end reasoning). One way to interpret the idea of improving design through the use of the flow diagrams amounts to the claim that if designers follow the prescriptions of these flow diagrams, they will solve a design task in a way that is, from a rational point of view, the most efficient; the same result (design as a noun) will be produced with less (or the least) amount of resources. But another way to interpret improvement in design concerns design as a noun and leads to the claim that the design process will lead to a better solution of the design problem, thus to a better design. If we take 'better' in a more or less technical sense, this may mean either that the design better (i.e., in a more effective and efficient way) fulfills the list of specifications, or that it fulfills a better list of specifications. This interpretation of improving design is also closely associated with design methodology. However, it raises an intriguing methodological problem: what, if any, is the relation between the rationality of a process and the quality of its outcome? Design methodology appears to be primarily concerned with process rationality, and its claim about improving the efficiency of design processes may make sense. But what about the idea that applying the results of design methodology will also lead to better designs? Since in general the criteria for evaluating a process and a product are very different, it is not self-evident at all that improvement of design as a process will automatically lead to improvement of design as a noun.

At this point our discussion of design as a process comes to a close. There is still one other feature of design as a process that needs to be addressed, namely the making of trade-offs, but we will do so when analyzing the role of values in design (Section 4).

21.3. Design as a Feature of a Technical Artifact

In a way, the title of this section already says it all: design as a noun refers to a feature of a technical artifact, and a crucial one, but what it is in more detail is hard to spell out. A technical artifact is more than just a physical object; it is the 'embodiment' of a design, which is a different way of saying more or less the same as that a technical artifact is a physical structure with a function. With reference to Hilpinen's (1993) definition of an artifact as an object with an author, an object may be said to be a technical artifact if and only if it has a design, which implies that it has a designer. It is the design or the function of a technical artifact that endows its physical object with its 'forness', which makes it into a means to a certain end. This is famously expressed by Paley (2006 [1802]: 14) who writes that in examining a watch what we see are:

> contrivance, design; an end, a purpose; means for the end, adaptation to the purpose. And the question, which irresistibly presses upon our thoughts, is, whence this contrivance and design. The thing required is the intending mind, the adapting hand, the intelligence by which that hand was directed.

But, of course, seeing this contrivance and design is something different from seeing, for instance, the color or form of the watch and its parts. That is precisely the reason why it is so difficult to specify in more detail this feature of technical artifacts.

Albeit implicitly, the notion of a design is contained in the description of the physical object that is the outcome of a design process. As we already pointed out, insofar as this description is necessary for making an instance of the technical artifact kind, it has normative force: it prescribes which physical features the object must have. The answer to the question why it has to have these physical properties is that they are necessary for performing its function. In the physical description, all the physical parts with their properties are (implicitly) arranged or organized in a purposeful, teleological way so that it is able to perform its function. The notion of design may be taken to refer to this purposeful arrangement or organization of the physical object and is therefore already implied by the normative physical description that is the outcome of a design process.

Still, what does it mean to say that a technical artifact is the actual realization of a design? One way to interpret this is to take a design to be something like a plan of a technical artifact. Of course, the notion of plan then cannot refer to a considered series of purposeful actions, since technical artifacts do not act. Instead, the notion of plan may be taken to refer to the idea of a purposeful arrangement or organization of physical parts so as to perform a certain function. In engineering it is common practice to describe such a plan by drawings (which brings us to the original meaning of the term 'design'). A given technical artifact is then the outcome of the execution or realization of a design as a plan and inherits the purposefulness of this plan. This interpretation fits well with the division-of-labor picture we presented earlier: the design phase is about conceiving an idea of or plan for a technical artifact and the making phase about realizing that plan.

These ways of interpreting the notion of design inevitably bring into play the designer (or, as Paley states, the 'intending mind, the adapting hand, the intelligence by which that hand was directed'): the designer prescribes, arranges or organizes. This may explain the origin of the teleological nature of technical artifacts, but not how we should interpret the teleological nature of the technical artifact itself. It is one thing to state that the technical artifact inherits its teleological nature from the designer, but another to interpret what this means in terms of the properties of a technical artifact. Earlier on, we have characterized technical artifacts as having two kinds of properties, namely physical and functional ones. From an ontological point of view, both are constitutive for being a technical artifact: an object with only physical properties is not a technical artifact, and neither is an object with only functional properties. With regard to the issue under discussion, this ontological characterization is far from innocent. The functional properties of a technical artifact are related to its design and its purpose and give it its teleological nature. This design and purpose, and therefore also the functional properties, are intimately related to human intentions. Thus, technical artifacts are constituted by physical properties on the one hand, and by intentions-related functional properties on the other. Moreover, these two kinds of properties are somehow related (constrain each other). This makes technical artifacts a kind of objects *sui generis*, different from physical and from social objects, with a hybrid or dual nature (Kroes 2012). It is because of their dual nature that an adequate conception (description) of technical artifacts cannot be achieved from a purely physical or from a purely intentional stance. This is the reason why the idea of a technical artifact as the embodiment of a design, attractive as it may sound, raises so many questions, not only because it is difficult to spell out more precisely what a design is, but also because the idea of embodiment is far from clear. With this observation we conclude our discussion of design as a noun and now turn our attention to the role of values in design.

21.4. Design and Values

As in any practice, values play an important role in engineering design practice. Here we will not be concerned with the role of values insofar as this design practice is a social practice of designers interacting with each other on the basis of a set of values. We will focus on the role of values with regard to what engineering design is about, its object matter. In terms of MacIntyre's (1984: 187) definition of practice, we are concerned with values in relation to producing the 'primary good' of the practice of engineering design, so with the role of values in producing a design. In the following we will briefly turn first to the variety of values that play a role in designing, then to the role of trade-offs and finally to the idea that technical artifacts, that is physical objects with a design, may embody values.

Engineering design is part of a process of value creation: producing a design for a technical artifact is producing something of value. Exactly what kind of value is primarily associated with a (design of) technical artifact depends on the point of view taken. From a design engineering point of view its value may be primarily associated with its innovative design, from a production engineering point of view with its ease of production, from a legal point of view with its patents, from an economic perspective with its monetary (exchange) value, from a user perspective with its (use) value for money, from a labor conditions perspective with its safety features, etc. Different stakeholders in the creation of a technical artifact bring into play a variety of values (and interests) that is reflected in the list of requirements for a design. Apart from the requirements that are directly related to its core technical function, that list may contain requirements that find their origin in conditions concerning available production facilities, or in labor, legal, environmental, financial or market conditions, or standards and norms, and so on. Behind each item on the list there are stakeholders that bring in their own values in the design of a technical artifact.

With such a variety of values that find their origin in different quarters, it is to be expected that it may be difficult, if not impossible, to satisfy all requirements. Often trade-offs between different values have to be made, for instance between safety and environmental friendliness in the design of cars (roughly, more safety implies more weight which in turn implies a less environmentally friendly car). Making trade-offs is indeed part and parcel of engineering design practice. Trade-offs involve the comparison of a gain in one value with the loss in another value. Under what conditions is such a comparison possible so that trade-offs between values may be made in a rational way? This not only involves (unresolved) issues about whether values are comparable and may be hierarchically ordered and if so in what kind of way (e.g., trade-offs between lexicographically ordered values appear to be out of the question) but also issues about what kind of measuring scales can be constructed for values and what kind of measuring scales are necessary for making trade-offs in a rational way (Van de Poel 2009).

Related to this issue is the question whether the notion of a 'best' design always makes sense. Suppose that, given a certain design problem, designers have to choose from a set of different design solutions and that these solutions score differently on various design criteria (values). So, the question arises, which option to choose from the proposed set. From a rational point of view it is obvious to go for the best option. This presupposes that it is possible to rank all options with regard to each other as being better (or worse) design solutions for the original design problem. However, it turns out that this is not always possible. It has been shown that a general procedure that satisfies certain very basic conditions for ranking design options as overall better or worse on the basis of how these options score on the various design criteria does not exist. In those cases there is no way to rank the design options such that the highest-ranking one may be considered the design that overall scores best on all relevant values (Franssen 2005).

Finally, we turn to some recent developments in engineering design that go under the banner of 'value sensitive design' or 'design for values' (see, for instance, the following websites: https://vsdesign.org, http://sciencepolicy.colorado.edu/about_us/find_us.html, http://ethicsandtechnology.eu/). They bring into focus an age-old problem about technology that is closely connected to the aforementioned issue about what it means to say that a technical artifact is the embodiment of a design. Design has always been practiced for (creating) values, but what is special about these developments is that they stress the role of moral values in design. The core idea appears to be roughly that taking into account moral values will lead to 'moral' specifications for the technical artifact to be designed, and that by satisfying these moral specifications, technical artifacts somehow will 'reflect', 'bear' or 'embody' the moral values involved. If moral values may indeed be built in, so to speak, in technical artifacts, this implies that technical artifacts are not morally neutral objects.[5] So these developments appear to reject in one form or another the moral neutrality thesis of technology and to take a definite position in the long-standing debate about the moral status of technical artifacts.

We will restrict ourselves here to the following comment. It is one thing to reject the idea that technical artifacts are simply morally neutral means and that therefore all moral issues about technology reside in the user context. It is another to make clear what it means that technical artifacts may harbor moral values. It is intimately related to the idea that technical artifacts are the embodiment of a design, and we have seen that it is difficult to spell out what this means. Several proposals have been advanced (see, e.g., Winner 1980; Latour 1992; Verbeek 2008; Kroes 2012), but at the moment a generally accepted, coherent explication of how technical artifacts may embody values is still missing.

21.5. Design and Creativity

Creativity is highly valued in design practice. Again, it is necessary to distinguish between creativity in relation to design as a noun and as a verb. For design as a noun creativity concerns features of a design. In those areas of engineering that border on the arts, such as architecture, creativity is often primarily associated with the aesthetic features of designs. But creativity in engineering design may be related also to other features. For instance, a design may be considered very creative from a technological point of view because it shows how a new technical function may be realized, or because it makes use of a new operational principle for realizing a function already performed by existing artifacts, or because it makes it possible to avoid certain trade-offs, etc.

Creativity is also strongly associated with design as a verb. In solving design problems, designers are supposed to think creatively, or 'out of the box'. One way to do so is to start the design process from a purely functional description of the artifact as a whole, instead of from some variation on an already existing design solution. So, the black box at the starting point is taken to be completely empty (for an example, see Otto and Wood (2001: 163). The underlying idea is that this will free the mind of the designers from existing design solutions so that they can come up with totally new, creative ones. But, of course, the creativity of designers may also extend to the ways they question, reconstruct and redefine design problems, in particular the ways they deal with the many, often conflicting constraints imposed on the object of design.

An interesting issue about creative thinking in design concerns its relation to rational thinking. Both creative thinking and rational thinking are highly valued in design. Underlying the notion of rational reconstruction of design processes discussed earlier is the idea that solving design problems can and ought to be done in a rational way. The question is whether creative and rational thinking can go hand in hand in design processes. It is often suggested that there is a tension between the two, in particular that creative thinking is somehow undermined by rational thinking. One way to argue that these two forms of thinking are more or less compatible in design is by stressing the role of creative thinking in the context of discovery and of rational thinking in the context of justification of design.

Leaving issues about how to enhance creativity in design aside, we will focus here on the notion of creativity itself. There are at least two (related) ways of interpreting this notion that are relevant for design practice, namely from an ontological and from what we will (for want of a better term) call a problem-solving perspective.

In order to clarify the ontological notion of creativity, we assume that technical artifacts, such as hammers, computers, cars, pencils, etc., are part of the 'furniture' of the universe; that is, are part of the ontology of the world (of what exists in the world).[6] This means that by creating technical artifacts, humans have in the course of time enriched the ontology of the world. This is a very basic, if not the most basic notion of creativity, of being creative: adding new things to what there is in the world. Now the question poses itself under what conditions an object made by humans is a technical artifact and is, as such, part of the ontology of the world. Many things made by humans (engineers) with the intention of creating a technical artifact end up as scrap in the junkyard; the things created turn out to be failures, not to be 'real' technical artifacts at all (e.g., the history of flight is rife with 'flying machines' that do not fly).

So the question arises under what conditions a human-made object is a technical artifact of some kind. For an answer we turn to Thomasson's (2003, 2007) theory of artifact kinds. According to this theory, an object is an instance of an artifact kind K (e.g., a steam engine) just in case it is the result of a largely successful execution of a largely correct substantive idea of what it means to be a K-er. For *technical* artifacts the notion of a largely correct substantive idea amounts to a largely correct *design* of the artifact kind under consideration. Note that this interpretation of making an artifact in general fits very well the earlier analysis of making a technical artifact by conceptually separating the intellectual and the physical aspects. Thomasson's theory imposes two success conditions on the making of a technical artifact: (1) the maker must have a largely correct substantive idea of a K-er; and (2) the maker must largely successfully execute this idea. For technical artifacts these two conditions together are intended to guarantee that the object made will be able to perform its function and will therefore be a 'real' technical artifact instead of just some human-made physical object. The most interesting case of creativity in engineering is the creation of a *new* kind of technical artifact; then the ontology of the world is enriched with new kinds of things. This case involves the conception or invention of a (largely correct) new design which is one of the most highly valued forms of creativity in engineering design.

For a brief clarification of the problem-solving notion of creativity, we turn to Briskman's (1980) analysis of creativity in science and art. He argues that the only way to analyze creativity in thinking is through analyzing the creativity of the product of that thinking. This means, in our case, that the creativity of a design has logical priority over the creativity of thinking. For Briskman, creativity of a product is a relative notion: something is creative only in relation to a given background. He claims that in order to qualify as creative relative to a background, the product has to satisfy four criteria. It has to: (1) be novel relative to a given background; (2) solve a problem emerging from that background; (3) conflict with parts of the background and (4) be favorably evaluated against standards that are part of the background. For Briskman creative products 'transcend' their background and often change that background by raising its standards by which products are evaluated. According to this view, newness (novelty) of a design is clearly not sufficient for being a creative design and therefore not for creative problem-solving or creative thinking in engineering design.[7] It remains to be seen whether Briskman's list of criteria captures all relevant aspects of the problem-solving notion of creativity in engineering design practice.

21.6. Conclusion: Engineering Design and the Social World

In this final section we will point to two particular ways in which the social world enters into engineering design. So far we have dealt with the design of physical technical artifacts. The social world enters in this activity in many different ways; not only is engineering design itself a social practice, but also the establishment of the list of requirements involves many social processes of negotiation between various stakeholders. Here, however, we are interested in the role of social factors in engineering design not as a process but in what is being designed. There are at least two different ways in which social factors may enter into the object of design. First, although many forms of engineering design exclusively deal with physical constructions that perform technical functions, there are forms in which alongside technical functions, symbolic functions may also be of central importance (such as in architecture and in industrial design). So, the object of design is a physical construction that has to perform not only a technical but also a symbolic function. So, we are dealing with techno-symbolic artifacts; the papal throne, e.g., is not just some chair to sit on, but also symbolizes the power of the Pope. Closely associated with the symbolic function of technical artifacts is the idea that technical artifacts may have meaning. Whereas the performance of technical functions is intimately related to the physical world, the performance of symbolic functions brings in the social world. Second, the social world enters engineering design in the design of sociotechnical systems. There is a growing awareness in the engineering world that many technical systems can perform their function only under appropriate social conditions. For instance, traffic lights can perform their function of regulating traffic only in conjunction with a set of social rules (such as, 'stop at a red light') that are enforced. So the traffic light system may be considered a sociotechnical system that has to be designed as such, which means that apart from its technical components, also its social components have to be designed and both have to be attuned to each other. Here, engineering the physical world and engineering the social world meet.

The growing importance of social elements in the object of design not only raises problems for engineering design practice itself, but also for analyzing it from a philosophical point of view. One of the core philosophical issues concerns how to interpret the notions of a techno-symbolic artifact and of a sociotechnical system. These issues bring us more or less back to our analysis of the notion of a technical artifact, since one of the main themes in dealing with all these issues appears to be how to interpret the hybrid nature of the things designed by engineers. From a philosophical point of view, technical and techno-symbolic artifacts as well as sociotechnical systems are not at home in the physical world, since they are more than simply physical constructions, nor are they at home in the social world, since more than social acceptance is involved in performing their function.

Related Chapters

Chapter 8: Prescriptive Engineering Knowledge (Sjoerd Zwart)
Chapter 10: Creativity and Discovery in Engineering (David H. Cropley)
Chapter 13: Systems Engineering as Engineering Philosophy (Usman Akeel and Sarah Bell)
Chapter 15: Artifacts (Beth Preston)
Chapter 16: Engineering Objects (Wybo Houkes)
Chapter 17: Use Plans (Auke Pols)
Chapter 18: Function in Engineering (Boris Eisenbart and Kilian Gericke)
Chapter 22: Values and Design (Ibo Van de Poel)

Further Reading

Meijers, A. (ed.) (2009). *Philosophy of Technology and Engineering Sciences*. Amsterdam: Elsevier. (Part III is devoted to philosophical problems about engineering design.)
Parsons, G. (2016). *The Philosophy of Design*. Cambridge: Polity Press. (An introduction to philosophical issues with regard to design in general.)

Notes

1. If pieces of art perform functions at all, they are usually taken to be different from practical functions.
2. From here on, the term 'structure' is to be taken as shorthand for structure or process.
3. In principle it states *all* physical properties of the technical artifact, whether or not relevant for performing its function. Here we will focus on the description of all physical properties that are relevant for performing its function.
4. In this respect there is a fundamental difference between technical and social artifacts; the latter perform their function on the basis of collective acceptance (see Searle 1995). Sociotechnical systems, to be discussed briefly

in the final section, are hybrid systems that perform their function on the basis of their physical structure and collective acceptance.

5. Note that this position will have far-reaching consequences for the moral responsibility of design engineers.
6. There are sparse ontologies for which this is not true, e.g., ontologies that only admit the existence of elementary particles of physics.
7. In patent law one of the basic criteria for granting a patent, apart from novelty, is non-obviousness; this is in line with the idea that newness does not imply creative thinking.

References

Black, M. (1964). The Gap Between "Is" and "Should". *The Philosophical Review,* 73(2), 165–181.

Briskman, L. (1980). Creative Product and Creative Process in Science and Art. *Inquiry,* 23(1), 83–106.

Dym, C.L. (1994). *Engineering Design; a Synthesis of Views.* Cambridge: Cambridge University Press.

Franssen, M. (2005). Arrow's Theorem, Multi-criteria Decision Problems and Multi-attribute Design Problems in Engineering Design. *Research in Engineering Design,* 16, 42–56.

Hilpinen, R. (1993). Authors and Artifacts. *Proceedings of the Aristotelian Society,* 93, 155–178.

Hubka, V. and Eder, W.E. (1996). *Design Science; Introduction to the Needs, Scope, and Organization of Engineering Design Knowledge.* London: Springer.

Hume, D. (1969 [1739–40]). *A Treatise of Human Nature.* Harmondsworth: Penguin Press.

Kroes, P. (2012). *Technical Artefacts: Creations of Mind and Matter.* Dordrecht: Springer.

Krohs, U. and Kroes, P. (eds.) (2009). *Functions in Biological and Artificial Worlds; Comparative Philosophical Perspectives.* Cambridge, MA: MIT Press.

Latour, B. (1992). Where Are the Missing Masses? The Sociology of a Few Mundane Artifacts. In W.E. Bijker, and J. Law (eds.), *Shaping Technology/building Society; Studies in Sociotechnical Change.* Cambridge, MA: MIT Press, pp. 225–258.

MacIntyre, A.C. (1984). *After Virtue: A Study in Moral Theory.* Notre Dame, IN: University of Notre Dame Press.

Meijers, A. (ed.) (2009). *Philosophy of Technology and Engineering Sciences.* Amsterdam: Elsevier.

Otto, K.N. and Wood, K.L. (2001). *Product Design; Techniques in Reverse Engineering and New Product Development.* Upper Saddle River, NJ: Prentice Hall.

Paley, W. (2006 [1802]). *Natural Theology: Or, Evidence of the Existence and Attributes of the Deity, Collected From the Appearances of Nature.* Eds. M. Eddy and D.M. Knight. Oxford: Oxford University Press.

Roozenburg, N.F.M. and Eekels, J. (1995). *Product Design: Fundamentals and Methods,* New York: John Wiley & Sons.

Searle, J. (1995). *The Construction of Social Reality.* London: Penguin Press.

Thomasson, A.L. (2003). Realism and Human Kinds. *Philosophy and Phenomenological Research,* 67(3), 580–609.

Thomasson, A.L. (2007). Artifacts and Human Concepts. In S. Laurence and E. Margolis (eds.), *Creations of the Mind: Essays on Artifacts and Their Representations.* Oxford: Oxford University Press, pp. 52–73.

Van de Poel, I. (2009). Values in Engineering Design. In A. Meijers (ed.), *Philosophy of Technology and Engineering Sciences.* Amsterdam: Elsevier, pp. 973–1006.

Verbeek, P-P. (2008). Morality in Design; Design Ethics and the Morality of Technological Artifacts. In P.E. Vermaas, P. Kroes, A. Light, and St. A. Moore (eds.), *Philosophy and Design: From Engineering to Architecture.* Dordrecht: Springer, pp. 91–103.

von Wright, G.H. (1963). Practical Inference. *The Philosophical Review,* 72(2), 159–179.

Winner, L. (1980). Do Artifacts Have Politics? *Deadalus,* 109(1), 121–136.

22

VALUES AND DESIGN

Ibo Van de Poel

22.1. Introduction

It is increasingly recognized that technological design expresses certain values and that it is therefore desirable to explicitly address values during the design process of new technology. One approach to do so is value sensitive design (VSD). The VSD approach was originally developed by Batya Friedman, and it has been further developed by her and her colleagues during the last decades (Friedman and Kahn 2003; Friedman et al. 2006; Davis and Nathan 2015). Also a range of similar approaches has been articulated and developed, like Values in Design (VID), Values at Play, and Design for Values (DfV) (e.g. van den Hoven et al. 2015; Flanagan et al. 2008). In addition, there are a range of approaches that focus more specifically on one value (or a limited number of values); for example, design for sustainability, inclusive design, and privacy-by-design (Keates and Clarkson 2003; Bhamra and Lofthouse 2007; Birkeland 2002).

The aim of this contribution is to discuss a number of general and more fundamental issues that should be addressed if one wants to systematically address values in the design of new technology. The focus is therefore not on one specific approach (like VSD) but on a number of more general conceptual and methodological issues that are crucial for systematically integrating values into technical design. These issues are: (1) what values are, and how to identify the values that should be integrated into a particular design; (2) the conceptualization, operationalization and specification of values which is required to make values operational in the design process; (3) issues of conflicting values and possible ways to address them and (4) the possibility of value change and options to address it during design. Before discussing these four issues, I start with a brief historical overview.

22.2. Historical Overview

It has long been recognized in the philosophy of technology that technology is value-laden (e.g. Heidegger 1962; Ellul 1964). The first to give a systematic account of how specific artefacts (rather than technology as such) and design choices may be value-laden was probably Langdon Winner (1980). This value-ladenness of technological artefacts is now more or less generally accepted (see also Chapter 53, 'Engineering and Contemporary Continental Philosophy of Technology' by Diane P. Michelfelder, this volume), although there is not yet agreement on how it can best be understood, as different philosophical accounts have been proposed (Kroes and Verbeek 2014).

Initial work on VSD did not originate from the philosophy of technology, but from the area of information systems in combination with social science. Since the early 1990s, Batya Friedman and colleagues have been working on proposing and further developing the VSD approach (Davis and Nathan 2015). Although this approach was initially developed for the design of information systems, it is more widely applicable.

VSD takes what Friedman et al. (2006) call an interactional stance with respect to how values can get embedded in technology. With this, they mean that which values are realized will depend on both how a technology has been designed and how it is implemented and operated; both technical and social factors thus play a role, and it is the interplay between them that makes technology value-laden.

A further core element of the VSD approach is a tripartite methodology consisting of an iterative combination of empirical, conceptual and technical investigations (Friedman and Kahn 2003). Empirical investigations, among others, involve empirically investigating relevant actors and their values; conceptual investigations involve a further conceptualization of these values and possible value trade-offs; technical investigations involve implementing values in technical design choices but also identifying ethical and value issues on basis of existing designs and technologies.

In addition to the tripartite methodology, Flanagan et al. (2008) propose three phases or activities for integrating values into design:

1. *Discovery*, resulting in a list of relevant values for the design project at hand.
2. *Translation*, i.e. "the activity of embodying or expressing . . . values in system design" (Flanagan et al. 2008: 338).
3. *Verification*, the assessment—e.g. through simulation, tests or user questionnaires—of whether the design indeed has implemented the values that were aimed at.

A wide range of more specific methods for VSD is now available. Some of these have been specifically developed for VSD, other are more generally applicable social science methods that are also useful for VSD. Friedman et al. (2017) list the following 14 methods: (1) direct and indirect stakeholder analysis; (2) value source analysis; (3) co-evolution of technology and social structure; (4) value scenarios; (5) value sketches; (6) value-oriented semi-structured interview; (7) scalable information dimensions; (8) value-oriented coding manual; (9) value-oriented mock-ups, prototypes, and field deployments; (10) ethnographically informed inquiry regarding values and technology; (11) model for informed consent online; (12) value dams and flows; (13) value sensitive action-reflection model and (14) envisioning cards.

A good overview of the developments and discussions in VSD is presented in Davis and Nathan (2015), while van den Hoven et al. (2015) provide an overview of the broader field of design and values. As said in the Introduction, my aim for the remainder of this chapter is to elaborate a number of general themes that are relevant to any attempt to systematically include values into design.

22.3. What Are Values?

In the literature on VSD, values are usually defined as referring "to what a person or a group of people consider important in life" (Friedman et al. 2006: 349).[1] Although this is helpful as a first-order characterization of values, it also raises questions. For example, is everything that a person or group concerns important in life also a value? What if people consider it important to treat people of a different race or ethnicity differently? It seems, then, that there are things people find important but which do not amount to values. For such reasons, it has been argued that values should be distinguished from mere preferences or interests of people, and that we should distinguish between what people find important in life and what they have good *reasons* for to find important in life.

In the psychological literature, somewhat more precise definitions of values can be found. Rokeach (1973), for example, defines values as "enduring beliefs that a specific mode of conduct is personally or socially preferable to an opposite or converse mode of conduct or end-state of existence" (Rokeach 1973: 5). According to Schwartz and Bilsky (1987) "[v]alues are (a) concepts or beliefs, (b) about desirable end states or behaviors, (c) that transcend specific situations, (d) guide selection or evaluation of behavior and events, and (e) are ordered by relative importance" (Schwartz and Bilsky 1987: 551).

There are no clear-cut definitions in the philosophical literature, but usually values are associated with what is *good* (Dancy 1993; Raz 1999). The notion of value can then refer to what is good (metaphysics), or what we believe (epistemology) or express (semantics) to be good (Hirose and Olson 2015). Different values may be understood as different varieties of goodness, i.e. as different ways in which a state of affairs (or an object) can be good (von Wright 1963). General as this philosophical characterization of values is, it nevertheless helps to distinguish values from a number of related concepts:

1. From *attitudes*. The difference is that values are more stable and enduring and are also more general and abstract than attitudes (Rokeach 1973; Hechter 1993);
2. From *interests and preferences*. While some authors have associated values with interests (Schwartz and Bilsky 1987), values are not just beliefs or expressions about what is in the interest of an agent, but rather expressions or beliefs about what is *good* either for an agent or more generally.
3. From *norms*. Norms are more specific than values and contain prescriptions (including recommendations, obligations, prohibitions, restrictions) for action often based on sanctions (Hechter 1993). They belong to the deontic domain of normativity, while values belong to the evaluative domain (Dancy 1993; Raz 1999). Values help to evaluate certain states of affairs in terms of goodness.
4. From *goals and aims*. Goals and aims are more specific and concrete than values. Values help to evaluate states of affairs and may thus suggest certain goals or aims to strive for, but they are not themselves goals.

Even if we have a general idea what values are, the question for VSD remains: how are we to determine what values are relevant and important for the design of a specific technology? Friedman et al. (2006) suggest a list of values—like human welfare, privacy, accountability, freedom from bias, trust and environmental sustainability—that are often implicated in system design, but they stress that this list is not exhaustive. Others have questioned the usefulness of such lists, as designers should be open to other relevant values and values should always be elicited in the context of specific design projects and require the input from users and stakeholders (Borning and Muller 2012; Dantec et al. 2009).

The question of how to determine the values that are to be taken into account in the design of technology has two aspects. One is what might be called the discovery of values. Discovery is aimed at finding the potentially relevant values for the design of a technology. A large range of methods and approaches is available for value discovery, including checklists, interviews, surveys, focus groups, (value) scenarios, discourse analysis, hermeneutics, ethnography, participatory research, living labs and Q-methodology. The other aspect is how to decide which of the values resulting from value discovery *should* be taken into account in a specific design process. This is a normative question that extends beyond listing the (potentially) relevant values. This may require a substantive ethical theory (Manders-Huits 2011) or at least a criterion to distinguish 'mere' values from 'real' values or, more important, from less important values. With respect to the latter, philosophers sometimes distinguish between intrinsic (or final) values, that are good for their own sake, and instrumental (or extrinsic) value, that are good insofar as they contribute to other intrinsically good values (Zimmerman 2004; Moore 1922), although such distinctions have also been criticized (Dewey 1922).

The normative question of what values to take into account also has a political dimension: Who gets to decide what values are to be included? Most of the VSD literature (and other literature on values and design) advocates an approach in which not just the designers, but also direct stakeholders (users) and indirect stakeholders (others affected by the technology) have a say in answering this question. However, this may itself raise further normative (and political) questions. For example, in the development of new encryption software, criminals are, strictly speaking, also an indirect stakeholder; but should their values also be taken in account?

22.4. The Conceptualization, Operationalization and Specification of Values

Identifying relevant values and selecting those on which the design process should focus is a first step in any systematic attempt to take values into account in the design of new technology. However, for these values to impact the actual design process, they need to be made more specific and operational so that they can guide design choices. I propose here to distinguish between three main activities that are important in order to make values bear on the actual design process of new technologies:

- *Conceptualization of values* is "the providing of a definition, analysis or description of a value that clarifies its meaning and often its applicability" (Van de Poel 2013: 261). Conceptualizations typically provide an explanation or reasons why a value is good or positive, as well as an interpretation of how to understand the value. Conceptualization of values is largely a philosophical activity that is independent from the specific context where the value is applied, although some conceptualizations may be more appropriate or adequate for certain contexts than for others. For example, privacy may be conceptualized in terms of the "right to be left alone" (Warren and Brandeis 1890), but also in terms of control over what information about one self is shared with others (cf. Koops et al. 2017). Although both conceptualizations are general, in the design of a specific information system, the second one may be more relevant than the first.
- *Operationalization of values* refers to the process of making values measurable, so that it becomes possible to measure to what degree a state of affairs or a certain design realizes (or meets) a certain value. Value measurement has similarities but also distinct differences with the measurement of physical concepts like temperature (Kroes and Van de Poel 2015). Unlike physical concepts, values are normative (and not descriptive) in nature. This makes it more difficult to measure values objectively, but it does not rule out the operationalization of values, nor does it make value measurements arbitrary, as I will discuss in greater detail.
- *Specification of values* refers to the translation of values into more specific norms and design requirements that can guide the design process of new technology. In specification, contextual information is added, which makes it more specific what it means to strive for (or respect or meet) a certain value in a certain context. For example, in the context of designing a chemical plant, safety may be specified in terms of minimizing explosion risks, as well as in terms of providing containment, so that in case of an accident, it is less likely that hazardous materials leak into the environment of the plant.

Next, I will further discuss and illustrate these three activities with the help of a number of examples.

22.4.1 Conceptualization

Concerning conceptualization, it is often the case that different philosophical conceptions of a value are available in the literature. Take for example well-being. In the philosophical literature, three main theories of well-being have been articulated (Crisp 2013):

- Hedonism, which understands human well-being in terms of pleasurable experiences.
- Desire satisfaction accounts which conceptualize well-being in terms of the fulfilment of people's desires.

- Objective list accounts which understand well-being in terms of a list of general prudential values (such as accomplishment, autonomy, liberty, friendship, understanding and enjoyment; cf. Griffin 1986).

Each of these conceptualizations suggests another approach to how to design for well-being. On a hedonistic account of well-being, design should be aimed at creating products that give users and other stakeholders pleasurable experiences. A number of design approaches are aimed at just that (e.g. Koskinen et al. 2003; Desmet and Hekkert 2007). On a desire-satisfaction account, people's (actual) desires should be the starting point of design, and a method like Quality Function Deployment (QFD) (e.g. Akao 1990; Hauser and Clausing 1988) that starts from user demands and translates these into engineering characteristics of the product-to-be-designed could be used. For objective list accounts, there are not many systematic methods, but a design approach is suggested in Van de Poel (2012).

The fact that there are various, often competing, philosophical conceptions of many values (like well-being, privacy, sustainability, justice) raises the question which conception practitioners who want to design for a certain value should adopt. There is no straightforward answer to this question, but it is important to be aware that two types of considerations are relevant in answering this question in a specific case. One type consists of more general philosophical considerations about the adequacy of certain conceptualizations. The mentioned philosophical theories of well-being have each been criticized, which in response has resulted in more sophisticated accounts that try to meet some of the raised criticism, and one might have good philosophical reasons to prefer one account over the others. In addition to such philosophical considerations, there are also more practical considerations. Not every technology will (potentially) affect the same dimensions of well-being, and depending on how a technology may or may not impact humans, there may be good reasons to focus on a specific notion of well-being in the design of a specific technology. For example, if a technology mainly affects human experiences and to a lesser extent other aspects of well-being (according to the other two accounts), it may be justified to focus on a more hedonistic notion of well-being, while in other cases such a notion may be too narrow, for example because a technology also affects friendship or personal relations.

22.4.2 Operationalization

Operationalization of values is the process of making values measurable. In some cases, it might be possible to directly measure values in a subjective way. We can, for example, ask users (or other stakeholders) to rank design options in terms of a value like safety; the result would be a subjective measurement of the value of safety.

In many cases, however, it is hard to measure values directly in a reliable way. In such cases, it may be more appropriate to first further operationalize the value at stake. Operationalization of values, in many cases, can be modelled as a two-stage process, namely: first, a translation of the value into a number of evaluation criteria that are important for judging whether the value is met (in the specific context considered); and second, the association of attributes, that can be more directly and objectively measured, with these criteria (Kroes and Van de Poel 2015). The attributes can, then, be seen as proxies for the attainment of the criteria. Figure 22.1 gives an impression of what the operationalization of values might look like in a concrete case (adopted from Kroes and Van de Poel 2015).

The case at hand is the design of alternative coolants for refrigerators. Values that are relevant in this case include environmental sustainability, safety and health. These values can be associated with certain evaluation criteria and attributes. For the attributes mentioned in the figure, there are measurement procedures and measurement scales that are, for example, laid down in relevant technical codes and standards.

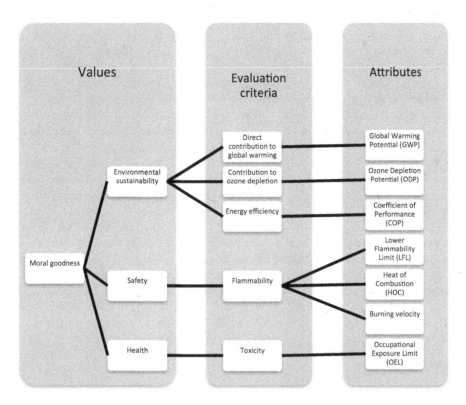

Figure 22.1   The operationalization of the values environmental sustainability, safety and health for the case of refrigerant coolants

Source: From Kroes and Van de Poel (2015).

A number of observations are in place with respect to this process of operationalization of values. First, it is context-specific. The operationalizations shown in Figure 22.1 are sensible for the case of coolants, but they would probably make no sense in the context of the design of another technology (like, for example, cars, where the same values would be relevant). Second, it involves so-called second-order value judgements, i.e. value judgements not about the technology designed (first-order value judgements), but about how to interpret and operationalize the relevant values in the specific context. Third, often more than one attribute will be associated with one value (as suggested in the figure). This means that these attributes are not only just proxies, but also that there is the question how to aggregate the different attributes into one measure for the relevant value. Obviously, any attempt to answer this aggregation question would involve a range of further value judgements (see also Chapter 7, 'Engineering Design and the Quest for Optimality' by Maarten Franssen, this volume).

Although the operationalization of values involves value judgements, and often different operationalizations are justifiable, it does not follow that operationalization is arbitrary. Not any operationalization will do, and in practice there are often operationalizations available for various technological domains, for which there is a degree of social consensus; for example, those laid down in technical codes and standards (see also Chapter 42, 'Standards in Engineering' by Paul B. Thompson, this volume). Of course, this does not necessarily imply that these operationalizations are also philosophically and morally justified, or are beyond debate. There may sometimes be good reasons to change an operationalization of a value in the light of new technological or social developments.

22.4.3 Specification

A third relevant activity to make values bear on design processes is specification. This process involves translating values into more concrete norms and design requirements that can guide the design of new technology. A useful tool here might the so-called values hierarchy (Van de Poel 2013). Figure 22.2 shows an example of a values hierarchy.

A values hierarchy consists of three main layers, i.e. values, norms and design requirements; each layer may in turn have a number of sublayers. A values hierarchy is held together by two relations. Top-down, this is the relation of specification, i.e. lower-level elements in the values hierarchy are specifications of higher-level elements. Specification is a non-deductive reasoning process, in which contextual information is taken into account in specifying what a higher level element (like a value) implies in a specific context. Bottom-up, the relation between the elements can be characterized as "for the sake of", i.e. the lower-level elements are strived for the sake of higher-level elements. For example in Figure 22.2, the design requirement that biofuels should (preferably) be based on non-edible crops is desirable for the sake of avoiding an increase in food process through competition, which, in turn, is desirable for the sake of intergenerational justice.

As suggested by these two relations, a values hierarchy can be construed top-down as well as bottom-up. In the first case, we start with values that are then specified in terms of norms and design requirements. In the second case, we can start, for example, with already formulated design requirements for a design task, and ask the question for the sake of what are these design requirements strived for, and so reconstruct the underlying values. In practice, the construction of a values hierarchy will usually be an iterative process, consisting of moving bottom-up as well as top-down. For example, if values are first reconstructed bottom-up (on basis of a set of given design requirements), we can then ask the question whether the set of design requirements is indeed the most appropriate specification of these values, or should perhaps be adapted.

Since specification is non-deductive, the translation of values into design requirements involves second-order value judgements (like in the case of value operationalization), and often different specifications of a set of values may be justifiable. It should be noted that a values hierarchy as such does not answer the question how to make such value judgements, or which specification to choose from a set of justifiable but competing specifications. Nevertheless, the values hierarchy might still be useful to locate where value judgements are to be made, and to track possible disagreements about the specification of values. As such, the values hierarchy is perhaps better seen as a deliberative tool, rather than a straightforward method to specify values.

As in the case of operationalization, not any value specification will do. In fact there are a number of criteria to judge the adequacy of a specification (Van de Poel 2013). For example, one crucial issue is whether meeting the design requirements would count as an instance of meeting the underlying value.

22.4.4 Value Conflicts

In most design processes, not just one but a range of values is relevant. Oftentimes, these values will conflict. Value conflicts can take many forms, but we can distinguish two main varieties:

1. Conflicts between values, i.e. situations in which it is not possible to fully realize the various relevant values at the same time in the design of a technology. In other words, the realization of one value comes at the cost of another value. For example, safety and sustainability may be conflicting in car design, because lightweight cars usually consume less energy and are therefore more sustainable. However, in order to make a really lightweight car, it may be necessary not to include certain safety features (that often add quite a lot of weight), so that sustainability is achieved at the cost of safety. We will call this types of conflicts multiple-value conflicts.

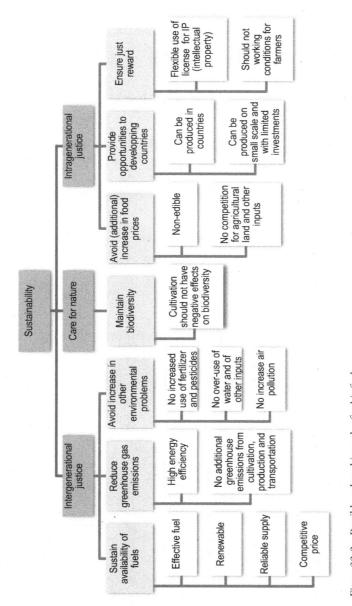

Figure 22.2   Possible values hierarchy for biofuels

Note: Values are dark gray; norms are light gray and design requirements are white

Source: From Van de Poel (2017b).

2. Conflicts between agents, i.e. situations in which agents disagree about the values. One form this may take is that agents attach different priorities (or weights) to the relevant values. In such cases, different agents may prefer differ solutions to multiple-value conflicts. However, agents may as well disagree about what values are relevant for a design, or about the conceptualization, operationalization and specification of these values. We will call this type of conflicts multiple-agent conflicts.

In reality, value conflicts will often be multiple-value as well as multiple-agent. Nevertheless, in some cases, it may be justified to treat them primarily as either multiple-value or multiple-agent conflicts. For example, if we consider the situation that one designer is designing for a multiplicity of values, it may be appropriate to treat this primarily as a multiple-value conflict. Similarly, if we take the perspective of the government, which has the responsibility to safeguard certain public values in the design of technology, it may also be proper to treat value conflicts primarily as multiple-value conflicts.[2] In other situations, it may be more appropriate to treat the conflict primarily as a multiple-agent conflict; for example, the design of large infrastructures that impact a large number of stakeholders, and for which the support of a wide diversity of agents is required for the success of the project.

There are many approaches to deal with value conflicts; overviews are presented in Van de Poel (2009, 2015). Here I will restrict myself to three main categories of strategies that can be applied to multiple-value as well as multiple-agent conflicts. These three strategies are: (1) creating a win/win option, (2) compromise and (3) finding an integrated solution.

22.4.5 Win/Win

In case of multiple-value conflicts, a win/win option is an option that improves the current situation in terms of all relevant values, and, similarly, in case of a multiple-agent issue, it is an improvement for all agents involved.[3] At the start of the design process, such win/win options may not exist, or at least not be evident. However, since design is a creative and innovative process that may result in *new options*, it may be seen as the challenge for the designer(s) to create such win/win options, in order to solve the value conflict. Van den Hoven et al. (2012) argue that designers have a moral obligation to look for such win/win options in particular if a multiple-value conflict amounts to a moral dilemma.

It should be noted that the creation of a win/win option for multiple-agent conflicts does not necessarily require that all the agents agree on the relevant values (or on the priority of these values, and their conceptualization, operationalization and specification). It 'only' requires that there is (at least) one option that, for all relevant agents, is an improvement compared to the current situation. The values, on basis of which the individual agents decide whether this is the case, may differ from agent to agent. Moreover, for each individual agent, the win/win option may still be a compromise, in the sense that it does not fully meet the values of that agent but still is considered an improvement (with respect to the current situation) for that agent. For that reason, a win/win option in terms of agents may not be a win/win option in terms of values.

Since win/win options for multiple-value conflicts do not coincide with win/win options for multiple-agent conflicts, designers who aim at creating a win/win option should first answer the question whether the value conflict at hand is best approached as a multiple-agent or a multiple-value conflict (or perhaps as both). Both approaches can be sensible, but each comes with its own pitfalls of which the designers should be aware. Generally speaking, the main pitfall of treating a value conflict solely as a multiple-value conflict is that, perhaps, an ethically acceptable solution (in terms of the relevant values) may be found, but that this solution is not accepted by the relevant agents, and therefore not executed. Conversely, if the conflict is solely treated as a multiple-agent issue, the resulting win/win solution may be accepted by all parties, but it may be ethically unacceptable, or it may neglect the legitimate interest of agents that cannot be involved (like future generations). Ideally a win/win solution, then, both implies an improvement in terms of the relevant values, as well as an improvement for all relevant agents, but obviously such ideal solutions may often be hard, if not impossible, to find even for the most creative designers.

22.4.6 Compromise

A compromise is a solution to a value conflict, in which either one (or more) of the relevant values is compromised (in case of a multiple-value conflict) or the perspective (or interest) of one (or more) of the agents is compromised (in case of a multiple-agent conflict). Because of this characteristic, compromises are often seen as less than ideal solutions to value conflicts. But, generally speaking, compromises are better than solutions, in which just one value or one agent dictates the solution, or in which a solution is reached by means of violence. So, even if compromises are less than ideal, they may the best that can be achieved.

There are many approaches to reach a compromise in a multiple-value or multiple-agent conflict. For multiple-value conflicts, compromising may often involve trading off some value(s) against one another, as is done, for example, in multiple criteria decision-making (MCDM) but also in many more specific engineering methods like Quality Function Deployment (QFD) or Pugh charts (e.g. Pugh 1991; Bogetoft and Pruzan 1991; Hauser and Clausing 1988). There are, however, also balancing methods for values that do not involve (direct) trade-offs between values (Van de Poel 2015). For multiple-agent conflicts, achieving a compromise will often involve some form of negotiation between the relevant agents.

Although compromises are sometimes unavoidable (or to be preferred to dominated or violent solutions), in general, compromises raise two further issues. One is that a specific compromise may be ethically unacceptable. In particular, negotiation in a multiple-agent conflict may result in solutions that are accepted by the most powerful agents but that either neglect important values or are achieved at the cost of less powerful agents. Another issue is that a compromise may be an unstable solution. The reason for this is that in case of a compromise, there is always at least one value or actor perspective compromised; however, the reasons for that value or agent perspective are typically not annulled (Chan and Protzen 2018). In other words, from the viewpoint of the value or agent that is compromised, there remain good reasons to reject the compromise, once there is the opportunity to do so.

22.4.7 Integrated Solution

An integrated solution will here be understood as a solution to a value conflict that does justice to all relevant values (in case of a multiple-value conflict) or to all relevant agents (in case of a multiple-agent conflict). So, contrary to a compromise, there is not a value or agent perspective compromised, and hence there is no remaining reason to undo the solution, as in the case of a compromise. At the same time, an integrated solution does not necessarily create a win/win situation, at least not in terms of the original value or agent perspectives that defined the original value conflict. The reason for this is that achieving an integrated solution involves *deliberation* about the values and the value conflict and may lead to a reformulation of the original value conflict, for example by reinterpreting or respecifying the relevant values, or because agents revise their perspective. This reformulation or reframing of the original value conflict may open the way to solutions that are acceptable in the light of all relevant values (in case of a multiple-value conflict) or to all agents (in case of a multiple-agent conflict), thus leading to a consensus solution rather than a compromise.

One possibility for the case of multiple-value conflicts is that deliberation results in a *respecification* of the relevant values in terms of design requirements. In multiple-value conflicts, often the values are not directly conflicting but, rather, the specifications of these values (in terms of design requirements) are conflicting. Because often different specifications of a value are rationally and morally defensible, it may be possible to respecify the values in such a way that the resulting design requirements are no longer conflicting, while the values are still all respected.

In case of multiple-agent conflicts, agents may need to adopt a solution-oriented attitude rather than a competing, or interest-maximizing, perspective in order to achieve an integrated solution (Chan and Protzen 2018; Carens 1979). It may particularly be helpful if agents come to a shared understanding of the value conflict, for example through reframing the problem, which may require agents to reflect on, and be willing to adapt, their own belief and value systems (Van de Poel and Zwart 2010). An integral solution, however, does not require that all agents adopt the same perspective. The result may also be what John Rawls has called an "overlapping consensus" (Rawls 2001). That is consensus (in this case about a solution to a value conflict) that each agent can morally justify in terms of his or her moral belief system, without all agents agreeing on the underlying moral belief system.

Achieving an integral solution may of course not always be possible. Nevertheless, the process of looking for one may trigger deliberation and lead to reflection on the relevant values and agent perspectives, which in itself may not only make it easier to achieve a compromise but also lead to better compromises (in terms of meeting a number of values and agent perspectives). At the same time, too much emphasis on consensus may result in an unjustified watering down of values, or may create too much pressure on agents to revise their perspective. It is important to be aware that (value) conflict can also be constructive in the sense that it may reveal hidden or marginalized values or perspectives on the issue (Rip 1987; see also Chapter 45, 'Engineering as a Political Practice' by Govert Valkenburg, this volume).

The three discussed strategies are ideal types, in the sense that real-world approaches and solutions to value conflicts will contain elements of all three approaches. Moreover, in different phases of the design process, different approaches may be more appropriate (Van de Poel 2017a). For example, once a first formulation of the design problem has settled and a value conflict surfaces, designers should first look for possible win/win solutions. This invites designers to think out of the box and to look for possibly better design options than they were initially considering. This may result in better but not yet ideal solutions, and a next phase may involve further deliberation to see whether an integrated solution can be found. When this is not possible, the designer may settle for a compromise. The point is that even if the eventual solution is a compromise it is likely to be a better compromise (in terms of values and agent perspectives respected) than if the designers had from the start aimed at a compromise.

22.5. Value Change

Studies in the philosophy of technology have shown that sometimes new technologies induce value change (Boenink et al. 2010; Swierstra 2013). The classical example is contraceptives that have triggered a change in sexual morality (Boenink et al. 2010; Swierstra 2013). New values may also emerge in order to deal with the sometimes unexpected consequences of human action and technology. An example is sustainability that may be seen as a response to the human endangerment of the natural environment.

Value changes after a technology has been designed may create a mismatch between the values for which a technical artifact or a sociotechnical system was originally developed and which are embedded in the technology, and the values we currently consider important. An example is that of energy systems, which according to many insufficiently reflect the value of sustainability, which was a less important, or even absent, value when these systems were originally developed. However, due to the large technological and institutional momentum of energy systems, these systems cannot easily be adapted to better meet the value of sustainability. This has led to discussions about the need for an energy transition, which not only is an economic, technical and institutional challenge but also requires a change in the values embedded in these systems (Demski et al. 2015).

To understand how values may change, we may understand values as emerging from earlier responses to moral problems. In line with pragmatist philosophers like Dewey, values can be seen as generalized responses to earlier moral problems. In many situations, existing values are adequate as a response to (morally) problematic situations people encounter. However, in new types of situations or due to new experiences, current values may no longer be adequate or sufficient. Such situations may require an adaptation of current values or the adoption of new values (Dewey 1922).

This suggests that values may change in response to new problematic situations or new experiences. One might, for example, think of the following possibilities:

- Technologies lead to new types of *consequences* that require new evaluative dimensions and therefore new values (e.g. privacy, sustainability) to evaluate sociotechnical systems.
- Technologies offer new *opportunities* (e.g. to protect homes against earthquakes) that lead to new moral obligations and therefore new values.
- Technologies create new *moral choices and dilemmas* where previously there were no choices (e.g. predictive genetics) that require new values.
- Technologies lead to new *experiences* (e.g. friendship online) that lead to new values or change existing values.

In the literature, there has until now been little attention for how to deal with value change, but one can think of different possible strategies. Anticipatory strategies try to predict or anticipate value changes. Currently most approaches do not try to predict value changes but rather develop, for example, technomoral scenarios to sketch different possible futures (Boenink et al. 2010). These can then be the basis for deliberation, and technologies may be so designed that they can deal with certain anticipated value changes or they may be designed in an attempt to steer value change in a certain direction. However, not all value changes can be predicted or even anticipated (Van de Poel 2017c). Not only may values change in other ways than anticipated, also how we perceive and morally evaluate value changes may be hard to anticipate beforehand. This means that anticipatory strategies alone are not enough to deal with value change. Partly we may deal with unanticipated value change by adaptive strategies. Possible adaptive strategies include the redesign of some of the technical parts of the sociotechnical system, redesign of some of the relevant institutions, adaptation of the operation of the sociotechnical system (operational strategies) or changing the behaviour of the actors in the system (behaviour strategies).

However, often it is better to be prepared for adaptation beforehand, even if the specific to-be-expected value changes cannot be anticipated. This may be done by employing a range of proactive design strategies, technical as well as institutional. One might think of design strategies or principles like robustness, flexibility, adaptability, and modularization.

22.6. Conclusions

That technology is value-laden is hardly controversial nowadays. This seems also to bring the obligation for engineers and designers to address values more systematically in the design processes of new technologies. I have discussed four (philosophical) challenges that such attempts face, i.e. (1) deciding what values to include in a design, (2) to make these values operational in the design, (3) to deal with conflicting values and (4) to deal with value change. For each, I discussed earlier work that has been done on the issue, and I presented some conceptual tools and approaches that might be useful for better dealing with these issues, theoretically as well as practically. It is, however, good to be aware that each of these issues cannot be completely solved in an abstract theoretical way. Each will require some form of moral judgement and (political) decision-making by those involved (i.e. the designers and relevant stakeholders) in the context of a concrete design process. Nevertheless, such judgement and decision-making are likely to be better if they are informed by some of the considerations offered in this contribution.

Acknowledgements

This publication is part of the project ValueChange that has received funding from the European Research Council (ERC) under the European Union's Horizon 2020 research and innovation programme under grant agreement No 788321.

Related Chapters

Chapter 7: Engineering Design and the Quest for Optimality (Maarten Franssen)

Chapter 20: Towards an Ontology of Innovation (Vincent Blok)

Chapter 21: Engineering Design (Peter Kroes)

Chapter 24: Human-Centred Design and Its Inherent Ethical Qualities (Marc Steen)

Chapter 25: Sustainable Design (Steven A. Moore)

Chapter 33: Values in Risk and Safety Assessment (Niklas Möller)

Chapter 34: Engineering and Sustainability: Control and Care in Unfoldings of Modernity (Andy Stirling)

Chapter 38: Health (Marianne Boenink)

Chapter 40: Ethical Considerations in Engineering (Wade L. Robison)

Chapter 43: Professional Codes of Ethics (Michael Davis)

Chapter 44: Responsibilities to the Public—Professional Engineering Societies (Joseph Herkert and Jason Borenstein)

Chapter 45: Engineering as a Political Practice (Govert Valkenburg)

Further Reading

Friedman, Batya and Kahn, Peter H. Jr. (2003). Human Values, Ethics and Design. In J. Jacko and A. Sears (eds.), *Handbook of Human-Computer Interaction*. Mahwah, NJ: Lawrence Erlbaum Associates, pp. 1177–1201. (Introduction to the VSD approach.)

Kroes, Peter, and Verbeek, Peter-Paul (eds.) (2014). *The Moral Status of Technical Artefacts*. Dordrecht: Springer. (Overview of different approaches how to understand the embedding of values in technology.)

Van den Hoven, Jeroen, Vermaas, Pieter E. and Van de Poel, Ibo (eds.) (2015). *Handbook of Ethics and Values in Technological Design. Sources, Theory, Values and Application Domains*: Springer. (Extensive overview of different approaches of designing for values for a variety of values in a multiplicity of domains.)

Winner, Langdon. (1980). Do Artifacts Have Politics? *Daedalus* (109), 121–136. (A classical on the value-ladenness of technology.)

Notes

1. One could interpret this as a definition of value in which there can be morally good and morally bad values; values are 'simply' what people consider important in life. In what follows, I will propose a definition of value that is more closely tied to what is good.
2. It should be noted that in such cases, distributive justice may be one of the values, so that ethical considerations about the distributions of advantages and disadvantages among stakeholders are taken into account even if the conflict is approached as a multiple-value conflict.
3. More formally, we may want to employ the notion of Pareto Improvement from economics and choice theory (see also Chapter 12, 'Engineering Design and the Quest for Optimality' by Christian Dieckhoff and Armin Grunwald, this volume). For a multiple-agent setting, a Pareto Improvement is defined as an

improvement that makes at least one agent better off and the other agents not worse off. Similarly for multiple-value issues, we may define it as an improvement in which at least one value is better achieved and no values are compromised.

References

Akao, Y. (ed.) (1990). *Quality Function Deployment. Integrating Customer Requirements Into Product Design*. Cambridge, MA: Productivity Press.

Bhamra, Tracy, and Lofthouse, Vicky (2007). *Design for Sustainability: A Practical Approach*. Aldershot: Gower.

Birkeland, Janis (2002). *Design for Sustainability. A Source Book for Ecological Integrated Solutions*. London: Earthscan.

Boenink, Marianne, Swierstra, Tsjalling and Stemerding, Dirk (2010). Anticipating the Interaction Between Technology and Morality: A Scenario Study of Experimenting With Humans in Bionanotechnology. *Studies in Ethics, Law, and Technology*, 4(2), 1–38.

Bogetoft, Peter, and Pruzan, Peter (1991). *Planning With Multiple Criteria: Investigation, Communication, Choice*. Amsterdam: Elsevier.

Borning, Alan, and Muller, Michael (2012). Next Steps for Value Sensitive Design. In *Proceedings of the SIGCHI Conference on Human Factors in Computing Systems*, Austin, TX, USA.

Carens, Joseph H. (1979). Compromises in Politics. *Nomos*, 21, 123–141.

Chan, Jeffrey Kok Hui, and Jean-Pierre Protzen (2018). Between Conflict and Consensus: Searching for an Ethical Compromise in Planning. *Planning Theory*, 17(2), 170–189. doi:10.1177/1473095216684531.

Crisp, Roger (2013). Well-Being. In Edward N. Zalta (ed.), *The Stanford Encyclopedia of Philosophy*, Summer 2013 ed. https://plato.stanford.edu/archives/sum2018/entries/it-privacy

Dancy, Jonathan (1993). *Moral Reasons*. Oxford: Blackwell Publishers.

Dantec, Christopher A. Le, Shehan Poole, Erika and Wyche, Susan P. (2009). Values as Lived Experience: Evolving Value Sensitive Design in Support of Value Discovery. In *Proceedings of the 27th International Conference on Human Factors in Computing Systems*, Boston, MA, USA.

Davis, Janet, and Nathan, Lisa P. (2015). Value Sensitive Design: Applications, Adaptations and Critiques. In Jeroen van den Hoven, Pieter E. Vermaas and Ibo Van de Poel (eds.), *Handbook of Ethics and Values in Technological Design*. Dordrecht: Springer, pp. 11–40.

Demski, Christina, Butler, Catherine, Parkhill, Karen A., Spence, Alexa and Pidgeon, Nick F. (2015). Public Values for Energy System Change. *Global Environmental Change*, 34, 59–69. doi:10.1016/j.gloenvcha.2015.06.014.

Desmet, Pieter, and Hekkert, Paul (2007). Framework of Product Experience. *International Journal of Design*, 1(1), 57–66.

Dewey, John (1922). *Human Nature and Conduct: an Introduction to Social Psychology*. New York: Holt.

Ellul, Jacques (1964). *The Technological Society*. Translated by John Wilkinson. New York: Alfred A. Knopf. Original edition, La Technique.

Flanagan, Mary, Howe, Daniel C. and Nissenbaum, Helen (2008). Embodying Values in Technology. Theory and Practise. In Jeroen Van den Hoven and John Weckert (eds.), *Information Technology and Moral Philosophy*. Cambridge: Cambridge University Press, pp. 322–353.

Friedman, Batya, Hendry, David G. and Borning, Alan (2017). A Survey of Value Sensitive Design Methods. *Found. Trends Hum.-Comput. Interact*, 11(2), 63–125.

Friedman, Batya and Kahn, Peter H. Jr. (2003). Human Values, Ethics and Design. In J. Jacko and A. Sears (eds.), *Handbook of Human-Computer Interaction*. Mahwah, NJ: Lawrence Erlbaum Associates, pp. 1177–1201.

Friedman, Batya, Kahn, Peter H. Jr. and Borning, Alan (2006). Value Sensitive Design and Information Systems. In Ping Zhang and Dennis Galletta (eds.), *Human-computer Interaction in Management Information Systems: Foundations*. Armonk, NY: M.E. Sharpe, pp. 348–372.

Griffin, James (1986). *Well-being: Its Meaning, Measurement, and Moral Importance*. Oxford [Oxfordshire]; New York: Clarendon Press.

Hauser, John R. and Don Clausing (1988). The House of Quality. *Harvard Business Review*, 66(3), 63–73.

Hechter, Michael (1993). Values Research in the Social and Behavioral Sciences. In R. Michod, L. Nadel and M. Hechter (eds.), *The Origin of Values*. Berlin: Aldine de Gruyter, p. 28.

Heidegger, Martin (1962). *Die Technik und die Kehre*. 7. Aufl. ed. Pfullingen: Neske.

Hirose, Iwao, and Olson, Jonas (2015). *The Oxford Handbook of Value Theory*. New York: Oxford University Press.

Keates, Simeon, and Clarkson, John (2003). *Countering Design Exclusion: An Introduction to Inclusive Design*. London: Springer.

Koops, Bert-Jaap, Bryce Clayton Newell, Tjerk Timan, Ivan Škorvánek, Tomislav Chokrevski, and Maša Galič (2017). A Typology of Privacy. *University of Pennsylvania Journal of International Law*, 38(2), 483–575.

Koskinen, Ilpo, Battarbee, Katja and Mattelmäki, Tuuli (eds.) (2003). *Emphatic Design. User Experience in Product Design.* Helsinki: IT Press.

Kroes, Peter, and Poel, Ibo van de (2015). Design for Values and the Definition, Specification, and Operationalization of Values. In Jeroen van den Hoven, Pieter E. Vermaas and Ibo Van de Poel (eds.), *Handbook of Ethics, Values, and Technological Design: Sources, Theory, Values and Application Domains.* Dordrecht: Springer, pp. 151–178.

Kroes, Peter, and Verbeek, Peter-Paul (eds.) (2014). *The Moral Status of Technical Artefacts.* Dordrecht: Springer.

Manders-Huits, Noëmi (2011). What Values in Design? The Challenge of Incorporating Moral Values Into Design. *Science and Engineering Ethics,* 17(2), 271–287. doi:10.1007/s11948-010-9198-2.

Moore, G.E. (1922). *The Conception of Intrinsic Value. Philosophical Studies.* London: Kegan Paul, Trench, Trubner, pp. 253–275.

Pugh, Stuart (1991). *Total Design: Integrated Methods for Successful Product Engineering.* Wokingham, England and Reading, MA: Addison-Wesley Pub. Co.

Rawls, J. (2001). *Justice as Fairness. A Restatement.* Cambridge, MA: The Belknap Press of Harvard University Press.

Raz, Joseph (1999). *Engaging Reason. On the Theory of Value and Action.* Oxford: Oxford University Press.

Rip, A. (1987). Controversies as Informal Technology Assessment. *Knowledge: Creation, Diffusion, Utilization,* 8(2), 349–371.

Rokeach, M. (1973). *The Nature of Human Values.* New York: The Free Press.

Schwartz, Shalom H. and Wolfgang Bilsky (1987). Toward a Universal Psychological Structure of Human Values. *Journal of Personality and Social Psychology,* 53(3), 550–562. doi:10.1037/0022-3514.53.3.550.

Swierstra, Tsjalling (2013). Nanotechnology and Technomoral Change. *Etica & Politica/Ethics & Politics,* XV(1), 200–219.

Van de Poel, Ibo (2009). Values in Engineering Design. In Anthonie Meijers (ed.), *Handbook of the Philosophy of Science. Volume 9: Philosophy of Technology and Engineering Sciences.* Oxford: Elsevier, pp. 973–1006.

Van de Poel, Ibo (2012). Can We Design for Well-being? In Philip Brey, Adam Briggle and Edward Spence (eds.), *The Good Life in a Technological Age.* London: Routledge, pp. 295–306.

Van de Poel, Ibo (2013). Translating Values into Design Requirements. In D. Mitchfelder, N. McCarty and D.E. Goldberg (eds.), *Philosophy and Engineering: Reflections on Practice, Principles and Process.* Dordrecht: Springer, pp. 253–266.

Van de Poel, Ibo (2015). Conflicting Values in Design for Values. In Jeroen van den Hoven, Pieter E. Vermaas and Ibo Van de Poel (eds.), *Handbook of Ethics, Values, and Technological Design.* Dordrecht: Springer, pp. 89–116.

Van de Poel, Ibo (2017a). Dealing with Moral Dilemmas through Design. In Jeroen van den Hoven, Seumas Miller and Thomas Pogge (eds.), *Designing in Ethics.* Cambridge: Cambridge University Press, pp. 57–77.

Van de Poel, Ibo (2017b). Design for Sustainability. In David M. Kaplan (ed.), *Philosophy, Technology, and the Environment.* Cambridge, MA: MIT Press, pp. 121–142.

Van de Poel, Ibo (2017c). Moral Experimentation with New Technology. In Ibo Van de Poel, Donna C. Mehos and Lotte Asveld (eds.), *New Perspectives on Technology in Society: Experimentation Beyond the Laboratory.* London: Routledge.

Van de Poel, Ibo, and Zwart, Sjoerd D. (2010). Reflective Equilibrium in R&D Networks. *Science, Technology & Human Values,* 35(2), 174–199.

Van den Hoven, Jeroen, Lokhorst, Gert-Jan and Van de Poel, Ibo (2012). Engineering and the Problem of Moral Overload. *Science and Engineering Ethics,* 18(1), 143–155. doi:10.1007/s11948–011–9277-z.

Van den Hoven, Jeroen, Vermaas, Pieter E. and Van de Poel, Ibo (eds.) (2015). *Handbook of Ethics and Values in Technological Design. Sources, Theory, Values and Application Domains.* Dordrecht: Springer.

von Wright, Georg Henrik (1963). *The Varieties of Goodness.* London: Routledge & Kegan Paul.

Warren, Samuel D. and Brandeis, Louis D. (1890). The Right to Privacy. *Harvard Law Review,* 4(5), 193–220. doi:10.2307/1321160.

Winner, Langdon (1980). Do Artifacts Have Politics? *Daedalus* (109), 121–136.

Zimmerman, Michael J. (2004). Intrinsic vs. Extrinsic Value. In Edward N. Zalta (ed.), *The Stanford Encyclopedia of Philosophy,* Fall 2004 ed. http://plato.stanford.edu/archives/fall2004/entries/value-intrinsic-extrinsic/.

23

DESIGN METHODS AND VALIDATION

Sabine Ammon

23.1. Introduction

Any handbook entry on design methods is bound to face a serious challenge given that the notion of design methods embraces a fundamental tension. A method, broadly understood, is a set of systematic rules that are applied in order to achieve a certain result. Given that artefacts and processes are the result of successful design activities, design methods must be defined as a set of systematic rules that are applied in design praxis in order to arrive at a new artefact or process. However, design is, above all, a creative activity: it involves intuition, spontaneity and subjectivity (see Chapter 10 "Creativity and Discovery in Engineering" by David H. Cropley, this volume). Methods, by contrast, need to be rational, controlled and objective. The very claim that there are or should be design methods implies a specific approach to design, namely, the assumption that the intuitive, spontaneous and subjective design process can somehow be restrained and transformed in a systematic way in order to arrive faster at better results. It comes as no surprise that this approach is controversial in the design disciplines and that the history of design is a history of the rise and fall of design methods.

The heyday of the Design Methods Movement in the 1960s continues to have an impact on discourse and practices even today. A product of postwar optimism and of a steadfast belief in science-based progress, the movement emerged from a visionary melting pot in which the design disciplines intermingled with cybernetics, computer science and artificial intelligence (Cross 2006; Langrish 2016). Its overambitious aims envisaged an intelligent computer system capable of solving design problems automatically. What was completely ignored, however, was the complexity of actual design practices. This heavily technicized approach led to the movement's precipitous decline just one decade later, triggered largely by its former proponents, many of whom had by then become its most eminent critics (Section 3).

The ramifications of this failure for the different design disciplines could not have been more diverse. Architecture in particular developed a profound mistrust of design methods. This 'no methods' backlash not only prompted an exaggerated emphasis on the creative act of the inspired star designer; it also strengthened the role of the user by introducing participatory elements. The latter response drew on the idea that good design can originate from almost anyone, provided they are given the right set of tools. Creative techniques supported this trend, which spread from product and communication design into other areas. Many of these techniques focus on the creative potential of groups to generate ideas collaboratively, especially in the early phases of a design. None of them, however, serves to manage the overall design process; rather, they support specific steps within the process.

It was primarily in engineering where comprehensive design methods gained ground again. A new generation of methods flourished in the 1980s and 1990s which proved to be far more successful than those of the movement's first generation and led to an influential research tradition in the German-speaking context especially. This second generation of methods adheres to the idea that a specific design process can be enhanced by employing certain generic methods, but it abandons the pretence of being able to do this automatically by means of an intelligent computer system. This novel approach has yielded so-called process models which structure the design process in a systematic way (Section 4) and stipulate that the quality and maturity of the (interim) results be examined using methodical testing, an approach discussed in terms of the concepts of verification and validation (Section 5). That it was engineering which fuelled the advance of process models can be explained by the prevalence of large and complex projects in this field. A framework that structures activities and reduces risks is clearly of much greater use to a group of collaborators striving to develop multiple components within a rigid set of time and cost restrictions than to an individual designer working on a clearly defined project.

The collapse of the Design Methods Movement's aspirations prompted a profound paradigm shift in computer systems: they were no longer meant to *replace* but to *support* the designer. The vision of a computer system functioning as an operator and being assisted by the designer switched to a principle that envisages the designer as an operator being assisted by a computer system. This principle still informs today's computer-based design tools and is a significant factor in their proliferation. Here, the relation between design tools and methods is less pronounced and yet effectual: design tools impose specific working routines on a given course of action. Black-boxed within the tool are constraints that structure the conceptual arena in which design thinking takes place. It is through the 'back door' of tool design, then, that methods enter creative processes (Section 6).

This leads ultimately to the question of where we stand right now, given the to-and-fro of the 'methods—no methods—minimal methods' swing (Section 7). However, before we can start to explore these issues in more detail, an implicit assumption in the concept of design methods needs to be discussed first. The concept of design methods implies an epistemology of design, namely, the claim that we gain insights through the praxis of designing, that this praxis is a way of reasoning and thinking, and that successful inquiry leads to knowledge. Only if one can reasonably argue that designing is not only intuitive and irrational but also contributes toward a given epistemic interest is the foundation laid for usefully applying (scientific) design methods (Section 2).

23.2. The Epistemology of Design

Whereas the primary aim of science is said to be that of generating new knowledge, the primary aim of design is usually described as that of creating new artefacts (such as products or processes). Just as scientific methods serve to make scientific endeavours as successful as possible by providing a rigorously reviewed, reliable and systematic course of action, design methods are similarly meant to contribute to successful design by offering a reliable and systematic course of action, although the element of review and the requirement of systematicity are usually less pronounced. Yet design is as much related to the generation of new knowledge as is science, albeit the epistemology of design is not immediately apparent. The design process is essentially about understanding a future, not-yet-existent artefact; this being so, designing itself must be conceptualized as an epistemic activity (Ammon 2017). A successful design process results in both specific and generic knowledge—about the artefact itself (specific artefact knowledge), about how to develop this kind of artefact (generic artefact knowledge) and about how to steer the process of development (processual knowledge). These insights, which remain implicit to a large extent, can be summarized as design knowledge that manifests subsequently in the completed product.

The question of what makes designerly activities distinctive from an epistemological point of view has been a key one in the emerging philosophy of the design sciences. Is design, which develops and shapes the components of new structures, fundamentally different from the research practices of scientists, which identify the components of existing structures? (Alexander 1964). Does design focus on inventing the not-yet-existent, whereas science focuses on problem-solving related to what already exists? (Gregory 1966). One of the most telling answers to this question was offered by Herbert Simon, a social and cognitive scientist who described the science of design as a "body of intellectually tough, analytic, partly formalizable, partly empirical, teachable doctrine about the design process" (Simon 1996 [1969]: 113). He goes on to outline a rough and simplified comparison which, nevertheless, works well as a heuristic for carving out important differences. According to Simon, the analytical approach (predominant in the natural sciences) is based on the epistemic activity of discovery; it explores how things are. Its formative element is the reduction of complexity, accompanied by practices of inquiry that are characterized by necessity on account of the researcher's subservience to natural laws (ibid.: xi). In contrast to this, the synthetic approach (predominant in the design sciences) is interested in how things might be or ought to be; it explores potential futures. Here, complexity is an integral ingredient, accompanied by practices of inquiry characterized by contingency due to the inherent malleability of the (design) environment (ibid.: xi; 3–5). What Simon's contrastive comparison makes clear is that the design disciplines—where the synthetic approach plays a fundamental role—constitute an independent domain of knowledge that has its own specific epistemic aims and methods.

The epistemological emancipation of the design sciences inevitably prompts the next question: what characterizes designing as a mode of thinking and reasoning? Often addressed as "design thinking",[1] the predominant paradigm used to describe this activity is problem-solving. However, a number of issues render this approach questionable. Design starts not with a well-structured problem, but rather with an indistinct set of aims and conditions. Hence, design is essentially a process of clarification in which the designers attempt to work out principles, drivers and constraints within a formative framework. It involves making trade-offs based on value judgements in order to respond to competing demands. Only from a retrospective point of view does the result of a successful design process appear as a solution to a problem—an effect often described as the co-evolution of problem and solution (Dorst and Cross 2001). However, if taken seriously, this interpretation would reduce the paradigm of problem-solving to absurdity.

With his focus on actual design practices, philosopher and urban planner Donald Schön established a new perspective on the epistemology of design. Based on ethnographic observations, he investigated designing in terms of "reflection-in-action" (Schön 1983). According to Schön, design artefacts—such as sketches, drawings, plans and models—enter into dialogue with the designer in the very process of their generation. They trigger a thinking process which in turn stimulates further design moves. In this way, a web of inferences unfolds which allows the designer to frame their design problem in a new way, to try out variations, to explore consequences, to engage in comparisons and to draw (virtual) conclusions.

This shift from normative to descriptive design research not only allows for a better understanding of the epistemology of design but also demonstrates the role of methods in knowledge generation. With explorations in design being the art of investigating the not-yet-existent, the methods used to do so need to ensure that it is possible to pursue and stabilize emerging knowledge without having a specific goal. Hence, knowledge generation in design is tied to 'substitute systems' which configure facets of the future artefact for exploration. Based on sketching, drawing, modelling and tinkering, these exploratory quests are embedded in milieus of reflection which entail specific constraints and affordances of the medium deployed (Ammon 2018; see also Ferguson 1992; Ammon and Capdevila-Werning 2017). Due to these restrictions, different milieus of reflection become interwoven with one another, and a plurality of design methods makes it possible to cross-check interim results. The outcomes of these tests are fed iteratively into the design process and thereby facilitate the growing maturity of the future artefact. Thus, methods contribute essentially to an epistemology of design and to emerging design knowledge.

23.3. A Brief History of Design Methods

The desire for systematized procedures of design and innovation is anything but new. Simple machines such as the lever, wedge, wheel and axle, inclined plane, pulley, and screw, out of which any machine could be constructed, can be traced as far back as Archimedes. In architecture, descriptions of the basic elements and rules of composition have a long-standing tradition that includes, not least, works by Vitruvius and Leon Battista Alberti. An early conveyor of a general theory was Gottfried Wilhelm Leibniz, who proposed an *ars inveniendi* based on a principle that remains in use to the present day. It begins with the instruction to break down a problem into its component parts in order to address them separately. On this basis it then becomes possible, using the laws of association, to fashion a systematic synthesis, one guided by general principles and performed using a set of symbols. Leibniz's approach highlights the challenges of any design method. Are complex design problems better dealt with when they are broken down into parts, or does this impede a holistic approach? Which symbol systems can be used to represent and work on design issues in a way that is both appropriate and satisfactory? To what extent can design methods be outlined in a general way? At the beginning of the 18th century, Enlightenment philosopher Christian Wolff set a first counterpoint. He maintained that design methods need to be domain-specific if they are to be applied at all. In order to show this, Wolff developed a proof of concept for the field of architecture. His *ars inveniendi speciales* introduced heuristic rules for linking general strategies to the requirements of a specific domain (Poser 2013: 138–146).

Another 160 years would pass until the first influential account of design methods was published for the field of engineering. Franz Reuleaux's *Kinematics of Machinery* set out a logic of construction that makes it possible to deduce new designs based on foundational elements such as kinematic pairs, chains and mechanisms (Reuleaux 1876). Increasingly complex patterns of movement and machine systems can thus be generated by applying rules of inference and drawing on a specific symbolic system. Reuleaux's highly theoretical approach prompted the question of the extent to which design methods should be formalized and rendered in scientific terms. This pivotal question culminated in a dispute at the *Königlich Technische Hochschule zu Berlin* (today's TU Berlin) about the right to confer doctorates. Reuleaux's opponent, Alois Rielder, likewise a professor of mechanical engineering, represented the "practitioner's movement", a group of influential construction engineers who opposed any overemphasis on theory, physics or mathematics in education. In the end, the "theorist" Reuleaux was forced to give way to the "practitioner" Riedler and the latter's research ideals, which consisted in practical, laboratory-based work with minimal methodic input: from 1899 onwards, engineering students became eligible to receive doctorates (König 2013: 420–430; see also Chapter 5, "What Is Engineering Science?" by Sven Ove Hansson, this volume).

Design methods began to gain broader recognition in the postwar period with the emergence of approaches intended to support the ideation process; many of these are still in use today. While Alex Osborne's *brainstorming* techniques make use of intuition to produce a wide variety of ideas, especially in teams (Osborn 1953), Fritz Zwicky's morphological analysis explores design in a more systematic way (Zwicky 1959, 1969). His matrix, the so-called *morphological box*, enables the designer to identify the dimensions of a design task (as separate features or parameters). Once a variety of instantiations have been identified for each dimension, the combination of selected instantiations leads the way toward a new design. An approach that goes further than this is *TRIZ*, the Theory of Inventive Problem Solving, by Genrich Altshuller and Raphael Shapiro (Altshuller and Shapiro 1956; Altshuller 1984). This works not just with a combinatorial approach on the basis of parameters but rather with *principles of invention*. From their review of numerous patents, Altshuller and Shapiro derived 40 principles (including division, taking out, merging and inversion) which enable contradictions rooted in the technical parameters of design to be resolved. Once resolved, these contradictions in turn lead toward a new solution.

Starting with a series of conferences in the 1960s, the Design Methods Movement triggered a rational turn in the design process (Jones and Thornley 1963; Archer 1965; Jones 1970). The aim of the movement was to achieve a scientization of design by adopting new approaches in problem-solving, cybernetics and systems theory. By highlighting the potential of computerization, especially the promise of the emerging field of artificial intelligence, proponents of the movement stoked high hopes regarding the automation of design processes. A key figure was Herbert Simon, mentioned earlier, whose approach grew out of his interest in complex systems and decision-making in organizations. Driven by the vision that there could be a general mechanism for solving problems, he framed designing as a form of problem-solving and developed, along with computer scientists John Shaw and Allen Newell, the computer programme "general problem solver" (GPS). Although this programme, which established an affinity between design methods and artificial intelligence, was able to solve simple problems, it failed to solve complex real-world problems. In the course of his work, Simon coined two influential concepts which were to prove fruitful for design theory and the development of design methods. The first of these, "bounded rationality", refers to decision-making in design processes being based on limited rationality due to the complexity of the problem at hand as well as individual cognitive limitations and available time. Second, his neologism "satisficing" means finding solutions that are good enough: instead of being optimal, these solutions both satisfy and suffice (Simon 1996 [1969]).

Less well known, but worth mentioning in the context of this *Handbook*, is the work of Johannes Müller who, after studying philosophy, became a leading methodologist for engineering design in the former East Germany. In the spirit of the Design Methods Movement, he envisioned an "ideal designer" in the form of a computer programme capable of representing the inner principles of the thinking process in the form of algorithms, unhampered by psychological, social or aesthetic limitations (Müller 1966; Heymann 2005: 278–280). Müller's disillusionment, after years of trying to implement his *systematic heuristic* in design praxis without any notable success, gave way to a new understanding of design methods. Instead of regarding design as fully computable and rationally reducible, Müller instead characterizes it in terms of incomplete information, approximation (due to the need to reconcile the diverse requirements of different design elements), value-based stipulations (in response to an ambiguous context) and a great variety of features (whose consequences can be predicted only imperfectly). Now, it is the human designer and not the machine who, in this diffuse situation, is able to think in holistic clusters and to switch between different levels and component problems (Müller 1990: 11–12).

Many first-generation methodologists in the movement radically transformed their views on design. A well-known example is mathematician and architect Christopher Alexander. In *Notes on the Synthesis of Form* (Alexander 1964) he was still concerned with defining independent design patterns by formal means, an approach he later disowned, thus distancing himself from the Design Methods Movement. His new approach resulted in a design manual entitled *A Pattern Language* (Alexander et al. 1977). Alexander and his team transform space into a complex network of 253 elementary patterns and their relations, which can be modified or enlarged in scope. The manual not only provides a language to describe potential towns, buildings and modes of construction but also serves as a design method at the same time. Sub- and superordinate patterns enable the reader to grasp each design task individually. As each pattern contains both a description of and a solution to the problem, laypeople too are enabled to participate actively in planning processes (see also Chapter 24, "Human-Centred Design and Its Inherent Ethical Qualities" by Marc Steen, this volume).

An influential new framing of design formed after the decline of the Design Methods Movement stems from design theorist Horst Rittel and urban designer Melvin Webber. Their critique takes its cue from "tame" or well-defined problems, which "are definable and separable and may have solutions that are findable", such as solving an equation in mathematics or analysing an unknown compound in organic chemistry (Rittel and Webber 1973: 160). These are set in contrast to "wicked problems", typically found in planning and design in general, which are not solvable as there is no exhaustive formulation of the problem itself (ibid.: 161): given that clarification of the problem is part of its solution, it is impossible to obtain the knowledge required to solve it. This is where contingency comes in: problem resolution rests on decisions in which values—technical as well as ethical—exert a major influence. Without any explicit natural end point, wicked problems need to be terminated for pragmatic reasons—limited resources such as time or money, for example. As Rittel and Webber point out, designers have no right to be wrong as they are responsible for the actions resulting from their design moves. Hence, planners need to make sure that the design works well—a demand that has lost none of its topicality (see Chapter 22, "Values and Design" by Ibo Van de Poel, this volume).

23.4. Contemporary Process Models

The failure of the Design Methods Movement's vision of an automated designer led to a reset in design methods. On the one hand, the focus shifted from the logic of design and design reasoning to new computer-based design tools as important conveyors of (implicit) design methods (see Section 6). On the other hand, methods now centred on the management of design processes. These so-called process models seek to provide a general description of the design process. Prescriptive in nature (Vermaas and Dorst 2007), they structure the design process in consecutive phases in a systematic and scientific way.

Current process models emerged during the 1980s and 1990s as a new generation of design methods that became influential in the engineering sciences. A remarkably strong community developed in the German-speaking context with works by design methodologists Vladimir Hubka and Wolfgang Eder, Gerhard Pahl and Wolfgang Beitz, as well as Klaus Ehrlenspiel (Hubka 1982; Hubka and Eder 1988; Pahl et al. 2007 [1988]; Ehrlenspiel 1995). Characteristic of this approach is the framing of the design process as a generic, overarching problem-solving activity. Industrial designers and design methodologists collaborated in devising the method guidelines eventually issued by the German Association of Engineers (VDI, *Verein Deutscher Ingenieure*).

Of particular significance is guideline VDI 2221, *Systematic Approach to the Development and Design of Technical Systems and Products* (Verein Deutscher Ingenieure 1993), which serves as an umbrella for a number of associated secondary guidelines. It subdivides the design process into several stages which, rather than proceeding in a purely linear way, also occur iteratively. Each stage is characterized by a specific activity leading to a result, as shown in the diagram of the process model (Figure 23.1). Despite being a highly idealized representation of the design process, this framing at least makes it possible to utilize checklists and to define gateways as means of supporting the management of the process.

Despite their proliferation, the practical relevance of generic process models has always been contested. Guidelines such as VDI 2221 suffer from being too problem-based rather than solution-focused (Cross 2008: 40) and in being too serialist and systematic; hence, "they have the tendency to put the intuitive and impulsive designer off" (Childs 2014: 16). In addition, the guideline's core assumption can be questioned, namely, in terms of whether design processes can be captured in a generic way at all—and if they can, to what extent this is so. Any generic description needs to abstract from specific contextual factors. However, for a process model to be applicable, it needs to

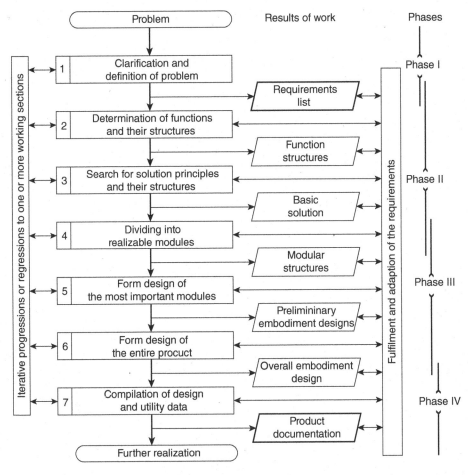

Figure 23.1 General procedure of systematic design according to VDI 2221

Source: Redrawn from Verein Deutscher Ingenieure (2004: 5).

consider the complex and specific contextual factors that are constitutive of design processes. Is it possible to achieve a sensible balance between a generic methodology and a specific application? Does the structure of design really rest on a generic methodical essence, or do design processes rather display 'family resemblances' (which would render the effort of an overall generalization futile)? The debate surrounding the latest revision of VDI 2221 is rather revealing in this respect. Stressing the uniqueness of each design process, the authors concede that the abstract model must be adapted to the needs of specific contextual factors in order to make it useful. Systematicity and the inner logic of the design process are no longer at the forefront. The overarching aim has become a modest one: the general model is to serve as a framework for orientation, to make the justification and documentation of process design easier, and to enable internal and external parties to develop a shared understanding (Verein Deutscher Ingenieure 2018: 4). Design researchers John Clarkson and Claudia Eckert offer an answer to this tricky conundrum. They suggest that design be characterized with regard to its inherent patterns which, in turn, are triggered by certain drivers. This would establish a comparative approach across domains, processes and levels of detail without presupposing the existence of a generic design schema (Clarkson and Eckert 2005: 18–21).

23.5. Testing as a Design Method: Verification and Validation

Just as designing explores the not-yet-existent, testing plays an essential role in stabilizing emerging knowledge about the future artefact. Testing makes it possible to check the consequences of design moves. A specific design constellation is subjected to careful interrogation; a series of tests enable methodical exploration of different design options. Test benches and their standardized environment allow a reduction of complexity and a systematic investigation of drivers and constraints. The test outcomes are fed iteratively into the overall design process, thereby enhancing the maturity of the product.

Different questions and design stages require specific testing methods. The V-chart (Figure 23.2) devised by software engineer Barry Boehm, illustrates the essential role of testing in design. This idealized model of design starts on the upper left arm of the V with conceptual explorations and a definition of the requirements. The procedure cascades downward, like a waterfall, in increasing detail until it reaches the bottommost point of the V. The right arm shows the corresponding testing phases, which proceed in reverse order, from detail to the finalized product in use. The graph assigns specific ways of testing to each phase. In the case of software development, the 'unit test' checks the 'code', the 'integration and system test' checks the detailed design and the 'acceptance test' checks the 'product design'—all summarized as modes of verification. Modes of validation take place when the phase of 'installation operational test and evaluation' checks 'plans and requirements' and 'usage and support' give feedback on the general concept.

The model highlights a demarcation between verification and validation, which is depicted in the graph by the requirements baseline. Below this line, tests are specified as a mode of verification; the question corresponding to this mode would be "Am I building the product right?" (Boehm 1979: 711). Above the line, tests are specified in terms of validation and correspond to asking "Am I building the right product?" (ibid.). Verification constitutes an internal error check as to whether the requirements (sometimes also referred to as specifications that clarify and define the task) have been

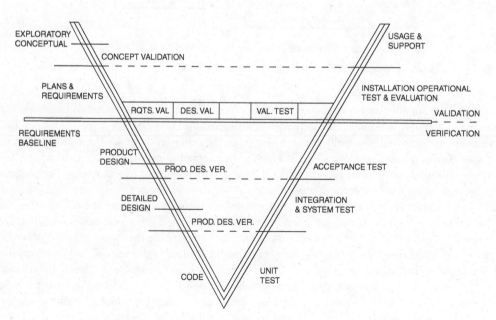

Figure 23.2   Testing as verification and validation in design processes

Source: Redrawn from Boehm (1979).

implemented correctly. Validation addresses the overall goals of the product, which are tested against real-world conditions (in this case, actual usage).

In practice, however, the demarcation between verification and validation is less obvious than the definition suggests. There are three reasons for this. First, contemporary process models do not wait until the very end of the design process to begin validation. Nowadays, users often enter into the design process early on, leading to a co-occurrence of verification and validation testing at the same time. The iterative development of products, which advances in an ongoing spiral-like loop, reinforces this effect. Second, verification and validation take place not only at the level of the process model but also at the level of the product model. To explore the product model, development processes draw on various forms of simulation as a means of generating knowledge about the future artefact and increasing its maturity. Simulation-based analysis draws, for example, on the finite element method (FEM), computer fluid dynamics (CFD), multibody simulation (MBS) and hardware- and software-in-the-loop (HiL/SiL) in order to gain knowledge about the mechanical and thermal behaviour of the future artefact, its interaction with liquids, gases or particles, the interplay between hardware and software and so on. These simulation models, in turn, draw on equations that require verification and validation. Whereas in this case verification is about solving the equations correctly, validation is about solving the right equations (Oberkampf and Roy 2010: xi). In other words, validation seeks to ensure that the model is properly set up so that it will deliver conclusions about the future artefact in a reliable way. Third, the issue of verification and validation returns at the level of the simulation tools. Black-boxed, the software manifests the results of previous verification and validation methods. As an analytic tool, it is itself a final product of a design process. Thus, the software artefact freezes implicit epistemic assumptions in the model setup which usually remain non-transparent during its use. Given this intricate interweaving of verification and validation in the design process, a clear-cut dividing line seems a dubious proposition. Nonetheless, the multilevel nested methods of verification and validation show the crucial importance of testing in design.

23.6. Design Tools—Design Methods Through the Back Door

Compared with the high-flying and yet vain vision of an intelligent design problem-solver (such as GPS or Müller's systematic heuristic; see Section 3), design tools turn methodological strategy on its head. Whereas the intelligent design problem-solver executes a generic design process that needs to be adapted to the conditions of the specific case and that *automates* decision-making and problem-solving, the generic design tool is employed in a specific design process and *supports* the designer's decision-making and problem-solving. In the first case, the operator is a computer system assisted by a designer; in the second, the operator is a designer assisted by a computer system. This inverted methodological strategy entails more than just a change of perspective; it is a veritable paradigm shift. Design tools bypass the unresolved dual challenge of automated design problem-solvers: they are created to support context-specific design praxis. Hence, design tools require neither an exhaustive explication of implicit rules and heuristics nor the prescription of a generic process.

Currently, there are three major families of tools used in processes of product development (Verein Deutscher Ingenieure 2018: 26). *Authoring tools* allow designers to work on the structure of core elements and their relations. Computer-aided design (CAD), or rather its domain-specific versions such as M-CAD in mechanical engineering, serve to shape processes of exploration. The resulting geometrical model can be enriched with information on dimensions, tolerances, materials and instructions for manufacturing. Likewise, electrical engineering and software development draw on E-CAD and CASE (computer-aided software design). *Analytic tools* provide feedback on a design-in-progress. Various simulation techniques facilitate the exploration and testing of mechanical and thermal behaviour, the flow of liquids and gases in an artefact, the effects of a given shape on the kinematics of tiny particles, and the consequences of software and hardware constellations. Nowadays, software packages of authoring tools integrate many analytic tools directly. As so many aspects of a future arte-fact need to be validated, design draws on multifarious tools, modelling and simulation techniques, which are often interrelated with each other. Finally, *data management tools* support the management of the design process itself. They interact closely with authoring tools (Team Data Management—TDM), supervise the overall generation of knowledge (Product Data Management—PDM), and consider the entire cycle of development, production, use and disposal of a product (Product Life-cycle Management—PLM).

By creating a framework for design praxis, tools contribute to specific milieus of reflection (Ammon and Capdevila-Werning 2017; Ammon 2018). Let us take the example of computer-aided design: the user chooses preconfigured basic elements and determines their relations and actions, thereby stipulating specific logics of design. For instance, 2D modelling draws on basic elements such as lines, splines (as free-form curves), circles and points. Selected attributes such as thickness, type, colour and offset are assigned to these elements, and they can be acted upon by translation, scaling, rotation, mirroring and shearing. By contrast, 3D modelling facilitates spatial thinking in a very different way. Wireframe modelling addresses shape as an edge, surface modelling as an envelope, solid modelling as a body, and parametric modelling as parameters. With advanced authoring tools, the design logic inherent in them profoundly affects the arena of designerly exploration. For instance, parametric modelling describes design objects according to constraints that incorporate simple rules within the model. These may be mathematical relations (the length of the object is three times its width) or logical operations (if the length of the object is more than 10 cm, its width is 3 cm, otherwise 2 cm). In this way, design heuristics enter explicitly into the modelling process. Scripting tools, which can automatically generate design variants, go even further. Thus, processes of reasoning increasingly enter the tool setup, integrating design methods into their system through the back door.

23.7. Conclusion

The overview provided in this chapter remains cursory—necessarily so, as design methods and the sciences of design are to a large extent *terra incognita* in the research literature: a challenging field which awaits further philosophical exploration. However, the overview clearly shows a great variety of design methods. Roughly speaking, they fall into two major categories. First, *partial methods* (used, for example, in ideation or validation) serve as temporary interventions enacted for a specific purpose within the design process. Second, *comprehensive methods* seek to address the overall process. They differ with respect to their point of departure, which can be in designerly thinking and reasoning, in the structure or logic of the design (as a future artefact) or in the management of the process of designing itself. Besides the process, recent approaches additionally highlight the composition of the design team (with regard to interdisciplinarity, suspension of hierarchies, participation of users and stakeholders) as well as the space where the creative process takes place.

What is all too easily overlooked here is a third category, which usually remains implicit: *meta methods* refer to the art of choosing the right method at the right time for a given challenge in design. Selecting an appropriate procedure out of a toolbox of methods and incorporating it in a fruitful way for the question at hand is essential in design—and perhaps the most important design method of all. Rarely taught explicitly, it is usually a variety of hands-on knowledge that helps the designer to deal with method pluralism and to apply the right method mix.

So where do we now find ourselves along the spectrum of 'methods—no methods—minimal methods'? With the rising popularity of *Design Thinking* and agile process management (such as *Scrum*), one might be inclined to conclude that we are seeing a decline in methods once again. However, are these modes of ideation and collaboration not also genuine methods, methods which have broken free from overregulation and have released creative intuition to flourish within a given framework? It looks as if the long-standing antagonism between methods-based and non-methods-based design is less about the pros and cons of methods; surely it is rather about the very nature of methods themselves. Should they be generic and scientific? Should they be informal and intuitive? It comes as no surprise that the decline in generic and scientifically oriented methods has been accompanied by a growing popularity of design tools. Their omnipresence provides a structuring framework with respect to both the logic of design and the process of designing. In the long run, this strategy has the potential to reconcile methodological requirements which minimize risk and maximize creativity in complex design tasks. If this endeavour is to be successful, however, the developers of design tools will need to make their methodological constraints transparent and to simplify their operability. It remains exciting to see where this movement will take us.

Acknowledgements

The research on this publication was funded by the Deutsche Forschungsgemeinschaft (DFG, German Research Foundation) under grant agreement AM 405/4–1 as well as under Germany's Excellence Strategy—EXC 2002 "Science of Intelligence"—project number 390523135.

Related Chapters

Chapter 2: A Brief History of Engineering (Jennifer Karns Alexander)

Chapter 5: What Is Engineering Science? (Sven Ove Hansson)

Chapter 6: Scientific Methodology in the Engineering Sciences (Mieke Boon)

Chapter 10: Creativity and Discovery in Engineering (David H. Cropley)

Chapter 21: Engineering Design (Peter Kroes)

Chapter 22: Values and Design (Ibo Van de Poel)

Chapter 24: Human-Centred Design and Its Inherent Ethical Qualities (Marc Steen)

Chapter 25: Sustainable Design (Steven A. Moore)

Chapter 28: Models in Engineering and Design: Modeling Relations and Directions of Fit (Michael Poznic)

Chapter 30: Computer Simulations (Hildrun Lampe)

Chapter 32: On Verification and Validation in Engineering (Francien Dechesne and Tijn Borghuis)

Further Reading

Alexander, C., Ishikawa, S., Silverstein, M. and Jacobson, M. (1977). *A Pattern Language: Towns, Buildings, Construction*. New York: Oxford University Press. (Classic pattern-based design methodology.)

Clarkson, J. and Eckert, C. (2005). *Design Process Improvement: A Review of Current Practice*. London: Springer. (An overview of today's design methods and practices.)

Pahl, G., Beitz, W., Feldhusen, J. and Grote, K.-H. (2007 [1988]). *Engineering Design: A Systematic Approach*, 3rd ed. London: Springer. (Influential account of a prescriptive process model.)

Schön, D. A. (1983). *The Reflective Practitioner: How Professionals Think in Action*. New York: Basic Books. (Classic on design thinking.)

Simon, H. A. (1996 [1969]). *The Sciences of the Artificial*, 3rd ed. Cambridge, MA: MIT Press. (Classic delineating of a theory of the design sciences.)

Note

1. Lawson (1980); Rowe (1987); Cross and Dorst (1992). The debate within design theory and philosophy of the design sciences should not be confused with the recent association of this concept with a user-centred method in product and organizational development (see e.g. Brown and Katz 2009).

References

Alexander, C. (1964). *Notes on the Synthesis of Form*. Cambridge, MA: Harvard University Press.

Alexander, C., Ishikawa, S., Silverstein, M. and Jacobson, M. (1977). *A Pattern Language: Towns, Buildings, Construction*. New York: Oxford University Press.

Altshuller, G.S. (1984). *Creativity as an Exact Science: The Theory of the Solution of Inventive Problems*. New York: Gordon and Breach.

Altshuller, G.S. and Shapiro, R.B. (1956). О Психологии изобретательского творчества (On the Psychology of Inventive Creation). (in Russian). *Вопросы Психологии (The Psychological Issues)* 6, 37–39.

Ammon, S. (2017). Why Designing Is Not Experimenting: Design Methods, Epistemic Praxis and Strategies of Knowledge Acquisition in Architecture. *Philosophy & Technology,* 30(4), 495–520.

Ammon, S. (2018). Drawing Inferences: Thinking with 6B (and Sketching Paper). *Philosophy & Technology,* 4(5), 345.

Ammon, S. and Capdevila-Werning, R. (eds.) (2017). *The Active Image: Architecture and Engineering in the Age of Modeling*. Cham, Switzerland: Springer, s.l.

Archer, L.B. (1965). *Systematic Method for Designers*. London: Council for Industrial Design.

Boehm, B. (1979). Guidelines for Verifying and Validating Software Requirements and Design Specifications. In P.A. Samet (ed.), *Euro IFIP 79*. Amsterdam: North-Holland, pp. 711–719.

Brown, T. and Katz, B. (2009). *Change by Design: How Design Thinking Transforms Organizations and Inspires Innovation*, 1st ed. New York: Harper Business.

Childs, P.R.N. (2014). *Mechanical Design Engineering Handbook*. Oxford: Butterworth-Heinemann.

Clarkson, J. and Eckert, C. (2005). *Design Process Improvement: A Review of Current Practice*. London: Springer-Verlag.

Cross, N. (2006). *Designerly Ways of Knowing*. London: Springer-Verlag.

Cross, N. (2008). *Engineering Design Methods: Strategies for Product Design*, 4th ed. Chichester: John Wiley.

Cross, N. and Dorst, K. (eds.) (1992). Research in Design Thinking. *Proceedings of a Workshop Meeting Held at the Faculty of Industrial Design Engineering*, Delft University of Technology, The Netherlands, May 29–31, 1991. Delft: Delft University Press.

Dorst, K. and Cross, N. (2001). Creativity in the Design Process: Co-Evolution of Problem—Solution. *Design Studies,* 22(5), 425–437.

Ehrlenspiel, K. (1995). *Integrierte Produktentwicklung: Methoden für Prozeßorganisation, Produkterstellung und Konstruktion*. Hanser, München.

Ferguson, E.S. (1992). *Engineering and the Mind's Eye*. Cambridge, MA: MIT Press.

Gregory, S.A. (1966). *The Design Method*. London: Butterworth.

Heymann, M. (2005). *"Kunst" und Wissenschaft in der Technik des 20. Jahrhunderts. Zur Geschichte der Konstruktionswissenschaft*. Zürich: Chronos Verlag.

Hubka, V. (1982). *Principles of Engineering Design*. Burlington, UK: Butterworth-Heinemann.

Hubka, V. and Eder, W.E. (1988). *Theory of Technical Systems: A Total Concept Theory for Engineering Design*. Berlin: Springer.

Jones, J.C. (1970). *Design Methods: Seeds of Human Futures*. London: Wiley Interscience.

Jones, J.C. and Thornley, D.G. (eds.) (1963). *Conference on Design Methods*. Oxford, UK: Pergamon Press.

König, W. (2013). Zwischen Algorithmus und Intuition: Ein Analogieangebot für die Architekturtheorie aus der Geschichte der Maschinenkonstruktion. In S. Ammon and E.M. Froschauer (eds.), *Wissenschaft entwerfen: Vom forschenden Entwerfen zur Entwurfsforschung der Architektur*. Paderborn: Fink, pp. 417–438.

Langrish, J.Z. (2016). *The Design Methods Movement: From Optimism to Darwinism*. www.drs2016.org/222. Accessed March 31, 2019.

Lawson, B. (1980). *How Designers Think*. London: Architectural Press.

Müller, J. (1966). *Operationen und Verfahren des problemlösenden Denkens in der technischen Entwicklungsarbeit—eine methodologische Studie: Habilitationsschrift*. Universität Leipzig.

Müller, J. (1990). *Arbeitsmethoden der Technikwissenschaften: Systematik, Heuristik, Kreativität*. Berlin: Springer.

Oberkampf, W.L. and Roy, C.J. (2010). *Verification and Validation in Scientific Computing* [Elektronische Ressource]. Cambridge: Cambridge University Press.

Osborn, A.F. (1953). *Applied Imagination: Principles and Procedures of Creative Thinking*. New York: Scribner.

Pahl, G., Beitz, W., Feldhusen, J. and Grote, K.-H. (2007 [1988]). *Engineering Design: A Systematic Approach*, 3rd ed. London: Springer.

Poser, H. (2013). Ars inveniendi heute: Perspektiven einer Entwurfswissenschaft der Architektur. In S. Ammon and E.M. Froschauer (eds.), *Wissenschaft entwerfen: Vom forschenden Entwerfen zur Entwurfsforschung der Architektur*. Paderborn: Fink, pp. 135–166.

Reuleaux, F. (1876). *The Kinematics of Machinery*.

Rittel, H.W.J. and Webber, M.M. (1973). Dilemmas in a General Theory of Planning. *Policy Sciences,* 4(2), 155–169.

Rowe, P.G. (1987). *Design Thinking.* Cambridge, MA: MIT Press.

Schön, D.A. (1983). *The Reflective Practitioner: How Professionals Think in Action.* New York: Basic Books.

Simon, H.A. (1996 [1969]). *The Sciences of the Artificial,* 3rd ed. [Nachdr.]. Cambridge, MA: MIT Press [u.a.].

Verein Deutscher Ingenieure (1993). *VDI2221 Methodik zum Entwickeln und Konstruieren technischer Systeme und Produkte (Englisch)* [Systematic Approach to the Development and Design of Technical Systems and Products]. Berlin: Beuth Verlag.

Verein Deutscher Ingenieure (2004). *VDI2223 Methodisches Entwerfen technischer Produkte Titel (Englisch)* [Systematic Embodiment Design of Technical Products]. Berlin: Beuth Verlag.

Verein Deutscher Ingenieure (2018). *VDI2221 Blatt1 Entwicklung technischer Produkte und Systeme—Modell der Produktentwicklung (Englisch)* [Design of Technical Products and Systems—Model of Product Design]. Berlin: Beuth Verlag.

Vermaas, P.E. and Dorst, K. (2007). On the Conceptual Framework of John Gero's FBS-model and the Prescriptive Aims of Design Methodology. *Design Studies,* 28(2), 133–157.

Zwicky, F. (1959). *Morphologische Forschung: Wesen und Wandel materieller und geistiger struktureller Zusammenhänge.* Winterthur: Buchdr. Winterthur in Komm.

Zwicky, F. (1969). *Discovery, Invention, Research Through the Morphological Approach.* Toronto: Palgrave Macmillan; New York: Collier-Macmillan.

24

HUMAN-CENTRED DESIGN AND ITS INHERENT ETHICAL QUALITIES

Marc Steen

24.1. Introduction

What do we need to know about human-centred design (HCD), about the practices of designers, developers and engineers, who help to shape our world? In everyday life, we often focus on the *output* of design processes; for example, when we interact with the digital devices or online services that were designed by these people—our smartphones, tablet computers or social networking services. Or, in the industry, for example, people focus on the *input* of a design process and are interested, for example, in the expertise and resources that are needed in a project. In this chapter, however, I will focus on the *process* of HCD and argue that HCD contains *inherent ethical qualities*—qualities which often remain implicit and unexamined (Bijker 1993; Winner 1993).

In this chapter I will present three specific HCD projects and the social processes happening in these projects, in order to discuss the ethical qualities inherent to HCD. This focus on the *specific* and the *social* follows from the character of design practices, which are specific in that they are concerned with developing specific solutions for specific problems, and social in that communication and cooperation are at the heart of design (Bucciarelli 1994; Devon 2004; Devon and Van de Poel 2004). This focus is in line with Van de Poel and Verbeek's (2006) proposal to "perform a context-sensitive form of ethics," to study the social practices of the people involved in specific projects.

24.2. Human-Centred Design

Human-centred design (HCD) emerged in the field of information and communication technology (ICT) as an approach to counter *technology push* (Cooper 1999; Thackara 1999, 2006), which can lead to products or services that people cannot or do not want to use (Nielsen 1993; Norman 1988). The term can be used as an umbrella to include diverse approaches (Steen 2011). Conveniently, there is an ISO standard for it: *Human-Centred Design for Interactive Systems* (ISO 2019). This standard describes the following key principles: to start with an explicit understanding of prospective users and their tasks and environments; to involve prospective users throughout the process of design and development; to involve prospective users in timely and iterative evaluations and to let these evaluations drive and refine the process of design and development; to organize an iterative process; to view the user experience holistically, e.g., not just as usability, but also as people's aspirations and emotions; and to organize a multidisciplinary project-team.

These principles help to understand HCD as a combination of a concern for understanding the *present*, e.g., the problem that is focused on, which is a *research* orientation, as in ethnography, and a concern for creating the *future*, e.g., the generating of possible solutions, which is a design orientation, as in co-design (Sanders and Stappers 2008); and as a combination of *researchers and designers moving* towards users and their daily lives, as happens in empathic design, and of *users moving* towards researchers and designers and their project, as in participatory design (Steen 2011).

HCD builds on the tradition of participatory design (PD). PD, in the Scandinavian tradition, is "an approach towards computer systems design in which the people destined to use the system play a critical role in designing it" (Schuler and Namioka 1993: xi). PD aims to empower putative or potential end users of ICT products and services to participate in the design process. HCD, however, is similar to PD but has less explicit political motivations. In practice, HCD tends to aim at supporting people in the industry to focus on end users, rather than at explicitly empowering end users (despite its commitment to user involvement).

Furthermore, HCD has informed value sensitive design (VSD) (see Chapter 22, "Values and Design" by Ibo Van de Poel, this volume). VSD "is a theoretically grounded approach to the design of technology that accounts for human values in a principled and comprehensive manner throughout the design process" (Friedman et al. 2006: 348); it aims to understand different stakeholders' values and interests, and to negotiate and combine these during the design process. In that sense, VSD has a broader scope than HCD, which focuses on one type of stakeholder, end users, and their values and interests.

Moreover, it is relevant to note the difference between user-centred design (UCD) (Nielsen 1993) and HCD. UCD tends to look at people in their role of *users*, whereas HCD aims to look at people *more holistically*, not only as users of a specific product or service, but also as citizens, as parents, as friends, as coworkers, etc.

24.3. Ethical Lenses to Look at Ethical Qualities

In this section, I will present three HCD projects, which I will discuss by adopting three different ethical lenses—lenses which complement each other:

- The first project (WeCare) will be discussed from the perspective of *virtue ethics*, in order to study the thoughts, feelings and actions of project-team members involved;
- The second project (FRUX) will be discussed from the perspective of *ethics-of-alterity*, in order to study the encounters between project-team members and prospective users;
- The third project (TA2) will be discussed from the perspective of *philosophical pragmatism*, in order to focus on the organization of collaborative and creative processes.

I selected these three ethical traditions because they are typically focused on *specific* and *social* practices, similar to the focus of design practices on the specific and social practices. Virtue ethics focuses on people in specific, concrete and social contexts and their thoughts, feelings and actions in these situations. Ethics-of-alterity views people as inherently social beings; one always finds oneself in specific relationships to others. And pragmatist ethics takes people's practices and experiences as a starting point for analysis and aims to deliver practical results. The lens through which we look thus matches the phenomena we study.

24.4. Virtues in Human-Centred Design

The WeCare project aimed to improve older people's well-being by enabling them to engage in online social networking, thereby promoting social interaction and participation, both online and in real life. The project consortium included industry partners (e.g. a supplier of online video communication), care or service providers (e.g. a provider of tele-care services for people in rural areas), organizations that represent older people and their interests, and research organizations in four countries (Finland, Spain, Ireland and the Netherlands). For each country, a HCD process was organized that involved older people and their family and friends in the design and evaluation of four online social networking services, one for each country. The services were developed as prototypes and evaluated in user trials, and they included tools for social communication, such as video communication and discussion forums, and for coordinating social activities, such as shared calendars and ways to request or offer support.

24.4.1 Virtue Ethics

I looked at this project via the lens of Aristotelian virtue ethics. This tradition focuses on cultivating virtues and enabling people to flourish (*eudaimonia*). Virtue ethics starts with an ultimate goal: the goal for people to flourish, to live the good life. Virtues are "dispositions not only to act in particular ways, but also to feel in particular ways. To act virtuously . . . is to act from inclination formed by the cultivation of virtues" (MacIntyre 2007: 149).

In virtue ethics, one aims for an appropriate *middle* between deficiency and excess, given the specific circumstances. For example, the virtue of courage requires striking an appropriate middle between cowardice and recklessness, and plays out differently for different people in different circumstances. Finding this middle "requires therefore a capacity to judge and to do the right thing in the right place at the right time in the right way" (*op cit.*, 150). Finding this middle is concerned with striving for excellence (*arete*), *not* with moderation or mediocrity, and with cultivating well-formed types of natural desires (*op cit.*, 160), *not* with countering desires. One can learn to think, feel and act virtuously by trying out virtuous behaviours or by observing people who behave virtuously.

In the following, I will argue that *promoting cooperation, collaborative curiosity, collaborative creativity* and *empowerment* are key virtues that are needed in HCD.

24.4.2 Promoting Cooperation

Cooperation is critical for HCD, and indeed, cooperation has been at the heart of one of its preceding traditions: participatory design (PD) (Bjerknes and Bratteteig 1995; Bratteteig and Stolterman 1997; Kensing and Blomberg 1998). Cooperation needs to be promoted carefully, with patience and attention for group dynamics, so that the people involved can engage in *cooperative curiosity* and *cooperative creativity* (see the following). Regarding cooperation, one will aim for a middle between the *deficiency of neglecting* the subtleties of group dynamics and cooperation, and the *excess of controlling* people and forcing them to cooperate. This virtue is especially needed in people in management or leadership roles.

One intervention of project manager Sharon can illustrate this virtue. Every couple of months, she organized a project-team meeting. Usually, in such meetings, people leave their laptop computers open and combine attending the meeting with reading and writing emails. Sharon, being aware of the need to promote cooperation, asked people to close their laptops and to pay full attention to the meeting and to the others. In addition, she organized relatively long lunch breaks to encourage project-team members to socialize and relax. Sharon understood that one needs to invest in such activities in order to promote cooperation. Such interventions helped project-team members to collaborate effectively throughout the project.

24.4.3 Cooperative Curiosity

The virtue of *cooperative curiosity* is a disposition of being open and receptive towards other people and one's own experiences. Typical methods to promote curiosity are mutual learning (Bødker et al. 1987; Bjerknes and Bratteteig 1987) or ethnography (Blomberg et al. 1993; Button 2000). Mutual learning was pioneered in the Utopia project, in which system developers cooperated with graphic workers to develop and evaluate information systems to support workers (Bødker et al. 1987). The developers and the workers had meetings in which the developers learned about the workers' ways of working, about skills and usage of tools and in which the workers learned about technologies. Another approach to foster curiosity is to draw from the tradition of ethnography, for example, by conducting all sorts of fieldwork to inform or inspire the design process. Ethnography can help one to focus on other people, rather than on one's own ideas about these people (Blomberg et al. 1993).

One needs to find a middle between the deficiency of too little sensitivity to other people's or one's own experiences, and the excess of too much receptiveness to other people's or one's own experiences. Jannie's actions can illustrate this virtue. Jannie worked for an organization that represents older people and their interests, and her role in the project was to promote a better understanding of older people. In several meetings, she noticed that people tend to use stereotypes when talking about older people. In order to counter that tendency, Jannie invited others to find out what older people actually *do* with computers, for example, by organizing workshops in which project-team members and older people met and exchanged knowledge and ideas, to promote cooperative curiosity.

24.4.4 Cooperative Creativity

The virtue of *cooperative creativity* is a disposition of jointly generating ideas, combining ideas of different people, and creating products or services. Typical methods to promote creativity are, e.g., *Future Workshops*, in which people engage in three collaborative and creative phases: *Critique*, of the current situation; *Fantasy*, about more desirable alternatives; and *Implementation*, articulating short-term actions (Kensing and Madsen 1991), or cooperative prototyping (Bødker et al. 1987; Ehn and Kyng 1991).

Cooperative prototyping, that is, the hands-on creation and evaluation of mock-ups and prototypes, was also pioneered in the Utopia project (Ehn and Kyng 1991; Bødker et al. 1987). In that project, mock-ups were sometimes as simple as a cardboard box with the text "laser printer" written on it; "everybody has the competence to modify [these mock-ups]; they are cheap, hence many experiments can be conducted without big investments in equipment, commitment, time, and other resources" (Ehn and Kyng 1991: 172–173).

One needs to find a middle between the deficiency of too little attention for other people's or one's own ideas, and the excess of too much realization of other people's or one's own ideas. Stefan's role can illustrate this virtue. Stefan was responsible for coordinating the project partners' activities of developing and combining software modules into working prototypes. This became critical when prototypes were going to be used by people in their daily lives. In one meeting, it became clear that specific modules were not delivered on schedule and did not meet the user requirements. Often, such a situation makes people look backward and blame others—not very productive for finding solutions. Instead, Stefan stayed calm and invited people to talk constructively with each other, to look ahead and to explore and develop practical solutions, to promote cooperative creativity.

24.4.5 Empowerment

One also needs the virtue of *empowerment*: the disposition to share power and agency with others, also with people outside the project, for example, the people who are supposed to be going to benefit from the project's results. One can do that by aiming for a middle between the deficiency of being passive and hesitant, for example, assuming that people will cope and thrive without help, and the excess of being patronizing and directive, for example, assuming that people will prosper if only they follow your advice. In the PD tradition, the *tool perspective* has been key to empowering workers: "The idea is that new computer-based tools should be designed as an extension of the traditional practical understanding of tools and materials used within a given craft or profession" (Ehn 1993: 57). The tool perspective respects people's tacit knowledge and skills and enables them to contribute to the development of the tools which they will be using. Moreover, it advocates developing tools that people can use actively and creatively, thus empowering them, rather than developing finished products that can only be used in predetermined and fixed ways, with the risk of making their users passive and disempowering them.

The virtue of empowerment can be illustrated with an example of John Thackara (1999), at that time project manager of the Presence project, which aimed to develop user-friendly Internet services for older people (similar to WeCare). This is what he wrote about the project-team members' first encounter with their so-called "target group":

> So we went and found some older people and told them how we had come to help them with the Internet, and they said, "Piss off! . . . We don't need your patronising help, you designers. If you've come here to help us, you're wasting your time; we don't want to be helped, thanks just the same. Yet we do have some interesting observations to make about our daily lives, about our lifestyles, about our communication, and about all of their attendant dysfunctions. If you could kindly change your attitude and help us explore how we will live, then perhaps we can do something together."

In other words, one needs to share power and agency with prospective users so that they can become active participants and creative contributors, rather than passive receivers.

In sum, we can view HCD as a *praxis* in which the people involved need to cultivate the virtues of cooperation, collaborative curiosity, collaborative creativity, and empowerment.

24.5. Human-Centred Design as a Fragile Encounter (FRUX Project)

The FRUX project aimed to develop two innovative mobile telecom services for two user groups, and to organize the design process in close cooperation with them: one for and with police officers, and another for and with informal caregivers. The projects combined a technology-centred approach (to develop telecom services) and a human-centred approach (to cooperate with prospective users).

The project-team members organized observations, interviews, workshops and field trials with prospective users, and designed and evaluated two prototypes, one for each target group: a mobile telecom service that helps different types of police officers to share information and to collaborate while they are out on the street, and an online social networking service that helps people to communicate and coordinate informal care for people with dementia, for example, sharing care and other tasks between family members who jointly provide care for one of their (grand)parents.

There were project-team members, with their experiences, knowledge and ideas to develop telecom services, and there were so-called "users," with their experiences, knowledge and ideas about their daily lives. The project aimed to bring these people together in face-to-face interactions.

24.5.1 Ethics-of-Alterity

I looked at this project through the lens of *ethics-of-alterity*; a type of ethics that takes the other and the relationships between other and self as a starting point, with Emmanuel Levinas (1906–1995) and Jacques Derrida (1930–2004) as key proponents. Levinas wrote extensively about the encounter between other and self, and Derrida about *différance* and otherness. In their *ethics-of-alterity* one always finds oneself within other-self relationships, which are inherently ethical relationships (not unlike Aristotelian virtue ethics, which views people as inherently social).

In a HCD project, people attempt to communicate and cooperate, which Levinas and Derrida would conceive of as encounters between other and self, and as situations that are inherently loaded with all sorts of ethical qualities. Let me attempt to deconstruct two key assumptions of HCD as a way to bring the ethical qualities of HCD to the fore, based on readings of Levinas and Derrida.

24.5.2 *Developing Knowledge and the Tendency to Grasp the Other*

A key assumption in HCD is that project-team members can jointly learn new things; that they can gather and develop new knowledge, for example, about prospective users and their needs and preferences (the HCD principles of involving and understanding prospective users, and of viewing their experiences holistically). It can be hard, however, for project-team members, to be *open* towards *others* and to learn new things, for example, when they interact with prospective users in interviews or workshops.

In his oeuvre, Levinas was concerned with the difficulties of encounters between people and with the violence that so often occurs in these encounters. He argued that one tends to *not* see the other as *other*, but as an object, and to reduce the other to concepts that one is already familiar with. Levinas put it as follows: "The foreign being . . . falls into the network of a priori ideas, which I bring to bear, as to capture it" (Levinas 1987: 48, 50). He characterized this tendency as the making of a grasping gesture (Levinas 1996: 152); one pulls the other into one's own way of thinking. In an attempt to develop knowledge, we will (unintentionally, inevitably) *grasp* the *other*, which makes it very difficult for us to learn anything new.

HCD practitioners cannot escape this tendency. Their ambitions, knowledge and ideas get in the way of their attempts to be open towards other people and their ambitions, knowledge and ideas. In the FRUX project, we conducted a series of four workshops with different groups of police officers; we discussed problems they experienced in their work and explored possible solutions for these problems. Based on the findings from each workshop, we gradually changed our project's focus and developed a mobile telecom application that promotes cooperation between police officers. It does so by automatically making suggestions to share implicit knowledge between police offices. In HCD, such learning, based on interactions with users, is considered good practice.

Nevertheless, we also missed several opportunities to learn from police officers and to let their ideas influence our project. In our interactions with police officers, we often privileged our own ideas. In the first workshop, for example, we jointly explored four areas that the police officers experienced as problematic. After the workshop, however, we chose to focus on the one area that was comfortably close to our ambition to develop an innovative telecom application. Consequently, we ignored other areas that were relevant to the police officers, such as their problems with their current systems for sharing and accessing information, or their struggles with their professional roles and with the police's organizational culture.

In order to counter this tendency to "grasp the other," Levinas envisioned an attempt to escape the gesture of grasping via a form of desire that is not aimed at satisfying the self and is respectful of the otherness of the other: "This desire without satisfaction hence takes cognizance of the alterity of the other" (Levinas 1987: 56).

24.5.3 *Making Decisions and the Tendency to Program Innovation*

Another key assumption in HCD is that the people involved can organize iterative phases of divergence, of research and exploration, towards openness, and phases of convergence, of evaluation and drawing conclusions, towards closure (the HCD principles of user involvement and of organizing an iterative process and multidisciplinary teamwork). Project-team members not only need to be open towards others and to explore; they also need to draw conclusions and to deliver results, to create closure and to make progress.

Regarding the process of decision-making, Derrida remarked that genuine decisions are "exceptional": "a decision that does not make an exception, that does nothing but repeat or apply the rule, would not be a decision" (Derrida 2001: 29), and that a genuine decision cannot be made by merely applying knowledge or following rules. A decision that is based on knowledge is "an application, a programming" (Derrida 1995: 147–148). Similarly, Derrida observed that people often attempt to *program* innovation and argued that this can lead to "the invention of the same" (Derrida 1989: 46, 55).

Because of this tendency to *program* innovation, one tends to stay within one's own comfort zone, which makes it hard to create anything new. In HCD, project-team members cannot escape this tendency. They bring their own backgrounds and methods to the encounters with other people, and these influence the balance between openness and closure, typically more towards closure.

In the FRUX project, we cooperated with primary informal caregivers; more specifically, we collaborated with people who provide informal care to people who suffer from dementia and who live at home; often the husband or wife of the person with dementia. Different project-team members followed different approaches to talk with them about their daily lives and their needs. Some project-team members, who were familiar with dementia and informal care and who worked in social science research roles, conducted a survey with hundreds of people with dementia and their primary informal caregivers. Other project-team members, for whom dementia and informal care were relatively new areas, and who worked in design roles, conducted informal interviews to inspire their creative process.

Both approaches were attempts to move toward openness, to learn from other people about their daily lives. However, they were also moves toward closure; to draw conclusions about other people's needs and creating products for them. The people doing the survey used a standardized questionnaire, so responses had to match its categories. The people doing the design-interviews wanted to create an innovative telecom application and were looking for inspiration, which influenced their interviews. Both groups brought their methods to the encounters with others as a way to focus and to move towards closure.

To escape these tendencies towards *closure* and *programming*, Derrida advocated welcoming the other: "To invent would then be to 'know' how to say 'come' and to answer the 'come' of the other" (Derrida 1989: 56). This would be an active form of passivity because it requires an effort to *not* make the other into a theme within one's own program.

In sum, we can view HCD as a fragile, face-to-face *encounter* between people, involving attempts to develop knowledge and being open towards others (and to counter the tendency to *grasp* the other), and attempts to make decisions and progress and to balance openness and closure (and to counter the tendency to *program* innovation).

24.6. Human-Centred Design as a Process of Joint Inquiry and Imagination (TA2 Project)

The TA2 project aimed to develop and evaluate a series of innovative telecommunication, multimedia and gaming applications, and to better understand how such technologies can help groups of people to engage in social communication when they are separated in space and in time, so that they can experience togetherness—TA2 stands for Together Anywhere, Together Anytime. The project involved a collaborative effort of approximately 40 researchers, designers and developers, with different backgrounds, such as technology, business and social science, from 14 organizations, ranging from international corporations and small enterprises to universities and research institutes.

The project delivered a series of prototypes for different target groups and usage contexts: *Space Explorers*, a game that combines TV-based video communication and a board game, which groups of friends can play from different locations; *Sixth Age*, a series of casual games for TV or tablet computer, which also facilitate social communication, for example, between grandparents and grandchildren; *Jump Style*, an application for creating, editing and sharing videos, which, for example, teenagers can use to create and share video clips of dance moves; *My Videos*, an application for creating and sharing video compilations of, for example, a school concert, based on footage shot by multiple people; and *Connected Lobby*, a social networking service that facilitates social communication by sharing status updates.

The project manager facilitated a HCD process in which diverse project-team members collaborated with each other and with people from different target groups, involving various methods: in-home interviews at the start of the project to learn about the daily lives of their inhabitants; creative workshops and discussions of ideas in iterative cycles throughout the project, to explore, discuss and improve ideas; and evaluations of prototypes, further on in the project, both in the lab and in people's daily lives.

24.6.1 Pragmatist Ethics

I looked at this project through the lens of philosophical pragmatism. This strand of philosophy emerged in the USA in the late 19th century, with key figures such as William James, C.S. Peirce and John Dewey. Here, I will focus on texts by Dewey (1859–1952) because his perspective is most relevant to discussions of technology (Hickman 1990), engineering (Emison 2004) and design (Dalsgaard 2009). A key theme in his work was the productive combination of practice and theory, and his advocacy for an empirical method of moving back and forth between practices (primary experiences) and reflections (secondary experiences) (Dewey 1965: 36). In contrast to mainstream views on science as a search for universal knowledge, Dewey contended that knowledge is always provisional, particular and contingent rather than universal and necessary (Dewey 1920: 78). Another key theme in Dewey's work was his meliorism: "the belief that the specific conditions which exist at one moment, be they comparatively bad or comparatively good, in any event may be bettered" (Dewey 1920: 178) and his advocacy for cooperation and empowerment. His concerns for practical experiences and for promoting positive change converged in his ideas concerning *inquiry* (Hickman 1998), which will be the basis of our discussion.

Dewey envisioned a process of joint inquiry and imagination in which people can better understand their current situations, imagine more desirable situations and develop ways to cooperate in their realization, so that they move from a situation of perplexity towards a resolution: "Inquiry is the controlled or directed transformation of an indeterminate situation into . . . a unified whole" (Dewey 1938: 104–105). HCD can be understood as a similar process, involving collaborative design thinking (Dorst 2011), collaborative problem-setting—and solution-finding (Lawson 2006: 125; Cross 2006: 80).

Dewey saw inquiry and imagination as processes with inherent ethical qualities. Moral experiences were his starting point, and empowering people to cope with moral questions was his primary goal (Stuhr 1998: 85). Similarly, HCD can be understood as a process of moral inquiry which proceeds "by dialogue, visualization, imagining of motor responses, and imagining how others might react to a deed done" (Hildebrand 2008: 77; also Lloyd 2008).

Dewey conceptualized this process of inquiry and imagination as consisting of different phases (1938: 101–119), which are ideally organized as an iterative process, moving from problem exploration and definition, via perceiving the problem and conceiving of possible solutions, to trying out and evaluating solutions.

24.6.2 Problem Exploration and Definition

At first, people experience a specific situation as problematic, without yet knowing what is precisely problematic about it. Dewey stressed that personal and subjective experiences are critical for the start of an inquiry process, to make the situation questionable. Expressing and sharing these experiences are critical: "inquiry is not a purely logical process—feeling is a useful and orienting presence throughout each phase" (Hildebrand 2008: 57). A provisional problem definition is formulated, which can later be restated and refined.

The ethics of HCD are enacted when participants express their experiences and empathize with others. In the TA2 project, workshops were organized to facilitate problem exploration and definition. Three months into the project, a workshop was organized in which key team members were invited to empathize with specific groups of people and to take them, and their experiences, as starting points for developing five scenarios: short narratives of people using the TA2 applications. Another example was a workshop, in the tenth month of the project, in which team members were invited to engage more personally, and morally, with the theme of togetherness, and the project's goal to promote togetherness. Such workshops helped project-team members to ground the project's problem definition in specific and moral experiences.

24.6.3 Perception of the Problem and Conception of Possible Solutions

In an iterative process, the problem and possible solutions are simultaneously explored and developed (Dewey 1938: 109). Dewey proposed that problems are best explored using *perception*, one's capacities to see, hear, touch, smell and taste, and that solutions are best developed using *conception*, one's capacities to imagine and envision alternative situations.

The ethics of HCD occur, for example, when participants use their capacities for perception and engage with visualizations of the problem (Sleeswijk Visser 2009) or their capacities for conception and engage in creative activities (Sanders 2000). Ideally, participants can imagine or rehearse current (problematic) situations or alternative (desirable) situations (see also "moral imagination" or "dramatic rehearsal" in Fesmire 2003: 55–91).

In TA2 this process was facilitated by creating and discussing five storyboards: for each of the TA2 applications, a series of five to ten drawings with accompanying narratives. These storyboards were developed in an iterative process between key project-team members and a professional illustrator. Creating these storyboards helped the people involved to discuss how the project's overall goal and ideas for specific solutions relate to each other. Moreover, the storyboards were discussed in a series of focus groups with different groups of people, which helped the project-team members to improve their ideas.

Creating and discussing these storyboards brought to the fore the ethics of HCD in that project-team members and users were able to jointly perceive a problem, in this case the current lack of togetherness between people; to jointly conceive of possible solutions, that is, specific features in one of the TA2 applications, and to move between perception and conception, for example, when project-team members listened to users talking about their problems and modified their prototypes accordingly.

24.6.4 Trying Out and Evaluating Solutions

In order to find out which solutions work, different possible solutions are tried out and evaluated, for example, in practical experiments. The project becomes more real and the stakes get higher. It may become clear, for example, that different participants or stakeholders have different interests. In such cases, the people involved need to negotiate carefully in order to bring the project to successful completion. They will need to find ways to combine their interests productively, in order to deal with even "deep-seated and fundamental value conflicts" (Keulartz et al. 2004) and develop solutions that work for all of them.

The ethics of HCD occur when the people involved are able to jointly create results and critically evaluate these, and to productively negotiate and combine their different interests.

In TA2 this process involved the development and evaluation of several prototypes, in cooperation with potential users, in laboratory experiments and in field trials in people's homes. The project-team members working on MyVideos, for example, cooperated with two groups of parents with children in two high schools. One group of parents made video recordings of a school concert in which their children performed, and evaluated a first prototype of MyVideos while viewing and editing the video material of that concert. They also participated in discussions about options for further development, which helped to steer the development of a second prototype.

In sum, we can view HCD as a process of *joint inquiry and imagination*, involving perception in problem-setting and conception in solution-finding; a process in which people are enabled to use "the power of intelligence to imagine a future which is the projection of the desirable in the present, and to invent the instrumentalities of its realization" (Dewey 1917: 69).

24.7. Conclusions

I studied three HCD, using different ethical lenses to look at different aspects of the HCD process, and its inherent ethical qualities:

- Virtue ethics helped to understand the dispositions of people who work in HCD projects and to argue that they need virtues related to cooperation, curiosity, creativity and empowerment. Ideally, they can cultivate these virtues, so that their thoughts, feelings and actions become aligned and help them in mutual learning or collaborative prototyping.
- Ethics-of-alterity helped to understand the ethical qualities of face-to-face encounters between people in HCD projects; for example, between team members and prospective users, in interviews or workshops. Ideally, they become aware of these ethics, so that they can balance concerns for other and self, and openness and closure.
- A pragmatist perspective helped to look at the ethics of organizing HCD projects: a collaborative and creative process of problem-setting and solution-finding. Ideally, the people involved can engage in a process of joint inquiry and imagination, for example, by organizing the project in iterative cycles of research, design and evaluation.

There are several recurring themes in these discussions of HCD: they are based on cooperation, and they require inwards-directed moves and outward-directed moves of the people involved—see Table 24.1.

Table 24.1 Ethical qualities inherent in human-centred design practices

Perspective	Virtue Ethics	Ethics-of-Alterity	Pragmatist Ethics
Focus	Participants' feelings, thoughts and actions	Face-to-face encounters and interactions	Managing a project and its iterative cycles
Cooperation as a Basis	Promoting cooperation and empowerment	Encounters between other and self	Process of collaborative design thinking
An Inwards-Directed Move	Cooperative curiosity: openness, empathy and joint learning	Developing knowledge: being open to the other (not grasp the other)	Joint inquiry: perception, empathy and problem-setting
An Outwards-Directed Move	Cooperative creativity: developing, realizing and trying out ideas	Making decisions: balancing openness and closure (not program)	Joint imagination: conception, creativity and solution-finding

HCD practices are based on *cooperation* between different people: on the virtues of cooperation and empowerment; on face-to-face encounters between diverse people; and on organizing collaborative problem-setting and solution-finding.

Participants need to allow for a move inwards: in cooperative curiosity and joint learning; when they develop knowledge and are open to other people; and when they engage in *joint inquiry* and use their capacities for perception and empathy in problem-setting.

Additionally, they need to allow for a move outwards: in cooperative creativity and joint development; when they make decisions and balance *openness* and *closure*; and in *joint imagination*, using their capacities for conception and creativity in solution-finding.

24.8. Discussion

We can relate human-centred design (HCD) to participatory design (PD) and to value sensitive design (VSD). As was mentioned in the introduction, HCD builds on the tradition of participatory design (PD). However, HCD aims to support people in the industry to *focus* on end users, rather than aiming to *empower* end users, as is done in PD. One might argue that HCD implies a *moral* appeal to designers to empathize with end users and work for their benefit, whereas PD is a *political* tool that aims to transfer power from designers to end users (Steen 2013b).

Regarding the relation between HCD and VSD, the introduction to this chapter also pointed out that VSD works with a broader scope than HCD and typically involves more and more diverse stakeholders than HCD. The chapter on values and design (Chapter 22, this volume) contains a discussion of conflicts that can occur between values, and between agents and various ways to handle and solve these conflicts (win/win, compromise and integration). These insights can also be useful in HCD, as there will also be conflicts between values and agents. If, for example, a HCD project is stuck because of some conflict, an exercise of identifying the values that are at stake, and of discussing these, can help to facilitate collaboration and creativity.

Finally, I would like to propose that people involved in HCD need to make the ethical qualities of HCD (more) explicit. These ethical qualities influence their practices anyway, either negatively—for instance, when they experience misunderstandings and frictions—or positively—for instance, when they experience the joys of learning and creating. In both cases, it would be productive when participants cope with these ethics more consciously, in order to more fully realize the transformative potential of HCD to make projects more participatory, human-centred and co-creative. The people involved in HCD can make these ethics explicit by embracing *reflexivity* (Rhodes 2009) or "professional self-awareness" (Stovall 2011). Reflexivity can help them to become more aware of their feelings, thoughts and acts, of their encounters and interactions, and of their processes of problem-setting and solution-finding.

Acknowledgements

The studies discussed were conducted in three projects: FRUX, which received funding from the Dutch Ministry of Economic Affairs (BSIK 03025); TA2, which received funding from the European Community's Seventh Framework Programme (FP7/2007–2013: ICT-2007–214793); and WeCare, which received funding in the European Ambient Assisted Living Joint Programme (AAL-2009–2–026). This chapter is partly based on Steen (2015), which in turn was based on Steen (2012, 2013a, 2013b).

Related Chapters

Chapter 3: Western Philosophical Approaches and Engineering (Glen Miller)
Chapter 10: Creativity and Discovery (David H. Cropley)

Chapter 13: Systems Engineering as Engineering Philosophy (Usman Akeel and Sarah Bell)
Chapter 22: Values and Design (Ibo Van de Poel)
Chapter 40: Ethical Considerations in Engineering (Wade L. Robison)
Chapter 43: Professional Codes of Ethics (Michael Davis)

Further Reading

Bowles, C. (2018). *Future Ethics*. Hove, UK: NowNext Press. (Excellent introduction to ethics for designers, with a focus on online/digital/data/algorithms, from a designer's perspective.)

Dorst, K. (2015). *Frame Innovation: Create New Thinking by Design*. Cambridge, MA: MIT Press. (Discussion of design thinking as a way to address wicked problems—implicitly related to Dewey's pragmatism.)

Hildebrand, D. (2008). *Dewey: A Beginner's Guide*. Oxford, UK: Oneworld Publications. (An accessible introduction to John Dewey's philosophical pragmatism.)

Peperzak, A., Critchley, S. and Bernasconi, R. (eds.) (1996). *Emmanuel Levinas: Basic Philosophical Writings*. Bloomington and Indianapolis, IN: Indiana University Press. (Some of Levinas's basic philosophical writings.)

Vallor, S. (2016). *Technology and the Virtues: A Philosophical Guide to a Future Worth Wanting*. Oxford, UK: Oxford University Press. (Excellent discussion of "technomoral" virtues that we need in order to flourish in the 21st century.)

Van de Poel, I. and Royakkers, L. (2011). *Ethics, Technology and Engineering: An Introduction*. Chichester: Wiley-Blackwell. (Thorough and practical treatment of ethical issues for engineers and engineering practice.)

References

Bijker, W.E. (1993). Do Not Despair: There is Life After Constructivism. *Science, Technology, & Human Values,* 18(1), 113–138.

Bjerknes, G. and Bratteteig, T. (1987). Florence in Wonderland: System Development with Nurses. In G. Bjerknes, P. Ehn and M. Kyng (eds.), *Computers and Democracy: A Scandinavian Challenge*. Aldershot: Avebury, pp. 279–296.

Bjerknes, G. and Bratteteig, T. (1995). User Participation and Democracy: A Discussion of Scandinavian Research on System Development. *Scandinavian Journal of Information Systems,* 7(1), 73–98.

Blomberg, J., Giacomi, J., Mosher, A. and Swenton-Hall, P. (1993). Ethnographic Field Methods and Their Relation to Design. In D. Schuler and A. Namioka (eds.), *Participatory Design: Principles and Practices*. Hillsdale, NJ: Lawrence Erlbaum Associates, pp. 123–155.

Bødker, S., Ehn, P., Kammersgaard, J., Kyng, M. and Sundblad, Y. (1987). A Utopian Experience: On Design of Powerful Computer-based Tools for Skilled Graphic Workers. In G. Bjerknes, P. Ehn and M. Kyng (eds.), *Computers and Democracy: A Scandinavian Challenge*. Aldershot: Avebury, pp. 251–278.

Bratteteig, T. and Stolterman, E. (1997). Design in Groups—and All That Jazz. In M. Kyng and L. Mathiassen (eds.), *Computers and Design in Context*. Cambridge, MA: MIT Press, pp. 289–315.

Bucciarelli, L. (1994). *Designing Engineers*. Cambridge, MA: MIT Press.

Button, G. (2000). The Ethnographic Tradition and Design. *Design Studies,* 21(4), 319–332.

Cooper, A. (1999). *The Inmates Are Running the Asylum: Why High-tech Products Drive Us Crazy and How to Restore the Sanity*. Indianapolis, IN: SAMS Publishing.

Cross, N. (2006). *Designerly Ways of Knowing*. London: Springer-Verlag.

Dalsgaard, P. (2009). *Designing Engaging Interactive Environments: A Pragmatist Perspective*. PhD dissertation, Aarhus University, Aarhus.

Derrida, J. (1989). Psyche: Inventions of the Other (Translated by Catherine Porter). In L. Waters and W. Godzich (eds.), *Reading de Man Reading*. Minneapolis, MN: University of Minnesota Press, pp. 25–64.

Derrida, J. (1995). Dialanguages. In *Points . . . Interviews, 1974–1994*. Stanford, CA: Stanford University Press, pp. 132–155.

Derrida, J. (2001). Deconstructions: The Im-possible. In S. Lotringer and S. Cohen (eds.), *French Theory in America*. New York and London: Routledge, pp. 12–32.

Devon, R. (2004). Towards a Social Ethics of Technology: A Research Prospect. *Techne: Research in Philosophy and Technology,* 8(1), 99–115.

Devon, R. and Van de Poel, I. (2004). Design Ethics: The Social Ethics Paradigm. *International Journal of Engineering Education,* 20(3), 461–469.

Dewey, J. (1917). The Need for a Recovery of Philosophy. In J. Dewey (ed.), *Creative Intelligence: Essays in the Pragmatic Attitude*. New York: Henry Holt and Co, pp. 3–69.

Dewey, J. (1920). *Reconstruction in Philosophy*. New York: Henry Holt and Co.

Dewey, J. (1938). *Logic: The Theory of Inquiry*. New York: Henry Holt and Co.

Dewey, J. (1965). *Experience and Nature*. La Salle, IL: Open Court Publishing.

Dorst, K. (2011). The Core of 'Design Thinking' and its Application. *Design Studies, 32*(6), 521–532.

Ehn, P. (1993). Scandinavian Design: On Participation and Skill. In D. Schuler and A. Namioka (eds.), *Participatory Design: Principles and Practices*. Hillsdale, NJ: Lawrence Erlbaum Associates, pp. 41–77.

Ehn, P. and Kyng, M. (1991). Cardboard Computers: Mocking-it-up or Hands-on the Future. In J. Greenbaum and M. Kyng (eds.), *Design at Work: Cooperative Design of Computer Systems*. Hillsdale, NJ: Lawrence Erlbaum Associates, pp. 169–196.

Emison, G.A. (2004). American Pragmatism as a Guide for Professional Ethical Conduct for Engineers. *Science and Engineering Ethics, 10*(2), 225–233.

Fesmire, S. (2003). *John Dewey and Moral Imagination: Pragmatism in Ethics*. Bloomington, IN: Indiana University Press.

Friedman, B., Kahn, P.H. and Borning, A. (2006). Value Sensitive Design and Information Systems. In P. Zhang and D. Galletta (eds.), *Human-Computer Interaction in Management Information Systems: Foundations*. New York: M.E. Sharpe Inc, pp. 348–372.

Hickman, L.A. (1990). *John Dewey's Pragmatic Technology*. Bloomington and Indianapolis, IN: Indiana University Press.

Hickman, L.A. (1998). Dewey's Theory of Inquiry. In L.A. Hickman (ed.), *Reading Dewey: Interpretations for a Postmodern Generation*. Bloomington and Indianapolis, IN: Indiana University Press, pp. 166–186.

Hildebrand, D. (2008). *Dewey: A Beginner's Guide*. Oxford: Oneworld Publications.

ISO (2019). *ISO 9241-210:2019: Ergonomics of Human-System Interaction—Part 210: Human-Centred Design for Interactive Systems*. Geneva, Switzerland: ISO.

Kensing, F. and Blomberg, J. (1998). Participatory Design: Issues and Concerns. *Computer Supported Cooperative Work, 7*(3–4), 167–185.

Kensing, F. and Madsen, K.H. (1991). Generating Visions: Future Workshops and Metaphorical Design. In J. Greenbaum and M. Kyng (ed.), *Design at Work: Cooperative Design of Computer Systems*. Hillsdale, NJ: Lawrence Erlbaum Associates, pp. 155–168.

Keulartz, J., Schermer, M., Korthals, M. and Swierstra, T. (2004). Ethics in Technological Culture: A Programmatic Proposal for a Pragmatist Approach. *Science, Technology, & Human Values, 29*(1), 3–29.

Lawson, B. (2006). *How Designers Think: The Design Process Demystified,* 4th ed. Amsterdam: Elsevier.

Levinas, E. (1987). Philosophy and the Idea of Infinity. Trans. Alphonso Lingis [original 1957]. In *Collected Philosophical Papers*. Dordrecht: Martinus Nijhoff Publishers, pp. 47–59.

Levinas, E. (1996). Transcendence and Intelligibility [original 1984]. In A. Peperzak, S. Critchley and R. Bernasconi (eds.), *Emmanuel Levinas: Basic Philosophical Writings*. Bloomington and Indianapolis, IN: Indiana University Press, pp. 149–159.

Lloyd, P. (2008). Ethical Imagination and Design. *Design Studies, 30*(2), 154–168.

MacIntyre, A. (2007). *After Virtue*, 3rd ed. London: Duckworth.

Nielsen, J. (1993). *Usability Engineering*. London: Academic Press.

Norman, D.A. (1988). *The Psychology of Everyday Things*. New York: Basic Books.

Rhodes, C. (2009). After Reflexivity: Ethics, Freedom and the Writing of Organization Studies. *Organization Studies, 30*(6), 653–672.

Sanders, E.B.N. (2000). Generative Tools for Co-designing. In S.A.R. Scrivener, L.J. Ball and A. Woodcock (eds.), *Collaborative Design: Proceedings of CoDesigning 2000*. London: Springer-Verlag, pp. 3–12.

Sanders, E.B.N. and Stappers, P.J. (2008). Co-creation and the New Landscapes of Design. *CoDesign, 4*(1), 5–18.

Schuler, D. and Namioka, A. (1993). *Participatory Design: Principles and Practices*. Hillsdale, NJ: Lawrence Erlbaum Associates.

Sleeswijk Visser, F. (2009). *Bringing the Everyday Life of People Into Design*. Doctoral Dissertation, Delft University of Technology, Delft.

Steen, M. (2011). Tensions in Human-centred Design. *CoDesign, 7*(1), 45–60.

Steen, M. (2012). Human-centred Design as a Fragile Encounter. *Design Issues, 28*(1), 72–80.

Steen, M. (2013a). Co-design as a Process of Joint Inquiry and Imagination. *Design Issues, 29*(2), 16–29.

Steen, M. (2013b). Virtues in Participatory Design: Cooperation, Curiosity, Creativity, Empowerment and Reflexivity. *Science and Engineering Ethics, 19*(3), 945–962.

Steen, M. (2015). Upon Opening the Black Box and Finding it Full: Exploring the Ethics in Design Practices. *Science, Technology, and Human Values, 40*(3), 389–420.

Stovall, P. (2011). Professional Virtue and Professional Self-awareness: A Case Study in Engineering Ethics. *Science and Engineering Ethics, 17*(1), 109–132.

Stuhr, J.J. (1998). Dewey's Social and Political Philosophy. In L.A. Hickman (ed.), *Reading Dewey*. Bloomington and Indianapolis, IN: Indiana University Press, pp. 82–99.

Thackara, J. (1999). An Unusual Expedition (Preface). In K. Hofmeester and E. De Charon de Saint Germain (eds.), *Presence: New Media for Older People*. Amsterdam: Netherlands Design Institute, pp. 7–9.

Thackara, J. (2006). *In the Bubble: Designing in a Complex World*. Cambridge, MA and London, England: MIT Press.

Van de Poel, I. and Verbeek, P.-P. (2006). Ethics and Engineering Design. *Science, Technology, & Human Values,* 31(3), 223–236.

Winner, L. (1993). Upon Opening the Black Box and Finding it Empty: Social Constructivism and the Philosophy of Technology. *Science, Technology, & Human Values,* 18(3), 362–378.

25

SUSTAINABLE DESIGN

Steven A. Moore

The topic of *sustainable design* combines a socially contested modern concept (sustainability) with an established ancient practice (design). To begin this essay it will be helpful to define both terms before they are tested as a combined concept in subsequent sections.

For designers (Chapters 21–26, this volume), *sustainability* (see also Chapter 34, "Engineering and Sustainability: Control and Care in Unfoldings of Modernity" by Andy Stirling, this volume) is best understood, not as an *algorithm*, but as a *storyline* (Dryzek 1997; Eckstein and Throgmorton 2003; Moore 2007). An algorithm is generally understood to be an automated path devised to solve a class of problems with a high degree of *certainty*. Our current global situation may make it rational to want such a tool to solve our problems, but it is unlikely, or even undesirable, that we will construct one. Although some enthusiasts (Kurzweil 2005) believe that artificial intelligence (AI) will soon have the capacity to make such calculations, others are highly skeptical that AI will ever capture the complexity of context-dependent problems like sustainability (Scott, 1998; Latour 2017). In contrast, a storyline can be understood to be a socially constructed path devised to tell a specific, open-ended story with a reasonable degree of *uncertainty* (see also Chapter 11, "Uncertainty" by William M. Bulleit, this volume). Although engineers are prone to prefer the certainty of algorithms, the concept of sustainable design is a problem-context in which the modern quest for scientific certainty simply cannot be achieved (Rittel and Webber 1973). By embracing uncertainty, the experimental construction of *storylines* may be more useful as a tool (Lanham et al. 2016; see also Chapter 12, "Scenarios" by Christian Dieckhoff and Armin Grunwald, this volume).

In the most basic or biological sense, a *sustainable* system is one that has the capacity to reproduce itself. In isolation, we imagine that a sustainable system would reproduce itself *ad infinitum*. Unfortunately (or perhaps fortunately for us), no system exists in isolation. Rather, all systems are nested within others. In other words, all systems are subject to unanticipated disturbances from outside the boundaries humans establish to understand and study them. The sustainability of any single system we might identify is, then, uncertain because the boundaries constructed by human are porous.

The second half of the topic, *design*, is an ancient social practice by which humans plan the construction of an *action*, *artifact* or *system*. Human actions can be designed to restore functionality to systems (intertidal zones, for example) that have been disrupted by higher-order systems (e.g., human fisheries) that have lost the ability to reproduce themselves. Other categories of human action can be designed to create novel systems (aquiculture) with predictable outcomes (tons of seaweed per year). In all cases, the practice of design can said to be hypothesis-making. A *design* is a prediction that *if you build in this particular way, we expect a, b or c to be the consequence.* Herbert Simon (1969: 55) put it this way:

> Every one designs who devises courses of action aimed at changing existing conditions into preferred ones. The intellectual activity that produces material artifacts is no different fundamentally from the one that prescribes remedies for a sick patient or the one that devises a new sales plan for a company of a welfare policy for the state. Design, so construed, is the core of all professional training: it is the principal mark that distinguishes the professions from the sciences.
>
> *(p. 55)*

Simon (1969: 56) also made a further distinction in arguing, "The natural sciences are concerned with how things are . . . Design, on the other hand, is concerned with how things ought to be." Given that the "existing condition" of the world is unsustainable, the project of design is to transform those conditions into preferred, or sustainable ones.

The danger here is to assume that the actions we design will accurately transform the highly complex relationships between existing *ecological*, *social* and *technical* variables into the conditions we prefer for the future. The pathway of conditions, from existing to preferred, is fraught with dynamic variables and epistemological uncertainties that require constantly shifting data and interpretations along the way. For these reasons, the practice of sustainable design is discussed in this chapter across four sections: (1) *precedents*, (2) *definitions*, (3) minor *critiques* and (4) major *challenges*. This historical and theoretical approach provides context for the reader to take considered action in the future.

25.1. Precedents of Sustainability

It is commonly assumed that the idea of sustainability emerged in response to the politically prompted global energy crises of 1973 and 1979. As one response to these events, in 1980 the International Union for the Conservation of Nature (IUCN 1980), published what is thought to be the first definition of the contemporary concept, which links "social, economic, and political progress" as does the three legs of a common stool (Figure 25.1). It is this *balanced* model of "sustainable development" that was adopted by the World Commission on Environment and Development (WCED 1987) a United Nations-sponsored deliberation chaired by Norway's former Prime Minister, Gro Harlem Brundtland. The final report of this three-year international negotiation, *Our Common Future*, argued that "Humanity has the ability to make development sustainable—to ensure that it meets the needs

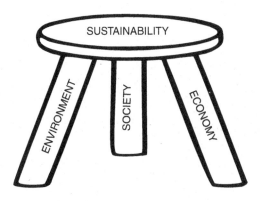

Figure 25.1   The three-legged stool of sustainability

of the present without compromising the ability of future generations to meet their own needs" (WCED 1987: 8). This definition was criticized upon its publication, and ever after, as being weak, indeterminate and impossible to measure (Dammann and Elle 2006). However, others argue that the *constructive ambiguity* of the meme is precisely what has nurtured a "pluralist" global dialog about the competing interests of *ecological preservation*, *economic development* and *social equity* (Guy and Moore 2005). To understand how these sets of interests came into conflict so aggressively at the end of the twentieth century it is helpful to consider deeper and older cultural precedents.

Premodern proposals for resource preservation (those in circulation prior to the mid-seventeenth century) were published in England by John Evelyn (1664), in Japan by Shogun's Proclamation (1666) and in Germany by Hans Carl von Carlowsky (1713) (Moore 2002; Moore and Wilson 2014; Diamond 2005). These early proposals were what we now call *sumptuary codes* (see also Chapter 42, "Standards in Engineering" by Paul B. Thompson, this volume, and Hunt 1996), or edicts that limit what specific social classes must, or must not consume. The regulation of consumption emerged in recognition that local forests, essential for development of any kind, could be depleted by unregulated human practices. In this premodern context, having to import wood over long distances was perceived as an economic burden to be avoided. However, an economic burden may be necessary, yet insufficient to constitute an *unsustainable* condition in the modern sense. It was not until the nineteenth century that modern knowledge in physics, biology and economics were available to envision the broader consequences of resource depletion at the core of sustainability as a modern concept.

In physics, Rudolph Claussius (1822–1888) developed the innovative concept of *entropy* about 1865, which posed that, in a closed system, certain changes are *irreversible*. If forests, mineral deposits, or water, for example, are understood as closed systems, it became possible, for the first time in history, for moderns to envision how human extraction practices could consume resources to a point of irreversibility—*where the system could no longer reproduce itself.*

Moderns conceptualized the cosmos as a giant clock in which the entropy-producing loss of each tree, mineral or jug of water represented one tick of the clock toward chaos and the end-time. The fundamental concept of a "cosmos" was itself popularized by the geographer and polymath Alexander Humboldt (1769–1859). He meant by this term not God or His realm of the heavens, but a complex yet rational and orderly system which was available for *human management*—a design process intended to slow the clock's ticking.

In biology, Charles Darwin (1809–1882) developed the concept of *evolution* some time before the 1859 publication of his historic work, *On the Origin of Species*. A key observation by Darwin was that diversity of genetic stock in species of any kind is a necessary, if insufficient, condition for successful reproduction. Should diverse genetic stock be depleted by processes of over-hunting, climatic events, or war, remaining species have reduced choices. Reducing the number and kinds of choices that any individual can make is to change the possible trajectories of its history. In other words, Darwinian history is not linear, as Christians had imagined. In the shadow of new knowledge moderns could conceptualize, for the first time, an inglorious end to human history.

At about the same time, Ernst Heinrich Philipp August Haeckel (1834–1919)—who studied the diverse fields of zoology, biology, medicine, politics, philosophy and art—built upon work of others to promote *a new concept which he called ecology*. Where most of his contemporary scholars studied species in isolation, Haekel argued for the indivisible relationship between species and their environmental context. One could not, Haekel argued, study the frog effectively without also studying the frog pond. Although this idea became popular amongst German romantics and racial *polygenists* (those who argued that the human "races" have not one, but multiple origins), it also contributed to transforming modern approaches to environmental development.

The term globalization was not coined until the 1970s (Smith 2013). Although the phenomenon of Euro-American spatial expansion began much earlier, radical transformation of transportation and communication technologies only became visible to historians in the modern era. Sailing ships in the Age of Discovery (the fifteenth and sixteenth centuries) opened the European world to the colonial exploitation of distant cultures, resources and the development of new investment opportunities by capital. These technologies and economic forms gave way to steam locomotives, the telegraph and the corporation in the nineteenth century. By the twentieth century, rapid technological development led to equally rapid ecological and cultural disruption. Although the banner of *international development* was initially welcomed in some places, it soon became clear to colonized societies that capitalist development would inevitably be "uneven" (Smith, 1982), leaving those inhabiting sites of short-term investment to deal with the ecological and cultural aftermath of extractive capital (Harvey 1996).

These five powerful ideas—*entropy, cosmos, evolution, ecology* and *globalization*—among other political and social changes—transformed how we modern humans conceptualize our situation. Sustainability is, then, an inherently modern meme of balanced, competing forces, not one appropriated from a romanticized or fabricated past. Although these prescient ideas had almost immediate consequences in a few realms, such as the emergent idea of *public health* (Melosi 2000) developed in the context of British industrialization and the American Civil War, they did not gain popular recognition or practical application in the US until the second or third decade of the twentieth century.

As suggested earlier, it was not only in science that major changes were afoot. English utilitarian philosophers Jeremy Bentham (1748–1832) and John Stuart Mill (1806–1873) had a strong influence not only in Britain, but in the American colonies, particularly upon the authors of the American Constitution (1776). Perhaps surprisingly, ideas incorporated there are those we now associate with *sustainability*. At the core of utilitarianism is "the Greatest Happiness Principle," or the idea that the experience of "pleasure and pain" constitute the standard by which choices, made by individuals or societies, maximize the greatest good. This seemingly hedonistic calculus was extended to practical considerations for *ethics* and *economics*. In this context the Greatest Happiness Principle translates into the difficulty of balancing benefits to individuals and society as a whole. For example, Bentham held that the poor do not suffer from God's punishment for living slothful lives, as was held by premodern religious superstition, but from living in unhealthful environmental conditions imposed on them by social and spatial hierarchy. Rather than ignore the suffering of the poor, he proposed that society should care for all citizens as potentially useful workers, which would, in turn, contribute to full employment and improve the "civic economy" (Chadwick 1965 [1842]). In Britain, the utilitarian project was most integrated into public works by such "sanitarian" engineers as Edwin Chadwick (1800–1890), who conducted the first empirical survey of health conditions in Britain and subsequently contributed to radical reform of the *Poor Laws*. In the US, the utilitarian project was advanced by Col. George Waring (1833–1898) who developed and realized innovative sanitary sewer technologies in Memphis (following a series of cholera epidemics, 1867–1879), and in squalid New York after 1895 (Melosi 2000). The nineteenth-century utilitarian conception of the social contract is a primary source of what we now call *social equity*, one dimension of sustainable development discussed below as an artifact of the twentieth century.

At the turn of the twentieth century the modern *professions* (see also Chapter 5, "What Is Engineering Science" by Sven Ove Hansson, this volume), as distinct from premodern guilds and arts, were struggling to define their social and economic interests and boundaries. In this competition to develop dominance in the embryonic areas of modern knowledge, new professional disciplines emerged through the cracks that have come to define the professions of architecture, engineering and medicine (see also Chapter 2, "A Brief History of Engineering' by Jennifer Karns Alexander, this volume). The discipline of *public health* emerged as a hybrid of engineering and medicine, just as *city planning* emerged as a hybrid of architecture and engineering (Brain 1994). Chadwick and Waring, discussed earlier, are exemplars of modern public health professionals in the Anglo-American tradition.

At this transitional moment in US history, sociologist Robert Brulle (2000) characterized three primary environmental discourses (the *conservation, preservation* and *reform* movements). These three competing ideas were the standard against which any resolution of the apparent conflict between humans and nature were tested. It was in this discursive context that Sir Patrick Geddes (1854–1932) and Lewis Mumford (1895–1990) developed prescient proposals for urban development that are the antecedents of modern sustainable design. Over a century of writing, planning and designing, Geddes and Mumford developed historical and empirical models of urbanism that they referred to, in developmental stages as *paloetechnic, neotechnic, geotechnic* and *ecotechnic*. Although Geddes and Mumford never influenced the practice of modern city-making as they had hoped, planner Robert Young (2017) has recently revisited their contribution with fresh insights. It is clear that both of these urbanists understood the inclusion of social equity as a necessary component of what we now call sustainable design.

It was, then, only in the late twentieth century that sustainability emerged, with the authorization of the UN WCED, as an "umbrella concept" (Allen and Sachs 1992; Thompson 2010) which aspired to provide a big tent under which many disparate discourses might roost. Sponsorship of the UN WCED also provided an opportunity for non-Western traditions to influence, or appropriate, this emergent set of ideas. Examples include: the cosmology of *Jainism* in the Indian subcontinent (which challenges the dominance of *homo sapiens* in the world order); the Confucian ideals of social sustainability (particularly with regard to seniors); and the adaptation of traditional craft approached (appropriated from Lapland or LA) as a model of *place-based production*. Non-Western cultural practices provide an almost endless supply of practical examples upon which moderns might draw, without embracing a romantic view of the past.

In sum, these pre- and early-modern precedents both expand, and limit, what the modern concept of sustainability would become. The same can be said of the new sub-discipline of sustainable design.

25.2. Defining Sustainability

The definition of "sustainable development" by the World Council on Economic Development (WCED 1987), cited in Section 1 can be said to be *seminal*. This term is generally reserved for those texts that both *reflect* and *contribute to* a "paradigm shift" (Kuhn 2012 [1962]) in how the world understands itself. After the publication of *Our Common Future*, it became increasingly difficult to isolate the 3 Es—*economy, ecology* and *equity*—from any public conversation about global development. The linkage of these terms became only stronger as the reality of *climate change* became more scientifically and publicly accepted. But, as argued in the introduction, the vague definition provided by WCED left the door open for others to add specificity.

One attempt to be clearly quantitative was by industrial and chemical engineer David Allen. Industrial engineering can, argues Allen, be understood as the "science of sustainability." As a demonstration he and his colleagues (Allen et al. 2010) invoked an equation initially developed by Ehrlich and Holdren to study energy use in the United States.

> The equation relates impact (I), which can represent energy consumption, materials use, environmental emissions, or other types of impacts, to population (P), affluence (A) and technology (T).
>
> $$I = P * A * T$$
>
> This conceptual relationship, often referred to as the IPAT equation, suggests that impacts are the product of the population (number of people), the affluence of the population (generally expressed as gross domestic product of a nation or region, divided by the number of

people in the nation or region), and the impacts associated with the technologies used in the delivery of the affluence (impact per unit of gross domestic product). If the IPAT equation were used to describe energy use in the United States, then I would represent energy use per year, P would represent the population of the United States, A would represent the annual GDP per capita, and T would represent the energy use per dollar of GDP.

(Allen et al. 2010: 106)

The result of Allen's investigation provides very useful insight, yet it is inconclusive in the end because the team itself recognized "there are clearly societal and cultural implications of mobility that influence design decisions" (Allen et al. 2010: 113). Like most, this highly talented group of engineers accepted the popular notion that *sustainability* is synonymous with *energy efficiency* because they didn't know how to calculate the complicated question of *social equity*. In other words, recommendations for action—which must be the goal of sustainability design—require measurement of the qualitative messiness of community engagement.

Perhaps the most influential model of sustainability has been the "Planners' Triangle," developed by Scott Campbell (1996) of the University of Michigan (Figure 25.2). What Campbell achieved was to maintain the clarity of the three competing forces identified by the IUCN in 1984 at the points of the equilateral triangle, and then adding a layer of information along the triangle's sides that identifies the nature of the conflict between those social forces. In referring to such conflicts as "rational" (Campbell 1996), he fabricated a platform for *conflict negotiation* between the competing interests of *economic development*, *environmental protection* and *social equity*.

In the design professions and building industry, an ever-growing number of not-for-profit certification organizations have subsequently emerged to provide designers with tools to calculate how balance between the three poles can be achieved (Conroy 2007; GBA 2018). Each organization, of course, developed scoring systems that reflect their own frame of interpreting and transforming conditions. For example, by far the largest of these certification organizations, the US Green Building Council (USGBC), developed the LEED (Leadership in Energy and Environmental Design) certification program. LEED tends to emphasize the interests of economic development by favoring innovative technologies that reduce energy consumption and avoiding the problem of social equity. In contrast, the SEED (Social, Economic, and Environmental Design) Network has developed an "an interactive software program that provides a protocol to help guide, document, [and] evaluate" design decisions in a manner that emphasizes social equity and "community values" (SEED 2018). A third certification organization, the Living Building Challenge, emphasizes the interests of environmental preservation by valuing "net zero energy and water" technologies over economic or social variables (LBC 2018). Each certification system provides an incentive for designers to make decisions that have concrete eco-sociotechnical consequences in the overlapping sectors of the Planners' Triangle. Sustainability certification systems are, then, an attempt by civil society groups to codify the built environment from outside the authority allowed to government or assumed by industry (Moore and Wilson 2014).

Upon the twentieth anniversary of Campbell's prescient publication of 1996, planner Lisa Schweitzer conducted a citation analysis and concluded,

> The article has been cited more than a thousand times according to Google Scholar, and a little more than 272 times according to Web of Science, which puts Campbell's contribution among the top most-cited articles ever to have appeared in JAPA.

(Schweitzer 2016: 374)

Her quantitative analysis also demonstrated the tremendous influence of Campbell's model beyond the profession of planning.

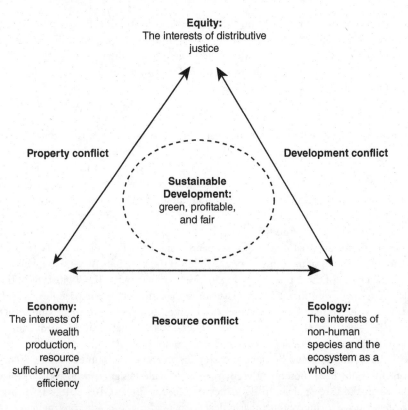

The Concept of Sustainability is inscribed within a triangle of
competing interests. In this construction, the concept is necessarily
discursive and democratic.
The Development Conflict sets those with an interest in protecting
the environment against those with an interest in distributing
available resources.
The Property Conflict sets those who control the means of
production against those with an interest in distributive justice.
The Resource Conflict sets those with an interest in economic
development against those with an interest in resource conservation.
The Sustainable City is one that negotiates and balances conflict and
each set of competing interests.

Figure 25.2   The Planners' Triangle

Source: Derived from Campbell (1996).

If the Planner's Triangle has become the dominant definition of sustainability for designers and
others, it has also been controversial. In the context of this chapter, it will be helpful to review the
most significant critiques because they will influence how the storyline of sustainability bends into
the future (see Chapter 55, "Reimagining the Future of Engineering" by Neelke Doorn, Diane P.
Michelfelder, et al., this volume).

25.3. Critiques of the Dominant Triangular Model

There are at least four "substantive critiques" (Moore 2016) of the dominant triangular model of
sustainability that deserve attention: *missing variables*, *the suppressed E*, *compromise versus transformation*
and *lack of complexity*.

25.3.1 Missing Variables

Sonia Hirt (2016: 383) argues that the model has omitted two important variables: "art and health, which have been with us since the 19th century." If we accept the hypothesis developed earlier, that the concept of *sustainability* is most useful as a storyline that convenes supporters and motivates action, and not as a formula for energy efficiency, Hirt's critique makes sense. With hindsight, omitting from the playing field for dominance (Owen and Dovey 2008) those actors first concerned with "beauty" (architects), and those first concerned with human "health" (medical professionals), seems to handicap the effort from the start. Had Campbell configured a more inclusive *pentagram*, rather than a *triangle*, the acceptance of sustainability into these domains might have been either faster (by attracting more adherents) or hybrid (attracting and integrating more ideas). The very nature of hindsight is, however, the impossible plot of substituting one historic context for another. If we could return to the 1980s, we would find two conditions that may explain Campbell's choices. First, architects of that era were primarily engaged in a highly abstract and isolated discourse about style (postmodern historicism) that made them unlikely receptors of sustainability. Second, with regard to health, Hirt observed that

> The original intersection of health and planning had to do with fears from the respiratory and contagious diseases that ravaged late-19th and early- 20th century U.S. cities as a result of widespread pollution and lack of sanitation. But these diseases had come largely under control by the mid-1900s.

In other words, reaching back to the then-dated origins of planning as a profession may have seemed more like going backwards to Campbell, rather than forward. The irony in such time-bound reasoning is that, by 2015, public health had again become a key variable in thinking about the built environment, but with regard to new maladies. In lieu of tuberculosis and typhoid (the urban diseases of the nineteenth century) asthma and obesity (which reflect the threats inherent in twenty-first-century cities) have come to take their place. The appearance of the Covid-19 pandemic, which ravaged the world beginning in January 2020, almost immediately challenged the norms of sustainable design. The good part of the substantive critique by Hirt is that it opens up possibilities for developing new models.

25.3.2 The Suppressed E

A second substantive critique is made by economist Michael Oden (2016). In this case the critique is aimed less at Campbell's logic (although Campbell's articulation of the concept could have been more robust) than at the public's tendency to avoid the topic of *social equity* as a necessary condition to achieve *ecological* and *economic* sustainability. This tendency persists in spite of the utilitarian foundation of the US Constitution discussed earlier (which argues for social equity) and tends to reduce the concept of sustainability to "greenwashing." In response, Oden identifies several "durable, conceptual and ideological divides that make serious discussion of equity highly charged and politically problematic." Upon that ground he develops the idea of "complex equity," initially proposed by Waltzer as a way forward. In elegant terms, Oden (2016: 34) argues: "*It is not inequalities within individual spheres that constitute the principal problem of equity, rather it is inequalities in one sphere spilling over and shaping distributions in another sphere with different intrinsic values and standards of distribution*" (italics original). To make Oden's logic concrete, one might argue that US citizens are likely to complain vociferously when ISIS mullahs in Syria, for example, intervene in the determination of what qualifies as art. However, the same US citizens are neither likely to notice nor complain when hedge fund managers sit on museum boards to make the same determination (Moore 2016). Both mullahs and hedge-fund managers operate beyond their social spheres. Oden's point is that such *spill-over effects* are ontologically inequitable and harmful to individuals *and* society in both contexts. *Social equity is not the uniform distribution of wealth; it is the dynamic recognition of value within each sphere of society.*

25.3.3 *Compromise Versus Transformation*

A third substantive critique observes the distinction between *compromise* and *transformation* (Moore 2016). As proposed by Campbell, "compromise" is the discursive process that moves the competing advocates of economy, ecology and equity to a shared position at the center of the triangle. In business it is a well-established truism that if no party is happy after a negotiation, it must have been a success. The problem with this assessment is that most analysts of our current predicament argue that we are so close to ecological collapse that *transformation*, not *compromise*, is required to avoid major economic and political disruptions (Rees 2004). In this view, even *good* business as usual will be inadequate to the task.

A more radical alternative is offered by those pragmatist philosophers (James 1977 [1899]; Barber 1984) and planners (Blanco 1994; Holden 2008) who advocate for *difficult* public discourse, in lieu of *easy* compromise, as a road forward. Pragmatists hold that compromise requires us only to be a bit less vehement in holding our ground. In contrast, the pragmatist proposal for "social learning" is "learning achieved through the practice of collective enquiry and public deliberation over public goals" (Minteer 2002: 3). Through compromise, individuals and society learn nothing, other than to give up as little as possible. But, through social learning, *individuals* learn to listen for unexpected moments of common experience, and *societies* learn to construct new epistemologies.

25.3.4 *Lack of Complexity*

A fourth substantive critique of the Planners' Triangle is offered by advocates of *complexity theory*, particularly Lanham et al. (2016). In simple terms, where advocates of the Planners' Triangle seek compromise and lasting *balance* at the center of any dispute, complexity theorists understand the cosmos, and any situation within it, to be *dynamic* and never balanced, never at rest—it is a cosmos where actors must constantly adapt to change. Another way to distinguish between these two cosmologies is to understand the Planners' Triangle to be a map, or snapshot of a particular conflict. In such a closed system, as defined only by three competing interests, it may be possible to achieve *certainty* of the outcome. In contrast, complexity theorists insist that the boundaries to any system, particularly one in conflict, are artificial intellectual constructs that only mask actual conditions. To help us appreciate our situation, they employ the graphic metaphor of a "fitness landscape" (Figure 25.3) in which the only certainty is *uncertainty*. In the dynamic landscape illustrated in this figure, one's point of view is constantly changing. At one moment a scientific observer may be on a peak where most data is visible, but in the next she might be in a valley where new data might surprise an observer from behind.

There are other critiques of the Planners' Triangle, but most of these are friendly amendments of a minor nature (Moore 2016). However, in addition to these minor and major critiques of sustainability as a three-sided arbitration process, there are more recent concepts that vie to replace sustainability altogether as a core modern concept.

25.4. Challenges to Sustainability

Like any powerful theory, *sustainable design* has been tested by multiple hypotheses from many disciplines, in this case three: *ecology*, *economics* and *philosophy*. Each of these deserve brief mention before tests from within the disciplines of design can be discussed.

From the discipline of ecology, three hypotheses have appeared that challenge the *balanced interests* seen by moderns as central to designing sustainable systems. The sub-discipline of *political ecology* has sought to politicize the idea that technocrats might quantitatively balance competing claims through the design of algorithms, certification criteria or codes as discussed earlier. As early as 1935, Frank Thone used the term "political ecology" to highlight the fundamental political conflicts and injustices embedded in the concept of ecological preservation. In contrast, the hypothesis of *civic*

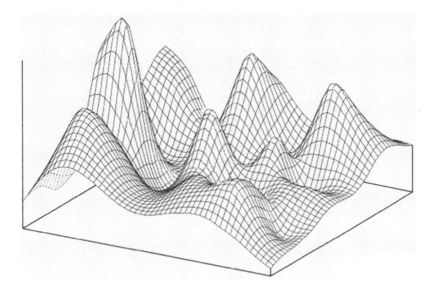

Figure 25.3   Example of a fitness landscape

Source: Adapted from a visualization created by Randy Olson, CC BY-SA 3.0.

environmentalism, initially developed by Dewitt John (1994), argues for a pragmatic rather than theoretical approach for "designing answers to environmental problems," by engaging "a critical mass of community leaders, local activists, and businesspersons [to] work with frontline staff of federal and state agencies" (John 1994: 219). Third, the more well-known hypothesis of *resilience* (see Chapter 35, "The Role of Resilience in Engineering" by Neelke Doorn, this volume), as initially developed by C.S. Holling (1973), has been key to understanding the dynamic capacity of systems to "bounce-back" or "bounce-forward" (Stevenson et al. 2016) after experiencing a disruption inflicted through a higher-order system. Paul B. Thompson (2010), however, has thoughtfully argued that each of these hypotheses may add nuance to the basic theory of sustainability developed in the 1980s, but none of the challenges introduced from the discipline of ecology are inconsistent with those proposed by the original framers.

From the discipline of economics, no fewer than two relevant hypotheses have emerged that challenge the dominant theory of sustainability. The literature regarding *degrowth*, as summarized by Schneider et al. (2010: 511), defines that idea

> as an equitable downscaling of production and consumption that increases human well-being and enhances ecological conditions at the local and global level, in the short and long term. The paradigmatic propositions of degrowth are that economic growth is not sustainable and that human progress without economic growth is possible.

Although the degrowth hypothesis has significant implications for design practice, as witnessed in the *tiny house* movement, it is more of a strategy to realize sustainable development than a new theory in itself. Likewise, *circular economics* is a term championed by some (Geissdoerfer et al. 2017) as a "new paradigm," but as discussed later in the context of "regenerative design," it is another addendum to sustainability theory than a paradigm shift.

If the disciplines of ecology and economics have not seriously challenged sustainable design theory, philosophy has. Earlier in this chapter, sustainability was characterized as an essentially *modern* idea. Bruno Latour has, however, proposed a cosmological shift away from modern scientific assumptions that defies sustainability theory in a substantive way. In his 2017 book, *Facing GAIA*, Latour reconstructs the concept of GAIA, originally authored by James Lovelock (1991) in terms that are clearer and more substantive than Lovelock's original text. In the simplest of terms, Latour argues that *modern* philosophies have characterized *nature* as a set of knowable systems that (a) infinitely interact with certain outcomes that are also (b) available for human agents to predict and control. But, in lieu of such immutable laws of nature, the Lovelock-Latour hypothesis argues that nature too is *uncertain*. The GAIA hypothesis holds that what moderns refer to as a singular *God*, or *Nature*, is better understood as related, yet distinct forces embedded in the interactive zone of earth's crust. Most important is that *these plural forces exert agency themselves in responding to geological history, and in constructing novel futures*. If the GAIA hypothesis has validity, *natural processes are not fixed* as moderns assumed.

This fundamental flaw in modern cosmology spills over into the methods we have used to study nature(s) (see also Chapter 6, "Scientific Methodology in the Engineering Sciences" by Mieke Boon, and Chapter 21, "Engineering Design" by Peter Kroes, this volume). For example, the logic of the *ceteris parabus* clause, which holds that in the space and time of experimentation (see also Chapter 31, "Experimentation" by Viola Schiaffonati, this volume) "all other things remain equal" outside the laboratory, is a reductive convenience that masks dynamic ontological conditions—the conditions of *being* in the world. In this radically altered context, *designing preferred conditions* becomes a very different practice.

Within the design disciplines, two hypotheses have attempted to embrace what Latour (1993) refers to as a *nonmodern* (or what Dewey long before him 2012 [1954]) referred to as *unmodern*) epistemology. These are the *smart city* and *regenerative sustainability*. The literature theorizing *the smart city* is large and growing from many disciplinary directions that cannot be easily summarized. Perhaps the best explanation of this pluralistic phenomenon is by planning theorist Antoine Picon (2015), who argues that globalizing cities (discussed earlier in this chapter and in Chapter 46, "Global Engineering Ethics" by Pak-Hang Wong, this volume) now compete for economic dominance and sustainable futures by employing interconnected information technologies in the construction of novel infrastructures. New infrastructures are, however, increasingly harder to predict because they emerge not only from material science and industry, but also from their interaction with "social imagination" or historical "imaginaries" (Picon 2018). Radical advocates of the smart city go so far as to propose that new human/digital "cyborgs" (Haraway 1995) will "transcend biology" (Kurzweil 2005) to become collective superorganisms. Others, however, envision such automated feedback systems to be new and dangerous forms of technological determinism.

In opposition to such deterministic design problems and design processes dominated by elites, the advocates of "regenerative sustainability" emphasize "the quality of social processes" as well as "technical competence" (Cole 2006: 304). The objective of these scholars is to design "a 'culture of assessment' that places enormous value in objective evaluations" (2006: 307) of historical, qualitative and quantitative data, not only by expert technicians but also *by citizens themselves*. In lieu of a smart city dominated by technical efficiency, the advocates of *regenerative sustainability* propose a messy, but deep democracy enabled by constantly triangulated data.

25.5. Conclusion

The requirement for rigorous assessment of our sustainability projects brings the reader back to a proposal made in the introduction to this chapter, and to the biggest question presented by the topic of *sustainable design*. If all designs are hypotheses, or storylines about how we might live better in the future, how will we judge *what* is better, and *who* will do so? These are, of course, political and philosophical questions, as well as methodological ones, that go well beyond the rigid boundaries established by the *modern concept of sustainability* and *the ancient practice of design*. The logic built here is not that designers must *also* be politicians or philosophers. Rather, it is that none of these traditional disciplines, *in isolation*, has quarantined knowledge adequate to make life-enhancing choices for our diverse communities. What we have to conjure up is how to build ever-evolving interdisciplinary knowledge groups with the capacity to materialize better, yet plural conditions in an uncertain world.

Related Chapters

Chapter 2: A Brief History of Engineering (Jennifer Karns Alexander)

Chapter 5: What Is Engineering Science? (Sven Ove Hansson)

Chapter 6: Scientific Methodology in the Engineering Sciences (Mieke Boon)

Chapter 11: Uncertainty (William M. Bulleit)

Chapter 12: Scenarios (Christian Dieckhoff and Armin Grunwald)

Chapter 20: Towards an Ontology of Innovation (Vincent Blok)

Chapter 21: Engineering Design (Peter Kroes)

Chapter 22: Values and Design (Ibo Van de Poel)

Chapter 23: Design Methods and Validation (Sabine Ammon)

Chapter 24: Human-Centred Design and Its Inherent Ethical Qualities (Marc Steen)

Chapter 31: Experimentation (Viola Schiaffonati)

Chapter 34: Engineering and Sustainability: Control and Care in Unfoldings of Modernity (Andy Stirling)

Chapter 35: The Role of Resilience in Engineering (Neelke Doorn)

Chapter 42: Standards in Engineering (Paul B. Thompson)

Chapter 46: Global Engineering Ethics (Pak-Hang Wong)

Chapter 55: Reimagining the Future of Engineering (Neelke Doorn, Diane P. Michelfelder, et al.)

Further Reading

Abendroth, Lisa M. and Bell, Bryan (eds.) (2015). *Public Interest Design Practice Guidebook*. New York: Routledge. (This compendium is a highly practical guide to community engagement in sustainable design.)

Baird Callicott, J. and Frodeman, Robert (eds.) (2009). *The Encyclopedia of Environmental Ethics and Philosophy*. New York: Gale Cengage Learning/Macmillan. (A very broad consideration of engineering design and environmental ethics.)

Guy, S. and Moore, S. A. (2005). *Sustainable Architectures: Natures and Cultures in Europe and North America*. London: Routledge/Spon. (A compendium that documents plural, place-based practices of sustainability.)

Mitcham, Carl Ed. (2014). *Ethics, Science, Technology, and Engineering: An International Resource*, 2nd edition. New York: Palgrave Macmillan. (This resource considers sustainable design in a broader ethical context.)

Robison, Wade, Raffelle, Ryne and Selinger, Evan (eds.) (2010). *Sustainability Ethics: 5 Questions*. Copenhagen: Automatic Press/VIP. (A compendium of philosophers who consider the same five basic questions concerning sustainability.)

References

Allen, D., Murphy, C.F., Allenby, Braden, R. and Davidson, Cliff I. (2010). Engineering Sustainable Technologies. In S.A. Moore (ed.), *Pragmatic Sustainability: Theoretical and Practical Tools*. New York and London: Routledge, pp. 106–116.

Allen, P. and. Sachs, C. (1992). The Poverty of Sustainability: An Analysis of Current Discourse. *Agriculture and Human Value*, 9(4), 30–37.

Barber, B. (1984). *Strong Democracy: Participatory Politics for a New Age*. Berkeley, CA, Univ. of California Press.

Blanco, H. (1994). *How to Think About Social Problems: American Pragmatism and the Idea of Planning*. Westport, CT, Greenwood Press.

Brain, D. (1994). Cultural Production as 'Society in the MAKING': Architecture as an Exemplar of the Social Construction of Cultural Artifacts. In D. Crane (ed.), *The Sociology of Culture*. Cambridge, MA, Blackwell.

Brulle, R.J. (2000). *Agency, Democracy and Nature: The US Environmental Movement From a Critical Theory Perspective*. Cambridge, MA: MIT Press.

Campbell, S. (1996). Green Cities, Growing Cities, Just Cities: Urban Planning and the Contradictions of Sustainable Development. *APA Journal*, 466–482, Summer.

Chadwick, Edwin (1965 [1842]). *The Poor Law Commissioners on an Inquiry Into the Sanitary Conditions of the Labouring Population of Great Britain* [online source]. London, 1842.

Cole, R. (2006). Building Environmental Assessment: Changing the Culture of Practice. *Building Research & Information*, 34(4), 303–307.

Conroy, M.E. (2007). *Branded: How the 'Certification Revolution' is Transforming Global Corporations*. Gibriola Island, BC: New Society Publishers.

Dammann, S. and Elle, M. (2006). Environmental Indicators: Establishing a Common Language for Green Building. *Building Research & Assessment*, 34(4), 387–404.

Dewey, J. (2012 [1954]). *Unmodern Philosophy and Modern Philosophy*. Edited and with an introduction by Phillip Deen. Carbondale, IL: Southern Illinois University Press.

Diamond, J. (2005). *Collapse: How Societies Choose to Fail or Succeed*. New York: Penguin Press.

Dryzek, J.S. (1997). *The Politics of the Earth: Environmental Discourses*. Oxford and New York: Oxford University Press.

Eckstein, B. and Throgmorton, J. (eds.) (2003). *Story and Sustainability*. Cambridge, MA: MIT Press.

Evelyn, John (1664). *Sylva: A Discourse of Forest, Trees, and the Propagation of Timber*. Teddington, Middlesex: Echo Library, 2009 London.

GBA (2018). *Green Building Alliance*. www.go-gba.org/resources/building-product-certifications/. Accessed October 30, 2018.

Geissdoerfer, Martin, Savaget, Paulo, Bocken, Nancy M.P. and Hultink, Erik Jan (2017). The Circular Economy—A New Sustainability Paradigm? *Journal of Cleaner Production*, 143, 757–768. doi:10.1016/j.jclepro.2016.12.048.

Guy, S. and Moore, S.A. (2005). *Sustainable Architectures: Natures and Cultures in Europe and North America*. London, UK: Routledge/Spon.

Haraway, D. (1995). Situated Knowledge: The Science Question in Feminism and the Privilege of Partial Perspective. In A. Feenberg and A. Hannay (ed.), *Technology & the Politics of Knowledge*. Bloomington, IN: Indiana University Press, pp. 175–194.

Harvey, D. (1996). *Justice, Nature & the Geography of Difference*. Cambridge, MA: Blackwell.

Hirt, Sonia A. (2016). The City Sustainable: Three Thoughts on "Green Cities, Growing Cities, Just Cities". *Journal of the American Planning Association,* 82(4), 383–384.

Holden, M. (2008). Social Learning in Planning: Seattle's Sustainable Development Codebooks. *Progress in Planning*, 69, 1–40.

Holling, C.S. (1973). Resilience and Stability of Ecological Systems. *Annual Review of Ecological Systems*, 4(1), 1–23.

Hunt, A. (1996). *Governance of the Consuming Passions: A History of Sumptuary Law*. New York: Macmillan Press.

IUCN (1980). *World Conservation Strategy*. Gland, Switzerland: International Union for the Conservation of Nature.

James, W. (1977 [1899]). On a Certain Blindness in Human Beings. In J. J. McDermott (ed.), *The Writings of William James: A Comprehensive Edition*. Chicago: University of Chicago Press, pp. 629–644.

John, D. (1994). Civic Environmentalism. *Issues in Science and Technology*, 10(4), 30–34.

Kuhn, Thomas S. (2012 [1962]). *The Structure of Scientific Revolutions*, 4th ed. Chicago, IL: University of Chicago Press.

Kurzweil, R. (2005). *The Singularity Is Near: When Humans Transcend Biology*. New York: Penguin Press.

Lanham, H.J., Jordan, M. and McDaniel, R.R. Jr. (2016). Sustainable Development: Complexity, Balance, and a Critique of Rational Planning. In S. A. Moore (ed.), *Pragmatic Sustainability: Dispositions for Critical Adaptation*. New York: Routledge, pp. 48–66.

Latour, B. (2017). *Facing Gaia: Eight Lectures on the New Climatic Regime*. Trans. C. Porter. Cambridge, UK: Polity Press.

Latour, B. (1993). *We Have Never Been Modern* (C. Porter, Trans.). Cambridge, MA: Harvard University Press.

LBC (2018). *Living Building Challenge*. https://living-future.org/lbc/. Accessed October 30, 2018.

Lovelock, J. (1991). *Gaia: The Practical Science of Planetary Medicine*. Oxford: Oxford University Press.

Melosi, M. (2000). *The Sanitary City: Urban Infrastructure in America from Colonial Times to the Present*. Baltimore: Johns Hopkins University Press.

Minteer, B.A. (2002). Deweyan Democracy and Environmental Ethics. In S.A. Moore and B. P. T. Minteer (eds.), *Democracy and the Claims of Nature: Critical Perspectives for a New Century*. Lanham, MD: Rowman & Littlefield, pp. 33–48.

Moore, S.A. (2002). Sustainability in History and at UT. *Platform: Journal of The University of Texas School of Architecture*. Fall, 4–5, 14–15.

Moore, S.A. (2007). *Alternative Routes to the Sustainable City: Austin, Curitiba and Frankfurt*. Lanham, MD: Lexington Books, Rowman & Littlefield.

Moore, S.A. (2016). Testing a Mature Hypothesis: Reflection on 'Green Cities, Growing Cities, Just Cities: Urban Planning and the Contradiction of Sustainable Development'. *Journal of the American Planning Association*, 82(4), 385–388. doi:10.1080/01944363.2016.1213655

Moore, S.A. and Wilson, B.B. (2014). *Questioning Architectural Judgment: The Problem of Codes in the United States*. New York: Routledge.

Oden, M. (2016). Equity: The Awkward E in Sustainable Development. In S.A. Moore (ed.), *Pragmatic Sustainability: Dispositions for Critical Adaptation*. New York: Routledge.

Owen, C. and Dovey, K. (2008). Fields of Sustainable Architecture. *The Journal of Architecture*, 13(1), 9–21.

Picon, A. (2015). *Smart Cities: A Spatialized Intelligence*. New York: Wiley.

Picon, A. (2018). "Urban Infrastructure, Imagination and Politics: From the Networked Metropolis to the Smart City." *International Journal of Urban and Regional Research* 42(2).

Rees, W. (2004). *Evolution and Human Mal-adaptation*. Discussion Paper. The University of Texas Center for Sustainable Development.

Rittel, H. and Webber, M. (1973). Dilemmas in a General Theory of Planning. *Policy Sciences*, 4, 155–169.

Schneider, F.G.K. et al. (2010). Crisis or Opportunity? Economic Degrowth for Social Equity and Ecological Sustainability. Introduction to this special issue *Journal of Cleaner Production*, 18(6), 511–518.

Schweitzer, L.E. and Link, T. (2016). Tracing the Justice Conversation After Green Cities, Growing Cities. *Journal of the American Planning Association*, 82(4), 374–379.

Scott, J.C. (1998). *Seeing Like a State*. New Haven: Yale University Press.

SEED (2018). *Social, Economy, Ecology, and Design Network*. https://seednetwork.org/. Accessed October 30, 2018.

Simon, H. (1969). *The Sciences of the Artificial*, Cambridge, MA, MIT Press.

Smith, K. (ed.) (2013). *Sociology of Globalization*. New York: Routledge.

Smith, N. (1982). Gentrification and Uneven Development. *Economic Geography*, 58(2), 139–155.

Stevenson, F., Baborska-Narozny, M. and Chatterton, P. (2016). Resilience, Redundancy and Low-carbon Living: Co-producing Individual and Community Learning. *Building Research & Information*, 44(7), 789–803. doi:10.1080/09613218.2016.1207371.

Thompson, P.B. (2010). What Sustainability is and What it Isn't. In S.A. Moore (ed.), *Pragmatic Sustainability: Theoretical and Practical Tools*. New York: Routledge, pp. 15–29.

Thone, Frank (1935). Nature Rambling: We Fight for Grass. *The Science Newsletter*, 27, 717, January 5: 14.

WCED (1987). *Our Common Future*. New York: United Nations World Council on Economic Development.

Young, R. (2017). 'Free Cities and Regions'—Patrick Geddes Theory of Planning. *Landscape and Urban Planning*, 166, 27–36.

26

MAINTENANCE

Mark Thomas Young

Maintenance is not a topic that has received much attention in the philosophy of technology.[1] This is surprising considering the importance of maintenance—both catastrophic failures of large structures such as the recent collapse of the Ponte Morandi bridge in Genoa, and what many perceive to be a crisis of aging infrastructure in the US and other parts of the world are often understood to reflect how maintenance can be ignored or deferred only at our own peril. With the exception of a brief section in Carl Mitcham's *Thinking Through Technology* (Mitcham 1994: 233), however, maintenance is a topic which is absent from nearly all volumes in the philosophy of technology. Yet far from being a trivial question, the nature of maintenance represents a difficult and important philosophical issue. For while we may have a clear understanding of the kinds of practices which are classified as maintenance, the question of what kinds of action we are performing when we maintain technologies is difficult to answer. When we engage in such practices, are we making or using the technologies in question? What does attending to maintenance practices reveal about the nature of technology itself?

The general neglect of such questions by philosophers of technology means that any philosophical analysis of maintenance will need to start at the very beginning—by asking what exactly *is* maintenance? One way we can begin to approach this question is by exploring the etymology of the word itself. The English word 'maintain' has two etymological roots; the first derives from the Latin *manu tenere*, a term used to denote the activity of providing stability to something by holding it in one's hand. The second etymological root derives from the Old French *maintenir* which reflects the temporal character of maintenance by capturing something which is performed continuously and habitually (Stevenson 2010: 1068). Combining these two ideas provides us with a preliminary conception of maintenance as an activity through which we provide stability to something over time. But exactly what this 'stability over time' involves in relation to technology, is a question which still awaits a thorough philosophical analysis.

The goal of this chapter is to prepare the ground for such an analysis, by exploring how the study of maintenance practices makes possible new ways of conceiving of technology. The first section aims to identify a deeply entrenched bias towards design in common conceptions of how artifacts are produced. According to this perspective, which I label 'the design paradigm', the production of artifacts occurs through the development and realization of designs. In addition to exploring how the design paradigm has informed conceptions of engineering practice, this section also examines how it has functioned to obscure the full significance of maintenance practices by understanding them to contribute only to the preservation and not the production of artifacts. The second section reviews some of the difficulties the design paradigm faces in providing an adequate understanding of the practices and products of engineering. Accordingly, this section argues for the need to reconceptualize engineering in a way that accords a more productive role to maintenance practices. The final section of this chapter will explore the consequences of such a shift for philosophical conceptions of technology. This section is more schematic and aims to identify areas for further study which emerge as a result of understanding maintenance to contribute to the production of artifacts.

Like many studies in the philosophy of engineering, this chapter will draw on a variety of disciplinary perspectives. For while maintenance has long been a neglected topic in philosophy, a small but growing number of studies in fields outside philosophy have attended to maintenance directly. This chapter aims to take advantage of this recent interest in maintenance by providing an overview of the central themes to emerge from this literature and exploring the extent to which insights gleaned from such studies can be applied to central questions within the philosophy of engineering.

26.1. Technology as Form: Maintenance as Conservation

It has long been common to understand the nature of engineering practice in terms of design. Nearly thirty years ago, Carl Mitcham noted how 'virtually all general articles on engineering and all introductory engineering textbooks identify designing as the essence of engineering' (Mitcham 1994: 220). While many aspects of engineering may have changed since then, the identification of engineering practice with design, it seems, has not. Even a cursory survey of current handbooks for engineering students reveals the extent to which design is still viewed as the essence of engineering. Furthermore, both educational and professional engineering institutions commonly reinforce this perception by privileging acts of innovative design over other forms of engineering practice (Russell and Vinsel 2019). Similarly, the way in which engineering design itself is understood does not appear to have changed much either. The definition of engineering design, offered by the influential engineering educator Ralph J. Smith in 1969, as the 'planning in the mind a device or process or system that will effectively solve a problem or meet a need' (Smith 1983: 160), continues to garner citations as a definitive statement of the essence of engineering practice. This particular way of understanding engineering design is widespread and corresponds to common conceptions of the nature of design in other fields. For our purposes, however, it is worth unpacking this claim in order to identify two components which will be relevant for our exploration of maintenance.

The *first* aspect of this definition that I want to draw attention to is the idea that the activity of engineering design represents a form of problem-solving. This reflects the common tendency to understand the products of design more generally in terms of solutions to existing problems. In his book *The Philosophy of Design*, Glenn Parsons promotes a similar conception of design as a problem-solving activity. According to Parsons, 'a design problem can usually be viewed as a problem of function . . . there is some task we want performed . . . and it is the designer's job to create an object that performs it well' (Parsons 2016: 46). Engineering design is often considered to be a special form of this problem-solving activity, which is distinguished by the methods it employs to develop solutions. According to Donald Schön, these methods reflect a model of technical rationality, which he described as 'the view that professional activity consists in instrumental problem-solving made rigorous by the application of scientific theory and technique' (Schön 1983: 21). Since the advent of modernity, Schön argues, engineering has been understood to represent one of the clearest examples of this ideal of technical rationality. It is this widespread confidence in the capacity of the methods of engineering to solve problems that gave rise to the idea of the 'technological fix', a term originally coined by Alvin Weinberg in 1969 to capture the possibility of solving complex social problems by reframing them as engineering problems (Scott 2012).

The *second* aspect of the definition of engineering offered earlier in this section that we will consider is the emphasis it places on intending—the activity of creating a plan for something in advance. Again this reflects common assumptions about the nature of design more generally; it is often noted, for example, that designing and intending are synonyms (Mitcham 1994: 220; Parsons 2016: 17), and that the association between making and intending captured by the term 'design' stems from at least the early modern period.[2] This association also often determines how the products of design, artifacts, are conceived (see also Chapter 15, "Artifacts" by Beth Preston, this volume). Among philosophers, for example, artifacts are nearly always understood as manifestations of the intentions of their creators. As Amie L. Thomasson notes, 'it seems to be part of the very idea of an artifact that it must be the *product* of human intentions' (Thomasson 2007: emphasis in original). This emphasis on intentions in discussions of artifacts reflects aspects of a broader philosophical theory concerning how technologies are made which originated in classical philosophy. According to Aristotle, artificial objects are created through a process in which a human agent imposes an antecedently existing form upon a material substrate. The sculptor, for example, begins by envisioning the form of the statue in her mind, and then proceeds to realize that form in the material they are working with by removing pieces of stone. As Beth Preston (2013) notes, because Aristotle considered the mental design to be completed before construction commences, his theory of making involved 'two clearly demarcated phases in the overall production process—an antecedent design phase and a subsequent construction phase' (18).

Smith's definition of engineering practice in terms of solving problems through the activity of planning can therefore be understood to reflect long-standing ideals about the way technology is made. Furthermore, in the history of engineering, it is not hard to see how these ideals have influenced the development of the profession itself. Engineering in the 19th century, for example, saw the institutionalization of Aristotle's separation of the phases of design and construction in different ways. The increasingly important role of technical drawing in the production of artifacts, for example, culminated in the emergence of the profession of draftsmanship, in which plans for structures were created in the form of blueprints which eventually acquired legal force by the latter half of the century (Sennett 2008: 41). This was complemented by the emergence of contracting firms whose job was primarily to execute finished designs, further consolidating the separation between design and construction.

Due to the influence it has exerted upon the history of engineering, it should therefore not surprise us that the practice of engineering design today is often understood in terms very similar to Aristotle's conception of making. Take the account of engineering design outlined by Peter Kroes (Chapter 21, this volume), for example, who suggests that;

> the mental activity involved in making a technical artefact of a particular kind consists of coming up with a largely correct design. The physical activity is referred to as 'a largely successful execution' of the design, that is, a largely successful material realization of the design. The definition suggests that the making of a technical artefact involves two steps, one involving mental and the other physical activity, and that these steps may be performed independently of each other and may therefore be separated in time. Each step has its own success criterion; the first mental step is successful if it produces a largely correct design and the following physical step is successful if it comes up with a physical object that is a largely correct execution of that design.
>
> *(Kroes 2012: 128)*

By suggesting making to be exhausted by the satisfaction of success criteria, Kroes draws attention to a feature of this account which often goes unremarked, and which is important for our purposes here. That is, understanding the production of a technology to consist in the formulation and realization of a design establishes the basis for a rigid dichotomy between the activities of making and using. It does so by positing a point at which an artifact can be considered finished. As Tim Ingold notes:

to draw a line between making and using means marking a point in the career of a thing at which it can be said to be finished . . . this point of completion can only be determined in relation to a totality that already exists, in virtual form, at the outset—that is, in relation to a design.

(Ingold 2013: 47)

In other words, framing processes of making around the formulation and realization of design functions to demarcate making from using by identifying a point at which production of an artifact is completed; the point at which the makers' intentions have been successfully realized in material form. This idea, that processes of making conclude with the material realization of a design, is common among philosophers of technology. Risto Hilpinen (Hilpinen 1992), Randall Dipert (Dipert 1993) and, more recently, Amie Thomasson (2007) and Peter Kroes (2012), for example, have all offered accounts of making which conclude with the successful execution of a design or intention. It should therefore not surprise us, then, that the distinction between making and using artifacts is rarely questioned in the philosophy of technology, where the two categories have often been considered to represent fundamentally different forms of technical activity (Winner 2004; Mitcham 1994: 210).

Yet it is worth noting that framing the project of making around the realization of designs establishes not only a dichotomy between making and using, but also a hierarchy. This is because understanding technological artifacts as realizations of the intentions of designers often encourages us to privilege the moment construction finishes as the point in an artifact's history where the intentions of the designer are most fully realized. Because the material world exists in a constant state of change, matter is always in the process of diverging from the designed forms that are imposed upon it. This divergence results in part from natural processes of change that all materials undergo over time. The materials from which a suspension bridge is constructed, for example, such as steel and concrete, are subject to corrosion over time which eventually weakens their structural integrity. Divergence also occurs as a result of human interaction—in the case of bridges, constant traffic degrades the bridge deck, and vehicle collisions can cause damage to different aspects of the structure. When artifacts are understood as realizations of a design, these different forms of change inevitably appear as processes which diverge from, and therefore obscure, the intentions of the designer.

It is from this way of thinking about technology that we derive our most common way of understanding the activity of maintenance: as conservation. From this perspective the different forms of maintenance suspension bridges constantly undergo, in order to slow or prevent corrosion or to repair damage caused by vehicles, appears as an attempt to either return the artifact to its designed form, or to delay its departure from it. This particular way of understanding maintenance, as preserving or conserving something that was previously created, is so common that it is rarely questioned. It informs definitions of maintenance offered by engineers themselves. For example, according to one prominent handbook for maintenance engineers, 'the act of maintaining is to keep in an existing state or preserve from failure or decline' (Smith and Mobley 2008: 5). I want to draw attention here to two distinguishing features of this conception of maintenance that will be important for our discussion in the rest of this chapter.

The first feature concerns the temporal orientation of the activity of maintenance. In attempting to hinder or reverse processes of change, this account understands maintenance to be directed backwards, against the passage of time. The idea that maintenance is directed against time is reflected in various cognate terms for maintenance in the English language, many of which share the same prefix; to *re*store, *re*pair, *re*place, *re*construct and *re*plenish. This conception of maintenance therefore exhibits a particular relationship to the flow of time in technical activity—as something to be resisted. In doing so, it reflects fundamental assumptions inherent in the conception of technology from which it derives. For Aristotle, as for Plato, the form a sculptor imposes upon matter does not itself change over time (Politis 2004: 233). When a sculpture undergoes a process of decay or destruction, it is the matter from which it is composed that changes, resulting in the loss of form. The longevity of an artifact therefore depends on the extent to which the union between matter and form can be preserved by slowing or curbing the processes of change the matter undergoes. Because the essence of an artifact, the form, is understood to be atemporal, time and change are cast as external influences against which a technology must be protected, giving rise to a conception of maintenance as conservation.

The second aspect of this conception of maintenance that I want to draw attention to is the derivative place it occupies within this particular understanding of how technologies are made. For understanding an artifact to be finished with the realization of a designer's intentions marginalizes the activity of maintenance at the outset. Instead of contributing to the efficacy of a technology itself, maintenance is instead relegated to preserving an efficacy which was achieved through the realization of a design. Understood in this way, design comes to be viewed as the sole source of the capacities of a technology. These kinds of assumptions are often apparent in the way engineers themselves describe the technologies they design. In *Why Buildings Stand Up*, the Italian engineer Mario Salvadori, for example, reflects on how;

> most laymen seldom look at architectural structures, or ask the simple question, 'What makes buildings stand up?' . . . A structure is an artifact expressing one of the many aspects of human creativity, but it is an artifact that cannot be created without a deep respect for the laws of nature. A beautiful structure is the concrete revelation of nature's laws.
>
> *(Salvadori: 1980: 25)*

Maintenance is largely absent from the account Salvadori provides of why buildings stand up, and this should not surprise us. By focusing entirely on design, Salvadori's account reflects the conception of the production of structures as the imposition of a designed form upon matter. According to this perspective, the building stands because the designed form conforms to correct principles of structural mechanics. Implicit in this view is a conception of maintenance not as responsible for enabling the building to stand, but rather for preserving the realization of this form in material which is constantly changing.

The idea that maintenance does not contribute to the production of a technology and is instead limited to conserving that which has already been created helps explain why maintenance has so often been neglected not only by philosophers, but also by historians. In the history of technology, it has rarely been considered necessary to include maintenance practices in the story of how technologies were developed (Edgerton 2010; Pursell 1995). Instead, historians have traditionally restricted their accounts of the development of technologies to the design and construction of artifacts. However, there is now a growing recognition among historians that this bias towards innovation has excluded aspects of technological practice which are important for explaining how technologies come to represent stable features of our environment. Central among these are practices of maintenance, performed to sustain the function of a technology over time. As David Edgerton notes, despite being 'the most widespread forms of technical expertise . . . we are not (even) in a position to give an overview of the main trends in the history of maintenance and repair' (Edgerton 2006: 80). In the next section, we will turn our attention to the few existing studies which have sought to address this lacunae, in order to gain insights which will allow us to take our first steps towards sketching the outlines for a philosophy of maintenance.

26.2. Limitations of the Design Paradigm: Bringing Maintenance Into View

The foregoing section described how the marginalization of maintenance emerges as a consequence of common assumptions concerning the nature of technology and the processes by which it comes into being. We identified these assumptions as a conception of design as a problem-solving activity, alongside an intentional theory of making which distinguishes production from use. While this way of understanding the practice of engineering is widespread, aspects of this picture are increasingly being drawn into question by scholars from a variety of fields, including science and technology studies, history of technology and material culture studies. These calls to attend to maintenance practices are often framed around criticisms of some of the basic assumptions underpinning the design paradigm. This section aims to provide an overview of these criticisms, with a view towards setting the stage for the preliminary philosophical analysis of maintenance in the final section.

The most common way of criticizing this approach to the study of technology problematizes the agency which is ascribed to the designers of a technology. According to this line of reasoning, understanding the production of an artifact to be framed around the intentions of the designer conflicts with the commonly accepted idea that engineering design always operates under conditions of partial uncertainty (see also Chapter 11, "Uncertainty" by William M. Bulleit, this volume). According to Billy Vaughn Koen, any engineer working on a design problem must contend with the fact that 'knowledge about the system before, during, and after the transition is incomplete, inconsistent, or would require more time to accumulate than the lifetime of the problem' (Koen 2003: 10). For this reason, Koen argues, 'the final state always has a reality that the engineer, situated at the initial state cannot anticipate' (Koen 2003: 12). This inherent uncertainty affecting the practice of engineering design is understood to occur as a result of various factors. Sometimes is it understood as a consequence of the complexity of a technology. As Nathan Rosenberg notes;

> For a range of products involving complex, interdependent components or materials that will be subject to varied or prolonged stress in extreme environments, the outcome of the interaction of these parts cannot be precisely predicted . . . the performance of these products, therefore, is highly uncertain.
>
> *(Rosenberg 1982: 122)*

Other times it is related to the inherent limitations faced in predicting how a technology will affect or be affected by the wider natural or social environment in which it will operate. In the philosophical literature on engineering design, these problems have often been considered to apply primarily or especially to complex sociotechnical systems,[3] such as electricity networks. Yet while these problems may be most apparent in such cases, there is no reason to think that they do not apply universally to all technologies. The design and construction of a suspension bridge, for example, is also subject to irreducible uncertainty concerning the effects it will have on the local ecosystem, or on life in the surrounding communities. Likewise, future weather conditions, or the manner in which the bridge is used, also introduces irreducible aspects of uncertainty which limit the information available during the design process. Unanticipated problems with the bridge which emerge after construction must be remedied on the fly—through corrective maintenance. These activities include the repair of corroded pillars or structural steel, the replacement of bearings or the fixing of impact damage (Alampalli 2014). Such activities represent a challenge to the design paradigm by revealing how problem-solving occurs not only during the production phase, but also after production has finished—during the operation of a technology. Furthermore, this is not limited to bridges. Problem-solving extends through the operational phases of a wide range of different technologies. In his study of the design and operation of aircraft engines, for example, Rosenberg illustrates how attaining desirable operating costs for a particular engine depends crucially on the remedying of design problems through maintenance practices after production (Rosenberg 1982: 131).

As a result of such considerations, there is an increasing recognition that the bias towards innovation in the history of technology often creates a misleading picture of the processes by which technologies come to represent stable and reliable features of our environment. According to Steven J. Jackson, for example, the emphasis on innovation has created:

> a false and partial representation of how worlds of technology actually work, when they work. . . . Against fans and critics of design alike, innovation rarely if ever inheres in moments of origination, passing unproblematically into the bodies of the objects and practices such work informs. For this reason, the efficacy of innovation in the world is limited—until extended, sustained, and completed in repair.
>
> *(Jackson 2014)*

For a vast range of technologies, reliability and efficiency depend on practices of maintenance. This applies as much as for automobiles as for large structures such as bridges. In order for a highway bridge to reach its desired life expectancy of 75 to 100 years, for example, both preventative and corrective maintenance is required. Recognizing the importance of these practices in sustaining the integrity of such structures over time allows us to return to the question Salvadori poses in *Why Buildings Stand Up*, with a different answer. Because like bridges, the life of a building also depends on maintenance. Without proper maintenance, the materials of a building can deteriorate over time, undermine the stability of the structure itself and, in some cases, lead to collapse (Panchdhari 2003). For many buildings, then, the question of why they stand up must be answered not only by looking at how their design correctly embodies principles of structural mechanics, but also by looking at the wide variety of maintenance practices which are performed in response to both predictable and unpredictable events and processes which occur after their construction.

The fact that much of this activity is performed and managed by engineers themselves reveals yet another way in which the design paradigm has been considered to cultivate a misleading impression of the nature of engineering practice. As historian David Edgerton notes, despite the common tendency for engineers to present themselves as designers, 'the majority have always been mainly concerned with the operation of things and processes; with the uses of things, not their invention or development' (Edgerton 2006). Furthermore, recent literature suggests this to be an adequate characterization of a wide variety of engineering professions, from civil engineering to software engineering (Russell and Vinsel 2019; Vinsel 2019).

Finally, the design paradigm has not only obscured the frequency of maintenance work amongst engineers, it has also obscured the nature of maintenance work itself. Against the common characterization of maintenance work as derivative, as concerned primarily with the preservation of the products of design and construction, studies of maintenance contend that such practices are not simply conservative—they are also deeply creative. As Steve Jackson argues;

> The remarkable qualities and energies that innovation names and unleashes—creativity, invention, imagination, and artfulness—are . . . distributed more broadly in the technology landscape than our dominant discourses of innovation . . . are keen to acknowledge.
>
> *(Jackson 2014)*

The inherent creativity of maintenance practices is reflected in the way a wide variety of different technologies continue to develop into new forms after construction. As an example, consider the evolution of the Bronx-Whitestone Bridge in New York City. Within the decade after construction was completed in 1939, cable stay ropes and stiffening trusses were added to the bridge, and sidewalks were converted into an extra lane of traffic in order to remedy design flaws revealed by the recent collapse of the Tacoma Narrows Bridge. The further addition of a tuned mass damper was made in 1985 to address this same problem. However, while improving the aerodynamic performance of the bridge, these modifications also increased the dead load on the cables—leading to deterioration of the cables themselves and their concrete anchorages. Further modifications were required to address these problems, including the installation of a new lighter bridge deck, a new anchoring system for the cables and the removal of the stiffening trusses. Beginning in 2008, the approach structures were also redesigned in order to meet current highway standards and tackle the increased weight of today's truck loads (Lorentzen et al. 2006). The forgoing description considers only major modifications; a more comprehensive survey would include drainage repair, replacement of the sealant surrounding expansion joints and repairing of cracks, alongside common preventative maintenance practices. The point, however, should nonetheless be clear. Not only does maintenance appear to require great amounts of labor, but this work also often transforms the structure itself beyond the form originally envisaged by the designer. This capacity to evolve after construction is illustrated by a wide range of different technologies, including cars which are often remodeled and equipped with new features or buildings which often expand over time through extensions or are converted to new uses. But the point is rendered perhaps most vividly by the example of modern battleships, which, as David Edgerton has shown, continue to evolve after construction through maintenance to such an extent that they are often fail to resemble their original forms in any way by exhibiting entirely different shapes, artillery and engines (Edgerton 2006: 92).

As we have seen in this section, studies of maintenance identify a variety of ways in which the design paradigm has failed to provide an adequate framework for understanding technological practices and objects. As we might expect, then, a common feature of such studies is the recognition of the need for new ways of thinking about technology. As Lee Vinsel and Andrew Russell note, the study of maintenance 'involves a turn, a switch, not only in regard to the object of study, but also in the underlying attitude toward technology' (Russell and Vinsel 2018). Edgerton, on the other hand, suggests that 'by thinking about technology-in-use a radically different picture of technology, and indeed of innovation and invention, becomes possible' (Edgerton 2006: xi).

In the next section, we will explore further what this radical rethinking of the nature of technology might involve. We will focus in particular on two aspects of technology; the first concerns the distinction between the production and use of technologies, and the way in which, as Carl Mitcham has noted, maintenance often presents a challenge to common ways of distinguishing between making and use.[4] The second aspect concerns the relationship between technology and time. For as we will see, the study of maintenance requires us to reflect over the temporal nature of technologies to a much deeper extent than is often required by other topics in the philosophy of technology.

26.3. Technology as Process: Maintaining as Making

The previous section noted how studies of maintenance often consider the traditional distinction between making and using to have prevented us from appreciating the true character of maintenance as a creative and transformative technical activity upon which the efficiency and reliability of technologies often depend. It should not surprise us, then, that calls to transcend this distinction are a common feature of studies which approach maintenance from a variety of disciplinary orientations, including material culture studies,[5] the history of technology[6] and science and technology studies.[7] In making such arguments, these studies remind us that despite forming a natural part of our modern technological vocabulary, there is nothing necessary about the distinction between making and using. Historical studies, for example, have revealed how before the development of distinct professional identities for makers and maintainers, societies of previous eras often failed to recognize meaningful distinctions between the activities of making and maintaining (Russell and Vinsel 2018).

A common argument motivating calls to transcend the distinction between making and using is that because the solving of problems appears to occur continuously throughout the life of an artifact, we should not understand an artifact to be finished upon the realization of a designer's intentions in material form (Brand 1994: 64; Ingold and Hallam 2014). Instead, studies of maintenance commonly imply that the completion of an artifact should be understood in relation to its use. From this perspective, an artifact can be considered finished only when it is no longer used, when it is used up or broken or perhaps even when we have simply lost interest in employing it. However, understanding production to continue throughout the operation of a technology in this way requires new ways of conceptualizing the practice of making. One suggestion as to how this might be done emerges from the work of anthropologist Tim Ingold, who challenges the widespread assumption that practices of making begin and end in relation to a designer's intention and instead outlines a view of making which involves 'no absolute distinction between making and growing . . . what we call 'making things' is, in reality not a process of transcription at all but a process of growth' (Ingold 2000: 88). According to this perspective, we should consider the life of an artifact not to be punctuated by separate phases of making and using, but rather as a continuous process of growth.

Reconceptualizing making as growing represents one way in which dominant modes of discourse on the development of technologies might be disrupted so that the creative potential of maintenance can come to light. For there are inherent similarities between the kinds of human practices that are often required for the cultivation of plants and those required for the production of artifacts. In both cases, human agents must manage processes of change and relationships between objects and their environments in order to achieve desirable results. So just as large structures such as bridges are coated in protective paint to prevent corrosion, trees can be wrapped in burlap to protect them from the cold. To help them withstand heavy wind, bridges can be stabilized with stiffening trusses, not unlike the way newly planted trees are stabilized with stakes driven into the ground before a storm. The growth of technologies, then, like the growth of plants, often depends on human practices geared towards making periodical adjustments based on a close observation of the changing context in which an artifact is situated. The idea that technologies grow as a result of human practices in this way is an insight which has emerged from the study of technology in various fields. Architectural theorist Stewart Brand, for example, applies this idea to the transformation of buildings after construction by noting how, over time,

> *all buildings grow*. Most grow even when they're not allowed to. Urban height limits and the party walls of row houses, for instance, are no barrier. The building will grow into the back yard and down into the ground—halfway under the street in parts of Paris.
>
> *(Brand 1994: 10, italics in original)*

Edgerton, on the other hand, notes how the global history of the automobile provides many examples of automobiles being embraced by consumers not as finished products but rather as seeds for development. These attitudes are visible in long-standing traditions of car modification around the world, in which cars are customized through remodeling, decoration or accessorization (Edgerton 2006: 97).

Yet challenging the distinction between making and using in this way raises a host of new questions concerning the nature of what is made. If what is being made is not an idea in the mind of a designer, then what is it? Instead of understanding technology as a material realization of an atemporal form, this perspective encourages us instead to understand the essence of technology as temporal, as processes which are extended and guided through time by the activity of human agents. From this perspective, the stability over time of technology results from human practices performed after construction; repairing damage and replacing parts, applying treatments to prevent processes of corrosion or decay and stabilizing, reinforcing or extending through the act of construction itself. Technologies are understood as fluid entities which constantly respond to the changing environments in which they exist, and which therefore require guidance through time in order to appear for us as stable entities. Learning to see not only cars but also bridges and buildings as processes in this way can be achieved only by paying adequate attention to the varied and continual forms of human practice upon which the smooth functioning of these technologies depend. However, paying attention to maintenance in this way reveals not only that our technologies are extended in time, but also *how* they are extended in time. As Jackson notes, 'bringing maintenance and repair work to the fore in our thinking about technology may help to . . . [offer] new insights and approaches to the understanding of technology as a timely or rhythmic phenomenon' (Jackson 2014).

The rhythmic character of technology is manifest first in the scheduling of certain forms of maintenance, such as preventative maintenance activities, which often occur at regular intervals. But a different rhythm is revealed by attending to practices of corrective maintenance. This second rhythmic character of technology is revealed in the work of sociologist Andrew Pickering, who proposes a general theory of technological practice based around the observation that human engagement with technology consists in alternating phases of activity and passivity. Utilizing the example of tuning equipment in experimental physics, Pickering notes how scientists engage in cyclical activities which involve making adjustments to machines, then stepping back to observe their performance, before making further modifications and stepping back again to observe the results. According to Pickering, tuning equipment requires human beings and the world to take turns expressing their agency, and he underscores the rhythmic character of this process by describing it as a 'dance of agency' (Pickering 1995: 21). Yet Pickering's account is not limited to science and is used to capture a wide variety of different human performances, including civil engineering. According to Pickering, the construction and maintenance of levees and weirs on the Mississippi River, for example, can also be considered:

> a 'dance of agency,' as I would call it, between the engineers and the river. The human agents, the engineers, try something—raising the levees, say—and then the nonhuman agent takes its turn by rising still higher and flooding New Orleans. In response the humans do something else—building the weir between the Mississippi and the Atchafalaya—to which the river does something else—ripping and tearing away at it. And so on, forever.
>
> *(Pickering 2008)*

The dance of agency represents a promising framework within which the significance of maintenance practices can be apprehended.[8] By explaining how the stability of technologies emerge precariously from such performances, it enables us to see how maintenance can contribute to the production of artifacts; how maintenance itself can be understood as a form of making. Not only does this reveal maintenance practices to be essential rather than derivative forms of technical activity, but it also alters our understanding of their temporal orientation. For instead of aiming at the conservation of something already created, maintenance appears instead as looking forwards in time; towards the next storm or heat wave and towards the various activities which we must perform in response, in order to carry our technologies into the future.

26.4. Conclusion

Before turning his attention to design in *What Engineers Know and How They Know It*, Walter Vincenti noted how a complete epistemology of engineering will need to address both the design and operation of a technology (Vincenti 1990: 7). Yet fifty years later, we have barely taken our first steps towards overcoming the bias towards design which will be necessary for the completion of this project. Change has been particularly slow in the philosophy of technology, which has yet to experience a turn towards maintenance similar to that which is slowly gaining traction among historians and sociologists of technology. However, because it attends to a profession deeply involved with maintenance practices, the philosophy of engineering represents perhaps the most promising subdiscipline within which such a turn can take place. But for this to be achieved, the way in which we often approach the study of technology will require reorientation; away from the traditional focus on innovative design and more towards the ongoing, ubiquitous and sometimes hidden practices of maintenance upon which the stability of our technologies often depend. For it is only by attending to such practices that we may come to see the artifacts around us not as objects that have already been created, but instead as processes, evolving under the guidance of human hands, in and with the world around them.

Acknowledgements

I would like to thank Tim Ingold, Hallvard Fossheim, Kristian Larsen, Neelke Doorn and Diane P. Michelfelder for helpful comments on earlier drafts of this chapter. Any remaining errors are my own.

Related Chapters

Chapter 11: Uncertainty (William M. Bulleit)
Chapter 15: Artifacts (Beth Preston)
Chapter 16: Engineering Objects (Wybo Houkes)
Chapter 19: Emergence in Engineering (Peter Simons)
Chapter 21: Engineering Design (Peter Kroes)

Further Reading

Brand, Stewart. (1994). *How Buildings Learn: What Happens After They're Built*. New York: Penguin Press. (Classic work of architectural history which traces how buildings are changed and adapted after construction, challenging the emphasis traditionally placed on original intent in assessing the meaning of a building or structure.)

Edgerton, David. (2006). *The Shock of the Old: Technology and Global History Since 1900*. London: Profile Books. (Important work which calls for rethinking distinctions between invention and use in order to subvert prevailing biases towards novelty and innovation in the history of technology. Features sustained discussions of maintenance and the continued development and significance of technologies after construction.)

Graham, Stephen and Thrift, Nigel (2007). Out of Order: Understanding Repair and Maintenance. *Theory, Culture and Society*, 24(3), 1–25. (Influential article drawing attention to processes of maintenance and repair of urban infrastructure that have been neglected by prior scholarship. Adopts a phenomenological approach which explores the visibility and hiddenness of these processes and why these processes are often overlooked.)

Jackson, Steven J. (2014). Rethinking Repair. In T. Gillespie, P. Boczkowski and K. Foot (eds.), *Media Technologies: Essays on Communication, Materiality and Society*. Cambridge, MA: MIT Press, pp. 221–240. (Argues for a reorientation of the study of technology towards processes of breakdown and decay and the different forms of human practices which arise in response. Relates this approach to various philosophical perspectives, including phenomenology, standpoint theory and the ethics of care.)

Russell, Andrew L. and Vinsel, Lee. (2018). After Innovation, Turn to Maintenance. *Technology and Culture*, 59(1), 1–25. (Summarizes the recent turn towards the study of maintenance in the history of technology by tracing the origins and agenda of the maintainers movement, a scholarly community devoted to the study of maintenance. Provides an overview of existing studies of maintenance, outlines different possible theoretical orientations and points towards fruitful areas for future research.)

Notes

1. Exceptions include Ravetz (2008), Vogel (2019) and Young (2020).
2. For a fascinating etymological analysis of the term engineer, see Sawday (2007).
3. See for example, Kroes (2012: 157).
4. According to Mitcham (1994), 'maintaining is in some sense intermediary between making and using and thus deserves special consideration'.
5. See for example, Colloredo-Mansfeld (2003) and Naji and Douny (2009).
6. SeeCowan (1983), Pursell (1995) and Edgerton (2006).
7. Oudshoorn and Pinch (2003).
8. For an application of this idea to the maintenance of Paris subway signs, see Denis and Pontille (2019).

References

Alampalli, Sreenivas (2014). Designing Bridges for Inspectability and Maintainability. In M. Frangopol and Y. Tsompanakis (eds.), *Maintenance and Safety of Aging Infrastructure*. London: CRC Press.

Brand, Stewart (1994). *How Buildings Learn: What Happens After They're Built*. New York: Penguin Press.

Colloredo-Mansfeld, R. (2003). Introduction: Matter Unbound. *Journal of Material Culture,* 8(3), 245–254.

Cowan, Ruth Schwartz (1983). *More Work for Mother: The Ironies of Household Technology From the Open Hearth to the Microwave*. New York: Basic Books.

Denis, Jérôme and Pontille, David (2019). The Dance of Maintenance and the Dynamics of Urban Assemblages: The Daily (Re)Assemblage of Paris Subway Signs. In I. Strebel, A. Bovet and P. Sormani (eds.), *Repair Work Ethnographies: Revisiting Breakdown, Relocating Materiality*. Singapore: Palgrave Macmillan, pp. 161–186.

Dipert, Randall R. (1993). *Artifacts, Art Works and Agency*. Philedelphia: Temple University Press.

Edgerton, David (2006). *The Shock of the Old: Technology and Global History since 1900*. London: Profile Books.

Edgerton, David (2010). Innovation, Technology or History: What Is the Historiography of Technology About? *Technology and Culture,* 51(3), 680–697.

Hilpinen, Risto (1992). On Artifacts and Works of Art. *Theoria,* 58, 58–82.

Ingold, Tim (2000). *Perception of the Environment: Essays on Livelihood, Dwelling and Skill*. New York: Routledge.

Ingold, Tim (2013). *Making: Anthropology, Archaeology, Art and Architecture*. New York: Routledge.

Ingold, Timothy and Hallam, Elizabeth (2014). Making and Growing: An Introduction. In Timothy Ingold and Elizabeth Hallam (eds.), *Making and Growing: Anthropological Studies of Organisms and Artefacts*. Surrey: Ashgate Publishing, pp. 1–24.

Jackson, Steven J. (2014). Rethinking Repair. In T. Gillespie, P. Boczkowski and K. Foot (eds.), *Media Technologies: Essays on Communication, Materiality and Society*. Cambridge, MA: MIT Press, pp. 221–240.

Koen, Billy Vaughn (2003). *Discussion of the Method: Conducting the Engineer's Approach to Problem Solving*. New York: Oxford University Press.

Kroes, Peter (2012). *Technical Artefacts: Creations of Mind and Matter*. Dordrecht: Springer.

Levy, Matthys and Salvadori, Mario (1987). *Why Buildings Fall Down: How Structures Fail*. New York: W.W. Norton.

Lorentzen, J. et al. (2006). Planning and Engineering for the Future: Capacity Increase and Cable Replacement at the Bronx-Whitestone Bridge. In Khaled Mahmoud (ed.), *Advances in Cable Supported Bridges*. London: Taylor & Francis, pp. 145–162.

Mitcham, Carl (1994). *Thinking Through Technology: The Path Between Engineering and Philosophy*. Chicago: University of Chicago Press.

Naji, M. and Douny, L. (2009). Editorial. *Journal for Material Culture,* 14(4), 411–432.

Oudshoorn, Nelly and Pinch, Trevor (eds.) (2003). *How Users Matter: The Co-construction of Users and Technology*. Cambridge, MA: MIT Press.

Panchdhari, A.C. (2003). *Maintenance of Buildings*. New Delhi: New Age International

Parsons, Glenn (2016). *The Philosophy of Design*. Cambridge: Polity Press.

Politis, Vasilis (2004). *Routledge Philosophy Guidebook to Aristotle and the Metaphysics*. New York: Routledge.

Pickering, Andrew (1995). *The Mangle of Practice: Time, Agency and Science*. Chicago: University of Chicago Press.

Pickering, Andrew (2008). New Ontologies. In Andrew Pickering and Keith Guzik (eds.), *The Mangle in Practice: Science, Society and Becoming*. Durham: Duke University Press, pp. 1–16.

Preston, Beth (2013). *A Philosophy of Material Culture: Action, Function and Mind*. New York: Routledge.

Pursell, Carroll (1995). Seeing the Invisible: New Perspectives in the History of Technology. *Icon,* 1, 9–15.

Ravetz, Jerome (2008). Maintenance as Morality. Paper presented at the 2nd WPE November 10, Royal Academy of Engineering, London.

Rosenberg, Nathan (1982). *Inside the Black Box: Technology and Economics*. Cambridge: Cambridge University Press.

Russell, Andrew L. and Vinsel, Lee (2018). After Innovation, Turn to Maintenance. *Technology and Culture,* 59(1), 1–25.

Russell, Andrew L. and Vinsel, Lee (2019). Make Maintainers: Engineering Education and an Ethics of Care. In M. Wisinoiski, E. Hintz and M. Kleine (eds.), *Does America Need More Innovators?* Cambridge, MA: MIT Press, pp. 249–272.

Salvadori, Mario (1980). *Why Buildings Stand Up: The Strength of Architecture*. New York: W.W. Norton.

Sawday, Jonathon (2007). *Engines of the Imagination: Renaissance Culture and the Rise of the Machine*. New York: Routledge.

Schön, Donald A. (1983). *The Reflective Practitioner: How Professionals Think in Action*. New York: Basic Books.

Scott, Dane (2012). Insurance Policy or Technological Fix? The Ethical Implications of Framing Solar Radiation Management. In Christopher Preston (ed.), *Engineering the Climate: The Ethics of Solar Radiation Management*. Plymouth: Lexington Books, pp. 151–168.

Sennett, Richard (2008). *The Craftsman*. New Haven: Yale University Press.

Smith, Ralph J., Butler, R. Blaine and Lebold, William K. (1983). *Engineering as a Career,* 4th ed. New York: McGraw Hill.

Smith, Ricky and Mobley, Keith R (2008). *Rules of Thumb for Maintenance and Reliability Engineers.* Oxford: Elsevier.

Stevenson, Angus (2010). *Oxford Dictionary of English,* 3rd ed. Oxford: Oxford University Press.

Thomasson, Amie L. (2007). Artifacts and Human Concepts. In E. Margolis and S. Laurence (eds.), *Creations of the Mind: Theories of Artifacts and Their Representation.* New York: Oxford University Press, pp. 52–74.

Vincenti, Walter G. (1990). *What Engineers Know and How They Know It: Analytical Studies From Aeronautical History.* Baltimore: Johns Hopkins University Press.

Vinsel, Lee (2019). Researching "What Do Engineers Do All Day?": Innovation, Maintenance and Everyday Engineering. *The Maintainters. Org.* http://themaintainers.org/blog/2019/5/21/researching-what-do-engineers-do-all-day-innovation-maintenance-and-everyday-engineering. Accessed November 8, 2019.

Vogel, Steven (2019). The Humanness of Infrastructure. Paper presented at the 7th PotC, October 5, University of Detroit Mercy.

Winner, Langdon (2004). Technologies as Forms of Life. In David Kaplan (ed.), *Readings in the Philosophy of Technology.* Maryland: Rowman Littlefield Publishing, pp. 103–113.

Young, Mark Thomas (2020). Now You See It (Now You Don't): Users, Maintainers and the Invisibility of Infrastructure. In Nagenborg, M et al. (eds.), *Technology and the City: Towards a Philosophy of Urban Infrastructure.* Switzerland: Springer International Publishing.

PART V

Engineering Activities and Methods

27

MEASUREMENT

Lara Huber

27.1. Introduction

Measurement is key in very different fields of research and application. Accordingly, various professional communities rely on the prospects of measurement. And, of course, laypeople are also familiar with these, given that measurement is a common practice in everyday life. Maybe this is why measuring, at first glance, seems to be a common and rather easy practice that not only provides us with secure knowledge but also is broadly applicable in the sciences and beyond. But then, as scientists and practitioners have come to understand, measurement is anything but an easy option if we take a closer look and, what is more, should never be treated lightly. There are two reasons for this:

To begin with, measurement requires a set of elaborate means, that is to say, reliable instruments, international standards ("measures"), and approved procedures. The latter includes all the preparatory steps, the actual process of measuring as much as the evaluation of outcomes. Taken together, they allow for the *practical* realization of measurement and accordingly represent measurement as a *practice*.

Secondly, scientists tend to achieve a certain kind of knowledge through measuring. Some might even say that true knowledge in the sciences can only be realized on the basis of measurement. Naturally, this issue deserves further study, for instance when we consider the critical acclaim this ideal of knowledge has caused, and is still causing, especially in the social and medical sciences.

Drawing upon these considerations, the purpose of this chapter is to review measurement as a practice. I shall present measuring as practical engagement with measuring instruments as well as with targets of measurement, the said "properties" of phenomena that could be assessed in a quantitative manner. Therefore, the chapter not only presents us with objectives of measurement from the perspective of measurement theory, but also analyses and discusses key challenges of measurement, from the perspective of its practical realization or application in science and engineering.

27.2. Measurement as a *Practice*

Measurement is commonly regarded as being a very distinguished scientific practice, differentiating itself from observation on the one side and experimentation on the other. Measurement and experimentation are regarded as defining two very different enquiries into physical realities: whereas experimentation aims to inform about the unknown by means of discovery, measurement provides us with certain and detailed information about objects ("targets of measurement"). Significantly, the essential features ("properties") of these objects have had already been identified in advance. So, let us not forget that this account paints a very narrow and therefore easily misleading picture of both these practices. Since Ian Hacking's landmark book about "representing and intervening" in the sciences (Hacking 1983), much has been said about the reality of scientific practices. For instance, it has been acknowledged that experimenting might be more independent of theorizing than earlier assumed. What is more, experimentation might not always be described as a linear process ("from hypothesis to testing") in the strictest sense, especially in the laboratory sciences (i.e., Knorr Cetina 1999). Against the backdrop of the practical turn in the philosophy of science, general assumptions about measuring and experimenting are revised. This happens not only with regard to the diversity of measuring instruments, but also concerning the heterogeneity of experimental designs. Measuring, more often than not, is a fundamental part of experimenting, and experimenting in turn is a key prerequisite of the process of measurement as such, if not a characteristic of measuring itself. Take, for instance, the case of assessing the precision of a measuring device through a series of measurements. Accordingly, the *International Vocabulary of Metrology* (VIM, third edition) defines measurement as an experimental process aimed at "obtaining one or more quantity values that can reasonably be attributed to a quantity" (JCGM 2012, Def. 2.1).

The classic distinction between experimenting and measuring is seemingly under threat—given that both approaches towards knowledge in science involve interactions with actual systems (whether physical, biological or purely technical in nature) with the aim of discovering and/or quantitatively representing properties of these very systems. This chapter refrains from exploring questions of mere theoretical interest, as for example, how different measurement theories interpret the scope or extent of numerical mapping. We shall also not consider questions such as how numerical mapping, conceptualization and/or empirical realization relate, and hence are justified.[1] The purpose is to consider the perspective of practitioners of measurement by introducing general concepts of measurement, specifying basic operations and reviewing practical challenges in measuring. Let us start by delineating and analysing key objectives of measurement.

27.3. Objectives of Measurement

Measurement is perceived as fostering genuine progress in science and therefore as coming with exceptional knowledge. In a lecture delivered at the London Institution of Civil Engineers in 1883, William Thomson (Lord Kelvin) quite famously introduced this ideal of knowledge through measuring, as follows (Thomson 1889: 81f):

> I often say that when you can measure what you are speaking about and express it in numbers, you know something about it; but when you cannot measure it, when you cannot express it in numbers, your knowledge is of a meagre and unsatisfactory kind; it may be the beginning of knowledge, but you have scarcely in your thinking advanced to the stage of science, whatever the matter might be.

We come to understand that measurement aims at specific ends, which in turn inform this very ideal of knowledge. Commonly, these ends are *certainty*, *precision* and *robustness*. Obviously, they not only inform knowledge through measurement but also orientate knowledge through scientific practices as such: Certainty is considered to be a general objective in science and engineering. Robustness is equally as important in measurement as it is in experimental trial design. As for precision, it might justify its claim as a genuine and specific status as regards measurement. Nonetheless, it is worth mentioning that all three ends receive specific attention in measurement theory.

Certainty in measurement refers to the question of whether a property that one intends to map numerically ("quantity") is distinctively and truthfully assessed through measurement. Hence, it is essential that a property intended for measurement has been thoroughly defined; that is to say, its distinction has been successfully disclosed ("concreteness"). Besides, the quest for certainty involves rather practical preconditions, including the exclusive use of scientifically approved means of measurement and other procedures for evaluating measurement outcomes. Therefore, certainty claims to relate to both the conceptual concreteness of a given property and the sound realization of a measurement procedure that reliably informs about this property and accordingly allows disclosing an essential feature of the phenomenon of interest.

Precision concerns the scope and range of a given measurement or measurement procedure, obtained by repeated measurement ("consistency" of measurement outcomes). This is why the quest for precision is attributable to technical invention and innovation in general and the respective scope/ range of measuring instruments in particular. In common language, "precision" is used as an umbrella term for addressing measurement outcomes as regards their representational scope or resolution in general. In measurement theory, precision is differentiated from a second concept, namely *accuracy*. The latter denotes how close a particular measurement outcome is to the "true value" of a given property (i.e., salinity of seawater). Accuracy presupposes that we are able to assess the "true value" of a property by means other than the measuring instrument at hand.[2] In measurement it is rather pragmatically achieved by means of comparing outcomes obtained by different measuring instruments and by defining "fixed points" through statistical approximation that serve for calibrating devices (i.e., portable salinometers) accordingly. Further practical issues worth considering as regards precision include, for instance, the need to ascertain that instruments operate properly and to define exclusively which measurement units are used, given the area of research and application. The latter refers to the genesis and implementation of "standards" of measurement, the former to issues of "calibration".

Robustness addresses the specific value of measurement outcomes. In the philosophy of science and beyond, it is an issue of great current interest and controversy, especially with regard to data analysis (i.e., Stegenga 2009). As regards measurement, the quest for robustness presupposes both scientifically approved means of measurement and internationally acknowledged ways of standardization and calibration. The latter concerns units, technical devices and measurement procedures. Accordingly, the robustness of outcomes could not be considered as an essential characteristic of a given property, but as the means by which it is assessed. These means are, for example, standardized measurement procedures, calibrated instruments and/or scientifically approved protocols for evaluating measurement data.

Before I evaluate on what grounds these ends are purported to be met, I shall introduce a more developed account of measurement. Measurement theory differentiates between an item or instrument that is used for measurement purposes ("measure"), a procedure that implements or realizes the very process of measurement ("measuring") and the given phenomenon and/or property of an object that is purported to be measured ("measurand").[3]

Accordingly, a *measure* could be a standard weight (also: "etalon") or reference material ("prototype"), and/or any auxiliary instrument, for example a balance, which is used for informing about quantity values of a phenomenon, body or substance in a given case ("measuring instrument"). The very process of measurement is defined and controlled on the basis of a given set of operations ("to measure"), referring to standard units, reference material and/or auxiliary instruments. The property of an object that one intends to measure, for example the weight of a chemical compound, is called *measurand*. The latter presupposes that the measurand could be reliably related to a given measurement procedure, that is, to a quantitative assessment in reference to a measurement standard and/or measuring instrument.

Ideally, an attribute of an object that is observable directly, for example the length of a table, could be identified and represented quantitatively, using a device, for example a folding carpenter's ruler. If no ruler or tape measure is at hand, we might try to assess the length of this table ("measurand") in reference to an object, the length of which is known to us (i.e., our feet = "measure"). Whatever "reference" or measuring instrument we choose impacts on the question if we are able to evaluate measurement outcomes. To cut a long story short: our feet are not standardized, whereas the ruler is. In order to ascertain the length of our feet, we sooner or later have to use an approved device for measuring, namely a foot-measuring device to assess shoe size. We learn two important issues through measuring feet: first, if we compare the foot length of different people, we arrive at very different lengths; secondly, if we proceed thoroughly, that is, if both feet of every person are measured consequently, we come to understand that feet are often not symmetrical in length. This characteristic is quite common throughout biological systems: symmetry in absolute terms could not be found in nature, but rather corresponds to our human, that is to say, technical interpretation of the world. In case I ignore measuring the length of both of my feet, I shall be simply unable to adjust my measurement outcome of the table (i.e., ten times the length of my right foot) to any other measurement outcome of this table, on the basis of other feet, or proper measuring instruments. Any measurement of tables or other objects on the basis of individual feet or other individual measures represents an epistemic challenge. To overcome this challenge, we seek both standardization of measures and international agreement on units of measurement, which provide us with defined (fundamental) magnitudes of a given set of quantities ("shared standards").

Standardization is a technical means of ensuring uniformity of objects, especially those that are used for measurement purposes (see also Chapter 42, "Standards in Engineering" by Paul B. Thompson, this volume). Every carpenter's ruler, in comparison to our feet, bears identical units. But again, there are differences: Common carpenters' rulers, or tape measures in the United States, bear customary units, quite similar but not identical to the imperial units in the United Kingdom (one yard or three feet, respectively, with 36 inches and 360 fractional inches). Rulers in the large majority of countries worldwide, where the metric system was adopted, bear metric units (one meter with 100 centimeters and 1000 millimeters). After all, the metric system is exemplary as regards an internationally agreed system, in which all base units interrelate and could be divided by an integer power of ten (decimal system). All rulers or other devices bearing metric units realize the "standard meter" (originally a unique artifact, nowadays a physical constant). Any realization, therefore, could be traced back to this genuine standard ("traceability").

From this point of view, we have to acknowledge that even a carpenter's ruler, a rather simple instrument from a technical perspective, is a quite elaborate means of measurement. The ability to measure length in reference to a carpenter's ruler or tape measure might be of minor interest, though, if we turn to the diversity and technical complexity of measuring instruments that inform today's practices of science and engineering: voltmeters, frequency and intensity meters, oscilloscopes, spectroscopes and chromatographs, spectral photometers, sensor technologies, X-ray diffraction and electron microscopes—just to name a few. In the majority of cases, scientists and engineers are interested in properties of objects that are not directly observable; in other words, properties that could be assessed only by means of measuring one or additional properties, if at all. This is why measurement theory differentiates between direct or fundamental measurements on the one hand, and indirect or derived measurements on the other: If we are interested in measuring an electrical charge, we have to rely on the measurement of time and electrical current. If we turn to measuring volume, we immediately and necessarily do so by referring to another fundamental or basic measure, namely length. Philosophical accounts on derived measurement often focus on thermometry, where we refer to changes in volume or pressure of physical materials, for example of liquids that expand with heat (i.e., mercury-in-glass-thermometer).[4] On a different level, this fundamental differentiation between "basic" and "derived" also informs the *International System of Units* (SI). In this case, though, the latter defines *base units*, which "are regarded as dimensionally independent by convention" (the metre, the kilogram, the second, the ampere, the kelvin, the mole, the candela) on the one side, and *derived units* that "are formed by combining the base units according to the algebraic relations linking the corresponding quantities" on the other.[5]

Thus, measurement theory is concerned with general assumptions about which property is measured by reference to a given measurement procedure, or whether a given property is necessary and/

or essential for informing about the phenomenon of interest. Even if we are able to identify a feature or property as a probable candidate for measuring, we are certainly faced with further preconditions of inventing and introducing a device that is able to measure this feature or property reliably. The latter draws attention to the question of on what grounds we commonly presume the stability and integrity of instruments. This is why this chapter, in what follows, focuses on preconditions and challenges in acknowledging measurement as a *practice*. In other words, measurement is scrutinized as an activity or set of operations which take place in the world and have an impact on the world. Measurement depends not only on the instrument(s) chosen, but also on further prerequisites, such as the know-how of practitioners, who are putting instruments to use, or scientists in general, who analyze measurement outcomes.

27.4. Preconditions of Measuring

Measurement in the sciences is often reduced to its impact on acquiring or ascertaining (true) knowledge. In engineering, measurement is as much a product of technical excellence as an auxiliary means to other ends of engineering, where they are not concerned with the intervention or refinement of the measuring instruments themselves. Thus, measurement not only aims at acquiring and improving knowledge about phenomena and their properties, but also presupposes a significant degree of knowing your way around measurement procedures—including cognitive, methodological and/or strictly technical "know-how". Some philosophers have stressed that technoscientific kinds of knowledge fundamentally differ from traditional accounts of knowledge in science and technology. Davis Baird addresses technoscientific knowledge as "instrumentally encapsulated knowledge" or "thing knowledge presented in a measuring device" (2004: 68); Alfred Nordmann, distancing himself from Baird's account, presents a more detailed description of technoscientific knowledge as "habit of action" in reference to Charles Sanders Peirce (2012: 27). For Nordmann as for Baird, technoscientific knowledge is objective due to the "transformation of available knowledge, technology, and skill into material processes and things" (ibid.: 26). But in contrast to Baird, this "objectivation" or "materialization" for Nordmann is but one aspect of technoscientific knowledge. It is the classical disciplinary sciences that "tacitly inform technoscientific action" by feeding their knowledge into an "elementary mastery of phenomena or systems" (ibid.: 27). As Nordmann further points out, this very knowledge actually "disappears in the formation of habits that are grounded in intimate familiarity with the behavior of a system" (ibid.). In a nutshell, "knowing that" becomes "knowing how"; theoretical knowledge is transformed into practical knowledge. In reference to Peirce, Nordmann specifies his reading of technoscientific action as "knowledge of control" as follows (ibid.):

> A habit of action, achieved capability, or knowledge of control guides us even in complex situations when we can safely rely on intersubjectively available causal relations and when we are cognizant immediately of this reliability and robustness rather than merely interpreting it as grounded in and derived from general laws.

Sociological and philosophical technoscientific accounts of measuring in science and engineering raise concerns about a common development in today's scientific research: the need to include only specific devices or numerical models in a given area of interest, and to, accordingly, exclude alternative designs/models. This mainly holds true for applied research, but also increasingly affects regimes in basic research—at least to a certain degree, depending on the given nature of inquiry. Research is a very demanding practice, as regards cognitive resources (scientific personnel, technical know-how, etc.) as much as laboratory infrastructure, including the development as much as the maintaining of technical equipment and other experimental preconditions. Both cognitive and technical resources have to be ensured prior to their being put to use. We should not forget that practitioners in science and engineering achieve experience through training with key procedures and standard tools in their area of interest (see also Chapter 5, "What Is Engineering Science?" by Sven Ove Hansson, this volume). They also have to be familiar with measurement protocols and the evaluation of outcomes. Accordingly, measuring as a practice addresses the following areas of applied knowledge:

- *which instrument* to choose, preferably, depending on the task or target of interest;
- *which operational steps* to consider, particularly, *to allow for a sound measurement process*, to minimize uncertainty and exclude known measurement errors; and—last but not least—
- *which strategies* to consult, *to respond to problems* that might arise in the course of measuring.

Each and every area of applied knowledge answers to practical preconditions in order to secure key objectives of measurement and also responds to challenges arising on the basic of given means of measurement. Take, for instance the case of temperature measurement: Depending on its area of application and the given system of interest, ensuring the accuracy of an instrument, from a technical point of view, might not suffice. In medicine, the clinical validity of an instrument presupposes that the medical staff is familiar with practical considerations of temperature assessment. This includes the choice of thermometer (e.g., infrared thermometer, liquid crystal thermometer) but also the proper assessment of body temperature, depending on the question of where (e.g., rectal, oral application of instruments) and when measurements are taken (timing and interval).

27.5. Challenges of Measuring

Practitioners in science and engineering are familiar with challenges of *measuring*, which address the technical refinement of devices, the revision of measurement procedures from a methodological point of view as much as the assessment and evaluation of measurement outcomes: What is measured or which property is assessed, respectively? Is a given measurement carried out correctly? Are measurement outcomes accurate? Drawing upon practical challenges, philosophers of science have stressed general concerns regarding *uncertainty* in measurement, *inaccuracy or imprecision* of devices and queries as concerns the *epistemic evaluation* of measurement results. The latter highlights the ability to verify observations as "phenomena" and to disqualify others as "artifacts", respectively.[6] In the following part, the chapter elaborates on these three major threats to measuring in science and engineering:

Uncertainty in measurement refers to conceptual as much as practical challenges, as was mentioned earlier. In the first case, very different scenarios are possible, covering the entire spectrum from conceptual fuzziness to ignorance. For example, a given property of a phenomenon of interest that one intends to measure has not been defined thoroughly. One might depict this kind of uncertainty as lack of conceptual knowledge. Hence, we are either unable to ascertain which property in comparison to other possible aspects or properties of a phenomenon is specifically assessed through measurement, or if a given property is assessed distinctly and truthfully. The latter might be described as lack of knowledge as regards the value of the measurand, for example concerning the length of a table. As a matter of consequence, our knowledge through measurement of a phenomenon might be hampered significantly. Some of these challenges are discussed in reference to Percy W. Bridgman's "operational analysis". This includes the position that our understanding of a concept (e.g., temperature) is correlated to the very method of measurement, given that each measurement presupposes a specified set of operations. According to "operationalism", we are confronted with as many concepts of temperature as there are methods of assessing it via measuring.[7] Another important issue of uncertainty in measurement concerns the question of whether measurement of a certain property informs us about the given phenomenon of interest and, if yes, to what degree, depending on defined parameters of testing, and/or measuring.[8] The latter corresponds to the common structural deficit of models in science, given that models are always "reduced" or "idealized" means in representing only a certain number of features or a given causal relation of a phenomenon they claim to model.

Secondly, we are also confronted with cases where practical preconditions of measurement are not sufficiently fulfilled, for example, if practitioners exclusively refer to instruments that have not yet been approved to serve as a means for a given purpose, from a scientific point of view. Besides technical challenges, there are structural ones. Take, for instance, the case of meteorology or climate research. Here scientists have to define reliable intervals of "observation" to allow a truthful assessment of all events that are relevant to map the phenomenon of interest, in order to develop advanced and precise models of weather forecasting or climate change (i.e., timing and interval of observation; spatial distribution of weather and environmental monitoring stations).

Inaccuracy in measurement is related to the scope and range of an instrument or measurement procedure, in comparison to the scope and range of other instruments or procedures that allow for assessing the "true value" of a property. It should be possible to overcome this, at least in principle, through further technical refinement and/or innovation. Any measurement outcome that could not be repeated, while using the very same device, would be regarded as *imprecise*. The latter might also be a result of mere day-to-day challenges of measuring, that is, of constant use or aging of instruments. In this case, the stability and integrity of measuring instruments are under threat, because they no longer operate properly due to deterioration, corrosion or structural damage. Besides, imprecision and inaccuracy in measurement relate to further questions—for example, whether instruments are calibrated correctly, if their use is sufficiently sound, whether practitioners are skilled enough or practiced in handling a given instrument. This includes the capacity of calculative error or other (external) effects that could have an impact on the measurement process—for example, any faulty operations in the course of measurement in general whatsoever, as well as any erroneous handling of instruments employed in particular, that might challenge the "proper" realization of a measurement procedure.

Epistemic evaluation: It is commonly assumed that if all precautions are fulfilled, a means truly and accurately informs about a given end in measurement. Another major precondition of measurement was only briefly mentioned before, namely the need to scientifically approve a measuring instrument or a formalized procedure, as for example, testing. In fact, a growing number of historical and philosophical accounts address the question of how instruments are approved to serve a given scientific purpose; in brief, how they are made "instrumental": For instance, consider Chang's *Inventing Temperature: Measurement and Scientific Progress* (2004), Nicolas Rasmussen's *Picture Control: The Electron Microscope and the Transformation of Biology in America* (1997) or William Bechtel's *Discovering Cell Mechanisms: The Creation of Modern Cell Biology* (2006). The question of how devices become "instrumental" (and are kept "instrumental") is tied up with fundamental challenges of measuring. This entails ascertaining a given measurement procedure as a technical and practical "arrangement", which includes said "instruments" (technical devices), defined steps of "operations" referring to one or more technical devices (technoscientific practices) as well as methodological regimes of evaluating and approving measurement data (scientific epistemology). Historical and philosophical accounts of a small number of selected scientific practices shed a light on general prerequisites in this very process of becoming "instrumental". Let me elaborate on this "arrangement" by referring to W. E. Knowles Middleton's *History of the Thermometer* (Middleton 1966): Here, Middleton informs us about the piecemeal process of scientific research into thermometric substances and the history of technical innovation and invention of scientific instruments, referring to differences in volume (from "thermoscopes" to "thermometers"). Against this background, he also discusses the need to identify valid thermometric benchmarks (i.e., the temperature of boiling water) and to ascertain that the defined fixed points of a calibrated temperature scale are truthfully "constant".[9] Depending on the instruments used, a given set of practical virtues in measuring emerges that has to be acknowledged to allow for epistemic success and reliability. With regard to thermometry, for example, it is mandatory to assess relevant changes due to temperature, which have to be defined in advance, rapidly and ideally with no delay. This presupposes that these changes are reversible, without substantially corrupting the physical property of a material. These practical virtues pave the way for a shared understanding of how a given measurement procedure is evaluated for specific purposes in science and engineering. Quite notably, this case study informs us about *structural* challenges of *scientific* epistemology, which—as these cases also illustrate—could not be overcome, because they are inherent to the very idea of science and are accustomed to practice. Thus, one may safely assume that these structural challenges of making something "instrumental" with regard to a given scientific purpose still impact on today's practices in science and engineering (see also Chapter 20, "Towards an Ontology of Innovation" by Vincent Blok, this volume). This regards the evaluation of outcomes as much as the question of on what grounds we define whether scientific objectives of measurement are fulfilled. Let us briefly revisit the historical case of a traveling meteorologist in the 19th century. A note added by Ludwig Schön, head of the Meteorological Institutes of the Great Duchy of Saxony-Weimar-Eisenach, to Johann Wolfgang Goethe's meteorological diary that he kept during a bathing sojourn in Bohemia in 1823, reads as follows (Goethe 2005: 148f):

> The meteorologist shall travel on foot with a messenger as his companion. The latter is to carry the following meteorological instruments that are well preserved and capable of being securely and comfortably transported, and easy to rapidly set up for means of observation, namely:
>
> 1. One portable barometer
> 2. Two thermometers
> 3. One portable hair hygrometer
> 4. One compass containing 22 wind segments
> 5. One waterproof umbrella.

Schön's note mentions in detail which specific instruments are of value for travelers such as Goethe, who were interested in meteorological events in an era when meteorology was only just becoming a scientific field of inquiry. He also lists an additional auxiliary instrument, namely the umbrella, which by necessity has to be waterproof. The umbrella is not regarded as a meteorological instrument *per se*, but as a necessary means to ascertain the measurement procedure. Schön distinctly states that the umbrella serves to protect the accurate technical performance of barometer and thermometer from objectionable effects due to sun and rain. Any account that addresses how instruments for the purpose of measurement are used appropriately must consider the issue of how known external influences, that otherwise would hamper the process of measurement, could be successfully excluded. To put it bluntly: it is only if measuring instruments are handled with care and foresight, and, if required, additional auxiliary means (i.e., a *waterproof* umbrella) are taken into account, that the measurement process could be regarded as correctly realized and valid in its outcome.

Goethe's meteorological diary with Schön's annotations is but one of many documents from the era of the advent of formalization and standardization in the natural sciences. But none, surely, covered more scientific fields of interest than the *Manual of Scientific Enquiry Prepared for the Use of Her Majesty's Navy and Adapted for Travellers in General*, edited by John F. W. Herschel in 1849. The manual examines metrical, technical as much as methodological issues of fields as varied as, for instance, astronomy, hydrography, botany, mineralogy, or statistics. The majority of papers were written by known experts in these given fields of inquiry, such as for example, Charles Darwin (botany), G. R. Porter (statistics) or Herschel (meteorology) himself. In his paper on meteorology, Herschel elaborates on technical preconditions, which instruments are preferred for given purposes and how these particular instruments have to be used (1849: 268–322). He also names general rules and precautions regarding the very process of measurement as much as the question of how travelers shall maintain a regular meteorological register. Furthermore, Herschel provides tables that define corrections to be applied to barometers, etc. (ibid.: 318–322).

These historical documents, Goethe's diary as much as Herschel's manual for travelers in general, allow us to reflect upon both the dimensions of epistemic evaluation of instruments of measurement and of measurement outcomes. They could be regarded as early accounts of how measurement necessitates a scientifically approved, but nevertheless social network of credibility and trust (i.e., Shapin 1994), to serve purposes in science and engineering: Credibility of methods and instruments used, reliance on units of measurement and a given metric system, as much as trust in practitioners who engage in measurement procedures and the evaluation of outcomes. Many scholarly accounts of scientific practices focus on the need to define or develop shared standards and to, correspondingly, "pacify" metrology (i.e., Mallard 1998). Within this debate, awareness is growing that this includes the necessity to re-evaluate and, if necessary, to revise shared accounts of scientific practices from the perspective of epistemic pluralism and epistemic injustice, respectively.[10]

Seemingly, several virtues stand out that correspond to key objectives of measurement, but also answer to challenges of day-to-day measurement practices. Virtues answer to the need to identify, and as a consequence exclude, corruptive practices of scientists that would degrade these practices as truly scientific ones. This, of course, concerns the peer group as much as the individual. As regards measurement, these virtues highlight the following needs, namely

- to *ascertain* a given measurement outcome, for example to verify that the visual representation of measurement data (i.e., graph) actually informs us about a "phenomenon" and does therefore not reflect an "artifact" (*awareness* of technical and methodological flaws);
- to *assess* which objectionable (external) factors might impact on a given measurement procedure, and to come up with strategies to successfully exclude these factors (*prudence* in identifying and eliminating objectionable effects);
- to *(re)evaluate* a given set of operations that render a said measuring device "instrumental" on a regular basis (*vigilance* to anticipate possible threats yet unknown);
- to *(re)evaluate* a said "standard" of measurement ("unit", "reference sample", etc.) due to its operational value as concerns a given purpose on a regular basis (*critical awareness* of the representational scope and extent of "references" in science);
- to *review* the use and handling of devices and resources with regard to scientific purposes (*appropriateness* of means to an end); and
- in depicting the day-to-day practice of measuring especially, to *review* and *critically reflect on* one's own performance as regards the rightful use and appropriate handling of measuring instruments and resources as regards a given scientific purpose (*self-awareness* as a practitioner in science and engineering).

This preliminary list of cognitive and practical virtues, which is all but exhaustive, informs measurement as a *practice*. Taken together, these virtues answer to known challenges in measurement. Some correlate with mere practical queries that, for instance, could be traced back to the technical scope or range of devices and procedure. Others precede mere practical considerations: if we aim to protect ourselves from avoidable ignorance, we have to acknowledge the limitations of measurement from a methodological point of view in general, and from the specific limitations of means of measuring from a technical perspective in particular. Consequently, this is about every practitioner's duty of taking a critical stance towards testing and measuring and of assessing the scientific value of such means that are purported to serve a given end. Lastly, this is also about engaging in the scientific debate on the means, as much as the ends, of measuring.

27.6. Conclusion

The chapter has presented key objectives of measurement (certainty, precision/accuracy, robustness) and also specified fundamental challenges of measurement (uncertainty, imprecision/inaccuracy, epistemic evaluation). Both objectives and challenges were addressed from the perspective of philosophy of science and epistemology, as much as from the practical realization or application of measurement in science and engineering. Measurement, very much like experimentation, is a key method of the sciences. It is not restricted to what is called the "natural" sciences or its application in engineering, but rather regarded as a fundamental strategy in a diversity of scientific fields of inquiry, including the "life" sciences, psychology, economics and the social sciences. It is these fields where Kelvin's ideal of *knowledge through measurement* has faced the most profound critique. What kind of knowledge is acquired through measurement? In how far is our knowledge restricted to a given scientific method, such as measurement? How do we acquire and ascertain knowledge of phenomena that cannot be assessed through measurement? How do we acknowledge and evaluate "uncertainty"? An inquiry into these mere general challenges of measuring may start by reflecting upon Bruno Latour's figurative analogy between the scientist and the termite. In *Science in Action: How to Follow Scientists and Engineers Through Society*, he compared what he calls the "enlightened networks" of scientists with the "obscure galleries" of termites (1987: 251): whereas the latter are built "with a mixture of mud and of their own droppings", the former reflect their given technical and methodological genesis. Consequently, as Latour claims, termites and scientists share a common trait: "they can travel very far without ever leaving home" (ibid.).

Measuring, in other words, not only comes with an exceptional perspective on target phenomena, it always provides us with an "exclusive" view in a double sense: Ideally, measurement allows for knowledge of a specific kind and therefore could be regarded as a unique prerequisite of ensuring scientific knowledge. Then again, it also comes with restrictions in perspective (*ex*clusive in contrast to *in*clusive). That way, even if practitioners in science and engineering are seemingly forced to hold on to the claim that progress in science is possible and best achieved through measurement and technical innovation, it might yet be of some advantage to also consider general boundaries of scientific methodology and critically assess possible costs that come with this singular but also inherently exclusive perspective measurement provides.

Related Topics

Chapter 11: Uncertainty (William M. Bulleit)
Chapter 20: Towards an Ontology of Innovation (Vincent Blok)
Chapter 31: Experimentation (Viola Schiaffonati)
Chapter 32: On Verification and Validation in Engineering (Francien Dechesne and Tijn Borghuis)
Chapter 36: Trust in Engineering (Philip J. Nickel)
Chapter 42: Standards in Engineering (Paul B. Thompson)

Further Reading

Gould, S. J. (1996) *Mismeasure of Man*, revised and expanded ed. New York and London: W. W. Norton. (A cultural history of the prospects and perils of intelligence testing and far-reaching social challenges of scientific misperceptions.)

Kula, W. (1986). *Measures and Men*. Trans. R. Szreter. Princeton: Princeton University Press. (The landmark book on the cultural history of measurement.)

Tal, E. (2020). Measurement in Science. In E. N. Zalta (ed.), *Stanford Encyclopedia of Philosophy*, Fall 2020 ed. https://plato.stanford.edu/entries/measurement-science/. Accessed November 4, 2020. (Introduces and discusses important strands of the philosophical discourse on measurement in science.)

Wise, N. (ed.) (1995). *The Values of Precision*. Princeton: Princeton University Press. (A collection of historical accounts of how values in the theory of measurement were perceived and evaluated.)

Zupko, R. E. (1990). *Revolution in Measurement: Western European Weights and Measures since the Age of Science*. Philadelphia: The American Philosophical Society. (A classical historical account of the scientific revolution in measurement and the genesis of the metric system.)

Notes

1. An account of a representational theory of measurement is given in Berka (1983).
2. For a critical discussion of "true value" in metrology, see Grégis (2015).
3. For an introduction, see VIM 3 (JCGM 2012).
4. For a critical assessment, see, for instance, Chang and Cartwright (2014).
5. This and further information could be found online, via www.iso.org.
6. An overview provides, for instance, the edited volume by Boumans et al. (2014).
7. As regards "operationalism", consult Chang (2019); for an analysis of the genesis of quantitative concepts in the natural sciences, see Schlaudt (2009).
8. "Uncertainty" in measurement and corresponding strategies to evaluating measurement data are addressed in the *Guide to the Expression of Uncertainty in Measurement* (JCGM 2008).
9. Chang (2004) reflects on benchmarks in thermometry and corresponding concepts of measurement.
10. See, for instance, Mitchell (2009) for an account of "pluralism" in science, and Fricker (2007) for her account of "epistemic injustice".

References

Baird, D. (2004). *Thing Knowledge: A Philosophy of Scientific Instruments*. Berkeley, Los Angeles and London: University of California Press.

Bechtel, W. (2006). *Discovering Cell Mechanisms: The Creation of Modern Cell Biology*. Cambridge: Cambridge University Press.

Berka, K. (1983). *Measurement: Its Concepts, Theories and Problems*. Translated from the Czech by A. Riska. Dordrecht, Boston and London: R. Riedel Publishing Company.

Boumans, M., Hon, G. and Petersen, A.C. (eds.) (2014). *Error and Uncertainty in Scientific Practice*. London: Routledge (Pickering & Chatto).

Chang, H. (2004). *Inventing Temperature: Measurement and Scientific Progress*. Oxford and New York: Oxford University Press.

Chang, H. (2019). Operationalism. In E.N. Zalta (ed.), *Stanford Encyclopedia of Philosophy*. Winter 2019 ed. https://plato.stanford.edu/entries/operationalism/. Accessed November, 4, 2020.

Chang, H. and Cartwright, N. (2014). Measurement. In S. Psillos and M. Curd (eds.), *Routledge Companion to Philosophy of Science*, 2nd and revised ed. London and New York: Routledge.

Fricker, M. (2007). *Epistemic Injustice: Power and the Ethics of Knowing*. Oxford and New York: Oxford University Press.

Goethe, J.W. von (2005 [1823]). Beobachtung und Beschreibung der atmosphärischen Phaenomene von Ende Juny bis den 18. Septbr. 1823 [Materialien 9.9]. In Ibid., *Die Schriften zur Naturwissenschaft*, Zweiter Band: Zur Meteorologie und Astronomie, Ergänzungen und Erläuterungen, bearb. v. Gisela Nickel (Goethe. Die Schriften zur Naturwissenschaft, Zweite Abteilung: Ergänzungen und Erläuterungen, Bd. 2). Weimar: Verlag Hermann Böhlaus Nachfolger.

Grégis, F. (2015). Can We Dispense with the Notion of 'True Value' in Metrology? In O. Schlaudt and L. Huber (eds.), *Standardization in Measurement: Philosophical, Historical and Sociological Issues*. London: Routledge (Pickering & Chatto).

Hacking, I. (1983). *Representing and Intervening: Introductory Topics in the Philosophy of Natural Science*. Cambridge: Cambridge University Press.

Herschel, J.F.W. (1849). *A Manual for Scientific Enquiry, Prepared for the Use of Her Majesty's Navy: And Adopted for Travellers in General*, published by Authority of the Lords Commissioners of the Admiralty. London: John Murray.

Joint Committee on Guides in Metrology (JCGM) (2008). *Evaluation of Measurement Data—Guide to the Expression of Uncertainty in Measurement* (GUM), 1st ed. (1995 version with minor revisions), Sèvres. www.bipm.org/en/publications/guides/gum.html. Accessed February 20, 2018.

Joint Committee on Guides in Metrology (JCGM) (2012). *International Vocabulary of Metrology—Basic and General Concepts and Associated Terms* (VIM), 3rd edn (2008 version with minor revisions), Sèvres. www.bipm.org/en/publications/guides/vim.html. Accessed February 20, 2018.

Knorr Cetina, K. (1999). *Epistemic Cultures: How the Sciences Make Knowledge*. Cambridge, MA and London: Harvard University Press.

Latour, B. (1987). *Science in Action: How to Follow Scientists and Engineers Through Society*. Cambridge: Harvard University Press.

Mallard, A. (1998). Compare, Standardize, and Settle Agreement: On Some Usual Metrological Problems. *Social Studies of Science,* 28, 571–601.

Middleton, K. (1966). *A History of the Thermometer and Its Use in Meteorology*. Baltimore, MD: Johns Hopkins University Press.

Mitchell, S.D. (2009). *Unsimple Truths: Science, Complexity, and Policy*, Chicago: University of Chicago Press.

Nordmann, A. (2012). Object Lessions: Towards an Epistemology of Technoscience. *Scientiae Studia,* 10, 11–31.

Rasmussen, N. (1997). *Picture Control: The Electron Microscope and the Transformation of Biology in America, 1940–1960*. Stanford: Stanford University Press.

Schlaudt, O. (2009). *Messung als konkrete Handlung: Eine kritische Untersuchung über die Grundlage der Bildung quantitativer Begriffe in den Naturwissenschaften*. Würzburg: Königshausen & Neumann.

Shapin, S. (1994). *A Social History of Truth: Civility and Science in Seventeenth-Century England*. Chicago: Chicago University Press.

Stegenga, J. (2009). Robustness, Discordance, and Relevance. *Philosophy of Science,* 76, 650–661.

Thomson, W. [Lord Kelvin] (1889). Electrical Units of Measurement [1883]. In Ibid., *Popular Lectures and Addresses*, in Three Volumes, Vol. 1: Constitution of Matter. London and New York: Palgrave MacMillan.

28

MODELS IN ENGINEERING AND DESIGN

Modeling Relations and Directions of Fit

Michael Poznic

28.1. Introduction

Models are abundantly used in engineering research, in engineering design, as well as in science. In this chapter, the urge to pose a generic question about models or engineering models will be contained. It is reasonable to assume that the generic question of what models are does not allow for a simple and uniform answer (cf. Frigg and Hartmann 2012; Poznic 2017), and neither does the less generic but still ambitious question of what engineering models are. Yet, attempts to answer specific questions of how particular engineering models are used seem to fare better. For example, one can ask for which purposes particular models are deployed. A further twist to the question about models may be to focus on the practice of modeling instead of regarding models as objects, in the first instance.[1] This will be the strategy of this chapter, namely to focus on the practice of modeling and to study particular instances of the uses of models as tools in different contexts of engineering.

Different purposes may direct the practice of modeling. In the literature on models, one finds many answers to the question about what their function is. Modeling may be practiced in order to explain (Bokulich 2017); models can be used as exploratory means (Gelfert 2016); modeling can be pursued to predict (Levins 1966); models can be used to conduct simulation studies (Winsberg 2010); models can be used as epistemic tools (Boon and Knuuttila 2009); or one can regard modeling as the practice of constructing or evaluating things (Eckert and Hillerbrand 2018).

This latter purpose of constructing seems to be one of the main functions of modeling practices in engineering design. Especially when one focuses on engineering as a practice that is primarily aimed at producing artifacts, this purpose of constructing something seems to direct the practice as the final end. Still, the other purposes may also be relevant for engineering. For example, one could make the point that, in many cases, engineering models are used to predict. Different interests may guide the search for predictions, e.g., the question of how artifacts will behave under different circumstances. Examples of such questions are whether a particular planned bridge will be resilient enough or not, or the question of how particular changes to a sociotechnical system affect different subsystems of this system or even society as a whole. However, the purpose to explain something may also drive modeling enterprises in engineering. For example, models may be used to study failures in constructions. A particular case is the study of constructed levees in the context of Hurricane Katrina in the US in 2005. Different groups of engineers tried to explain why the levees did not withstand large amounts of water due to this weather phenomenon (cf. Pirtle et al. 2018).

In many contributions to philosophy of technology, it is stated that engineering is rather about how things *should be* in contrast to how things *are*. Let us call this the basic assumption of philosophy of engineering and technology.[2] In this chapter, I aim to question whether this assumption is indeed correct when spelled out in terms of modeling enterprises. I apply a distinction of *directions of fit* that is commonly employed in philosophy of mind and philosophy of language to the relation between vehicles of modeling and targets of modeling.[3] Pictured simply, this is about the relation between model and the thing that is to be modeled. According to this distinction, there are two relations with different direction of fit: One relation is about how the target should be according to the vehicle of modeling, and the other relation is about how the target is according to the vehicle. Examples for vehicles are computational models of buildings, scale models of a bridge, mathematical models specified by equations, simulation models, etc. But also, model descriptions such as sentences in a technical language, descriptions in ordinary language, strings in programming codes, and mathematical equations can be seen as vehicles, among other examples. Examples for targets are the things that are modeled with the help of the vehicles, things such as buildings, bridges, human organs, jet engines, etc.

The result of the application of the distinction of directions of fit is that there are two different modeling relations. With the help of this result I will show, on the one hand, what is correct about the basic assumption that engineering modeling is rather about how things should be in contrast to how things are. On the other hand, I will also show that this assumption is too simplistic, because engineering modeling is also about how things are and not only about how things should be. Beyond that, some enterprises encompass modeling relations of both sorts.

In the next section, I will introduce the distinction between two directions of fit and apply it to discussions about models (§2). Three examples of modeling enterprises will be discussed. One example is a model that was developed with the goal of creating a technological surrogate system for a human organ. This surrogate system may function as an epistemic tool in drug research (§3.1). Another example is the design of a weekend cottage, which is constructed with the aid of an architectural model (§3.2). A third example is a model that is used by designers who aim to develop solutions to a problem in the context of producing jet airplane engines (§3.3). In a following section (§4), I will discuss open questions concerning the interpretation of the models from the three different contexts of engineering research, artistic design, and engineering design in terms of modeling relations with the directions of fit. In the last parts of the chapter, the following three questions will be posed: the question concerning indirect views of modeling applied to design models such as the weekend cottage (§ 4.1), concerning an alternative interpretation of models as props in games of make-believe (§ 4.2) and concerning the difference between product and process models studied in design research (§ 4.3). The chapter will close with some concluding remarks (§ 5).

28.2. Directions of Fit in Modeling Relations

Let me elaborate on an example in which one can easily distinguish two different ways in which vehicles are related to their corresponding targets. In an influential monograph from 1957, Elizabeth Anscombe gives the following vignette. Imagine a customer going to the supermarket. His wife wrote a shopping list for him the other day, and he now has the plan of buying the things that are written down on the list, the vehicle. The customer enters the supermarket with the shopping list and starts to put things in the shopping basket according to the list. There is a supermarket detective spying on the customer. The detective is writing down a record of all the items the customer collects in his basket and so creates a second list of the items, another vehicle. After finishing the shopping, the customer carries his basket to the cash point. In an ideal situation, his list agrees with the items in the basket. The same goes for the detective's list. However, it may be that the lists and the items do not agree, and in case they do not agree, they may do so for different reasons.

> It is precisely this: if the list and the things that the man actually buys do not agree, and if this and this alone constitutes a mistake, then the mistake is not in the list but in the man's

performance . . . whereas if the detective's record and what the man actually buys do not agree, then the mistake is in the record.

(Anscombe 1957: 56)

In case that lists and items do not agree, shopper and detective would correct the apparent mistake differently. The shopper would adjust the items in the basket according to the list, whereas the detective would adjust the list according to the items. So, there are different *directions of fit* in the relation between the shopper's list and the items and the relation between the detective's list and the items. Expressed with the terminology of this chapter, the first relation involves a "target-to-vehicle" direction of fit, and the second one involves a "vehicle-to-target" direction of fit (cf. Poznic 2016a). Let us call the first relation the *design* relation and the second relation the *representation* relation. This analysis can be combined with a perspective on modeling relations in science and engineering. Given that modeling not only involves vehicles such as models and targets but also users and purposes, the full picture from philosophy of science applied to the shopping example is the following one. In the instance of the customer who has the purpose to buy particular items, the relation between the vehicle (here, the list) and the target (here, the basket) has a target-to-vehicle direction of fit. In this case, the customer is the user and the wish to buy the items is the purpose. In the instance of the detective, who follows the purpose to produce a record of the chosen items in the basket, the relation has a vehicle-to-target direction of fit. Here, the detective is the user and the purpose is documenting, i.e., to describe which items are in the basket. With the help of this distinction between different directions of fit one can interpret the relation between many models and targets as a representation relation with a vehicle-to-target direction of fit. Especially in science, where oftentimes the goal is to study natural phenomena that exist independently of users or models, the establishment of representation relations between models as vehicles and natural phenomena as targets is the prevailing aim of modeling practices. In the first example of a model from an engineering research context, we will see that the relation of representation is relevant for this particular modeling task in engineering as well.

28.3. Three Examples of Models

28.3.1 *A Bioengineering Surrogate Model: The Lung Chip*

Let us start with the first example of a particular model in an engineering research context. To be more precise, the context is bioengineering, and applications of this model are utilized in medical studies and drug research. The example is a device that is meant to mimic core functions of a human lung. It is called "lung-on-a-chip" or "lung chip" for short. It involves not only technologically constructed materials but also living human cells. These cells are cultivated on a silicon chip that has a specific architecture. Different compartments are built on the chip, and the two central ones stand for air sacs and blood vessels in the human lung. Endothelial and epithelial tissues are each placed on one and the other side of a membrane that separates the two central compartments. Mechanical forces are applied to the cells so that, for example, the breathing motions of a lung can be simulated. Biochemical influences can be imposed by inducing particles or fluids. Particles can be introduced in the alveolar channel, the compartment standing for the air sac. Microfluidic technology is applied to steer fluids through the vascular channel, the compartment standing for the blood vessel.

In one particular representational use of this lung chip, fluid accumulation in the lung, a disorder called *pulmonary edema* is modeled (cf. Huh et al. 2012). There is a modeling relation between the chip as vehicle and the respective target of modeling.[4] This relation is a relation where the vehicle is adjusted to the target. Should the vehicle not accurately model the target, one would correct such an inaccuracy by changing the vehicle. It is presupposed that the vehicle is used to adequately represent the target. Taking the terminology from the previous section, we could say that the relation between vehicle and target is a relation of representation, a relation with a vehicle-to-target direction of fit.

So, the lesson is that this particular bioengineering model has a representational function. The representation relation is one modeling relation that is relevant for the uses of the lung chip. One could ask whether the design relation may also be relevant for modeling practices related to the lung chip. Indeed, the construction of the lung chip with the help of a so-called protocol can be reconstructed as a *modeling* of a human organ chip. This task involves using "the protocol" of the chip as a vehicle in order to establish a relation with a target-to-vehicle direction of fit (cf. Poznic 2016a). To produce such a lung chip involves following a complicated procedure, which Dongeun Huh and colleagues have described in detail (Huh et al. 2013). The protocol of this procedure is split into separate parts that involve numerous steps. The whole process of following the protocol takes at least several days before the lung chip with its full functionality is completed. The full description of this protocol works like a recipe. In addition, critical steps are explicitly mentioned, and extensive advice for troubleshooting is provided.

The full picture of the lung chip as a representational model of the human lung involves not only a representation relation between the chip and the target but also a design relation between the protocol and the chip. The main function of the lung chip, however, is to represent particular targets.

The basic assumption seems to be falsified by this example. At least, it is falsified if one interprets the assumption to be about all cases of engineering models. Because the modeling involves as its pivotal relation a representation relation between lung chip and target of modeling, the use of the lung chip is mainly to inquire about what is the case, and not about what should be the case.

In contrast, there is the option to object to this interpretation of the model, falsifying the assumption. One could argue that the example partly confirms the first half of the assumption, because the modeling of the chip also involves reasoning about how the lung chip should be constructed.

In order to mediate between these two ways of relating the example to the assumption, a modest reaction may be to say that it challenges the assumption. One potential response from a proponent of the assumption may be to stress that the model is used in a research context. Because the model is from a research context rather than a context of design, the basic assumption may not apply to this example comprehensively. In any case, this example shows that the basic assumption is not a valid claim when interpreted as applying to all models in engineering.

The design relation that was used in a preliminary step of the modeling with the lung chip has a more prominent place in the next example. We will see that the model discussed in that example is used as a generative model of a target, i.e., a model used to construct a target.

28.3.2 An Architectural Model of a Building: The Weekend Cottage

For the second case, we will introduce a model from a different, non-research context. The model is an architectural model and it is from a rather artistic context. It is an architect's model, but probably a model of a more mundane than fancy design. The following instance inspired by Galle (1999) gives a simple story of a possible situation in which such a model is used as a means of communication. It is an ideal-type situation involving three individuals who fill the roles of designer, client and builder, respectively. In simplified form, the narrative goes as follows. The school teacher asks her neighbor, the architect, whether she could design a weekend cottage for her. The architect agrees, and she asks the retired carpenter to build the cottage. So, the teacher fills the role of the client, the neighbor fills the role of the designer and the carpenter fills the role of the builder.

To give a bit thicker description of this illustration, the architect starts designing such a building and, after a short while, she presents a first model to the teacher. After some hours of discussion, they agree that the architect as designer should refine the model in specific ways, and they make a new appointment. At the next meeting, the architect presents a revised model and client and designer again start to discuss. This process may continue for a while until they agree that the designer will make a few final revisions. The designer revises the model accordingly and, after showing the next version to the teacher, they agree on the design of the cottage that is embodied in the final model, the "design model" from now on. The architect gives the design model together with other things such as drawings, sketches and maps to the carpenter to build the cottage. Here in this second step, there may also occur some rounds of discussion among designer and builder, further revisions and so on. For reasons of simplicity, let us assume that the builder just constructs the cottage according to the design the teacher and neighbor settled upon.

The design model functions as a central tool of communication in this example. The architect as designer and the teacher as client use it in order to reach an agreement about the design of the cottage. To be more precise, they use the different versions of the model in order to reach this agreement that is then, in the end, embodied in the final version of the design model. This final version, the model of the cottage, is used by designer and client to check whether they in fact have the same understanding of the design. In the second step, the design model is used by the designer to communicate to the builder how the cottage should be structured in the end. The design model can also be used by the builder to guide construction. In all of these instances, the final model is standing to the building in a relation with a target-to-vehicle direction of fit. The building as the target is adjusted to the model as vehicle. Here, the model stands in a design relation to its target.

Until now, we have not said what the model is made of. It may be a complicated material object made out of cardboard, plastic or other materials.[5] Or, it may be constructed with the help of a particular computer software. Building information modeling is a technique that many architects use to design buildings (cf. Eastman et al. 2011).[6] Maybe the neighbor is used to using only paper and pencil, however, and does not use a computer. In any case, whether the model is a concrete physical model or a computational model, it is standing to its target within a design relation.

Here, the basic assumption seems to be confirmed. The model is used as a tool to foster reasoning about how something should be. Designer and client use the design model to reach an agreement about how the cottage should look like in the end. And, of course, the builder is also engaged in this type of reasoning when using the model to build the cottage.

28.3.3   A Designer's Product Model: The Jet Engine

So far, we have discussed two different cases. One modeling task—the lung chip—in which the main relation is a modeling relation with a vehicle-to-target direction of fit exemplifies the first case, and another task—the cottage model—in which the main modeling relation is a design relation with a target-to-vehicle direction of fit exemplifies the second case.

The third example is a modeling enterprise followed by designers to foster an agreement about solutions to a particular task in the context of jet engine design. The overarching design problem is to maximize the combustion temperature of a jet engine. Within this overarching problem there is the subproblem of designing particular components of the engine that are supposed to withstand high temperatures. The crucial issue is that not every material can withstand the high temperatures that are needed in order for the engine to work efficiently. The geometry of the structural composition of the jet engine is crucial as is the materials to be used in the engine. Within this third model example, the non-rotary engine components are particular objects of study.

In the context of jet engine design, usually one company, the manufacturer, is responsible for the control of the main combustion process and the production of the complete jet engine. The manufacturer sits at the top of a supply chain, with various suppliers working on subproblems of the design of the jet engine. For the particular subproblem of designing the non-rotary components of the engine, one parameter is the central variable, namely the optimal combustion temperature of the jet engine. As the knowledge about the final combustion temperature is uncertain at the beginning of the design process, one has to start with only presumed combustion temperatures. There are many design teams working in parallel on the task of designing the components. For example, let us imagine that there are three different design teams. Team 1 uses an established material that can meet a presumed combustion temperature which is not that high. Team 2 studies an experimental material. This innovative material might allow for a temperature increase that would give the manufacturer some leeway that allows for a temperature that is much higher than the presumed temperature. However, this material is very costly to test, and it is not certain that the use of this material will in fact work out. Finally, team 3 develops a solution that involves air-cooling. Under this measure, the components can withstand very high temperatures. Here, the downside is constituted by the high costs for manufacturing this air-cooled solution (cf. Eckert et al. 2018).

In this example, the targets do not yet exist while the modeling is being done. On the one hand, the models may be standing in a design relation to their targets. In the case that the manufacturer and the different design teams agree on a solution to the problem of the engine components, the models of this non-rotary component are standing in a design relation to the components. The models then may be used to construct the targets, as in the example of the model of the cottage. On the other hand, the models may also be standing in a representation relation to their targets. The models are used to make predictions about the system, how well the jet engine will work given the specific solution to the problem of the non-rotary components that would be chosen in the end.

Here, both relations seem to be equally relevant in the modeling task. The same may be said about the cottage model, and this will be discussed further in the next section.

28.4. Open Questions

28.4.1 Indirect Views of Modeling and the Weekend Cottage

In philosophy of science, indirect views of representation are widely discussed. According to such views, modeling tasks have a tripartite structure consisting of model descriptions, model systems and target systems (cf. Giere 1988; Weisberg 2007; Frigg 2010). For example, mathematical models consist of mathematical equations and interpreted structures. The mathematical equations of such a mathematical model constitute the model descriptions, and the interpreted structures that are specified by the equations are the model systems of the modeling task. The phenomenon to be studied with the help of the modeling is the target system. The equations specify structures and so are standing in a particular relation to the structures. The structures, the model systems, are standing in a further relation to the target system of the modeling. The relation between the model system and the target system is usually interpreted to be a representation relation, a modeling relation with a vehicle-to-target direction of fit. In Poznic (2016a), I argued that the modeling of the human lung with the help of a lung chip can be reconstructed with such an indirect view of representation.

The model of the weekend cottage may also be related indirectly to its target, the cottage. The reasoning goes via two steps. First, there are also representational uses of the architectural model. Second, to think of the model standing in a direct modeling relation to the building may be problematic. Because of that, the model may be conceptualized to be indirectly related to the building.

Now, to a more detailed line of reasoning. Just as the modeling of the jet engine components involves the two modeling relations of design and representation, the modeling with the help of the cottage model may also involve the representation relation in addition to the design relation. The model of the cottage indeed seems to be used with the aim of representing something. For example, in the case that a lamp is used as a stand-in for the sun, one can use the model in order to predict how the planned building creates shadows in its surroundings. It may even be possible to predict how the sunlight will fall into the building through windows or other openings. In this case, the model will be changed when one is not happy with the content of these predictions; for example, if the shadows cast are too big, or too little light is shining into the rooms. So, the model may be standing in a representation relation to something.

In the case that this target is interpreted to be the planned building, there is a worry, though. The target is—at this stage—only a possible building, because the building is only planned and not yet constructed. The plan may be changed in the end and so the actually built cottage would be a different object from the possible object. And, if the building is never constructed, there would be no actual object (cf. Poznic 2018).[7]

One may claim that the model represents a *plan* of the building rather than the building itself. This plan comprises all properties that the designer and the customer together determined the potential building would have. All the tools that were produced—sketches, drawings, computer models, etc.—jointly determine the plan of the building that comprises all properties the building should have in the end. The plan may be seen as the product of the practice of designing. Model and plan would then stand in the representation relation to each other. What about the claimed design relation to the building? Taking an inspiration from indirect views of representation, one may argue that model and building do stand in a relation to each other, but that this relation is indirect. The model stands in a representation relation to the plan, and the plan stands in a design relation to the building. If the target of the representation relation is not the building, the worry of the possible but not actual cottage can be answered. The model is still related to the building, even though in an indirect way. Because the model is firstly related to the plan, the representational uses of the model can be made sense of, and the modeling relation of design connects the plan and the building (cf. Poznic 2018).

In the next section, a different interpretation of a model apparently involving a design and a representation relation will be discussed. The modeling of the jet engine components may be interpreted with the help of the concept of a game of make-believe.

28.4.2 *Games of Make-Believe and the Jet Engine*

As we saw in §3.3, the modeling of the jet engine components also seems to involve both modeling relations, the design relation and the representation relation. It may be that a similar story about indirect modeling as in the preceding section could be told about the modeling of the jet engine components. Here, however another option of developing the narrative about the jet engine will be discussed. There is an alternative way of reconstructing this case of modeling with the help of notions such as props, principles and games of make-believe. This is based on prevalent accounts of make-believe in the arts and sciences (Walton 1990; Frigg 2010; Toon 2012) but also goes beyond these known applications and addresses the issue of how certain rules of the game can be changed while maintaining a set of central rules of the game (Eckert et al. 2018).

According to the theory of make-believe, the phenomenon of make-believe is a fundamental activity of any human being. In children's games everyone learns about this practice, and in adult life the appreciation of films, literature, computer games, etc., involves games of make-believe as well.

In any game of make-believe, there is an agreement by players of the game that certain objects can be used to imagine particular propositions that do not have to be true but still are agreed upon by the players. An easy example is a children's game where the participants treat broomsticks as horses. If one player takes a broomstick for a horse and the community of players agrees about this as a legitimate move in the game, then the proposition that this player is riding a horse is made "fictional" in the respective game of make-believe. The constituents of the game are so-called "props," in this case the broomsticks, and so-called "principles of generation," rules to imagine something that does not have to be true, for example, that someone is riding a horse, in case that she is in fact riding a broomstick. One general claim of the theory of make-believe is that "props and principles generate fictional propositions" (cf. Walton 1990). The prop, the broomstick and the rule to imagine that someone is riding a horse if she is riding a broomstick generates the fictional proposition that the player is riding a horse. The theory of make-believe was adopted by philosophers of science in order to make sense of modeling enterprises that involve idealizing and other false assumptions. In the case of scientific models, it is also claimed that the models are props in specific games of make-believe and that the models together with certain principles generate fictional propositions (Frigg 2010; Toon 2012). The particular accounts of make-believe in philosophy of science may be criticized (cf. Poznic 2016b), but their application of the theory of make-believe to modeling is a fruitful idea, and it may be transferred to engineering models.

The case of the models of jet engine components seems to be especially well suited for such an application of the account of make-believe. Different design teams use models of the components, and they have to base their models on potential combustion temperatures about which they cannot be sure whether they are feasible in the end. The targets of many of the models seem to be only possible and not actual objects. The advantage of choosing a make-believe interpretation of the models is that one does not have to think of model-target relationships in the first instance. As the targets of the models may be non-existing or possible objects, the relation to such objects may be difficult to account for. According to the theory of make-believe, the models are not related to targets; the models are props, and it is their function to generate fictional propositions that every player of the game should agree upon. With the help of this move one can circumvent worries about non-existing or possible objects. It may even be the case that the approach of make-believe can tackle the problem of the absent artifact of the cottage model. The advantage of a make-believe approach is that one can give an account of modeling without having to posit targets as relata of modeling relations (cf. Toon 2012; Eckert et al. 2018).

Here, we have seen that there is the option to interpret practices of modeling as not involving targets that correspond to models. In the next section, though, we will return to modeling relations between models and targets.

28.4.3   *Process Models and Product Models*

Engineering practice is about reaching effective and efficient designs of technical artifacts such as buildings, cars, aircraft, etc., or components of these artifacts such as window frames, steering wheels and landing gears. A more specific goal is to manage the process of designing itself. It is a reflexive task of *designing* the design process. Important tools in this regard are models of the design and development process, so-called "process models." These process models are often distinguished from *product* models, i.e. models of the products of such processes (cf. Eckert and Stacey 2010). The examples of models discussed in the previous sections may all be regarded as product models. The process models seem to be of a different nature from that of the product models, and some scholars have proposed to analyze them with the help of a number of different categories. One proposal is to distinguish four categories: (i) analytical models on a micro-level, (ii) abstract models on a meso-level, (iii) procedural models on a macro-level and (iv) management science and operations research models (Wynn and Clarkson 2018). To manage a design and development process is a complex task that involves iterations on different levels. Engineers have to cope with challenges of novelty, and they have to coordinate many teams and activities. David Wynn and John Clarkson state that some of the process models are "representations of an emerging design" (ibid., 160). How is this notion of representation related to the notion of representation or the generic notion of a modeling relation discussed in the previous sections? The process models have different functions. Some are used to document the design and development process. They may be interpreted to be related to the process in a representation relation. Some are used to guide future and present design and development processes as best practice models. Here, the design relation may be used to interpret the relation between models and targets. Other models are used as a means of communication among different stakeholders, e.g., multiple engineering teams, individual engineers, clients, etc. It is up to further research to study the connection between these different types of process models and modeling relations such as design and representation relations or even the relation of make-believe if one thinks that make-believe is indeed a modeling relation.[8]

28.5. Concluding Remarks

This chapter distinguished design relations and representation relations as modeling relations between models and other vehicles on one hand, and different target systems on the other. The relations are distinguished by the different directions of fit. Three examples of modeling enterprises were discussed—the lung chip, the weekend cottage and the jet engine; the two modeling relations with different directions of fit were analyzed in the three examples. The lung chip is standing in a representation relation to its corresponding target, and the modeling of the cottage involves a design relation to the cottage. Both examples also involve the complementary other modeling relation: the protocol of the chip is standing in a design relation to the chip, and the cottage model is standing in a representation relation to the design plan. The modeling of the jet engine components involves both modeling relations as well.

What does this tell us about the basic assumption? The examples of the jet engine and the weekend cottage do not disconfirm that engineering modeling is about how things should be rather than about how things are. However, the first example of the lung chip cannot be accommodated so easily. It shows that a representation relation is prominently involved in one of the models. This modeling involves reasoning about how things are rather than about how things should be. So, one modeling enterprise seems to be rather about what is than about what should be. The other two examples may be seen as confirming evidence for the basic assumption.

Although the basic assumption was not sufficiently disconfirmed, the first example challenges the assumption when interpreted to be a valid claim about all engineering models. It is at least questionable whether one can speak about all engineering models as being used to study what should be rather than to study what is.

What is common to all three models is that they involve design *and* representation relations. In order to make sense of this involvement of design and representation relations, an indirect view of the modeling of the lung chip and of the weekend cottage was discussed. Alternatively, the interpretation of the modeling of the jet engine as a game of make-believe did not make it necessary to posit an actual target system next to particular propositions.

Beside product models, which were those mainly discussed in this chapter, there are also process models in engineering design. Whether these process models also stand in design or representation relations to targets is a question for future research. Another open question is whether the process model can be accounted for with the Waltonian concept of a game of make-believe.

Related Chapters

Chapter 6: Scientific Methodology in the Engineering Sciences (Mieke Boon)
Chapter 21: Engineering Design (Peter Kroes)
Chapter 29: Scale Modeling (Susan G. Sterrett)

Further Reading

Pincock, C. (2012). *Mathematics and Scientific Representation*. New York: Oxford University Press. (Important book about mathematical models as representational tools.)

Weisberg, M. (2013). *Simulation and Similarity: Using Models to Understand the World*. New York: Oxford University Press. (Very influential publication about computational and other kinds of models.)

Wimsatt, W. C. (2007). *Re-Engineering Philosophy for Limited Beings: Piecewise Approximations to Reality*. Cambridge, MA: Harvard University Press. (Engineering applied to philosophy and philosophy applied to engineering; contains a seminal chapter about "false models.")

Notes

1. In the philosophy of science, some scholars want to speak about modeling without acknowledging that there are models as genuine objects. One slogan is accordingly "modeling without models" (see, for example the title of Levy 2015). In contrast, I will assume that there are models. However, I want to stay uncommitted concerning the question of what their exact ontological status is. As the focus of this chapter is rather on the practice of modeling and on the question of how models are used, the ontological status of models seems to be not directly relevant in this present context. For an overview of positions concerning the ontology of models, see Gelfert 2017.

2. For example, in Franssen et al. (2015), claims along these lines are mentioned in the context of a discussion of the difference between technology and science or the different attitudes of engineers and scientists, respectively. Some chapters of this *Handbook* make the claim that engineering is about how things should be (Sjoerd Zwart does so explicitly in Chapter 8, and Peter Kroes does so more implicitly in Chapter 20).

3. The terminology of directions of fit was introduced by John Searle (1983), but the idea behind it dates back to an earlier discussion by Elizabeth Anscombe (1957). I will elaborate on it in more detail in the next section.

4. The target may be conceptualized as the human lung, pulmonary edema in a human lung or the treatment of pulmonary edema with a specific chemical substance.

5. In such a case one may interpret the model to be a scale model (see also Chapter 29, "Scale Modeling" by Susan G. Sterrett, this volume).

6. In fact, these so-called BIMs are used not only by architects or engineers to design something. They are also used by contractors, owners or facility managers to build and, then, maintain the designed artifacts.

7. Per Galle calls this worry the problem of the absent artifact and proposes to interpret the target of the model to be not the building but an "idea" of the designer (cf. Galle 1999).

8. For example, props and fictional propositions of games of make-believe may be interpreted to be vehicles and targets of modeling, respectively.

References

Anscombe, G.E.M. (1957). *Intention*. Oxford: Basil Blackwell.

Bokulich, A. (2017). Models and Explanation. In L. Magnani and T.W. Bertolotti (eds.), *Springer Handbook of Model-Based Science*. Cham, Switzerland: Springer, pp. 103–118.

Boon, M. and Knuuttila, T. (2009). Models as Epistemic Tools in Engineering Sciences. In A. Meijers (ed.), *Philosophy of Technology and Engineering Sciences*. Amsterdam: Elsevier, pp. 693–726.

Eastman, C.M., Eastman, C., Teicholz, P., Sacks, R. and Liston, K. (2011). *BIM Handbook: A Guide to Building Information Modeling for Owners, Managers, Designers, Engineers and Contractors*, 2nd ed. Hoboken, NJ: Wiley.

Eckert, C. and Hillerbrand, R. (2018). Models in Engineering Design: Generative and Epistemic Function of Product Models. In P.E. Vermaas and S. Vial (eds.), *Advancements in the Philosophy of Design*. Cham, Switzerland: Springer, pp. 219–242.

Eckert, C., Hillerbrand, R., Poznic, M. and Stacey, M. (2018). Jet Engines, Design Teams and the Imagination: Designing as Playing Games of Make-believe. Paper presented at *Forum on Philosophy, Engineering and Technology (fPET) 2018*, University of Maryland, College Park, MD, USA, May 30–June 1, 2018.

Eckert, C.M. and Stacey, M.K. (2010). What is a Process Model? Reflections on the Epistemology of Design Process Models. In P. Heisig, J. Clarkson and S. Vajna (eds.), *Modelling and Management of Engineering Processes*. London: Springer, pp. 3–14.

Franssen, M., Lokhorst, G.-J. and Van de Poel, I. (2015). Philosophy of Technology. In E.N. Zalta (ed.), *The Stanford Encyclopedia of Philosophy*, Fall 2015 ed. http://plato.stanford.edu/archives/fall2015/entries/technology/

Frigg, R. (2010). Models and Fiction. *Synthese*, 172(2), 251–268.

Frigg, R. and Hartmann, S. (2012). Models in Science. In E.N. Zalta (ed.), *The Stanford Encyclopedia of Philosophy*, Fall 2012 ed. http://plato.stanford.edu/archives/fall2012/entries/models-science/

Galle, P. (1999). Design as Intentional Action: A Conceptual Analysis. *Design Studies*, 20(1), 57–81. doi:10.1016/S0142-694X(98)00021-0

Gelfert, A. (2016). *How to Do Science With Models: A Philosophical Primer*. S.l.: Springer.

Gelfert, A. (2017). The Ontology of Models. In L. Magnani and T.W. Bertolotti (eds.), *Springer Handbook of Model-Based Science*. Cham, Switzerland: Springer, pp. 5–23.

Giere, R.N. (1988). *Explaining Science: A Cognitive Approach*. Chicago: The University of Chicago Press.

Huh, D., Kim, H.J., Fraser, J.P., Shea, D.E., Khan, M., Bahinski, A., Hamilton, G.A. and Ingber, D.E. (2013). Microfabrication of Human Organs-on-chips. *Nature Protocols*, 8(11), 2135–2157. doi:10.1038/nprot.2013.137

Huh, D., Leslie, D.C., Matthews, B.D., Fraser, J.P., Jurek, S., Hamilton, G.A., Thorneloe, K.S., McAlexander, M.A. and Ingber, D.E. (2012). A Human Disease Model of Drug Toxicity-Induced Pulmonary Edema in a Lung-on-a-Chip Microdevice. *Science Translational Medicine*, 4(159), 159ra147. doi:10.1126/scitranslmed.3004249

Levins, R. (1966). The Strategy of Model Building in Population Biology. *American Scientist*, 54(4), 421–431.

Levy, A. (2015). Modeling Without Models. *Philosophical Studies*, 172(3), 781–798.

Pirtle, Z., Odenbaugh, J., Hamilton, A. and Szajnfarber, Z. (2018). Engineering Model Independence: A Strategy to Encourage Independence Among Models. *Techné: Research in Philosophy and Technology*, 22(2), 191–229.

Poznic, M. (2016a). Modeling Organs With Organs on Chips: Scientific Representation and Engineering Design as Modeling Relations. *Philosophy & Technology*, 29(4), 357–371. doi:10.1007/s13347-016-0225-3

Poznic, M. (2016b). Make-Believe and Model-Based Representation in Science: The Epistemology of Frigg's and Toon's Fictionalist Views of Modeling. *Teorema: Revista internacional de filosofía*, 35(3), 201–218.

Poznic, M. (2017). *Models in Science and Engineering: Imagining, Designing and Evaluating Representations*. Delft: Delft University of Technology. doi:10.4233/uuid:a1bec569-8d24-45ea-9e7e-63c0b900504e

Poznic, M. (2018). Architectural Modeling: Interplay of Designing and Representing. Paper presented at *Models and Simulations 8*, University of South Carolina, Columbia, SC, USA, March 15–17.

Searle, J.R. (1983). *Intentionality: An Essay in the Philosophy of Mind*. Cambridge, Mass: Cambridge University Press.

Toon, A. (2012). *Models as Make-Believe: Imagination, Fiction, and Scientific Representation*. Basingstoke: Palgrave Macmillan.

Walton, K.L. (1990). *Mimesis as Make-Believe: On the Foundations of the Representational Arts*. Cambridge, MA: Harvard University Press.

Weisberg, M. (2007). Who Is a Modeler? *British Journal for the Philosophy of Science*, 58(2), 207–233.

Winsberg, E. (2010). *Science in the Age of Computer Simulation*. Chicago: University of Chicago Press.

Wynn, D.C. and Clarkson, P.J. (2018). Process Models in Design and Development. *Research in Engineering Design*, 29(2), 161–202. doi:10.1007/s00163-017-0262-7

29

SCALE MODELING

Susan G. Sterrett

29.1. Role of Scale Models in Engineering Practice

Scale models are used in engineering design and analysis today, and they have been used in the profession of engineering for well over a century. The methodology of scale modeling is at least potentially applicable to any field of engineering, technology, or science. It is thus a puzzle that many discussions about models in philosophy of science have (mistakenly) assumed that scale modeling is an obsolete methodology that has been replaced by computer models (e.g., Oreskes 2007). For not only is experimentation using scale models still employed in many fields of engineering (Sterrett 2017b), but many of the computer programs used in building and analyzing computer models in engineering rely crucially on data that was generated by extensive scale model experiments set up and performed specifically for the purpose of generating data needed to write those computer programs. So scale modeling is an essential part of much engineering work, even though its involvement in engineering practice is not always obvious. In addition to the scale models used for research, analysis, and design, there are also configurable scale models that are constructed specifically for educational use in engineering curricula. Such configurable models provide students the opportunity to design, set up, and carry out model experiments.[1]

Scale models have become much more sophisticated in recent decades due to significant advancements in measurement technologies (e.g., lasers for measuring distances) and the development of advanced materials (Sterrett 2017b). These recent advancements have been incorporated into the design of engineering scale model experiments, with the result that some of the scale modeling practiced today was not possible, or even imagined, a hundred years ago.

29.2. Scale Models in Philosophy

Most current discussion in philosophy about models has excluded philosophical treatment of accurate accounts of scale modeling used in engineering.[2] For instance, even though Weisberg's widely read *Simulation and Similarity: Using Models to Understand the World* featured a scale model constructed and used by the Army Corps of Engineers on the cover and in the text, his "weighted feature matching" discussion of similarity is an extension of a psychologically based conception of similarity (Weisberg 2013), and does not provide a scientific explanation of how and why the methodology of scale modeling worked for that model. As this *Handbook* goes to press, the tide is turning, though, and some recent publications hint at future work underway that may help to rectify the current situation that, other than the few individuals mentioned earlier, scale modeling is not appropriately recognized in philosophy of science (Sanchez-Dorado 2019; Bokulich and Oreskes 2017; Pincock 2020).

Due to the current lack of engagement with the methodology of scale modeling in the philosophy of science literature, there are not really current debates in the field. There were certainly debates within the profession of engineering about the foundations, merits, and applicability of scale modeling in previous eras, but not within the past half-century. Inasmuch as differences of opinion about scale modeling currently exist in the philosophical community, they are attributable to misconceptions about scale modeling. Thus, this chapter on scale modeling does not address current debates *per se* but aims to provide an introduction to the foundations of the methodology and identify misconceptions that currently exist about it in philosophy of science.

29.3. What Are Scale Models?

Scale models, as the term is used in engineering, are physical objects or situations, usually specially constructed for the purpose, that are employed experimentally to learn about another imagined or existing physical object or situation. Scale models in engineering are usually constructed by humans, though it's possible to use the methodology of scale modeling to interpret naturally occurring objects or situations as scale models, too. The scale model experiment generally includes the surroundings that influence the behavior of the model, e.g., forces and ambient conditions, and these are designed to be analogous to (i.e., to correspond to) those in the surroundings of what it is intended to model. Construction of the scale model includes determining not only ratios of distances, but ratios of other measurable quantities such as various material properties and forces. Not just any ratio will be of significance in building a model that is informative about the thing it is supposed to model. Which ratios of measurable quantities to use in specifying the model, and how they are used to construct and interpret the model, are determined by employing the theory of dimensions.

After the scale model is constructed, its behavior can be observed, and the observations and measurements made in the model, suitably interpreted, are informative about the object or situation modeled. The formal methodology of scale models provides not only some prescriptions as to how the model is to be constructed, but a quantitative translation of the measurements made in the model to the corresponding measurable quantities associated with what it is intended to be a model of. Engineering knowledge is then used to make sense of the results regarding the problem or question being investigated, a process often referred to as "interpretation" of the model experiment.

This methodology is distinctively different from the kind of model-building in which the modeler starts from a mathematical equation describing the model or its behavior (Sterrett 2002, 2017a). It's a significant philosophical difference, as models in science have generally been associated with scientific equations (Bailer-Jones 2009). Further, epistemological issues in modeling also differ for scale models. This is because issues important in epistemology associated with scale models, such as evaluation of the external validity of the model, and analyses of how fundamental laws and experimental data on which the model is based are employed in modeling, differ from those that arise in the usual approach on which a model is a mathematical equation. Hence, most current philosophical accounts of how models manage to inform us about the world, and what we can conclude from them, are not applicable to scale modeling. They could, however, become enriched by adapting to incorporate the methodology of scale modeling.

29.4. Scale Models in Practice—Unique Challenges, Unique Versatility

The scale model can then be used in other experimental tests. Often, a scale model is useful when we are interested in understanding behavior that results from some unusual event or environmental change: the observations and measurements are informative about how the object or situation modeled will be affected by the corresponding modeled event or changes. Thus, although an experimental test might be designed to model expected normal operation in order to observe the overall behavior of the object or situation modeled (as when used in the pre-construction phase of the design of chemical processing plants), it can likewise involve subjecting the model to the application of an environmental factor such as heat, a temperature difference, a flow process such as an wind, river flow, or wave motion, or some event (e.g., an impact force, a periodic or nonperiodic motion, the initiation of a landslide, to give a few examples). After the results from the measurements taken in the model have been mapped, i.e., transformed or translated to the object or situation modeled, it is possible to produce tables or graphs of how the object or situation responds to various events or changes, according to the model.

The materials used in a scale model are generally not exactly the same materials that occur in the actual situation that the scale model experiment aims to simulate, for even material properties must be properly scaled. In the kind of model shown in Figure 29.1, which is used in educational institutions, there might be scaling of material particle size and intergranular friction in order for the model to provide the kind of behavior of interest, such as the progression of material dispersal over long time spans in the river modeled.[3] A common example where the material used in the model can differ from the material in the system modeled is flow in piping systems; water is sometimes used to model a more viscous fluid, such as oil. An example of the kind of difficulties encountered and care taken in getting the crucial material properties right is the case study of modeling ocean cable using plasticized PVC (polyvinyl chloride) piping to get the proper modulus of elasticity in the small scale model—then, in order to get the appropriate density in the model, the material was impregnated with powdered lead (Herbich 1998: 331).

Figure 29.1 Scale modeling in an educational setting

Source: Configurable (interactive) scale model from Little River Research and Design. "Em4" model. Descriptions and videos of these models are at http://emriver.com/models/em4/. Used with permission.

Sometimes several different scale models of a given object or situation are built in the course of designing it, as several different scale model experiments are needed to predict the several different kinds of behaviors of an object that are of interest to a modeler, or the different behaviors that are dominant at different scales. The scale model an experimentalist builds to predict the diffusion of heat in a given structure might not work well as an experimental model for predicting other kinds of behavior of the same structure, such as mechanical responses to earthquakes. Also, different phases of the situation modeled, such as the different stages in the life of a volcanic eruption and its aftermath, might require separate scale models, as the behavior of interest to the researcher might differ at different stages as the eruption progresses, and different phenomena will be dominant at different stages.

Scale models are so called because the models usually happen to be built according to a scale that indicates how one should translate measurements of distances from the model to what is modeled. To take a familiar example, a 1/8-scale model of a car would mean that any distance you pick out on the model car corresponds to a distance 8 times as long in the car it is intended to be a model of; thus in the case of geometrical scale, it is easy to comprehend that a 1/8-scale model car that is 1 foot long would indicate that the car it models is 8 feet long. Architectural models of buildings or building complexes are generally scale models in which distances are the *only* thing precisely scaled. However, for the more general notion of scale model used in engineering, other quantities such as velocity and current are generally scaled as well, and more than one quantity is scaled concurrently. Comprehending how scaling works in such complex cases is much more involved.

Scale models are often thought of by the layperson as being constructed as if the model were made by shrinking an object to a smaller size. If only geometrical similarity is to be achieved, rather than, for example, dynamic similarity (in which forces in the model correspond to forces in the situation modeled), that is not inappropriate. However, if similarity of physical behavior (bending, vibrating, buckling, stretching, expanding, cooling, etc.) is desired in a model, then the important interrelationships between all the quantities involved in that physical behavior must change in a coordinated manner. Then, the values of the quantities in the model are related to the quantities in what is modeled in very complex ways. Distance might be translated according to one scale factor, time according to another, and mass according to yet another. Thus, translating a quantity like velocity in a scale model to velocity in the situation it models is not as straightforward as it is for a hobbyist building a 1/8-scale model car where only geometrical similarity is of interest. The quantities in the environment acting on the scale model need to be scaled as well. Thus it is more appropriate to speak of a *physical system*, rather than a physical object such as a ship or plane, when discussing model experiments and the practice of scale modeling.

We then say that, ideally, we aim for the model and what it models to be *physically similar systems*, and we say that we construct a model to be physically similar to what it models. For two things to be "physically similar" or not always needs to be qualified (whether explicitly so stated or not) as similar *with respect to* some behavior considered within the realm of physics. For example, the behavior might be the magnitude of a liquid flowrate, electrical charge, or stress in a structural element; or it might be the existence of turbulence, the existence of buckling, or the existence of a phase change.

Most scale models are smaller in size than what they model. There is no reason in principle why a scale model cannot be made on a larger scale than the object or situation it models, though, and in fact some of them are. The advantage of making a scale model is to be able to experiment on a model of something, as a proxy for experimenting on something that cannot itself be experimented upon. Some scale models are tabletop models, as pictured in Figure 29.1, but there are also some large testing facilities, such as wind tunnels, models of river basins (LSU Center for River Studies), and volcanoes (Sterrett 2017b). These are seldom easily accessible to the public, but there are a few retired models that are accessible. The San Francisco Bay Model discussed in Weisberg (2013) is one such model. Another place to view scale models is the early facility for testing proposed ship designs that has since been replaced by the current David Taylor Model Basin; it is shown in Figure 29.2. In

Figure 29.2 Experimental Model Basin, Washington Navy Yard, Washington, DC—interior view, c. 1900; this was the first model basin (towing tank) for the US Navy.

Source: US Navy (www.dt.navy.mil/div/about/galleries/gallery1/012.html), public domain.

such testing facilities, the scale models of ships can be quite large, on the order of 20 feet long. The first experimental facility for testing ships built there was constructed at the very end of the nineteenth century, in 1896. The current facility on that site contains a shallow water basin, a deep water basin, and a high speed basin. (ASME 1998: 2)

One of the largest scale models, perhaps the largest ever built, is the scale model of the Mississippi River Basin, called the Mississippi Basin Model, or MBM (Figure 29.3). Like the David Taylor Model Basin, it holds a special place in the US history of scale models: at 40 acres in size, it is known as "the largest small-scale model" in the world. Many other hydraulic models were built by the same facility (The Waterways Experiment Station). Historical research into that facility's establishment reveals that there were debates as to the validity of the method of using scale models at that time, around the 1930s (Manders 2011: p. 56); the subsequent investment in and use of the MBM reflects the eventual outcome of that debate. The MBM model has not been preserved, in spite of its significance as a cultural icon, but this is not due to the technology of scale modeling itself becoming obsolete. A new indoor model of part of the basin, costing $4 million and requiring a quarter acre of space, the "Lower Mississippi River Physical Model," has recently been built in a new facility (LSU Center for River Studies).

Before the MBM was retired, data were collected from experiments that were specifically designed and carried out to provide data for use in computer programs in the 1970s. The data was incorporated into computer programs used to simulate the flow of water in the Mississippi River Basin (Foster 1971: vii). Thus the computer model that was used in lieu of the physical MBM scale model after its retirement was not independent of the scale modeling work. When it was in service, the MBM model was used to make predictions, most famously during the 1952 Missouri River flood. Predictions could be generated from the scale model by controlling water levels in it to correspond

Figure 29.3 Postcard of Waterways Experiment Station

Source: Personal collection of the author. Text on reverse: "WATERWAYS' EXPERIMENTAL STATION, VICKSBURG, MISSISSIPPI. The most unique Educational attraction in this part of the world is the U. S. Waterways' Experimental Station, located on a reservation four miles south of the city. It employs about one hundred graduate engineers and maintains the largest and best equipped laboratory of its kind in the world. Weighty problems concerning our vast waterways system are under constant study and miniature, scale-built models of our most temperamental streams, have been built for study. Ektachrome by Woody Ogden MADE BY DEXTER, WEST NYACK, NY Pub. by Jackson News Co., Jackson, Miss."

with real-time inputs of actual river level measurements. Time in a scale model goes faster; the events of an entire day in the actual river system took only a few minutes in the model (Foster 1971: 21–27). Likewise, the Center for River Studies housing the current basin model reports that in its model, "one year of the Mississippi River is simulated in one hour" (LSU Center for River Studies).

The scale factors that map, or translate, quantities in the model (including the quantities of the modeled environment), to quantities in whatever it is that is modeled, are determined by the ratios used to design the engineering scale model experiment (Pankhurst 1964). The selected ratios are kept invariant between the scale model and what it models. (That is the aim, at least.) It is in this sense that these ratios are called invariants. The key to understanding how scale model experiments are designed, and why model experiments that work well do so, when they do, is understanding the role of invariants and similarity in the practice of scale modeling.[4] We begin with an extremely simple case in order to make the ideas clear.

29.5. Scale Models, Invariants, and Similarity: The Basic Ideas

To illustrate the basic ideas behind scale modeling, i.e., the ideas of physical similarity and physically similar systems, we will first make the basic concepts involved clear for the simpler case of geometric similarity. Geometric similarity is generally easy to understand, because we can easily grasp the idea of two figures having the same geometric shape. A major misconception that abounds in philosophy about scale models is that the methodology of scale models is geometric similarity. It is not. The (correct) statement, often found in textbooks on the topic, is that the method of scale models is a *generalization* of geometric similarity (Sedov 2014: p. 43). This statement seems to have been grossly misunderstood in philosophy, and the misunderstanding is widespread. In the sections that follow, I hope to show the deep analogy between geometric similarity and the kinds of similarity used in scale modeling that are specific instances of physical similarity: kinematic similarity, dynamic similarity, hydrodynamic similarity, and thermal similarity, to name a few. Hence we begin by explicitly setting out the logical structure of reasoning about similarity already familiar from geometry, so as to see how to extend reasoning about similarity to physics.

29.6. The Logical Structure of Arguments From Geometric Similarity

One of the simplest examples of geometric similarity is the circle; all circles have the same geometric shape. Any two circles of different sizes are geometrically similar to each other, *in spite of the fact that none of the individual measurements* made on one circle (diameter, area, circumference) will be the same in another circle of a different size. Recall that the ratio of the circumference of a circle to its diameter is invariant no matter how small you shrink a circle in size, nor how large you expand it in size: so long as the figure keeps its shape, i.e., so long as it is a circle, *this* ratio will be the same. Many other ratios of geometrical quantities of a circle are *not* invariant between circles of different sizes: the ratio of circumference to area is *not* the same for all circles, for instance—*that* ratio *will* vary depending on the size of the circle. Not so for the ratio of circumference to diameter; it's invariant among all circles. We don't even have to know the numerical value of that ratio in order to make the statement that the ratio of the circumference of Circle #1 to the diameter of Circle #1 is equal to the ratio of the circumference of Circle #2 to the diameter of Circle #2. We can say that, whatever that ratio is, *it doesn't vary* between circles; whatever it is, it is *the same* for every circle. It is *invariant* from any circle to any other circle.

What is required to establish that two things have the same geometric shape? First, they must be the same *sort of* thing; for example, they must both be closed curves, or both be three-dimensional solids. Secondly, they must be geometrically similar. One way to ensure that two figures are geometrically similar to each other is to *construct* a figure that is similar to a given one. That's the general approach taken in scale modeling: to *construct* something that is similar in the relevant ways. However, as we shall see later, the *analogous notion of similarity in scale modeling* has to be generalized quite a bit from the case of *geometrical similarity*.

When there is at least one ratio that is invariant between all geometrical figures of a certain shape, i.e., between all figures that are geometrically similar to a certain figure, and to each other, that invariant ratio can be used to find the value of some distances that are not directly measurable. The method is an extremely simple example of scale modeling: construct a figure that is geometrically similar to one that involves the distance one wishes to know the length of. Then, using the ratio that is invariant between all figures of that shape, construct a proportional equation by equating the ratio expressed in terms of the line segments for one of the figures to the ratio expressed in terms of the line segments for the other figure. If the length of the line segments in the figure you have constructed are known or can be measured, this may allow solving for the distance one wishes to know.

The method is used in a common middle school exercise asking students to determine the height of a tall object such as a tree or flagpole on a bright day, by measuring its shadow and the shadow of their own body (Figure 29.4). It will be helpful to identify the structure of the reasoning here, for later use. So long as the area in which the tree and child is sufficiently flat, the right triangle formed by the student, her shadow, and the line connecting them is geometrically similar to the right triangle formed by the tree, its shadow, and the line connecting them. The ratio of [Height of Tree]/[Length of Tree's Shadow] is the same as the ratio of [Height of Student]/[Length of Student's Shadow]. A worksheet prepared for use by middle school teachers illustrates the sun-object-shadow situations in which we find these two similar triangles.

A. The *Terrific Tree*

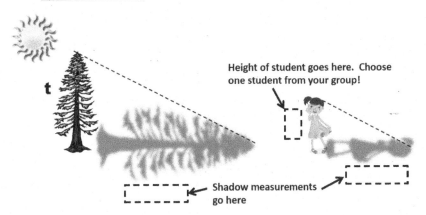

Height of student goes here. Choose one student from your group!

Shadow measurements go here

Figure 29.4   A worksheet designed for use with middle school students showing how to use your own body to determine the height of a tree, from shadow measurements

Source: "Similar Figures and Indirect Measurement: The Outdoor Lesson," Barry Schneiderman, TeachersPayTeachers.com, 2014. Used with permission.

If the height of the student and the lengths of both shadows can be obtained by measurement, the height of the tree can be determined by equating these ratios expressed as follows:

(Height of Tree/Length of Tree's Shadow)
= (Height of Student/Length of Student's Shadow)

Stated in more general terms, the knowledge that this ratio is invariant between the two (sun-object-shadow) situations allows us to equate the ratios. The proportion that results then provides the means to determine the height of the tree, as follows:

t = Height of the Tree =
[Height of Student/Length of Student's Shadow] × Length of Tree's Shadow

One way to look at what we are doing when we indirectly measure the height of the tree this way is that the student-sun-shadow situation has served as a model of the tree-sun-shadow situation, with respect to height.

Note that the criterion of similarity in use here is objective. In spite of the fact that the situations compared have aesthetic aspects and that human cognition is involved in apprehending the two triangular figures associated with the two physical situations, the criterion of geometrical similarity between the two triangular figures indicated in Figure 29.4 is completely objective. The question of whether two plane triangles are geometrically similar is settled here by the fact that the two triangles are right triangles and the angle at the top of the tree and the angle at the top of the student's head are formed by rays of the sun in the sky hitting them at the same angle. That angle need not even be known in order to conclude that the triangles indicated in Figure 29.4 are similar triangles. The reasoning from geometric similarity is objective, too, i.e., the *consequence* of the fact that these two triangles have the same shape, i.e., are geometrically similar, is that ratios between corresponding sides are the same. The reasoning from geometric similarity is straightforward reasoning according to the methods of Euclidean geometry. In Euclidean geometry, what's similar are two dimensional closed curves (figures), or, if three dimensional, solid figures.

29.7. Generalizing Similarity in Geometry: What's Analogous in Physics?

Progressing now from the simple case of geometrical similarity to the more complex case of physical similarity: what could be analogous to geometric shape, for physically similar systems? There isn't really a term for it, but we can explain such a concept in terms of the invariant ratios that remain the same between physically similar systems (Sterrett 2017a). That is the proper way to think about an analogue of shape in physics: just as we explain geometric shape in terms of the invariant ratios that remain the same between geometrically similar figures, so we conceive of something like shape of a physical system in terms of the value of the invariant ratios that remain the same between physically similar systems. There is a difference, though: geometric shape of closed plane figures is uniquely determined, whereas there are different kinds of similarity in physics. For the more complex kinds of similarity, similarity of physical behavior under gravitational forces, or heating, or cooling, or being set in motion by an earthquake, or undergoing pressurization, and so on, the invariants are certain dimensionless ratios composed of quantities used in physics. Which dimensionless ratios are relevant depends upon what behavior the modeler is interested in modeling. Dimensionless ratios will be explained next; for now, we want to state the concept of *physically similar systems* on analogy to *geometrically similar figures*.

To say that two physical configurations or situations S-one and S-two are "physically similar" with respect to a certain kind of behavior (rather than just geometrically similar), is to say that System S-one and System S-two have the same values of the (dimensionless) ratios that determine that kind of behavior. That is, we are considering a case in which a system S-one can have the same ratios of the dimensionless quantities that are relevant to a given behavior as another system S-two has—for example, the same ratios of certain forces—even though the values of some or all of its measurable quantities may not be the same system in S-one as the values of the corresponding quantities in system S-two (Buckingham 1914). So long as the values of the dimensionless ratios are the same, the specific behavior of the two systems on which the similarity of systems was drawn is the same. Specific numerical values of quantities in the model system and in the system it models will differ, of course; these values are related by a scale factor, which is recoverable from the dimensionless quantities.

What is so philosophically significant about scale modeling is that, unlike many other philosophical accounts of models, the methodology of scale models provides a scientific basis for determining that correlation (i.e., the correlation between a certain quantity in the model and a quantity in what it is intended to model (the "target" system, or any other physically similar system). This is so *even in cases in which the modeler does not know of an equation describing the behavior of the system.* One way to put it is that the method of dimensionless parameters provides: (i) a way to construct a model system that is physically similar to the thing of interest; and (ii) a means of interpreting the behavior of the model system in a way that is informative about what it models. Putting the point in terms of the terminology of a "key" as recently employed in philosophy of scientific representation (Frigg and Nguyen 2017), scale modeling *provides its own* "key" by which the results of experimentation on that constructed model are to be interpreted to give quantitative values for the quantities in the thing it models. (Pankhurst 1964) That is truly philosophically significant. It is the holy grail that many other current philosophical accounts of models seek. Often philosophical accounts of modeling leave that aspect to the judgment or knowledge of the modeler, or to experiment. Hence my claim that philosophical accounts of modeling stand to gain much by taking account of how the method of using scale models manages to be as successful as is.

Thus, a close study of scale modeling methods allows us to answer the following question: "In what way can the behavior of two systems be said to be the same, if none of the quantities measured in them is the same?" This question often arises in explaining the practice of scale modeling, since the quantities with which physics is concerned will not have the same values in the model as they do in the object or situation modeled. The answer is: the behavior is said to be the same in the model as in the situation modeled, on analogy to the way that two geometrical figures of different size are said to have the same shape. That is, for two figures to have the same geometrical shape, certain ratios of lengths in the figures are the same in both figures. Analogously, for two physical systems to be the same with respect to a certain kind of physical behavior, certain ratios of (measurable) quantities must be the same in both systems. This can be thought of as an analogy between geometric similarity and physical similarity, or as a generalization of geometrical similarity to physics, as shown in Table 29.1.

To offer an example that is easy to grasp visually, one kind of similarity that may hold between two physically similar systems is kinematic similarity. When kinematic similarity holds between two systems, the paths of the particles or bodies in the system trace out figures of the same shape. The paths are said to be homologous, which means that the particles of the two systems have corresponding velocities at corresponding times. (In the simple case where the particles have uniform velocity, the velocities and times will scale linearly between the two systems.) However, not all kinds of similarity in physics lend themselves to such visualization.

Another common way to grasp the physical significance of the nature of the similarity that holds between physically similar systems is to conceptualize the crucial dimensionless ratio or ratios in terms of a ratio of two kinds of forces. Thus the Froude number Fr, which is often expressed in terms of the quantities of velocity, length, and the gravitational constant, is commonly thought of as the ratio of a fluid's inertial force to its gravitational force. The Reynolds number, Re, which is often expressed in terms of the quantities of fluid density, velocity, length, and fluid viscosity, is commonly thought of as the ratio of a fluid's inertial force to its viscous force. The kind of behavior of interest in constructing the model determines which ratios one chooses to keep invariant between the model and the situation or object one wishes to model: the Froude number is used to construct a scale model when wave and surface behavior are important in a situation, as in designing ships for sea travel, while the Reynolds number is used to construct a scale model for a variety of phenomena associated with turbulent flow (examples are flows in piping systems, the response of buildings to high winds, and high speed travel in the atmosphere such as aircraft and projectiles).

Table 29.1 Generalization of similarity in Euclidean geometry to similarity in physics (mechanics, including heat, fluids, etc.)

(Geometrically) Similar Figures	*(Physically) Similar Systems*
Certain *ratio(s)* of quantities (lengths) are the same in both figures. A proportion holds.	Certain *dimensionless ratio(s)* involving quantities used in physics are the same in both systems. So proportion(s) hold.
To *establish geometric similarity*: *Construct a figure* so that it is geometrically similar to a given figure; or *deduce* that two figures are geometrically similar.	To *establish that two systems are physically similar systems*: *Construct a system* so that it and the given system are physically similar systems with respect to a certain behavior; or *deduce* that two systems are physically similar systems with respect to a certain behavior.
To reason from geometric similarity of two figures: From the knowledge of the equality of certain ratios in the given and constructed figures, knowledge of all the quantities in the constructed figure that occur in those ratios, and of some of the quantities in the given figure, *deduce the value of previously unknown quantities in the given figure*. (Proportional reasoning.)	*To reason from the fact that two systems are physically similar systems: From the knowledge of the equality of certain dimensionless ratios in the given and constructed systems*, knowledge of all the quantities in the constructed system that occur in those ratios, and of some of the quantities in the given system, *deduce the value of previously unknown quantities in the given system*. (Proportional reasoning.)

Table 29.2 illustrates how the Reynolds number would be used to construct a model of an object or situation such that the Reynolds number is invariant between the model and what it models:

Table 29.2 Using dimensionless ratios in scale modeling

What You Want a Model Of:	The Model You Construct:
A system with density $\rho 1$, *velocity u1, some characteristic length L1, and dynamic viscosity $\mu 1$. Also, some fixed ratios reflecting the physical configuration.* The Reynolds number characterizing the system, which can be thought of as the ratio of the *inertial force* to the *viscous/frictional force*, is expressed as: $Re1 = \rho 1 \, u1 \, L1 \, / \, \mu 1$	The model is a system with density $\rho 2$, *velocity u2, some characteristic length L2, and dynamic viscosity $\mu 2$. Also, it has the same fixed ratios reflecting the physical configuration as the system you want a model of has.* How to design the model you construct: choose a fluid velocity and a fluid with fluid properties (density and dynamic viscosity) such that *the Reynolds number in the model equals the Reynolds number in the system you are modeling*, i.e.: Choose $Re2 = \rho 2 \, u2 \, L2 \, / \, \mu 2$ such that $Re2 = Re1$. This will result in a model in which the ratio of the inertial force to the frictional force is the same in the model as it is in the given system.

29.8. Selection of Invariants: Which Dimensionless Parameters Matter?

Since the dimensionless parameters that are kept invariant between the model and what it models are so crucial to the method and its success, the question of where they come from deserves at least a brief answer here. There are two main analytical means of determining the dimensionless parameters relevant to a certain behavior. One method, nondimensionalizing the governing equation to identify the dimensionless parameters that can play the role of invariants for the behavior governed by the equation, relies upon knowing the differential equations governing the phenomenon. The second method, using dimensional analysis and applying the principle of dimensional homogeneity, does not require knowing the actual equation or equations; it requires only knowing what quantities are involved (Sterrett 2017b).

In the aforementioned example, using the Reynolds number, the equality of the Reynolds number in the model and what it models was sufficient to establish that the model and what it models were physically similar systems. However, in many cases, what is required to establish that two systems are physically similar systems is to show that a certain set of two or more dimensionless parameters has the same value in the model as in the system it models. The theory upon which a set of dimensionless parameters sufficient to establish that two systems are physically similar systems is dimensional analysis, or the theory of dimensions (Buckingham 1914; Pankhurst 1964; Sterrett 2006, 2009, 2017a, 2017b). The set of parameters is not unique; what is determined is *how many* dimensionless parameters are required to establish the physical similarity of two given systems with respect to a certain kind of behavior.

In practice, modelers do not usually derive the relevant dimensionless parameters anew for each experiment, or, even, for each kind of experiment. Rather, which dimensionless parameters are appropriate to select as invariant(s) to guide construction of a model is often already established by the community of researchers in which the experimenter is working. Dozens of dimensionless parameters have been identified and given proper names. Though there are by now canonical formulations of each named dimensionless parameter in the community of researchers and practitioners who use that dimensionless parameter, there is not even one unique expression of every dimensionless parameter. Likewise, though there are by now established choices of which set of dimensionless parameters to use to establish similarity of a model to what it models for a certain kind of behavior, there are many different sets of dimensionless parameters that are equivalent in terms of establishing that two systems are physically similar, i.e., that can play the role of invariant(s).

New kinds of experiments are constantly being conceived, too. For these, a combination of analytical approaches and experimentation is used, to determine the appropriateness of the choice of invariant in capturing the kind of behavior one wishes to study, using the kind of model and testing conditions employed. While it is true that experimenter knowledge and practical experience are involved in carrying out these kinds of investigations, the criterion of similarity is still an objective matter; similarity is a matter of a set of relevant invariant dimensionless parameters[5] (the set of parameters might not be unique) being the same between model and situation modeled, a matter of the two systems being physically similar systems. Changes to model materials, testing conditions, and other features of the experiment are evaluated as well, in tandem with the choice of invariants, in order to obtain a sufficiently effective model experiment protocol.

29.9. Inherent Limitations of the Method of Physical Similarity

In practice, exact similarity of systems in physics is not always achievable. In particular, for dynamic similarity of physically similar systems, exact similarity is in general unachievable unless the model is a full-size scale model.[6] However, most scale models are not full-size scale models, and most scale models are only approximately physically similar to what they model. In this section, we briefly indicate the reasons for this.

The reason that it is not in general possible to achieve complete dynamic similarity with a model that is not full-size is that the problem of ensuring that the model and what it models are dynamically (physically) similar systems is *overconstrained*. Solving that problem requires finding a combination of values of all the quantities that appear in all the dimensionless parameters that one needs to keep invariant. It is a simple mathematical matter, an application of linear algebra, to show that in the general case, there are so many constraints that a general solution to the problem of dynamically similar systems is not possible except with a full-size model. Thus, in practice, modelers compromise and construct a model that is only *approximately similar*, rather than *exactly similar*, to what it is a model of. When the scale model is not exactly similar to what it models, scale effects can arise. Part of the modeler's job is to quantify scale effects and design the scale model so that the kind of scale effects that arise in that model are not important to the kind of behavior the model is being used to investigate. As noted earlier, often several different scale models of the same object or situation are made, each one designed to investigate a different kind of physical behavior.

Another reason that exact similarity is not achieved in practice is more deliberate. Most models of large bodies of water, such as large lakes or rivers, are distorted: the vertical dimension of the model corresponding to the depth of the river or lake uses a scale much larger than the horizontal dimensions of the model corresponding to the earth's surface. This is because if the same scale were used, the depth of the water in the model would be impracticably shallow, and the effects of the river bottom or lake bottom would be much exaggerated in the model behavior as a result. Engineering experience gained from experimentation is involved in the process of arriving at a good selection of dimensionless parameters, and engineering expertise is involved in deciding which trade-offs to make in constructing a model that is only approximately similar. However, the criterion of exact similarity is still well defined, even if seldom obtained. Exact similarity consists in the values of the dimensionless parameters that characterize the model being equal to the values those dimensionless parameters have in the situation it models.

29.10. Misconceptions: A Brief List

Common misconceptions about scale modeling include: (i) confusing the sense in which invariant is used in scale modeling with the sense of invariant used to denote an "invariant of nature"; (ii) that all dimensions must be expressed in terms of a particular set of base dimensions, such as {[M], [L], [T]}; (iii) that the "generalization of geometrical similarity" is another kind of geometrical similarity; (iv) that anything that can be achieved with a scale model can be achieved with a computer simulation; and (v) that scale modeling requires more information than using an equation. (In fact, scale modeling requires less information than numerical simulation, as it does not require that the modeler be in possession of an equation describing the behavior of interest (Sterrett 2002, 2017b).

29.11. Concluding Remarks

Scale modeling is essential in current engineering practice, both in the building of scale models to investigate behavior and in providing empirical data for use in the design of software for computational simulations. The basis for scale modeling is known as physically similar systems, and can be thought of on analogy to geometrically similar figures, as explained in this chapter. Instead of ratios of like geometrical quantities, it is dimensionless ratios consisting of ratios of products of physical quantities that play the role of invariants in the theory of scale modeling. In practice, it is often not possible to achieve exact (or full) similitude in a scale model, and empirical investigations are often carried out to make informed judgments about the best compromises to make in the design of a model experiment using an approximately similar scale model.

Related Chapters

Chapter 8: Prescriptive Engineering Knowledge (Sjoerd Zwart)
Chapter 15: Artifacts (Beth Preston)
Chapter 16: Engineering Objects (Wybo Houkes)
Chapter 21: Engineering Design (Peter Kroes)
Chapter 23: Design Methods and Validation (Sabine Ammon)
Chapter 28: Models in Engineering and Design: Modeling Relations and Directions of Fit (Michael Poznic)
Chapter 30: Computer Simulations (Hildrun Lampe)
Chapter 31: Experimentation (Viola Schiaffonati)

Further Reading

Becker, Henry A. (1976). *Dimensionless Parameters: Theory and Methodology*. Wiley. (An extremely concise, insightful, philosophically minded introduction to the topic with a variety of examples, including some from chemical engineering.)

Buckingham, E. (1914). On Physically Similar Systems; Illustrations of the Use of Dimensional Equations. *Physical Review*, 4, 345. (This is the classic paper, tightly written by a philosopher-physicist and worth a close study; even the more well-known P. Bridgman's *Dimensional Analysis* credits Buckingham as its main resource.)

David, F. W. and Nolle, H. (1982). *Experimental Modeling in Engineering*. Butterworths. (Excellent textbook that has stood the test of time as a reference for fundamental principles as well as advanced applications. Re-issued in ebook format in 2013.)

Pankhurst, R. C. (1964). *Dimensional Analysis and Scale Factors*. Butler and Tanner. (A small, short, clear, concise and comprehensive work covering many kinds of similarity, by the Superintendent of the Aerodynamics Section of the National Physical Laboratory.)

Sterrett, S. G. (2017). Experimentation on Analogue Models. In L. Magnani and T. Bertolotti (eds.), *Springer Handbook of Model-Based Science*. Springer Handbooks. Cham, Switzerland: Springer. (Discusses the bases for scale modeling as compared to other methods for using concrete physical models in science.)

Notes

1. Figure 29.1 shows a configurable model developed for educational use. URLs of time-lapsed videos of the model illustrating its use are provided in the caption.
2. Kroes (1989), Zwart (2009), and Sterrett (2006) are some of the few exceptions known to this author.
3. The spectacular visual effects of using color-coded particles can be seen in the videos of progression of sediment transport in the river model, which are available online at the manufacturer's website: http://emriver.com/models/em4/.
4. For a deeper explanation of why keeping these invariants the same actually results in the model behavior reflecting the behavior of what it models, see Sterrett (2009, 2017a).
5. The set of parameters is not unique, as explained in Buckingham (1914) and Sterrett (2006, 2009).
6. Even when full-size scale models are used, the practice is still referred to as scale modeling, for the principles of scale modeling described earlier are still involved in setting up the experiment. For example, the same principle is adhered to in determining the fluid properties (density, viscosity) and conditions (temperature, pressure, velocity) to use for the fluid in a flow channel or wind tunnel. The crucial thing is still to keep the appropriate dimensionless parameter(s) (ratio(s)) the same in the model as in what it models.

References

ASME International (1998). *The David Taylor Model Basin*. West Bethesda, MD: ASME International.

Bailer-Jones, Daniela (2009). *Scientific Models in Philosophy of Science*. Pittsburgh, PA: University of Pittsburgh Press.

Bokulich, Alisa and Oreskes, Naomi (2017). Models in the Geosciences. In Lorenzo Magnani and Tommaso Bertolotti (eds.), *Springer Handbook of Model-Based Science*. Cham: Springer, pp. 891–911.

Buckingham, E. (1914). On Physically Similar Systems; Illustrations of the Use of Dimensional Equations. *Physical Review*, 4, 345.

Foster, J.E. (1971). *History and Description of the Mississippi Basin Model*. Vicksburg, MS: Waterways Experiment Station.

Frigg, R. and Nguyen, J. (2017). Models and Representation. In L. Magnani and T.W. Bertolotti (eds.) *Springer Handbook of Model-Based Science*. Cham: Springer.

Herbich, John B., Editor (1998). *Developments in Offshore Engineering: Wave Phenomena and Offshore Topics*. Houston, TX: Gulf Publishing Company.

Kroes, Peter (1989). Structural Analogies Between Physical Systems. *The British Journal for the Philosophy of Science*, 40(2), 145–154.

Louisiana State University (LSU) Center for River Studies (website). https://lsu.edu/river/. Accessed March 10, 2019.

Manders, Damon (2011). Research and Development in the U.S. Army Corps of Engineers: Improving the Common Stock of Knowledge. *U. S. Army Corps of Engineers, Omaha District*. Paper 145. http://digital commons.unl.edu/usarmyceomaha/145/

Oreskes, Naomi (2007). From Scaling to Simulation: Changing Meanings and Ambitions of Models in Geology. In Creager, Angela, Elizabeth Lunbeck and M. Norton Wise (eds.), *Science Without Laws: Model Systems, Cases, Exemplary Narratives*. Durham, NC: Duke University Press.

Pankhurst, R.C. (1964). *Dimensionless Parameters and Scale Factors*. London, UK: Chapman and Hall.

Pincock, C. (2020). Concrete Scale Models, Essential Idealization, and Causal Explanation. *British Journal for the Philosophy of Science*.

Sanchez-Dorado, Julia (2019). *Scientific Representation in Practice: Models and Creative Similarity*. Doctoral Thesis, University College, London. http://discovery.ucl.ac.uk/10064872/

Sedov, L.I. (2014). *Similarity and Dimensional Methods in Mechanics*. Reprint of the 1959 edition. Academic Press.

Sterrett, S.G. (2002). Physical Models and Fundamental Laws: Using One Piece of the World to Tell About Another. *Mind and Society*, 5(3), 51–66.

Sterrett, S.G. (2006). *Wittgenstein Flies a Kite: A Story of Models of Wings and Models of the World*. New York: Pi Press.

Sterrett, S.G. (2009). Similarity and Dimensional Analysis. In Anthonie Meijers (ed.), *Handbook of the Philosophy of Science, Volume 9: Philosophy of Technology and the Engineering Sciences*. Amsterdam: Elsevier Science and Technology.

Sterrett, S.G. (2017a). Physically Similar Systems: A History of the Concept. In Lorenzo Magnani and Tommaso Bertolotti (eds.), *Springer Handbook of Model-Based Science*.

Sterrett, S.G. (2017b). Experimentation on Analogue Models. In Lorenzo Magnani and Tommaso Bertolotti (eds.), *Springer Handbook of Model-Based Science*. Cham, Switzerland: Springer.

Weisberg, Michael (2013). *Simulation and Similarity: Using Models to Understand the World*. Oxford Studies in Philosophy of Science. Oxford University Press.

Zwart, Sjoerd D. (2009). Scale Modeling in Engineering: Froude's Case. In Anthonie Meijers (ed.), *Handbook of the Philosophy of Science, Volume 9: Philosophy of Technology and the Engineering Sciences*. Amsterdam: Elsevier Science and Technology.

30

COMPUTER SIMULATIONS

Hildrun Lampe

30.1. Introduction

Computer simulation (CS) is a relatively new phenomenon in the history of science and engineering with a fundamental impact on research and engineering practice as well as on political decisions. A new and pervasive technique like CS gives rise to questions of how it relates to established methods and existing knowledge, to questions about its capability and limitations and about best practices and appropriate standards. Questions of this sort are traditionally of interest for the philosophy of science and have led to important debates about understanding and "legitimating simulation" (Morrison 2015: 248) during the last decades.

This chapter is an introduction to the philosophical discussion of CS. It is dedicated to the characterization of CS as a method. Whereas this might seem trivial at first glance, it has important consequences for the kind of knowledge gained and the justification of the results. These topics belong to the epistemology of CS and concern questions like the following: What are the ingredients of a simulation model? Does CS allow to set aside laboratory experiments? Which conditions render the results of CS reliable? What is the role of the computer in the simulation process? What do we learn about an object of inquiry using CS? One purpose of such a general consideration is to complement disciplinary accounts. Whether CS is a powerful method of investigation depends on features of the task, e.g., the data basis, validation possibilities or the explanatory power of the mathematical model. But it depends on general constraints and possibilities of the method as well.

In this chapter, the characterization of CS follows four interconnected routes, separated into sections mainly for the sake of presentation. The first route is a discussion of existing definitions, followed by a methodological reconstruction of the simulation process. For the third route (Section 4), accounts to clarify the relation between CS and other methods, namely thought experiments and laboratory experiments, are presented. The fourth route is a characterization of CS in terms of complexity and epistemic opacity of their underlying mathematical models. These features justify the view that CS is a novel and distinct method in the scientific toolbox.

30.2. Characterization of CS: Definitions

The use of the term "simulation" is highly diverse. It is even deemed "promiscuous", since "a simulation is almost anything from a training exercise to an algorithm" (Durán 2017: 175, adopting Nelson Goodman's description of the usage of "model"). The general term "simulation" is commonly used, thereby concealing whether it refers to the method of computer simulation, the computational model, a simulation run, a computer experiment, a simulation study or possibly even programming languages (Saam 2015: 82). Typically, "simulation" on the one hand refers to the whole process of forecasting, reenacting or reconstructing a certain scenario, and on the other hand, it merely refers to the central step of the process, i.e. the actual calculation (Bungartz et al. 2013: 1).

In the course of philosophical reflection, several definitions have been presented. Much-quoted definitions have been introduced by Humphreys (1991) and Hartmann (1996). Humphreys defined simulation as "any computer-implemented method for exploring the properties of mathematical models where analytic methods are not available" (Humphreys 1991: 500), whereas Hartmann proposed that a (computer) simulation "imitates one process by another process" with "process" referring to "some object or system whose state changes in time" (Hartmann 1996: 83). Thus, the former emphasizes the usage of CS in connection with analytically non-tractable models, whereas the latter emphasizes the dynamical perspective introduced by CS. The implementation step seems to be marginal to Hartmann while it is central to Humphreys. Each of these working definitions disqualifies cases classified as CS by the other, e.g., CS in cases where analytic models are available and CS that investigate a model without imitating a process just because analytic methods are not available (Grüne-Yanoff and Weirich 2010: 22–23). Grüne-Yanoff and Weirich themselves make a distinction between calculation and simulation to categorize CS. According to their distinction, Monte Carlo simulations are considered as a method of calculation, despite their name, because they construct a probabilistic analogy of a deterministic system in order to calculate approximations of the system's properties. The probabilistic approach is considered as alternative computation of deterministic properties and not as a simulation of the system, since it is done without a mimetic purpose.[2] Thus, to represent a system in the form of a surrogate or stand-in is regarded as the characteristic feature of CS (Grüne-Yanoff and Weirich 2010: 29–30). While mimesis or imitation is central to both Hartmann and Grüne-Yanoff and Weirich, the former specifically addresses an imitation of the dynamics.

Whether or not CS are taken to be novel is influenced by the way simulation is defined (Grüne-Yanoff and Weirich 2010: 29). While the novelty of CS as a scientific technique is uncontroversial, its novelty regarding issues of the philosophy of science, especially epistemology, is contested. Novelty in a philosophical sense means that issues brought up and questions raised in connection with CS are not specific to the method but are only a variant of problems that have been discussed before in other contexts. An account of simulation as not strongly differentiated from models is naturally sceptical towards the novelty claim (Frigg and Reiss 2009), whereas an account focusing on computation (Humphreys 2009) lends itself to such a claim.

Definitions of simulation vary, depending on whether or not a distinction between analogue simulations and digital (computer) simulations is made. While it may be elaborated explicitly (e.g., Durán 2017), it is often consciously ignored in the debate on the characterization of simulation for the sake of generally discriminating simulation from other methods. Simulations in a broad sense "include studies of computer models, model organisms in laboratories, and model airplanes in wind tunnels" (Parke 2014: 517) vis-à-vis CS in the narrow sense. We find the broad sense of simulation used predominantly in philosophical and historical accounts (Gramelsberger 2011 about the history of fluid dynamics being one example), as well as in biological and medical contexts.

In general, the philosophical discussion about CS is shaped by certain scientific disciplines. Saam (2015: 83) assumes that different paradigmatic examples partly explain why philosophical views on CS are heterogeneous or contradictory. Yet, in order to reflect the method appropriately, it seems advisable not to refrain from case studies too soon because they illustrate the diversity of scientific practice. Authors who acknowledge this diversity emphasize context sensitivity with respect to the epistemic value of CS (Parke 2014) or claim to be "careful not to abstract away from the details of scientific practice" (Barberousse and Vorms 2014: 3596). However, Resch (2017) warns that the choice of case studies—in respect of a general understanding of CS—may be biased: Due to a lack of technical insight,

> [s]ocial scientists, humanities scholars, and philosophers prefer to focus on simulations of climate or astrophysics, where simulation exhibits both its greatest strengths and its most profound shortcomings. Although understanding such shortfalls may offer productive results, it does not address the true "essence" of simulation.

Furthermore, to address relevant issues "clear communication" between technical scientists and others is required and "vague descriptions of technology" must be avoided because they are misleading (Resch 2017: 29).

30.3. Characterization of CS: Methodological Reconstruction of the Simulation Process

In the introductory chapters of student textbooks (e.g., Bungartz et al. 2013), the simulation process is presented by way of the so-called "simulation pipeline", also termed "technical-scientific conception of simulation" (Resch 2017: 24). It is meant to illustrate the diversity of tasks and their sequence.

Typically, the simulation pipeline consists of the following steps (Bungartz et al. 2013: 3):

1. Mathematical modelling of the target
2. Treatment of the mathematical model for computer processing
3. Suitable implementation and visualization (data exploration in general)
4. Reliability tests of the results
5. Embedding into a context

A mathematical model of the target, step 1, is the basis of any CS. While this is new territory in some sciences and therefore a challenge for simulation practice, mathematical modelling is well established in others. CS clearly differ by means of their underlying mathematical model. Typically, CS based on a system of partial differential equations and CS based on cellular automata ("agent-based") are distinguished. Since most of the philosophy of science literature is dedicated to equation-based simulations—thereby addressing their pervasiveness in engineering—this chapter does so as well.[3] Concerning equation-based simulations, the possibility of numerically solving analytically non-tractable systems of partial differential equations is essential. Before the rise of CS, in order to derive solutions, theoretical models remained restricted to ideal cases far simpler than actual applications (Gramelsberger 2011: 131).

Step 2, the treatment of the model in order to allow numerical calculation, involves discretization and a choice of efficient algorithms. Beside classical implementation, i.e. code programming, step 3 increasingly involves skilful software development on a grand scale, including, e.g., parallelization (Bungartz et al. 2013: 3). Data exploration is the task of generating a "model of the phenomenon" (Winsberg 2010: 17, 19) out of a deluge of numbers. As easy as it may be in some cases, in others it is "a science in itself" (Bungartz et al. 2013: 3). The data has to be interpreted and relevant information extracted, visualization being "by far the most effective means" in order to identify characteristic qualitative features within large and complex dynamical data sets (Winsberg 2010: 22).

To investigate the reliability of the results, step 4, is very important since the whole process "is ripe with all sorts of uncertainties that need to be managed" (Winsberg 2010: 19). The terms "verification" and "validation" embrace various methods developed for this end (e.g., Morrison 2015: 248–286). Comparisons among different models, algorithms or codes as well as comparisons of simulated results with experimental data play a prominent role. One example of a context is a manufacturing process. Here, the task of step 5 is to integrate the simulation, e.g., to define interfaces or to create test beds.

It has been claimed that every explicit pipeline formulation is "incomplete" (Resch 2017: 25). What is regarded as the core of a simulation varies from the choice of initial and boundary conditions, to numerical issues and hardware aspects, depending on what is considered essentially fixed.

Although the pipeline metaphor may suggest that the steps are executed separately from each other by very different experts in the sense of a production line, the steps are in fact closely entangled. The simulation process is remarkably complex and involves feedback loops, i.e., the steps are typically iterated several times (Bungartz et al. 2013: 2–3).

The danger of taking the pipeline metaphor too literally has been a topic of philosophical debate. It is of concern because an important consequence would be to misinterpret CS as a pure calculation process where the theoretical rigour of the underlying mathematical model automatically justifies the results of the computation (given that it is executed properly). The "idealized view of computer simulation" depicted by Barberousse and Vorms (2014: 3605–3607) can be interpreted as an instance of such an interpretation. It conceives the simulation process as being "a deductive one applied to an initial set of input values for variables and parameters in a given set of equations". According to this view, the implementation does not change the original content, practical issues can be ignored as being irrelevant, and only the input and the mathematical model are deemed to be responsible for the simulation result. Barberousse and Vorms criticize this view because "[m]any other elements from empirical origin force their way to the process at virtually every step of the program writing and checking", and "there is hardly anything as messy as a simulation program" ("messy" referring to lurking bugs and errors and fine-tuning).[4]

A similar account is given by Winsberg (2010). According to him, the ideal consists in the belief that model construction is governed or determined by theory. By means of several examples, he demonstrates instead a "motley methodology" (Winsberg 2010: 19), whereby the mathematics of a simulation model stems from three sources, theory being only one of them. The other sources, or principles of model building and kinds of background knowledge, respectively, are physical intuition (i.e., intuitive or speculative acquaintance with the system of interest) and "model-building tricks" to counterbalance computational limitations (Winsberg 2010: 19; 65). The most prominent "model-building tricks" are statistical models based on empirical data, so-called parametrization schemes. They are designed to represent the dynamics of subgrid processes, small-scale physical processes that occur at length scales that cannot be adequately resolved on the computational grid. Although small-scale, they may be important and thus have to be accounted for. Parametrization schemes describe their aggregated effect over a larger scale. In contrast to idealizations, these "cooked-up techniques" (Winsberg 2010: 15) are substitutes or additions to mitigate what is lost during discretization. Winsberg's examples are eddy viscosity (designed to restrict dissipative effects to short length scales in simulations of turbulent fluids) and the Arakawa operator (designed to overcome non-linear instability in global circulation models) (Winsberg 2010: 13–15).[5] These additional techniques designed to enhance realism of the simulation results in the face of discretization and restricted computational power are salient elements of empirical origin.[6] Most remarkably, the choice of additional techniques to improve the simulation results is not made in relation to theory but in relation to prior results (simulationists "go back and tinker with their choice"; Winsberg 2010: 16). The modelling is guided by the outcome and not by theoretical rigour, in contrast to what the pipeline may suggest. What counts is the "best solution set" (Winsberg 2010: 9), i.e., the outcome that uncovers the relevant features of the system. Because it is not included in the picture of a pipeline, the important role of this feedback may be underestimated.

According to Kaminski et al. (2016), the guiding idea of the "pipeline" is that of a formal translation where the formal structure is preserved during the transfer from one step to the next. This guidance is limited by ruptures between the different model types involved (mathematical, numerical, implemented, etc.), resulting in "as if" translations. The transfer between step 1 and step 2 is the most comprehensible instance of an "as if" translation. The created numerical model is not equivalent to the mathematical model, but instead a functionally adequate approximation. Although original properties get lost (which is generally risky), it renders the simulation possible. Similar to Winsberg, the authors emphasize the technical tricks enabling the transfer because these tricks handle these ruptures. Kaminski et al. (2016) call means "tricks" in relation to an end if the means are an improvement in reaching the end (e.g., in terms of speed), if the means open up the possibility of the end (e.g., enables the calculation), or if the end is substituted (e.g., the numerical imitation of a solution while analytical non-tractability endures).

Consequently, in contrast to the idea of seamless translation, every transfer from one step to the next may be called a simulation of one type of model by another type of model. The challenge is then to ensure that "this imposing sequence of transfers" finally simulates the target system "and not something else" (Hubig and Kaminski 2017: 125). However, if both the model and the numerical method are regarded as fixed, simulation is reduced to a computational solution process (Resch 2017: 26). This perspectivity may explain the two contrasting accounts. To reveal the modelling decisions and technical tricks of each step may provide reasons for viewing each step as a simulation in itself.

To sum up, simulation is a process where feedback loops are of great importance. Because of the loops, there is a significant orientation towards the overall performance of the model at the expense of theoretical rigour. This leads to the applicability of theory.[7] Empirical elements and additional decisions accompany each step and lead to a certain independence of the implemented model from the theoretical underpinning. This is significant for the justification of the results, and one reason why the transitions and relationships between the chosen steps are of special philosophical interest. Regarding CS as a deductive calculation process is clearly a misunderstanding of the procedure.

30.4. Characterization of CS: Relation to Other Methods

Generally, when CS is considered in relation to other methods, the discussion takes place on "the usual methodological map" (Galison 1996: 120). This figurative expression relates to the two classical pillars of scientific inquiry, namely theory and experiment (see Chapter 31, "Experimentation" by Viola Schiaffonati, this volume). There is remarkable disagreement in the philosophy of science literature in locating CS on this map (Saam 2015: 66). According to a widely shared intuition, a CS "is taken to be a hybrid of the commonly distinguished practices of theorizing and experimenting: it is like theory in terms of what is manipulated by the scientist but like experiment in terms of how this manipulation is being done" (Petersen 2000: 270). Sometimes the intermediate position here referred to as a "hybrid" is expressed as lying "in between" theory and experiment. Lenhard (2004) depicts the idea as the "Janus-faced character" of CS: it has properties of both experiment and theory.[8] However, CS is also classified as a "tertium quid" vis-à-vis theory and experiment (Saam 2017: 304), i.e., as a method *sui generis* demanding an explication of special features in order to be distinct. Advocates of this view may still acknowledge a clear linkage to methodological features of both experimenting and the theoretical activity of modelling. Yet, due to the uniquely generated type of knowledge CS provide, which is claimed to raise its own epistemological concerns, CS "ought not to be assimilated with one or the other, nor should it be considered a hybrid"; instead, simulation data should be "taken in their own right" (Morrison 2015: 250–251).

Either way, it is not denied that CS are associated with experimental features. Thus, the relation of simulation and experiment has been a subject of debate within the philosophy of simulation. In the following, the advocated positions are presented, mainly based upon the overview provided by Saam (2015, 2017), respectively. The aim is to clarify the relation between CS and two established methods, namely thought experiments (TE) and laboratory experiments (LE), one being theoretical and the other empirical.[9]

In part, the intended purpose of the debate on the relation of simulation to TE and LE is to claim an epistemic privilege for LE over simulations. This purpose is led by the intuition that, apart from the question of relevance of the knowledge gained for applications, in contrast to LE, CS may present fictional rather than just alternative realities.

It has been a long tradition to link CS to TE.[10] TE are considered as templates for possible worlds. By means of TE "what if" questions are investigated, and it is examined whether a certain situation is possible given what we know, i.e., TE reveal implicit consequences of our theories about the world. A comparison with TE helps to elaborate properties of CS (Saam 2015: 69). CS seem to be appropriate instead of TE whenever a scenario is too complex to be processed in the mind only. In contrast to TE, CS require explicit assumptions to be made on the one hand, but force approximations to be made on the other. TE are ascribed to initiate conceptual changes (such as, e.g., the famous TE conducted by Galileo and Einstein), whereas CS seem to be part of "normal science" (in the sense of Thomas Kuhn). Both TE and CS generate new knowledge because both can be reconstructed as arguments.[11] CS therefore seem to continue TE with other means and different properties (cf. Humphreys 2004: 115: "Indeed, many simulations are examples of what would have been, in technologically more primitive times, thought experiments").

With a common ground between TE and CS having been acknowledged, it is also fruitful to emphasize their differences. As a consequence of complexity of the simulation model, CS may be referred to as opaque TE (Paolo et al. 2000). In the case of CS, we cannot know why certain conclusions can be deduced from given premises, even if the deduction can be shown. Consequently, a simulation model has to be made understandable by appropriate methods, e.g., by the systematic isolation of potentially explaining factors. Again, CS seem to carry on TE, because they promise progress where TE fail. They come into operation where the constraint of epistemic transparency is unrealizable.

Two links to the second method for comparison with CS, namely LE, have been mentioned already. One is the lack of transparency of the model, which has to be explored, similar to an experimental setting. The other is the manner in which this exploration is done, namely by controlled manipulation. Thereby, varieties of the notion of experiment become apparent. Different positions concerning CS result from whether an experiment is characterized primarily by materiality (e.g., Parke 2014), by exploration of an unfamiliar or opaque entity (e.g., Dowling 1999; Lenhard 2004) or by control and repeatability (e.g., Hubig 2010). Lenhard points to the fact of the controversial notion of experiment in the philosophy of science. Although it seems therefore defective as a basis for an analysis of CS, he still advocates "a heuristic use of 'experiment'" (Lenhard 2004: 96–97). To him the features of observing a behaviour by visualization and the capability of surprise are salient.

Comparisons of CS and LE either invoke their common features or emphasize differences between them. One common feature is the fact that CS are used like experiments to gather novel data about the target of inquiry (Saam 2015: 75). It has even been argued that the results of some CS are interpretable as measurements instead of calculations, because both rest on models (Morrison 2009). The motivation behind this claim is to broaden the scope of what is regarded as experimental knowledge. The argument builds on an investigation of modern experimental practice, revealing how heavily measurement relies on models. Thus, the seemingly obvious gap between empirical measurement and theoretical calculation is narrowed down.

Models (or theories) also play a prominent role within the views of CS as experiments on theories and CS as models of experiments. Within the view of CS as experiments on theories[12] (Dowling 1999), the simulation model is strategically taken as a black box, opaque and unpredictable, in order to creatively and experimentally play around with it. It may be objected that the mathematical modification of equations is typical of traditional paper-and-pencil calculations as well without the need for calling such a procedure an experiment (Saam 2015: 78). However, a point brought up by the debate on epistemic opacity (portrayed in the next section) is that the black box situation is not a matter of choice but a given essential feature of complex models. If this were the case, the use of "experiment" may still be justified against the previous objection because the "playing around" functions differently.

According to the view of CS as models of experiments (Beisbart 2018), a CS represents a detailed model of a potential experiment with the target system (see Chapter 28 "Models in Engineering and Design: Modeling Relations and Directions of Fit", Michael Poznic, this volume). Virtual intervention and virtual observation replace intervention and observation within the experiment, i.e. they are modelled within the simulation. This conception is compatible with the view that CS can be reconstructed as arguments (Beisbart 2012) and that they function like TE in this respect.

Others have focused on a fundamental difference between simulation and LE. This seems reasonable since even advocates of CS as experimental practice concede, e.g., that CS cannot be used to determine the existence of physical phenomena (Morrison 2015: 245). Saam (2015) identifies two different justifications for making a clear distinction. The first is the so-called materiality thesis. In experiments, there is a material similarity between object and target, while in simulations there is only an abstract, formal similarity.[13] Within the second account, different beliefs are claimed as demarcating CS and LE (Winsberg 2009). In the case of experiments, it is the belief in material similarity, whereas in the case of CS, the model building knowledge is relied upon. Whether or not the respective belief is warranted decides the epistemic value of each method. Thus, the inferential power is context-dependent (Parke 2014). The context includes the validation of CS, the practical possibility of conducting an experiment as well as the dependence of the research question on "physical, physiological or phylogenetic object-target correspondence" (Parke 2014: 534). Obviously, the two justifications identified by Saam (2015) are meant to serve different goals. The first aims to show that LE are epistemically privileged over CS, while the second seeks to demonstrate that the quality of the model-building knowledge is crucial, thus denying any claim of a general epistemic privilege.

Yet, there is a sense in which the intuition of an epistemic privilege of LE can be kept also within the scope of the second justification: However reliable the model building knowledge is in the case of CS, empirical data are fundamental: "We know how to do good computer simulations precisely because we have gained knowledge about the world through observation and experiment" (Parke 2014: 518). CS assume "that we already know a great deal" about the interesting features of the target system in order to build good models, and this knowledge depends on what one has learned "from a long history of experiment and observation": "we do not commit ourselves to the reliability of model building principles unless they have been tested against experiments and observations" (Winsberg 2009: 591). The intuition is elaborated in the over-control argument (Beisbart 2018). The term "over-control" refers to the results of CS being already entailed by the assumptions that define the set-up of the simulation. As a consequence of over-control there is "no space left for a possible answer by nature". The answer, i.e. the result, is determined by the questions, i.e. the model and the input data. The existence of over-control distinguishes CS from experiments.

One may wonder how to evaluate this long-lasting debate. Two aspects may be evaluated separately. The first concerns the topic of the debate, the second its results. Starting with the first, the debate has proceeded by using traditional notions to describe new phenomena. Resch et al. (2017) take this procedure as one possible strategy of dealing with transformations of science (which is assumed to be underway by means of CS). The advantage is that presuppositions about what is essentially new are not required. However, that the strategy can only diagnose (dis)similarity (by this framing the reflection of the new) is a disadvantage. Although the strategy contributes to the continuity of discourse, it does not lead to deeper insight and can only be a starting point (Resch et al. 2017: 1–2). In contrast, Frigg and Reiss (2009) warn that the emphasis on novelty "prevents philosophers from appreciating what a discussion of simulations could contribute to ongoing debates" (Frigg and Reiss 2009: 596), in this case the debates about models and experiments. With respect to results, the second of the two evaluation aspects, the present situation has been characterized as a "battlefield". Although Saam (2017) refers to this result as "state of the art", she herself deems the debate as being fruitful because it had led to other debates.

30.5. Characterization of CS: Complexity Barrier and Epistemic Opacity

This section is dedicated to a characterization of CS by means of complexity as the salient feature of the underlying mathematical model and its consequences for the justification of CS results and the kind of knowledge generated by CS. Complexity arises because the numerous calculations involved are highly interdependent (Lenhard 2011). The increased complexity of models and a lack of analytical tractability are, aside from calculation speed, the reason to make use of the computer as a modelling instrument. Advocates of the epistemological novelty of CS put forward the epistemic opacity thesis. As has been mentioned before, it has been questioned that the properties of CS require a novel epistemology. Admittedly, Frigg and Reiss (2009) argue mainly against the hype of novelty and strive for soberness. But even they recognize novelty "owed to the fact that models are more complex than traditional philosophy of science allows". In relation to CS, they furthermore acknowledge that the treatment of mathematical equations for calculation on a computer "fundamentally change[s] their mathematical characteristics" (Frigg and Reiss 2009: 601). In the following, the epistemic opacity thesis and its implications are presented.

To Humphreys, who was among the first to introduce CS into the philosophy of science, a new perspective is required: "For an increasing number of fields in science, an exclusively anthropocentric epistemology is no longer appropriate because there now exist superior, non-human epistemic authorities" (Humphreys 2009: 617). Humphreys also introduced the notion of epistemic opacity as a central aspect of the novelty of CS: "[A] process is epistemically opaque relative to a cognitive agent X at time t just in case X does not know at t all of the epistemically relevant elements of the process" (Humphreys 2009: 618). Epistemic opacity may be removable or non-removable ("relative" and "essential" in Humphreys's terms). Humphreys claims two sources of epistemic opacity. First, the calculation process is necessarily too fast to be followed, and second, "computationally irreducible processes" are concerned (Humphreys 2004: 147–148), meaning there is no simpler description available than the actual computation nonexecutable for humans.

Kaminski et al. (2018) seek to clarify the notion of epistemic opacity. They discern three aspects, namely social, technical and mathematical opacity. Their thesis is that although the first two are highly relevant to CS, only the last is special for CS in the sense Humphreys introduced it. Social opacity has to do with the dependence on one's own knowledge claims on others and with the social organization of the epistemic process. Technical opacity is a feature owing to the fact that techniques may be used without knowledge of either their context of discovery or their context of justification. Mathematical opacity can occur in different modes (being proven, unclear, of practical necessity and terminable). The reason why the computational speed finally is of importance is that the calculation is long relative to a human lifetime and there is no shortcut ("computational irreducible processes"). A calculation of future states is only possible along intermediate steps.[14] Thus, the computation becomes opaque.

A consequence of epistemic opacity of complex simulation models is that analytic insight is blocked (Kaminski 2018: 14) and "traditional transparency" cannot be achieved (Humphreys 2004: 148). As a consequence, CS cannot be classified as models in the "classical" sense implying insight and transparency.[15] Transparency and the justification of results are closely related. However, step-by-step insight is the basis only of internalistic justification strategies (argument and mathematical proof being the prototype). Externalistic justification strategies resort to reliability and are guided by criteria of performance and success. To CS the latter are deemed to be of greater importance due to the lack of insight (Kaminski et al. 2018).

Epistemic opacity provides an explanation why CS appear as experiments, because their outcome is unforeseeable and there is no shortcut as well (Kaminski 2018). One possibility to deal with the situation is the use of trial-and-error procedures, i.e., empirical methods to explore the simulation (Humphreys 2004: 150–151).

The role of exploration in CS has been found to be different from exploration in TE (Lenhard 2011). In TE, the role of exploration is temporary, while it is permanent in CS. The reason for the difference is the complexity of the simulation model due to the multitude of calculation steps and their interdependencies. In the case of TE, repeated exploration investigates the premises from which a conclusion can be deduced. TE aim at a continuous, complete, consistent line of argument such that further iterations become redundant. While the outcome of a TE is initially opaque (which is the reason to call it "experiment" in the first place), it turns into a transparent argument. This final epistemic transparency is the precondition for a TE to be acceptable. In other words, before a situation is thought through, i.e., before the TE starts, it is not clear which premises lead to a given result. The experiment consists in repeated trials of different premises in an argument. If the occupation with the situation is successful, the result is transparency concerning the relevance or logical consistency of the premises. At the same time, the exploration is over. CS, however, do not turn into obvious arguments. Even after many simulation runs, the exploration is not over. Despite algorithmic transparency, the initial opacity is not eliminated in the course of iteration. The reason has to do with the complexity of the simulation model, i.e., the large number of calculation steps and the interdependencies among them. CS compensate for the absent transparency between premises (input values) and conclusion (output values) in that they systematically pile single results and document them, e.g., by means of parameter variations. Each change of the model requires a new calculation.

Exploration as an experimental strategy is a means to cope with opacity in CS. It results in a certain kind of knowledge, as has been demonstrated by Lenhard by means of case studies from nanoscience (Lenhard 2004, 2005, 2006). With respect to the purposes of mathematical modelling, complexity implies that one has a mathematical model, but it does not help. CS bypass the problematic situation by means of an imitation of the mathematical model, the model thereby becoming computationally tractable. Even though the imitation of the mathematical model and its computational exploration do not allow for theory-based insight, they make the opaque model assessable. (Thus, "simulation" refers to both the imitation of the complex opaque model and the experimental strategy to cope with the model; cf. Saam 2015: 82.) CS overcome the "complexity barrier" (Lenhard 2006: 609) for theory-based insight of the simulation model by allowing for prediction and control and creation of phenomena. They provide "a kind of understanding that manipulation becomes possible" (Lenhard 2004: 98). This kind of understanding, termed "pragmatic", aims at handling skills and stable design rules in contrast to mechanistical explanations. It is based primarily on a specific combination of an experimental handling and visualizations. As a consequence, researchers are not dependent on an insight into the details of the complex interaction of the theoretical components. They interact with the model behaviour instead. In contrast to the traditional model as an instrument for insight (although into an idealized model world), the simulation model is an instrument to explore and control model behaviour. As another consequence, the assessment of a CS takes place with regard to the overall behaviour of the model and not by a reconstruction of the steps (thus, by externalistic justification according to Kaminski et al. 2018).

It is questionable whether "we must abandon the insistence on epistemic transparency for computational science" (Humphreys 2004: 150). At least it is "crucial to explicate just what is and what is not understood when only the results of changes in the initial conditions and parameter values are known" (Kuorikoski 2012: 177). In particular, it is not understood why the manipulations have the observed effects. It is even more important to clarify the epistemic gain, because an "illusion of understanding" (Kuorikoski 2012: 173) is likely, especially when scientific results are presented to a non-expert audience. Understanding proper, according to Kuorikoski, includes counterfactual inferences and means that new "what-if-things-had-been-different" questions can be answered. These are virtues of classical models. Examples of an illusion of understanding are to conflate visualization and insight, i.e. knowledge of dynamics and knowledge of underlying causes, and to conflate a manipulation of model dynamics (knowledge of dependencies) with an understanding of the mechanism (knowledge of the reasons of dependency).

To sum up, CS are based on complex mathematical models. This complexity is enhanced further by social and technical aspects, together leading to epistemic opacity. With epistemic opacity, analytic insight and transparency are blocked. As a consequence, justification strategies based on reliability guided by criteria of performance and success are of major importance. CS are a means to cope with the complexity. They are instruments to explore and control model behaviour, thereby providing handling skills.

30.6. Conclusions

Within the previous sections, four interconnected routes to characterize CS have been presented, in order to approach CS philosophically. Each of them adds to an understanding of the method.

Working definitions of CS have contributed to their characterization. Despite many guises of CS, two threads can be discerned. One emphasizes simulation (in the broad sense, including CS) as a method of imitating, either a model or a process. The other emphasizes the use of the computer which enables CS as a method of exploring a mathematical model. In the following sections of this chapter, the two threads reappear. Within the simulation pipeline, certainly the step of discretizing the mathematical model can be recognized as an imitational activity. When focusing on the properties of the simulation model, simulating as exploring activity comes to the fore. The two threads are interconnected: the imitation enables the exploration, and the results of the exploration feed back into the design of the imitation. Following one thread or the other, a CS is at least not a simple calculation process but instead an autonomous modelling activity, where the theoretical underpinning is just one ingredient of the final simulation model.

Whether the exploration is an experimental access, and to what, has been one of the issues of the second route of characterizing CS. CS exhibit features similar to thought experiments, e.g., both are reconstructable as arguments, but complexity and approximations, among other characteristics, distinguish CS. CS are similar to experimental laboratory settings in how they represent target systems; under certain conditions their results may even be interpreted as measurements instead of calculations. However, remarkable differences remain. In contrast to thought experiments, the experimental feature of CS is permanent and does not lead to argumentative transparency. In contrast to laboratory experiments, CS are "over-controlled" (Beisbart 2018), meaning that their results are preset by the assumptions involved. The views of CS as experiments on theories (or explorations of models) and of CS as models of experiments (or virtual experiments) emphasize CS as a method of theory without neglecting its strong empirical character at the same time.

The necessity of exploration may be explained by epistemic opacity, i.e. properties of the simulation model hindering or excluding theory-based insight and mechanistic transparency. CS as an exploring activity seems to lead primarily to handling skills. The virtual handling is intended to correspond to the handling of complex systems in reality.

From an engineering perspective it may seem rather natural that simulationists "go back and tinker with their choice" (Winsberg 2010: 16). The simulation model then is an object of design, to be judged by its ability to perform a special task related to the realistic imitation of system behaviour. Such a pragmatic attitude suggests itself in the light of this chapter. However, at least two implications are challenging. First, if performance concerning a special task is crucial, simulation results have to be tested. Second, if failure occurs, it has to be traced back. What seems to be straightforward turns out to be tricky in the case of CS because these requirements are often difficult to achieve. For example, CS are often used in cases where comparisons with experimental data and test situations are out of reach or only partially feasible, and epistemic opacity is clearly disadvantageous for the detection of failures.

Related Chapters

Chapter 23: Design Methods and Validation (Sabine Ammon)

Chapter 28: Models in Engineering and Design: Modeling Relations and Directions of Fit (Michael Poznic)

Chapter 31: Experimentation (Viola Schiaffonati)

Chapter 32: On Verification and Validation in Engineering (Francien Dechesne and Tijn Borghuis)

Chapter 36: Trust in Engineering (Philip J. Nickel)

Further Reading

Morrison, M. (2015). Legitimating Simulation: Methodological Issues of Verification and Validation. In M. Morrison (ed.), *Reconstructing Reality: Models, Mathematics, and Simulations*. Oxford: Oxford University Press, pp. 248–286. (About the testing of simulation results: an introduction to the methods of verification and validation.)

Lenhard, J. and Winsberg, E. (2010). Holism, Entrenchment, and the Future of Climate Model Pluralism. *Studies in History and Philosophy of Modern Physics*, 41, 253–262. (Why it is often difficult to trace failure back: thesis of epistemological holism of highly complex simulation models, illustrated by a climate model case study.)

Notes

1. The abbreviation CS is used for both singular and plural.
2. The denied simulating property of the Monte Carlo method has also been discussed elsewhere, e.g., in terms of saving the label "simulation" of a mere numerical technique by emphasizing the dynamization of the task (Hartmann 1996). Galison (1996) provides a study of the history of the Monte Carlo method which includes differing views on its simulating property.
3. Please note that this is not a statement on how the issues discussed are relevant for agent-based simulations as well.
4. However, the authors admit that "some rare cases of simulations comply with it" (with the idealized view). These are cases "when the computer model is analytically solvable".
5. For the original detailed presentation of the Arakawa operator, see Küppers and Lenhard (2005).
6. "Empirical" here is equivalent to "non-theoretical". It refers to modelling knowledge gained in the course of the modelling process, in contrast to modelling knowledge deduced from theory (cf. the distinction of a priori justification and empirical support by Barberousse and Vorms 2014: 3598).
7. The history of fluid and aerodynamics (Gramelsberger 2011) illustrates this point. A schism between theory and applications dominated research in the 19th century. It consisted in the fact that only highly idealized models could be solved analytically and thus investigated theoretically, while real problems had to be investigated experimentally. The schism has been overcome first by analogue simulation and later on by CS accompanied by even more advantages (e.g., in terms of accuracy).
8. Note that although this feature may be acknowledged, a corresponding novelty claim may still be rejected: In an examination of the idea of in-betweenness, Frigg and Reiss argue that thought experiments and mathematical models are hybrids as well (first metaphorical reading), while scientific models and experiments have also been regarded as mediators (second metaphorical reading) (Frigg and Reiss 2009: 607–611).
9. "Laboratory" is not necessarily meant literally. Rather, it expresses a type of investigation, with the laboratory representing ideal working conditions.
10. However, Saam (2015: 69) critically remarks that such a linkage usually does not take up the discussions within the philosophy of TE. Thus, linkage claims might be superficial.
11. Within the philosophy of thought experiment it has been debated if every TE is reconstructable as an argument. There seems to be no 1:1 relation of an argument and a TE since there are cases where one TE serves different arguments and may be reused in a different context (Saam 2015: 67–68). For the reconstruction of CS as arguments see Beisbart (2012).
12. The expression "experimenting on theories" has been criticized for conceptual reasons by both Winsberg and Parker. Being abstract entities, neither models nor theories can be experimented on. Thus, Winsberg concludes, "we are forced to take seriously the material characteristics of computation" since something is

in fact manipulated (Winsberg 2009: 578). This request should be distinguished from attempts to push the materiality issue further and to argue that by CS one experiments with the material programmed digital computer (Parker 2009).

13. Note that this is only one interpretation of the "materiality thesis". The alternative reading emphasizes the material dimension of CS, especially in terms of the potential malfunctioning of the machine (Barberousse and Vorms 2014: 3607–3608).

14. This is not to be confused with an acceleration of computation, e.g., by parallel processing (Kaminski et al. 2018).

15. "Classical models" refer to mechanistical models which became important during the 19th century. They permit retracing of the causal relationships and interdependencies among the model components and deliver mechanistic explanations for input-output relations.

References

Barberousse, A. and Vorms, M. (2014). About the Warrants of Computer-based Empirical Knowledge. *Synthese*, 191(15), 3595–3620.

Beisbart, C. (2012). How Can Computer Simulations Produce New Knowledge? *European Journal for Philosophy of Science*, 2(3), 395–434.

Beisbart, C. (2018). Are Computer Simulations Experiments? And if Not, How Are They Related to Each Other? *European Journal for Philosophy of Science*, 8(2), 171–204.

Bungartz, H.-J., Zimmer, S., Buchholz, M. and Pflüger, D. (2013). *Modellbildung und Simulation. Eine anwendungsorientierte Einführung*, 2nd ed. eXamen.press. Berlin, Heidelberg: Springer Spektrum.

Dowling, D. (1999). Experimenting on Theories. *Science in Context*, 12(2), 261–273.

Durán, J.M. (2017). Varieties of Simulations: From the Analogue to the Digital. In M.M. Resch, A. Kaminski and P. Gehring (eds.), *The Science and Art of Simulation 1. Exploring—Understanding—Knowing*. Cham, Switzerland: Springer, pp. 175–192.

Frigg, R. and Reiss, J. (2009). The Philosophy of Simulation: Hot Issues or Same Old Stew? *Synthese*, 169, 593–613.

Galison, P. (1996). Computer Simulations and the Trading Zone. In P. Galison and D. Stump (eds.), *The Disunity of Science. Boundaries, Contexts, and Power*. Stanford: Stanford University Press, pp. 118–157.

Gramelsberger, G. (2011). From Computation With Experiments to Experiments on Computation. In G. Gramelsberger (ed.), *From Science to Computational Sciences. Studies in the History of Computing and its Influence on Today's Sciences*. Zürich, Berlin: Diaphanes, pp. 131–142.

Grüne-Yanoff, T. and Weirich, P. (2010). The Philosophy and Epistemology of Simulation: A Review. *Simulation & Gaming*, 41(1).

Hartmann, S. (1996). The World as a Process: Simulations in the Natural and Social Sciences. In R. Hegselmann, U. Mueller and K. Troitzsch (eds.), *Modelling and Simulation in the Social Sciences From the Philosophy of Science Point of View*. Dordrecht: Kluwer, pp. 77–100.

Hubig, C. (2010). Leistungen und Grenzen der Virtualität beim Wissenserwerb. In K. Kornwachs (ed.), *Technologisches Wissen. Entstehung, Methoden, Strukturen*, acatech diskutiert. Berlin, Heidelberg: Deutsche Akademie der Wissenschaften, pp. 211–225.

Hubig, C. and Kaminski, A. (2017). Outlines of a Pragmatic Theory of Truth and Error in Computer Simulation. In *The Science and Art of Simulation I. Exploring—Understanding—Knowing* (pp. 121–136). Springer.

Humphreys, P. (1991). Computer Simulations. *PSA: Proceedings of the Biennial Meeting of the Philosophy of Science Association*, 1990, 497–506.

Humphreys, P. (2004). *Extending Ourselves. Computational Science, Empiricism, and Scientific Method*. Oxford University Press.

Humphreys, P. (2009). The Philosophical Novelty of Computer Simulation Methods. *Synthese*, 169, 615–626.

Kaminski, A. (2018). Der Erfolg der Modellierung und das Ende der Modelle: Epistemische Opazität in der Computersimulation. In A. Brenneis, O. Honer, S. Keesser and S. Vetter-Schultheiß (eds.), *Technik—Macht—Raum*. Wiesbaden: Springer.

Kaminski, A., Resch, M. and Küster, U. (2018). Mathematische Opazität—Über Rechtfertigung und Reproduzierbarkeit in der Computersimulation. In A. Friedrich, P. Gehring, C. Hubig, A. Kaminski and A. Nordmann (eds.), *Arbeit und Spiel*, Vol. 4 of *Jahrbuch Technikphilosophie*. Baden-Baden: Nomos, pp. 253–277.

Kaminski, A., Schembera, B., Resch, M. and Küster, U. (2016). Simulation als List. In G. Gamm, P. Gehring, C. Hubig, A. Kaminski and A. Nordmann (eds.), *List und Tod*, Vol. 2 of *Jahrbuch Technikphilosophie*. Zürich and Berlin: Diaphanes, pp. 93–121.

Kuorikoski, J. (2012). Simulation and the Sense of Understanding. In P. Humphreys and C. Imbert (eds.), *Models, Simulations and Representations*. New York and London: Routledge, pp. 168–187.

Küppers, G. and Lenhard, J. (2005). Validation of Simulation: Patterns in the Social and Natural Sciences. *Journal of Artificial Societies and Social Simulation*, 8(4), 3.

Lenhard, J. (2004). Nanoscience and the Janus-faced Character of Simulation. In D. Baird, A. Nordmann and J. Schummer (eds.), *Discovering the Nanoscale*. Amsterdam: IOS Press, pp. 93–100.

Lenhard, J. (2005). Kreation oder Kontrolle von Phänomenen? Computersimulationen in der technikorientierten Wissenschaft. In G. Abel (ed.), *Kreativität: XX. Deutscher Kongress für Philosophie. Sektionsbeiträge I*. TU Berlin Universitätsverlag.

Lenhard, J. (2006). Surprised by a Nanowire: Simulation, Control, and Understanding. *Philosophy of Science*, 73(5), 605–616.

Lenhard, J. (2011). Epistemologie der Iteration. Gedankenexperimente und Simulationsexperimente. *Deutsche Zeitschrift für Philosophie*, 59(1), 131–145.

Morrison, M. (2009). Models, Measurement and Computer Simulation: The Changing Face of Experimentation. *Philosophical Studies*, 143(1), 33–57. doi:10.1007/s11098-008-9317-y

Morrison, M. (2015). *Reconstructing Reality. Models, Mathematics, and Simulations*. Oxford Studies in Philosophy of Science. New York: Oxford University Press.

Paolo, E.A.D., Noble, J. and Bullock, S. (2000). Simulation Models as Opaque Thought Experiments. In M. Bedau, J. McCaskill, N. Packard and S. Rasmussen (eds.), *Artificial Life VII: The Seventh International Conference on the Simulation and Synthesis of Living Systems*. Cambridge, MA: MIT Press, pp. 497–506.

Parke, E.C. (2014). Experiments, Simulations, and Epistemic Privilege. *Philosophy of Science*, 81(4), 516–536.

Parker, W.S. (2009). Does Matter Really Matter? Computer Simulations, Experiments, and Materiality. *Synthese*, 169(3), 483–496.

Petersen, A.C. (2000). Philosophy of Climate Science. *Bulletin of the American Meteorological Society*, 81(2), 265–272.

Resch, M.M. (2017). On the Missing Coherent Theory of Simulation. In M.M. Resch, A. Kaminski and P. Gehring (eds.), *The Science and Art of Simulation 1. Exploring—Understanding—Knowing*. Cham, Switzerland: Springer, pp. 23–32.

Resch, M.M., Kaminski, A. and Gehring, P. (eds.) (2017). *The Science and Art of Simulation I. Exploring—Understanding—Knowing*. Springer.

Saam, N.J. (2015). Simulation in den Sozialwissenschaften. In N. Braun and N.J. Saam (eds.), *Handbuch Modellbildung und Simulation in den Sozialwissenschaften*. Wiesbaden: Springer Fachmedien, pp. 61–95.

Saam, N.J. (2017). What is a Computer Simulation? A Review of a Passionate Debate. *Journal for General Philosophy of Science*, 48(2), 293–309.

Winsberg, E. (2009). A Tale of Two Methods. *Synthese*, 169(3), 575–592.

Winsberg, E. (2010). *Science in the Age of Computer Simulation*. Chicago and London: University of Chicago Press.

31

EXPERIMENTATION

Viola Schiaffonati

31.1. Introduction

Experiments play a central role in the scientific method. Part of modern science after the scientific revolution in the 17th century has developed in an experimental direction, and the concept of experimentation has been shaped in accordance with the development of science. To define an experiment is a complex task, as experiments cover many roles in science and they work with different purposes. In a very general sense, an experiment is a set of observations and actions, performed in a controlled context, to test a given hypothesis. This expresses the original idea of experiment as *controlled experience*, consisting in the production of controlled circumstances, where the phenomenon under investigation is treated as an isolated object and it is assumed that other factors not under investigation do not influence the investigated object. "An experiment, in the standard scientific sense of the term, is a procedure in which some object of study is subjected to interventions (manipulations) that aim at obtaining a predictable outcome or at least predictable aspects of the outcome" (Hansson 2016: 613). This is the working definition of experiment I will use in this chapter and on which I will further elaborate in Section 2. This definition emphasizes the centrality of manipulation that is usually expressed with the label "controlled" experiment. In a controlled experiment, *independent variables* are manipulated to test the effects on the *dependent variable* which is the variable being tested in the experiment.

Experiments are usually characterized with respect to general criteria. *Reproducibility* is the possibility of verifying, in an independent way, the results of a given experiment. It refers to the fact that independent experimenters, different from the one claiming for the validity of some results, are able to achieve the same results, by starting from the same initial conditions, using the same type of instruments, and adopting the same experimental techniques. *Repeatability* concerns the fact that a single result is not sufficient to ensure the success of an experiment. A successful experiment must be the outcome of a number of trials, performed at different times and in different places. These requirements guarantee that the result has not been obtained by chance, but is systematic. From these, other criteria flow, such as *comparison*, that is the knowledge of what has already been done in the past within a field and the possibility to accurately compare new results with the old ones; and *justification*, that is the capacity to derive well-justified conclusions starting from the experimental data.

In the discussion about experimentation in engineering, this canonical view of experiment has been mostly adopted. For many years, the engineering disciplines have been conceptualized as applied sciences with little attention to their peculiarities and differences from the natural sciences. Accordingly, experimentation has been considered along the same lines. As the debate on the status of engineering and engineering knowledge progressively developed, and engineering (or at least engineering research and the engineering sciences) gained an ontological autonomy from the sciences, the nature and role of experiments in engineering have been reconsidered. However, notwithstanding the differentiation of engineering from science on the basis of their different methodologies (Staples 2014), the methods of engineering have been mostly equated to the experimental approaches in the natural sciences (Boon 2012b; Staples 2015). Thus, it is not surprising that a general account of engineering experiments is also missing from those works that recognize the peculiarity of the engineering sciences within the framework of the so-called philosophy of engineering (Van de Poel and Goldberg 2010; Michelfelder et al. 2013; Michelfelder et al. 2017).

In this chapter I will focus on experiments in the engineering sciences (and not engineering in general), defined as scientific research in the context of technological application (Boon 2012a), leaving aside industrial contexts. Moreover, and more importantly, I will discuss the case of computer engineering[1] and, in particular, the field of experimental computer science and engineering, defined as the building of, or the experimentation with or on, non-trivial hardware and software systems.

The reason for this latter delimitation is twofold. First, a general account of the experimental method in the engineering sciences is at risk of being generic and without any real insight about the multifaceted practice of engineering. Second, notwithstanding the focus of the engineering disciplines, including computer engineering, on technical artifacts,[2] the artifacts of computer engineering present some peculiarities worth investigating in more detail. Computers are complex and *malleable* artifacts. The concept of logical malleability (Moore 1985) expresses their universality: there is no a priori limit on their functionality, which raises expectations about their performances and makes the consequent evaluation more difficult. Moreover, these artifacts share a different relationship with their theories. Indeed, computer engineering does not depend on elaborate theoretical foundations in the same way that, for example, experimental physics draws on theoretical physics. This is not to deny the roots of computer science and engineering in mathematics and the theoretical import of theoretical computer science. Rather, it is to stress that experiments are used to validate some informal thesis not rigorously specified by theory (National Research Council 1994). On the contrary, the artifacts of other engineering disciplines are often constrained by well-defined physical phenomena (e.g., gravity, compressibility of gases), thus limiting their variety and presenting clearer criteria to experimentally evaluate them.

So, in this chapter I will discuss how experiments in computer science and engineering have much in common with traditional sciences, such as physics and biology, but also present significant differences due to their aim of producing technical artifacts (i.e., computing systems, programs) that are complex, malleable, and not directly descending from theories. Computer engineers have practically used the term experimentation as a "search to find implementations of systems that worked" (Tedre and Denning 2017: 9), and in methodological discussions the adoption of the same experimental criteria of the natural sciences has been emphasized.

This chapter is organized as follows. In Section 2, I will review how experiments have been conceived and used in computer science and engineering. In Section 3, I will present different views of experiments in this field, by stressing how experiments can be considered in different ways. In Section 4, I will focus on experiments in software engineering and autonomous robotics to highlight some of their peculiarities. In Section 5, I will discuss what can be learned from these case studies and whether some claims can be generalized to the whole of computer science and engineering. Finally, in Section 6, I will conclude the chapter by addressing some open issues and by advancing a proposal for a novel conceptualization of experiments in computer science and engineering.

31.2. Experimental Computer Science and Engineering

Already in 1967, George Forsythe, when discussing the foundations of computer science, pointed out: "good experimental work in computer science must be rated very high indeed . . . Computer science is in part a young deductive science, in part a young experimental science, and in part a new field of engineering design" (Forsythe 1967: 4). The acknowledgement of computer science as an experimental science is essential to overcome the idea that formal verification is sufficient to prove that the behavior of a program corresponds to how the program was intended to behave, without the need of any experimental procedure, where debugging can be substituted by proving that a program meets its specification (McCarthy 1962) and demonstrations can take the place of experiments (Hartmanis 1994).

In the attempt to move away from the idea of computer science as equated to the natural sciences (and from the conception of experiments presented accordingly), the acknowledgment of computer science as a new field of engineering is an essential step. The following view of engineering that perfectly fits with computer science can be of some help for the present analysis:

> Although engineering often draws on science, it is not science, and is not merely applied science. . . . What distinguishes engineering from technology is methodology—a systematic approach for the use and growth of objective knowledge about how the physical world can be made to meet requirements.
>
> *(Staples 2014: 2276)*

This view is in accordance with that provided by Walter Vincenti in his seminal book *What Engineers Know and How They Know It*, in which the idea of engineering as distinct from a merely applied science has already been defended: "Engineers use knowledge primarily to design, produce, and operate artifacts, goals that can be taken to define engineering" (Vincenti 1990: 226).

The emphasis on requirements, and on artifacts as well, is evident in the very famous paper written by Allen Newell and Herbert Simon, after they received the Turing Award (Newell and Simon 1976), which offers a view of computer science as a particular form of experimental science:

> Computer science is an empirical discipline. We would have called it an experimental science, but like astronomy, economics, and geology, some of its unique forms of observation and experience do not fit a narrow stereotype of the experimental method. None the less, they are experiments. Each new machine that is built is an experiment. Actually constructing the machine poses a question to nature; and we listen for the answer by observing the machine in operation and analyzing it by all analytical and measurement means available. Each new program that is built is an experiment. It poses a question to nature, and its behavior offers clues to an answer. Neither machines nor programs are black boxes; they are artifacts that have been designed, both hardware and software, and we can open them up and look inside. We can relate their structure to their behavior and draw many lessons from a single experiment. We don't have to build 100 copies of, say, a theorem prover, to demonstrate statistically that it has not overcome the combinatorial explosion of search in the way hoped for. Inspection of the program in the light of a few runs reveals the flaw and let us proceed to the next attempt.
>
> *(Newell and Simon 1976: 114)*

This long quote shows the originality of this approach in between the traditional view of experiment as a question to nature and the new idea that computers and programs are artifacts and, therefore, construct a different relationship with the surrounding natural world. So, for example, experiments in this field do not need to respect the repeatability criterion, but one single experiment is sufficient to draw many lessons. Whether or not the renunciation of the traditional standards of experimentation, such as repeatability, is fully justified in Newell and Simon's framework, it is important to note that this is a first attempt to not fully equate computer science with the natural sciences and to reason about the nature of its experiments accordingly. This attempt was part of Simon's more general project on the foundations of the sciences of the artificial (Simon 1969), although the view on experiments has not been further developed within this context. An interesting example in the same line of Newell and Simon's approach can be found in Khalil and Levy (1978), where the role of programming in computer science is equated to that of laboratories in the physical sciences.

The same attention to the peculiarities of the experimental method within computer engineering has not been reached in the debate following the crisis of what was then called *experimental computer science* that took place in the 1980s. From its very beginning this debate has been characterized by the importance of measurement and testing of computing algorithms and systems (Feldman and Sutherland 1979) that should bring major advantages to the different fields of computer engineering (McCracken et al. 1979). At the same time it was suggested that experimental computer science should be rejuvenated by adopting the same traditional standards of science, because in the absence of these standards, it risks not being taken seriously (Denning 1980). On the one hand, this proposal tries to go beyond the view advanced by Newell and Simon to emphasize some peculiarities of experimentation in computer science, and it advocates a more rigorous approach to experimentation by adopting the same experimental criteria, such as reproducibility and repeatability, of the natural sciences. On the other hand, the necessity to adjust to the same experimental standards as the natural sciences is taken for granted and the overall picture of science is often a simplified one, as the process of classifying knowledge derived from observations.

Whether reductive or not, the invitation to apply the same experimental criteria of the natural sciences to computer engineering has remained mostly unattended, notwithstanding some efforts over the years (Denning et al. 1989; Tichy 1998; Langley 1988). Recently, the discussion on the experimental method in computer engineering has reemerged and has involved different research areas. Experiments are considered essential for assessing computing systems' performances and for triggering new developments (Denning et al. 1989; Freeman 2008; Morrison and Snodgrass 2011). Experimentation plays an essential role in use-inspired research and product design (Snir 2011), and, in general, in understanding computations too complex for a purely mathematical analysis, in proving their correctness, in checking consistency with hypotheses, in uncovering performance constraints, and in showing whether original goals are met (Denning and Freeman 2009).

Despite the increasing interest in more rigor in approaching experiments in computer engineering, many lament that the current methodology is inadequate and that, in comparison with other fields (e.g., physics, biology), computer scientists should experiment more (Denning 2005). However, the invitation to experiment more and better is often based on general statements, such as the acceleration of progress (Tichy 1998) and the tremendous success of scientific reasoning (Morrison and Snodgrass 2011), without any specific analysis of the concrete advantages of this approach.

Two constant patterns can be identified within this discussion. First of all, the emphasis on the necessity to adopt the same experimental criteria as used for the natural sciences is a sort of leitmotiv justified by the importance to assess the scientific status of the discipline against the opposite view of computer engineering as *hacking science*.[3] Secondly, the notion of experimentation to be adopted is, in most cases, oversimplified and the traditional view of controlled experiment taken for granted without any further discussion about its applicability to the engineering sciences.

31.3. Different Views on Experimentation in Computer Science and Engineering

Starting from the so-called Feldman report (Feldman and Sutherland 1979), the label *experimental computer science and engineering* has become quite popular in at least one part of the computing literature. However, its clear meaning has remained unexplained, and although experimentation has been discussed in Feldman's report in terms of exploration, construction and testing, hypothesis testing, demonstration, and modeling, different interpretations have since been advanced. These interpretations include experimental computer science as opposed to theoretical computer science, the view emphasizing its being a hypothesis-driven discipline, and also the idea that experimentation could cover all kinds of empirical data collection, demonstrations of technology, and building of things (Denning 1980).

Over the years experimental computer science has been used as a label to indicate at least three different views (Feitelson 2006): an engineering field devoted to the demonstration of feasibility, the mathematical modeling of the behavior of complex systems whose properties have to be empirically tested, and the evaluation of computer systems by using the same experimental means—such as, for instance, reproducibility—of the natural sciences.

A finer-grained taxonomy of the different meanings of experiments, strongly rooted in a careful historical analysis, is presented in (Tedre 2014). In this analysis the focus is on how the term experiment has been used, and how these different uses can be traced to different traditions, and not on how it should be used. Matti Tedre claims that at least five views of experiment can be found in the history of computer science and engineering.

Feasibility experiment. In this case the term is used for those efforts aimed at empirically demonstrating the proper development and working of a technology. It should not be surprising that in this context *demonstration* and *experiment* are used as synonyms, and experimental computer engineering is the one producing techniques, insights, and understanding that come from building and using computer systems. Other terms used in this case are *proof of concept* and *proof of principle* (Hartmanis 1994).

Trial experiment. This is the case in which a system is evaluated, using predetermined variables, usually in a laboratorial context. Trial experiments go further than feasibility experiments because they aim at evaluating how well a system works with respect to criteria such as fastness, memory, and robustness. Here experiment is synonymous with test and is designed to evaluate the quality of a system, being mostly focused on measurements rather than hypothesis testing (McCracken et al. 1979).

Field experiment. This is the type of experiment aimed at evaluating the aspects of a system outside the laboratory, namely in the real world, as many of the requirements of the system are not internal to it but relative to its surroundings. Similar to trial experiments, field experiments are not usually meant to generalize outside the particular system. The term "field experiment" is commonly used in computer science and engineering, for instance in information systems (Palvia et al. 2003) or in robot car races (Freeman 2008).

Comparative experiment. This form of experimentation is devoted to comparing different solutions, in a search for the best one for a specific problem, an approach that is surely applicable to many areas of experimental computer science and engineering. Comparative experiments are performed to measure and compare one solution with a competing solution with the same dataset and parameters (Basili and Briand 1996).

Controlled experiment. This type of experiment is considered the gold standard of science and can be subsumed into the idea of rigorously testing hypotheses under controlled circumstances while typically enabling both generalization and prediction. These controlled experiments have been promoted in some areas of computer engineering, from computer security to software development (Morrison and Snodgrass 2011; Feitelson 2007).

From a normative point of view (in the sense of analyzing the label *experiment* not just in terms of its actual use, but in terms of its correct use), it is worth considering a report on experimentation in computer science and engineering dating back to 1994 which, surprisingly, has not received much attention within the community (National Research Council 1994). Experimental computer science and engineering (ECSE) is firstly defined as a *synthetic discipline*: the computational artifacts that are its objects of investigation have been created by humans rather than being given by nature. Computational artifacts are processes, algorithms, mechanisms, robots, and the like that manipulate and transform information. The synthetic nature of the discipline is essential to reconsider the role of experimentation that, in the tripartite model (theory, abstraction, design) advocated by Denning

(1980), is ignored. Here, experimentation is meant as a performance evaluation of the computational artifacts that have been constructed. These artifacts are too complex to be understood by direct analysis, where complexity refers not only to the total number of constituent parts, but also to the interaction among these parts and with the surrounding environment. They must be implemented so that the behavior of a system and the interaction of its parts can be observed in practice. This view of experimentation stands apart from that of the natural sciences and the idea that ECSE should conform to the traditional standards of natural sciences is rejected: "Experimentation carries different connotations in ECSE than it has in physics, biology, or medicine" (National Research Council 1994: 17). Accordingly, artifacts cover at least three roles in the report: proof of performance, proof of existence, and proof of concept.

Proof of performance. An artifact that acts in this role offers an apparatus or testbed for measurement or experimentation, and the findings are usually quantitative. An example is an operating system designed to prove the improvement of the techniques adopted to implement parallelism and distribution.

Proof of concept. An artifact that acts in this role demonstrates by means of its behavior that a complex configuration of elements can perform these types of activities. This behavior cannot be simply argued for by logical reasoning or abstract demonstration. An example is represented by the so-called experimental computers used for understanding the effects of different design philosophies, for instance, for evaluating the best solutions to avoid the long delays of processors in accessing memory in parallel computing.

Proof of existence. An artifact that acts in this role conveys a new phenomenon. Because computation is synthetic, it is possible to produce phenomena that have not been previously imagined, and these phenomena are often better explained by demonstration than by description. A well-known example is the computer mouse, whose use as a pointing device in human-computer interaction is better conveyed by showing it in action, rather than by verbal description.

These three roles offer a view of experimentation as implementation, a necessary step to evaluate the ideas and the theoretical models behind them:

> an experiment in ECSE does not verify a prediction from theoretical computer science or rely heavily on a model developed by theoreticians, although . . . good experimental work is grounded in testable models and hypotheses. Experiments are most often conducted to validate some informal thesis derived from a computational model that is informed but not rigorously specified by theory and that may have been developed expressly for the experiment.
>
> *(National Research Council 1994: 15)*

In this section I have shown how the notion of experiment can be differently conceptualized and practically intended within computer science and engineering. It is true to say that none of these notions prevailed over the others but that, in most of the cases, they coexist. In the next section I will present examples from different research areas to evidence in greater detail some of the essential features of experimentation within the field of computer engineering.

31.4. Experimentation: The Cases of Software Engineering and Autonomous Robotics

In this section I discuss in greater detail two fields of computer engineering in which the debate on the conduction of experiments has recently taken center stage: software engineering and autonomous robotics. By presenting these two cases, I do not mean to say that the attention to experimentation has not been raised in other fields. Indeed, significant examples are represented by machine learning

(Langley 1988; Drummond 2009), signal processing (Barni and Perez-Gonzalez 2005; Vandewalle et al. 2009), artificial intelligence and intelligent agents (Hanks et al. 1993), theoretical computer science (Johnson 2002), and heuristic methods (Barr et al. 1995). However, software engineering and autonomous robotics are particularly representative not only in showing how different conceptualizations of the experimental method can coexist in the same field (see Section 3), but also in evidencing some peculiarities and, thus, the specificity of computer engineering with respect, on the one hand, to the natural sciences and, on the other, to other engineering disciplines.

Software engineering is a field of computer science and engineering aimed at the study and application of techniques for the design, development, operation, and maintenance of software. In the debate about the nature and role of experiments in this field, a very influential paper was published in 1997 (Zelkowitz and Wallace). The authors examine 612 papers published in the journals *IEEE Transactions of Software Engineering* and *Software* and in the *Proceedings of the International Conference on Software Engineering* in the years 1985, 1990, and 1995 to establish a taxonomy of experimental data collection techniques, ranging from "no experimentation" at all to "replicated experiments". This work concludes that, although the analysis shows that the trend is moving in the direction of a greater attention toward rigorous experimental methodology, about one third of the analyzed papers had a weak form of experimentation (called "assertion" by the authors) that favors the technology or tool described in the paper over other alternatives. Moreover, by comparing these results with works published in journals of other disciplines, the authors show that experimental rigor in software engineering is lower than in other scientific disciplines. According to the authors, some elements show that software engineering has not yet reached the same standards in experimentation as other disciplines (for instance, physics), and thus lacks experimental rigor. These standards are: originality, careful literature survey, canonical methods to collect data, standard format in the organization of technical papers (e.g., presentation of the hypothesis, development of the concept, and experimental validation), and experimentations rather than simple assertions. Along the same line, Tichy et al. (1995) finds that only a small part of the published papers in the field validate the claimed hypotheses through experimentation.

The lack of a well-developed experimental methodology in software engineering can be seen as the driving force in the foundation of the journal *Empirical Software Engineering* (Basili and Briand 1996) and of the first *International Symposium on Empirical Software Engineering* (IEEE 2007). Notwithstanding these efforts, Juristo and Moreno (2001) provide a list of caveats to show how software engineering is still immature with respect to these methodological issues. In particular, it emphasizes: a lack of training in assessing the importance and meaning of the scientific method, a lack of statistical training to understand how to analyze experimental data, and a lack of interest in publishing empirical studies conducted to check the ideas of others.

In general, experiments in software engineering are aimed at establishing criteria that help make the best choices regarding all the factors involved in the creation of software, such as the requirements specification language, the teams of programmers, and the programming paradigm and language. All the relevant activities are heavily human-centered, and following a rigorous method to do research on this kind of practice becomes particularly critical. However, notable exceptions can be found, like the one described in Cepeda Porras and Guéhéneuc (2010), where the performance of human subjects is measured with chronometers and eye-tracking devices to verify which of four competing graphic notation standards provides the best support to software developers in the task of recognizing design patterns in the architecture of a program. Such experimentation has a twofold purpose: investigating the effects of certain information visualization techniques on humans, and analyzing the search for the best practice in one aspect of the complex software development process. The merit of this kind of experimentation is not to show the superiority of one solution over another (in the case just discussed, which type of diagram software engineers should use for the aforementioned tasks), but to illustrate in detail an experimental procedure that software engineers may follow when such decisions need to be taken.

Whereas the main focus of this methodological debate in software engineering has concentrated almost exclusively on the implementation of traditional experimental categories, in particular the reproducibility of experiments (see Juristo and Gomez 2012 for a comprehensive and influential work on this topic), more recent attempts have put these discussions into the wider framework of the philosophy of science and the philosophy of technology. In this context, reflection on experiments is accompanied by the analysis of the disciplinary status of computing. For instance, Schiaffonati and Verdicchio (2016) analyzed the 50 most cited papers (according to Scopus) from *Empirical Software Engineering: An International Journal* in the 2003–2012 decade. The authors observe that, besides case studies, reviews, empirical analyses, comparative analyses, and field studies, more than one fifth of the articles present a form of controlled experimentation (even though not all of them explicitly refer to the presented work as a controlled experiment). Indeed, the articles pay attention to reproducibility and present results derived from repeated rounds; they carefully design the experimental setting, with an informed reflection on the choice and variation of controlling factors; they devote attention to the issue of considerably enlarging the number of subjects involved in the experimentation; they present in most of the cases a considerable statistical analysis of their results; they are interested in both the internal and external validity of their results, thus making it possible to generalize them. When considering those works that present the results of controlled experiments performed with human subjects, it is interesting to note how the most rigorous experimental procedures (see, for instance, the reproducibility criterion) are applied in a context where the human factor plays a primary role and brings in its typical and unforeseeable effects.

In recent years a similar discussion on experiments, and on the effort made to develop good experimental methodologies, has also received increased attention in autonomous robotics. This field is oriented towards the development of robotic systems that are autonomous, meaning they have the ability to operate without continuous human intervention, in order to work in places hardly accessible by humans or in cooperation with humans in common environments. In autonomous robotics, human operators evolve from being active controllers of the robot systems to being more passive supervisors of the same robotic systems. There are different reasons for the recent interest in experimentation in the field: from a scientific perspective, it concerns the desire of this rather novel community to adopt the same methodological standards of other scientific disciplines; from a more practical perspective, it deals with the possibility of measuring some parameters (e.g., safety, efficiency, etc.) in a standard way and of having rigorous benchmarks to compare and evaluate different products. The attention to benchmarks, in particular, plays a central role in the debate about good experimental methodologies in robotics (Amigoni et al. 2015) and in the way the notion of progress has been conceptualized in the discipline (Müller 2020).

The interest in solidly based experimental research in autonomous robotics has increased progressively over the last 20 years. However, it is only recently that this interest has been coupled with a careful analysis on how the concept of experimentation should be translated in the practice of autonomous robotics, also giving rise to a debate about the status of the discipline itself. To this end, both the creation of the EURON Special Interest Group on Good Experimental Methodology in Robotics Research (Bonsignorio et al. 2018a) and the series of workshops concerning replicable experiments in robotics (Bonsignorio et al. 2018b) have been playing a decisive role. The workshop series, in particular, has contributed in raising several issues and in increasing the sensibility of the community on these topics (Bonsignorio and del Pobil 2015).

In an effort to improve the quality of experimental activities, some attempts (Amigoni et al. 2009) have been made to take inspiration from how experiments are performed in the natural sciences, by trying to translate into the practice of autonomous robotics their general experimental criteria (e.g., comparison, repeatability, reproducibility, etc.). However, from a more recent analysis (Amigoni et al. 2014), it emerges that these criteria are not yet fully part of the current research practice. Notwithstanding the emphasis on the importance of reproducibility, for instance as a way to increase the experimental level of the field, good practices to promote it, such as the availability of shared data and code, are still not very common, and the attempts to critically analyze how the aforementioned criteria should be attained in experiments with autonomous robots are largely unexplored. From this systematic analysis (based on the papers presented in the first 10–15 years of the 21th century at the *International Conference on Autonomous Agents and Multiagent Systems*), it emerges that none of the experiments considered in autonomous robotics can be properly labeled as controlled. The increasing use of public data over which different systems can be run and compared is surely a sign of how comparison is acquiring a crucial importance in this field, as well as the recent trend toward the development of comparable implementation of systems, starting from their descriptions in papers and reports, and using the same code that was adopted in previous experiments. Moreover, the public distribution of code and/or problem instances (data sets) is a positive sign that experimentation is moving toward the reproducibility of its results. However, experiments involving several data sets referring to different environments (indoor or outdoor) are still not so common; hence, the difficulties in implementing similar experiments in order to understand, for instance, which parameters influence a robotic system. Also, reports of performance anomalies, which could help in the detection of issues deserving further attention, are rare. Finally, little attention is given to statistical analysis of results, thus compromising the possibility to justify and explain them.

It is true that, dealing with technical artifacts, robotics cannot be straightforwardly assimilated to a traditional scientific field, where experiments are generally conducted for hypotheses-testing purposes and with a strong theoretical background. At the same time, the type of experiment to which autonomous robotics aspires is the controlled one. However, the reality of autonomous robotics practice is rather different from that of the natural sciences, as robotic systems are human-made artifacts. Accordingly, the goal of experimentation is to demonstrate that a given artifact is working in accordance with a reference model (e.g., its design requirements or its expected behavior) and, possibly, that it works better than other similar artifacts with respect to certain metrics, thus making experiments closer to tests typical of engineering disciplines. In this sense, experiments in autonomous robotics have a different aim than experiments in the natural sciences because in the evaluation of technical artifacts, normative claims are introduced (see Section 5 for a more detailed discussion on this). At the same time, the most advanced robot systems are extremely complex, and their behavior is hardly predictable, even by their own designers, especially when considering their interactions with the natural world that are difficult, if not impossible, to model in a fully satisfactory way. In this sense, experiments in autonomous robotics also aim at understanding how these complex systems work and interact with the world. Moreover, they aim at generalizing these results to a broad class of artifacts that have similar features and behaviors with respect to that which is the object of the experimental evaluation.

31.5. Discussion

When Vincenti discusses the anatomy of design knowledge, and in particular the different types of knowledge-generating activities, he introduces experimental engineering research and claims that "a great deal of engineering experiment has a character very much its own" (Vincenti 1990: 232). As Vincenti discusses the peculiarities of experimentation in aeronautical engineering, in the present chapter I analyze the specifics of experimentation in computer engineering. This does not mean that some of the features that characterize its experimental approach are not present in other fields of engineering, only that I do not discuss them here, not even in this section in which I elaborate on the results of the previous cases (software engineering, autonomous robotics) to put forward some more general considerations.

I have already argued elsewhere that, although there is general agreement on the fundamental features of the scientific method, the straightforward application of such a framework to computer engineering can lead to oversimplifications that prevent achieving a complete picture of the practice in this field (Schiaffonati and Verdicchio 2014). Experiments in computer engineering have much in common with those in the natural sciences, but they also present significant differences due to their aim to produce technical artifacts (i.e., computing systems, programs) and the key role played by human actors and organizations (e.g., programmers, project teams, software houses) in such a production. These artifacts are of a special kind: they are malleable and extremely complex, both in their structure and in their interaction with the environment. Moreover, experimentation is not always performed to test theories or in a theory-based framework.

A closer look at the debate on reproducibility in the field may offer some insights. First of all, it presents a concrete example of the tendency to conform to traditional experimental criteria and, at the same time, of the differences computer engineering has in respect to these criteria. Notwithstanding a general call for reproducible results that should guarantee the rigor of experimental results in computer engineering, many authors have failed to justify why computer engineers should aspire to work like other scientists, given that their subject matter differs (Tedre 2011). Moreover, few publications seem to be interested in investigating what reproducibility means in the practice of computer engineering, not to mention the fact that several of them do confuse reproducibility and repeatability.

A notable exception is represented by Vandewalle et al. (2009). This work distinguishes three levels of reproducibility, namely: reproducibility of the *algorithm*, the *code*, and the *data* to conclude that algorithms are generally well described and compared with each other, whereas implementation details (e.g., the code) and data are provided only in a very small number of cases. This concrete analysis is paired with more general considerations on how the relevant research community should be reframed to promote reproducibility, from the publication of negative results to the organization of special sessions in journals and at conferences.

More recently, Collberg and Proebsting (2016), while recognizing reproducibility and repeatability as cornerstones of the scientific method, have focused on repeatability (meant as re-running the researcher's experiment using the same method in the same environment and obtaining the same results). From an analysis of 601 papers from the ACM (Association for Computing Machinery) conferences and journals, they show how repeatability (even a weak version of it) is scarcely achieved in the whole community and propose recommendations to try to avoid these pitfalls. One example is represented by the suggestion of funds for the so-called *repeatability engineering*, including hiring programming staff to document and maintain code, to do release management, and to assist other research groups wanting to repeat published experiments. Given the level of detail of the analysis, it is surprising that this paper offers no insight into why repeatability should be promoted and why it is difficult to achieve, along, for example, the same lines adopted in Drummond (2009), in which a distinction between reproducibility and replicability is advanced, and the mere idea of replicability, as achieved by the replication of the code, is highly criticized as a scientific goal.

There are various limits in taking inspiration from experimental criteria at work in scientific disciplines that have different goals from computing. Moreover, these criteria are not applicable in general, but they need to be specifically characterized in the different fields of computing. Consider the examples of experiments in software engineering (Section 4). Here, experimental results may have been obtained under conditions that are difficult if not impossible to reproduce (e.g., the persons involved are different) and, more importantly, that are not always those in which researchers and practitioners are currently interested (e.g., the software to work on is different). The extreme specialization in software engineering is driven by the requirements to be met, and the peculiarity of their features, together with the pragmatic constraints imposed by the circumstances, compel the designers to quickly make decisions based on common sense and their experience, or to promote further experimental activity to obtain results that are locally relevant. This shows how the human factor plays an important role in these experiments.

The notion of technical artifact is also important to illustrate why adherence to experimental criteria, as conceived in the natural sciences, is not always possible in computer engineering. Consider for instance robotic systems. These are technical artifacts, meaning physical objects with a *technical function* and a *use plan*[4] deliberately designed and made by humans. There is a notable difference between research on naturally occurring phenomena and on artificial (human-made) phenomena (Tedre 2011), and this difference should be taken into account when analyzing the nature of experiments in the engineering sciences in general and in computer engineering in particular. Experiments evaluate technical artifacts according to whether and to what extent the technical function for which they have been designed and built is fulfilled. Hence, normative claims are introduced in engineering experiments. A technical artifact, such as a robotic system, can be "good" or "bad" (with respect to a given technical function reference model), whereas a natural object, such as an electron, whether existing in nature or produced in a laboratory, can be neither "good" nor "bad". In fact, it is analyzed without any reference to its function and use plan, and it is free from any normative constraints regarding its proper working mode. In this context, experiments are performed to test[5] how well an artifact works with respect to a reference model and a metric. Moreover, these experimental results do not necessarily tell anything new about the world; they tell us more about the people that have done the job: think for example of algorithm behavior or usability of software.

Another important element, worth taking into account in this discussion, is related to the low precautions that experimenters in computer engineering take against experimental bias (Fletcher 1995). Although in the traditional protocol of experimental sciences, a researcher should be an outsider with respect to the phenomenon to be explained,[6] it is not clear how much a computer engineer can be an outsider to a phenomenon he/she has created (Tedre 2011). This evidences a common practice: experiments in computer engineering are often conducted by the same researchers who designed the artifact. One could ask why researchers should need to test artifacts that they have themselves designed and created. The answer is that experiments are essential due to the complexity of the artifact and of the environment (including humans) surrounding it. This complexity makes the predictability of the artifact's behavior very difficult, as already explained: the complexity of "the systems built in ECSE and of the underlying models and theories means that experimental implementation is necessary to evaluate the ideas and the models or theories behind them" (National Research Council 1994: 15). This is also extremely important when reaffirming that physics is not a good model to describe the relationship between experimentation and theory in computer engineering. An experiment in ECSE does not verify any prediction from theoretical computer science; and it does not even rely heavily on a model developed by theoreticians. An experiment is usually conducted to validate some informal thesis derived from a computational model not fully specified by a theory.

31.6. Conclusions

In this chapter I discussed experimentation in computer engineering. What emerged is a plurality of approaches to experimentation, ranging from trial experiments to controlled ones. A careful analysis of software engineering and autonomous robotics revealed some peculiarities in experimentation: experiments are performed on technical artifacts rather than on natural objects, and the role of the designers is often overlapped with that of experimenters. These features emerged in the discussion of the two cases, but it is plausible to say that they can be seen as characterizing the whole field of computer engineering, where experiments have different objects of investigation (technical objects rather than natural phenomena) and other aims (to test the proper working of the artifact rather than a theory) with respect to experiments in the sciences. Whether these elements can be extended to engineering sciences in general is not the topic of this chapter, even though some of these elements are general enough to characterize engineering sciences in general.

While recognizing that the traditional notion of controlled experiment, as devised in the natural sciences, is applicable only to a limited number of situations in computer science and engineering, other conceptions of experimentation are worth taking into account, such as the notion of *explorative experiment* (Schiaffonati 2016), elaborated in the field of autonomous robotics as a technological form of experimentation in line with the notion of directly action-guiding experiment (Hansson 2016). Explorative experiments test technical artifacts in a style of investigation not guided by a theoretical background. The traditional experimental criteria (e.g., reproducibility and repeatability) are no longer considered as standard and universally accepted criteria, and the distinction between designers and experimenters is eliminated, given that experiments are performed by the designers themselves. Also, the traditional idea of experimental control, meant to decide which experimental factors are to be manipulated, is revised, as control cannot take place already at the beginning, but only after the artifacts are inserted into their environment. The notion of explorative experiment requires further attention to be fully developed in the specific context of autonomous robotics and to be evaluated in the larger framework of engineering sciences, but it seems promising in illustrating the differences between experimentation in the natural and in the engineering sciences.

Related Chapters

Chapter 1: What Is Engineering? (Carl Mitcham)

Chapter 5: What Is Engineering Science? (Sven Ove Hansson)

Chapter 6: Scientific Methodology in the Engineering Sciences (Mieke Boon)

Chapter 21: Engineering Design (Peter Kroes)

Chapter 23: Design Methods and Validation (Sabine Ammon)

Chapter 54: Engineering and Contemporary Continental Philosophy of Technology (Diane P. Michelfelder)

Further Reading

Tedre, M. (2014). *The Science of Computing*, Boca Raton, FL: CRC Press. (Extensive introduction to the mathematical, engineering, and scientific traditions of computing.)

Notes

1. In this chapter I will mostly use the term "computer science and engineering" to label a field that has both a scientific characterization and a strong engineering component, without entering into the long-standing debate about the nature of this discipline. For the purpose of this chapter, it suffices to say that computing is today an engineering science. Matti Tedre (2014) offers in his book an extended reconstruction of this debate.
2. By technical artifacts in this chapter I adopt the definition provided in Vermaas et al. (2011): technical artifacts are material objects deliberately produced by humans in order to fulfill some practical functions.
3. The mention of computer science as "hacking science" is discussed, among others, in Denning (1980). It appeals to the view of computer science as an art, more than a science, which is more attractive to hackers (the term was originally used to name elite programmers).
4. According to Vermaas et al. (2011) the technical function specifies what the artifact is for, whereas the use plan is a series of actions to be met by the user to ensure that the function of the artifact is realized.
5. An analysis of the differences between tests and experiments in engineering contexts is surely of valuable interest, in particular because tests and experiments can be seen as having fuzzy boundaries. However, this distinction is out of scope in this chapter; here it suffices to say that experiments have a more general goal than tests and that they aim at some form of generalization. For further discussion, see Ammon (2017).
6. This must be true at least in principle, even if then in practice, the prevention of this specific bias is not always taken seriously.

References

Amigoni, F., Bastianelli, E., Berghofer, J., Bonarini, A., Fontana, G., Hochgeschwender, N., Iocchi, L., Kraetzschmar, G., Lima, P., Matteucci, M., Miraldo, P., Nardi, D. and Schiaffonati, V. (2015). Competitions for Benchmarking: Task and Functionality Scoring Complete Performance Assessment. *IEEE Robotics & Automation Magazine*, 22(3), 53–61.

Amigoni, F., Reggiani, M. and Schiaffonati, V. (2009). An Insightful Comparison Between Experiments in Mobile Robotics and in Science. *Autonomous Robots*, 27(4), 313–325.

Amigoni, F., Schiaffonati, V. and Verdicchio, M. (2014). Good Experimental Methodologies for Autonomous Robotics: From Theory to Practice. In F. Amigoni, V. Schiaffonati (eds.), *Methods and Experimental Techniques in Computer Engineering*. Springer Briefs in Applied Sciences and Technology. New York: Springer, pp. 37–53.

Ammon, S. (2017). Why Designing Is Not Experimenting: Design Methods, Epistemic Praxis and Strategies of Knowledge Acquisition in Architecture. *Science and Engineering Ethics*, 30(4), 495–520.

Barni, M., Perez-Gonzalez, F. (2005). Pushing Science Into Signal Processing. *IEEE Signal Processing Magazine*, 120, 119–120.

Barr, R., Golden, B., Kelly, J., Resende, M., Stewart, W. Jr. (1995). Design and Reporting on Computational Experiments With Heuristic Methods. *Journal of Heuristics*, 1, 9–39.

Basili, V.R., Briand, L.C. (eds.) (1996). *Empirical Software Engineering: An International Journal.* Springer.

Bonsignorio, F. and del Pobil, A. (2015). Toward Replicable and Measurable Robotics Research. *IEEE Robotics & Automation Magazine*, 22(3), 32–35.

Bonsignorio, F., Hallam, J. and del Pobil, A. (2018a). *GEM Guidelines.* www.heronrobots.com/EuronGEMSig/downloads/GemSigGuidelinesBeta.pdf. Accessed March 2018.

Bonsignorio, F., Hallam, J. and del Pobil, A. (2018b). *Special Interest Group on Good Experimental Methodologies.* www.heronrobots.com/EuronGEMSig/gem-sig-events. Accessed March 2018.

Boon, M. (2012a). Scientific Concepts in the Engineering Sciences: Epistemic Tools for Creating and Intervening With Phenomena. In U. Feest, F. Steinle (eds.), *Scientific Concepts and Investigative Practice*. Berlin: De Gruyter, pp. 219–224.

Boon, M. (2012b). Understanding Scientific Practices: The Role of Robustness Notions. In L. Soler, E. Trizio, T. Nickles, W. Wimsatt (eds.), *Characterizing the Robustness of Science After the Practical Turn of the Philosophy of Science*, Boston Studies in the Philosophy of Science, Springer, pp. 289–315.

Cepeda Porras, G., Guéhéneuc, Y.G. (2010). An Empirical Study on the Efficiency of Different Design Pattern Representations in UML Class Diagrams. *Empirical Software Engineering*, 15, 493–522.

Collberg, C. and Proebsting, T. (2016). Repeatability in Computer Systems Research. *Communications of the ACM*, 59(3), 62–69.

Denning, P.J. (1980). What Is Experimental Computer Science. *Communications of the ACM*, 23(10), 543–544.

Denning, P.J. (2005). Is Computer Science Science? *Communications of the ACM*, 48(4), 27–31.

Denning, P.J. and Freeman, P. (2009). Computing's Paradigm. *Communications of the ACM*, 52(12), 28–30.

Denning, P.J. et al. (1989). Computing as a Discipline. *Communications of the ACM*, 32(1), 9–23.

Drummond, C. (2009). Replicability Is Not Reproducibility: Nor Is It Good Science. In *Proceedings of the Evaluation Methods for Machine Learning Workshop at the 26th ICML*, Montreal, QC.

Feitelson, D.G. (2006). *Experimental Computer Science: The Need for a Cultural Change.* Unpublished manuscript. www.cs.huji.ac.il/~feit/papers/exp05.pdf. Accessed March 2018.

Feitelson, D.G. (2007). Experimental Computer Science. *Communications of the ACM*, 2(9), 497–502.

Feldman, J.A., Sutherland, W.R. (1979). Rejuvenating Experimental Computer Science. *Communications of the ACM*, 22(9), 497–502.

Fletcher, P. (1995). The Role of Experiments in Computer Science. *Journal of Systems Software,* 30, 161–163.

Forsythe, G.E. (1967). A University's Educational Program in Computer Science. *Communications of the ACM*, 10(1), 3–11.

Freeman, P. (2008). Back to Experimentation. *Communications of the ACM*, 51(1), 21–22.

Hanks, S., Pollack, M. and Cohen, P. (1993). Benchmarks, Test Beds, Controlled Experimentation, and the Design of Agent Architectures. *AI Magazine*, 14(4), 17–42.

Hansson, S.O. (2016). Experiments: Why and How? *Science and Engineering Ethics*, 22, 613–632.

Hartmanis, J. (1994). Turing Award Lecture on Computational Complexity and the Nature of Computer Science. *Communications of the ACM*, 37(10), 37–43.

IEEE (2007). *Proceedings of First International Symposium on Empirical Software Engineering and Measurement* (ESEM 2007). Madrid.

Johnson, D. (2002). A Theoretician's Guide to the Experimental Analysis of Algorithms. In *Proceedings of the 5th and 6th DIMACS Implementation Challenges*, American Mathematical Society, pp. 215–250.

Juristo, N. and Gomez, O. (2012). Replication of Software Engineering Experiments. In B. Mayer, M. Nordio (eds.), *Empirical Software Engineering and Verification*. Oxford: Springer, pp. 60–88.

Juristo, N. and Moreno, A.M. (2001). *Basics of Software Engineering Experimentation*. Dordrecht: Kluwer.

Khalili, H., Levy, L.S. (1978). The Academic Image of Computer Science. *ACM SIGCSE Bulletin*, 10(2), 31–33.

Langley, P. (1988). Machine Learning as an Experimental Science. *Machine Learning*, 3, 5–8.

McCarthy, J. (1962). Computer Programs for Checking Mathematical Proofs. *American Mathematical Society Proceedings of Symposia in Pure Math*, 5.

McCracken, D.D., Denning, P.J., Brandin, D.H. (1979). An ACM Executive Committee Position on the Crisis in Experimental Computer Science. *Communications of the ACM*, 22(9), 503–504.

Michelfelder, D., McCarthy, N., Goldberg, D.E. (eds.) (2013). *Reflections on Practice, Principles, and Process*. Dordrecht: Springer.

Michelfelder, D., Newberry, B. and Zhu, Q. (eds.) (2017). *Philosophy and Engineering. Exploring Boundaries, Expanding Connections*. Dordrecht: Springer.

Moore, J.H. (1985). What Is Computer Ethics? *Metaphilosophy*, 16(4), 266–275.

Morrison, C. and Snodgrass, R. (2011). Computer Science Can Use More Science. *Communications of the ACM*, 54(6), 38–43.

Müller, Vincent C., 2020, Measuring Progress in Robotics: Benchmarking and the 'Measure-Target Confusion'. In Elena Messina, Angel P. del Pobil, and John Hallam (eds.), *Metrics of Sensory Motor Coordination and Integration in Robots and Animals*, Fabio Bonsignorio (Cognitive Systems Monographs 36). Cham: Springer International Publishing, 169–179. doi: 10.1007/978-3-030-14126-4_9

National Research Council (1994). *Academic Careers for Experimental Computer Scientists and Engineers*, Washington, DC: The National Academies Press.

Newell, A. and Simon, H. (1976). Computer Science as Empirical Inquiry: Symbols and Search. *Communications of the ACM*, 19(3), 113–126.

Palvia, P., Mao, E., Salam, A.F. and Soliman, K.F. (2003). Management Information Systems Research: What's There in a Methodology? *Communications of the Association for Information Systems*, 11(16), 1–32.

Schiaffonati, V. (2016). Stretching the Traditional Notion of Experiment in Computing: Explorative Experiments. *Science and Engineering Ethics*, 22(3), 647–665.

Schiaffonati, V. and Verdicchio, M. (2014). Computing and Experiments. *Philosophy & Technology*, 27(3), 359–376.

Schiaffonati, V. and Verdicchio, M. (2016). Rethinking Experiments in a Socio-Technical Perspective: The Case of Software Engineering. *Philosophies*, 1(1), 87–101.

Simon, H. (1969). *The Sciences of the Artificial*, MIT Press.

Snir, M. (2011). Computer and Information Science and Engineering: One Discipline, Many Specialties. *Communications of the ACM*, 54(3), 38–43.

Staples, M. (2014). Critical Rationalism and Engineering: Ontology. *Synthese*, CXCI (10), 2255–2279.

Staples, M. (2015). Critical Rationalism and Engineering: Methodology. *Synthese*, CXCI (1), 337–362.

Tedre, M. (2011). Computing as a Science: A Survey of Competing Viewpoints. *Minds and Machines*, 21, 361–387.

Tedre, M. (2014). *The Science of Computing*, Boca Raton: CRC Press.

Tedre, M. and Denning, P. (2017). Shifting Identities in Computing: From a Useful Tool to a New Method and Theory of Science. In H. Werthner, F. van Harmelen (eds.), *Informatics in the Future*. Dordrecht: Springer, pp. 1–16.

Tichy, W. (1998). Should Computer Scientists Experiment More? *IEEE Computer*, 31(5), 32–40.

Tichy, W., Lokowicz, P., Prechelt, L. and Heinz, E. (1995). Experimental Evaluation in Computer Science: A Quantitative Study. *Journal of Systems and Software*, 28(1), 9–18.

Van de Poel, I., Goldberg, D.E. (eds.) (2010). *Philosophy and Engineering: An Emerging Agenda*. New York: Springer.

Vandewalle, P., Kovacevic, J. and Vetterli, M. (2009). Reproducible Research in Signal Processing. *EEE Signal Processing Magazine*, 37, 37–47.

Vermaas, P., Kroes, P., Van de Poel, I., Franssen, M. and Houkes, W. (2011). *A Philosophy of Technology: From Technical Artefacts to Sociotechnical Systems*. Williston, VT: Morgan & Claypool Publishers.

Vincenti, W. (1990). *What Engineers Know and How They Know It*. Baltimore and London: The John Hopkins University Press.

Zelkowitz, M.V., Wallace, D.R. (1997). Experimental Validation in Software Engineering. *Information and Software Technology*, 39(11), 735–743.

32

ON VERIFICATION AND VALIDATION IN ENGINEERING

Francien Dechesne and Tijn Borghuis

32.1. Introduction

Loosely speaking, engineering is about creating systems that fulfill certain needs. According to Wasson (2006: 705), there are two ways to go about this:

- "Option 1: Employ the hobbyist approach based on the BUILD, TEST, FIX paradigm '*until we get it right*' philosophy.
- Option 2: Do the job RIGHT the first time."

Verification and validation are needed to *do the job right*. They are epistemic engineering activities: they produce knowledge about quality and value. For example, in a commercial setting, stakeholders want to *know* "up front" that resources are invested *efficiently* and *effectively* towards achieving the goal, within constraints of resources (such as time and budget). Verification and validation are reflective processes throughout the systems engineering life cycle that are aimed at providing insights about the effectiveness of the process and the product. With respect to efficiency, there is a trade-off between the cost of these reflective processes, and the value of the insights that they can produce, in terms of creating optimal user value and avoiding costly fixes and retrofits later in the product life cycle.

Verification and validation aim to contribute to the reliability of the systems that are being developed, and to ensure that stakeholder value has been realized by checking specifications and requirements. There can be different purposes for the evaluative actions, depending on the perspectives of different stakeholders, including the developers. Throughout the engineering life cycle there are different combinations of what it is that is evaluated against what (e.g. a mathematical model against a formalized requirement, or a finished product against user satisfaction), for which purpose and how.

We write this contribution coming from a background in mathematical modeling and formal verification in computer science. We observe that much of the theory about verification and validation comes from computer science, where model based engineering is common practice. Models play a central role in the activities of verification and validation, for example as proxies for systems that are not physically available (yet). It is important to always be aware that the purpose of the modeling activity is essential for the modeling choices that are made in relation to the engineering process, product and requirements.

In this chapter, we will first describe the evaluative engineering processes of verification and validation and how they differ from each other (Section 2). We then present categories of methods for verification and/or validation (Section 3). We reflect on the practices by discussing some operational and theoretical challenges and limitations (Section 4) and then conclude with a look on the role of verification and validation in the light of changing engineering practices.

32.2. Verification and Validation in General Engineering

Verification and validation are often contrasted using aphorisms as

> Verification: Are we building the system right?
> Validation: Are we building the right system?

However, giving a more comprehensive account of their differences and resemblances is difficult, not only because of the large variety in definitions of verification and validation that are in use inside and across different branches of engineering, but also because of different ways of structuring engineering processes and the varying levels of detail at which engineering activities are described in the literature.

Verification and validation are mainly associated with the development phase of the system life cycle, as in the standard ISO/IEC 15288:2008,[1] summarized in Table 32.1. We will take this phase as the starting point for our exploration of the contrasts.

32.2.1 Elementary Development

We will first look at the simplest possible case, depicted in Figure 32.1, where development is assumed to be an atomic activity, at the end of which a complete system is available for verification and validation.

Verification (the lower half of the figure) is the confirmation that the system satisfies the requirements imposed at the beginning of the development phase. Validation (the upper half of the figure) is the confirmation that the system meets the needs of the stakeholder(s).

Table 32.1 System life cycle processes according to ISO/IEC 15288

Phase	Activities
Concept	Stakeholder needs
	Explore concepts
	Propose viable solutions
Development	Refine system requirements
	Create solution description
	Build System
	Verify and validate
Production	Produce systems
	Inspect and test
Utilization/Support	Operate system to satisfy needs/Provide sustained system capabilities
Retirement	Store, archive or dispose of system

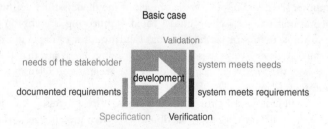

Figure 32.1   Verification and validation for the elementary case

Given this simple picture, we can already see a number of respects in which verification and validation differ.

Documentation. To be able to compare properties or behaviors of the system against requirements that were imposed before the start of the development process, these requirements have to be documented. For this reason verification is preceded by the activity of *specification* in which the requirements for the system are documented for future reference, often according to specific technical standards. For validation, prior documentation is not a necessary condition. Typically, development will somehow be informed by the stakeholders' needs, but in principle one could develop a system without consulting them and, once finished, determine if the system satisfies their needs. This may seem like an academic concern, but apparently stakeholders' needs can go undocumented in actual system development. Wasson (2006, author's note 53.7) mentions that it is good practice to document the mutual understanding of the users' needs because "People tend to change their minds about WHAT they said or intended to say."

Evidence. Both verification and validation require evidence, but the *nature of the evidence* differs. Definitions of verification typically require that evidence be *objective*, e.g. ISO/IEC 2008 (section 4.38): "Confirmation, through the provision of objective evidence, that specified (system) requirements have been fulfilled."

Definitions of validation mostly just require "evidence," like IEEE 1012–2004, (3.1.35): "The process of providing evidence that the software and its associated products satisfy system requirements allocated to software at the end of each life cycle activity, solve the right problem, and satisfy intended use and user needs." Some even leave room for *value* judgments whether or not to accept the obtained result, like ISO/IEEE 2015 (1, Section 6.4.8): "The set of activities ensuring and gaining confidence that a system is able to accomplish its intended use, goals and objectives (i.e., meet stakeholder requirements) in the intended operational environment. The right system was built."

Participants. A further contrast is brought out by PMI (2013) which defines verification as: "The evaluation of whether or not a product, service, or system complies with a regulation, requirement, specification, or imposed condition. It is often an *internal* process," and validation as: "The assurance that a product, service, or system meets the needs of the customer and other identified stakeholders. It often involves acceptance and suitability with *external* customers." Hence, there can be a difference in *who* participates in the activity: the members of engineering team carry out verification by themselves, but validation requires the involvement of stakeholders, and perhaps other external actors (such as experts). Note that for both verification and validation, it is sometimes recommended to assign tasks to engineers not directly involved in the development.

Environment. That verification and validation can differ in *where* the activity takes place is brought out by ISO/IEEE 2015 quoted earlier which states that validation activities should take place in the "intended operational environment" of the system. This puts the stakeholders in the best possible position to judge whether the system is able to accomplish its "intended use, goals and objectives."

Definitions of verification don't specify where the activities have to be carried out. Any environment that allows for the required comparisons between the expected outcome (documented in the specification) and the obtained outcome (performance of the system) will do.

32.2.2 Multi-Stage Development

In the elementary case, we assumed that development is an atomic activity. In fields like software engineering and systems engineering the development phase is considered to consist of multiple discernable stages. The 1985 DOD-STD-2167, Military Standard Defense System Software Development, prescribes a development cycle that includes the phases: Preliminary Design, Detailed Design, Coding and Unit Testing, Integration, and Testing. The 2005 IEEE 1220–2005, Standard for Application and Management of the Systems Engineering Process, which has wider scope, distinguished four consecutive stages: System Definition, Preliminary Design, Detailed Design and FAIT (Fabrication, Assembly, Integration and Test). With this more fine-grained view of the development cycle come strongly structured approaches to engineering design, like the Waterfall model (Royce 1970; Bell and Thayer 1976) and the V-model (Forsberg and Mooz 1991).

To explore the contrast between validation and verification in this context, we start from the following schematic multi-stage development process (see Figure 32.2).

For validation, the presence of multiple stages makes little difference. The aim is still to determine that the system meets the needs of the stakeholders, once it is completed after the final stage (*development_n*). Verification happens at the end of each development stage (*Verification_1–Verification_n*).

As described for the elementary case, verification can be performed only if documented requirements are present as well as an artifact that can be inspected. For the final stage (*development_n*), the artifact is the complete system. In the preceding stages, it is some kind of intermediate product. For the first stage (*development_1*), the initial requirements have been documented (*Specification_1*), before each of the following stages (*development_2–development_n*) there is a dedicated activity (*Specification_2–Specification_n*) to document the requirements, in which the results of the previous stage(s) are taken into account. In this way a chain of alternating development and verification activities is created which links the system back to the initial specification. The IEEE 829–2008 (3.1.54) standard emphasizes this chain building by defining verification in the context of software engineering as:

> [the] process of providing objective evidence that the software and its associated products comply with requirements for all life cycle activities during each life cycle process, satisfy standards, practices, and conventions during life cycle processes, and successfully complete each life cycle activity and satisfy all the criteria for initiating succeeding life cycle activities.

The staged development process brings out two further aspects on which verification and validation can be contrasted.

Scope. Validation always regards the *system as a whole*, but verification can also pertain to *components*, as in IEEE 1012–2004 (3.1.36): "The process of evaluating a system or component to determine whether the products of a given development phase satisfy the conditions imposed at the start of that phase."

Realization. Validation takes place once the system is *realized*. In earlier development stages the system is not yet (fully) realized, and hence not available for inspection. Verification can then be performed on *proxies* such as models and simulations, or a combination of realized components and models (Braspenning 2008). We will return to this topic in Section 3.

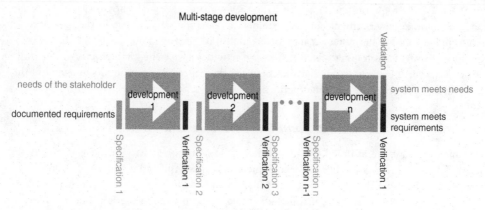

Figure 32.2   Verification and validation for multi-stage development

32.2.3 Systems of Systems

Engineering solutions are becoming increasingly complex, and ways of managing this complexity are developed and studied in the field of systems engineering (INCOSE 2000; SEBoK n.d.; see also Chapter 13, "Systems Engineering as Engineering Philosophy" by Usman Akeel and Sarah Bell, this volume). Where we assumed for the case of elementary development that a single system is developed, it is more common nowadays to think in terms of "systems of systems," or to assume a hierarchy of system levels in the development of complex solutions.

Figure 32.3 shows a schematic multilevel development process, where on a given level multiple systems are developed in parallel, typically involving multiple engineering disciplines. At the next level up, these systems are integrated into larger systems. The procedure is repeated until at the top level, the complete system (of systems) has been realized. This brings out a final further aspect of the contrast between verification and validation.

Level. The activities can differ in the *system level* at which they take place. For verification, the aim is as before to determine that each element of the system meets the requirements of a documented specification. In a system of systems, this implies that verification is performed at *each level* of the system hierarchy. To determine whether the entire system meets the needs of the stakeholders, validation for a system of systems has to occur *only at the top level* of the systems hierarchy (SEBoK, Implementing and Proving the Solution).

32.2.4 Evolutionary Development

From our short exploration of only one phase of the system life cycle, development, it will be clear that large numbers of validation and verification activities need to be performed over the course of an engineering project. To keep track of requirements and verification and validation outcomes across multiple multi-stage, multilevel development stages alone is a complex task that involves the

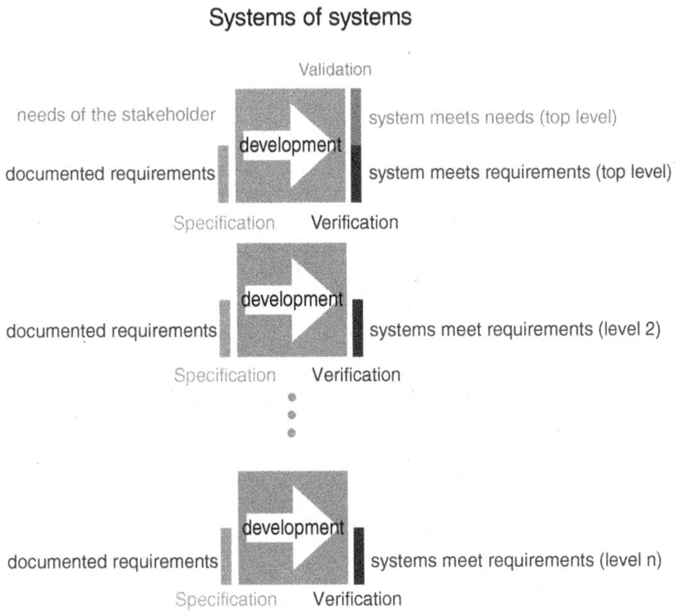

Figure 32.3 Verification and validation for systems of systems

production of large amounts of documentation. The SEBoK (system validation) recommends the development of validation and verification strategy as part of project planning.

In particular in the field of software engineering, these "heavyweight" development processes were criticized as being overly bureaucratic. A number of "lightweight" alternatives evolved which are now commonly referred to as "agile," after the *Manifesto for Agile Software Development* (Beck et al. 2001). They have also found some use in fields outside of software engineering. Agile methods are iterative, incremental and evolutionary. Development takes place in short iterations, called "sprints," each of which results in the release of working pieces of software. Iterations include verification and validation, as described earlier for elementary development. There are documented requirements, which are checked against the produced software, and at the end of a sprint, the working software is validated by demonstrations to stakeholders. Multiple iterations may be needed to complete a product release, but since there are few predetermined dependencies between sprints, there is flexibility to deal with changes in the understanding of the problem and in the stakeholder priorities.

32.3. Verification and Validation Methods

The evaluative engineering activities that constitute verification and validation processes can be summarized under the common term "testing": a test, according to the *Oxford English Dictionary*, is "that by which the existence, quality, or genuineness of anything is or may be determined; 'means of trial.'" In engineering, however, the notion of "test" is often defined a bit more specifically as *measuring* the *response* of a system to inputs *in a controlled environment*:"[t]he act of executing a formal or informal scripted procedure, measuring and recording the data and observations, and comparing to expected results for purposes of evaluating a system's response to a specified stimuli in a prescribed environment with a set of constraints and initial conditions" (Wasson 2006: 693).

Due to the wide range of engineering practices, a *general* standard categorization of verification and validation methods does not seem to exist. Across engineering disciplines, a myriad of different types of evaluations exist that are and can be performed to achieve the goals of verification and validation. It may depend on the discipline and purpose of the engineering system under evaluation which types of tests are performable, most efficient and/or most effective. Some types of tests can be both part of verification and validation, some are specific to one of them.

Verification tasks and actions enable technical programs to evaluate risk assessments, product and process capabilities, compliance with requirements, proof of concept; they include analysis, modeling and simulation, demonstrations, and tests. Validation methods are aimed at establishing confidence that the system developed meets user needs, and include user interviews, prototyping, demonstration, qualification tests, test markets, and field trials (Wasson 2006: 694, 699).

We discuss some relevant classes of evaluation methods, including the activities mentioned as the standard verification methods in the *Systems Engineering Handbook* of the International Council on Systems Engineering: inspection, analysis, demonstration and test (INCOSE 2000: 275). The different methods contribute different types of knowledge to the verification and validation processes.

Inspection is a (visual) examination of the system (hardware, software, associated documentation) for compliance with standards and requirements, without the use of special laboratory equipment or procedures.

Demonstration is a set of test activities with system stimuli selected by the system developer, that may be used to show that system or subsystem responses to stimuli are suitable. It is a qualitative exhibition of functional performance, usually accomplished with no or minimal instrumentation.

Testing (*measurements*) consists of a formal or informal scripted procedure to measure and record outputs of the system in response to specified stimuli in a controlled environment (with a set of constraints and initial conditions). These outcomes are then compared to the expected results. For example, technical performance measurement is the verification of the degree of anticipated and actual achievement of technical parameters. It involves identification of critical parameters and their relationships, and selection of critical requirements and objectives for tracking, before entering the development phase.

Analysis is the use of analytical data or simulations under defined conditions to show theoretical compliance. Analysis is used where testing to realistic conditions is impossible or not cost-effective.

Analytical evaluation methods are distinct from empirical evaluation methods, which perform observations or measurements to compare results with expectations. Rather than on a physical artifact, analytical methods operate on formalized representations of (parts of) the system, and they are aimed at testing for overall logical system conditions such as consistency, completeness and satisfiability of requirements. Analysis also plays an important role in "requirements engineering": requirements analysis is a verification of the system requirements from a system-designer perspective, checking them for adequacy, redundancy and completeness (ISO/IEC/IEEE 29148, 2018).

A specific and pure example of an analytical method is formal verification. This method assesses, usually through *computational proof*, whether a formalized representation of the system satisfies formalized representations of the requirements.

Formal verification seems particularly applicable in software engineering where the system or artifact already is a formal object, so one might expect that no formal representation is necessary. We can express the requirements in principle in the same formalism as the systems behaviors. However in practice, the complexity of software engineering systems requires abstraction into a model of lesser complexity, and in established formalisms for verification in order for computational techniques to be applicable.

From an epistemic point of view, it is important to realize that the translations from the system and requirements into formal representations are steps whose validity must also be critically evaluated to qualify the outcome of a formal verification process. The validity of that outcome, or its meaning in terms of the quality of the actual system, depends highly on the adequacy of the abstractions made in the formalization of the system into a model and its requirements into properties to be checked on the model. Checking the validity of these translations is *not* a process that can itself be formally verified.[2]

The following two "methods" *delegate* the actual verification or validation activities (consisting of inspection, demonstration, measurement and analysis) to previous instances of validation and verification. In that sense, they can be seen as second order: the verification or validation is either derived from the check of a product that is *sufficiently similar*, or it is executed by an *outside authority* against legal or technical standards.

Similarity as a verification or validation method allows to use the verification or validation of a previous system design that is sufficiently similar, as the verification or validation of a newly developed system (Wasson 2006: 377). Similarity as a verification activity then boils down to verifying that the current system is *sufficiently similar* to the verified or validated previous system for the verification or validation to be transferable. Purpose also plays a role here, indirectly through the requirements (verification) that also must be similar for system_old and system_new, for validation in that the stakeholder_old that was satisfied with system_old should also be satisfied with system_new, and vice versa. Put differently, not only the systems but also problems should be sufficiently similar. This method therefore depends on providing a valid measurement for similarity of systems, and an adequate notion of similarity. It is therefore dependent on context whether this method is considered to be sufficient.

Certification is also mentioned as a verification technique (INCOSE 2000: 275), typically for commercial contexts (think of CS certification in Europe and UL certification in North America). This method refers to the checking of compliance with legal and/or industrial standards by an outside authority. Here it should be up to the authority to determine how the requirements are to be verified—through inspection, analysis, demonstration, testing or otherwise (see Chapter 42, "Standards in Engineering" by Paul B. Thompson, this volume).

32.3.1 The Use of Simulation Models in Validation and Verification

An interesting analytical engineering practice relevant for verification and validation is the use of computational simulation models. Physical systems or products are not always accessible or controllable enough to provide actual data for testing, inspection or demonstration for verification purposes. In the development phase, engineers employ models and simulations to gain insights into the system interactions for a prescribed set of operating scenarios and conditions. With the abundance of computational power and data to build models effectively simulating the behavior of complex systems or the postulated behavior of physical products still under construction, simulation models have become richer and more adequate for the (tentative) evaluation of system behavior; such models make "verification by simulation" a possible method (Wasson 2006: 651, 699).

Note that there is an important difference between simulation models and the models used in the formal verification process described earlier. Computational simulation models can be used to generate and inspect a finite set of possible behaviors of the system, while formal models are designed to generate and inspect *all* possible executions of the system (see Chapter 30, "Computer Simulations" by Hildrun Lampe, this volume).

The use of simulation models as proxies for the realized system raises interesting philosophical questions. As in the case of formal verification, the validation of the model with respect to the realized system (*Is it the right model?*) becomes a crucial part in the process. Is the simulation *sufficiently similar* in the relevant characteristics of its behavior to be used for validating user needs?[3]

32.4. Challenges and Limitations

Verification and validation are mechanisms to test and confirm the "rightness" of a system, in different ways as explained earlier in the chapter. Of course, there are limitations and challenges, on the operational as well as on the fundamental level.

32.4.1 Operational Challenges and Limitations

The activities of verification and validation are subject to a number of operational issues. In commercial systems engineering, there is the prominent trade-off for the system acquirer to pay for verification and validation on the one hand and possibly investing in more functionalities on the other. Note that in principle, the need for verification and validation increases only when choosing for extra functionalities. Furthermore, it may be hard to strike a balance between the necessary independence of the verifying/validating team and the knowledge they have available of how the systems work or should work (Wasson 2006: 704).

Verification activities are testing systems against the requirements. This makes the process dependent on the quality of formulation of the requirements. The requirements need to satisfy certain conditions and quality criteria themselves. The INCOSE Systems Engineering handbook lists a number of these conditions, such as clarity (of terminology and formulation), consistency with standards, achievability, completeness and necessity (INCOSE 2000: 223).

In verification, requirements will always contain underlying assumptions about the context of use of the system that is developed. Many of them will be implicit; for instance, the assumption that all staff necessary to operate the system are present. If such an assumption is violated in the context of use (as happened with the luggage handling system at Heathrow's Terminal 5),[4] a situation may arise where validation of the system failed (the system did not satisfy stakeholder needs) while verification succeeded (the system was built according to the requirements).

It is significant to note that the shift towards evolutionary development paradigms has an impact on all of these operational challenges to verification and validation. In Agile methods, the balance of the trade-off tips towards more functionalities, and with respect to the quality criteria on the requirements, the standard to strive for in Agile is "just barely good enough" (Ambler n.d.).

32.4.2 Fundamental Challenges and Limitations

Validation assumes some kind of stability in the needs of the stakeholders. This assumption is not unproblematic. Under the aspect Documentation in Section 2, we already quoted Wasson's personal note that for validation, documenting user needs is good practice, because people tend to change their mind about what they said or intended to say. The fact that they do may be due to more than just carelessness or inattention. Steve Jobs famously held the opinion that users are not sufficiently aware of their needs: "A lot of times, people don't know what they want until you show it to them."[5] The possibility of experiencing the artifact at the end of the development process may bring about changes in the very user needs the development was intended to satisfy.

In general, the relation between development contexts and users' practice is complex and unpredictable: technologies are frequently used in different ways from what were initially intended. According to Don Ihde's phenomenological analyses of human-technology relations this ambiguity is fundamental. Technology does not have an essence or basic meaning apart from the use contexts it enters into, what he calls "multi-stable" (Ihde 1977, 1993). Hence, effects of mediation on stakeholder needs cannot be avoided, but some development processes provide greater flexibility than others: evolutionary methods, such as agile development, that go through a series of working versions of the product can adapt to changing stakeholder needs more easily than strongly structured approaches, such as the Waterfall and V-model (cf. Section 2).

In verification, the evidence that a system satisfies a requirement has to be objective. Anticipating on the obligation to provide objective evidence can lead to "expressibility bias," where requirements that can be stated in commonly used mathematical terms and parameters for which measurement procedures are readily available are preferred over requirements for which these terms and procedures would have to be researched or developed. When the easily expressible requirements get most of the attention during development, verification can become "myopic," especially in multi-stage or multi-level development, where there can be long chains of specification and verification activities before the completed system is available for validation.

A limitation in the case of formal verification already touched upon in Section 3 is that the computational complexity of checking all possible behaviors of a system against the requirements can be prohibitive. Although possible in principle, completing the process would take tens, hundreds or even thousands of years in such cases, and can hence not contribute to the development of the system. One then has to settle for the verification of weaker statements (pertaining to a simplified model of the system and/or a subset of the requirements) that can be performed within the development timeframe.

Verification and validation are processes that aim to contribute to the reliability of the systems that are being developed and to ensure that stakeholder value has been realized. This presupposes that the outcomes of these processes can be trusted. There are two fundamental challenges here: the complexity of the processes, and the objectivity of the persons carrying them out.

Regarding the latter point, we already mentioned the practice to let verification/validation be carried out by a separate sub-team, or to delegate it to an outside authority. However, the challenge is to avoid not just conflicts of interest, but also the systematic errors introduced by human cognitive biases. Mohanani et al. (2018) summarized the small body of literature on the effects of these biases in software engineering, mapping these biases to knowledge areas of the SWEBoK (n.d.). Confirmation bias is related to software testing, where it manifests itself as 'positive test bias': "developers tend to test their program with data that is consistent with 'the way that the program is supposed to work', and not with any inconsistent data." On the basis of only positive tests a developer would not be justified in claiming that the program satisfies a certain requirement.

When the processes involve software, as in the case of formal verification, trusting the outcome of the process can boil down to trusting a system that may be as complex as the system under development. However, there are approaches to formal verification that allow for a form of delegation in which the proof constructed by a complex program can be checked by a much simpler independent program that a skeptical user could easily write him/herself. Such approaches are said to satisfy the De Bruijn criterion, after the Dutch mathematician N.G. de Bruijn, who was the first to notice the possibility of reducing the problem of trusting the outcome of a complex computational process to trusting the outcome of a much simpler process (Barendregt and Geuvers 2001).

32.5. Conclusions and Looking Ahead

In engineering standards and literature, the evaluative activities of verification and validation are identified as relevant throughout the system life cycle. We have distinguished the two activities in different aspects in the development phase of the life cycle: documentation, evidence, participants, environment, scope, realization and level. Standard types of methods are inspection, demonstration, testing and analysis, and delegated approaches are similarity and certification.

We have not found a well-established body of literature and active research *across* the engineering disciplines on the *theory* of verification and validation. Theory seems to be developed mostly in the context of software engineering, including requirements engineering, and systems engineering. We speculate that the sense of necessity for such theory is strongest for software engineering for the following reasons. In software engineering, the design space is bounded by logic, rather than the laws of physics. This makes it much larger, requirements have to be more comprehensive, and eliciting needs is more difficult. In physical engineering, the engineers have a shared understanding of the laws of physics, and a catalog of standard solutions within their disciplines. For software development and virtual engineering this understanding is *under construction*. Since the "systems of systems" that are the subject of systems engineering typically involve software, systems engineering will have inherited some of its need to develop theories of validation and verification from software engineering. A further reason for the existence of a body of literature on verification and validation in this field is its inherently wide scope, the definition of "engineered system" includes sociotechnical elements (SEBoK, glossary, engineered system).

Due to developments such as the virtualization of engineering (cf. Chapter 55, "Reimagining the Future of Engineering" by Neelke Doorn, Diane P. Michelfelder, et al., this volume), we expect the use of proxies, such as simulation models, in verification and validation to increase. This will lead to techniques that make a wider range of problems amenable to computational verification. On the other hand, while systems become more and more "cyberphysical" where the human is considered to be a part of the system, models and theories from the social sciences and humanities, such as psychology, sociology and ethics, may lead to different types of verification and also new kinds of validational proxies (Chapter 55, this volume).

Acknowledgements

We would like to dedicate this chapter to Kees van Overveld (1957–2014), as his approach to engineering design and modeling has been a main inspiration for us in writing this chapter.

Related Chapters

Chapter 13: Systems Engineering as Engineering Philosophy (Usman Akeel and Sarah Bell)
Chapter 30: Computer Simulations (Hildrun Lampe)

Chapter 42: Standards in Engineering (Paul B. Thompson)

Chapter 55: Reimagining the Future of Engineering (Neelke Doorn, Diane P. Michelfelder, et al.)

Further Reading

Magnani, Lorenzo and Bertolotti, Tommaso (eds.) (2017). *Handbook of Model-Based Science*. Springer. (Addresses novel issues surrounding the use of computational models for doing science, in particular the use of simulations, and how this affects processes of verification and validation—cf. Part A, chapter 5.)

SEBoK Guide to the Systems Engineering Body of Knowledge (v.2.0) www.sebokwiki.org/wiki/Guide_to_the_Systems_Engineering_Body_of_Knowledge_(SEBoK) (Compendium of the key knowledge sources and references of systems engineering organized and explained to assist a wide variety of users. It is a living product, accepting community input continuously.)

Wasson, Charles S. (2006). *System Analysis, Design, and Development—Concepts, Principles, and Practices*. Hoboken, NJ: John Wiley & Sons, Inc. (A classic practitioners textbook written to advance the state of practice in system engineering.)

Notes

1. ISO/IEC 15288:2008 has been revised by ISO/IEC/IEEE 15288 (2015), which no longer prescribes a system life cycle in view of the variety of life cycle models used in practice. Instead, it provides a set of *life cycle processes* which can be used in defining a system's life cycle.
2. A very interesting observation is that formal verification methods may turn into an engineering rather than an evaluative activity, as the proof of a requirement may produce the procedure (algorithm) to realize the requirement (Paulin-Mohring 1989; Ramadge and Wonham 1987).
3. In fact, Winsberg has argued that in the context of computational simulation validation and verification are too much interwoven to be distinguishable (Winsberg 2010).
4. One of the factors contributing to the initial failure of the new luggage handling system was that airport staff arriving for work could not find their way to the staff car park due to unclear signage. *What did go wrong at Terminal 5?* BBC, 27 March 2008, http://news.bbc.co.uk/2/hi/uk_news/7318568.stm
5. BusinessWeek, 25 May 1998: www.businessweek.com/1998/21/b3579165.htm.

References

Ambler, Scott W. (n.d.). *Just Barely Good Enough Models and Documents: An Agile Best Practice*. www.agilemodeling.com/essays/barelyGoodEnough.html

Barendregt, H. and Geuvers, H. (2001). Proof Assistants Using Dependent Type Systems. In A. Robinson and A. Voronkov (eds.), *Handbook of Automated Reasoning*, Vol. 2. Amsterdam: Elsevier, Chapter 18, pp. 1149–1238.

Beck, Kent, Grenning, James, Martin, Robert C., Beedle, Mike, Highsmith, Jim, Mellor, Steve, van Bennekum, Arie, Hunt, Andrew, Schwaber, Ken, Cockburn, Alistair, Jeffries, Ron, Sutherland, Jeff, Cunningham, Ward, Kern, Jon, Thomas, Dave, Fowler, Martin and Marick, Brian (2001). *Manifesto for Agile Software Development*. http://agilemanifesto.org/ *Agile Alliance*. Accessed March 17, 2019.

Bell, Thomas E. and Thayer, T.A. (1976). Software Requirements: Are They Really a Problem? *Proceedings of the 2nd international conference on Software engineering*. IEEE Computer Society Press, Stanford, CA, USA.

Braspenning, N.C.W.M. (2008). *Model-based Integration and Testing of High-tech Multi-disciplinary Systems*. IPA-dissertation series 2008–05, Eindhoven: Technische Universiteit Eindhoven. doi:10.6100/IR632482

DOD-STD-2167 (1985). Military Standard Defense System Software Development.

Forsberg, Kevin and Mooz, Harold (1991). The Relationship of System Engineering to the Project Cycle. In *Proceedings of the First Annual Symposium of National Council on System Engineering*, October, pp. 57–65.

IEEE (2004). *IEEE Standard for Software Verification and Validation*. Institute of Electrical and Electronics Engineers (IEEE) Standards Association: IEEE 1012–2004.

Ihde, Don (1977). *Experimental Phenomenology: An Introduction*. New York: Putnam.

Ihde, Don (1993). *Postphenomenology*. Evanston: Northwestern University Press.

INCOSE (2000). International Council on Systems Engineering, Systems Engineering Handbook Version 2.0, July 2000.

ISO/IEC/IEEE 15288 (2015). *Systems and Software Engineering—System Life Cycle Processes*. Geneva, Switzerland: International Organization for Standardization (ISO)/International Electrotechnical Commission (IEC)/Institute of Electrical and Electronics Engineers (IEEE).

ISO/IEC/IEEE 29148 (2018). *Systems and Software Engineering—Life Cycle Processes—Requirements Engineering.* Geneva, Switzerland: International Organization for Standardization (ISO)/International Electrotechnical Commission (IEC)/Institute of Electrical and Electronics Engineers (IEEE).

Mohanani, R., Salman, I., Turhan, B., Rodriguez, P. and Ralph, P. (2018). *Cognitive Biases in Software Engineering: A Systematic Mapping Study.* https://arxiv.org/abs/1707.03869

Paulin-Mohring, Christine (1989). *Extraction de programmes dans le Calcul des Constructions* [Program Extraction in the Calculus of Constructions]. Paris Diderot University, France. https://tel.archives-ouvertes.fr/tel-00431825

PMI (2013). *A Guide to the Project Management Body of Knowledge (PMBOK® Guide).* 5th ed. Newtown Square, PA: Project Management Institute (PMI).

Ramadge, P. and Wonham, M. (1987). Supervisory Control of a Class of Discrete Event Processes. *SIAM J Control Optim*, 25(1), 206–230.

Royce, Winston (1970). Managing the Development of Large Software Systems. *Proceedings of IEEE WESCON*, 26, 1–9, August.

SEBoK (n.d.). Guide to the Systems Engineering Body of Knowledge (v.2.0). www.sebokwiki.org/wiki/Guide_to_the_Systems_Engineering_Body_of_Knowledge_(SEBoK)

SWEBoK (n.d.). Guide to the Software Engineering Body of Knowledge (v. 3.0), IEEE. www.computer.org/education/bodies-of-knowledge/software-engineering

Wasson, Charles S. (2006). *System Analysis, Design, and Development—Concepts, Principles, and Practices.* Hoboken, NJ: John Wiley & Sons, Inc.

Winsberg, E. (2010). *Science in the Age of Computer Simulation.* Chicago, IL: University of Chicago Press.

PART VI

Values in Engineering

33

VALUES IN RISK AND SAFETY ASSESSMENT

Niklas Möller

33.1. Introduction

From the cave paintings of prehistoric times to detailed accounts of historians, it is evident that in its broadest form, humans have been engaged in risk and safety assessment right from the start. This is hardly surprising, since successfully avoiding danger in any other way than by sheer luck means that some sort of assessment of the risks has to be made, whether the issue is trusting a tree branch enough to climb it, or constructing the roof of a new dwelling.

In the more specific sense in which the term is used in professional contexts, however, risk and safety assessment is a phenomenon of modernity. Here, a risk and safety assessment[1] is a structured activity conducted by scientists and other experts of the field in question. The aim of a risk assessment is to give an answer to the question of what may go wrong, how likely it is to go wrong and what the consequences are if something would, in fact, go wrong (Apostolakis 2004; Kaplan and Garrick 1981).

Risk assessment on this understanding should be distinguished from the *whole* process of handling risk and safety. Let 'risk analysis' refer to the entire process from the start of an investigation of the risk involved in a specified system or activity (such as a power plant or airplane flight) or substance (such as a new drug), to making a decision about this risk. On the traditional understanding, risk analysis can be divided into two distinct stages (NRC 1983; EC 2003). In the first stage, we investigate the world in order to estimate, to the best of our ability, what the risks *are*. This is the risk assessment stage. In a second stage, we use this data as input for our decision what to do: fundamentally whether to accept these risks or not. This stage is typically called *risk management* (Figure 33.1).

The aim of this chapter is to discuss values in risk assessment. Do values enter into risk assessment, and if so, in which way? A traditional answer to this question is in the negative. Risk assessment is value-free. The general strategy in arguing for this claim is to utilize the division of labour between the two stages of risk analysis. Admittedly, a decision to act—in this case, whether to accept or allow an activity, system or substance—is always based on values. Deciding that a chemical plant is safe enough, or that a medicine has a sufficiently low risk, will always be a value decision. Consequently, risk analysis on the aforementioned interpretation cannot be completely value-free. But that does not necessarily mean that *both* stages of risk analysis are value-laden. On the traditional argument against values in risk assessment, only the risk management stage is fraught with values. The risk assessment stage itself is value-free.

This traditional answer typically relies on the idea that risk assessment is the scientific stage of risk analysis, and that since science is value-free, so is risk assessment. In what follows, we will first

Figure 33.1   The traditional two-stage picture of risk analysis

discuss the idea that risk assessment is a science. We will then apply several objections to the idea that science is value-free to the risk assessment context. Lastly, we will discuss one source of values that is quite specific to the very role of risk assessment in risk analysis, before ending with some concluding thoughts.

33.2. Risk Assessment and Science

A risk assessment is a structured activity where experts in the field in question estimate the potential threats and dangers involved in a societal activity, system or substance. In essence, a risk assessment attempts to answer, for a given situation, the question of what may go wrong, and if so, what the consequences would be (Apostolakis 2004; Kaplan and Garrick 1981). While there is a continuum of approaches to risk assessment, two broad categories are typically given. If the assessment involves explicit estimates of probabilities for harmful events as well as specifications of the magnitude of such harms, the risk assessment is called *quantitative risk assessment* (Aven 2011); if the probabilities and harms are not explicitly assessed, but rather given a qualitative estimate, such as 'high', 'medium', and 'low', the assessment is called *qualitative risk assessment* (Tiusanen 2018).

The grounding thought for risk assessment as value-free is that it is a purely scientific endeavour, an investigation into properties of the world. In other words, in risk assessment, we are searching for facts of the world in the most structured and reliable way available to us. We are looking at bridges and power plants and complex new medicines, and we attempt our best at predicting how they may harm us. Surely, such investigations are fraught with uncertainties. When complex systems are built and new substances developed, even the best team of experts often lack a certain answer to every negative thing that may happen, how likely it is that it will happen and what exactly the consequences would then amount to. Still, the systematic investigations of risk assessment are performed in accordance with the best scientific methods and standards we have developed. As such, the argument goes, it should be considered a scientific process (Hansson and Aven 2014).

Whether risk assessment is a science naturally depends on what should be understood by 'science'. In a recent defence of the scientific status of risk assessment, Sven Ove Hansson and Terje Aven utilise a broad understanding of science as

> the practice that provides us with the most reliable (i.e., epistemically most warranted) state-
> ments that can be made, at the time being, on subject matter covered by the community of
> knowledge disciplines, i.e., on nature, ourselves as human beings, our societies, our physical
> constructions, and our thought constructions (Hansson and Aven 2014:
> 1175. Cf. also Hansson 2013).

This definition explicitly aims to cover not only natural science, but also all other academic fields, including the humanities, thus coming much closer to the German 'Wissenschaft' (with analogues in other Germanic languages) than the more restricted English usage of the term 'science' (Hansson and Aven 2014: 1174).[2] Given this definition, Hansson and Aven argue, risk assessment should indeed be considered a science, since risk assessment provides the most reliable statements of nature and our constructions that can be made, also when those statements are fraught with uncertainty.[3] Indeed, the characterizations of these uncertainties in risk assessment represent the most well-developed models and other tools we have at our disposal (Hansson and Aven 2014: 1180).

The conclusion that risk assessment is scientific is indeed controversial, but for our purposes in this chapter, it provides a useful starting point, since even granted this assumption, there are potential 'intrusions' of values, as will soon become clear.[4]

33.3. Science and Values

The argument that science is free from values typically starts out with a central distinction: the distinction between epistemic and non-epistemic values. The claim that science is value-free should be understood not as the claim that science is independent on *any* kind of value, but as the claim that science is independent from *non-epistemic values*.

Epistemic values are those which help us with central scientific tasks such as to formulate, reject or accept a hypothesis, choose between rival theories, and assess the virtues of a suggested methodology. Although the exact list is debated, typical values often mentioned include reliability, testability, generality and simplicity (McMullin 1982; Kuhn 1962; Lakatos and Musgrave 1970). Since whether or not to accept a hypothesis given a certain set of data is ultimately a matter of which standards of inference we have, we need epistemic values in science. Consequently, science is not completely free from values. Still, epistemic values on this picture are a science-internal affair. They are the kind of values that the scientific experts are competent to apply in their scientific enterprise. They relate only to what we have *reasons to believe*, to the central scientific task of establishing sufficiently secure knowledge. As such, epistemic values are not the kind of values that the 'value-free ideal' refers to.

The values *not* part of science on this picture are the *non-epistemic values*, particularly practical values such as political or ethical values, or prudential values such as well-being. These values have no role in establishing scientific knowledge, proponents of the value-free ideal argue. If ethical or political values factored into what the scientists took as corroborated in their studies—whether the objects of study are electrons in the lab or people in society—the result would not, and should not, be counted as scientific knowledge.

33.4. External Influences

One potential objection to science as free from (non-epistemic)[5] values comes from the presence of seemingly obvious counter-examples. To start with, the history of science contains many examples of then-respectable sciences which have been exposed as value-laden. Phrenology, popular as a science in the 1800s, was famously based on the idea that measuring the human skull could determine the character and temperament of people, and many proponents of the discipline conveniently concluded that European superiority over other 'inferior races' was demonstrated by their difference in skull sizes to the (ideal) Caucasian skull. But these 'demonstrations' were in fact based on explicit and implicit values, and the phrenology predictions that did turn out to have some correspondence to facts were mere 'very lucky guesses' (Young 1970: 12; cf. also Cooter 1985). Similarly, there have been many cases of scientific hoaxes and cases where scientists have published 'scientific results' which turned out to be false (Sokal 2010). Often, these results have been motivated by values, such as ideological values (e.g. Holocaust denial) and instrumental values (e.g. the scientist's attempt to receive fame and fortune—or at least funding).

At this stage it is important to keep two distinctions in mind. The first distinction is between *actual* scientific practice and the *norm or ideal* of scientific practice. The second is the distinction between *internal* and *external* value-ladenness. Starting with the former distinction, few, if any, would deny that as a matter of empirical fact, there have been many cases where what has been labelled as science has been value-laden in the way described in the previous paragraph. But that does not show that science as a norm or (reachable) ideal is value-laden. Phrenology was not exactly a science in this normative sense since it did not provide us with the most reliable statements that could be made on the subject matter (to use Hansson and Aven's definition) or properly utilize the scientific method of hypothesis testing (to use a common alternative). It did, in other words, not properly utilize the *epistemic values* central to science. As long as it is *possible* to conduct scientific investigation in a value-free way, cases of misconduct are neither here nor there.

The second distinction aims to capture the difference between the process and content of science—scientific method and knowledge—with the external context in which scientific is conducted. As with any activity, scientific practice is part of society. That means that there are many practical values—moral, political and instrumental—which play a part in the scientific enterprise. Societal decision-makers select which research projects get funded, and society (if not all scientists) puts ethical restrictions on how scientific investigations may be conducted. Likewise, there may be personal values involved in the decisions of an individual scientist to study a particular subject matter, as well as ethical values in how science may be conducted (e.g. related to living subjects). But these are all *external values*, proponents of the value-free ideal of science argue. The actual standards for accepting a hypothesis as scientific knowledge may—and should—still be free from values.

Of course, it is often relevant to ask who is funding a particular scientific activity, since external values may put (explicit or implicit) pressure on experts to modify their scientific standards (Ziman 1996). The scandals of tobacco company-funded research are perhaps among the most famous examples. A more recent case is asbestos, which is interesting also since the external pressure here is displayed in both directions. While historically, it has come from the asbestos industry in the 'default' form of risk denial, recent studies clearly indicate a reversed phenomenon: experts hired by lawyers attempting to receive compensation have made biased assessments of the lungs of the claimants, judging them to be harmed by asbestos to a much larger degree than subsequent independent assessments (Gitlin et al. 2004; Janower and Berlin 2004).[6]

The danger of external pressure is particularly relevant in an applied science such as risk assessment, which is often commissioned by stakeholders with a particular agenda. Consequently, in *practice*, societal mechanisms such as regulations and the possibility of independent assessments are vital, especially in applied sciences such as risk assessment. But as a *principled objection* to the value-free ideal of science, the existence (and potential pressure) of external values does not suffice.[7]

33.5. Value Boundary

Whereas the objection from both revealed cases of scientific misconduct and the psychological plausibility of scientific biases from external values should cure us from too naive a notion of scientific practices, more is needed for a principled threat to the ideal of science as value-free.[8] In this section, we will look, in the context of risk assessment, at the first of two principled objections to the idea of science as value-free, the *value boundary objection*.

The central idea that epistemic values can (and should) be included in science whereas non-epistemic values can (and should) be excluded is premised on the possibility of properly distinguishing between the two. However, it has been argued that the list of values dubbed 'epistemic' is both arbitrary and contains some values that are directly questionable.[9] Looking at historical cases, religious and other cultural non-epistemic values have been influencing what has been accepted and rejected as hypotheses in physics and other sciences (Rooney 1992). For example, a male perspective and often also assumptions of male superiority have influenced the biological sciences, such as interpreting male behaviour as active and female as passive on very unclear evidential bias (Hubbard 1982), and assuming male-female differences in biological theories of higher cognitive functioning (Fausto-Sterling 1985).

While historical cases may—at least to an extent—be met with the aforementioned strategy of distinguishing actual misconduct from the normative ideal of science, in the case of risk assessment, there is a strong case to be made that there are underlying values intrinsic to the process which are not merely epistemic. Indeed, we need not go beyond the very core concepts of risk and safety to find a strong indication of this. Whereas there are parts of risk assessment that consist solely of 'classic' scientific activities such as investigating potentially toxicological effects of a certain medical substance, the aim of risk and safety assessment is, as the name suggests, to assess what the *risks* are. Even under the presumption that there are purely epistemic values guiding potential toxicological effects of inhaling a certain chemical substance or a probability distribution for structural damage of a bridge given an assumed load distribution, the inference to assessments of risk and safety are far from given by these results. Crudely put, although we all have a broad notion of what a risk may be, there is no uncontroversial yet precise conception for scientists to adhere to. And when scrutinizing the candidates, we note that several potential non-epistemic considerations enter into the deliberation (Hansson 2010; Möller 2012).

To see this, we need go no further than to the classic notion of risk as the 'expected value of harm' (Möller et al. 2006; Rechard 1999). Here, risk is understood as a probabilistic quantitative measure, where the severity of a possible harmful outcome is weighted by its probability. To take a simple example: if one harmful outcome of an activity would result in the death of 100 people, and the probability for that event is 0.01, then the expected value of harm, measured in loss of life, is 1 person. Already in this simple example, however, the potential problem is evident. 'Harm' is, in itself, not a well-defined scientific concept. Looking only at harm to humans, it seems clear that what is harmful depends on what we value, in a non-epistemic sense of value. In the example, we used the common measure of a loss of human life. But a harmful consequence may not necessary mean death: both losing an arm and becoming allergic to a substance are certainly harms even if neither is deadly (Möller 2012). Measures such as QALY (quality-adjusted life-years) try to accommodate many of these additional complexities (Nord 1999). Still, this measure is not uncontroversial. And even if it were, its status as non-epistemic seems to retain: how we should value and compare different harms is a normative question over and above that which is settled by epistemic values.

In addition to harm, other aspects have been pointed to as adding a non-epistemic value dimension to risk assessment. Here, the expectation value of harm conception of risk has often been the target. It has been debated, for example, whether the uncertainty that is an inherent aspect of risk considerations may be sufficiently captured by the probability notion underlying the expected value conception of risk. Defenders argue that this uncertainty aspect, however it should be understood, is epistemic in nature and thus does not constitute an objection to the ideal of value-free science. Others have added that it is in any case not the role of risk assessment to take a stand in what constitutes a risk or what is safe, but to deliver the best statements of facts possible (Möller 2012; Hansson and Aven 2014).

The success of these strategies depends to a large extent on how to evaluate *the argument from inductive risk* and *the handover dilemma*, to which we now turn.

33.6. Inductive Risk

The next objection to the idea of science as value-free goes under many names, but a common contemporary label is the *argument from inductive risk*, a term coined by Carl Hempel (1965) and later put back on the terminological map by Heather Douglas (2000).[10] The argument from inductive risk is grounded in the basic feature of empirical science that there is always in principle a chance that the researcher is wrong in accepting (or rejecting) a scientific hypothesis. (The term 'inductive risk' refers directly to that chance—or rather risk—of error.) Since the practical consequences of error may vary greatly, it does not seem to be merely a matter of internal epistemic values to set the bar for acceptance or rejection of a hypothesis, in particular if the consequences of error may be significant. This problem is further highlighted by the empirical fact that scientific contexts vary greatly within as well as between different scientific disciplines, and that there is no 'fixed epistemic level' for when a hypothesis should be accepted or rejected (Wilholt 2013). As Douglas puts it: 'Where non-epistemic consequences follow from error, non-epistemic values are essential for deciding which inductive risks we should accept, or which choice we should make' (Douglas 2000: 565).

Proponents of the value-free ideal in science have not denied the phenomenon of inductive risk (which would be futile). Instead, the main strategy has been to insist that the practical consequences of error neither *should* nor *need to* play any role in deciding which hypothesis to accept or reject. The role of science is to build on our base of secure knowledge. Consequently, the epistemic bar should always be high for accepting a hypothesis. How the scientific results can be used for good or bad may affect practical issues such as whether the knowledge should be freely available or (as with some potentially dangerous results in atomic physics) restricted; but it should not in any way affect our evidentiary standards, which should stay purely epistemic (Levi 1960; Laudan 1984).

This defence relies on an asymmetry between two types of error: false positives (type I error) and false negatives (type II error). A false positive is the error of inferring that there is an effect—typically, accepting a hypothesis—when in fact there is no effect, whereas a false negative is the error of inferring that there is no effect when in fact there is. For example, if our hypothesis is that a certain alloy may rust when exposed to oxygen and moisture, wrongly accepting the hypothesis would constitute a false positive, whereas wrongly rejecting the hypothesis constitutes a false negative.

Using this terminology, the main defence of the value-free ideal translates to prioritizing the avoidance of false positives. In advancing science, we should at virtually any cost avoid situations where we *wrongly* take something to be scientific knowledge. Consequently, we must avoid endorsing a phenomenon as real when in fact it is not. Not endorsing a phenomenon when in fact it is real, on the other hand, is an unfortunate but unavoidable consequence of our epistemic priorities.

The argument for keeping a 'high entry fee' to the scientific knowledge base seems reasonable indeed. Science enjoys a high status in society, and if something is 'scientifically demonstrated',[11] we expect to be able to rely on it as a fact. Consequently, scientists have increasingly been participating with their knowledge in societal decision-making processes (Lentsch and Weingart 2009). If we were to accept that the epistemic standards should be a function of the practical consequences of the scientific research, we could no longer be as secure in trusting the scientific corpus as containing facts.

For some scientific activities, the demand for high epistemic standards to avoid false positives may be perfectly unproblematic. In particle physics, for example, accepting a result that is later proven to be false would arguably have little effect on society, beyond a potential waste of money for embarking on a wrong research direction until the mistake is revealed (Douglas 2016: 615). The very motivation for risk assessment, however, is fundamentally practical. In risk assessment, our investigation of the potential effects of a substance, system or activity is not motivated by sheer scientific curiosity,[12] but by the societal goal of avoiding harm. Consequently, it has been argued that even if in basic (or 'pure') science, conservative epistemic standards to avoid false positives are justified, in an applied science such as risk assessment, avoiding false negatives might be equally, if not more, important (Hansson 2007; Hansson and Aven 2014).

The reason for this is that in risk assessment, the attitude towards potential threat and dangers is clearly different from that in science in general. Not only is a mere suspicion—as opposed to solid evidence—of potential harm sufficient for demanding further investigation, but even when there is no such prior suspicion, the goal of avoiding harm entails that risk assessment must actively try to search for ways in which harm may occur (Taleb 2007). This calls not only for other selective criteria and methods in risk assessment than in much of basic science, but for other treatment of the results as well. Even if, for monetary or other reasons, a suspicion has not been corroborated by the risk assessment process, it may still be a relevant and interesting result to put forward.

That even the suspicion of potential causes of harm are relevant for risk management is especially clear today, where the precautionary principle is gaining influence in many domains such as environmental risk management (Munthe 2011). The exact formulations (and thus strength) of the principle vary greatly (Sandin et al. 2002), but the underlying intuition that the various versions try to capture is the idea that we should take far-reaching steps to ensure that the system and substances we permit in society *do not* cause harm, even when that means to reject systems and substances which later turn out not to entail the potential harmful effects we feared (ibid.). In other words, we should prioritize the avoidance of false negatives over the avoidance of false positives.

To conclude, it seems safe to say that regardless of the force of the argument for a uniform standard in science proper that is independent of potential consequences of error, risk assessment would transform to another activity completely if it were to adhere solely to such a standard. That is not to say, of course, that risk assessment does not contain steps which fit perfectly with the epistemic standards thusly understood. As previously mentioned, methods of investigating potential toxicological effects in a medicine, for example, may be conducted in perfect alignment with established scientific methods identical to those used outside of risk assessment. But in order to supply relevant information to decision-makers, additional concerns, such as those sketched earlier, need to influence both what enters into the risk assessment process—how the process is performed—and the output of the process in the form of information to the decision-makers and other stakeholders.

33.7. The Handover Dilemma

So far, we have looked at potential value objections which primarily have been directed at science *per se*, and discussed its implications for values in risk assessment. In this penultimate section, we will look at a complication that directly has to do with the role of risk assessment in relation to the entire risk analysis process, viz. the boundary between risk assessment and risk management.

As mentioned in the introduction, the process of risk analysis is typically divided into two separate stages, risk assessment and risk management, where the first step signifies the scientific stage in which the risk and safety is assessed, and the second is the stage where it is decided how the risks should be handled, which centrally includes decision-making as to whether the risk should be considered acceptable or not. The extent to which values enter into risk assessment depends on where the line between these two main steps of risk analysis is drawn. In the traditional picture, the output of a risk assessment is a knowledge base containing the scientific results of the processes, based on which the risk decision is made.

In the previous sections, we have seen arguments to the effect that this knowledge base is not wholly 'free' from non-epistemic values. The very concepts used in the evaluation of risks are value-laden, according to the *delimitation argument*, and precautionary principle and other concerns relating to how the *inductive risk* should be handled entails non-epistemic concerns also for which epistemic standards to pursue in risk assessment. But regardless of the success of these objections, there is an additional value complication emanating from the fact that the knowledge base is actually playing merely an *indirect* guiding role in risk management. Given the ideal of a division of labour between the scientific stage (risk assessment) and the decision and handling stage (risk management), the knowledge gained from the risk assessment should be used by the decision-makers to make an informed decision. But these results are typically too complicated and complex to simply 'hand over' to the decision-makers in the risk management stage. The *direct* guiding role is played by evaluations and summaries based on the knowledge base. This is the source of what we may call *the handover dilemma*.

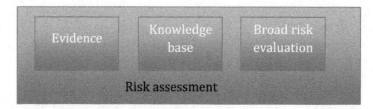

Figure 33.2   Hansson and Aven's model of the risk assessment process
Source: Hansson and Aven (2014: 1177).

Sven Ove Hansson and Terje Aven have recently provided a clear analysis of the problem (Hansson and Aven 2014). The interpretation of the knowledge base, they point out, is a complicated affair which is dependent on a substantial amount of background (scientific) knowledge:

> We may have tested the vaccine extensively and studied its mechanism in great detail, but there is no way to exclude very rare effects or effects that will only materialize 20 years into the future. Although the decision to disregard such possibilities is far from value-free, it cannot, in practice, be made by laypeople since it requires deep understanding of the available evidence, seen in relation to our general biological knowledge.
>
> *(Hansson and Aven 2014: 1177)*

Hansson and Aven suggest that risk assessment must be understood as including an additional step after the knowledge base, which they call 'broad risk evaluation' (Figure 33.2).

In this final step of risk assessment, the scientific experts make an evaluation of the knowledge base and produce a summary judgement on the risks and uncertainties involved in the case at hand. In order for such an evaluation to contain useful information for the decision-makes, non-epistemic values must be utilized. Examples of value questions include which notion(s) of risk to employ, what to use as appropriate levels of risk to judge an outcome or event to be safe enough, values as to where the break-off point for a risk should be—e.g. minimal level of risk or uncertainty for an alternative or event (Hansson and Aven 2014; Hansson 2008; Pigliucci and Boudry 2014). Hansson and Aven argue that the values that should be used in the broad risk evaluation are those of the risk managers, in their role as decision-makers (Hansson and Aven 2014: 1177). However, these value questions are typically interconnected with scientific issues, and it is not uncommon for risk managers to rely (partly or entirely) on the experience of the scientific experts in order to make the value judgements necessary to produce a risk evaluation that the decision-makers may use.

The dilemma is thus as follows. It may indeed be considered perfectly reasonable to rely on the experts also when it comes to many non-epistemic aspects of evaluation. After all, the experts involved in risk assessment are typically more experienced than the decision-makers in specifying a concept of risk and weighing different harmful outcomes and uncertainties. But by doing so, scientists are then performing a role 'outside of their jurisdiction' in the sense that they are acting as evaluators also of non-scientific considerations. If, on the other hand, experts were not to be involved in evaluating the results, the non-experts would in practice not be able to make an informed decision at all.

33.8. Conclusion

For good reason, risk assessment is often assumed to be the scientific stage of risk analysis. Risk assessment uses established scientific methods and is typically performed by scientific experts of the relevant fields. Even so, we have in this chapter seen several potential sources of (non-epistemic) values entering the process. First, as a matter of empirical fact, there have been many cases where biases of scientific experts have been revealed. Risk assessment, being to a significant extent a 'consulting business', is surely not immune to such dangers. However, while the potential presence of this kind of value-influence actualizes the importance of control measures such as regulation and ethical codes, it does not in itself indicate any principled way in which risk assessment is value-laden.

Other potential sources of values in risk assessment treated in the chapter are harder to dismiss, however. While the distinction between epistemic and non-epistemic values is relatively uncontroversial as such, the boundary between them is not; consequently, there is a risk that some of what is taken as epistemic values should rather be seen as non-epistemic. In a domain such as risk research, where the fundamental concepts utilized concern potential harm, it seems clear that what may seem as epistemic concern ('How large is the risk?') in fact depends also on non-epistemic values.

The practical nature of risk considerations also complicates the matter of potential error. In core science, avoiding false positives has long been prioritized. But in risk assessment, avoiding false negatives is arguably just as important: claiming that there is no (demonstrated) negative effect of a chemical substance when in fact there is, is often worse from a risk assessment perspective than claiming that there is a negative effect when in fact there is not.

The last source of non-epistemic values discussed in this chapter concerns the fact that the scientific output, the so-called knowledge base resulting from the scientific investigations and reviews of risk assessment, in practice can be properly interpreted only by scientific experts. The scientific output must thus in turn be summarized and evaluated by scientists, and this is a process which cannot be done without non-epistemic values.

In sum, the case is strong for the conclusion that risk assessment is not value-free but includes values whose content ultimately depends on which society we want to promote—in other words, ethical, political and other values. Is that, then, something that should worry us? Well, some theorists do worry. Critics argue that unless risk decisions are completely based on fully scientifically established facts, the role of science is marginalized (Gray and Bewers 1996: cf. also Durodié 2003). Today, however, most contemporary theorists seem to accept that an applied science such as risk assessment neither should nor can wholly avoid non-epistemic values. As Hansson puts it:

> Practical rationality demands that the required weight of evidence in practical matters be influenced by our evaluations of the possible outcomes. When there are strong but contestable indications that a volcano may erupt in the next few days, we evacuate its surroundings, rather than waiting for full scientific evidence that the eruption will take place. A principle that we should act against other types of possible dangers only if we have full scientific proof would be difficult to defend.
>
> *(Hansson 2007: 265)*

Indeed it would. Moreover, the decision to wait for 'full scientific proof' until we act would itself be a value-decision. Since the information decision-makers are in need of depends on our (non-epistemic) values, rather than an 'unfortunate circumstance', it is in fact *central* that risk assessment can provide, as far as possible, the relevant value-laden information. To the degree that we want to take also non-confirmed suspicions into consideration, risk assessment must aim to provide that information. If we are primarily interested in avoiding fatalities but less interested in minor injuries, risk assessment should arguably take that into consideration in its broad risk evaluation.

Two final comments are here in order. First, that risk assessment cannot be 'insulated' from values does not mean that there is no reason to differentiate between different levels of information. Whether a piece of information is a fact that is scientifically established through the 'high bar' epistemic values which prioritize avoidance of false positives, or whether it is an unconfirmed suspicion is naturally relevant for risk decisions. How to distinguish between different levels of a scientific knowledge base is a complex question to which we do not have a general or uncontroversial answer.[13] Still, it is important to keep in mind that an influx of values in risk assessment does not that mean 'everything goes'.

Secondly, what the presence of values in risk assessment does call for is *transparency*. Since the evaluation of the risk at hand depends on non-epistemic factors such as which alternatives were investigated, which methods were used and which notion of risk were employed, the entire risk analysis process, risk assessment as well as risk management, must be as upfront as possible in declaring with transparency every assumption and value that may be controversial. Whereas condensed claims such as 'this power plant is safe' may have clear rhetorical benefits, in reality any such claim is a simplification that must not only be backed up with science but also with transparent information about how it is even supposed to be understood. While we cannot expect to always agree on matters of risk and safety, we should be able to ensure that we are not talking past each other. That alone is quite a task for risk analysis, but a task on which we should never give up.

Related Chapters

Chapter 6: Scientific Methodology in the Engineering Sciences (Mieke Boon)
Chapter 35: The Role of Resilience in Engineering (Neelke Doorn)
Chapter 36: Trust in Engineering (Philip J. Nickel)
Chapter 39: Philosophy of Security Engineering (Wolter Pieters)
Chapter 40: Ethical Considerations in Engineering (Wade L. Robison)

Notes

1. For simplicity, the shorter term 'risk assessment' will be used in what follows.
2. It can be argued that this definition does not distinguish science from other knowledge-gathering activities such as, say, the knowledge gained from counting the number of shoes in the hallway. The latter activity may be ever so reliable, yet it typically is not considered scientific. In order to make this further distinction between these two types of knowledge production, some sort of reference to 'the scientific method' is typically made (Newton-Smith 1981). Still, that distinction is arguably not central for the task in the main text, since the aim is to distinguish science from pseudo-science or other value-laden activities.
3. It should be noted that while Hansson and Aven ask the broader question of whether *risk analysis* is scientific, their answer in the affirmative applies to risk assessment rather than risk management (Hansson and Aven 2014: 1181–1182).
4. For examples of the literature discussing the scientific status of risk assessment, see e.g. Weinberg (1981); Ricci and Molton (1981); Scheines (2008); Hansson (2010).
5. For matters of convenience, when it is clear that we refer to non-epistemic values, we will often talk simply of 'values'.
6. In one study, independent experts found abnormalities in only 4.5% of the 500 claimant X-rays, compared to 96% for the consultant experts (Gitlin et al. 2004).
7. That is, at least up to a point. For a comparison, take the nuclear safety debate. When there has been a nuclear plant incident, it is a common objection from defenders of nuclear power that since it typically is possible to pinpoint a number of human-made mistakes in design or execution that could have been avoided, the incident does not indicate that nuclear power is unsafe *as such*, but instead that the power plant in question has been poorly designed or managed. While that argumentative strategy is not without merit, it only goes so far. Eventually, if the 'human factor' cannot in practice be factored out from the equation, it must be included in our risk and safety estimations (cf. Möller and Wikman-Svahn 2011).
8. It may be objected that this conclusion is premature. There are more sophisticated versions of the above-mentioned charges arguing that even when conducted in perfect agreement with established epistemic values, non-epistemic values present in the scientists made them oblivious to alternative explanations for the phenomenon investigated, resulting in biased outcomes (Harding 1986; Haraway 1989; Longino 1990).

Whether the cases pointed to what should be best understood as improper applications of the scientific method/existing epistemic values, or as an indication of a systematic—and thus deeper—problem, is unclear. For space reasons that debate is excluded in this chapter, but see, e.g., Brown (2013), Elliott (2011) or Intemann (2005).

9. Cf. Longino (1995) and Douglas (2009). For example, 'simplicity' seems to be an aesthetic value rather than an epistemic one. As Heather Douglas (2009) notes, sometimes the world may be messy, and the complex may do a better explanatory job than the simple.

10. It has also been called the 'error argument' (Brown 2013; Elliott 2011) and 'the normative challenge' (Douglas 2016). Cf. also Churchman (1948), Rudner (1953), Biddle (2013), Elliot and Richards (2017).

11. Which should not be interpreted as, strictly speaking, '(logically) proven'. As is well known, the scientific method, while able to provide strong corroboration for a hypothesis, is fallible, relying on bridge premises which may, in principle, be false (Newton-Smith 1981).

12. The motivation of individual scientists may, of course, be just that (or any other potential motivation).

13. Some authors, e.g. Dobbins (1987) and Cranor (1990), have suggested what Hansson calls an 'applications-adjusted corpus', where evidential requirements should be a function of the practical decision at hand (Hansson 2007: 265). Others, e.g. Hansson himself, suggests an 'all-purpose corpus' to be used for theoretical and practical purposes alike, and a supplement which may vary in accordance with the evidential requirements of the practical problem at hand (Hansson 2007: 266–67).

References

Apostolakis, G.E. (2004). How Useful Is Quantitative Risk Assessment? *Risk Analysis,* 24(3), 515–520.

Aven, T. (2011). *Quantitative Risk Assessment: The Scientific Platform.* Cambridge: Cambridge University Press.

Biddle, J.B. (2013). State of the Field: Transient Underdetermination and Values in Science. *Studies in History and Philosophy of Science Part A,* 44(1), 124–133. doi:10.1016/j.shpsa.2012.09.003.

Brown, M.J. (2013). Values in Science Beyond Underdetermination and Inductive Risk, *Philosophy of Science,* 80(5), 829–839.

Churchman, CH.W. (1948). *Theory of Experimental Inference.* New York: Macmillan.

Cooter, R. (1985). *The Cultural Meaning of Popular Science: Phrenology and the Organisation of Consent in Nineteenth Century Britain.* Cambridge: Cambridge University Press.

Cranor, C.F. (1990). Some Moral Issues in Risk Assessment. *Ethics,* 101, 123–143.

Dobbins, J.G. (1987). Regulation and the Use of 'Negative' Results from Human Reproductive Studies: The Case of Ethylene Dibromide. *American Journal of Industrial Medicine,* 12, 33–45.

Douglas, H. (2000). Inductive Risk and Values in Science, *Philosophy of Science,* 67(4), 559–579.

Douglas, H. (2009). *Science, Policy, and the Value-Free Ideal.* Pittsburgh, PA: University of Pittsburgh Press.

Douglas, H. (2016). Values in Science. In Paul Humphreys (ed.), *Oxford Handbook of Philosophy of Science.* Oxford: Oxford University Press, pp. 609–630.

Durodié, B. (2003). The True Cost of Precautionary Chemicals Regulation. *Risk Analysis,* 23(2), 389–398.

EC (2003). *Technical Guidance Document in Support of Commission Directive 93/67/EEC on Risk Assessment for New Notified Substances,* Joint Research Centre, EUR 20418 EN, Office for Official Publications of the EC, Luxembourg.

Elliott, K.C. (2011). *Is a Little Pollution Good for You? Incorporating Societal Values in Environmental Research.* New York: Oxford University Press.

Fausto-Sterling, A. (1985). *Myths of Gender: Biological Theories About Women and Men.* New York: Basic Books.

Gitlin, J., Cook, L., Linton O. and Garrett-Mayer, E. (2004). Comparison of "B" Readers' Interpretations of Chest Radiographs for Asbestos Related Changes. *Academic Radiology,* 11(8), 843–856.

Gray, J.S. and Bewers, M. (1996). Towards a Scientific Definition of the Precautionary Principle. *Marine Pollution Bulletin,* 32(11), 768–771.

Hansson, S.O. (2007). Values in Pure and Applied Science. *Foundations of Science,* 12, 257–268.

Hansson, S.O. (2008). Regulating BFRs—From Science to Policy. *Chemosphere,* 73, 144–147.

Hansson, S.O. (2010). Risk: Objective or Subjective, Facts or Values. *Journal of Risk Research,* 13(2), 231–238.

Hansson, S.O. (2013). Defining Pseudoscience and Science. In M. Pigliucci and M. Boudry (eds.), *Philosophy of Pseudoscience.* Chicago: University of Chicago Press.

Hansson, S.O. and Aven, T. (2014). Is Risk Analysis Scientific? *Risk Analysis,* 34, 1173–1183. doi:10.1111/risa.12230.

Haraway, D.J. (1989). *Primate Visions: Gender, Race, and Nature in the World of Modern Science.* New York and London: Routledge.

Harding, S.G. (1986). *The Science Question in Feminism*. Ithaca, NY: Cornell University Press.

Hempel, C.G. (1965). Science and Human Values. In C.G. Hempel (ed.), *Aspects of Scientific Explanation*. New York: Free Press, pp.81–96.

Hubbard, R. (1982). Have Only Men Evolved? In S. Harding and M.B. Hintikka (eds.), *Discovering Reality*. Dordrecht: Riedel, pp. 45–69.

Intemann, K. (2005). Feminism, Underdetermination, and Values in Science, *Philosophy of Science*, 72(5), 1001–1012.

Janower, M. and Berlin, L. (2004). "B" Readers' Radiographic Interpretations in Asbestos Litigation. *Academic Radiology*, 11(8), 841–842.

Kaplan, S. and Garrick, B.J. (1981). On the Quantitative Definition of Risk. *Risk Analysis*, 1(1), 11–27.

Kuhn T (1962). *The Structure of Scientific Revolutions*. Chicago: University of Chicago Press.

Lakatos, I. and Musgrave, A. (eds.) (1970). *Criticism and the Growth of Knowledge*. London: Cambridge University Press.

Laudan, L. (1984). *Science and Values: The Aims of Science and Their Role in Scientific Debate*. Oakland: University of California Press.

Lentsch, J. and Weingart, P. (eds.) (2009). *Scientific Advice to Policy Making: International Comparison*. Farmington Hills, MI: Barbara Budrich Publishers.

Levi, I. (1960). Must the Scientist Make Value Judgements? *Journal of Philosophy*, 57(11), 345–357.

Longino, H.E. (1990). *Science as Social Knowledge: Values and Objectivity in Scientific Inquiry*. Princeton: Princeton University Press.

Longino, H.E. (1995). Gender, Politics, and the Theoretical Virtues. *Synthese*, 104(3), 383–397.

McMullin, E. (1982). Values in Science. *PSA: Proceedings of the Biennial Meeting of the Philosophy of Science Association*, 2, 3–28.

Möller, N. (2012). The Concepts of Risk and Safety. In Sabine Roeser, Rafaela Hillerbrand, Martin Peterson, and Per Sandin (eds.), *Handbook of Risk Theory*. Dordrecht: Springer, pp. 55–85.

Möller, N., Hansson, S.O. and Peterson, M. (2006). Safety is More Than the Antonym of Risk. *Journal of Applied Philosophy*, 23, 419–432.

Möller, N. and Wikman-Svahn, P. (2011). Black Swans and Black Elephants of Nuclear Safety. *Ethics, Policy & Environment*, 14(3), 273–278.

Munthe, C. (2011). *The Price of Precaution and the Ethics of Risk*, Dordrecht: Springer.

National Research Council (1983). *Risk Assessment in the Federal Government Managing the Process*. Washington, DC: National Academy Press.

Newton-Smith, W.H. (1981). *The Rationality of Science*. Boston: Routledge & Kegan Paul.

Nord, E. (1999). *Cost-Value Analysis in Health Care: Making Sense out of QALYs*, Cambridge: Cambridge University Press.

Pigliucci, M. and Boudry, M. (2014). Prove it! The Burden of Proof Game in Science vs. Pseudoscience Disputes. *Philosophia*, 42(2), 487–502.

Rechard, R.P. (1999). Historical Relationship Between Performance Assessment for Radioactive Waste Disposal and Other Types of Risk Assessment. *Risk Analysis*, 19, 763–807.

Ricci, P.F. and Molton, L.S. (1981). Risk and Benefit in Environmental Law. *Science*, 214, 1096–2001.

Rooney, P. (1992). On Values in Science: Is the Epistemic/Non-Epistemic Distinction Useful? In K. Okruhlik, D. Hull and M. Forbes (eds.), *PSA: Proceedings of the Biennial Meeting of the Philosophy of Science Association*, Vol. 1. Chicago: Philosophy of Science Association, pp. 13–22.

Rudner, R. (1953). The Scientist Qua Scientist Makes Value Judgments. *Philosophy of Science*, 20, 1–6.

Sandin, P., Peterson, M., Hansson, S.O., Rudén, C. and Juthe, A. (2002). Five Charges Against the Precautionary Principle. *Journal of Risk Research*, 5, 287–299.

Scheines, R. (2008). Causation, Truth, and the Law. *Brooklyn Law Review*, 73(2), 625–650.

Sokal, Alan (2010). *Beyond the Hoax: Science, Philosophy and Culture*. Oxford: Oxford University Press.

Taleb, N.N. (2007). *The Black Swan: The Impact of the Highly Improbable*. New York: Random House.

Tiusanen, R. (2018). Qualitative Risk Analysis. In N. Möller, S.O. Hansson, J-E. Holmberg and C. Rollenhagen (eds.), *Handbook of Safety Principles*. Hoboken, NJ: Wiley, pp. 463–492.

Weinberg, A.M. (1981). Reflections on Risk Assessment. *Risk Analysis*, 1, 5–7.

Wilholt, T. (2013). Epistemic Trust in Science. *The British Journal for the Philosophy of Science*, 64(2), pp. 233–253. doi:10.1093/bjps/axs007.

Young, R. (1970). *Mind, Brain and Adaptation in the Nineteenth Century. Cerebral Localization and Its Biological Context from Gall to Ferrier*. Oxford: Clarendon Press.

Ziman, J. (1996). 'Postacademic Science': Constructing Knowledge with Networks and Norms. *Science Studies*, 9, 67–80.

34

ENGINEERING AND SUSTAINABILITY

Control and Care in Unfoldings of Modernity

Andy Stirling

34.1. Sustainability, Engineering and Control

Sustainability is about reversing the negative impacts of contemporary Modernity on social equity, human well-being and ecological integrity (Brundtland 1987; Holden et al. 2014). Here, engineering of many kinds has undoubtedly helped enable many material (and some wider) benefits. But it remains a driving concern of Sustainability[1] (Stirling 2009), that diverse kinds of technological, environmental and institutional engineering have also interacted with other factors to bring a host of (ostensibly unintended) adverse social and ecological consequences (Holling and Meffe 1996). Accordingly, the Sustainable Development Goals (UN 2015) offer a crucial contribution in contemporary high-level global governance, towards actions to address the resulting compelling imperatives around inequality (Piketty 2014), oppression (UN 2015), climate disruption (IPCC_WGII 2014), ecological destruction (UNEP 2014), toxic pollution (Harremoës et al. 2002), nuclear risks (UNSCEAR 2016) and the perennial scourge of war (Mueller 2001).

As such, Sustainability addresses an 'Achilles' heel' at the core of Modernity—the vulnerability of engineering cultures and practices to so many kinds of fixations, fallacies and failures of *control* (Stirling 2019a). Depending on political perspective, examples might include fossil fuel dependencies (Unruh 2000), high-input agriculture (IAASTD 2009), pharmaceutical based healthcare (Davies 2011; GFHR 2004), automobile-based transport infrastructures (Arthur 1989), high-rise housing (Ward 1976), technocratic urban design (Cecla 2012), top-down development programmes (Parfitt 2017), agrarian collectivisation (Viola 1996), re-engineering of rivers (Zavialov 2005), large hydro-electric schemes (McCully 2001), nuclear power (Collingridge 1983), nuclear weapons (ICAN 2015), and the litany of repeatedly disastrous preparations for war (Blue et al. 2014). Historically, all these infrastructures have privileged instrumental-style engineering thinking and expert technical interventions involving various kinds of control rhetoric (Jasanoff 2005; Stirling 2015). In each field, claims are routinely made that engineered 'solutions' offer means of control, by which to achieve declared aims. But again and again, though these narratives and performances of control persist, the successful realisation of control remains elusive.

So, a lens of control offers a potentially useful means to examine relations between Sustainability and engineering. For, engineering is not only deeply implicated *empirically* in so many of the persistent social and environmental problems that Sustainability seeks to address. In a "traditional" view (Allenby 2005), engineering is also seen quite fundamentally in its own right to embody—indeed, be shaped and driven by—the underlying formative aspirations and assertions of Modernity itself: to control nature and society. So it is in this sense that Sustainability movements find themselves also wrestling with some of the deepest commitments of Modernity in the constituting *ideologies* of engineering around control (Millar and Mitchell 2017).

Associated dilemmas have long been well recognised. For instance, Heidegger explored the tragic paradox that the instrumental engineering vision of technology as an ostensibly consummate "means to an end", should itself be so evasive of human "mastery" (Mumford 1934; Ellul 1964; Winner 1977; Feenberg 1999; Verbeek 2007). In wrestling with this, many contrasting modalities of control have since been distinguished around engineering—for instance distinguishing diverse gradations of 'hard' or 'soft' 'mastery' (Williams and Edge 1996). Associated implications are variously viewed in favourable (Wolpert 1992) or unfavourable (Wynne 1993) ways. But across the enormous resulting scope for debate and ambiguity, the fundamental aspirations of engineering are again frequently held to coalesce around different kinds and contexts of *control* over society and nature (Ison 2010; Fuller and Collier 2004; Aronowitz 1988).

Accordingly, contrastingly conceived detailed notions of control are also central preoccupations in technical discussions more specifically in the *philosophy* of engineering. Indeed, discussing "ideals for engineering" in a recent key text in this field, Pirtle and Szajnfarber envisage a quite existentially controlling role for engineering in that "[h]umanity's ability to exist in the long-run will depend on our being able to continue controlling and influencing ourselves, our environment, and the systems which we have embedded into it" (Michelfelder et al. 1998). It is hardly surprising, then, that various principles around control feature prominently in many contemporary specific contributions to this field (Floridi in RAEng 2011; Demir 2012). Control forms a defining focus in philosophical examinations of systems engineering (Van de Poel et al. 2010). Likewise, control is especially salient in understandings of responsibility in engineering (Fischer and Ravizza 1998; Van de Poel et al. 2010. In different senses, notions of control remain central discussions in the philosophy of engineering over "social control" (Collingridge 1980) and "rational acceptance" (Goldberg et al. 2008).

So, before considering in detail the implications of all this for relationships between engineering and Sustainability, it is worth asking (for present purposes) exactly 'what is control?' (Stanovich et al. 2014). For such an apparently simple and familiar idea, notions of control are actually quite tricky to define (Leigh 2004). How do they differ, for instance, from related ideas concerning contrasting degrees and modalities for the exercise of agency over a given process of change—variously by 'influencing', 'effecting', 'impacting', 'modifying', 'conditioning' or 'managing' formative processes? (Bandura 1989). Here, a typical dictionary gloss of the distinctive colloquial meaning of 'control' refers to a capacity to "exercise power or authority . . . to determine . . . behaviour or action" (OED 2013). So, a crucial ingredient is the reference to the rather ambitious aim of '*determination*'. But if such a strong aspiration is to hold, then associated agency needs to be both *effective* and *specific* in quite demanding ways (Stirling 2019b).

In these terms, then, 'control' implies an *efficacy* of agency in a given context, sufficient fully to realise the particular aim(s) in view (rather than achieving this only partially or conditionally). This, of course, also implies a faith that it is possible to resolve and aggregate such a requisitely precise definition of purpose in the first place. If an intervention succeeds in effecting large-scale consequences, but these are not aligned with any original purpose, then what has occurred is more like 'impact' than 'control'. So, the idea of control also implies a high degree of *specificity* in the extent to which the available instrumental means are held to make it possible to avoid inadvertent side effects beyond given intended ends. This means that those outcomes that are realised are restricted only to ends that were initially intended, as distinct from other possible alternative or collateral consequences. After all, the wider and deeper the unintended consequences, the less any given intervention can properly be held to involve control (rather than some other kind of intervention). So, in their most idealised form, then, any aims or claims involving control (disaggregated as appropriate with respect to specific ends), inherently require *realising fully and solely, a prior set of intended end(s), with no unintended effects*.

Although demanding, there is no shortage of examples of situations in which this strict concept of control can be recognised to be workably appropriate. A simple paradigmatic instance of control in this precise sense might be recognised in many aspects of the everyday exercise of agency through a mechanical device such as an automobile. Among the multiple immediate aims in play, in this kind of confined setting, are various aspects of the process of moving from one location to another. In one such aspect of agency, the normal practice of steering (for instance) involves an experience of moving in one particular direction at any given moment—and only in this way. Wobbling between orientations (even if the average remains around the intended orientation), would not properly count as control. Similarly, the modulation of speed, must—to be held as control—result in a precise rate of movement, and not some kind of lurching fast and slow around this rate. Likewise, control over other functions like windscreen wipers or heating should not interfere with other aspects of control, nor result in unintended results like engines cutting out or wheels falling off.

What is striking about general relationships between engineering and Sustainability, then, is the extension of this familiar (when operational) experience of control *within* a given *engineered system*, to control *by* engineering interventions of wider *encompassing circumstances*. But as depths, degrees and dimensionalities of non-linear complexity extend beyond circumscribed domains of mechanical linearity, ideas of control in this instrumental engineering sense quickly become highly questionable. And there are, as we have seen, plenty of other possible concepts than 'control', for addressing the kinds of formative relationships that actually pertain between entire engineered infrastructures or technological systems and the plural social contexts and operational environments within which these are located (Hommels et al. 2014).

In efforts to achieve (for example) provision of food, shelter, energy, livelihoods, mobility, communications or security, Sustainability debates teach an important lesson. Engineering interventions in such contexts cannot typically reasonably be held fully to deliver on the more hubristic aims or claims of 'control'. In areas like these, too much policy discourse around engineering takes the form of 'solutionist' promises, emphasising supposedly singular and complete 'magic bullet' responses to global challenges (Strebhardt and Ullrich 2013). Rather than these kinds of controlling visions, what the considerable potentialities of engineering can instead be relied upon to achieve is, more modestly, to 'influence', 'affect', 'impact', 'modify', 'condition' or 'modulate' the realisation of some part of the intended ends. What is required, then, is a degree of *humility* around conventional modernist instrumental imaginations of control.

For if the function of engineering were distinctively to count as *control* (rather than some other among these more modest kinds of formative relation), then delivery of stated aims would not incur any of the many adverse collateral impacts that it is the defining purpose of Sustainability to address. It is the totalising self-confidence of idealised imaginations of control—and associated propensities to sideline emerging criticism or concern—that serve as much as the material workings of the engineered systems, to exacerbate the negative impacts of so many instrumental engineering interventions. And it is the resulting mismatch between modernist idealisations of control and the messy practical actualities of engineering responses to real-world challenges that dominate the negative side of relations between engineering and Sustainability.

This problematic mismatch is further exacerbated when it is considered that formal frameworks like the Sustainable Development Goals represent a rather incomplete and idiosyncratic—if not downright misleading (Farley and Smith 2013)—reflection of the actual interests, values and drivers that really tend to motivate large-scale institutional and infrastructural interventions around the world (Stirling 2009). Governance discourse around Sustainability involves powerful actors ostensibly subscribing exclusively to values advancing human well-being, social equity and ecological integrity (Voss et al. 2006). But that these declared aims of Sustainability are so prominent in elite discourse, does nothing to diminish the hard realities of the actual purposes that are typically far more influential in shaping and driving large-scale engineering endeavours around the world (Parkinson and Spedding

2001). Despite the undoubted efforts, commitments and integrity of individual engineers, it implies no cynicism to observe that the most formatively important forces shaping the configuration and orientation of much of the world's engineering more often lie in values like individual careers, disciplinary reputation, private profit, sectional privilege, commercial interest, national prestige or military domination (Stirling 2018). None of these highly formative drivers of engineering appear anywhere in the SDGs. Whilst isolated serendipitous alignments with wider aims of Sustainability can occur under favourable circumstances, it would be eccentric simply to assume these. So, recognising these further uncomfortable realities underlying the driving purposes of efforts at control, puts relations between engineering and Sustainability in an even more fraught light.

34.2. Modernity, Control and Care

So far, attention has focused on failures in conventional large-scale, engineering-style efforts at control. Maybe the 'solution' lies within engineering professions, institutions and cultures themselves? (see Chapter 42, "Standards in Engineering" by Paul B. Thompson, Chapter 43, "Professional Codes of Ethics" by Michael Davis, and Chapter 44, "Responsibilities to the Public—Professional Engineering Societies" by Joseph Herkert and Jason Borenstein, this volume). If engineering practice were to reform within its wider social context, in ways that make it more 'ethical' or 'responsible', perhaps these negative syndromes might be solved? Such impulses may be understandable and positive as human reactions. But great caution is required over any idea that such circumscribed interventions alone might in themselves constitute a fully satisfactory response to the dilemma. For there are sadly many deeper dimensions to the encompassing modernist culture of control, which arguably render such programmes themselves, often to be little more than naïvely instrumental 'solutionism'. If these dimensions are not also addressed, then efforts at alleviation that are too restricted in scope may actually inadvertently exacerbate the problems.

It is also across this deeper-seated, more pervasive and long-standing 'infraculture' of Modernity as a whole, then (Hoffman 2015), rather than merely inside the more visible but circumscribed cultures around particular existing practices, disciplines or institutions, that the real challenge lies for resolving relations between engineering and Sustainability. For it is centrally around refractions of this same impulse to control, that the entire condition of Modernity—in all its worldwide forms (Eisenstadt 2000; Mignolo 2007; Seth 2016; Kiely 2005; Santos 2002)—has itself been arguably most deeply constituted for several hundred years. In foundational analyses like that of Condorcet (Condorcet 1979), for instance, particular 'enlightenment' notions of Modernity (see Chapter 53, "Engineering and Contemporary Continental Philosophy of Technology" by Diane P. Michelfelder, this volume) have since their inception centred around "the belief that technological progress will give humanity increasing control over nature" (Inglehart and Welzel 2005; see also Chapter 51, "Engineering and Environmental Justice" by Benjamin R. Cohen, this volume). Likewise, at least since Weber (Scott 2006), Modernity has been deeply associated in philosophies of society, politics, culture and history, with a range of contrastingly diagnosed modalities of control (Gregory et al. 2009). Primarily, these include: rationalisation, individualisation, capitalisation, industrialisation, nation-forming, bureaucratisation, democratisation and disenchantment from tradition (Stirling 2019c). It is in these terms that Modernity in itself and as a whole, across its many diverse forms and aspects, might be characterised in quite deep general terms as an 'infraculture', a more generally formative cultural configuration lying beneath and beyond otherwise disparate manifestations across cultures—and conditioning their evolution over long time periods (Hoffman 2015). If so, the most prominent single feature of this underlying 'infraculture' of Modernity might be identified to be imaginations of control. Unless this pervasive constituting characteristic of the encompassing infraculture of Modernity is not addressed, then it must be expected that little that can be done by interventions implicating engineering alone.

This point may warrant some elaboration, for it may not be immediately obvious how so many of the widest and deepest constituting foundations of Modernity itself are so centrally about romanticisations of control (Beck et al. 1994). To take each of the above widely discussed aspects in turn, then: *rationalisation* (for instance), involves aspired control over legitimate ways of understanding the world (in science) (Hegel 1974) and justifying action (in public affairs) (Habermas 1985). *Individualisation* is about imaginations of greater control by individuals over the course of their own lives (Durkheim 1984)—disembedding from constraining traditional norms (Comte 2009). *Capitalism* involves efforts by particular political-economic interests to concentrate control over wider means of production (Marx 1976). *Industrialisation* (socialist or capitalist) involves attempts at enhanced control over productive activities (Arendt 1959). *Nation-forming* was notionally about more clearly demarcating particular geographical domains of political control (Gellner 1983)—or (as in more contemporary 'Brexit' parlance) "taking back control" (VLC 2016). *Bureaucratisation* involves the performance of tightened control over organisational processes (Weber 1947). Many varying notions of *democratisation* (for all their flaws) centre around ideas of more distributed social control over unfolding politics (Tocqueville 2003).

Yet even this formidable list of the constituting control dynamics of Modernity, as identified in conventional academic accounts, is still arguably seriously incomplete. Further to these familiar, deeply structured processes, there is an important additional way in which aspirations to control have exercised formative effects on prevailing cultures and infrastructures of engineering. This concerns the profoundly shaping effects of ever-present cross-cultural forces of *coloniality* (Mignolo 2007; Quijano 2007). Also arguably present before and outside Modernity (Dietler 2010; Halperin 2007; Deringil 2003; Buzon 2008; Latvus 2006; Larsen 1974; Barrett 2003) this most pervasive and intractable modality for the hegemonic entrenchment of power has taken especially devastating forms in successive waves of European *colonialism* that attended the advent of Modernity (Trigger and Washburn 1996; Mignolo 1995; Said 1994) (as well as their various non-European eddy currents (Chen 1970; Mwangi 2001; Halperin 2007; Gladney 2004; Yi-chong 2014)). Involving particularly extensive and acute forms of violence (Churchill 2002; Jacobs 2009), Euro-American colonialism has proven arguably more important than any other single factor in shaping the underlying infrastructures and imaginations that condition engineering around the world (Grove 1997; Churchill 2002; Hudson 1972; Harvey 2003; Ahmad 2004).

Currently entrenching globalised architectures for production, investment, trade, mobility, innovation, knowledge accreditation and military domination were all originally engineered under intense pressures of coloniality—and continue to reproduce these conditions today (Bukharin 1917; Wallerstein 2006). Interacting with all the other aspects of Modernity discussed earlier, these each display their own fixations, fallacies and failures of control. Behind well-meaning individual motivations and justificatory rhetorics, then, the hierarchical structures and centralised orderings underlying all these engineered systems, and their embedded artefacts, continue (despite countervailing intentions among some embedded actors) to facilitate various kinds of global extraction and appropriation (Hildyard 2016). So, perhaps more than any other among the conventionally recognised constitutive dynamics of Modernity, it is arguably the under-acknowledged relations and processes of coloniality that most drive cumulative challenges of social inequality and environmental degradation at the heart of relations between engineering and Sustainability (see also Chapter 50, "Engineering and Social Justice" by Caroline Baillie, this volume).

As if this were not enough, the picture so far has still only centred on the more material expressions in Modernity of this deep-seated 'infraculture' of what Nehru called "controlism" (Jones 2010). Underlying and shaped by these constituting practices, there is also the dimension of ideology and imagination. For control is also central to the engineering of the distinctive *ways of knowing* the world, that are variously seen to characterise contemporary global Modernity (Misa et al. 2003; Stirling 2015). Recognising that "knowledge itself is power", Francis Bacon famously articulated in the

seventeenth century a defining theme in the then-erupting coloniality of the European Enlightenment (McGovern 2005). He went on to urge that, in order to achieve 'progress', science must "put
nature on the rack and torture her secrets out of her" (Keller 1984). Especially significantly, Bacon
anticipated an "empire of man over nature" (Morgenthau 1947). There could hardly be a more
graphic expression in the foundational empiricism of the ostensibly neutral world of science, showing
how the controlism of Modernity entangles with the distinctive violence of coloniality.

In all these ways, instrumental-style engineering imaginations pervade not only the material
structures of globalisation, but also their driving imaginations and ways of knowing (see Chapter
8, "Prescriptive Engineering Knowledge" by Sjoerd Zwart, this volume). Bodies like the World
Bank, World Trade Organisation, International Monetary Fund, UN development frameworks and
multilateral networks for regulating global environmental and security issues—as well as numerous
transnational corporations—are as intrinsic to the controlling orientation of contemporary engineering as the disciplines and cultures of engineering itself as a discipline. All alike reproduce Baconian
efforts to: reduce complexities; standardise disparities; aggregate variabilities; integrate ambiguities;
homogenise pluralities; domesticate risks; externalise uncertainties and deny indeterminacies—and
so discipline (if not "torture") diversity and change (Jasanoff 2004). In seeking to address relations
between engineering and Sustainability, then, it is again essential to go beyond claims-making about
professional ethics or responsibility *within* disciplinary cultures, to also challenge the constituting
controlling obsessions of Modernity itself.

So what is the alternative? Amidst so much deeply embedded complexity, what can be the practical implications for relations between engineering and Sustainability? It is here that there comes most
strongly to the fore once again, the human and environmental values of long-standing Sustainability
movements. For, as set out by Brundtland and detailed further in the carefully negotiated global aims
encoded in the Sustainable Development Goals (SDGs), Sustainability presents arguably the most
important general challenge to the encompassing *controlling* ambitions of Modernity. It does this at a
level that is appropriately much deeper than purely expert interventions or technical reforms in or by
particular disciplines or institutions. Fulfilling ambitious aims around social equity, ecological integrity and human well-being require overtly political engagements addressing the most general arenas
of international governance—in the most central traditions of progressive social movements, addressing the most entrenched forms of global power. For all the many serious flaws, compromises and
civilising hypocrisies (Elster 1999) in the present form of the SDGs themselves, then, it is this explicit
framework of diverse normativities that helps shift attention away from the closed, instrumentalised technocratic idealisations of *control* towards a more openly political and potentially empowering
idiom of caring—for the neglected unintended consequences of just this kind of control for: people,
societies, ecologies and the future of the world (Bellacasa 2017; Robinson 2011; see also Chapter 52,
"Beyond Traditional Engineering: Green, Humanitarian, Social Justice and *Omnium* Approaches" by
George D. Catalano, this volume).

Whether in terms of the Brundtland values of human well-being, social equity and ecological
integrity, then, or the many more specific (if often compromised) Sustainable Development Goals,
the crucial contribution made by this more 'caring' style of Sustainability discourse and practice, is to
help open up narrow, vertical, technical notions of control (in which driving ends remain instrumentally invisible) (Preston and Wickson 2016). It is in the resulting onward political, cultural and discursive contexts that emerge, that more opportunities can arise for currently conventional controlling
styles of engineering to be balanced with wider and more horizontal—more overtly political (and so
accountable)—ideas and practices of care (Wickson et al. 2017). In this way, the driving imperatives
of governance around engineering can deepen and expand from narrow expert wrangles over asking
merely "How fast?"; "What risk?"; or "Who leads?" in some presumed single trajectory controlled in
each sector by incumbents as 'inevitable' (Stirling 2014b). In ways elaborated in the following, when
the pluralities of these more caring values and dispositions of Sustainability are taken seriously, discussion

far more plural and fundamental questions may emerge for engineering over "Which way?"; "Who says?" or "Why?" (Stirling 2008).

But this prompts difficult questions. What then characterises more concretely and exactly, this contrast between care and control in relationships between engineering and Sustainability? It is here that long-standing feminist literatures around the ethics and politics of care offer many practical insights over how to shift attention away from the deeply embedded modernist imaginations of control (Plumwood 1993). Of course, some strands of feminism display their own fixations with control (Cuboniks 2015). And much formative thinking about care has also gone on outside feminism (Pellizzoni 2004; Hamilton 2013; Groves 2015; Schweitzer 1947; Pope_Francis 2015). But it is in deep engagements with the comprehensive subalternities of gender, race and other intersectionalities (May 2015) that some of the hardest thinking has been done about the kinds of enhancements of social agency that Sustainability is striving for, which feature more as 'matters of *care*' (Bellacasa 2017; Fisher and Tronto 1991) than as aspiring control. It is only at this level of political-cultural depth and pervasiveness that the most intractably hardwired effects of coloniality on engineering can arguably be addressed.

It is due to feminism, for instance, that a further analogy may be helpful in illuminating the control/care distinction advanced here (see also Chapter 48, "Feminist Engineering and Gender" by Donna Riley, this volume). To focus on this distinction should not be taken to imply the kinds of hierarchical binaries so well critiqued in a related strand of feminist thought (Taylor et al. 2010). To contrast one idea with another, rather than merely assert a single (ostensibly disembodied) notion on its own is, after all, a way to be more explicit and accountable—caring for the enabling of critique. So, in relation to the example of driving an automobile given earlier, then, a contrasting analogy might focus on society's relationships with children (Gilligan 1993; Held 2005). After all, many different kinds of societies all care very much about the ways in which children develop. But (however it may manifest in some settings), this concern is typically only very poorly (arguably dysfunctionally) describable as ambitions to *control*.

Key in the example of child-rearing (as in Sustainability) is that care is typically greatly more flexible, open and plural than is captured by the idea of control (Gilligan 1993; Held 2005). Intentions and ends at issue are generally more diverse and carefully deliberated between subjects and objects of care. The means are more mutualistic than deterministic. There is a more relational than categorical stance, with greater humility and reflexivity over complexities, uncertainties, and other (third) loci of agency. Deeper precaution is adopted towards adverse effects, than is captured in controlling notions of 'trade-offs'. It is these kinds of differences that lead efforts to exert general controlling relationships over children (for instance) to be widely regarded (across different societies) as a pathology. So, it is distinctions like these that make relations with children (like the concerns of Sustainability) better envisaged to be more about 'care' than 'control' (Davies 2012; Arora 2017; Groves 2013). It is on this basis that the implications of Sustainability can be considered, for the development of a kind of engineering that is more caring than controlling.

34.3. From Control to Care in Engineering for Sustainability

So, what might be the practical lessons for relations between engineering and Sustainability, of this analysis of the contrast between control and care in response to challenges of Modernity? Before turning to more specific repercussions for engineering, it is worth reflecting on some important general implications for discourse and practice around Sustainability itself. For the deep-seated hegemonies of control, discussed earlier, have not remained inactive in the face of Sustainability challenges. On issues like heavy metals, chlorine chemicals, ionising radiation, pesticides, carcinogens and toxic wastes, decades of values-based environmental campaigning achieved repeated successes through applications of values-based—'caring'—actions, contrasting markedly with the more controlling imaginations of government, business and mainstream academia (Gee et al. 2013; Harremoës and European Environment 2001).

In the face of unrelenting scientistic, expert-led, 'risk-based', 'no alternatives' rhetorics from elite institutions, this struggle involved broad political mobilisations emphasising uncertainties and alternative political visions (Grove-White 2001). But these earlier articulations of Sustainability are now under pressure. On high-profile issues like climate change, for instance, a main emphasis is now on exactly the kinds of control rhetoric that were in the past deployed *against* Sustainability concerns. Scientistic analysis increasingly substitutes for explicitly values-based arguments (Hulme 2009). Policy debates become dependent on complex risk-based modelling (Shackley et al. 1996). Primary agency is afforded to exactly the kinds of incumbent science and policy interests that in the past most resisted Sustainability concerns (Wynne 2010).

And in struggles to protect the earth's climate from rising emissions of climate-forcing gases, there is an especially disturbing way in which a modernistic infraculture of control is warping what were originally the more caring sensibilities of environmentalism (Bloom 2003; Lannoo 2010; Yearley 2005; Pepper 1996). Under a controlling imagination, this challenge is undertaken in a very different way to previous successful environmental struggles to curb uncontrolled ecological impacts (Harremoës et al. 2001; EEA 2013). Instead of hope-driven, values-based actions, what increasingly hold sway are fear-laden, expert-defined visions of control over the earth's climate as a whole—modulating mean temperature within pre-set model-defined limits.

Under the burgeoning instrumentalism of this planetary engineering imagination, it is as if the 'natural' state of the climate were somehow static; that this condition is self-evidently desirable; that it is 'change' itself that is negative; and that what should be controlled for are the 'optimal Holocene conditions' of a 'stabilized earth' (Steffen et al. 2018; Sterner et al. 2019). In other words, the aim shifts from protecting other—largely unknown—kinds of agency on the earth, to asserting with renewed vigour the particular controlling singularity of agency associated with exactly the presumptively all-knowing modernity that is causing the problem in the first place (Stirling 2019c).

A caring approach to the earth's climate, by contrast, might be about more politically curbing 'climate disruption', than technically controlling 'climate change' (Pidgeon et al. 2016). Driven by acknowledged uncertainties over uncontrolled impacts, rather than asserted certainties towards planetary control, this would focus not so much on controlling global temperature, but more directly on substituting the polluting practices. And a caring approach would recognise that the problem is caused not indiscriminately, 'anthropogenically' by humanity *per se*, but by particular *modernistic* kinds of society (Lövbrand et al. 2015). Indeed, it is this very possibility of political alternatives in ways of being human, that makes any struggle worth pursuing at all.

This pattern reflects a wider shift in many other areas of environmental struggles, as these impinge on high-level governance processes around the world. With the caring multiplicities of Sustainability decried as 'Stupid Development Goals' (Economist 2015), repeated efforts are made to control the focus of attention down to a supposedly singular and more technical 'nexus' (Beddington 2009) of far narrower 'grand challenges' (CEC 2009). A growing mood of 'environmental authoritarianism' (Beeson 2010) argues that the democratic arenas on which environmental struggles of the past depended so much, might actually be seen as an 'enemy of nature' (Euractive 2010). Leading environmentalists proclaim that "it may be necessary to put democracy on hold for a while" (Hickman 2010). Attempts are made to reduce the vibrant political arena of Sustainability into an inaccessible technical discourse around a narrow set of 'planetary boundaries' as "control variables" (Rockström et al. 2009) about which "non-negotiable" expert pronouncements brook "no uncertainty" (Rockström 2010).

All this is especially relevant to engineering, because momentum is accelerated by a new wave of unprecedentedly ambitious new infrastructures promising planetary-scale control (Cairns and Stirling 2014). A host of emerging 'geoengineering' technologies help drive the momentum (Shepherd et al. 2009), in which current global political—and especially military (Brzoska et al. 2012)—incumbencies openly position themselves for a mission of 'planetary management' (Newton 1999). So powerful has this modernist control-delusion become, that a demonstrably massive failure on the part of Modernity even to be able to control itself, has been spun into an even more inflated (self-appointed) mandate to control the whole 'earth system' (Lövbrand et al. 2009). In an oddly under-discussed confusion of 'impacts' with 'control', a newly burgeoning ideology of 'the Anthropocene' (Economist 2011) hubristically misnames a fleeting geological instant of destruction as a presumed entire new prospective epoch of notional human control. As foundational scientific texts put it, the Anthropocene is a coming epoch in which "a self-conscious control force that has conquered the planet" (Schellnhuber 1999), with a homogenized and depoliticized humanity "taking control of nature's realm" (Crutzen and Schwagerl 2011). Just as diagnoses burgeon of the possible end of Modernity, so Modernity hysterically intensifies the prescriptions that so helped make the problem in the first place: impelling, in the name of Sustainability itself, a new wave of fixations, fallacies (and inevitable further failures) of control.

So, how to defend past hard-fought political gains by Sustainability from this currently resurgent modernist hegemony? How to reverse erosions of scope for democratic struggle? How to resist wider authoritarian appropriations of the institutions and practices of Sustainability itself and their subversion to technical agendas of control, rather than more emancipatory politics of care? Central as it is to visions of Anthropocene geoengineering, the engineering community arguably has a particular responsibility (see Chapter 44, this volume).

A start might lie in reaffirming where Sustainability movements came from in the first place—and the nature of the political forces that have most supported them. In keeping with Brundtland's emphasis on themes of participation and democratic struggle (Brundtland 1987), Sustainability discourse was historically pressured onto high level international agendas (like the World Commission on Environment and Development) through emergent waves of collective action and "uninvited" public engagements in pursuit of social justice and environmental protection (Saille 2014). Just as in other earlier movements for the emancipation of subjugated classes, ethnicities, slaves, workers, colonies, women, young and disabled people and diverse sexualities, Sustainability was only pressured to become a focus for bodies like the Brundtland Commission, through diverse, protracted, radically challenging and overtly political agonistic struggles by subaltern social movements (Mouffe 1999).

Mediated often in new musical and artistic forms, these movements acted through the deepest and broadest political spaces of culture as a whole. These unruly processes were a far cry from the kinds of expert-led 'integrated assessment' or 'evidence-based' control highlighted in contemporary elite planetary management. Far removed from the 'cockpitism' of transitions processes like that so unsuccessfully presided over by the IPCC, transformations of this kind are more about a mutually choreographed rhizomic 'culturings' of change. Here knowledge and action are not forced into vertical separation by elite institutional etiquettes and divisions of labour, but are freely combined, with new configurations of practice, identity, values and interests horizontally shaping each other. It is mainly by such means, after all (rather than by more controlling interventions), that other great progressive gains of history have also been achieved—against slavery (Davis 2014) and serfdom (Leonard 2011); oppressed labour (Thompson 1966) and colonised people (Fanon 1967); subjugated women (Paletschek and Pietrow-Ennker 2004) and disabled people (Block et al. 2016); and minority ethnicities (Sudbury 1998) and sexualities (Giffney and O'Rourke 2009).

In keeping with the aphorism that 'the medium is the message', then, these ways in which Sustainability movements came into being and have thrived so long also strongly resonate with their normative content (Stirling 2016). Again, this key strand of Sustainability arose historically more through a messy mutualistic politics of care, than by orderly engineering of hierarchical instruments of control. Take, for instance the development of issues around occupational hazards, resource degradation, consumer chemicals, ionising radiation, atmospheric pollution, water contamination and climate change (Gee et al. 2001; EEA 2013). All were typically pioneered by subaltern communities of workers or affected people, then picked up by the social movements who cared for these interests. In each case, it was caring recognition of uncertainties that most strongly advanced progressive causes, not the controlling assertions of 'uncompromising', 'non-negotiable' certainties now redolent of controlling Anthropocene discourse.

Indeed, these imperatives were at each stage strongly contested by precisely the authoritarian controlling language of risk, now used by mainstream science and engineering and high-level governance institutions seeking to champion the reframing of Sustainability as 'planetary management'. The kinds of control-style aggregating analysis, optimising models and categorical boundaries that now structure 'global assessments', for 'earth systems governance', were—like earlier notions of 'assimilative capacities'—all used in efforts to resist Sustainability movements. The formative kinds of concrete action that most grew momentum around Sustainability issues were less about quantitative expert control of risks, and more about qualitative values of care for fellow people and their environments.

Turning from the motivating problems to aspiring engineering 'solutions', similar general patterns can be seen. Innovations such as wind turbines, ecological farming, super-efficient buildings, and green chemistry all owed their pioneering origins and early development to subaltern social movements (Garud and Karnøe 2012; Smith et al. 2013). All were systematically marginalised (if not actively suppressed), by incumbent interests in science, government and industry. Again, these transformative responses were nurtured not by controlling management, but by mutualistic—caring—struggle. That so many of these innovations have now become central elements in prospective transformations to Sustainability, is despite—rather than because of—'sound scientific', evidence-based elite policy discourse. Again, it was the politics of care, that brought these presently growing signs for hope from engineering into being, far more than the currently celebrated technical imaginations of control.

So at the end, we come to the question of how engineering can best assist in struggles towards Sustainability: to rebalance modernist technical fixations with control, by helping in the culturing of a newly vibrant mutualistic politics of care (Stirling 2014). And here, for the purpose of drawing practical conclusions, it might be useful to think, as a heuristic, about two ideal-typical imaginations of radical social change that pervade current global policy debates around the implications of Sustainability (Stirling 2011).

On the one hand, are what might called *sustainable transitions* (Markard 2017; Rotmans et al. 2001; Schot and Kanger 2018; Geels 2005): directed under incumbent structures by means of orderly management according to tightly disciplined technical expertise and technology-based innovations (Shove and Walker 2007; Meadowcroft 2009; Smith and Stirling 2010; Stirling 2011). Often driven especially by fear, these focus largely on presumptively well-known and singular means, with details of the wider driving ends remaining relatively less questioned. Notionally rigid categories are emphasised over inconveniently open relations. This typically emphasises integrated multidisciplinary science directed at controlling management through formal procedures in hierarchical organisations centred around technical infrastructures sponsored by the convening power of government. Exemplified in a range of currently widely propounded frameworks, this is the mode of change most characteristic of instrumental-style engineering under modernist hegemonies of control.

On the other hand, there are *transformations to Sustainability* (Brand et al. 2013; Temper et al. 2018; Stirling 2014a, 2011; Hölscher et al. 2018), involving more diverse, emergent and unruly political alignments. Best driven more by hope than by fear, these reflect less disciplined (incommensurable, tacit and embodied) knowledges and social (more than technical) innovation (Scoones et al. 2015). They involve pursuit of contending (sometimes even presently unknown) means towards contrasting ends that remain hotly contested. The focus is on open-ended relations rather than supposedly fixed categories. So, in this more political mode, space opens up for subaltern interests, social movements and civil society to struggle for in ambiguous (sometimes invisible) ways to orient the broader normative and cultural climates in which all explicit structures are set. Challenging incumbent interests associated with the modernist hegemony of control, this idea is more congruent with the mutualistic pluralism of caring imaginations of Sustainability. It is these features that might help characterise a more caring style of engineering.

Although taking contrasting forms in different contexts, it is arguably the more caring dynamics of *transformation* (more than controlling transition in these senses) that are most in keeping with the original driving values and practices of the collective action movements that brought the Brundtland Report into being. And it is associated qualities described here as care for (rather than control of) fellow people and the earth, that feature most strongly (for all their flaws) in frameworks like the SDGs. These kinds of social change are not best enacted through rigidly categorical hierarchies of deterministic cause and effect. They do not depend on incumbent interests commissioning justificatory 'evidence' or 'research' from elite institutions of academic or policy actions. They thrive instead in myriad, ostensibly small-scale, mutually shaping actions, flowing together in 'murmurations' that condition transformative social change not through rigid hierarchies of control, but through more dynamic and mutualistic relations of culture.

Referring to the often rapid changes of direction seen in flocking behaviours in nature, the idea of murmurations shows how ambitious social transformations are arguably only truly achieved through horizontal relations of care, rather than the vertical structures of control. Interestingly, the colloquial English word for such processes—murmurations—refers both to the sense of unstructured mutual coordination and subaltern critical dissent (Stirling 2016). Here, knowledge and action need not be treated as separate and sequential as prescribed in the rigid controlling protocols of 'evidence-based policy' discussed earlier. Instead, they can be recognised to be deeply entangled and mutually shaping—into multiple kinds of 'knowing doings' of kinds that shaped the culturing of Sustainability movements themselves (Stirling 2014a, 2014b).

Choreographed in this way more by distributed autonomous normative compasses than by centralised, instrumentally coercive grids, this kind of politics of transformation defies not only any prior controlling orchestration but also the imposition of such expedient storylines after the fact. Formative phenomena are not neatly nested, but rhizomically entangled across all 'phases', 'scales' and 'levels'. Key generative processes eschew the conveniently tractable logistic curves so beloved of expert diagrams, instead surging to and fro in non-monotonic waves, where it may be unclear throughout exactly which is the direction of change.

If it is to become more aligned to these potential caring dynamics of transformation, as distinct from the more controlling hierarchies of control, then engineering needs to engage more with these more relational, processual, non-linear and rhizomic forms of understanding and action. First, of course, there are the well-known values of responsibility (see also Chapters 43 and 44, this volume). And in the spirit of 'first do no harm', there are obvious implications for some of the most self-evidently unsustainable applications of engineering—for instance, in the military (Barry 2012). Beyond this, emphasis might shift from technical means to political ends—with the latter addressed more directly and in ways not seen through such instrumental lenses (Stirling 2016). This means recognition for the irreducible ambiguities and pluralities of politics such that even the idea of a singular engineering 'solution' in any given context is acknowledged to be untenable—and even such ambitions as downright damaging (Morozov 2013). Where engineering communities find themselves—as is often the case in international affairs—sitting at the "top tables" of global governance, they might try harder to avoid not only the assertion of narrow institutional agendas, but also the frequently somewhat wishful conceit of representing uninvited interests (Felt et al. 2008). More caring efforts might be directed towards opening up of political space, such that these other voices can speak directly for themselves (Stirling 2008). In short, engineering imaginations should accept the role of servant, rather than master—in the service of democratic struggles rather than seeking to control them.

Either way, the point of this distinction between 'transition' and 'transformation' is not to insist on terminology. These words are often entirely reasonably used in interchangeable ways. Nor is the point to argue that only one dynamic can be historically effective. In reality, any major political change will require interactions between each. Nor is the message that one is always more positive under a progressive view or the other necessarily negative. These are dynamics for radical social change of any normative orientation. So real-world diversities and complexities give many examples where either process spans the political spectrum. What is argued instead, are the following three—simple but potentially crucial—points.

First, it is practically useful to distinguish these two processes. Exclusion of this distinction, or failures to make it in particular settings (with whatever words), will likely lead to a default situation in which vulnerabilities are reinforced to the modernist fixation with control discussed in this chapter. On the basis of evidence documented here, this powerful current hegemony can reasonably be expected to respond to Sustainability pressures by emphasising expedient technical mechanisms of transition and sidelining the inconvenient (more care-focused) politics of transformation. Even without any countering bias, simply making the distinction is a precondition for rebalancing this political pressure.

Second, there is the point, made in this section, that it is the caring politics of transformation (rather than the more controlling management of transition), that has proven more crucially formative in the histories of emancipatory movements that have always driven Sustainability. This remains so, despite mixes of both processes being salient in different stages of complex real-world histories (and each dynamic being potentially regressive as well as progressive). This history is important in reminding how the unruly generative vibrancy of care presents a direct challenge to more controlling environmental authoritarianism currently consolidating around initiatives like 'the Anthropocene' and 'the nexus', as well as manipulative 'nudge' strategies and 'planetary management'. Under these incumbent pressures, 'transition' processes thus tend to look after themselves. Serious progressive actions towards Sustainability should therefore emphasise complementary processes of 'transformation'.

Third, there is the point that process matters. Although either political dynamic can lead to different normative ends, they are not simply 'plug and play'. Just as means are not entirely divorced from ends, so the contrasting attributes of each kind of process may be expected to leave sticky imprints on the consequences towards which they lead. No matter how well-intended they may be as means to challenge particular forms of control, interventions that are also controlling in different ways will likely reinforce this shared hegemony. Perhaps details will be reoriented. But underlying cultures of control remain intact. An extreme case is especially pertinent to historical use of controlling violence as a means towards 'revolutionary' visions of transformative change. No matter how progressive the revolutionary visions, such controlling violence can easily serve merely to perpetuate (in new modes) the oppressive relations it was ostensibly intended to control. Reflecting long-standing recognition for this point in subaltern progressive social movements, this is why Brundtland emphasised the caring imperatives of participatory process and democratic struggle.

If all this seems quite abstract and conceptual, it is not hard to see the practical salience of these contrasting faces of transformation for current high-stakes Sustainability politics. Consider, for example, the radically contrasting orientations of early moves (in alternative potential forms of Sustainability 'transitions' or 'transformations') currently underway in different sectors around the world. Entirely plausibly imaginable worldwide transformations to low-input, care-intensive agroecology may be sidelined by transitions to high-technology synthetic biology driven by intensive industrial agriculture. Transformations towards distributed community-based renewable energy and interactive energy services will be suppressed by transitions highlighting centralised grids for 'small modular' nuclear power reactors, interlinked with military propulsion and weapons infrastructures. Climate disruption may be alleviated by transformative improvements in political economy, resource efficiency and lifestyle change, or by transitions to global climate management using geoengineering technologies and their associated institutions of planetary control.

Each of these radically contrasting possibilities—and many more—are currently loudly propounded under different political views as (sometimes the 'only') possible means to find Sustainable resolutions to particular challenges in different sectors. In terms of their material feasibility, each is equally plausibly realisable in an appropriate economic context. But commitments in one direction foreclose chances of realising the other. Many well-known processes of lock-in help shape single track 'race-like' modernist visions for technology, infrastructure and institutional change. As time goes on, path dependencies increasingly do their job, and entirely feasible alternative possibilities find themselves crowded out. Although these examples are stylised, and diverse permutations are possible in this picture, the stakes between transition and transformation are clearly very high.

34.4. Practical Implications for Sustainability and Engineering

In order to draw practical conclusions, the key themes of this analysis are readily summarised. For all its progressive aspects, Modernity continues to shape and condition engineering of many kinds through a range of fixations, fallacies and failures of control. Despite specific gains under particular conditions and notions of social advance, these syndromes tend to impact unacceptably on prevailing levels of inequality, injustice and environmental degradation. Against this backdrop, Sustainability can (by contrast) be seen to be about caring for these neglected harms to people, societies, nature—and their implications for the future of the world. This compels that radical improvements must be sustained in human well-being, social equity and ecological integrity. If engineering is to assist in this, it must not only be reformed in its own practices and priorities, but play its part in aiding the wider necessary political transformations.

Here is not the place to detail the diversity of practical actions that engineers themselves have conceived and undertaken, broadly (if not explicitly) away from controlling transitions and towards more caring transformations to Sustainability. In particular, engineers repeatedly join scientists and other technical specialists around the world, variously organising in favour of 'sustainable engineering' (Davidson et al. 2010; Jonker and Harmsen 2012) or related efforts at 'engineering without borders' (Helgesson 2006) or for 'global responsibility' (INES 2019) or striving for particular visions for engineering that are (for instance) 'ethical' (Bown 2009), 'humanitarian' (Mitcham and Munoz 2010), 'responsible' (Woodhouse 2001), 'holistic' (Buch 2016) or even 'caring' (Lucena 2013).

Beyond the influence of these kinds of initiatives, further effects may also be exercised by prospective engineers 'voting with their feet'—or, more accurately, their career choices (Parkinson and Spedding 2001). This is an especially important dynamic in areas of engineering which ethical or sustainability values most obviously disfavour—like the military or nuclear industries (Blue et al. 2014; Langley 2005). Frequent lamentations in such areas over the challenges presented by a "skills gap" (HoC 2009) is an illuminating reflection of the ways in which the prevailing orientations for engineering as conditioned by incumbent structures of Modernity may sometimes be in tension with everyday normativities of ordinary people (and prospective engineering students). Mismatched "prosocial motivations" of many otherwise would-be engineers (especially women) are often observed to be a key reason for the large rates with which prospective students choose other courses (Miller et al. 2000)—or accredited engineers actually leave the profession (Rulifson and Bielefeldt 2017). Were movements towards a greater emphasis on Sustainability to be successful in rebalancing patterns of engineering activity away from areas like military and nuclear and towards (say) renewable energy or closed cycle production (CAAT 2014), then such impacts on the engineering profession might be expected to be lessened. To this extent, it is arguable that the 'murmurating' effect of cumulative career choices may be one among many cultural factors exercising convivial pressures for moves away from control to more caring engineering for Sustainability.

The implication here is not that initiatives confined to disciplines and institutions of engineering—no matter how ambitious—can completely eliminate the controlling hegemony of Modernity. Indeed, such ambitions would be ironically self-refuting in their own controlling aims. The vision is, that relatively circumscribed technical movements within and around the engineering professions must, to be successful, be accompanied by parallel action in wider public arenas—including (indeed, especially) by engineers. Synergising with movements within engineering disciplines, such complementary outside action can challenge the overarching imaginations and structures of control, within which the instrumentalised practices and institutions of engineering are continuously reproduced.

Even here, the aim cannot be eradication of all notions of control within engineering. As has been discussed, these do retain more circumscribed and conditional applicabilities in spatially and temporally confined operational contexts *within* (rather than *around*) engineered systems. The aim of the present analysis is rather towards a more nuanced and re-balancing of attention to caring and controlling idioms in engineering, reflexively countering the inherent biases of encompassing modernist cultures. And (in multiple ostensibly minor forms), this kind of effort in everyday practice can, in itself, be an example of the murmurating dynamics of transformation. Even (perhaps especially) if they are below the radar of established canons of practice, such 'political jujitsu' or 'Trojan horse' moves can *broaden out* apprehensions within engineering of the different values, understandings and possibilities that constitute the emancipatory core of Sustainability (Stirling 2016).

When articulated by engineers towards patrons and sponsors of wider infrastructures (or in horizontal actions that are uninvited and unauthorized by incumbents), there is much potential for such moves. They can help *open up* wider political appreciations in society at large, for the multiplicity of alternative pathways that exist around the world for realising—iteratively and carefully—concrete moves towards Sustainability in particular settings. And the cumulative effect of such moves, even within the traditional heartland of engineering cultures and practices, may help in processes of eroding the overbearing hegemony of Modernity itself—aiding in the vital process of *letting go*—to subvert, dissolve and pivot prevailing idioms of control, without reproducing them in self-negating attempts at countervailing domination.

If global societies are serious about enabling people to care better for each other and for the earth, then it is these kinds of actions (not countervailing efforts at control) that may most help in murmurating away from the fixations, fallacies and failures of controlling Modernity. And in seeking to advance this imperative for transformative struggles towards more caring possibilities of Sustainability, arguably no community is more important (nor their actions potentially more influential) than the professions, institutions and practices of engineering.

Acknowledgements

I owe a great debt to my SPRU colleague (former engineer and extraordinary scholar) Saurabh Arora, especially in relation to the discussion of coloniality, but also ranging far more widely.

Related Chapters

Chapter 8: Prescriptive Engineering Knowledge (Sjoerd Zwart)

Chapter 25: Sustainable Design (Steven A. Moore)

Chapter 42: Standards in Engineering (Paul B. Thompson)

Chapter 44: Responsibilities to the Public—Professional Engineering Societies (Joseph Herkert and Jason Borenstein)

Chapter 48: Feminist Engineering and Gender (Donna Riley)

Chapter 50: Engineering and Social Justice (Caroline Baillie)

Chapter 51: Engineering and Environmental Justice (Benjamin R. Cohen)

Chapter 52: Beyond Traditional Engineering: Green, Humanitarian, Social Justice and *Omnium* Approaches (George D. Catalano)

Chapter 53: Engineering and Contemporary Continental Philosophy of Technology (Diane P. Michelfelder)

Further Reading

Hommels, A., Mesman, J. and Bijker, W. (eds.) (2014). *Vulnerability in Technological Cultures: New Directions in Research and Governance*. Cambridge, MA: MIT Press. (A state-of-the-art summary of major strands in the exploring of relations between technological cultures and societal challenges.)

Jasanoff, S. (ed.) (2004). *States of Knowledge: The Co-production of Science and Social Order*. London: Routledge. (A useful outline in the broad spirit of the present analysis, concerning how political economy, state politics and the politics of knowledge are all entangled.)

Jordan, A. and Adger, N. (eds.) (2009). *Governing Sustainability*. Cambridge: Cambridge University Press. (A rich survey of the politics of sustainability, including attention to pressures for subversion.)

Meadowcroft, J. and Lunghelle, O. (eds.) (2019). *What Next for Sustainable Development? Our Common Future at Thirty*. Oxford: Oxford University Press, pp. 1–22. (An up-to-date review of the current status of several key strands in the politics of sustainable development.)

Scoones, I., Newell, P. and Leach, M. (2015). *The Politics of Green Transformations*. Edited by I. Scoones, M. Leach, and P. Newell. London: Earthscan. (An engaging account of the politics of green transformations from several diverse political and academic perspectives.)

Note

1. Later in this chapter, a series of serious political pressures will be discussed that act to appropriate and subvert Sustainability institutions and discourse in order to legitimize unrelated values or interests. The fact that the verb 'sustain' can be applied in principle in English to refer to any imaginable object, value or interest further exacerbates vulnerability to these political pressures. To help resist such misuse, reasons are elaborated in (Stirling 2009) for using a capital 'S' when using the term Sustainability in relation to the *particular* publicly deliberated values and interests around human well-being, social equity and ecological integrity made explicit and accountable in Sustainable Development processes since Brundtland (Brundtland 1987) and on through the Sustainable Development Goals (UN 2015). The capacity to maintain any *unspecified* (even entirely undeclared) object, value or interest, can then be referred to as sustainability with a small 's'.

References

Ahmad, A. (2004). Imperialism of Our Time. *Socialist Register*, 43–62.

Allenby, B. (2005). *Reconstructing Earth: Technology and Environment in the Age of Humans*. Washington, DC: Island Press.

Arendt, H. (1959). *The Human Condition, American Sociological Review*. Chicago: University of Chicago Press. doi: 10.2307/2089589.

Aronowitz, S. (1988). *Science as Power: Discourse and Ideology in Modern Society*. Minneapolis: University of Minnesota Press.

Arora, S. (2017). *Defying Control: Aspects of Caring Engagement between Divergent Knowledge Practices*. Brighton: STEPS Centre.

Arthur, W. B. (1989). Competing Technologies, Increasing Returns, and Lock-in by Historical Events. *The Economic Journal*, 99(394), 116–131.

Bandura, A. (1989). Human Agency in Social Cognitive Theory. *American Psychologist*, 44(9), 1175–1184.

Barrett, J. H. (ed.) (2003). *Contact, Continuity and Collapse: The Norse Colonization of the North Atlantic*. Turnhout.

Barry, J. (2012). *The Politics of Actually Existing Unsustainability: Human Flourishing in a Climate-Changed, Carbon-Constrained World*. New York: Oxford University Press.

Beck, U., Giddens, A. and Lash, S. (1994). *Reflexive Modernisation: Politics, Tradition and Aesthetics in the Modern Social Order*. London: Polity.

Beddington, J. (2009). *Food, Energy, Water and the Climate: A Perfect Storm of Global Events?* London.

Beeson, M. (2010). The Coming of Environmental Authoritarianism. *Environmental Politics*, 19(2), 276–294.

Bellacasa, M. P. de la (2017). *Matters of Care: Speculative Ethics in More than Human Worlds*. Minneapolis: University of Minnesota Press.

Block, P. et al. (eds.) (2016). *Occupying Disability: Critical Approaches to Community, Justice, and Decolonizing Disability*. Dordrecht: Springer.

Bloom, A. (2003). *'Takin "It to the Streets": A Sixties Reader'*. Edited by A. Bloom and W. Breines. Oxford: Oxford University Press. Available at: http://www.ncbi.nlm.nih.gov/pubmed/21928481.

Blue, E., Levine, M. and Nieusma, D. (2014). *Engineering and War: Militarism, Ethics, Institutions, Alternatives*. San Rafael: Morgan and Claypool.

Bown, W. R. (2009). *Engineering Ethics: Outline of an Aspirational Approach*. Berlin: Springer.

Brand, U. et al. (2013). Debating Transformation in Multiple Crises. In *Wotld Social Science Report 2013: Changing Global Environments*. New York: UNESCO.

Brundtland, G. H. (1987). *Report of the World Commission on Environment and Development: Our Common Future*. Oxford: Oxford University Press.

Brzoska, M. et al. (2012). Geoengineering: An Issue for Peace and Security? *Security and Peace*, 4, 29–31.

Buch, A. (2016). Ideas of Holistic Engineering Meet Engineering Work Practices. *Engineering Studies*, 8(2), 140–161. doi: 10.1080/19378629.2016.1197227.

Bukharin, N. (1917). *Imperialism and the World Economy*. London: Martin Lawrence.

Buzon, M. R. (2008). A Bioarchaeological Perspective on Egyptian Colonialism in Nubia during the New Kingdom *. 94, 165–182.

CAAT (2014). *Arms To Renewables*. London.

Cairns, R. and Stirling, A. (2014). "Maintaining Planetary Systems" or "Concentrating Global Power?" High Stakes in Contending Framings of Climate Geoengineering. *Global Environmental Change*, 28(1). doi: 10.1016/j.gloenvcha.2014.04.005.

CEC (2009). *The Lund Declaration Europe Must Focus on the Grand Challenges of Our Time*. Brussels.

Cecla, F. La (2012). *Against Architecture*. Oakland, CA: PM Press.

Chen, E. I. (1970). Japanese Colonialism in Korea and Formosa: A Comparison of the Systems of Political Control Source. *Harvard Journal of Asiatic Studies*, 30, 126–158.

Churchill, W. (2002). *Struggle for the Land: Native North American Resistance to Genocide, Ecocide and Colonization*. San Francisco: City Lights.

Collingridge, D. (1980). *The Social Control of Technology*. Milton Keynes, UK: Open University Press.

Collingridge, D. (1983). *Technology in the Policy Process: Controlling Nuclear Power*. London: Frances Pinter.

Comte, A. (2009). *A General View of Positivism*. Cambridge: Cambridge University Press. doi: 10.1017/CBO9780511692888.

Condorcet, J.-A. (1979). *Sketch for a Historical Picture of the Progress of Human Mind*. Westport, CT: Hyperion Press.

Crutzen, P. J. and Schwagerl, C. (2011). Living in the Anthropocene: Toward a New Global Ethos. *Yale Environment*, 360(24 January), 6–11.

Cuboniks, L. (2015). *Xenofeminism: A Politics for Alienation*. London: Verso.

Davidson, C. I. et al. (2010). Preparing Future Engineers for Challenges of the 21st Century: Sustainable Engineering. *Journal of Cleaner Production*. Elsevier Ltd, 18(7), 698–701. doi: 10.1016/j.jclepro.2009.12.021.

Davies, Sally C. (2011). *Infections and the Rise of Antimicrobial Resistance - Annual Report of the Chief Medical Officer, Volume II*. London: Department of Health.

Davies, G. (2012). Caring for the Multiple and the Multitude: Assembling Animal Welfare and Enabling Ethical Critique. *Environment and Planning D: Society and Space*, 30, 623–638. doi: 10.1068/d3211.

Davis, D. B. (2014). *The Problem of Slavery in the Age of Emancipation*. New York: Knopf.

Demir, H. (ed.) (2012). *Luciano Floridi's Philosophy of Technology: Critical Reflections*. Berlin: Springer.

Deringil, S. (2003). "They Live in a State of Nomadism and Savagery": The Late Ottoman Empire and the Post-colonial Debate. *Comparative Studies in Society and History*, 45(2), 311–342.

Dietler, M. (2010). *Archaeologies of Colonialism: Consumption, Entanglement and Violence in Ancient Mediterranean France*. Berkeley: University of California Press.

Durkheim, E. (1984). *The Division of Labour in Society*. London: MacMillan.

Economist (2011). The Anthropocene: A Man-made World - Science is Recognizing Humans a Geological Force to be Reckoned With. *Economist* (26 May), 1–5.

Economist (2015). The 169 Commandments: The Proposed Sustainable Development Goals Would Be Worse Than Useless. *Economist*, 1, 2015–2016. doi: 10.1017/CBO9781107415324.004.

EEA (2013). *Late Lessons from Early Warnings: Science, Precaution, Innovation—Summary*. Copenhagen, Denmark: European Environment Agency.

Eisenstadt, S. N. (2000). Multiple Modernities. *Daedalus*, 129(1), 1–29. doi: 10.2307/20027613.

Ellul, J. (1964). *The Technological Society*. New York: Alfred A. Knopf.

Elster, J. (1999). *Alchemies of the Mind: Rationality and the Emotions*. Cambridge: Cambridge University Press.

Euractive (2010). "Guilt Card" to Force Green Behaviour on Consumers? *Euractive* (4 June), 1–2.

Fanon, F. (1967). *Toward the African Revolution: Political Essays*. New York: Grove Press.

Farley, H. M. and Smith, Z. A. (2013). Defining Sustainability: Refocusing a Distorted Concept Heather. In, pp. 1–34.

Feenberg, A. (1999). *Questioning Technology*. London: Routledge.

Felt, U. et al. (2008). *Taking European Knowledge Society Seriously: Report of the Expert Group on Science and Governance to the Science, Economy and Society Directorate, Directorate-General for Research, European Commission*. Edited by U. Felt and B. Wynne. Brussels: European Commission.

Fischer, J. M. and Ravizza, M. (1998). *Responsibility and Control: A Theory of Moral Responsibility*. Cambridge: Cambridge University Press.

Fisher, B. and Tronto, J. (1991). Toward a Feminist Theory of Care. In E. Able and M. Nelson (eds.), *Circles of Care: Work and Identity in Women's Lives*. Albany, NY: SUNY Press.

Fuller, S. and Collier, J. H. (2004). *Philosophy, Rhetoric, and the End of Knowledge: A New Beginning for Science and Technology Studies*. Hahwah, NJ: Lawrence Erlbaum.

Garud, R. and Karnøe, P. (2012). Path Creation as a Process of Mindful Deviation. In R. Garud and P. Karnøe (eds.), *Path Dependence And Creation*. Hove, East Sussex: Psychology Press, pp. 1–38.

Gee, D. et al. (eds.) (2001). *Late Lessons from Early Warnings: the precautionary principle 1896–2000*. Copenhagen: European Environment Agency.

Geels, F. (2005). *Technological Transitions and System Innovations: A Co-evolutionary and Socio-technical Analysis*. Cheltenham, UK; Northampton, MA: Edward Elgar.

Gellner, E. (1983). *Nations and Nationalism*. Ithaca: Cornell University Press.

GFHR (2004). *The 10/90 Report on Health Research: 2003-2004*. Geneva: Global Forum for Health Research.

Giffney, N. and O'Rourke, M. (eds.) (2009). *The Ashgate Research Companion to Queer Theory*. Farnham, Surrey: Ashgate.

Gilligan, C. (1993). *In a Different Voice: Psychological Theory and Women's Development*.

Gladney, D. C. (2004). Internal Colonialism and the Uyghur Nationality: Chinese Nationalism and its Subaltern Subjects. *Cahiers d'Etudes sur la Méditerranée Orientale et le monde Turco-Iranien*, 25, 3–12.

Goldberg, C. D. E. et al. (2008). *Workshop on Philosophy & Engineering*. London.

Gregory, D. et al. (eds.) (2009). *The Dictionary of Human Geography*. London: Wiley Blackwell.

Grove, R. H. (1997). *Ecology, Climate and Empire: Colonialism and Global Environmental History, 1400-1940*. Cambridge: White Horse Press. doi: 10.2307/3985186.

Grove-White, R. (2001). The Rise of the Environmental Movement. In T. C. Smout (ed.), *Nature, Landscape and People since the Second World War*.

Groves, C. (2013). Horizons of Care: From Future Imaginaries to Responsible Innovation. *Shaping Emerging Technologies: Governance, Innovation, Discourse*, 185–202.

Groves, C. (2015). Logic of Choice or Logic of Care? Uncertainty, Technological Mediation and Responsible Innovation. *NanoEthics*, 9, 321–333. doi: 10.1007/s11569-015-0238-x.

Habermas, J. (1985). In Foreste. In G. S. Habermas (ed.), *Critical Theory and Public Life*. Cambridge, MA, pp. 414, 415.

Halperin, S. (2007). Re-envisioning Global Development: Conceptual and Methodological Issues. *Globalizations*, 4(4), 543–558. doi: 10.1080/14747730701695810.

Hamilton, J. T. (2013). *Security: Politics, Humanity, and the Philology of Care*. Princeton: Princeton University Press.

Harremoës, P. et al. (eds.) (2001). *Late Lessons from Early Warnings: The Precautionary Principle 1896-2000*. Copenhagen: European Environment Agency.

Harremoës, P. et al. (2002). *The Precautionary Principle in the 20th Century: Late Lessons from Early Warnings*. London: Earthscan. Available at: http://books.google.co.uk/books?id=JSBjiEUPyJkC.

Harremoës, P. and European Environment, A. (2001). *Late Lessons from Early Warnings: The Precautionary Principle 1896-2000*. Luxembourg: Office for Official Publications of the European Communities. Available at: http://reports.eea.eu.int/environmental_issue_report_2001_22/en/Issue_Report_No_22.pdf.

Harvey, D. (2003). *The New Imperialism*. Oxford: Oxford University Press.

Hegel, G. W. F. (1974). *Hegel's Lectures on the History of Philosophy*. London: Humanities Press.

Held, V. (2005). *The Ethics of Care: Personal, Political and Global*. Oxford: Oxford University Press. doi: 10.1093/0195180992.001.0001.

Helgesson, C. I. (2006). Engineers Without Borders. *IEEE Engineering in Medicine and Biology Magazine*. IEEE, 25(June), 32–35. doi: 10.1109/MEMB.2006.1636348.

Hickman, L. (2010). James Lovelock: Humans are Too Stupid to Prevent Climate Change. *Guardian* (29 March), 2–5.

Hildyard, N. (2016). *Licensed Larceny Infrastructure, Financial Extraction and the Global South*. Manchester: Manchester University Press.

HoC (2009). *Engineering: Turning Ideas into Reality*. London.

Hoffman, K. M. (2015). Connecting People – An Evolutionary Perspective on Infraculture. In E. A. Picot et al. (eds.), *The Economics of Infrastructure Provisioning: The (Changing) Role of the State*. Cambridge, MA: MIT Press, pp. 1–23.

Holden, E., Linnerud, K. and Banister, D. (2014). Sustainable Development: Our Common Future Revisited. *Global Environmental Change*. Elsevier Ltd, 26, 130–139. doi: 10.1016/j.gloenvcha.2014.04.006.

Holling, C. S. and Meffe, G. K. (1996). Command and Control and the Pathology of Natural Resource Management. *Conservation Biology*, 10(2), 328–337. doi: 10.1046/j.1523-1739.1996.10020328.x.

Hölscher, K., Wittmayer, J. M. and Loorbach, D. (2018). Transition versus Transformation: What's the Difference? 27(April 2017), 1–3. doi: 10.1016/j.eist.2017.10.007.

Hommels, A., Mesman, J. and Bijker, W. (eds.) (2014). *Vulnerability in Technological Cultures: New Directions in Research and Governance*. Cambridge, MA: MIT Press.

Hudson, M. (1972). *Super Imperialism: The Origins and Fundamentals of US World Dominance*. London.

Hulme, M. (2009). *Why We Disagree About Climate Change*. Cambridge: Cambridge University Press.

IAASTD (2009). *Agriculture at a Crossroads: International Assessment of Agricultural Knowledge Science and Technology for Development (IAASTD)*. Washington, DC: Island Press.

ICAN (2015). *Catastrophic Humanitarian Harm*. Geneva.

INES (2019). *Mission Statement of the Internation al Network of Engineers and Scidntsts for Global Responsibility*. Available at: http://inesglobal.net/ineshome/mission-statement/. Accessed January 17, 2019.

Inglehart, R. and Welzel, C. (2005). *Modernization, Cultural Change and Democracy: The Human Developmente Sequence*. Cambridge: Cambridge University Press.

IPCC_WGII (2014). Climate Change 2014: Impacts, Adaptation, and Vulnerability. (March), 1–44.

Ison, R. (2010). *Systems Practice: How to Act in a Climate-change World*. London: Springer.

Jacobs, M. D. (2009). *White Mother to a Dark Race: Settler Colonialism, Materialism, and the Removal of Indigenous Children in the American West and Australia, 1880-1940*. Lincoln, NB: University of Nebraska Press.

Jasanoff, S. (ed.) (2004). *States of Knowledge: The Co-production of Science and Social Order*. London: Routledge. doi: 10.4324/9780203413845.

Jasanoff, S. (2005). *Designs on Nature: Science and Democracy in Europe and the United States*. Princeton: Princeton University Press. doi: 10.1017/CBO9781107415324.004.

Jones, M. (2010). *After Hiroshima: The United States, Race and Nuclear Weapons in Asia, 1945-65*. Cambridge: Cambridge University Press.

Jonker, G. and Harmsen, J. (2012). *Engineering for Sustainability: A Practical Guide for Sustainable Design*. Amsterdam: Elsevier.

Keller, E. F. (1984). Science and Power for What. In E. Mendelsohn and H. Nowotny (eds.), *Nineteen Eighty-four: Science between Utopia and Dystopia*. Dordrecht: Reidel.

Kiely, R. (2005). *The Clash of Globalisations: Neoliberalism, the Third Way and Anti-globalization*. Leiden: Brill.

Langley, C. (2005). *Soldiers in the Laboratory: Military Involvement in Science and Technology - and Some Alternatives*. Folkestone: Scientists for Global Responsibility.

Lannoo, M. J. (2010). *Leopold's Shack and Ricketts's Lab: The Emergence of Environmentalism*. Berkeley: University of California Press.

Larsen, M. T. (1974). [review of] Louis L. Orlin, The Old Assyrian Colonies in Anatolia. *Journal of the American Oriental Society*, 94(4), 468–475.

Latvus, K. (2006). Decolonizing Yahweh: A Postcolonial Reading of 2 Kings 24–251. In R. S. Sugirtharajah (ed.), *The Postcolonial Biblical Reader*. London: Blackwell.

Leigh, J. R. (2004). *Control Theory*. Stevenage: Institute of Electrical Engineers.

Leonard, Carol S. (2011). *Agrarian Reform in Russia: The Road from Serfdom*. Cambridge: Cambridge University Press.

Lövbrand, E. et al. (2015). Who Speaks for the Future of Earth ? How Critical Social Science Can Extend the Conversation on the Anthropocene Who Speaks for the Future of Earth ? How Critical Social Science Can Extend the Conversation on the Anthropocene. *Global Environmental Change*, 32, 211–218.

Lövbrand, E., Stripple, J. and Wiman, B. (2009). Earth System Governmentality. *Global Environmental Change*, 19(1), 7–13. doi: 10.1016/j.gloenvcha.2008.10.002.

Lucena, J. (ed.) (2013). *Engineering Education for Social Justice: Critical Explorations and Opportunities*. Berlin: Springer.

Markard, J. (2017). Sustainability Transitions: Exploring the Emerging Research Field and its Contribution to Management Studies. In *Proceedings of the 33rd EGOS Colloquium, Copenhagen, July 6-8*. Copenhagen: ETH Zurich.

Marx, K. (1976). *Capital: A Critique of Political Economy - Volume 1*. Harmondsworth: Penguin.

May, V. M. (2015). *Pursuing Intersectionality, Unsettling Dominant Imaginaries*. London: Routledge.

McCully, P. (2001). *Silenced Rivers: The Ecology and Politics of Large Dams*. London: Zed Books.

McGovern, U. (ed.) (2005). *Chambers Dictionary of Quotations*. Edinburgh: Chambers Harrap Publishers. doi: 10.5005/jp/books/10552_1.

Meadowcroft, J. (2009). What about the Politics? Sustainable Development, Transition Management, and Long Term Energy Transitions. *Policy Sciences*, 42(4), 323–340. doi: 10.1007/s11077-009-9097-z.

Michelfelder, D. P., Newberry, B. and Zhu, Q. (eds.) (1998). *Responsibility and Control: A Theory of Moral Responsibility*. Cambridge: Springer.

Mignolo, W. D. (1995). *The Darker Side of the Renaissance: Literacy, Territoriality, and Colonization*. Ann Arbor: University of Michigan Press.

Mignolo, W. D. (2007). Delinking: The Rhetoric of Modernity, the Logic of Coloniality and the Grammar of De-coloniality. *Cultural Studies*, 21(2–3), 1–75.

Millar, S. W. S. and Mitchell, D. (2017). The Tight Dialectic: The Anthropocene and the Capitalist Production of Nature. *Antipode*, 49, 75–93. doi: 10.1111/anti.12188.

Miller, P. H. et al. (2000). A Desire to Help Others: Goals of High-achieving Female Science Undergraduates. *Women's Studies Quarterly*, 28(1–2), 128–142.

Misa, T. J., Brey, P. and Feenberg, A. (eds.) (2003). *Modernity and Technology*. Cambridge, MA: MIT Press.

Mitcham, C. and Munoz, D. (2010). *Humanitarian Engineering, Synthesis Lectures on Engineers, Technology and Society*. San Rafael: Morgan and Claypool. doi: 10.2200/S00248ED1V01Y201006ETS012.

Morgenthau, H. (1947). *Scientific Man vs Power Politics*. London: Latimer House.

Morozov, E. (2013). *To Save Everything, Click Here: The Folly of Technological Solutionism*. London: Allen Lane.

Mouffe, C. (1999). Deliberative Democracy or Agonistic Pluralism. *Social Research*, 66(3), 745–758.

Mueller, J. (2001). *Retreat from Doomsday: The Obsolescence of Major War*. New York: Basic Books.

Mumford, L. (1934). *Technics and Civilization*. London: Routledge and Kegan Paul.

Mwangi, W. (2001). Of Coins and Conquest: The East African Currency Board, the Rupee Crisis, and the Problem of Colonialism in the East African Protectorate. *Comparative Studies in Society and History*, 43(4), 763–787.

Newton, P. (1999). A Manual for Planetary Management. *Nature*, 741, 1998.

OED (2013). *Oxford English Dictionary – Online*. Oxford University Press. Available at: http://www.oed.com/.

Paletschek, S. and Pietrow-Ennker, B. (eds.) (2004). *Women's Emancipation Movements in the Nineteenth Century: A European Perspective*. Standford: Stanford University Press.

Parfitt, T. (2017). Inhuman Development ? Technics as Enframing or Poiesis? *Third World Quarterly*. Routledge, 6597(April), 1–19. doi: 10.1080/01436597.2016.1229565.

Parkinson, S. and Spedding, V. (2001). *An Ethical Career in Science and Technology?* London.

Pellizzoni, L. (2004). Responsibility and Environmental Governance. *Environmental Politics*, 13(3), 541–565. doi: 10.1080/0964401042000229034.

Pepper, D. (1996). *Modern Environmentalism: An Introduction*. London: Routledge.

Pidgeon, N. et al. (2016). *European Perceptions of Climate Change (EPCC): Sociopolitical Profiles to Inform a Cross-national Survey in France, Germany, Norway and the UK*. Oxford: Climate Outreach.

Piketty, T. (2014). *Capital in the Twenty First Century*. Cambridge, MA; London, UK: The Belknap Press of Harvard University Press.

Plumwood, V. (1993). *Feminism and the Mastery of Nature*. London: Routledge.

Pope_Francis (2015). *Laudato Si - Praised Be: Encyclical on Care of Our Common Home*. Rome: The Vatican.

Preston, C. J. and Wickson, F. (2016). Broadening the Lens for the Governance of Emerging Technologies: Care Ethics and Agricultural Biotechnology. *Technology in Society*. Elsevier Ltd, 45, 48–57. doi: 10.1016/j.techsoc.2016.03.001.

Quijano, A. (2007). Coloniality and Modernity / Rationality. *Cultural Studies*, 2386. doi: 10.1080/09502380601164353.

RAEng (2011). *Philosophy of Engineering - Volume 2 of the Proceedings of a Series of Seminars Held at the Royal Academy of Engineering*. Edited by K. Guy. London: Royal Academy of Engineering.

Robinson, F. (2011). *The Ethics of Care: A Feminist Approach to Human Security*. Philadelphia: Temple University Press.

Rockström, J. (2010). *Let the Environment Guide Our Development*. TED Talks. Available at: http://www.ted.com/talks/johan_rockstrom_let_the_environment_guide_our_development.html. Accessed January 18, 2019.

Rockström, J. et al. (2009). A Safe Operating Space for Humanity. *Nature*, 461(24 September), 472–475.

Rotmans, J., Kemp, R. R. and Asselt, M. van (2001). More Evolution Than Revolution: Transition Management in Public Policy. *Foresight*, 3(1), 15–31.

Rulifson, G. and Bielefeldt, A. (2017). Motivations to Leave Engineering: Through a Lens of Social Responsibility. *Engineering Studies*, 9(3), 222–248. doi: 10.1080/19378629.2017.1397159.

Said, E. W. (1994). *Culture and Imperialism*. New York: Vintage Books.

Saille, S. de (2014). Dis-inviting the Unruly Public. *Science as Culture*, 24(1), 99–107. doi: 10.1080/09505431.2014.986323.

Santos, B. de S. (2002). The Proicesses of Globalisation. *Časopis za književnost i kulturu, i društvena pitanja*, 68(14), 67–131.

Schellnhuber, H. J. (1999). "Earth System" Analysis and the Second Copernican Revolution. *Nature*, 402 SUPP(2 December), C19–C23.

Schot, J. and Kanger, L. (2018). Deep Transitions: Emergence, Acceleration, Stabilization and Directionality. *Research Policy*. Elsevier (March), 1–15. doi: 10.1016/j.respol.2018.03.009.

Schweitzer, A. (1947). *Albert Schweizer: An Anthology*. Edited by C. R. Joy. Boston: Beacon Press.

Scoones, I., Newell, P. and Leach, M. (2015). *The Politics of Green Transformations*. Edited by I. Scoones, M. Leach, and P. Newell. London: Earthscan.

Scott, J. (2006). *Social Theory: Central Issues in Sociology*. London: Sage.

Seth, S. (2016). Is Thinking with "Modernity" Eurocentric? *Cultural Sociology*, 10(3), 385–398. doi: 10.1177/1749975516637203.

Shackley, S., Wynne, B. and Waterton, C. (1996). Imagine Complexity: The Past, Present and Future Potential Complex Thinking. *Futures*, 28(3), 201–225.

Shepherd, J. et al. (2009). *Geoengineering the Climate: Science, Governance and Uncertainty*. London: The Royal Society.

Shove, E. and Walker, G. (2007). CAUTION! Transitions Ahead: Politics, Practice, and Sustainable Transition Management. *Environment and Planning A*, 39(4), 763–770. doi: 10.1068/a39310.

Smith, A., Fressoli, M. and Thomas, H. (2013). Grassroots Innovation Movements: Challenges and Contributions. *Journal of Cleaner Production*. Elsevier Ltd, pp. 1–11. doi: 10.1016/j.jclepro.2012.12.025.

Smith, A. and Stirling, A. (2010). The Politics of Social-ecological Resilience and Sustainable Socio-technical Transitions. *Ecology and Society*, 15(1).

Stanovich, K. E. et al. (2014). Dual-process Theories of the Social Mind. *Dual-Process Theories of the Social Mind*, 642. doi: 10.1080/00224545.2014.953874.

Steffen, W. et al. (2018). Trajectories of the Earth System in the Anthropocene. *PNAS*, 1–8. doi: 10.1073/pnas.1810141115.

Sterner, T. et al. (2019). Policy Design for the Anthropocene. *Nature Sustainability*. Springer US, 2(January). doi: 10.1038/s41893-018-0194-x.

Stirling, Andrew (2008). "Opening Up" and "Closing Down": Power, Participation, and Pluralism in the Social Appraisal of Technology. *Science, Technology and Human Values*, 23(2), 262–294.

Stirling, Andrew (2009). Participation, Precaution and Reflexive Governance for Sustainable Development. In A. Jordan and N. Adger (eds.), *Governing Sustainability*. Cambridge: Cambridge University Press. doi: 10.1017/CBO9780511807756.011.

Stirling, Andrew (2011). Pluralising Progress: From Integrative Transitions to Transformative Diversity. *Environmental Innovation and Societal Transitions*, 1(1), 82–88. doi: 10.1016/j.eist.2011.03.005.

Stirling, Andrew (2014). *Emancipating Transformations: From the Controlling 'the Transition' to Culturing Plural Radical Progress*. Brighton: STEPS Centre, University of Sussex.

Stirling, Andy (2014a). From Sustainability to Transformation: Dynamics and Diversity in Reflexive Governance of Vulnerability. In *Vulnerability in Technological Cultures: New Directions in Research and Governance*. Cambridge, MA: MIT Press, pp. 1–61.

Stirling, Andy (2014b). Towards Innovation Democracy? Participation, Responsibility and Precaution in Innovation Governance. In *Annual Report of the Government Chief Scientific Adviser 2014, Innovation: Managing Risk, Not Avoiding It. Evidence and Case Studies*. London: UK Government, pp. 49–62.

Stirling, Andy (2015). Power, Truth and Progress: Towards Knowledge Democracies in Europe. In J. Wilsdon and R. Doubleday (eds.), *Future Directions for Scientific Advice in Europe*. Cambridge: Cambridge University Press, pp. 133–151.

Stirling, Andy (2016). Knowing Doing Governing: Realising Heterodyne Democracies. In J.-P. Voß and R. Freeman (eds.), *Knowing Governance: The Epistemic Construction of Political Order*. Basingstoke: Palgrave MacMillan.

Stirling, Andy (2018). *How Deep is Incumbency? Introducing a 'Configuring Fields' Approach to the Distribution and Orientation of Power in Socio-technical Change*. 2018–23. Brighton.

Stirling, Andy (2019a). Engineering and Sustainability: Control and Care in Unfoldings of Modernity. 06.

Stirling, Andy (2019b). How Deep is Incumbency? A "Configuring Fields" Approach to Redistributing and Reorienting Power in Socio-material Change. *Energy Research & Social Science*. Elsevier, 58(July), 101239. doi: 10.1016/j.erss.2019.101239.

Stirling, Andy (2019c). Sustainability and the Politics of Transformations from Control to Care in Moving Beyond Modernity. In J. Meadowcroft and O. Lunghelle (eds.), *What Next for Sustainable Development?: Our Common Future at Thirty*. Oxford: Oxford University Press, pp. 1–22.

Strebhardt, K. and Ullrich, A. (2013). Paul Ehrlich's Magic Bullet Concept: 100 Years of Progress. *Nature Reviews Cancer*, 8(July 2008), 473–480. doi: 10.1038/nrc2394.

Sudbury, J. (1998). *Other Kinds of Dreams: Black Women's Organisations and the Politics of Transformation*. London: Routledge.

Taylor, Y., Hines, S. and Casey, M. E. (eds.) (2010). *Theorizing Intersectionality and Sexuality*. London: Palgrave Macmillan.

Temper, L. et al. (2018). A Perspective on Radical Transformations to Sustainability: Resistances, Movements and Alternatives. *Sustainability Science*. Springer Japan, 0(0), 0. doi: 10.1007/s11625-018-0543-8.

Thompson, E. P. (1966). *The Making of the English Working Class*. New York: Knopf.

Tocqueville, A. de (2003). *Democracy in America*. Harmondsworth: Penguin Classics.

Trigger, B. G. and Washburn, W. E. (eds.) (1996). *The Cambridge History of the Native Peoples of America*. Cambridge: Cambridge University Press.

UN (2015). *Transforming Our World: The 2030 Agenda for Sustainable Development*. New York: United Nations.

UNEP (2014). *Assessing Global Land Use: Balancing Consumption with Sustainable Supply. A Report of the Working Group on Land and Soils of the International Resource Panel*. New York: United Nations Environment Programme.

Unruh, G. C. (2000). Understanding Carbon Lock-in. *Energy Policy*, 28(2000), 817–830.

UNSCEAR (2016). *Sources, Effects and Risks of Ionizing Radiation: UNSCEAR 2016 Report: Report to the General Assembly*. New York: United Nations Scientific Committee on the Effects of Atomic Radiation.

Van de Poel, I. et al. (eds.) (2010). *Philosophy and Engineering: An Emerging Agenda*. Berlin: Springer.

Verbeek, P.-P. (2007). Morality in Design: Design Ethics and the Morality of Technological Artifacts. In P. E. Vermaas et al. (eds.), *Philosophy and Design: from Engineering to Architecture*. Berlin: Springer, 91–103.

Viola, L. (1996). *Peasant Rebels Under Stalin: Collectivization and the Culture of Peasant Resistance*. Oxford: Oxford University Press.

VLC (2016). *Taking Back Control from Brussels*. London.

Voss, J., Bauknecht, D. and Kemp, R. (eds.) (2006). *Reflexive Governance for Sustainable Development*. Cheltenham: Edward Elgar.

Wallerstein, I. (2006). *World Systems Analysis: An Introduction*. London: Duke University Press.

Ward, C. (1976). *Housing: An Anarchist Approach*. London: Freedom Press.

Weber, M. (1947). *The Theory of Social and Economic Organization*. New York: Free Press.

Wickson, F. et al. (2017). Addressing Socio-economic and Ethical Considerations in Biotechnology Governance: The Potential of a New Politics of Care. *Food Ethics*. Food Ethics, (June). doi: 10.1007/s41055-017-0014-4.

Williams, R. and Edge, D. (1996). The Social Shaping of Technology. *Research Policy*, 25(6), 865–899. doi: 10.1016/0048-7333(96)00885-2.

Winner, L. (1977). *Autonomous Technology: Technics Out of Control as a Theme in Political Thought*. Cambridge, MA: MIT Press.

Wolpert, L. (1992). *The Unnatural Nature of Science - Why Science Does Not Make (Common). Sense*. London: Faber and Faber.

Woodhouse, E. (2001). Curbing Overconsumption: Challenge for Ethically Responsible Engineering. *IEEE Technology and Society* (Fall), 23–30. doi: 10.1109/44.952762.

Wynne, B. (1993). Public Uptake of Science: A Case for Institutional Reflexivity. *Public Understanding of Science*, 2, 321–337. doi: 10.1088/0963-6625/2/4/003.

Wynne, B. (2010). *Rationality and Ritual: Participation and Exclusion in Nuclear Decision-Making*. 2nd edn. London: Earthscan.

Yearley, S. (2005). *Cultures of Environmentalism: Empirical Studies in Environmental Sociology*. London: Palgrave Macmillan. doi: 10.1057/9780230514867.

Yi-chong, X. (2014). Chinese State-owned Enterprises in Africa: Ambassadors or Freebooters? *Journal of Contemporary China*. Taylor & Francis, 23(89), 822–840. doi: 10.1080/10670564.2014.882542.

Zavialov, P. (2005). *Physical Oceanography of the Dying Aral Sea*. Berlin: Springer Verlag.

35

THE ROLE OF RESILIENCE IN ENGINEERING

Neelke Doorn

35.1. Introduction

Resilience is a value that has only recently entered the domain of engineering in its present meaning. In a general sense, resilience might be seen as a value or ideal to which engineering should aspire or contribute. Thus, similar to, for example, engineering for health (see also Chapter 38, "Health" by Marianne Boenink, this volume) or engineering for sustainability (see also Chapter 25, "Sustainable Design" by Steven A. Moore, and Chapter 34, "Engineering and Sustainability: Control and Care in Unfoldings of Modernity" by Andy Stirling, this volume), one would expect to find under the heading of "engineering for resilience" approaches that focus on specific engineering strategies or activities that aim to strengthen resilience. However, in engineering, the term "resilience" is more commonly used to refer to a particular approach within safety management, usually denoted as "resilience engineering."

This chapter will primarily focus on *resilience engineering* and less on *engineering for resilience*. The reason for this is that the term "resilience" is interpreted quite differently across disciplines and contexts. Since the purpose of engineering for resilience is very much dependent on how resilience is interpreted, it is difficult to find one specific approach that can be captured under the heading of *engineering for resilience*.

The outline of this chapter is as follows. Section 2 presents a brief discussion of the history of the term "resilience" and its use in different disciplines. In this section, we will also look at how *resilience engineering* and *engineering for resilience* are related. Section 3 presents the Resilience paradigm that forms the basis for resilience engineering, followed by a discussion of the four cornerstones of resilience engineering (Section 4) and its implementation in practice (Section 5). Finally, the measures to implement resilience are compared with traditional safety measures (Section 6). The concluding Section 7 summarizes the main findings.

35.2. History and Use of the Term in Different Disciplines

Much of the literature on resilience refers to the work of the ecologist Holling, who introduced the term in the context of ecosystems (Holling 1973). However, the history of the term goes back even further. The first use in an engineering context can be found at the beginning of the nineteenth century, where resilience was used to describe the property of certain types of wood that could accommodate sudden and severe loads without breaking (McAslan 2010). In the mid-nineteenth century, the naval architect of the British Admiralty, Robert Mallet, used the phrase *modulus of resilience* as a measure for assessing the ability of materials to withstand severe conditions in warship design (Mallet 1862).

Thus, although not the source of the first use of the term, Holling's paper is probably one of the first in which an attempt is made to provide a more precise description of "resilience." Drawing on examples from ecology, Holling distinguished resilience from stability, defining the latter as "the ability of a system to return to its equilibrium state after a temporary disturbance" (p. 14). He described resilience, as "the persistence of systems and their ability to absorb change and disturbance and still maintain the same relationships between populations or state variables" (ibid.). Hence, resilience does not require a system to return to an equilibrium state. Resilience, according to Holling, is primarily about being able to absorb and accommodate unexpected future events.

After its introduction in ecology in the 1960s and 1970s, the term "resilience" became popular in other domains as well, entering the field of safety management around 2000 (Woods and Wreathall 2003) as well as other fields such as psychology (Connor and Davidson 2003; Southwick et al. 2005), disaster management (Adger 2000; Paton and Fohnston 2001; Walker et al. 2004), and even business (Hamel and Valikangas 2003).

With the use of the same term in different domains, different definitions and interpretations emerged (Doorn 2017). In the last decade, several studies have been published that refer to various meanings and definitions of resilience. An often-cited taxonomy is that provided by Folke (2006), who distinguishes between three notions of resilience, ranging from a narrow interpretation ("engineering resilience"), to ecological/ecosystem and social resilience, and an even broader social-ecological interpretation of resilience. The narrow interpretation refers to the capacity to return to a stable equilibrium and focuses on recovery and constancy, similar to what Holling referred to as "stability." The second notion of resilience originates from ecology and follows Holling's idea of resilience as persistence. It can be applied to ecosystems, but also to social systems, and it refers to a system's buffer capacity and ability to withstand shocks and maintain its functions. The third notion refers to social-ecological resilience, which incorporates ideas about adaptation, learning, and self-organization, in addition to the general ability to recover from a disturbance.

Reference to "engineering resilience" or the "engineering interpretation of resilience" is misleading in the sense that the field of "resilience engineering," which emerged at the end of the twentieth century, represents a distinctly different idea from that of the "modulus of resilience" mentioned earlier, and it has a much broader sense than a mere capacity for recovery or return to stable equilibrium. Resilience engineering concerns a particular approach to safety management that actually has more in common with the social-ecological interpretation of resilience than with the narrow engineering interpretation of resilience as stability.

In resilience engineering, the meaning of the term has moved in the direction of social and socio-ecological resilience. Thus, the term no longer simply refers to a return to some stable equilibrium, but rather denotes a new way of looking at how complex sociotechnical systems work. In the context of healthcare engineering, for example, Fairbanks et al. (2013) argue that resilience should not be seen as a dichotomous outcome (a system being resilient or not resilient or being safe or unsafe), but rather as a measure that indicates performance. Based on this view, resilience is a measure of the potential of a system or an organization to perform in a certain way.

This broader conception of resilience engineering prompts the question already raised by Carpenter et al. (2001) in the context of ecological resilience: resilience *of* what *to* what? If resilience is best seen as the potential of a system to perform in a certain way, this system would seem the obvious answer to the "resilience *of* what" question. The Resilience paradigm is typically applied to *socio*technical systems, which suggests that the potential for resilience is not restricted to physical artifacts but also looks at how humans function within that system.

The "resilience *to* what" question refers to the type of threats to which the system should be able to respond. Typical examples in the context of engineering include natural hazards, such as earthquakes, flooding, and volcanic eruptions, but they may also include human-induced threats, such as security attacks, or demographic trends, such as urbanization and aging. These processes may put pressure on a system in various ways.

The view of resilience as the potential of a sociotechnical system to perform in a certain way suggests that resilience itself is not the ultimate aim, but that it is a formal concept that should be elaborated in terms of what is considered desirable performance (Doorn et al. 2019). This desirable performance is ultimately linked to the goal of engineering, which is often described in terms of its contribution to specific values; for example, human well-being or safety (cf. NSPE 2007; Royal Academy of Engineering 2014), or, more broadly, to the promotion of human flourishing (Bowen 2014). Here, resilience, as a value that engineering aims to achieve (*engineering for resilience*), also comes into play as an aspect of human welfare; for example, a person's psychological or physiological resilience. This resilience, however, is different from the characteristic by which we describe the performance of the sociotechnical system as a whole. When discussing resilience in engineering, it is therefore important to keep in mind whether resilience is used to denote the value that a sociotechnical system should contribute to (*engineering for resilience*), or whether it is used to characterize the performance of the system as a whole (*resilience engineering*). The role of humans in the two approaches also differ. In *engineering for resilience*, humans are the beneficiaries, in the sense that the engineering activities contribute to human resilience. In *resilience engineering*, humans are part of the sociotechnical system, and they thereby contribute to the system's resilience and, although not the focus, as part of the system, they could also be the beneficiaries of the system's resilience.

One domain of application in which we can recognize both approaches is in the field of urban resilience (Wardekker forthcoming). In analogy with the resilience engineering approach, and with a strong basis in systems dynamics, we see approaches to urban resilience that define it as "the ability of the city to maintain the functions that support the well-being of its citizens" (Da Silva et al. 2012), conceptualizing cities as systems with components, functions, and flows of resources, materials, and people, among other elements (e.g., Wardekker et al. 2010; Meerow et al. 2016). Parallel to this systems framing of urban resilience, we also see a community framing of urban resilience, which has its roots in the fields of disaster preparedness and psychology, and which focuses on how communities are impacted by disturbances. Here, the focus is on fostering community characteristics that are considered intrinsically valuable, such as urban life, community bonds, and self-sufficiency (e.g., Leichenko 2011; Berkes and Ross 2013).

In engineering, the focus will more often be on the systems approach to urban resilience. In the remainder of this chapter, we will look in more detail at what resilience engineering entails and how it is different from traditional approaches to safety management, leaving aside the question of how to interpret resilience in *engineering for resilience*.

35.3. Resilience Engineering: A New Paradigm of Safety Management

Resilience engineering is about the functioning of *sociotechnical systems*, where such systems should be seen as heterogeneous engineering systems that consist not only of technical infrastructure but also of people and institutions (Kroes et al. 2006). A flood defense system, for example, is composed of material objects (storm surge barriers, closeable gates, dikes, and levees), people operating the system, and rules or "institutions" which stipulate the conditions under which a barrier or gate needs to be closed. These different elements are essential for the functioning of the system as a whole. The focus on *sociotechnical systems* unavoidably prompts questions about the interaction required between human beings and the technology. Looking at these human factors is therefore indispensable for resilience engineering. In the remainder of this section, when discussing systems, I explicitly mean sociotechnical systems and not pure technological systems.

The earlier discussion reveals that resilience is not an easy term to define. In resilience engineering, as mentioned, the term reflects a specific view on safety management and the functioning of systems. Resilience engineering is therefore unavoidably linked to a particular safety paradigm. This section will thus begin by discussing the safety paradigm that underlies "resilience engineering."

The origin of the resilience engineering paradigm is often linked to the first symposium of the Resilience Engineering Association (REA), which was held in 2004 and which has since been organized once every two to three years (Patriarca et al. 2018). The discussions and afterthoughts of the first four symposia were documented in a series of volumes dedicated to both theoretical and conceptual work as well as more practical applications. Especially in the early years of REA's existence, some important conceptual work was done to clarify the term "resilience engineering" and its relation to more traditional approaches to safety management.

Erik Hollnagel, one of REA's co-founders and a leading scholar in resilience engineering, describes it as part of a Safety-II paradigm, while Safety-I is used to refer to traditional approaches to safety management (Hollnagel 2014). Although Safety-I and Safety-II should not be seen as entailing incompatible or conflicting views, for the sake of clarity it is worth discussing the main differences. The following discussion is largely based on Hollnagel's deconstruction of the two paradigms in terms of their phenomenology, etiology, and ontology. For the sake of readability, I use the term "Resilience" to refer to the Safety-II paradigm and "Safety" to refer to the Safety-I paradigm.

35.3.1 Phenomenology

The phenomenology of the two paradigms refers to how safety and resilience manifest themselves in terms of observable characteristics. When do we call systems safe or resilient? Something "being safe" refers to a situation in which the outcome of an action is as expected (Hollnagel 2014: p. 3). For example, a railway track is considered safe if the train can be driven on the track without derailing or crashing into a car at a crossing. Although this sounds trivial and a definition that few people would oppose, this is not how safety is usually assessed or measured. Safety is rarely assessed in terms of instances of success or tasks that are performed without incident or accident. Rather, safety is more often assessed in terms of things that have gone *wrong*. The safety of a railway track is expressed in terms of the number of accidents that have happened on or related to the track. Thus, in the Safety paradigm, the implicit definition of safety is a condition in which "as few things as possible go wrong" (ibid.: p. 147). The phenomenology of safety is therefore concerned with the occurrence of adverse outcomes.

While Safety is focused on failure, the Resilience paradigm focuses on success; for example, a system working normally or a person succeeding in doing a task. These good outcomes can be found in daily work routines. Although it may be difficult to perceive processes that go right—where the fact that something goes right often means that it went as expected—learning to understand how things are normally done will provide insight into the best possible conditions to ensure that the processes continue to succeed (ibid.: pp. 135–136). Thus, from a resilience perspective, it may, for example, be more important to look at how a train driver adequately responds to signals that indicate a reduction in speed is needed to prevent the train from derailing as a result of an unknown object on the track.

35.3.2 Etiology

Etiology is the study of causes and the reasons why things happen, and it describes the "mechanisms" behind how systems function. The Safety paradigm is very much based on the assumption that the behavior of a system can be explained in a linear causal way. Since the Safety paradigm focuses on what goes wrong, the etiology of this paradigm is about the probable causes of these adverse outcomes and the mechanisms that produce them. A typical example of linear causal thinking in the Safety paradigm is Failure Mode and Effect Analysis (FMEA), which investigates, for each component in a system, what the modes of failure of this component might be and how any failure would propagate through the system (Dhillon 1992). In the example of a crash between two cars, the Safety paradigm will probably identify one of the cars driving in the wrong lane as the cause. It may identify underlying causes, such as slippery roads and poor vision due to bad weather conditions, or the driver being distracted due to long working hours, etc., but ultimately, the accident will be causally reconstructed.

In the Resilience paradigm, adverse outcomes are seen as emergent rather than caused by some malfunctioning component in the system. Hollnagel speaks of "elusive causes," which may be transient phenomena, combinations of conditions, or conditions that exist only at a particular point in time (Hollnagel 2014: p. 131). Combined, these conditions and phenomena may account for the adverse outcomes, but they cannot be conceived of as the "cause" in a deterministic sense. They should be seen as *patterns* rather than causal mechanisms. If we conceive of emergent outcomes as arising from unexpected combinations of performance variability, while all variability might be within a small magnitude, an outcome itself might be large enough to be noticeable. There is no proportionality between the variability in performance and the emergent outcome, and the pattern by which certain outcomes emerge should therefore be seen as non-linear. In the Safety paradigm, a logical response would be to constrain variability if an undesired outcome does emerge from performance variability. In the Resilience paradigm, however, this same performance variability also leads to *resilient*—that is, desirable—performance, and variability should therefore not be constrained as such. Instead, efforts should be put into monitoring when variability becomes critical (see Section 4.2).

35.3.3 *Ontology*

Ontology is the study of the nature of that "which is." In this context, it refers to the essential characteristics of safety and resilience in the respective paradigms. In the Safety paradigm, a system consists of different components that are connected. This system can be in one of two modes: the system either fails or functions correctly. A case of failure will be traced back to an underlying cause, which will then be eliminated or neutralized; for example, by replacing or redesigning the corresponding component of the system. The underlying assumption is that outcomes can be understood as effects that follow deterministically from prior causes, and given complete knowledge, we can also predict how a system will perform (Hollnagel 2014: pp. 96–104). In the example of the car crash, the solution will possibly be sought in putting a guardrail between the lanes, because that would prevent a situation in which a car ends up in the wrong lane.

In the Resilience paradigm, predictability is replaced by intractability. Systems have become so complex that the conditions under which a system needs to perform will almost always be different from what has been specified or prescribed. Consequently, people—both individually and collectively as an organization—continuously need to adjust their behavior (action, tools, methods) so that it corresponds to the situation. The performance of the system can therefore not be described in the bimodal way that characterizes Safety (functioning properly or failure) but rather in terms of variability. This performance variability is the very essence of the Resilience paradigm and should therefore not be interpreted negatively as "performance deviation" that should be constrained (ibid.: p. 127). In the example of the car crash, the solution will probably not be sought in constraining driver behavior, but rather in monitoring and feedback mechanisms that indicate when a situation becomes critical.

35.4. Four Cornerstones of Resilient Performance

The deconstruction of the Resilience paradigm into its phenomenology, etiology, and ontology suggests that resilience should not be seen as a single state that is acceptable or not, but rather as a quality of functioning, with respect to the potential for resilient performance. Resilient performance is something that occurs in degrees and is never completely absent.

The Resilience paradigm assumes that regardless of whether things go wrong or right, they are all the result of the same underlying process. It therefore makes little sense to constrain performance variability in order to avoid adverse outcomes, since that would also constrain the possibilities to achieve the desired acceptable outcomes. The starting point for improving resilient performance is therefore to investigate the situation in which things went right and to determine where performance variability is moving in the right direction and where it is moving in the opposite direction. This requires that one recognizes performance variability in the first place. Only when we have a good understanding can we attempt to adjust its functioning if necessary.

Thus, where do we start when investigating performance variability? For Hollnagel, a resilient system can be defined as one that "is able effectively to adjust its functioning prior to, during, or following changes and disturbances, so that it can continue to perform as required after a disruption or a major mishap, and in the presence of continuous stresses" (Hollnagel 2009: 117). In a subsequent publication, Hollnagel refers to "sustain[ing] required operations under both expected and unexpected performance" (Hollnagel 2017). Four potentials or abilities can be derived from this definition, which have since become the defining characteristics of a resilient system in resilience engineering.

35.4.1 Ability to Respond: Dealing With the Actual

A resilient system is one that is able to respond to the demands of the current situation (Pariès 2011a). This "current situation" may comprise both regular and irregular events, disturbances, and opportunities. The ability to respond requires that the system is able to assess the situation, knows what to respond to, and subsequently knows what to do and when to do it. For example, an anesthesiologist is confronted with an urgent operation on a patient who never underwent general anesthesia before. Usually, relevant risk factors are identified during a pre-anesthetic visit. Because the operation is unplanned, no pre-anesthetic visit has taken place. When intubation of the patient turns out to be problematic, the anesthesiologist should be able to deal with this situation and find a workaround (Cuvelier and Falzon 2011).

35.4.2 Ability to Monitor: Dealing With the Critical

While the ability to respond refers to "knowing what to *do*," the ability to monitor refers to "knowing what to *look for*" (Hollnagel 2011b: xxxvii). Although the ability to respond also requires that the system is able to assess the situation, the ability to monitor goes further in that it not only considers the overall state of a system, but also monitors that which could become a threat. This enables the system to change from a state of normal operation to a state of readiness when the conditions indicate that a critical disturbance or failure is imminent (Hollnagel 2009: 124–125). For example, in aviation industry, fatigue is known to be a major risk for safety. Fatigue can occur when there is inadequate time to rest and recover from the work periodic (acute fatigue) and it can result from insufficient recovery from acute fatigue over time (chronic fatigue). This chronic fatigue is less studied and people are often not aware of their negative impacts on performance. Here, resilience requires that the aircrew is monitored not only for acute fatigue but also for chronic fatigue (Cabon et al. 2011).

35.4.3 Ability to Anticipate: Dealing With the Potential

The ability to anticipate is about knowing what to expect, such as future threats and their consequences, but also the future capacity of the system to deal with these threats. While anticipation in the traditional approach to safety management attempts as much as possible to anticipate possible threats and to avoid them (Wildavsky 1988), anticipation in resilience engineering is also—or perhaps even primarily—about the adaptive capacity of the system and the question of whether this adaptive capacity is adequate to address the threats it could encounter in the future (Woods 2011: 121). For example, a resilient system in a hospital is able to recognize in time the danger of "bed crunch" in intensive care units (Cook 2006).

35.4.4 Ability to Learn: Dealing With the Factual

The ability to learn in resilience engineering is probably most distinct from its equivalent in traditional safety management approaches. While learning in the traditional safety management approach focuses on learning from mistakes, learning in resilience engineering refers to learning from what goes right. As stated earlier, the Resilience paradigm builds on the idea that failure or success both follow from performance variability. It therefore makes little sense to limit learning to situations where things have gone wrong, as these are not only less frequent, but probably also more extreme and less similar and, thus, more difficult to generalize. The idea is that precisely the frequent and rather similar situations will provide the most effective learning environments (Hollnagel 2011a).

35.5. Implementing Resilience Engineering

The four abilities discussed in Section 4 describe a resilient system as one that is able to respond to the actual, to monitor the critical, to anticipate the potential, and to learn from the factual. While they do not tell us how this resilience can be engineered into a system, these abilities provide clues for implementing resilience in a system.

With regard to the ability to respond, some responses might be implemented in the system beforehand; for example, by anticipating potential disruptive situations and implementing ready-for-use solutions (e.g., specific procedures for abnormal and emergency operation, or specific reaction skills and cognitive competencies (Pariès 2011b). For more unpredictable and irregular situations, it is often impossible to anticipate what a potentially threatening situation will require. In these situations, resilience entails that a system is able to adapt its ongoing functioning to the new situation (Hollnagel 2017). There is a potential tension between both conditions, especially when anticipation strategies reduce the system's ability to cope with unanticipated events. Many approaches to resilience engineering therefore distinguish between situations that fall within some foreseeable range of variability of a system and those situations that can be described as "unthought-of" variability (Cuvelier et al. 2012). The former situations can often be dealt with by relying on technical, domain-specific skills and more prescriptive emergency procedures (Bergström et al. 2009). The "escalating" situations, in contrast, require flexible, adaptive, and more generic competencies so that people can deal with "fundamental surprise" situations, or, phrased more paradoxically, are "prepared for the unprepared" (Pariès 2011a: 6–7).

The ability to respond thus builds on the ability to monitor, which is specifically aimed at distinguishing between these two situations. Building on the idea that performance is based on a combination of internal processes and external drivers, it is crucial to monitor both the intermediate processes and activities in the system and what is happening in the environment in which a system functions (Wreathall 2011). If the intermediate processes and activities cannot be monitored directly, indicators can be used as proxy measures for items identified as important in the underlying models of safety.

As discussed earlier, the ability to anticipate includes both the anticipation of future threats and keeping an eye on the system's adaptive capacity to be able to respond to future threats. Regarding the latter, it may be good to identify signs that would indicate when the system is exhausting its adaptive capacity (Woods 2011). An increase in recovery time after disruption may indicate that a system is nearing a tipping point, after which it will collapse (Scheffer et al. 2009), while the erosion of buffers and reserves may indicate that a system will soon not be able to deal with another disruption (Cook and Rasmussen 2005). Thus, to keep an eye on the system's adaptive capacity, indicators are needed that provide a warning when this capacity is approaching its limits, beyond which it will no longer be able to adapt. Another aspect of adaptive capacity is the room to make trade-offs between goals. A resilient system is able to shift priority from, for example, economic goals to safety goals, if the situation so requires (Tjørholm and Aase 2011).

In a case study of Airline Operations Contingency Planning, Richters et al. (2016) found that rotating roles may increase the level of shared situation awareness, while at the same time contribute to greater efficiency by identifying opportunities for trade-offs on the least critical aspects. Similar to the competencies required for the ability to respond, this specific aspect of the ability to anticipate is strengthened by providing training that goes beyond specific disciplines and professions (Tjørholm and Aase 2011: pp. 169–170). Adaptive capacity can also be strengthened by avoiding interdependencies in the system that have a positive, and thereby escalating, feedback mechanism; for example, by implementing damping mechanisms that prevent variability within the system that could lead to undesirable outcomes (Hollnagel 2012). With respect to the ability to anticipate future threats, it is important not to merely extrapolate from existing conditions but also to use imaginative tools that are able to identify more surprising and unexpected future threats (Adamski and Westrum 2003).

As stated in Section 4.2, the basis for learning in resilience engineering is normal operation and not only through accidents or failure. Whereas learning in the traditional Safety paradigm is focused on finding underlying root causes, learning in resilience engineering aims at understanding dependencies among functions and the typical variability of these functions (Hollnagel 2009). This suggests that learning should also continuously take place and not be limited to specific events. Preferably, learning should be based on both qualitative and quantitative data (Herrera et al. 2009). Examples of how learning can be implemented in an organization include: daily team reflections, in which team members discuss what went well that day, thereby transforming implicit knowledge into explicit knowledge (Siegel and Schraagen 2017); and improving both individual and institutional knowledge (Hollnagel 2009).

35.6. Resilience Engineering and Traditional Safety Management Compared

The previous two sections discussed the basic assumptions underlying the Resilience paradigm and the four cornerstones of resilience engineering. By doing so, the difference between resilience engineering and the traditional approaches to safety management were emphasized. Nevertheless, resilience engineering should not be seen as a simple replacement for traditional safety management, but rather as broadening the latter's scope.

Depending on the complexity of the system that is the locus of resilience engineering, resilience engineering has more to add to traditional approaches to safety management. As mentioned in Section 2, resilience engineering is targeted at sociotechnical systems. For a very simple artifact that has few interdependencies with the environment in which it is used, resilience engineering probably has little to add. However, it is questionable whether there are many such examples in reality, if any.

The more complex the system, the more important it will be to focus on the four abilities in resilience engineering. Compared to traditional approaches to safety management, there is much greater emphasis on learning, while the basis for learning is also different. Although safety management has shifted its focus from merely technological failure to human factors, organizational factors, and safety culture (Hale and Hovden 1998), safety management is still based on a linear causal view of how systems fail. As explained earlier, resilience engineering takes as its premise that, in our contemporary complex systems, there is no fundamental difference between performance that leads to failure and performance that leads to success, which is why we should attempt to understand performance in general.

However, some measures to improve the four abilities in the Resilience paradigm may be quite similar to those in traditional safety management. For example, the use of safety reserves (also referred to as safety factors or safety margins) is a well-established measure in engineering to protect against a particular integrity-threatening mechanism (Doorn and Hansson 2011). Although formulated in terms of failure, these safety reserves are intended to build some redundancy into the system to account for imperfect theory and unknown failure mechanisms, thereby recognizing the limitations of the linear causal model of failure. This same redundancy may also provide the room to maneuver that is necessary for resilience (Woods and Branlat 2010). The application of safety reserves is not limited to physical systems but can equally apply to sociotechnical systems; for example, the availability of several roads rather than just one to leave a dangerous area, or backup facilities in case one communication service breaks down (Doorn and Hansson 2017).

When dealing with complexity, there may be a tension between the measures taken within the traditional Safety paradigm and those taken within the Resilience paradigm. Whereas traditional safety management measures attempt to ensure safety by reducing complexity through standardization, the Resilience paradigm appears to embrace complexity by stimulating flexible policies, behavior, and technologies (Guédon et al. 2017). In the field of patient safety in the operating room, for example, standards and guidelines aimed to reduce complexity have significantly benefited patient safety in relation to anesthesia (Gaba 2000; Klein 2008) and surgical practice (Haynes et al. 2009; De Vries et al. 2010; Chera et al. 2015). However, these same guidelines may also result in unsafe practices due to a mismatch with existing work practices (Klein 2008; Urbach et al. 2014). This difference between what is implemented at the "blunt end" and what happens at the "sharp end"—or between *work-as-imagined* and *work-as-done* (Hollnagel 2014: 39–41)—is precisely the reason why resilience engineering also focuses on adaptive capacity so that work practices can be attuned to the context at hand. However, relying only on local, self-organized regulations may not lead to the desired performance (Nyssen 2008). To avoid the loss of valuable safety management measures when the Resilience paradigm is fully embraced, the challenge for resilience engineering is to look for tools that are adaptive to day-to-day variability (Guédon et al. 2017).

35.7. Conclusions

The discussion of resilience engineering shows that this approach to safety management starts from a completely different view of how systems function. It builds on the idea that both successful or desired performance and unsuccessful or undesired performance emerge from the same functional processes. Resilience should therefore not be seen as the flipside of failure; rather, it indicates levels of performance on a scale from very poor to extremely strong.

When it comes to operationalization and implementation, resilience engineering builds on four abilities: the ability to respond, the ability to monitor, the ability to anticipate, and—perhaps most important—the ability to learn. In combination, these four abilities indicate how well a system is able to deal with and adapt to changing circumstances.

Many of the safety measures that follow from traditional approaches to safety management are still relevant, as long as they do not limit a system's room for maneuver. This applies especially to the human or organizational side of the system. Procedures intended to improve safety may be counterproductive if they reduce the system's necessary flexibility. However, practical case studies in resilience engineering show that too little guidance may also lead to suboptimal solutions. The future of resilience engineering may well be one in which standardized measures in the Safety paradigm are transformed into flexible measures that are adaptive to the specific context at hand.

Related Chapters

Chapter 19: Emergence in Engineering (Peter Simons)
Chapter 33: Values in Risk and Safety Assessment (Niklas Möller)
Chapter 36: Trust in Engineering (Philip J. Nickel)
Chapter 38: Health (Marianne Boenink)

Further Reading

Hollnagel, E. (2014). *Safety-I and Safety-II: The Past and Future of Safety Management*. Farnham, UK: Ashgate. (An in-depth comparison of the traditional safety paradigm and the resilience paradigm.)

Hollnagel, E., Woods, D.D. and Leveson, N. (eds.) (2006). *Resilience Engineering: Concepts and Precepts*. Boca Raton, FL: CRC Press/Taylor & Francis. (The first comprehensive volume on resilience engineering.)

Nemeth, C.P., Hollnagel, E. and Dekker, S. (eds.) (2008). *Resilience Engineering Perspectives. Volume 2: Preparation and Restoration*. Farnham, UK: Ashgate Publishing, Ltd. (Volume on resilience engineering, also including the public policy and organizational aspects of resilience.)

Woods, D.D. and Wreathall, J. (2003). *Managing Risk Proactively: The Emergence of Resilience Engineering*. Columbus, OH: Institute for Ergonomics, The Ohio State University (The first book on resilience engineering.)

References

Adamski, A.J. and Westrum, R. (2003). Requisite Imagination: The Fine Art of Anticipating What Might Go Wrong. In E. Hollnagel (ed.), *Handbook of Cognitive Task Design*. Boca Raton, FL: CRC Press, pp. 193–220.

Adger, W.N. (2000). Social and Ecological Resilience: Are They Related? *Progress in Human Geography*, 24, 347–364.

Bergström, J., Dahlström, N., Van Winsen, R., Lützhöft, M., Dekker, S. and Nyce, J. (2009). Rule- and Role-Retreat: An Empirical Study of Procedures and Resilience. *Journal of Maritime Research*, 6, 75–90.

Berkes, F. and Ross, H. (2013). Community Resilience: Toward an Integrated Approach. *Society & Natural Resources*, 26, 5–20.

Bowen, W.R. (2014). *Engineering Ethics: Challenges and Opportunities*. Heidelberg: Springer.

Cabon, P., Deharvengt, S., Berecht, I., Grau, J.Y., Maille, N. and Mollard, R. (2011). From Flight Time Limitations to Fatigue Risk Management Systems: A Way Toward Resilience. In E. Hollnagel, J. Pariès, D.D. Woods and J. Wreathall (eds.), *Resilience Engineering in Practice: A Guidebook*. Boca Raton, FL: CRC Press/Taylor & Francis, pp. 69–86.

Carpenter, S., Walker, B., Anderies, J.M. and Abel, N. (2001). From Metaphor to Measurement: Resilience of What to What. *Ecosystems*, 4, 765–781.

Chera, B.S., Mazur, L., Buchanan, I. et al. (2015). Improving Patient Safety in Clinical Oncology: Applying Lessons From Normal Accident Theory. *JAMA Oncology*, 1, 958–964.

Connor, K.M. and Davidson, J.R.T. (2003). Development of a New Resilience Scale: The Connor-Davidson Resilience Scale (CD-RISC). *Depression and Anxiety*, 18, 76–82.

Cook, R.I. (2006). Being Bumpable: Consequences of Resource Saturation and Near-saturation for Cognitive Demands on ICU Practitioners. In D.D. Woods and E. Hollnagel (eds.), *Joint Cognitive Systems: Patterns in Cognitive Systems Engineering*. Boca Raton, FL: CRC Press/Taylor & Francis, pp. 23–35.

Cook, R.I. and Rasmussen, J. (2005). "Going Solid": A Model of System Dynamics and Consequences for Patient Safety. *Quality & Safety in Health Care*, 14, 130–134.

Cuvelier, L. and Falzon, P. (2011). Coping with Uncertainty: Resilient Decisions in Anaesthesia. In E. Hollnagel, J. Pariès, D.D. Woods and J. Wreathall (eds.), *Resilience Engineering in Practice: A Guidebook*. Boca Raton, FL: CRC Press/Taylor & Francis, pp. 29–43.

Cuvelier, L., Falzon, P., Granry, J.C. and Moll, M.C. (2012). Managing Unforeseen Events in Anesthesia: Collective Trade-off Between "Understanding" and "Doing". *Work*, 41, 1972–1979.

Da Silva, J., Kernaghan, S. and Luque, A. (2012). A Systems Approach to Meeting the Challenges of Climate Change. *International Journal of Urban Sustainable Development*, 4, 125–145.

De Vries, E.N., Prins, H.A., Crolla, R.M.P.H., Den Outer, A.J., Van Andel, G., Van Helden, S.H., Schlack, W.S., Van Putten, M.A., Gouma, D.J., Dijkgraaf, M.G.W., Smorenburg, S.M. and Boermeester, M.A. (2010). Effect of a Comprehensive Surgical Safety System on Patient Outcomes. *New England Journal of Medicine*, 363, 1928–1937.

Dhillon, B.S. (1992). Failure Mode and Effects Analysis: Bibliography. *Microelectronics and Reliability*, 32, 719–731.

Doorn, N. (2017). Resilience Indicators: Opportunities for Including Distributive Justice Concerns in Disaster Management. *Journal of Risk Research*, 20, 711–731.

Doorn, N., Gardoni, P. and Murphy, C. (2019). A Multidisciplinary Definition and Evaluation of Resilience: The Role of Social Justice in Defining Resilience. *Sustainable and Resilient Infrastructure*, 4(3), 112–123. doi: 10.1080/23789689.2018.1428162.

Doorn, N. and Hansson, S.O. (2011). Should Probabilistic Design Replace Safety Factors? *Philosophy & Technology*, 24, 151–168.

Doorn, N. and Hansson, S.O. (2017). Factors and Margins of Safety. In N. Möller, S.-O. Hansson, J.-E. Holmberg and C. Rollenhagen (eds.), *Handbook of Safety Principles*. Hoboken, New Jersey: John Wiley & Sons.

Fairbanks, R.J., Perry, S., Bond, W. and Wears, R.L. (2013). Separating Resilience From Success. In E. Hollnagel, J. Braithwaite and R.L. Wears (eds.), *Resilient Health Care*. Farnham, UK: Ashgate.

Folke, C. (2006). Resilience: The Emergence of a Perspective for Social–ecological Systems Analyses. *Global Environmental Change*, 16, 253–267.

Gaba, D.M. (2000). Anaesthesiology as a Model for Patient Safety in Health Care. *BMJ*, 320, 785–788.

Guédon, A.C.P., Spruit, S.L., Wauben, L.S.G.L., van der Elst, M., Doorn, N., Dankelman, J., van den Dobbelsteen, J. and Klein, J. (2017). Delicate Balance: Adaptive Support to Improve Patient Safety. *BMJ Innovations*, 3, 1–6.

Hale, A.R. and Hovden, J. (1998). Management and Culture: The Third Age of Safety. A Review of Approaches to Organizational Aspects of Safety, Health and Environment. In A.M. Feyer and A. Williamson (eds.), *Occupational Injury: Risk Prevention and Intervention*. London: Taylor and Francis.

Hamel, G. and Valikangas, L. (2003). The Quest for Resilience. *Harvard Business Review*, 81, 52–65.

Haynes, A.B., Weiser, T.G., Berry, W.R., Lipsitz, S.R., Breizat, A.-H.S., Dellinger, E.P., Herbosa, T., Joseph, S., Kibatala, P.L., Lapitan, M.C.M., Merry, A.F., Moorthy, K., Reznick, R.K., Taylor, B. and Gawande, A.A. (2009). A Surgical Safety Checklist to Reduce Morbidity and Mortality in a Global Population. *New England Journal of Medicine*, 360, 491–499.

Herrera, I.A., Nordskag, A.O., Myhre, G. and Halvorsen, K. (2009). Aviation Safety and Maintenance Under Major Organizational Changes, Investigating Non-existing Accidents. *Accident Analysis & Prevention*, 41, 1155–1163.

Holling, C.S. (1973). Resilience and Stability of Ecological Systems. *Annual Review of Ecology and Systematics*, 4.

Hollnagel, E. (2009). The Four Cornerstones of Resilience Engineering. In C.P. Nemeth, E. Hollnagel and D.S.W.A. (eds.), *Preparation and Restoration*. Aldershot, UK: Ashgate, pp. 117–134.

Hollnagel, E. (2011a). To Learn or Not to Learn, That is the Question. In E. Hollnagel, J. Pariès, D.D. Woods and J. Wreathall (eds.), *Resilience Engineering in Practice: A Guidebook*. Boca Raton, FL: CRC Press/Taylor & Francis, pp. 193–198.

Hollnagel, E. (2011b). Prologue: The Scope of Resilience Engineering. In E. Hollnagel, J. Pariès, D.D. Woods and J. Wreathall (eds.), *Resilience Engineering in Practice: A Guidebook*. Boca Raton, FL: CRC Press/Taylor & Francis, pp. xxix–xxxix.

Hollnagel, E. (2012). *FRAM: The Functional Resonance Analysis Method for Modelling Complex Socio-technical Systems*. Farnham, UK: Ashgate.

Hollnagel, E. (2014). *Safety-I and Safety-II: The Past and Future of Safety Management*. Farnham, UK: Ashgate.

Hollnagel, E. (2017). Resilience Engineering and the Future of Safety Management. In N. Möller, S.-O. Hansson, J.-E. Holmberg and C. Rollenhagen (eds.), *Handbook of Safety Principles*. Hoboken, NJ: John Wiley & Sons.

Klein, J. (2008). Multimodal Multidisciplinary Standardization of Perioperative Care: Still a Long Way to Go. *Current Opinion in Anesthesiology*, 21, 187–190.

Kroes, P., Franssen, M., Van de Poel, I.R. and Ottens, M. (2006). Treating Socio-technical Systems as Engineering Systems. *Systems Research and Behavioral Science*, 23, 803–814.

Leichenko, R. (2011). Climate Change and Urban Resilience. *Current Opinion in Environmental Sustainability*, 3, 164–168.

Mallet, R. (1862). *Great Neapolitan Earthquake of 1857*. William Clowes and Sons: London.

McAslan, A. (2010). *The Concept of Resilience. Understanding Its Origins, Meaning and Utility*. Adelaide, Australia: The Torrens Resilience Institute.

Meerow, S., Newell, J.P. and Stults, M. (2016). Defining Urban Resilience: A Review. *Landscape and Urban Planning*, 147, 38–49.

NSPE (2007). *NSPE Code of Ethics for Engineers*. Alexandria, VA.

Nyssen, A.S. (2008). Coordination in Hospitals: Organized or Emergent Process? Towards the Idea of Resilience as the Agents', Groups', Systems' Capacity to Project Themselves into Future. In E. Hollnagel, F. Pieri and E. Rigaud (eds.) *3rd International Symposium on Resilience Engineering*. Antibes—Juan-les-Pins, France, pp. xxix–xxxix.

Pariès, J. (2011a). Resilience and the Ability to Respond. In E. Hollnagel, J. Pariès, D.D. Woods and J. Wreathall (eds.), *Resilience Engineering in Practice: A Guidebook*. Boca Raton, FL: CRC Press/Taylor & Francis, pp. 3–8.

Pariès, J. (2011b). Lessons from the Hudson. In E. Hollnagel, J. Pariès, D.D. Woods and J. Wreathall (eds.), *Resilience Engineering in Practice: A Guidebook*. Boca Raton, FL: CRC Press/Taylor & Francis, pp. 9–27.

Paton, D. and Fohnston, D. (2001). Disasters and Communities: Vulnerability, Resilience and Preparedness. *Disaster Prevention and Management*, 10, 270–277.

Patriarca, R., Bergström, J., Di Gravio, G. and Costantino, F. (2018). Resilience Engineering: Current Status of the Research and Future Challenges. *Safety Science*, 102, 79–100.

Richters, F., Schraagen, J.M. and Heerkens, H. (2016). Balancing Goal Trade-offs When Developing Resilient Solutions: A Case Study of Re-planning in Airline Operations Control. In P. Ferreira, J. Van Der Vorm and D.D. Woods (eds.), *Proceedings 6th Symposium on Resilience Engineering: Managing Resilience, Learning to be Adaptable and Proactive in an Unpredictable World*. Lisbon, Portugal: Resilience Engineering Association, pp. 34–39.

Royal Academy of Engineering (2014). *Statement of Ethical Principles for the Engineering Profession*. London.

Scheffer, M., Bascompte, J., Brock, W.A., Brovkin, V., Carpenter, S.R., Dakos, V., Held, H., van Nes, E.H., Rietkerk, M. and Sugihara, G. (2009). Early-warning Signals for Critical Transitions. *Nature*, 461, 53–59.

Siegel, A.W. and Schraagen, J.M.C. (2017). Beyond Procedures: Team Reflection in a Rail Control Centre to Enhance Resilience. *Safety Science*, 91, 181–191.

Southwick, S.M., Vythilingam, M. and Charney, D.S. (2005). The Psychobiology of Depression and Resilience to Stress: Implications for Prevention and Treatment. *Annual Review of Clinical Psychology*, 1, 255–291.

Tjørholm, B. and Aase, K. (2011). The Art of Balance: Using Upward Resilience Traits to Deal with Conflicting Goals. In E. Hollnagel, J. Pariès, D.D. Woods and J. Wreathall (eds.), *Resilience Engineering in Practice: A Guidebook*. Boca Raton, FL: CRC Press/Taylor & Francis, pp. 157–170.

Urbach, D.R., Govindarajan, A., Saskin, R., Wilton, A.S. and Baxter, N.N. (2014). Introduction of Surgical Safety Checklists in Ontario, Canada. *New England Journal of Medicine*, 370, 1029–1038.

Walker, B., Holling, C.S., Carpenter, S.R. and Kinzig, A. (2004). Resilience, Adaptability and Transformability in Social-ecological Systems. *Ecology and Society*, 9, 5–13.

Wardekker, A. (forthcoming). Framing 'Resilient Cities': System versus Community Focussed Interpretations of Urban Climate Resilience. In L.J. Herrero, O.G. Castillo, J. Pont Vidal and E. Santibanez (eds.), *Urban Resilience: Methodologies, Tools and Evaluation*. Berlin: Springer.

Wardekker, J.A., De Jong, A., Knoop, J.M. and Van der Sluijs, J.P. (2010). Operationalising a Resilience Approach to Adapting an Urban Delta to Uncertain Climate Changes. *Technological Forecasting & Social Change*, 77, 987–998.

Wildavsky, A.B. (1988). *Searching for Safety* New Brunswick: Transaction Books.

Woods, D.D. (2011). Resilience and the Ability to Anticipate. In E. Hollnagel, J. Pariès, D.D. Woods and J. Wreathall (eds.), *Resilience Engineering in Practice: A Guidebook*. Boca Raton, FL: CRC Press/Taylor & Francis, pp. 121–125.

Woods, D.D. and Branlat, M. (2010). Hollnagel's Test: Being 'in Control' of Highly Interdependent Multi-layered Networked Systems. *Cognition, Technology & Work*, 12, 95–101.

Woods, D.D. and Wreathall, J. (2003). *Managing Risk Proactively: The Emergence of Resilience Engineering*. Columbus, OH: Institute for Ergonomics, The Ohio State University.

Wreathall, J. (2011). Monitoring: A Critical Ability in Resilience Engineering. In E. Hollnagel, J. Pariès, D.D. Woods and J. Wreathall (eds.), *Resilience Engineering in Practice: A Guidebook*. Boca Raton, FL: CRC Press/Taylor & Francis, pp. 61–68.

36

TRUST IN ENGINEERING

Philip J. Nickel

36.1. Introduction

According to one account, after the (first) Quebec Bridge Disaster in 1907, "it was immediately recognized that a serious blow had been struck to public confidence in the whole engineering profession" (Roddis 1993: 1544). Responsibility for the Quebec Bridge Disaster is largely ascribed to the consulting engineer, Theodore Cooper. In his design, Cooper extended cantilevered bridge construction beyond its earlier scope of application without sufficient testing. He was out of touch with the on-site engineers and workers who observed the bridge showing increasing signs of strain and material failure during construction. In the bridge failure dozens of workers died, and after the collapse of the structure, the two sides of the Saint Lawrence remained unbridged near Quebec City until 1919 (except when the river was frozen over).

In Canada, this disaster and a second that followed it at the same location led to a new awareness of professional responsibilities of engineers, as embodied in the Ritual of the Calling of the Engineer (ibid., 1545). Even today, many newly certified Canadian engineers pledge awareness of their responsibility towards society and receive a special ring to wear as a reminder of this professional responsibility.[1] One plausible explanation of this practice is that it instills and expresses the value of trustworthiness within the engineering profession, in much the way that the Hippocratic Oath and its associated symbolism functions within the medical profession.

Trust and trustworthiness are complementary attitudes. Trust, roughly speaking, is the expectation of one person or entity, that a second person or entity will uphold their commitments and meet certain standards. Trustworthiness, on the other hand, is the disposition of a person or entity to perform in the way that others reasonably expect it to perform, given relevant commitments and standards. For engineering to be trustworthy means that those designated as professional engineers can be reasonably expected to carry out certain tasks such as bridge construction to a high standard. When these trust expectations are disappointed, as in the Quebec Bridge Disaster, the salience of trust and trustworthiness suddenly become obvious. In the words of Annette Baier, one of the founding figures in bringing philosophical attention to trust, "We inhabit a climate of trust as we inhabit an atmosphere and notice it as we notice air, only when it becomes scarce or polluted" (Baier 1986: 234).

In this chapter, my agenda is threefold. First, I consider the traditional notion of engineers as trustworthy professionals with particular competencies, in order to see what this implies for the ethical orientation of engineers. Second, I consider how engineers try to "engineer trust" in contexts where agents interact, so as to achieve a certain desired form of interaction. This form of social engineering raises epistemic and moral questions about whether trust within these designed contexts is correctly based on the reality of these contexts, and about the unintended consequences (such as "filter bubbles") that sometimes result. Third, I consider trust in engineered systems themselves, especially "smart" systems that collect data and use it to make automated decisions (or to advise humans how to do so) and take over human tasks. I argue that there are important epistemic and moral questions about what type and amount of evidence is needed when relying on such systems. In addition, philosophical questions are raised by the very fact that researchers are inclined to call such reliance on automation "trust".

Because of this manifold agenda, the notion of trust and trustworthiness that I consider in relation to engineering is multifaceted. I do not expect to develop an overall definition of trust covering all these cases.[2] It will be useful at various points to consider the explanatory work being done by the concept of trust as contrasted with other possible attitudes of reliance. It is especially useful to have on hand a contrasting notion of *strategic reliance*, in which one person or entity relies on a second person or entity on a purely pragmatic basis, without the first at any time thinking that the second *should* take this reliance into account, or that the second is committed to behaving in accordance with particular norms. When I drive over a bridge, what is the difference, after all, between saying that I *trust* it to hold the weight of my vehicle, and saying that I merely rely on it to hold that weight? Arguably, trust plays no distinctive explanatory role in such situations.[3] When we talk about trust as an attitude distinct from such strategic reliance, we have in mind something affectively and morally loaded. Such an attitude does distinctive explanatory work in accounting for cooperation and the nature of relationships between individuals.

36.2. Trust in Engineers as Professionals: Linear but Indirect

The first thing we often think of when we consider trust in engineering, is trust toward engineers as professionals. Engineers take on special responsibilities toward clients, colleagues, technology users, and society, in virtue of representing themselves as professionals (see Chapter 44, "Responsibilities to the Public—Professional Engineering Societies" by Joseph Herkert and Jason Borenstein, this volume). Engineers are held to have a responsibility of "non-maleficence" (non-harm) to society and to users of technology. More broadly, engineers are committed to carrying out work on the basis of up-to-date scientific and technical expertise, in line with best practices for the relevant engineering discipline. These responsibilities and stringent standards are explicitly stated in nearly every code of professional ethics of the various engineering disciplines (see Chapter 43, "Professional Codes of Ethics" by Michael Davis, this volume, for additional discussion). When taken at face value, such explicit commitments signal the trustworthiness of engineers and provide a strong interpersonal basis for trust in engineering as a profession.

The sociology of professions contains trust as a significant (Evetts 2006; Brown and Calnan 2016) but limited (Adams 2015) theme. One of the main purposes of professional designations is to provide a readily accessible reason for trusting those persons on which they are conferred. The professional designation works as follows: "Education, training and experience are fundamental requirements but once achieved (and sometimes licensed) . . . the exercise of discretion . . . based on competencies is central and deserving of special status. . . . Because of complexity it is often necessary to trust professionals' intentions" (Evetts 2013: 785).

However, compared with the study of trust in, and within, other professional disciplines such as medicine and law (e.g., Brown and Calnan 2012; O'Neill 2002; Hall 2002), there has been very little empirical and ethical research on trust in professional engineering. One reason for this is that engineers do not interact directly with members of the public the way that physicians, nurses, and lawyers do. In other professions, direct personal trust in known persons often remains essential to the delivery of service. By contrast, most people's reliance on engineering is impersonal. Users interact with cars, bridges, and heart monitors, and unidentified engineers are presumed to have designed these technological artifacts with suitable care. It is a *linear* trust relationship (USER → TECHNOLOGY → ENGINEER), but an *indirect* one.

Despite the lack of scholarly study of trust in the engineering profession, we can still find important clues about how this *linear-but-indirect* trust relationship has become more complex over time. Emotional reactions such as blame toward engineers and others in the aftermath of accidents reveal the occasion and the object of distrust, and by extension the people and entities that one was inclined to trust in the first place. In its historical context, the Quebec Bridge Disaster caused distrust in Cooper and in engineers more broadly. However, in more recent engineering failures such as Dieselgate and the bridge failure in Genoa, the failure of a technology seems to have caused distrust in corporate entities (e.g., Volkswagen, Autostrade) and government oversight bodies, instead of or in addition to distrust in engineering as a discipline. This historical shift in the object(s) of distrust following cases of technological failure is partly due to a diffusion of responsibility. One important factor is the state's assumption of responsibility for technological risk. According to Beck's "risk society" theory, in the twentieth century, government entities assumed greater responsibility for the overall balance of technological hazards, compared with the "limited state" of prewar Western modernity (Beck 1992). One scholar describes Beck's theory as follows:

> With the beginning of societal attempts to control, and particularly with the idea of steering towards a future of predictable security, the consequences of risk become a political issue. . . . It is societal intervention—in the form of decision-making—that transforms incalculable hazards into calculable risks.
>
> *(Elliott 2002: 295)*

When the state assumes responsibility for the unwanted effects of technological development, responsibility for technological risks, as well as trust and distrust, are diffused away from engineers (Renn 2008: 28).

Managerial changes over the last forty years have further diffused both responsibility for and trust in the engineered world around us. The advent of the "audit society" has placed some of the responsibility for technological risks with large companies and non-state governance bodies that rationalize technological risks from the perspective of management (Power 1997). Practices of *internal control* have also emerged in which the task of oversight is partly delegated by the government to organizations themselves and accountancy firms (Power 2007). These practices have made trust in engineering, which was already indirect and impersonal, more diffuse, complex, and abstract (see Figure 36.1). They may also lead to a paradoxical yearning for more personal forms of trust (Brown and Calnan 2012) and the need, mentioned by Evetts above, to "trust professionals' intentions" in situations of complexity (*op cit.*). Perhaps the yearning for personal trust in our impersonal engineered environment leads us to seek channels of focused expression for trust or distrust, for example

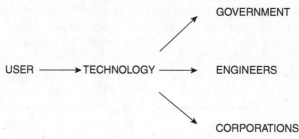

Figure 36.1 Diffusion of linear-but-indirect trust in engineering

in the "smart" technological products and services we use (e.g., Apple devices, Facebook) or in the celebrity technology entrepreneurs we associate with them (e.g., Steve Jobs, Mark Zuckerberg).

Figure 36.1 suggests that the immediate object of trust in engineering is an *artifact or technology that we use practically* in some way. The linear-but-indirect relationship has become more complex because what stands behind our immediate relationship to an artifact or technological system is a whole array of people and institutions.

By way of contrast, it is interesting to compare trust in engineering with trust in science, where science is thought of as a body of knowledge and a way of interpreting experience. We might say that trust in science is not usually directly manifested in acts of *practical reliance*, but rather in acts of *believing in* certain established scientific claims, as well as adopting certain interpretations of experience. When scientific knowledge is crystallized into an act of concrete practical reliance (e.g., having a medical device installed, or trying out a new way of fertilizing the soil), we are inclined to redescribe it as trust in engineering. In this way, even though science and engineering are inherently related, trust in engineering remains conceptually independent of trust in science because it is (in philosophical terminology) practical rather than theoretical.

36.3. Engineering Interpersonal and Interagential Trust

There is widespread awareness that technology shapes human relations. Engineering sometimes has the explicit design goal to shape or increase trust relationships between people. Architecture, urban planning, web design, industrial engineering, and industrial design are all fields that can take this kind of "social engineering" as a goal. In what follows, I focus on the design of digital environments as a way of illustrating a general trend toward reflexivity in the practice of (social) engineering that influences trust relationships. For example, in scholarly literature about online interactions starting about twenty years ago, concerns were voiced about the sustainability of trust relationships online, where the interacting parties are anonymous or pseudonymous, and not answerable for their words and deeds (Pettit 2004). Anonymity and the inability to rely on reputation in online environments have been used to explain fraud in one-off market exchanges in the digital sphere, as well as trolling behavior. Fraud within market exchanges has been empirically linked to the logic of one-off transactions between rationally self-interested agents with no reputation to protect (Ba et al. 2003). Concerns about trolling have also been empirically grounded in an "online disinhibition effect" in which people experience less inhibition to behave in impulsive or antisocial ways when online (Suler 2004).

These effects were originally unintended side effects rather than desired effects of participation in online platforms, but as they became increasingly well understood, they have been taken into account in engineering design (Resnick and Zeckhauser 2002). Indeed, since these effects are predictable and empirically grounded, engineers of networked environments have a responsibility to take them into account (Friedman et al. 2000). Similarly, those who design environments for offline interaction, such as architects, urban planners, and industrial engineers, must keep in mind similarly relevant social scientific knowledge about the predictable effects of their interventions.

Recently, other effects on trust in digital environments and social media have been studied, related to the digital transformation of society itself. For example, trust and distrust have been linked to pervasive personalization and network effects. Turcotte et al. (2015) found that recommendations of news items by Facebook "friends" perceived as "opinion leaders" increase trust in the source of these items. One can easily imagine that with billions of people reading news primarily on or via social media, such trust phenomena could strongly influence people's worldviews and political beliefs, leading to phenomena such as "fake news".

In design practice, the implications of such findings depend on many additional premises about the ecology of social media. For example, if opinion leaders tend to have larger networks than other users, then they should have outsized effects on trust in news sources. Furthermore, if people tend not to have *contradictory* opinion leaders within their social networks, these trust effects would likely be one-sided, leading most individuals to trust only a select (and potentially biased) set of sources. It is tempting to explain political polarization partly in terms of such effects. Recently, the dark side of social engineering of social media has captured public attention, associated with concepts such as "filter bubbles" and "fake news". Social media has been held accountable for manipulation. This type of explanation may be elusive, however, because the social media ecology is constantly changing with technological, economic, and social forces. When digital natives who have had no experience of political and social life before social media become the norm, emergent phenomena observed within social media may change fundamentally and will also be evaluated differently (for example, because norms of privacy have evolved).

As a result, the effects of design on social interactions, as well as how these interactions are experienced and evaluated by participants, are not easily predictable. They depend to a large extent on conventions, culture, and user practices. In order to anticipate and plan for these effects, it is useful to combine empirical, technical, and ethical inquiry during the design process in order to achieve the right balance of trust within designed social environments. Such combined inquiry has come to be called "value sensitive design", "design for values", or "responsible innovation", and has been applied explicitly to the value of trust on a number of occasions (Friedman et al. 2000; see examples in Nickel 2014; see also Chapter 22, "Values and Design" by Ibo Van de Poel, this volume, for a broader discussion of such methods).

Innovation that takes the value of trust into account must face the complexity of the concept and its value. Trust has two dimensions of value: practical and epistemic. On the practical dimension, trust is constitutive of healthy relationships. Trusting and being trusted have an intrinsic practical value for friends, neighbors, and co-citizens. They also have instrumental value in the sense that they help to obtain the benefits of cooperation and compliance. On the epistemic dimension, the value of trust is determined by whether it is well grounded. One person should trust a second person just to the extent that that second person really is trustworthy or reliable: there should not be over- or under-trust. Furthermore, in some contexts it is important for the first person to have sound *reasons* for believing the second person is reliable, because not having such reasons would be irresponsible or reckless (Manson and O'Neill 2007; Voerman and Nickel 2017). This is particularly relevant when engineers are designing for other professionals, who have an obligation to act reasonably when trusting others. (The reasons at stake are often implicit, as van Baalen and Carusi [2017] argue in the case of multidisciplinary clinical decision-making.)

36.4. Trust in Automation

The previous sections looked at two important kinds of trust in relation to engineering practice: first, trust in engineers themselves; and second, trust in other people whom we encounter within engineered environments. In this section I turn to a third kind of trust in engineering: trust in automated technologies. These technologies are often designed to invite trust, and the language of trust is used to talk about how we rely on them. In this section, I will assess whether this talk of trust is superficial or serious. If it is merely superficial, then we can replace it with talk of strategic reliance without any explanatory loss. I argue that trust in automation should be taken seriously.

"Automation" here refers to technological systems that take over human tasks requiring some intelligence and skill. Examples include trust in a robot to take over parts of a shared assembly task, and trust in artificial intelligence to make decisions in a shared medical task such as a surgery, but also more "casual" artificial agents such as those that assist in everyday decisions (e.g., about what music playlist to put on at a party).

A useful point of reference is a recent review article synthesizing a model of trust in automation on the basis of over a hundred selected empirical studies of the subject (Hoff and Bashir 2015). The model distinguishes two phases of trust: the "initial learned trust" that one develops prior to interacting with the automated system, and the "dynamic learned trust" that one develops subsequently during one's interaction with the automated system. The overall picture of trust and its determinants can be seen in Figure 36.2.

Hoff and Bashir follow Lee and See's (2004: 54) definition of trust as "the attitude that an agent will help achieve an individual's goals in a situation characterized by uncertainty and vulnerability". This definition involves a belief (or related representation) that automation will have *instrumental utility* within a context of uncertainty and vulnerability.

The trouble with this definition, and the scholarship from which it is derived, is that we do not know whether to take the talk of trust seriously. The literature does not clearly distinguish between trust and strategic reliance, defined earlier as calculative, instrumental reliance. When I rely strategically on somebody or something, this does not carry the implication that the relied-upon entity *should* take my reliance into account, nor does it carry the thought that the relied-upon entity *is committed to* or *is responsible for* behaving in accordance with any particular norms.

We must therefore ask, *what is gained by thinking of reliance on automation in terms of trust?* Answering this question is a serious challenge because it requires us to consider what we expect from automation itself, such that it, unlike ordinary artifacts such as bannisters and buses, might be the object of genuine trust. One possible answer is that there is nothing to be gained. Scientists are simply using the word "trust" in an extended and casual sense. They have no interest in a distinctively normative notion of trust, nor in an explanatory contrast between trust and strategic reliance.

This answer is unsatisfying. Automation has been introduced in areas of life such as medicine that are full of ethically weighty decisions. Inviting trust in that automation carries significant ethical consequences. It is doubtful that the word "trust" has been carelessly chosen; it seems to invite rather than avoid these ethical implications. For the sake of charitable understanding, we must at least try to find a different answer to our question.

Another possible answer is that trust in automation is a distinctive normative attitude, but an indirect one. We saw earlier that trust in engineering is linear but indirect. The engineer is an indirect

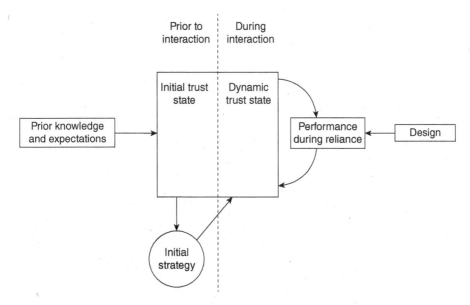

Figure 36.2 Trust in automation

Source: Simplified version of figure from Hoff and Bashir (2015).

object of the attitude of trust, standing behind the technological artifact with which users interact. In a stage play, the director and the author are off stage, but they take most of the responsibility—and the criticism—for the performance. Analogously, one might maintain that the reason we talk about trust in automation is because engineers (or companies) stand in appropriate relations to that automation.

However, there are two problems with this account. The first is that automation itself is not a real object of trust according to this story. The engineer, alongside other institutional entities, is the object of genuine trust. But we do not actually rely on engineers for the particular "performances" that we require during the activity of carrying out the shared task. Hence, their agency is insufficiently related to what we actually rely on to make it count as trust (Nickel 2013a). The second problem is that this story holds for buses, bridges, and bannisters just as well as for automated systems. The notion of trust at stake is therefore just as thin as for other everyday artifacts. In that case, there is nothing special about automation.

A third possible answer holds that the perceptible features of automation technology *invite* distinctive trust attitudes. There are several features of automation technology that make it psychologically a suitable object of trust. Some of these are discernible in the scientific literature. Hoff and Bashir summarize advice in the literature for "creating trustworthy automation", such as: "increase the anthropomorphism of automation" taking into account "age, gender, culture, and personality"; use "gender, eye movements, normality of form, and chin shape of embodied computer agents to ensure an appearance of trustworthiness"; and "increase politeness" (Hoff and Bashir 2015: 425). From these strategies it is clear that many automation technologies take on human social, bodily, and communicative characteristics that make them appear different from dumb artifacts or mere things in our environment. Anthropomorphic robots are among the most striking examples of automated systems that elicit interpersonal trust attitudes because of their human form and characteristics (Coeckelbergh 2012).

On its own, however, this third answer seems shallow. Although we might be happy to talk about trust in automation because the design of automation exploits human trust cues, on further reflection, talk of trust should be seen as a useful but insignificant manner of speaking, at most an explanatory stance (Dennett 1989). Since we *know* that anthropomorphic characteristics that invite trust are design add-ons and reflect no underlying personhood, they do not ultimately support robust talk about trust in automation. They may help us understand the phenomenology of the technology user's reliance on automation (and the cunning of the designer!), but they do not reveal a genuine trust relationship. In a recent article, Tallant (2019) puts forward a view along these lines.

The final answer to be considered looks at the fact that automation technologies are complex sociotechnical systems with a high degree of intelligence and rationality built into them both functionally and contextually. Functionally, they are capable of perceiving and responding to the environment. We evaluate them not just as being either reliable or unreliable (like a bridge), but as having correct or incorrect representations and as drawing correct or incorrect inferences. These functional aspects of sophisticated automation make them subject to some of the same kinds of normative evaluation that also underwrite our trust attitudes. For example, an artificial agent such as Watson that gives answers in response to questions invites many of the kinds of normative evaluation of its speech—whether it has made a relevant and sincere assertion, whether it has interpreted a question correctly, etc.—as a human speaker does. It is not at all surprising that we would describe our attitude toward such an agent in terms of trust (Nickel 2013a). We do not need to countenance automation as having free will or moral capacities in order for some of these normative evaluations to be appropriate. Such a notion of trust may be thinner than interpersonal trust, but it is not obviously thinner than the trust that we take towards institutions or the roles within them, for example.

Conversely, it is likely that automation technologies will also begin to assess the reliability of humans, or, as it were, the quality of the working relationship between automation and human, when the functioning of the automation depends on humans doing their part within that working relationship. Whether we call such a faculty a capacity for *trust*, or not, will depend on both the specific character that the capacity takes, and the matter of what we can gain, explanatorily, from using trust concepts to describe it.

In sum, it is plausible that reliance on automation involves a distinctive attitude of trust, contrasted with strategic reliance. The three factors that support talking about complex automation in terms of trust—the indirect role of the human engineer, the designed social cues that invite trust in the automated systems themselves, and the sociotechnical complexity of these systems (including representational states and perceptual and inferential processes)—are mutually reinforcing of one another.

Suppose we grant that real trust in automation is possible in cases where the aforementioned factors are present to a high degree. There are still open questions concerning whether and when we *should* trust automation. Philosophically, we must take a critical epistemological and ethical view of trust practices. Trust is not a goal to be unreflectively increased or a purely psychological given. People have *reasons* for trusting, both contextual reasons (e.g., there are few better options) and broadly evidential reasons (e.g., the trusted entity is seen to have the trustor's interests at heart). These reasons can be evaluated for their epistemological and ethical adequacy.

Engineers can design automated systems to manipulate our trust. Suppose an artificial agent is made to mimic a user's facial appearance and voice in order to gain their trust (Verberne et al. 2015). Such a strategy for increasing trustworthiness counts as *deceptively manipulative* in circumstances where increases in perceived trustworthiness do not correspond with good epistemic reasons for trust or greater usability (Spahn 2012; Smids 2018).

A theory of justified trust in automation would be useful as a starting point for ethical and epistemological evaluation, but currently no account of it exists. At a minimum, we want two things from such an account. First, we want a criterion for over-trust and under-trust, something that tells us when trust is appropriate. Such a criterion or norm concerns the would-be *trustor*, the one who is put in a position to trust. Let us call this the "user norm". Second, we want an account that illuminates the obligations and responsibilities of the engineer, the designer of automation. We want, for example, to know what would count as wrongful manipulation and exploitation of trust. Let us call this the "design norm". (Simon's 2015 identification of "epistemic responsibility" and "governance/design" as two relevant aims of an account of epistemic trust in the digital sphere is analogous to the distinction I draw here between user norms and design norms. Simon's account is not directly applicable to trust in automation because her account is explicitly tailored to epistemic trust involving belief or knowledge, rather than acts of practical reliance.)

To spur our imagination a bit, consider a not-too-improbable case of trust in automation (based on Google's service *Hire*). Suppose the human resources department of a company or university begins to use artificial intelligence to filter job applications and reduce the "grunt work" of recruitment officers who previously sifted through hundreds or thousands of files. Suppose that the recruiters care professionally about the values of fairness and diversity, and suppose that this AI tool has built-in features that are intended to increase fairness in recruitment, such as ignoring sex, age, and name when identifying salient candidates. Under what conditions should a recruiter trust such automation to suggest salient candidates?

I propose two main conditions for well-grounded trust of a user U in an automated agent A. Because of the lack of scholarly literature on the topic, these conditions are meant as hypotheses for discussion rather than a battle-tested theory. The first condition is meant to track the idea that the automation does not operate on reasons that are completely irrelevant or contradictory to the reasons that the user has (or should have). For example, is ignoring demographic characteristics sufficient to secure the values of fairness and diversity in recruitment? The recruitment officer might reasonably hold that affirmative action is required to secure these values, which would imply being *aware* of demographic characteristics rather than ignoring them. In that case, trusting this form of automation might lead to a sort of betrayal of the recruiter's values.

The second condition is meant to track the idea that the user must have sufficient reason to believe that the actual performance of the automation does or will satisfy standards of competence at carrying out the tasks at hand while not undermining U's interests. This sufficient reason can come from various sources depending on whether, as in Figure 36.2, the user is in an "initial state" prior to actually interacting with the automation technology, or is in a "dynamic state" in which there is some experience with the performance of the system. For example, before relying on the automation technology, a particular recruiter might hear from a trusted colleague that the system has improved the quality and diversity of candidates, while freeing up valuable time for recruiters to communicate quickly with those candidates.

Accordingly, let us then hypothesize the following conditions on warranted trust in automation:

User U's trust in automated agent A to do task T is reflectively warranted if and only if the following two conditions are met:

(Hypothetical Transparency Condition)

If U knew and rationally reflected upon the reasons that influenced A in relation to T, U would recognize the relevance and strength of these reasons from the perspective of U's own interests.

(Internal Adequacy Condition)

U has either experienced A doing T adequately or has other relevant support for her expectation that A can T adequately, such as the judgment of epistemic peers to the effect that A can T adequately, or the normative commitment of those who designed and deployed A to take U's interests into account when doing T.

Annette Baier (1986) proposes a similar hypothetical transparency test of the moral adequacy of trust. However, she rejects anything like the internal adequacy condition as a general criterion for the moral adequacy of trust, because she points out that children are often perfectly warranted in their trust even though they do not have any such reasons (ibid.). By contrast, I do adopt such a condition here because *reflective warrant* is meant to be appropriate for sovereign adults relying on automation. Particularly for professionals like those in our example who are in a position to trust automation in the workplace, it is plausible that they need to have some kind of positive reason for doing so. The idea of reflective warrant is meant to capture this idea. When children rely on automation, a different standard applies.

When coupled with the claim that users often have a strong (moral) interest in having a warrant for trust in automation, it follows that engineers who design automation have an obligation, other things equal, not to deceive people about the grounds for trust, and that they may have a positive obligation to give users access to information that helps fulfill the conditions of the warrant. In this way, we derive a relevant "design norm" against deception and exploitation from the "user norm". This should remind us of the point we started with in this chapter, that there is always a connection between trust and trustworthiness, between the user's trust and the engineer's trustworthiness. It is not a shallow connection, but a deep and ethically significant one.

Such an account of warranted user trust in automation also carries implications for governance, suggesting a new interest at stake in Europe's General Data Protection Regulation (GDPR). That law asserts that individuals have a right to obtain "meaningful information about the logic involved" that is used in automatically processing their personal data (Regulation 2016/679, 2016: 2.15.1.h), and data controllers have a corresponding duty to provide such information (ibid., 2.13.2.f). Most people have understood the aim of the GDPR as protecting one's interest in confidentiality and control over one's personal data. But there is in fact a second interest at stake here: trust in automation. When personal data processing is being carried out in order to provide automated services (whether at work or at home), the user has an interest in "meaningful information about the logic involved" *in order to safeguard the Hypothetical Transparency Condition*. This would include information about *the reasons for* using data inputs of a certain type and the *design thinking* behind the automation: reasons that help explain why the automation behaves as it does. These reasons should normally be conveyed during the mundane process of training a person to work with an automated system.

36.5. Conclusion

The opening question of the chapter can now be examined from a new perspective: What will trust in the engineer and the engineering profession look like in the future? Engineers may need to rethink how they want to be trusted, and to regard not just reliability and trustworthiness, but *trust* as an important commodity in itself. Brown and Calnan (2012) and Evetts (2013) emphasize the way in which trust is used by people as a way of navigating institutional and organizational complexity. Trust will be in ever greater demand in a future technological environment that is ever more complex and abstract. With engineers at the center of some of the most important technological changes in our future, the question of trust will likely arise for them in a pointed way.

Another way in which the future of engineering will not be similar to its past is that engineering will apply automation to itself. Calculative and mechanical tasks that used to form the core competences of engineering will be carried out by computers and robots. Engineers will design and care for sociotechnical systems including both human and mechanical components using a broad set of skills: computer programming and data science, interdisciplinary communication, lifelong learning, ethical awareness, and critical design thinking. Designing for interpersonal trust, and for trust in automation, will be core competences. So although trust and engineering have many strands—as exhibited in the preceding sections—these look increasingly as if they will be woven together in the future of engineering. (For more on this topic, see Chapter 55, "Reimagining the Future of Engineering" by Neelke Doorn, Diane P. Michelfelder, et al., this volume.)

Related Chapters

Chapter 22: Values and Design (Ibo Van de Poel)
Chapter 43: Professional Codes of Ethics (Michael Davis)
Chapter 44: Responsibilities to the Public—Professional Engineering Societies (Joseph Herkert and Jason Borenstein)
Chapter 55: Reimagining the Future of Engineering (Neelke Doorn, Diane P. Michelfelder, et al.)

Further Reading

Classics in the philosophical literature on trust include Annette Baier's paper "Trust and Antitrust" (1986) and Richard Holton's "Deciding to Trust, Coming to Believe" (1994). For contributions to the recent philosophical literature on trust, see Faulkner and Simpson (eds.), *The Philosophy of Trust* (Oxford 2017). For an earlier treatment of challenges to the idea of trust in technology and the relationship of trustworthiness to reliability in engineering, see Nickel et al. (2010).

Notes

1. "The Ritual of the Calling of an Engineer" (University of Guelph Department of Engineering, www.uoguelph.ca/engineering/events/2018/03/ritual-calling-engineer, Accessed September, 28 2018).
2. How trust is theorized depends on one's explanatory agenda. For example, in behavioral economics and political theory, trust is often theorized as a mechanism through which cooperation emerges from interactions

between individuals or state actors. In the theory of child development, by contrast, trust is theorized as a normal psychological disposition of the child to rely comfortably on the primary carer(s). I borrow elements from these diverse conceptions of trust as the context demands. For philosophical approaches to the concept of trust, see Nickel (2017), Simpson (2012), and McLeod (2002).

3. But see Nickel (2013b).

References

Adams, T.L. (2015). Sociology of Professions: International Divergence and Research Directions. *Work, Employment, and Society,* 29, 154–165.

Ba, S., Whinston, A.B. and Zhang, H. (2003). Building Trust in Online Auction Markets Through an Economic Incentive Mechanism. *Decision Support Systems,* 35(3), 273–286.

Baalen, S. van and Carusi, A. (2017). Implicit Trust in Clinical Decision-making by Multidisciplinary Teams. *Synthese.* doi:10.1007/s11229-017-1475-z

Baier, A. (1986). Trust and Antitrust. *Ethics,* 96, 231–260.

Beck, U. (1992). *Risk Society—Towards a New Modernity.* London: Sage.

Brown, P. and Calnan, M. (2012). *Trusting on the Edge: Managing Uncertainty and Vulnerability in the Midst of Serious Mental Health Problems.* Bristol: The Policy Press.

Brown, P. and Calnan, M. (2016). Professionalism, Trust, and Cooperation. In M. Dent, I.L. Bourgeault, J.-L. Denis and Kuhlmann (eds.), *The Routledge Companion to the Professions and Professionalism.* Routledge, pp. 129–143.

Coeckelbergh, M. (2012). Can We Trust Robots? *Ethics and Information Technology,* 14, 53–60.

Dennett, D.C. (1989). *The Intentional Stance.* Cambridge, MA: MIT Press.

Elliott, A. (2002). Beck's Sociology of Risk: A Critical Assessment. *Sociology,* 36, 293–315.

Evetts, J. (2006). Trust and Professionalism: Challenges and Occupational Changes. *Current Sociology,* 54, 515–531.

Evetts, J. (2013). Professionalism: Value and Ideology. *Current Sociology Review,* 61, 778–796.

Friedman, B., Kahn, P.H. Jr., Howe, D.C. (2000). Trust Online. *Communications of the ACM,* 43, 34–40.

Hall, M.A. (2002). Law, Medicine, and Trust. *Stanford Law Review,* 55, 463–527.

Hoff, K.A. and Bashir, M. (2015). Trust in Automation: Integrating Empirical Evidence on Factors That Influence Trust. *Human Factors,* 57, 407–434.

Lee, J.D. and See, K.A. (2004). Trust in Automation: Designing for Appropriate Reliance. *Human Factors,* 46, 50–80.

Manson, N.C. and O'Neill, O. (2007). *Rethinking Informed Consent in Bioethics.* Cambridge: Cambridge University Press.

McLeod, C. (2002). *Self-Trust and Reproductive Autonomy.* Cambridge, MA: MIT Press.

Nickel, P.J. (2013a). Artificial Speech and Its Authors. *Minds and Machines,* 23(4), 489–502.

Nickel, P.J. (2013b). Trust in Technological Systems. In M.J. De Vries, S.O. Hansson, and A.W.M. Meijers (eds.), *Norms in Technology: Philosophy of Engineering and Technology,* Vol. 9. New York: Springer, pp. 223–237.

Nickel, P.J. (2014). Design for the Value of Trust. In J. van den Hoven, I. Van de Poel, P. Vermaas (eds.), *Handbook of Ethics, Values and Technological Design.* Berlin/ Heidelberg: Springer-Verlag.

Nickel, P.J. (2017). Being Pragmatic About Trust. In P. Faulkner and T. Simpson (eds.), *The Philosophy of Trust.* Oxford: Oxford University Press, pp. 195–213.

Nickel, P.J., Franssen, M. and Kroes, P. (2010). Can We Make Sense of the Notion of Trustworthy Technology? *Knowledge, Technology and Policy,* 23, 429–444.

O'Neill, O. (2002). *Autonomy and Trust in Bioethics.* Cambridge University Press.

Pettit, P. (2004). Trust, Reliance and the Internet. *Analyse & Kritik,* 26, 108–121.

Power, M. (1997). *The Audit Society: Rituals of Verification.* Oxford University Press.

Power, M. (2007). *Organized Uncertainty.* Oxford University Press.

Regulation (EU) 2016/679 of the European Parliament and of the Council (2016). *Official Journal of the European Union.* L 119/1-L199/88

Renn, O. (2008). *Risk Governance: Coping With Uncertainty in a Complex World.* London: Earthscan.

Resnick, P. and Zeckhauser, R. (2002). Trust Among Strangers in Internet Transactions: Empirical Analysis of eBay's Reputation System. In Michael R. Baye (ed.), *The Economics of the Internet and E-commerce (Advances in Applied Microeconomics, Volume 11).* Bingley: Emerald Group Publishing Limited, pp. 127–157.

Roddis, K. (1993). Structural Failures and Engineering Ethics. *Journal of Structural Engineering,* 119, 1539–1555.

Simon, J. (2015). Distributed Epistemic Responsibility in a Hyperconnected Era. In L. Floridi (ed.), *The Onlife Manifesto.* Springer. doi:10.1007/978-3-319-04093-6_17.

Simpson, T.W. (2012). What is Trust? *Pacific Philosophical Quarterly,* 93, 550–569.

Smids, J. (2018). *Persuasive Technology, Allocation of Control, and Mobility: An Ethical Analysis.* Doctoral dissertation, Technische Universiteit Eindhoven, Eindhoven.

Spahn, A. (2012). And Lead Us (not) Into Persuasion . . . Persuasive Technology and the Ethics of Communication. *Science and Engineering Ethics,* 18, 633–650.

Suler, J.R. (2004). The Online Disinhibition Effect. *CyberPsychology and Behaviour,* 7, 321–326.

Tallant, J. (2019). You *Can* Trust the Ladder, but You Shouldn't. *Theoria.* doi:10.1111/theo.12177.

Turcotte, J., York, C., Irving, J., Scholl, R.M. and Pingree, R.J. (2015). News Recommendations from Social Media Opinion Leaders: Effects on Media Trust and Information Seeking. *Journal of Computer-Mediated Communication,* 20, 520–535. doi:10.1111/jcc4.12127.

Verberne, F.M.F., Ham, J. and Midden, C.J.H. (2015). Trusting a Virtual Driver that Looks, Acts, and Thinks Like You. *Human Factors,* 57, 895–909.

Voerman, S.A. and Nickel, P.J. (2017). Sound Trust and the Ethics of Telecare. *Journal of Medicine and Philosophy,* 42, 33–49.

37

AESTHETICS

Stefan Koller

37.1. Introduction: Aesthetics Beyond Art and Beauty

Two breakthroughs in recent aesthetic philosophy make aesthetics relevant to philosophical reflection on engineering more than ever. One breakthrough divorces aesthetics from its historical preoccupation with the study of artworks, and it grounds the field instead on a particular, namely aesthetic, mode of experience promiscuous in its range of objects (Nanay 2018: 4). Aesthetics thus reconfigured becomes the study of that experiential mode: its nature, value, and impact on human beings. The other breakthrough extends inquiry of aesthetic value (and of aesthetically valuable experience) beyond beauty (Levinson 2016: chs. 3–4), for instance by querying how some artifacts manage "to vivify, deepen or, exceptionally, modify our understanding" of the world around us (Kieran 2013: 295). Aesthetics divorced from fixation on art and beauty widens the scope of aesthetic inquiry—and opens up questions of how we understand, design, and experience engineered artifacts. This is quite independent of whether engineered artifacts have lofty artistic ambitions or aspire to beauty—while some of them, like the Ferrari F1 displayed in New York's Museum of Modern Art, or indeed that museum itself (as a blend of structural engineering and aesthetic façade design) arguably do (Fisher 2015), many others do not. Artistic ambition and beauty are no longer necessary for artifacts to become salient objects of aesthetic inquiry and valuation.

Nor is that all. Design in the arts and in engineering nowadays overlap in practice, if less so in method. Many products from smartphones to cars are hybrids of engineering and aesthetic design, the end product of multidisciplinary design teams. Management of such products draws widely on all facets of design, from prototyping to patent filing and market distribution. Philosophical reflection on engineered products, by contrast, lags behind, preferring to compartmentalize analysis into the hard sciences and leaving aesthetics as an afterthought—a packaging. Yet, just as designers in engineering and the arts draw freely on each other's design philosophies, philosophers reflecting on engineering design will find untapped riches in recent aesthetic philosophy. Or so this chapter demonstrates. To that end, the chapter scrutinizes a subset of philosophical debates in aesthetics particularly pertinent to a philosophy of engineering, with particular emphasis on the creation of aesthetic and moral value (cf. Chapter 22 "Values and Design" by Ibo Van de Poel, this volume). The overall goal is to highlight avenues of future research by which next-generation scholars will surmount present limitations.

Given spatial limitations, this chapter cannot do justice to the methodological pluralism and diversity that characterizes recent aesthetic work at its best. Instead, the chapter focuses on key debates in analytic aesthetics, a type of aesthetics that prominently features the application of logic, conceptual analysis, and a quasi-scientific method relying on hypothesis, counter-example, and modification (Lamarque and Olson 2019: xvi). As such, analytic aesthetics may be well suited to complement the engineering sciences.

At the same time, the chapter's analytic focus unfortunately marginalizes other eras and traditions in philosophy, leaving plenty of room for disagreement on points of principle and detail. Work in phenomenological and feminist aesthetics is critical of several assumptions made in this chapter, not least its assumptions about value, ontology, and lack of emphasis on (and with) the social identity of those creating and experiencing engineered artifacts (Cahn and Meskin 2009: 327–389; Korsmeyer 2004; Devereaux 2005; cf. Chapter 48, "Feminist Engineering and Gender" by Donna Riley, this volume; on a more reconciliatory note, see Smith 2018: 46). Inquiries into how art relates to contemporary engineering, such works argue, should focus on innovative processes giving rise to new artifacts—and not focus on artifacts on their own, as under the analytic approach probed in this chapter (Coeckelbergh 2018: 503). Non-Western traditions offer similar critiques (Chakrabarti 2016). In addition, analytic philosophers are latecomers to debates on how aesthetics, engineering, and new technologies interface. By contrast, their Continental colleagues have contributed to debates on all three at a pace contemporaneous with contemporary advances in technology. For instance, Jean Baudrillard explored the interdependency of man and machine in fractal and virtual realities, and the very phenomenon of the "extended mind" (expanded scope of non- and aesthetic experience by cognitively enhancing implants) decades before such reflections entered analytic philosophy (Baudrillard 1989; Menary 2010). Since anyone drawing on analytic aesthetics must be as keenly aware of its limitations as its potential, the chapter will recurrently address both.

This in place, we can start mapping the terrain ahead. After categorizing central debates under three headings—experience, ontology, and value—the chapter scrutinizes the nature and content of aesthetic experience and looks at frameworks for the moral appraisal of aesthetic as well as non-aesthetic engineered artifacts. Where these opening sections probe the applicability of aesthetics to engineering, the concluding sections probe how recent technology has altered aesthetic experience and aesthetic environments. Digital renderings and geoengineering give rise to phenomena that cry out for philosophical treatment but also challenge assumptions present in contemporary aesthetics. Hence, instead of simply applying analytic philosophy *to* new technologies, philosophical attention to new technologies may well recast foundational issues in aesthetics. A Coda reflects on how greater attention to analytic aesthetics may affect philosophy of engineering not just at the level of content but irrevocably reform its method—hopefully for the better.

37.2. Key Issues in Contemporary Aesthetics: Experience, Ontology, and Value

Key debates in recent aesthetic philosophy have focused on the proper characterization of aesthetic experience, the ontological nature of aesthetically appraisable artifacts (in short, aesthetic artifacts), and the moral appraisability of such artifacts. All three debates can easily be recast from aesthetic to engineered artifacts—where an artifact qualifies as engineered if its causal history and design align with a full account as to what engineering is. (This chapter is system-neutral as regards such accounts but presupposes that the set of engineered artifacts is non-empty; cf. Chapter 15, "Artifacts" by Beth Preston, and Chapter 16, "Engineering Objects" by Wybo Houkes, this volume).

Such recasting gives rise to three (sets of) questions:

> *Experience.* Can engineered artifacts be experienced in an aesthetic mode? Should they? If so, how? And should that experiential mode inform the design of engineering artifacts, or inform it to a greater degree than it currently does?

Ontology. Which properties do engineered artifacts have, and how much overlap is there between
those properties and the kinds of properties aesthetic artifacts (typically) have?

Value. To the extent engineered artifacts can be experienced in an aesthetic mode, which values
do these artifacts disclose that they otherwise would not? In particular, do engineered artifacts
have moral value accessible only in the aesthetic mode?

These three areas of experience, ontology, and value are interrelated: how we ideally experience and
interact with engineered artifacts should, at least partially, drive how we design and engineer these
artifacts—what properties they (ought to) have. An aesthetic stance on engineering design, then,
just like aesthetic philosophy itself, will have a dual focus on the creation and reception of specific
artifacts. Given the focus on aesthetic experience, our inquiry naturally takes us there first. We begin
with a recent case study, and then delve into analytic frameworks.

37.3. Aesthetic Experience: The New Environment

Although many aesthetic experiences such as a trip to the museum are voluntary and expected,
some are not. Some, indeed, occur as a mix of the two, and the surprise is not always welcome. In
2003 pre-screenings of the computer-animated comedy *Shrek*, children started to cry at the uncanny
appearance of one character. "Princess Fiona" (later voiced by Cameron Diaz) was so life-like, the
animation nearly indistinguishable from an actual human actress, that children experienced not sympathy but alienation and even fear. Cognitive scientists studied the phenomenon and found a correlation of audience reactions to a phenomenon that early computer scientists had termed the "uncanny
valley" (Gray and Wegner 2012). Humans react favorably to machines or artificial representations
that are like humans—especially robots. Indeed, they react more favorably the more these machines
resemble actual humans. Yet, there is a sharp decline in emotive responses to robots and virtual ani-
mations after a certain degree of semblance—at which point a feeling of uncanniness sets in.

Without having a name or empirical study to the phenomenon, the animators in *Shrek* began to
dial back the lifelikeness of the Fiona character, making its figure and animation more cartoony. This
resulted in warmer audience receptions and ultimately helped the film towards commercial success.
In the time since *Shrek*, other empirical studies of digital animation have confirmed the phenomenon, pinpointing the onset of its receptivity in childhood development (Brink et al. 2017) and
demonstrating that we are not remotely in a position to avoid (or ignore) it altogether. Engineers,
it appears, are not merely limited in their current means to prevent involuntary and unpleasant aesthetic audience reactions when confronted with novel technologies: they remain improperly trained
to predict such reactions before they jeopardize multimillion-dollar global entertainment projects.

As technological engineering transforms the world around us, it does not always do so in predictable or intended ways—and sometimes, the reactions to such transformations evoke aesthetic
responses that are not readily understood or categorized in frameworks that date to the eighteenth
century. Not coincidentally, Gray and Wegner relied on an aesthetic category of much later progeny:
the notion of uncanniness, in German *Unheimlichkeit* (of feeling not quite at home). But what exactly
does it consist in? Which components does such experience have—and if our engineers can deliberately elicit it, can they equally well design ways around it? Precisely because engineering conjures
up novel aesthetic experiences every day, and because these phenomena remain so little understood
beyond empirical correlations, we need sophisticated frameworks that help deliver that understanding one piece at a time.

Enter analytic aesthetics. As a first step, the field imposes a heuristic distinction, or rather, recognizes two categories of components to aesthetic experience: one emotive, the other representational.
That is, aesthetic experience arguably blends feelings and emotional reactions with cognitive uptake,
where the latter can be represented (if imperfectly) at a quasi-propositional level. For instance, looking

at a Van Gogh landscape, we experience a variety of sensations all at once. We may, for instance, experience elation at the vibrancy of color, while taking in the representation of landscape before us. Analytic aesthetics attempts to move from this commonplace to greater precision and consistency: we want a theoretic understanding not only of this particular experience, but of any other sufficiently similar to it. The divide of experiential qualities into emotive and representational, while artificial, helps to attend to that task in a more piecemeal fashion.

Philosophers divide on the question whether or not to countenance a *distinctive* range of emotional states associated with aesthetic experience—whether, for instance, such experiences contain emotions we do not experience outside them (Ismeninger 2004: 99). In the following, we shall raise the question whether engineered experiences and landscapes can successfully simulate their 'natural' counterparts. Here, by contrast, the question rather arises whether engineering could actually expand the range of aesthetic emotion by creating a wholly unprecedented range of sensations. Resolving this question has barely begun, and will (like much else covered in this chapter) likely require attention not only of aesthetic philosophy but the cognitive sciences as well.

A second matter on which philosophers divide concerns not just the qualitative nature of aesthetic emotions, but also their cognitive value (if any). Where emotions are purely associative—such as, becoming nostalgic upon listening to a Christmas song—philosophers regard them as extrinsic, and not essential, to the artifacts that elicit them (Robinson 2019: 583). Of the emotions that remain, and are suitably related to the works that arouse them in us, philosophers query whether such emotions hinder or aid our comprehension of artifacts (cf. Mole 2015: 75–77). Robinson's main example is musical works that rely on creating a structured (temporally extended) experience, comprising feelings of surprise and relief elicited by tonal-harmonic structures. By commanding (and deliberately eliciting) these feelings in turn, listeners hone in on the piece's other musical features, and become more attuned to the progression of melody and theme. Robinson's empirical work here builds on that of established musicologists like Krumhansl and Meyer (Robinson 2019: 593).

A framework like this raises questions regarding the scope of its application beyond music. Naturally, any artifact capable of eliciting temporally extended series of emotional reactions can tap into such cognitive stimulation (or manipulation)—and that potential does not seem limited to artifacts traditionally categorized as artistic. That is doubly so in times where the boundaries between artistic and engineered artifacts have become increasingly porous:

> Whether or not something is art might determine whether or not it is eligible for an award from a government arts agency or whether its sale or import should be taxed. For example, a question arose over the importation of Brancusi's abstract sculpture *Bird in Flight* as to whether it was an artwork or a collection of industrial metal. If industrial tubing, a customs fee would have to have been paid; but if art, it could enter the U.S. duty free.
>
> *(Carroll 1999: 207)*

Where artworks increasingly rely on ready-mades and (relatively unmodified) components of industrial engineering, such works' expressive and emotive potential seems to remain in place quite regardless of how we—or indeed, the US Internal Revenue Code—would classify such objects. This, for now, shall suffice to discuss cognitive value and the dimension of aesthetic experience's emotive component.

On a competing framework, we sidestep questions of emotive quality altogether, and limit characterizations of aesthetic experience to their cognitive and representational features (Ismeninger 2004: 99). This framework is likely better suited for an account of engineered artifacts, given that many of them are so ingrained in our lives that our emotional reactions are confined to moments where they operate in unintended ways, or cease to operate at all. In an example made famous by Heidegger, primal reactions to engineered artifacts occur in moments of frustration, such as a hammer falling apart, or in a moment of the artifacts' own physical absence, as when I mislaid the hammer. Since these emotions can hardly be said to characterize (a typical experience of) the artifact, a deflationary account that regards emotional reactions as optional may fit engineered artifacts more frequently. (This 'side-stepping' of emotive quality is naturally open to challenge. Technical artifacts give rise to a wide spectrum of emotional responses even in the absence of breakdown. In fact, positive emotional responses—such as the psychological phenomenon of flow—very much assume the opposite.)

In addition, Carroll has a uniquely fitting characterization of the representational content of such experience, orienting it to the artifact's function. In particular, he probes an account whereby appreciation of an artifact may have less to do with liking it than with 'appreciating' it for its accomplishment of a given task (Carroll 1999: 150, 2016: 4). This accomplishment typically derives from attention to an artifact's primary function(s), where appreciation focuses on the designerly elegance at accomplishing a set task, or on doing so in a novel way—see Section 4 below on the functionalist stance.

If this is so, then engineered artifacts on Carroll's account either try to work so flawlessly and effortlessly as to not draw our emotional attention to themselves—or on the few occasions when they do, aesthetic appreciation is deeply intertwined with functionalist concerns that already (or at least, should) inform engineering design. While now a historical rather than contemporary experience, anyone's first Internet browsing experience with a lightweight iPad is likely of this variety: aesthetic enjoyment here simultaneously focuses on the effortless delivery of a high-resolution screen and a second-order awareness of the design feat of delivering just that experience through a novel combination of haptic and other features.

If ease of access and use typifies modern engineering design—especially in the consumer realm—then accessibility too should be a focal point in anyone's research into engineering aesthetics. Here, analytic philosophers contribute relevant frameworks through their recent study of so-called mass art, that is, art both mass produced and produced for mass audiences. Historically, most aesthetic philosophers denigrated this range of aesthetic experience, based on elitist conceptions of what range of objects are suitable for such experience. Quite a few in the Continental tradition (Adorno, for instance) were proponents of avant-garde art, leading them to consign mass art to the realm of pseudo-art (Carroll 2019: 617). That is, if only avant-garde art is true art, and mass art lacks the features that typify the avant-garde, then mass art is of necessity an unsuitable object for those preoccupied with true art.

Recent work in analytic aesthetics rejects those assumptions (and in doing so, incidentally following Benjamin's work a century earlier). First, as we saw above, recent work rejects an exclusive focus on art, let alone true art. Secondly, to the degree it studies artworks, analytic aesthetics is interested in a wider range of objects than the avant-garde. Thirdly, analytic aesthetics, instead of denigrating mass art, is interested in understanding the aesthetic and commercial success of the engineered product that typifies mass art.

Three features, it is claimed, particularly encapsulate the essence of mass art (Carroll 2019: 608–615). Firstly, mass art is made for mass consumption—that is, a wide demographic, without reliance on demographic or class distinctions (much less, elitism). Scale of audience is critical to success. To attain mass consumption, the production of such artifacts requires the other features. Secondly, mass consumption requires a mass medium that eases the work's distribution and consumption, with film (over individualized stage productions) being a case in point. Thirdly, mass consumption requires, indeed presupposes, mass accessibility. This, in turn, has given rise to analytic inquiry into just what makes some engineered products accessible where others fail to do so. On one account, accessibility requires that the design considerations going into such works' manufacture be "such that they can be grasped and understood almost on first contact" (Carroll 2019: 611). Meaning, key design features have to be "legible to the average untutored audience" so as to facilitate "ease of comprehension and consumption," and skirt over nuanced cultural background knowledge that may be accessible only to a small demographic (Carroll 2019: 612).

While mass accessibility is perhaps a necessary condition for commercial success, it does not guarantee it. Instead, philosophers argue, a full account of mass art design must consider—and explain—how some works rely on a mass medium in a way that instantly "command[s] the attention of mass audiences" (Carroll 2019: 615). It is here, above all, that analytic aesthetics taps into cognitive science and studies how aesthetic experience works at a quasi-mechanical level. For instance, the success in the late 1990s and early 2000s of MTV music clips critically relies on a cut-frame rate so high as to stimulate humans' "involuntary tendency of our attention to reawaken (for sound adaptive reasons) upon the onset of new stimuli" (Carroll 2019: 616).

In the aggregate, these reflections have taken analytic aesthetics far away from the rarified world of elite artworks—and in doing so, have rendered aesthetics more readily relevant to reflection on the commercial and designerly success of some engineered artifacts as opposed to others. And where artifacts intended for mass consumption do fail to be successful, the answer will likely lie in the type of aesthetic response the artifact elicited, and the type of response it failed to elicit: to what degree did engineers assume background knowledge? How far did they project their own expertise onto the realm of consumption—and is there a way back, by tapping more directly into untutored responses?

37.4. The Functionalist Stance

Another breakthrough friendly to reflection on engineered artifacts is recent aesthetics' take on the place of functionalist considerations in aesthetic experience and valuation. Reflections in this area influence what we earlier called questions of ontology: the proper characterization of artifacts, and questions of whether or not to countenance artifacts thus characterized as fundamental to a philosophical account of what, ultimately, exists in the world around us.

Traditionally, aesthetic judgments—and what informs them—were taken to be antithetical to utilitarian considerations of any sort. Since (it was believed) the ideal or prototype of aesthetic experience was of an artwork, and an artwork was best experienced for its own sake ('art pour art'), it followed that considerations of the artwork's economic value or other usefulness in the affairs of daily ('mundane') life were irrelevant at best and distracting at worst. The key philosopher associated with this anti-functionalist stance is Immanuel Kant, who expressed it as (attention to) "purposiveness without purpose":

> Judgments of value are related to purpose. If I want to know whether an [artifact] X is a good X, I need to know what Xs are for—that is how I tell what makes a good knife, or a good plumber, and so on. Judgments of perfection are similar: I cannot know what is a perfect X without knowing what is the function of an X. Judgments of beauty, however, cannot be quite like this, since [on Kant's view] they do not bring their objects under any concept X.
>
> *(Kenny 2007: 252)*

By contrast, recent analytic aesthetics has made room for reflections on type and utility, opening up its frameworks to reflections on engineered artifacts—artifacts we cannot properly experience, and evaluate for adequacy or 'perfection' in the absence of classification, as pointed out by Kenny apropos knives. Carroll calls this the functionalist stance, because in tying aesthetic experience to aesthetic form, and aesthetic form to an artifact's function, we correlate the very essence of aesthetic experience to experience and "designerly appreciation" of functional features (Carroll 1999: 152).

It is obvious how such identification of the representational or 'content' level of aesthetic experience closes the gap between so-called artistic and engineered artifacts. Recall also the earlier example whereby industrial tubing features in artistic installations, blurring the same boundary at the material level. Increasingly, it is not material or institutional features of an artifact, but our response to it, that classify it as suitable for aesthetic engagement—and thus, for aesthetic inquiry (Carroll 1999: 148–152). This presents an inversion of twentieth-century aesthetics which had put categorization ahead of (indeed, informing the very content of) response:

> [H]ow we respond to an object . . . depends decisively upon whether or not we categorize it as an artwork. Suppose we come across a living, breathing couple seated at opposite sides of a wooden table, starting intently at each other. Ordinarily we might pay no attention to them at all. . . . [b]ut if we categorize the situation as . . . a performance piece . . . our response will be altogether different.
>
> *(Carroll 1999: 6)*

By instead preserving the openness of aesthetic experience to artifacts regardless of their institutionalized type, contemporary aesthetics allows us to scrutinize a much wider range of artifacts. Instead of, for instance, only scrutinizing the physical, particularized token of a certain design, say an Italian sports car, we can also aesthetically appreciate that design's intangible type, as in a drawing or an abstract type. A patent engineer or appraiser can appreciate the conceptual and 'designerly' elegance of this or that design solution, even if it were not of a designed object but considerably more abstract like a mathematical formula (Breitenbach 2013; Fenyvesi and Lähdesmäki 2017). Similar considerations extend aesthetic appreciation to non-visual and visual computer art (Gaut 2009).

With this openness of aesthetic experience at the content level in place, we turn to normative issues: firstly, whether aesthetic experience, and the object of such experience, can be suitably moralized, and secondly, whether the transformation of such experience at the hand of engineering is something to be welcomed or be critical of (readers interested in further normative inquiries, such as how beauty intersects with epistemic norms in mathematics, are directed to McAllister 1999).

37.5. Moral Value in Aesthetic Design

Much recent work on analytic aesthetics has focused on morally appraising artworks. To appreciate its subtle abstractions, we begin with a concrete example. In late 2018, Volvo ran an advertisement for its newest car, reminding readers that, just as Volvo "had not lied to" its customers when promising to install seatbelts in an ad back in 1967, so Volvo was honest in 2018 about its emission reporting to purchasers of its cars. As is well known, the patent for the seat belt belongs to a Volvo engineer, Nils Bohlin, and the company gave it away for free to the public domain in 1959, well before the US Congress made the use of a seat belt mandatory in 1968 in response to public pressure (Nader 1965). Indeed, much of the public misgiving about seat belts targeted the very *availability* of the patent in the absence of adoption by industry and commerce. Volvo's emphasis on designing cars with proper emission reporting, by contrast, is meant to contrast the scandal surrounding Volkswagen's cars being designed rather differently. Ongoing litigation over emission reporting targets both environmental degradation as well as the erosion of consumer trust (see Chapter 36, "Trust in Engineering" by Philip J. Nickel, this volume).

What exactly is at issue in the 2018 Volvo ad? Is it the moral promise of a company, of an engineer, of a car, or indeed all three? We are accustomed to pinning moral value attributes on people, or large groups thereof (corporations)—less so when it comes to cars. Is a car morally virtuous while a Volkswagen is corrupt and deceptive? Do our smartphones spy on us and betray the trust we place in them—or are they mere extensions of (broken) trust between a company and the end consumer?

Surprisingly, philosophers of engineering have shirked these questions, preferring to moralize human reactions or conduct instead. Aesthetic philosophers, by contrast, have recently tackled these questions head-on, and their efforts deserve closer study.

As stated earlier, much recent work on analytic aesthetics has focused on morally appraising artworks. Accounts of such appraisal should minimally meet two constraints: the appraisal should literally target the artifact (engineered or aesthetic), and the appraisal should be genuinely moral. While these constraints border on truisms, it is surprising how few theoretic models ultimately escape their scrutiny (Koller 2015: 28–32). That is, on a *prima facie* plausible set of expectations, moral appraisals of artifacts seem to warrant being taken at face value. Yet most theories proclaim that such appraisals are ultimately not about the artifacts themselves. What underlies that sentiment is the assumption, frequently un-argued, that only persons, or person-like artifacts, are morally appraisable, and artifacts are by definition not persons (e.g., Carroll 2012: 191). Instead, aesthetic philosophers have tried to find suitable substitutes of moral appraisal, and attributed person-like features to aesthetic artifacts— for instance, the implied (not necessarily authorial) point of view a work of fiction conveys (Carroll 2012: 192). One off-shoot of this view is that engineered artifacts, say buildings, are no longer intrinsically moral but only derivatively so—and statements that ostensibly attribute moral features to artworks are infelicitous and have to be reinterpreted, for instance, as judgments about actions an artist performs through her work (Sauchelli 2012: 139).

On another proposal, the derived relationship is best explained in terms of symbolization, such as a dictator's palace (like Ceausescu's in Bucharest) being "a symbol of everything that is evil" (Illies and Ray 2006: 1225). Of course, what is alleged to be morally evil here is not the building but what it houses—the person inhabiting it. Appeal to symbolism is necessarily too narrow to serve as a general theory of artifacts' moralization: various consumer products (from processed coffee beans to shower gel) praise themselves as ethically responsible for being properly sourced. It is not plausible to recast these examples in terms of the symbolization of political (or other) regimes. That is why certain writers reject artifacts' moralization entirely, and prefer to regard such moralizations as committing a category mistake, as literally meaningless, and best delegated to a theory whereby such statements are to be eliminated, or reduced to their emotive content, expressing how people feel rather than describing what the world is like (Schroeder 2015).

As should be apparent from this brief survey, analytic aesthetics offers a variety of views to moralize artifacts—some of them more critical of our appraising practices than others. It should be equally apparent that few of these frameworks translate straightforwardly to ordinary engineered artifacts. A pipeline or computer hard drive will rarely assume or convey a point of view or other suitably anthropomorphic feature; nor will it readily symbolize a political or moral stance we can praise or condemn by virtue of association. The eliminativist strategy to regard all moral appraisal of artifacts as empty noise, however, may seem too radical. In that regard, engineered artifacts are not so much an illustration as a challenge to existing theoretical paradigms in analytic aesthetics.

A perhaps more promising avenue focuses on the causal effects of engineered artifacts, such as the psychophysical harm they (foreseeably) create or avert in their users (Illies and Ray 2006: 1222). The problems with this approach are less principled (ontological and semantic) than epistemic. Outside exceptional cases (leading to a manufacturer's recall of a faulty product), the causal progeny of psychophysical harm is rarely traceable to the artifact itself—as opposed to a complex mix of social milieu, user idiosyncrasy, and communication failure between product and consumer. (Were it otherwise, commercial tort litigation would hardly produce multimillion-dollar settlements in cases like Volkswagen and beyond.) While the totality of such circumstances may be morally appraisable, this moves us very far away from appraising artifacts themselves, in isolation. Moreover, once we divert attention from artifacts to their use context or use plan (see Chapter 17, "Use Plans" by Auke Pols, this volume), the initial ambition of evaluating artifacts has dropped by the wayside. Like aesthetic philosophers, engineering philosophers perennially prefer changing the subject (literally) to admitting that existing frameworks have yet to provide an account that treats moralizations of artifacts at face value (Koller 2015: 56). Instead, a future framework should ground moral appraisals in the kind of properties such artifacts uncontestably have—for instance, (so-called) categorial and intrinsic properties. Such a framework would have to establish the kind to which the artifact belongs, an observation applicable to artworks and engineered artifacts alike, and is significantly informed by the (primary) use to which such objects are put (Thomasson 2014, 2019). Needless to say, most work of establishing such frameworks remains to be done.

We now move beyond topics firmly entrenched within the triad of experience, ontology, and value. Instead of applying aesthetics to engineering, we ask what challenges contemporary engineering poses to reflections in analytic aesthetics.

37.6. Engineering Aesthetic Experience

Engineering has not always been viewed as a benign influence on the arts and aesthetic experience. Walter Benjamin (2008), for one, argues that the onset of phenomena made possible by engineering's success story—industrialization and mass production—have undercut several aesthetic values, including the uniqueness and distinct "aura" of an artwork. It is one thing to behold Da Vinci's *Mona Lisa* in its carefully orchestrated set-up in the Paris Louvre, another to look at the painting's photographic reproduction in the confines of one's home or computer screen. The same applies to religious artworks from prior eras, such as frescos or sculptures. As modern engineering evolves and new technologies such as 3D printing emerge and become commercially available, philosophical questions of artworks' uniqueness and accessibility—and the relative value thereof—will if anything become even more disputable.

While new technologies may in principle render such artworks experientially available to larger (especially economically disadvantaged) audiences, Benjamin's challenge stands: is the aesthetic experience of such works the same as that afforded by the original artwork? Modern artworks that rely on mechanical duplication to begin with—such as serial art, whether of paintings or digital music—all but eradicate the historic distinction of original to copy (cf. Danto 1983). In that vein, present and future advances of engineering may be vehicles to techno-aesthetic change, that is, of change in our aesthetic sensitivities and value judgments.

Some recent work challenges this line of argument. Contemporary fascination with technology, so far from eradicating pre-modern aesthetic sensibilities, taps into them to endow contemporary engineered artifacts with aesthetic qualities quite foreign to their gestation. Robots, for instance, and fascination with the autonomy and automation of bodies of human semblance, have given rise to a new romanticism of technology (Coeckelbergh 2017). This is but one instance where the aesthetic experience of appropriated engineered artifacts may gain increasing social traction, and the old dichotomy of artistic and engineered artifacts may become increasingly institutionally and commercially obsolete. For many a sales display area of contemporary engineered artifacts (say, an Apple Store) deliberately engineer spaces of heightened aesthetic experience through artful use of digital lighting, much like religious spaces in earlier centuries tried to accentuate the value of artifacts through lighting effects. As before, contemporary philosophers may here well draw on Benjamin's (2002) reflections on aesthetic space and its constant redesign through commerce, an inquiry that readily lends itself to contemporary engineering and architecture (Koller 2016).

Some philosophers see art and technology as not necessarily inimical, but as potentially mutually beneficial. Roeser (2013), for instance, argues that contemporary understanding, design, and interaction with new technologies all benefit if engineers and consumers alike draw on emotive responses. Such responses carry cognitive and moral value not easily replicated in stances bereft of emotional input. Contrariwise, such responses may alert us when value is lacking, as it well may be where engineered artifacts like buildings give rise to unpleasant visual experiences. Just how to demarcate valuable from harmful aesthetic experience, however, is an open issue. Recent work on that question increasingly relies on the cognitive sciences and suggests that we consider redrawing the very lines of (valuable) aesthetic experience (Nanay 2018). This is particularly true for the study of emotive content and emotive quality, where brain scans reveal highly divergent patterns between non-aesthetic and aesthetic types of pleasure:

> Even though our pleasure funnel through the same deep brain systems, to say that our pleasure in art is the same as the pleasure we receive from tasting sugar would be silly. Of course aesthetic encounters are more complicated. Aesthetic experiences go far beyond the simple pleasures of our basic appetites. The emotional rewards of aesthetic encounters are more nuanced, and the experiences are more modifiable by our cognitive systems.
>
> *(Chatterjee 2014: 181)*

Such input from the cognitive sciences modifies and challenges earlier work on aesthetic pleasure, specifically regarding its idiosyncrasy compared to other types of pleasures. In that regard, new work in aesthetics is more open to the hard sciences and engineering disciplines as an input to century-old debates on aesthetic value and experience, debates largely uninformed by (if not deliberately insulated from) empirical input. A particularly critical ingredient is the study of evolutionary cognition: if, as Chatterjee believes, our cognitive systems modulate our experience of aesthetic pleasure, and those systems evolve over time, it stands to reason that aesthetic beauty, informed as it is by aesthetic pleasure, itself evolves over time (Chatterjee 2014: 46–47).

This research paradigm is highly pertinent to the design and experience of engineered artifacts: how exactly do our brains and cognitive systems process pleasant interaction with such artifacts? Is it possible to use brain scans to map out changes over time, or perhaps even predict such changes? If so, groundbreaking work in cognitive aesthetics presents an untapped resource for engineers and philosophers of engineering. Philosophers must be wary of such forays, however. Invoking the cognitive sciences cannot conclusively answer normative and conceptual questions in aesthetics (Levinson 2016: 9–10). Traditionally, philosophers termed this risk the naturalistic fallacy: you cannot infer how things *ought to be* given how they (empirically) happen to be (for a relatively small sample of the population). Forays into cognitive aesthetics, then, ought to be cognizant of such fallacies.

37.7. The Aesthetics of Geo-Engineering

As we saw earlier, contemporary aesthetics offers several frameworks to rethink the moral impact of engineered artifacts. In addition, aesthetics also offers a means to more discriminately evaluate the normative rights and wrongs implicated in engineering's impact. Technology's impact on the natural environment has reached a scale no longer confined to the life world of humans but extending to the very integrity of the natural environment. Whether directly (as explored here), or indirectly as in geo-engineering seeking to undo the effects of climate change, contemporary engineering may soon transform—for better or worse—the very face of the earth. That is, we have yet to see whether fundamental alterations to our micro-climates in the atmosphere, stratosphere, and perhaps oceanic environment will not fundamentally alter the macroscopic appearance that characterizes the natural environment around us and informs our aesthetic appreciation of it.

To fully assess such impacts, environmental aesthetics presents some pressing questions as to which values are implicated. Importantly, the field recently shifted its focus from aesthetic beauty—the scenic quality of our environment—to aesthetic integrity, the wholesomeness of nature untouched or not irremediably damaged by human impact (see, e.g., Brady 2019; Carlson 2019; Saito 2007: 31–41). Even supposing that geo-engineering effected *no* drastic visual impact immediately observable to the human eye, we could likely not look at the landscape in the same way. For, as Brady (2019: 676) argues, proper environmental aesthetic appreciation is not insulated from "scientific fact" but very much informed by it.

To illustrate her point, imagine future engineers were equipped to artificially recreate an entire mountain landscape that prior environmental degradation had altered beyond recognition. While visually indiscernible from its (original) natural counterpart, and perhaps of equal beauty, the engineered copy lacks natural-historical properties that contribute to the very value comprising the landscape's integrity—and our enjoyment of nature (Brady 2019: 681). For such enjoyment draws not on a landscape frozen in time and place, but on narrative qualities having to do with the landscape's longevity and history—in fact, it draws quite a lot from "[t]he fact that nature is natural—not our creation," in marked contrast to a geo-engineered scenery (Carlson 2019: 670).

The example of an engineered mountain range is less futuristic than it may initially seem. A recent study reflects on the re-creation of the Swiss Matterhorn in Disneyland and ponders the relative aesthetic value of original and copy (Ballantyne 2012). While marveling at ever new wonders that engineering bestows on human enjoyment, aesthetics is uniquely positioned to differentiate types of enjoyment—say, between the tourist and the conservationist (Ballantyne 2014). In addition, philosophical aesthetics draws on a wide range of ontological concepts, from intrinsicality to supervenience, to properly differentiate various relations of imitation to copy. By drawing distinctions overlooked by other disciplines, environmental aesthetics furnishes us with indispensable means to a comprehensive understanding of engineered nature and its impact on future generations.

37.8. Art as Technology

Instead of conceiving of technology as art, we may look at art as a technological enterprise, or, as myriad variants of technological or engineering engagements. This would deploy the philosophy of engineering—its positions, tools, distinctions—so as to shed greater light on questions of experience, ontology, and value in the arts. Historically, this proposal has met with considerable resistance—but recent analytic work invites optimism to the contrary.

Aristotle famously distinguishes the practical from the theoretical disciplines (or lines of inquiry). In theory, we aim at the truth—and depending what kind of truths we are after, depending what the objects of inquiry are, Aristotle classifies the pursuit as mathematical, physical, or metaphysical. By contrast, practical inquiries do not solely aim at unearthing theoretic truths but at changing the world, one way or another. One domain of *praxis*, for Aristotle, is human action, which in turn comprises several kinds, such as moral action. Moral action has no end (or goal) external to itself and strives to internalize its own excellence or *telos*. Aristotle differentiates action of this kind from production, from an activity that produces a distinct external product, such as a chair—much less so when, e.g., a flute player 'produces' a song (*Poetics* 1). In contemporary jargon, technology deals with causally distinct means and ends where the arts and moral action do not (or, at least, less obviously so). (See also Chapter 8, "Prescriptive Engineering Knowledge" by Sjoerd Zwart, this volume.)

If philosophers of engineering adhere to this broad outline of Aristotle's philosophy and its taxonomy of action and so on, they cannot deploy philosophy of engineering to study the arts or vice versa. For, as just established, aesthetic activity as a kind is radically at odds with goal-driven activity directed at the making of a (physically) distinct, external object—such as the making of an engineered artifact. As a result, philosophical reflection on aesthetics and engineering would enjoy limited, perhaps no, mutual relevance. And that would benefit neither sub-discipline.

Fortunately, there exist alternatives to Aristotle's framework. Instead of seeing art and engineering as mutually exclusive practices, current aesthetics asks us to redefine art, technology, and human action in novel ways—ways remote from Aristotle but open to their mutual relevance:

> [The arts] aim at making things . . . [and] participate in crafts and technologies. They are
> tied to the manual and the constructive. . . . There is an intimate link between technology

and organized activities. Roughly, a tool (such as a hammer or a computer) is the hub of an organized activity. Technology is not mere stuff. It is the equipment with which we carry on our organized activities. Technologies organize us; properly understood, they are *evolving patterns of organization*.

(*Noë 2015: 19, emphasis in original*)

By developing a nuanced account of *organized activity*, Noë is able to illustrate what Aristotle could not: that we can not only study engineering from an aesthetic point of view, but instead view the arts, in their attempt to engender aesthetic experience, as technology. It remains to be seen how this research project unfolds, and whether philosophers of engineering will significantly contribute to revising our (future) understanding of the arts.

37.9. Coda: Aesthetics as Methodological Lode Star? Lessons From Contemporary Philosophy of Architecture

As this chapter demonstrated, analytic aesthetics is an untapped conceptual reservoir to clarify open philosophical issues in engineering. In addition, aesthetics should serve another role for future work on philosophy of engineering. Like the latter, aesthetics must satisfy two audiences at once: practitioners and philosophers. The concrete reality of aesthetic production and reception keeps aesthetic philosophers in and prevents aloof reflections untethered from practical reality. Analytic philosophers, in turn, make sure that those reflections attain a level of systematicity, rigor, and coherence that typifies their discipline elsewhere. Aesthetics, here, has become a "special case for metaphysics, ontology, epistemology, theory of meaning, value theory, and social or political philosophy" (Lamarque and Olson 2019: xvi).

The same cannot be said of philosophy of engineering: after its salutary "empirical" turn—a turn *away* from purified philosophy, if towards concrete engineering practice—the field is now ripe for having its findings checked more profoundly against analytic metaphysics, epistemology, and the philosophy of language (Koller and Franssen 2016). The success story of aesthetics, having gone through an empirical turn of its own before returning to the analytic fold, ought to provide ample inspiration and encouragement.

Yet, having extolled aesthetics' applicability to engineering, it would be remiss to not single out some pitfalls likely awaiting us on this journey. One of them is analytic philosophy's alleged "emptiness"—that is, the charge that analytic philosophy's contributions have offered "next to nothing about the character of concrete reality" (Unger 2017) and are thus unlikely to unearth novelties about engineering and engineered reality. This charge is particularly true where philosophers' arguments nominally engage a practice, such as art or engineering, but do so without engaging its everyday particulars or history.

For instance, recent philosophical work on architecture raises philosophical issues that arise *once* we view buildings as instantiating a particular philosophical category—categories such as 'artifact with significant relation to function or form,' 'abstraction rather than material concretion,' or 'virtual (rather than real) thing' (Fisher 2015; Grosz 2001). While the philosophical questions raised by such classification are interesting, these questions can be resolved without much regard to the practice and history of the discipline that gave rise to it—architecture. Identical claims could be probed (and reached) if we asked: what result, philosophically, if we viewed bananas as abstract or virtual?

Incidentally, the more interesting labor may concern not what *follows* such classification but what *merits* it. Why should we philosophically think of architecture as abstract? Is it really true that architectural practice is (or should be) more informed by the building's blueprint than its built reality?

On such matters, architecture has produced two thousand years of recorded reflection that could aid analytic philosophical practice. In reality, it is often ignored. Reviewing a particularly well-regarded sample of the genre (Scruton 2013), an architecturally trained reader found the entire book set up

> as a take-it-or-leave-it argument, almost like a mathematical proof that sets out to validate the correctness of a formula by deriving it from previously proven formulas or from axioms. If, as a reader, one disagrees with only one step in the reasoning, that is enough to render the whole argument—and especially its conclusions—unconvincing.
>
> *(Heynen 2017: 210)*

Architectural readers are particularly prone to 'disagree' with philosophers' first premises where these are set up, not by proof but by unsubstantiated invitation—to view architecture as this, that, or the other (abstract, virtual, etc.). As Aristotle pointed out long ago, a philosopher's arguments are particularly "dull" where they "ask us to grant one absurd premise, and the rest will follow—no grand accomplishment" (Aristotle 1988: 82, modified). Similarly, a philosophical "method of 'postulating' what [premise] we want has many advantages; they are the same as the advantages of theft over honest toil" (Russell 1920: 71). One man's postulate is another man's schemata.

This is not to say that all philosophical work on architecture is of necessity schematic, or schematic to the same extent; nor is it to excuse architects from engaging analytic philosophy schematically, or ornamentally only (a Wittgenstein quote here, a Dummett quote there). It is just to be alive to the challenge, and to distrust work that skips over key premises.

It would be naïve to regard the philosophy of engineering as completely immune from such concerns, when many of them are in view in this chapter. Present sections invited readers to apply this or that debate in analytic aesthetics to engineering instead of architecture, often on empirically thin grounds. Similar worries arise for other research frameworks. While intellectually exciting to view contemporary engineering practice through the lens of contemporary systems theory or feminism, do such views furnish novel claims distinctly true of contemporary engineering? Or could their results be equally at home in a Guidebook to the Philosophy of Bananas?

While not immune, the philosophy of engineering is likely better poised than aesthetic philosophy to stave off the worst excesses of schematism—for, its empirical turn is more recent, perhaps more prolonged too. Furthermore, schematism is unavoidable in reference works like the present, works that map out terrains readers are encouraged to traverse in greater detail on their own. If subsequent journeys show up dead ends and unexplored vistas, this chapter and its siblings will have served their function.

Related Chapters

Chapter 8: Prescriptive Engineering Knowledge (Sjoerd Zwart)
Chapter 9: The Art of Engineering and Engineering as Art (Lara Schrijver)
Chapter 15: Artifacts (Beth Preston)
Chapter 16: Engineering Objects (Wybo Houkes)
Chapter 17: Use Plans (Auke Pols)
Chapter 22: Values and Design (Ibo Van de Poel)
Chapter 36: Trust in Engineering (Philip J. Nickel)
Chapter 48: Feminist Engineering and Gender (Donna Riley)

Further Reading

Carroll, N. (1999). *Philosophy of Art. A Contemporary Introduction*. New York: Routledge. (A classic overview to analytic aesthetics, geared at undergraduate students and newcomers.)

Kieran, M. (2005). *Contemporary Debates in Aesthetics and the Philosophy of Art*. New York: Routledge. (An introduction to key debates in analytic aesthetics by matching contemporary practitioners arguing against each other; recommended for upper-division undergraduates and above.)

Lamarque, P. and Olson, H. S. (eds.) (2019). *Aesthetics and the Philosophy of Art: The Analytic Tradition*. London: Blackwell. (Anthology of key papers in analytic aesthetics supplemented by short introductions: for ease of reference, the present chapter frequently refers to papers anthologized here.)

Levinson, J. (ed.) (2005). *The Oxford Handbook of Aesthetics*. Oxford: Oxford University Press. (Authoritative reference work on key movements and key figures.)

References

Aristotle (1988). *A New Aristotle Reader*. Ed. by J. Ackrill. Princeton: Princeton University Press.

Ballantyne, A. (2012). Remaking the Matterhorn. *Nordic Journal of Architecture,* 3(2), 34–39.

Ballantyne, A. (2014). Trying to Think in a Connected Sort of Way. Andrew Ballantyne in Conversation With Stefan Koller. *Architecture Philosophy,* 1(1), 119–125.

Baudrillard, J. (1989). Videowelt und fraktales Subjekt. In *Ars Electronica: Philosophien der neuen Technologie*. Berlin: Merve, pp. 113–131.

Benjamin, W. (2002). *The Arcades Project*. Cambridge, MA: Harvard University Press.

Benjamin, W. (2008). The Work of Art in the Age of Mechanical Production. In Cahn and Meskin (2009), pp. 327–343.

Brady, Emily (2019). Aesthetic Character and Aesthetic Integrity in Environmental Conservation. In Lamarque and Olson (2019), pp. 684–694.

Breitenbach, A. (2013). Beauty in Proofs: Kant on Aesthetics in Mathematics. *European Journal of Philosophy,* 23(4), 955–977.

Brink, K.A., Gray, K. and Wellman, H.M. (2017). Creepiness Creeps. In Uncanny Valley Feelings Are Acquired in Childhood. *Child Development* (unspecified volume), 1–13. doi:10.1111/cdev.12999.

Cahn, S.M. and Meskin, A. (eds.) (2009). *Aesthetics: A Comprehensive Anthology*. Malden, MA: Blackwell.

Carlson, A. (2019). Appreciation and the Natural Environment. In Lamarque and Olson (2019), pp. 670–683.

Carroll, N. (1999). *Philosophy of Art. A Contemporary Introduction*. New York: Routledge.

Carroll, N. (2012). *Art in Three Dimensions*. Oxford: Oxford University Press.

Carroll, N. (2016). Art Appreciation. *Journal of Aesthetic Education,* 50(4), 1–14.

Carroll, N. (2019). Defining Mass Art. In Lamarque and Olson (2019), pp. 607–622.

Chakrabarti, A. (2016). *The Bloomsbury Research Handbook of Indian Aesthetics and the Philosophy of Art*. London: Bloomsbury.

Chatterjee, A. [Anjan] (2014). *The Aesthetic Brain. How We Evolved to Desire Beauty and Enjoy Art*. Oxford: Oxford University Press.

Coeckelbergh, M. (2017). *New Romantic Cyborgs. Romanticism, Information Technology, and the End of the Machine*. Cambridge, MA: MIT Press.

Coeckelbergh, M. (2018). The Art, Poetics, and Grammar of Technological Innovation as Practice, Process, and Performance. *AI & Society*, 33, 501–510.

Danto, A. (1983). *The Transfiguration of the Commonplace*, Cambridge, MA: Harvard University Press.

Devereaux, M. (2005). Feminist Aesthetics. In Levinson (2005), pp. 647–665.

Fenyvesi, K. and Lähdesmäki, T. (eds.) (2017). *Aesthetics of Interdisciplinarity: Art and Mathematics*, Cham, Switzerland: Birkhäuser/Springer.

Fisher, S. (2015). The Philosophy of Architecture. In E. Zalta (ed.), *Stanford Encyclopedia of Philosophy* (online).

Gaut, B. (2009). Computer Art. *American Society for Aesthetics Newsletter*, 29(3), 1–4.

Gray, K., Wegner, D.M. (2012). Feeling Robots and Human Zombies: Mind Perception and the Uncanny Valley. *Journal of Cognition*. doi:10.1016/j.cognition.2012.06.007.

Grosz, E. (2001). *Architecture From the Outside. Essays on Virtual and Real Space*. Cambridge, MA: MIT Press.

Heynen, H. (2017). Review of Scruton 2013. *Architecture Philosophy,* 2(2).

Illies, C. and Ray, N. (2006). *Philosophy of Architecture*. Cambridge, MA: Cambridge Architectural Press.

Ismeninger, G. (2004). The Aesthetic State of Mind. In M. Kieran (ed), *Contemporary Debates in Aesthetics and the Philosophy of Art*. London: Routledge, pp. 98–110.

Kenny, A. (2007). *A New History of Western Philosophy. Volume IV. Philosophy in the Modern World*. Oxford: Oxford University Press.

Kieran, M. (2013). Value of Art. In B. Gaut and D. Lopes (eds.), *The Routledge Companion to Aesthetics,* 3rd ed. New York: Routledge, pp. 289–298.

Koller, S. (2015). *The Ethics of Tectonics*, PhD Thesis, Delft, The Netherlands.

Koller, S. (2016). Visual Vertigo, Phantasmagoric Physiognomies. In P. Healy and A. Radman (eds.), *Constellation of Awakening: Benjamin and Architecture*, published as *Footprint,* Vol. 18. Delft: NaI Publishing.

Koller, S. and Franssen, M. (2016). Philosophy of Technology as a Series Branch of Philosophy: The Empirical Turn as a Starting Point. In M. Franssen et al. (eds.), *Philosophy of Technology Beyond the Empirical Turn*. Cham: Springer, Synthese Library, pp. 31–61.

Korsmeyer, C. (2004). *Gender and Aesthetics: An Introduction*. London: Routledge.

Lamarque, P. and Olson, H. S. (eds.) (2019). *Aesthetics and the Philosophy of Art: The Analytic Tradition*. London: Blackwell. (Anthology of key papers in analytic aesthetics supplemented by short introductions: for ease of reference, the present chapter frequently refers to papers anthologized here.)

Levinson, J. (2016). *Aesthetic Pursuits. Essays in the Philosophy of Art*, Oxford: Oxford University Press.

McAllister, J. (1999). *Beauty and Revolution in Science*. Ithaca, NY: Cornell.

Menary, R. (2010). Introduction: The Extended Mind in Focus. In Richard Menary (ed.), *The Extended Mind*. Cambridge, MA: MIT Press, pp. 1–26.

Mole, C. (2015). *Attention as Cognitive Unison*. Oxford: Oxford University Press.

Nader, N. (1965). *Unsafe at Any Speed: The Designed-In Dangers of the American Automobile*. New York: Grossman Publishers.

Nanay, B. (2018). *Aesthetics as Philosophy of Perception*, Oxford: Oxford University Press.

Noë, A. (2015). *Strange Tools. Art and Human Nature*. New York: Hill and Wang.

Robinson, J. (2019). Listening With Emotion: How Our Emotions Help Us to Understand Music. In Lamarque and Olson (2019), pp. 583–599.

Roeser, S. (2013). Aesthetics as a Risk Factor in Designing Architecture. In C. Basta and S. Moroni (eds.), *Ethics, Planning, and the Built Environment*. New York: Springer, pp. 93–106.

Russell, B. (1920). *Introduction to Mathematical Philosophy,* 2nd ed. London: Unwin.

Saito, Y. (2007). *Everyday Aesthetics*. Oxford: Oxford University Press.

Sauchelli, A. (2012). Functional Beauty, Architecture, and Morality. *Philosophical Quarterly,* 62, 128–147.

Schroeder, M. (2015). *Expressing Our Attitudes: Explanation and Expression in Ethics*. Oxford: Oxford University Press.

Scruton, R. (2013). *The Aesthetics of Architecture,* revised ed. Princeton: Princeton University Press.

Smith, D. (2018). *Exceptional Technologies: A Continental Philosophy of Technology*, Oxford: Bloomsbury.

Thomasson, A. (2014). Public Artifacts, Intentions and Norms. In P. Vermaas et al. (eds.), *Artefact Kinds: Ontology and the Human-Made World*. Cham: Springer, Synthese Library, pp. 45–62.

Thomasson, A. (2019). The Ontology of Art and Knowledge in Aesthetics. In Lamarque and Olson (2019), pp. 108–116.

Unger, P. (2017). *Empty Ideas. A Critique of Analytic Philosophy*. Oxford: Oxford University Press.

38

HEALTH

Marianne Boenink

When asking people what they value in life, almost everyone includes health on the list. If you would ask the same people what this value actually entails, you would probably be confronted with a variety of answers, punctuated by hesitation and silence. A frequently heard answer is that health is the 'absence of disease'—which of course begs the question what 'disease' is. Are you diseased when you experience complaints? Or only when you received a medical diagnosis? What about the woman without complaints, but with a positive mammography? Or the diabetic who engages in high-level sports and enjoys life to the full? And the person experiencing chronic fatigue for which doctors cannot find an explanation? As these examples show, defining health by referring to disease is not necessarily helpful, since defining 'disease' also raises a lot of controversies.

Moreover, many people after some reflection argue that 'absence of disease' does not automatically imply 'health'—it seems more is required. Demonstrating this way of reasoning, the World Health Organization (WHO) in 1948 defined health as "a state of complete physical, mental and social well-being and not merely the absence of disease or infirmity" (WHO 1948). Many critics since then have wondered, however, whether this definition isn't erring on the other side. When is well-being ever 'complete'? Should lack of social well-being (because of experienced inequality or discrimination, for example) be seen as a determinant of health? The WHO definition seems to equate health not only with well-being, but even with happiness. But can't you be healthy and unhappy, or happy while unhealthy?

There may actually be more truth in the saying that health is the absence of disease than most people would realize at first sight. The German philosopher Gadamer has argued that health is an elusive phenomenon, calling it an "enigma" (Gadamer 1996), because health does not present itself to us in any direct way. The French philosopher Canguilhem in a similar vein pointed out that health actually comes to the fore when it is absent, and for that reason is hard to study or even define (Canguilhem 1991). So when reflecting on health, we cannot even avoid reflecting on its opposite at the same time, and this chapter will be no exception.

This chapter reviews different ways of thinking about health as well as disease, to facilitate the competence of engineers and those engaged and interested in technology development to reflect on the potential health implications of engineering and technology. Technologies actually play an important role in the way we pursue individual health or (more often) try to avoid its opposite. Importantly, technologies also shape what we experience as health or the lack of it. A better understanding of different ways to conceptualize health and disease may therefore help engineers to better anticipate some of the ethical and social implications of their work. Although technology also plays a major role in public health, discussion of this would require a separate analysis. The chapter therefore focuses on the health of *individuals*.

The chapter starts with an exploration of the ways in which 'health' actually can be at stake in engineering. We then disentangle different meanings assigned to 'disease' and, by implication, to 'health'. The next two sections discuss how technological developments have co-shaped the historical evolution of our understanding of health and disease and how specific technologies mediate what is perceived as health or disease. Finally, we review several philosophical theories of disease and health, all of which aim to revise our use of these concepts to make it more coherent, and we conclude with a summary of our findings.

38.1. Health as a Value in Engineering

What role does 'health' actually play in engineering? First of all, health is a central value in all engineering activities, even when it is not the main aim pursued. It is part of the professional responsibility of engineers that their inventions and designs, whatever aims these pursue, should not harm the health of users, producers and other parties that may be affected. This is visible in professional codes of ethics for engineering, where the very first statement usually includes the norm that engineers should "hold paramount the safety, health and welfare of the public" (this phrasing comes from the code of the National Society of Professional Engineers, www.nspe.org/resources/ethics/code-ethics; similar statements are included in, among others, the codes of the American Society of Mechanical Engineers and the Institute of Electrical and Electronics Engineers). Avoiding harm to health also is a no-brainer in engineering textbooks and training. To be able to fulfil this professional obligation, engineers should be aware 'health' can have different meanings.

Secondly, 'health' obviously plays a large role in the domain of biomedical engineering. When developing diagnostic, prognostic, therapeutic or rehabilitation technologies, 'health' functions not only as a negative requirement ('first do no harm'), but also as a more positive goal. At the same time, most biomedical engineers would probably shy away from the claim that their technologies actually fully realize 'health'. Their work helps reduce disease and suffering and, as such, indirectly contributes to the realization of health. Even if we grant that 'health' in biomedical engineering is often narrowed down to 'absence of disease', this still leaves open the question what counts as 'disease' and how it is distinguished from 'health'.

Finally, an increasing amount of engineering activity focuses on developing 'health technology', aiming to contribute to health in a variety of ways. For example, sensoring technologies are developed that monitor healthy individuals' bodily and mental functioning, with the aim to identify deviations early on and to enable interventions (whether technological or otherwise) to sustain health. Technologies that aim to optimize or explicitly enhance functioning of healthy individuals also attract a lot of attention: possibilities in robotics, biomaterials and theranostics (combining in vivo sensoring and drug delivery), among others, are explored with this type of aim in mind. In all these cases, engineers need to have some understanding of what 'health' is above and beyond the absence of disease, and how technology can help to realize such a positive conception of health.

So health plays a role in almost any form of engineering, albeit in different ways. The relation between technology and health is usually thought of in instrumentalist terms: technology is a means, an instrument, to realize more or less health. This suggests that there is a clear-cut view of what health entails, and that technological instruments can help to achieve this pre-set goal. Moreover, what counts as 'health' or as (absence of) 'disease' is often thought of as a factual issue. Both assumptions, however, have been thoroughly criticized by philosophers. Philosophers of technology have argued that *technologies are never neutral instruments*; they co-shape the goals they are thought to achieve (Ihde 1990; Verbeek 2005; Hofmann 2001). Many philosophers of medicine have argued, moreover, that health and disease are inherently *normative* concepts (Murphy

2015). The distinction between health and disease describes not only what is normal or not, it is also about what is (not) desirable/acceptable, and any attempt to express this distinction in neutral terms is doomed to fail. These concepts are, moreover, not innocent: they have very real implications for the way we organize society and assign rights, duties and responsibilities, and for the way individuals experience their life. So when engineers develop technologies intended to contribute to (or at least not jeopardize) health, they are engaging in a highly normative activity: their work may help determine what counts as health or disease. If we then realize that there are many different understandings of health and disease, and of the boundary between the two, it should be clear that technology developers had better be aware of what view of health and disease they are actually contributing to.

38.2. Health as Opposed to Disease, Illness and Sickness

As mentioned in the introduction, the phenomenon of health is hard to grasp, since it is hardly directly experienced. We tend to experience and get to know what health is via the experience that we lack it. For this reason, I will start exploring the different ways we label the lack of health, before turning to health. In the English-speaking world, the phenomena associated with lack of health are actually described with different terms: *disease*, but also *illness* and *sickness*. All three labels refer to reduced functioning of an individual, but each has a particular meaning (Twaddle 1994; Hofmann 2002). 'Illness' refers to reduced personal functioning: the subjective experience of feeling unwell (bodily, mentally, socially or a combination), experiencing pain, suffering and/or not being able to do what you are used to or would like to do. This implies that whether or not you are ill can be decided only by the subject undergoing these experiences. 'Disease' refers to reduced *bodily* functioning, something usually established by experts, like a doctor. In contemporary Western medical practice such an assessment will usually be based on the insights from the biomedical sciences and thus boil down to reduced biological functioning. 'Sickness', finally, refers to reduced *social* functioning: for example when you are not able to go to work or care for your family. In this case, social actors (i.e. employers, parents, insurance companies) determine whether you are indeed sick, and if so, you usually are absolved from certain social obligations (like going to work or to school), or even entitled to certain rights (like compensation of the costs made to get better). The three terms thus not only refer to different types of functioning but also point to different sources and authorities for determining whether someone does deserve the label: the person him/herself, an (often biomedical) expert or one's social environment. Please note that the distinction between the three labels is not meant to distinguish mental from physical disease. It can actually be applied to both 'physical' and 'mental' manifestations of dysfunctioning; in both cases, the dysfunctioning can be described with reference to experiential, biological and/or social terms.

Obviously, the three labels can (and often do) overlap to a large extent. Advanced colon cancer, appendicitis and Alzheimer's disease, for example, usually deserve all three terms: people suffer from it, it can be localized in the body by experts and it is considered a valid reason to stop working or to be referred to a hospital. If the three labels do not converge, however, confusion and social controversy may arise. Chronic fatigue, for example, is experienced by a substantial number of people, but experts have not been able to localize a bodily correlate; and partly as a result, societies are at a loss of whether or not to acknowledge it as sickness. Many screening technologies, to mention another example, identify as 'diseased' individuals who do not feel ill themselves, raising again the question of what social consequences this should have. A final example: many societies continued to label 'homosexuality' as sickness, long after doctors stopped labelling it as a disease, and while individuals were not necessarily suffering from the experience. In all these cases of divergence, the question of who is allowed to pronounce that someone is not healthy becomes a major issue. Should the (non-)sufferer, the expert or society have the last say?

It is important to note, however, that the three labels are not fully independent. For example, whether or not a phenomenon is labelled as a 'disease' can strongly influence what rights one has in society, and if something is considered 'sickness' in society, this could actually cause individual suffering and thus contribute to 'illness'. The distinctions also help to realize that knowledge of disease starts with and depends on the experience of illness. It is because individuals feeling unwell at some point visited a doctor that diseases are identified and distinguished. The illness experience is, as Canguilhem said, "epistemologically prior" to the phenomenon of disease (Canguilhem 1991). Without relating them to collected evidence about feelings of subjective lack of well-being, the meaning of more objective observations of bodily functioning is never fully clear: are they just variations, or signs of pathology?

If we now shift back to 'health', we can see that this concept can have at least three different meanings:

1. the subjective experience of functioning well, whether bodily, psychologically and/or socially (when I say that I feel healthy);
2. good bodily or biological functioning (for example when a doctor tells the parents their newborn baby is healthy);
3. good social functioning (when you are for example able to fulfil your duties as an employee or to live up to social expectations of a parent).

These distinctions help to see why conceiving of health as the absence of disease is often considered a 'reductionist' move: a rich, multifarious phenomenon is reduced to only one of its constitutive elements, bodily functioning. The WHO definition of health referred to in the introduction ("a state of complete physical, mental and social well-being and not merely the absence of disease or infirmity") was a deliberate attempt to do justice to the other dimensions of health. Notwithstanding this attempt, in contemporary Western societies, health is thought of most often in opposition to disease. This is the result of developments in medicine and healthcare, in which technology played a crucial role.

38.3. Technology and the Shift From Illness to Disease

Historians looking back at the history of Western medicine and healthcare observe that the emergence of medical science and the accompanying technologies have brought about a gradual shift of attention from illness to disease. Jewson (1976) shows that European doctors at the end of the 18th century vied for patients (usually from social elites) in private practices by putting the needs and subjective assessments of these patients centre stage. Each patient was thought to present a unique constellation of events, and illness experiences were framed in terms of a disturbed balance between different elements in an individual's situation. It was the doctor's task to sort out how the equilibrium (and thus health) could be restored, and in doing so, anything in a person's life could be relevant and a clue for intervention. Rosenberg (2002; 2007) presents similar findings for 18th-century medicine in the USA. In both cases, doctors worked with their senses, listening to a patient's complaints and life story, and observing his or her body and bodily fluids via visual inspection, smelling, hearing and touching. Since every patient was considered to be unique, a treatment that worked with a former patient need not be effective in the current situation. Ultimately, the patient's subjective assessment of his or her functioning determined whether a medical intervention was or was not successful.

Stanley Reiser (1978) discusses how since the 19th century an increasing number of instruments have entered medical practice, from the stethoscope, the thermometer, the microscope and X-ray machines to the electrocardiogram and the computer. In his view such instruments led the physician's attention away from the patient's subjective report to the objective observation of bodily characteristics. Compared to physical examination by the doctor's senses, the results these instruments produced were considered both more reproducible (because they were less dependent on the individual doctor's skill and memory) and more standardized (comparable among different settings). According to Reiser these instruments alienated doctors from their patients; having a personal relationship with the patient became less important, and the meaning of illness experiences for a patient received less attention.

This last claim is too generalizing, since some technologies, rather than ignoring subjective complaints, reshape the way these are experienced, presented to and discussed with the doctor; see, for example, Willems (1995, 2000) on self-monitoring technologies for asthma and Pols (2012) on telecare technologies for monitoring heart disease and COPD. Moreover, Reiser's thesis neglects that technologies can still approach disease in different ways, as we will discuss later. However, the observation that biomedical technologies have put 'disease' centre stage in medicine and have fostered a view of disease as a phenomenon that is somehow independent of the patient at hand is widely shared (Rosenberg 2002). Whereas earlier medical practices focused on subjective symptoms, technologies opening up the (deceased, but later also the living) body, reframed these symptoms as only secondary signs of an underlying disease. These diseases were thought of as *species* (to be distinguished and categorized like plant or animal species), and patients became cases exemplifying a specific disease. An ever-increasing number of tools ("instruments of precision"; Rosenberg 2002: 243) was developed to support the diagnostic work. The search for specific diseases also stimulated thinking about therapies as 'magic bullets' (a term coined by Paul Ehrlich; see Mulinari 2014) that would precisely counter the cause of the disease without negatively affecting other parts of the body. Developing such therapies has proven challenging. It still is a major aim driving contemporary visions of 'precision medicine', however.

Overall, then, the introduction of diagnostic technologies in medical practice has contributed to the shift from a phenomenological, illness-oriented approach to a more objective, but also more narrow disease-oriented approach. What implications did this have for the way health was conceptualized and experienced? Again, it is challenging to come up with direct evidence for what was historically seen as 'health'. The shifts described earlier, however, suggest that health in the 18th century was more like 'absence of illness' than like 'absence of disease'. Moreover, whereas human beings probably always have experienced health as a fragile phenomenon that can never be fully controlled, health in the era of technoscientific medicine has become something that can be diminished without your knowing it. This view has become so widespread, moreover, that most people tend to be careful when claiming they are healthy. When asked how their health is, they qualify their response: "Good—at least, as far as I know . . ."

38.4. Technological Mediations of Disease and Health

As indicated previously, the increasing focus on 'disease' can actually still take different forms, partly depending on the technologies used, and with different implications for how 'health' is conceptualized. When technologies enabled doctors to open up bodies (literally, by knife, from the 17th century onwards, and more metaphorically by all kinds of imaging technologies from the 19th and 20th centuries), 'disease' was first localized as an abnormal entity in bodily space. An abscess, a tumour, clotted arteries, but also the plaques visible on brain scans of some individuals with dementia: in all these cases, disease is first and foremost conceived of as a deviant entity in (or on) the body that should first be localized and then somehow removed. This *ontological* conception of disease, as it is called (Temkin 1977; Rosenberg 2003), views both disease and health as qualitatively different states. Moreover, health and disease have a mutually exclusive, binary relationship: you are healthy or you are not, and if you are not healthy, you are diseased. Imaging technologies in particular (looking for disease 'lesions') contribute to this way of approaching disease, and surgical techniques seem to be its preferred way of treatment. The same conception is also visible in microbiology, however, where disease is thought to be caused by a pathogen that should be eradicated or countered. Medical genetics prior to the Human Genome Project also framed genetic mutations in spatial terms, as just waiting to be discovered and, if feasible, to be removed by gene therapy (Boenink 2016).

Starting already with the emergence of different types of biochemical technologies in the 19th century, however, an alternative way of conceptualizing disease has come along. The *physiological* concept of disease frames disease as a dynamic process in time, rather than as a state in space (Rosenberg 2003). The biological processes continuously going on in the living body sometimes go awry, resulting in a higher or lower production of certain substances (proteins, neurotransmitters, RNA, etc.). Disease and health according to this way of thinking can be identified by measuring the absence/presence or even the exact quantities of such substances, and the main challenge for doctors is to re-establish normal quantities. This way of thinking has been visible, among others, in diabetes, rheumatoid arthritis and many neurodegenerative diseases. It also is widely prevalent in most of contemporary molecular medicine, which is keen to measure all kinds of 'biomarkers' to identify disease, as well as to reconstruct pathological mechanisms. Canguilhem has actually labelled this view the 'quantitative view of health and disease' (Canguilhem 1991), indicating that disease according to this view is a matter of producing too much or too little of one or more specific bodily substances, compared to the population average. Functioning is healthy, in contrast, as long as levels remain between the boundaries of what is considered 'normal'. Canguilhem also pointed out that the interpretation of quantitative measurements of bodily functioning still depends on epidemiological research correlating these findings with clinical diagnoses, and ultimately with subjective experiences of what is or is not acceptable personal functioning. The criteria for what is or is not normal are therefore always laden with subjective evaluations.

The physiological model of health and disease, in contrast to the ontological model, conceives of health and disease as located on a continuous scale. They are only quantitatively different, and there can be a substantial grey zone between them. Health and disease are matters of gradation. What counts (literally!) as a disease is determined by cut-off points that seem epistemologically arbitrary. This way of thinking often leads to a 'cascade model of disease' (Boenink 2010). The image of the cascade is often used to label hypothesized disease mechanisms, like the 'rheumatoid arthritis cascade' and the 'amyloid cascade hypothesis' (supposedly reconstructing the process leading to Alzheimer's disease). Such a model has important implications for the way we approach health. Cascades are usually conceived of as something that you hardly notice at first: they start really small (on the molecular level). The process that follows is thought of in a mechanistic or even deterministic way: each step tends to trigger the next one, unless an external factor intervenes. Moreover, the process is usually conceived of in a unidirectional way—there is no turning back, although blocking the cascade may be possible. It is not hard to understand that such a cascade model invites proposals for early detection and early treatment. The assumption is that the earlier you intervene, the easier it should be to counter the disease process. Doing so requires that you are able to identify the disease process in its early stages ('downstream'). This way of thinking underlies, among others, screening programmes for breast and colon cancer, but also current proposals to diagnose Alzheimer's disease at an earlier stage by way of molecular biomarkers. An extreme but striking illustration of the 'early detection logic' driven by the cascade model is the response of some women who carry a BRCA mutation. The mutation means these women are at substantially increased risk of hereditary breast and ovarian cancer, which is why they sometimes decide to have their healthy breasts and/or ovaries removed via preventive surgery. Some of these women call themselves 'pre-vivors' (Force 2018): they did not survive disease and suffering, but proactively intervened in the supposed cascade to prevent the disease from reaching the stage in which it causes bodily complaints and suffering.

It should be noted that the ontological and physiological concepts of disease and health are ideal types, which can be more or less present in the way we approach a specific disease. It is nonetheless helpful to distinguish them, because they have different implications regarding the way disease is best diagnosed and treated, and because each one tends to overlook aspects of health and disease that do not fit its respective way of thinking. An ontological concept of disease seems to neglect, among other things, that disease always develops in time. The physiological concept of disease has difficulty justifying what the difference between health and disease is and tends to ignore that this distinction is never made on the basis of numbers alone. Reflection on which type of conceptualization of health and disease is implied by specific biomedical technologies thus can help to become aware of the potential limitations of these technologies. Distinguishing these different concepts of disease also helps to recognize shifts in the way emerging technologies tend to frame disease (see, for example, Boenink 2016).

38.5. Philosophical Accounts of Health and Disease

Even if one agrees with most historians that medical practice, partly because of the ever-increasing role of technology, has tended to focus on disease rather than illness, this still leaves open the question whether this development should be applauded. *Should* the question whether a specific person is healthy or not be answered by referring to bodily phenomena to be identified and measured by others than the person at hand? Or *should* it rather be answered by the person him/herself, referring to subjective experiences that may be hard or impossible to assess from an external perspective? Several philosophers argue that a coherent theory of health and disease requires that we give precedence to either one or the other. These theories are not meant as empirically adequate descriptions of how we currently use the concepts of health and disease; they rather prescribe how to revise our use of them if we want to be coherent. This means that examples of different types of use do not necessarily falsify these theories; they should be judged on their consistency. Philosophical views on what is the most coherent theory of health and disease vary, however.

The theory that comes closest to the disease-oriented, quantitative approach so prevalent in modern medicine has been formulated by Christopher Boorse (1975, 1977, 1997). He has put forward the 'biostatistical' theory of disease and health, according to which someone is diseased if his or her biological functioning objectively and substantially deviates from what is standard for persons similar in gender and age, in a way that ultimately decreases one's chances of survival and reproduction. Health, in contrast, is statistically normal functioning of an individual's body, supposed to contribute to reproduction and survival. What is health and what is disease in this view can be identified only by doing longitudinal population research, establishing which biological characteristics do or do not contribute to reproduction and survival. Once we know which characteristics are normal and which ones are deviant in certain populations, we can identify 'diseased' individuals.

According to Boorse, diagnosing 'disease' along the lines of his theory would be an objective activity, based on statistical knowledge of biological characteristics. A person displaying all the symptoms associated with, say, lung infection is diseased, ultimately not because of the symptoms, but because experts know these symptoms are associated with increased mortality. 'Illness' is a sub-class of 'disease', indicating those cases where the biological abnormal functioning has led to subjective experiences of not feeling well. A person with osteoporosis or high blood pressure, even while feeling healthy, is diseased because it is known that these characteristics predispose to an early death. To be sure, the concepts of disease and health proposed by Boorse, even while claiming objectivity, are not free from normativity. The values of reproduction and survival are central to his theory; these determine what should (not) be considered 'normal'. Boorse argues, however, that the prevalence of these norms in life (not only in human life, by the way) can be established in a value-free manner. Their validity does not depend on subjective views of what is a good life for a human being. For this reason, his theory is often called a 'naturalist' theory of health and disease.

Many critics have taken issue with Boorse's proposal; I will focus here on three types of criticism. First, some have argued that the norms of reproduction and survival may be constitutive for 'evolutionary fitness', rather than for 'health' (Methot 2011). Although fitness may be part of or contribute to health, it does not cover all aspects of health. Because of this, the biostatistical theory repeats the reductionist move of modern medical practice. Secondly, other critics have argued that Boorse's theory is much more normative then he suggests. The identification of what is normal and abnormal functioning for a specific individual depends on the choice of relevant reference groups. These are usually defined in terms of sex, age and increasingly on biological characteristics measured by biomarkers. As Kingma (2007) argues, however, one could imagine other criteria to determine reference groups, like 'sexual orientation'. In Boorse's view, homosexuality interferes with reproduction and for that reason should be considered as a disease (he argues that this does not automatically mean it is undesirable or should be medically treated). If we distinguish reference groups with different sexual orientations, however, it would not be a disease. Kingma points out that the choice to identify certain criteria as relevant and others as irrelevant for identifying reference classes is a normative endeavour and that Boorse's theory thus is less objective than he suggests. Finally, some critics have pointed out that what are relevant reference classes may depend on the context. What is normal in one situation can be abnormal in another one. Whether a certain blood cell count is normal or not depends, for example, on altitude. Boorse realizes this and argues that, for example, infections that are typical reactions to pathogens present in a specific environment should be seen as normal, not as a disease. However, his critics argue that these may still cause suffering, which this theory cannot acknowledge (Nordenfelt 1995; Musschenga 1997). In sum, the biostatistical theory of health and disease does not seem to do full justice to the richness of the phenomenon of health, it is more normative than Boorse claims and it insufficiently recognizes the relevance of the illness experience.

Opposed to naturalist theories are normativist theories, which claim that the concept of health and disease ultimately should depend on the values and norms regarding what is good for an individual. Havi Carel (2008), for example, has argued that health should be conceptualized in terms of one's lived, subjective experience. Neither the doctor nor technology should determine whether or not one is unhealthy. It should be the person him/herself, because people experience bodily deviations in very different ways. Human beings can give different meanings to phenomena that seem objectively 'the same'. Whether or not a biological disease affects our functioning depends on how we construct meaning for ourselves, whether or not we feel at home in our bodies, maybe even in the world. This implies that one can be healthy while living with a chronic or terminal disease, or with a mutation predisposing one to genetic disease. Carel's approach is phenomenological; it gives precedence to the lived experience of individual persons.

The normativity implied in such a phenomenological approach has been further elaborated by Lennart Nordenfelt (1995, 2007). He also argues that defining health and disease ultimately should be a subjective activity, which depends on the values and views of the good life of the individual involved. Experiences like the incapacity to hear, which can be very debilitating for someone living in a world in which speech and sound are crucial, need not be problematic for someone born in a family of deaf people and living in an environment in which sign language is widely used. Along similar lines of reasoning, carrying a genetic mutation for breast cancer may be much more burdensome to someone who has seen her mother, an aunt and a sister die of this disease, than for someone lacking such experiences. Nordenfelt therefore would like to reserve 'disease' for those elements of one's constitution that hinder someone in achieving his/her goals in life. He also goes beyond disease in the biological sense. The causes of the problematic constitution may, but need not, be biological.

Critics have argued that making 'disease' dependent on subjective views of what is important opens the door to counterintuitive results. Reasoning along Nordenfelt's lines, someone who is training to become a great cyclist but does not succeed in winning the Tour de France could say that he is diseased (and, for that matter, would have a perfect excuse to use drugs, which would then not be seen as 'doping'). It also is questionable whether Nordenfelt's approach allows for criticism of cultural practices in which, for example, women are satisfied with low levels of bodily functioning, simply because they are used to these. The lack of more objective standards for what is a good life for a human being makes this theory vulnerable for accusations of relativism.

Revising our understanding of health and disease along the lines of what either Boorse or Carel and Nordenfelt propose would have pretty radical implications. However, it is questionable whether any of these two revisionary proposals could ever lead to a fully coherent use of the terms 'health' and 'disease', because the subjective and objective phenomena involved may be unavoidably inter-woven. The French philosopher Canguilhem (1991/1966) already stressed that the subjective norms can never be really separate from the objective numbers. What is made 'objectively' visible in the clinic depends on subjective evaluations (people deciding when to see a doctor), but these 'subjective' experiences in turn are mediated by the technologies available. As discussed earlier, since imaging technologies became available, people are aware that their body may harbour a tumour even without their noticing. As a result they will experience their body in a different way from that of people in the pre-imaging era. In a similar vein contemporary developments in self-tracking technologies shape the way people experience their bodily and possibly also their psychological and social functioning.

Canguilhem actually proposed a third way of approaching health and disease, stressing the dynamic equilibrium between a living human being and its material and sociocultural environment. He argued that health is not a permanent state, and neither is 'normal' functioning as defined by population averages. In both cases, the focus is too much on the body. An organism (whether human or non-human) does not live in a vacuum, however. Whether or not it flourishes depends on how well the organism succeeds in addressing the challenges posed by the environment (Canguilhem 1991). Health, in this way of thinking, is a form of *resilience*: being able to adapt to the changing environment in such a way that subjective functioning is not diminished. (Interestingly, this way of thinking shows many analogies with the shift from 'risk' to 'resilience' in philosophical reflection on engineering and safety, which is discussed in Chapter 35, "The Role of Resilience in Engineering" by Neelke Doorn, this volume.) This implies that disease does not occur when subjective functioning is diminished, but only when an organism is insufficiently capable of *restoring* subjective functioning. As in Carel's and Nordenfelt's theories, health for Canguilhem depends on subjective norms and ide-als rather than on objective numbers. These norms, however, are dynamic themselves and interact with the (also changing) environment. This means that if you were an active cyclist who becomes asthmatic, and you succeed in adapting to the circumstances by learning to love Nordic walking or chess even though the thought of such activities would have appalled you in the past, Canguilhem would consider you healthy. In contrast, if you stick to your earlier aims and never succeed in accom-modating your coughing, wheezing and shortness of breath, even while these may be diminished by taking drugs, Canguilhem would say you are diseased. Technology can play a major role in facilitating adaptation not just by curing (stopping or reversing) bodily processes, but also by enabling people to make the disease burden bearable.

This approach fits quite well with more recent attempts to explain that health is more than absence of bodily 'disease', while at the same time steering away from the all-encompassing defini-tion of the WHO. Huber and colleagues (2011) argue that ageing populations and changing patterns of disease (less acute and lethal, more chronic diseases) give reason to update the WHO definition of health. In 2011 they proposed to define 'positive health' as "the ability to adapt and self-manage in the face of social, physical and emotional challenges" (Huber et al. 2011). Recently, they also opera-tionalized this health concept by distinguishing and developing measurement tools for six dimen-sions: daily functioning; bodily functioning; mental functioning and perception; spiritual/existential dimension; quality of life; social and societal participation. On this basis they developed a tool to chart an individual's functioning on each of these dimensions (Huber et al. 2016). Interestingly, their definition entails that health is not so much measured by one's one-time scores on this combination of dimensions, but by the changes in functioning over time after repeated measuring: stability of or an increase in scores indicates health, whereas a decrease may point out diminished health. This way of measuring tries to do justice to the highly individual, the multiple as well as the dynamic character of the health phenomenon and is an interesting attempt to escape the ubiquity of the 'health as absence of disease' approach. However, critics have pointed out, among others, that this conception of health has difficulties to account for the difference between 'normal' (for example due to pregnancy or ageing) and abnormal variation, and may erroneously imply that any adaptation is good. In addition, it may open the door for a huge extension of the healthcare domain and possibly a medicalization of human life (Kingma 2017). Such arguments make us aware that there are also advantages of a more limited or even reductionist approach to health and disease. Ultimately, they also show that philosophical theories of health and disease may have more added value as tools to better grasp and locally adjust our evolving practices, than as blueprints for a radically revised way of dealing with health and disease. Incoherence and tensions are likely here to stay: what is the best or most convincing approach of health for some purposes may not be the best one for others.

38.6. Conclusion

The phenomenon of health proves to be quite elusive, not only in our personal experience, but also in philosophical understanding. In particular, in a world in which biomedical technology plays a major role in the way we experience and deal with health, grasping what health is seems to require that we first pay attention to situations in which it is absent. The distinction between disease, illness and sickness helps to understand how the emergence of biomedical science and technologies have led to a reframing (and narrowing) of health as the absence of objective pathology, rather than of subjective complaints. It can also help to clarify the controversies regarding what should be recognized as health or the lack of it, and which and whose criteria should determine this. The further distinction between ontological and physiological conceptualizations of disease and health helps to analyse how specific technologies direct how we experience health and disease, and what limitations or blind spots may be involved.

Philosophical theories of health and disease claim to offer more coherent frameworks to approach the phenomena of disease and health than we tend to find in everyday practice. Both naturalist and phenomenological or normativist theories, however, may ignore the interdependency of subjective experiences and objective observations of health and disease, in which technology actually plays an important role as mediator. Because of this mediating role and the important personal and societal implications of how we understand and deal with health and disease, engineers working on technologies that should avoid harm to health, help reduce disease or actively contribute to positive health would do well to reflect on what kind of health they are actually about to realize and whose purposes this actually serves.

Related Chapters

Chapter 35: The role of Resilience in Engineering (Neelke Doorn)
Chapter 45: Engineering as a Political Practice (Govert Valkenburg)

Further Reading

Boorse, C. (1977). Health as a Theoretical Concept. *Philosophy of Science,* 44, 542–573. (The classic example of a naturalist approach of health.)
Canguilhem, G. (1991/1966). *The Normal and the Pathological.* New York: Zone Books (An early formulation of the dynamic view of health.)

Huber, M. et al. (2011). How Should We Define Health? *BMJ*, 3(43), d4163. (The first publication about the concept of positive health as "the ability to adapt and self-manage".)

Nordenfelt, L. (1995). *On the Nature of Health: An Action-Theoretic Perspective*. Dordrecht: Kluwer (A classic example of a normativist approach of health.)

Rosenberg, C.E. (2007). *Our Present Complaint: American Medicine, Then and Now*. Baltimore: Johns Hopkins University Press. (A collection of historical essays on the evolution of US healthcare with ample attention for the underlying views of health and disease.)

References

Boenink, M. (2010). Molecular Medicine and Concepts of Disease: The Ethical Value of a Conceptual Analysis of Emerging Biomedical Technologies. *Medicine, Health Care and Philosophy*, 13(1), 11–23.

Boenink, M. (2016). Disease in the Era of Genomic and Molecular Medicine. In J. Marcum (ed.), *The Bloomsbury Companion to Contemporary Philosophy of Medicine*. London: Bloomsbury.

Boorse, C. (1975). On the Distinction Between Health and Illness. *Philosophy and Public Affairs*, 5, 49–68.

Boorse, C. (1977). Health as a Theoretical Concept. *Philosophy of Science*, 44, 542–573.

Boorse, C. (1997). Rebuttal on Health. In J.M. Humber and R.F. Almeder (eds.), *What Is Disease?* Totowa: Humana Press, pp. 1–134.

Canguilhem, G. (1991/1966). *The Normal and the Pathological*. New York: Zone Books.

Carel, H. (2008). *Illness: The Cry of the Flesh*. London: Routledge.

Force (2018). www.facingourrisk.org/understanding-brca-and-hboc/publications/newsletter/archives/2009winter/what-is-previvor.php. Accessed July 13, 2018.

Gadamer, H.G. (1996). *The Enigma of Health. The Art of Healing in a Scientific Age*. Cambridge: Polity Press.

Hofmann, B. (2001). The Technological Invention of Disease. *Journal of Medical Ethics: Medical Humanities*, 27, 10–19.

Hofmann, B. (2002). On the Triad Disease, Illness and Sickness. *Journal of Medicine and Philosophy*, 27(6), 651–673.

Huber, M. et al. (2011). How Should We Define Health? *BMJ*, 3(43), d4163.

Huber, M., van Vliet., M., Giezenberg, B., Winkens, Y., Heerkens, P., Dagnelie, C. and Knottnerus, J. A. (2016). Towards a 'Patient-centred' Operationalization of the New Dynamic Concept of Health: A Mixed Methods Study. *BMJ Open*, 5, e010091.

Ihde, D. (1990). *Technology and the Lifeworld: From Garden to Earth*. Bloomington, IN: Indiana University Press.

Jewson, N.D. (1976). The Disappearance of the Sick-man from Medical Cosmology, 1770–1870. *Sociology*, 10(2), 225–244.

Kingma, E. (2007). What Is It to Be Healthy? *Analysis*, 67(294), 128–133.

Kingma, E. (2017). Kritische Vragen bij Positieve Gezondheid. *Tijdschrift voor Gezondheidszorg en Ethiek*, 3, 81–83.

Methot, P-O. (2011). Research Traditions and Evolutionary Explanations in Medicine. *Theoretical Medicine and Bioethics*, 32, 75–90.

Mulinari, S. (2014). The Specificity Triad: Notions of Disease and Therapeutic Specificity in Biomedical Reasoning. *Philosophy, Ethics and Humanities in Medicine*, 9, 14.

Murphy, Dominic (2015). Concepts of Disease and Health. In Edward N. Zalta (ed.), *The Stanford Encyclopedia of Philosophy*, Spring 2015 ed. https://plato.stanford.edu/archives/spr2015/entries/health-disease/

Musschenga, A.W. (1997). The Relation Between Concepts of Quality-of-Life, Health and Happiness. *The Journal of Medicine and Philosophy*, 22(1), 11–28.

Nordenfelt, L. (1995). *On the Nature of Health: An Action-Theoretic Perspective*. Dordrecht: Kluwer.

Nordenfelt, L. (2007). The Concepts of Health and Illness Revisited. *Medicine, Health Care and Philosophy*, 10(1), 5–10.

Pols, J. (2012). *Care at a Distance: On the Closeness of Technology*. Amsterdam: Amsterdam University Press.

Reiser, S.J. (1978). *Medicine and the Reign of Technology*. Cambridge: Cambridge University Press.

Rosenberg, C.E. (2002). The Tyranny of Diagnosis: Specific Entities and Individual Experience. *Milbank Quarterly*, 80(2), 237–260.

Rosenberg, C.E. (2003). What Is Disease? In Memory of Owsei Temkin, *Bulletin of the History of Medicine*, 77(3), 491–505.

Rosenberg, C.E. (2007). *Our Present Complaint: American Medicine, Then and Now*. Baltimore: Johns Hopkins University Press.

Temkin, O. (1977). The Scientific Approach to Disease: Specific Entity and Individual Sickness. In *The Double Face of Janus and Other Essays in the History of Medicine*. Baltimore: Johns Hopkins University Press, pp. 441–455.

Twaddle, A. (1994). Disease, Illness and Sickness Revisited. In A. Twaddle and N. Nordenfelt (eds.), *Disease, Illness and Sickness: Three Central Concepts in the Theory of Health*. Linköping: Studies on Health and Society, 18, 1–18.

Verbeek, P.P. (2005). *What Things Do: Philosophical Reflections on Technology, Agency and Design*. University Park, PA: Penn State University Press.

WHO (1948). *Constitution of the World Health Organization*. Geneva: World Health Organization.

Willems, D. (1995). *Tools of Care. Explorations into the Semiotics of Medical Technology*. PhD dissertation, University of Maastricht.

Willems, D. (2000). Managing One's Body Using Self-management Techniques: Practicing Autonomy. *Theoretical Medicine and Bioethics*, 21, 23–38.

39

PHILOSOPHY OF SECURITY ENGINEERING

Wolter Pieters

39.1. Introduction

Technology can contribute to undesirable events, or prevention thereof, in many ways. Technology can be a causal factor in floods, weaponized attacks, nuclear incidents, and cybercrime. Dikes, security cameras, nuclear safety systems, and data encryption are technologies aimed at protection against such incidents. The associated events are labeled undesirable because they involve some kind of harm to some (human or non-human) actors or systems.

In this context, security is generally understood as different from safety, but there are various interpretations of this distinction. The most common definition appears to be that security is the protection against deliberately caused events (e.g. Burns et al. 1992; Piètre-Cambacédès and Chaudet 2009). This implies that there are actors (adversaries) that have an interest in causing these events. In the aforementioned examples, this would apply to weaponized attacks and cybercrime, as in both cases harm is caused intentionally. In financial cybercrime, there is monetary benefit for the adversaries associated with the harm caused to others; in weaponized terrorist attacks the benefits to the adversaries are immaterial (although material harm may be caused).

Technology and engineering can play at least two different roles in security against such adversarial acts. Firstly, the design and deployment of technologies may have implications for security, in the sense that they may increase (or decrease) the risk of deliberately caused harmful events. Weapons are an obvious example, but also technologies that at first sight seem to benefit society may be used to cause harm. An important example is the impact of computer networks and the Internet ("cyberspace") on new forms of crime and even terrorism (Koops 2010). When security is taken into account in the design of such technologies, engineers may want to think of ways to reduce such undesirable forms of use. Secondly, to prevent security threats such as crime and terrorism, both online and offline, technologies are often designed specifically to mitigate these threats by preventing, detecting, or responding to them. For example, physical fences as well as digital firewalls may be deployed to limit access of adversaries to facilities. Surveillance cameras or Internet monitoring systems may be put in place to detect threats. It may be hard to assess the effectiveness of such technologies against security threats, and they may also have side effects on other values (Hildebrandt 2013), of which privacy is often mentioned, in particular concerning surveillance technologies.

In this contribution, we cover the foundational concepts of security risk, and how this materializes (or fails to materialize) in engineering practices. To this end, we draw on literature from philosophy, criminology, and engineering disciplines, with a focus on computer science and cybersecurity. The key philosophical topics that we address are (1) ontologies of security and security risk in relation to technology, (2) conceptual challenges in embedding security in engineering practices, (3) issues in evaluating the effectiveness of engineered security, and (4) the ethics of security engineering. While outlining these topics, we will touch upon several key debates in the field, such as secret versus transparent security, and responsibility of users versus responsibility of designers.

39.2. Ontologies of Security

In describing harmful events in order to analyze and mitigate the associated risk, it is often said that a *threat agent acts against an asset* (Jones 2005). In safety, threat agents such as earthquakes are assumed to "act" randomly. There is no plan, strategy, or intentionality behind the events. This leads to methods of risk analysis based on probabilities of occurrence of such events. It is often assumed that security events are instead dependent on strategic decisions by adversaries, who adapt to whatever the defender does and vice versa, justifying the application of game theory (Cox Jr. 2009). In this case, the probabilities of occurrence of events are an *output* rather than an input of the risk analysis process, as they depend on strategic choices of agents. So, rather than historic earthquake frequencies being an input to the risk analysis, the probability of attack is an output that is calculated based on what attackers would be interested in and capable of.[1]

Both safety and security generally refer to protection against harm. This harm can materialize in entities that are of value to the stakeholders concerned, often called *assets*. These can be material entities, but also immaterial ones such as activities or reputation. Assets may also have value for adversaries, for the same or for different reasons. For example, transactions are of value to banks as part of their core business, but they are also of value to adversaries as a possible target of fraud. When adversaries go after such value, the value of the assets for the original stakeholders may diminish or disappear entirely: harm may be caused.

Threat agents, adversaries, or attackers are those actors who have an interest in initiating the events that may cause harm to other stakeholders. In order to decide what needs to be done to make a system secure, one needs to understand the type of threat agents that one is facing. Several taxonomies of relevant properties have been proposed, which typically distinguish between the *motivation* of the threat agents and their *capabilities* (cf. Casey 2007; De Bruijne et al. 2017). Motivation determines which events are of interest to the adversaries; capabilities determine to what extent they are able to initiate those events.

The concept of *vulnerability* is also central to the field, but it can mean different things. It can either refer to a specific weakness in a (computer) system, enabling access to unauthorized parties (countable noun), or to a general level of susceptibility to harmful events (uncountable noun). In both meanings, (a) vulnerability is what makes it possible for a threat agent to reach/act against an asset. Vulnerabilities may reside in different "places" from those of assets, for example when an employee falls for a phishing scam (vulnerability), giving cyber spies access to company secrets (asset). In practice, an attack often consists of a sequence of events with associated vulnerabilities, the so-called "kill chain", rather than a single event (cf. Ayrour et al. 2018).

In the view outlined earlier, security means the reduction of harm caused by (uncertain) intentional events. Such harm can be reduced by preventing the events, or by mitigating the harm when the events do occur. While originally this was very much an art of keeping the bad guys out at all cost, it can easily be argued that security is essentially a risk management endeavor, with associated acceptance of residual risk levels (Blakley et al. 2001).

A risk management approach to security is closely connected to the possible actions of adversaries. In particular, the ability of adversaries to cause harm depends largely on what they can access and how easy/difficult or cheap/costly this is. For example, the easier it is for terrorists to acquire weapons, the more harm they can cause, and similarly for spies and the ease of access to sensitive information.

In the philosophical literature, such access relations have been conceptualized in different ways. From the perspective of information ethics, Floridi (2005) speaks about "ontological friction" when discussing the extent to which actors have to spend effort to access information. Ontological friction represents resistance against the flow of (personal) information, thus reducing accessibility and improving privacy. From the perspective of systems theory, Pieters (2011) uses the concept of "causal insulation" from Niklas Luhmann (1993) to discuss the extent to which access is limited. In this perspective, the focus is on restrictions on the ability of a system to influence the environment, and vice versa, thereby also restricting action possibilities for adversaries. Vuorinen and Tetri (2012) conceptualize security as an "order machine" that blocks chaotic forces from the outside, but thereby also disrupts the inside, i.e. organizational practices. This highlights the limitations that increased security and limited accessibility impose on the regular use of a system, in addition to restricting interference by adversaries. In all of these approaches, there is some kind of "distance metric" suggesting that certain undesirable influences need to be kept at a minimum "distance": access to assets needs to be difficult for those adversarial actors/forces.

Because of the importance of access relations, it matters a lot for security what access adversaries are assumed to have initially. This leads to a distinction between "insider threat" and "outsider threat" (Probst et al. 2010). Many security events are in fact caused by insiders, for example, employees of the victim organization. This distinction is not as binary as it first appears, as adversaries can have any level of "insiderness" (Coles-Kemp and Theoharidou 2010) depending on the initial access they have (e.g. subcontractors).

In relation to the "kill chain" discussed earlier and the associated steps in an attack, the notion of reachability is an important concept: it represents how easy it is for a specified adversary to reach a particular goal predicate (for example, having access to a particular asset). Note that this depends on the type of adversary: for a more powerful adversary, or an adversary with more initial access, it may be easier to reach a particular asset than for others. Therefore, specifying a desirable security level (acceptable risk) inevitably involves an adversary specification.

An example here is the notion of burglar resistance (European ENV 1627:1999; Jongejan and Woldendorp 2013). Burglar resistance levels of doors and windows are specified in terms of a burglar type, including equipment, and a minimum time that this burglar needs to spend to obtain access by force. In fact, what is happening here is that by making the adversary model explicit, the required security level becomes independent from the threat environment. Böhme (2010) discusses this in terms of a security production function, which is decomposed in two steps: first from cost to security level, and then from security level to benefits. The first step is independent from the threat environment, while the second step incorporates the threat environment to assess benefits (in terms of the reduction of harm).

This view suggests that security risk may be mitigated by increasing the "distance" between potential threat agents and assets. In fact, perimeters such as (fire)walls are indeed often employed. However, because of the simultaneous need for authorized actors to pass those perimeters, and because of the possibility of other weaknesses in them, multiple perimeters or, more generally, compartmentation of a system is often used (defense-in-depth). In addition, detective and responsive controls may help in mitigating harm when threat agents do pass through the perimeters. More generally, the approach of situational crime prevention highlights how environmental design can contribute to reducing crime, in addition to focusing on the criminals. In this approach, there are five key classes of controls by which designs can help to reduce crime: increase the effort for the threat agents, increase the risks for the threat agents, reduce the rewards for the threat agents, reduce provocations, and remove excuses (Cornish and Clarke 2003).

At the same time, a perspective focusing on individual human threat agents seems rather limited in relation to security and technology. From the point of view of actor-network theory, it can be argued that key security problems emerge because of the *composition* of human and non-human actors in a human–technology network that is capable of actions causing harm (Van der Wagen and Pieters 2015). This point of view does not take the human adversary as a starting point, but rather the way in which programs of action are being translated in new human–technology configurations. This also brings up the question of how prosecuting the perpetrators and fixing the system should be balanced in high-tech crime.

Security thus involves protection against harm caused by opponents or adversarial courses of action. In particular, we want to prevent harm caused by undue access by such adversaries. In order to embed security in engineering practices, a design should assist in increasing the difficulty of adversarial behavior that may cause harm, in terms of the *use* that may be made of a design.

39.3. Security Engineering

Security engineering (Anderson 2010) involves defining security requirements for the design of a technology, and employing techniques to meet those requirements in the design. In the sense of preventing adversarial use of a technology, security requirements state what a technology should *not* do rather than what it should do, making security a non-functional requirement (Glinz 2007). In the terminology of the previous section, the technology should *not* support adversaries in gaining access that enables them to cause harm. Therefore, in order to show that a design meets a security requirement, one should ideally be able to show that there are no exceptions, no black swans, but the possible exceptions cannot be enumerated. In other words, security is a falsifiable property (if a weakness is found), but insecurity is not (Herley 2016).

Although theoretically always imperfect, it is nevertheless argued that some level of security engineering is possible. In the requirements engineering process, this means that rather than imagining possible uses of a technology, one should imagine possible *misuses*, or misuse cases (Sindre and Opdahl 2005). In this context, there are several options for misuse. A technology could be used as a weapon (e.g. using nuclear waste for a dirty bomb), as a target (e.g. disrupting critical electricity infrastructure), or both (using weaknesses in the Bitcoin system to commit transaction fraud; Pieters and Dechesne 2017). Misuse cases and security requirements, if deemed sufficiently critical, should then lead to specific controls in the design aimed at making such misuse less attractive (fences, cameras, tamper-proof seals, encryption, denaturation of alcohol, etc.). Obviously, such controls need to be designed as well, in which case security is the main design goal.

A key debate, particularly prominent in information security, is whether security designs should be made public or kept secret. In cryptography, the so-called Kerckhoffs principle (Petitcolas 2011) states that the security of a cryptographic algorithm should not depend on the algorithm being kept secret. Instead, a secret cryptographic key is used as input to the algorithm. In cryptography, but also in other areas of security, keeping designs secret is referred to as "security-by-obscurity". Whether or not obscurity contributes to security is subject to debate (Pavlovic 2011). On the one hand, like in breaking other security measures, adversaries will (at least initially) need to spend effort to discover the secret design. On the other hand, security mechanisms that rely on secrecy carry the risk of becoming useless when the design becomes public. Being transparent about security has another benefit: it enables users to take security into account when choosing a product or service. This prevents information asymmetry leading to a so-called market of lemons, which would be dominated by less secure but cheaper offerings (Anderson 2001).

Beyond transparency of the design, it is also being discussed to what extent publishing the implementation in a computer program (source code) contributes to security, or rather jeopardizes it (Hoepman and Jacobs 2007; Schryen 2011). On the one hand, open source software could be checked for security problems by anyone; on the other hand, attackers could also use the source code to spot vulnerabilities that they could exploit to gain access. The question of open design and open source also connects to the notion of trust in technology: to what extent can people trust a system—for example, an electronic voting system—with respect to security? (Nickel 2013; Pieters 2006). It could be argued that open source increases trust, in the sense that people can rely on independent experts to check the programs for issues. It is non-trivial, though, to make sure such checks are done sufficiently in practice, and to establish whether the published code is actually used in a system.[2]

Another key debate relates to whether security measures that do not increase attacker effort or risk a lot can still be meaningful. Such "security theatre" (Schneier 2008) could be beneficial in two ways: it may deter attackers (if they believe things *did* change), and it may reassure the public. However, as it doesn't contribute to "actual" security, it could also be seen as fake or deceptive security. If deceiving attackers is seen as a legitimate security measure, then security theatre may make a lot of sense, but when the public is deceived at the same time, this raises ethical concerns. In the end, to what extent a strict separation between actual and perceived security is meaningful is a question that is tightly connected to core philosophical perspectives, in particular realism versus constructivism.

As outlined earlier, security engineering adds specific considerations to the engineering process regarding misuse in addition to use of a technology. In general, it is often argued that engineers need to employ a security-by-design approach (cf. Siriwardena 2014), rather than adding security as an add-on to a system, which often leads to either weak security or high cost. The issue with a security-by-design approach is that it suggests a neutral position regarding security, which can be conceived as problematic, given that there are always assumptions regarding security of what against whom. In addition, the inherent limitations in planning and verifying security in the design stage provide arguments to consider security in later stages as well. We will come back to these issues later, but let us first discuss how security can be measured and tested for a designed technology.

39.4. Security Metrics and Testing

In order to evaluate how secure a particular design is, one needs metrics and associated testing methods. In safety science, the safety contribution of controls can be evaluated in terms of the extent to which they reduce the probability and/or impact of undesirable events. In security, the behavior of the adversaries influences such metrics, for example, because they may simply circumvent a control. In addition, secrecy often hides data on security incidents. Because of these issues, security metrics may be difficult to obtain and often lack scientific rigor (Herley and Van Oorschot 2017; Sanders 2014; Spring et al. 2017).

One can distinguish between two basic types of security metrics (Herley and Pieters 2015). In this classification, type 1 security metrics would involve "what-if" reasoning: if an attacker would try to gain access, how difficult would it be? By contrast, type 2 metrics would gather data on expected or actual incidents when a system is deployed. This means that type 1 metrics are independent of the threat environment, whereas type 2 metrics are dependent on it. For example, the security of the same type of electronic voting system deployed in a stable versus an unstable country would be identical under type 1 metrics, but not under type 2 metrics. Type 1 metrics are therefore useful if one wants to describe properties of the device itself; type 2 metrics are useful if one wants to study the security risk of the device deployed in its environment.

Type 1 security metrics typically involve a specification of how easy it is to obtain a certain kind of access for a certain threat agent, as in the burglar resistance example. In fact, a very detailed metric would map each possible threat agent against a metric of ease of access, for example, in terms of invested time and budget and/or probability of success. Framed differently, this would entail a mapping of the "force" exerted against the probability of success. In safety engineering, this relation is often referred to as fragility, represented in so-called fragility curves (cf. Shinozuka et al. 2000).

In practice, it is often only possible to test a system against a limited set of threat agents. For example, in so-called penetration testing (Bishop 2007), one can hire ethical hackers, burglars, etc. to test how easy it is for them to gain access to some kind of system. If multiple types of threat agents are involved, one could then interpolate the results, but this is often not the case. Also, many tests report only on whether access was obtained, without much detail on the level of effort needed.

In the space of security, data available about a specific technology or a specific organization is often very limited. In this situation, it becomes crucial to obtain information about the level of security from others. However, sharing security information is often seen as particularly sensitive. To move this forward, organizations often set up sector-specific sharing initiatives, or issue "bug bounties" (Maillart et al. 2017) to incentivize hackers to report security problems rather than sell them on the black market.

When more exact metrics are lacking, maturity levels (Le and Hoang 2016) are often defined as an approximation. Maturity levels are ordered levels describing a step-wise improvement from ad-hoc to structured processes in an organization, with each higher level imposing additional requirements. In a security context, it is then assumed that the actual level of security in a design or organization is correlated with the implementation of certain security practices. Whether this is adequate in the highly dynamic and non-linear domain of cyberspace is subject to debate. In fact, a single weakness in a computer program can cause havoc even in mature organizations or technologies. Still, maturity levels (and associated certification practices) provide some kind of indication of the knowledge and effort that is being put into security engineering processes.

When testing security, it is crucial to search for the boundaries of system operation. Many security weaknesses surface only when exceptional input is given to a system, for example, a name of 10,000 characters rather than 10. Since security is not a functional property, things that the system should *not* do should be searched primarily in such exceptional circumstances. Formal verification may help with some of these (see Chapter 32, "On Verification and Validation in Engineering" by Francien Dechesne and Tijn Borghuis, this volume). However, there are security protocols that were proven correct by formal methods under certain assumptions, and they still had flaws when those assumptions were changed (e.g. the Needham-Schroeder protocol; Lowe 1996). The issue is that we often fail to identify types of exceptional circumstances that have not yet caused security problems (Pieters and Consoli 2009). Only when new types of issues become generally known do they become part of security verification and testing. An example are the many types of "side channels" that can leak confidential information, such as power consumption or computation time in a computer system (Standaert et al. 2009). Thinking "outside the box" might be facilitated by advanced penetration testing services by ethical hackers.

As exceptional conditions are easily overlooked, we cannot assume that security testing can be done exhaustively during the design stage. This means that we may deploy security-sensitive technologies in society subject to uncertainty about possible security weaknesses. In this sense, this constitutes a social experiment (Van de Poel 2013). In contrast to technologies with uncertain safety implications, the behavior of possible adversaries plays a crucial role here: whether security issues materialize depends on whether adversaries have a motivation to address the technology, and whether they are capable of doing so (Pieters et al. 2016). The bigger the scale at which a technology is deployed, the more interesting it becomes for adversaries. Think, for example, about Internet voting technology in an election. Therefore, consciously scaling up is even more important in this case.

Finally, security is not the only property that should be tested, even for technologies aimed specifically at security. When higher levels of security lead to much greater effort by benevolent users, for example because of bad usability, this may lead to loss of productivity/time wasted. In addition, users may decide to circumvent the technology altogether, for example by using commercial file storage services rather than the secure ones offered by their employer (shadow security; Kirlappos et al. 2014). There is a big debate in this field on whether users need to be educated to use technologies securely, or whether the technology should be designed such that it encourages or enforces secure use (Adams and Sasse 1999). In particular when secure behavior conflicts with social norms, it is hard to enforce secure behavior via training (Pieters and Coles-Kemp 2011). There is also a link here with the notion of meta-task responsibility: it could be argued that designers should enable users to use their technology responsibly, which also means securely (Ahmed and Van den Hoven 2010).

Compared to safety science, a scientific approach to security is much more controversial, precisely because the resulting risk is partly determined by adversarial behavior rather than by probabilistic events. This creates confusion about what needs to be measured to say something about security, which in turn makes it difficult to test security properties of a design. In addition, as discussed in the next section, security is a heavily politicized concept, meaning that stakeholders may have reasons for advancing a particular interpretation of security and associated metrics that helps them meet their own goals.

39.5. Ethics of Security Engineering

If security is about regulating access, questions can be asked about the responsibility for ensuring proper controls, responsibility for responding to discovered security weaknesses in those controls, as well as responsibility for incidents caused by undue access. In addition, regulating access in or through technological design may have side effects on values other than security.

A key question in the ethics of security engineering is who is responsible for misuse. Because of the role of adversaries, it is easy to blame the attackers when things go wrong, as opposed to harmful events that happen by accident. For example, compare the situations of a chemical factory exploding by accident, versus a chemical factory being blown up by terrorists. How much responsibility for safety/security can be attributed to the designers/operators in those cases? This example illustrates that ethical concepts such as the precautionary principle (Pieters and Van Cleeff 2009) or duty of care (Tikk 2011) may need a different interpretation in the presence of adversarial threats. Rather than designers or operators being the primary responsible agents in case of an accident, the question now becomes to what extent they have *contributed* to the malicious acts of others.

There may be many reasons why security is not addressed in a design, or fails in practice. For example, market pressure requires a piece of software to be published as quickly as possible, making it more difficult to do proper security testing, and high costs may limit the time during which security updates are supplied. This can lead to legacy software that cannot be updated for security anymore, for example software embedded in devices in the Internet of Things. There is a meta-design question here, namely how the sociotechnical system in which designers do their work can be designed such that it incentivizes designers to pay sufficient attention to security (Bauer and Van Eeten 2009). Policy and legislation may play a role here, implying a responsibility for the meta-designers as well, but there are debates in this space too. For example, would insurance coverage for security breaches improve or reduce overall security? On the one hand, insurance may diminish the need to implement controls, as damage is covered anyway (so-called moral hazard; Gordon et al. 2003). On the other hand, when premiums depend on the controls in place, this may indeed incentivize clients to implement better security.

A second key ethical question for security engineering is under what conditions engineers should embed security concerns in their designs. Security concerns are always stakeholder-specific, in the sense that they reflect potential harm affecting those stakeholders, while at the same time representing other stakeholders as threats that may cause such harm. Security always distinguishes between inside and outside and therefore includes and excludes stakeholders. It also does so in a very imprecise way, enabling more access than necessary, while at the same time disabling more access than necessary. These concerns may therefore be very political in nature, requiring careful consideration of what is being included or excluded in these concerns. In fact, stakeholders may use security as an argument to safeguard other interests, in particular to justify claims related to access (e.g. to data), often referred to in terms of *securitization* (McDonald 2008). When implemented in a design, security concerns can materialize in the form of granting certain types of access to some stakeholders, while withholding certain types of access from others. This may contribute to a shift in power balances (Brey 2008). It can therefore be argued that engineers should be aware of the *politics* or *framing* behind security concerns (Fichtner et al. 2016; Fichtner 2018). If they are not, they may unknowingly contribute to the embedding of specific interests in their designs.

A third key question is how to handle security weaknesses that are discovered after deployment, so-called "responsible disclosure" (Matwyshyn et al. 2010). This holds both for the researcher that finds the vulnerability and for the person or organization that it is reported to (if at all). The researcher may decide to report it to the problem owner (software developer), make it public, misuse it, or sell it on the black market, or even to an intelligence agency. If the issue is reported, the software developer may decide to fix the issue (and possibly pay the researcher), ignore it, or even sue the researcher. All these choices could be said to involve different consequences, moral imperatives, and virtues, and can be analyzed from the corresponding ethical theories, with possibly different results. That there lies a moral responsibility with both agents would be acknowledged by most of us. In particular, for the police or intelligence agencies, is it acceptable to employ newly found vulnerabilities for investigation or intelligence gathering rather than reporting them, at the expense of many computer systems around the world remaining vulnerable?

Fourthly, technologies aimed specifically at improving security may also have side effects on other values. This often involves technologies focused on detection of security threats, such as surveillance cameras and online surveillance. Proving the effectiveness of such technologies is often difficult (Cayford and Pieters 2018), and at the same time the question is raised whether the impact on privacy is proportional to the claimed effect. The fundamental debate in this context is whether the tension between security technologies and privacy always involves a trade-off, or whether there are solutions possible that support both values (Solove 2011; Valkenburg 2015). For security engineering, the issue of side effects means that supporting the value of security by means of a technology can never be treated in isolation from other values. In particular, the deployment of security technologies may redistribute power, and without checks and balances, such redistribution may affect other values, including but not limited to privacy.

In short, the practice of security engineering requires awareness of the interaction between security requirements, stakeholder interests, and other values. Security engineering does not stop after deployment, and the handling of security updates, discovered security issues, and also incidents imposes many moral choices as well. Responsibility for security lies not only with the security engineers but also with those providing the context in which they do their work.

As a final observation, harm in security does not only occur via the direct damage done by security incidents. Typically, successful attacks provide monetary or other value to the attackers, which means that even if the damage on the defenders' side is limited, a fairness argument applies: is it acceptable to design something that provides opportunities for illegitimately obtained benefits to those who seek these, giving them resources that can be used to cause additional harm? Designing insecure technologies facilitates unfairness in the sense of some adversaries running off with benefits that belong to all.

39.6. Conclusion

In this chapter, we have covered the major philosophical concepts describing the field of security engineering as well as questions and debates that the theory and practice of security engineering have sparked. We have seen that security distinguishes itself as a topic in engineering because it entails a combination of (a) dealing with adversaries and adversarial risk, (b) the intricacies of nonfunctional requirements and associated verifiability problems, (c) poorly established guidelines for a "science of security" and associated metrics, and (d) a heavily politicized context. While we can expect a gradually increasing maturity of this field of engineering, the inherent normative tension in distinguishing between good and bad guys will sustain many of the controversies for years to come.

Acknowledgements

The research leading to these results was conducted while the author was employed at Delft University of Technology, and has received funding from the European Union's Horizon 2020 Research and Innovation Programme, under Grant Agreement no 740920 (CYBECO). This publication reflects only the authors' views, and the Union is not liable for any use that may be made of the information contained herein. The author wishes to thank Martijn Warnier for useful comments.

Related Chapters

Chapter 32: On Verification and Validation in Engineering (Francien Dechesne and Tijn Borghuis)
Chapter 33: Values in Risk and Safety assessment (Niklas Möller)
Chapter 35: The Role of Resilience in Engineering (Neelke Doorn)

Further Reading

Freund, J. and Jones, J. (2014). *Measuring and Managing Information Risk: A FAIR Approach*. Oxford: Butterworth-Heinemann. (A detailed approach to information [security] risk.)
Meyer, T. and Reniers, G. (2016). *Engineering Risk Management*. Walter de Gruyter GmbH & Co KG. (An overview of engineering aspects of risk management, including safety and security.)
Van den Hoven, J., Blaauw, M., Pieters, W. and Warnier, M. (2018). Privacy and Information Technology. In Edward N. Zalta (ed.), *The Stanford Encyclopedia of Philosophy*, Summer 2018 ed. https://plato.stanford.edu/archives/sum2018/entries/it-privacy/. (More on the philosophy of privacy, technology, and security.)
Vuorinen, J. (2014). *Parasitic Order Machine. A Sociology and Ontology of Information Securing*. PhD thesis, University of Turku. www.utupub.fi/handle/10024/99059. (PhD thesis on the ontology of information security.)

Notes

1. In practice, there is a spectrum of security incidents, ranging from more or less random (e.g. computer virus infections) to targeted (Sony hack; Sullivan 2016), and how adversarial threats can be distinguished from accidental ones can be problematized (cf. Campbell 1997).
2. A related debate concerns the issue of whether monocultures (lots of systems using the same software implementation) are dangerous for security (Schneier 2010). On the one hand, the more effort put into and experience gained with an implementation, the more likely that flaws will be fixed; however, if a flaw is found in a widely used implementation, it affects loads of systems worldwide.

References

Adams, A. and Sasse, M.A. (1999). Users Are Not the Enemy. *Communications of the ACM*, 42(12), 40–46.
Ahmed, M.A. and van den Hoven, J. (2010). Agents of Responsibility—Freelance Web Developers in Web Applications Development. *Information Systems Frontiers*, 12(4), 415–424.
Anderson, R.J. (2001). Why Information Security is Hard—An Economic Perspective. In *Proceedings 17th Annual Computer Security Applications Conference, ACSAC 2001*. IEEE, Stanford, CA, USA, pp. 358–365.
Anderson, R.J. (2010). *Security Engineering: A Guide to Building Dependable Distributed Systems*. Hoboken, NJ: John Wiley & Sons.
Ayrour, Y., Raji, A. and Nassar, M. (2018). Modelling Cyber-attacks: A Survey Study. *Network Security*, 2018(3), 13–19.
Bauer, J.M. and Van Eeten, M.J. (2009). Cybersecurity: Stakeholder Incentives, Externalities, and Policy Options. *Telecommunications Policy*, 33(10–11), 706–719.
Bishop, M. (2007). About Penetration Testing. *IEEE Security & Privacy*, 5(6).

Blakley, B., McDermott, E. and Geer, D. (2001). Information Security Is Information Risk Management. In *Proceedings of the 2001 Workshop on New Security Paradigms*. ACM, pp. 97–104.

Böhme, R. (2010). Security Metrics and Security Investment Models. In *International Workshop on Security*. Berlin, Heidelberg: Springer, pp. 10–24.

Brey, P. (2008). The Technological Construction of Social Power. *Social Epistemology*, 22(1), 71–95.

Burns, A., McDermid, J. and Dobson, J. (1992). On the Meaning of Safety and Security. *The Computer Journal*, 35(1), 3–15.

Campbell, R. (1997). Philosophy and the Accident. *Clio Medica*, 41, 17–34.

Casey, T. (2007). Threat Agent Library Helps Identify Information Security Risks. *Intel White Paper*.

Cayford, M. and Pieters, W. (2018). The Effectiveness of Surveillance Technology: What Intelligence Officials Are Saying. *The Information Society*, 34(2), 88–103.

Coles-Kemp, L. and Theoharidou, M. (2010). Insider Threat and Information Security Management. In *Insider Threats in Cyber Security* (pp. 45–71). Springer, Boston, MA.

Cornish, D.B. and Clarke, R.V. (2003). Opportunities, Precipitators and Criminal Decisions: A Reply to Wortley's Critique of Situational Crime Prevention. *Crime Prevention Studies*, 16, 41–96.

Cox, Jr., L.A. (2009). Game Theory and Risk Analysis. *Risk Analysis,* 29(8), 1062–1068.

De Bruijne, M., van Eeten, M., Gañán, C.H. and Pieters, W. (2017). *Towards a New Cyber Threat Actor Typology: A Hybrid Method for the NCSC Cybersecurity Assessment*. Delft University of Technology—Faculty of Technology, Policy and Management, WODC. www.wodc.nl/binaries/2740_Volledige_Tekst_tcm28-273243.pdf

Fichtner, L. (2018). What Kind of Cyber Security? Theorising Cyber Security and Mapping Approaches. *Internet Policy Review*, 7(2). doi:10.14763/2018.2.788.

Fichtner, L., Pieters, W. and Teixeira, A. (2016). Cybersecurity as a Politikum: Implications of Security Discourses for Infrastructures. In *Proceedings of the 2016 New Security Paradigms Workshop*. ACM, pp. 36–48.

Floridi, L. (2005). The Ontological Interpretation of Informational Privacy. *Ethics and Information Technology*, 7(4), 185–200.

Glinz, M. (2007). On non-functional Requirements. In *Requirements Engineering Conference, 2007. RE'07. 15th IEEE International*. IEEE, Stanford, CA, USA, pp. 21–26.

Gordon, L.A., Loeb, M.P. and Sohail, T. (2003). A Framework for Using Insurance for Cyber-risk Management. *Communications of the ACM*, 46(3), 81–85.

Herley, C. (2016). Unfalsifiability of Security Claims. *Proceedings of the National Academy of Sciences*, 113(23), 6415–6420.

Herley, C. and Pieters, W. (2015). If You Were Attacked, You'd Be Sorry: Counterfactuals as Security Arguments. In *Proceedings of the 2015 New Security Paradigms Workshop*. ACM, pp. 112–123.

Herley, C. and Van Oorschot, P.C. (2017). Sok: Science, Security and the Elusive Goal of Security as a Scientific Pursuit. In *Security and Privacy (SP), 2017 IEEE Symposium on*. IEEE, Stanford, CA, USA, pp. 99–120.

Hildebrandt, M. (2013). Balance or Trade-off? Online Security Technologies and Fundamental Rights. *Philosophy & Technology*, 26(4), 357–379.

Hoepman, J.H. and Jacobs, B. (2007). Increased Security Through Open Source. *Communications of the ACM*, 50(1), 79–83.

Jones, Jack A. (2005). *An Introduction to Factor Analysis of Information Risk (FAIR)*. Risk Management Insight.

Jongejan, A. and Woldendorp, T. (2013). A Successful CPTED Approach: The Dutch 'Police Label Secure Housing'. *Built Environment*, 39(1), 31–48.

Kirlappos, I., Parkin, S. and Sasse, M.A. (2014). Learning from "Shadow Security": Why Understanding Noncompliance Provides the Basis for Effective Security. In *Proceedings of the Workshop on Usable Security*. USEC '14. 23 February 2014, San Diego, CA, USA. Internet Society, ISBN 1-891562-37-1.

Koops, E.J. (2010). The Internet and its Opportunities for Cybercrime. In *Transnational Criminology Manual*. Nijmegen: WLP, pp. 735–754.

Le, N.T. and Hoang, D.B. (2016). Can Maturity Models Support Cyber Security? In *Performance Computing and Communications Conference (IPCCC), 2016 IEEE 35th International*. IEEE, Stanford, CA, USA, pp. 1–7.

Lowe, G. (1996). Breaking and Fixing the Needham-Schroeder Public-key Protocol Using FDR. In *International Workshop on Tools and Algorithms for the Construction and Analysis of Systems*. Berlin, Heidelberg: Springer, pp. 147–166.

Luhmann, N. (1993). *Risk: A Sociological Theory*. New Brunswick: Transaction Publishers.

Maillart, T., Zhao, M., Grossklags, J. and Chuang, J. (2017). Given Enough Eyeballs, All Bugs are Shallow? Revisiting Eric Raymond With Bug Bounty Programs. *Journal of Cybersecurity*, 3(2), 81–90.

Matwyshyn, A.M., Cui, A., Keromytis, A.D. and Stolfo, S.J. (2010). Ethics in Security Vulnerability Research. *IEEE Security & Privacy*, 8(2).

McDonald, M. (2008). Securitization and the Construction of Security. *European Journal of International Relations*, 14(4), 563–587.

Nickel, P.J. (2013). Trust in Technological Systems. In *Norms in Technology*. Dordrecht: Springer, pp. 223–237.

Pavlovic, D. (2011). Gaming Security by Obscurity. In *Proceedings of the 2011 New Security Paradigms Workshop*. ACM, New York, NY, USA, pp. 125–140.

Petitcolas, F.A.P. (2011). Kerckhoffs' Principle. In H.C.A. van Tilborg and S. Jajodia (eds.), *Encyclopedia of Cryptography and Security*. Springer, Boston, MA

Pieters, W. (2006). Acceptance of Voting Technology: Between Confidence and Trust. In *International Conference on Trust Management*. Berlin, Heidelberg: Springer, pp. 283–297.

Pieters, W. (2011). The (Social) Construction of Information Security. *The Information Society*, 27(5), 326–335.

Pieters, W. and Coles-Kemp, L. (2011). Reducing Normative Conflicts in Information Security. In *Proceedings of the 2011 New Security Paradigms Workshop*. ACM, New York, NY, USA, pp. 11–24.

Pieters, W. and Consoli, L. (2009). Vulnerabilities and Responsibilities: Dealing With Monsters in Computer Security. *Journal of Information, Communication and Ethics in Society*, 7(4), 243–257.

Pieters, W. and Dechesne, F. (2017). Adversarial Risks in Social Experiments With New Technologies. In *Experimentation beyond the Laboratory: New Perspectives on Technology*. London: Routledge.

Pieters, W., Hadžiosmanović, D. and Dechesne, F. (2016). Security-by-experiment: Lessons From Responsible Deployment in Cyberspace. *Science and Engineering Ethics*, 22(3), 831–850.

Pieters, W. and van Cleeff, A. (2009). The Precautionary Principle in a World of Digital Dependencies. *Computer*, 42(6).

Piètre-Cambacédès, L. and Chaudet, C. (2009). Disentangling the Relations Between Safety and Security. In *Proceedings of the 9th WSEAS International Conference on Applied Informatics and Communications (AIC '09)*. www.wseas.us/e-library/conferences/2009/moscow/AIC/AIC27.pdf

Probst, C.W., Hunker, J., Gollmann, D. and Bishop, M. (2010). Aspects of Insider Threats. In *Insider Threats in Cyber Security*. Boston, MA: Springer, pp. 1–15.

Sanders, W.H. (2014). Quantitative Security Metrics: Unattainable Holy Grail or a Vital Breakthrough Within Our Reach? *IEEE Security & Privacy*, 12(2), 67–69.

Schneier, B. (2008). The Psychology of Security. In *International Conference on Cryptology in Africa*. Berlin, Heidelberg: Springer, pp. 50–79.

Schneier, B. (2010). The Dangers of a Software Monoculture. *Information Security Magazine*.

Schryen, G. (2011). Is Open Source Security a Myth? *Communications of the ACM*, 54(5), 130–140.

Shinozuka, M., Feng, M.Q., Lee, J. and Naganuma, T. (2000). Statistical Analysis of Fragility Curves. *Journal of Engineering Mechanics*, 126(12), 1224–1231.

Sindre, G. and Opdahl, A.L. (2005). Eliciting Security Requirements with Misuse Cases. *Requirements Engineering*, 10(1), 34–44.

Siriwardena, P. (2014). Security by Design. In *Advanced API Security*. Berkeley, CA: Apress, pp. 11–31.

Solove, D.J. (2011). *Nothing to Hide: The False Tradeoff Between Privacy and Security*. New Haven: Yale University Press.

Spring, J.M., Moore, T. and Pym, D. (2017). Why a Science of Security Is Hard: A Philosophy of Science Perspective. In *Proceedings of the 2017 New Security Paradigms Workshop (NSPW2017)*. ACM, New York, NY, USA.

Standaert, F.X., Malkin, T.G. and Yung, M. (2009). A Unified Framework for the Analysis of Side-channel Key Recovery Attacks. In *Annual International Conference on the Theory and Applications of Cryptographic Techniques*. Berlin, Heidelberg: Springer, pp. 443–461.

Sullivan, C. (2016). The 2014 Sony Hack and the Role of International Law. *Journal of National Security Law & Policy*, 8, 437.

Tikk, E. (2011). Ten Rules for Cyber Security. *Survival*, 53(3), 119–132.

Valkenburg, G. (2015). Privacy Versus Security: Problems and Possibilities for the Trade-off Model. In *Reforming European Data Protection Law*. Springer, Dordrecht, pp. 253–269.

Van de Poel, I. (2013). Why New Technologies Should Be Conceived as Social Experiments. *Ethics, Policy & Environment*, 16(3), 352–355.

Van der Wagen, W. and Pieters, W. (2015). From Cybercrime to Cyborg Crime: Botnets as Hybrid Criminal Actor-networks. *British Journal of Criminology*, 55(3), 578–595

Vuorinen, J. and Tetri, P. (2012). The Order Machine-The Ontology of Information Security. *Journal of the Association for Information Systems*, 13(9), 695.

PART VII

Responsibilities in Engineering Practice

40

ETHICAL CONSIDERATIONS IN ENGINEERING

Wade L. Robison

Ethical considerations permeate engineering practice, especially its intellectual core, solving design problems. No one could tell that by looking at the Accreditation Board for Engineering and Technology's (ABET) definition of engineering design:

> It is an iterative, creative, decision-making process in which the basic sciences, mathematics, and engineering sciences are applied to convert resources into solutions . . . For illustrative purposes only, examples of possible constraints include accessibility, aesthetics, codes, constructability, cost, ergonomics, extensibility, functionality, interoperability, legal considerations, maintainability, manufacturability, marketability, policy, regulations, schedule, standards, sustainability, or usability.
>
> *(Criteria 2018–2019)*

Ethics merits not a mention, even as a possible constraint. That may reflect the view that engineering is a purely quantitative enterprise with an objectivity not supposedly found in ethics. That view is mistaken: ethical considerations permeate engineering, and given the importance of ethical considerations to engineering, it is surprising that ethical constraints are not among the examples listed in ABET's definition.

40.1. Ethics Enters the Design Process

It should come as no surprise, however, that ethical considerations permeate engineering or even that they enter into the design process itself. Five unsurprising truths make it clear that ethical considerations not only can but do enter the intellectual core of engineering (Robison 2014: 1–4).

40.1.1 No Design Problem Necessitates a Particular Solution

The relationship between the statement of a design problem and any solution is contingent. We cannot deduce a solution from the design problem statement no matter how detailed or how conditioned by constraints. If we could, a design problem would have the same solution no matter which engineer tried to solve it—assuming the engineers did not make a mistake in their deductions. We owe the wide variety of solutions to any one particular problem to the contingency of the relation between the statement of the design problem and a solution. That is why we have different cell phones to choose from, different engines in different cars, different operating systems in different computers.

That is not to deny that engineers might converge on the same solution, but as in evolution, convergence is no proof of necessity. In evolution it is only evidence that the initial conditions were similar enough to produce roughly the same effect. In engineering it is only evidence that one solution appears 'obvious'—perhaps because it is similar to the way engineers have solved similar problems. Convergence is consistent with contingency.

The contingent relation between the statement of a design problem and a solution leaves space for creativity, for an imaginative engineer to solve a design problem in an utterly novel way. It also leaves conceptual space for ethical considerations to enter. Indeed, as we shall see, ethical considerations enter into how the design problem is framed, and those considerations can play themselves out in that conceptual space.

40.1.2 An Engineer Could Intentionally Choose a Solution Which Will Unnecessarily Cause Harm

We can imagine an engineer unintentionally choosing a design solution which will cause unnecessary harm. That happens perhaps too frequently, unfortunately. But we can also imagine an evil genius of an engineer who intentionally chooses solutions which cause unnecessary harm—just because they do. Such an engineer would indeed be evil, and were the engineer truly a genius, we would be in for some diabolical design solutions.

That we can imagine such an engineer is proof that between the statement of a design problem and a solution, there is room for ethical considerations to enter. The conceptual space created by there being a contingent gap between the problem statement and a solution can be filled with an ethical judgment—or an unethical one, as in the case of the evil genius of an engineer.

We would criticize such an engineer, as we should, but we should be thankful that no example quickly comes to mind—except, perhaps, Osama bin Laden. Some engineer may be an evil genius, that is, but that would be the exception. The point remains, however, that ethical considerations can enter the gap between a design problem and a solution—as the intentional choice of an evil genius of an engineer illustrates.

40.1.3 A Person Can Cause Harm Without Intending to Cause Harm

We try to teach this to children early on when, for instance, we say to our child, 'Don't swing that bat around like that! You are going to hurt somebody.' We are teaching children to be responsible by telling them it is wrong to cause harm, and we are teaching them that they can cause harm even without intending to cause harm if they are careless, for example.

We have all said at one time or another, 'But I didn't mean to do it!' We say that because we want to take off some of the ethical pressure we feel and deflect the ethical criticism we think others are directing at us for what we did. We do make ourselves look better and deflect some ethical criticism when we point out that we did not intend to cause whatever harm we caused. What we are saying is that we are not the evil person we may appear to be. We caused the harm without any evil intent, that is.

But causing harm and intending to cause harm are distinct. It is ethically wrong to intend to cause harm whether we fail or succeed. When we succeed in causing harm we intend to cause, we are subject to ethical criticism both for intending to cause harm and for succeeding. But we could also cause harm without any intent to cause harm. That is wrong, even without the relevant intent, if it comes about because we were careless, for instance. We thus hold others ethically responsible for the harms they bring about when they cause an accident because they were for some reason not on their game.

A 20-year-old driver who crossed the median strip and killed 13 in a bus said, 'I'm sorry, I'm sorry. "I was texting."'—as though that explained and so excused it all. We do not accept as an excuse, let alone a justification, 'I was texting' (Lozano 2017). Driving is an act of skill, requiring careful attention to the vehicle and to the road conditions. A failure to drive skillfully can be an ethical failure for any of a number of reasons—being unfit to drive because drunk or high, being distracted by talking and gesturing to a passenger, texting or using your cell phone. In such cases, the ethical problem arises because we are not exercising carefully the skill we should be using to drive, not because we intend to cause harm.

40.1.4 *In Solving a Design Problem, Engineers Are to Use Their Special Knowledge and Skills Competently and With Due Care*

Engineers might be negligent—as were some of the engineers responsible for the Hyatt Regency disaster in Kansas City (Petroski 1992: 87; Levy and Salvadori 1994: 221–230). They might be careless—as were the engineers who failed to check what unit of measurement was being used for the Mars Orbiter (Mars 1999) or the French engineers who discovered too late that the new trains they designed were too wide for many platforms (Great Miscalculations 2014). They might just be incompetent—as were, perhaps, the engineers on the BP oil platform who failed to have an agreed-upon procedure for the temporary abandonment of the deepwater well that blew and killed many of them as well as badly polluted the Gulf and the shores of the states bordering it (*Deep Water* 2011: 104). A failure of engineers to use properly the special knowledge and skills they have learned to become practicing engineers makes them ethically at fault, whatever their intent. They are in that way no different from the young man who killed 13 people while texting. They are ethically responsible because they failed to exercise properly the skills they were supposed to have learned to become engineers.

40.1.5 *They Are* **Prima Facie** *Acting Ethically When They Make Use of the Knowledge and Exercise Properly the Skills They Learned to Become Engineers*

What makes the news are instances of unethical conduct—bombing civilians, torturing animals—and instances where people go out of their way to help others. Instances of ethical behavior that are not in any way extraordinary do not make it into the news. They hardly register in our consciousness as ethical. I drive to and from work, within the lane I am supposed to be in, using turn signals to change lanes, stopping at stop lights, maintaining whatever regular speed others are maintaining so as not to be a risk or an impediment for others. We do not think of such things as ethically admirable or even worth noting. But noted or not, they are ethical—even though I no longer have to think about staying in my lane or keeping up with traffic. I do it all by habit.

There is one ethical theory that requires that for us to be ethical, we have to think about what we ought to do before doing it and make a conscious decision to do the right thing. Such an ethical theory leaves us with a world in which only few actions, if any, are ethical and most of what we do does not deserve being called 'ethical' because we did not go through the process of thinking about what we ought to do before doing it. On this theory, most of what we do could not be ethical. We make decisions too quickly, as Daniel Kahneman would point out, without much if any thought (Kahneman 2011: 13, 50ff.). When I swerve my car to avoid a child running into the street, I am saving the child's life perhaps, but if I did not first think about whether I should swerve or not and then decide to swerve, my avoiding the child would have no ethical content on this theory.

It seems unlikely that this philosopher's theory—Immanuel Kant's—has so wormed its way into our collective consciousness as to cause us not to think of ordinary ethical acts as ethical. But for some reason we do tend to put to one side our ordinary day-to-day ethical behavior as not worth remarking on. It would be a mistake, however, to treat what engineers do on a regular basis as not worth noting. Some have certainly made mistakes and rightly been held to account, but what engineers do daily as they solve design problems and carry on their other activities is *prima facie* ethical.

We thus have the following commonplace observations:

1. No design problem necessitates a particular solution.
2. An engineer could intentionally choose a solution that causes unnecessary harm.
3. A person can cause harm without intending to cause harm.

These three observations tell us that it is perfectly possible for an engineer to go from a design problem to an unethical solution, all without intending such an outcome. An engineer could intend to solve a design problem in an ethically admirable way and fail. If so, the engineer would be ethically accountable for that failure—if, for instance, the engineer failed by not using the knowledge engineers learn or by not properly using the skills they have learned to become engineers.

These are not the only ways in which a design solution could go wrong, of course. That is why engineers are only *prima facie* acting ethically when they properly use their special knowledge and special skills. We judge them to have acted ethically in such cases, that is, until it is proven otherwise. It may turn out that unbeknownst to them, the software generally used is flawed. An engineer could then produce unintended harms through a design solution without any intention to cause harm.

In any event, when engineers use that knowledge and those skills properly, they are at least *prima facie* acting ethically. That is what the next two points tell us:

4. In solving a design problem, engineers are to use their special knowledge and skills competently and with due care.
5. They are *prima facie* acting ethically when they make use of the knowledge and exercise properly the skills they learned to become engineers.

So students are not just replacing their well-entrenched habits and forms of thought with a rigorous and measured response to the problems they will meet as engineers. They are also learning how to respond in an ethical way. That is one way in which ethical considerations enter into the intellectual core of engineering.

40.2. The Character of an Engineer

Engineers should get ethical credit for doing what, as engineers, they ought to do. If that seems odd, think of what would happen if an engineer failed to calculate the stress properly for a bridge that is to carry semi-trailer trucks. The mistaken calculations may show more support is needed than necessary, in which case the bridge would never be stressed, but would be more expensive to build—an unnecessary harm. Or the calculations may show that less support is needed than necessary, presumably saving money, but risking failure. Getting the calculations right avoids potential harms, and that is enough to make the calculations of ethical concern. Getting the calculations right is thus the right thing to do. Our failure to make much of engineers doing what they ought to do does not preclude its being the right thing to do.

It is the right thing to do not just because failing to do it could cause unnecessary harms and thus be an ethical failure. It is the right thing to do because it is what engineers are trained to do: it is a function of their having become engineers, of their gaining special knowledge and becoming adept at special skills before they can graduate. They graduate *as* engineers, entitled to represent themselves as competent to use the skills they have acquired with the knowledge they have gained. They graduate embodying the accumulated wisdom of those who came before them and having honed the knowledge and skills necessary to be an engineer. That legacy will continue to be refined and perfected as working engineers find problems with what was the standard way of doing things and work to ensure new standards that will not have such problems. That is, what the students have learned are the norms of the profession, what someone who is an engineer ought to do. Though the norms may change over time, at any one time engineers have an obligation to follow the norms.

So in teaching students the knowledge and skills they need to solve design problems, faculty are teaching them ethical behavior—what they are to do if they are to be engineers, doing what engineers ought to do. Faculty are trying to change not just the way students think, but the students themselves. They are trying to turn someone who used to say, 'That's good enough,' into someone who says, 'We can do better.' They are trying to instill a combination of qualities that make for an admirable engineer, one who can be trusted to determine with regularity what is reasonable for any design solution. Becoming adept at what is needed to graduate as an engineer is no guarantee, obviously, that a student is an ethical engineer. An evil genius is as capable of learning the knowledge and skills as anyone else, but the aim is to graduate students who are honest, creative and risk-averse, meticulous but able to see the big picture, team players able to communicate clearly, and set and meet goals—all the traits we should expect in an engineer. It is not likely that students who enter engineering already possess all the characteristics of an engineer, and so, in teaching them how to think in a new way about solving problems and so rooting out, when necessary, old habits of thought, faculty are molding their characters.

It is no small task, but certainly an admirable one to root out embedded character traits that may be contrary to those necessary for an engineer and replace them with the traits that make for a trustworthy engineer, capable of furthering the profession's mission of serving the public good. That is a challenge with ethical import and so another way in which ethical considerations enter into engineering.

40.3. The First Step in the Decision-Making Process: 'What's the Problem?'

In solving design problems, engineers thus display their professional character by their ethical behavior in using properly the knowledge and skills they have gained in becoming engineers. Ethical considerations thus permeate the decision-making process. But each step in that process also requires ethical judgment. Ethical considerations enter the very first step, 'What's the problem?'

Here are some different projects:

- Design a gear shift lever for a lawn tractor that is as strong as a steel lever, but made of high-impact plastic.
- Design a ladder that can extend to 24 feet, will support 250 pounds, and weighs no more than 20 pounds.
- Develop a test that will instantaneously tell us whether water has lead in it (Wamsley 2017).
- Develop a small portable water-purification device that makes any water safe to drink (Adler 2007).
- Design a land mine that is undetectable by any known technology, is extremely sensitive so that it will explode upon the slightest pressure within a radius of three or four feet, and spews out bomblets upon detonation—a cluster land mine.

This is an odd collection of projects, but it illustrates what is common to all design problems: an engineer's choice of a design solution is an ethical choice, whether good or bad, for at least two different reasons

An engineer might choose the first project, a bit dull and run-of-the-mill but perhaps paying well. Or the second, less run-of-the mill and much more of a challenge since it requires a solution which would cut in half the usual weight of such ladders. An engineer would need a creative solution for this problem. In either case, and in all the others, the engineer will make a design choice that, first, reflects some value judgments and, second, will, once realized in an artifact and sent out into the world, have effects, some good and some harmful.

We can readily imagine a shift lever that fails to provide a firm grip, for instance, the engineer having preferred the aesthetics of a smooth surface to the ridges or bulges that would provide something to grab and hold onto. That preference of values will mark the design, an aesthetic consideration outweighing, consciously or subconsciously for the engineers, a concern to make the lever easy to grab and hold. The engineer may have carefully considered the choice made or made it without thinking about it at all. It does not matter. The lever will display the aesthetic value chosen and the consequent failure to have any ridges or bulges whether the engineer gave the choice any serious thought or not. The choice will also have effects once realized into a lever attached to a lawn tractor. The harms may well be minor, a failure to shift properly, for instance, until we get accustomed to grabbing the lever tighter than we normally would. But they are harms nonetheless and come into existence because of an ethical choice to value a particular aesthetics over making the lever easier to grab and hold onto.

Ethical considerations enter in another way in regard to design problems. When an engineer accepts such an assignment, the engineer is making an ethical commitment, accepting an assignment that is to be completed by such-and-such a date. When the person or company comes to see the prototype and the engineer says, 'Oh, I decided not to do it,' the engineer has both legal and ethical troubles. Accepting a project is the equivalent of making a contract, or making a promise. Failing to do what you have committed to do is neither legal nor ethical.

The same problems arise if you complete the project, but it fails to come anywhere near meeting the specifications. The ladder weighs 60 pounds, for instance, nowhere near 20, or the gear shift lever breaks on demonstration. When you make a commitment to do a project, you are committing yourself to make use of the skills and knowledge you possess as an engineer to do the best job you can. That is an ethical commitment. If the best you can do for a ladder is worse than ladders that are already available, you have failed to fulfill your promise in accepting the assignment.

That is an ethical failing as well as, perhaps, a legal one, and it has other harmful consequences. You will earn a lousy recommendation as an engineer—'less than mediocre'—and a bad reputation as someone who cannot be trusted to do even simple engineering projects, let alone more complicated ones like the lightweight ladder. Your attempts mark an engineering as well as an ethical failure.

If we look back at the list of projects, the last three fall into a different category from that of the first two: they are themselves ethically charged. Those who successfully take on the third and fourth, about ensuring water is safe to drink, deserve ethical praise for trying to do something that helps keep individuals healthy and so significantly furthers the public good. If successful, they did something admirable and so deserve two accolades.

The fifth project, the land mine, is also ethically charged—negatively rather than positively. Any engineer who decides to take on the project needs at least to ask whether the world would be a better or worse place with such a land mine. Whatever the answer, the reasons would depend upon some claim about what is right and what is wrong.

The last three examples go to show that choosing a design problem can be an ethical matter and that the design problem itself reflects ethical considerations. Asking the question 'Can we design a straw that will purify water wherever the water comes from?' is an indication of an ethical concern that individuals should not have to drink water with Guinea worm larvae or any other pathogens. Some design problems may be ethically neutral, but some at least are not. In either case, as we saw with the lever, ethical considerations enter into the nature of the design problem itself and cause great harm.

40.4. The Second Step in the Decision-Making Process: 'What Solutions Are Possible?'

Ethical considerations also enter in what follows in the process of solving design problems, the determination of what solutions are possible. Engineers are to brainstorm possible solutions, thinking as best they can 'out of the box' to come up with ingenious and elegant ways of handling a problem. Many possibilities are tossed as non-starters because they are too expensive, too idiosyncratic to risk the time and money to work them through, too complicated and thus too prone to problems when they are used, and so on.

One criterion for tossing out possible solutions is that they cause unnecessary harms. The principle that we not cause unnecessary harm is a minimal ethical constraint, and no one ought to object if an engineer tosses out a possible solution because it would cause unnecessary harms. It is not possible to introduce an artifact into the causal stream of the world without changing the flow, producing effects, some good, some harmful.

So when the engineers at GM designed a mercury light switch for trunk lids, they introduced a welcome benefit—a light to see what is inside the trunk when the lid is opened. But they also introduced an unnecessary harm: mercury in vapor form is extremely toxic to our central nervous system, among other things. The harm they introduced was unnecessary because there were other ways to design a trunk light switch that did not require a toxic substance. We can also find necessary harms, features of design solutions that are likely to cause harm but are necessary to achieve the proper effects. Computer screens emit blue light. It is not possible to have computer screens without blue light emissions, but too much can cause macular degeneration.

I think engineers generally toss out possible solutions that cause unnecessary harms all the time without thinking too much about the ethical reason for doing it. They just do not see any good reason to go with a design that would have unnecessary harmful features.

It is, unfortunately, relatively easy to find possibilities that reached fruition and should have been tossed out when they were mere possibilities. In retrospect, it is easy to see that this ethical constraint, however minimal, was not at work—probably for a variety of reasons.

Some unnecessary harms are missed because engineers fail to consider how their design choices will be used. When you close the door in the Subaru SVX, a harness automatically fastens across your chest, but not the seat belt. You need to reach down and fasten that separately. The oddity is that there was no need for a motorized harness that closes when the door closes if you still have to go to the trouble of fastening your seat belt, and there is enough of a risk of significant harm that there is a warning on the harness and in the instructions that you should be sure to fasten your seat belt because otherwise you risk 'severe head trauma.'

What is likely to happen is that when you have an accident and your body begins to move forward, the upper torso will be constrained by the harness, but the lower part of your body will move forward and draw you down so that your neck will meet the harness. Your head may escape undamaged, but it may no longer be attached to your body since the harness may slice right through it. The warning is not quite accurate, but Subaru was lucky that so few SVX's were sold that no one lost their head, as it were, driving them (Robison 2016: 171–172).

We have to wonder what the engineers were thinking when they chose this design. They clearly recognized that it could cause great harm since they provided warnings, but if so, we have a puzzle as to why they stuck with that solution for the problem they were trying to solve. Drivers and passengers alike are likely to put themselves at risk of having their necks severed without realizing it.

Not all the harms engineers should avoid are as bad as the risk of severed heads, but are evidence that they did not consider how the design chosen could cause harm. Drivers in the Mazda RX-8 may find themselves unable to shift the car because the edge of the side wall next to the clutch pedal forms a lip at the same height as the pedal. Unless the driver has very narrow feet, the lip will catch their shoe, preventing them from pushing the clutch pedal down—annoying, to say the least, and dangerous as well (Robison 2016: 158–159).

We can find a raft of other examples where engineers have apparently failed to consider how their design solutions will end up being used. One of the easiest ways to find examples is to go to Amazon and read the reviews of various toasters. I recommend the Sunbeam 3910–100 2-Slice Wide Slot Toaster. This toaster 'invokes feelings of rage and hatred' for one reviewer (Carrier 2017). It has many problems. For instance, the toaster cord comes out of the front of the toaster where the controls are and is only 12 inches long. Somehow we are supposed to plug it into a wall outlet and still be able to set the dial and push down the lever. To put it mildly, this is not an optimal design solution. The harm of not being able to adjust the degrees of heat without unplugging the toaster is relatively minor as harms go, but it is a harm. Even worse, the toaster is notorious, according to the reviews, for toasting bread unevenly even when left on the same setting. Having to toss a piece of toast because it is burnt is another relatively minor harm, made worse by the realization that unplugging the toaster and adjusting the setting is not likely to solve the problem since uneven toasting occurs at all settings. We need only imagine coming into the kitchen in the morning with dread, knowing that the likelihood of getting a proper piece of toast is minimal. The toaster's design flaws turn what ought to be a relatively easy job into a stressful chore, one that can indeed invoke 'feelings of rage and hatred.' As another reviewer put it, 'How difficult is it to design a toaster that toasts'? (mwreview 2014).

It is not clear how many of the difficulties with this toaster are the result of engineers not doing a proper job and how many are due to a manufacturer's cutting corners, but a clear example of a problem caused by engineers was the design of the original airbags. The crucial design problem for the original airbags was that the airbags exploded at about 180 mph while the person being protected was being thrown forward towards the airbag. Those being protected would risk severe injury if not death if they were too close to the airbag and it struck them at 180 mph, but would hit the airbag and risk injury if they were still moving forward when the airbag had fully inflated. The engineers' solution was to design an airbag that would work for a 'normal' person, one of exactly the right height and weight to be exactly the right distance away from the airbag to be caught and cradled, not crushed, by it. The engineers chose as their 'normal' person the 50th percentile of men, thus putting at risk those who were at either end of the bell curve for weight and height. Short, petite women would be sitting closer to the airbags and thus would receive the full force of its exploding at 180 mph.

The engineers had to draw a line somewhere. No design would protect every single driver given the wide variety of drivers, some tall, some short, some heavy, some petite. So no matter where they drew the line, some drivers would be significantly better protected than others. What makes the original airbag a good illustration of how ethical considerations enter into design problems is not just that some choice had to be made about whom to protect the best and whom to put at risk, but that the engineers chose as their norm the normal man. As I have pointed out elsewhere, their choice gives new meaning to the phrase, 'Women and children first' (Robison 2016: 159–164).

Here is one last example of how engineers fail to consider how a design solution will play out when it is realized in a product. In the 1990s, Cadillac introduced cars with trunk lids that would close themselves—provided that someone closed them by hand down to about 12 to 15 inches where the motor would take over. It does not take much imagination to understand what harms will result from someone who knows how to close a trunk, having been doing that for years. They will assume the trunk works, well, the way trunks have always worked. You push it down, and it latches. Unfortunately, if you push the Cadillac trunk lid down past the point where the motor would take over, you break the motor. If you break the motor, the trunk will not latch. So the result of someone's doing what they always have done to close trunk lids means you will need to use a bungie cord to hold your trunk down until you can get the motor replaced. Replacing the motor requires removing the back seat—at significant additional cost. What is worse is that it is easy to break the motor. There is no catch to prevent you from pushing the trunk down. No warning about what you need to do. No pressure pushing back that might signal to you that things are not as they should be (Robison 2016: 142).

No one will be hurt or injured because of this design, but the harms are significant, and they were completely unnecessary. There was not much of a gain in adding a motor to the trunk since the trunk had to be lowered by hand anyway to the point where the motor took over, and drivers would always have to be on guard for someone's pushing the trunk closed—a bellhop at the hotel, a child, a helper at a grocery store. In short, the gains made by adding a motor do not compensate for the harms the design is likely to produce.

One possible reason for the failure to consider whether a design solution will produce unnecessary harms when realized in an artifact is that senior design students generally consider only a narrow set of risks. As a study of senior design courses shows, 'Risk management includes identification, analysis, and mitigation of risks that, should the risks materialize, have the potential for negatively impacting the completion of a project' (Groth and Hottell 2007: 45–54). The students' concern, quite understandably, is to ensure that they get their project finished, and so they consider such risks as one of their team falling ill or free riding, or 'mission creep' as they try to satisfy a client, or an inability to obtain in a timely manner some component they need.

The mindset of focusing on what risks could prevent completion of a project is presumably not limited to students. When engineers take on a project, they have committed themselves to a goal, finishing something within a certain time frame. That goal becomes the primary driver for the engineers, and so it is only natural to focus on the risks that they might face in reaching that goal.

Those risks fall into two categories, those that are internal to the project—e.g., a team member's falling ill—and those external to the project—e.g., a change in the amount of money the contracting party is willing to provide for a prototype (Block 1983: 2–7). The possibilities for either type of risk are numerous and varied, and the engineers need to fasten on what mitigation strategy to use. Are they to prevent the risk? Reduce it? Transfer it? Or just accept it? That is a time-consuming process, but it has to be done for the engineers to have some confidence that they can complete a project on time.

But if engineers are considering only what risks may prevent them from completing their project, they are not focusing on harms that can occur downstream from the project's entry into the world or what effects it will have upstream. There is some evidence that when students are asked to expand the horizons of risk beyond the project and consider the harms that they may introduce into the world, they are capable of articulating a wide variety, both downstream and upstream (DeBartolo and Robison 2018). All engineers ought to ensure that the risks they consider, and thus the potential harms taken into account, include more than just those that will impede the completion of their project.

The reason is obvious. Once a design solution is realized in an artifact and introduced into the world, it will have effects. Some of these effects will flow from the artifact's introduction, and some will be produced because of what is necessary for the artifact's introduction. What materials need to be mined for the artifact? What chemical compounds need to be produced, with what side effects, to complete the product? What will happen when people try to use the product? What will happen when the product ends its useful life?

Consider in this regard the mercury light switches we mentioned. As we noted, the engineers did not have to use switches with mercury in them. Other solutions would have worked, and yet they chose a solution which left us with an enormous amount of mercury that would need to be collected and then disposed of safely when the vehicles had reached the end of their lives. A program to recover that mercury began in 2005, and before General Motors went bankrupt, 2.5 million switches had been recovered, containing 6500 pounds of mercury. That might seem like a lot, but the switches were used in 36 million vehicles built prior to 2000, leaving 87,100 pounds unaccounted for. Since it is toxic, that puts many at risk—an unnecessary risk (Robison 2016: 176–179).

The mercury switches had effects upstream and downstream. The mercury first had to be secured, subjecting the individuals who secured it to potential harm, and when it is released into the environment, it poses a hazard that needs to be mitigated.

No artifact is without harmful effects, and any engineer would be hard-pressed to design something that could be introduced into the causal stream without any harmful effects. But they are not being asked to do that. They are being asked to do what they already generally do, minimize the harms that do occur by ensuring that none are unnecessary. It is on the ground that they produce unnecessary harmful effects that possible solutions to design problems ought to be tossed without further consideration, and it is in tossing them that engineers display how ethical considerations play a role in considering possible solutions.

40.5. The Third Step in the Decision-Making Process: 'Which Solution Is Best?'

As we saw, ABET's definition of engineering design lists a variety of constraints—from accessibility to usability—but makes no mention of any ethical considerations. Not only is there no mention in the central description of the process. There is no mention in the list of nineteen constraints ABET takes the time to list: 'accessibility, aesthetics, codes, constructability, cost, ergonomics, extensibility, functionality, interoperability, legal considerations, maintainability, manufacturability, marketability, policy, regulations, schedule, standards, sustainability, or usability.' Very few of these constraints have ethical aspects. A design solution that meets the constraint of sustainability has, presumably, fewer harmful effects than one that does not, and if a solution is ergonomically designed, presumably those who use it will not somehow harm themselves in using it.

I find the exclusion of ethical considerations stunning because it is, I think, obvious that engineers do not exclude them when solving design problems, and especially when choosing the best design solution. We would otherwise live in a world filled with harmful artifacts, designed by engineers without any concern at all about flooding the world with what will harm us. We can certainly use artifacts to harm ourselves—driving vehicles too fast, not being careful enough with a chainsaw—but were engineers not cognizant of potential harms and careful to avoid them, we would live in a far worse world.

Imagine a world in which all engineers are evil and intend to cause harm through their diabolical design solutions. Imagine another world in which no engineer intends to cause harm but pays no attention to whether a design solution causes harm. Now imagine our world. It is a very different world from the first two possible worlds. Engineers do sometimes opt for designs that cause unnecessary harm. The mercury light switches are an example of that. They even opt for designs that will ensure harm because even the most intelligent, informed, and motivated will be led by the design into making a mistake. I call these design choices error-provocative, and they occur far more often than we would like. In one toaster, for instance, bread is lowered by pushing the lever down, but if you pull the lever up to get the toast out, you will break the toaster. It cannot be repaired for less than $35, which is more than the cost of the toaster—which means you have lost $35 and the landfill has one more useless object in it (Robison 2016: 3). But, all in all, the artifacts that realize the design solutions engineers opt for do more good than harm and do not generally cause unnecessary harms.

I am not arguing, that is, that engineers ought to make decisions that reflect ethical considerations, but that they cannot help but make such decisions and that they generally make the right decisions. They generally have no trouble identifying potential harms and mitigating the risks of their occurring. They generally have no trouble weighing harms and benefits. They thus already incorporate ethical considerations into their design solutions.

So, my concern is not that engineers make unethical decisions. Few do. My concern is that engineers make value-laden decisions without necessarily thinking about whether their decisions are ethical. Making explicit the ethical considerations that already enter into the design process should encourage a better understanding of how such considerations could enter into the process and broaden the kinds of ethical considerations engineers already take into account, whether they realize it or not. It also encourages engineering schools to expand the risk analysis that their students do in senior design to include external as well as internal risks and risks that arise upstream as well as downstream. That would encourage a new generation of ethical engineers to do better what the present generation generally does well.

Related Chapters

Chapter 21: Engineering Design (Peter Kroes)
Chapter 22: Values and Design (Ibo Van de Poel)
Chapter 44: Responsibilities to the Public—Professional Engineering Societies (Joseph Herkert and Jason Borenstein)
Chapter 55: Reimagining the Future of Engineering (Neelke Doorn, Diane P. Michelfelder et al.)

References

Adler, Margot (2007). Functional Designs that Change Lives. *NPR*, June 17.

Block, Robert (1983). *The Politics of Projects*. New York: Yourdon Press.

Carrier, Toby (2017). September 20. www.amazon.com/Sunbeam-3910-100-2-Slice-Toaster-Black/product-reviews/B0007Y17WE/ref=cm_cr_getr_d_paging_btm_next_3?ie=UTF8&filterByStar=one_star&reviewerType=all_reviews&pageNumber=3#reviews-filter-bar. Accessed March 20, 2018.

Criteria for Accrediting Engineering Programs (2018–19). www.abet.org/accreditation/accreditation-criteria/criteria-for-accrediting-engineering-programs-2018-2019/#definitions. Accessed November 3, 2018.

DeBartolo, Elizabeth A. and Robison, Wade L. (2018). *Risk Management and Ethics in Capstone Design*. American Society for Engineering Education, Salt Lake City, Utah, June 24–27.

Deep Water: The Gulf Oil Disaster and the Future of Offshore Drilling (2011). Washington, DC: National Commission on the BP Deepwater Horizon Spill and Offshore Drilling.

Great Miscalculations: The French Railway Error and 10 Others (2014). *BBC Magazine*, 22 May. www.bbc.com/news/magazine-27509559. Accessed March 14, 2018.

Groth, D.P. and Hottell, M.P. (2007). How Students Perceive Risk: A Study of Senior Capstone Project Teams. *20th Conference on Software Engineering Education & Training (CSEET'07)*, Dublin.

Kahneman, Daniel (2011). *Thinking, Fast and Slow*. New York: Farrar, Straus and Giroux.

Levy, Matthys and Salvadori, Mario (1994). *Why Buildings Fall Down: How Structures Fail*. New York: W. W. Norton.

Lozano, Juan A. (2017). Witness: Truck Driver in Texas Crash that Killed 13 was Texting. *USA Today*, March 13.

Mars Climate Orbiter Mishap Investigation Board Phase I Report, 6. ftp://ftp.hq.nasa.gov/pub/pao/reports/1999/MCO_report.pdf. Accessed March 19, 2018.

mwreview (2014). *Toaster Causes Emotional Distress*, April 16. www.amazon.com/Sunbeam-3910-100-2-Slice-Toaster-Black/product-reviews/B0007Y17WE/ref=cm_cr_getr_d_paging_btm_next_4?ie=UTF8&filterByStar=one_star&reviewerType=all_reviews&pageNumber=4#reviews-filter-bar. Accessed March 20, 2018.

Petroski, Henry (1992). *To Engineer Is Human: The Role of Failure in Successful Design*. New York: Vantage Books.

Robison, Wade L. (2014). Ethics in Engineering. *Ethics in Science, Technology and Engineering*. 2014 IEEE International Symposium on Ethics. ISBN 978-1-4799-4992-2.

Robison, Wade L. (2016). *Ethics Within Engineering: An Introduction*. New York: Bloomsbury Academic Publishing.

Wamsley, Laurel (2017). Troubled By Flint Water Crisis, 11-Year-Old Girl Invents Lead-Detecting Device. *NPR*, October 20.

41

AUTONOMY IN ENGINEERING

Eugene Schlossberger

41.1. Introduction

Because the notion of autonomy plays an important role in ethical deliberation, it impacts many of the decisions engineers must make. For example, the importance of autonomy underlies the recognition that engineers generally have greater reason to refrain from imposing an involuntarily risk than an otherwise similar risk that is voluntarily borne (Schlossberger 1993). More broadly, some hold that Kant's imperative to show respect for persons as autonomous agents plays the central (Donagan 1977) or a very important (Frankena 1986) role in moral decision-making of any kind. Both Rawls (1971)[1] and Kant (1797) located autonomy at the heart of a just political theory.

In addition to the role that autonomy plays in moral thinking generally, at least nine issues specifically pertinent to engineering ethics focus on questions of autonomy. After examining the nature and value of autonomy, this chapter will discuss each of the nine issues in turn.

A. Engineers' need to *have* autonomy:

    1. Regarding the role of the engineer within the organization (firm/agency/corporation), and
    2. Regarding the role of the engineer within the state or society, e.g., with respect to regulation in managing risk.

B. Engineers' need to *respect* autonomy:

    3. Issues of informed consent,
    4. Respecting the autonomy of individual stakeholders (e.g., consumers or clients), and
    5. Respecting the autonomy of society in decisions that affect it (e.g., nuclear power plants).

C. Autonomous technologies:[2]

    6. Ethical issues regarding the use of autonomous technologies (e.g., self-driving cars and autonomous lethal combat devices), and
    7. Questions about when autonomous machines gain moral status (e.g., merit rights).

D. Technologies that foster/impair autonomy:

    8. Bioengineering technologies (e.g., human cloning and genetic engineering) that may affect human autonomy, and, more generally,

9. Whether we have become too dependent on technology (and how this question should affect engineering decisions).

41.2. Nature and Value of Autonomy

At its most basic, autonomy is the ability of agents to make their own decisions about their own lives. The term "autonomy" is used in more than one way. It is used to designate our capacity to make deliberate choices as well as to indicate a sphere of freedom within which to make such choices. It is because small children are held to have limited autonomy in the first sense that limiting the autonomy of small children in the second sense is justified. Raz (1986) views autonomy broadly as self-creation and partly authoring one's life. Christman (2015), commenting on Feinberg's (1989) identification of four meanings of autonomy,[3] suggests that "central to all of these uses is a conception of the person able to act, reflect, and choose on the basis of factors that are somehow her own (authentic in some sense)."

There are numerous ways of grouping competing accounts of autonomy. Generally, there are three elements of autonomy: the *ability* to select, the *independence* of selection, and the *efficacy* of selection. My neighbor and a prisoner may both choose to spend their weekend camping on a beach (selection), but my neighbor can actually implement her decision while the prisoner cannot. My neighbor's choice has an efficacy that the prisoner's choice lacks. Both my neighbor and a randomizing program in a 1990 PC have the ability to select a number, but my neighbor's selection will have a kind of independence that the PC will not. Different authors call for different conditions on these three aspects for a decision to count as autonomous. *Thin theories* of autonomy place relatively undemanding constraints on these three conditions, while *thick theories* place robust constraints that build in fairly rich demands. For example, Narayan advocates a relatively thin view of autonomy in saying a "person's choice should be considered autonomous as long as the person was a 'normal adult' with no serious cognitive or emotional impairments and was not subject to literal outright coercion by others" (Narayan 2002: 429). Views of autonomy can be relatively thick or relatively thin about each of the three elements of autonomy. A thin theory of selection might require only that two or more options are available, while a rich theory might require that a wide range of reasonably desirable options exist.[4] A thin theory of efficacy might require only that the agent is not formally barred (e.g., by law) from pursuing (not necessarily obtaining) the selected option, while a richer theory might require that the agent have a reasonable chance of succeeding at a feasible cost.

Most controversial are conditions on independence, either from external influences (such as manipulation) or from internal obstacles (the ability to self-govern). Buss (2013) distinguishes between internalist and externalist accounts of autonomy. Externalist accounts regard decisions, desires, and so forth as autonomous only when they have an appropriate relationship to something external to them, such as what caused them. Internalism assesses our autonomy only by the nature of our states themselves. For example, coherence conditions on autonomy require that the person's choice fits some element of his or her outlook, such as the agent's highest-order desires (Frankfurt 1988), judgments about what is most worth doing (Watson 1975), or that agent's "authentic" self. Coherence conditions are thus internalist. Reason-responsiveness conditions require that the choice results from, or that deliberation would be appropriately influenced by, relevant reasons for or against the choice (Fischer and Ravizza 1998). For example, if I autonomously choose to eat a piece of pie but, as I lifted the fork to my mouth, I was told the pie is poisoned, I would appropriately respond to that piece of information. Causation conditions require that the choice not be determined by certain sorts of causes, such as manipulation, threat, or, for incompatibilists such as van Inwagen (1983), any cause other than the agent's free will. (Reason-responsivist and causation conditions are "externalist.") What counts as manipulation or threat is a difficult question. Almost all choices bear some cost. The threat of your not laughing if I don't tell you a joke doesn't defeat the autonomy of my choosing not to tell the joke. Is it the magnitude of the threat that determines whether a choice is autonomous, and, if so, is the criterion the objective harm of what is threatened or the subjective degree to which the agent fears the harm? How large is the threat, objectively or subjectively, of having a spider placed beside one to an arachnophobiac? Epistemological conditions require that the choice be, to some degree, an informed choice, e.g., that the agent be aware of all information that might be relevant to making a rational choice. Epistemological conditions present some of the same problems as manipulation and threat. No human being is omniscient. Which kinds of knowledge deficiency (and to what degree) defeat autonomy?

Some feminists have advocated "relational autonomy," in which social connections are incorporated into the framework of autonomy, either negatively, as internalized oppressive socialization, or positively, as an essential part of being a person. "If we ask ourselves what actually enables people to be autonomous," says Nedelsky, "the answer is not isolation, but relationships—with parents, teachers, friends, loved ones" (Nedelsky 1989: 12). "Autonomy," adds Oshana, "must be regarded, in part, as constituted by social relations that are extrinsic to facts about the psychological states of the individual" (Oshana 2006: 20). (See also Christman 2004.)

The choice between different accounts of autonomy is morally important because each places different conditions on respecting or violating autonomy. If the independence conditions for autonomy require freedom from any form of pressure by others, then Robin Morgan may be correct that, because married partners sometimes had sex when they "wished just to sleep instead . . . Most of the decently married bedrooms across America are settings for nightly rape" (Morgan 1977: 166). Moderately strong epistemological conditions for autonomous choice mean that Ford violated the autonomy of Pinto purchasers and users when withholding from the public information about the vulnerability of the Pinto gas tank.[5] Much stronger epistemological conditions might entail that lying about one's income to a prospective sexual partner would make sex nonconsensual. The difference between thin and thick selection conditions is critical to determining whether engineers have work autonomy, since almost every engineer working on a project has some options but operates within some constraints.

In addition, autonomy, however defined, can be categorized by the nature of what is to be chosen. One important distinction is between mere-choice autonomy, proclamative autonomy, and high-impact autonomy. A mere-choice is making any selection, regardless of the content, such as choosing between chocolate and vanilla ice cream. That mere-choice as such is not morality's highest concern is shown by the fact that trivial choices not tied to moral or personal values may be overridden for a significant end. For example, we would not hesitate, if we can do so without causing pain or frustration, to fool (and thus violate the autonomy of) an alien organism that spends its days aimlessly but freely chooses to pick up and drop either a red or a blue button. The alien chooses freely, but its choice is trivial (Schlossberger 1992). By contrast, a choice that proclaims a value and announces what one stands for, such as choosing to die rather than betray one's country, is a proclamative choice. Insofar as moral agents are worldviews in operation (a lived stance giving meaning to the world), proclamative choices are central to what it is to be a moral agent. A high-impact choice, such as choosing between surgery and chemotherapy, significantly affects the ability to make proclamative choices. Respecting proclamative and high-impact choices carries more moral weight than respecting mere-choices (Schlossberger 2013).

Ethicists differ about the moral importance of respecting autonomy. For some, such as Nozick (1974), it is an inviolable side-constraint on other moral considerations. Some regard autonomy as merely one moral consideration among others. Conly (2012) argues against the priority of autonomy since, because of cognitive bias, people do not reason well about their own goals. Some utilitarians claim autonomy is of only derivative value, important only insofar as respecting autonomy increases utility, perhaps because autonomy is seen as an element of well-being (Mill 1859). Most strict utilitarians insist autonomy (as such) is not a significant moral factor at all. Gewirth (1978) argues that because we pursue purposes, we must value that which makes such pursuit possible, namely freedom. Since we recognize others have the same reason to value their freedom, we must value the freedom of others.

41.3. Engineers' Need to Have Autonomy

Questions arise about the degree to which engineers have (Meiksins and Watson 1989) or should have autonomy within their organizations and within the larger society (including its laws). Freidson (1986) argues that constraints placed on professionals by the organizations in which they work are, by and large, unobtrusive, a conclusion with which Meiksins and Watson (1989) do not fully concur. (In particular, they found that engineers report constraints on autonomy regarding time and record keeping as well as resources and project choice.) Bailyn (1985) distinguishes between design and implementation freedom. Individual engineers might need autonomy concerning their choice of project, design parameters, and constraints (for example, the strict limits on cost, weight, and product development time imposed on the engineering team developing the Ford Pinto resulted in an unsafe gas tank), methods and processes, and deciding when success is achieved and implementation feasible (e.g., the engineering team at Morton Thiokol, who initially decided it was not safe to launch the space shuttle *Challenger* under prevailing conditions, was pressured to reverse their stance).[6] Boisjoly insists engineers should not, under any circumstances, sign off on projects about which they have safety reservations (Boisjoly 1995, 2006), while Lynch and Kline (2000) suggest the emphasis should be on shaping engineering practice to avoid the need for heroic moments when engineers must assert their autonomy against management. For engineering as a whole, Kasher distinguishes between "framework autonomy, conceptual autonomy, and social autonomy": autonomy in determining the nature of a profession, engineering's particular vocation, and how engineers internalize social norms in their practice (Kasher 2005: 89).

Davis (1991) argues that, generally, engineers' duty is to follow their profession's code of ethics (as opposed to deferring to their individual moral judgment), while Schlossberger suggests, "many of the most crucial moral issues are either untouched by or not clearly settled by professional codes" (Schlossberger 1993: 167), although engineers have a special (but not absolute) duty to obey the law (Schlossberger 1995). Luegenbiehl suggests the Western ideal of autonomy may not be cherished elsewhere in the world. He argues that the autonomy of the engineer from managerial orders is conceived as necessary to protect society (guarantee public safety), because managers are concerned with profit and rely on regulation to safeguard society. However, he avers, in parts of Asia there is no history of professionalism or of capitalist greed. Autonomy is not valued as much as flexibility and fitting in. "The identity of the engineer is found in the corporation as a whole" (Luegenbiehl 2004).

Coeckelbergh (2006) argues that the dichotomy between external regulation and increased personal autonomy of engineers is misleading because the relation between regulation and autonomy is complex: regulations are only the outcome of a process involving public debate; implementation requires engineers to resolve within themselves the tension between a professional community lacking social expertise and those outside the profession lacking technical expertise; legal regulation is supplemented by peer pressure, organizational culture, and marketplace demands; and legal requirements, which on the surface limit engineers' autonomy, may enhance an engineer's autonomy in resisting managerial pressure.

Finding an appropriate balance between (a) the autonomy of consumers and society, (b) the legitimate needs and expectations (including marketplace realities) of management in directing the organization that employs engineers, (c) the professional, legal, and institutional ideals to which engineering as a profession is dedicated, and (d) the personal conscience and preferences of individual engineers requires complex weighing on a case-by-case basis. Similarly, legal regulators must steer a course between allowing too much individual discretion (which can result in unfair, biased, and dangerous choices) and rules that are too narrow and unresponsive to differences in actual cases.

41.4. Engineers' Need to *Respect* Autonomy

Experimenting on human subjects requires some degree of informed consent, although ethicists disagree about what counts as "informed" and what counts as "consent" (Eyal 2012). Most ethicists regard the primary justification for requiring informed consent as respecting the autonomy of the patient, although Taylor (2004) argues otherwise. Informed consent is undermined by lies, deception, or withholding of critically pertinent information, by subjects' inability to understand, by pressure, and/or by subjects' lack of opportunity to reflect, ask questions, object, or otherwise withhold consent. Because blind testing requires withholding some information from subjects, a balance must be found between being forthcoming and invalidating the experiment. Respecting autonomy plays a role in deciding when to impose and when to publicize risks, e.g., risks knowingly and voluntarily assumed are less ethically challenging, other things being equal, than risks involuntarily imposed (Schlossberger 1993). While most engineering codes of ethics prioritize or hold paramount safety, almost any process can be made safer at additional cost, inconvenience, or inefficiency. Engineers' professional commitment to safety must be weighed against affording consumers, clients, and other stakeholders some autonomy in choosing how to balance safety with other concerns (Schlossberger 2013). As Martin and Schinzinger (1983) point out, engineers are neither mere order-takers carrying out the dictates of management or society nor all-wise saviors. Rather, the autonomy of engineers is best balanced against the autonomy of society regarding engineering decisions that affect society when engineers function as social enablers and catalysts, working with society and helping society understand the ramifications of engineering decisions. As a result, engineers generally have a collective duty to speak up about engineering issues that affect society (Schlossberger 2015). For example, before a nuclear power plant is built near a residential area, engineers should help the public understand the potential benefits (such as reducing carbon footprint) and risks, both long and short term (such as accidents and the problem of storing nuclear waste), of nuclear power, and participate in (and listen to) the public debate. If, after a thorough and informed public debate, the residents who most bear the brunt of the risks are opposed to the construction of the plant, the fact that the construction of the plant violates their autonomy is a major factor in deciding whether it is unethical for engineers to work on the plant's construction.

41.5. Autonomous Technologies

Autonomous technologies, such as self-driving automobiles and autonomously operating weapons, raise a variety of ethical issues. Self-driving vehicles, which are never tired, impaired, or subject to human shortcomings, can potentially decrease the number of accidents. Lethal weapon systems are generally cheaper, faster, longer lasting, insusceptible to biological or chemical weapons, and more precise (Guetlein 2005). However, Charisi et al. (2017) point out that earlier intelligent machines, such as automated household appliances, functioned within a segregated environment or "working envelope" (e.g., the actions of an automated coffeemaker affect only a small and limited domain) or possessed little capacity to harm the environment. Autonomous cars and lethal weapons break both barriers. Moral theories, the authors claim, are context dependent, and AI systems have trouble with ambiguity. Conforming such systems to societal values, they assert, is the central focus of machine ethics.

Many of the relevant issues are raised by a much-discussed, updated version of the Trolley Problem, made famous by Foot (1967). A self-driving car completing a mountainous turn must decide between hitting and killing two pedestrians or driving off the road and killing the passenger. System designers must balance several potentially conflicting aims: insuring the vehicle's responses to such problems are morally defensible; consistent; situation-sensitive enough to address morally relevant information and uncertainty (risk assessment); transparent; socially acceptable without dampening buyer demand; and reliable and efficient, yet complex enough to accomplish all that is required. Is killing (the passenger) morally worse than allowing (the pedestrians) to die?[7] Should system makers prioritize passenger safety over the safety of others?

> Should [autonomous vehicles] account for the ages of passengers and pedestrians? If a manufacturer offers different versions of its moral algorithm, and a buyer knowingly chose one of them, is the buyer to blame for the harmful consequences of the algorithm's decisions?
> *(Bonnefon 2016: 1576)*

Who, ask Birnbacher and Birnbacher (2017), should decide such questions (whose values should govern), and who is responsible for the consequences?

Some reason exists for thinking that system makers have a fiduciary duty to prioritize passenger safety over the safety of others. Passengers, after all, are being asked to purchase and use the vehicles in replacement of human-driven vehicles. A human driver has the opportunity and the legal right to choose to avoid serious harm to him/herself and vehicle passengers in these situations, especially when the vehicle user is not at fault. It is unclear on what basis designers may take away that right. At the very least, designers would need to respect user autonomy by making the lack of priority for passengers known to users. Users are not in a position to examine and understand the vehicle's programming and must rely upon designers' and manufacturers' commitment to their welfare. Many users would not knowingly and voluntarily sacrifice themselves by using such vehicles, if given a choice (Bonnefon et al. 2016). For these reasons, it seems reasonable to hold that designers have a fiduciary duty to prioritize avoiding serious harm to passengers. It does not follow, however, that absolute priority must be given to avoiding any harm at all to passengers. A driver who killed a pedestrian to avoid incurring a small scratch would be subject to moral censure and legal peril.

Similarly, system design for autonomous weapons must address the moral issues raised by collateral damage and harm to noncombatants. Tonkens (2013) suggests creating autonomous lethal machines is wrong because warfare itself is wrong. Non-pacifists must decide under what conditions autonomous systems may knowingly harm or impose the risk of harming human combatants and non-combatants. An extensive literature, both moral and legal, addresses such questions, from just war theory based on Augustine to contemporary pacifism (Walzer 1977; Allhoff et al. 2013; Fiala 2014). The doctrine of double effect, employed by Aquinas to justify self-defense, is often invoked to defend collateral damage in war. It is sometimes permissible, the doctrine states, to bring about a bad effect (such as the death of a civilian) as a foreseeable consequence as long as one does not intend to bring about that effect. While the doctrine itself is a matter of controversy (McIntyre 2014), applying it to a machine is especially problematic. Does a lethal machine have intentions? Is what counts the intention of the designer, a rational observer, or the deployer of the machine? If the relevant intention is that of the deployer, how high up the chain of command must one search? Arkin (2010) suggests that under battle conditions, robotic systems may ethically outperform human agents for six reasons: robots can act conservatively, as they do not need to protect themselves; are free from judgment-clouding emotions such as anger and fear; can monitor and report ethical violations by others; are in an epistemologically superior position because their sensors are more accurate; can more rapidly integrate information; and are not subject to "scenario fulfillment" (filtering out information and options that do not fit prematurely fixed beliefs). These advantages must be weighed against concerns about ceding responsibility for killing to machines, e.g., creating an accountability gap because robots cannot be held responsible (Sparrow 2007), and making it easier to kill by distancing ourselves from the act.

Additional issues raised by autonomous machines include the resulting human loss of jobs (truckers, bus drivers, chauffeurs, etc.) and the extent to which autonomous machines should deliberately behave imperfectly in order to accommodate human imperfections. For example, driverless cars strictly obeying the speed limit might prompt road rage or dangerous driving on the part of impatient human drivers behind them. Woods (2016) points to the "brittleness problem": because autonomous systems are complex networks of multiple algorithms, feedback loops, and human operators interacting under varying conditions, normal reliability engineering may be insufficient.

Finally, machine ethics must address when autonomous machines attain moral status and should be accorded rights (see, for example, Neely 2014). While AI systems approaching the boundary of moral personhood seem remote at the current time, this question will likely become an urgent one in the future. Extending the debate about the nature of moral personhood beyond animal rights and abortion may well challenge our conception of what it is to be a moral agent.[8]

41.6. Technologies That Foster/Impair Autonomy

Bioengineering in general, and genetic engineering in particular, has fostered concerns of impairing human autonomy, from the unsupported insistence by Kass (1997) that human cloning will produce automatons unworthy of the designation "moral agent," to arguments by Habermas and others, described (and opposed) by Mameli (2007), that "cloning and genetic engineering should be prohibited because these biotechnologies would undermine the autonomy of the resulting child" (p. 87). By contrast, Savulescu (2009) defends genetic engineering because of the enhanced well-being and moral superiority (including greater autonomy in the thick sense—see Schaefer et al. 2014) of the offspring and because moral enhancement is required if humanity is to survive (Persson and Savulescu 2014). Tiefel (1982) argues that, since risks are not definitively known and future offspring cannot give consent, use of technologies such as IVF, cloning, and (Tiefel would surely add) genetic engineering is unethical. But, as Edwards (1974) points out, consent from future offspring cannot be obtained when administering acetaminophen (or pre-natal vitamins) to a pregnant woman. Moreover, if it is unproblematic for a parent to give proxy consent for a toddler, on what grounds is parental consent illicit in the case of genetic engineering or reproductive technology? (Schlossberger 2008). To those who oppose cloning and genetic engineering because they believe it violates autonomy, Michael Sandel answers that "the language of autonomy . . . is ill equipped to address the hardest questions posed by genetic engineering," because "none of us chooses his genetic inheritance" and the autonomy argument fails to address those who choose genetic enhancement for themselves (Sandel 2004: 51; Sandel opposes designer babies in particular on other grounds).

More generally, some fear that increasing dependence on technology will undermine or supersede human autonomy (see, for example, Seely 2013). The term "autonomous technology" is sometimes used by those who believe technology has gotten out of control and has taken on a life of its own (Winner 1977). A 2010 Rasmussen Reports poll indicates 70% of respondents to a telephone poll agreed that Americans are too dependent on technology. Others point to ways in which diverse technologies, such as information and communication technology (Kline and Stokes 2017) and medical enhancement (Eysenbach et al. 2017), increase autonomy. Sussman (1989) believes technology will undermine totalitarian attempts to shackle information, leading to greater political autonomy. Marx[9] appears to hold that, eventually, advances in technology will both eliminate or vastly reduce scarcity and largely free humanity from undesirable tasks, which will be performed by intelligent machines. Technology will thus enable human beings to engage in activities expressing their nature as creative producers, make the state unnecessary, and, in sum, permit a system in which each takes according to his or her needs (since scarcity is abolished) and contributes according to his or her creative abilities. For Marx, then, technology is the ultimate key to unlocking true human autonomy.

Perhaps all of these generalizations are too sweeping. Social media helped undermine totalitarianism during the Arab Spring but, arguably, weakened informed debate during the 2016 US elections. Technologies such as washing machines have freed human beings from some undesired tasks, but other technologies create undesirable tasks human beings must perform. An assembly line worker's task is not necessarily more enjoyable than the work of a skilled craftsperson the assembly line replaced. What can be said is that engineers have some responsibility to consider the ramifications of the projects on which they work (Schlossberger 1997), and the extent to which society and individuals are replaced by, become dependent upon, or are liberated by a technology is one such ramification to consider when deciding to work on that technology.

41.7. Conclusion

Autonomy, itself a complex notion, enters into engineering decisions in a wide variety of nuanced and sophisticated ways. One possible (but controversial) lesson to be taken from this discussion is that ethical decisions in engineering should generally be made by examining how all the relevant factors apply to the particular decision to be made and giving reasons for thinking that some factors may be more pertinent or central to that specific decision, and that a particular option represents the resultant that best balances the relevant ethical vectors in that particular case (Schlossberger 1992). If so, diverse considerations of autonomy are important factors to be balanced in a large range of engineering decisions.

Related Chapters

Chapter 33: Values in Risk and Safety Assessment (Niklas Möller)
Chapter 36: Trust in Engineering (Philip J. Nickel)
Chapter 38: Health (Marianne Boenink)
Chapter 40: Ethical Considerations in Engineering (Wade L. Robison)
Chapter 48: Feminist Engineering and Gender (Donna Riley)

Notes

1. A vast literature debates exactly what Rawls's view of autonomy was and what role it played in his writing. The issue is complicated by the fact that Rawls clarified and/or revised his view several times.
2. The term "autonomous technology" is used here to refer to examples of self-directed technology, such as self-driving cars. The term has also been used to describe "the belief that technology has gotten out of control and follows its own course, independent of human direction by those who see technology in general as out of control" (Winner 1977: 13). This latter issue is discussed in Section 6.
3. Namely, (1) the capacity to govern oneself (impaired, for example, when an agent is drunk), (2) the actual condition of ruling oneself (which, for example, a monk might partially surrender to his superiors), (3) an ideal of character (as when someone is exhorted to be or become more autonomous), and (4) the absolute right or authority to be the sovereign of a certain domain (such as one's own self-regarding decisions).
4. In this regard, Anatole France's reminder that "In its majestic equality, the law forbids rich and poor alike to sleep under the bridges of Paris" (*The Red Lily*, 1894, Chapter 7), sometimes re-parsed as "rich and poor are equally free to sleep under the bridge," is pertinent.
5. For information about the Ford Pinto, see Birsch and Fielder (1994).
6. For information about the *Challenger* disaster, see Boisjoly et al. (1989).
7. There is a vast literature on whether killing is morally worse than allowing to die. For an overview, see Woollard and Howard-Snyder (2016). Brody (1975) viewed the distinction as so stringent that, he believed, it is wrong to abort a fetus (killing) even to save the mother's life (not allowing to die). Rachels (1975) attacked the tenability of the distinction itself.
8. A large literature contains many conflicting accounts of the basis or grounds of moral agency and its relation to moral considerability. See, for example, Jaworska and Tannenbaum (2018), Kittay (2005), and Schlossberger (1992).

9. Controversies abound concerning how to interpret Marx. Key texts on these points are *The Critique of the Gotha Program* (1875) and the *Grundrisse* (1857), especially the portion sometimes referred to as "the fragment on machines."

References

Allhoff, Fritz, Evans, Nicholas G. and Henschke, Adam (eds.) (2013). *Routledge Handbook of Ethics and War*. New York: Routledge.

Arkin, Ronald (2010). The Case for Ethical Autonomy in Unmanned Systems. *Journal of Military Ethics*, 9(4), 332–341.

Bailyn, Lotte (1985). Autonomy in the Industrial R & D Lab. *Human Resource Management*, 24, 129–146.

Birnbacher, Dieter and Birnbacher, Wolfgang (2017). Fully Autonomous Driving: Where Technology and Ethics Meet. *Intelligent Systems, IEEE*, 32(5), 3–4.

Birsch, D. and Fielder, J. (eds.) (1994). *The Ford Pinto Case: A Study in Applied Ethics, Business, and Technology*. Albany: SUNY Press.

Boisjoly, Robert (1995). Commentary on 'Technology and Civil Disobedience: Why Engineers Have a Special Duty to Obey the Law. *Science and Engineering Ethics*, 1(2), 169–171.

Boisjoly, Robert (Updated 2016). Ethical Decisions—Morton Thiokol and the Space Shuttle Challenger Disaster— Index. *Online Ethics Center for Engineering* (orig. May 15, 2006). www.onlineethics.org/CMS/profpractice/ppessays/thiokolshuttle.aspx

Boisjoly, Robert, Curtis, E.F. and Mellican, E. (1989). Roger Boisjoly and the Challenger Disaster: The Ethical Dimensions. *Journal of Business Ethics*, 8(4), 217–230.

Bonnefon, Jean-Francois, Shariff, Azim and Rahwan, Ivad (2016). The Social Dilemma of Autonomous Vehicles. *Science*, 352(6293), 1573–1576.

Brody, Baruch (1975). *Abortion and the Sanctity of Human Life: A Philosophical View*. Cambridge, MA: MIT Press.

Buss, Sarah (2013). Personal Autonomy. *Stanford Encyclopedia of Philosophy*. https://plato.stanford.edu/entries/personal-autonomy

Charisi, V., Dennis, L.A., Fisher, M., Lieck, R., Matthias, A., Slavkovik, M., Sombetzki, J., Winfield, A.F.T. and Yampolskiy, R. (2017). Towards Moral Autonomous Systems. arXiv:1703.04741 [cs.AI] October. https://arxiv.org/pdf/1703.04741.pdf

Christman, J. (2004). Relational Autonomy, Liberal Individualism, and the Social Constitution of Selves. *Philosophical Studies*, 117, 143–164.

Christman, J. (rev. 2015). Autonomy in Moral and Political Philosophy. *Stanford Encyclopedia of Philosophy*. https://plato.stanford.edu/entries/autonomy-moral/#ConVar

Coeckelbergh, Mark (2006). Regulation or Responsibility? Autonomy, Moral Imagination, and Engineering. *Science, Technology, & Human Values*, 31(3), 237–260.

Conly, Sarah (2012). *Against Autonomy: Justifying Coercive Paternalism*. Cambridge, MA: Cambridge University Press.

Davis, Michael (1991). Thinking Like an Engineer: The Place of a Code of Ethics in the Practice of a Profession. *Philosophy and Public Affairs*, 20(2), 150–167.

Donagan, A. (1977). *The Theory of Morality*. Chicago: University of Chicago Press.

Edwards, R.G. (1974). Fertilization of Human Eggs in Vitro: Morals, Ethics, and the Law. *Quarterly Review of Biology*, 40(3), 3–26.

Eyal, Nir (2012). Informed Consent. *Stanford Encyclopedia of Philosophy*. https://plato.stanford.edu/archives/fall2012/entries/informed-consent.

Eysenbach, Gunther, Flott, Kelsey, Crooks, George, Phanareth, Klaus, Vingtoft, Søren, Christensen, Anders Skovbo, Nielsen, Jakob Sylvest, Svenstrup, Jørgen, Berntsen, Gro Karine Rosvold, Newman, Stanton Peter, and Kayser, Lars (2017). The Epital Care Model: A New Person-Centered Model of Technology-Enabled Integrated Care for People With Long Term Conditions. *JMIR Research Protocols*, 6(1), 1–15.

Feinberg, Joel (1989). Autonomy. In John Chrsitman (ed.), *The Inner Citadel: Essays on Individual Autonomy*. New York: Oxford University Press, pp. 27–53.

Fiala, Andrew (2014). Pacifism. *The Stanford Encyclopedia of Philosophy*. https://plato.stanford.edu/archives/win2014/entries/pacifism/

Fischer, J. and Ravizza, M. (1998). *Responsibility and Control: A Theory of Moral Responsibility*. Cambridge: Cambridge University Press.

Foot, Philippa (1967). The Problem of Abortion and the Doctrine of the Double Effect. *Oxford Review*, 5, reprinted in Foot, Philippa. (1978). *Virtues and Vices*. Oxford: Basil Blackwell.

Frankena, W.K. (1986). The Ethics of Respect for Persons. *Philosophical Topics*, 14, 149–167.

Frankfurt, Harry G. (1988). Freedom of the Will and the Concept of a Person. In Frankfurt (ed.), *The Importance of What We Care About*. Cambridge: Cambridge University Press, pp. 11–25.

Freidson, Eliot (1986). *Professional Powers*. Chicago: University of Chicago.

Gewirth, Alan (1978). *Morality and Freedom*. Chicago: University of Chicago.

Guetlein, M. (2005). Lethal Autonomous Systems—Ethical and Doctrinal Implications. *Naval War College Joint Military Operations Department Paper*. Cited by (Arkin 2010).

Jaworska, Agnieszka and Tannenbaum, Julie (2018). The Grounds of Moral Status. *The Stanford Encyclopedia of Philosophy*. https://plato.stanford.edu/archives/spr2018/entries/grounds-moral-status/.

Kant, Immanuel (1797). *The Doctrine of Right*, reprinted in Mary Gregor (ed.) (1996). *The Metaphysics of Morals*. Cambridge: Cambridge University Press.

Kasher, A. (2005). Professional Ethics and Collective Professional Autonomy: A Conceptual Analysis. *Ethical Perspectives*, 12(1), 67–97. doi:10.2143/ep.12.1.583363. PMID: 16619429.

Kass, Leon (1997). The Wisdom of Repugnance: Why We Should Ban the Cloning of Humans. *New Republic*, 2, 17–26, reprinted in Glenn McGee (ed.) (2000). *The Human Cloning Debate*, 2nd ed. Berkeley, CA: Berkeley Hills Books, pp. 68–106.

Kline, Howard, and Stokes, Peter (2017). The Dynamic of ICT and Smart Power: Implications for Managerial Practice. In Varda Muhlbauer and Wes Harry (eds.), *Redefining Management: Smart Power Perspectives*. Dordrecht: Springer, pp.31–46.

Kittay, Eva Feder (2005). At the Margins of Moral Personhood. *Ethics*, 116, 100–131.

Luegenbiehl, Heinz C. (2004). Ethical Autonomy and Engineering in a Cross-Cultural Context. *Techne: Research in Philosophy and Technology*, 8(1), 57–78. https://scholar.lib.vt.edu/ejournals/SPT/v8n1/luegenbiehl.html

Lynch, William T. and Kline, Ronald (2000). Engineering Practice and Engineering Ethics. *Science, Technology, & Human Values*, 25(2), 195–225.

Mameli, M. (2007). Reproductive Cloning, Genetic Engineering and the Autonomy of the Child: The Moral Agent and the Open Future. *Journal of Medical Ethics*, 33(2), 87–93.

Martin, Mike and Schinzinger, Roland (1983). *Ethics in Engineering*. New York et al.: McGraw-Hill Higher Education.

Marx, Karl (1857). *Grundrisse: Foundations of the Critique of Political Economy*, reprinted in David McLellan (ed.) (1977). *Karl Marx: Selected Writings*. Oxford: Oxford University Press.

Marx, Karl (1875). *Critique of the Gotha Program*, reprinted in David McLellan (ed.) (2000). *Karl Marx: Selected Writings*, 2nd ed. Oxford: Oxford University Press.

McIntyre, Alison (2014). Doctrine of Double Effect. *The Stanford Encyclopedia of Philosophy*. https://plato.stanford.edu/archives/win2014/entries/double-effect/

Meiksins, Peter F. and Watson, James M. (1989). Professional Autonomy and Organizational Constraint: The Case of Engineers. *The Sociological Quarterly*, 30(4), 561–585.

Mill, John Stuart (1859). *On Liberty*, reprinted in John M. Robson (ed.) (1963–91). *The Collected Works of John Stuart Mill*. Toronto: University of Toronto Press, London: Routledge and Kegan Paul

Morgan, Robin (1977). *Going Too Far: The Personal Chronicle of a Feminist*. New York: Random House.

Narayan, U. (2002). Minds Of Their Own: Choices, Autonomy, Cultural Practices and Other Women. In L. Antony and C. Witt (eds.), *A Mind of One's Own. Feminist Essays on Reason and Objectivity*. Boulder, CO: Westview, pp. 418–432.

Nedelsky, J. (1989). Reconceiving Autonomy: Sources, Thoughts and Possibilities. *Yale Journal of Law and Feminism*, 1, 7–36.

Neely, Erica (2014). Machines and the Moral Community. *Philosophy & Technology*, 27(1), 97–111.

Nozick, Robert (1974). *Anarchy, State, and Utopia*. New York: Basic Books.

Oshana, M. (2006). *Personal Autonomy in Society*. Hampshire: Ashgate.

Persson, Ingmar and Savulescu, Julian (2014). *Unfit for the Future: The Need for Moral Enhancement*. Oxford: Oxford University Press.

Rachels, James (1975). Active and Passive Euthanasia. *New England Journal of Medicine*, 292(2), 78–80, reprinted in J.M. Humber and R .F. Almeder (eds.) (1979). *Biomedical Ethics and the Law*. Boston: Springer.

Rasmussen Reports (2010). *70% Concerned Americans Have Become Too Dependent on Computers, Electronic Devices*. www.rasmussenreports.com/public_content/lifestyle/general_lifestyle/september_2010/70_concerned_americans_have_become_too_dependent_on_computers_electronic_devices

Rawls, John (1971). *A Theory of Justice*. Cambridge, MA: Harvard University Press.

Raz, J. (1986). *The Morality of Freedom*. Oxford: Clarendon Press.

Sandel, Michael (2004). The Case Against Perfection. *The Atlantic Monthly*, 51–62. www.theatlantic.com/magazine/archive/2004/04/the-case-against-perfection/302927/

Savulescu, Julian (2009). The Moral Obligation to Create Children With the Best Chance of the Best Life. *Bioethics*, 23(5), 274–290.

Schaefer, G. Owen, Kahane, Guy, and Savulescu, Julian (2014). Autonomy and Enhancement. *Neuroethics,* 7, 123–136.

Schlossberger, Eugene (1992). *Moral Responsibility and Persons.* Philadelphia: Temple University Press.

Schlossberger, Eugene (1993). *The Ethical Engineer.* Philadelphia: Temple University Press.

Schlossberger, Eugene (1995). Technology and Civil Disobedience: Why Engineers Have a Special Duty to Obey the Law. *Science and Engineering Ethics,* 1(2), 163–168.

Schlossberger, Eugene (1997). The Responsibility of Engineers, Appropriate Technology, and Lesser Developed Nations. *Science and Engineering Ethics,* 3(3), 317–326.

Schlossberger, Eugene (2008). *A Holistic Approach to Rights: Affirmative Action, Reproductive Rights, Censorship, and Future Generations.* Lanham, MD: University Press of America.

Schlossberger, Eugene (2013). The Right to an Unsafe Car? Consumer Choice and Three Types of Autonomy. *Journal of Applied Ethics and Philosophy,* 5, 1–9.

Schlossberger, Eugene (2015). Engineering Codes of Ethics and the Duty to Set a Moral Precedent. *Science and Engineering Ethics,* 22(5), 1333–1344.

Seely, Mark (2013). *Anarchist by Design: Technology and Human Nature.* Surrey: Old Dog Books.

Sparrow, R. (2007). Killer Robots. *Journal of Applied Philosophy,* 24(1), 62–77.

Sussman, Leonard R. (1989). *Power, the Press and the Technology of Freedom: The Coming Age of ISDN.* Washington, DC: Freedom House.

Taylor, J.S. (2004). Autonomy and Informed Consent: A Much Misunderstood Relationship. *The Journal of Value Inquiry,* 38, 383–391.

Tiefel, H.O. (1982). Human in Vitro Fertilization: A Conservative View. *Journal of the American Medical Association,* 247, 3235–3242.

Tonkens, Ryan (2013). Should Autonomous Robots Be Pacifists. *Ethics and Information Technology,* 15, 109–123.

Van Inwagen, Peter (1983). *An Essay on Free Will.* Oxford: Clarendon Press.

Walzer, Michael (1977). *Just and Unjust Wars: A Moral Argument With Historical Illustrations.* New York: Basic Books.

Watson, G. (1975). Free Agency. *Journal of Philosophy,* 72, 205–220, reprinted in Watson, Gary (2004). *Agency and Answerability.* Oxford: Oxford University Press, 13–32.

Winner, Langdon (1977). *Autonomous Technology.* Cambridge, MA: MIT Press.

Woods, David D. (2016). The Risks of Autonomy: Doyle's Catch. *Journal of Cognitive Engineering and Decision Making,* 10(2), 131–133.

Woollard, Fiona and Howard-Snyder, Frances (2016). Doing vs. Allowing Harm. *The Stanford Encyclopedia of Philosophy.* https://plato.stanford.edu/archives/win2016/entries/doing-allowing/

42

STANDARDS IN ENGINEERING

Paul B. Thompson

The term "standard" covers an array of conventions that specify, determine and oversee engineering practice in virtually all domains of application. In its original meaning, the word "standard" referred to the flag or banner carried to identify the forces of a given chieftain or leader. A silversmith, cooper or other craftsman might analogously apply a distinctive mark to identify the products of his labor. A maker's mark came to convey a number of things. It might signify the quality of workmanship in the artifacts bearing the mark, but standards of quality are in many respects different from the standards that govern engineering practice today. Each of the craftsman's artifacts were one-off works of individual skill, and even products that were quite similar in appearance had many unique qualities: minor differences in size, shape or composition of materials. When craft industries began to utilize common marks to indicate the purity or mixture of metals being used, they took a major step toward contemporary practices of standardization.

There is also a fairly obvious connection between the general idea of standards and morality. Dictionary definitions of the word "standard" include morals, ethics and habits established by authorities, custom or individuals. The National Society of Professional Engineers Code of Ethics comprises twenty written principles divided into three main categories: fundamental canons, rules of practice and professional obligations, with the latter categories including many sub-rules (NSPE 2017). In contrast, technical standards are documents that establish criteria for products, material processes and systemic interaction of devices, appliances or techniques. Since engineering codes of conduct are covered elsewhere, this chapter will concentrate on the ethical implications of technical standards, including non-material and social practice standards such as workplace rules or contractually enforced requirements.

Virtually every aspect of contemporary engineering practice is governed by standards. Some standards dictate performance criteria for engineered systems and their components, such as the tensile strength or conductivity of a particular material. Other standards may apply to processes. They can apply either to material processes such as those being used in manufacturing (e.g. the temperature of a process must not fall below a given level) or to social processes, such as workflow management or required rest periods for key workers. Standards can be created to ensure technical interoperability of components in a single appliance or among a number of different devices having multiple applications. Indeed, the range of activities subject to standardization defies easy summary or classification.

42.1. Technical Standards

The most common models for engineering standards specify properties or characteristics that must be exhibited in a product, work process or design. Properties are specified through measures that include physical objects (such as a ruler for defining length) and experimental operations that bear some defined relationship to the performance of artifacts and engineered systems. Criteria for strength or a material's ability to carry loads are critical features of products and designs of built structures and most technical artifacts. Such criteria can be specified as standards that specify ability to resist breakage, crumbling or deformation under stress. In other areas, standards may specify features crucial for basic functionality of a device. For example, standards for signal transmission continue to play a crucial role in the development of telephones, television and, more recently, all manner of electronic communication devices. Whether devices allow two-way exchange (as in the telephone as compared to television), and whether information identifying the type of signal (e.g. audio, video) can be imbedded within the signal itself (as it is for computers) as opposed to being designed into the sending and receiving artifact (television) all depend upon malleable properties of signal transmission. The creation of the Transmission Control Protocol (TCP) standard allowed signal transfer for innumerable distinct applications (e.g. video, text, audio, computation) to be distributed across an indefinitely large number of physical media, as distinct from the dedicated lines that characterized early signal transfer standards for telephony.

In any case, performance criteria become technical standards when they can be associated with measurable indicators or tests. As such, the standard may be literally *defined* in terms of testable quantities, such as pulses per millisecond or units of force. While engineering involves considerable creativity in development of technical designs, virtually every aspect of engineering practice is circumscribed by tests that certify compliance with prior criteria for devices and processes of a given type. Although engineers' training equips them to work to standard, and practicing engineers rapidly become knowledgeable about the standards relevant to their professional specialization, the source of the particular criteria that govern their activity may be largely unknown to them. Engineers who simply work *to* the tests that determine standards of engineering practice may presume that the development and adoption of a standard is itself a purely technical matter, devoid of normative or ethical significance.

However, all standards involve significant human judgment and social coordination at the time of their creation, and in some cases they are highly contingent upon values and interests of the people who participate in the standard setting process. In the model that has come to govern a large part of engineering practices, standards are created by independent *standards developing organizations* (SDO) such as the American National Standards Institute (ANSI), ASTM International (originally named the American Society for Testing and Materials) and the International Organization for Standardization (ISO). TCP, for example, is currently administered through the Internet Society, a non-governmental membership organization, through a subgroup known as the Internet Architecture Board. There are thousands of non-governmental SDOs operating in the United States alone, and they operate under many different rubrics. Alternatively, standard setting may be done by a governmental agency such as the Standards Administration of China (SAC). In some cases, standards development may be conducted under the aegis of a professional society such as the Institute of Electrical and Electronics Engineers (IEEE), which provides a basis of support and oversight for standards development. Trade associations also sponsor standards development, and in some cases development of standards occurs on a self-funding or for-profit basis. When the SDO is not a governmental entity, standards are often proprietary forms of intellectual property that must be licensed for use in engineering or other commercial practice.

Standards produced by SDOs are written documents developed and adopted through a formal process that is defined by the organization itself. A formal standard includes defined indicators or tests for compliance. Some SDOs also function as *standards setting organizations* (SSO), that unify practice in standards for a given technical domain and that include users in the process of settling on a standard that will have widespread (especially international) application. The difference between an SDO and an SSO may seem subtle, but when two or more incompatible standards have been developed and implemented, there may be a need for a coordinating body to resolve incompatibilities and establish a single standard. The *Codex Alimentarius*, for example, is a body at the Food and Agriculture Organization (FAO) of the United Nations that develops uniform standards for food safety. ISO functions as both a standards development organization and as a body for setting international standards. In its capacity as an SDO, ISO has generated standards for engineering and business practices. One of the best known is ISO 9000, actually a series of specific standards intended to standardize quality control and customer service across a broad range of industries. But ISO also serves as an international coordinating body where competing standards can be renegotiated and harmonized, especially when incompatible standards have been developed or implemented in different countries.

Technical standards are also developed *within* for-profit firms and may attain widespread influence through entirely informal practices. Electronic playback or display formats (such as Sony's Blu-Ray or Adobe's PDF) and computer operating systems (such as Microsoft Windows or Apple OS) are developed within firms as part of the normal course of product development, but as the products attain widespread use, they attain the status of standards that are applied throughout an industry. Such informal standards can become pervasive when a particular company's product so thoroughly dominates a given market that competitors must emulate specific design or performance characteristics in order to participate. Although there may be no written or expressed statement, these informal conventions become *de facto* standards. Failure to conform or comply is generally sanctioned through success or failure in markets, but non-compliance can occasionally be ruled unlawful. In fact, legal standards governing such things as product liability include many informal criteria that are applied through the judgment of the courts.

In some domains of practice, certification that a standard has actually been met is done by the SDO itself, but in many cases certification is contracted to third-party firms. The nature of certification varies according to the industry and the standard in question. Certification may include physical testing of products, an audit of documents (such as engineering reports or internally produced test results) or an on-site inspection of facilities, work processes or other evidence pertinent to the standard in question. Third-party certifiers are often paid by the firm being certified, creating an incentive for laxity in the certification process that undermines the integrity of the standard. Thus, in most jurisdictions, certifiers are themselves overseen by an accreditation body that reviews their activity. Accreditation is often done by government or quasi-governmental agencies. The framework of SDOs, certifiers and accreditation creates what is known as a tripartite standards regime (Loconto et al. 2012).

42.2. What Standards Do

Both formal and informal standards are used to accomplish designated ends that SDOs and for-profit firms were *trying* to achieve, but standards also have unintended consequences. Most obviously, standards are intended to ensure that a technical artifact or practice possesses a desired performance attribute. Safety is foremost among the attributes sought by standard setting. Fire safety, structural integrity and toxic load are among the qualities governed by technical standards. Standards that have been developed to promote safety are the most likely to become legal requirements. In the parlance of civil, chemical and mechanical engineering practice, such legally mandated standards are referred to as *codes*. Similarly, other public goods may be sought by developing tests intended to ensure the quality of a given practice. The so-called "stress tests" applied to financial institutions illustrate how technical standards are being applied to more and more areas of everyday life (Quagliariello 2009).

Beyond various safety and protective goals, the main objective for standardization is to ensure the interoperability of technical artifacts and technical systems. Interoperability was the key to the manufacture of clocks, firearms and other artifacts using interchangeable parts—an innovation apocryphally attributed to Eli Whitney (Woodbury 1960). It is also crucial to supply relationships in which articles manufactured by one firm are used in the products of another. In the contemporary world, standards such as TCP, PDF and the operating systems for personal computers have created platforms that permit the interoperability of an astounding array of devices and applications. In this respect, standards for the functionality of technical systems have the unintended effect of creating new potential for innovation, including the invention of the novel devices and uses we see with smartphones and the Internet.

The standardization associated with interchangeable parts also had the effect of creating enormous cost savings in the manufacturing process. Efficiencies of many kinds can be both a goal and a byproduct of standardization. Under competitive economic conditions, savings in the cost of materials, energy or labor will lead to price reductions, making technology affordable to a larger number of potential users. These efficiencies thus extend the benefits of technology more broadly and allow consumers to spend money that might have gone to purchasing a high-cost good on other things. This increase in consumption in turn fuels the process of economic growth, creating markets for entirely new types of goods (as well as the employment needed to produce those goods). It is thus not an exaggeration to say that technical standards have played a significant role in improving quality of life across the board, fueling processes of economic development that are central to the economies of industrialized societies.

At the same time, standards can stifle competition because the licensing can create barriers to entry for new firms. By the same token, dominant standards may present disincentives for technical innovation. Standardized technology may achieve consistency in product attributes while driving higher-quality products out of the marketplace (Ronnen 1991). When standards are coupled with automation, they can de-skill manufacturing processes, allowing employers to substitute low-wage workers or even replace them entirely with mechanized assembly lines. Standards are also applied *to* workers in the form work rules and manufacturing design. Frederick Winslow Taylor (1856–1915) developed time and motion studies in order to standardize work routines and eliminate wasted movements on assembly lines, but the putatively optimal way to perform a physical task was rapidly deployed in service to workplace discipline (Kanigel 2005).

While standards can be the source of efficiencies, simply having a standard does not imply that the standard is being met. Standards are thus accompanied by audits and testing procedures that constitute sometimes elaborate and costly certification processes (Loconto et al. 2012). Michael Power has argued that while there are clearly cases where verified standards do indeed deliver benefits discussed previously, there is an ongoing expansion of standards and auditing into many domains where the cost of compliance exceeds the benefits, where certifications become ritualized charades that do not, in fact, assure quality and where tests and indicators are both vague and inadequately tied to the desired characteristics to be meaningful (Power 1997).

42.3. The Logical Properties of Technical Standards

Evaluating standards begins by examining what a given standard does relative to the type of judgment or purpose it was intended to serve. To this end, it is worth noting some very general features of technical standards. Lawrence Busch argues that a standard can reflect one of four distinct types of normative judgment: filters, optima, ranks and divisions. Each of these forms reflect general goals or purposes for which more specific technical standards might be developed. Busch's classification system produces the easy-to-remember acronym FORD (though Busch referred to optima as "Olympic" standards). He argues that all technical standards can be placed into one of these four categories (Busch 2011).

A filter articulates a threshold or baseline that a product or process must meet in order to be deemed acceptable. For example, a toxicological standard may specify the parts per million (or billion) of a hazardous substance that can be permitted in air, water or another fluid medium.

Samples that exceed this level fail to meet the standard. The goal for developing the standard is thus to ensure some minimum or maximum of some measurable quantity exists, but the standard does not address quality differences above or below the specified threshold. Many health and safety regulatory standards function as filters. Indeed, filters are probably among the most common type of standard.

As explained by Busch, optima articulate the specific criteria that will be used to determining the best or "the winner" among possible candidates for judgment. The winner of the race is "the first one to cross the finish line," assuming other criteria (such as staying in one's lane and following the rules) are met. There, of course, many ways to understand "the best" or optimality, and the point of optimization standards is to define the criteria that are to be deployed in a particular case. Optima emerge when performance criteria need to conform to quite narrow tolerances. For example, standards for signal transmission often need to be quite specific in order to assure interoperability; hence, specifying a winner among alternatives becomes a practical necessity. As in sporting competitions, optimal standards may be specified in advance of any actual design or process. Dava Sobel's book *Longitude* discusses how the British government's Board of Longitude established standardized criteria for solving a technical puzzle in navigation, with John Hamilton's chronometer being the first design to claim the prize offered to the first person who could meet the standard. Alternatively, an existing device that embodies a number of different features may function as an optimum, creating a *de facto* standard. However, optima may evolve into filters as a competitive space for performance develops. Sony's Blu-Ray standard for high resolution DVDs eventually became the single winning standard, dominating the marketplace. One area where optima are commonplace in engineering practice is in developing tests for certifying compliance with other standards (be they filters, ranks or divisions). The standard being met may be a threshold, but only one testing method is deemed to be superior to others in ascertaining whether or not that threshold has been met.

Ranks categorize candidates for judgment into ascending and descending categories of quality. Divisions also categorize candidates, though with no corresponding judgment of better or worse. Grading schemes are a common instance of ranks, where individual items will be classified (A, AA, AAA) in a manner that allows them to be sorted by whatever level of quality or performance may be requisite for a task at hand. Bolts and fasteners are graded by tensile strength, for example, allowing engineers to weigh cost savings against functional needs in a given application. Alternatively, ranks may be intended to incentivize higher performance while still permitting lower levels. The widely known LEED standards for architectural design and performance of built structures provide a clear example of ranks, where "silver," "gold" and "platinum" represent ascending categories of compliance with a complex set of characteristics intended to reduce the environmental impact of built infrastructure (Cidell 2009).

Divisions arise when there is a need to compartmentalize phenomena but no reason to apply criteria for ranking one category over another. One of the most interesting studies on standards concerns the development of standards for disease classification. For reasons that involve diagnostics, allocation of healthcare resources and billing for services, standards for categorizing patients according to particular disease classifications began to emerge in the 1950s, becoming pervasive throughout healthcare in the industrialized world during the 1970s and 1980s. The development of these divisions involved extensive scientific and ethical debate and had a significant impact on healthcare delivery. Patients who did not fit well into the classification schemes arguably suffered from these efforts. Yet there was never any sense in which being classified into one disease category was a respective quality judgment (Bowker and Starr 1999).

Disease classification illustrates how the borderline case was a problem with ethical significance for divisions. Borderline cases, however, are endemic to all four types of standardized judgment. There are always individual items that either "nearly" or "just barely" meet a standard; hence, there is always the potential for classifications that do not conform to the intent of a standard. In the technical world, this problem is dealt with both by developing specific *tests* that answer the question of whether or not a standard has been met in a particular case, as well as *tolerances* that attempt to specify the degree of accuracy that must be achieved in applying the standard (Busch 2011). While a clinician may *treat* the disease in a non-standardized way, Bowker and Starr argue that standardized classification systems designed to serve administrative or insurance purposes can constrain more artful practices, sometimes to a patient's detriment. Tests and tolerances are, in effect, standards within standards: they dictate how a given standard is to be used by substituting technical procedures for human judgment.

42.4. Ethical Significance of Standards

Many standards are introduced in order to satisfy ethically significant dimensions of technical performance. Buildings and bridges are expected to bear specified loads not only to satisfy technical needs, but because builders, building owners and municipalities have an ethical responsibility to prevent foreseeable injuries and loss of life. Similar rationales apply to standards in toxicology, fire safety, engineering control systems and, indeed, many of the technical domains in which engineers practice. However, the efficiencies brought about through standardization are also supported by a broad ethical rationale. When standards reduce the cost of manufacturing or using a technical product, resources can be expended on other goods. Technical efficiency is, in general, supported by the utilitarian maxim to achieve "the greatest good for the greatest number" of affected parties.

However, as already indicated, developing and implementing standards itself comes with a cost. There are, first, the direct costs of convening experts to specify the criteria and measures for a given standard, and, second, the costs that users incur both in licensing the standard and in undertaking whatever is needed to be in compliance with it. The costs for standards development influence who has a say in the process. Compliance costs include both the provision of key elements (e.g. a fire code may require that fire extinguishers or sprinklers be installed, both of which entail costs) and also the expense of inspection, certification and record keeping needed to demonstrate that compliance has been achieved. As such, application of the utilitarian ethical rationale for technical efficiencies requires one to consider whether the cost reductions achieved through standardization are, in fact, greater than the additional costs associated with developing standards and then demonstrating compliance with them. As Michael Power (1997) has argued, the enthusiasm for regulatory oversight can lead us to neglect such trade-offs.

The tripartite standards regime itself is a complex institutional arrangement for addressing potential conflicts of interest that are inherent in technical standards. SDOs aim to create standards that satisfy demand for quality assurance, and firms that license a standard benefit from the confidence that such assurance lends to their products and services. But firms also have incentives to cut costs by marketing substandard products, and they may conspire with SDOs or certifiers to deceive buyers by implying that standards are either more rigorous or more adequately met than they actually are. This conflict of interest illustrates one way in which standards must be informed by ethics in order to achieve their intended result. Accreditation agencies may be brought in to add an another layer of oversight, but at the end of the day, the layering of monitoring activity depends upon trust relationships. The ethics of standards are thus embroiled in the problem of *Quis custodiet ipsos custodes*, or Who will guard the guardians? (Hackman and Wageman 1995).

Beyond these straightforward trade-offs and conflicts of interest, standardization is subject to classic ethical problems in the distribution of burdens. One way of conceptualizing the problem emphasizes fairness in distributing benefits and burdens across the population of affected parties. Standardization could, on this view, be ethically problematic when losers are not compensated. For example, imagine an international negotiation to reconcile competing technical standards for some kind of signal transmission. Suppose that a relatively small company from a newly industrializing country has entered the market first and had some initial success, but they must now negotiate with engineers from a very large and well-capitalized technology multinational. Though our case is described as a thought experiment, it is a situation that is not unusual in standards harmonization. The small company will be at both a financial and a political disadvantage and will need to shoulder comparatively larger burdens simply to participate in the standards process.

An alternative way to approach the ethical question of distributing the burdens of standardization is to identify a moral criterion that is independent of the standardization process, such as individual rights. If the effect of standardization is to deprive individuals of a moral right, then efficiencies achieved through standardization would become irrelevant: the entire process of standardization would be judged morally unjustifiable. For example, in recent years, many technical standards that denied access to people with physical handicaps have been overturned, sometimes at significant cost to the efficiency of a technical process. An example would be standards for curb height and access by wheelchairs. Cases where standards are alleged to violate individual rights are likely to involve controversy over whether the alleged right is, in fact, morally actionable. Standardization can achieve technical efficiencies that lead to loss of jobs or deskilling of the workplace, for example, but the question of whether these outcomes violate individual rights is hotly disputed (Wickramasekara 2009).

42.5. Standards, Power and Social Justice

Technical standards become implicated in power relations in two ways. First, standards are like any set of rules or laws in terms of their power effects. The role that standards play in ensuring health and safety, and the consequences they can have for workers and for firms' ability to enter markets, means that those who control the development, implementation and verification of standards have direct influence over what other people can do, and whether they are exposed to risk of harm or economic loss. The principal difference between technical standards, on the one hand, and laws or policies, on the other, is that technical standards are often developed and adjudicated by engineers as part of their professional capacity. They give power to engineers in virtue of previously discussed aspects, and hence engineering judgment merges with the criteria of justice in standard setting.

This implies that principles of justice that are commonly applied to governmental decision-making are at least potentially applicable within the process of standard setting. For example, standards should be developed and implemented so as to avoid undue interference in the liberty of affected parties. They should not be used so as to place individuals at an unfair disadvantage. In the past, civil engineers may have developed curb and gutter standards without much thought to their impact on the mobility of people who are confined to wheelchairs. They have been focused solely on flow rates and average rainfall. But the infrastructure created in compliance with these standards was judged to be unfair, unjust, by at least one standard of justice. When technical standards are recognized to have many of the same impacts as laws and social policies, it becomes feasible to ask standard setting groups to include justice in their deliberations (Thompson 2012).

The second way in which standards raise questions of justice is with respect to network power. Network power arises unintentionally when a large number of individual decisions coalesce in a pattern that has coercive effects on others. For example, writers who wish to have their work published today must submit texts that conform to one of the standard formats for electronic files: docx, PDF, or TXT. There is no law that demands this, but the mere fact that publishers and the writing community have adopted these formats so widely means that someone who wished to avoid complying with the standards for electronic texts would be barred from access to virtually all publishing opportunities. Network power is thus a form of coercion associated with social relations, rather than a form of power wielded by individuals, organizations or offices. David Grewal has argued that both formal and *de facto* standards can be associated with network power. He cites the use of English as a *de facto* international standard that has significant discriminatory effects on non-native speakers, who must bear significant costs to participate in many international business and technical forums (Grewal 2008)

None of this is to imply that firms or groups can never develop standards with the intention of making profits, gaining an economic edge over competitors or fulfilling a cultural or aesthetic value. However, standards should neither cause systemic harm to third parties nor have discriminatory effects on groups defined by race, gender, ethnicity or religious affiliation. Indeed, balancing the intended goals of a standard against criteria of just power distributions is itself an instance of the need for justice in deliberative judgment. The qualifications noted in this section are doubly true for standards developed and administered by government authorities, where rulemaking for technical efficiencies is conjoined with the coercive authority of the state.

42.6. The Ethical Evaluation of Standards

Technical and quasi-technical standards are not always recognized to have ethical significance. Once adopted, they structure the work activity of engineers so thoroughly that it can be difficult to imagine how things might have been different. Yet technical standards have ethical implications for three broad reasons that comprehend a number of more parochial concerns. First, they are developed, adopted and enforced in order to achieve certain goals, and the purposes those goals serve may be ethically legitimate or illegitimate. They may furthermore be laudable and progressive, or unambitious and even regressive, when viewed in light of larger ethical aspirations. Second, the processes for developing and implementing technical standards can themselves be judged according to criteria of fairness and justice. Finally, technical standards produce consequences or outcomes for the way that technology affects the health, wealth and well-being of its users, as well as third parties. Those outcomes can be praiseworthy, blameworthy or ethically neutral. Simply acknowledging these implications is an important element of engineering ethics.

However, engineering ethics should take further steps to include discussion of and debate over the strengths and weaknesses of particular standards. Do they, in fact, accomplish the health, safety and public interests they may have been intended to protect, and do they do so in ways that exceed the inconvenience and transaction costs that were the focus of Powers's critique? Are standards being deployed in order to achieve cost efficiencies at the expense of worker well-being? If so, are these impacts justified by some off-setting benefit to the public at large? While a line engineer may not be in any position to question or resist standards that are already in place, engineering ethics educators must note that it is often senior engineers who, with attorneys and management, are in a position to develop and set standards. The appropriate forum for ethical evaluation of standards is thus not in the work practice of the entry-level engineering professional, but in the role that technical expertise plays in the development and implementation of new technical standards.

The legal scholar Lawrence Lessig has undertaken an extensive evaluation of the TCP standard and its implications for access and control of the Internet. While the standard was developed with the goal of making the entire system of interconnected computers less vulnerable to disruption or failure through physical causes (such as a power failure), the TCP also had the feature of making it relatively difficult to control or even monitor signal traffic on the system in a centralized way. Lessig argues that this feature of TCP lay at the heart of subsequent ethical and political debates about control of the Internet. Those who saw the Internet as a place for free and unfettered communication and creative expression found the openness and uncontrollability of the early Internet as an ethically positive feature. Those who wanted to monitor the activity of citizens or employees were less supportive of this feature. What Lessig refers to as the democratic features of the Internet also presented early challenges to the development of security features that were crucial for commercial use, such as the financial transfers associated with buying or selling merchandise (Lessig 1999).

This history of the Internet since Lessig wrote in the waning years of the last century demonstrates how innovation and refinement of technical standards can achieve a wide variety of performance objectives. Internet commerce is now commonplace, and both firms and repressive governments have achieved regulatory control over significant components of Internet use. At the same time, the emergence of platforms such as Facebook, Twitter and YouTube have effectively layered new standards-based capabilities onto the basic systems for transferring files among interlinked computers, in some cases even increasing users' ability to develop unmonitored and difficult-to-trace applications. However, Lessig's more general point still holds: these ethically controversial uses and forms of control are among the various affordances we now associate with networked computers. The underlying architecture that either creates or frustrates the emergence of a given capacity is a function of standards—both formal standards such as TCP and *de facto* standards such as PDF. The engineers responsible for developing these standards may or may not have appreciated the ethical significance of their work, but future engineers should understand that in creating a technical standard, they are undertaking an engineering practice of enormous importance and with significant ethical implications.

Indeed, at this writing, the developers of social media platforms are struggling with the realization that their internal computer codes may have left the platform vulnerable to manipulation. They have had to modify systems after they have become tools recruiting terrorists, and have learned that the democratic value of openness also carries the risk of sensationalizing acts of bullying, sexual abuse and murder. In similar fashion, developers of operating systems face continued need for modification and updating as flaws and vulnerabilities in the systems are discovered by hackers, some with malicious intent (Shao et al. 2017). Computer code may or may not rise to the level of a formal or *de facto* standard, though it is difficult to deny that the Windows and OS systems marketed by Microsoft and Apple, respectively, function as standards for the interoperability of many otherwise distinct technical applications. Yet the controversies over whether platforms can or cannot be controlled, and when the technical staff in charge of a platform have an ethical responsibility to monitor use, are emblematic of the ethical questions that can and should be raised in the course of developing many technical standards.

42.7. Ethics, Standards and Engineering Education

The ethical dimensions of standardization notwithstanding, standards and standard setting are not at this time featured elements of engineering ethics education. Given the role of technical standards in structuring the development and applications of modern technology, education about standards and standard setting should arguably become a component of *general* education, a competency related to effective understanding of the sociotechnical environment in which citizens currently live. However, one study indicates that administrators of most educational programs in both arts and sciences are themselves unaware of how standards come to embody and express power relationships, or of how they shape the potential policy responses of governments. Although engineering faculty discussed in the study demonstrated cognizance of the significance of standards and SDOs in their profession, they stated that education about these elements of engineering were not part of current curricula (Thompson et al. 2015).

The features of standards and the ethical issues discussed in this chapter suggest that standards development and harmonization is actually one of the most important forums in which engineering can have morally significant and enduring impact. SDOs and SSOs are a key forum for the practice of engineering ethics, and it seems appropriate that the ethical aspects of standards should be a component of engineering ethics.

Related Chapters

Chapter 27: Measurement (Lara Huber)
Chapter 32: On Verification and Validation in Engineering (Francien Dechesne and Tijn Borghuis)
Chapter 36: Trust in Engineering (Philip J. Nickel)
Chapter 50: Engineering and Social Justice (Caroline Baillie)
Chapter 51: Engineering and Environmental Justice (Benjamin R. Cohen)

Further Reading

Busch, Lawrence (2011). *Standards: Recipes for Reality*. Cambridge, MA: MIT Press. (Busch's book is the only comprehensive overview of standardization practices and processes and of the organizational infrastructure that supports them. Not limited to engineering standards, the book places the standardization processes that support engineering in a broader social context.)

Lessig, Lawrence (1999). *Code and Other Laws of Cyberspace*. New York: Basic Books. (Lessig's widely read study examines the social implications of file transfer protocol (FTP) as well as how alternative technical standards for constructing information systems serve different social ends. Though technically outdated, the book remains a model for ethical analysis of a putatively technical standard.)

Stone, John V., Loconto, Alison and Busch, Lawrence (2012). Tri-partite standards regime. *Wiley-Blackwell Encyclopedia of Globalization*. Online. doi:10.1002/9780470670590.wbeog919. (This short essay provides a clear explanation of the relationship between standards development, certification and the accreditation of non-governmental standardization processes. It explains how governments can either control standard setting or allow competition among standard settings to flourish.)

Thompson, P. B. (2012). "There's an App for That": Technical Standards and Commodification by Technological Means. *Philosophy and Technology*, 25, 87–103. (This article argues that processes of commodification—restructuring goods, services and relationships so that they can be bought and sold—are sometimes achieved by creating technical standards. In addition, standard setting is identified as the forum in which many of the politically most significant aspects of technological innovation are decided.)

References

Bowker, G.C. and Starr, S.L. (1999). *Sorting Things Out: Classification and Its Consequences*. Cambridge, MA: MIT Press.

Busch, Lawrence (2011). *Standards: Recipes for Reality*. Cambridge, MA: MIT Press.

Cidell, J. (2009). A Political Ecology of the Built Environment: LEED Certification for Green Buildings. *Local Environment*, 14(7), 621–633.

Grewal, D. (2008). *Network Power: The Social Dynamics of Globalization*. New Haven, CT: Yale University Press.

Hackman, J.R. and Wageman, R. (1995). Total Quality Management: Empirical, Conceptual, and Practical Issues. *Administrative Science Quarterly*, 309–342.

Kanigel, R. (2005). *The One Best Way: Frederick Winslow Taylor and the Enigma of Efficiency*. Cambridge, MA: MIT Press.

Loconto, A., Stone, J.V. and Busch, L. (2012). Tri-partite Standards Regime. *Wiley-Blackwell Encyclopedia of Globalization*. Online. doi:10.1002/9780470670590.wbeog919.

NSPE (National Society of Professional Engineers) (2017). *Code of Ethics*. www.nspe.org/resources/ethics/code-ethics. Accessed September 14, 2017.

Power, M. (1997). *The Audit Society: Rituals of Verification*. Oxford: Oxford University Press.

Quagliariello, M. (ed.) (2009). *Stress-testing the Banking System: Methodologies and Applications*. Cambridge, MA: Cambridge University Press.

Ronnen, U. (1991). Minimum Quality Standards, Fixed Costs, and Competition. *The RAND Journal of Economics*, 490–504.

Shao, C., Ciampaglia, G.L., Varol, O., Flammini, A. and Menczer, F. (2017). The Spread of Fake News by Social Bots. *arXiv preprint arXiv:1707.07592*.

Sobel, D. (1995). *Longitude: The True Story of a Lone Genius Who Solved the Greatest Scientific Problem of His Time.* New York: Walker.

Thompson, P.B. (2012). "There's an App for That": Technical Standards and Commodification by Technological Means. *Philosophy and Technology,* 25, 87–103.

Thompson, P.B., Stone, J.V. and Busch, L.M. (2015). Standards Education in the Liberal Arts: Curricular Materials and Educational Strategies. *Standards Engineering,* 67, 12–15.

Wickramasekara, P. (2009). Development, Mobility, and Human Rights: Rhetoric and Reality. *Refugee Survey Quarterly,* 28(4), 165–200.

Woodbury, R.S. (1960). The Legend of Eli Whitney and Interchangeable Parts. *Technology and Culture,* 1(3), 235–253.

43

PROFESSIONAL CODES
OF ETHICS

Michael Davis

Engineering has had formal codes of ethics for at least a century. Some are professional codes (or codes of professional ethics). Some are not. Among the earliest was the "Code of Principles of Professional Conduct" of the American Institute of Electrical Engineers (AIEE), March 8, 1912. While codes of ethics for engineers still seem to be more common in North America than elsewhere, they did appear, within a decade or two of the AIEE's, in Europe and Asia. For example, Norwegian engineers adopted the "Etiske Retningslinjer" ("Ethical Guidelines") in 1921; the Chinese Institute of Engineers did something similar in 1933. Today, such codes can be found from Brazil to Canada, from Japan to Germany, from Mexico to India.[1]

A code of ethics may appear in engineering under a variety of names, not only "code of ethics," "ethical guidelines," or "principles of professional conduct," but also "rules of practice," "ethical canons," "standards of conduct," and so on. A professional code of ethics for engineers may, or may not, include the word "professional," "ethics," or "engineer." The name of a document does not settle whether it is a professional code, a code of engineering ethics or, indeed, even a code at all. We must, then, say more about "profession," "ethics," "code," and "engineer" before saying anything helpful about the design, use, interpretation, or future of professional codes of ethics for engineers.

43.1. Ethics

"Ethics" has many senses in English. I shall now distinguish the five most likely relevant here and explain why I think one of them better than the others for understanding expressions like "professional code of ethics." This explanation should help avoid the confusion common when "ethics" is left undefined in discussions of engineering ethics.

In one sense, "ethics" is a mere synonym for ordinary morality (those universal standards of conduct that apply to moral agents simply because they are moral agents). Etymology fully justifies this first sense. The root of "ethics" (*ēthos*) is the Greek word for habit (or character) just as the root of "morality" (*mores*) is the Latin word for it. Etymologically, "ethics" and "morality" are twins (as are "ethical" and "moral," "ethic" and "morale," and "etiquette" and "petty morals"). In this first sense of "ethics," codes of ethics are just statements of ordinary morality; they add nothing to what engineers are obliged to do. There is no point to speaking of "ethics" rather than "morality." This does not seem to be the sense of "ethics" in expressions like "code of ethics." A code of ethics seems to be more than a simple moral code. That, perhaps, is why there does not seem to be any official code of engineering ethics with "morality" in the title.

In at least four other senses of "ethics," "ethics" differs from "morality." For philosophers, ethics is the attempt to understand morality as a reasonable practice. That is the sense it had, for example, in Aristotle's *Nicomachean Ethics* and typically has today in philosophy courses with "ethics" in the title. But it is not the sense "ethics" has in "code of ethics." A code of ethics is not (or, at least, typically is not) an attempt to understand morality but an attempt to document existing standards of conduct or establish new ones.

In a third sense, ethics consists of those standards of conduct that moral agents *should* follow ("critical morality"); morality, in contrast, is said to consist of those standards that moral agents generally do follow ("positive morality"). In this sense, "morality" is very close to its root *mores*; it can be unethical (in the first sense of "ethics"). What a certain group treats as morally right (racial segregation, forced female circumcision, or the like) can be morally wrong. "Morality" (in this sense) has a plural. There can be as many morali*ties* as there are moral agents. Even so, ethics (in this sense) can be a standard common to everyone. Hence, this third sense of "ethics" is as irrelevant here as the first two, since our subject is *professional* codes, not codes that apply to everyone.

In a fourth sense, "ethics" is contrasted with "morality" in another way. *Morality*, then, consists of those standards applying to every moral agent. Morality is a universal minimum, the standard of moral right and wrong. Ethics, in contrast, is concerned with moral good, with whatever is beyond the moral minimum. This is another sense that seems not to fit professional codes of ethics—for at least two reasons. First, this ethics-of-the-good is still universal, applying outside professions as well as within. Second, codes of ethics consist (in large part, at least) of *requirements*, the right way to conduct oneself rather than just a good way to. Typically, a code of ethics does not sound like mere recommendation, aspiration, or ideal. Any sense of "ethics" that does not include the right cannot be the sense relevant to "codes of ethics."

"Ethics" can also be used in a fifth sense: it can refer to those *morally permissible standards of conduct applying to members of a group simply because they are members of that group*. In this sense, engineering ethics is for engineers and no one else; business ethics, for people in business and no one else; and so on. Ethics—in this sense—is "relative" even though morality is not. Ethics (in this sense) resembles law and custom, both of which can also vary from place to place, time to time, and group to group. But ethics (in this sense) is not mere *mores*. By definition, ethics (in this sense) must be at least morally permissible. There can be no Mafia ethics or Nazi ethics, except with scare quotes around "ethics" to signal an ironic or analogical use.

A code of ethics, though not a mere restatement or application of ordinary morality, can be—and typically is—morally binding on those to whom it applies (that is, a code of ethics typically imposes moral obligations or requirements on those to whom it applies). How is that possible? Some codes of ethics are morally binding (in part at least) because of an oath, promise, or other expression of consent (for example, one's signature on a contract that makes following an employer's code of ethics a condition of employment). In general, though, codes of ethics bind in the way rules of a (morally permissible) game bind when one is a voluntary participant. While one voluntarily receives the benefits of a code of ethics, one has a moral obligation, an obligation of fairness, to do what the code says. If cheating consists of violating the requirements of a voluntary, morally permissible, cooperative practice, such as a game, then doing anything the rules of such a practice forbid is violating the moral rule, "Don't cheat."

Since law applies to those subject to it whether they wish it or not, law (as such) cannot bind in the way a code of ethics can. Since a code of ethics applies only to voluntary participants in a special practice, not to everyone, a code (if generally followed) can create trust beyond what ordinary moral conduct can. It can create a special moral environment. For example, a code of ethics can justify trust in the claims of an engineer beyond the trust those claims would deserve if the engineer were just an ordinary decent person. But the engineer must voluntarily enter the "game" by identifying him/ herself as an engineer. The engineer must "profess" his/her profession.

43.2. Profession[2]

The word "profession" resembles "ethics" in having several legitimate senses. "Profession" can be a mere synonym for "vocation" (or "calling"), that is, for any useful activity to which one devotes (and perhaps feels called to devote) much of one's life, even if one derives no income from so doing (or, at least, does not engage in the activity to earn a living). If the activity were not useful, it would be a hobby rather than a vocation. In this sense of "profession," even a gentleman, in the sense of "gentleman" still current early in the nineteenth century, could have a profession, even if only public service or private charity. The opposite of profession in this sense is trade ("a mere money-making calling").

"Profession" can also be a synonym for "occupation," that is, for any typically full-time activity defined in part by a "discipline" (an easily recognizable body of knowledge, skill, and judgment) by which its practitioners generally earn a living. In this sense, we may, quite properly, speak of someone being a "professional thief" or "professional athlete." The opposite of "professional" in this sense is "amateur" (one who engages in the activity "for love," not to earn a living).

"Profession" can, instead, be used for any occupation one may openly declare or profess, that is, an honest occupation: though athletics can be a profession in this sense, neither thieving nor being a gentleman can be. Thieving cannot because it is dishonest. Being a gentleman cannot because, though an honest way of life, it is not a way to *earn* a living. A gentleman is supposed to live off his wealth, not work for it. This seems to have been the primary sense of "profession" outside English-speaking countries until quite recently. It certainly is the sense it had, for example, when Durkheim (1957) and Weber (1978) wrote about "professions" early in the twentieth century, making their writing about "professions" more or less irrelevant to understanding professions in the following sense, the one of interest here.

"Profession" can also be used for a special kind of honest occupation. This is the sense of a "profession" that allows us to distinguish between professional soldiers and mere mercenaries. There are at least two approaches to defining this sense of "profession." One, what we may call "the sociological," has its origin in the social sciences. Its language tends to be statistical; that is, the definition does not purport to state necessary or sufficient conditions for an occupation to be a profession but merely what is true of "*most* professions," "the *most* important professions," "the *most* developed professions," or the like.

Generally, sociological definitions understand a profession to be an occupation the practitioners of which have high social status, high income, advanced education, important social function, or some combination of these or other features easy for the social sciences to measure. For social scientists, there is no important distinction between what used to be called "the liberal professions" (those few honest *vocations* requiring a university degree in most of early modern Europe) and today's professions (strictly so called). Plumbers cannot form a profession (in the sociological sense) because both the social status and education of plumbers are too low. Medicine certainly is a profession (in this sense) because physicians have relatively high status, high income, advanced education, and important social functions. Business managers are also a profession (in this sense) because they too have high income, high status, advanced education, and an important social function. Nurses may only be "quasi-professionals" because they typically have only a bachelor's degree and less income and social status than physicians, managers, and the like. For most sociologists, it is obvious that professions have existed in Europe and America for many centuries.

Refuting a sociological definition is not easy. Because its claim is stated in terms of "most," a few counter-examples do not threaten it. If the counter-examples seem to grow more numerous than the professions fitting the definition, the defenders of a sociological definition can distinguish "true professions," "fully developed professions," "the ideal type of profession," or the like from those not fitting the definition ("pseudo-professions," "quasi-professions," "emerging professions," and so on). The only professions that seem to be on every sociologist's list of "true" or "fully developed" professions are law and medicine.

The other approach to defining "profession" is philosophical. A philosophical definition attempts to state necessary and sufficient conditions for profession. While a philosophical definition may leave the status of a small number of would-be professions unsettled, it should at least be able to explain (in an intuitively appealing way) why those would-be professions are neither clearly professions nor clearly not professions. A definition covering "most professions" is not good enough.

Philosophical definitions may be developed in one of (at least) two ways: the "Cartesian" and the "Socratic" (as we may call them). The Cartesian way tries to make sense of the contents of one person's mind. That person develops a definition by asking him-/herself what he/she means by a certain term, setting out that meaning in a definition, testing that definition by counter-examples and other considerations, revising whenever a counter-example or other consideration seems to reveal a flaw, and continuing to examine his/her beliefs until satisfied with the order into which he/she has put them.

In contrast, the Socratic way of defining a term seeks common ground between one or more philosophers and the "practitioners" (those who normally use the term in question and are therefore expert in its use). Thus, Socrates would go to the poets to define "poetry," to politicians to define "good government," and so on. A Socratic definition typically begins with what a practitioner offers. A philosopher responds with counter-examples or other criticism, inviting practitioners to revise. Often a philosopher will help by suggesting revisions. Once the practitioners seem satisfied with the revised definition, a philosopher may again respond with counter-examples or other criticism. And so the process continues until everyone is happy with the result. Instead of the private monologue of the Cartesian, there is public conversation, "a dialogue," between philosophers and all those with some experience using the term. The process restarts as soon as a philosopher or practitioner expresses doubt about what has been achieved.

The Socratic way of defining has yielded this definition of "profession:" *a number of individuals in the same occupation voluntarily organized to earn a living by openly serving a certain moral ideal in a morally permissible way beyond what law, market, morality, and public opinion would otherwise require.*

According to this definition, a profession is a collective term—like army, family, or club. There can be no profession with just one member—though there can be just one expert, inventor, or teacher of a certain kind. This may seem a small point, but it has the immediate consequence of disqualifying "oath" (such as "the Engineer's Hippocratic Oath" or "the Engineer's Creed") as a *professional* code. Since a single person can take an oath whatever anyone else does, an oath cannot, as oath, be a professional code (though its contents may correspond to the contents of a professional code). To be a professional code, the oath would have to bind each and every member of the profession, taking effect only when all have so sworn and remaining in effect only while each new member of the profession also takes the oath and none of the old oath-takers renounces it. An oath can be an (actual) *professional* code of ethics only in the context of a complex practice—a practice so complex it probably could not be maintained for long in any large group not otherwise organized. Engineers who appeal to an oath for their professional ethics are simply confused. They should appeal to a professional code.

To be a profession (according to the Socratic definition), the members of the group in question must have an occupation. Mere gentlemen cannot form a profession. Hence, members of any of the traditional "liberal professions" (lawyers, physicians, clergy, and so on) could not form a profession until quite recently—that is, until they ceased to be gentlemen, began to work for a living, and recognized the change of circumstance. I think that happened sometime after 1850. So, if law and medicine were the first professions, they must (according to this definition) have become professions less than two centuries ago.

The members of a would-be profession must *share* an occupation. A group consisting of, say, lawyers and engineers cannot today be a profession, though lawyers can be one profession and engineers another. The two occupations cannot form a single profession because they lack a common discipline. Hence, they have separate schools, degrees, licenses, libraries, and technical organizations.

Each profession is designed to serve a certain moral ideal, that is, to contribute to a state of affairs everyone (every reasonable person at his/her most reasonable) can recognize as good to achieve, perhaps even good to approach. So, physicians have organized to cure the sick, comfort the dying, and protect the healthy from disease; lawyers, to help people obtain justice within the law; engineers, to improve the material condition of humanity; and so on. But a profession does not just organize to serve a certain moral ideal; it organizes to serve it *in a certain way*, that is, according to standards beyond what law, market, morality, and public opinion would otherwise require. A would-be profession must, then, set *special* (morally permissible) standards. Otherwise it would remain nothing more than an honest occupation. Among those special standards may be requirements for education, character, experience, and skill, but inevitably some of the special standards will concern conduct. The special standards of conduct will be ethical (in our fifth sense of "ethics"). They will govern the conduct of all the group's members simply because they are members of that group—and govern no one else.

These ethical standards may constitute a code of ethics, depending on what "code" means.

43.3. Code

The word "code" comes from Latin. Originally, it meant any wooden board, then boards covered with wax used to write on, then collections of papers or parchment bound between such boards, and then any book ("codex"). This last sense is the one it had when first applied to the book-length systemization of Roman law that the Emperor Justinian enacted in 529 AD. Justinian's book-length systemization differed from an ordinary compilation of law in at least one important way: it received the legal authority that Justinian had as Emperor. His systemization of law was authoritative, replacing all that preceded it. A "codification" is a new beginning.

Since 529 AD, any document much like Justinian's *Code* could be called "a code." Sometimes the analogy with Justinian's *Code* is quite close, as it is, say, in the *Code Napoleon* or the *Illinois Criminal Code*. A spy's "secret code" (cipher) is a code in a more distant sense. While a spy's code is an authoritative system of written rules (and thus resembles Justinian's *Code* in that way), the rules of a spy's code concern only converting one set of symbols into another (ciphering and deciphering); a spy's code does not constitute a code of *laws*. Sometimes, however, the analogy is not close at all. For example, "genetic code" is a code only in a rather distant sense: an arrangement of molecules that yields useful information when treated as if it were a spy's code.

One important feature of Justinian's *Code* is that it was written. Could a code nonetheless be *unwritten*? Since the point of codification (strictly speaking) is to give law (and, by analogy, any similar system of guidance) an authoritative formulation, an unwritten code might seem to be no code at all, an unformulated formulation. Nonetheless, there seem to be at least three senses in which a code might be said to be "unwritten" without losing all connection with the fundamental idea of code. First, a code might, though not in writing, have an authoritative *oral* formulation, for example, a short list of slogans like "Safety first." The absence of a written formulation would then seem a mere technicality—easily overcome by writing down the oral formulation. Second, a code, though unformulated, might be so obvious to those who know the practice that they need only put it into words and write the words down to have the written formulation accepted as "the code" in question. Though some fields of engineering may today have a few rules unwritten in one of these two ways, no field of engineering seems to have enough such rules to constitute an unwritten code. Nor is it likely that any would. How, in fields where so much changes so quickly, are individuals separated by age, experience, and immense distances to reach agreement without putting the agreement in writing?

Third, an unwritten code may be written but not written as a formal code. It might, for example, exist only *implicitly* in a large body of technical documents. The structure, language, or effect of such documents may be seen as expressing certain standards of conduct every engineer (at his most reasonable) accepts. In this sense, engineering's technical documents may have served as an "unwritten" code of ethics for perhaps a century or two before engineering first adopted a formal code of ethics—much as the British are said, even today, to have an "unwritten constitution" that includes such documents as the "Magna Carta," "The Bill of Rights of 1689," and various statutes and judicial opinions. The content of such an unwritten code or constitution will, of course, be controversial in part—or, at least, open to interpretation.

Whether written or unwritten, a code of ethics will typically belong to one of three categories. First, a code may be *professional*, that is, apply to all, and only, members of a certain profession. The AIEE Code is a professional code in this sense, however "profession" was understood in 1912. The code applies to "engineers" as such (and to no one else). Second, a code can be *organizational*. An organizational code applies to members of an organization, for example, to the members of a technical or scientific society enacting it, whatever their profession. An organizational code can also (like the code of ethics of a government, business, or not-for-profit) apply to an organization's employees, contractors, or the like. The IEEE Code of Ethics (2006) is an organizational code in this sense (though the IEEE is the AIEE's descendant). Thus, that code begins: "In recognition of the importance of our technologies in affecting the quality of life throughout the world . . . [we] do hereby commit ourselves to the highest ethical and professional conduct and agree . . ." Because the IEEE's "we" (its members) include computer scientists, mathematicians, and technical managers, as well as engineers, the IEEE's code applies to many non-engineers. Because the code applies to non-engineers of several occupations (as well as to many engineers), it cannot be a professional code. Third, a code can be *institutional*; that is, it can apply to a number of individuals involved in a certain activity whatever their education, experience, or organizational connections. The Computer Ethics Institute's "Ten Commandments of Computer Ethics" is an institutional code in this sense. It applies to anyone using a computer.

A code of ethics may include ordinary moral rules (such as "Don't steal," "Keep your promises," or "Help the needy"), but it must include more than such rules to be a code of *ethics*. A code of ethics includes standards beyond what law, market, morality, or public opinion would otherwise impose.

A code of ethics may nonetheless be incorporated into law. For example, Illinois's Professional Engineering Practice Act of 1989 (225 ILCS 325) includes several sections of "Violations" that may appear in a code of ethics properly so called, for example, that engineers are not to "practice, attempt to practice, or offer to practice professional engineering without a license as a licensed professional engineer" (Sec. 39 (4)). But a code incorporated into law is not a code of ethics unless those it governs can (at their most reasonable, at least) recognize it as consisting of standards of conduct they want to govern those it purports to govern so much that they are willing to be so governed even when the standards cannot be legally enforced.

An engineer may, then, be subject to more than one code of ethics, for example, a professional code (such as that of the National Society of Professional Engineers), the engineer's employer's code of conduct, and the Ten Commandments of Computer Ethics. These may conflict. An engineer may also find that one or more of the codes of ethics governing him/her may conflict with law. For example, a city's building code may require materials (such as asbestos) that have proved unsafe. Which standards, if any, take priority in such a conflict? That depends in part on what "engineer" means.

43.4. Engineer

What, then, is an engineer here? That is not a question about how the term "engineer" is typically used, a question for the dictionary. In some languages, there is not even a word corresponding to the English "engineer." One then needs a phrase to distinguish an engineer (or a certain kind of engineer) from other "technologists," such as architects, industrial designers, electricians, and so on.

Even in English, which has a word for "engineer," there is some confusion. A linguistically proper use of the term "engineer" (or "engineering") is no guarantee that what is in question is an engineer (or engineering) in the sense relevant here. The custodian of my apartment building has only a high school education but has an "engineer's license" that entitles him to oversee operation of the building's two boilers. A few miles south of those boilers is the local office of the Brotherhood of Locomotive Engineers and Trainman. Neither my building's "licensed engineer" nor those "locomotive engineers," though properly called "engineers," are engineers in the sense relevant to engineering's professional codes of ethics. They are engineers only in an old sense indicating a connection with engines; that is, they are technicians much like mechanics, trainman, or bus drivers.

Then there are the knockoffs of engineering proper: "genetic engineering," "social engineering," "climate engineering," "financial engineering," "re-engineering," and so on. Engineers (in the sense relevant here) are notably absent from these activities. Those engaged in them might more accurately be called "geneticists," "social tinkers," "climate experimenters," and so on.

Equally confusing is that there are fields of engineering *not* called "engineering." One is naval architecture. In the United States, programs in naval architecture are accredited as engineering programs, describe themselves in those terms, and in fact have much the same requirements as other engineering programs. Naval architects are called "architects" only because there was once a tradition of English shipbuilding in which gentlemen, not trained as engineers, used detailed drawings to instruct tradesman how to build large sailing ships. Having a classical education, they knew Greek and preferred to use the Greek term "architect" rather than the English "master shipwright"—a term analogous to "master carpenter," "master builder," and the like, suggesting a tradesman with dirt under his fingernails who worked his way up until he could take responsible charge of the shipbuilding. The name outlived the tradition. Applied physics (sometimes called "engineering physics") is another of those misnamed fields; it is not physics but engineering. "Rocket science" would be another—if the term ever appeared anywhere but in the humorous observation, "It's not rocket science." The only "rocket science" is engineering (aeronautical, mechanical, and so on).

What, then, distinguishes engineers, in the sense appropriate here, from other technologists—assuming, for the moment, that architects, computer scientists, and so on are technologists but not engineers? The answer cannot be the *function* of engineers. Like other technologists, engineers design, manage the building, and otherwise contribute to the life (and death) of technology (that is, useful artifacts embedded in a social network that designs, builds, distributes, maintains, uses, improves, and disposes of them). *Equating* engineers with other designers, builders, or the like promises at least two bad consequences.

The first is that study of differences between engineers and other technologists becomes conceptually impossible. Architects, computer scientists, and the like also design, build, and so on. Much may be learned about technology from study of the ways in which engineers work differently from other technologists. Such a study is defined out of existence if engineers just are technologists (or technologists just are engineers).

Second, and equally important, is that equating engineers with designers, builders, or the like gives a misleading picture of what engineers in fact do. Though some engineers do design, build, or the like, many, in addition or instead, control quality, write regulations, evaluate patents, sell complex equipment, reconstruct equipment failures, or teach engineering. Whether all of these activities are properly engineering—indeed, whether any of them are—is a question for the philosophy of engineering. My point now is that quality control, writing regulations, evaluating patents, and so on are functions that engineers routinely perform not only in the sense that they are functions some engineers *happen* to perform, but in the philosophically more interesting sense that they are functions that some engineers are supposed to perform *as engineers*. Employers sometimes advertise for engineers rather than for other "technologists" to perform these functions. I agree that design (engineering design) is central to understanding engineering, but I do not see how designing, building, or the like can be *the* defining function of engineering, or even *a* defining function, because, as I see it, there is no function that engineers, and only engineers, seem to perform (except, of course, engineering itself, which is what we are trying to define).

If function cannot define engineering, what can? For this question, I have a double answer. One is that if "define" means giving a classic definition (say, by genus and species), there are only practical definitions, useful for a particular purpose. There can be no philosophical definition, that is, a formula that captures the "essence" of engineering, because engineering has no essence. All attempts at a philosophical definition of "engineering" will: (a) be circular (that is, use "engineer" or a synonym or use an equally difficult term, such as "technical"); (b) be open to serious counter-example (whether because the definition excludes some activity clearly belonging to engineering or includes some activity clearly not belonging); (c) be too abstract to be informative; or (d) suffer a combination of these errors. The same is true of "engineer."

The other answer, the one that in part explains this first, is that engineering, like other professions, is self-defining but not in the classic sense of "define." Engineering's definition is not (primarily) a matter of words. There is a core of people, more or less fixed by history at any given time, who are undoubtedly engineers. This core constitutes engineering in two ways. First, it decides what is within the joint competence of those people (engineers) and what is not. Second, that core of people also constitutes engineering insofar as it admits or rejects candidates for the title "engineer" (in our sense), using criteria such as similarity with themselves in education, method of work, experience, and product. Often, applying these criteria is routine. So, for example, an ordinary lawyer clearly is not an engineer (that is, competent do engineering), while the typical graduate of an ABET-accredited engineering program with a few years' experience successfully working as an engineer clearly is an engineer. But sometimes the criteria cannot be applied without exercising judgment. Does someone with a degree in chemistry who, say, has successfully managed a large chemical plant for five years, count as an engineer because what he/she has been doing is, in effect, "chemical engineering"?

Since I have argued for this double answer elsewhere (Davis 1996, 2009a, 2010), I shall not repeat the arguments here. Instead, I shall simply point out one consequence: The philosophy of engineering is *not* simply the philosophy of technology (or even the philosophy of technologists). It is in part at least the philosophy of a profession. Each profession is an historic individual, not the realization of an essence or Platonic form. The philosophy of engineering should attempt to understand that historically given practice, not attempt to understand what some theory of technology says engineering should be. The key to understanding engineering is its history. But the key to understanding its history is conceiving that history as the biography of a self-conscious group constructing a discipline and passing it on from teacher to student.

Engineering was a discipline a century or two before it was a profession. But long before that discipline appeared, there were people who *functioned* as engineers (more or less) as engineers now do. Whoever built Skara Brae on Orkney 5,000 years ago or the Great Pyramid at Giza 4,000 years ago was not an engineer, though what he, she, or they did somewhat resembles the work of modern engineers. We know he, she, or they were not engineers because there is no chain of teachers and students to connect them with today's engineers. They do not share a discipline with today's engineers. On the other hand, if we start with today's engineers, we can find such a chain leading back to the officers of the French Army's *corps du génie*, disappearing a few decades before 1700. Though soldiers called "engineers" existed a century or two before that, they did not share a discipline. There were then no schools of engineering, no common curriculum, nothing to connect them to today's engineers but a name, one today's engineers still must share with boiler supervisors, train drivers, and soldiers with shovels.

43.5. Design and Use of Codes

Attempts have been made to distinguish between short, general, or uncontroversial codes of ethics ("code of ethics" proper) and longer, more detailed, or more controversial codes ("code of conduct," "guidelines," or the like). While some such distinction may sometimes be useful in practice, it is hard to defend in theory. A "code of conduct" is as much a special standard as a "code of ethics" is (except where the "code of ethics," being a mere restatement of morality, is just a moral code). "Codes of conduct" are also (typically) as morally binding as "codes of ethics." They are more than mere suggestions, aspirations, or hopes. Short codes of ethics, such as the IEEE's 2006 code, are about 250 words long; long codes are at least four times that. There are also a few codes, surprisingly few, that are shorter than a long code but longer than a short code. For example, the IEEE's 1979 code was about 550 words. In what follows, I ignore the distinction between short and long, general and detailed, and controversial and uncontroversial codes.

Codes of ethics—whether professional, organizational, or institutional—can have at least four legitimate uses in engineering: First, and most important, a code of ethics can establish special standards of conduct where experience has shown common sense to be inadequate. Second, a code of ethics, being an authoritative formulation of the rules governing a practice, can help those new to the practice to learn how to act. Third, a code can remind those with even considerable experience of details they might otherwise forget. Fourth, a code can provide a framework for settling disputes, even disputes among those with considerable experience.

A code of ethics can also be used to justify formal discipline or legal liability, but either of these uses turns a code of ethics into (something like) law. When, then, someone complains that a code of ethics is "mere window dressing" unless it is "enforced" by penalties or unless violators are "held legally accountable," that person has made a mistake. That person has assumed that punishment is the only, or at least the primary, way to guide conduct and has forgotten about social pressure, conscience, and (most important of all) good will. For those who will the good, to know what they should do is reason enough to do it. For them, "enforcement" is primarily a matter of learning what is justifiably expected of them—and of being reminded of that now and then. They hold themselves accountable.

A *professional* code of ethics has at least three other legitimate uses in engineering. First, it can help those outside engineering ("the public") understand what may reasonably be expected of any engineer. The code tells those outside the profession what engineers expect of themselves, indeed, what they invite the world to expect. Second, a professional code can provide a vocabulary for interpreting engineering's technical standards, for example, "If we accepted your interpretation of this specification, we'd be putting the public safety at risk. But, as engineers, we agree that the public safety should be paramount." Third, a professional code can help engineers resist pressure from client, employer, or other superior to do what they should not. An engineer can say, "You hired an engineer. Engineers are hired, in part, to maintain certain standards. If you wanted someone to do whatever you want, you should not have hired an engineer."

While I believe an engineer's professional code always pre-empts the wish, rule, or practice of a client, employer, or superior, I know of only one code of engineering ethics that makes that explicit. The "Fundamentals of Engineering Ethics" (2002) of the German Federation of Engineers (VDI) contains the following provision: "such professional regulations have priority over individual contracts." Some such provision should be more common in engineering's professional codes, since it is already implicit in all of them.[3]

When writing a professional code of ethics, one should consider the purposes that the code should serve. When using a professional code, one should consider what purposes it was meant to serve. The primary purpose of any code is to provide a *systematic* statement of relevant standards of conduct. So, when using a code of ethics, one should treat it as a single document (a system), not as a collection of independent sentences. For example, suppose a code has (as the AIEE code of 1912 does) a first General Principle stating that "In all of his relations the engineer should be guided by the highest principles of honor." Suppose too that the code has this seemingly contradictory principle (as the AIEE code also does): "The engineer should consider the protection of a client's or employer's interests his first professional obligation, and therefore should avoid every act contrary to this duty." It would be wrong to interpret the second principle as requiring the engineer to do *whatever* protects the client's or employer's interest. The "highest principles of honor" limit what the engineer may do for a client or employer to what is honorable. The second principle must be read in the context of the first (Davis 1999). Though "honor" probably meant more in 1912 than it does today (justified good reputation), it is, I think, clear even today that the "highest principles of honor" must condemn lying, cheating, and similar questionable conduct even when such questionable conduct would serve a client or employer. But, in case there was any doubt of that, there is the second General Principle in the AIEE's code:

> It is the duty of the engineer to satisfy himself to the best of his ability that the enterprises with which he becomes identified are of legitimate character. If after becoming associated with an enterprise he finds it to be of questionable character, he should sever his connections with it as soon as practicable.

The interests of client or employer are an engineer's "first *professional* obligation" only if what the client or employer is doing is of a "legitimate character." The "professional obligation" takes effect only within an organization that is acting as it should.

Oddly, this interpretation of the AIEE code has proved controversial. Relying on Sarah Pfatteicher's work, Carl Mitcham has objected that this interpretation ignores

> the conflicting influences active in the emergence of the [AIEE] code and the particular language used in internal discussions that extended from the 1870s to the early 1900s . . . [For example,] in a proposal leading up to adoption of the AIEE code, it is clearly stated that "the electrical engineer should consider the protection of his client's interests as his first obligation [not just his first *professional* obligation].
>
> *(Mitcham 2009: 39–40)*

This objection makes two mistakes. First, it ignores the fact that the first-obligation language cited was eventually amended to insert "professional" before obligation (and to delete "electrical"), allowing for honor to pre-empt the client's or employer's interest (and for the Code to treat engineering as a whole, not just electrical engineering, as the profession in question). The original first-obligation language turned out not to be what the AIEE wanted in its code. No doubt Mitcham is right about the swirl of influences that resulted in the code. He is wrong only in *not* giving priority to the code that actually resulted. What the code says is also history. Second, what was said before adoption of the code does not have the same standing as the Code itself. The Code is what codes typically are, a new beginning, not what was or might have been but is not. Its actual language is what binds engineers, not earlier authoritative language, much less language proposed and rejected.

Though there has been some criticism of engineering codes of ethics as such (e.g. Ladd 1980), the number of such codes seems to be rising as engineering societies in non-English-speaking countries adopt their own codes, and new fields of engineering in English-speaking countries feel the need to adopt codes designed for their special needs. These new codes, or the discussions surrounding them, suggest at least three (not necessarily inconsistent) directions of change in codes of engineering ethics.

The first, and most obvious, direction of change is the addition of explicit language imposing some sort of obligation on engineers for "sustainable development." For example, the Code of Ethics of the American Society of Civil Engineers (2006) requires engineers to "strive to comply with the principles of sustainable development in the performance of their professional duties." The requirement merely to "strive" is unusually weak for any engineering code of ethics, though still more than mere "aspiration." The requirement should become stronger as engineers gain more experience in trying to follow the principles of sustainable development.

The second direction of change that codes of engineering ethics may take are provisions to guide engineers in "other cultures," that is, parts of the world that may seem to want engineers not to maintain the same standards of conduct engineers typically maintain "at home." What engineering standards should be universal? What exceptions to engineering standards are permissible under specified conditions?

The third direction of change in codes of engineering ethics may be the addition of provisions, or the writing of special codes, to guide the conduct of engineers when they have "human subjects" or patients. While engineers have long "experimented on human subjects" in the sense that they have let the products of their ingenuity loose on the world without fully knowing what would happen, engineers are increasingly involved in research involving "human subjects" in the sense in which that term is used in medicine. This relationship, much more personal than the relationship of engineers to the public, has little precedent in engineering. No major code of engineering ethics has yet to deal with it, even though many engineers in chemical, electrical, and mechanical engineering are now involved in biomedical research in which they work beside physicians who have patients before them. The Biomedical Engineering Society Code of Ethics (2004), though the work of a small society, is nonetheless worth reading. It suggests the radical changes the rise of biomedical engineering may lead to. Consider, for example, its requirement that biomedical engineers involved in healthcare activities:

1. Regard responsibility toward and rights of patients, including those of confidentiality and privacy, as their primary concern.
2. Consider the larger consequences of their work in regard to cost, availability, and delivery of healthcare.

Rule 1 seems to put patient welfare ahead of the public welfare. Rule 2 reinforces that interpretation.

Related Chapters

Chapter 1: What Is Engineering? (Carl Mitcham)

Chapter 2: A Brief History of Engineering (Jennifer Karns Alexander)

Chapter 3: Western Philosophical Approaches and Engineering (Glen Miller)

Chapter 21: Engineering Design (Peter Kroes)

Chapter 33: Values in Risk and Safety Assessment (Niklas Möller)

Chapter 36: Trust in Engineering (Philip J. Nickel)

Chapter 40: Ethical Considerations in Engineering (Wade L. Robison)

Chapter 41: Autonomy in Engineering (Eugene Schlossberger)

Chapter 42: Standards in Engineering (Paul B. Thompson)

Chapter 44: Responsibilities to the Public—Professional Engineering Societies (Joseph Herkert and Jason Borenstein)

Chapter 53: Engineering and Contemporary Continental Philosophy of Technology (Diane P. Michelfelder)

Chapter 54: Engineering Practice From the Perspective of Methodical Constructivism and Culturalism (Michael Funk and Albrecht Fritzsche)

Notes

1. I draw these conclusions from an examination of the Engineering section of the Ethics Code Collection, http://ethics.iit.edu/ecodes/ethics-area/10, accessed August 28, 2018, maintained by the Center for the Study of Ethics in the Professions, Illinois Institute of Technology. All codes referred to in this chapter can be found at this site. The Codes Collection also includes, beside nearly four thousand non-engineering codes, a bibliography for codes of ethics at: http://ethics.iit.edu/ecodes/bibliography. Accessed August 28, 2018.
2. For a fuller argument for this definition, see Davis (2009b).
3. For a somewhat different (but equally useful) list of uses of codes of ethics in engineering, see Martin and Schinzinger (2005: 44–46).

References

Davis, M. (1996). Defining Engineering: How to Do It and Why It Matters. *Journal of Engineering Education,* 85(April), 97–101.

Davis, M. (1999). Professional Responsibility: Just Following the Rules? *Business and Professional Ethics Journal,* 18(Spring), 65–87.

Davis, M. (2009a). Defining Engineering—From Chicago to Shantou. *Monist,* 92, 325–339, July.

Davis, M. (2009b). Is Engineering a Profession Everywhere? *Philosophia,* 37(June), 211–225.

Davis, M. (2010). Distinguishing Architects From Engineers: A Pilot Study in Differences Between Engineers and Other Technologists. In I. Van de Poel and D. Goldberg (eds.), *Philosophy and Engineering: An Emerging Agenda.* Dordrecht: Springer, pp. 15–30.

Durkheim, E. (1957). *Professional Ethics and Civic Moral.* London: Routledge.

Ladd, J. (1980). The Quest for a Code of Professional Ethics: An Intellectual and Moral Confusion. In R. Chalk, M.S. Frankel, and S.B. Chafer (eds.), *AAAS Professional Ethics Project: Professional Ethics Activities in the Scientific and Engineering Societies.* Washington, DC: AAAS, pp.154–159.

Martin, M.W. and Schinzinger, R. (2005). *Ethics in Engineering,* 4th ed. Boston: McGraw Hill.

Mitcham, C. (2009). A Historico-ethical Perspective on Engineering Education: From Use and Convenience to Policy Engagement. *Engineering Studies,* 1(1), 35–53.

Weber, M. (1978). *Economy and Society.* Eds. G. Roth and C. Wittich. Berkeley: University of California Press.

44

RESPONSIBILITIES TO THE PUBLIC—PROFESSIONAL ENGINEERING SOCIETIES

Joseph Herkert and Jason Borenstein

44.1. Introduction

Engineers both individually and collectively have responsibilities to the public. This is due in part to the specialized expertise and training they possess. Few, if any, other persons or entities have the relevant knowledge to inform certain types of critical decisions that directly impact the "health, safety, and welfare" of the public. For example, making a determination about whether a bridge is structurally sound requires the expertise of civil engineers and transportation engineers; demonstrating that an airplane is safe enough for passengers requires the expertise of aeronautical engineers, mechanical engineers, materials engineers, and electronics engineers.

Individual engineers can have a range of responsibilities in part because they serve in various roles. Not only do they have formal obligations to the public associated with the requirements of their job, but they can also participate as citizens in discussions and activities beyond the workplace (Woodhouse 2001). For example, an engineer might serve as a *pro bono* expert on matters related to the public's well-being, such as by testing air or water quality. Or, an engineer might participate in politics and governance, which could be accomplished by providing advice to a city planning board. In this chapter, however, the main focus will be on the collective responsibility of the engineering profession, particularly professional engineering societies.

What is meant by "collective responsibility" entails more than the ever-growing practice of an engineer working in teams with other engineers (and often with members of other fields). The collective responsibility of the engineering profession extends to organizations of engineers at the local, national, and international levels. There are many different types of engineering societies, and these societies carry with them ethical responsibilities. For example, an engineering society may have the requisite knowledge to inform public policy debates on such issues as whether water is clean enough for public consumption. Alternatively, a society might advise government agencies regarding which factors contributed to a disaster and whether it is safe to rebuild in the relevant region (e.g., ASCE 2007).

When considering the roles of engineers and of engineering societies, a fundamental concept to take into account is "social responsibility." Individual engineers can exhibit social responsibility both in their engineering work and outside of their employment as an engineer; social responsibility, however, is particularly relevant to the role of professional societies, both in terms of the numbers of engineers they represent and the position and authority these societies have in society. Although the concept is difficult to define precisely, social responsibility normally refers to ethical obligations that go beyond the rules, guidelines, and norms that are minimally required to operate within one's profession. For example, running a computer simulation correctly and then honestly reporting the resulting data to one's employer or client are responsibilities internal to the profession. Yet, how the results of such simulations are used to benefit or harm the public are generally considered external social responsibilities of the profession.

44.2. Traditional Professional Engineering Societies

44.2.1 Types of Traditional Professional Engineering Societies

Engineers can belong to many types of societies. In this chapter, a distinction will be drawn between traditional professional engineering societies, which usually have direct ties to a specific engineering discipline or the engineering profession as a whole, and what will be referred to as "quasi-professional" engineering societies, which usually focus on a specific issue or kind of activity. The latter type of society is normally organized outside of traditional professional institutions.

Most traditional professional engineering societies are organized by discipline. For example, the most prominent in the United States are the five "founder" societies, which are as follows:

- American Institute of Chemical Engineers (AIChE)—established in 1908.
- American Institute of Mining, Metallurgical and Petroleum Engineers (AIME)—established in 1871.
- American Society of Civil Engineers (ASCE)—established in 1852.
- American Society of Mechanical Engineers (ASME)—established in 1880.
- Institute of Electrical and Electronics Engineers (IEEE)—created in 1963 by a merger of the American Institute of Electrical Engineers (established in 1884) and the Institute of Radio Engineers (established in 1912).

The membership of these societies includes both engineers who are faculty at academic institutions and those who are practitioners. Furthermore, many of these societies have members worldwide and profess to represent the engineering profession globally. Yet other countries are also home to traditional professional engineering societies, including the Japan Society of Civil Engineers (established in 1914), the Indian Institute of Chemical Engineers (established in 1947), and the South African Institute of Electrical Engineers (established in 1909).

Discipline- or multi-discipline-based societies have also emerged for more specialized or newer branches of engineering. This would include, for example, the American Society of Safety Engineers (established in 1911), the Association for Computing Machinery (ACM) (established in 1947), the Biomedical Engineering Society (established in 1968), and the Institution of Engineering and Technology in the United Kingdom, which formed in 2006 through a merger of two organizations dating to the late 19th century.

Other US-based traditional professional engineering societies cut across technical disciplines while focusing on a particular aspect of engineering. These include the National Society of Professional Engineers (NSPE) (established in 1934), which represents licensed professional engineers, and the American Society of Engineering Education (ASEE) (established in 1893). Similarly, societies may serve a particular engineering constituency such as the Society of Women Engineers (established in 1950) and the National Society of Black Engineers (established in 1975). Parallels exist in other countries, such as the Indian Society for Technical Education (established in 1988) and the Women's Engineering Society in the UK (established in 1919).

Moreover, some organizations bring together different types of engineering societies such as the American Association of Engineering Societies (established in 1979). The World Federation of Engineering Organizations (WFEO) (established in 1968) transcends the boundaries of any particular country or engineering discipline and seeks to represent the profession globally.

44.2.2 Traditional Professional Engineering Societies and Social Responsibility

Engineering societies engage in many activities, including sponsoring technical conferences and publishing journals. Historically, and even to some degree today, engineering societies have tended not to address the social responsibilities of engineers directly. Yet their views on social responsibilities of engineers are often implicitly made manifest in their formulation and consideration of public policy including issuance of formal position statements on policy issues. Engineers Australia (2018) (established in 1919), for example, has "five public policy priority issues," which are the future workforce, energy, infrastructure, the defense industry, and skilled migration. The ASCE (2015) has issued policy statements, for example, on the civil engineering profession's response to the impact of climate change. Similarly, Engineers Canada (established in 1936) issues national position statements that "represent the collective position of the engineering profession" on such issues as diversity/inclusion and infrastructure. Policy matters and concerns for social responsibility also arise indirectly in other activities of traditional professional engineering societies, especially in the promulgation and enforcement of codes of ethics and in the creation of technical standards.

44.3. Codes of Ethics

In 1910, England's Institution of Civil Engineers was the earliest engineering society to establish a code of ethics. The first engineering code of ethics in the US was adopted in 1912 by the American Institute of Electrical Engineers (AIEE). The underlying motivation for codes of ethics has evolved over time. Though codes in the first half of the 20th century stressed obligations to employers and clients, most contemporary codes include a "paramountcy clause," i.e., a provision that pledges engineers to "hold paramount" public health, safety, and welfare.

A profession promulgates a code of ethics for a variety of reasons:

- It provides a minimum threshold of behavior that practitioners should not fall below.
- It provides a standard against which someone could in principle be held accountable.
- It articulates what is important to the profession.
- It can help to reassure the public of the profession's commitment to a social good.

While codes can serve an important function by articulating and codifying the values that a profession espouses, they have their share of critics (e.g., Ladd 1980).

44.3.1 Limits and Critiques of Codes

A variety of limitations and criticisms can afflict engineering codes of ethics. For one, they can be difficult to enforce; the organization that promulgates the code may be unable or unwilling to hold members accountable to the code. Also, professional societies can expel or limit activities of members who violate their code, but they do not have a direct way of controlling the behavior of non-members (and many engineers are not part of a professional society). Codes are seen by some as self-serving (i.e. enhancing the profession's reputation) and may not be taken seriously by members of the profession (Ladd 1980).

Although a code often includes provisions aimed at assisting engineers when they are making a good faith effort to follow the code, professional societies rarely take action to protect engineers whose employment is threatened in this type of circumstance (Chertow et al. 1993). In particular, societies have typically been reluctant to support "whistleblowers" who go outside normal reporting channels within or external to their place of employment. Much of this reluctance can be explained by tensions arising in the engineering societies between "business" interests and "professional" interests (Layton 1971).

Furthermore, the codes could be perceived as being too prescriptive, undermining the notion of moral autonomy (Ladd 1980; Coeckelbergh 2006). It could also be claimed that the guidance the codes provide is too general and deliberately avoids specific, controversial issues (such as the appropriateness of weapons development). Some of these issues are ones that have a direct bearing on the public's well-being, and yet the codes might be silent on such matters (Riley and Lambrinidou 2015).

44.3.2 The Codes and Social Responsibility

Despite the perceived weaknesses of codes of ethics noted in the previous section, the codes remain the most visible sign of the engineering profession's moral commitments. The codes have generally been interpreted as applying to the behavior of individual engineers (i.e., responsibilities internal to the engineering profession). The paramountcy clause in the codes, however, could be read as a call for engineers to uphold the tenets of social responsibility. Additionally, some codes have explicit provisions that collectively apply to engineers and the engineering profession. The IEEE code (2020), for example, pledges its members "to improve the understanding by individuals and society of the capabilities and societal implications of conventional and emerging technologies, including intelligent systems." The code from the Japan Society of Civil Engineers (2014) pledges engineers to "be pro-active in sharing their expertise and knowledge in their endeavors and communicate in an open exchange of views with the people." The Engineers Ireland (2018) (established in 1835) states in its code that "members shall foster environmental awareness within the profession and among the public."

By emphasizing public health, safety, and welfare and, in some cases, the societal implications of technology, engineering codes of ethics indicate the importance of an engineer's and the engineering profession's social responsibilities, including as they pertain to matters of public policy. Until recently, however, the professional societies did not consistently include considerations of the social and ethical responsibilities of the engineering profession when formulating policy statements; such statements have largely been seen as a lobbying function, usually national in scope. Policies advocated by professional societies on product liability, for example, sometimes seem at odds with ethical mandates to protect public health, safety, and welfare (Herkert 2003). As will be discussed later, this gap has been filled somewhat by the "quasi-professional engineering societies" and, more recently, by initiatives within some of the traditional engineering societies.

44.4. Standards

National standards organizations began to emerge in the late 19th century, including the German Imperial Institute of Physics and Technology (established in 1887), the British National Physical Laboratory (established in 1989), and the American National Bureau of Standards (NBS) (established in 1901; today known as the National Institute of Standards and Technology). Around that time, the US engineering societies began to engage in standardization efforts, and in 1918, the founder societies partnered with the government to form what would later become the American Standards Association (ASA; and still later, the American National Standards Institute or ANSI). NBS standards (government driven) and ASA standards (voluntary consensus) set up a dynamic tension that exists to this day (Russell 2005). Individual professional societies soon became involved in promulgating standards; historian Andrew Russell (2005) notes that "the establishment of technical standards occurred within the broader context of engineers struggling to define their profession."

One of the earliest individual professional society initiatives was

[T]he ASME Boiler and Pressure Vessel Code (B&PVC) . . . conceived in 1911 out of a need to protect the safety of the public. This need became apparent shortly after the conception of the steam engine in the late 18th century. In the 19th century there were literally thousands of boiler explosions in the United States and Europe, some of which resulted in many deaths.

(ASME 2011)

Though safety is often a primary focus of standards, for the most part professional societies have traditionally not directly linked them to professional ethics or social responsibilities of the profession. On occasion, societies have encountered ethical scrutiny and legal liability for their standards, such as when ASME settled an anti-trust law suit involving its boiler standard for $4.75 million USD (May 1982).

44.5. Policy Statements by Professional Engineering Societies

Many professional engineering societies issue position statements on matters of public policy. The reasons for this activity vary; the stated purpose is generally to inform members and policymakers of the organization's position on matters ranging from the status of the profession to important technological issues such as energy policy, environmental policy, and climate change. ASME (n.d.), for example, states the following: "Position statements are effective for communicating the independent views of various segments of the Society—or the Society as a whole—to policymakers who are confronted with decisions on a wide variety of technical issues." Engineers Canada (n.d.) states that it issues "National Position Statements" in order to:

- represent the collective position of the engineering profession
- influence public policy
- facilitate discussion with government
- provide information for our members and those of the engineering profession

In some cases, societies delineate the values underlying their policy positions and statements. For example, the Engineers Ireland (n.d.) identifies five core values, including Sustainability and Trustworthiness.

Specific position statements may also speak directly to issues of ethics and social responsibility, as in the following case (ASCE 2013):

> The American Society of Civil Engineers (ASCE) actively participates in and strongly supports the involvement of civil engineers, as well as individual citizens and coalitions, in the legislative and regulatory decision making processes at the local, state and national levels.

Position statements are rare at the international level, though the WFEO issues Declarations, usually in conjunction with a particular conference or event.

44.6. Quasi-Professional Engineering Societies

In addition to the traditional engineering societies, there are many "quasi-professional" engineering societies that are not directly tied to a traditional society or institution. The quasi-professional societies are organized in a number of ways. Some focus on humanitarian or social justice concerns; others serve as honorary societies and often have a service mission. Traditional societies may have specialized divisions that have a similar mission such as a section on humanitarian engineering. Quasi-professional societies may have the advantage of not being subject to control by the professional societies or influenced by their internal politics.

44.6.1 Humanitarian Engineering Societies

Several engineering societies have a humanitarian mission, usually oriented toward engineering projects in less developed nations. The members of such organizations are usually engineering practitioners or engineering educators and students, or in some cases, a hybrid of both. In the US, a prominent humanitarian society is Engineers Without Borders-USA (EWB-USA), incorporated in 2002. The mission of EWB-USA (n.d.) is to build "a better world through engineering projects that empower communities to meet their basic human needs and equip leaders to solve the world's most pressing challenges." EWB-USA has thousands of volunteers working on sanitation, water, transportation, energy, and other projects in over forty countries. EWB-USA is affiliated with several other national humanitarian organizations through an umbrella group, Engineers Without Borders-International.

Ingenieurs Sans Frontieres (ISF) in France (established in 1982) is the oldest national society with the title "Engineers Without Borders," though it is not a member of EWB-International (Paye 2010). The society has evolved over time from a focus on technical intervention in development efforts by amateur volunteers including students, to professionalization of the membership and engagement with major funders, and more recently, a return to a focus on student involvement under the slogan "the socially-responsible engineer."

Belgium has three humanitarian engineering organizations, all members of EWB-International (Meganck 2010). Ingenieur Assistance Internationale (IAI-ISF) is located in the Walloon region, the French-speaking part of the country, and its members are primarily practicing and retired engineers. The other two are in the Flemish region of the country. Ingenieurs zonder Grenzen (established in 1992), has a makeup similar to IAI-ISF. The other organization, part of a larger group called Ex-Change, provides expert consultations on sustainable development in underdeveloped nations.

44.6.2 Engineering Societies Oriented Toward Social Responsibility and Social Justice

While humanitarian engineering societies often attend to issues of social responsibility and social justice, other groups have formed that more directly identify social responsibility as part of their mission. For example, the Union of Concerned Scientists (UCS), founded in 1969, arose out of concern its founders had over the misuse of science in the Vietnam War and in contributing to environmental pollution. The current mission statement of the UCS (n.d.), both reflects and expands on the initial mission:

> The Union of Concerned Scientists puts rigorous, independent science to work to solve our planet's most pressing problems. Joining with people across the country, we combine technical analysis and effective advocacy to create innovative, practical solutions for a healthy, safe, and sustainable future.

Broader in scope and geographic reach than UCS, an international network of academics, practitioners, and activists formed Engineering, Social Justice and Peace (ESJP) in 2004. ESJP (2010) is committed to:

> envisioning and practicing engineering in ways that extend social justice and peace in the world. This commitment manifests in two major areas: First, by understanding how technology and society are co-constructed, we are committed to identifying and dismantling specific occurrences of injustice related to engineering and technology. Second, in collaboration with community groups facing specific structures of injustice, we are committed to devising and developing technologies and other engineering solutions (broadly conceived) to the problems they face.

44.6.3 Honorary/Public Service Engineering Societies

Many different honorary engineering societies exist around the globe. The US National Academy of Engineering (NAE) (n.d.), which was founded in 1964, elects individuals from the engineering profession to serve as members. The NAE convenes expert committees to help advise on a variety of issues of interest to the profession and the broader public. It is not a government agency, but it is often asked to provide advice to the US federal government. The Royal Academy of Engineering (2018) (established in 1976) plays a parallel role in the UK. Other countries with national academies of engineering include China, India, and Australia.

Sigma Xi (n.d.), founded in 1886 by engineering students and faculty at Cornell University, is an international scientific research honorary society with "over 500 chapters in the U.S., Canada and other countries, including Switzerland, Thailand, Lebanon, New Zealand and Australia." Among other activities, Sigma Xi is active in the areas of research ethics and science communication.

The Corporation of the Seven Wardens (n.d.) is a Canadian organization that coordinates the Ritual of the Calling of an Engineer, in which engineering students recite their Obligation to follow engineering duties and responsibilities and are honored with an iron ring. A similar practice, based on the Canadian precedent, takes place in the US and is referred to as the Order of the Engineer (n.d.).

Some engineering societies directly mention public service goals as part of their mission. For example, the VDI Association of German Engineers is an interdisciplinary group of experts; the VDI (n.d.) states that its experts "share their technological knowledge with others for the general good."

44.7. Selected Issues of Concern to Engineering Societies that Highlight Social Responsibility and Public Policy

44.7.1 Responsibilities to Research Subjects

Martin and Schinzinger (2005: 88–95) argue that engineering can be thought of as a form of social experimentation. This is due in part because the consequences of introducing new technology into society can be difficult to predict; a variety of unforeseen benefits and harms can emerge. Martin and Schinzinger's notion is intriguing to consider. But in this section, the narrower, traditional sense of the phrase "human subjects research" as defined by federal or other policies will be the focus of the discussion. Researchers also have ethical obligations, which in part may be encapsulated in national laws or regulations, when conducting experiments on non-human animals, but that topic will not be discussed here.

Those who conduct research on human beings are expected to uphold fundamental ethical principles. In the US and many other regions of the world, the principles that guide engineers or other researchers working with human subjects are as follows: *respect for persons, beneficence, non-maleficence,* and *justice* (National Commission 1979).

In the context of research, respect for persons refers to whether individuals have a meaningful opportunity to make a voluntary choice about participation in a study. Interconnected with the principle of respect for persons is informed consent, a process that is supposed to ensure potential research subjects are participating voluntarily. Beneficence is the ethical obligation that engineers and other researchers have to do as much good for research subjects, or for the larger population that the subjects are part of, as possible. The principle of non-maleficence is the obligation to "do no harm." Within the context of research, justice pertains to how the benefits and risks of a study are distributed; the principle also encompasses whether individuals or groups have a meaningful opportunity to participate in research studies.

Given the broad scope and effects of human subjects research, the resulting ethical challenges may need to be addressed by both individual engineers and engineering organizations. Engineering organizations may need to be involved in informing deliberations about when, and under which specific conditions, it is appropriate to perform research on human beings. For example, vast amounts of data are being collected by social media companies about Internet users, and from that data, much could be learned about the users' beliefs and behaviors. In fact, Internet users may be research subjects in countless studies without even being aware that such activities are occurring. For example, over 680,000 people were part of an experiment on Facebook that involved manipulating the content of their newsfeeds, and they were not directly informed that the experiment was taking place (Kramer et al. 2014). Engineering organizations should be involved in deliberations regarding what constitutes proper research practice, including when personally identifiable information is being collected from human subjects.

44.7.2 Privacy and Security

Perhaps the defining ethical, social, and legal issue of the digital age is privacy and its counterpart, security. In the digital age, countless technologies raise privacy concerns, ranging from personal computers to the Internet to mobile applications embedded in smartphones. Emerging technologies such as the Internet of Things and robotics only serve to heighten fears that privacy may be diminishing or eroding to such a degree that it might not be recoverable. Protecting private information from unauthorized business uses, unjustified government surveillance, and criminal activity has been discussed for decades now (e.g., Forester and Morrison 1991). Yet, revelations of system vulnerabilities seem to be almost a daily topic in the news, a recent example being that most of the world's microprocessor chips have flaws that hackers can exploit (Neuman 2018).

One approach to these issues is "Privacy by Design" (PbD), a concept developed by Ann Cavoukian, the former Information and Privacy Commissioner of Ontario, Canada. According to Cavoukian and Winn (2012), "PbD is a proactive approach to privacy protection. It seeks to avoid data breaches and their consequential harm, thereby being preventative in nature." Originally developed for large-scale data systems, PbD has been applied to such technologies as smart power meters (Cavoukian and Winn 2012). Engineers have begun to embrace the concept of "ethical design" which, in addition to privacy, could apply to a broad range of issues including safety and the Digital Divide (Baldini et al. 2018). Professional societies could add to this momentum by articulating technical standards for the privacy-related features of technology.

44.7.3 Social Justice

Like social responsibility, it can be difficult to arrive at a precise, consensus definition of social justice. Yet the concept seeks to capture the ethical dimensions of how individuals and groups are treated in society. Social justice is integrally interwoven with notions of fairness and oftentimes equality. Roughly stated, upholding social justice can entail whether each person is protected by the same rights and can meaningfully pursue the same opportunities as other members of society. Social justice scholars seek to investigate whether, how, and to what degree segments of society have historically and/or presently suffered from unfair discrimination. Integrally tied to that type of inquiry is determining the kinds of changes in behavior and policy that could potentially mitigate or remedy unjust practices and contribute to a more just future.

A frequent topic at the intersection of social justice and engineering is the distribution of a technology's benefits and risks. For example, if a city's investment in an infrastructure project disproportionately improves the lives of the wealthy over those who are poor, that can be a social justice concern; depending on the engineers' role in the design and implementation of the project, they may bear at least some moral responsibility for the project's asymmetrical impacts on the public. Another example illustrating the concept of social justice is creating a potentially lifesaving medical device, which due to its high price, would only realistically be available to affluent customers. Engineers can have a decisive role in a technology's availability to the public. In this specific instance, if an engineer deliberately chooses to use more expensive materials in a medical device, rather than using less costly alternatives of comparable quality and reliability, that could directly impact the device's accessibility. In effect, this and other design decisions could prevent some portions of the public from being able to use the device.

44.7.4 Environmental Justice

A subset of social justice concerns intersects with environmental justice. The realm of environmental justice often, but not always, focuses on the harmful effects of the policy decisions related to the placement of hazardous facilities and waste or other forms of pollution in a community. For example, a series of reports by the United Church of Christ (UCC 2007) suggests that the placement of solid waste facilities in the United States is correlated to regions densely populated with people of color. While the existence and nature of a causal relationship in this case is difficult to ascertain, the UCC reports indicate that people of color are disproportionally exposed to the risks of living in close proximity to hazardous waste.

Professional societies can have a role in addressing the issue of environmental justice by, for example, seeking to inform regulatory entities that determine policies on land and water use. Engineering codes of ethics are with greater regularity emphasizing an engineer's ethical obligations to the environment, which could include a responsibility to address environmental justice concerns and sustainability issues. The latter topic will be the focus of the next section.

44.7.5 Sustainability and Sustainable Development

The foregrounding of discussions about sustainable development within professional and other communities traces back to the United Nations World Commission on Environment and Development's 1987 report entitled *Our Common Future*, which is also referred to as the Brundtland Report. According to the Report, sustainable development is "development that meets the needs of the present without compromising the ability of future generations to meet their own needs" (WCED 1987: 43). Although it arguably needs to be further specified and operationalized, the definition serves as the starting point and foundation for the field of sustainability.

Historically, engineering societies did not directly mention the environment within their codes of ethics. However, over the last few decades, obligations to the environment, often including an overt statement about the importance of sustainable development, have begun to appear. Within the US engineering societies, ASCE was the first to mention sustainable development. The current ASCE code (2017) states that engineers "shall strive to comply with the principles of sustainable development in the performance of their professional duties." According to ASCE (2017), "Sustainable Development is the process of applying natural, human, and economic resources to enhance the safety, welfare, and quality of life for all of the society while maintaining the availability of the remaining natural resources."

ASME (2011) states in its code of ethics that "Engineers shall consider environmental impact and sustainable development in the performance of their professional duties." IEEE (2017) revised the paramountcy clause in its code to state that engineers are "to hold paramount the safety, health, and welfare of the public, to strive to comply with ethical design and sustainable development practices, to protect the privacy of others, and to disclose promptly factors that might endanger the public or the environment."

The Code of Ethics of Engineers Australia (2013) has an entire section entitled "promote sustainability"; it includes subsections entitled "engage responsibly with the community and other stakeholders," "practice engineering to foster the health, safety and wellbeing of the community and the environment," and "balance the needs of the present with the needs of future generations." As with other components of the codes, the tenets related to the environment are often directed toward the individual engineer. Yet the environmental challenges humanity faces are frequently so complex and large scale that they may require the expertise and resources of engineering societies.

Even though a settled scientific consensus has been reached on the existence of climate change (e.g., IPCC 2014), it can of course be politically contentious regarding how to react to the resulting problems. Arguably, the engineering profession needs to take the potential impacts of climate change into account when making decisions (e.g., transportation infrastructure planning) (Douglas et al. 2017). This can include how assumptions are formulated when creating computer models or simulations. The effects of climate change can raise social justice concerns (e.g., disproportionate levels of flooding in poor regions), and engineers can play a role in the distribution and management of the negative consequences of climate change. In addition to mitigating climate change's effects, engineers have the expertise to devise solutions that lessen the magnitude of climate change, including improvements in energy efficiency and development of energy resources that do not emit greenhouse gases (Pacala and Socolow 2004).

44.7.6 Dual Use

Traditionally, "dual use" is defined as information or technology that has both civilian and military applications (Stowsky 1997). The US NIH (n.d.) discusses a similar concept referred to as "dual use research of concern"; the phrase captures the notion that certain types of research may have been intended to serve a beneficial purpose but could be maliciously misused to cause widespread harm. Much of the conversation on dual use research has emerged in the life sciences, particularly in connection with biological weapons potentially deployed by nations, terrorist groups, or even disgruntled individuals. For example, global governments and organizations considered blocking the publication of avian flu studies, or at least redacting essential details about them, in anticipation of malicious actors trying to exploit the research findings (Yong 2012).

The engineering community, including its professional societies, have important ethical responsibilities in the "dual use" realm. This can include making recommendations on whether to place publication restrictions on research findings that have a high potential for causing large-scale harm to the public. For example, a paper that reveals the vulnerabilities in the power grid could serve as an impetus for making security improvements, but it could arguably provide the foundation for a malevolent attack (Markoff and Barboza 2010). A similar type of debate could ensue related to sharing information about the aforementioned security vulnerability in widely used computer processes (Neuman 2018). Engineering societies can actively participate in deliberations, and author policy statements, on the sharing of dual use research with the public.

44.7.7 Weapons Development

The history of engineering is often intertwined with weapons research and development (Riley 2008). The term "civil engineer" was coined to distinguish engineers who work on civilian projects from those engaged in weapons-related projects. Much of the modern conversation about the ethics of weapons development was sparked by the Manhattan Project and the resulting dropping of the two atomic bombs on Japan. Debates ensued about the appropriateness of continuing to pursue nuclear weapons after World War II. The issue created quite a divide in the scientific and engineering world, a divide often filled by quasi-professional engineering and science societies. For example, one of the core tenets of the Bulletin of the Atomic Scientists (n.d.), which made its first appearance in 1945, is to advocate for the abandonment of nuclear weapons development. The UCS has a similar goal of trying to prevent the proliferation of nuclear weapons.

Reasonable people can disagree about whether it is ethically acceptable for engineers to partici-pate in weapons development (even if there is widespread agreement that certain types such as chem-ical and biological weapons should be forbidden). Nevertheless, professional engineering societies have a role to play in considering the ethical and public policy implications of weapons development. As will be discussed later in this chapter, an area of increasing concern to both quasi-professional and traditional professional engineering societies is the design and deployment of military robots, includ-ing autonomous weapons systems.

44.7.8 Communicating With the Public

Given how many facets of human life are intertwined with engineering, an engineer's ability to communicate clearly, competently, and honestly with the public takes on crucial importance. For example, engineers could help inform the public on matters ranging from how to operate a particular device and the impact the device may have on daily life, to how to prepare for a natural disaster and whether evacuating one's home is necessary. Obviously during the digital age, many methods are available for obtaining information, and countless sources seek to provide it. Correspondingly, engi-neering communities need to find mediums that the public uses and trusts (for example, television shows, podcasts, blogs, etc.) and convey information in a way that the public is likely to understand. A related challenge is whether to work with the popular press, science journalists, or other interme-diaries to convey the findings competently, or whether reaching out to the public more directly is a better approach.

In addition to informing the public, engineers have an obligation to *listen* to and understand the concerns of members of the public, especially when they are potentially put at risk by engineering decisions. Social scientists have found that public perceptions of risk, while different from those of technical experts, can provide valuable information to the risk assessment process that is often over-looked by experts (Slovic 1987). While a two-way model of risk communication has long been the expected norm, many technical experts still cling to the traditional model of experts "informing" the public about risk (Árvai 2014). Engineering societies can play a key role in structuring more effective and ethical risk assessment processes. Instead of merely seeking to "inform," societies can listen and participate in meaningful dialogue with the public.

44.8. Conclusion: New Roles for Professional Engineering Societies

Over the last several years, engineers, both individually and collectively, have been involved in efforts to articulate and clarify what their social responsibilities are in connection with a range of emerging technologies, such as artificial intelligence (AI), robotics, autonomous vehicles (AVs), the Internet of Things, and human enhancement technologies. Contrary to the historical disconnect between ethics and policy discussed earlier, professional engineering societies are beginning to exhibit leadership in terms of integrating ethical and public policy considerations in all three of the aforementioned activ-ity areas—codes of ethics, standards, and policy positions or initiatives.

Part of what sparked debate about the responsibility of scientists and engineers at the turn of the century was an article in *Wired* written by noted computer scientist and entrepreneur Bill Joy (2000). He warned of the dangers of genetics, nanotechnology, and robotics (GNR), going so far as to ask whether a moratorium on GNR research may be necessary. Joy's argument was roundly criticized in computing and engineering communities (e.g., Kurzweil 2001; Dyson 2003). Yet less than two decades later, many distinguished technical leaders have made statements to the effect that the devel-opment of AI, particularly with respect to military applications, may pose an existential threat to humanity (Hawking et al. 2014; Eadicicco 2015).

Appreciating the impact that AI may have on countless facets of human life, IEEE (2019) commissioned a series of reports entitled *Ethically Aligned Design* as part of its Global Initiative on Ethics of Autonomous and Intelligent Systems. The intent of the reports is to provide a comprehensive overview of ethical issues related to AI and robotics, including how computing devices may affect human well-being, how they should adhere to differing cultural norms, and the ethical appropriateness of their attempt to nudge human behavior. Interconnected with this effort, IEEE has also formed a collection of working groups that seek to develop standards for ethical design of AI systems (Bryson and Winfield 2017). Also, as previously noted, IEEE (2020) highlights the importance of "ethical design" and singles out "intelligent systems" as a particular area of interest in its Code of Ethics.

Among the emerging technologies that are expected to have the greatest effect on the public's well-being are autonomous vehicles. Concerned about the effect that autonomous vehicles may have on public safety, especially if insufficient testing has been conducted and if a licensed engineer is not directly involved in the design process, NSPE has released relevant policy statements (NSPE 2017; Kaplan-Leiserson 2016). NSPE has sought to inform the decision-making of the US Congress while it is in the process of crafting federal transportation legislation. This illustrates that the professional society sees itself as having a responsibility to proactively address ethical issues related to the design and deployment of the technology.

The use of robots, drones, and other autonomous or semi-autonomous technologies for military purposes is another focal point that highlights the responsibilities of the engineering profession. The Future of Life Institute (2015), for example, issued an open letter on the use of autonomous weapons. Signatories of the letter seek to condemn the development and use of "offensive autonomous weapons beyond meaningful human control." Though not an engineering society *per se*, many engineers are affiliated with the Institute. In 2017, a group of prominent leaders in the AI and robotics industries sought to advise the United Nations to consider a ban on autonomous weapons (Gibbs 2017). Such statements made by engineering organizations could have global implications.

Historically, professional engineering societies have tended not to directly address the collective social responsibilities of the profession. While this void has been partially filled by humanitarian and social justice-oriented quasi-professional societies such as EWB and UCS, these groups often lack the influence of the larger, more mainstream societies. As the examples in this chapter have shown, social and ethical concerns regarding emerging technologies highlight the need, and opportunities, for traditional societies to assume leadership roles in addressing such concerns.

Related Chapters

Chapter 25: Sustainable Design (Steven A. Moore)
Chapter 33: Values in Risk and Safety Assessment (Niklas Möller)
Chapter 34: Engineering and Sustainability: Control and Care in Unfoldings of Modernity (Andy Stirling)
Chapter 39: Philosophy of Security Engineering (Wolter Pieters)
Chapter 41: Autonomy in Engineering (Eugene Schlossberger)
Chapter 42: Standards in Engineering (Paul B. Thompson)
Chapter 43: Professional Codes of Ethics (Michael Davis)
Chapter 45: Engineering as a Political Practice (Govert Valkenburg)
Chapter 46: Global Engineering Ethics (Pak-Hang Wong)
Chapter 47: Engineering Practice and Engineering Policy: The Narrative Form of Engineering Policy Advice (Natasha McCarthy)
Chapter 49: Socially Responsible Engineering (Jennifer Smith and Juan Lucena)
Chapter 50: Engineering and Social Justice (Caroline Baillie)

Chapter 51: Engineering and Environmental Justice (Benjamin R. Cohen)
Chapter 52: Beyond Traditional Engineering: Green, Humanitarian, Social Justice and *Omnium* Approaches (George D. Catalano)

References

Árvai, J. (2014). The End of Risk Communication as We Know It. *Journal of Risk Research*, 17(10), 1245–1249. doi:10.1080/13669877.2014.919519.

ASCE (2013). *Policy Statement 139—Public Involvement in the Decision Making Process*. American Society of Civil Engineers. www.asce.org/issues-and-advocacy/public-policy/policy-statement-139-public-involvement-in-the-decision-making-process/. Accessed February 2, 2018.

ASCE (2015). *Policy Statement 360—Impact of Climate Change*. American Society of Civil Engineers. www.asce.org/issues-and-advocacy/public-policy/policy-statement-360—impact-of-climate-change/. Accessed January 23, 2018.

ASCE (2017). *Code of Ethics*. American Society of Civil Engineers. www.asce.org/code-of-ethics/. Accessed January 23, 2018.

ASCE Hurricane Katrina External Review Panel (2007). *The New Orleans Hurricane Protection System: What Went Wrong and Why*. Reston, VA: American Society of Civil Engineers.

ASME (2011). *The History of ASME's Boiler and Pressure Vessel Code*. American Society of Mechanical Engineers. www.asme.org/engineering-topics/articles/boilers/the-history-of-asmes-boiler-and-pressure. Accessed February 6, 2018.

ASME (n.d.). *Policy Publications*. American Society of Mechanical Engineers. www.asme.org/about-asme/advocacy-government-relations/policy-publications. Accessed February 2, 2018.

Baldini, G., Botterman, M., Neisse, R. et al. (2018). Ethical Design in the Internet of Things. *Science and Engineering Ethics*, 24, 905–925. doi:10.1007/s11948-016-9754-5.

Bryson, J. and Winfield, A. (2017). Standardizing Ethical Design for Artificial Intelligence and Autonomous Systems. *Computer*, 50(5), 116–119.

Bulletin of the Atomic Scientists (n.d.). *Background and Mission: 1945–2018*. https://thebulletin.org/background-and-mission-1945-2018. Accessed January 23, 2018.

Cavoukian, A. and Winn, C. (2012). *Applying Privacy by Design Best Practices to SDG&E's Smart Pricing Program*. Information & Privacy Commission, Ontario. www.sdge.com/sites/default/files/documents/pbd-sdge_0.pdf. Accessed January 30, 2018.

Chertow, M., Lenz, J. and Plourde, R.P. (1993). Whistle-Blowing—A Professional Responsibility. *Journal of Professional Issues in Engineering Education and Practice*, 119(1), 27–30.

Coeckelbergh, Mark (2006). Regulation or Responsibility? Autonomy, Moral Imagination, and Engineering. *Science, Technology, & Human Values*, 31(3), 237–260.

The Corporation of the Seven Wardens (n.d.). *The Calling of an Engineer*. www.ironring.ca/. Accessed January 23, 2018.

Douglas, Ellen, Jacobs, Jennifer, Hayhoe, Katharine, Silka, Linda, Daniel, Jo, Collins, Mathias and Alipour, Alice et al. (2017). Progress and Challenges in Incorporating Climate Change Information Into Transportation Research and Design. *Journal of Infrastructure Systems*, 23(4), 04017018.

Dyson, F. (2003). The Future Needs Us! *New York Review of Books*, 50(2).

Eadicicco, Lisa (2015). Bill Gates: Elon Musk Is Right, We Should All Be Scared of Artificial Intelligence Wiping Out Humanity. *Business Insider*, January 28. www.businessinsider.com/bill-gates-artificial-intelligence-2015-1. Accessed January 23, 2018.

Engineering, Social Justice and Peace (ESJP) (2010). *Our Commitments*. http://esjp.org/about-esjp/our-commitments. Accessed June 17, 2018.

Engineers Australia (2013). *Our Code of Ethics*. www.engineersaustralia.org.au/sites/default/files/2017-05/Engineers%20Australia%20Code%20of%20Ethics.pdf. Accessed January 23, 2018.

Engineers Australia (2018). Government and Policy. www.engineersaustralia.org.au/Government-And-Policy. Accessed January 23, 2018.

Engineers Canada (n.d.). *National Position Statements*. https://engineerscanada.ca/public-policy/national-position-statements. Accessed January 23, 2018.

Engineers Ireland (2018). *Code of Ethics*. www.engineersireland.ie/getattachment/About/Code-of-Ethics-and-Bye-laws/Revised-Code-of-Ethics.pdf.aspx. Accessed January 23, 2018.

Engineers Ireland (n.d.). *Policy*. www.engineersireland.ie/communications/policy.aspx. Accessed February 2, 2018.

Engineers Without Borders-USA (EWB-USA) (n.d.). *Mission & History*. www.ewb-usa.org/about-us/mission-and-history/. Accessed January 23, 2018.

Forester, Tom and Morrison, Perry (1991). *Computer Ethics: Cautionary Tales and Ethical Dilemmas in Computing.* Cambridge, MA: MIT Press.

Future of Life Institute (2015). *Autonomous Weapons: An Open Letter from AI & Robotics Researchers.* https://futureoflife.org/open-letter-autonomous-weapons/. Accessed January 23, 2018.

Gibbs, Samuel (2017). Elon Musk Leads 116 Experts Calling for Outright Ban of Killer Robots. *The Guardian,* August 20. www.theguardian.com/technology/2017/aug/20/elon-musk-killer-robots-experts-outright-ban-lethal-autonomous-weapons-war. Accessed February 2, 2018.

Hawking, Stephen, Russell, Stuart, Tegmark, Max and Wilczek, Frank (2014). Stephen Hawking: 'Transcendence Looks at the Implications of Artificial Intelligence—but are We Taking AI Seriously Enough?' *The Independent,* May 1. www.independent.co.uk/news/science/stephen-hawking-transcendence-looks-at-the-implications-of-artificial-intelligence-but-are-we-taking-9313474.html. Accessed January 23, 2018.

Herkert, J.R. (2003). Professional Societies, Microethics, and Macroethics: Product Liability as an Ethical Issue in Engineering Design. *International Journal of Engineering Education,* 19(1), 163–167.

IEEE (2019). *Ethically Aligned Design: A Vision for Prioritizing Human Well-being with Autonomous and Intelligent Systems.* http://standards.ieee.org/content/dam/ieee-standards/standards/web/documents/other/ead1e.pdf. Accessed October 31, 2020.

IEEE (2020). *IEEE Code of Ethics.* Institute of Electrical and Electronics Engineers. www.ieee.org/about/corporate/governance/p7–8.html. Accessed October 31, 2000.

Intergovernmental Panel on Climate Change (IPCC) (2014). *Climate Change 2014: Synthesis Report.* Contribution of Working Groups I, II and III to the Fifth Assessment Report of the Intergovernmental Panel on Climate Change [Core Writing Team, R.K. Pachauri and L.A. Meyer (eds.)]. IPCC, Geneva, Switzerland.

Japan Society of Civil Engineers (2014). *Code of Ethics for Civil Engineers.* www.jsce-int.org/about/p_engineer. Accessed January 23, 2018.

Joy, B. (2000). *Why the Future Doesn't Need Us.* https://www.wired.com/2000/04/joy-2/. Accessed November 1, 2020.

Kaplan-Leiserson, Eva (2016). Driving the Future. *PE Magazine,* January/February. www.nspe.org/resources/pe-magazine/january-2016/driving-the-future. Accessed February 2, 2018.

Kramer, A.D.I., Guillory, J.E. and Hancock, J.T. (2014). Experimental Evidence of Massive-scale Emotional Contagion through Social Networks. *PNAS,* 111(24), 8788–8790.

Kurzweil, Ray (2001). *Promise and Peril.* www.kurzweilai.net/promise-and-peril. Accessed January 23, 2018.

Ladd, John (1980). The Quest for a Code of Professional Ethics: An Intellectual and Moral Confusion. In Rosemary Chalk, Mark S. Frankel and S.B. Chafer (eds.), *AAAS Professional Ethics Project: Professional Ethics Activities in the Scientific and Engineering Societies.* Washington, DC: AAAS, pp. 154–159.

Layton, E.T. (1971). *Revolt of the Engineers: Social Responsibility and the American Engineering Profession.* Cleveland, OH: Case Western Reserve University Press.

Markoff, John and Barboza, David (2010). Academic Paper in China Sets Off Alarms in U.S. *The New York Times,* March 20.

Martin, M.W. and Schinzinger, R. (2005). *Ethics in Engineering,* 4th ed. New York: McGraw Hill.

May, L. (1982). Professional Action and the Liabilities of Professional Associations: ASME v. Hydrolevel Corp. *Business & Professional Ethics Journal,* 2(1), 1–14.

Meganck, M. (2010). From Boy Scouts and Missionaries, to Development Partners. *IEEE Technology and Society Magazine,* 29(1), 27–34.

National Academy of Engineering (NAE) (n.d.). *About the National Academy of Engineering (NAE).* www.nae.edu/About.aspx. Accessed January 23, 2018.

National Commission for the Protection of Human Subjects of Biomedical and Behavioral Research (1979). *The Belmont Report: Ethical Principles and Guidelines for the Protection of Human Subjects of Research.* www.hhs.gov/ohrp/regulations-and-policy/belmont-report/index.html. Accessed February 2, 2018.

National Institutes of Health (NIH) (n.d.). *Dual Use Research of Concern.* https://osp.od.nih.gov/biotechnology/dual-use-research-of-concern/. Accessed January 23, 2018.

National Society of Professional Engineers (2017). *NSPE Position Statement No. 1772 Autonomous Vehicles.* www.nspe.org/resources/issues-and-advocacy/position-statements/autonomous-vehicles. Accessed February 2, 2018.

Neuman, Scott (2018). Intel Acknowledges Chip-Level Security Vulnerability in Processors. *NPR: The Two-Way,* January 4. www.npr.org/sections/thetwo-way/2018/01/04/575573411/intel-acknowledges-chip-level-security-vulnerability-in-processors. Accessed January 23, 2018.

Order of the Engineer (n.d.). www.order-of-the-engineer.org/. Accessed January 23, 2018.

Pacala, S. and Socolow, R. (2004). Stabilization Wedges: Solving the Climate Problem for the Next 50 Years with Current Technologies. *Science,* 305(5686), 968–972.

Paye, S. (2010). Ingénieurs Sans Frontières in France: From Humanitarian Ideals to Engineering Ethics. *IEEE Technology and Society Magazine,* 29(1), 20–26.

Riley, D.M. (2008). *Engineering and Social Justice*. Williston, VT: Morgan and Claypool Publishers.

Riley, D.M. and Lambrinidou, Y. (2015, June). Canons against Cannons? Social Justice and the Engineering Ethics Imaginary. Paper presented at *2015 ASEE Annual Conference & Exposition*, Seattle, Washington. 10.18260/p.23661

Royal Academy of Engineering (2018). *40 Years of the Academy*. www.raeng.org.uk/about-us/what-we-do/40-years-of-the-academy. Accessed January 23, 2018.

Russell, A.L. (2005). Standardization in History: A Review Essay With an Eye to the Future. *The Standards Edge: Future Generations*, 247–260.

Sigma Xi (n.d.). *Come Up Higher: The History of Sigma Xi*. www.sigmaxi.org/about/history. Accessed January 23, 2018.

Slovic, P. (1987). Perception of Risk. *Science*, 236(4799), 280–285.

Stowsky, Jay (1997). The Dual-Use Dilemma. *Issues in Science and Technology*, 13(2) (Winter).

Union of Concerned Scientists (UCS) (n.d.). *Union of Concerned Scientists Mission Statement*. www.ucsusa.org/about/mission.html#.WnS3k6inFPY. Accessed February 2, 2018.

United Church of Christ (UCC) (2007). *Toxic Wastes and Race and Toxic Wastes and Race at Twenty (1987–2007)*. www.ucc.org/environmental-ministries_toxic-waste-20. Accessed January 23, 2018.

VDI Association of German Engineers (n.d.). *VDI Societies*. www.vdi.eu/engineering/. Accessed January 23, 2018.

WCED (1987). *Our Common Future. Report of the World Commission on Environment and Development*. Oxford: Oxford University Press.

Woodhouse, E.J. (2001). Curbing Overconsumption: Challenge for Ethically Responsible Engineering. *IEEE Technology and Society Magazine*, 20(3), 23–30.

Yong, Ed (2012). The Risks and Benefits of Publishing Mutant Flu Studies. *Nature News*. www.nature.com/news/the-risks-and-benefits-of-publishing-mutant-flu-studies-1.10138. Accessed January 23, 2018.

45

ENGINEERING AS A POLITICAL PRACTICE

Govert Valkenburg

45.1. The Technological and the Political

Making technology has consequences for the social world in which those technologies will be used: they open up new possibilities for action, close off others, and continue to have this influence once their makers have long left the scene. That is to say, technologies have agency: the capacity to act, and to effectuate change—whether we see this as an extension of what is still ultimately human agency, or as an agency beyond the human and something technology has in and of itself.[1] If we think of politics as a range of ways people arrange the organization of their societies—whether democratically, by the exercise of power, or any other form of rule—then it is easy to see how technologies, when seen as having agency, are relevant to politics: they directly bring influence to bear on the composition of the world, and what it affords and constrains. If we grant technologies a place in our thinking about politics, we branch out from what is conventionally thought of as politics: the institutional arrangements, i.e. *human-social relations*, that people make to resolve differences, conciliate interests, and make collective decisions (Crick 2004). This chapter explores the connection between engineering as the making of technology, and specifically how this making of technology is ultimately a political affair.

The exploration of technologies as political artifacts has a long history. One seminal example is the analysis by Langdon Winner (1988), who explained how bridges were purposely built low, such that only cars could pass under them, not buses.[2] By consequence, the area secluded by those bridges, a recreational area with beach access, could be negotiated only by rich people travelling by car, not poor people travelling by bus. Importantly, the rich-poor divide coincided with racial divides. This is thus an example of an infrastructural arrangement that is instigated by (disputable) social ideals, and which in turn reproduces social relations.

In addition, Bruno Latour (1987, 1988, 1993) explored the problematic nature of the distinction between technological and political relations and proposed that we should think of political and technological relations as essentially of the same kind, with distinction between them being as itself a matter of negotiation rather than given *a priori*. Are trains and railways political technologies? Not if we just think of them as means of transportation, and they are indeed usually staged as such. But the role of railways in sociopolitical history is hardly overestimated, and they have been very much political, for example in European integration (Schot and Vleuten 2004). It depends on the exact relations in which technologies end up, whether we can easily recognize them as political and make a convincing case for them to be seen as political.

Finally, Andrew Feenberg (1991, 2002, 2005, 2017) expanded upon the Marxist insight that technological relations pre-structure socioeconomic relations. Notably, he argues that scientific rationality pretendedly settles technological design choices in a fair and neutral way, while in fact it *underdetermines* technologies, leaving the implementation in the context of application vulnerable for colonizing by dominant interests. On the one hand, we see that indeed the underdetermination of the Internet and social media allows for them to be captured by great neoliberal interests. At the same time, the Internet in its underdetermination allows for marginalized groups to organize and emancipate themselves (Feenberg and Friesen 2012).

Zooming in from the political character of technology towards the narrower thematic of how we can think politically about the practice of *engineering*, or *making technologies*, leads to three sub-questions. The first is what it means to think about engineering as a *practice*. What happens in engineering, if there is a distinct activity by that name in the first place? Is engineering a practice in the sense of a collection of mainly human relations, where members pursue some shared goals? What is and is not part of engineering? (see also Chapter 55, "Reimagining the Future of Engineering" by Neelke Doorn, Diane P. Michelfelder, et al., this volume). Who does engineering? What is *specific* about engineering, innovation and technology design practices compared to other practices?

The second question is how politics unfolds in the *practice itself*. In its most mundane form, this question is about whether and how what happens in the design laboratory, from the workbench to the coffee corner, is subject to political relations and even biases such as sexism, classism and racism (see also Part IV of this *Handbook* on engineering design processes). In a more fundamental sense it is about how particular methodologies, epistemologies (theories of knowing) and problem definitions gain dominance over others (see also Part II of this *Handbook* on engineering reasoning). While these appear as strictly methodological concerns that are the domain of 'the engineering way-of-knowing', they can be very political: they ultimately determine what kind of problems can or cannot be addressed in engineering and what solutions may or may not emerge.

The last question is how practices of engineering and innovation are political in an *external* sense, i.e. how they relate to social and political processes outside the engineering practice, in the broader social world. This relation is dialectical. On the one hand, the boundary between the inside and outside of engineering is permeable: things go in and out all the time, and these include technologies just as much as political content such as ideologies, regulations and conflicts—or at least, so I assume for the argument of this chapter. On the other hand, it should be assumed that the boundary of the practice itself is perpetually kept in place, renegotiated and policed. One central endeavour in this 'boundary work' (Gieryn 1983) is to perform engineering as politically neutral, so as to exempt it from politics and leave it to autonomous governance by engineers themselves. That is to say, boundary work is to keep the politics out. Another claim that enjoys continual and active support is that engineering and innovation, notwithstanding some obvious perversities, in general contribute to the common good and the prosperity of humankind; and therefore, it should not be made subject to interference from politics (which here refers to the more conventional, narrower understanding of politics limited to the aforementioned institutional arrangements). Again, an argument to keep the politics out.

Opposition to those claims is found in long-standing traditions of *technology assessment* and *constructive technology assessment* (Rip et al. 1995), and in perspectives on the ramifications of technology (including technologies in the making) into the spheres of ethics and morality (Swierstra and Rip 2007). A recent branch of this concern is found in the discourse of *responsible research and innovation* (Owen, Macnaghten et al. 2012; Schomberg 2011; Stilgoe, Owen et al. 2013), which seeks to democratize not only the use of technologies but also their development and the setting of broader innovation agendas.

All this starts, however, with the question of what makes something 'politics' or 'political', and whether there is a difference between the two. This will be the topic of the next section. After that, sections will be devoted to the respective questions: how engineering can be thought of as a practice, how this practice is internally political, and how the practice relates in a political sense to the outside world.

45.2. To Call Something Political

Notions of politics and the political abound. The point of this chapter is not to conclusively settle what politics is. But we do need a heuristic conception of politics that helps us identify the phenomena we are looking for. In fact, Mark Brown (2017) has pointed out that calling something 'political' is in itself a potentially ambiguous attribution. On the one hand, it may refer to something being relevant to politics; i.e. it may have political origins, implications, or effects. We could call this political in an 'object sense': we do politics *about* it. On the other hand, calling something political may refer to the thing itself becoming connected to and implicated in political processes. This we could call political in a 'subject sense': the thing joins in the process of doing politics. But these are only preliminary steps in thinking about what could be political about engineering practices.

Intuitively, one might think of politics as what concerns the *state* (or more generally, the *polity* if we also want to include forms of organization that differ from present-day states). Indeed, a key political philosopher such as John Rawls is primarily concerned with what he calls 'the basic structure of society', which he aspires to arrange in such a way that political decisions can be made in a way that is fair to all members of society, irrespective of the religion or *comprehensive doctrine* they abide by (Rawls 1971, 1993). Also, politics can indeed be thought of as a specific, *institutionalized* activity (Crick 2004). When referring to this narrower idea of politics, I will speak of *conventional politics*.

However, if we talk about politics, we often also talk more generally about some process of decision-making of which the outcomes are somehow collectively binding (Luhmann 2000; Peters 2004; Weale 2004). It is the natural consequence of people being social beings: from couples to entire cities and nations, many of the actions people commit are either done in concert with others, or have consequences for others. Hence, those actions will be subject to collective decision-making. Politics is what emerges if something is at stake, notably in relation to social values. This is not to say that this politics is always democratic, or somehow justified in how (the majority of) the populace believes things should be arranged. Politics can also be hegemonic, dictatorial, or conflictual, and it has even been argued that contestation is crucial to any functioning form of politics (Mouffe 2005, 2013). What unites these notions of politics is that any member of the community is bound by the emerging rules. Politics is, for now, best understood as the pursuit of collectively binding decisions in a context of power differences and conflicts of value interests.

Following Brown (2014), two questions emerge here. First, how does something create *matters of concern*, dealing with which must be reckoned as political? This is the object sense of politics: it is something we do politics on. One example is that any modern society will have members that want to have a smaller state and lower taxes, as well as members that want a larger state and higher taxes. Decisions on what should or should not be the concern of the state are thus clearly matters of concern that require politics. Alternatively, one could think of the influx of African and Middle-Eastern immigrants that Europe is currently facing. Regardless of how small or big immigration and possible problems in its wake really are, the issue sparks concern among many and is readily mobilized by some politicians and media. This makes it political, which in this case is not clearly connected to how things are in any objective sense.

On a side note, an important aspect of matters of concern, elaborated by Noortje Marres (2007), is that these matters of concern unite and even create publics, and thus form a vital element in the emergence of politics. This means that even if we see an object as something on which politics is 'done', it is never *just* something on which politics is done, but always something on which we came to do politics through specific histories, fraught with interests and power relations.

The second question that emerges is how something potentially becomes a *site of politics* in the sense that the prioritization of alternatives becomes a matter of negotiation, power, and possibly conflict. This is the subject sense of politics: something comes to play a part in the process of politics, or the process of collective world making. Again, a straightforward example is the fact that parliaments and ministries take and enforce decisions that matter to the public at large, and produce the world in which the public lives. An example that often stays under the radar is the fact that history education in primary and secondary school usually appears as a mere conveyance of facts regarding the past, while in fact it serves an important role in the production of nationhood and citizenship. Thus, history education is quintessentially political in the subject sense.

The potential to create publics (Marres 2007) already hints at the subject and object senses sometimes being hard to distinguish. Yet, things can also contribute to politics without becoming an issue, or a matter of concern. For example, the aforementioned example of social media that shape how certain forms of information (and a lot at that!) comes to us, also entails that the information on which we base our political positions might somehow be skewed. Politics is 'done' *in* those media which thereby become a site of politics, without their being explicitly *objects* of politics.

Posing these questions specifically about engineering will be central to the following sections. The reason we have to engage with these questions is that engineering practices are not necessarily open to participation by all. If we accept that engineering and its products might be political in either an object or a subject sense or both, and we observe that not all people are members of the engineering practice, than we have *de facto* something political that is not generally accessible. This means that a certain realm of politics is kept outside of public reach, entailing that it cannot (straightforwardly) be called to account, and that the public cannot (straightforwardly) participate in decision-making processes that concern them. In short, we might be facing a *democratic deficit*; a place where politics fails to be democratic politics. This is the whole point in exploring engineering as a political practice. Things that engineering practices may impose on the broader world may call for some form of democratic, political control.

45.3. Engineering as a Practice

To think through engineering as a political practice, the notion of practice needs to be developed first. Various notions of practice exist (MacIntyre 1981; Schatzki 2012), and a conclusive definition is beyond the scope of this chapter. A number of shared properties, though, can easily be distilled. To begin with, a practice is essentially social: the practice consists of human members. These members engage in relations and interactions. Second, these relations and interactions are structured according to norms: some things are accepted and others are not. Norms are taken in a broad sense here, ranging from unreflexive routines to deep moral concerns, from mundane habits to heartfelt traditions, and including agreements, standards, etc. This is not to say that all members of the practice will agree on each and every norm, but there is sufficient shared normativity to bind the practice together.

While these first two properties of being 'social' and 'norm-based' could also be said about society at large, the third one sets practices apart as a more specific level of aggregation and analysis: practices are defined as sharing an activity or set of activities that are oriented to some commonly understood good. This means the celebration of religious beliefs for church communities, the pursuit of elegant and winning play in football, and the creation of working, efficient, and elegant technologies in engineering. This shared notion of a specific good entails that the practice has more or less clearly defined boundaries. This notion of practice is mostly associated with MacIntyre (1981) but also resonates well with how De Vries (2007) conceives more specifically of technoscientific practices (see later in this chapter). Also, it needs to be said that with Schatzki (2012), a practice is principally open-ended, meaning that it consists of the things people actually do; it is not something that is closed and scripted *a priori*.

What exactly circulates in engineering practices merits further attention, as it helps explain how it is or is not political (in either subject or object sense). The *goods* pursued in engineering practices are informative here. While the question of what makes good engineering can be expanded *ad infinitum*, some core concerns are accepted across large swaths of engineering ethics. The first virtue of engineers is generally taken to consist of making contributions to the common good, and more specifically, engineers are tasked with *solving problems*: 'resolving an undesirable condition through the application of technologies' (Sheppard et al. 2006). Note the convergence with the earlier observation that engineering is kept exempt from politics by appeal to its arguable contribution to the common good. The point here is that this is not only an argument mobilized instrumentally to the outside of the practice, but also something that we may presume to be heartfelt among engineers themselves. (See also Part VI of this *Handbook* about norms and values in engineering.)

Another generally recited virtue of engineers is the safeguarding of the public from harm (Harris 2008). As this connects to the obligation to deliver good and working technologies, it follows that engineers should master relevant parts of mathematics and natural and engineering sciences and develop a sensitivity to the risks that may emerge as the consequence of the technologies they build (Harris 2008). Also related is the norm expressed by many engineering codes that stipulates that an engineer shall never accept an assignment for which s/he is not qualified (e.g. IEEE 2006, art. 6, KIVI Engineering Society 2018, art. 6; see also Chapter 43, "Professional Codes of Ethics" by Michael Davis, this volume). In fact, this constitutes a direct connection to conventional politics outside the practice: in many countries, the state has the possibility to attach consequences to a violation of these principles, up to the level of withdrawing licences.

So far, we have been talking of the norms that *guide* engineering. That is still something different from what *happens* in engineering. What is it that circulates in engineering practices? What is engineering done upon, and what means does it use? If we want to further understand how engineering is or is not political, following the things that circulate in engineering provide insight into how the processes may be political: what kind of interactions and negotiations take place. This offers a more dynamic and conflictual view, and more politically substantial at that, than the guiding values that only abstractly provide the direction for engineering practices.

The very first thing that should draw our attention is the *knowledge* engineers have. Obviously this consists of a good dose of natural-scientific knowledge, but it also extends well beyond that: knowledge about solving design problems, about technical norms and standards, about economics, about legal matters, and about translating clients' desires into technical specifications (Dias de Figueiredo 2008; Meijers and De Vries 2013). Indeed, one of the key reasons we need the whole notion of practice lies in the observation that the activity of *knowing* is not merely an individual affair, but something located at a higher level of organization, namely the practice (Reich et al. 2014).[3] (See also various chapters in Part II of this *Handbook* on engineering reasoning.)

Shifting our understanding of knowing to the higher level of practice allows for it to include non-human parts. In particular, the role played by material entities in practices has been articulated (Shove 2017). Models, even if their 'materiality' in a very strict sense is limited when they exist only as mathematical equations, perfectly fit this idea. The use of models is deeply ingrained in the engineering way of knowing. Models are to be understood broadly here (Frigg and Hartmann 2018): not only literally scale models of envisioned or existing technologies, but also theoretical models, sets of mathematical equations that represent systems, and computer programmes that simulate technologies. Models are utterly diverse in both form and application. They are used to represent reality, to represent theories, to test designs at a smaller scale, and to partially test theories while leaving out part of reality (idealization), to mention a few applications. People might also learn from models and conduct experiments on them.

One thing that models have in common is the fact that models allow for certain ways of thinking and solution-finding, and eliminate others. Much like how problem definitions determine what is and is not part of engineering, models pre-structure the kind of outcomes engineering can produce. Hence, the selection of models is ultimately not a politically neutral affair. Also, models amount to both an abstraction and a simplification of the real or envisioned system they represent. As will be argued later, this contributes to the *decontextualization* of technology design, which serves its positioning as non-political.

One other important class of things that circulate is made up by the *problem definitions* of engineering. We may have an intuitive notion of what typical engineering problems are, and within this notion, a virtually endless range of problems is likely to fit. Yet, problem definitions are subject to more criteria and negotiations than only the fact that 'they can likely be solved through technological innovation'. For example, it is controversial whether engineers should lend their efforts to further exploiting fossil-fuel resources. Some will say this is 'just' a technical issue that is secondary to decisions made regarding fossil-fuel use in political institutions (in the conventional sense), and hence unproblematic for engineers to take up. Others will argue in contrast that this challenge should not be considered a qualified engineering problem, because fossil-fuel solutions should (arguably) not be further pursued. The politics of engineering already looms here, in the sense that negotiating what world we want to build is operationalized into what kind of engineering we want to do.

Another thing is that engineering practices are typically characterized by division of labour (Sheppard, Colby et al. 2006). An engineer working in the automobile industry will never 'design a car', but rather develop a new head rest, or decrease the susceptibility to wear for a specific part of the engine. This means that the problem definitions an actual engineer faces are in fact partial problem definitions. This entails that the individual engineer will have only a limited perception on how their work relates to the context of use.[4] This contributes further to the decontextualization that is characteristic of engineering ways of thinking.

This decontextualization complements the following. Engineering is of course not done only for its own, internal sake. Engineers produce technologies for the broader world. This connectedness to the outside is not just some boundary phenomenon at the fringes of the engineering practice, but definitive of the practice as a whole. Engineering is essentially what in common parlance would be termed an *applied* form of knowledge work. At the same time, this connection to the outside world is anything but trivial: where the boundary exactly is remains contestable, and what crosses the boundary in terms of problems, solutions, tenders, payments, contracts, is highly heterogeneous. Indeed, a sensitivity towards this social connectedness is considered one of the goods that circulates in the practice of engineering (Harris 2008).

45.4. Internal Politics: The Objects of Engineering

We need the notion of practice from Section 3 because it allows us to direct our attention to how the content that circulates may or may not be political. From the two senses of political discussed earlier—something being an object we do politics on, or something being a subject that acts in the process of doing politics—we can already glean how what circulates in engineering practices is potentially political. In the object sense, the aforementioned problem definitions, models, and epistemologies that circulate are political. They are the result of choices, and as such are potentially subject to negotiations and power relations. Also, they have potentially ground-shaking consequences, as they lead to inclusion and exclusion, to redistribution of power and wealth, and at the fundamental level to which conceptions of the good life are and are not facilitated by the technologies we build.

In the subject sense, all parties that contribute to setting the problem definition are involved in this politics. These include engineers but also potentially all other actors that matter to the issue. Ultimately, they may also include non-human entities, such as laws, institutions, and technologies; these may have far-reaching consequences for what kind of technological world we ultimately build. While this may appear far-fetched in many cases, it is less so in others. For example, standards, once they have been set, have a strong influence on inclusion and exclusion of what innovations are possible. And, as standards themselves may require revision, they determine at which pace innovations are even possible (Egyedi and Blind 2015).

Regarding this very political content itself, De Vries (2007) has argued that much of the writing about politics and science has too unreflexively talked about the political as something just being there, as though it were available from some reservoir of political problems that we can pick and choose to engage with. Rather, De Vries argues, we should think of the political as itself being constructed and continually modified, and constituted through its very interactions with members of a practice. The question in relation to engineering practices then becomes whether and how what circulates in engineering practices becomes *constructed as* political through the interactions in the practice.

In the Section 3 discussion of engineering as a practice, a number of circulating things were mentioned: epistemologies, models, and problem definitions, to mention the three taken as central. However, contrary to the plea by De Vries (2007), we have just assumed these to be political from an external perspective. While this may rightly articulate the political *potential*, it does not say anything about these circulating things being (constructed as) political in its real context. If we accept the idea of things being constructed as political, then what is it more specifically that helps construct epistemologies as political? Or models, and problem definitions?

The point is: they are usually *not* constructed or enacted as political, but rather the contrary. Most of these things are made to circulate under guises other than 'politics'. Problem definitions are staged as 'what the client desires', or as 'optimal solutions' in the light of specific criteria. In the selection and development of models, choices are necessarily made, but these are not likely to be connected explicitly to ideologies and power relations, but rather to notions of accuracy, elegance, and usability. And epistemologies are not even explicitly constructed in the engineering practice itself, but are largely thought of as taught in engineering schools, and moreover related to some objective methods of truth-finding.

Interestingly, the only thing that seems constructed 'as political' within the engineering practice, is 'the politics that is kept out'. Thus, the only thing explicitly political is the notion of ethics and politics itself. It serves as a rhetorical counterpoint to what engineering is. If things are politics, they are not the work of engineers. What is more, the aforementioned codes of conduct may potentially serve as a device to pacify issues: once values are condensed into seemingly unambiguous principles, it may become harder to discuss the ways in which they are equivocal (Martin and Schinzinger 2010 [2000]). This means that moral considerations are in some sense put outside the realm of negotiation. At least in appearance, they become 'fixed'.

This approach to the internal politics of engineering as constructed indeed suggests that the objects may also be constructed as 'a-political'. This is compatible with De Vries's point, if we take it to be symmetrical. Indeed, even if defining something as non-political is essentially a political affair, this does not render it any less a 'construction of the a-political'. That this is perhaps the most relevant part of the politics of engineering will indeed be discussed in the penultimate section of this chapter.

45.5. External Politics: Engineering as Society-Building

The potential political character of engineering is fundamentally connected to what happens *outside* the very practice of engineering itself. As argued earlier, if we want to understand how engineering is political, we need a broad understanding of politics as the progressive composition of the world (Latour 2004; Luhmann 2000). Before engaging in the next section with how problematic this demarcation between inside and outside of the engineering practice is in relation to the distribution of politics, it first makes sense to articulate how engineering is political in the broader world, regardless of whether or not that broader world can be distinguished from engineering in any meaningful way.

Even starting from a comparably narrow understanding of politics as pertaining to arranging the basics of society, it is only one step to seeing how engineering contributes to basics that matter: modern life is unthinkable without infrastructures, production technologies, communication and information technologies, etc. While much of this is formally privatized in capitalist societies and hence outside the reach of what is generally called politics, this is not at all self-evident. Endeavours of engineering have contributed to the normalization of large-scale animal farming, and the mobility paradigm based on cars that are now inextricably tied up with the right to eat and the right to move—to mention only two things that can straightforwardly be reckoned as objects of politics. Mind that there are also modes of transport that are in fact *hampered* by the automobility paradigm—all the more reason not to think of the latter too simplistically as a matter of free choice or free markets.

In line with the general idea of structuration (Giddens 1984), which holds that social structures are at once determining our actions and themselves the consequences of our actions, it can be seen how technologies are both human made and have important consequences for human life. The moment that cars became available, they emerged as an add-on to existing mobility paradigms. But as they gradually became pervasive throughout society, they became the new standard of personal mobility, and what is today for example expected in terms of home–work commuting at least puts some social pressure on the arguably free choice of where to live and where to work. Similarly, the argument has been made that the mere existence of *preimplantation genetic diagnostics*—indeed, a human-made technology—may lead to social pressure on prospective parents to eliminate in specific cases the possibility of giving birth to a child with a disability, which is thus displaced from the realm of 'chance' to the realm of 'choice' (Buchanan et al. 2000; Cameron and Williamson 2003; King 1999; Scannell 2002).

What is more, technologies keep carrying the traces of the context in which they are made: the normativities that rule the context of design are *inscribed* in those technologies (Akrich 1992). For example, if diversity is lacking in the context in which technologies are made—which has traditionally been the case in engineering practices (Franzway et al. 2009)—it is likely that technologies are designed with imagined users in mind that are only representative for part of society, not its whole. This happens, for example with male-dominated Silicon Valley (Corbyn 2015, 2018), where technologies are created that are allegedly based on a too-narrow experience, and that provide a poor match with the society they are to serve (Wajcman 2016). Here, a first hint at a *democratic deficit* emerges: those close to the fire of technology-making have a better shot of getting their normativities inscribed, which may lack justification for them to be imposed on the broader society.

At the same time, the ability of technologies to convey normativities outside the engineering practices should not be treated as a fixed given, but as something that is up for negotiation. As Orlikowski (2000) argues, fixation of social structures through technologies only works if people in fact abide with those structures. Instead, they can also opt for insubordination and seek to alter the structures into which they are seemingly forced. Insofar as we are concerned with the responsibilities of engineering practices to provide society with technologies that are 'good' against some criterion of societal desirability, this concern should also include the negotiations that are still possible *after* a technology design has seemingly been completed. For example, people today can still live a life without a car or without preimplantation genetic diagnostics. Hackers come up with solutions to use technologies, such as allegedly privacy-invasive social media, in different and more privacy-compliant ways. And in every workplace, one can find laptops with sticky notes covering their webcams, because their owners refuse to live by imposed paradigms of visibility.

These connections to the world outside engineering practices potentially substantiate the aforementioned *democratic deficit*: issues that should somehow be contained by properly organized political processes, but are not. In an extreme form, this could be reckoned a *technocracy*: a political system in which decisions are determined by expert knowledge and the position of experts in a technocratic system, where moral worth is replaced by calculation, and where important social relations are fixed by technological relations (cf. Collins and Evans 2017; Feenberg 1999; Habermas 1968). In a less extreme form, such considerations underlie approaches such as *responsible research and innovation* (to be discussed in the concluding section), *constructive technology assessment* (Rip et al. 1995), and *scenario studies* (Wright et al. 2013), each of which in their own way aspire to connect technoscientific knowledge processes to sociopolitical knowledge processes (where, again, the boundary between those classes is itself a topic worth studying).

Once we understand that engineering knowledge is essentially incomplete regarding the application context of technologies, and that this incompleteness can only to some extent be remedied by the aforementioned strategies of democratization, it follows that engineering is fundamentally an experimental attitude to the broader world (Martin and Schinzinger 2010 [2000]; Van de Poel 2016). This experimental nature is even more strongly felt if the technology is itself inherently risky and uncertain, as Van de Poel (2011) discusses in the case of nuclear energy technology.

45.6. The Non-Politics of Engineering

Given the clear political implications of engineering in the previous section, and the political activities within engineering in the section before that, one might wonder why it is the case that we do not naturally think of engineering as a political affair. We do not typically find engineering matters in the politics sections of newspapers. People studying political science or political philosophy do not naturally think of engineering and technology as their object of analysis. And when asked what a politician does, most people will not answer anything along the lines of 'well, building bridges and powerlines, or anything else an engineer does, too'. Why is this?

The first answer here is that it seems not entirely nonsensical to think of engineering as the practice of putting nature's laws to work. The question of *what exactly* technology is, is often answered with 'tools that support human action'. This thought is also known as *technological instrumentalism* (Feenberg 1995). Also, the thought of leaving engineering to engineers and politics to politicians seems in itself not a nonsensical thought—even though it has extensively been shown in the scholarly literature that such a strictly neutral vision of technology is not tenable. In fact, most people can easily come up with examples of technologies indeed not being innocent, and technologies not being exactly neutral with respect to the kinds of lives people can live with them. Still, the idea that natural laws are independent of what we do politically seems hard-wired in society.

A second answer is that many parties involved have an interest in acting *as though* engineering were something radically separate from politics. Appeals to technological instrumentalism often surface when questions of the responsibility of engineers emerge. Engineers can arguably do their work best when it is not impeded upon by complex, political considerations. This suits engineers fine, and it is also convenient for politicians that they can exclude a certain practice from their realm of concern.

Yet, there is nothing natural or logically inevitable about this division of labour, of engineers doing engineering and politicians doing politics. In fact, the boundary where engineering stops and the social, political world begins is a construct, and the process of constructing the boundary is itself essentially a political process. One of the things that happens in such processes is that the genesis of technologies tends to become invisible once the technology moves from the design table to the context of use: it becomes a black box (Latour 1987, 1993). And if all the human relations that in fact went into the construction of the technology have become invisible, then indeed it requires a fair amount of critical thinking to remain aware of how the technology can be seen as political, rather than 'natural laws in a box'.

One further step of inference is that if we accept that the boundary between the inside and the outside of the practice of engineering is not self-evident, and if we accept that relevant actors on both sides of the divide have an interest in the boundary appearing as self-evident, it becomes clear that it is in fact a political move to define exactly *what* engineering is in relation to non-engineering.

Ironically, this boundary is then used as a foundation of defining what is neutral. At the same time, this neutrality is used to justify the existence of the boundary.

45.7. Towards a Good Politics of Engineering

Notwithstanding the persistence of instrumental visions of technology and the ensuing vision of engineering as—neutral—applied natural science, the political aspects have clearly been articulated and attempts have been made to reconnect engineering and the political. For example, new social movements have played a crucial role in politicizing technical domains that liberal discourse had formerly isolated from the scope of politics (Thorpe 2008).

Over recent years, the notion of *responsible research and innovation* (RRI) has gained currency. The most commonly referenced definition is the one by René von Schomberg (2011), which holds that RRI is a process by which societal actors and innovators become mutually responsive with a view to the ethical acceptability, sustainability, and societal desirability of the innovation process and its products. This conception reflects the very idea that science and technology cannot be seen apart from the effects they have outside the practices of research and development. Therefore, the external political relevance must be somehow connected to the inner content of those very practices.

A fundamental challenge to overcome here, which looms more clearly if we see engineering as a *practice*, is the fact that a practice comes with a particular *epistemology* or theory of knowledge. This means that knowledge alien to that practice is likely to be dismissed as irrelevant, unsound, or even irrational. Thus, if we want to connect the outside and the inside of the practice to aim at mitigating any democratic deficit, then considerable thought will have to be spent at connecting incompatible knowledges from different epistemologies. Concretely, how will a conversation take shape between an engineer thinking in terms of design criteria, and a politician thinking in terms of interests and ideologies?

One of the more practical ideas is to democratize the *agenda* of programmes of innovation. This would democratically provide direction to innovation practices. However, here the same incompatibility of knowledges potentially plays out. If the broader public is invited to contribute to the agenda, then it matters who makes the invitation, as a way of putting it. If the invitation comes from the engineering practice itself, it is likely that engineering epistemologies will provide the framework in which the agenda is discussed. This renders the broader audience at a disadvantage at best, and at worst puts it in a *de facto* excluded position. If the process of democratic agenda setting is initiated from elsewhere, for example from political institutions narrowly defined, then it risks remaining anathema to engineering practices. In a general sense, the process of agenda setting is not by itself neutral with respect to different sorts of knowledge. Such processes thus require experts in transdisciplinary research and efforts of translation.

It is vital for members of the engineering practice themselves to remain aware of these differences, and of the fact that their position comes with specificities that have sociopolitical consequences. The nature and boundaries of the practice of engineering cast a specific light on the ethical considerations typically reflected in professional codes: realizing the relevant values is not only a matter of individual responsibility, but also a matter of dealing with collective identities, collectively enacted separations between the inside and the outside of the practice, and with ensuing power differences between inside and outside. Insofar as engineers are already held to a *noblesse oblige* because of their mere privileged knowledge position, they are even more held to this because of the social structures that carry this position.

The concerns raised in this chapter connect to broader societal problems such as racism, sexism, and classism. Engineering as a practice is actively constructed, with its boundaries and its internal and external politics. Not the least, this is to construct and reproduce 'who gets to do engineering'. This means that the correction of many societal problems and many emancipations of subordinate groups will come with making technologies such that they counter those problems, which in turn requires that engineering itself becomes inclusive and receptive to those groups.

This chapter has discussed engineering at the level of practice. It could be argued that engineering should rather be discussed at the level of specific practices, rather than the somewhat abstract and rather generalized way we have done here. True enough, engineering practices are in fact highly diverse. However, most of these practices are in fact also highly internally heterogeneous in terms of the science and engineering disciplines that populate them and the problems they revolve around. Also, they will in practice differ widely on the external, sociopolitical influence they have. Yet, the framework in this chapter offers a good perspective to study them. On the one hand, the abstract properties discussed here are most likely to pertain to each of them, at least to a certain extent. On the other hand, this is only a starting point, and what these properties mean for concrete practices remains to be further specified, and continues to be a valuable object of empirical analysis.

Related Chapters

Chapter 40: Ethical Considerations in Engineering (Wade L. Robinson)

Chapter 43: Professional Codes of Ethics (Michael Davis)

Chapter 50: Engineering and Social Justice (Caroline Bailly)

Chapter 51: Engineering and Environmental Justice (Benjamin R. Cohen)

Chapter 54: Engineering Practice From the Perspective of Methodical Constructivism and Culturalism" (Michael Funk and Alfred Fritzsche)

Further Reading

Feenberg, A. (2017). A Critical Theory of Technology. In U. Felt et al. (eds.), *The Handbook of Science and Technology Studies*. Cambridge, MA: MIT Press, pp. 635–663. (An overview of critical theory of technology and the critique of rationality in modern culture.)

Leftwich, A. (2004). *What Is Politics? The Activity and Its Study*. Cambridge: Polity. (An edited volume providing a comprehensive introduction to various theories of politics and the political.)

Martin, M. W. and Schinzinger, R. (2010 [2000]). *Introduction to Engineering Ethics,* 2nd ed. Boston: McGraw-Hill Higher Education. (An introduction to the ethical problems that are typical for engineers to encounter, with approaches to dealing with them.)

Schomberg, R. (2011). *Towards Responsible Research and Innovation in the Information and Communication Technologies and Security Technologies Fields*. Brussels: Directorate General for Research and Innovation. (A manifesto on how innovation should be positioned not as something separate from society, but as something intrinsically connected to society and its democratic governance.)

Notes

1. A narrower definition of agency, strictly linking it to *intention* and thereby limiting it to *human* agency, is central to debates in philosophy of mind, starting with Anscombe (1957) and Davidson (1980). This is, however, beyond the scope of this chapter, and the current broad understanding of agency is apt for the present argument.

2. The example centres on low bridges that only let through rich, white people travelling by car, and not poor, black people travelling by bus. The example has later been invalidated (Joerges 1999) and so must not be taken as an historical account, but as an illustrative exploration of how material, social, and political worlds can be thought to conflate.

3. See also Meijers and De Vries (2013), though the point is not couched in terms of practice there.

4. Mind that this is in fact not a matter of scale, and of engineers working at higher levels of aggregation (entire devices or even composite systems and global networks) having a better view of the context of operation. Rather, each view is partial because of the networkedness of engineering. For example, engineers working at the systems level will have a poorer view of the micro-interactions between individual persons and system elements than human-technology interface designers have.

References

Akrich, M. (1992). The De-scription of Technical Objects. In W.E. Bijker and J. Law (eds.), *Shaping Technology/ Building Society: Studies in Sociotechnical Change*. Cambridge, MA: MIT Press, pp. 205–224.

Anscombe, G.E.M. (1957). *Intention*. Oxford: Basil Blackwell.

Brown, M.B. (2014). Politicizing Science: Conceptions of Politics in Science and Technology Studies. *Social Studies of Science,* 45(1), 3–30. doi:10.1177/0306312714556694.

Brown, M.B. (2017). Re: Not Everything Political is Politics. Reflection on the March for Science. *Public Seminar [online]*. Accessed September 20, 2017.

Buchanan, A. et al. (2000). *From Chance to Choice: Genetics and Justice*. Cambridge: Cambridge University Press.

Cameron, C. and Williamson, R. (2003). Is There an Ethical Difference Between Preimplantation Genetic Diagnosis and Abortion? *Journal of Medical Ethics,* 29, 90–92.

Collins, H.M. and Evans, R. (2017). *Why Democracies Need Science*. Cambridge, UK: Polity.

Corbyn, Z. (2015). Silicon Valley Is Cool and Powerful. But Where are the Women? *The Guardian*.

Corbyn, Z. (2018). Why Sexism Is Rife in Silicon Valley. *The Guardian*.

Crick, B. (2004). Politics as a Form of Rule: Politics, Citizenship and Democracy. In A. Leftwich (ed.), *What Is Politics?* Cambridge: Polity Press, pp. 67–85.

Davidson, D. (1980). *Essays on Actions and Events*. Oxford: Clarendon Press.

De Vries, G. (2007). What Is Political in Sub-politics? How Aristotle Might Help STS. *Social Studies of Science,* 37(5), 781–809.

Dias de Figueiredo, A. (2008). Toward an Epistemology of Engineering. *Workshop on Philosophy and Engineering*. The Royal Academy of Engineering, London, November 10–12.

Egyedi, T. and Blind, K. (eds.) (2015). *The Dynamics of Standards*. Cheltenham, UK: Edward Elgar.

Feenberg, A. (1991). *Critical Theory of Technology*. Oxford: Oxford University Press.

Feenberg, A. (1995). *Alternative Modernity*. Berkeley, Los Angeles and London: University of California Press.

Feenberg, A. (1999). *Questioning Technology*. Oxon: Routledge.

Feenberg, A. (2002). *Transforming Technology: A Critical Theory Revisited*. New York: Oxford University Press.

Feenberg, A. (2005). Critical Theory of Technology: An Overview. *Tailoring Biotechnologies,* 1(1), 47–64.

Feenberg, A. (2017). A Critical Theory of Technology. In U. Felt et al. (eds.), *The Handbook of Science and Technology Studies*. Cambridge, MA: MIT Press, pp. 635–663.

Feenberg, A. and Friesen, N. (eds.) (2012). *(Re)inventing the Internet: Critical Case Studies*. Rotterdam: Sense Publishers.

Franzway, S. et al. (2009). Engineering Ignorance: The Problem of Gender Equity in Engineering. *Frontiers: A Journal of Women Studies,* 30(1), 89–106.

Frigg, R. and Hartmann, S. (2018). Models in Science. In E.N. Zalta (ed.), *Stanford Encyclopedia of Philosophy*. Stanford: Stanford University Press.

Giddens, A. (1984). *The Constitution of Society: Outline of the Theory of Structuration*. Berkeley and Los Angeles: University of California Press.

Gieryn, T.F. (1983). Boundary-work and the Demarcation of Science from Non-science: Strains and Interests in Professional Ideologies of Scientists. *American Sociological Review,* 48(6), 781–795.

Habermas, J. (1968). *Technik und Wisschenschaft als Ideologie*. Frankfurt am Main: Suhrkamp Verlag.

Harris, C.E., Jr. (2008). The Good Engineer: Giving Virtue Its Due in Engineering Ethics. *Science and Engineering Ethics,* 14(2), 153–164.

IEEE (2006). *IEEE Code of Ethics*. New York: IEEE.

Joerges, B. (1999). Do Politics Have Artefacts? *Social Studies of Science,* 29(3), 411–431.

King, D. (1999). Preimplantation Genetic Diagnosis and the 'New' Eugenics. *Journal of Medical Ethics,* 25, 176–182.

KIVI Engineering Society (2018). *KIVI Ethische Code 2018*. The Hague: Koninklijk Instituut van Ingenieurs.

Latour, B. (1987). *Science in Action: How to Follow Scientists and Engineers Through Society*. Cambridge, MA: Harvard University Press.

Latour, B. (1988). How to Write The Prince for Machines as Well as for Machinations. In B. Elliott (ed.), *Technology and Social Change*. Edinburgh: Edinburgh University Press, pp. 20–43.

Latour, B. (1993). *La clef de Berlin et autres leçons d'un amateur de sciences*. Paris: Éditions la Découverte.

Latour, B. (2004). *Politics of Nature: How to Bring the Sciences into Democracy*. Cambridge, MA: Harvard University Press.

Luhmann, N. (2000). *Die Politik der Gesellschaft*. Frankfurt am Main: Suhrkamp Verlag.

MacIntyre, A. (1981). *After Virtue: A Study in Moral Theory,* 2nd ed. Notre Dame, IN: University of Notre Dame Press.

Marres, N. (2007). The Issues Deserve More Credit: Pragmatist Contributions to the Study of Public Involvement in Controversy. *Social Studies of Science,* 37(5), 759–780.

Martin, M.W. and Schinzinger, R. (2010 [2000]). *Introduction to Engineering Ethics,* 2nd ed. Boston: McGraw-Hill Higher Education.

Meijers, A.W.M. and De Vries, M.J. (2013). Technological Knowledge. In J.K.B.O. Friis, S.A. Pedersen and V.F. Hendricks (eds.), *A Companion to the Philosophy of Technology.* Chichester: Wiley-Blackwell, pp. 70–74.

Mouffe, C. (2005). *On the Political. Thinking in Action.* London and New York: Routledge.

Mouffe, C. (2013). *Agonistics. Thinking the World Politically.* London and New York: Verso.

Orlikowski, W.J. (2000). Using Technology and Constituting Structures: A Practice Lens for Studying Technology in Organizations. *Organization Science,* 11(4), 404–428.

Owen, R., Macnaghten, P. and Stilgoe, J. (2012). Responsible Research and Innovation: From Science in Society to Science for Society, With Society. *Science and Public Policy,* 39(6), 751–760.

Peters, B.G. (2004). Politics Is About Governing. In A. Leftwich (ed.), *What Is Politics? The Activity and its Study.* Cambridge, UK: Polity Press, pp. 23–40.

Rawls, J. (1971). *A Theory of Justice,* revised ed. Cambridge, MA: Harvard University Press.

Rawls, J. (1993). *Political Liberalism.* New York and Chichester: Columbia University Press.

Reich, A. et al. (2014). Engineers' Professional Learning: A Practice-theory Perspective. *European Journal of Engineering Education,* 40(4), 366–379.

Rip, A., Misa, T.J. and Schot, J. (1995). *Managing Technology in Society—The Approach of Constructive Technology Assessment.* London: Pinter.

Scannell, K. (2002). To Be or Not to Be—Preimplantation Genetic Diagnosis. *The Permanente Journal,* 6(4).

Schatzki, T.R. (2012). A Primer on Practices. Theory and Research. In J. Higgs et al. (eds.), *Practice-Based Education: Perspectives and Strategies.* Rotterdam, Boston and Taipei: Sense Publishers, pp. 13–26.

Schomberg, R. (2011). *Towards Responsible Research and Innovation in the Information and Communication Technologies and Security Technologies Fields.* Brussels: Directorate General for Research and Innovation.

Schot, J. W. and Vleuten, van der, E. B. A. (2004). Transnational Infrastructures and the Rise of Contemporary Europe. In *Proceedings for SHOT 2004, TIE Working Documents Series,* Vol. 4.

Sheppard, S. et al. (2006). What Is Engineering Practice? *International Journal of Engineering Education,* 22(3), 429–438.

Shove, E. (2017). Matters of Practice. In A. Hui, T.R. Schatzki and E. Shove (eds.), *The Nexus of Practices. Connections, Constellations, Practitioners.* London and New York: Routledge.

Stilgoe, J., Owen, R. and Macnaghten, P. (2013). Developing a Framework for Responsible Innovation. *Research Policy,* 42(9), 1568–1580.

Swierstra, T.E. and Rip, A. (2007). Nano-ethics as NEST-ethics: Patterns of Moral Argumentation About New and Emerging Science and Technology. *NanoEthics,* 1(1), 3–20.

Thorpe, C. (2008). Political Theory in Science and Technology Studies. In E.J. Hackett et al. (eds.), *The Handbook of Science and Technology Studies.* Cambridge, MA and London: MIT Press, pp. 63–82.

Van de Poel, I. (2011). Nuclear Energy as a Social Experiment. *Ethics, Policy & Environment,* 14(3), 285–290.

Van de Poel, I. (2016). An Ethical Framework for Evaluating Experimental Technology. *Science & Engineering Ethics,* 22(3), 667–686.

Wajcman, J. (2016). *Pressed for Time: The Acceleration of Life in Digital Capitalism.* Chicago: The University of Chicago Press.

Weale, A. (2004). Politics as Collective Choice. In A. Leftwich (ed.), *What Is Politics?* Cambridge: Polity Press, pp. 86–99.

Winner, L. (1988). Do Artifacts Have Politics? *Daedalus,* 109(1), 121–136.

Wright, G., Cairns, G. and Bradfield, R. (2013). Scenario Methodology: New Developments in Theory and Practice. *Technological Forecasting and Social Change,* 80(4), 561–565.

46

GLOBAL ENGINEERING ETHICS

Pak-Hang Wong

Global engineering ethics is the engineering ethics' response to *globalization*. Globalization, defined as "a multidimensional set of social processes that create, multiply, stretch, and intensify worldwide social interdependencies and exchanges while at the same time fostering in people a growing awareness of deepening connections between the local and the distant" (Steger 2003: 13), plays a major role in the received narrative about the need for a global engineering ethics.[1] The received narrative is often illustrated by stories of some engineers *A* (of culture *X*) who interact with people or organizations of culture *Y*, and as a result encounter conflicts between their (i.e. culture *X*'s) ethical values and culture *Y*'s ethical values that generate ethical conundrums to the engineers (see, e.g. Luegenbiehl 2004; Downey et al. 2007; Harris et al. 2014; Luegenbiehl and Clancy 2017). Global engineering ethics is thus needed to help engineers to navigate through these ethical conundrums. While the received narrative is useful to raise awareness of the global dimension in engineering ethics, it is insufficient in attending to the different *nature* and *scope* of ethical challenges for engineers and engineering practices in a globalized context.

To understand these differences, we can clarify the presuppositions underlying the need of a global engineering ethics. Identifying these presuppositions and their different interpretations can also enable us to better understand current approaches to global engineering ethics. Hence, I will first describe three basic presuppositions that create the need for global engineering ethics and explain their various interpretations. These presuppositions will then form the basis of my discussion of current approaches to global engineering ethics. As I shall argue, global engineering ethics is not merely *reactive* but also *proactive*. However, the existing approaches have paid insufficient attention to the proactive dimension of global engineering ethics. So, I will end this chapter by arguing for the importance of the proactive dimension and exploring what it demands from engineers and engineering practices.

46.1. Three Presuppositions of Global Engineering Ethics

Three basic presuppositions underlie the ethical conundrums in global engineering ethics. These presuppositions, however, can be interpreted differently, thereby leading to varying concerns and emphases in different approaches to global engineering ethics. By elaborating these presuppositions and their different interpretations, we can understand better the different approaches to global engineering ethics. The three basic presuppositions I shall elaborate are *pluralism of values*, *values in engineering and technology*, and *globality of engineering*.

Let's start with the presupposition about *pluralism of values*. It refers to the idea that there exist diverse and conflicting sets of (ethical) values in different cultures, and that these sets of values are equally valid; or, in Charles Taylor's (1992: 66) view, there is a presumption of equal worth of cultures.[2] Pluralism of values can come in *radical* and *moderate* versions. The radical version can be understood as a form of ethical relativism: there is, and can be, no common ground for different cultures to decide on value-related issues, and every culture requires decisions to be made based on their *own* sets of values.[3] The more moderate version agrees that there are diverse and conflicting sets of (ethical) values, but it also argues that there are, or can be, some values, interests, or ethical standards shared by different cultures.

One formulation of moderate pluralism of values is John Rawls's fact of reasonable pluralism and overlapping consensus. Rawls suggests that we are living in a world characterized by "a pluralism of comprehensive religious, philosophical, and moral doctrines . . . a pluralism of incompatible yet reasonable comprehensive doctrines" (Rawls 1993: xvi). In such a world, individuals have distinct views on value-related issues, and they can justifiably disagree with one another on those issues so long as the values they invoke to support their views are *reasonable*. Despite the incompatibility between different sets of reasonable values, Rawlsian pluralism suggests that mutual agreements can be achieved among different *reasonable* cultures through overlapping consensus, that is—different cultures *can* agree on some values, interests, or ethical standards with their own reasons and from their own point of view (cf. Rawls 1993: 134). So construed, Rawlsian pluralism accepts that there are diverse and conflicting sets of values, but, at the same time, it uses *reasonableness* as a normative standard for the validity of those sets of values. We should note that Rawlsian pluralism can only arrive at a minimal moral denominator, as it explicitly precludes people's conceptions of the good in the course of overlapping consensus (Wong 2009). Yet, moderate pluralism of values can also be formulated with more substantive values, e.g. minimum requirement for well-being (see, e.g. Kekes 1996).

The second presupposition is *values in engineering and technology*. The idea that engineering and technology are not value-neutral is a major lesson we have learned from philosophy of technology and science and technology studies (see, e.g. Winner 1986; Latour 1992; Verbeek 2011). Value-ladenness of engineering and technology amounts to the claim that values—ethical or otherwise—are built into the end products at various stages of design and production by opening up new options or by making some options more salient and accessible to users, and *vice versa*. Or, simply, as the goal of engineers is to change people's lives for the better with their products, they achieve this goal by embedding values in their products that influence, constrain, or shape users' decisions and behaviors.[4] We can call this the *material interpretation* of values in engineering and technology.

Alternatively, the presupposition about values in engineering and technology can also have a practice-based understanding: the idea that engineering practices are not value-free activities. Engineering practices are guided by specific sets of norms and values (Davis 2015). Ibo Van de Poel (2015) argues that societal values, such as safety, health, human well-being, etc., are either internalized as guiding standards of engineering practices, or these societal values justify the guiding role of some of the values internal to engineering practices, e.g. technological enthusiasm, effectiveness, efficiency, etc. So construed, the values that guide engineering practices are derived from societal values. We can call this the *practice interpretation* of values in engineering and technology.[5]

Finally, the presupposition about *globality of engineering* is the descriptive claim that engineering practices and their end products have a global reach. For instance, people from different nations work together in engineering projects, bringing their values with them in their works. Countries also import and export engineering products and technologies, thereby introducing values embedded in engineering and technology to their importers or introducing exporters' values into their home countries.

Globality of engineering is, perhaps, most pertinent to the need of global engineering ethics: engineering projects and their products are where different—and, possibly conflicting—sets of values meet. Engineers, while working with people from a different culture, or in a company in the host country, can encounter foreign norms and values. Or, the engineers have to reconsider the values (to be) embedded in the end products, as people in the host country might not share with them the same sets of societal values. In other words, if engineering practices and their end products are merely *local* and do not involve global encounters, then there is not a meeting of values, and thus not a need for a global engineering ethics.

46.2. The Nature and Scope of Global Engineering Ethics

We can view pluralism of values as the source of ethical conundrums for engineers in a globalized context, i.e. the existence of different sets of equally valid values of different cultures make it ethically problematic for engineers to simply apply their own values and ignore the values of the host country, especially when their values come into conflict with the host country's values. Yet, at least in the case of moderate pluralism, pluralism of values also suggests a potential answer to these ethical conundrums through some shared values, interests, or ethical standards. Here, the shared values, interests, or standards can be understood in either *thin* or *thick* terms (Walzer 1994; cf. Rawls 1971: 395–399): that is, it can either be a minimal moral denominator that informs us what is *right*, where rightness is arrived at by consensus independent of people's substantive ethical outlooks; or it can go beyond the minimal moral denominator and include a more substantive view of human well-being. As such, global engineering ethics can be construed as a quest of normative guidelines for engineers and engineering practices, or as an attempt to promote well-being via engineers' works.

The presupposition of values in engineering and technology also carries two subject matters for global engineering ethics. On the material interpretation of values in engineering and technology, engineers should consider the potential conflicts between the values embedded in the engineering products during their design and production and the values in host countries, as well as whether and how these conflicts of values ought to be resolved. On the practice interpretation, engineering practices carry with them the home (societal) values that might be in conflict with the values in host countries, and the challenges again are to navigate through these conflicting values ethically.

We can thus distinguish at least three different, non-exclusive tasks of global engineering ethics:

1. Global engineering ethics concerns establishing and defending some normative guidelines for engineers and engineering practices in a cross-cultural context.
2. Global engineering ethics concerns promoting well-being in a cross-cultural context via engineers' works.
3. Global engineering ethics concerns scrutinizing the values embedded in engineering products and/or practices and their ethical implications in a cross-cultural context.

So far, the exercise of elaborating the basic presuppositions is to remind us of the multiplicity in global engineering ethics, i.e. there are distinct tasks to be tackled by global engineering ethics. Moreover, if my characterization of the tasks of global engineering ethics is broadly accurate, this exercise should also enable us to observe which, if any, areas of global engineering ethics are currently underexplored.

Before turning to the different approaches to global engineering ethics, I want to reject two ways to engage global engineering ethics for their failure to take seriously the 'global' in global engineering ethics; they are radical pluralism of values and a rejection of pluralism of values respectively. Both radical pluralism and the rejection of pluralism eliminate the need for a genuine global engineering ethics, the goal of which is to help engineers to navigate the ethical conundrums that arise from cross-cultural encounters in a globalized world. My argument goes as follows: both radical pluralism and the rejection of pluralism supply *definitive* answers to engineers' cross-cultural ethical encounters, and thus no ethical conundrum shall arise for the engineers. In the case of radical pluralism, engineers will be advised to act according to the host culture's values for their acts to be considered as ethical; whereas in the case of the rejection of pluralism, engineers' acts will be informed by some universal ethical norms and values. Either way, engineers will face *no* ethical conundrum so long as they are equipped with the host culture's values or with a set of universal ethical truths.[6]

More importantly, neither radical pluralism nor the rejection of pluralism can be genuinely 'global' in responding to the ethical problems for engineers and engineering practices in a globalized context. While the ethical relativism inferred from radical pluralism requires one to familiarize oneself with the host culture's values, the ethical decisions are ultimately local, i.e. it requires one to consider *only* from the ethical viewpoint of the host culture. Similarly, the rejection of pluralism is, at best, to globalize a specific account of (engineering) ethics, where some norms and values are to be considered as universal, and thus they will be applicable regardless of any cultural background. At its worst, an attempt to globalize a specific account of engineering ethics will encourage us to ignore local norms and values, and it might even lead to a form of technological imperialism (Wong 2016).

46.3. Current Approaches to Global Engineering Ethics

Qin Zhu and Brent K. Jesiek (2017a) have recently summarized four approaches to global engineering ethics and examined their limitations. Their taxonomy of approaches offers a convenient starting point to investigate different ways of doing global engineering ethics and their relations to the basic presuppositions and the tasks of global engineering ethics I have identified. The four approaches they have summarized are: *global ethical codes, functionalist theory, cultural studies*, and *global ethics and justice*.

Zhu and Jesiek characterize the global ethical codes approach as one that aims to "build up a code of ethics that is expected to be applicable across cultures" (Zhu and Jesiek 2017a: 3). They point out that many professional engineering societies and organizations around the world already have their own codes of ethics; the goal of this approach, therefore, is to agree on a common code of ethics among the different professional engineering societies and organizations. Technical codes and standards might also be used in regulating engineers and engineering practices as well (Pritchard 2009), thereby contributing to the global ethical codes approach. For instance, a set of *internationally* enforced technical codes and standards can sanction engineers when their products and practices fail to satisfy the values those technical codes and standards exemplify. These technical codes and standards, therefore, could serve as a shared normative standard for engineers from different cultures.

Yet, Karim Jamal and Norman E. Bowie (1995) point out that professional codes of ethics can be inattentive to—or even mistaken about—public interest. For instance, they argue that the provisions designated as obligations of professional courtesy, e.g. "promoting public understanding of the profession", are often about the interests of the profession but not the interests of the public. Similarly, they argue that codes of ethics often represent the moral views of the profession, which can contradict the moral position of the public. Professional codes of ethics, therefore, can be considered by the public ethically dubious, especially when the provisions do not explicitly prioritize the interests of the public over the interests of the profession.

The global ethical codes approach also faces a practical challenge. Zhu and Jesiek rightly note that a global ethical code requires a supra-national institution to coordinate, govern, and enforce it, which could be practically difficult to achieve. Moreover, the approach seems to face a dilemma in grounding its normative authority, namely the normative ground of the global ethical codes is either from the institutional arrangement(s) of professional engineering societies and organizations without further normative justification, which will then be viewed as arbitrary, or it is grounded on a fundamental set of norms and values, which will make the global ethical codes unnecessary as the more fundamental set of norms and values can better ground the normativity of global engineering ethics (Davis 2015).

A more serious problem to the global ethical codes approach appears when it begins to limit the globality of engineering. Since the acceptance and compliance of profession's ethical codes are usually a prerequisite of professional membership, global ethical codes can serve to delimit who are entitled to be part of the engineering community and to perform engineering and related services. In accordance with the global ethical codes approach, those who disagree with, or refuse to accept, the global ethical codes would be excluded from the global engineering community. It thus limits the 'globality' of engineering to cultures and societies with shared norms and values. The result is often harmful to developing nations, who are newcomers in the engineering community; they will be forced to accept the norms and values that are long established by developed nations in order to be recognized as *de jure* engineers and allowed to *legally* perform engineering and related services.[7]

The second approach summarized by Zhu and Jesiek is the functionalist theory, which is based on "some fundamental, shared characteristics internal to the engineering profession that apply globally" (Zhu and Jesiek 2017a: 5). Both Michael Davis (2015) and Heinz C. Luegenbiehl (2009; also see Luegenbiehl and Clancy 2017) have argued that there are norms and values internal to the engineering culture and its community that are independent of people's cultural backgrounds. Since these norms and values are culture-independent and shared by engineers, they can offer a common normative standard to engineers from different cultures. The argument from the functionalist theory is primarily teleological in nature: one starts with a *teleological* definition of engineering, e.g. engineering is a set of activities that respond to the needs and interests in a society via transforming nature, and then asserts that engineers are the agents to realize the end of engineering; the purpose of engineering thus provides a normative standard for engineers and engineering practices.[8]

There are two ways to develop and defend the functionalist theory. Earlier, I argued in my discussion of values in engineering that the values (or, the purposes) of engineering are not independent of people's cultures, as they are drawn from the broader societal values. So construed, the first way to advance the functionalist theory is to sever the link between engineering's norms and values and societal values, where the societal values can vary from one culture to another. In other words, the first way to advance the functionalist theory requires its proponents to construct and defend a set of norms and values that are unique to engineering culture and shared by its community; both Davis (2015) and Luegenbiehl (2009; also, Luegenbiehl and Clancy 2017) seem to be taking up this line of argument. There is a second way to advance the functionalist theory, however. Instead of suspending the link between engineering's norms and values and the societal values, one can argue that there are some societal needs and interests that are shared by people in different cultures, and the norms and values in engineering are derived from these *shared* societal values.

Yet, those who opt for the second way are required to answer the question of how shared societal values are determined (or, discovered). Here, Noreen M. Surgrue and Timothy G. McCarthy (2015) suggest that a possible solution is to return to the idea of overlapping consensus proposed by Rawls: if different societies and/or cultures can agree on some needs and interests, even for different reasons, then the shared needs and interests can serve as the foundation of the functionalist theory. In real-world situations, the search for overlapping consensus will likely require performances of stakeholder analysis to discern the needs and interests of different parties affected by the engineering products and practices.[9] One should note, however, that the second way to advance the functionalist theory is no longer *purely* functionalist with the addition of overlapping consensus and/or other similar procedures because it is now grounded on societal values that go *beyond* those that are internal to engineering, i.e. the norms and values are not only derived from the function(s) of engineers and engineering practices.

The last two approaches examined by Zhu and Jesiek are the cultural studies approach and the global ethics and justice approach. Whereas the former focuses on "the importance of cultural differences in formulating effective ethical decisions in the global context" (Zhu and Jesiek 2017a: 2), the latter emphasizes the significance of a universal normative standard for *all* cultures and is "inspired by the idea of *minimal moral realism* (Zhu and Jesiek 2017a: 8). As I have already discussed the potential dangers of either approach in the previous section, I shall only reiterate that an excessive focus on cultural differences might lead to ethical relativism, and thus overlooking the role of cross-cultural reflections in global engineering ethics. And, a strong form of global ethics and justice approach that asserts there is *only* one true, objective, universal normative standard will leave no room for local norms and values. In other words, in the strong formulations, both approaches risk removing the need of a genuine global engineering ethics, for there are no unique ethical issues that arise from cross-cultural encounters.

There are surely less radical formulations of cultural studies approaches that do not insist on irreconcilable differences of norms and values in different cultures, e.g. a pragmatic approach to culture differences (Zhu and Jesiek 2017b), and weaker formulation of global ethics and justice approaches that maintain people's cultures do play an essential role in arriving at the shared, common normative standard. In order for them to be feasible, however, proponents of these approaches must formulate strategies that enable engineers and engineering communities from different cultures to communicate norms and values to each other, to respect—or, at least, tolerate—their differences, and to come to terms with each other on a common normative foundation. Here, Rawls's idea of overlapping consensus might again prove to be useful in constructing a *shared* normative foundation without asserting there is only one 'true' reason to justify it.

46.4. From Global Engineers' Ethics to Engineering Ethics for Global Problems

The task of global engineering ethics in the preceding discussion has been understood as searching and building a common normative standard that helps engineers to think through when some engineering practices are ethically acceptable in their cross-cultural encounters. This understanding of global engineering ethics can be labeled as, to borrow Josep M. Basart and Montse Serra's (2013, italics added) term, a "global *engineers'* ethics", i.e. a paradigm that views engineering ethics to be something internal to engineers' works.[10] However, recall that there are other tasks we can expect from global engineering ethics as well. Recall the second and third tasks listed in Section 2:

2. Global engineering ethics concerns promoting well-being in a cross-cultural context via engineers' works.
3. Global engineering ethics concerns scrutinizing the values embedded in engineering products and/or practices and their ethical implications in a cross-cultural context.

The second task has to do with the potentials of engineers to improve human well-being with their skills and knowledge and with the questions about their positive duty to help and contribute to the *global* common good. The third task calls for a deeper analysis and reflection on the values of engineering products and practices with reference to the values in other cultures. In the final part of this chapter, I shall briefly illustrate what they demand from engineers and engineering practices, and why they are important in a global engineering ethics.

Here, Basart and Serra's (2013) insight that engineering ethics should go *beyond* engineering activities performed by engineers has already opened up the ethical questions concerning the roles and impacts of engineering to the society more broadly. Yet, it is Carl Mitcham and David Muñoz's (2010) idea of 'humanitarian engineering' that exemplifies a distinct task of global engineering ethics. Defined as "design under constraints to *directly improve* the well-being of under-served populations" (Mitcham and Muñoz 2010: xi; my emphasis), humanitarian engineering is not about avoiding ethical pitfalls in engineers' works, but about *proactively* promoting well-being of the population in developing nations. Similarly, Charles E. Harris (2015) proposes an 'aspirational ethics' in global engineering ethics that focuses on designing for well-being in both developing and developed societies. One immediate argument in favor of this *proactive* dimension in global engineering ethics, of course, is the global interconnectivity and interdependency, which entail engineers' works are likely to have global implications, and thus it is prudent to take this opportunity to transform the world for better.[11] Unfortunately, *promoting* human well-being is ostensibly more contentious than *safeguarding* justice or *avoiding* harm.[12] We are, therefore, in need of more research on theories of well-being (or, theories of the good life) that inform us the plausibility of this task and how to best proceed (see, e.g., Brey 2012).

Related to the proactive dimension of global engineering ethics is the importance of reflecting on the values embedded in engineering and their manifestation in cross-cultural encounters. Particularly, for engineering to work—in developing nations, or in a foreign culture—the values embedded in engineering and technology and the host's values must be carefully scrutinized (Oosterlaken 2011, 2012; also see Akubue 2017). In other words, mismatches between values in engineering and technology and in societies where they are introduced must be resolved if they are to contribute *positively* to the society. Luckily, there is abundant research in philosophy and ethics of technology that offers theoretical and practical tools to engineers who are interested in this endeavor.[13]

To conclude, global engineering ethics does not, and indeed should not, *only* be concerned with establishing a common normative standard among engineers that could apply across different cultures. In this respect, the received narrative is insufficient to capture the various tasks of global engineering ethics. Global engineering ethics is also about the promotion of well-being via engineering and related services, as well as critically reflecting on the values embedded in the end products and practices. All these tasks require, I believe, the means to understand, communicate, and respect the norms and values of foreign cultures. This, in turn, suggests that global engineering ethics will be an interdisciplinary endeavor that includes philosophical study of ethical theories and theories of well-being in cross-cultural contexts, empirical investigation of the values and moral viewpoints in the different societies, and practical work to translate the insights from philosophy and social sciences into engineering products, and thus requires collaboration between philosophers, social scientists, and engineers, who shall examine the normative grounds of global engineering ethics, the values embedded in engineering, the values held by locals, and the practical questions of implementing values in engineering.

Related Chapters

Chapter 3: Western Philosophical Approaches and Engineering (Glen Miller)
Chapter 4: Eastern Philosophical Approaches and Engineering (Glen Miller, Xiaowei (Tom) Wang, Satya Sundar Sethy, and Fujiki Atsushi)
Chapter 40: Ethical Considerations in Engineering (Wade L. Robison)
Chapter 43: Professional Codes of Ethics (Michael Davis)
Chapter 44: Responsibilities to the Public—Professional Engineering Societies (Joseph Herkert and Jason Borenstein)
Chapter 52: Beyond Traditional Engineering: Green, Humanitarian, Social Justice and *Omnium* Approaches (George D. Catalano)

Further Reading

Ladikas, M., Chaturvedi, S., Zhao, Y. and Stemerding, D. (eds.) (2015). *Science and Technology Governance and Ethics: A Global Perspective from Europe, India and China.* Cham, Switzerland: Springer. (A recent comparative analysis of the different understandings of the role and values of science and technology between European countries and non-European countries.)

Luegenbiehl, H. C. and Clancy, R. (2017). *Global Engineering Ethics.* Kidlington, England: Butterworth-Heinemann. (A significant, and perhaps the first, textbook on engineering ethics from a global perspective.)

Murphy, C., Gardoni, P., Bashir, H., Harris, Jr., C.E. and Masad, E. (eds.) (2015). *Engineering Ethics for a Globalized World*. Cham, Switzerland: Springer. (An edited volume that examines various facets of global engineering ethics.)

Oosterlaken, I. (2015). *Technology and Human Development*. London: Routledge. (An extended discussion on the role of technology in the context of global human development.)

Notes

1. It is necessary to point out that globalization is a complex, multifaceted phenomenon, and that the precise characterization of 'globalization' remains very much in question; see e.g. Axford (2013). However, the debate on how globalization is to be characterized should have no direct implication to my discussion of global engineering ethics in this chapter. My account only requires one to recognize the growing interactions between different nations and/or cultures, which is shared by various understandings of globalization.

2. Pluralism of values, as I formulated here, is both *empirical* and *normative*, that is—there are *in fact* different sets of (ethical) values in different cultures, and they are *justified* by and within the different cultures. Pluralism of values is related to, but different from, value pluralism, which is a meta-ethical view that asserts (ethical) values are incommensurable, and that they are not reducible to one supervalue; see Mason (2018)

3. For an example of how ethical relativism can be derived from a *radical* pluralism of values, see Moore (2009).

4. Of course, engineers and their choices cannot dictate how users will be using their products (see e.g. Albrechtslund 2007), but the influence of designs and production processes on users' decisions and behaviors should be properly acknowledged.

5. Interestingly, Van de Poel does not consider whether these societal values could vary in different cultures, and he seems to assume that these values are shared globally. Yet, if pluralism of values is correct, then there are good reasons to suppose that different societies may have different sets of societal values, and, in turn, different sets of values in engineering practices.

6. Here, this claim should not be mistaken as one that there will be no ethical questions for engineers and in engineering practices in a globalized context. Even if radical pluralism is true, or if pluralism of values is successfully rejected, engineers are still required to make ethical decisions, but radical pluralism and universalism of values shall enable them to respond to the ethical questions rather straightforwardly assuming that they can identify the host culture's values in the case of radical pluralism, or they have a satisfactory account of universalism of values. Surely, one can question the feasibility of identifying appropriate values in the host culture amidst its complexity, or question the plausibility of universalism of values, but these questions aim at whether radical pluralism or universalism of values can be true, but not their implications to global engineering ethics.

7. I have discussed in more detail the imperialistic concerns from global standards in Wong (2016), but see AlZahir and Kombo (2014) for a more optimistic view.

8. Davis's (2015) argument is built on the idea of 'profession' as inherently normative, and that engineering as a profession can serve as a common cultural backdrop for engineers from different cultures; whereas Luegenbiehl's (2009) argues that six ethical principles, i.e. (i) the principle of public safety, (ii) the principle of human rights, (iii) the principle of environmental and animal preservation, (iv) the principle of engineering competence, (v) the principle of scientifically founded judgment, and (vi) the principle of openness and honesty can be derived from the definition of engineering as "Engineering is the transformation of the natural world, using scientific principles and mathematics, in order to achieve some desired practical end" (Luegenbiehl 2009: 153) via reason. Both Davis's and Luegenbiehl's accounts of global engineering ethics thus rely on a teleological understanding of engineering.

9. It is beyond the scope of this chapter to explore the prospects and problems of stakeholder analysis; see Freeman et al. (2010) for an overview of stakeholder theory.

10. My use of the expression 'engineers' ethics' is slightly different from the use in Basart and Serra (2013). Basart and Serra use the expression to highlight the fact that engineering depends on actors and activities *beyond* engineers and engineering practices. My use of the expression, on the other hand, intends to highlight that global engineering ethics can engage with normative issues that go beyond *regulating* engineers and engineering practices and therefore is not internal to engineers' works. In this respect, my use of the expression shares with Basart and Serra's the importance of considering actors and activities *beyond* engineers and engineering.

11. However, for a more cautious note on 'engineering to help', see Schneider et al. (2009).

12. Although one can argue that even in the case of safeguarding justice or avoiding harm, with different understandings of 'justice' and 'harm', it can be controversial as well.

13. For a recent overview of different approaches to the ethics of technology, see Hansson (2017).

References

Akubue, A.I. (2017). International Technology Transfer. In S.O. Hansson (ed.), *The Ethics of Technology: Methods and Approaches*. New York: Rowman & Littlefield, pp. 35–49.

Albrechtslund, A. (2007). Ethics and Technology Design. *Ethics and Information Technology*, 9, 63–72.

AlZahir, S. and Kombo, L. (2014). Towards a Global Code of Ethics for Engineers. In *Proceedings of the IEEE 2014 International Symposium on Ethics in Engineering, Science, and Technology*. https://ieeexplore.ieee.org/document/6893407

Axford, B. (2013). *Theories of Globalization*. Oxford: Wiley.

Basart, J.M. and Serra, M. (2013). Engineering Ethics Beyond Engineers' Ethics. *Science and Engineering Ethics*, 19(1), 179–187.

Brey, P.A.E. (2012). Well-Being in Philosophy, Psychology, and Economics. In P.A.E. Brey, A.R. Briggle and E.H. Spence (eds.), *The Good Life in a Technological Age*. New York: Routledge, pp. 15–34.

Davis, M. (2015). "Global Engineering Ethics": Re-inventing the Wheel? In C. Murphy, P. Gardoni, H. Bashir, C.E. Harris and E. Masad (eds.), *Engineering Ethics for a Globalized World*. Cham, Switzerland: Springer, pp. 69–78.

Downey, G.L., Lucena, J.C. and Mitcham, C. (2007). Engineering Ethics and Identity: Emerging Initiatives in Comparative Perspective. *Science and Engineering Ethics*, 13(4), 463–487.

Freeman, R.E., Harrison, J.S., Wicks, A.C., Parmar, B.L. and De Colle, S. (2010). *Stakeholder Theory: The State of the Art*. Cambridge: Cambridge University Press.

Hansson, S.O. (2017). *The Ethics of Technology: Methods and Approaches*. New York: Rowman & Littlefield.

Harris, C.E. (2015). Engineering Responsibility for Human Well-Being. In C. Murphy, P. Gardoni, H. Bashir, C.E. Harris and E. Masad (eds.), *Engineering Ethics for a Globalized World*. Cham, Switzerland: Springer, pp. 91–107.

Harris, C.E., Pritchard, M.S., Rabins, M.J., James, R. and Englehardt, E. (2014). *Engineering Ethics: Concepts and Cases*, 5th ed. Boston, MA: Cengage.

Jamal, K. and Bowie, N.E. (1995). Theoretical Considerations for a Meaningful Code of Professional Ethics. *Journal of Business Ethics*, 14(9), 703–714.

Kekes, J. (1996). *The Morality of Pluralism*. Princeton: Princeton University Press.

Latour, B. (1992). Where Are the Missing Masses? The Sociology of a Few Mundane Artifacts. In W.E. Bijker and J. Law (eds.), *Shaping Technology/Building Society*. Cambridge, MA: MIT Press, pp. 225–259.

Luegenbiehl, H.C. (2004). Ethical Autonomy and Engineering in a Cross-Cultural Context. *Techne*, 8(1), 57–78.

Luegenbiehl, H.C. (2009). Ethical Principles for Engineers in a Global Environment. In I. Van de Poel and D. Goldberg (eds.), *Philosophy and Engineering: An Emerging Agenda*. Dordrecht: Springer, pp. 147–159.

Luegenbiehl, H.C. and Clancy, R. (2017). *Global Engineering Ethics*. Kidlington, England: Butterworth-Heinemann.

Mason, E. (2018). *Value Pluralism*. In E.N. Zalta (ed.), *The Stanford Encyclopedia of Philosophy (Spring 2018 Edition)*. https://plato.stanford.edu/archives/spr2018/entries/value-pluralism/

Mitcham, C. and Muñoz, D. (2010). *Humanitarian Engineering*. San Rafael, CA: Morgan & Claypool.

Moore, M.J. (2009). Pluralism, Relativism, and Liberalism. *Political Research Quarterly*, 62(2), 244–256.

Oosterlaken, I. (2011). Inserting Technology in the Relational Ontology of Sen's Capability Approach. *Journal of Human Development and Capabilities*, 12(3), 425–432.

Oosterlaken, I. (2012). Inappropriate Artefacts, Unjust Design? Human Diversity as a Key Concern in the Capability Approach and Inclusive Design. In I. Oosterlaken and J. van den Hoven (eds.), *The Capability Approach, Technology and Design*. Dordrecht: Springer.

Pritchard, M.S. (2009). Professional Standards in Engineering Practice. In A.W.M. Meijers (ed.), *Handbook of the Philosophy of Technology and Engineering Sciences*. Boston: Elsevier, pp. 953–971.

Rawls, J. (1971). *A Theory of Justice*. Cambridge, MA: Harvard University Press.

Rawls, J. (1993). *Political Liberalism*. New York: Columbia University Press.

Schneider, J., Lucena, J. and Leydens, J.A. (2009). Engineering to Help. *IEEE Technology and Society Magazine*, 28(4), 42–48.

Steger, M.B. (2003). *Globalization: A Very Short Introduction*. New York: Oxford University Press.

Surgrue, N.M. and McCarthy, T.G. (2015). Engineering Decisions in a Global Context and Social Choice. In C. Murphy, P. Gardoni, H. Bashir, C.E. Harris and E. Masad (eds.), *Engineering Ethics for a Globalized World*. Cham, Switzerland: Springer, pp. 79–90.

Taylor, C. (1992). The Politics of Recognition. In A. Gutmann (ed.), *Multiculturalism: Examining the Politics of Recognition*. Princeton: Princeton University Press, pp. 25–73.

Van de Poel, I. (2015). Values in Engineering and Technology. In W.J. Gonzalez (ed.), *New Perspectives on Technology, Values, and Ethics*. Cham, Switzerland: Springer, pp. 29–46.

Verbeek, P.-P. (2011). *Moralizing Technology*. Chicago: University of Chicago Press.

Walzer, M. (1994). *Thick and Thin: Moral Argument at Home and Abroad*. Notre Dame, IN: University of Notre Dame Press.

Winner, L. (1986). Do Artifacts Have Politics? In *The Whale and the Reactor*. Chicago: University of Chicago Press, pp. 19–39.

Wong, P.-H. (2009). What Should We Share? Understanding the Aim of Intercultural Information Ethics. *SIG-CAS Computers and Society*, 39(3), 50–58.

Wong, P.-H. (2016). Responsible Innovation for Decent Nonliberal Peoples: A Dilemma? *Journal of Responsible Innovation,* 3(2), 154–168.

Zhu, Q. and Jesiek, B.K. (2017a). Engineering Ethics in Global Context: Four Fundamental Approaches. In *Proceedings of the 124th ASEE Annual Conference and Exposition.* https://peer.asee.org/28252

Zhu, Q. and Jesiek, B.K. (2017b). A Pragmatic Approach to Ethical Decision-Making in Engineering Practice: Characteristics, Evaluation Criteria, and Implications for Instruction and Assessment. *Science and Engineering Ethics,* 23(3), 663–679.

47

ENGINEERING PRACTICE AND ENGINEERING POLICY

The Narrative Form of Engineering Policy Advice

Natasha McCarthy

47.1. Introduction

Why does engineering have a role in policy—what are the particular skills of engineers and the contribution of engineering knowledge and experience in policy? This chapter argues that there are particular roles for engineering in developing and delivering policy, in distinction (but not in isolation) from other disciplines and professions, giving engineers a specific and essential role in policymaking. It seeks to argue that this role can be best explained by examining what I would argue is the *narrative structure* of engineering policy advice, drawing on a range of accounts of narrative explanation set out in Roth (1989) and Morgan and Wise (2017).

Understanding this narrative structure depends on first examining philosophical reflections on the nature of engineering, as well as engineers' own reflections on the nature of engineering policy advice. Then, considering a number of examples of engineering policy, or policy advice where engineers play a major role, I will suggest that the proper and appropriate account of how engineers can be involved in policymaking is dependent on a specific narrative structure that is more closely akin to engineering than to other disciplines that play complementary and equally critical roles in policy. Engineers' contribution to such narratives depends on the elements of design and development of solutions to practical challenges in engineering, which are carried out by virtue of engineers' ability to use knowledge and theory as 'tools' to bring about practical outcomes in the face of contingencies and unpredictable events. As a result, engineers' role in policymaking has a particular value, and the engineering profession has a responsibility to use their engineering expertise to address major policy challenges. This is not to say that engineering's contribution to policy is necessarily superior to that of other disciplines—but that it often forms a necessary part of the interdisciplinary evidence base for policy.

47.2. Preliminary Considerations on Types of Engineering Policy Advice

There is an often-made distinction in discussions of science policy between science for policy and policy for science—and a similar distinction could be made in engineering. This is broadly a distinction between policy developed to support engineering practice and the use of engineering evidence in wider policy. The reality is more complex—for example, Pielke (2007) develops four different roles for science in policy, depending on the contribution the scientist makes and in what circumstances. Focusing on engineering in policy specifically, the following are four forms of engineering policy advice which provide useful context for this chapter, and which focus on the purpose of the policy advice and the nature of the policy activity in question. This is not an exhaustive list but gives a sense of the range of ways that engineering contributes to policy in practice.

47.2.1 Policy for Engineering

Policy for engineering is the endeavour of making the case to governments regarding the conditions that enable academic and industrial engineering to flourish, and the ways that government can aid in bringing about those conditions. These kinds of arguments were set out famously over half a century ago by Vannevar Bush, in his *Science: The Endless Frontier* (Bush 1945), at a critical time for research in industry and universities in the United States. Bush argues:

> The most important ways in which the Government can promote industrial research are to increase the flow of new scientific knowledge through support of basic research, and to aid in the development of scientific talent. In addition, the Government should provide suitable incentives to industry to conduct research.
>
> *(Bush 1945)*

Such cases are made and remade as economic and social environments change. They can relate to science and engineering in the broad sense espoused by Bush, or in relation to specific areas where there is a significant need for engineering research and development.

47.2.2 Engineering for Policy

Scientists and engineers have a clear role in arguing for the conditions for their disciplines to flourish but also apply their expertise to a wider range of policy challenges. For example, responding to the threats and risks from climate change no doubt has a clear role for engineers both in designing systems which will enable us to produce less carbon and meet the Paris Agreement expectations in cutting carbon emissions; and also to adapt to the impacts of climate change. Thus, engineering has a role in influencing policy at national and international levels.

The particular role of engineering in this policy context is set out well by Pirtle (2013). He sees the role of engineering in climate policy as sitting between science and politics. Engineering policy can focus on the means by which we can reduce carbon emissions, or adapt to the effects of climate change, rather than examining the evidence of climate change or the politics of the kind of response required. Pirtle suggests:

> The recent shift toward energy innovation on the global warming debate may help to create technologies that can reduce the cost that society needs to pay to address global warming, which may make political agreement on climate change policy easier to attain.
>
> *(Pirtle 2013: 378)*

Pirtle's argument is against the linear model of innovation where results flow from the application of basic research, arguing that innovation can be through the direct development of technology. He therefore also suggests that in the response to climate change, policy can helpfully "support late-stage development and demonstration projects . . . Successful demonstrations reduce uncertainty in a new technology, which can enable adequate technologies to develop and receive more investment." Engineering has a key role in achieving carbon reduction targets, through the strategies of low carbon energy production, reducing the demand for carbon through vehicles and buildings and ensuring resilient infrastructure systems designed to withstand climatic change. But in turn, government has a role in supporting engineering in creating these solutions so that it can achieve positive impact on climate change—thus in this case, engineering for policy and policy for engineering are closely related.

47.2.3 Governance of Engineering and Technology—Regulators, Professional Bodies and Standards

Another key role of engineering in policy goes beyond either the support of engineering research and practice, or developing the responses to policy challenges external to that practice. It is an activity that, to an extent, connects the two, by focusing on the ways that engineering and technology development are carried out. The design, development and implementation of technologies and engineering systems is not a process that is independent of policy and politics, as it is often highly dependent on systems of governance and regulation, intended to ensure that technologies are developed in such a way as to maximise social and economic benefits while limiting harms to individuals, groups and societies.

Engineers are key to understanding practical governance arrangements in order to create approaches that enable safe and rapid implementation of new technology—understanding that technology changes can bring with it great benefit as well as potential harm, and that in fact delaying these benefits can constitute harm. As stated in Craig (2018), and drawing on Walport and Craig (2014):

> Risks are popularly considered to be negative but many professional fields speak of "upside", as well as downside, risks. Risk and innovation are intertwined in this way. All innovation carries risk but, at least in the sense of change and growth, innovation is essential. So both acting and not acting carry risks.
>
> *(Craig 2018: 18)*

This intertwining of upside and downside risks is demonstrated in the development of artificial intelligence (AI) technologies, in relation to which new governance frameworks are being developed to enable their safe uptake. For example, the British Standards Institute (BSI) issued in 2016 its standard BS 8611 on the 'ethical hazards of robots' which "gives guidelines for the identification of potential ethical harm arising from the growing number of robots and autonomous systems being used in everyday life" and which covers "safe design, protective measures and information for the design and application of robots" (BSI 2016). Such standards and other forms of technology governance are important in building trust among users, giving assurance that the technologies have been developed and used appropriately, and in this case, ethically.

47.2.4 Engineering in Policy

Engineering is central to developing large-scale systems that deliver critical public services. Building smart electricity grids, creating resilient infrastructure and addressing flood risk are all engineering projects that interface with policy either because they are designed to deliver a policy outcome—for example, managing energy demand in order to lower carbon emissions—or they may be procured by public money; or they may require public or democratic decision-making for them to go ahead.

In the UK the National Infrastructure Commission was established in order to create long-term, stable policy for the planning of infrastructure. Led (at the time this chapter was written) by the engineer Sir John Armitt, who had successfully led the engineering of London's Olympic Stadium, the Commission states: "The NIC aims to be the UK's most credible, forward-thinking and influential voice on infrastructure policy and strategy."[1] Commissioners are from a range of disciplinary backgrounds; however, engineering expertise plays a key role. In areas of infrastructure delivery, which can take many years to complete, cause disruption and impact on communities in many ways, engineering and policy decisions are closely coupled. Creating systems that deliver policy objectives of sustainability and efficiency—e.g. through the better use of data and digital technologies—depends on engineering insight into how systems can be designed to practically achieve these ends.

There are, therefore, many areas of engineering practice where engineering and policy are inextricably woven together, albeit often with the essential involvement of other disciplines such as the natural sciences, economics, the social sciences and humanities, such as ethics and philosophy. Throughout all of these areas there are commonalities in the ways that engineering advice provides input, best understood in terms of its narrative form, underpinned by an understanding of the nature of engineering knowledge.

47.2.4.1 The Nature of Engineering Knowledge—Tool Use, Cookbook Engineering and Design

A number of different reflections on the nature of engineering, and on how engineers work, are based on distinctions made between engineering and science, both between how they are practised and the knowledge that they produce. A basic aspect of this is the idea that engineering knowledge is based on *know how*, rather than *know that*, or that the distinction between engineering and scientific knowledge can be traced back to the distinction between *techne* and *logos*, according to Davis (1997). The idea that engineering revolves centrally around this pragmatic form of knowledge means that engineers have a particular relationship with the formal methods and theories that apply.

Walter Vincenti's seminal book on engineering knowledge suggests that this idea of engineering as being rooted in practical knowledge has a very literal basis: "By the very nature of things, some of what had been learned about flush riveting could only remain in the neuromuscular skills of the workers and the intuitive judgement of the engineers" (Vincenti 1990: 188).

Of course this focuses on the skills that lie at the practical extreme of what is a very broad spectrum of engineering knowledge. However, there is an element of 'tool use' that is common to the ways that engineers make use of a diversity of theoretical knowledge as it is to how they make use of riveting techniques. For example, I argued (McCarthy 2009: 65):

> Mathematics has the status of a tool in engineering, a method that engineers use to support their work in design and testing . . . It is as important for an engineer to know where the mathematics breaks down and applies only approximately as it is for them to know where it applies at all.

This idea of engineering knowledge as tool use draws on comments from Goldman (2004).

Joseph Pitt (2001) considers the nature of engineering knowledge in relation to scientific knowledge. Examining Vincenti's account of engineering knowledge, he highlights the fact that unlike scientific knowledge, engineering knowledge is "task-specific and aims at the production of an artefact to serve a predetermined purpose" (Pitt 2001: 25). The nature of this knowledge is distinct from science in that engineering knowledge is systematised, in terms of establishing a set of methods to achieve the tasks that engineering knowledge is focused on: "The solution to specific *kinds* of problems ends up catalogued and recorded in the form of reference works which can be employed across engineering areas" (Pitt 2001: 25). He describes this as "cookbook engineering," an idea of engineering knowledge which should not be seen as underplaying the complexity of engineering knowledge but actually showing its value. Engineering knowledge is a set of systematized recipes, and the skill of the engineer is to see when a situation exemplifies a certain type of problem in need of a certain type of solution. This can be seen as choosing the right recipe to use, or, as I suggested earlier, knowing the right tool that will work in a given situation. And Pitt suggested that a great benefit of engineering knowledge is that "cookbook" knowledge can work across fields, in a wide range of contexts.

This should not lead us to believe that engineers merely implement preconceived solutions. Consider another important characterisation of engineering—in which design is key to the nature of engineering. Writing as an engineer reflecting on engineering practice, David Andrews (2010: 35) states: "I would contend that it is largely the design element in engineering practice that distinguishes engineering as an activity from the sciences." Andrews argues that this design role is met by engineers in a pragmatic fashion, making use of relevant knowledge—not only mathematics—as an appropriate tool:

> Much of what an engineer does in the process of design, which distinguishes engineering design from most other design endeavours, is applying scientific analysis using scientific knowledge. But of course, in using the engineering sciences, we as engineers do not do so in a pure scientific manner. We are pragmatists, we want answers and we will use what knowledge there is to best progress our analysis . . . the engineering designer has to be much more than just an applied physical scientist. He or she also has to be skilled in the human sciences and all of these topics seem to require ever greater breadth in the designer's awareness.
>
> *(Andrews 2010: 35)*

The design role of engineers highlights not only the pragmatic, tool-using nature of engineering, but its focus on some external outcome. This idea is set out for example in Schmidt (2013). In "Engineering as Willing," Schmidt states:

> Science is widely perceived as an especially systematic approach to knowing: engineering could be conceived as an especially systematic approach to willing . . . [in engineering] although the intellect is implicitly involved, the will is primary, because the goal is pragmatic: some outcome that is usually subjective.

While Schmidt is arguing for an understanding of engineering in relation to a particular theory of insight which I am not specifically endorsing here, the essential point in Schmidt's argument is the focus on a *goal* and *outcome* in engineering practice—a goal that is often externally set.

47.3. How Engineers Talk About Engineering in Policy

How do these characterisations lead us to understand the role of engineering in policy advice? A challenge here is that one might be tempted to assume that engineering knowledge reduces to the following simplistic structure: 'You need to achieve this, you apply tool X, bearing in mind considerations ABC.' It is an instructional format. It is practical and outcome focused, and in Schmidt's case he suggests that there is an element of the end goal being subjective and imposed from without. One might assume that the key role of engineering knowledge is in bringing about a pre-ordained outcome, based on knowledge of the situation and a suite of exceptionally complex tools, and by applying a set of skills from the well codified to the craft skill.

The risk with engineering in policy is that this conception of engineering can create a role for engineers at the implementation and delivery end of the policy cycle. Policymakers choose an outcome—even a technology—and consider the role of engineering to be finding a means of implementing that outcome.

However, engineers might argue that this is the wrong conception of how pragmatic engineering knowledge can inform policy. The following argument was made by a group of engineering organisations in response to a parliamentary inquiry into the role of engineering in government.

> Engineering is concerned with solving practical problems and in changing the physical world, using scientific, technical and business skills. Science, on the other hand, is principally about understanding the nature of the world. The practical nature of engineering means that engineering advice and expertise is of great value in developing policy and delivering projects. For example, the need for engineering advice is particularly pertinent in the area of climate change. The big challenge is no longer the search for evidence for climate change but rather the search for means of avoiding its advance and mitigating its effects, many of which will be matters of engineering and technology.
>
> *(Royal Academy of Engineering 2008: 4)*

A key reason why engineers have this broader role in policy is due to the fact that engineering involves systems approaches:

> There are few areas of government policy that do not have an engineering dimension to their delivery. This strategic capacity is therefore critical when commissioning engineering consultancy, designing major engineering projects and receiving engineering advice relevant to policymaking. The experience of Chartered Engineers in delivering projects and their ability to think at a systems level mean that engineers in the civil service can make valuable contributions right through the policymaking and policy delivery cycles.
>
> *(Engineering the Future 2011: 5)*

To understand what this means for engineering policy and the nature of the narratives of engineering policy, we can revert back to characterisations of the distinction between engineering and science set out earlier. The major differentiating factor between engineering and science, according to many engineers, is the fact that engineers do not take the role of gathering evidence and constructing narratives that explain causal relationships and processes to give an account of the evidence observed. Rather, they deal in the art of the possible rather than the actual, and engage in the process of design rather than (primarily) description. This gives them—or engineering evidence—a particular role in policy. Before showing this role in action, and analysing its form, I will set out some aspects of the idea of narrative explanation.

47.4. Narrative Explanation

In order to set out the nature of engineering in policy, I will draw on the idea of narrative explanation explored in depth through the collection by Morgan and Wise (2017) and set out in Roth (1989). Narratives are familiar from fiction and have the form of stories which "set out how things happen and why things happen" (Morgan 2017: 2). They are most closely associated with explanations in history, because they are thought to be of greatest relevance in setting out relationships between contingent events rather than those governed by deterministic laws, making sense of a flow of chance-like events and giving understanding as to why a certain outcome came about. Morgan and Wise, among others, seek to demonstrate that narrative explanations, 'stories' that chart a path through different events and seek to set out how they could be connected, also have a significant role in the sciences as well as in history and social sciences.

There are different forms of narrative explanation—a key one being narratives that set out how events 'unfold in time.' This is set out by Morgan and Wise in their introductory paper to their *Science Narratives* volume:

> The second prominent way in which many of our papers show narrative functioning in the sciences, concerns its use in making things known and understandable by revealing how, like a story, they "unfold" in time. . . . Both forward following, and backward tracing reveal narrative paths that are neither fully predictable nor fully explainable, for their ingredient of randomness, variety, and unpredictability, create narrative paths that are themselves a source of illumination to the scientist.
>
> *(Morgan and Wise 2017: 2)*

A key feature of narratives according to Morgan and Wise is that:

> a narrative account makes it easy to think not just about contingencies, but also about possibilities and counterfactuals. Narratives deal in these characteristics routinely. . . . The explanatory power of narratives lies in being able to chart a satisfactory path not just through contingencies, possibilities and alternatives, but do so by making active use of those features.
>
> *(ibid.)*

Narratives therefore have a form different from that of a logical argument where laws and generalisations play an overt or explicit role and are in contrast to the kind of deductive-nomological form of explanation assumed by Hempel. Importantly, narratives involve a "forward or backward understanding of [a] path of events" (Morgan and Wise 2017: 1), deal in the relationships between particular events and enable the identification of points in time that are critical to bringing about certain outcomes—even when there were many potential outcomes that might have arisen. Narratives do call upon various mechanisms that might connect those points, including causal or theoretical relations, but they are various in form and bring coherence to a set of events in different ways. The explanatory nature of the narrative is not specifically reducible to a given form but rather:

> What counts as an explanation, and understanding, within a science depends less on a universal ideal, than on what satisfies the scientific norms and values and shared knowledge set of a community.
>
> *(Morgan and Wise 2017: 4)*

Morgan explains the purpose of narrative as follows:

> Scientific narratives focus on the reasons how and why things happen, whether these are ordered through time, or along some other perspective. Thus, it is the ability of the narrative scientist (as for the novelist or historian) not merely to order their materials., but to do so in answering how or why questions that lies at the heart of narrative.
>
> *(Morgan 2017: 87)*

Roth (1989) uses the idea of narrative explanation as a challenge to the idea that explanations have at root the logical form of an argument, and sets out the idea of a form of explanation that "invokes no laws or even probabilistic generalisations" (Roth 1989: 450). In setting out how narratives can provide explanation, Roth argues that "Narratives explain, on this account, by providing stories as solutions to problems" (Roth 1989: 469). That is to say, a narrative takes a situation and problematizes it to some degree—sets it up as a puzzle—and gives a narrative of that event which provides a solution to that puzzle. What makes a narrative such a solution, and hence enables it to serve as an explanation, is an invocation of paradigmatic narrative forms which are—in the manner of the paradigms in science described by Thomas Kuhn—accepted within a given knowledge community as explanations.[2] To serve in such a way, narratives have to be sufficiently generalizable to have paradigmatic status:

> Paradigms are solutions to problems posed by concrete phenomena; they are answers to puzzles. As solutions, they become paradigmatic insofar as they are sufficiently flexible to allow of extension to related phenomena, or phenomena perceived as related.
>
> *(Roth 1989: 468)*

Morgan (2017) explains this further by suggesting that the 'paradigmatic' nature of narrative explanations (a term that Roth uses carefully and Morgan treats with further caution) relates to the fact that such explanations "enable extension to related phenomena" (p. 94)—that is, despite being ostensibly about one particular situation, it can be generalised out to 'explain' related situations. Roth describes the narratives that he refers to as treating narrated events as *tokens* of event *types*.

I will argue that a narrative form of explanation also has a crucial role in engineering policy advice, due to the nature of the situations to which engineering policy pertains, and also due to the nature of engineering knowledge and evidence. I will set out some examples of engineering policy, and show how they exhibit a narrative form which I argue bears a relation to the idea of narrative explanation set out by Morgan, Wise and Roth.

47.5. Engineering (Narratives) in Policy—a Series of Case Studies

All of the following examples of engineering policy advice are linked in that they focus in particular on engineering policy advice which is future-focused in a variety of ways, which we can consider perhaps as *strategic* engineering policy advice. They do not include the latter two of my types earlier—where engineering policy is focused on regulating technology or in which policy advice is tightly woven into procurement processes—but there is a case for saying that the narrative structure I will identify applies to these forms of policy advice too, as I will explain at the end of the chapter.

47.5.1 *Internet of Things: Realising the Potential of a Trusted Smart World*

First is the 2018 report from the Royal Academy of Engineering and the PETRAS research consortium which is focused on issues relating to security in the Internet of Things. The report *Internet of Things: Realising the Potential of a Trusted Smart World* sets out its purpose thusly:

> This report examines the policy challenges for the Internet of Things (IoT), and raises a broad range of issues that need to be considered if policy is to be effective and the potential economic value of IoT is harnessed.
>
> *(Royal Academy of Engineering and Petras 2018: 5)*

The report sets out a statement of promise or vision in broad terms relating to best utilisation of IoT technologies: "[the] IoT is an enabling technology that has the potential to fundamentally change society and business processes within and across sectors" (ibid.: 4). The aim of this report, therefore, might be seen as how we bring about a future where the IoT is put to beneficial societal and business uses, bringing about positive economic impact.

The report sets out a series of recommendations on how we can achieve that outcome. These include recommendations about how policy should be shaped:

> Recommendation 1: In developing policy for IoT, the Department for Culture, Media and Sport (DCMS), the Department for Business, Energy and Industrial Strategy (BEIS) and other government departments should distinguish the differing policy outcomes for industrial, public space and consumer applications of IoT.
>
> *(ibid.: 6)*

The authors also include recommendations about regulation:

> There will need to be new mechanisms, and perhaps new regulatory frameworks, for cooperation between sector regulators and a genuine systems approach to policymaking.
>
> *(ibid.: 22)*

And recommendations about technology:

> Recommendation 2c Government should continue to facilitate the development and deployment of standards for IoT where needed, building on progress to date.
>
> *(ibid.: 8)*

What is interesting here is the focus on the need for roadmaps and strategies. The report is inherently forward looking, and while it highlights risks—"There are substantial and interdependent issues around privacy, ethics, trust, reliability, acceptability, safety, and security for the systems that are created" (ibid.: 4)—the focus of the advice is not on analysis of risk and failure but on building a system that works—and the steps needed to put that system in place.

47.5.2 Greenhouse Gas Removal (GGR)

The next example is about a much more fine-grained and explicit goal that most members of the United Nations are bound to and the means by which it can be reached. The goals in question relate to the goals for limiting global temperature rise as set out in the Paris Agreement in 2015, and the report sets out the roles of technologies for GGR in meeting those goals. This report sets out the need for GGR, then looks at methods, the means for implementing them and then explicitly focuses on two scenarios, which are in turn considered because they are potential means to achieve the climate targets in the Paris Agreement. The purpose of looking at these scenarios is as follows:

> Two example scenarios have been developed to examine how individual greenhouse gas removal methods and cross-cutting issues might apply in the practice of meeting real-world targets . . . they present plausible, though not necessarily optimal or desirable, future worlds in which ambitious efforts are made to limit climate change using GGR.
>
> *(Royal Academy of Engineering and Royal Society 2018: 90)*

This report importantly brings together scientists and engineers alongside other disciplines, and sets out the science of greenhouse gas removal methods alongside with the steps needed to *use* them to bring about a desired scenario. The report considers two specific scenarios: "achieving net-zero emissions in the UK in 2050; and limiting the global temperature rise on pre-industrial levels to 1.5°C as of 2100" (ibid.: 9).

The report highlights the significant challenges in reaching these scenarios:

UK net-zero in 2050: In the UK, reducing greenhouse gas emissions to the greatest degree considered feasible would leave remaining emissions of around 130 MtCO2 pa by 2050. Off-setting these emissions with GGR to reach 'net-zero' for the UK is possible, but very challenging. It involves deployment of many different GGR methods, and import of biomass.

(ibid.: 9)

Global cumulative GGR compatible with 1.5°C by 2100: Integrated assessment models provide evidence that a cumulative GGR of around 810 GtCO2 is expected to be required from now until 2100 to limit the rise in temperature to 1.5°C on pre-industrial times. This is the equivalent to about 15 years of 2017 greenhouse gas emissions.

(ibid.)

47.5.2.1 The Recommendations

This report gives fine-grained recommendations about how to achieve these specific outcomes, based on the examinations of the methods in the report. These include things that physically need to be done specifically in order to reach the net zero scenario:

Pursue rapid ramp-up of forestation, habitat restoration, and soil carbon sequestration, across large UK land-areas.
   Grow and import sustainable biomass at large scale to meet the need for both energy and GGR demands.

(ibid.)

They also include the mechanisms that might encourage or enable these steps to be taken:

Establish an incentive or subsidy system to encourage changes of land practice, particularly for soil carbon sequestration . . .

(ibid.)

And methods that allow indication that the right approaches have been taken to achieve these outcomes:

Develop monitoring and verification procedures and programmes to track the effectiveness of GGR delivered by each method. . . .
   This should include assessment of the co-benefits, social and environmental risks, monitoring and evaluation, and include field-based pilot demonstrations.

(ibid.: 106)

The report also makes broader recommendations about the research and engineering system that needs to be in place to enable this; e.g.:

RECOMMENDATION 3 Build Carbon Capture and Storage infrastructure. Scenario-building indicates that substantial permanent storage, presently only demonstrated in geological reservoirs, will be essential to meet the scale required for climate goals.

RECOMMENDATION 4 Incentivise demonstrators and early stage deployment to enable development of GGR methods. This allows the assessment of the real GGR potential and

of the wider social and environmental impacts of each method. It would also enable the
process of cost discovery and reduction.

(ibid.: 114–115)

This report therefore identifies the technical, policy and research steps that need to be taken—
through the use of GGR methods—that get us to the scenarios set out, which are in turn essential
to the overall goal of meeting the commitments in the Paris Agreement.

47.5.2.2 Building Evidence for Policy

Also interesting to note here is a common form of recommendation in engineering policy text—
which is the need for demonstration that a system will work at scale (recommendation 4). The report
summary gives more detail to this:

> Some GGR methods are already in use today, while others require significant development
> and demonstration before they can remove emissions at scale. When considered at the scale
> required, none of the methods have been fully evaluated across their life cycle.

(ibid.: 8)

The significance of this is that the nature of engineering predictions is not deterministic; it is not
possible always to generalise from things working in one situation to their working in another; or for
processes and systems that work at one scale necessarily scaling up.

47.5.3 Space Weather

The Royal Academy of Engineering's policy report *Extreme Space Weather: Impacts on Engineered
Systems and Infrastructure* is similarly future-focused, though in this case the scenario is a risk that we
need to avert. This is an archetypal challenge in engineering in that it deals with low probability and
high-impact scenarios and involves expert knowledge of the circumstances in such a scenario and
how we could change situations so that this did not come about.

> How often superstorms occur and whether the above are representative of the long term
> risk is not known and is the subject of important current research. The general consensus
> is that a solar superstorm is inevitable, a matter not of 'if' but 'when?'
>
> *(Royal Academy of Engineering 2013: 5)*

What is better understood and set out in the report, however, is what would happen if there were a
superstorm. The report sets out likely impacts on engineered infrastructure both terrestrial—i.e. the
electricity grid—and in space—in terms of satellite systems that provide essential services. In both
cases these systems are likely to suffer damage that can put critical infrastructure at threat. The challenge for this area of policy advice is that such events are so rare that they have not been seen in the
context of the current complex power and communications infrastructure.

The report sets out a series of actions that will mitigate the effects of such a storm, based on engineering knowledge of the effects of such a storm on engineered infrastructure and the technological
steps needed to enable recovery. These include:

> Electricity grid: The current National Grid mitigation strategy should be continued. This
> strategy combines appropriate forecasting, engineering and operational procedures. It
> should include increasing the reserves of both active and reactive power to reduce loading

on individual transformers and to compensate for the increased reactive power consumption of transformers.

Global navigation satellite systems (GNSS): All critical infrastructure and safety critical systems that require accurate GNSS derived time and or timing should be specified to operate with holdover technology for up to three days.

High frequency (HF) communications: The aviation industry and authorities should consider upgrades to HF modems (similar to those used by the military) to enable communications to be maintained in more severely disturbed environments. Such an approach could significantly reduce the period of signal loss during a superstorm and would be more generally beneficial.

(ibid.: 6)

47.6. The Narrative Form of Engineering Policy Advice

Implicit in, or underpinning, the logic in these examples of engineering policy advice is a narrative of one of the following forms:

- If we want to achieve situation X, we must do ABC;
- If situation X occurs, we must prepare by doing ABC, which will enable us to reach situation Y (where disastrous outcomes are averted); or
- The way the world is now, we will be in situation X (unless we do ABC).

That is to say, narratives of these forms can easily be imposed on the key messages in these pieces of policy advice.

I argue that there is a narrative structure to these examples of policy advice for two reasons. First is that the structures here are essentially temporal, due to the fact that the account links a desired outcome with the set of preceding events and activities that need to be put in place in order to reach that outcome. These cases of policy advice set out potential future paths that events might follow if certain key steps are taken. They are predictive in nature but rest on *narratives of possible worlds* in which steps are taken and the situation is achieved, or steps are taken and it is not. They can be seen as explaining how some outcome can come about in those possible worlds.

Second, the nature of links being made are based on practical actions and their outcomes. The relationship is not one of logical implication, e.g. through appeal to application of deterministic laws, but is based on managing a series of contingent events in the 'real world.' The advice emphasises the need for testing, for scaling up and responding to contingent events. These are future-facing stories about how real world events can unfold rather than descriptions of events under controlled conditions, or following predictable laws.

Why are these *engineering* narratives? After all, much of policy has the form of setting out how the world could be, or is hoped to be, based on taking certain steps, and policy rests on much more than just engineering evidence and expertise. There is a specific aspect of these policy reports which I would argue rests on the particularly distinct input of engineering.

First is a general argument having to do with the nature of engineering practice. Because engineering is not simply descriptive of events that have happened but is about drawing out the possibility space, in order to say what could or should happen, there is a clear contrast with historical or natural science narratives which describe actual processes that have taken place or models/simulations of such processes. Science is of course also predictive and therefore is also future focused, but the key difference for engineering is that it is focused on actively *bringing about* a future situation or *changing* the context in which it will take place. This is the essential design element to engineering, set out by Andrews and Schmidt.

Second, the nature of the story that engineers tell about what is possible, or about what has to be the case in the possible world where this comes to be, is based not on strict laws but on contingencies, on the actions of people and their interactions with natural systems and technologies. This is where we return to the nature of engineering knowledge and its practical purpose. First, engineers work not on stories of idealised possible worlds, but on possible worlds that are linked through practical steps to our current situations. And engineers identify the tools that can be applied, in certain ways, to bring about that situation.

It is this latter point that makes Roth's (1989) account of narrative explanation relevant here. The engineering advice being given may be providing a route through a set of contingencies, but there are established rules for this, not guesswork. These tools take the form of paradigmatic solutions to problems, ones which engineers will recognise as providing the appropriate solution. Or, in Morgan's framing, these are solutions that can extend from one case to another. In the engineering sense they are the *tools* and approaches that are known to work in one kind of situation and which are therefore invoked in order to explain how to address the new challenges set out in a related or comparable policy area.

This narrative idea is also reminiscent of the "cookbook engineering" knowledge that Pitt discussed. The kinds of narratives that Roth discusses are well-accepted stories that provide solutions to specific problems or puzzles. This is akin to providing a given recipe or solution to bring about a token of an event or design type in engineering, as per the "cookbook" approach. This is also seen in the policy examples, where specific sorts of steps are presented as means to reach a goal which is set up as a puzzle to be solved. (How do we get to this outcome? How do we avert this potential situation?)

This is, I believe, the place where engineering makes a specific contribution to policy, where it draws on the practical experience of engineers to set out how we can use theoretical evidence and technological solutions to bring about a specific outcome in messy, real-life circumstances. In many cases, though, that engineering contribution will be embedded and implicit and entwined with other disciplines, all of which have a critical role. Indeed, one of my key case studies was a joint report between the Royal Society and the Royal Academy of Engineering—and it makes critical use of the *science* that underpins GGR. But the nature of the engineering knowledge and expertise that adds into this policy advice suggests that there is a specific engineering contribution to the narrative that runs through these case studies. This is telling a story about a possible world, full of contingency, choice and well-designed interventions which lead us to a point where we achieve our stated goal; or we avert a potential disaster of failure or respond to that disaster or failure if and when it comes. A key role for engineers and their experience is in knowing how to find a tool that will work, and how to use it, in these complex situations.

47.7. Discussion

The kinds of policy advice discussed here are strategic policy advice and fit into the categories of 'policy for engineering' and 'engineering for policy' in the taxonomy that I presented earlier. However, there are similarities in the form of engineering advice in regulation and procurement. In each case, a link is being made between certain steps, whether these steps adhere to technology standards or business practices, or invest in given systems and technologies, and bringing about a desired outcome. More examination of such forms of policy would be fruitful.

There are a number of further questions to discuss in order to best understand the role of engineering in policy advice. First is further isolating or extracting the engineering narrative and those of other disciplines that make an essential contribution to policy advice. It may be that the nature of engineering's role in policy is easier to distinguish from the sciences than it is from other areas of practical knowledge that inform policy. There is also value in better understanding how policymaking methods compare with the design cycles of engineering.

However, what this initial consideration of the narrative of engineering policy does is to clearly establish a role for engineers in the policy process which is distinct from that of closely related disciplines and shows that the uniquely practical and pragmatic nature of engineering makes it highly valuable in the policy process. Policy challenges are complex and require a complex set of disciplines to inform them, but understanding the nature of the engineering contribution can show why engineers are able to contribute much more than implementing policies once those policies are already set.

Acknowledgements

I would like to thank the organisers and participants in the LSE Science Narratives workshop on Expert Narratives in December 2018 for the opportunity to present and hear feedback on an earlier version of this chapter, in particular Mat Paskins and Mary Morgan.

Related Chapters

Chapter 8: Prescriptive Engineering Knowledge (Sjoerd Zwart)
Chapter 12: Scenarios (Christian Dieckhoff and Armin Grunwald)
Chapter 13: Systems Engineering as Engineering Philosophy (Usman Akeel and Sarah Bell)
Chapter 21: Engineering Design (Peter Kroes)
Chapter 45: Engineering as a Political Practice (Govert Valkenburg)

Further Reading

Pielke, R. A. Jr. (2007). *The Honest Broker, Making Sense of Science in Policy and Politics.* Cambridge: Cambridge University Press. (For a broader exploration of the roles that science and engineering can play in policy.)

Notes

1. www.nic.org.uk/what-we-do/; retrieved 28 February 2019.
2. Roth cites Kuhn (1975, 1977).

References

Andrews, D. (2010). Philosophical Issues in the Practice of Engineering Design. In *Philosophy of Engineering: Volume 1 of the Proceedings of a Series of Seminars held at The Royal Academy of Engineering.* London: The Royal Academy of Engineering, pp. 35–44. www.raeng.org.uk/publications/reports/philosophy-of-engineering-volume-1. Accessed February 27, 2019.

British Standards Institute (2016). BS8611 (short description). https://shop.bsigroup.com/ProductDetail/?pid=000000000030320089. Accessed February 25, 2019.

Bush, V. (1945). *Science: The Endless Frontier.* Washington, DC: United States Government Printing Office. https://nsf.gov/od/lpa/nsf50/vbush1945.htm. Accessed February 24, 2019.

Craig, C.H. (2018). *How Government Listens to Scientists.* Cambridge: Palgrave Macmillan.

Davis, M. (1997). *Thinking Like an Engineer.* Oxford: Oxford University Press.

Engineering the Future (2011). *Response to the House of Commons Science and Technology Select Committee Engineering in Government Inquiry.* London: Engineering the Future. www.raeng.org.uk/publications/responses/engineering-in-government-inquiry-response. Accessed February 28, 2019.

Goldman, S.L. (2004). Why We Need a Philosophy of Engineering. In Howard Cattermole (ed.), *Interdisciplinary Science Reviews,* 29(2).

Kuhn, T. (1975). *The Structure of Scientific Revolutions.* Chicago: University of Chicago Press.

Kuhn, T. (1977). *The Essential Tension.* Chicago: University of Chicago Press.

McCarthy, N. (2009). *Engineering: A Beginner's Guide.* Oxford: One World.

Morgan, M. (2017). Narrative Ordering and Narrative Explanation. In Morgan and Wise (2017).

Morgan, M. and Wise, N. (eds.) (2017). Narrative Science. *Studies in History and Philosophy of Science*, 62.

Pielke, R.A., Jr. (2007). *The Honest Broker*. Cambridge, MA: Cambridge University Press.

Pirtle, Z. (2013). Engineering Innovation: Energy, Policy, and the Role of Engineering. In D. Michelfelder, N. McCarthy and D. Goldberg (eds.), *Philosophy and Engineering: Reflections on Practice, Principles and Process. Philosophy of Engineering and Technology*, Vol. 15. Dordrecht: Springer.

Pitt, J.C. (2001). What Engineers Know. *Techne,* 5(3), 18–30.

Roth, P.A. (1989). How Narratives Explain. *Social Research*, 56(2), 449–478.

Royal Academy of Engineering (2008). *Engineering in Government: Response to House of Commons IUSS Select Committee*. London: Royal Academy of Engineering. www.raeng.org.uk/publications/responses/engineering-in-government. Accessed February 24, 2019.

Royal Academy of Engineering (2013). *Extreme Space Weather: Impacts on Engineered Systems and Infrastructure*. London: Royal Academy of Engineering.

Royal Academy of Engineering and PETRAS (2018). *Internet of Things: Realising the Potential of a Trusted Smart World*. London: Royal Academy of Engineering.

Royal Academy of Engineering and Royal Society (2018). *Greenhouse Gas Removal*. London: Royal Society.

Schmidt, J.A. (2013). Engineering as Willing. In D. Michelfelder, N. McCarthy and D. Goldberg (eds.), *Philosophy and Engineering: Reflections on Practice, Principles and Process. Philosophy of Engineering and Technology*, Vol. 15. Dordrecht: Springer.

Vincenti, W. (1990). *What Engineers Know and How They Know It*. Baltimore: Johns Hopkins Press

Walport, M. and Craig, C. (2014). *Annual Report of the Government Chief Scientific Advisor 2014. Innovation and Risk: Managing Risk, Not Avoiding It*. London: Government Office for Science.

PART VIII

Reimagining Engineering

48

FEMINIST ENGINEERING AND GENDER

Donna Riley

48.1. Feminist Engineering Ontologies

Ontology is the study of what exists, or what is real, and the relationships among those things. Engineering has, for the most part, been rooted in materialist presumptions, whether or not it is aware of any ontological alternatives or the ways in which emerging technologies might challenge these assumptions (Goldberg 2010). In materialist ontological frameworks, it can be difficult to introduce notions of socially constructed phenomena such as identities around gender, race, disability, or sexual orientation, due to a tendency to root reality in observable biological or chemical phenomena such as anatomy, DNA, brain chemistry, and the like (Haslanger 1995).

Engineering ought not to be considered only through a materialist ontological lens. One can readily identify anti-materialist (including, but not limited to, idealist) ontologies in engineering topics, both historically and in the present day. For example, at least one early exploration of the conservation of energy by Julius Meyer was rooted in a belief that there was something more to existence than solely the material (von Baeyer 1999). Meyer's experiments helped to challenge the conventional wisdom of the time that heat was a material substance (the caloric). While energy ultimately has become part of the materialist schema, we are now confronted with new challenges to materialism in engineering ontologies in digital settings (Goldberg 2010; see also Chapter 55, "Reimagining the Future of Engineering" by Neelke Doorn, Diane P. Michelfelder, et al., this volume). Brey (2003) offers a helpful ontological analysis of virtual reality using Searle's (1995) notion of social ontology. He comments on a 1990s instantiation of virtual reality in the early days of the Internet, where gamers created multi-user dungeons (MUDs), spaces in which users engaged in interactive role-playing games, assuming characters represented by programmed objects. Brey (2003: 280) characterizes MUDs as sites of "ontological uncertainty":

> A regular topic of discussion among MUD users is the status of violence and sexual assault in them. If MUDding is to be understood as playing a game, then perhaps violence and sexual assault are permissible, because they can be introduced as elements of normal play. However, if MUDding is to be understood as built up out of real social interactions, then perhaps violence and sexual assault in MUDs should be understood as really happening, and should be treated as such.
>
> *(Brey 2003: 281)*

Kendall (1996) illustrates through her personal account of MUDs the complexities of gender relations online, and the profound implications engineered worlds have for our lived experience of gender. In documenting the hostilities enacted against female personae in MUDs, regardless of the user's gender "in real life," she raises ontological questions about construction and reproduction of gender and power in virtual worlds, which engineers ignore at their (and society's) peril.

48.1.1 Engineering as a Gendered Existence

There is a robust empirical literature documenting through observation and the application of feminist analysis the ways in which the engineering profession is gendered. At the heart of this work lies an assumption that gender is socially constructed, as is engineering as a discipline, a body of knowledge, a profession, and a set of activities.

Pawley (2012) describes engineering as a block of Swiss cheese, in which women's contributions to the field have been omitted in the very definition of the field. She analyzes faculty narratives about engineering to identify locations of field boundaries that implicitly include certain things and exclude others, and which actors are considered to belong or not belong in the spaces set by these boundaries. In this way a gendered picture of engineering emerges, one that benefits male actors in masculine spaces.

One historical example that illustrates Pawley's "Swiss cheese" metaphor is documented in the work of Amy Sue Bix (2002); home economics developed in parallel with engineering, even as female students learning household equipment design used the same principles as those found in chemical, mechanical, and electrical engineering, with the same preparation in math and physics. Because the equipment was used by women in the home, this part of what might otherwise be considered engineering was treated separately as home economics and thus became the purview of women.

Cindy Foor and Susan Walden (2009) investigate the field of industrial engineering (sometimes derogatorily referred to as "Imaginary Engineering") as a marginal and feminized space within engineering. Through ethnographic interviews, they found that industrial engineering was able to depart at times from engineering gender norms and thus become more gender inclusive, but at the same time there remained proscribed gender roles for men and women to fulfill, with women matching heteronormative feminine business-woman ideals, and with men reimagining masculinity in ways paired with this normative female archetype, drawing contrasts with stereotypical narrow engineering geeks and making room for males as nurturers.

Wendy Faulkner (2000, 2007, 2009a, 2009b) documents in a series of workplace studies the ways in which engineering practice encodes gender. Her ethnographies illustrate the complex interactions of engineering and gender identities, practices, and cultures. She argues that there is a mutually reinforcing relationship between narrowly technical engineering identities and hegemonic masculinities that emphasize hands-on orientations, while women are more readily identified with the broader sociotechnical and people-oriented reality of engineering work, setting up conditions in which they must continually prove both that they are "real engineers" and "real women" (Faulkner 2007).

Other accounts of gendered workplace norms come from Paulina Borsook (1996), who describes the sexism of Silicon Valley tech culture through her career as a journalist for *Wired*. Prominent females like Esther Dyson were scrutinized based on whom she was dating or what she wore, a standard not applied to male tech leaders. When Borsook spoke to her editor about this sexism, he remarked that he saw Dyson as neither male nor female; in other words, she was invisible as a woman to him. Other scholars have observed invisibility as a strategy some female engineers consciously take on to avoid negative or unwanted attention from male peers; both Seymour (1995) and Hacker (1989) describe women seeking to blend in through androgynous dress.

Lisa Frehill (2004) provides an account of the establishment of a masculine engineering profession in the United States in the late 19th and early 20th centuries, as the nation moved to mass production and increasing urbanization. Frehill draws on Oldenziel (2000) to show how the profession sought to establish itself in terms of a white, male, middle-class norm, modeling engineering education in the image of Theodore Roosevelt, known for his rugged character and love of the wild outdoors. This leads quite directly, Frehill argues, to the development of a rigorous weed-out culture, in which men proved themselves as they would in military boot camp.

Years earlier, Hacker (1989) employed autoethnography to describe her experiences of a deeply masculinized and militarized engineering curriculum, in which students must physically regulate their bodies in deprivation of food, sleep, and even pleasure in a disciplined regimen directed toward success on high-stakes tasks such as exams or problem sets.

Karen Tonso (1996) offers a discourse analysis of the gender norms articulated in an engineering design class. Through the use of sexual innuendo, profanity, and violent metaphors, male students and the professor communicated a presumption of male professional ability linked to male sexual desirability, as well as an assumed incompatibility between engineering ability and sexual desirability in females. More recent work by Akpanudo et al. (2017) identifies hegemonic masculine social norms operating in engineering, characterized by "emotional regulation, the exertion of significant dominance over others, and the desire to win."

Some of the descriptions rendered by Tonso and Hacker in particular note incidents of violence embedded in engineering. Riley (2008) documented the use of a mnemonic device in electrical engineering for resister color codes (Black Boys Rape Our Young Girls But Violet Gives Willingly) that reveals the kind of overt racism, misogyny, and threat of violence that can appear as "situation: normal" in engineering classrooms. Indeed, there is documentation of prevalent sexual violence in engineering from the earlier sexual harassment studies of Lafontaine and Tredeau (1986) and McIlwee and Robinson (1992) to broader present-day studies documenting gender-based harassment and assault across STEM disciplines, such as that conducted by Clancy et al. (2014). The dominant expected response is to ignore, grin, and bear this violence is to develop a trait once characterized as persistence, but now rebranded or "improved" as grit (Slaton et al. 2019).

48.1.2 Intersectionality and Ontology

While the previous section focused on engineering as a gendered existence, the notion of intersectionality (Anzaldúa 1987; Crenshaw 1989; Hill Collins 1998) indicates that the gendering of engineering cannot be understood in isolation. While engineering is a cis-masculine profession, it is simultaneously also a white, straight, able-bodied, professional class activity. The effects are not merely additive, requiring focused study of intersectional experience. Some of the work referenced in the previous section explicitly considers this multiplicity (see, e.g., Frehill 2004), while for others it is implicit, or even rendered invisible as whiteness, straightness, able-bodiedness, and middle-class existence are the presumed normative state of being (Hammonds 1994). Engineering education research is beginning to narrate intersectional experiences, particularly for women of color, but also considering intersectionality in the lived reality of LGTBQ students (e.g., Leyva 2016; Camacho and Lord 2013; Revelo et al. 2017; Cech and Rothwell 2018).

Carbin and Edenheim (2013) describe an ontological slippage in feminist work with intersectionality, noting that its origins in Black Feminist Thought are structuralist, while its popularity has caused it to be adopted and adapted to fit poststructuralist as well as liberal feminisms, without a careful unpacking of its ontological presumptions. In particular, they articulate a concern that "the basic premise of poststructuralism is missing in this inclusion: that of the fundamental impossibility of accurately representing the world" (243). Further, they are concerned that intersectionality's loose inclusion of both structures and agents allow for intersectionality to exist both in the classic structural sense as presented in Black Feminist Thought, and operationalized at the individual and agential level, but never questioning that very binary, which is, they argue, a central project of poststructuralist feminism.

Slaton and Pawley (2018) also note this problematic and note that "we are a long way from being sure that these two analytic objects—individual and collective experiences of identity—even represent two distinct ontological categories" (147). They propose using queer theory to embrace these incompatibilities with critical reflexivity, allowing ontology to inform epistemology, acknowledging that, "if it does not question our sense of what counts as progress, or fails to treat privileged researchers' ideas of progress as themselves continuously productive of privilege, research is unlikely to produce social change" (148).

48.1.3 *Feminist Ontologies and the Reconstruction of Engineering*

The previous discussion on intersectionality demonstrated one way in which feminists engage their work from a variety of ontological perspectives and presumptions. Haslanger and Sveinsdóttir (2017) provide a helpful framework for understanding the range of feminist ontologies by outlining three key interrelated themes in feminist metaphysics: (1) the social construction of gender (and race, class, and other categories); (2) the social relations in which, and against which, selves are defined; and (3) critiques of dualisms. In the first case, gender is not a natural category but rather a social one. In the second case, feminists challenge the ideas that gender is intrinsic, that selves are atomistic, independent, rational beings, and that the category of woman is defined against that of a man, where woman is the other defined against a presumed normative subject, generating a particular social order around gender.

In the third case, Haslanger and Sveinsdóttir (2017) note, feminist thought challenges the notion of binaries by revealing how a category, such as man, which appears to merely describe some objective reality, is in fact non-substantive in nature and sets up a normative state of existence. One way of challenging the gender binary is to point out that there exist not only women, but perhaps everything in between man and woman, and other sexes or genders entirely, and people who exist outside of sex or gender altogether (e.g., Butler 1990). Another approach (see, e.g., Frye 1996) is to seek to define the category of woman on its own terms, not in relation to men or any kind of binary, and to, among other things, focus on the differences among women.

Applying these ideas to engineering, then, we have already seen how applying ideas about the social construction of gender can help make visible the ways in which engineering is gendered. But what would a feminist engineering look like that moves beyond the critique?

Following Maruska's (2010) fashioning of feminist ontologies of international relations, we might employ relational ontology to examine the ways in which engineers operate in the social, political, and economic relations of engineering. To this end, Wendy Faulkner (2007), for one, envisions a sociotechnical engineering. On the one hand, she argues that professional engineers are already navigating the sociotechnical reality of engineering work. At the same time, the refusal to acknowledge this reality and the elevation of a masculinist, purely technical "real engineering" strains gender relations in the engineering workplace. She envisions a world in which the sociotechnical reality of engineering is acknowledged and accepted by engineers, and she argues that this would transform gender relations in engineering.

Pat Treusch (2017) and Laura Forlano (2017), in two separate pieces in a special issue of *Catalyst* on remaking feminist science, posit new human-machine relations that enact (and ontologically engineer) feminist futures. Treusch (2017) seeks to account for affective labor and co-production between humans and machines in the work of robotics development. Queering notions of success and failure in this work makes possible a transformation of capitalist notions of automation into a process that reflectively and consciously constructs human-machine relations. Forlano (2017) critically reflects on her own use of smart technologies to manage diabetes as an intimate human-machine relationship, from which a lived experience of feminist data practice emerges, contributing to new understandings of crip time specifically and ontologies of difference more broadly. Both of these works inhabit new ways of *being engineering*.

Riley et al. (2009) provide some seeds of ideas of how engineering education might need to be in order to prepare engineers to enact these kinds of new relational realities. Pawley (2017) proposes that act on the presumption of diversity as the normative state of existence in engineering education. Kehdinga Fomunyam (2017) has begun a conversation envisioning a decolonized engineering education curriculum.

The Engineering, Social Justice, and Peace community has sought to envision new realities for engineering in a justice frame (e.g., Riley 2008; Baillie and Catalano 2009; Nieusma and Blue 2012; Leydens and Lucena 2017; Siller and Johnson 2018). While some accounts are explicitly feminist, others are compatible or may be adaptable with some additional analysis or critique.

Much of this work points to limited engineering imaginaries. It can be difficult to conjure exactly how engineering can be different. One of the most critical failures of engineering imaginary at present is knowing when not to engineer (Slaton et al. 2019).

Ultimately, expanding engineering imaginaries is not merely an ontological project and must also simultaneously entail knowing and doing. Thus, we turn now to the potentialities of feminist epistemologies in engineering, and finally to developing feminist engineering ethics.

48.2. Feminist Engineering Epistemology

48.2.1 *Feminist Philosophy of Science and Ways of Knowing*

If engineering existence is gendered, it may stand to reason that engineering knowledge might also be gendered. However, it is the normative masculinity of engineering that obscures this very reality. Waller (2006) noted that research projects in engineering disciplines are predominately rooted in positivist assumptions about knowledge and sought to make these epistemic assumptions visible to colleagues. Montfort et al. (2014), in a study of civil engineering professors' personal epistemologies, identified ontologies that affirmed the existence of objective reality, the truth of which can be ascertained through empirical observation, even as some affordances were made for the complexity of the world.

Engineering's epistemologies are traditionally based on that of the physical sciences and thus are subject to the broad critiques offered by feminist philosophers of science. Central to feminist critiques is science's (or engineering's) presumed objectivity, characterized by reductionism and abstraction, logical positivism, and a "view from nowhere" that evacuates emotion and normative values (Anderson 2017; Haraway 1991; Harding 1991; Keller 1985; Longino 1990). Scientific objectivity cleanly separates knowers from their subjects of scientific inquiry, observing a presumed objective reality that does not change based on the observer, eliding numerous choices made by researchers that impact how reality is conceived and represented (Longino 1990, in Anderson 2017). However, feminist ontologies would require that some realities are socially constructed, and not accounting for this possibility causes scientists to mistakenly naturalize gendered or raced power structures (e.g., Fausto-Sterling 1986; Proctor 2006).

Harding (2006) questions the universalism of masculinist epistemology in science for its epistemic hubris, observing that "the ideal of one true science obscures the fact that any system of knowledge will generate systematic patterns of ignorance as well as of knowledge" (125). These patterns, Harding notes, privilege male perspectives and interests, as well as White perspectives and interests, and colonial interests of the global North.

In addition to feminist critiques of science's epistemic frames, it is important to note engineering's pragmatic and instrumentalist knowledge presumptions. Montfort et al. (2014) observed that civil engineering professors' personal epistemologies as valuing knowledge for its truth as well as its practical usefulness. Slaton (2012: 7) notes that

> such instrumentality systematically displaces attention from the actors involved: those who bring about engineering (engineers and their patrons and clients) and those who live with its benefits (material comfort and well-being, profit, lucrative and secure employment, etc.) and costs (to health, safety, national and personal security, labor conditions, etc.).

Another central concern in feminist epistemology revolves around who has the authority to know and be believed, and the unequal distribution of knowledge and expertise in society. Miranda Fricker (2007) has furthered the concept of epistemic injustice to characterize the effects of unjust social relations on who is able to claim or wield epistemic authority, and whose knowledge is considered suspect or unreliable.

48.2.2 *Critiques of Rigor, Depoliticization, and Meritocracy in Engineering*

Amy Slaton's (2010) history of racial inequality in American engineering education reveals the close relationship between conceptualizations of rigor and ideological commitments to meritocracy. She shows how the two work together to create standards that, while appearing objective or value-neutral, in fact have served to limit access to the profession for African Americans in particular. Even where educational administrators were well-meaning (or convinced by workforce needs) and sought to improve access for African Americans, a commitment to "high standards" and concerns about "diminished rigor" led to the perpetuation of underrepresentation of African Americans in the field.

In an experiment that revealed sexism in admissions standards for computer science, Carnegie Mellon showed how the relaxation of one seemingly objective and meritocratic but in fact gender-biased admissions criterion, that of high school programming experience, led to a dramatic demographic shift as the female population jumped from 7% to 42% over seven years (Margolis and Fisher 2002).

Feminist epistemologies undergird critiques of prevailing practices of power/knowledge in engineering. Riley (2017) builds on feminist epistemologies as she attempts a jujitsu move in a critical participation piece in *Engineering Studies*; using the same forms of commonplace sexual innuendo and puerile humor that readily mark engineering culture as masculine, she seeks to address head-on (pun intended) the phallic implications of rigor, and its functions as an epistemic gatekeeper and disciplinarian.

Erin Cech (2013) shows how the technical-social dualism in engineering forms the basis of an ideology of depoliticization that renders concerns about equity or justice as lying beyond the field's scope. She further argues that meritocratic thinking naturalizes inequities as the logical result of a rewards system assumed to operate on a level playing field, privileging only those who deserve it due to their talent or effort. These inequities, then, need no correction, stymieing efforts toward diversity in engineering or broader engagement of the profession with social justice issues.

Cech's critique echoes earlier findings by Winner (1990) and Zussman (1985) regarding the lack of concern engineers take for the ends of their work. Similarly, Carl Mitcham (2009; see also Chapter 1 "What Is Engineering" by Carl Mitcham, this volume) regards it a "philosophical inadequacy of engineering" that, while the profession purports to serve societal ends (Mitcham, after Thomas Tredgold, chooses the phrasing "human use and convenience"):

> there is nothing in engineering education or knowledge that contributes to any distinct competence in making judgments about what constitutes "human use and convenience."

Engineering as a profession is analogous to what medicine might be if physicians had no expert knowledge of health or to law if attorneys knew nothing special about justice.

(339)

48.2.3 Epistemologies of Ignorance

This knowledge gap that Mitcham identifies points to one particularly fruitful arena in feminist epistemology that holds promise for engineering: epistemologies of ignorance. Tuana and Sullivan (2006) argue for the importance of ignorance studies to feminist epistemology, as a site for understanding how practices of ignorance both stem from and give rise to or sustain unjust social relations. Tuana (2006) provides a useful taxonomy of ignorance that explores key differences among, for example, being not *interested* to know vs. not knowing what you don't know.

Epistemologies of ignorance hold promise for explaining the recalcitrance of engineering to recognize the social as part of its purview, to acknowledge the political interests inherent in the profession and its applications, or to deal directly with inequity in engineering education and the workplace. Riley (2016, 2019) has employed epistemologies of ignorance to explore the denial of gender-based violence in STEM and the systematic exclusion of sexual violence as an explanatory factor in women's underrepresentation in engineering.

48.2.4 Constructing Feminist Epistemologies of Engineering

Drawing on feminist ontologies that position our knowledge of existence amidst social relations, feminist epistemology is commonly characterized by situated knowledges, shaped by the positionality of the knower (Haraway 1988). Acknowledging situated knowledges provides various ways forward toward engaging feminist science, or feminist engineering.

One approach is to reform empirical research methods to minimize sexist bias, developing new procedures for feminist scientific research. One set of corrections focuses on eschewing value neutrality and detachment for what Keller (1985) terms "dynamic objectivity," in which one's relationship to the subject of study is a wholly relevant, integral, and healthy part of knowing.

Harding's (1993) "strong objectivity" calls for a reflexivity that is transparent about the social relations of knowledge: its biases, assumptions, and interests. While representations will be necessarily partial truths, by triangulating among multiple representations, one can build a stronger, more valid representation than any single representation alone. Using feminist standpoint theory, Harding further argues that representation of perspectives from marginalized groups will strengthen the reflexive positioning and build a more complete and more objective picture of reality.

Helen Longino (1990) also uses multiple accounts to arrive at a feminist rendering of objectivity, but her process is one of democratic discussion that does not rely on standpoint epistemology. Through a cooperative process of public participation, criticism and evaluation, and response to criticisms, more objective knowledge may result. This is predicated on all participants having equal epistemic authority.

In the engineering setting, Gwen Ottinger (2017) explored the hermeneutic injustices experienced by communities located near energy facilities struggling to gain the knowledge resources necessary to collect, analyze, and interpret air quality data. She explores the promises and limitations of using storytelling as an alternative resource for meaning making related to air quality. These narratives, while incomplete in some ways, can nonetheless provide a more holistic accounting of harms to health and the disrespect encountered by communities than methods afforded by strictly scientific frameworks.

The presumption of equal epistemic authority is not easily achieved, as recognized by Patty Lather (2007) in *Getting Lost*. Among other explorations, she considers the loss of authority and expertise as an epistemic method that seeks to reckon honestly with the power/knowledge relations inherent in research.

Some other poststructuralist approaches follow a similar theme; Donna Harraway (1992) seeks full accountings of non-linearities in scientific concepts in a process she calls diffraction, following their convoluted histories complete with twists, turns, gaps, slippages, and unknowings. Fortun and Bernstein (1998) seek to account for feminist ontologies in their conception of reality, which seeks to reflexively represent the subjective relationships between scientific practitioners and their observations.

Riley (2017) has proposed vigor as a replacement for rigor that responds to Hacker's (1989) call for embodiment in engineering practice, re-eroticizing engineering in an inclusive rather than dehumanizing set of relations among pleasure, power, and technology.

48.3. Feminist Engineering Ethics

48.3.1 Critique of Masculinist Ethics

Riley (2013) provides an overview of feminist ethics for the purposes of introducing a set of possibilities for feminist approaches to engineering ethics; the arguments from that piece are summarized in this section. As with each of our previous sections on ontology and epistemology, feminist ethics begins with a critique of masculinist ethics. Perhaps the clearest synthesis of these critiques is found in Margaret Urban Walker's (1989) thematic analysis of failures of masculinist ethics.

As with feminist ontology and epistemology, feminist ethics finds fault in an abstract, decontextualized, and depersonalized positionality in masculinist ethics, arguing that feminist ethics must be relational, and contextualized through the use of narrative. As with scientific knowledge, feminists caution against universals that set aside people and their relationships. Walker borrows from the feminist slogan "the personal is political" to argue that the personal is moral, and the moral is personal. This indeed suggests that the moral, then, is political; Walker urges our attention to the politics of language, and the politics of social institutions in our formulation of ethics.

As with feminist approaches to scientific knowledge, relational approaches get messy, and Walker notes that this results in the acknowledgement of "moral remainders"—ethical decision-making is imperfect, and there are inevitable consequences to relationships. In this way, communication becomes a crucial element of ethical action.

In the context of feminist critiques of the failure of masculinist ethics to attend to relationships, Nell Noddings (1984) developed the "Ethic of Care," employing Carol Gilligan's (1982) research identifying differences in discourse used by women and men in moral justifications, where men tended to speak more in terms of rights or justice, while women spoke more in terms of care or relationships (see also Chapter 34, "Engineering and Sustainability: Control and Care in Unfoldings of Modernity" by Andy Stirling, this volume).

Since that time, feminists have questioned whether adopting care plays into gender stereotypes and the oppression of women (Bartky 1995), or whether women care in response to a set of power relations that limit women's range of possible action (Card 1995). Uma Narayan (1995) shows how the concept of care justified patronizing colonial power relations, denying the agency of colonial subjects in the service of imperial interests. In response to these critiques, care ethicists such as Joan Tronto (1987, 1993) and Karen Warren (2000) incorporate considerations about power. Some limitations of the ethic of care are based in enduring questions around what to do with those who are uncaring, and how to address the fact that many remain uncared for.

At the same time, other feminist ethicists have sought to reform masculinist justice traditions. Iris Marion Young (1990) sought to broaden notions of distributive justice to a framework that could account for difference and counter oppression and domination. Susan Moller Okin (1989) furthered theories of justice within the family in order to address gender inequality in those settings. Black womanist ethics identifies justice as a central theme in black women's experiences recorded in Black history and Black literature (Cannon 1988). Feminist theories of justice have also been applied to development and global justice for women (Verma 2004; Jaggar 2009a).

Finally, at the heart of feminist ethics lie questions about who counts as a moral agent, and how social structures and power relations impede or support that agency. For example, Hilde Lindemann Nelson (2001) analyzes a case in which a nurse and a doctor experience a conflict over whether or not to tell a patient that he has leukemia. When the nurse expresses her professional ethical judgement that he ought to be told, the doctor dismisses this position as overly emotional, undermining her moral agency. Nelson illustrates how the nurse can work with her peers to reclaim moral agency through the development of a counter-narrative to the doctor's dismissive account. Transnational feminist ethicists also employ concepts of moral agency in order to empower actors resisting globalization, attending to the power relations of nation-states under global capitalism, and the ways in which race, gender, and sexuality interact on this stage (Harding and Narayan 1998; Narayan and Harding 1998; Jaggar 2009b).

48.3.2 Feminist Ethics and Ways of Doing and Making

Feminist ethical practices constitute engineering processes of iterative design (Whitbeck 1998) or design as care (Pantazidou and Nair 1999), and open a space for reimagining, rethinking, remaking, and redoing engineering (Riley and Longmaid 2013). Engineering and ethics are both normative activities, ways of seeking the world as it could be. Sites of exploration for feminist ethics as engineering include the following questions posed by feminist ethics that can generate new forms of engineering.

Who is a moral agent? Who has power to act, and on whose behalf? Questions of power, agency, and identity help us imagine an engineering accessible to marginalized communities, where problems are identified and defined not by disconnected experts or prevailing interests but by those whose lives are most impacted. How can attention to power and agency in design produce different engineering processes, opportunities, and results? What structures facilitate or inhibit agency, and how can these be better designed? Caroline Whitbeck (1995) transformed engineering ethics case studies in order to emphasize students as engaged moral agents rather than passive judges. Bauschpies et al. (2018) reimagine an engaged, community-led engineering that dismantles "the flaw of the awe"—the unjust power relations and community disempowerment that extends from an un-listening wielding of engineering expertise.

What new knowledge emerges from lived experience? As feminist ethics has emerged from women's lives where intersectional specificities matter and resist universalization as both unattainable and undesirable (Walker 1989), an emergent engineering will attend to specific persons rather than abstractions. This attendance to specific persons requires we shape a relational practice of engineering.

What narratives can we construct that provide new contexts for engineering? What new forms of communication come into being and serve as new methods and methodologies of engineering? Riley and Lambrinidou (2015) engaged in a thought experiment in which engineering's ethics canons were arranged as if people mattered. Using social work's ethical commitments as a guide, the new ethics canon centered on helping others; structurally addressing injustice; honoring all people; centering human relationships; and seeking peace.

How can alternative epistemologies of feminist ethics construct unreasonable or empathic engineering? Feminist ethics claims a role for affect and emotion, ways of knowing that are outside or beyond objectivity. This begs a question in response from traditional engineers: Can the bridge still stand up, or the plane stay in the sky (if we would still engineer bridges or planes), without universals and absolutes, without depersonalization and abstraction? Feminist ethics deals with the complexities of imperfect solutions, with moral remainders and associated relational fallout. How would using feminist philosophers' alternate constructions of validity or robustness (e.g., Haraway, Harding, Lather) change engineering? Slaton et al. (2019) posit a queering of engineering's desires, bending away from gritty capitalist strivings and embracing the strange.

Who benefits, and who loses? Who cares for and about whom? (Noddings 1984). Feminist ethics has been a generative site for doing ethics differently, offering both ethics of care and feminist conceptions of justice as alternatives to masculinist ethics.

How does attention to these central questions reshape the content and boundaries of engineering knowledge and practice? Of the various questions posed from feminist ethics, this site is perhaps the one best explored by engineering colleagues to date, with engineering design processes reconceived along both lines of care and empathy as well as lines of justice, with worked and lived examples (Catalano 2006; Pantazidou and Nair 1999; Hess et al. 2017; Nair and Bulleit 2019; see also Chapters 49–52, this volume). While the feminist origins and influence is not always explicitly acknowledged in this work, it is ever present (Riley 2013). This re-engineering effort is responsive to previous critiques that have, for example, identified engineering's cultures of depoliticization and meritocracy (Cech 2013), or the field's existence as a "war-built" discipline with militaristic epistemic foundations pervading all engineering thought, culture, and practice (Nieusma and Blue 2012). While at this time these reconceptions of feminist ethics as engineering have yet to be made fully concrete and embodied, part of the work of this chapter is to imagine these possibilities and map out their realization.

48.4. Conclusion

This chapter has explored what might be entailed in feminist engineering, reviewing possible transformations in engineering's ways of being, knowing, and doing. Among these, it is perhaps the ontological questions that prove most challenging: Can there even *be* a feminist engineering? When engineering begins to adopt feminist ways of being, knowing, or doing, it often ceases to be recognized *as* engineering. For there to be a feminist engineering, engineering must be (or become) the thing it cannot be. Resolving this contradiction is ultimately a relational project, taking care and caring for others in co-constructing new engineering knowledge, practice, and existence.

Related Chapters

Chapter 1: What Is Engineering? (Carl Mitcham)
Chapter 34: Engineering and Sustainability: Control and Care in Unfoldings of Modernity (Andy Stirling)
Chapter 49: Socially Responsible Engineering (Jennifer Smith and Juan Lucena)
Chapter 50: Engineering and Social Justice (Caroline Baillie)
Chapter 51: Engineering and Environmental Justice (Benjamin R. Cohen)
Chapter 52: Beyond Traditional Engineering: Green, Humanitarian, Social Justice and *Omnium* Approaches (George D. Catalano)
Chapter 55: Reimagining the Future of Engineering (Neelke Doorn, Diane P. Michelfelder et al.)

Further Reading

Faulkner, W. (2007). 'Nuts and Bolts and People' Gender-Troubled Engineering Identities. *Social Studies of Science*, 37, 331–356. (Wendy Faulkner's ethnography of engineers in the workplace contrasts purely technicist and sociotechnical characterizations of "real engineering," and explores the implications of each for gender and power in the workplace.)

Hacker, S. (1989). *Pleasure, Power & Technology: Some Tales of Gender, Engineering, and the Cooperative Workplace*. Boston: Unwin Hyman. (Sally Hacker's action research and participant observation of engineering education offer incisive critiques of exclusionary engineering culture, using social theory to explain the manifestation of engineering's rigors through mind-body dualism, as well as gender, race, and class.)

Pantazidou, M. and Nair, I. (1999). Ethic of Care: Guiding Principles for Engineering Teaching and Practice. *Journal of Engineering Education,* 88, 205–212. (In this piece, Marina Pantazidou and Indira Nair explore how feminist care ethics might be applied in engineering education and practice.)

Pawley, A. L. (2012) What Counts as 'Engineering'? Towards a Redefinition. In C. Baillie, A.L. Pawley and D. Riley (eds.), *Engineering and Social Justice: In the University and Beyond.* West Lafayette, IN: Purdue University Press, pp. 59–85. (Alice Pawley reveals what is included and excluded through our gendered definitions of engineering, and poses more inclusive alternatives.)

Riley, D. (2013). Hidden in Plain View: Feminists Doing Engineering Ethics, Engineers Doing Feminist Ethics. *Science and Engineering Ethics,* 19(1), 189–206. (Donna Riley reviews prior feminist work in engineering ethics, noting a distinct reluctance to use the "f-word.")

Whitbeck, C. (1998). *Ethics in Engineering Practice and Research.* New York: Cambridge University Press. (Caroline Whitbeck builds on feminist philosophy in this engineering ethics text, casting the work of ethics as more akin to iterative design processes in engineering than deterministic problem-solving.)

References

Akpanudo, U.M. Huff, J.L., Williams, J.K. and Godwin, A. (2017). Hidden in Plain Sight: Masculine Social Norms in Engineering Education. *2017 IEEE Frontiers in Education Conference (FIE)*, Indianapolis, IN, USA, pp. 1–5.

Anderson, E. (2017). Feminist Epistemology and Philosophy of Science. In Edward N. Zalta (ed.), *The Stanford Encyclopedia of Philosophy*, Spring 2017 ed. https://plato.stanford.edu/archives/spr2017/entries/feminism-epistemology/. Accessed April 2, 2018.

Anzaldúa, G. (1987). *Borderlands/La Frontera: The New Mestiza.* San Francisco, CA: Aunt Lute Books.

Baillie, C. and Catalano, G.D. (2009). *Engineering and Society: Working Toward Social Justice.* San Rafael, CA: Morgan and Claypool.

Bartky, S.L. (1995). Feeding Egos and Tending Wounds: Deference and Disaffection in Women's Emotional Labor, In Mary Rogers (ed.), *Power, Dignity and Social Structure: Readings in Multicultural Social Theory.* New York: McGraw Hill.

Bauschpies, W., Douglas, E.P., Holbrook, J.B., Lambinidou, Y. and Lewis, E.Y. (2018). Reimagining Ethics Education for Peace Engineering. Paper presented at *WEEF-GEDC 2018*, Albuquerque, New Mexico. https://weef-gedc2018.org/wp-content/uploads/2018/11/47_Reimagining-Ethics-Education-for-Peace-Engineering.pdf.

Bix, A. (2002). Equipped for Life: Gendered Technical Training and Consumerism in Home Economics, 1920–1980. *Technology and Culture,* 43, 728–754.

Borsook, P. (1996). Memoirs of a Token. In L. Cherney and E.R. Weise (eds.), *Wired Women: Gender and New Realities in Cyberspace.* Seattle, WA: Seal Press, pp. 24–41.

Brey, P. (2003). The Social Ontology of Virtual Environments. *American Journal of Economics and Sociology,* 62(1), 269–282.

Butler, J.P. (1990). *Gender Trouble: Feminism and the Subversion of Identity.* New York: Routledge.

Camacho, M.M. and Lord, S.M. (2013). *The Borderlands of Education: Latinas in Engineering.* Boulder, CO: Lexington Books.

Cannon, K.G. (1988). *Black Womanist Ethics.* Oxford: Oxford University Press.

Carbin, M. and Edenheim, S. (2013). The Intersectional Turn in Feminist Theory: A Dream of a Common Language? *European Journal of Women's Studies,* 20(3), 233–248.

Card, C. (1995). Gender and Moral Luck [1990]. In V. Held (ed.), *Justice and Care: Essential Readings in Feminism.* Boulder, CO: Westview Press.

Catalano, G.D. (2006). *Engineering Ethics: Peace, Justice, and the Earth.* San Rafael, CA: Morgan and Claypool.

Cech, E.A. (2013). The (Mis)Framing of Social Justice: Why Meritocracy and Depoliticization Hinder Engineers' Ability to Think About Social Injustices. In J. Lucena (ed.), *Engineering Education for Social Justice: Critical Explorations and Opportunities.* New York: Springer, pp. 67–84.

Cech, E.A. and Rothwell, W. (2018). LGBTQ Inequality in Engineering Education. *Journal of Engineering Education,* 107(4), 583–610.

Clancy, K.B.H., Nelson, R.G., Rutherford, J.N. and Hinde, K. (2014). Survey of Academic Field Experiences (SAFE): Trainees Report Harassment and Assault. *PLoS ONE,* 9(7), e102172. doi:10.1371/journal.pone.0102172.

Crenshaw, K. (1989). Demarginalizing the Intersection of Race and Sex: A Black Feminist Critique of Anti-discrimination Doctrine, Feminist Theory, and Antiracist Politics. *University of Chicago Legal Forum, 1989,* 139–168.

Faulkner, W. (2000). Dualisms, Hierarchies and Gender in Engineering. *Social Studies of Science,* 30, 759–792.

Faulkner, W. (2007). 'Nuts and Bolts and People' Gender-Troubled Engineering Identities. *Social Studies of Science,* 37, 331–356.

Faulkner, W. (2009a). Doing Gender in Engineering Workplace Cultures. I. Observations from the Field. *Engineering Studies,* 1, 3–18.

Faulkner, W. (2009b). Doing Gender in Engineering Workplace Cultures. II. Gender In/authenticity and the In/visibility Paradox. *Engineering Studies,* 1, 169–189.

Fausto-Sterling, A. (1986). *Myths of Gender: Biological Theories about Women and Men.* New York: Basic Books.

Foor, C.E. and Walden, S.E. (2009). "Imaginary Engineering" or "Re-imagined Engineering": Negotiating Gendered Identities in the Borderland of a College of Engineering. *NWSA Journal,* 21, 41–64.

Forlano, L. (2017). Data Rituals in Intimate Infrastructures: Crip Time and the Disabled Cyborg Body as an Epistemic Site of Feminist Science. *Catalyst: Feminism, Theory, Technoscience,* 3(2), 1–28. https://catalystjournal.org/index.php/catalyst/article/view/28843/pdf_13

Fomunyam, K.G. (2017). Decolonising Teaching and Learning in Engineering Education in a South African University. *International Journal of Applied Engineering Research,* 12(23), 13349–13358.

Fortun, M. and Bernstein, H.J. (1998). *Muddling Through: Pursuing Science and Truths in the 21st Century.* Washington, DC: Counterpoint.

Frehill, L.M. (2004). The Gendered Construction of the Engineering Profession in the United States, 1893–1920. *Men and Masculinities,* 6(4), 383–403.

Fricker, M. (2007). *Epistemic Injustice.* Oxford: Oxford University Press.

Frye, M. (1996). The Necessity of Differences: Constructing a Positive Category of Women, *Signs,* 21(4), 991–1010.

Gilligan, C. (1982). *In a Different Voice: Psychological Theory and Women's Development.* Cambridge, MA: Harvard University Press.

Goldberg, D.E. (2010). Why Philosophy? Why Now? Engineering Responds to the Crisis of a Creative Era. In I. Van de Poel and D.E. Goldberg (eds.), *Philosophy and Engineering: An Emerging Agenda.* New York: Springer, pp. 255–264.

Hacker, S. (1989). *Pleasure, Power & Technology: Some Tales of Gender, Engineering, and the Cooperative Workplace.* Boston: Unwin Hyman.

Hammonds, E. (1994). Black (w)holes and the Geometry of Black Female Sexuality. *Differences: A Journal of Feminist Cultural Studies,* 6, 126–145.

Haraway, D. (1988). Situated Knowledges: The Science Question in Feminism and the Privilege of Partial Perspective. *Feminist Studies,* 14, 575–599.

Haraway, D. (1991). *Simians, Cyborgs, and Women: The Reinvention of Nature.* New York: Routledge.

Haraway, D. (1992). Otherworldly Conversations, Terran Topics, Local Terms. *Science as Culture,* 3(1), 59–92.

Harding, S. (1991). *Whose Science? Whose Knowledge?* Ithaca: Cornell University Press.

Harding, S. (1993). Rethinking Standpoint Epistemology: 'What is Strong Objectivity?' In L. Alcoff and E. Potter (eds.), *Feminist Epistemologies,* New York: Routledge, pp. 49–82.

Harding, S. (2006). *Science and Social Inequality: Feminist and Postcolonial Issues.* Champaign, IL: University of Illinois Press.

Harding, S. and Narayan, U. (1998). Border Crossings: Multicultural and Postcolonial Feminist Challenges to Philosophy (Part II). *Hypatia,* 13(3), 1–5. See entire special issue.

Haslanger, S. (1995). Ontology and Social Construction. *Philosophical Topics,* 23(2), 95.

Haslanger, S. and Sveinsdóttir, Á.K. (2017). Feminist Metaphysics. In E.N. Zalta (ed.), *The Stanford Encyclopedia of Philosophy,* Fall 2017 ed. https://plato.stanford.edu/archives/fall2017/entries/feminism-metaphysics/. Accessed April 1, 2018.

Hess, J.L., Beever, J., Strobel, J. and Brightman, A.O. (2017). Empathic Perspective-taking and Ethical Decision-making in Engineering Ethics Education. In D. Michelfelder, B. Newberry and Q. Zhu (eds.), *Philosophy and Engineering: Exploring Boundaries, Expanding Connections.* Dordrecht: Springer, pp. 163–179.

Hill Collins, P. (1998). The Social Construction of Black Feminist Thought. In K.A. Myers, B.J. Risman and C.D. Anderson (eds.), *Feminist Foundations: Towards Transforming Sociology.* Thousand Oaks, CA: Sage Publications, pp. 371–396.

Jaggar, A. (2009a). Transnational Cycles of Gendered Vulnerability: A Prologue to a Theory of Global Gender Justice. *Philosophical Topics,* 37(2), 33–52.

Jaggar, A. (2009b). The Philosophical Challenges of Global Gender Justice. *Philosophical Topics,* 37(2), 1–15.

Keller, E. (1985). *Reflections on Gender and Science*. New Haven: Yale University Press.

Kendall, L. (1996). MUDder? I Hardly Know 'Er! Adventures of a Feminist MUDder. In L. Cherney and E.R. Weise (eds.), *Wired Women: Gender and New Realities in Cyberspace*. Seattle, WA: Seal Press, pp. 227–223.

Lafontaine, E. and Tredeau, L. (1986). The Frequency, Sources, and Correlates of Sexual Harassment Among Women in Traditional Male Occupations. *Sex Roles*, 15(7–8), 433–442.

Lather, Patti (2007). *Getting Lost: Feminist Efforts Toward a Double(d) Science*. Albany: SUNY Press.

Leyva, L.A. (2016). An Intersectional Analysis of Latin@ College Women's Counter-stories in Mathematics. *Journal of Urban Mathematics Education,* 9(2), 81–121.

Leydens, J.A. and Lucena, J.C. (2017). *Engineering Justice: Transforming Engineering Education and Practice*. New York: Wiley IEEE Press.

Longino, H.E. (1990). *Science as Social Knowledge: Values and Objectivity in Scientific Inquiry*. Princeton: Princeton University Press.

Margolis, J. and Fisher, A. (2002). *Unlocking the Clubhouse: Women in Computing*. Cambridge, MA: MIT Press.

Maruska, J.H. (2010). Feminist Ontologies, Epistemologies, Methodologies, and Methods in International Relations. In R.A. Denemark and R.Marlin-Bennett (eds.), *The International Studies Encyclopedia*. New York: Wiley-Blackwell.

McIlwee, J.S. and Robinson, J.G. (1992). *Women in Engineering: Gender, Power, and Workplace Culture*. Albany: SUNY Press.

Mitcham, C. (2009). A Philosophical Inadequacy of Engineering. The *Monist*, 92(3), 339–356.

Montfort, D., Brown, S. and Shinew, D. (2014). The Personal Epistemologies of Civil Engineering Faculty. *Journal of Engineering Education*, 103(3), 388–416.

Nair, I. and Bulleit, W.M. (2019). Pragmatism and Care in Engineering Ethics. *Science and Engineering Ethics*. doi:10.1007/s11948-018-0080-y

Narayan, U. (1995). Colonialism and Its Others: Considerations on Rights and Care Discourses. *Hypatia,* 10(2), 133–140.

Narayan, U. and Harding, S. (1998). Border Crossings: Multicultural and Postcolonial Feminist Challenges to Philosophy (Part I). *Hypatia,* 13(2), 1–6. See entire special issue.

Nelson, H.L. (2001). Identity and Free Agency. In P. DesAutels and J. Waugh (eds.), *Feminists Doing Ethics*. Lanham, MD: Rowman & Littlefield.

Nieusma, D. and Blue, E. (2012). Engineering and War. *International Journal of Engineering, Social Justice, and Peace*, 1(1), 50–62.

Noddings, N. (1984). *Caring: A Feminine Approach to Ethics and Moral Education*. Berkeley, CA: University of California Press.

Okin, S. (1989). *Justice, Gender, and the Family*. New York: Basic Books.

Oldenziel, R. (2000). *Making Technology Masculine: Men, Women, and Modern Machines in America, 1870–1945*. Amsterdam: Amsterdam University Press.

Ottinger, G. (2017). Making Sense of Citizen Science: Stories as a Hermeneutic Resource. *Energy Research & Social Science*, 31, 41–49.

Pantazidou, M. and Nair, I. (1999). Ethic of Care: Guiding Principles for Engineering Teaching and Practice. *Journal of Engineering Education,* 88, 205–212.

Pawley, A.L. (2012). What Counts as 'Engineering'? Towards a Redefinition. In C. Baillie, A.L. Pawley and D. Riley (eds.), *Engineering and Social Justice: In the University and Beyond*. West Lafayette, IN: Purdue University Press, pp. 59–85.

Pawley, A.L. (2017). Shifting the "Default": The Case for Making Diversity the Expected Condition for Engineering Education and Making Whiteness and Maleness Visible. *Journal of Engineering Education*, 106, 531–533. doi:10.1002/jee.20181

Proctor, R. (2006). *Racial Hygiene: Medicine Under the Nazis*, 2nd ed. Cambridge, MA: Harvard University Press.

Revelo, R.A., Mejia, J.A. and Villanueva, I. (2017). *Who Are We? Beyond Monolithic Perspectives of Latinxs in Engineering*. ASEE Annual Conference and Exposition, Conference. https://peer.asee.org/29125.

Riley, D. (2008). *Engineering and Social Justice*. San Rafael, CA: Morgan and Claypool.

Riley, D. (2013). Hidden in Plain View: Feminists Doing Engineering Ethics, Engineers Doing Feminist Ethics. *Science and Engineering Ethics*, 19(1), 189–206.

Riley, D. (2016). *Why Deny? The Willful Ignorance of Gender-Based Violence in STEM*. Gender, Bodies, and Technology, April.

Riley, D. (2017). Rigor/Us: Building Boundaries and Disciplining Diversity with Standards of Merit. *Engineering Studies*, 9(3), 249–265. doi:10.1080/19378629.2017.1408631.

Riley, D. (2019). Pipelines, Persistence, and Perfidy: Institutional Unknowing and Betrayal Trauma in Engineering. *Feminist Formations,* in press.

Riley, D. and Lambrinidou, Y. (2015). Canons Against Cannons? Social Justice and the Engineering Ethics Imaginary. Paper presented at *2015 ASEE Annual Conference & Exposition*, Seattle, Washington, DC, June. doi:10.18260/p.23661

Riley, D., Pawley, A.L., Tucker, J. and Catalano, G.D. (2009). Feminisms in Engineering Education: Transformative Possibilities. *National Women's Studies Association Journal* (now *Feminist Formations*), 21(2), 21–40.

Riley, D. (with Emily Rider-Longmaid) (2013). *How Feminist Ethics Brings Pluralism to Engineering: The Prison-Industrial Complex and Drone Warfare*. Feminist Ethics and Social Theory Conference, Tempe, AZ, October 17–20.

Searle, J. (1995). *The Construction of Social Reality*. Cambridge, MA: MIT Press.

Seymour, E. (1995). The Loss of Women From Science, Mathematics, and Engineering Undergraduate Majors: An Explanatory Account. *Science Education*, 79(4), 437–473.

Siller, T.J. and Johnson, G. (2018). Just Technology: The Quest for Cultural, Economic, Environmental, and Technical Sustainability. *Synthesis Lectures on Sustainable Development*, 1(1), 1–93.

Slaton, A.E. (2010). *Race, Rigor, and Selectivity in U.S. Engineering: The History of an Occupational Color Line*. Cambridge, MA: Harvard University Press.

Slaton, A.E. (2012). *The Tyranny of Outcomes: The Social Origins and Impacts of Educational Standards in American Engineering*. American Society for Engineering Education Annual Conference 2012, Austin, TX, June 10–13.

Slaton, A.E., Cech, E.A. and Riley, D.M. (2019). Yearning, Personhood, and the Gritty Ontologies of American Engineering Education. In Steve Fifield and Will Letts (eds.), *STEM of Desire: Queer Theories in Science Education*. Brill | Sense Publishers, pp. 319–340.

Slaton, A.E. and Pawley, A.L. (2018). The Power and Politics of Engineering Education Research Design: Saving the 'Small N'. *Engineering Studies*, 10(2–3), 133–157. doi:10.1080/19378629.2018.1550785

Tonso, K.L. (1996). The Impact of Cultural Norms on Women. *Journal of Engineering Education*, 85, 217–225.

Treusch, P. (2017). The Art of Failure in Robotics: Queering the (Un)making of Success and Failure in the Companion-Robot Laboratory. *Catalyst: Feminism, Theory, Technoscience*, 3(2), 1–27. https://catalystjournal.org/index.php/catalyst/article/view/28846/pdf_16

Tronto, J.C. (1987). Beyond Gender Difference to a Theory of Care. *Signs*, 12(4), 644–663.

Tronto, J.C. (1993). *Moral Boundaries: A Political Argument for an Ethic of Care*. New York: Routledge.

Tuana, N. (2006). The *Speculum of Ignorance*: The Women's Health Movement and Epistemologies of *Ignorance*. *Hypatia*, 21(3), 1–19.

Tuana, N. and Sullivan, S. (2006). Introduction: Feminist Epistemologies of Ignorance. *Hypatia*, 21(3), vii–ix.

Verma, S.K. (2004). Protecting Traditional Knowledge. *The Journal of World Intellectual Property*, 7, 765–805. doi:10.1111/j.1747-1796.2004.tb00228.x

Von Baeyer, H.C. (1999). *Warmth Disperses and Time Passes: A History of Heat*. New York: Modern Library.

Walker, M.U. (1989). Moral Understandings: Alternative "Epistemology" for a Feminist Ethics. *Hypatia*, 4(2), 15–28.

Waller, A.A. (2006). *Special Session—Fish Is Fish: Learning to See the Sea We Swim In: Theoretical Frameworks for Education Research*. Frontiers in Education Conference Proceedings, San Diego, CA.

Warren, K. (2000). *Ecofeminist Philosophy: A Western Perspective on What It Is and Why It Matters*. Lanham, MD: Rowman & Littlefield.

Whitbeck, C. (1995). Teaching Ethics to Scientists and Engineers: Moral Agents and Moral Problems. *Science and Engineering Ethics*, 1(3), 299–308.

Whitbeck, C. (1998). *Ethics in Engineering Practice and Research*. New York: Cambridge University Press.

Winner, L. (1990). Engineering Ethics and Political Imagination. *Broad and Narrow Interpretations of Philosophy and Technology*, 7, 53–64.

Young, I.M. (1990). *Justice and the Politics of Difference*. Princeton: Princeton University Press.

Zussman, R. (1985). *Mechanics of the Middle Class: Work and Politics Among American Engineers*. Berkeley: University of California Press.

49

SOCIALLY RESPONSIBLE ENGINEERING

Jessica M. Smith and Juan C. Lucena

Any effort to define and teach engineering ethics which does not produce a vital, practical, and continuing involvement in public life must be counted not just as failure, but as betrayal as well.

(Winner 1990: 64)

49.1. Introduction

Attempts to cultivate socially responsible engineers in the United States have largely focused on teaching undergraduate students engineering codes of ethics. Yet in their professional practice, none of our graduates or the engineers we came to know through our research turned to these codes when confronted with a range of ethical dilemmas in the context of their working lives—a finding that resonates with other surveys (e.g. Luegenbiehl and Puka 1983). In this chapter, we propose a new framework for teaching social responsibility to engineering undergraduates, as this educational experience is foundational to their professional identities. We begin by tracing the historical emergence of codes of ethics in the US and how they became stabilized in undergraduate engineering education. This history shows why codes align concepts of social responsibility with business interests while preventing engineers from being responsible for other interests. We then draw on scholarship in science and technology studies to propose a new framework for social responsibility that can enhance and help to critically analyze the norms that shape engineering education and practice (e.g., codes of ethics, ISO standards, corporate policies), and we propose strategies to make those pillars actionable through five steps which we illustrate through the corporate career of an engineer. We conclude by proposing a critical take on corporate social responsibility as a crucial arena of practice for socially responsible engineers.

49.2. Emergence of the Codes of Ethics in the US: Historical Background

Many US engineering ethics courses, introductory textbooks to engineering, and professional engineering (PE) exams take the codes of ethics for granted as a straightforward statement of engineers' social responsibilities. But codes of ethics emerge in specific historical, political, and sociocultural contexts. In this section we trace how codes of ethics emerged as a marker of professional identity for engineers, established the locus of their loyalty and the limits of their social responsibility, and became the basis for the ethical dimensions of accreditation. While the account in this chapter focuses exclusively on the US, for a brief comparative analysis of how the emergence of engineering ethics happened in other national contexts, see Downey et al. (2007). For more in-depth treatments of the history of engineering ethics, see Didier (1999) for France, Huning and Mitcham (1993) for Germany, and Luegenbiehl (2004) for a comparison of the US and Japan.

49.2.1 From Progressivism to Business as Usual

We begin at the time when engineers practiced engineering without the guidance of codes of ethics. During the Progressive Era (1880–1920), some engineers began to question the negative impact of industrial capitalism on society, specifically of smoke from coal burning on human health. These concerns emerged from interactions with ladies' clubs and physicians who were concerned about the aesthetic and health impacts of smoke on urban environments and populations. These engineers saw the problem of smoke as a problem of inefficient combustion and were able to demonstrate to wealthy industrialists that reducing smoke meant saving money in coal purchases (Stradling 2002). These engineers came from local engineering societies and, like other examples at the time such as ventilation engineers concerned about children's respiratory health in public schools (Baker 2012), represented a group of professionals with social responsibility that cared for the well-being of others who were not their employers and without the guidance of professional codes of ethics.

Interestingly, the Progressive Era also coincides with the largest percentage increase of engineers in American history. According to Edwin Layton,

> The rising demand for engineers by industry began the second stage of the emergence of the engineering profession. The golden age for the application of science to American industry came from 1880 to 1920, a period which also witnessed the rise of large industrial corporations. In these forty years, the engineering profession increased by almost 2,000 percent, from 7,000 to 136,000 members. The civil engineer was overshadowed by the new technical specialists who emerged to meet the needs of industry: the mining, metallurgical, mechanical, electrical, and chemical engineers.
>
> *(Layton 1986: 3)*

As Layton shows, this proliferation in numbers and disciplines challenged engineers with questions of *identity* (What is an engineer? How is one engineer different from another?), *loyalty* (Who are engineers loyal to? Their employers? Themselves as professionals? The public at large?), and *social responsibility* (What are engineers responsible for?).

Most of these questions would be settled during the initial third of the 20th century with the professional codes of ethics in what Mitcham (2009) calls the phase of "ethics as loyalty." But during the Progressive Era, engineers had their moment in history where they flirted with the causes of the Progressive movement, as seen earlier, and with the creation of the Federation of American Engineering Societies (FAES), which Layton labels "the high tide of engineering progressivism" (Layton 1986). The first president of FAES, Herbert Hoover, came to embody the ideals of those engineers seeking autonomy: "professionalism, progressivism and the engineering approach to social problems" (p. 189). "Extolling the superiority of engineering to other professions," Hoover "praised the unique potentialities of his profession for public service." But he positioned them outside of political and economic pressures by corporations when he wrote

> Engineers were through the nature of their training, used to precise and efficient thought; through the nature of their calling, standing midway in the conflicts between capital and labor; and above all, being in their collective sense independent of any economic or political interest, they comprise a force in the community absolutely unique in the solution of many national problems.
>
> *(Hoover quoted in Layton 1986: 189–190)*

The demise of the FAES in the mid-1920s marked the beginning of the dominance of conservative pro-business ideology over the engineering profession, leaving behind any serious attempt for social reform. FAES was replaced by the American Engineering Council (AEC), which expressed "conservative, business-oriented view of the engineers' social role and responsibilities . . . and became a defender of the business system and of *laissez faire*" (p. 208). This change in the organization that was supposed "to bring the engineering viewpoint to bear in the problems of the nation" was actually deemed to exemplify "the paralysis of the engineering profession" as the engineering profession was split in the 1930s. The progressive members of FAES were replaced by an "ultraconservative minority" who swayed the AEC "to take a position more conservative than that of the business community and unrepresentative of the engineering profession" (p. 230). This split was represented, on one hand, by engineers who called for licensing and collective bargaining but were careful not "to mix with nonprofessionals [technicians] . . . as amalgamation with labor threatened to cut off engineers not only from their professional brethren but also from management" (p. 241); and, on the other hand, by ultraconservative pro-business engineers who now controlled the AEC. Unable "to advance the status and the welfare of engineers," the AEC was abolished in 1940s when the American Society of Civil Engineers (ASCE) and American Institute of Electrical Engineers (AIEE) withdrew their support. Since then, this ideological fracture has taken many forms during World War II, the Cold War, and thereafter (see Lucena 2005; Wisnioski 2012) yet one commonality emerged between the two camps: allegiance to the codes of ethics.

Commitment to the codes accomplished three political results: it (1) provided a way for engineers to exhibit professionalism and differentiate themselves from technicians; (2) exempted engineers and their employers from the big problems that Americans witnessed in the Great Depression (e.g., industrial waste, urban pollution, poverty, hunger, homelessness, etc.) while claiming commitment to safety and welfare of an abstract "public"; and (3) safeguarded the interests of their employers (pro-business). Hence by the 1950s, the big questions about engineers' social responsibility were reduced to the canons in the codes of ethics. Mitcham shows alignment among the major professional societies (AIEE, ASME, ASCE) around the idea that "the engineer should consider the protection of a client's or employer's interest his first professional obligation" (Mitcham 2009: 39). This alignment led Layton to conclude that

> As America lost interest in reform, engineers either rejected the idea of social responsibility or gave it a conservative interpretation [in codes of ethics]. The close alliance between engineering and business that developed in the 1920s brought many material benefits. But the profession lost much of its precious independence. The studies of waste and the twelve-hour day demonstrated the sort of contributions an autonomous engineering profession might have made to national life . . . There can be little doubt that engineers derived substantial benefits from their alliance with business. But there was a danger that in gaining worldly things the engineering profession might have lost its own soul.
>
> *(Layton 1986: 201, 218)*

Around that time, this pro-business orientation to social responsibility also entered the engineering curriculum. David Noble (1979) shows that the interests of corporations and the military had been entering engineering education since World War I, but the creation of the Engineering Council for Professional Development (ECPD) in 1932 "signaled the complete triumph of corporate engineers and their particular brand of professionalism" (p. 243). Representing all the engineering societies at the time, ECPD

> became recognized at the central agency for all matters relating to the engineering profession, including college accreditation, professional standard of ethics, determination of

competence for practice, professional recruitment from high schools, and the general definition of what it means to be a professional engineering in America.

(Noble 1979: 243)

Before ECPD evaluated its first engineering degree program in 1936, there was no mechanism for
the engineering profession to encourage the teaching of ethics. ECPD accreditation became such a
tool, and it carried with it the pro-business orientations of its leaders. As Mitcham shows, in the first
four decades of the 20th century, "engineering ethics education was primarily an issue for professional societies and learning through apprenticeship. There was little by way of explicit promotion of
engineering ethics at the college level [and] certainly nothing by way of required courses" (Mitcham
2009: 44).

D. C. Jackson, dean of engineering at MIT and consultant to electric utilities, chaired the Committee on Principles of Ethics of ECPD from 1940 to 1951, which developed "a wide-ranging
Canon of Ethics and spent the rest of his time as chairman persuading various regional and national
engineering organizations to adopt it" (Stephan 2002). Jackson was also one of the main proponents
of the engineer as businessman and the businessman as an engineer—the side who ultimately won
the battle in which engineers sided with corporations. Pointing to the corporation as the locus of
engineers' social responsibility, "Jackson maintained that the modern corporation was the product of
an evolutionary law discovered by Spencer and, therefore, that it was the engineer's duty to defend
such corporation from government regulation" (Layton 1986: 72). Jackson's work on ethics in ECP
came to a halt because of World War II, when engineering schools had to align themselves with the
needs of the war. Yet, it is likely that Jackson's leadership left a lingering influence on how the committee advocated teaching ethics, and that influence was sympathetic to corporations and the codes
of ethics that espoused social responsibility that protected their interests. In sum, Jackson's leadership
reinforced the stabilization of codes of ethics as social responsibility both in engineering practice and
education.

49.2.2 From Three Mile Island to New Textbooks

The 1970s and 1980s witnessed a number of technological accidents and disasters like Three Mile
Island (1979), the Hyatt Regency walkway collapse (1981), Bhopal (1984), Chernobyl (1986), *Challenger* (1986), and Exxon Valdez (1989) which put technology and its consequences on human populations and the environment on center stage. In the historical development of engineering codes of
ethics, these catastrophes came at the end of what Mitcham (2009) calls the phase of "public safety,
health, and welfare." Beginning in the years after World War II, professional societies called attention to the complexities of technological systems, questioned engineers' abilities to control them,
and pushed safety, health, and welfare of the public (the "paramountcy principle") to the top of the
professional codes. The National Society of Professional Engineers (NSPE), a bastion of professional
autonomy among the larger pro-business engineering societies, was the first one to bring the paramountcy principle to the top of their list (1981), followed by IEEE (1990) and others.[1]

These highly publicized catastrophes and the ascendance of the paramountcy principle played
out in different ways in higher education. One way was to highlight them through campus lectures
such as Choices and Challenges at Virginia Tech, inviting experts such as Roger Boisjoly, the famous
whistleblower in the *Challenger* disaster, to interact with graduate students in STS and giving extra
credit to engineering undergraduates to attend. Another way was to become subjects of analysis in
STS scholarship for scholarly audiences (e.g., Vaughan 1997) such as those found in STS graduate
seminars and conferences. A third way was the publication of engineering ethics textbooks (see
Mitcham 2009: 44–45 for a review of these textbooks). While variation exists among those textbooks, general trends emerged: (1) content focused on engineers as individual actors, oftentimes in

extraordinary circumstances; (2) case studies as the dominant method for teaching; (3) ethical reasoning as the epistemic approach for reasoning through the dilemmas presented in the case; and (4)
the codes of ethics as the normative framework to guide behavior. While these prominent disasters
spurred increased reflection on engineering ethics, serving as flashy examples in textbooks and as
points of reference for professors and student audiences alike, they left intact the overarching business
orientation in the code of ethics and the way they were taught.

By the mid-1980s ECPD had changed its name to the Accreditation Board for Engineering and
Technology (ABET). The ethics committee disappeared, but ABET did not turn its back entirely on
the issue of engineering ethics. In 1984, two sentences which were to remain substantially unchanged
until the issuance of EC 2000 appeared in the criteria for the first time:

> An understanding of the ethical, social, economic, and safety considerations in engineering
> practice is essential for a successful engineering career. Course work may be provided for
> this purpose, but as a minimum it should be the responsibility of the engineering faculty to
> infuse professional concepts into all engineering course work.
>
> *(Stephan 2002: 13)*

With the end of the Cold War in the late 1980s and the emergence of economic competitive-
ness as the main preoccupation for engineering educators, the tensions inside the curriculum
were those between engineering science, as the crowning achievement of the Cold War, and
engineering design/manufacturing, as the new requirements for post-Cold War engineers to
compete in the global economy. Although the locus of attention changed, from defending the
US from Soviet aggression and communism to competing with Japan and other emerging Asian
economies, this change did not bring any substantial changes to engineering ethics instruction
(Lucena 2005). In the humanities and social sciences curriculum for engineers, it brought attention to cross-cultural education, foreign languages, and study abroad as responses to the new need
to compete in a global economy. The stabilization of engineering ethics education as the learning of the codes and case studies had survived even the most significant geopolitical event of the
20th century: the end of the Cold War. Even with new available textbooks in engineering ethics,
Mitcham concludes that

> teaching has remained to a significant degree focused on what may be described as a largely
> internalist and individualist emphasis—that is, on individual professional responsibility to
> promote public safety, health, and welfare—using a mix of analytic ethics and case studies
> with some modest introduction of social implications, always with explicit reference to the
> ethical codes of various professional engineering societies.
>
> *(Mitcham 2009: 46)*

49.2.3 From Sustainable Development to ABET 2000

The emergence of sustainable development in the early 1990s as a normative concept for engineers
did not fundamentally change engineering ethics instruction. In that decade, most of the engineering professions adopted the Brundtland definition of sustainable development as that which "meets
the needs of the present without compromising the ability of future generations to meet their own
needs". Yet Herkert, writing almost two decades after the Brundtland report and one decade after
the emergence of ABET EC 2000, concludes that

> engineering ethics and research to date [2008] have for the most part focused on microanal-
> ysis of individual ethical dilemmas in such areas as health and safety issues in engineering

design, conflict of interest, representation of test date, whistle blowing, accountability to clients and customers, quality control, trade secrets, gift giving, and bribes.

(Herkert 2008: 58)

All of these, of course, reflect the main points in the codes of ethics.

In spite of Herkert's call for engineers to embrace macroethics (i.e., social and environmental responsibility towards societal issues such as sustainable development, product liability), engineering ethics education has remained stable as the teaching of codes, often through case studies. In a systematic review of engineering ethics interventions published in peer-reviewed journal articles between 2000 and 2015, Hess and Fore (2018) found that 85% of their sample used codes of ethics or standards, and 81% used case studies. Similarly, in a comprehensive study of the teaching in the professions by the Carnegie Foundation for the Advancement of Teaching, Colby and Sullivan (2008) focused on engineering ethics in undergraduate engineering education and discovered that engineering codes are still the dominant normative framework of engineering ethics instruction.

Not surprisingly, Colby and Sullivan discovered that macroethics education, as proposed by Herkert, "did not appear to be a common practice in the engineering programs we reviewed. Even though we did encounter some attention to macroethical issues, few schools had instituted systematic programs to educate for this broad sense of professional responsibility" (p. 330). Colby and Sullivan take for granted that "ethics codes originate from within the profession [hence] provide a good sense of the kinds of ethical issues practicing engineers in various specialties are likely to confront" before concluding that "they can inform engineering educators about the kinds of issues and dilemmas graduates should be prepared to handle" (p. 328). This approach to understanding the place of codes in engineering education loses sight of the importance of questioning the power relations and struggles between professional autonomy and corporate interests in the historical development and current practice of codes. Given the historical evolution of those codes and their entrance into accreditation, mainstream engineering ethics education can reproduce a latent pro-business paradigm while ostensibly teaching about social responsibility.

49.3. Three Pillars of Social Responsibility

The previous section illustrated how professional codes of ethics became the basis for undergraduate engineering students' ethical education in the US.[2] The dominant pedagogy for teaching these codes became and remains the case study method, though the case study method can also be used to teach engineering ethics outside of the codes. Based on US engineering schools, the comprehensive Colby and Sullivan (2008) survey points to the limitations of relying so heavily on the codes: they are not sufficient to "frame a broad conception of goals" but instead "point to a wide-ranging set of understandings and competencies students need to develop" (p. 335). Moreover, they argue that the dominant case study method for teaching students to apply those codes "does not require students to struggle with the trade-offs involved in actual engineering decisions or with the fact that the consequences of those decisions become clear only in retrospect" (p. 331).

What Colby and Sullivan point to as the "trade-offs" of engineering decisions opens space to consider the inherently political dimensions of engineering, which scholars in science and technology studies argue is necessary for a more robust sense of professional social responsibility. Scholars such as Joe Herkert, Carl Mitcham, Langdon Winner, and the contributors to the *Science, Technology & Human Values* special issue on STS, ethics, and engineering design (Van de Poel and Verbeek 2006) critique these codes and the case studies used to teach them for making structures of power invisible and thus hidden from the kind of critique required to consider the broader context of engineering and its imbrication with social, political, and economic structures of power. In particular, these scholars argue that discussion of case studies, the dominant teaching pedagogy, contributes to this foreclosure of structural critique by positioning engineering ethics as a matter of individual engineers engaging in heroic actions of whistleblowing. This approach fails to consider the everyday, more mundane ways in which engineers encounter ethical dilemmas and are constrained by their corporate and organizational structure (Lynch and Kline 2000) and how deviance can be normalized, such as in Vaughan's (1997) historical ethnography of the routine work practices that contributed to the *Challenger* disaster.

Even more significantly, we argue, the combination of codes of ethics, taught through case studies, often fails to encourage students to question or contest *why* engineering is being used to support particular ends and with what implications for *particular* groups of people rather than an amorphous notion of society at large. Winner (1990) illustrates this difference using a hypothetical case study of an engineer discovering that the paint used on the shell of a cruise missile emits toxic fumes that could be dangerous to workers in the assembly plant. The locus of "ethics" in this case study is whether the engineer should go against his boss's desires for the project to stay on schedule, thus avoiding risk to his own job and career, rather than the decision to work for a defense contractor that builds weapons (pp. 53–54). The dominant focus in engineering ethics courses on issues of right and wrong in personal conduct, while important dimensions of professional ethics, does not open up crucial questions of social structure and history; as Winner writes, we must "move beyond questions of individual conduct to consider the nature of human collectivities and our membership in them" (p. 57). As summarized by Mitcham (2016: 71), "ethics is not enough"—what is required is a political theory of engineering to shift focus from personal heroism to the "particular place of engineering in human affairs and the political order."

In place of engineering ethics that legitimate and reinforce the status quo, Winner (1990) envisions "the genuine promise of the exercise of responsibility among scientific and technical professionals" to be "a gradual reorientation of patterns of research, development, and application of emerging and already existent technologies to accord with our civilization's higher principles" (p. 63). This vision is premised on *three pillars*. The first pillar proposed by Winner is the *responsibility of dialogue*: "engaging others in the difficult work of defining what the crucial choices are that confront technological society and how intelligently to confront them" (p. 62). Instead of assuming that options are inevitable, engineers should question why they are proposed, why they might not be opposed, and why alternatives are not presented. This responsibility is predicated on a second pillar not explicitly called out by Winner but implicit in his framework, what we call the *responsibility of awareness*: becoming aware of the structures of power that shape the possibilities of what is engineered and for whom. This capacity underlines Winner's vision for more radical notions of engineering responsibility. The third pillar is *responsibility of citizenship*: to raise public consciousness about key political issues and act on them.

Winner argues that these pillars are not just absent from the dominant case study method for teaching about social responsibility in the US, but that the case study method itself encourages engineers to be complacent. Case studies, in his view, tacitly presume and convey that "as one enters a profession, one simply embraces the existing commitments, institutional patterns, and power relationships the profession contains" (Winner 1990: 56). Colby and Sullivan (2008: 328) signal similar concerns in their survey, writing, "The ethical codes themselves are worthy of attention in the curriculum, but using them to frame a broad conception of goals goes well beyond telling students to 'learn the code and follow it.'" Instead, Winner calls for engineering students and educators to analyze and critique the social, political, and economic contexts giving rise to those cases, seeing how and questioning why engineering projects serve some economic or social interests and not others.

49.4. Making the Three Pillars Actionable: Engineering for Social Responsibility Framework

In this section we propose a framework with five steps to help engineering educators, students, and practitioners to do this work of critical questioning to arrive at more robust considerations of their social responsibilities. We do not seek to replace codes of ethics or other normative frameworks such as ISO standards, but to empower engineers to: (1) view those standards with a critical lens to determine the opportunities and limitations they offer; and (2) imagine new ways of practicing social responsibility.

This framework, like all intellectual projects, has its own history. Jon Leydens and Juan Lucena (2017) developed criteria for engineering for social justice: listening contextually; identifying structural conditions; acknowledging political agency/mobilizing power; increasing opportunities and resources; reducing imposed risks and harms; and enhancing human capabilities. While useful for engineers practicing community development from positions inside of NGOs, universities, the Peace Corps, and other government entities, they are difficult to translate to the for-profit companies that constitute the primary workforce destination of engineers. To translate and adapt these steps to a context with different constraints, we draw on research done by Jessica Smith, who used ethnographic methods to understand how the concept of corporate social responsibility shapes engineers' work, including how they navigate tensions among their employers' policies, their professional notions of social responsibility, and their own personal senses of right and wrong (Smith 2021).

Drawing on a literature review and the key insights from those interviews, she created the following steps toward practicing engineering for social responsibility and vetted them with other engineering educators and practicing engineers in workshops. We propose them here as a framework for engineering for social responsibility.

1. Understanding structural conditions and power differentials among specific stakeholders of an engineering project.
2. Contextually listening to all stakeholders, especially those who are marginalized, to understand their worldview and to grasp their needs, desires, and fears surrounding a specific project, decision, etc.
3. Collaboratively identifying opportunities and limitations of creating shared social, environmental and economic value for all stakeholders, especially those who are marginalized. This requires acknowledging when "value" is differently defined by stakeholders.[3]
4. Adapting engineering decision-making to promote those shared values, acknowledging situations in which this is not possible and engineering projects should not move forward.
5. Collaboratively assessing activities and outcomes with those stakeholders.

One key way in which this framework differs from codes of ethics is that it identifies specific groups of people to whom engineers are responsible rather than "society at large" or "the public." Johnson (2017) critiques the paramountcy principle for treating the public as a "black box . . . the term hides all the complexity and diversity in the public" (p. 93). In "Ethics is Not Enough," Mitcham (2016) attributes this focus on impossibly large social groups to engineers' comfort with idealized abstractions. Here we acknowledge an array of groups to whom engineers are responsible (the corporations employing them, employees, shareholders, users of products) but focus specifically on the people who are often negatively affected by engineering projects but not empowered to shape or contest them. The second key difference is that this framework, like Winner's, treats engineering as a sociotechnical practice that is always embedded in relationships of power. The responsibility of engineers is to acknowledge those power structures and their own roles within them, and then to find ways to direct their professional practice to mitigating those asymmetries.

49.5. Enacting Social Responsibility: Pillars + Framework

What would it look like for Winner's three pillars of social responsibility, enhanced with our concrete framework for making them actionable, to be put into action by practicing engineers? We do not advocate for engineers to stop learning about professional codes of ethics or to throw out the case study method for teaching them. Instead, we call for engineers and engineering students to wrestle with those codes in a way that is informed by the more robust sense of social responsibility we propose here. For example, engineering students should learn how to "hold paramount the safety, health and welfare of the public" when evaluating decisions. But they should be doing so while simultaneously *becoming aware* of how those decisions are inflected by asymmetrical structures of power, engaging in *dialogue* with the specific people affected by the decision, and enacting *citizenship* by sparking discussions about engineering projects and making decisions that create shared value for the people affected by that decision, especially those who are marginalized by structures of power.

The work experiences of one of the engineers in Smith's research project shed light on what these five steps could look like for those who choose corporate careers. After working in environmental remediation and compliance projects in the mining and oil and gas industries, this engineer went on to lead a major multinational company's social responsibility programs at strategic, new development projects around the world. One of the largest projects, totaling a multibillion-dollar investment, involved planning and permitting new oil and gas facilities, including new production fields and transportation facilities. The engineer integrated elements of all five steps to different degrees, in his social responsibility work on this project.

1. He emphasized the importance of *disaggregating the "local community"* to identify tensions within and among the groups closest to their operations. This required working with local residents to *identify power differentials* and then seeking out those whose interests were not necessarily aligned with those of the dominant political or family leaders.

2. The centerpiece of the community engagement he put into practice was *listening* to all of the project's stakeholders, including those who were in marginalized positions. Done in a culturally appropriate manner, listening provided him a technique to "triangulate . . . what the powerful were saying versus really what the mass of people in the community were voicing in terms of their concerns and needs."

3. With a grasp on the hopes, needs, and concerns of all of the stakeholders, he was then able to *direct collaborative community development projects*, ensuring to include the well-being of the most marginalized identified in #2. For example, he and his team devised training opportunities for local women to build on their existing capacities to create new small enterprises. Women were "eager to improve their economic situation," he said, as they were primarily responsible for raising children and did not have access to the same economic resources that men did. Working directly with the women to understand the intricacies of local kinship and economic systems, they found creative ways to enhance their economic earnings while guarding against those earnings being claimed by powerful male kin or instigating retaliation or increased surveillance by those male kin. "You had to find ways [of supporting the women] that fit within the cultural pattern while not exposing them to an increased amount of risk," he explained.

4. Detailed personal and cultural knowledge also allowed him to *inform engineering decision-making*. The original pipeline path that had been designed back in the company's US offices did not account for the households that would have to be resettled because they were located in the pipeline's path. The engineer and his team advised project managers to relocate it to minimize those resettlements, but the company went forward with the plan and encountered months of costly delays due to the lengthy processes of determining land rights and engaging in consultations over compensation. Learning from that experience, the engineering team then followed his advice and "took great pains to avoid residences and gardens and similar things wherever they could." His team advised their coworkers on more culturally appropriate routes.

5. To *assess* their activities, he and his team recognized the strategic importance of quantifying information for internal buy-in at the same time as they acknowledged the limitations of trying to put numbers on qualitative information. He explained,

> As an oil and gas firm with a bunch of engineers and scientists, they're forever pushing us get more metrics. And there's this belief that if you can measure it you can manage it, right? When you're dealing with people, it's not always that concrete. I'm a firm believer the fewer and the simpler metrics are probably better.

Instead, he drew attention to tying metrics to concrete outcomes. He wanted to know, "Did it make a measurable improvement in people's lives?" He explained,

> So you could say, "Okay, success: we delivered 50 ovens." Or, "One half of our target group were able to raise their incomes." Which is the success? I would prefer being able to say 50% of our target group were able to raise their incomes.

This engineer was able to enact all three pillars of social responsibility through his work—not through turning to his professional codes of ethics but through the kind of critical assessment we advocate through our five steps.

49.6. Conclusion: A Space for CSR in Engineering Education

The engineer profiled in Section 5 was working inside a corporation, as will the majority of US engineering graduates (Downey 2007; Lamancusa et al. 2013). That context provides particular kinds of opportunities and constraints for practicing socially responsible engineering. On the one hand, companies have substantial economic and political power that can be harnessed to promote goals such as sustainable community development, as long as engineers recognize and mitigate as possible the power differentials between them and the communities they seek to serve. On the other, for-profit companies are large, internally diverse organizations that are ultimately accountable to a financial bottom line, sometimes expanded to a "triple bottom line" that includes social and environmental performance. To manage these competing responsibilities, this engineer and others interviewed turned to their companies' corporate social responsibility (CSR) policies and World Bank/IFC performance standards. Tellingly, *none* of the over 70 engineers interviewed for the CSR project referred to their professional codes of ethics or remembered ethics case studies as influencing their decision-making.

What this engineer and others did was reference their employer's CSR policies in order to generate internal support for social responsibility work; in this way, community development projects could be framed as supporting the company's own policies rather than representing an externally imposed obligation. We therefore underline that while CSR is a highly controversial concept that, depending on how it is enacted, can extend corporations' moral authority and economic well-being while not delivering benefits for its intended recipients (Kirsch 2010; Rajak 2011), it is not simply an exercise in greenwashing. Done well, CSR is a crucial field of practice for engineers that should not be ignored in the teaching of engineering ethics (Smith 2021; Smith et al. 2016; Smith et al. 2018). A critical approach to CSR can open up macro-ethical questions about who engineering benefits and disadvantages, and how engineers can leverage their professional practices to enhance human capabilities.

Truly robust corporate social responsibility policies in the extractive industries, for example, must tackle the contribution of these industries to climate change and include the opportunity for communities to say "no" to natural resource development rather than simply being consulted on how it should happen (Owen and Kemp 2017). The engineer briefly profiled here wrestled with the question of whether companies should be operating in places where human rights were abused by people in political and familial positions of power. "That's a very valid question: 'By operating there are you implicitly endorsing such behavior?'" he said. "I would argue that if you're not some sort of agent for change, even a slow and gentle, then things may never improve." This pragmatic approach was shared by many of the engineers interviewed for the project (Smith 2021). While noting that there are many opportunities to serve communities' interests and the company's financial bottom line—the "win-win" scenario that academics refer to as the "business case" for CSR—they also recognized that sometimes those interests cannot be reconciled. These cases bring to the fore the need for engineers to be able to navigate multiple normative frameworks—employers' CSR policies, external performance standards, professional codes of ethics, etc.—while enacting *awareness, dialogue,* and *citizenship.*

Therefore while CSR should not be the only definition of engineers' social responsibilities, it should be integrated into undergraduate engineering curriculum so that engineers can approach this crucial field of practice with a critical eye, rather than simply learning codes of ethics that do not provide clear guidance on how to balance their responsibilities to their employers alongside their wider responsibilities. While acknowledging that mastering the "wide-ranging set of understandings and competencies" that comprise professional responsibility takes on-the-job experience, Colby and Sullivan (2008: 335) underline the importance of undergraduate education: "the undergraduate years represent an important opportunity to establish trajectories that will lead eventually to this broad range of competencies, understandings, and inclinations."

In our own Humanitarian Engineering program, our *Corporate Social Responsibility* course develops students' abilities to critically assess this concept and its potential to transform business to serve not just society in general, but the particular people who shoulder the largest burdens of its harms. Throughout the course, students read peer-reviewed articles and ethnographies that provide a critical view on the concept by analyzing how it is enacted in actual projects. In the students' semester-long project, they complete a series of scaffolded assignments that help them propose a critical intervention in a real-world engineering project. The students begin by mapping the project's stakeholders and relationships among them (criterion 1) and then research their worldview and particular concerns, desires, perceptions, and needs, drawing on peer-reviewed literature as well as sources such as blogs, newspaper articles, and films. This is an imperfect approximation of listening to stakeholders, given the limitations of the course (criterion 2). Using that information, they identify opportunities and limitations for the project to create social, environmental, and economic value, though again, without being able to truly do so collaboratively (criterion 3). They then propose how they will adapt specific engineering decisions—such as the size of a manufacturing plant, the location of a tailings pond, or the materials used in a particular product—to support those values, prioritizing the needs of the most marginal stakeholders (criterion 4). Sometimes students recommend that the project should not move forward at all, such as in the case of some controversial oil and gas development projects. While the students write a collaborative assessment plan (criterion 5), they are not able to enact it in practice. The assignments do, however, push students to ask *why* engineering is being used to support particular ends and with what implications for particular groups of people—the foundation of truly socially responsible engineering.

Related Chapters

Chapter 2: A Brief History of Engineering (Jennifer Karns Alexander)
Chapter 40: Ethical Considerations in Engineering (Wade L. Robison)
Chapter 41: Autonomy in Engineering (Eugene Schlossberger)
Chapter 43: Professional Codes of Ethics (Michael Davis)
Chapter 44: Responsibilities to the Public—Professional Engineering Societies (Joseph Herkert and Jason Borenstein)
Chapter 45: Engineering as a Political Practice (Govert Valkenburg)
Chapter 50: Engineering and Social Justice (Caroline Baillie)

Notes

1. Although not the focus of this chapter, for an in-depth account of how the *Columbia* disaster challenged organizational culture, decision-making, and risk management, see Starbuck and Farjoun (2005). These disasters also sparked new scholarly research and writing on safety culture, and risk and crisis management (Pidgeon 1991).
2. For a review of non-US initiatives in engineering ethics, most of which do not rely on the teaching of codes, see Zandvoort et al. (2013).
3. While the term "shared value" originally stems from strategic management theory (Porter and Kramer 2011), we expand the term to signal that "value" takes different form for people who live in, understand, and judge the world in diverse ways. In other words, values emerge from specific contexts and cannot be assumed to be shared, requiring practitioners to contextually listen to how their interlocutors understand and judge the world in addition to what they think about a particular engineering project.

References

Baker, L. (2012). *A History of School Design and Its Indoor Environmental Standards, 1900 to Today*. National Clearinghouse for Educational Facilities. https://eric.ed.gov/?id=ED539480

Colby, A. and Sullivan, W.M. (2008). Ethics Teaching in Undergraduate Engineering Education. *Journal of Engineering Education*, 97(3), 327–338.

Didier, C. (1999). Engineering Ethics in France: A Historical Perspective. *Technology in Society*, 21, 471–486. doi:10.1016/S0160-791X(99)00029-9

Downey, G.L. (2007). Low Cost, Mass Use: American Engineers and the Metrics of Progress. *History and Technology*, 23(3), 289–308.

Downey, G.L., Lucena, J.C. and Mitcham, C. (2007). Engineering Ethics and Identity: Emerging Initiatives in Comparative Perspective. *Science and Engineering Ethics*, 13(4), 463–487.

Herkert, J.R. (2008). Engineering Ethics and STS Subcultures. In *Research in Social Problems and Public Policy*, Vol. 16. Bingley: Emerald (MCB UP), pp. 51–69. doi:10.1016/S0196-1152(08)16003-3.

Hess, J.L. and Fore, G. (2018). A Systematic Literature Review of US Engineering Ethics Interventions. *Science and Engineering Ethics*, 24(2), 551–583. doi:10.1007/s11948-017-9910-6.

Huning, A. and Mitcham, C. (1993). The Historical and Philosophical Development of Engineering Ethics in Germany. *Technology in Society*, 15(4), 427–439. doi:10.1016/0160-791X(93)90014-F.

Johnson, D. (2017). Rethinking the Social Responsibilities of Engineers as a Form of Accountability. In D.P. Michelfelder, B. Newberry and Q. Zhu (eds.), *Philosophy and Engineering: Exploring Boundaries, Expanding Connections*. Switzerland: Springer, pp. 85–98.

Kirsch, S. (2010). Sustainable Mining. *Dialectical Anthropology*, 34, 87–93.

Lamancusa, John S., Zayas, Jose L., Soyster, Allen L., Morell, Lueny and Jorgensen, Jens (2013). 2006 Bernard M. Gordon Prize Lecture*: The Learning Factory: Industry-Partnered Active Learning. *Journal of Engineering Education*, 97(1), 5–11. doi:10.1002/j.2168-9830.2008.tb00949.x

Layton, E.T. (1986). *The Revolt of the Engineers: Social Responsibility and the American Engineering Profession*. Baltimore, MD: Johns Hopkins University Press.

Leydens, J.A. and Lucena, J.C. (2017). *Engineering Justice: Transforming Engineering Education and Practice*, 1st ed. Hoboken, NJ: Wiley-IEEE Press.

Lucena, J.C. (2005). *Defending the Nation: U.S. Policymaking to Create Scientists and Engineers From Sputnik to the "War Against Terrorism"*. Lanham, MD: University Press of America.

Luegenbiehl, H.C. (2004). Ethical Autonomy and Engineering in a Cross-Cultural Context. *Techné: Research in Philosophy and Technology*, 8(1).

Luegenbiehl, H.C. and Puka, B. (1983). Codes of Ethics and the Moral Education of Engineers. *Business and Professional Ethics Journal*, 2(4), 41–66.

Lynch, W.T. and Kline, R. (2000). Engineering Practice and Engineering Ethics. *Science, Technology, & Human Values*, 25(2), 195–225.

Mitcham, C. (2009). A Historico-ethical Perspective on Engineering Education: From Use and Convenience to Policy Engagement. *Engineering Studies*, 1(1), 35–53. doi:10.1080/19378620902725166.

Mitcham, C. (2016). Ethics is Not Enough: From Professionalism to the Political Philosophy of Engineering. *Leadership and Personnel Management: Concepts, Methodologies, Tools, and Applications*, 1350–1382. doi:10.4018/978-1-4666-9624-2.ch060.

Noble, D.F. (1979). *America by Design: Science, Technology, and the Rise of Corporate Capitalism*. Oxford: Oxford University Press.

Owen, J.R. and Kemp, D. (2017). *Extractive Relations: Countervailing Power and the Global Mining Industry,* 1st ed. London: Routledge.

Pidgeon, N.F. (1991). Safety Culture and Risk Management in Organizations. *Journal of Cross-Cultural Psychology,* 22(1), 129–140. doi:10.1177/0022022191221009

Porter, M. E., and Kramer, M. R. (2011). Creating Shared Value. *Harvard Business Review, January–February 2011,* 1–17. https://hbr.org/2011/01/the-big-idea-creating-shared-value

Rajak, D. (2011). *In Good Company: An Anatomy of Corporate Social Responsibility.* Palo Alto: Stanford University Press.

Smith, J.M. 2021. *Extracting Accountability: Engineers and Corporate Social Responsibility.* Cambridge, MA: The MIT Press.

Smith, J.M., McClelland, C.J. and Smith, N.M. (2016). Engineering Students' Views of Corporate Social Responsibility: A Case Study from Petroleum Engineering. *Science and Engineering Ethics,* 1–16. doi:10.1007/s11948-016-9859-x

Smith, Nicole M., Smith, Jessica M., Battalora, Linda A. and Teschner, Benjamin A. (2018). Industry—University Partnerships: Engineering Education and Corporate Social Responsibility. *Journal of Professional Issues in Engineering Education and Practice,* 144(3), 04018002. doi:10.1061/(ASCE)EI.1943-5541.0000367

Starbuck, W. and Farjoun, M. (eds.) (2005). *Organization at the Limit: Lessons From the Columbia Disaster,* 1st ed. Malden, MA: Wiley-Blackwell.

Stephan, K.D. (2002). All This and Engineering Too: A History of Accreditation Requirements. *IEEE Technology and Society Magazine,* 21(3), 8–15.

Stradling, D. (2002). *Smokestacks and Progressives: Environmentalists, Engineers, and Air Quality in America, 1881–1951,* revised ed. Baltimore, MD: Johns Hopkins University Press.

Van de Poel, I. and Verbeek, P.-P. (2006). Editorial: Ethics and Engineering Design. *Science, Technology, & Human Values,* 31(3), 223–236. doi:10.1177/0162243905285838

Vaughan, D. (1997). *The Challenger Launch Decision: Risky Technology, Culture, and Deviance at NASA,* 1st ed. Chicago: University Of Chicago Press.

Winner, L. (1990). Engineering Ethics and Political Imagination. In *Broad and Narrow Interpretations of Philosophy of Technology.* Dordrecht: Springer, pp. 53–64. doi:10.1007/978-94-009-0557-3_6

Wisnioski, M. (2012). *Engineers for Change: Competing Visions of Technology in 1960s America.* Boston, MA: MIT Press.

Zandvoort, H., Børsen, T., Deneke, M. and Bird, S.J. (2013). Editors' Overview Perspectives on Teaching Social Responsibility to Students in Science and Engineering. *Science and Engineering Ethics,* 19(4), 1413–1438. doi:10.1007/s11948-013-9495-7

50

ENGINEERING AND SOCIAL JUSTICE

Caroline Baillie

50.1. Introducing Engineering and Social Justice

When I first used the term *engineering and social justice* in 2003, it was met with confusion: "Surely that's an oxymoron". I was in the midst of exploring why I was being told, ever increasingly, that "Of course engineering is about making a profit". Funny that. I always assumed it was about serving peoples' needs. I had recently moved to Canada to take up my position as Chair of Engineering Education at Queens University, and I truly believed I was moving to a social justice haven. But within days of my arrival, the venture capitalists were at my office door (literally) offering to help turn my work on 'green' materials into profit—this would apparently help me 'prove' my worth as an engineer. Giving my ideas away for free, so they spread further and helped save the planet, was clearly not clever. Coming from the UK, I knew about nationalized engineering services (owned by the government and non-profit-making) such as British Gas and National Rail, and I didn't equate the profession to capital creation as an *inevitability*. I started to ask questions of political economists and sociologists and eventually understood. We had stopped putting people first. This began a life-long task which stemmed from asking a simple question: what would engineering look like if we put people first (and worried afterwards if we could afford it), rather than profit (and worrying afterwards if we had harmed anyone)? Would we engineer different things? Certainly. Would we engineer in a different way? Absolutely.

As time moved on, the Engineering, Social Justice and Peace organization (ESJP) was founded, and as people learned about our work, more groups became interested. As is necessary when facts evolve—the notion of what the term "engineering and social justice" meant became confused. Many would conflate and confuse the term with corporate social responsibility, ethics, humanitarian engineering, and sustainability. Even within social justice-oriented groups there will be a difference in focus about what they mean by the term 'justice'. In Riley's book *Engineering and Social Justice*, she introduces the novice engineer to a wide range of perspectives on social justice (Riley 2008). Several of these are represented in this book—the chapter by Riley herself on feminist perspectives and Ben Cohen on environmental justice, for example. To help us frame the arguments about social justice used in this chapter, selected commitments of the Engineering, Social Justice and Peace organization will be adopted (ESJP 2018), and used as a launching pad for comparing with other key "engineering and society" frameworks and for pointing out key differences among them. Because this chapter is being framed through a social justice perspective, it will show where other framings are not doing so. This is not to say that such framings are not *necessary* work, to the contrary, but that from a social justice perspective, they might be considered *insufficient*. Marullo and Edwards (2000) speak well of the difference between *charity* and *social justice* in relation to student community service programs and which explains such insufficiency. They refer to the "teach a person to fish" adage which is often used to help people understand sustainability of charity work. A charity perspective might be: "Give a person a fish and they have one meal, teach a person to fish and they can feed themselves". A social justice perspective would question why there wasn't enough fish in the lake to sustain the community.

There is no doubt that there are connections between all such frameworks and issues that overlap and extend arguments. Alongside this opportunity for collaboration, lies the real need to demonstrate the difference between them. Variation theory suggests that we understand the essence of something by beginning to see *what it is not* (Marton and Booth 1997). It is important for us to understand critical differences if we are to learn from and respect them. This chapter is framed around that notion of critical difference.

50.2. Engineering and Social Justice (ESJ) Commitments

1. We are committed to identifying and dismantling specific occurrences of injustice related to engineering and technology. Second, in collaboration with community groups facing specific structures of injustice, we are committed to devising and developing technologies and other engineering solutions (broadly conceived) to the problems they face (ESJP 2018).

In the first part of this commitment, ESJ work would entail a critical analysis or dismantling of unjust systems. An example might be work to explore suspected pollutants in an affected mining community's water system. This might result in evidence being brought against the mining company which would assist a community in resisting the mine from going ahead. The second part of the commitment might involve helping such a community create a filtration system to enable its members to drink already-contaminated water. In this second area, but rarely in the first, we would find ourselves in the company of many engineers working within sustainability or humanitarian engineering frameworks.

2. We are committed to resisting injustice in its many forms through promotion of diversity and inclusivity, and by working towards fair, equitable, and sustainable treatment of people and their environments. We are critical of structures of thought conducive to injustice, including the reductionism and positivism prevalent in engineering. We oppose globalized economic policies that lead to the breaking of local networks of labor, production, and food provision (ESJP 2018)

The first part of this commitment could align with all other frameworks listed earlier, depending on their interpretation of the terms, but not the second or third. We do not usually see engineers from an ethics stance, for example, questioning structures of thought or the dominant ideologies behind Western engineering practices, which are based on particular values and norms. We rarely see engineers working for sustainability questioning neoliberal economic policies.

3. While social justice is our goal, we do not adhere to a singular definition of social justice or a static notion of what it entails. We are committed to continually redefining social justice alongside those who experience injustices. We recognize that social justice is contextual, attaining significance within particular times, places, and social locations. Thus, our expression of shared commitments is necessarily always incomplete, and needs to change to reflect our growing consciousness and the shifting urgency of different forms of injustice. We look to social and community movements around the world and in our own locales to guide our commitments (ESJP 2018).

The shifting definition of social justice which is familiar to ESJ workers could be uncomfortable to many engineers from other frameworks. It is difficult for some engineers to accept a lack of precision in definition. It is assumed that if we cannot define it, then we cannot identify it, measure it, or address it. The idea of a *definition* of injustice coming from those negatively affected by an engineering practice would be considered too inexact and "subjective" to many.

4. We are committed to reflexivity—to resisting injustice even as we recognize our complicity with it. As members of different engineering communities, we recognize the structural forces impinging on the profession that perpetuate and reinforce problematic forms of power and privilege. Our participation in these structures necessarily affects how we view social justice and peace. We therefore seek to work critically and inter-disciplinarily to interrogate these structural forces and our relationships to them (ESJP 2018).

All the framings mentioned thus far will have a similar positioning regarding the desire for equity and fairness to all people. Groups working in CSR, for example, might take part in a community meeting to mediate a mining company's position on provision of a trust fund from royalties to provide for local affected community members. However, they may not acknowledge the power dynamic in such meetings or question who is selected to "represent" the community.

The commitments discussed earlier begin to demonstrate how ESJ work applies critical lenses to understand the role and function of engineers and engineering practice, and to use this knowledge to transform such practice to that which increasingly enhances justice.

In the following sections, we will deepen our discussion by comparing an engineering and social justice framing with three other key "engineering and society" framings: engineering ethics, humanitarian engineering and corporate social responsibility.

50.3. Engineering Ethics

Engineers produce codes of ethics, largely based on utilitarianism, Kantian ethics and virtue theory, and they often teach the subject as an "ethics of duty" using case studies of whistleblowers and engineering disasters as discussed in Chapter 49, "Socially Responsible Engineering" by Jennifer Smith and Juan Lucena, this volume (see also Goreman et al. 2000). An example is introducing students to the notion of "voluntarily undertaken risks"—in which someone might choose to ride a motorbike or climb a treacherous mountain, and contrast this with "involuntary risks" such as living next to a mine tailings dam. This approach might teach students, for example, that informed consent is required by local traditional land owners, before a development can happen on Indigenous land, and that such consent must not be *coerced*. This is known as "FPIC"—Free Prior and Informed Consent, the ruling for Indigenous communities within Convention 169 of the International Labor Organization. Teaching FPIC within an ethics framework (without any further analysis) might persuade an engineer that it is *necessary* for them to require informed consent by those potentially impacted by any engineering system. A social justice stance would suggest, however, that this approach is not *sufficient* to ensure justice is done.

A recent examination of the relationship between engineering ethics and social justice (Baillie and Levine 2013) maintains that adopting an ethical standpoint does not necessarily mean the same thing as being a socially just engineer. This is a complicated thing for novice engineers to grasp. Baillie and Levine note that

> Engineering students are rarely asked to reflect upon what they do, why they do it, and what the implication are within their own culture, let alone in relation to anyone else's.
>
> *(Baillie and Levine 2013: 4)*

An engineer following a good example of an ethics code (such as Engineers Australia 2018) might consider the needs of people different from themselves in their designs and implementation using a *user-centred design* framing, but they will rarely be asked to question why what they are doing is necessary, or question the assumptions underlying the actions of their organization. Baillie and Levine explain this as follows:

> The profession as a whole takes major aspects of the dominant discourse for granted—a discourse that is intrinsically linked to corresponding ingrained modes of practice and serves to reify such practice. How then, can we expect engineering students to learn how to make choices, or to even know that there are choices—many of which are important, to be made?
>
> *(Baillie and Levine 2013: 4)*

Ethics codes, just like laws, are predominantly interpreted by the dominant wisdom of the country within which the engineer is professionally registered. Precedent is created when these are applied and this is hard to shift. What we *mean* by this code or that law becomes *common sense*. To understand what this looks like in practice and in relation to our current topic, Baillie and Levine ask us to look at the case of a remote marginalized community which has been approached by a mining prospector. This community is not in a strong negotiating position. Its members have suffered years of oppression and need money and food.

> Is their agreement to the development of a new mine on their land, when they do not necessarily understand the implications, even if they are "told," a free and un-coerced agreement? Is it an "agreement" at all? A non-coerced, non-manipulative, and non-deceptive "agreement" may not be possible for those who have been socialized into accepting handouts as their traditional means of living. The agency and personal decision making capacity of some indigenous people may have been gradually eroded by Western urbanization, subjugation and indeed oppression. They are being asked, unjustly on some accounts, to play a game (follow a procedure) which they do not fully understand by means of rules they may not accept.
>
> *(Baillie and Levine 2013: 6)*

Our ethical codes will reflect the dominant paradigm in which they were created and are now being employed. Assumptions around such negotiations might be shared by both the mining engineers and the community members—that money from royalties will bring a better life, even when there are multiple examples for this not being the case. During interviews that my team undertook from 2014 to 2017 on the impact of mining, an art teacher in the outback town of Wiluna, Western Australia, told us she was very much against the sudden influx of money to communities from mining:

> I've got no idea of figures or anything like that but you know, it's just dreadful what happens if people drink it (the money) away and disastrous things happen in the community just through alcohol. The mines can't stop the alcohol but if they could maybe make payment in some other way . . . The violence wouldn't be as bad through you know, if people get a lot of money in a dry community it mightn't be as violent as in a wet community . . . couldn't say it's the mine's fault but they probably just provide more money to create more problems. But the pattern was already set, we set the pattern . . . there will be money for now but what's going to happen in two or three generations you know? Nothing will be left, or little will be left.
>
> *(Art teacher, Wiluna, 2016)*

An engineer in Australia would have fulfilled their ethical obligations to 'promote sustainability' if any agreement is made with the community. They are asked to:

> Engage responsibly with the community and other stakeholders promote the involvement of all stakeholders and the community in decisions and processes that may impact upon them and the environment.
>
> *(Engineers Australia 2018)*

The agreement made may be ethical according to dominant interpretations, but it may not ultimately bring justice.

50.4. Humanitarian Engineering

Mitcham and Munoz (2010: 27) define humanitarian engineering as the

> artful drawing on science to direct the resources of nature with active compassion to meet the basic needs of all—especially the powerless, poor, or otherwise marginalized.

Many engineering schools and programs such as EWB (Engineers Without Borders) have established humanitarian programs, and these often provide very important services for communities in support of their sanitation, water, energy, building and infrastructure needs. Students enjoy their involvement in such programs, especially if they involve international placements. They also gain both institutional credit for such activities as well as important additions to their resumes as employers often seek experience of this kind in differentiating between otherwise equally brilliant scholars.

However, rarely do students or professionals involved in such programs undertake important preparatory training before their visit, such as proposed by Michelfelder and Jones in their 'ethics of care' model of working with engineering students (Michelfelder and Jones 2015). They describe student teams who were asked to prepare for a visit to Guatemala by researching beforehand the socioeconomics of the area, as well as the problem context and their own engineering tasks. In most cases, however, even if students are given an introduction to the culture and history of the local people, this does not usually include a critical analysis of domination, colonialization and the effects of past "development" failures. In preparing my students at the University of Western Australia to work with the Noongar people of Western Australia, board members from NIWA (Noongar Institute of Western Australia) presented their experience of the "invasion" by white people and of their experience of being the "stolen generation" (stolen from their families and put into missions). They warn schools not to teach students about their culture through *Indigenous artifacts* but through a serious analysis of the history of colonization. At first this is very hard to swallow for some local students, but ultimately it is these students who experience the largest transformative experience when taking such classes as they begin to take on a responsibility for their own peoples' past actions. The problem of a lack in such preparation is that students inadvertently continue the cycle of domination and exploitation—this time for knowledge and cultural experience. In the name of community service, students gain, but at what cost to communities?

Mitcham and Munoz (2010) address criticisms levelled at humanitarian engineering by those who resist what might be considered contemporary colonialism. They refer to Ivan Illich's talk "To Hell With Good Intentions" which was directed at American students working to "help" communities in Mexico. His critique is aimed at an era of aid programs and what he calls *Mission-vacations* for rich white American students.

Mitcham and Munoz acknowledge this concern but note that

> The most profound theoretical challenge is thus to come to terms with our own interests and intentions—our praise and our criticism—and to consider carefully our understandings of what we are about, insofar as we engage in humanitarian engineering, its associated educational programs, and its criticism.
>
> *(Mitcham and Munoz 2010: 37)*

To do this, however, students need to be much more informed than they are. Being able to question one's assumptions and motivations requires students to learn to look through multiple critical lenses and reflect on their proposed actions. Humanitarian engineering should involve *praxis*—value and theory—informed practice.

In a study that views humanitarian engineering through a social justice lens, Vandersteen and Baillie (2009) conducted interviews with engineers and international professionals in Canada and in Ghana to discover "who pays and who benefits" using a framework from Ursula Franklin (Franklin 1999). Although there were a number of quoted benefits mostly to students, the list of liabilities discussed in the data is cause for concern (Vandersteen and Baillie 2009: 41):

1. Cross cultural social structures are difficult to understand if students have only been exposed to one way of living before their visit.
2. The creation of dependency can be very damaging. Communities become ever more used to Westerners supplying 'aid'—which leads to a lack of self-sufficiency and creativity in response to their own situation.
3. Placements can undermine others' ability to care for themselves or to demand good governance for themselves. Local governments may not ever improve local conditions if they think that foreigners will do this for them.
4. Adjusting cultural practices via external influence may not be in the best interest of the community. The influence of Westerners on lifestyle, motivation and assumptions of what is needed to have a 'good life' can change rapidly when outsiders visit.
5. Technology transfer is extremely difficult and often not appropriate.
6. Outsiders create unequal power relationships.

Despite the concerns, Vandersteen and Baillie conclude that student placements in particular can create opportunities to work for greater social justice and can inspire a lifetime of working with marginalized communities *given appropriate preparation and match between the engineers and the community*. As with ethical engineering, humanitarian engineering can be socially just—however, it is not necessarily the case without such preparation and reflection.

50.5. Corporate Social Responsibility (CSR)

Engineers learn about mining as a process of extraction. As mechanical, chemical, materials, civil and electrical engineers, we might be interested in the processing of these minerals and oils, or we might just use them to make our smart materials, our bridges, our cars and our soaps. We learn about health and safety issues and risk assessment. We also learn a little about ethics and environmental impact. However, few engineering students are exposed to the really difficult connection that mining can have with the people who live in close proximity with the mine site. In this section, we present the views of various stakeholders associated with the mining sector as it relates to social justice and corporate social responsibility, in a series of interviews which our team conducted during 2014–2017, together with Glevys Rondon of LAMMP (Latin America Mine Monitoring Program). (You can also read more about CSR and mining practices within Chapter 49, "Socially Responsible Engineering" by Jennifer Smith and Juan Lucena, this volume.)

Bruce Harvey, formerly Global Practice Leader—Communities and Social Performance at Rio Tinto, opens the conversation for us:

> The irony of course is that we would never dream of developing a mineral resource without getting a comprehensive picture of the resource, drilling it out at a requisite drill spacing, understanding the chemical composition, understanding the material handling characteristics of the ore itself. We would never dream of building a large coal washery without actually doing a lot of very expensive bench testing on the wash characteristics of the coal. Yet regularly we go into places and spend billions of dollars without understanding the social fabric in which we're working in which we now know is the most important fabric, right?
>
> *(BH)*

Mining companies know full well the problems that arise by not engaging properly with local communities. Tony Hodge, former President of the International Council for Mining and Minerals (ICMM 2017), notes:

> There has been a shift in values inside mining companies now . . . what the senior executives of mining companies are really looking for in young engineers for example that has changed dramatically, hugely. You can get design specialists, you can get computer wonks, you can get all of that, what they really need are people that understand all of that and know how to engage effectively and build relationships and that's a tough sell, tough ask . . . what our mining companies need most now are people that know how to build relationships . . . you have to be able to respect people, you have to understand how to interact effectively with people because companies are working in different cultures. . . . You've got to have the skill at it, the sensitivity of it. Have to learn just not to listen, but hear, and those are skills that are not typical in our engineering faculties.
>
> *(AH)*

However, Harvey is quite clear here—we don't need to become social scientists to have this new skill, but we do need to be aware enough of the issues to know when we need expert advice and to understand this advice when it is given:

> Step one, acquire a comprehensively verified validated understanding of the social landscape in which we're developing. Then step two is form partnerships with people who know how to do the work that we don't know how to do . . . There's another whole set of data that's more qualitative, and we're talking the data that anthropologists would collect, you know—how do people live their lives around here? Where does power come from? How are decisions made? How are the checks and balances in civil society around here undertaken? What role do cultural or religious authorities have versus civic authorities? What are the customary norms that prevail? . . . understanding the norms of people's lives in households and in extended families is a very important part of understanding how we will achieve societal stability in order for us to run a multigenerational mine without it becoming a victim of chaos or anarchy or civil discontent.
>
> *(BH)*

Such partnerships are supported through what is often called Corporate Social Responsibility, or CSR.

Current models of CSR distinguish between the state's *duty* to protect and the corporation's *responsibility* to protect (UNGP 2011). CSR frameworks centre around "responsible mining", "sustainability" and "social acceptance" and refer to the corporate commitment to doing something for host-communities such as investing in local education and infrastructure (Broad 2014). Typically, this does not involve change in the mining process itself but, rather, aims to offset or "correct" any injustices that result from exploitation activities within an overall and relatively recent framing of "Mining for Development" (Australian Government 2011; Newmont Mining Corporation 2014). Those promoting a Mining for Development framing, including government agencies that formerly funded aid work, utilize indicators such as GDP and economic growth to describe community benefits and poverty alleviation (Broad 2014; Graulau 2008). However, the nature and distribution of these benefits has been questioned where social conflict and environmental degradation have arisen (Dajer 2015; Zwitter et al. 2014). Heightened awareness of these issues has exposed miners to a new critique by consumers, driving a business case for Corporate Social Responsibility and "social license to operate" (Harvey 2013; Kemp et al. 2012).

The case for mining bringing "ultimate benefits" to the communities is a utilitarian perspective in that focus is given to maximizing utility benefits rather than their distribution (Bebbington 2012). Orihuela and Thorp (2012: 26) describe the constrained distribution of benefits of mining as self-perpetuating:

> the very nature of extractives both generates and sustains interpersonal and interregional inequality, with huge implications for political choices and structures.

Harvey describes clearly how a mining company presents itself to a community in providing ultimate benefits:

> I mean, societies of people are inclined to believe that the value of the mineral resource is the rock in the ground, hence they're going to tax us and rent us, and you know because that's the value. Well, the rock in the ground is a rock on the ground, it has no value. The real value proposition from a mining perspective is the discovery, the evaluation, the extraction, the beneficiation, the marketing. The real value is the intellectual value that goes into that value chain of converting a rock in the ground, an endowment, a latent endowment in the ground into some form of intellectual and financial and human and capital down the track.
>
> *(BH)*

Clearly, then, this investment of time and money expects a return. Interest-based negotiation is used to help communities decide what they want in return for this "valueless rock". The problem comes with the obvious power imbalance in such negotiations. Here's what Harvey has to say on this:

> We're coming in and we're saying we want to develop this major mine or we want to build this great, big minerals processing facility—and it's almost inevitably in a frontier by the way— you know, I mean major urban first world social communities don't want a mine. London is not going to host a new mine nor is Melbourne, nor is Perth for that matter. So, as a matter of, sort of, almost by definition, we're always going to be working at the frontier . . . and we're saying to people we're going to build this plant, this facility, this mine, and it is going to rapidly change and affect your social and physical landscape forever.
>
> *(BH)*

The value proposition which gets explored involves this "change of the physical landscape". It is not in fact the rock that they wish to buy, but compensation for this "change". Harvey goes on to tell us:

> Things are changing anyway, but this is going to exonerate it, and we're getting better at cleaning up. Nevertheless, let's be very clear and very honest that things are going to change around here and it's largely your choice. Now if you choose to support us in that endeavour, then we will help you build an economy. That's the value proposition and if people believe what we're saying and we do it consistently and we do it well, then they will buy into that value proposition.
>
> *(BH)*

The negotiation needed to ensure that communities know what they are letting themselves in for, and to get adequate and fair compensation, is how CSR personnel undertake community engagement. However, Tony Hodge, former Director of the International Council on Mining and Minerals ICMM (ICMM 2017), notes that even with the best will in the world, company CEOs may not in fact be in control of how community engagement is enacted, or of how promises are kept:

> In fact most people don't understand the mining industry. As best we can estimate, there's 21 companies on our council, there are about 6,000 mining companies in the world . . . our 21 companies employ about a million people out of two point five million in the formal sector, that doesn't include the miners who can range in numbers from 30 to a hundred million depending on what's going on in the world and nobody really knows . . . So you've got a little corporate office and you've got operations all over the world and . . . it's probably the greatest challenge to bringing a lot of these ideas into real practice than any, any other issues that arise.
>
> *(AH)*

If operations get so large that even those companies who claim to wish to do the right thing are not in control of their own staff, and in countries where governments do not protect the rights of their own citizens, conflicts will and do arise. CSR personnel are aware of this and work with the idea of a "cost of conflict" framework (Davis and Franks 2014). They support communities in the process of engagement with the clear appreciation that conflict, which might arise if community negotiations are not well done, would indeed cost the company's profit. Harvey tells us:

> I've never come across a situation yet where there is, where people don't want that (value proposition) to happen. Unless they (the community) have already been damaged or unless they've already been exposed to other deleterious developments or things that have been very unpleasant for them.
>
> *(BH)*

Mining companies may not even envisage a situation where communities *veto* the project unless there has been past experience of conflict. In fact, amongst mining and government personnel in many countries, there is a serious effort to suppress the idea that *free prior and informed consent* could mean that communities can actually give their *consent* to a mine moving ahead—i.e. that they could potentially say yes, *or say no* and veto the project as discussed also by Smith and Lucena in this volume. When there are differences of viewpoint amongst the community members, between men and women, between those who would get jobs and those who live next door to the mine and farm, the dissenter's voice is often marginalized and suppressed. In some contexts, those who voice opposition against the mine have frequently been harassed and turned into "enemies of the state". A recent example of this is found in the context of the Mina Conga, launched in 2004 as an expansion of the company's Yanacocha mine, in the Cajamarca region of Peru. Serious community resistance over many years lead to the official suspension of the Conga project and re-categorization as a mineral reserve in 2016. Newmont still hopes to mine this resource, and the battle continues. Prime Minister Oscar Valdes described protesters opposing the Newmont Conga extension in Peru as "anti-investment, anti-development and opposed to national interest" (Poole and Rénique 2012).

Those who oppose the mine are de-legitimized; they are also branded as terrorists. This rhetoric is powerful, denoting malicious connotations for people merely exercising their right to say no to a project that radically impacts their lives. An interviewee in a recent study of ours noted that in Cajamarca, members of the community viewed a protester as someone who is against the wealth of Cajamarca (Baillie et al. 2020).

A human rights framework can be seen as a "minimum floor" of requirements for social justice (Moyn 2015; Neier 2015). In 2016, my team undertook a study of five female human rights defenders who have been impacted by mining (Baillie et al. 2020) (see Figure 50.1).

- **Aura Lolita Chavez**, leader of the Council of K'iche People for the Defense of Life, Mother Nature, Land and Territory of Guatemala.
- **Lina Solano**, HRD from Ecuador, coordinator of the Women's Defenders of Mother Earth, President of Latin-American Union of Women (ULAM).
- **Lorena Cabnal**, self-described community feminist, co-founder of the Association of Indigenous Women of Santa Maria Xalapan.
- **Elizabeth Cunya**, Peruvian HRD and founding member The Association of Women Protectors of the Wetlands and Watersheds (AMUPPA).
- **Bertha Cáceres**, Honduran environmental activist.

The women were interviewed and their stories assessed through the lens of the UN Declaration of the Rights of Indigenous People (UNDRIP, UN 2008). We discovered human rights abuses in many

Figure 50.1 Human rights defenders interviewed by our team

Note: (L–R) Elizabeth Cunya, Lorena Cabnal, Berta Zúñiga Cáceres, Aura Lolita Chavez and Lina Solano Ortiz.

Source: Photo courtesy Eric Feinblatt (2016).

stages of the relevant mining development and for all of the women interviewed. These included the initial scoping stages where the EIA (Environmental Impact Assessment) developers had failed to consult with key stakeholders (UNDRIP Article 32.2), through development (Article 19) and into production where influxes of security forces had resulted in violence, abuse and even murder of activists (Articles 5 and 7) (Baillie et al. 2020). These cases are all instances where the women are saying "no" to a mine site, as they believe it will harm their land or water system, but they are being turned into "enemies of the state" and made to suffer for the care they have for home and family. This is a clear case of social injustice which CSR programs can do nothing about.

We might conclude that a social justice perspective to mining engineering would support the rights of *minorities* who claim their land and livelihoods are being destroyed, whilst frameworks such as CSR which are driven by the corporate agenda would try to ensure that the mining happens by ensuring that there is a good "interest based" negotiation process, leading to a "win-win" for the *majority*. In the former case, the right thing to do would be to stop the mine from going ahead if it would result in social or environmental injustice (as defined by the affected people) in the short or long term, even to a small group of people. In the latter case, the right thing to do would be to ensure that the negotiation goes ahead in such a way as to avoid "the cost of conflict", that the majority are satisfied by the compensation that they get, and others mollified enough to not protest.

In a case where CSR practice works to support community engagement practices which are equitable, with fair representation and strongly democratic mechanisms of negotiation, when a mine is wanted in a local area for the benefits it brings, and with full understanding of the costs of such a mine on the environment, then CSR might be considered socially just. However, we might also argue that the disparity of power in such negotiations is too high for them to ever be considered completely fair. Few wealthy white individuals would want a mine in their backyard, and they can *afford* to say no, which is rarely the case for mining communities in the Global South. Added to this is the problem that CSR practices can (and often intend to) sugar coat the real negatives of a mine site being developed. As with our previous arguments relating to humanitarian engineering and engineering ethics, CSR is not *necessarily* socially just and is often quite the opposite.

50.6. Moving Towards a Social Justice Framing for Engineering

Whether engineering ethics, humanitarian engineering or CSR, or any of the numerous other approaches we might select to frame our work, those which do not draw on critical theories to question underlying assumptions will not assist in transforming practice away from the dominant paradigm. Social justice requires such transformation, a profound shifting in the distribution of wealth and pathways to healthy and fulfilling lives. Those seeking to enhance engineering which promotes social and environmental justice must therefore question any chosen approach and scrutinize how it might inadvertently result in supporting the very thing it claims to want to change. Gruenewald considers that "decolonizing" our minds might be a necessary precondition for "reinhabitation" of new ideas where decolonization

> can be seen as a metaphor for the process of recognizing and dislodging dominant ideas, assumptions and ideologies and externally imposed.
>
> *(Smith and Katz 1993: 9 in Gruenewald 2003)*

Many years ago, Fleck (1979) argued that stable thought collectives form around organized social groups (such as professional engineers), and that thought styles of such collectives get fixed and formalized in structure if the group is large and lasts long enough. The longer the thought style is around, the more certain it appears. Engineering may be considered a particular community of practice, with an associated common sense and thought collective. If engineers blindly accept, and do not question the thought style that they work within, they will be part of a thought collective that they were not even aware of. All too often engineers are not in a position to do this critical questioning, as they did not learn the skills in school. Sustainability, ethics, CSR and humanitarian engineering will have nothing to do with social justice if there is an unquestioning acceptance of a dominant paradigm which is focused on an inherently unjust system. If we truly want to see justice in our engineering practice, we must get outside our thought collectives and question our assumptions.

50.7. Summary Thoughts

In this chapter, we have examined engineering in relation to social justice. We have explored this through the concept of *variation theory*—whereby understanding comes through varying around critically important aspects of a concept. Engineering and social justice has been compared with three framings typically used by engineers to describe their work when it relates to "engineering and society": engineering ethics, humanitarian engineering and corporate social responsibility. By exploring these framings as used by engineers, we note a key difference to the approaches taken by a social justice perspective—they do not necessarily question assumptions or common sense about the way the world around them works—a requirement for justice work. Ultimately we argue that seeing engineering from a social justice perspective requires us to break out of what Fleck (1979) calls "thought styles" and begin to question what we see around us and how to do it differently.

Acknowledgements

I would like to thank the multiple contributors to this chapter whose ideas, framings and thoughts are profoundly drawn upon, through the work we have done together. These include Kylie Macpherson, Jordan Aitken, Chloe Hewitt, Glevys Rondon, Rita Armstrong and Eric Feinblatt, as well as all those interviewed for our studies.

Related Chapter

Chapter 49: Socially Responsible Engineering (Jennifer Smith and Juan Lucena)

References

Australian Government. Department of Industry, Innovation and Science (2011). *Leading Practice Sustainable Development Program for the Mining Industry*. https://industry.gov.au/resource/Documents/LPSDP/guideLPSD.pdf

Baillie, C., Feinblatt, E., Mejia, J. and Rondon, G. (2020). *Mining and Social Justice: Rhetoric or Reality?* Engineers, Technology and Society Series. Ed. C. Baillie. Williston, VT: Morgan and Claypool.

Baillie, C. and Levine, M. (2013). Engineering Ethics From a Justice Perspective: A Critical Repositioning of What It Means to Be an Engineer. *International Journal of Engineering, Social Justice, and Peace*, 2(1), 10–20.

Baillie, C. and Male, S. (2019). Assisting Engineering Students Along a Liminal Pathway and Assessing Their Progress. *Australian Journal of Engineering Education*, 24(1), 25–34.

Bebbington, A. (ed.) (2012). *Social Conflict, Economic Development and Extractive Industry*. New York: Routledge.

Broad, R. (2014). Responsible Mining: Moving from a Buzzword to Real Responsibility. *The Extractive Industries and Society*, 1, 4–6.

Dajer, T. (2015). High in the Andes, a Mine Eats a 400-Year-Old City. *National Geographic*. http://news.nationalgeographic.com/2015/12/151202-Cerro-de-Pasco-Peru-Volcan-mine-eats-city-environment/. Accessed October 9, 2017.

Davis, R. and Franks, D. (2014). *Costs of Company-Community Conflicts in the Extractive Sector*. Business and Human Rights Resource Centre. https://www.business-humanrights.org/en/latest-news/pdf-costs-of-company-community-conflict-in-the-extractive-sector/

Engineers Australia (2018). *Code of Ethics*. www.engineersaustralia.org.au/ethics. Accessed March 1.

ESJP (2018). *ESJP Commitments*. http://esjp.org/about-esjp/our-commitments. Accessed March 1, 2018.

Fleck, L. (1979). *Genesis and Development of a Scientific Fact*. Chicago: University of Chicago Press.

Franklin, U. (1999). *The Real World of Technology*. Toronto: House of Anansi Press.

Goreman, M., Mehalik, M. and Werhane, P.H. (2000). *Ethical and Environmental Challenges to Engineering*. Upper Saddle River, NJ: Prentice Hall.

Graulau, J. (2008). Is Mining Good for Development? *Progress in Development Studies*, 8(2), 129–162.

Gruenewald, D. (2003). The Best of Both Worlds: A Critical Pedagogy of Place. *Educational Researcher,* 32(4), 3–12.

Harvey, B. (2013). Social Development Will Not Deliver Social License to Operate for the Extractive Sector. *The Extractive Industries and Society*, 1(1), 7–11. doi:10.1016/j.exis.2013.11.001

ICMM (2017). *International Council of Mining & Minerals*. www.icmm.com/. Accessed September 10, 2017.

Kemp, D., Owen, J.R. and Van de Graaff, S. (2012). Corporate Social Responsibility, Mining and the Audit Culture. *Journal of Cleaner Production*, 24, 1–10.

Marton, F. and Booth, S. (1997). *Learning and Awareness*. Mahwah, NJ: Lawrence Erlbaum Associates.

Marullo, S. and Edwards, R. (2000). From Charity to Justice: The Potential of University-Community Collaboration for Social Change. *American Behavioral Scientist,* 43(5), 895–912.

Michelfelder, D. and Jones, S. (2015). *From Caring About Sustainability to Developing Careful Engineers*. 7th International Conference on Engineering Education for Sustainable Development, Vancouver, Canada, June.

Mitcham, C. and Munoz, D. (2010). *Humanitarian Engineering*. Engineers, Technology and Society Series. Ed. C. Baillie. Williston, VT: Morgan and Claypool.

Moyn, S. (2015). Human Rights and the Age of Inequality. In D. Lettinga and L. van Troost (eds.), *Can Human Rights Bring Justice?* Amsterdam: Amnesty International Netherlands, pp. 13–18.

Neier, A. (2015). Human Rights and Social Justice: Separate Causes. In D. Lettinga and L. van Troost (eds.), *Can Human Rights Bring Justice?* Amsterdam: Amnesty International Netherlands, pp. 47–52.

Newmont Mining Corporation (2014). *Beyond the Mine—Our 2014 Social and Environmental Performance*. http:// sustainabilityreport.newmont.com/2014/_docs/newmont-beyond-the-mine-sustainability-report-2014. pdf. Accessed September 10, 2017.

Orihuela, J.C. and Thorp, R. (2012). The Political Economy of Managing Extractives in Bolivia, Ecuador and Peru. In A. Bebbington (ed.), *Social Conflict, Economic Development and Extractive Industry*. New York: Routledge, pp. 27–45.

Poole, D. and Rénique, G. (2012). Peru: Humala Takes Off His Gloves. *NACLA Report on the Americas,* 45(1), 4–5.

Riley, D. (2008). *Engineering and Social Justice*. Engineers, Technology and Society Series. Ed. C. Baillie. Williston, VT: Morgan and Claypool.

Smith, N. and Katz, C. (1993). Grounding Metaphor: Toward a Spatialized Politics. In M. Keith and S. Pile (eds.), *Place and the Politics of Identity*. London: Routledge, pp. 67–83.

United Nations (2008). *United Nations Declaration on the Rights of Indigenous Peoples*. www.un.org/esa/socdev/ unpfii/documents/DRIPS_en.pdf. Accessed October 9, 2017.

United Nations. Office of the High Commissioner (2011). *United Nations Guiding Principles on Business and Human Rights: Implementing the United Nations "Protect, Respect and Remedy" Framework*. www.ohchr.org/ Documents/Publications/GuidingPrinciplesBusinessHR_EN.pdf. Accessed March 1, 2018.

Vandersteen, J.D.J., Baillie, C. and Hall, K.R. (2009). International Humanitarian Engineering Placements: Who Benefits and Who Pays? *IEEE Technology and Society*. Special issue on "Volunteerism and Humanitarian Engineering" Part 1, 28(4), Winter.

Zwitter, A.J., Prins, A.E. and Pannwitz, H. (2014). *State of Emergency Mapping Database*. University of Groningen Faulty of Law Research Paper Series.

51

ENGINEERING AND ENVIRONMENTAL JUSTICE

Benjamin R. Cohen

51.1. Introduction

Engineering and environmental justice (EJ) fit together as a topic of community engagement. EJ addresses a number of situations. Among them are the ways environmental harms and benefits are distributed inequitably across different communities and the ways different populations have the opportunity to contribute to choices about the effects of environmentally intensive systems (energy, waste, water, etc.). No matter the diversity and complexity of EJ scenarios, they all have something to do with the ways communities seek to live in healthy ecosystems. Engineering is certainly too complex to summarize capably (see also Chapter 1, "What Is Engineering?" by Carl Mitcham, this volume). This chapter treats it as a form of trained professional expertise, a set of activities, and a way of configuring relationships through technological systems. No matter the intense history of engineering and the various ways modern engineered systems influence community choices, engineering always has some relationship with community settings.

EJ has not received significant attention within traditional engineering. Over the past half century, it has not been a prominent topic in engineering education, nor has it been a guiding factor in engineering practice. With the rise of modern environmental regulation after the 1960s, engineers increasingly responded to the culture of environmental impact assessments, but attention to the ways the results of engineered works affect different communities in different ways has yet to become a common feature of technological development. Since the turn of the century, an increasing number of people have begun to pay attention, some within science and technology studies (STS), engineering education reform, and engineering ethics; some within programs seeking to integrate environmental studies and engineering; and some within newer programs in sustainable design that strive to escape strictly technical metrics of sustainability to consider justice and equity as well (see e.g., Chapter 25, "Sustainable Design" by Steven A. Moore, this volume).

At the outset, there are two broad ways to position the identity of the engineer in the field of EJ. One is to recognize the ways engineers are members of communities and advocates for certain values that can foster a more just relationship between human and non-human nature. In this framing, engineers as people—and engineering as a professional activity—are shaped by social, environmental, and economic conditions that precede them (see Chapter 44, "Responsibilities to the Public—Professional Engineering Societies" by Joseph Herkert and Jason Borenstein, this volume). A second way to position the relationship of the engineer to EJ is to consider engineers as external agents, standing apart as experts and contributing their technical expertise to address an unjust situation. The ethics of relevance in the first circumstance are about political conditions and social questions, about how the engineer's options, choices, and actions are produced by a context larger than the individual (see Chapter 45, "Engineering as a Political Practice" by Govert Valkenburg, this volume). The ethics of relevance in the second circumstance are more about an individual's moral standing *as* an individual. What ought I to do as an engineer when faced with a situation of environmental injustice?

This chapter considers the various ways scholars and practitioners have sought to bring engineering and EJ together, with deference to the dynamic changes occurring in the fields of EJ and engineering practice and reform. The two ways of seeing the relationship amongst and between engineers and EJ just noted—engineers as members of communities experiencing environmental injustice and engineers as outside experts called on for their input—play out differently at different points in a technological life cycle, one that begins with design and development, goes through implementation and use, and continues with maintenance and disposal. Those different points in a technological life cycle also relate to different ways of understanding injustices, a point explored in this chapter.

The next part of the chapter (Section 2) reviews the terms and theoretical shape of environmental justice as a field of scholarship, a social movement, and a framework for redressing inequitable relationships between people and their environments. The section is meant to illustrate the ways environmental injustices are likewise produced at different points along the technological life cycle. Some are the consequence of design and development; some result from implementation use, maintenance and disposal. Section 3 of this chapter seeks to understand engineers as people and engineering as a professional activity in a dynamic, process-based manner. It articulates various ways one can understand the relationships between engineering and EJ before turning in the final section to possible future directions for holding the two together.

51.2. Environmental Justice

Environmental racism, environmental justice, and environmental injustice are complicated terms. The common point of interest among the three related terms concerns the ways human activity about the non-human world affects people in different ways. Although the terms have different historical valances, this chapter uses environment justice (EJ) to refer to the concerns raised with each label. At the start, EJ is more an argument about the environment that requires attention to how people live in the world than an environmental argument about nature or the natural. It places the questions of environmental ethics as a matter of political human interaction and manipulation rather than ideal categories about what counts as "nature." In a subsequent section, this chapter will return to the matter of interaction and manipulation with more direct discussion of engineering and technology. First, though, is an overview of the basic terms and commitments of EJ.

EJ is a field of study, a social movement, a set of methods and theories, and an ethic of engagement between humans and their non-human surroundings. It emerged with that array of identities in the later twentieth century (Cole and Foster 2000; Pellow and Brulle 2005; Holifield Smith et al. 2018). To be sure, the issues the field calls out have extensive and rich histories well preceding this more recent vintage. The damages of dispossession, colonialism, and hazardous working conditions have long been features of human activity in the non-human landscape. And they have long led to the inequitable sharing of environmental risks and burdens. The creation of national parks that depended on the removal of native populations; child labor in resource extractive vocations (textiles and other mill work, for example) in the early industrial years; and access to healthy water resources and waste disposal systems in quickly urbanizing cities, all provide historical examples of what later would be called cases of environmental injustice along with the social injustice. These are not limited to late-twentieth-century phenomena (Boone and Buckley 2018). The actual study of such phenomena, their theoretical framing, and the political organization developed to address them, though, are relatively recent.

EJ is and has since its beginning been a political project that challenges the kind of mainstream environmentalism that emerged in the 1960s and 1970s. Sze and London (2008) write that it "developed in an explicit reaction to the lack of adequate attention to race and class issues by the mainstream environmental movement" (1334). As a movement, EJ took hold in the United States more fully than elsewhere, a tendency that, as discussed in subsequent sections, recent work has been striving to overcome. In the 1970s, for instance, cases like Love Canal near Buffalo, New York, brought such issues to light. There, the remnants of an abandoned toxic waste dump polluted a school and neighborhood, leading to greater anti-toxics attention within the environmental movement (Blum 2008). Later that decade and into the 1980s, cases like that of a protest against toxic waste dumps in North Carolina, and housing inequities in Houston where city officials disproportionately placed landfills in predominantly Black neighborhoods, formed core examples for what would be called environmental racism. Work by Robert Bullard (1990, 2005; Bullard and Wright 1989), at the head of the efforts, made clear the striking racial component to the distribution of environmental goods and bads. Other cases informed a roster of issues about the ways the health of a community was intertwined with the health of an ecological system in ways that benefitted some and harmed others. Reactions to waste siting in predominantly Spanish-speaking areas of California or an incinerator in lower-income Chester, Pennsylvania, for example, brought the issues to greater light. To quote scholars and activists Luke Cole and Sheila Foster (2000), "Environmental hazards are inequitably distributed in the United States, with poor people and people of color bearing a greater share of pollution than richer people and white people" (10).[1]

With such cases adding to a growing catalog of scenarios, EJ as a field of study and a social movement first took shape by the 1980s through a framework of environmental racism. Robert Bullard's leadership in this regard was important as attention to environmental racism took initial form with his work. Further leadership by Rev. Benjamin Chavis in North Carolina and coalescence around basic principles in the First National People of Color Environmental Justice Summit in 1991 added to the clarity that an EJ framing offered something more politically attuned to different populations than mainstream environmentalism (United Church of Christ 1987, 1991). In 1994, the Clinton Administration gave EJ a federal platform for the first time, signing Executive Order 12898, on Environmental Justice "Federal Actions to Address Environmental Justice in Minority Populations and Low-Income Populations."[2]

By the twenty-first century, EJ offered a relatively common language and set of references for those working in communities. It also offered a viable field of study for scholars seeking to engage with those communities, either as members of or researchers examining and helping them. Still, the proliferation of efforts and methods attached to the EJ label have warranted analysis for the ways different ethical commitments and senses of "justice" play out under the now-broad label.

David Schlosberg's (2013) review of the theoretical contours within EJ is most helpful for the sake of articulating the common ground of engineering and EJ later in this chapter (also cf. Schlosberg 2007). He summarizes four forms: distributive, procedural, recognition, and capabilities. This quartet of approaches to issues of injustice, while being about diverse and pluralist communities, is itself a commitment to pluralist analysis.

Distributive justice deals with the ways environmental benefits and harms are distributed to different people. This follows from general distributive ethical concerns, that "The economic, political, and social frameworks that each society has—its laws, institutions, policies, etc.—result in different distributions of benefits and burdens across members of the society" (Lamont and Favor 2017). The environmental injustice in question has often been about equity, whether, for instance, people have equal access to clean water, air, soil, energy, and food. "The concept was used to illustrate that some communities received more environmental risks than others," writes Schlosberg (2013: 38). "Those environmental bads were simply another example of social injustice." Such distribution questions were prominent as the field developed in the 1980s through an environmental racism theme.

A second form has been procedural. Rather than the ways some risks flow to certain lower income or communities of color, this is about who gets to contribute to the procedures of rulemaking. Imagine there is a public hearing for a proposed toxic waste dump in a small, predominantly Spanish-speaking community. Anyone could theoretically attend. It is technically open to the public. In practice, though, the presentations are in English, the supporting documentation is a doorstop of a binder, also in English, and the time for public discussion is limited to a few questions. Even if there were more time, the attendees are all working early in the morning and have come to the forum after a long day already. Their opportunity to meaningfully participate is limited. These are the bare bones of procedural injustice.[3]

Thinking in terms of temporal patterns helps differentiate the two kinds of injustice. Matters of procedural justice are often (although not always) *a priori*. They generally occur before the toxic waste siting, before the new power lines go through a neighborhood, before diesel trucks idle for hours on roads in front of a house or nearby parking lots. Who had a say in making the decision to authorize these activities? The consequences of those decisions will affect some people more than others (*after the fact*). But the decision process before the fact was itself unjust. It limited contributions from some people over others in disproportionate ways. Matters of distributive justice are perhaps more about the ways environmental risks flow to different peoples once those environmental activities are underway. This sense of justice is at play after the fact. Thus, the two most prominent kinds of justice in EJ, distributive and procedural, fit together at different points in a timeline rather than compete for attention. They are complementary.

Two other aspects have helped the EJ movement grow into a fuller, more theoretically grounded, and socially attentive field. These are expansions beyond procedures and equities in distribution to what Schlosberg captures as recognition and capabilities. Recognition justice begins with the premise of giving people a seat at the table while involving more explicit and intentional attention to affected users in other contexts. Rather than just stakeholders, who should be gathered around the table to contribute to decision-making, a demand for recognition as a form of justice asks as well for attention to a few other aspects (cf. Iser 2013). One is the range of peoples not often included but affected by the decisions (communities of color, indigenous peoples, etc.). Another is the peoples beyond an immediate area still affected by ecological changes. In a word, those upstream. Decisions in Northeast Pennsylvania about natural gas drilling, to take a case close to my home in the Delaware River Watershed, have ramifications downstream. Thus, although I live outside the Marcellus Shale where rampant fracking has occurred, members of my community are connected to the decisions of others because of the broader ecology of the watershed.[4] These situations present legal difficulties, to be sure, which helps explain why environmental law attends to the overlap of political jurisdictions by recognizing ecological boundaries (jurisdictions) instead. In the case of EJ, recognizing the status of affected users and considering the voices of those who cannot speak, as it were, would follow from recognition justice. These are matters of identifying and including the previously voiceless, not only in general but also for the purpose of noting that access to environmental decision-making and healthy air, land, and water are matters of individual and community well-being.

The capabilities approach to justice draws from foundational work by Amartya Sen and Martha Nussbaum that focuses on what people are capable of being and doing in their given communities. Since ecological integrity is crucial for human well-being, then, as Rosie Day (2018) writes, "all individuals have a right . . . to live in an environment that provides the necessary resources and services to enable their other essential capabilities. Any contravention of this would constitute an injustice" (128). A capabilities approach to EJ is thus in essence about how well an ecosystem can maintain its capacity to function in a healthy manner so that all of its member can thrive. Undermining that capacity has differential effects on various communities. In the capabilities approach, to quote Schlosberg (2013), "the central issue continues to be the interruption of the capabilities and functioning of living systems—what keeps those living systems from transforming primary goods into the functioning, integrity, and flourishing of those that depend on them" (44). Further:

> When we interrupt, corrupt, or defile the potential functioning of ecological support systems, we do an injustice not only to human beings, but also to all of those non-humans that depend on the integrity of the system for their own functioning. It is the disruption and increasing vulnerability of the integrity of ecosystems that is at the heart of the injustice of climate change, for example, both in terms of its impact on vulnerable human communities and non-human nature. The treatment—or abuse—of human and non-human individuals and systems is based on the same loss of the ability to function.
>
> *(Schlosberg 2013: 44)*

Thus has EJ evolved in complexity and nuance in recent decades, with at least four overlapping ways to understand the "justice" part of environmental justice: as related to distribution, procedures, recognition, and capability.

If that was not enough, Sze and London (2008) articulate an additional sense of change and evolution in EJ by noting spatial expansion. Spatial expansion can refer to two things. One is the greater attention to global contexts, especially as scholars move beyond North American cases. This can mean attending to cases in areas outside the United States. It can also mean attending to injustices that are not limited to regions or nation-states, such as climate change and the attendant concerns for climate justice (Shepard and Corbin-Mark 2009; Roberts 2009). That second sense is a spatial expansion in its scalar form, that the injustice—be it distributive, procedural, or other—is at a scale of vastly greater proportions than a neighborhood or town.

All of this together—the different forms of justice, the different scales, the proliferation of cases—suggests that in the twenty-first century, the field has evolved considerably in theoretical scope, case study sensitivity, and sheer number of scholars and advocates investing time and energy to the issues. In fact it now seems clearer, historically, that if there was a first wave of environmentalism attached to a significant degree to national parks and wildlife preservation and fidelity to some lingering concept of wilderness—think John Muir and Gifford Pinchot, dams and Yosemite—and there was a second wave arising from the counter-cultural developments of the 1960s and 1970s that worked toward policy, governmental, organizational, and economic means for protecting, using, or impacting the environment—think EPA and Clean Water Acts, Endangered Species and Rachel Carson—then environmental justice is a third wave. The three waves accrue, rather than replace each other. New eras demand greater attention to previously overlooked considerations (Sandler and Pezzullo, eds. 2007).

At the end of the second decade of the century, EJ has developed enough that its scholars and practitioners argue over the boundaries of subfields. One finds in the journal *Environmental Justice*, for example, cases for energy justice (Hernández 2015), water justice (Butler et al. 2016), food justice (Gottlieb 2009), and climate justice (Shepard and Corbin-Mark 2009).[5] In the main, the boundary arguments consist of staking claims on which element should be at the center of EJ theory and practice—climate, food, energy, etc.—or, in other words, whether something like climate justice, for instance, is really a subfield, or the main one within which others sit. Related subfields like disaster studies similarly draw from core EJ theory to understand the differential responses to hurricanes, heat waves, and other "natural" disasters (Fortun et al. 2017). Regardless of the potential for articulating more refined focal areas within a broader EJ heading, the existence of such arguments demonstrates the value of EJ methods and their contributions to modern environmental discourse.

51.3. Engineering and Technology

In all of the cases, and no matter the kind of justice under discussion, two features remain consistent. One is that technological modernity has something to do with the issues captured under the label EJ. The other is that engineers play some role in the perpetuation of environmental injustice or the lack of attention to issues of EJ in technology design. The top of this chapter noted that engineering and EJ fit together as a topic of community engagement. That is because community health lies at the center of questions along each axis of EJ theory—procedures, distribution, recognition, and capabilities. Each element likewise requires attention to the increasing scale and spatial extension of the issues, both to different parts of the world and within global circulations. The technological products of engineering are involved in each area.

In turn, as EJ scholarship and attention to community health through an EJ lens have grown, so too has attention to the ways engineers, technology, and EJ come together. The early years of EJ were more prone to focus on how technological systems were the cause of injustices. That kind of appraisal was part of a broader set of social critiques about technological society that emerged in the decades after World War II. They were also bound up in larger social critiques about corporate capitalist and industrial metrics of progress. As Ottinger and Cohen (2011) put it:

> The increase in production-consumption cycles throughout the twentieth century has been the source for the environmental hazards that communities, including those instrumental in the rise of the EJ movement, have organized to confront. The polychlorinated benzene (PCB)-contaminated soil that African-American residents of Warren County, North Carolina, fought to keep out of their community in 1982, for example, originated as a by-product of industrial processes to make components for electrical systems (such as insulators and heating fluids) and plasticizers for a generation of new polymeric materials. The leaking drums of chemical wastes that a white community in Love Canal, New York, fought to have cleaned up in the late 1970s had their genesis thirty years prior with the closing of a chemical production facility. The construction and expansion of petrochemical facilities in the latter half of the century had ramifications for communities, as in the case of Norco, Louisiana, where a company's decision to build a new facility in the 1960s sandwiched a historic African-American community between a chemical plant and an oil refinery.
>
> (4)

Engineers were implicated in the critiques. As with the broader cultural criticism around industrial society rising in the postwar era, many cultural observers saw engineering as a culprit in the rise of unjust polluting systems (Wisnioski 2012).

Scholarship, advocacy, and community engagement in recent years have expanded our understanding of engineering-technology-EJ relationships. In keeping with a "risk society" thesis that argues we live in an era where science and engineering are both the font of environmental problems and the means to redress those problems, the expansion of work on the engineering-technology-EJ relationship has moved beyond causal arrows pointing in only one engineering-to-injustice direction (Beck 1986; Giddens 1999). Engineers can and have also worked to redress injustices, often as a matter of more equitable distribution of environmental goods and bads. They have also worked to design technologies with principles of justice in mind, often through a design approach that includes community members in the design process.

A few prosaic background features about engineering set the stage from which to broach their combinations with EJ. One is that engineering practice is and has always been evolving. It is historically dynamic. Over the past half century, let alone the last few centuries, what it means to be an engineer, who gets to be one, how one is trained and vetted as having status as an engineer, where one works, and what one works on have all evolved. A second is that engineering practice is an active process in its current state, in the present. It is less of a thing and more of a practice. Thus, even when we recognize the historically contingent identity of engineering, we also need to recognize its dynamism as practiced in the moment. A third background feature is that the process of engineering is, to be sure, a human activity, a point that is mundane but meant to front the value-laden status of all engineering practice and, thus, of all engineered products.[6] Although it would not be suitable to claim that engineering is solely responsible for the final form and use of a given technology—as users, regulators, technicians, and others have a role too—in this case it helps to note that engineering, like the technologies engineers craft and maintain, is a function of value-laden decisions (Oudshoorn and Pinch, ed. 2003). Put plainly, engineers are human; both engineering and technology are human activities. And, like science, engineering "is not only made, it is constantly in the process of being re-made in response to shifts in cultural terrain" (Ottinger and Cohen 2011: 10). EJ is one of those shifts.

A series of studies help elucidate the various ways engineering and EJ can fit together in a productive manner. Work by Kim Fortun (2001), Barbara Allen (2003), David Pellow (2004), and Julie Sze (2007) helped build a foundation for analysis about technology and EJ. Studies of disaster in Bhopal, Cancer Alley in Louisiana, urban waste in Chicago, and the politics of pollution in New York laid bare the differential effects of engineered environments. They addressed the relationships of expert knowledge and industrial consequences for communities, touching on the outcomes of engineering though not analyzing the engineering profession itself. Vallero and Vesilind (2007) were some of the first authors to provide attention to the ways EJ fit the profile of a socially responsible engineer. Their work provided a lens through which engineering students could recognize the virtues of "good" engineering as those that promote environmental health.

Other work has focused not on engineering as a profession, but on technology as a feature of community infrastructure. Environmentally just technologies, as an example in this vein, are those that strive to support an EJ agenda in their use and outcomes. Instrumentation to measure air quality and tools for assessing water health provide direct examples of technologies designed to support communities in their quests for environmental justice (Ottinger 2011; Kinchy et al. 2014). They are often lower cost, more adaptable and user-friendly, and tailored to specific contexts rather than mass-produced without recognition of place-based dynamics. Research programs in citizen science, participatory design, popular epidemiology, and community mapping seek to achieve such goals. These types of programs have seen increased attention over the last decade in ways that bring engineering and EJ closer together (Wilson et al. 2018; Haklay and Francis 2018).

Many of these cases are about how engineers and scientists influence EJ. Adding to this work, contributors to Ottinger and Cohen (2011) showed not how science and engineering shaped matters of justice or injustice, but how the EJ movement itself was influencing professional practice in science and engineering (also cf. Ottinger and Cohen 2012). Further work has proceeded in that vein, explaining how engineers as social agents inform their designs and professional activities through a set of values aimed at addressing questions of justice.

These examples among others are in large part about reaching from the side of EJ to that of engineering. Further work in engineering education reform speaks to the connection in the other direction, from the side of engineering to EJ. Recognizing that direction highlights the point that bringing engineering and EJ together requires engineering education reform, not just changes in professional practice once the engineer is in the field. Such potential educational changes rest on the view developed in citizen science, participatory design, popular epidemiology, and community mapping, namely that the values and concerns of non-engineers need to be part of, not post-hoc additions to, the creation of research agendas, technological design, and maintenance plans.

Community-centric design in engineering aims at that outcome in ways that may not have yet used EJ frameworks but could. For example, Lucena et al. (2009) articulate community-centric framings with attention to global development projects, showing how engineers should include community principles into their research agendas in ways that make the sustainability of the community the main goal. A lentil de-husker in Senegal, a grain crusher in Mali, a windmill in central India, and an irrigation project in Honduras all require attention to the assets and contours of prevailing community practice before the technology design can proceed. Maintaining the health of community practices needs to be a priority for engineering practice because doing so allows for the possibility of EJ principles. Community-centric design also allows for the possibilities of greater procedural and recognition justice.

Work in social justice and engineering similarly provides room for the questions of EJ (Schneider 2010). Donna Riley (2005) and Caroline Baillie (Chapter 50, "Engineering and Social Justice," this volume) have written more broadly about social justice and engineering, providing the conceptual space within which elements of an *environmental* justice-based agenda can grow. By considering questions of engineering ethics beyond the individual scope of justice, that is, a social justice framing helps show how to include environmental questions like healthy air, water, energy, and food, too. Nieusma and Riley (2010) bring together the global development analysis of Lucena et al. (2009) with social justice principles in ways that also encourage greater attention to EJ. As an argument about engineering to support EJ, they show how engineering for community requires recognition of political patterns—in the sense of community governance and social relations—and not just technologies that work or do not pollute. The considerations are farther upstream in the design process, in other words. Thus a second point to insist upon when bringing engineering and EJ together is that the ethical questions are not downstream, once the technology is in place. It is as much about process as product.

Global development examples are not necessarily or only about justice, but they could and perhaps should be. Just as well, the lessons of community engagement apply for any example of engineering and EJ, as they all rely upon engineers attending to community values first, not after the fact. In the language of EJ theory, this means advocating for procedural justice in the engineering design process. "Local knowledge and experiential knowledge often serve as the basis for community-initiated investigations of pollution and health issues," write Ottinger et al. (2017). Those community members "frequently ask different questions, looking for ways to represent ongoing, systemic hazards rather than assessing regulatory compliance or identifying the root cause of a particular [problem]" (1034). What is more, such a view can promote procedural justice in ways that could alleviate distributive injustices. Thus, the programs noted earlier for citizen science, etc.—those that begin from the community side and reach out to the engineering profession—can also frame the ways engineers structure their work to reach from their expert-based training out to the community.

The community-centric approach leads to a payout for later questions of distributive justice in a more specific way. It means the answer is not simply technical, nor does the answer lie simply with more "green" technology. More green technology, by whatever definition one uses to define it, follows from rather than stands in for issues of EJ. In other words, instrumental answers are insufficient and likely ineffective.

In summary, then, the four-fold framework of distribution, procedures, recognition, and capabilities offers a way to see where and how engineers have or could contribute to mitigating or avoiding environmental injustices. The spatial expansion into global scales and multinational contexts likewise offers a basis from which engineers can work.

To wit, many of the studies of toxins and pollution that provided early cases for the EJ movement show the problems of distribution. They are related to the ways technological systems—for energy production, manufacturing, and waste management—lead to the inequitable distribution of the hazardous effects of the industries. This can happen because engineers did not consider questions of equitable distribution or, just as likely, because the pressures of economic considerations rendered such questions less relevant for decision-makers.

For engineering, the question of procedural justice pushes designers and developers to include more voices in the design process. Many of the examples that would fit this category have to do with siting—the placement of waste facilities, the location of power lines and drilling derricks, the path for trucks, railways, and other forms of conveyance. Typically decisions about where to site such things have been made by business or government entities who do not live in the affected areas and thus do not recognize the need to include local and regional voices. Instead, many forms of community involvement are about listening sessions over concerns, not for the sake of designing the system, but to manage the worries and adjust parts of the system piecemeal. Newer approaches for just science and engineering seek to overcome those limitations in design practice.

Engineers and recognition justice relate to the approach of procedural justice just noted. If matters of procedural justice temporally precede distributive justice—attending to the process *before* the toxins and pollution are distributed—then recognition is even further upstream temporally. It asks engineers to do more work with communities to identify affected users even before convening public forums or building research programs with various non-engineers who live in the relevant communities. Again, community-based research and participatory methods fit here. They operate under the assumption that recognizing community members and affected users is a necessary and helpful condition for designing and implementing new systems.

The capabilities approach is the least explicated of the four and thus requires more research. If the premise of this approach is to attend to what it means for people to assert their standing in a community, seek well-being, and foster self-determination, then fostering capabilities EJ would mean engineers might prioritize technologies that help one another care for each other, attend to health and everyday life, and replace innovation as the paradigm of engineering quality with well-being.

Taken together, the payout for greater integration of EJ principles with engineering education and practice is to back up and attend to upstream issues. This happens in one way in how we educate engineers and redefine an engineering fundamental; it happens in another way in the cultural and political conditions that make possible certain forms of research and design; and it happens in a third way in the historical sense of understanding how these systems got here, where the flex points are, and how the momentum of the systems precludes some options but allows for others. A few opportunities canvassed in the final section show how the future connection of engineering and EJ might come about.

51.4. Future Directions

The entanglements of engineering and EJ require further research, analysis, and argument. Research and practice over the last decade has brought the two arenas closer together, but work on putting sustainable and just technologies more to the center of engineering identity remains limited. Four areas offer helpful opportunities for future work.

The first and perhaps most immediate area comes from engineering education reform. It is telling, for instance, that the *Journal of Engineering Education* does not host conversations about the topic. In its 107 volumes, only one article mentions EJ (Swearengen et al. 2002). *Science and Engineering Ethics* is similarly light on attention to EJ, with scattered articles referring to EJ as part of a discussion about the ethical dynamics of a given technology (cf. Duvall et al. 2002; Epting 2016). There is space for such work, as prior studies have increasingly brought attention in engineering education to issues of diversity, gender, and social justice.

For the undergraduate engineering curriculum, one specific action is to consider that environmental justice provides an important lens through which to bridge environmental ethics and engineering ethics. Given the degree to which engineering work is implicated and embedded in ecosystem health, lacing together engineering and environmental ethics is not only helpful but necessary. In such a case, the ethics of EJ take shape as the integration of more traditional engineering and environmental ethics. They are less about the application of moral philosophies, deontological, utilitarian, virtue, or otherwise, and more about the recognition of an individual's participation in keeping up the livelihood of ecosystem health. Given the influence of technologies to foster (or undermine) ecosystem health, considerations of appropriate and morally justifiable engineering practice should thus include questions not only of justice in general but environmental justice more specifically. Questions of distribution, procedures, recognition, and capabilities can inform the terms of such a curriculum. This kind of intervention takes place before engineering practice itself.

A second area for future work points to the better integration of EJ principles into professional codes of ethics once engineers are out in the profession, beyond the undergraduate classroom (see Chapter 43, "Professional Codes of Ethics" by Michael Davis, this volume). This would follow from work by Michelfelder and Jones (2011) and Brauer (2013) that urges greater attention to sustainability and environmental ethics in engineering codes of ethics. Extending their arguments to fold in the specific theories of EJ and the community engagement principles that drive them can help. On this count, consideration of environmental impact statements (EIS) is a start, but it is not enough. The environmental impacts of technologies and other engineered systems are only one element (a distributive one) of a larger portfolio of EJ considerations. Thus, a future direction for engineering and EJ is to include the necessary attention to environmental impacts while broadening the purview to include questions about equitable participation and beyond.

A third area involves more research into the global networks of injustice. This prong follows from the spatial expansion of EJ over the past decade. That has been an expansion into the global circulation of burdens and benefits, plus more attention to cases in countries beyond the West. More research into the global networks of injustice related to engineering and technology asks for work in engineering ethics and education to likewise develop spatial analysis at the global scale. Here, it would be productive to follow work that binds the first direction above, on education, with global issues of EJ. A starting point for such efforts would be to pull together work on global engineering ethics (Wang and Thompson 2013; Murphy et al. 2015; see also Chapter 46, "Global Engineering Ethics" by Pak-Hang Wong, this volume), comparative studies that consider engineering ethics in different national contexts (Downey et al. 2007), and other analysis in global engineering studies (Downey et al. 2006) with global EJ studies. Pairing the spatial expansion of EJ with spatial expansions of engineering would take seriously the point by Ottinger et al. (2017) that "global inequities are an equally important category of environmental injustice" and "global capital is implicated in the economic marginalization and environmental despoliation of developing countries" (1031). At that scale, "more work is needed to formulate an expanded science and engineering ethics that incorporates technical practitioners' responsibilities to environmental justice" (1046).

The first two areas mentioned move out temporally from earlier points in engineering life (education) to later professional practice (codes of ethics). The third moves out spatially to global scales. Finally, a fourth area for future work, integrating engineering and EJ, underlies each of those three directions. This last one encourages further attention away from the specific locus of engineering practice and toward the cultural and political conditions that give engineers standing in the first place. This is more the domain of philosophers of technology and engineering, STS scholars, and engineering education researchers where such scholars attend to the cultural conditions of engineering— to include economic, political, and moral aspects—that give meaning to engineering and technology in society. Incorporating further attention to EJ in those studies can inculcate a next generation of scholarship and encourage further emphases in engineering ethics to make matters of environmental injustice core elements of engineering education, practice, and identity.

Related Chapters

Chapter 25: Sustainable Design (Steven A. Moore)
Chapter 34: Engineering and Sustainability: Control and Care in Unfoldings of Modernity (Andy Stirling)

Chapter 38: Health (Marianne Boenink)

Chapter 40: Ethical Considerations in Engineering (Wade L. Robison)

Chapter 43: Professional Codes of Ethics (Michael Davis)

Chapter 44: Responsibilities to the Public—Professional Engineering Societies (Joseph Herkert and Jason Borenstein)

Chapter 45: Engineering as a Political Practice (Govert Valkenburg)

Chapter 46: Global Engineering Ethics (Pak-Hang Wong)

Chapter 49: Socially Responsible Engineering (Jennifer Smith and Juan Lucena)

Chapter 50: Engineering and Social Justice (Caroline Baillie)

Chapter 52: Beyond Traditional Engineering: Green, Humanitarian, Social Justice and *Omnium* Approaches (George D. Catalano)

Further Reading

Bullard, Robert (1990). *Dumping in Dixie: Race, Class, and Environmental Quality*. Boulder, CO: Westview Press. (One of the first influential treatments from a pioneer in the field.)

Cole, Luke and Foster, Sheila (2000). *From the Ground Up: Environmental Racism and the Rise of the Environmental Justice Movement*. New York: New York University Press. (The standard background text at a moment when the field of EJ had first developed clarity.)

Holifield, Ryan, Chakraborty, Jayajit and Walker, Gordon (eds.) (2018). *Handbook of Environmental Justice*, London: Routledge. (The most informed volume covering the nuances and complexity of EJ as scholarship and a social movement.)

Ottinger, Gwen (2013). *Refining Expertise: How Responsible Engineers Subvert Environmental Justice Challenges*. New York: New York University Press. (A trenchant analysis of the nuances between engineering and EJ.)

Notes

1. Cole and Foster's summary has become a classic account of several decades of work under the labels environmental racism and environmental justice. They use the metaphor of a river and tributaries to summarize the early work that came together by the end of the century. They identify "civil rights, grassroots anti-toxics, academic, labor, indigenous" and environmental movements as the main six.
2. The full text of this Order is available at www.archives.gov/federal–register/executive-orders/pdf/12898.pdf.
3. They are the terms summarized by Cole and Foster (2000) with respect to Kettleman City, California.
4. In this case, I mean "upstream" in the conceptual sense, while the environmental reference is also about the actual physical upstream flow of water in a river.
5. Of these, food justice is the most vibrant. See Gottlieb and Joshi (2010); Alkon and Agyeman (2011); Alkon (2018).
6. This touches on a host of works examining technology and the human values embedded in technological systems, beginning with sociological, historical, and philosophical work in STS (e.g., MacKenzie and Wajcman 1999; Sismondo 2004; Felt et al. 2017) and continuing with increasing attention from philosophers of engineering and technology, this volume included (e.g., Van de Poel and Goldberg 2010; Michelfelder et al. 2016).

References

Alkon, Alison (2018). Food Justice: An Environmental Justice Approach to Food and Agriculture. In *Routledge Handbook of Environmental Justice*. London: Routledge, pp. 412–424.

Alkon, Alison and Agyeman, Julian (eds.) (2011). *Cultivating Food Justice: Race, Class, and Sustainability*. Cambridge, MA: MIT Press.

Allen, Barbara (2003). *Uneasy Alchemy: Citizens and Experts in Louisiana's Chemical Corridor Disputes*. Cambridge, MA: MIT Press.

Beck, Ulrich (1992 [1986]). *Risk Society: Towards a New Modernity*. New Delhi: Sage.

Blum, Elizabeth (2008). *Love Canal Revisited: Race, Class, and Gender in Environmental Activism*. Kansas University Press.

Boone, Christopher and Buckley, Geoffrey (2018). Historical Approaches to Environmental Justice. In *Handbook of Environmental Justice*, 222–230.

Brauer, C.S. (2013). Just Sustainability? Sustainability and Social Justice in Professional Codes of Ethics for Engineers. *Science and Engineering Ethics,* 19, 875.

Bullard, Robert (1990). *Dumping in Dixie: Race, Class, and Environmental Quality.* Boulder, CO: Westview Press.

Bullard, Robert (2005). *The Quest for Environmental Justice: Human Rights and the Politics of Pollution.* San Francisco: Sierra Club Books.

Bullard, Robert, and Wright, B.H. (1989). Toxic Waste and the African American Community. *The Urban League Review,* 13, 67–75.

Butler, Lindsey, Scammell, Madeleine K. and Benson, Eugene B. (2016). The Flint, Michigan, Water Crisis: A Case Study in Regulatory Failure and Environmental Injustice. *Environmental Justice,* 9(4), 93–97.

Cole, Luke and Foster, Sheila (2000). *From the Ground Up: Environmental Racism and the Rise of the Environmental Justice Movement.* New York: New York University Press.

Day, Rosie (2018). A Capabilities Approach to Environmental Justice. In *Handbook of Environmental Justice,* 124–135.

Downey, G.L., Lucena, J.C., Moskal, B.M., Parkhurst, R., Bigley, T., Hays, C., Jesiek, B.K., Kelly, L., Miller, J., Ruff, S., Lehr, J.L. and Nichols-Belo, A. (2006). The Globally Competent Engineer: Working Effectively With People Who Define Problems Differently. *Journal of Engineering Education,* 95, 107–122.

Downey, Gary, Lucena, Juan and Mitcham, Carl (2007). Engineering Ethics and Identity: Emerging Initiatives in Comparative Perspective. *Science and Engineering Ethics,* 13(4), 463–487.

Duvall, Tim, Englander, Fred, Englander, Valerie, Hodson, Thomas J. and Marpet, Mark (2002). Ethical and Economic Issues in the Use of Zero-emission Vehicles as a Component of an Air-pollution Mitigation Strategy. *Science and Engineering Ethics,* 8(4), 561–578.

Epting, Shane (2016). A Different Trolley Problem: The Limits of Environmental Justice and the Promise of Complex Moral Assessments for Transportation Infrastructure. *Science and Engineering Ethics,* 22(6), 1781–1795.

Felt, Ulrike, Fouche, Rayvon, Miller, Clark and Doer-Smith, Laurel (eds.) (2017). *The Handbook of Science and Technology Studies,* 4th ed. Cambridge, MA: MIT Press.

Fortun, Kim (2001). *Advocacy After Bhopal: Environmentalism, Disaster, New Global Orders.* Chicago, IL: University of Chicago Press.

Fortun, Kim, Knowles, Scott, Murillo, Louis Felipe, Jobin, Paul, Liboiron, Max, Torre, Pedro de la, Choi, Vivian and Matsumoto, Miwao (2017). Researching Disaster From an STS Perspective. In Ulrike Felt, Rayvon Fouché, Clark A. Miller and Laurel Smith-Doerr (eds.), *Handbook of Science and Technology Studies.* Cambridge, MAMIT Press, pp. 1003–1028.

Giddens, Anthony (1999). Risk and Responsibility. *Modern Law Review,* 62(1), 1–10.

Gottlieb, Robert (2009). Where We Live, Work, Play . . . and Eat: Expanding the Environmental Justice Agenda. *Environmental Justice,* 2(1), 7–8.

Gottlieb, Robert and Joshi, Anupama (2010). *Food Justice.* Cambridge, MA: MIT Press.

Haklay, Muki and Francis, Louise (2018). Participatory GIS and Community-based Citizen Science for Environmental Justice Action. In *Routledge Handbook of Environmental Justice.* London: Routledge.

Hernández, D. (2015). Sacrifice Along the Energy Continuum: A Call for Energy Justice. *Environmental Justice,* 8(4), 151–156. doi:10.1089/env.2015.0015.

Holifield, Ryan, Chakraborty, Jayajit and Walker, Gordon (eds.) (2018). *Handbook of Environmental Justice.* London: Routledge.

Iser, Mattias (2013). Recognition. In Edward N. Zalta (ed.), *The Stanford Encyclopedia of Philosophy,* Fall 2013 ed. https://plato.stanford.edu/archives/fall2013/entries/recognition/. Accessed August 15, 2018.

Kinchy, Abby, Jalbert, Kirk and Lyons, Jessica (2014). What Is Volunteer Water Monitoring Good for? Fracking and the Plural Logics of Participatory Science. In Scott Frickel, David J. Hess (ed.), *Fields of Knowledge: Science, Politics and Publics in the Neoliberal Age.* Emerald Group Publishing Limited, pp. 259–289.

Lamont, Julian and Favor, Christi (2017). Distributive Justice. In Edward N. Zalta (ed.), *The Stanford Encyclopedia of Philosophy,* Winter 2017 ed. https://plato.stanford.edu/archives/win2017/entries/justice-distributive/. Accessed August 15, 2018.

Lucena, Juan, Schneider, Jen and Leydens, Jon (2009). *Engineering and Sustainable Community Development.* Morgan and Claypool Publishing.

MacKenzie, Donald and Wajcman, Judy (eds.) (1999). *The Social Shaping of Technology,* 2nd Edition. London: McGraw Hill.

Michelfelder, Diane and Jones, Sharon (2011). Sustaining Codes of Ethics for the Twenty-first Century. *Science and Engineering Ethics,* 19(1), 237–258.

Michelfelder, Diane, Newberry, Byron and Zhu, Qin (eds.) (2016). *Philosophy and Engineering: Exploring Boundaries, Expanding Connections.* Springer.

Murphy, Colleen, Gardoni, Paolo, Bashir, Hassan, Harris, Charles E., Jr. and Masad, Eyad (eds.) (2015). *Engineering Ethics for a Globalized World*. Springer.

Nieusma, Dean and Riley, D. (2010). Designs on Development: Engineering, Globalization, and Social Justice. *Engineering Studies*, 2, 29–59.

Oudshoorn, Nelly and Pinch, Trevor (ed.) (2003). *How Users Matter: The Co-Construction of Users and Technologies*. Cambridge, MA: MIT Press.

Ottinger, Gwen (2011). Environmentally Just Technology. *Environmental Justice*, 4(1), 81–85.

Ottinger, Gwen, Barandiaran, Javiera and Kimura, Aya (2017). Environmental Justice: Knowledge, Technology, and Expertise. In Ulrike Felt, Rayvon Fouche, Clark Miller, and Laurel Doer-Smith (eds.), *The Handbook of Science and Technology Studies*, 4th ed. Cambridge, MA: MIT Press, pp. 1029–1058.

Ottinger, Gwen and Cohen, B.R. (eds.) (2011). *Technoscience and Environmental Justice: Expert Cultures in a Grassroots Movement*. Cambridge, MA: MIT Press.

Ottinger, Gwen and Cohen, B.R. (2012). Environmentally Just Transformations of Expert Cultures: Toward the Theory and Practice of a Renewed Science and Engineering. *Environmental Justice*, 5(3), 158–163.

Pellow, David (2004). *Garbage Wars: The Struggle for Environmental Justice in Chicago*. Cambridge, MA: MIT Press.

Pellow, David and Brulle, Robert (2005). *Power, Justice and the Environment: A Critical Appraisal of the Environmental Justice Movement*. Cambridge, MA: MIT Press.

Riley, Donna (2005). *Engineering and Social Justice*. Williston, VT: Morgan & Claypool Publishing.

Roberts, J. Timmons (2009). The International Dimension of Climate Justice and the Need for International Adaptation Funding. *Environmental Justice*, 4, 185–190.

Sandler, Ronald and Pezzullo, Phaedra C. (eds.) (2007). *Environmental Justice and Environmentalism: The Social Justice Challenge to the Environmental Movement*. Cambridge, MA: MIT Press.

Schlosberg, David (2007). *Defining Environmental Justice: Theories, Movements, and Nature*. Oxford: Oxford University Press.

Schlosberg, David (2013). Theorising Environmental Justice: The Expanding Sphere of a Discourse. *Environmental Politics*, 22, 37–55.

Schneider, Jen (2010). Engineering and the Values of Social Justice. *Engineering Studies*, 2(1), 1–4.

Shepard, Peggy and Corbin-Mark, Cecil (2009). Climate Justice. *Environmental Justice*, 2(4), 163–166.

Sismondo, Sergio (2004). *An Introduction to Science and Technology Studies*. Oxford: Blackwell.

Swearengen, J., Barnes, S., Coe, S., Reinhardt, C. and Subramanian, K. (2002). Globalization and the Undergraduate Manufacturing Engineering Curriculum. *Journal of Engineering Education*, 91, 255–261.

Sze, Julie (2007). *Noxious New York: The Racial Politics of Urban Health and Environmental Justice*. Cambridge, MA: MIT Press.

Sze, Julie and London, Jonathan (2008). Environmental Justice at the Crossroads. *Sociological Compass*, 2(4), 1331–1354.

United Church of Christ (1987). *Toxic Waste and Race in the United States of America*. New York: Committee for Racial Justice.

United Church of Christ (1991). *Principles of Environmental Justice*. www.ejnet.org/ej/principles.pdf. Accessed August 20, 2018.

Vallero, Daniel and Aarne Vesilind, P. (2007). *Socially Responsible Engineering: Justice in Risk Management*. Hoboken, NJ: John Wiley.

Van de Poel, Ibo and Goldberg, David (eds.) (2010). *Philosophy and Engineering: An Emerging Agenda*. Springer.

Wang, G. and Thompson, R.G. (2013). Incorporating Global Components Into Ethics Education. *Science and Engineering Ethics*, 19(1) 287–298.

Wilson, Sacoby, Aber, Aaron, Wright, Lindsey and Ravichandran, Vivek (2018). A Review of Community-engaged Research Approaches Used to Achieve Environmental Justice and Eliminate Disparities. In *Routledge Handbook of Environmental Justice*. London: Routledge.

Wisnioski, Matthew (2012). *Engineers for Change: Competing Visions of Technology in 1960s America*. Cambridge, MA: MIT Press.

52

BEYOND TRADITIONAL ENGINEERING

Green, Humanitarian, Social Justice, and *Omnium* Approaches

George D. Catalano

52.1. Introduction

Engineering is the application of knowledge in the form of science, mathematics, and empirical evidence to the innovation, design, construction, operation, and maintenance of structures, machines, materials, devices, systems, processes, and organizations. According to various codes of ethics adopted by professional engineering societies, members of this profession, i.e. engineers, are expected to exhibit the highest standards of honesty and integrity, as the practice of engineering has a direct and vital impact on the quality of life for all people (see Chapter 43, "Professional Codes of Ethics" by Michael Davis, this volume). In fact, the first fundamental canon in the National Society of Professional Engineers (an umbrella professional engineering society) code of ethics requires that *engineers, in the fulfillment of their professional duties, shall hold paramount the safety, health, and welfare of the public.* It is the breadth of the word "public" which serves as the focus of the present work. How is "public" to be interpreted, that is, how expansive a view should be taken? Does "public" refer only to the members of society that are directly linked to the technology at the time that technology is developed? Or, when considering the health and welfare of the public, does the ethical code require a longer view in which the impact(s) of the technological advances upon the environment are also considered? And what of the parts of the world—the regions marked by poverty—that advancing technology rarely if ever touches at all? Does the engineer's ethical code have any concern for the millions of species that inhabit the rest of our planet, or is it simply all about one species, our species—humankind?

In response to these questions, this chapter provides an overview of several, relatively new ideas that offer a more expansive view of the role of engineering and the extent of its responsibilities. One such development in the practice of engineering seeks to broaden the professional duties of engineers to include an explicit commitment to the health of the natural environment, while another focuses upon the well-being of people in the parts of the world that endure poverty and injustice. A related concept to this concern for the impoverished is encapsulated in engineering for social justice. This movement seeks understanding of how technology and society are co-constructed, to identifying and dismantling specific occurrences of injustice related to engineering and technology, and to devising and developing technologies and other engineering solutions to the problems confronted. Lastly, another approach proposed is the notion of *omnium* engineering. A literal translation of the Latin word *omnium* is "all" or "all beings." In the context of engineering, the term is extended to more fully

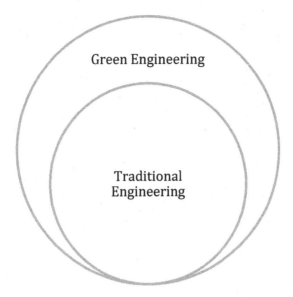

Figure 52.1 Expanding wave from traditional to green engineering

describe an engineering profession that considers the wants and needs of all life forms, not only that of the human species. Such an engineering takes at its root the ideas offered by the Lakota phrase: *Mitákuye Oyás'iŋ* ("All Are Related") (François 2007). It is a statement of values that seeks "oneness and harmony with all forms of life: other people, animals, birds, insects, trees and plants, and even rocks, rivers, mountains and valleys" (Maroukis 2005).

The present chapter shall describe each of these expansive approaches to engineering. For each approach, guiding principles are provided as well as a brief historical review, respective strengths, and possible limitations. In addition, the areas of knowledge needed for professional engineers using the different approaches shall also be discussed. Metaphorically, the direction of the expansive approaches from traditional engineering to *omnium* engineering discussed is analogous to the propagation of waves emanating from the impact of a stone in a still pond. Traditional engineering corresponds to the first wave generated when the impact occurs. Moving out from the first wave, there is an expansion to include concerns about the health of the environment. This is followed by further expansion first to include consideration of the poor and impoverished regions of the world, and then to broaden the boundaries even more to include the common good. The largest and final expansion corresponds to a consideration of all species, not only humankind.

52.2. Green Engineering

Green engineering encompasses all of the engineering and science disciplines, focusing on the design and synthesis of materials, processes, systems, and devices with the objective of minimizing overall environmental impact (including energy utilization and waste production) throughout the entire life cycle of a product or process. Green engineering can be considered environmentally conscious attitudes, values, and principles combined with sound science, technology, and engineering practice and is inherently inter- and cross-disciplinary in nature.

As can be seen in Figure 52.1, green engineering extends the boundaries beyond the traditional sense of professional responsibility. According to the Environmental Protection Agency, green engineering is the design, commercialization, and use of processes and products that minimize pollution, promote sustainability, and protect human health without sacrificing economic viability and efficiency (EPA 2019). The concept of green engineering grew out of environmental engineering which Buescher described as "one of the world's oldest professions" (Buescher 2018). He details the instances of the practice of environmental engineering from the early Egyptian dynasties to the classical Greek and Roman civilizations, as well as instances described in the *Oursuta Sanghita* and the *Susuruta Samhita* (Buescher 2018). It was after the Second World War that environmental concerns became increasingly important in the United States. With the emergence of synthetic chemicals, many of the nation's waterways began to foam as many of these new agents were not biologically biodegradable and existing water treatment plants could not break down these wastes (Buescher 2018). In 1962, Rachel Carson published *Silent Spring*, which alerted a large audience to the environmental and human dangers of indiscriminate use of pesticides, spurring revolutionary changes in the laws affecting air, land, and water (Carson 1962). The Love Canal case in upstate New York among others further raised awareness of the assault on the environment (Neuman 2016). In the summer of 1978, residents of Love Canal, a suburban development in Niagara Falls, New York, began protesting against the leaking toxic waste dump in their midst—a 16-acre site containing 100,000 barrels of chemical waste that anchored their neighborhood. Initially seeking evacuation, area activists soon found that they were engaged in a far larger battle over the meaning of America's industrial past and its environmental future. The Love Canal protest movement inaugurated the era of grassroots environmentalism, spawning new anti-toxics laws and new models of ecological protest. Today's issues of hydro-fracking and renewed concern about nuclear waste disposal continue to bring attention to the harmful impacts of technology upon the natural world (see Chapter 51, "Engineering and Environmental Justice" by Benjamin R. Cohen, this volume). Green engineering is a response to those concerns.

Several authors have identified core principles of green engineering (Allen 2002; Jimenez-Gonzales and Constable 2011). One set which is particularly encompassing has been offered by Anastas and Zimmerman (2003). The 12 principles of green engineering are: (1) ensure that all material and energy inputs and outputs are as inherently non-hazardous as possible; (2) strive to prevent waste rather than to treat or clean up after it is formed; (3) develop separation and purification operations that minimize energy consumption and materials use; (4) ensure that products, processes, and systems are designed to maximize mass, energy, space, and time; (5) minimize the amount of resources consumed to transform inputs into the desired outputs; (6) strive to minimize complexity; (7) target durability, not immortality, (8) strive to minimize excess; (9) minimize materials diversity to promote disassembly and value retention; (10) integrate material and energy flows using existing assets or energy sources with an emphasis on developing products, processes, and systems that require local materials and energy resources; (11) design products, processes, and systems for performance in a commercial "afterlife"; and (12) choose materials and energy that are from renewable sources rather than finite reserves.

Green design principles most commonly use an eco-efficient design methodology. That is, the goal of green design and green engineering primarily can be summed up as being "less bad." It is a "cradle-to-grave" methodology which seeks to reduce toxic wastes, reduce energy consumption, and reduce the demand on various natural resources, and it is subject to rules, regulations, and limitations in its attempts to be "less bad." In developing a different green design approach, McDonough and Baumgartner (2013) begin by asking the question, "When is being less bad not good enough?" Their approach is termed "cradle-to-cradle" and uses an eco-effective design methodology. It requires, for example, that processes purify waste water and actually produce drinking water, produce more energy than is consumed, and produce more and new materials for human and natural consumption, and it focuses upon creating abundance rather than adhering to limits.

Green engineering makes important contributions in promoting environmental awareness and sustainability. Training in green engineering requires a much greater emphasis on chemistry and materials science as well as rigorous engineering basics. Green engineering also requires an understanding of life cycles and ecosystems, particularly how such systems are impacted by various technologies. Green engineering is an approach that is particularly focused on issues related to respect for biodiversity and in stark opposition to what McDonough and Braumgart refer to as "attack of the one size fits all" design response and its attendant de-evolution or simplification on a mass scale. Issues related to the challenges of poverty and underdevelopment are not addressed, though there is reference to the production of an abundance of both natural materials and energy sources. Several questions remain unanswered. Who will in fact have access to this abundance? Furthermore, how will this abundance be distributed throughout the world? How ultimately will that distribution impact questions of peace and security? And what about the rest of the species and ecosystems with whom we share the earth? Where exactly do they fit in? What are the responsibilities of engineering to them? Such questions seem of pre-eminent importance, now more than ever before. A different kind of engineering, humanitarian engineering, addresses at least some of the questions raised particularly with respect to poverty.

Green engineering is growing in acceptance throughout the engineering profession. The NSPE Code of Ethics now encourages engineers to adhere to the principles of sustainable development (NSPE 2018). The Code defines "sustainable development" as the "challenge of meeting human needs for natural resources, industrial products, energy, food, transportation, shelter, and effective waste management while conserving and protecting environmental quality and the natural resource base essential for future development." The goal is to design projects in a manner that reduces their environmental impact during construction activities as well as during the life cycle of the project. Engineers have, for the most part, considered the environmental impact of their design solutions; what's new is that clients and others are now more interested in the environmental impact (see Figure 52.2).

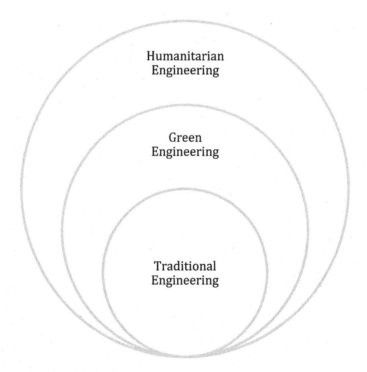

Figure 52.2 Expanding wave from green to humanitarian engineering

52.3. Humanitarian Engineering

Humanitarian engineering, a relatively new discipline in engineering, points the engineering profession in a much different direction than had been previously the case. Humanitarian engineering is defined as "the application of engineering to improving the well-being of marginalized people and disadvantaged communities, usually in the developing world" with an emphasis upon sustainability, low costs, and locally available resources for the solutions advanced (Mitcham and Munoz 2012). Sustainability herein is linked to maintaining technology locally linked to the state of the natural world. A different description of humanitarian engineering offered by Ohio State University is that it is "the creation of technologies that help people" (Humanitarian Engineering, Ohio State, 2018) while the University of Warwick states that "Humanitarian Engineering is defined as the use of science and engineering to invent, create, design, develop, or improve technologies which promote the well-being of communities which are facing grand humanitarian challenges (fast growing populations, poor, disaster-hit, marginalized, or underserved communities)" (Humanitarian Engineering, Warwick). To meet these challenges, humanitarian engineering offers a way to bring forth a new engineering that is:

> appropriately influenced by the humanitarian ideal, thus giving rise to opportunities for new understanding and an historically distinct practice of this technical profession . . . something called humanitarian engineering may be described in general terms as the artful drawing on science to direct the resources of nature with compassion to meet the basic needs of all persons irrespective of national or other distinctions—an artful compassion that, in effect, is directed towards the needs of the poor, powerless or otherwise marginalized persons.
>
> *(Mitcham and Munoz 2012)*

Humanitarian engineering offers, then, a view more expansive than the traditional view. That is, it is not enough to consider the notion of *public* in a very narrow sense but rather, it forces an integration of compassion towards our fellow human citizens. Passino has offered a list of the principles of humanitarian engineering (Passino 2015). Such stated principles focus on empathy, compassion, and building trust. It puts emphasis upon developing communities and being aware of the cultural and societal differences that exist. The focus remains firmly affixed to human dignity, human rights, and fulfillment. Humanitarian engineering focuses the skill and capabilities of engineering theory and practice toward aiding the greater good of humanity by offering stakeholder-centric solutions to medical and disaster relief, global outreach, human displacement, human safety, food security, cultural awareness/sensitivity, and economic development.

Training in humanitarian engineering must necessarily include history, politics, economics, sociology, language, as well as rigorous engineering basics. An interesting example arising from the author's personal experience working with the Onondaga Nation points to a range of different skills and expertise required in humanitarian engineering. The Onondaga (*Onönda'gaga'* or "Hill Place") people are one of the original five constituent nations of the Iroquois (Haudenosaunee) Confederacy in northeast North America (History of Onondaga 2018). Their traditional homeland is in and around present-day Onondaga County, New York, south of Lake Ontario. A group of clan mothers from the nation had expressed concern to the tribal leadership that with the advancing age of many of the women, it may soon become difficult if not impossible to continue to prepare corn soup in the traditional way. A design team of engineering students were then tasked with meeting the needs of the clan mothers as part of their capstone design experience. Their response required the students to immerse themselves in the rich culture of the Haudenosaunee in which the preparation of the corn soup is sacred. The proper preparation remains an important link back to the generations that have

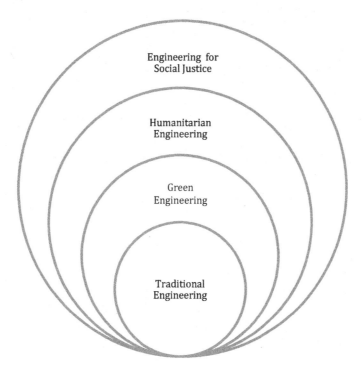

Figure 52.3 Expanding wave from humanitarian to engineering for social justice

come before. Working with the Haudenosaunee also forced the students to deal with the issue of trust. The team's clients in this case are members of a culture who have been subject to repeated lies, disappointments, and sometimes violence. In addition, the students were confronted further with very different notions of time and the importance of efficiency compared to what is understood by these terms in European cultures.

Humanitarian engineering does offer the engineering profession a chance to consider the plights of those who traditionally not been included in engineering classrooms—the poor in this country as well as overseas, and the native peoples who were subject to colonialization throughout the Americas. This potential is to be applauded, and nurtured. Compassion towards those whose concerns and presence too often have been absent from the engineering classroom seems arbitrarily limited if it is not extended beyond economic class, country, and the world. It falls short in the acknowledgement that the earth is inhabited with countless other species (see Figure 52.3).

52.4. Engineering and Social Justice

At the forefront of the engineering and social justice movement is the Engineering, Social Justice and Peace (ESJP) organization which began at Queens University, Kingston, Ontario, in 2004, founded by Baillie (Baillie et al. 2012). This effort challenges the notion of an engineering profession either as value-free or one that is firmly entrenched in capitalism. A brief recounting of several of the commitments stated by ESJP includes: (1) envisioning and practicing engineering in ways that extend social justice and peace in the world; (2) resisting injustice in its many forms through promotion of diversity and inclusivity, and by working towards fair, equitable, and sustainable treatment of people and their environments; (3) redefining social justice alongside those who experience injustices; (4)

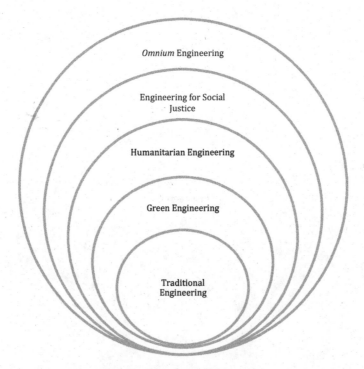

Figure 52.4 Expanding wave from social justice to *omnium* engineering

working for peace and nonviolence; (5) resisting injustice even as complicit with it; (6) translating ideas into action and recognizing their interconnectedness; and (7) committing to equity and sharing where they support justice goals (ESJP 2018).

Training in engineering for social justice must necessarily include history, politics, economics, sociology, and language, as well as rigorous engineering basics. An important additional subject matter is critical theory following the school of thought beginning with Horkheimer and Adorno (2007) and continuing through Marcuse (2002). The core concepts of critical theory are as follows: (1) critical social theory should be directed at the totality of society in its historical specificity (i.e. how it came to be configured at a specific point in time); and (2) critical theory should improve understanding of society by integrating all the major social sciences, including geography, economics, sociology, history, political science, anthropology, and psychology (ESJP 2018).

Interestingly, the organization's website further states that the group is committed to the fundamental principles first described by Berry: (1) respecting and promoting differentiation of all species; (2) recognizing the intrinsic worth of each element of the Universe, that is, each is a subject in its own right; and (3) nurturing the sense of community that binds each element together in the tapestry (Berry 2009). In fact, it is these three fundamental principles—differentiation, subjectivity and community that have led to the development of the final approach to engineering described in the present chapter: *omnium* engineering (see Figure 52.4).

52.5. *Omnium* Engineering

With rapidly advancing technology and science, the understanding of ethical responsibilities, both as a profession and as individual practitioners, has struggled to keep pace with those advancements. A literal translation of the Latin word *omnium* is "all" or "all beings"; yet, in the context of engineering, it could be extended to more fully describe an engineering profession that considers the wants and needs of all life forms, not only that of the human species. Such an expansion of what counts morally needs careful underpinning from an ethical paradigm that extends our understanding of our ethical responsibilities.

This new approach to such questions of responsibility is based on advances in quantum mechanics, eco-philosophy, and complex systems. From quantum mechanics, rather than a Universe described as being deterministic, in a quantum world the Universe can best be described as a rich tapestry of potentialities; and rather than a collection of disparate objects, the Universe is more accurately described as one, connected system. From eco-philosophy, the ideas of an evolving, dynamic Universe; governed by the principles of differentiation, communion and subjectivity are included. Lastly, from complex systems, the concepts of non-linearity, self-organization, and emergence are incorporated into this new approach. With respect to emergence and self-organization, emergence suggests the appearance of behavior that could not be anticipated from knowledge of the parts of the system alone, and self-organization means that there is no external controller or planner "engineering" the appearance of these emergent features—they appear spontaneously.

Omnium engineering would then:

- Recognize the Universe as being filled with infinite potentialities and that decisions that limit the various life-forms from reaching their unique potentials must be carefully considered;
- Recognize the Universe as complete and connected throughout; be cognizant of the various filters through which each of us experiences what we claim to be our reality; forgo the notion of linearity and its attendant principle of superposition, that is, the whole being simply the sum of its parts;
- Understand that a totally unexpected result may emerge from systems no matter how carefully and tightly controlled the systems and the inputs might seem;
- Recognize that various elements in a system may have desire to organize themselves in ways that cannot be anticipated form the outside;
- Promote diversity rather than seeking to reduce differentiation and move towards homogenization;
- Recognize the intrinsic value of all life forms rather than seeking to establish a hierarchy of life forms both across and within species; and
- Nurture the sense of a community rather than seeking to destroy those communities, societies, or cultures.

These principles suggest the following reflections on the practice of engineering. Are the totality of life forms, their intrinsic worth, and their desire to fulfill their individual and unique potentialities considered in the formulation of an engineering design response? And what of the local communities and their culture and their very way of life—are they considered in an engineering design response? Ultimately recalling the ideas of Popper (2002), can there be acceptance that engineering does not occur in a linear and deterministic Universe in which past, present, and future exist with certainty, much like scenes frozen for all time on video tapes or burned on DVDs? Moreover, can there be a moment to pause and reflect on the fact that the principle of superposition may not work unfailingly? Arrogance needs to be replaced by humility.

Training in *omnium* engineering again includes history, politics, economics, sociology, language, rigorous engineering basics, and critical theory. Important additional subject matters include the philosophical implications of quantum mechanics as well as the science of complex systems and eco-philosophy. Of particular relevance in eco-philosophy are the works of Teilhard de Chardin (1960), who described the unfolding of the cosmos and the evolution of matter to humanity; Berry (2011), who pointed the way to an ecological spirituality attuned to humankind's place in nature and giving rise to an ethic of responsibility; and Swimme (2017), who has written extensively on evolutionary cosmology.

52.6. Final Thoughts

The purpose of this chapter has been to discuss engineering approaches that extend the boundaries of the responsibilities of the engineering profession. Green engineering brings into focus concerns about the plight of the environment under the onslaught of advancing technology. Humanitarian engineering challenges the profession to consider that segment of society that is far too often neglected in modern engineering—the poor and the marginalized. Engineering for social justice calls upon engineers to cultivate a passion for social justice and peace and to develop the skill and knowledge set needed to take practical action for change within the profession. *Omnium* engineering, addresses both the well-being of these often invisible communities as well as that of the planet. Each of these constituencies consists of a communion of subjects rather than a collection of objects, and both in turn consist of individuals as well as systems from micro- to macro-level. Expanding the definition and thus understanding of the profession's responsibilities in engineering to the public is incredibly important at this time. This chapter seeks to push the ethical boundaries for the practice of modern engineering a bit further to include "all my relations," a communion or community of all forms of life: other people, animals, birds, insects, trees and plants, and even rocks, rivers, mountains and valleys, to borrow from the Lakota.

This call for an extension seems analogous to the present-day, highly contentious discussion of expanding the sphere of moral concern as a society to refugees and asylum seekers seeking protection in the West who often find themselves confronted by hostile attitudes and increasingly restrictive asylum policies. Tejero (2013) explored the moral underpinnings of asylum-giving in the West and evaluated how people's solidarity with refugees and asylum seekers can be expanded. Her work integrated two philosophical approaches for eliciting greater understanding and sympathy for the claims of refugees and asylum seekers from the public: the Kantian approach which emphasized rights and obligations; and the Rortian approach which highlighted the importance of sympathy. Tejero's analysis suggested that the two approaches are complementary and mutually reinforcing, with each necessary for the creation of true sanctuaries for refugees and asylum seekers. Just as the sense of duty and empathy compels society to consider the plight of those who have fled to the United States as their last refuge, clinging to the vision of America put forward by Abraham Lincoln who stated, "We shall nobly save or meanly lose the last best hope of earth" (1862). So too does Tejero's analysis challenge the engineering profession to meet its commitment to the safety and well-being of "all my relations": all peoples, animals, birds, insects, trees and plants, and even rocks, rivers, mountains, and valleys.

Related Chapters

Chapter 43: Professional Codes of Ethics (Michael Davis).

Chapter 44: Responsibilities to the Public—Professional Engineering Societies (Joseph Herkert and Jason Borenstein).

Chapter 49: Socially Responsible Engineering (Jennifer Smith and Juan Lucena)

Chapter 50: Engineering and Social Justice (Caroline Baillie)

Chapter 51: Engineering and Environmental Justice (Benjamin R. Cohen)

Further Reading

Baillie, Caroline and Riley, Donna. (2012). *Engineering and Social Justice: In the University and Beyond*. West Lafayette, IN: Purdue University Press. (A review of the growth of the social justice movement in engineering.)

Berry, Thomas. (2015). *Dream of the Earth*. Berkeley, CA: Counterpoint Press. (Description of the evolving Universe from an eco-theological perspective.)

Cuthbert, A. (1999). *More on Trash*. Oxford: Oxford University Press. (An extended treatment of garbage.)

Mitcham, Carl and Munoz, David. (2012). *Humanitarian Engineering* (Synthesis Lectures on Engineers, Technology, and Society). San Rafael, CA: Morgan Claypool Press. (A review of the growth of humanitarian engineering in engineering profession and education.)

Rankin, B. (1984). Where to Put the Hors Texte? In A. Cuthbert and N. Bloggs (eds.), *Garbage Deconstructed*. Cambridge: Cambridge University Press. (The classic deconstructive treatment of the waste disposal industry.)

References

Allen, David (2002). *Green Engineering: Environmentally Conscious Design of Chemical Processes*. Upper Saddle River, NJ: Prentice Hall.

Anastas, P.T. and Zimmerman, J.B. (2003). Design Through the Twelve Principles of Green Engineering. *Environmental Science and Technology*, 37(5), 94A–101A.

Baillie, Caroline, Pauley, Alice and Riley, Donna (2012). *Engineering and Social Justice: In the University and Beyond*. West Lafayette, IN: Purdue University Press.

Berry, Thomas (2009). *The Sacred Universe: Earth, Spirituality, and Religion in the Twenty-First Century*. New York: Columbia University Press.

Berry, Thomas (2011). *The Great Work: Our Way into the Future*. Phoenix, AZ: Crown Press.

Buescher, Charles (2018). *Environmental Engineering "One of World's Oldest Professions"*. Energy, Environmental & Chemical Engineering, Washington University in Saint Louis. https://eece.wustl.edu/eeceatwashu/about/Pages/environmental-engineering-history.aspx

Carson, Rachel (1962). *Silent Spring*. Boston, MA: Houghton-Mifflin.

Engineering, Social Justice, and Peace (2018). *A Network of Activists, Academics, and Practitioners Dedicated to Social Justice and Peace*. http://esjp.org/

Environmental Protection Agency (2019). *Design for the Environment Programs, Initiatives, and Projects*. www.epa.gov/saferchoice/design-environment-programs-initiatives-and-projects

François, Damien (2007). *The Self Destruction of the West: Critical Cultural Anthropology*. Saint-Denis: Publibook, pp. 28–29.

History of Onondaga Nation (2018). *People of the Hills*. www.onondaganation.org/history/

Horkheimer, Max and Adorno, T.W. (2007). *Dialectic of Enlightenment (Cultural Memory in the Present)*. Stanford: Stanford University Press.

Humanitarian Engineering. University of Warwick. https://warwick.ac.uk/fac/cross_fac/iatl/humanitarianengineering/

Humanitarian Engineering Center (2018). Ohio State University. https://osuhe.engineering.osu.edu/

Jimenez-Gonzales, C. and Constable, J.C. (2011). *Green Chemistry and Engineering: A Practical Design Approach*. Hoboken, NJ: Wiley.

Lincoln, Abraham (1862). *Annual Message to Congress*. www.abrahamlincolnonline.org/lincoln/speeches/congress.htm

Marcuse, Herbert (2002). *One-Dimensional Man: Studies in the Ideology of Advanced Industrial Society* (Routledge Classics), Vol. 78. London: Routledge.

Maroukis, Thomas Constantine (2005). *Peyote and the Yankton Sioux: The Life and Times of Sam Necklace*. Norman, OK: University of Oklahoma Press, p. 16.

McDonough, W and Baumgartner, M. (2013). *The Upcycle: Beyond Sustainability—Designing for Abundance*, 1st ed. New York: North Point Press.

Mitcham, Carl and Munoz, David (2012). *Humanitarian Engineering* (Synthesis Lectures on Engineers, Technology, and Society). Williston, VT: Morgan Claypool Press.

Newman, Richard (2016). *Love Canal: A Toxic History From Colonial Times to the Present*. Oxford: Oxford University Press, pp. 99–100.

NSPE (2018). *Code of Ethics*. www.nspe.org/resources/pe-magazine/client-push-sustainability

Passino, Kevin (2015). Humanitarian Engineering Creating Technologies That Help People. London: Routledge, pp. 670–672.

Popper, Karl (2002). *The Logic of Scientific Discovery* (Routledge Classics), Vol. 56. London: Routledge.

Swimme, Brian (2017). *The Hidden Heart of the Cosmos*. Maryknoll: Orbis Books.

Teilhard de Chardin, Pierre (1960). *The Phenomenon of Man*. Dublin: Collins Press.

Tejero, Debora (2013). Expanding the Sphere of Moral Concern: How City of Sanctuary Seeks to Create Solidarity and Protection for Refugees and Asylum Seekers. *Oxford Monitor of Forced Migration*, 3(2).

53

ENGINEERING AND CONTEMPORARY CONTINENTAL PHILOSOPHY OF TECHNOLOGY

Diane P. Michelfelder

53.1. Introduction

In a short article published two years before the inaugural meeting of the Society for Philosophy and Technology in Bad Hamburg, Germany, the Continental philosopher Hans Jonas raised a pointed and provocative question:

> To put it bluntly, if there is a philosophy of science, language, history, and art: if there is social, political, and moral philosophy, philosophy of thought and action . . . how could there not be a philosophy of technology, the focal fact of modern life?
>
> *(Jonas 1979: 34)*

Moving on to identify the elements that could be involved in such a philosophy, Jonas singled out three promising focal points for lines of investigation: (1) the dynamic forces present within technology itself, understood as an "abstract whole"; (2) the different kinds of modern technologies and the influences that they have on human life; and (3) the ethical questions raised by the abstract whole of technology and modern technologies themselves.

As the philosophy of technology began to take shape as a separate philosophical sub-discipline in the latter third of the 20th century, its development largely followed along the lines that Jonas had suggested. Still, during this time, the positive attention paid to the first of these directions by philosophers working within the Continental tradition tended to be identified and strongly critiqued as reflection on "Technology with a capital T" (Brey 2010). The shift away from such reflection, commonly called the "empirical turn" in the philosophy of technology, has been marked by an emphasis on the concrete analysis of specific technologies, technological systems, and engineering practices (Brey 2010; Kroes and Meijers 2000) rather than the consideration of technology as an abstract and autonomous force shaping the lifeworld of human experience. In recent years, however, there has been a resurgence of interest within Continental philosophy in thinking about technology as a whole but without a capital "T". This resurgence involves a combination of new research, new perspectives on previous contributions to the field, and new translations increasing the visibility of work in both categories.

This aim of this chapter is to explore what the implications of this resurgence might be for engineering—in particular engineering centered around the design of data-driven and interconnected everyday technologies within the context of a market-propelled economy (see also Chapter 55, "Reimagining the Future of Engineering" by Neelke Doorn, Diane P. Michelfelder, et al.).

Section 2 offers some background before considering two perspectives on "technology as a whole" found within this resurgence, with an emphasis on the ideas of what it is to be a human self that these perspectives disclose. Building on a distinction between the subject *of* engineering and the subject *for* engineering, Section 3 looks at two responses to the question *Who is the subject of engineering design?* prevalent within professional engineering practices today—the "experimental" and the "datafied" subject. It suggests that these subjects show how the subject *of* engineering is increasingly becoming the subject *for* engineering, argues why this is problematic, and traces how the views spelled out in the previous section disclose a constructive alternative from which engineering could learn, given its commitment to progress and the design of technologies and technological systems that can enhance the conditions under which human life can thrive. In short, the chapter aims to show that reflection on technology as a whole can offer a source of inspiration to engineering for rethinking the subject *of* engineering as the whole human self: dynamic, open, and life-affirming. The chapter wraps up with a brief look at some implications for future research.

53.2. Rethinking Technology as a Whole

53.2.1 Background

The philosophy of technology traces its origins to the influential work of a number of 20th-century, predominately European philosophers who, retrospectively, have been credited with shaping what is widely called "classical" (see for instance Brey 2010; Verbeek 2005) or "first-generation" (Achterhuis 2001: 3) philosophy of technology, but who did not consciously think of themselves as working within an area of inquiry by that name. Prominent among them is Martin Heidegger (1977), generally perceived as holding the view that the essence of modern technology could be characterized as "Enframing": an all-encompassing and self-distorting mode of revealing both artifacts and ourselves as resources, primed to be "on call" to deliver up what is stored within in the context of an order of things whose essence is "orderability" itself. Jacques Ellul (1964, see also 1990), another representative of this trajectory of thinking, famously argued that "technological society" was fundamentally dominated by a framework of rationality which he dubbed "technique", marked by its unswerving commitment to the principle of efficiency in production. As he saw it, the unstoppable momentum to identify more and more efficient means of production leading to an increasingly mechanized world came at the cost of remembering what ends are important to try to attain that would help us lead freer and more human lives. In the eyes of Peter-Paul Verbeek (2005), such "transcendental" approaches, by abstracting from specific technological artifacts and practices and looking at technology in terms of its conditions of possibility, end up becoming philosophically blinded to the very perspectives that would offer a less dystopian, less totalizing, and consequently more promising outlook on the potential that technologies can bring to the enhancement of human life.

Reacting to these perceived shortcomings of classical philosophy of technology, a number of philosophers working within the Continental tradition have developed theoretical lines of reflection that cast a deeply skeptical eye on the idea that technical rationality, no matter how pervasively it is reflected within society and its institutions, represents an all-pervasive and controlling force largely immune to public intervention. For instance, inspired primarily by the work of Herbert Marcuse and other Frankfurt School critical theorists, as well as work in science and technology studies (STS), Andrew Feenberg has proposed that the "technosystem"—taken as the technical framework joining technologies, markets, and administrations (Feenberg 2017a: x)—does not reflect a single rationality but rather multiple rationalities embedded in social contexts. As Feenberg sees it through the lens of what he calls critical constructivism, this makes it possible for the technosystem to be challenged politically by social movements organized around perspectives which were not represented in its original design (see also Feenberg 2017b). By a similar token, the design of particular technologies such as the Internet often reflect "layers" of technical functions (Feenberg 2017a: 99) stemming from the different and frequently competing political interests of different social actors (see also Chapter 45, "Engineering as a Political Practice" by Govert Valkenburg). It is this "internal" competition among different models of the purpose of particular technologies that, as Feenberg sees it, can open up a space for democratic values to be realized within them (Feenberg 1999).

Such "layering" can be compared to the idea of "multistability": a seminal concept in the work of the postphenomenologist Don Ihde (see for instance Ihde 2001, 2012a, 2012b). As multistable, technologies have a range of freedom embedded within them so that they do not necessarily have to be taken up in a set or prescribed manner. They can be used in a variety of ways to accomplish different ends. Because of its claims that there is an overarching "essence" to technology, classical philosophy of technology overlooks this range of freedom exhibited by the mobile phone and other new technologies and so, in Ihde's view, is completely "outdated" when it comes to being able to disclose new insights about them (Ihde 2012b).

But while the ongoing developments in this empirical trajectory within Continental philosophy of technology are sourced in questioning the idea of technical rationality as an all-encompassing force, in recent years emerging work in this tradition has come to the fore that challenges the irrelevance, "single-note", and "monolithic" reading of classical philosophy of technology. For instance, in their introduction to a collection of essays on Jacques Ellul marking the 100th anniversary of his birth, Helena Jerónimo et al. (2013) affirm that his work is "classical" not because of its early position in a temporal sequence, but because there is still much that can be learned from it. His view that technology brings about the "de-symbolization" of sociocultural practices and heritages foreshadows the contemporary erosion of not-easily-renewable social resources such as home maintenance skills (Cérézuelle 2013) and the creation of new digital objects such as the metadata triggered by interactions with social media (Hui 2013). And Søren Riis (2018) has taken a fresh, penetrating look at Heidegger's writings on technology, arguing that he does not tell a single story about modern technology, but rather situates it within multiple genealogies, only one of which characterizes the essence of modern technology as "Enframing" or standing-reserve.

Along with new readings of classical philosophy of technology has come a renewed interest in thinking technology as a whole on the part of some philosophers in the Continental tradition. A look at a "cluster" of work by two of these philosophers: Gilbert Simondon—whose work was largely not known outside of Europe until the 21st century—and Bernard Stiegler, over whose work he exerted a profound influence—highlights how out of such thinking new ways of looking at the self emerge from which contemporary engineering might have something to learn. This is not to say that the latter should take these theoretical perspectives in hand in order to apply them, but rather that it could benefit from taking seriously what they proffer as a reminder of the need to be attentive to the interests of the whole self.

53.2.2 Gilbert Simondon, "Open" Technology, and Concretization

On the Mode of Existence of Technical Objects (2016, originally published in French in 1958), a dissertation written by the French philosopher Gilbert Simondon (1924–1989), has received a good deal of attention in recent years. In it, Simondon committed himself to calling into question the belief that there was no inherent relationship between the sphere of technology and the sphere of what it is to be human. So critical did he think it was for philosophers to call into question the absolute difference between human reality and technical reality that he compared the task of uncovering the true value of the technical object to that of "the abolition of slavery and the affirmation of the value of the human person" (Simondon 2016: 15). The true value of the technical can appear once the development of technology and the development of the human self—as well as society as a whole—are understood as being tied together through the process of joint evolution. Key to this view is Simondon's belief that technical objects can be best understood through the lens of biology, particularly human life. In a human being, when life is going well, the systems making up the physical body (cardiovascular, nervous, digestive, etc.) function on their own individually and together without the need for any external support. The same is true when it comes to the whole self, which thrives when its physical, mental, and emotional lives are integrated with one another and can stand on their own.

A similar kind of rich integration can be found in technical objects. While all individual technologies have something in common, what they have in common resists being identified as a "transcendental essence". What they share is participation in what Simondon referred to as the process of "concretization", in which a technical object, initially an outcome of human creativity, "individuates" itself through becoming more like human life in having the ability to function as an organic whole. As an example, Simondon offered· the (then) contemporary combustion engine, which he saw as having been "concretized" to such an extent that the individual components exchange energy at the appropriate moment so that a cycle of causal relationships can be fired up. (2016: 26). This contrasts with the abstraction of the earlier thermal engine, in which each component could be said to wrap up its business after doing its part rather than being part of an integral whole, and which was more dependent on humans for being maintained (see also Novaes de Andrade 2008). By a similar token, a technical object such as a train becomes further concretized when it is assisted by the careful placement of tracks to be responsive to conditions of terrain and climate, and so to bond with the environment it travels through to form a unified whole or "milieu" (Novaes de Andrade 2008: 13).

When technologies continue in such a way to be "open" to new forms of concretization, they also, as it were, flow back on human life, "re-seeding" its potential and opening it up to new forms for individuals to develop in concert with others in a process that Simondon referred to as collective or "transindividuation". Once technology is understood as being integrally connected to the domain of the human rather than being consigned to an entirely separate domain of the "artificial", it becomes possible to set aside the belief that the purpose of technology, and the creative processes of engineering that lead to it, is to gain control over the forces of nature. What appears in its place is a dynamic "partnership" between technology, the whole human self, and society, geared toward the future and new possibilities for ways of being. Absent robust concretization in a technical object, however, these new ways of being cannot materialize. The case that Novaes de Andrade makes (2008: 11) for taking the contemporary automobile to be an example of this is even stronger today, as features such as infotainment systems and health-monitoring systems are simply "added on" or "assembled together" inside it. The same would hold true for thought or for the self. Interdisciplinary work which is an "assemblage" of different perspectives would count as another example, as would a self who is anchored primarily in the past. As Simondon would see it, a healthy self points toward the future and has energy to supply the "yeast" for new and more progressive social formations.

53.2.3 Bernard Stiegler and Individuation

Beginning with the three-volume set *Technics and Time*, published between 1994 and 2001, the vast output of work by the French philosopher Bernard Stiegler (1952–2020) has been shaped by many influences, including Marx and Heidegger. Above all, though, it reverberates with echoes of Simondon's ideas and a focus on critically interpreting and developing his thoughts of the evolutionary relations between humans and technological objects. Much as Simondon deplores how philosophy encouraged the belief that there was a firm split between the realm of biology and the realm of the technical, Stiegler opens the first pages of *Technics and Time* by noting that the dawn of Western philosophy is marked by an split between living beings that contain their own dynamics within them and technical objects that do not. The former are granted ontological privilege over the latter, which become relegated to the realm of the mechanical (Stiegler 1998: 2). For Stiegler, as for Simondon, this state of affairs is a sign that philosophically technics remains "the unthought", a situation that takes on a particular urgency to address, given that technics comprises the "most dynamic factor" in an era of unprecedented and furious change. (p. ix). In the word "dynamic", deep resonances can be heard to Simondon, as Stiegler adopts his general idea that individuation occurs through the concretization process (see, for example, Crogan 2010) involved in co-evolution. More simply put: "human and technics compose together a dynamic of mutual becoming" (Crogan 2010: 137). His adoption of Simondon is also an adaptation. In particular, Stiegler (1998) stresses the way that technologies, rather than being tools for us to shape a natural world that stands in extrinsic relation to human beings, serve as "prostheses" for them, pushing the idea of what humans are beyond the biological while at the same time opening up the technical for being more than the expression of a "forming intention" on the part of the designer.

In an intriguing reflection on World Wide Web founder Tim Berners-Lee's phrase "philosophical engineering"—the use of philosophy to create a new world—Stiegler (2014b: 189–190) emphasizes that the fundamental knowledge of how to live, act, and think rationally cannot be equated with forms of cognition, as arises in a joint process of psychological and technological individuation, ultimately allowing for individual knowledge to become the knowledge of a community. What most worries Stiegler about contemporary technologies—particularly those related to media and cultural industries (what Stiegler calls the "programming industries") is their potential to subvert the process of individuation by taking and shaping consciousness as a "raw material" for consumer markets. This not only works to the detriment of the full human self with respect to expanding its own singular capacities for desire, affect, memory, and creativity, but also with respect to trans-individuation: to the development of forms of political community based on a shared sense of history, a common good, and caring feeling toward one another (Stiegler 2014a). If the "I" and the "we" are best thought of a "process and not a state", and if they mutually constitute one another, then inhibiting the individuation of one will do the same to the other (Stiegler 2014a: 50, 61). The use of the programming industries to market technical prostheses to consumers means that what someone sees, be it publicity on television for streaming services or something on streaming services themselves, is not differentiated from one person to the next, which inhibits these processes of individuation.

In short, for Stiegler, the political stakes are extremely high if the process of individuation is impeded by technologies, to which he grants a special place (for reasons beyond the scope of this chapter) to programming industries. With this in mind, it is understandable why he (2010) expresses particular concern about ways in which these industries work to impede individuation by undermining the ability to pay attention to something. An attentive self is arguably a microcosm of a whole self, as to fully pay attention to something involves the integration of sense-perception, understanding, affect in the form of care, and imagining possibilities for co-being between oneself and what is being attended to.

53.3. Two Subjects for Contemporary Engineering

The preceding section offered a brief overview of the work of two philosophers in contemporary Continental philosophy of technology to highlight how its reflective orientation toward technology as a whole lets a view of the self as a dynamic and life-affirming whole come into relief. With this section, we turn to draw a contrast between this particular view and two other, contrasting views of the self or subject on which the world of contemporary engineering increasingly, if implicitly, relies.

To talk about the "world" of engineering is to speak in very general terms about a bounded space belonging to trained practitioners who have passed explicit "vetting" requirements that authorize their membership within it. Arie Rip once noted (2000) how the engineering world as seen from the perspective of professional engineers can be shown as a series of concentric circles, with engineers working within the inner realm, surrounded by other circles representing the social and cultural contexts for whom engineers design but which are taken to fall outside engineering spaces.

Such a view, grounded on the assumption that the results of engineering run linearly from the workplace to the world at large, ignores the many ways in which interactions among those in these circles shape these results of what Rip has named the "innovation journey". Between the time that paper was published and now, the boundaries separating the sphere within which engineers work and the public have become even more porous, in part due to the push toward open innovation that can bring new perspectives from non-experts (Fritzsche 2016) and introduce disequilibriating ideas (Michelfelder 2018) into the development of new technologies. The "world" of contemporary engineering is also increasingly caught up in two sets of sociopolitical dynamics that contribute to this porosity. One dynamic relates to economic stimulus: engineers are increasingly called upon to come up with innovative products that can serve as drivers to accelerate growth on both a local and a global scale. This dynamic is itself caught up within the dynamics of political regimes, with one outcome being the development of Internet-embedded digital engineering products that can support what has come to be known as surveillance capitalism (Zuboff 2019) It is within this complex environment of entanglements that two perspectives on the human subject to which we now turn have become increasingly important in the course of engineering work.

53.3.1 Engineering and the "Experimental Subject"

In order for self-driving vehicles to be successfully integrated into an environment marked by different modes of transportation, they will need to be able to predict the movements of pedestrians in an accurate manner. In 2019, researchers at the University of Michigan positioned self-driving vehicles at various Ann Arbor intersections in order to gather data using the vehicles' sensors and cameras on the motion of pedestrians engaged in routine activities involving walking. Those whose motions were registered during these "in the wild" sessions, as researchers described them, provided training data for a neural network that could be used to predict how pedestrians might move at traffic intersections (Du et al. 2019). In these sessions, the pedestrians were unaware that their movements were being observed for this purpose. They were subjects in a research study about which they had no knowledge and to which they had not given consent,

The idea that consent is unnecessary to be a research subject, at least in the United States, reaches back to the Second World War, where conscientious objectors were "drafted" as control subjects in biomedical research studies without their consent, the argument being that the military did not need to get permission from citizens before they were drafted (McCarthy 2008). While a similar justification for not asking for consent in the study just mentioned could be based on the idea that the behaviors being recorded are public and available for anyone to see, a key difference between the two is that the latter study is arguably embedded within the larger context of a large-scale "social experiment" (Van de Poel 2013) in which driverless cars are disruptively introduced into established transportation systems without a clear sense of what the positive and negative impacts to the individual and social good might be. Seen from the angle of a social experiment, akin to an experiment in science (see Chapter 31, "Experimentation" by Viola Schiaffonati), who counts as an "experimental subject" in the above situation is considerably more difficult to determine than it might initially seem.

Although the coherency of extending the notion of an experiment outside of a scientific setting is not without its challengers (for example, Peterson 2017), this section focuses on a context where the connection of social experimentation to identifiable experimental subjects is fairly clear-cut. The example given here is just one instance of how, in the fast expansion of the use of machine learning to advance engineering projects, reliance on experimental subjects has accelerated. The turn toward predictive algorithms in social media contexts, such as ones intended to identify those who are at risk for self-harm, has been marked by a conspicuous lack of transparency, as those whose posts offer training data for the AI involved are generally not aware that they are being used for this purpose (Marks 2019). Concerns over how the subjects of social experiments can be treated in an ethically responsible manner so that their interests could be well protected have been raised and responded to by Martin and Schinzinger (1996) and more recently by Van de Poel (2016). No matter what the nature of the particular social experiment is—from digital currency or social robotics—experimental subjects would share an interest in being treated well.

Looked at from another angle, another commonality among experimental subjects in the context underscored here is the fact that what is learned from them is derived from occurring actions and behaviors or ones that have already taken place. Translated into the distinction between the subject *of* engineering—the subject whose life and well-being engineering is intended to benefit—and the subject *for* engineering—the subject in whom engineering takes an interest as a means of benefiting engineering and a corporate bottom line—the experimental subject is a partial self, looked at mostly for what it has to offer in terms of data, but not necessarily with regard to how engineering could better care for it to realize its future potential so it could flourish and thrive. As engineering design processes become more entangled with AI and machine learning, increasing the reliance on experimental subjects, the question comes up of whether the subject *for* engineering will begin to displace the subject *of* engineering, so that engineering will more and more be focused on designs that will advance the interests of engineering to the detriment of the self as a whole.

The fact that experimental subjects have data collected about them and that they are often unaware that this collection is taking place are features they share with another subject *for* engineering, the datafied subject, to which we turn next. Unlike datafied subjects, however, the practice within engineering of collecting data from experimental subjects is not intended to directly reinforce the status of these subjects as experimental; the interest in experimental subjects extends only up to the point where their actions present data that can be analyzed and used within engineering projects. It is easy to imagine though voluntarily taking on the role of an experimental subject to raise awareness of a vulnerability in a new technology, as, say, in the case of informal experiments done on Teslas to get them to drive fully autonomously. When it comes to the datafied subject, though, the role engineering plays in its formation takes on increased intentionality.

53.3.2 Engineering and the "Datafied Subject"

With the term "datafication", Cukier and Meyer-Schoenberger (2013: 29) called attention to how the rise of "Big Data" marked the "ability to render into data many aspects of the world that have never been quantified before". A job candidate responding online to questions posed by the AI-based video interviewing system HireVue, which makes a hiring recommendation based on analyzing massive amounts of data related to facial expression, tone of voice, and the like, participates in datafication as a data subject, someone who provides data for a particular purpose. Such a person can also though be said to be a "datafied subject". While a data subject offers up data for collection and analysis, a datafied subject can be said to shape their selves in accordance with the values involved with a particular technology of datafication. When a job candidate attempts to tailor their interview "performance" based on what they believe HireVue is looking for, they can be said in this context to be a "datafied subject".

More generally, as the subject *of* engineering is displaced by the subject *for* engineering, the values of efficiency, control, productivity, and the like that pervade contemporary engineering practices are then taken up as primary goods to be pursued within a wide swath of other activities, including the "activity" of everyday life itself. In turn, this trajectory can serve to normalize a particular concept of what it is to be a human subject. Fitbits and other self-tracking devices used by many to gain knowledge about their physical and mental well-being, while not without some ability to augment personal autonomy (Sharon 2017), provide a case in point. Based on the values mentioned previously, they promote the impression that the constant, easy, uncritical, and productivity-oriented monitoring of bodily related data is a sure pathway for individuals to take greater control over their health (see Baker

2020), even though, for example, the popular productivity measure of taking ten thousand steps a day has its roots in an advertising slogan rather than in empirical research (McGlinty 2020). This particular ensemble contributes to the formation of the human subject as a datafied self whose information about decisions is fed back to companies to be used in additional product improvement and development that serve to reinforce the same values and to continue framing the subject in the same way. Such an iterative pattern can be found in many other instances, from social media to sociable robots, where Internet-enabled innovations are involved.

Seen through a temporal lens, much as in the case of the experimental subject, the datafied subject is a subject held to information provided at a point in time; that is to say, it is held to the past. Raphael Capurro observes: "Algorithms implement digital reifications of who we are and what roles we play in the drama of life" (Capurro 2019: 133). In describing what kind of vigilance would be required in order to escape such reifications, Capurro suggests using the "obfuscation" tactics proposed by Finn Brunton and Helen Nissenbaum, in which users of digital devices provide false or confusing data about themselves in order to take an end run around algorithms that aim to reify. Apart from the ethical questions involved in, say, trying to confound Amazon's algorithms by intensively searching for products that one has no intention of purchasing, such behaviors of resistance would in essence acknowledge and accept the "behavior" of algorithms to produce reifications of the self and so could implicitly serve to double down on the construction of the datafied self. Bruno Gransche (2020: 77) adds:

> The more things know about us and the more they proactively pre-arrange our world, the fewer opportunities we have to differ from our past digital selves. . . . *things get to know who we are and tie us down to who we were.*

Another outcome of the engineering of material objects that present a relatively unassuming façade of functionality to the user, while not calling attention to how it acts as an "interface" (Gransche 2020) to convey data to a wide array of others, is to change the lived experience of navigating spaces such as the home (Aydin et al. 2018). That such a change results in a narrowing of possibilities for navigating space is less evident than it is with regard to the datafied subject's experience of time.

53.3.3 *Implications*

It is widely accepted among engineers and the general public alike that engineering as a professional activity makes a contribution to advancing the social good because of its foundational character as a problem-solving enterprise. In particular, engineering innovation opens up new "possibility spaces" where familiar but intransient problems can be solved and even "wicked" ones can be mitigated. There is much, though, that this conventional view hides and occludes. It helps to sustain the impression that a technological solution to a problem is always the most appropriate solution, and so arguably works to discourage other approaches that might produce better results. In addition, the view that engineering is about problem-solving can serve to deflect attention from how a particular problem it is interested in solving is framed in the first place, with the result being that the adequacy of that frame might escape needed critical attention. When engineering responsibilities are framed solely as technological in character, as Erin Cech (2012), for instance, has pointed out, the role that engineering could play in alleviating social injustice risks going unconsidered. The focus of this section has been on another aspect of engineering that the conventional view assists in covering over: namely, how engineering can help to form and reinforce particular notions of the human subject or self.

In short, as subjects *for* engineering—and so as contributing to engineering as a driver and stimulus for corporate profitability and economic growth—both experimental subjects and datafied subjects become sites for the replication of particular engineering values. This puts professional engineering, particularly engineering in a difficult, if not paradoxical, position. The user-generated knowledge that engineers bring to the design process involves data that subjects directly give or indirectly give off (Susser 2016). Well before the concepts of the experimental and the datafied subject became a focus of academic investigation, the Czech phenomenologist Jan Patočka vigorously affirmed that we are more than the "sheer objectivity of massive givens" (Patočka 1998: 127). While what Patočka had in mind by "givens" was "sense data" and not data in the form of information, his observation is still a fitting one. Looked at from within a temporal frame, both the experimental and the datafied subject are anchored in what *has already* happened, as the result of ways of cognition such as decision-making, rather than in the space of what *may be*, the space of the *possible*. When it comes, though, to "leading a life" and to self-realization, it is the latter space that dominates (Patočka 1998: 129). Being in that space involves drawing on not only cognition but also other aspects of ourselves such as sensibility, affect, creativity, and imagination. By losing sight of the subject *of* engineering by attending less to the human subject as a whole self, engineering creativity and innovation risks undermining its aim of helping to better the conditions under which individuals can enjoy more flourishing lives.

In short, the resurgence of interest in thinking technology as a whole points to a need for engineering to more critically reflect on the role that it plays in promoting ideas of what it is to be a human subject, and how these ideas might be better aligned with the work that engineers do in the interests of supporting the whole, thriving human self.

53.4. Conclusion

This chapter highlights some work in contemporary Continental philosophy of technology that has taken up thinking about technology as a whole in a fresh way. It explores this work to clarify how it could lend its voice in order to contribute to how professional engineers might understand the scope of their practices, specifically by raising awareness among them as to how the decision-making activities of engineering design do not just lead to new or improved technologies and technological systems and to the deployment of increasingly engineered institutions and social practices. They also simultaneously work to form and sustain general models of the self or human subjectivity. If it is the case that the utmost challenge facing engineering today is the acquisition of a more critical understanding of what it is to be an engineer (Mitcham 2014), then part of facing up to this challenge is to understand the role that engineering plays, and the responsibilities that it bears, with respect to these model formations.

The sense of self-wholeness that is threatened by the subject *of* contemporary engineering becoming the subject *for* contemporary engineering is arguably given most directly in the feeling of being alive (Weber 2019). While nature plays a big role in nourishing this sense of wholeness (Weber 2019), it is not alone. Things can do it as well (Levinas 1991). Unless engineering takes steps to be vigilant when it comes to thinking about the question of how the things it makes can help to nourish the self as a whole, it can be said to disrupt and undermine its own potential for contributing as much as it could to the well-being of those to whom it bears professional responsibilities.

From this angle, particularly with respect to how raising awareness among engineers might result in increased attention to the whole self in the design process, the emergence within contemporary Continental philosophy of new ways of thinking about engineering as a whole can be seen as complementary to the two other trajectories of inquiry mentioned at the start of this chapter—inquiry into specific technologies and into technological ethics—which Jonas hoped would be represented in any philosophy of technology deserving of the name. A significant amount of newer work by philosophers in both of these areas reflect how Continental contributions to the field have gone beyond the "tag-teaming" focus on the analysis of the ethical, social, and political impacts of engineering objects which Carl Mitcham (1994) identified as a hallmark of the humanities approach to philosophy of engineering. The move away from "tag teaming" can be seen in the way that postphenomenologists have turned to analyze specific technologies in order to influence design and public policy (see Van Den Eede 2020). It comes out as well in the involvement of Continental philosophers in groups dedicated to teasing out fundamental ethical principles that could be used to support the development of responsible AI (see for example the European Commission High Level Expert Group on Artificial Intelligence 2019). In a similar way, reminders by post-classical philosophers of technology to engineers that their practices not only result in the generation of new artifacts but help to shape the self as well are attempts to get out in front of the design process rather than responding to its results once these results have entered societies and the consumer marketplace. Such recognition poses a challenge to think more intentionally about what kind of self a particular engineering design process is likely to support.

The perspectives explored in this chapter lead to a number of questions. How might they be directly applied to thinking about the design of specific technologies, as in Felix Hun Tan Lo's (2019) investigation, informed by the work of Simondon, into the issue of whether social robots should be built to resemble human beings? How might these general perspectives be brought to influence the design process by working in tandem with postphenomenological theory and its more specific focus on how, in the context of human-technology relations, particular technologies shape subjectivity in certain ways? How do these perspectives relate to thinking about engineering as an art (see also Chapter 9, "Engineering as Art and the Art of Engineering" by Lara Schrijver) or engineering as a form of handicraft? Might increased attentiveness within engineering design processes to the whole, dynamic, and open human self inspire new, and possibly more humble, narratives about engineering, ones that question the dominant narrative of engineering mentioned earlier? How might a shift in the preparation of professional engineers toward educating the "whole person" (see Chapter 55, "Reimagining the Future of Engineering" by Neelke Doorn, Diane P. Michelfelder, et al.) lead to caring more about the "whole self" in engineering design? It is hoped that these questions, and others inspired by them, might serve as promising vectors for future research.

Related Chapters

Chapter 1: "What Is Engineering?" (Carl Mitcham)
Chapter 9: "Engineering as Art and the Art of Engineering" (Lara Schrijver)
Chapter 31: "Experimentation" (Viola Schiaffonati)
Chapter 45: "Engineering as a Political Practice" (Govert Valkenberg)
Chapter 55: "Reimagining the Future of Engineering" (Neelke Doorn, Diane P. Michelfelder, et al.)

Further Reading

Campbell, T. (2011). *Improper Life: Technology and Biopolitics From Heidegger to Agamben*. Minneapolis, MN: University of Minnesota Press. (An intriguing defense of the "philosophical merits of play", understood as a practice of attention and *bios*.)

Hayward, M. and Geoghegan, B. (eds.) (2011). *SubStance* 41(3) on Gilbert Simondon. (A good source for "catching up", as the editors put it, with the work of Simondon, including how it relates to the thought of Bernard Stiegler.)

Smith, D. (2018). *Exceptional Technologies: A Continental Philosophy of Technology*. London: Bloomsbury Academic. (An insightful exploration of how Continental philosophy of technology would benefit from a combination of empirical and transcendental inquiry.)

Weber, A. (2019). *Enlivenment: Towards a Poetics for the Anthropocene*. Cambridge, MA: MIT Press. (A well-articulated proposal for replacing Enlightenment ways of thinking, including the belief in progress through the control of nature, with an "enlivenment" approach that acknowledges our intertwining with the natural world.)

References

Achterhuis, H. (2001). Introduction: American Philosophers of Technology. In H. Achterhuis (ed.), *American Philosophy of Technology: The Empirical Turn*, trans. Robert P. Crease. Bloomington, IN: Indiana University Press, pp. 1–11.

Aydin, C., Woge, M. and Verbeek, P.-P. (2018). Technological Environmentality: Conceptualizing Technology as a Mediating Milieu. *Philosophy & Technology*, 32, 321–338.

Baker, D.A. (2020). Four Ironies of Self-quantification: Wearable Technologies and the Quantified Self. *Science and Engineering Ethics*, 26, 1497–1498.

Brey, P. (2010). Philosophy of Technology After the Empirical Turn. *Techné: Research in Philosophy and Technology*, 14(1), 36–48.

Capurro, R. (2019). Enculturating Algorithms. *Nanoethics*, 13, 131–137.

Cech, E. (2012). Great Problems of Grand Challenges: Problematizing Engineering's Understandings of Its Role in Society. *International Journal of Engineering, Social Justice, and Peace*, 1(2), 85–94.

Cérézuelle, D. (2013). Technological Acceleration and the 'Ground Floor of Civilization'. In H. Jerónimo, J.L. Garcia and C. Mitcham (eds.), *Jacques Ellul and the Technological Society in the 21st Century*. Dordrecht: Springer, pp. 63–72.

Crogan, P. (2010). Bernard Stiegler: Philosophy, Technics, and Activism. *Cultural Politics*, 6(2), 133–156.

Cukier, K. and Meyer-Schoenberger, V. (2013). The Rise of Big Data: How It's Changing the Way We Think About the World. *Foreign Affairs*, 92(3), 28–40, May/June.

Du, X., Vasudevan, R. and Johnson-Robertson, M. (2019). Bio-LSTM: A Biomechanically Inspired Recurrent Neural Network for 3-D Pedestrian Pose and Gait Prediction. *IEEE Robotics and Automation Letters*, 4(2), 1501–1508, April.

Ellul, J. (1964). *The Technological Society*. Trans. John Wilkinson. New York: Vintage Books.

Ellul, J. (1990). *The Technological Bluff*. Trans. Geoffrey W. Bromiley. Grand Rapids, MI: Wm. B. Eerdmans Publishing Company.

European Commission High-Level Expert Group on AI (2019). *Ethics Guidelines for Trustworthy AI*. Brussels: European Commission, April.

Feenberg, A. (1999). *Questioning Technology*. New York: Routledge.

Feenberg, A. (2017a). *Technosystem: The Social Life of Reason*. Cambridge, MA: Harvard University Press.

Feenberg, A. (2017b). A Critical Theory of Technology. In U. Felt, R. Fouché, C.A. Miller, and L. Smith-Doerr (eds.), *Handbook of Science and Technology Studies*. Cambridge, MA: MIT Press, pp. 635–663.

Fritzsche, A. (2016). Open Innovation and the Core of the Engineer's Domain. In D. Michelfelder, B. Newberry, and Q. Zhu (eds.), *Philosophy and Engineering: Exploring Boundaries, Expanding Connections*. Dordrecht: Springer, pp. 255–266.

Gransche, B. (2020). Handling Things That Handle Us: Things Get to Know Who We Are And Tie Us Down to Who We Were. In *Relating to Things: Design, Technology, and the Artificial*. London: Bloomsbury Visual Arts, pp. 61–80.

Heidegger, M. (1977). *The Question Concerning Technology, and Other Essays*. Trans. William Lovitt. New York: Harper.

Hui, Y. (2013). Technological System and the Problem of Desymbolization. In H. Jerónimo, J.L. Garcia, and C. Mitcham (eds.), *Jacques Ellul and the Technological Society in the 21st Century*. Dordrecht: Springer, pp. 73–82.

Ihde, D. (2001). *Bodies in Technology*. Minneapolis, MN: University of Minnesota Press.

Ihde, D. (2012a). *Experimental Phenomenology: Multistabilities*, 2nd ed. Albany: SUNY Press.

Ihde, D. (2012b). Can Continental Philosophy Deal With the New Technologies? *Journal of Speculative Philosophy*, 26(2), 321–332.

Jerónimo, H., Garcia, J.L. and Mitcham, C. (2013). *Jacques Ellul and the Technological Society in the 21st Century*. Dordrecht: Springer.

Jonas, H. (1979). Toward a Philosophy of Technology. *Hastings Center Report*, 9(1), 34–43.

Kroes, P. and Meijers, A.W.M. (eds.) (2000). *The Empirical Turn in the Philosophy of Technology*. Amsterdam: JAI-Elsevier.

Levinas, E. (1991). *Totality and Infinity*. Dordrecht: Kluwer.

Lo, F.T.H. (2019). The Dilemma of Openness in Social Robots. *Techné: Research in Philosophy and Technology*, 23(3), 342–365.

Marks, M. (2019). Artificial Intelligence Based Suicide Prediction. *Yale Journal of Health Policy, Law, and Ethics*, 18(3), 98–121.

Martin, M.W. and Schinzinger, R. (1996). *Ethics in Engineering*, 3rd ed. New York: McGraw-Hill Higher Education.

McCarthy, C.R. (2008). The Origins and Policies that Govern Institutional Review Boards. In E.J. Emanuel et al. (eds.), *The Oxford Textbook of Clinical Research Ethics*. Oxford: Oxford University Press, pp. 541–551.

McGlinty, J.C. (2020). 10,000 Steps a Day? Fewer May Be Fine. *The Wall Street Journal,* June 13–14, A2.

Michelfelder, D.P. (2018). Risk, Disequilibrium, and Virtue. *Technology in Society,* 52, 32–38, February.

Mitcham, C. (1994). *Thinking Through Technology: The Path Between Engineering and Philosophy*. Chicago: The University of Chicago Press.

Mitcham, C. (2014). The True Grand Challenge for Engineering: Self-Knowledge. *Issues in Science and Engineering,* 31(1), Fall.

Novaes de Andrade, T. (2008). Technology and Environment: Gilbert Simondon's Contributions. *Environmental Sciences,* 5(1), 7–15.

Peterson, M. (2017). What Is the Point of Thinking of New Technologies as Social Experiments? *Ethics, Policy, and Environment,* 20(1), 78–83.

Patočka, J. (1998). *Body, Community, Language, World*. Trans. Erazim Kohák. Chicago: Open Court.

Riis, S. (2018). *Unframing Martin Heidegger's Understanding of Technology: On the Essential Connection Between Technology, Art, and History*. Trans. R. Walsh. Lanham, MD: Lexington Books.

Rip, A. (2000). There's No Turn Like the Empirical Turn. In P. Kroes and A. Meijers (eds.), *The Empirical Turn in the Philosophy of Technology*. Amsterdam: JAI, pp. 3–17.

Sharon, T. (2017). Self-Tracking for Health and the Quantified Self: Re-articulating Autonomy, Solidarity, and Authenticity in an Age of Personalized Healthcare. *Philosophy and Technology,* 30(1), 93–121.

Simondon, G. (2016). *On the Mode of Existence of Technical Objects*. Trans. C. Malaspina and J. Rogove. Minneapolis, MN: Univocal Publishing.

Stiegler, B. (1998). *Technics and Time, 1: The Fault of Epimetheus*. Trans. R. Beardsworth and G. Collins. Palo Alto, CA: Stanford University Press.

Stiegler, B. (2010). *Taking Care of Youth and the Generations*. Trans. S. Barker. Stanford, CA: Stanford University Press.

Stiegler, B. (2014a). *Symbolic Misery, Vol. 1: The Hyperindustrial Epoch*. Trans. B. Norman. Cambridge: Polity Press.

Stiegler, B. (2014b). Afterward: Web Philosophy. In Harry Halpin and Alexandre Monnin (eds.), *Philosophical Engineering: Toward a Philosophy of the Web*. Wiley-Blackwell, pp. 187–198.

Susser, D. (2016). Information Privacy and Social Self-Authorship. *Techné: Research in Philosophy and Technology,* 20(3), 216–239.

Van de Poel, I.R. (2013). Why New Technologies Should Be Conceived as Social Experiments. *Ethics, Policy, and Environment,* 16(3), 352–355.

Van de Poel, I.R. (2016). An Ethical Framework for Evaluating Experimental Technology. *Science and Engineering Ethics,* 22, 667–686.

Van den Eede, Y. (2020). The Purpose of Theory: Why Critical Constructivism Should 'Talk' and Postphenomenology Should 'Do'. *Techné: Research in Philosophy and Technology*. doi:10.5840/techne202027115.

Verbeek, P-P. (2005). *What Things Do: Philosophical Reflections on Technology, Agency, and Design*. State College, PA: Pennsylvania State University Press.

Weber, A. (2019). *Enlivenment: Toward a Poetics for the Anthropocene*. Cambridge, MA: MIT Press.

Zuboff, S. (2019). *The Age of Surveillance Capitalism*. New York: PublicAffairs.

54

ENGINEERING PRACTICE FROM THE PERSPECTIVE OF METHODICAL CONSTRUCTIVISM AND CULTURALISM

Michael Funk and Albrecht Fritzsche

54.1. Introduction

During the past decades, engineering has spread out into many new domains (Neely et al. 2018). It has become involved in a variety of problems which go far beyond the classical notion of industry and reach into the wider reality of human life (Fritzsche and Oks 2018). In the course of this process, engineering has become more diverse and split into specific fields which develop with their own dynamics and involve different stakeholders (Michelfelder 2017). At the same time, the philosophy of engineering has taken strong efforts to gain a deeper understanding of engineering practice and become engaged in a dialogue with the engineers themselves (Franssen et al. 2016; Kroes and Meijers 2000). An important motivation for these efforts is the insight that engineering cannot be reduced to an applied science (Michelfelder 2010); it is more deeply related to human thinking about problem-solving and requires therefore a deeper consideration of the world in which it takes places and its challenges for people living in it (e.g. Pitt 2000; Mitcham 1994).

Driven by the emergence of new technologies and the social phenomena related to them as causes or effects, the philosophy of engineering has rapidly advanced and developed its own repertoire of theories, concepts and methods of investigation which is addressed in various chapters of this volume (for instance, see Chapter 6, "Scientific Methodology in the Engineering Sciences" by Mieke Boon, Chapter 7, "Engineering Design and the Quest for Optimality" by Maarten Franssen, and Chapter 10, "Creativity and Discovery in Engineering" by David H. Cropley). From a wider perspective, it is also part of a larger development in philosophy which turns the attention increasingly to the actual conditions under which human life takes place. The roots of this development can be traced back at least to the late nineteenth century, a time which is not by coincidence characterized by the experience of rapid industrial growth and a changing social and material environment. This is reflected in various philosophical movements from pragmatism to phenomenology, which have left their mark on the philosophy of technology, but also on other streams of research. Among them, the German schools of methodical constructivism and methodical culturalism are of particular interest for the current discourse on engineering, as they put the observation of human practice in the center of the investigation. Moreover, the epistemic and methodical interest is not limited to scientific practice. It involves a wide range of cultural expressions of human activity. The schools offer a unique argumentative pattern from the combination of often separated philosophical approaches like phenomenology, hermeneutics and pragmatism on the one side, and formal logics and analytic philosophy of language on the other. Due to a lack of English translations—two important exceptions are Kamlah and Lorenzen (1984) and Janich (1997)—the schools have so far received fairly little attention internationally, in strong contrast to the considerable influence which they had on the German discussion. Nevertheless, they offer a lot of interesting arguments and points of view which can enrich the philosophical discussion and open up new directions for research on engineering.

This chapter gives a brief introduction to the genesis of the schools of methodical constructivism and culturalism and describes the characteristic elements of their approach to human practice. It then explores what the methodical schools, as we will address them here, can add to the current discussion of engineering and names some potential application areas.

54.2. Foundations of Methodical Constructivism and Culturalism

The methodical schools emerged as a distinctive approach in the philosophy of sciences, language, and technology in the middle of the twentieth century. Their main protagonists were Wilhelm Kamlah, Paul Lorenzen, Jürgen Mittelstraß, Kuno Lorenz (not to be confused with Paul Lorenzen), Friedrich Kambartel and Peter Janich, who first worked in Erlangen and later on in Constance and Marburg. Methodical constructivism and culturalism build upon four foundational elements which are directly linked to philosophical works from the early twentieth century in the wake of the scientific and technological developments at that time. In the following paragraphs these four foundational elements will be summarized, so that the works of the methodical schools can be more easily set in relation to other approaches which have proven to be highly influential in the philosophy of engineering and technology.

54.2.1 *Phenomenology and Hermeneutics*

The first element to be named here is the epistemic foundation of embodied lifeworld practice in Husserl's phenomenology and hermeneutics in his book *The Crisis of European Sciences and Transcendental Phenomenology: An Introduction to Phenomenological Philosophy* (published 1936; see Husserl 1970) and Heidegger's *Being and Time* (published 1927; see Heidegger 2010). Husserl argued that modern sciences had lost their capabilities to challenge real life. They had become abstract, mathematically formal, theoretically idealized far away from human reality in everyday life and therefore lacked social significance. Scientific worlds are theoretical worlds but not practical. In consequence, Husserl introduced the term "lifeworld" with at least two motivations. Firstly, it was to show how far scientific rationality had drifted away from people's real problems, as they presented themselves in the lifeworld as a practical sphere of human engagement, which should be meaningfully emphasized by scientists instead of useless theoretical argumentation. In the presence of a theoretical transition to independence, the meaning of practice in real life plays a particularly important role in Husserl's later approach. He contributed to a practical turn in twentieth-century philosophy which also moved the focus to technical actions.

Secondly, and maybe more importantly, Husserl demonstrated how the lifeworld itself was not only an object of scientific investigations but also its methodical precondition. Scientific inquiry is performed by scientists who grew up in real life, practically create experimental instruments, talk to each other and therefore could not, methodically speaking, banish their practical engagements from their theoretical investigations. The world is not a mathematical model; it is empirically given in everyday life, relative to a personal point of view, and before scientific explanations start. The art of metrology—the skill of measurement—is the lifeworld axiom for geometry. It serves as methodical foundation for the operational deduction of ideal geometrical forms by using basic elements. The basic elements are practically given in the succeeding actions of measurement (Husserl 1970). It was not only the relevance for real-life problems in scientific results that was overlooked, but also the real-life foundation of its result-creating processes. Since, due to a growing complexity of measuring instruments, metrology became an engineering skill itself, an intimate linking between scientific research and the engagement of technical experts was uncovered.

Heidegger also refers to everyday life practice as the foundation of human existence. Tool use is not only handcraft on an epistemic lower level than sciences. It is seen as the initial as well as the very intimate physical relation between humans, other humans and natural environments. Without implicit knowledge, craftsmanship and virtuosity in using instruments, there would be no human existence, neither technology nor sciences (Heidegger 2010). We reveal and interpret phenomena within a lifeworld of shared physical actions. What Husserl presented as a critique of useless scientific abstraction became a fundamental existential claim in Heidegger's analysis. Although to a lesser extent than Husserl, Heidegger's views also exerted some influence on methodical constructivism and culturalism, in particular through Kamlah's works. The discussion between methodical constructivism, culturally oriented philosophy and phenomenology is still going on today and illustrates even stronger conceptual links between the approaches (Janich 1999). Although still rather vague, a foundational element for a philosophy of engineering can be found already in this root: related with the lifeworld emphasis and focus on everyday life goes a unique revaluation of epistemic practice. This revaluation captures especially in the fundamental ontological analysis of Heidegger all areas of human behavior including technical practice, whereas Husserl presented claims about measurement as a link between sciences and technology. Implicit pre-understandings and body actions received particular attention from the methodological perspective as the starting point of epistemic reasoning.

54.2.2 *Methodology and Pragmatism*

The second element is the epistemic foundation of scientific operations by Dingler (1881–1954) in his *Essays Concerning Methodology* (1936, 1942, estate; see Dingler 1987) and other writings. What methodical constructivism and culturalism have in common with pragmatism is the intention to relate science to practical experience. Just like Husserl, Dingler has been intellectually inspired by geometry, mathematics and physics of atoms and quanta. Methodical geometry, as Dingler states, contains sentences and definitions that characterize the ideal basic elements/forms and at the same time serve as manufacturing instructions of how to realize them practically. The logical function of theoretical axioms is replaced by practical directions (Dingler 1987: 20, 27; see also Dingler 1933). In other words: there is no scientific ideal of an object without *knowing how to create it*. In a methodically explicit and systematic way, Dingler developed operational foundations of scientific research like "methodical order" (Wolters 2016) and "proto-theories" (Janich 2016). His influence on later developments of methodical constructivism and culturalism was high, and his ideas still motivate debates in philosophy of technologies and sciences (Janich 1984, 2006b).

With respect to engineering practice, the replacement of a logical truth criterion by a pragmatic one receives outstanding importance: repeatedly successful practice stands for truth both in linguistic and non-linguistic actions (see Chapter 8, "Prescriptive Engineering Knowledge" by Sjoerd Zwart, this volume). The knowing-how aspect of scientific ideals directly touches what today is also discussed under the label of technoscience—the intimate amalgamation of technologies and sciences. However, the current discussion seems rather related to Husserl, whose analysis of Galileo and his telescope serves, for example, as a paradigmatic case in Ihde's approach of technical mediation (Ihde 1990). It can be argued that the methodical schools addressed here are in some points close to current approaches like technoscience, mediating technologies or instrumental realism. But they also provide a much more systematic focus on language and the construction of scientific knowledge that goes far beyond technical mediation (Haas 2010; Mainzer 2010). Dingler, for instance, takes such an approach in showing how scientific terms and ideals are practically engineered. He suggests a technoscience interpretation in which engineering has a major impact on theoretical conceptualizations. This point of view is not so far away from a dictum attributed to Richard Feynman (e.g. Way 2017; Nordmann 2014): "What I cannot create, I do not understand." Consequently, engineering is not the application of scientific theory, but vice versa: scientific theory depends on engineering skills and the related capabilities to produce scientific measuring technologies. Here we can see the conceptual root for a point of view, where engineering is not reduced to the application of scientific theories only.

54.2.3 Analytic Philosophy and Philosophy of Mathematics

The third element is the epistemic foundation of formal language and scientific theories, as it has been elaborated by analytic philosophy. Formal language became a genuine topic of Frege's (1848–1925) works like *On Sense and Reference* (1892; see Frege 2008: 23–46) and others. Scientific sentences and theories have been logically analyzed and systematically investigated regarding sense and meaning. In his book released in 1928, *The Logical Structure of the World*, Carnap (1891–1970) proposed an approach of logical and formal reconstruction of reality (see Carnap 1998). Although Carnap's definition of "world" is abstract and different from Husserl's understanding of lifeworld, Lorenzen and others have recognized that it has relevance for the research program of the methodical schools. Methodical constructivism and culturalism are characterized by the combination of elements from phenomenology and analytic philosophy, which was in particular achieved by Kamlah and Lorenzen in the 1960s. The analytic approach also shaped later works by Lorenzen and Lorenz on game semantics and dialogical logics (Lorenz 2009; Lorenzen and Lorenz 1978; Mittelstraß and von Bülow 2015). This unique combination of phenomenology and analytic philosophy justifies a certain interest for the philosophy of engineering practice as it creates various access points from the Anglo-American traditions, initially in the theory of science and later in the theory of cultural and technical practice as well. Language and appropriate formulations in scientific sentences or theories are as important as experiments and other instrumental operations. The same is true for engineering. The methodical approach includes precise logical reconstructions that find its roots in contingent everyday life actions. Here also, the theoretical truth criterion related to the formal consistency of theoretical claims finds its place.

54.2.4 Critique of Language and Transcendental Grammar

The fourth element to be named is the epistemic foundation of socially shared language practice by Wittgenstein (1898–1951). Wittgenstein had a stronger influence on the second and third generation of the Constance constructivism (Kambartel and Rentsch) with his analyses of ordinary language and methodical foundations of transcendental grammars. In his *Philosophical Investigations* (1953 estate; see Wittgenstein 2009) he argued that meaning is created by language practice, not by theoretical definitions. Thereby, our use of language follows a non-semantic grammar which structures succeeding actions and how we are used to talking about it. It is a grammar of performances, not of formal signs. With his ideas, Wittgenstein influenced the concept of "transsubjectivity" which has been developed in methodical constructivism and culturalism by Lorenzen, Mittelstraß, Kambartel and Rentsch in order to describe the socially shared constitution of subjectivity: We are practically always individuals in a society and can only become individual persons in a shared (life)world of shared language, meaning, habitualizations, norms, etc. (Kambartel 1989, 1996; Rentsch 2000, 2003; Schwemmer 2010). Moreover, Wittgenstein developed a notion of pragmatic truth including tool use and bodily and linguistic actions which show generally strong similarities to the conceptualizations of Dingler and other constructivists (Funk 2018a). Especially his late writings are traversed by examples of technical practice which enable further approaches in the philosophy of technology and language (Coeckelbergh and Funk 2018). Wittgenstein's works provide important starting points for the philosophy of engineering practice since he adds another facet to the analysis of human practice (close to Husserl and Heidegger) with a unique focus on the generation of meaning while practically using language: how we are used to talk about how we use a certain technology (technical situation), and how we are used to talk about engineering.

54.3. The Evolution of the Methodical Schools Since the Late Twentieth Century

In the previous subsection, four foundational elements of the methodical schools in philosophy and their impact on the philosophy of engineering practice have been presented. This subsection gives a short overview of the genesis of methodical constructivism and culturalism with a particular focus on the unique points of view and argumentative patterns. Historically, the intellectual center was located at the institute of philosophy at Friedrich-Alexander University at Erlangen. For this reason, the first period in the 1960s is called the Erlangen School (Kambartel and Stekeler-Weithofer 2005). Here Kamlah and Lorenzen took on the development of what is also called a methodically ordered approach in the theory of science, from the combination of existing research lines. The principle of methodical order, as we will see in the next section, is one of the most important concepts with a high significance for the philosophy of engineering practice today. Kamlah was influenced by phenomenology, hermeneutics and anthropology, whereas Lorenzen's background was in logic, analytic philosophy, philosophy of language and mathematics. It was one of the seldom fruitful meetings between the so-called continental and analytic traditions of twentieth-century philosophy. A unique argumentative pattern of the methodical schools can be found in the combination of practice-oriented philosophy of tool use with analytic philosophy of language (Janich 1991; Mittelstraß 2008, 2012). In their 1967 published book *Logical Propaedeutic* (see Kamlah and Lorenzen 1967; English translation: Kamlah and Lorenzen 1984; see also Lorenzen 1968, 1974, 2000), results of this cooperation have been summarized, and it also became a programmatic manifesto for works in the field. Janich, Mittelstraß, Lorenz, and others picked up on this methodical approach to release their own contributions to methodical philosophy of science. For instance, the notion of proto-theory describes how concrete everyday practice—e.g. of using metrological tools—is turned into a prototype of a related scientific theory (Janich 1969, 1980, 2016). Therefore, the principle of homogeneity regulates how to set normalized aims for the production of indicated values of measurement (Janich 2008).

The second period started in the 1970s with the first generation of students (Janich, Mittelstraß, Lorenz, etc.) who went to Constance and continued their epistemological investigations there together with Kambartel. This marks the time of the Constance School. During this time, the movement became more and more diverse. While still connected by the methodical approach, most members also worked in other disciplines such as ethics, anthropology or logics, so that the whole movement turned into a general constructive philosophy (Mittelstraß 2016; Thiel 2010). Also due to different interests beyond the methodical manifesto, the close cooperation between Kamlah and Lorenzen, the original Erlangen pioneers, came to an end. In the 1970s, Mittelstraß started editing the overarching *Encyclopedia of Philosophy of Epistemology* (published since 1980 in two editions with four and eight books; see Mittelstraß et al. 1980ff.), which was written by members of the school and their students. Beyond being an encyclopedia, it went on to become a general conceptualization of terminology and definitions of methodical philosophy.

In the 1970s, a general extension of the focus of methodical philosophy started. The approach was applied to many fields of human lifeworld practice including ethics and technical knowledge. *Constructive Logics, Ethics and Philosophy of Sciences* is the title of another programmatic book by Lorenzen, released in 1973 with Schwemmer, which also expresses the new interest in ethics of technology in the German-speaking community since the 1970s (Lorenzen and Schwemmer 1975).

Since the 1980s, the Marburg School of constructivism relates to the works of Janich and colleagues who continued their research in Marburg, after Janich became a professor at this university. Overall, Janich is one of the very few persons—if not the only one—who has been actively involved in all different schools over the years, which are all reflected in different ways by approaches taken in this third period. From Janich's work, methodical constructivism was established as an umbrella term for the diverse investigations of the several generations of the Erlangen, Constance and Marburg Schools, which took place in parallel and in close cooperation with one another. Since the 1990s, Janich, together with Hartmann and others, further elaborated the methodical constructivist views into a culturalistic approach, which has since then received considerable attention in Germany (Janich 1996, 2006a, 2010; Hartmann and Janich 1996, 1998).

54.4. Applications to the Philosophy of Engineering

In the previous sections we briefly reconstructed four conceptual foundations of the methodical schools and described their historical evolution over time. This section discusses key claims of the schools and their relevance for engineering practice. In the true spirit of methodical constructivism and culturalism, we start with a practical illustration, as it was often used by Peter Janich in his lectures and writings (Janich 2006a: 27, 58, 74, 83 et passim)—manuals and recipes. We use this to elaborate the basic pattern of how the methodical schools further developed their four roots and how this applies to the philosophical conceptualization of engineering practice.

54.4.1. Uncovering the Richness of Sequential Order

Husserl already relates the analysis of lifeworld to everyday practice—an insight that later on became a key concept of the culturally sensitive approach (Janich 2006a: 73 et passim; Welter 2010). Argumentative patterns of the methodical schools include a lifeworld apriori (Gethmann 2005) as a presuppositionless systematic starting point. The methodological foundation addressed here can be observed in a simple experience that nearly everybody had in childhood: cooking. How do we learn cooking? Of course, the concrete story and cuisine differs, but what remains the same in all different cultures of food preparation are certain sequences that lead to successful results. Cooking is usually characterized as an art. It involves tacit knowledge (Polanyi 2009), in which daily experience plays a major role. What matters in cooking is not so much a scientific model of a dish, but the pragmatic success of a certain operational order in food preparation—by the way, also scientific models can be seen as mere means to fulfill a certain end (Janich 2006a: 24). The key question is: Am I able to create in a limited timeframe with my resources and tools, also in cooperation with other persons, a tasty and healthy meal that fulfills the needs of good nutrition? In other words: Am I able to produce with the appropriate means the intended aim? Usually, we learn cooking from our ancestors, family members or friends who share sequences of achievement with us, but not from scientific theories of thermodynamics. This sharing involves details which are likely to be discarded as trivial or obvious in a merely theoretical approach, but still play a decisive role in the process. When we want to create a soup, for instance, in our sequence of preparation, the herbs are placed at a later point than the preparation of the soup stock. We learn to first create a basis and then enrich it with additional flavors. If we put parsley, salt and pepper in an empty pot and activate the cooktop without any fluid inside the pot, we simply burn and destroy it—or at least, we create a culinary disaster. In order to prevent mistakes like this, we are taught to follow a certain sequence. We operate step by step. Therefore, recipes contain clear instructions in which order to proceed (see Figure 54.1).

This is a simple illustration of what has been called the principle of methodical order or pragmatic order in the works of the methodical schools (Lorenz 2016; Wolters 2016; Janich 2006a: 26–28, 40, 58, 80 et passim). The historical amount of sequentially ordered succeeding actions creates a

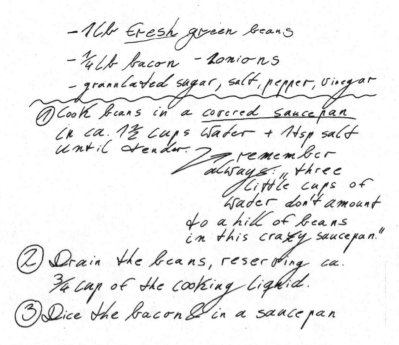

Figure 54.1 Illustration of a family recipe for a common German bean dish

Note: English translation, excerpt; dish will not taste nice when prepared without further treatment.

concrete irreversible cultural state ("Kulturhöhe"). Progress is not a matter of theoretical sentences, but of gradually synthesizing levels of succeeding constructive procedures. For instance, Euclidean geometry finds its roots in handcraft actions: the application of drawing instruments and rulers for geometrical investigations already requires the cultural state of being able to produce the drawing instruments and rulers (Janich 2006a: 37 et passim). Therefore, structures of non-verbal actions that lead to the construction and manufacturing of facts and circumstances, including technical products, are emphasized by the methodical schools with respect to the poiesis-paradigm (Janich 2006a: 22–28, 34–35, 39–44, 54–57 et passim).

Cooking is a praxis performed on a poiesis-related cultural state, dependent on the concrete manufacturing actions that lead to the means which are supposed to be used—this precondition is usually not mentioned explicitly in recipes. Instead, cooking instructions contain information about the amount of condiments. While cooking, we learn how to use a scale or measuring cup in order to gauge the appropriate amount of ingredients. A precondition of this procedure is the existence of normalized measuring units—like kilogram or liter—and a calibrated scale or cup. These units, of course, are a matter of convention, as their still existing diversity in different countries illustrates. The convention depends on the possibility of providing standardized artefacts to refer to lengths, volumes, etc. Both the convention and the infrastructure which allows for the recreation of calibrated references shape pragmatic preconditions which are also at least implicitly present while following the instructions of recipe books (poiesis-paradigm). In terms of engineering, more complex metrological tools fulfill the same function and require even more effort to be maintained (see also Fritzsche 2010, 2009). What the cooking example also illustrates very well is that measurements serve as a pragmatic idealization, a reference which we aspire to meet, without fully being able to achieve it (see also Wittgenstein 1984: 218; Funk 2018a: 362–363). As Poser (2012) shows, similar forms of vagueness play a huge role in the general context of engineering as well.

On the other hand, in safety-relevant technical applications, vagueness is supposed to be avoided. Forms of compliance are applied, based on experiences and tests that pragmatically justify concrete standardized gauges. Therefore e.g. fire extinguishers are not only distinguished into safety classes, accordingly produced and controlled at the point of sale. In order to ensure appropriate function, after a certain period of time the normalized features of extinguishers of a certain class need to be checked, which happens again in a concrete procedural order. The sequence of the procedures adds a structural component which is not present in the mere collection of properties which extinguishers are expected to possess. Here one can see that the concept of methodical order applies not only to the production or application of artifacts, but also to their maintenance.

Operational sequences embedded in wider lifeworld environments can also be found in user and construction manuals, as well as in many other forms of technical documentations that play an important role in engineering practice. In contrast to older forms of apprenticeship, training and education in the field of engineering takes a step away from actual learning on the job and requires the reference to formal descriptions of certain operations. A wider hermeneutical task of engineering is to translate the intended use of a technical artifact into an explicit sequential process description that can be understood by users. "Explicit" includes several meanings: words and sentences as well as images. Once more, cookbooks can serve as a popular example which shows how such descriptions are embedded in the lifeworld with pictures of the final dishes, physical movements in handling tools and ingredients, and other additional information which departs from scientific descriptions of cooking and creates references to further aspects which are relevant for a practice.

54.4.2. The Role of Practice for the Understanding of the Subject Matter

Further investigations into the visual representation of methodical orders can also provide important insights in advanced fields of engineering like currently emerging nanotechnologies or synthetic biology. Here, the situation is further complicated by the fact that the objects of operation, single-cell organisms or nanoparticles, are too small to be seen with the naked eye. Visualizations therefore must find a different basis to describe the treatment of the subject matter in research laboratories as well as industrial production facilities. Mere scientific models are insufficient to do so, as they depend themselves on experimentation to evolve over time. Visualizations therefore rely on other sources, such as the imagery from standardized routines and communicable practices from other fields of activity, where the application of common instruments can serve as a shared reference. This is illustrated, for example, by the notion of scissors in genetic engineering, which are even often depicted as such in publications (see Figure 54.2).

Images such as this one draw on practical experience with successful technical action and at the same time express the operational order related to this action that leads to the intended result. Similar observations can be made for words and other carriers of meaning in communication as well. They

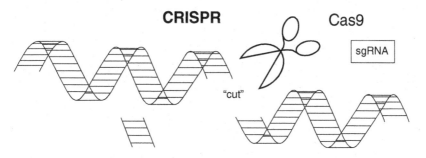

Figure 54.2 Lecture script illustrating genetic editing with scissors

emerge as vocabulary in the attempt to gain access to and control over the subject matter in question, such that repetitive operation, knowledge exchange and learning over time can take place. At the same time, they are constantly actualized by practical experience, such that the understanding of the subject matter is a floating entity, which can change over time (Fritzsche 2018).

The application of normalized components—like standardized gears in mechanical engineering or bio-bricks in synthetic biology (Bölker et al. 2016: 31–32)—is one basic element of engineering practice. The operative conceptions of engineering practice follow a similar dynamic. Engineers are able to act only if the pragmatic success of their actions is predictable in any situation. Due to this demand, not only are the components and operations normalized, but also the language in which the operations are described. For this reason, not only units of measurement, but entire terminologies in different fields of work often show a large diversity. Furthermore, the success of technical operations is a matter of convention as well, enabled by the possibility to attribute errors either to the design and soundness of implementation of technology or to the users handling it (see e.g. Fritzsche 2009). The criteria for this attribution are not inherent in any functional concept of a device, but added from the outside.

54.4.3. Engineering Knowledge, Scientific Knowledge and Teaching

Having clarified the importance of engineering practice and its influence on the understanding of the subject matter, the implications for engineering knowledge can now also be discussed in a broader sense. As other contributions to this volume show (for instance, see Chapter 5, "What Is Engineering Science?" by Sven Ove Hansson, Chapter 8, "Prescriptive Engineering Knowledge" by Sjoerd Zwart, Chapter 9, "Engineering as Art and the Art of Engineering" by Lara Schrijver, and Chapter 20, "Engineering Design" by Peter Kroes, this volume), one of the key questions of the philosophy of engineering is: What is engineering knowledge, and how does it relate to handcraft skills as well as (natural) scientific theories? Although these questions have already received considerable attention in the philosophy of engineering, the emphasis of methodical constructivism and culturalism on methodical order—including the poiesis-paradigm—may still be able to add a new perspective to the phenomenological, hermeneutical and analytic approaches which have been explored so far.

A suitable point of access is the concept of proto-theories, which emerge in the gradual development of theoretical knowledge from a pragmatic lifeworld basis (Janich 2006a: 73 et passim). In many respects, proto-theories can be described as mediators between abstraction and practical experience, which are accessible from both directions. As theories, they are still in a nascent and dynamic stage. They can therefore give insight into the efforts necessary for the formation of a scientific theory. At the same time, they mark the departure from individual experience in the search for common references within a group of experts. Studies like those performed by Janich (1969, see also Janich 1980) in the context of physics show how evidence about proto-theories can be collected. Again, the field of metrology can serve as an example in the way how simple practical procedures like hand-crafting measuring instruments step by step lead to the idealization of gauges and scales: the terms that describe and therefore also normalize the procedures that result in the measurement of numeric values.

Early work by Dingler (1933) and, to some extent, also Husserl (1970), Heidegger (2010), and Wittgenstein (2009) also serve as an inspiration for an etymological approach to the study of transitions from practice to theory in engineering knowledge. Looking at the origins of modern scientific terminology in ancient Greek, it turns out that abstract objects, such as lines or circles in geometry, are described by terms which have their origin in the poietic work of craftsmen, like the usage of a potter's wheel or drawing equipment (Janich 2006a: 27–28, 37, 41–42 et passim). As craftsmen were a common sight for thousands of years, it seems justified to assume that this sight left its mark in the understanding of abstractions. A similar effect can be assumed to take place today as well in teaching science and engineering. Although craftsmen might not be at the first glance present any more, abstract terms are still learned from the observation of gestures and the usage of tools and not from formal definitions (on the level of a concrete cultural state). This adds a richness to knowledge which is still very little understood. For engineers, however, who are constantly forced to relate abstract models with practical application, these aspects of knowledge can be expected to play an important role.

54.5. Summary and Outlook

In the course of the ongoing exchange among philosophers over the past decades, many thoughts expressed by the schools of methodical constructivism and culturalism have already found their way into the international discussion of engineering and technology. Their full body of work, however, has received very little attention. With this chapter, we have made an attempt to give some insight into the points of view and argumentative patterns of the methodical schools. As the previous examples have shown, methodical constructivism and culturalism can enrich current debates with a new perspective and complement existing research programs with another view on engineering practice and knowledge generation. Although details have not been elaborated, we hope that the chapter can serve as a resource for scholars who are looking for different access points for a philosophical study of engineering and possibilities to relate it to other philosophical works from the twentieth century.

The approach of the methodical schools may prove to be particularly informative in the discussion of emergent new fields of engineering and the digital transformation, which leads to a stronger reliance on formal-symbolic representations of lifeworld phenomena in the work of engineers. Based on the considerations in the previous chapters, it seems quite clear that this cannot mean that practical experience becomes obsolete. Instead, one must expect that digital technology increases the variety of possibilities to transform practice into abstract terminology as material constraints disappear. This makes it even more important to keep track of the development from the origins to the eventual outcome and the numerous choices that are taken on the way. It is an essential to avoid a reductionism which makes certain developments look natural and without alternatives, although they are, in fact, culturally informed. As an illustration, the analysis of dialogical logics in the works of Lorenzen, Lorenz and Mittelstraß provide effective instruments for analyzing the dialogue structures in interdisciplinary working groups and their impact on the results that are produced, as happens today, for example, in the fields of social robotics, persuasive technology, etc.

Furthermore, the methodical schools can also contribute to the discourse of engineering ethics. Especially in the linguistic reconstruction of moral terms that relate to technical practice, methodical constructivism and culturalism offer useful tools to critically reconstruct the impact of new technical actions on our common forms of expression (in particular, Lorenzen and Schwemmer 1975 and following works). References to methodical order can also be used to provide alternate and more precise models for ethical decision-making. This issue is of high relevance for the engineering of so-called "moral" machines or "autonomous robots"—fields where terms are taken out of the context of human social life without a methodical proof of the adequacy of their meaning in a different application scenario. Other examples in this context are terms such as "life" in synthetic biology or "information" and "communication" in the computer sciences. Moreover, on the basis of lifeworld foundation, pragmatic truth and pragmatic order, it can be shown that moral reasoning and the rational reflection of morality—what we are used to calling "ethics"—has its foundation in everyday practice, not in abstract disembodied passive theories. In consequence, it can be gradually reconstructed (a) that and (b) how top-down approaches to so-called "moral machines" must fail. Ethical codices like Asimov's three/four laws might be formulated in a correct grammar. But their application will fail in robotic systems due to a lack of pragmatic order and the foundations involved in everyday social life (Funk 2021).

Since engineering remains one of several cultural practices—like natural sciences, music, theater, etc.—the methodical reconstruction of the pragmatic genesis of claims made by engineers can additionally reveal the limitations when it comes to social life or the technical recreation of social interaction in social robotics. Engineering is already based on the preconditions of social life without explicitly emphasizing them. In conclusion, social practice as such cannot only be engineered in a limited range. Here are many points where the concept of transsubjectivity (Janich 2006a: 49, 75, 81–86 et passim) plays a significant role: we are always socially embedded, and the shared sphere of human interaction makes us individual—not vice versa. In consequence, transsubjectivity can be used not only to better understand why and how the engineering of morality and social life remains limited (because it is a methodical presupposition rather than an object of technical practice); it can also contribute to methodical reconstructions of collective actions—for instance in robotics or genetics (Funk 2018b). The division of labor is addressed in the concepts of inter- and transdisciplinarity, which also appear to become more and more important issues in the philosophy of engineering today. The methodical reconstruction of transdisciplinary and transsubjective lifeworld actions can contribute to a better understanding of both the epistemology and ethics of specialized technical actions.

In this sense, many opportunities for further work exist which now have to be explored.

Related Chapters

Chapter 5: What Is Engineering Science? (Sven Ove Hansson)
Chapter 6: Scientific Methodology in the Engineering Sciences (Mieke Boon)
Chapter 7: Engineering Design and the Quest for Optimality (Maarten Franssen)
Chapter 8: Prescriptive Engineering Knowledge (Sjoerd Zwart)
Chapter 9: The Art of Engineering and Engineering as Art (Lara Schrijver)
Chapter 10: Creativity and Discovery in Engineering (David H. Cropley)
Chapter 21: Engineering Design (Peter Kroes)

Further Reading

Janich, P. (1997). Methodical Constructivism. In D. Ginev and R.S. Cohen (eds.), *Issues and Images in the Philosophy of Science*. Boston Studies in the Philosophy of Science, Vol 192. Dordrecht: Springer. (A short English overview of the history, key figures and claims of the methodical schools—written by Peter Janich, one of the main figures in the field.)

Janich, P. (2006). *Kultur und Methode, Philosophie in einer wissenschaftlich geprägten Welt*. Frankfurt am Main: Suhrkamp. (A summary of Peter Janich's most important papers about methodical culturalism, including conceptual foundations and areas of application like chemistry, ICT and genetic knowledge—in the German language.)

Kamlah, W. and Lorenzen, P. (1984). *Logical Propaedeutic. Pre-school of Reasonable Discourse*. Lanham, MD: University of America Press. (English translation of the classical manifesto of the Erlangen School, originally published in German in 1967—a unique combination of so-called Continental and analytic philosophy.)

References

Bölker, M. et al. (2016). Synthetic Biology. Diverse Layers of Live. In M. Engelhard (ed.), *Synthetic Biology Analysed. Tools for Discussion and Evaluation. Ethics and Technology Assessment,* Vol. 44. Dordrecht: Springer, pp. 27–50.

Carnap, R. (1998). *Der logische Aufbau der Welt*. Hamburg: Felix Meiner.

Coeckelbergh, M. and Funk, M. (2018). Wittgenstein as a Philosopher of Technology. Tool Use, Forms of Life, Technique, and a Transcendental Argument. *Human Studies,* 41(2), 165–191, June. doi:10.1007/s10746-017-9452-6.

Dingler, H. (1933). *Die Grundlagen der Geometrie, ihre Bedeutung für Philosophie, Mathematik, Physik und Technik*. Stuttgart: Ferdinand Enke.

Dingler, H. (1987). *Aufsätze zur Methodik*. Hamburg: Felix Meiner.

Franssen, M., Vermaas, P., Kroes, P.E. and Meijers, A.W.M. (eds.) (2016). *Philosophy of Technology After the Empirical Turn*. Dordrecht: Springer.

Frege, G. (2008). *Funktion, Begriff, Bedeutung. Fünf logische Studien*. Göttingen: Vondenhoeck & Ruprecht.

Fritzsche, A. (2009). *Schatten des Unbestimmten. Der Mensch und die Determination technischer Abläufe*. Bielefeld: transcript.

Fritzsche, A. (2010). Engineering Determinacy. The Exclusiveness of Technology and the Presence of the Indeterminate. In I. Van de Poel and D. Goldberg (eds.), *Philosophy and Engineering. Philosophy of Engineering and Technology*, Vol. 2. Dordrecht: Springer, pp. 305–312.

Fritzsche, A. (2018). Dancing the Device: A Translational Approach to Technology. In J.C. Pitt and A. Shew (eds.), *Spaces for the Future: A Companion to Philosophy of Technology*. New York and London: Routledge.

Fritzsche, A. and Oks, S.J. (2018). Translations of Technology and the Future of Engineering. In A. Fritzsche and S.J. Oks (eds.), *The Future of Engineering: Philosophical Foundations, Ethical Problems and Application Cases*. Cham, Switzerland: Springer, pp. 1–12.

Funk, M. (2018a). Repeatability and Methodical Actions in Uncertain Situations. Wittgenstein's Philosophy of Technology and Language. In M. Coeckelbergh, M. Funk and S. Koller (eds.), *Wittgenstein and Philosophy of Technology. Techné: Research in Philosophy and Technology. Special Issue, Volume 22, Issue 3 (2018)*, pp. 351–376. doi:10.5840/techne201812388.

Funk, M. (2018b). Paleoanthropology and Social Robotics: Old and New Ways in Mediating Alterity Relations. In J. Aagaard, Jyn Kyrre Berg Friis, J. Sorenson, O. Tafdrup and C. Hasse (eds.), *Postphenomenological Methodologies: New Ways in Mediating Techno-Human Relationships*. Lanham, MD: Rowman & Littlefield/ Lexington Books, pp. 125–149.

Funk, M. (2021). *Roboter- und Drohnenethik. Eine methodische Einführung*. Wiesbaden: Springer.

Gethmann, C.F. (2005). Apriori, lebensweltliches. In J. Mittelstraß et al. (eds.), *Enzyklopädie Philosophie und Wissenschaftstheorie. Band 1: A-B. 2. neubearbeitete und wesentlich ergänzte Auflage*. Stuttgart and Weimar: J.B. Metzler, p. 185.

Haas, G. (2010). konstruktiv/Konstruktivität. In J. Mittelstraß et al. (eds.), *Enzyklopädie Philosophie und Wissenschaftstheorie. Band 4: Ins-Loc. 2. neubearbeitete und wesentlich ergänzte Auflage*. Stuttgart and Weimar: J.B. Metzler, pp. 313–314.

Hartmann, D. and Janich, P. (eds.) (1996). *Methodischer Kulturalismus. Zwischen Naturalismus und Postmoderne*. Frankfurt a.M.: Suhrkamp.

Hartmann, D. and Janich, P. (eds.) (1998). *Die Kulturalistische Wende. Zur Orientierung des philosophischen Selbstverständnisses*. Frankfurt a.M.: Suhrkamp.

Heidegger, M. (2010). *Being and Time*. New York: State University of New York Press.

Husserl, E. (1970). *The Crisis of European Sciences and Transcendental Phenomenology: An Introduction to Phenomenological Philosophy*. Evanston: Northwestern University Press.

Ihde, D. (1990). *Technology and the Lifeworld. From Garden to Earth*. Bloomington and Indianapolis, IN: Indiana University Press.

Janich, P. (1969). *Die Protophysik der Zeit*. Mannheim: Bibliographisches Institut.

Janich, P. (1980). *Die Protophysik der Zeit. Konstruktive Begründung und Geschichte der Zeitmessung*. Frankfurt a.M.: Suhrkamp.

Janich, P. (ed.) (1984). *Methodische Philosophie, Beiträge zum Begründungsproblem der exakten Wissenschaften in Auseinandersetzung mit Hugo Dingler*. Mannheim a.o.: Wissenschaftsverlag.

Janich, P. (ed.) (1991). *Entwicklungen der methodischen Philosophie*. Frankfurt a.M.: Suhrkamp.

Janich, P. (1996). *Konstruktivismus und Naturerkenntnis. Auf dem Weg zum Kulturalismus*. Frankfurt a.M.: Suhrkamp.

Janich, P. (1997). Methodical Constructivism. In D. Ginev and R.S. Cohen (eds.), *Issues and Images in the Philosophy of Science. Boston Studies in the Philosophy of Science*, Vol. 192. Dordrecht: Springer.

Janich, P. (ed.) (1999). *Wechselwirkungen, Zum Verhältnis von Kulturalismus, Phänomenologie und Methode*. Würzburg: Königshausen & Neumann.

Janich, P. (2006a). *Kultur und Methode. Philosophie in einer wissenschaftlich geprägten Welt*. Frankfurt a.M.: Suhrkamp.

Janich, P. (ed.) (2006b). *Wissenschaft und Leben. Philosophische Begründungsprobleme in Auseinandersetzung mit Hugo Dingler*. Bielefeld: Transcript.

Janich, P. (2008). Homogenitätsprinzip. In J. Mittelstraß et al. (eds.), *Enzyklopädie Philosophie und Wissenschaftstheorie. Band 3: G-Inn. 2. neubearbeitete und wesentlich ergänzte Auflage*. Stuttgart and Weimar: J.B. Metzler, pp. 432–433.

Janich, P. (2010). Kulturalismus, methodischer. In J. Mittelstraß et al. (eds.), *Enzyklopädie Philosophie und Wissenschaftstheorie. Band 4: Ins-Loc. 2. neubearbeitete und wesentlich ergänzte Auflage*. Stuttgart and Weimar: J.B. Metzler, pp. 408–409.

Janich, P. (2016). Prototheorie. In J. Mittelstraß et al. (eds.), *Enzyklopädie Philosophie und Wissenschaftstheorie. Band 6: O-Ra. 2. neubearbeitete und wesentlich ergänzte Auflage.* Stuttgart and Weimar: Verlag J.B. Metzler, pp. 478–479.

Kambartel, F. (1989). *Philosophie der humanen Welt. Abhandlungen.* Frankfurt a.M.: Suhrkamp.

Kambartel, F. (1996). Transsubjektiv/Transsubjektivität. In J. Mittelstraß et al. (eds.), *Enzyklopädie Philosophie und Wissenschaftstheorie. Band 4 der ersten Auflage.* Stuttgart and Weimar: J.B. Metzler, p. 332.

Kambartel, F. and Stekeler-Weithofer, P. (2005). Erlanger Schule. In J. Mittelstraß et al. (eds.), *Enzyklopädie Philosophie und Wissenschaftstheorie. Band 2: C-F. 2. neubearbeitete und wesentlich ergänzte Auflage.* Stuttgart and Weimar: J.B. Metzler, pp. 390–391.

Kamlah, W. and Lorenzen, P. (1967). *Logische Propädeutik. Vorschule des vernünftigen Redens.* Mannheim: Bibliographisches Institut.

Kamlah, W. and Lorenzen, P. (1984). *Logical Propaedeutic. Pre-school of Reasonable Discourse.* Lanham, MD: University of America Press.

Kroes, P. and Meijers, A. (eds.) (2000). *The Empirical Turn in the Philosophy of Technology.* Dordrecht: Springer.

Lorenz, K. (2009). *Dialogischer Konstruktivismus.* Berlin and New York: de Gruyter.

Lorenz, K. (2016). Prinzip, methodisches. In J. Mittelstraß et al. (eds.), *Enzyklopädie Philosophie und Wissenschaftstheorie. Band 6: O-Ra. 2. neubearbeitete und wesentlich ergänzte Auflage.* Stuttgart and Weimar: J.B. Metzler, pp. 434–435.

Lorenzen, P. (1968). *Methodisches Denken.* Frankfurt a.M.: Suhrkamp.

Lorenzen, P. (1974). *Konstruktive Wissenschaftstheorie.* Frankfurt a.M.: Suhrkamp.

Lorenzen, P. (2000). *Lehrbuch der konstruktiven Wissenschaftstheorie.* Stuttgart: J.B. Metzler.

Lorenzen, P. and Lorenz, K. (1978). *Dialogische Logik.* Darmstadt: WBG.

Lorenzen, P. and Schwemmer, O. (1975). *Konstruktive Logik, Ethik und Wissenschaftstheorie.* Mannheim: Bibliographisches Institut.

Mainzer, K. (2010). Konstruktion. In J. Mittelstraß et al. (eds.), *Enyzklopädie Philosophie und Wissenschaftstheorie. Band 4: Ins-Loc. 2. neubearbeitete und wesentlich ergänzte Auflage.* Stuttgart and Weimar: J.B. Metzler, pp. 310–312.

Michelfelder, D.P. (2010). The Philosophy of Technology When 'Things Ain't What They Used to Be'. *Techné: Research in Philosophy and Technology*, 14(1), 60–68.

Michelfelder, D.P. (2017). Risk, Disequilibrium, and Virtue. *Technology in Society*, 52, 32–38.

Mitcham, C. (1994). *Thinking Through Technology: The Path Between Engineering and Philosophy.* Chicago: University Of Chicago Press.

Mittelstraß, J. (ed.) (2008). *Der Konstruktivismus in der Philosophie im Ausgang von Wilhelm Kamlah und Paul Lorenzen.* Paderborn: Mentis.

Mittelstraß, J. (ed.) (2012). *Zur Philosophie Paul Lorenzens.* Münster: Mentis.

Mittelstraß, J. (2016). Philosophie, konstruktive. In J. Mittelstraß et al. (eds.), *Enzyklopädie Philosophie und Wissenschaftstheorie. Band 6: O-Ra. 2. neubearbeitete und wesentlich ergänzte Auflage.* Stuttgart and Weimar: J.B. Metzler, p. 276.

Mittelstraß, J. and von Bülow, C. (eds.) (2015). *Dialogische Logik.* Münster: Mentis.

Mittelstraß, J. et al. (eds.) (1980ff.). *Enzyklopädie Philosophie und Wissenschaftstheorie.* Stuttgart and Weimar: Metzler.

Neely, A., Fell, S. and Fritzsche, A. (2018). Manufacturing With a Big M—The Grand Challenges of Engineering in Digital Societies From the Perspective of the Institute for Manufacturing at Cambridge University. In A. Fritzsche and S.J. Oks (eds.), *The Future of Engineering.* Cham, Switzerland: Springer, pp. 191–200.

Nordmann, A. (2014). Synthetic Biology at the Limits of Science. In B. Giese, C. Pade, H. Wigger and A. von Gleich (eds.), *Synthetic Biology: Character and Impact.* Berlin: Springer, pp. 31–58.

Pitt, J. (2000). *Thinking About Technology. Foundations of the Philosophy of Technology.* New York and London: Seven Bridges.

Polanyi, M. (2009). *The Tacit Dimension. With a New Foreword by Amartya Sen.* Chicago and London: University of Chicago Press.

Poser, H. (2012). Knowledge of Ignorance. On the Problem of the Development and the Assessment of Technology. In G. Abel and J. Conant (eds.), *Rethinking Epistemology*, Vol. 1. Berlin and Boston: De Gruyter, pp. 369–391.

Rentsch, T. (2000). *Negativität und praktische Vernunft.* Frankfurt a.M.: Suhrkamp.

Rentsch, T. (2003). *Heidegger und Wittgenstein. Existenzial- und Sprachanalysen zu den Grundlagen philosophischer Anthropologie.* Stuttgart: Klett-Cotta.

Schwemmer, O. (2010). Intersubjektivität. In J. Mittelstraß et al. (eds.), *Enzyklopädie Philosophie und Wissenschaftstheorie. Band 4: Ins-Loc. 2. neubearbeitete und wesentlich ergänzte Auflage.* Stuttgart and Weimar: J.B. Metzler, pp. 47–49.

Thiel, C. (2010). Konstruktivismus. In J. Mittelstraß et al. (eds.), *Enzyklopädie Philosophie und Wissenschafts-theorie. Band 4: Ins-Loc. 2. neubearbeitete und wesentlich ergänzte Auflage*. Stuttgart and Weimar: J.B. Metzler, pp. 314–319.

Way, M. (2017). What I Cannot Create, I Do Not Understand. *Journal of Cell Science*, 130, 2941–2942.

Welter, R. (2010). Lebenswelt. In J. Mittelstraß et al. (eds.), *Enzyklopädie Philosophie und Wissenschaftstheorie. Band 4: Ins-Loc. 2. neubearbeitete und wesentlich ergänzte Auflage*. Stuttgart and Weimar: J.B. Metzler, pp. 482–487.

Wittgenstein, L. (1984). *Werkausgabe Band 2. Philosophische Bemerkungen. Aus dem Nachlass herausgegeben von Rush Rhees*. Frankfurt a.M.: Suhrkamp.

Wittgenstein, L. (2009). *Philosophische Untersuchungen. Philosophical Investigations*. Chichester: Blackwell Publishing.

Wolters, G. (2016). Prinzip der pragmatischen Ordnung. In J. Mittelstraß (eds.), *Enzyklopädie Philosophie und Wissenschaftstheorie. Band 6: O-Ra. 2. neubearbeitete und wesentlich ergänzte Auflage*. Stuttgart and Weimar: J. B. Metzler, pp. 436–437.

55

REIMAGINING THE FUTURE OF ENGINEERING

*Neelke Doorn, Diane P. Michelfelder, Elise Barrella, Terry Bristol,
Francien Dechesne, Albrecht Fritzsche, Gearold Johnson, Michael Poznic,
Wade L. Robison, Barbara Sain, Taylor Stone, Tonatiuh Rodriguez-Nikl,
Steven Umbrello, Pieter E. Vermaas, Richard L. Wilson*

Reimagining suggests the idea of opening up new, unconventional spaces of possibilities for an activity or an entity that already exists. At its most transformative, the activity of reimagining develops spaces of possibilities that alter the very definition of that activity or entity. What, then, would it be to reimagine the future of engineering?

An exploration of such a topic cannot be done well by a single individual but rather requires the combined perspectives and insights of a number of people. The thoughts presented in this chapter had their beginnings in a workshop on this topic which took place at a meeting of the Forum on Philosophy, Engineering and Technology (fPET) at the University of Maryland, College Park, in 2018. Because participants in the workshop came from the fPET community, they included philosophers and engineers from both inside and outside the academy. On this account, reimagining the future of engineering is a matter of reimagining and redrawing the spaces of engineering itself: spaces for designing, action, problem framing, professional and disciplinary identity, and for the training of future engineers.

55.1. The Virtuality of Future Engineering

A concrete example of one new space in engineering is digital space. Digital technology permeates engineering work, just as it does all parts of human life. In cyber-physical architectures, digital representations are closely associated with the physical systems to which they refer, such that both are treated as a unity. Comprehensive simulations are used to support the design of such systems, which provide digital representations of physical phenomena that include user behavior to get to know the workings of engineered systems. All this has given engineers better access to the subject matter of their work, but it has also changed the points of orientation that help them find direction in their activities. Digital models allow for new types of benchmarks in the ideation, design, and evaluation of engineering products. So, apart from the extended affordances digitalization may provide to products of engineering, it fundamentally changes the engineering process.

Epistemological and normative implications of digitalization for the engineering process are extensively explored in many different domains. What remains unclear is whether digitalization also changes the notion of engineering in its core. At first glance, it would seem to be the opposite. For a long time, engineering has been related to the application of abstract mathematical methods to designing, building, creating, operating, and maintaining engineering products and systems, e.g. in the specification of the general desiderata for an end product into operationalizable requirements and measurable goals. The use of digital technology goes along with a formalization of the subject matter, which relates practice even more than before to symbolic representations such as models and constructs. This makes digitalization fit very well into already-existing approaches to engineering. Moreover, it might explain why the digital transformation is strongly embraced by so many engineers: it fulfills the wildest dreams about changing the world by systematic design, enframed in symbolic structures of interrelated and controllable objects and properties. One could therefore argue that digital technology makes current practices of engineering more powerful as it grants access to any field of human experience through bits and bytes. Doesn't that strengthen the position of engineering as it is, underlining its leading force in progress and development?

At the same time, however, one could put forward an argument pointing in the opposite direction: in expanding its scope to comprehensive representations of the lifeworld, engineering cannot stay as it is. As the contribution of engineering to the solution of human problems changes, the foundations of its operative principles need a careful review to redefine its role in society.

Support for this argument can be taken from simple combinatorial considerations of the size of the design spaces which digital technology opens up for engineering. Digital technology has increased the range of options for engineering design in so many different directions that the fixed points of reference that have so far given orientation and direction disappear, creating the need for the choice of artificial replacements which give the design space a completely new texture.

The automotive industry is one domain where the changing conceptions of engineering under the influence of digitalization can clearly be seen (e.g. Kirk 2015; Hars 2015; Gusikhin et al. 2007). Auto manufacturers are turning into IT companies and data brokers who do not primarily take physical improvements in materials, mechanics, or aerodynamics as drivers of innovation, but IT. More and more layers of IT are created between users and the effects of their actions in the physical world, as well as between auto mechanics and the car as a physical object.

Going a step further, controls and displays in automobiles themselves are already mostly digitally operated, even if the driver does not perceive it. Analog displays (if still existing) are superficial, and behind them is digital data; the resistance of the brake and gas pedal is also created artificially, as there is no physical connection to other parts of the car, just a digital recognition of the signals created by the driver which are sent to the respective automotive controls.

Digitalization also shows up in the process of automotive design. The techniques for shaping and making automotive parts are in their own way also cyber-physicalized via additive manufacturing. It gives endless opportunities to create new parts with properties never seen before based on digital designs, making them lighter and better resistant to pressure and heat, and allows changing aesthetics.

Beyond the engineering products and processes, digitalization impacts the business logic behind automotive engineering, driving innovation in unexpected directions. New profit margins are created by assistance systems (parking, navigation, accident prevention) and service provisions enhancing the driving experience without being related to motion in space at all (entertainment systems, telematics insurance schemes, car sharing, and mobility solutions). Add to that the potential of reusing the data gathered in the use of such services, giving those data exploitable value, and we can see how other models of the functionality of the car emerge in larger economic structures.

Like the smartphone, the car has become its own kind of platform onto which many different aspects of the user's life can be linked. Whereas the smartphone is carried close to one's skin and taken everywhere, the car creates a sphere of intimacy where we are nurtured by technology. This does not require the classical strengths of automotive engineering: powerful engines, smooth riding, and full control.

All this gives the impression that the car as an object of engineering, although remaining present in a physical embodiment, becomes virtual in different ways. The experience of driving is simulated, while the mechanics and parts of the control happen behind the user interface, both inside the car and beyond it. Also, the actual instance of the car loses importance, as is demonstrated, for example, by how car-sharing, etc., provide mobility-as-a-service, where the material objects are totally replaceable. And with the services it uses from outside, by being with connected, and with continuous updates on controls and displays, the car as an engineering product extends beyond its former spatial and temporal limits.

Digital representations thus become the predominant subject matter of engineering with endless new possibilities to be divided into pieces and set together again in a new way. This has consequences for the design space: it practically explodes to form a whole new universe. Furthermore, even the notion of a singular artifact itself becomes blurred: In the case of the car, one does not know exactly where its design space ends and the design space of one of the other objects (interconnected devices, cameras, traffic control, insurance) starts. Formerly discrete design spaces become inherently related, creating additional degrees of freedom for engineering.

This explosion of engineering design space in endless and often opaque options for future development has consequently a radical impact that needs to be addressed on operative-practical, epistemic, and ethical levels. Who in this design space can be said to be designers? Can engineers continue to be seen as the "authors" of artifacts? When (if ever) do we consider an engineering product a finished artifact? What is the artifact's function or the problem that it solves? In addition, as basic needs are increasingly satisfied, technological innovation will take on an increasingly totemic flavor, and engineers who work in areas that most appeal to rapidly changing consumer preferences will increasingly find themselves working on curiosities for consumerist consumption. How will this change the engineer, who is, at least in a nominal sense, a problem-solver interested in the benefit of humankind? Will the engineer, by and large, participate in this self-referential dynamic of technological innovation for the sake of technological innovation? To what extent will this participation be conscious? How will the profession define itself? Even engineers who "remain behind" to work on fundamental problems will find challenges, as they discover that the technical contribution to the solution pales in comparison to the contributions of public policy and other fields. How will the engineer and the profession respond as the innovative aspects of engineering work fade in importance or become subsumed by meaningless consumption?

These questions indicate that there is yet another design process to be addressed, since it does not seem possible to answer them convincingly based on mere factual information. With its expansion through digital technology, engineering itself becomes the subject of a design process.

55.2. The Self-Awareness of Future Engineers

Another example of a new kind of space within which modern engineering operates is the space of open systems, where design and research problems exist within social and political contexts. In open systems, engineering is more about framing problems and viewing the solution space in different ways than "solving" the problem or finding *the* solution. Open system problems are typified by the Millennium Development Goals, UN Sustainable Development Goals, or the notion of a "grand challenge" or "wicked problem".

This space is connected to an emerging ("new") self-awareness in the modern engineering community and in the community at large. The "problem" (agenda) of engineering is (traditionally) the problem of design. Herbert Simon characterized engineering design problem-solving as attempting to move from a current state of affairs to a future, more desirable state of affairs. By being perhaps a definitional tautology of all problem-solving, Simon's characterization will also apply to the work of the future engineers. But unlike the older imagined, value-neutral scientific spectator on reality, the future engineer is an embodied and active participant in the historical emergence of reality and possesses a value agenda. The engineer of the future will be more attuned to how engineering design fits within this emergence where values, materiality, culture, and politics are inseparably interconnected.

From here, engineers will be naturally led to design not just technical solutions, but many possible solutions reflecting different values and valuation. Samuel Florman pointed out that we are naturally creative "existential" engineers in that we are embodied with the ability to act in the world, but, in important ways, without a defining script (i.e., inherent uncertainty) (Florman 1994 [1976]). We are not "scientifically determined". Rather, we are constrained with an irreducible component of creative freedom. We each choose how to design our lives—e.g., how much time to spend with family, how much time to spend at work. And we are all naturally concerned with how to treat/relate to others as well as, self-reflexively, ourselves. This points to the idea that engineering is about the development of how we should live. And "how we should live" is the defining question of morality. Here, one might claim that the fundamental framework of engineering is concerned with morality.

With the new, general conception of engineering as a moral framework, engineering would re-unify the idealized "scientific" and "humanities" cultures (Grasso and Burkins 2010; Petroski 2016). The engineer of the future asks "*for whom* are we designing and *how can* this be designed?" How such an engineer approaches or responds to those fundamental questions will be influenced by a number of factors: personal values, professional values, societal values, education and experience, position and authority, etc. In order to consider possible responses, and based on a common practice in engineering as a means of exploring a problem-solution space, the following section proposes some scenarios and uses them as a lens to look at the question of *who* an engineer is. The personas presented, and the virtues or vices that define them,[1] may typify different engineering practitioners or researchers in future scenarios connected to open system spaces.

55.3. The Personas of Future Engineers

Addressing tasks and problems in open systems will strengthen the requirement that engineers will work in teams (with other engineers and non-engineers) when taking up projects. This raises the question of what roles engineers can take up in these teams. We discern several possibilities in terms of personas that are not mutually exclusive: an engineer may take up a role characterized by two or more personas. These personas are simplified descriptions of aspects of real-life engineers' personalities. Furthermore, the personas are intertwined with moral and epistemic virtues of engineers understood as relatively stable character traits. The virtues reflect the general values of engineering such as collaboration and teamwork as well as efficient problem-solving. In the following, we will propose an ordering of personas along the two analytic dimensions of systems orientation and openness to society.

The first persona concerns the existing role of engineers to deliver expert knowledge and skills about technology and design to a project. This persona of *technology expert* ignores the emergence of open systems and their societal aspects. This persona is inclined to leave choices about developing systems and bringing in societal considerations to other team members. The technology expert needs to develop epistemic virtues such as open-mindedness and creativity but may lack important moral virtues such as empathy and care.

Two other personas do address the system character of open systems but might not take the societal aspects on board. The *systems expert* brings in expertise about systems design. This persona is still subservient on the team but takes up broader responsibility to implement changes and developments. The *moderator* takes up a coordinating role on the team by using systems expertise to align the work of team members and stakeholders and translate it into opportunities for and developments of engineering systems. These two personas embody central moral and epistemic virtues that are important traits of any good, responsible engineer.

Still, in these two scenarios, all is not "business as usual". For instance, the moderator shows the virtues of humility and empathy and thereby is already more open to society by also being oriented to stakeholders as clients, users, and others involved or affected by a project. This openness is central to two further personas. The *social engineer* is committed to issues in society such as the UN Sustainable Development Goals and initiates and participates in projects to address them. The *visionary* is focused on society as well, not on specific issues but by envisaging innovative ways to address existing or future aims. Creativity is a virtue especially relevant to the visionary engineer; it can be said to be *the* character trait of engineers that embody a visionary persona. In general, the social and visionary engineer both take up responsibility, as do the systems expert and moderator, but in their best implementation they also display the virtues of justice and wisdom.

Two more personas, already becoming visible today and likely to remain so in the future, stand somewhat outside the sequence of the five described so far. These personas are characterized by certain vices rather than virtues. The first is the *disrupter*. Addressing tasks and problems in open systems may require overhauling current technologies, infrastructures, and societal arrangements, and for this a more disruptive approach can be chosen by hacking systems and by introducing game changers. In this manner, vices such as negligence and carelessness may become apparent. The second persona is the one of the *compromised engineer* whose contributions to tasks and problems are eventually assessed as unprofessional, unethical, or illegal. In the language of virtues and vices, one may call such a persona the "vicious" engineer who acts irresponsibly as, for example, was seen in the cases of the Volkswagen emissions scandal or the Cambridge Analytica incident. Vices of moral and epistemic injustice may be further character traits possessed by a compromised engineer.

As said, these personas are not mutually exclusive, and they are not descriptions of real-life persons. For instance, the visionary engineer is typically one who thinks in terms of technological possibilities. The engineer who moderates in a team can do so by taking a systemic perspective on the team's project. And the engineer who ends up into a compromised position can have gotten there through a series of virtuous contributions to regular technological developments within teams.

55.4. Implications for Future Engineering Education

This section presents a reflection on the implications for engineering education—*how should* we "design" engineers, and *for whom* are we designing the engineering education system?

Idealistically (and at the risk of being reductive), engineering is problem-oriented. It takes up challenges with a range of complexity and moral weight (i.e., creating a means of transport across this body of water, improving the well-being and comfort of a patient, etc.), and seeks to provide practical, workable, and often innovative solutions. Many of the problems that we see emerging, and that will in all likelihood continue to define the 21st century, are large, complex, systemic, multifaceted, etc. As already mentioned, these are often referred to as "wicked problems" or, more recently, "grand challenges". A paradigmatic example would be climate change, which has far-reaching social, political, and technical causes and ramifications.

As modern engineering practice shifts from contained, well-defined technological challenges to open, ambiguous societal challenges, an "engineer" becomes a broader label. In the exploration just given, the notion of "engineer" as a one-size-fits-all label built upon a fixed body of disciplinary knowledge and professional identity receded, and multiple scenarios (or personas) that reflect future engineers' different roles, values, and relationships with the sciences, humanities, and society came into view. Rather than offering a definitive answer to the question "Who is an engineer?" a multidimensional framing of a diverse engineering profession was presented.

Instead of defining engineers by what they "do" or "know", the engineering profession in the future will be defined by who engineers are as people. What characters and cultures do people who do engineering represent? In 2015, a social media and marketing campaign using #ILook-LikeAnEngineer presented a wider view for the public of who is an engineer and what engineers do, beyond the typical stereotypes and buzzword industries (Guynn 2015). The types of people who do engineering in an open-space problem environment is broader than the traditional 20th-century engineering stereotype of someone who applies math and science to solve a problem. Perhaps an engineer is someone who uses engineering tools, processes, and mindsets to move from a current state to a better state. Or perhaps engineers will be defined by their character and striving for virtues, both ethical (e.g., honesty, courage) and epistemic (e.g., wisdom, creativity).

This way of thinking suggests moving beyond defining engineers by their disciplinary body of knowledge, which is common with professional organizations, licensure, and college accreditation. It might, for example, suggest a reorientation and reorganization of engineering, from sub-disciplines to challenges (i.e. a climate change engineer instead of a civil or aerospace engineer, etc.).

Would such a reorientation be more effective in addressing and solving grand challenges like climate change? What is the "minimum" or "shared" knowledge and skills across engineering disciplines? (Vincenti 1990). Does it differ by career track? Is education more about ways of thinking and doing than specific knowledge and skills? (cf. Neely et al. 2018). "Engineer" already refers to diverse people and roles, and so engineering education in the future will need to reflect more diverse ways of learning and evaluating success. If engineering education is to remain relevant in the future, it needs to adapt and educate the whole person. Based on the scenarios, an engineer is defined by knowledge/skills, processes, relationships, character, and worldview. Thus, engineering education and training should also focus on the whole person.

In the future, will engineering graduates identify as engineers, and will they find pleasure in the work of engineering? Perhaps the answer primarily depends on whether engineers find themselves in the right engineering track to address the societal challenges that matter to them.

55.5. Future Spaces of Engineering Responsibility

Questions of social, moral, and legal responsibility permeate the philosophy and ethics of engineering (including presentations at fPET). Responsibility is a key issue in engineering ethics education, and foundational to well-established theoretical topics such as risk. And, responsibility (broadly conceived) is an underlying foundation for more recent developments in philosophy of technology, such as value-sensitive design, responsible research and innovation, and questions of agency in postphenomenology/mediation theory. Thus, what seems undeniable is that engineers—and engineering as a whole—carry a heavy burden of responsibility. Yet this burden also creates new opportunities, for finding creative, innovative, and workable solutions is then also the responsibility of engineering. Importantly, it raises questions of how to (continue to) take ownership of this "responsibility for responsibility" and further incorporate it into the demarcated *moral* (not just legal) space of engineers and their toolbox. This also requires new approaches to engineering ethics education.

55.6. New Approaches to Case Studies

Historically, case studies have been a valued component of engineering ethics education. For instance, in their textbook *Engineering Ethics*, Charles Harris, Michael Pritchard, and Michael Rabins state that "the importance (of cases) cannot be overemphasized. It is by studying cases that a person can most easily develop the abilities necessary to engage in preventive ethics" (Harris et al. 1995; p. 12). Indeed, well-designed cases can enable students to see how engineering design is embedded within a complex framework of cultural, ethical, political, financial, and social factors. If, however, this complex framework remains *simply* as the unreflected-upon context for a case study, the study fails to capture important lessons for engineering education. In this regard, Donna Riley and Yanna Lambrinidou take this criticism even further (Riley and Lambrinidou 2015), worrying that the typical focus of case studies on "micro-ethical dilemmas" serves to do more harm than good by obscuring and rendering invisible the larger social and structural contexts which are the real root of the dilemmas in the first place. Without speaking of personas *per se*, they also make the point that the case study approach often gives the persona of *technological experts* (the first persona mentioned earlier) to the engineers involved. This effectively legitimizes the boundary between engineering understanding and the understandings of the publics affected by the decisions of engineers. For instance, in a case study examining the devastating 2010 earthquake in Haiti and considering what engineers should do looking forward as Haiti is rebuilt, Charles Fleddermann (2012: 117) emphasizes the need to use "state-of-the-art earthquake resistant design" and (in the case of engineers from outside Haiti) to "help support Haitian engineers in learning appropriate design and construction techniques", without giving consideration to public input and perhaps looking to rectify injustices perpetrated by existing institutions.

Another difficulty with using case studies to teach engineering ethics is that students may be left with the impression that ethical challenges are easy to identify, that there is always a "smoking gun", or that the ethical problems should be left to the ethical experts (and not engineers). If students are not adequately engaged, they may leave the discussions about ethics without having developed their own abilities to identify and examine novel problems in engineering in their full sociohistorical context. In addition, when faculty select cases, they may be tempted to pick ones that are canonical and much discussed already, and so replicate the persona of the *technological expert* within the classroom, inhibiting the process of discovery in their students.

One approach to consider in mitigating these challenges is to have students themselves identify current cases relevant to their discipline for which there is not yet an authoritative "right" answer to the ethical problems. Allowing students to work through the messiness of a case, perhaps over the length of an entire course, could help them develop valuable competencies for identifying, examining, and addressing ethical issues.

55.7. Dealing With the Hidden Curriculum

A focus of the previous section was the often-hidden context of case studies. We turn now to another challenge of integrating ethics more effectively into engineering curricula, namely, the "hidden curriculum". For Van de Poel et al. (2001: 278), the "hidden curriculum" of engineering studies is largely tied to attitudes that students acquire informally in the course of their program which "leads to an a priori skeptical attitude on the part of engineering students towards 'soft' disciplines like ethics". For instance, students may experience a disparity between explicit claims about the importance of ethics and the reality in which ethical implications of student projects are considered to lie beyond the scope of engineering. The formation of the "hidden curriculum" may also be seen as a matter not only of attitudes but also personal epistemologies which contribute to the belief that ethics is "a matter of personal opinion". As Tormey et al. (2015) suggest, if students' first encounters with engineering involve a strong focus on fundamental, agreed-upon scientific principles, they may end up unconsciously adopting a personal epistemology which "may lead students to come to see problems which do not have a single correct answer as being 'simply a matter of opinion' rather than an opportunity to engage in a different way of constructing their understanding of the world" (Tormey et al. 2015: 4). This personal epistemology can also feed into and reinforce a formal classroom emphasis on producing artifacts that work rather than reflecting on the ethical implications of the artifacts to be created (Tormey et al. 2015; Van de Poel et al. 2001).

Any effort at reimagining the future of engineering ethics education needs to directly open up educational spaces in order to confront this hidden curriculum. Educators need to carefully consider the messages that they may unintentionally be sending about the value of ethics and professionalism while designing their pedagogical approaches, and maybe even discuss the very existence of this hidden curriculum. A potential benefit of revealing the hidden curriculum to students is that it can encourage them to critically analyze the social institutions in which they are embedded. This may make students less likely to simply accept workplace norms that are ethically dubious once they join the profession.

55.8. Concluding Thoughts

By design, the thoughts presented here are inconclusive. They take up and address the question of reimagining the future of engineering, but are far from giving the question a definitive answer. Pitt and Shew (2017) have turned to the idea of "space"—specifically ethical, political, virtual, personal, and inner and outer space—as a way of organizing the areas within which philosophers of technology now and in the future will be likely to work. The word "space" figures prominently in the thoughts presented in this chapter as well, as a device for exploring how the spaces of engineering might be reconfigured. We have deliberately not given any indication of how the spaces highlighted here might be prioritized or ordered. For instance, one might naturally think that reimagining the future of engineering is fundamentally grounded in reimagining the future of engineering education. However, the space for changing engineering education is also dependent on the space provided by society for making such a change. Likewise, the various engineering personas presented in this chapter are not just a matter of engineers assuming these roles but also of society allowing engineers to adopt them. We hope the reader will see the inconclusiveness of this chapter as a virtue rather than a negative characteristic and that it will serve to inspire future dialogue between philosophers and engineers.

Note

1. Preliminary discussion of the scenarios and personas presented here took place in a subgroup at the fPET 2018 workshop.

References

Fleddermann, Ch.B. (2012). *Engineering Ethics,* 4th ed. Upper Saddle River, NJ: Prentice Hall.
Florman, Samuel C. (1994 [1976]). *The Existential Pleasures of Engineering,* 2nd ed. New York: St. Martin's Press.
Grasso, D. and Burkins, M. (eds.) (2010). *Holistic Engineering Education: Beyond Technology.* New York: Springer.
Gusikhin, O., Rychtyckyj, N. and Filev, D. (2007). Intelligent Systems in the Automotive Industry: Applications and Trends. *Knowledge and Information Systems*, 12(2), 147–168.
Guynn, J. (2015). #ILookLikenEngineer Challenges Stereotypes. *USA Today*, August 4. www.usatoday.com/story/tech/2015/08/03/isis-wenger-tech-sexism-stereotypes-ilooklikeanenginer/31088413/
Harris, Ch.E., Pritchard, M.S. and Rabins, M.J. (1995). *Engineering Ethics: Concepts and Cases.* Belmont, CA: Wadsworth.
Hars, A. (2015). Self-driving Cars: The Digital Transformation of Mobility. In *Marktplätze im Umbruch.* Berlin, Heidelberg: Springer, pp. 539–549.
Kirk, R. (2015). Cars of the Future: The Internet of Things in the Automotive Industry. *Network Security,* 2015(9), 16–18.
Neely, A., Fell, S. and Fritzsche, A. (2018). Manufacturing With a Big M—The Grand Challenges of Engineering in Digital Societies from the Perspective of the Institute for Manufacturing at Cambridge University. In A. Fritzsche and S.J. Oks (eds.), *The Future of Engineering: Philosophical Foundations, Ethical Problems and Application Cases.* Cham, Switzerland: Springer, pp. 191–200.
Petroski, H. (2016). Refractions: Feeling Superior. *Prism,* September. www.asee-prism.org/refractions-sep-2/
Pitt, J.C. and Shew, A. (eds.) (2017). *Spaces for the Future: A Companion to Philosophy of Technology.* London: Routledge.
Riley, D.M. and Lambrinidou, Y. (2015). *Canons against Cannons? Social Justice and the Engineering Ethics Imaginary.* Presented at the 122nd Annual Conference and Exposition, June 14–17. Seattle. American Society for Engineering Education. Paper ID #12542.

Tormey, R., LeDuc, I., Isaac, S., Hardebolle, C. and Voneche Cardia, I. (2015). The Formal and Hidden Curricula of Ethics in Engineering Education. In *Proceedings of the 43rd Annual SEFI Conference*. www.sefi.be/wp-content/uploads/2017/09/56039-R.-TORMEY.pdf

Van de Poel, I.R., Zandvoort, H. and Brumsen, M. (2001). Ethics and Engineering Courses at Delft University of Technology: Contents, Educational Setup and Experiences. *Science and Engineering Ethics,* 7(2), 267–282. This refers to their discussion on p. 278.

Vincenti, W. (1990). *What Engineers Know and How They Know It.* Baltimore and London: The Johns Hopkins University Press.

INDEX

Note: numbers in **bold** indicate a table. Numbers in *italics* indicate a figure.

3-Es (economy, ecology, equity) 344, 346, *348*, 350
5-key-term approach 256–257, *257*

Accreditation Board for Engineering and Technology
  (ABET) 44, 547, 556, 587, 665–666
accuracy: and attributes 199, 613; in Big Data 202; in
  critical thinking 144, 147; of measurements 373,
  376, 380; observational 267; thermodynamics 14;
  tolerances 574; *see also* inaccuracy
Ackoff, Russell L. 179
Adorno, T.W. 706
Advanced Driver Assistance Systems (ADAS) 161
aesthetic experience 508–515
aesthetic philosophy 506–510, 513
aesthetics: architectural 12, 16; contemporary
  507–508; of design 296, 512–514; of engineering
  21, 128–135, 506–520; feminist 507; of geoengineering 515–516; limitations 319; scientific
  76; of technology 127
aesthetic value 459n9, 515–516, 576
affordances 236, 239–240, 251–252, 317, 577, 736
Africa 34; immigrants from 609; *see also* South Africa
African Americans 652, 692
agency (institutional) 540; accreditation 574;
  engineer's role inside 558; government 98,
  570–571, 592, 598, 681
agency (will or volition) 1, 741; and automation 500;
  efficacy of 462; of colonized peoples 654, 677;
  exercise of 462–463; and empowerment 331–332;
  Gaia forces 352; human 52, 617n1; moral 655;
  political 668; primary 468; social 467; and
  technology(ies) 361, 365, 607
Age of Discovery 345
Agricola 15
agriculture 26–27; advent of 213; in China 52; and
  Green Revolution 55; high input 461; in India 54;

industrial 472; as a practical art 57; and Systematic
  Parameter Variation 120
air traffic controller 226
Akeel, Usman 176–190, 439
Aken, Joan E. van 116, 118–119, 122
Akpanudo, U.M. 649
Alberti, Leon Battista 130, 318
Alexander, Christopher 319
Alexander, Jennifer Karns 25–37, 129–130, 345
Alexander, Samuel 264
algorithm engineering 33
algorithms 87–88, 104–106, 194–200, 270;
  aggregation 105; Arrowian criteria for 108;
  automated 194; autonomous systems as
  563–564; "behavior" of 717; biases of 198;
  classifier 196; as computational artefact
  425; cryptographic 536; design of 350; and
  experimental computer science 424; feedback
  loops produced by 195; as "ideal designer"
  319; moral 563; optimization models 169;
  predictive 715; probabilistic techniques 197;
  reproducibility of 430; simulation as 408, 410,
  416; sustainability and 342
Allenby, Braden 22
Allen, Barbara 693
Allen, D. 347
alterity *see* ethics-of-alterity
Altshuller, Genrich 318–319
Amer, M. 166
American Association of Engineering Societies 593
American Engineering Council (AEC) 663
American Institute of Chemical Engineers 33
American Institute of Electrical Engineers (AIEE)
  33, 663
American National Bureau of Standards *see* National
  Bureau of Standards (NBS)

American National Standards Institute (ANSI) 570, 595

American Society for Testing and Materials (ASTM) 570

American Society of Civil Engineers (ASCE) 593–594, 596, 663

American Society of Engineering Education (ASEE) 593

American Society of Mechanical Engineers (ASME) 593, 595–596; code of ethics 600, 663

American Society of Safety Engineers 593

American Standards Association (ASA) 595

American Telephone & Telegraph (AT&T) 142, 178

Ammonites 281

Ammon, Sabine 315–327

analytic ontology *see* ontology

analytic philosophy 21

analytic tools 323

analysis 47; consequentialist 45; cost-benefit 45; economic 17; engineering 2, 14, 80; of engineering design 71; of engineering practice 39; by engineering technologies 55; gap 46; geometrical 29; graphical means of 15; moral 43; risk 45; *see also* logic; evaluation

analytics 191–192, 194, 198, 201

alchemy 30

Ananny, M. 197

Anastas, P.T. 702

Andreasen, M.M. 253, *254*, 258

Andrews, David 634, 641

animal rights 564

Anscombe, G.E.M. (Elizabeth) 48n1, 384–385, 392n3, 616

Anthropocene 47, 61, 135; discourse 470; emergent 20; ideology of 469

Apollo 11, 178, *183*

applied natural science 74–75

arbitrariness 227–228; of disease 526; of epistemic values 452; of global ethical codes 623; of operationalization 305; of value measurements 303

Archimedes of Syracuse 29, 318

architecture: Ancient world 129–131; contemporary philosophy of 517–518; and engineering 130–131; Modern 131–132; industrial age 132

Aristotelian philosophy 42; and virtue ethics 39–42, 55, 330, 333

Aristotelians 38; realists 213

Aristotle 41, 517; on artifacts 289; on artificial objects 358; concepts, analysis of 22; on design 358–359; on disciplines, practical or theoretical 516; on emergence 264; essence by analogy 23; on existence and natural things 209–211; and mereology 217; *Metaphysics* 223, 231; *neotherizein* 284n2; *Nichomachean Ethics* 581; on philosophical premises 518; as Plato's student 130; on political order 279–280; on virtue 16; on the whole being greater than sum of its part 183

Arkin, Ronald 563

Armstrong, Neil 20

Arrow's Theorem 103–105, 108

art: and beauty 506–507; and engineering 127–135; as technology 516–517; *see also* aesthetics; architecture

artes 66–67

artifact kinds 214–217

artifacts 209–219; design 234–235; epistemic utility of category of 213–216; metaphysics of 211, 214, 217; parts 217–218; technical 289–290, 293–294, 300

artificial intelligence (AI) 270, 315, 319, 602–603; accountability in 201; agents (robotic) 500–501, 508; experimentation in 427; governance frameworks for 632, 719; Partnership on AI to Benefit People and Society 201; and sustainability 342; transparency, lack of 715; trust in 498; *see also* Big Data

artificial photosynthesis 84–86

artificial, the 116, 211, 713; sciences of 133–134, 424

artisanal knowledge 130

artisanship and craftsmanship 12–13, 15, 39, 223; Japan 58, 61; objects in relationship to 231; philosophical considerations of 225

Asimov's three/four laws 731

Association for Computing Machinery (ACM) 201, 430, 593

Association for the Advancement of Science (Britain) 69

asylum seekers 708

Atsushi *see* Fujiki, Atsushi

Augustine 563

Austin, J.L 222

Australia: Banj Bim 26–27; black swans 152; national academy of engineering 598; Woomera 139; *see also* Engineers Australia

authoring tool 323–324

AUTOCAD 121

autoethnography 649

automation and automata 131; cellular 410; trust in 498–503

automotive industry 161–162, 737

autonomous agents (robotics) 428–429, 431–432, 731; ethics of 603

autonomously driving car 160–161

autonomy 19; in engineering 558–567; Kant's views on 43; ontological 422; rise of 21

Aven, Terje 450, 452, 456, *456*

aviation industry 487, 641; *see also* civil-aviation system

awareness 671; contextual 202; cultural 704; designer's 634; environmental 595, 702; ethical 503; fairness through 198–199; lack of 195; raising 620, 716, 718; responsibility of 667; second-order 510; in security engineering 540; self- 338; situation 489; of technology 497; unawareness 199

Aydede, M. 48n5

B&K method 80, 84–91

Bacon, Francis 13, 16, 280, 282–283, 465–466

Baier, Annette 494, 502
Bailer-Jones, Daniela M. 92
Baillie, Caroline 465, 674–686, 694; Engineering, Social Justice and Peace 705
Bailyn, Lotte 561
Baird, Davis 375
Baker, Lynne Rudder 210
Bak, P. 152
Barberousse, A. 411, 418n6
Barrella, Elise
Basart, J.M. 625, 627n10
Baudrillard, Jean 507
Bechtel, William 377
Beck, U. 496
Beitz, Wolfgang 320
belief: in automation 499, 501; in computer simulation versus laboratory experiment 414; dogmatic 45; ethical 38; justified and effective 2, 123; in modernity 20; moral 43–44, 310; political 497; prejudice as 195; in rational thought 131; religious 199, 610; right 57; in science 315; in the supernatural 216; superstitious 55; in technological progress 414; and truth values 122; and use plans 235–238; values as 301–302
belief consistency 238
belief system 181
Bell, Sarah 176–190, 439
beneficence 44, 598
Benjamin, Walter 510, 514
Bentham, Jeremy 39, 48n7, 345
Berger, Gaston 162
Bernstein, H.J. 654
Betz, G. 164, 171–172
Bhabha, Homi 56
Bharatitya Janata Party (BJP) 56
Bhopal gas tragedy 56, 664, 693
bias 191–202, 608; cognitive 560; cultural 474; evidential 452; experimental 431; expressibility 443; and fairness, methods to assess 195–200; freedom from 302; gender 652; in ideal theory 242; scientific 452, 457; selective 498; sexist 653; structural 240; towards innovation 360–361, 365; unbiasedness 172
Big Bang 75
big-bang controllers 228
Big Data 191–202; bias and discrimination 194–198; data chain 192–194; fairness 198–200; provenance 194–195
Biomedical Engineering Society 593
Birnbacher, Dieter 563
Birnbacher, Wolfgang 563
Bishop, P. 166
black box 193, 197, 245–248, *246*, 290, 296; paramountcy principle and the public as 668; simulation 413; technology as 615
Black Death 139
Black Feminist Thought 650
black holes 75
black market 538, 540

black swan event 150, 152, 156–157, 536
Black womanist ethics 655
Blok, Vincent 53, 273–286, 378
Boehm, Barry 322
Boenink, Marianne 521–532, 482
Bohlin, Nils 512
Böhme, R. 535
Boisjoly, Robert 561, 664
Boltzmann, L. 266
Bonnefon, Jean-Francois 563
Boon, Mieke 80–94, 352, 722
Boorse, Christopher 527–529
Borenstein, Jason 32, 495, 592–606, 687
Borghuis, Tijn 194, 435–446
Börjeson, L. 163, 166
Borsook, Paulina 648
Boumans, M. 89
Bowie, Norman E. 623
Bowker, G.C. 574
Boyle's law 87
Bradfield, R. 162
Brady, Emily 515
Brand, Stewart 364
Bratman, M. 233
Brauer, C.S. 696
Brey, P. 647
Briskman, L. 297
Bristol, Terry 736–744
brittleness problem 564
Broad, C.D. 264
Brotherhood of Locomotive Engineers and Trainmen 586
Brown, Mark 609
Brown, P. 503
Brulle, Robert J. 346
Brundtland, Gro Harlem 343, 466, 469; Brundtland Report 471–472, 665
Brunel, Isambard Kingdom 32
Brunelleschi, Filippo 131
Brunton, Finn 717
Bucciarelli, L.L. 108
Buchanan, Richard 16–17
Buddhism 52, 54–56, 59, 61; stupa of Duthagamini 28
Buescher, Charles 702
Buhl, H. R. 138, 141, 144, **146**
building blocks 210, 284; epistemic 86; physical 86
Bullard, Robert 689
Bulleit, William M. 47, 149–159
Bunge, Mario 14, 47, 111, 113; and "groundedness" 14, 118, 125n23; technical rules 118, 121
bureaucratization 464–465
Burghardt, M.D. 138–139
Burke, Edmund 132
Busch, Lawrence 572–573
Bushi-do 59
Bush, Vannevar 19, 631
business and businesspeople 161, 275; and engineering 594, 635, 663–666; ethics 581, 585; female 648; and innovation 274, *283*; and internet

of things 637–638; privacy and security issues 599; as profession 582; risk assessment as 457; social responsibility in 661; standards for 571, 576, 642
business-as-usual scenario 166, 739
business cycles *275*
Buss, Sarah 559
butterfly effect 152

Cabnal, Lorena 683
Cáceres, Bertha 683
Cajamarca 683
Calder, Alexander 134
Calders, T. 196
Campbell, Colen 15
Campbell, Scott 347, 349–350
Canada 594, 679; codes of ethics 580; Sigma Xi 598; Queen's University 674; *see also* Cavoukian, Ann; Engineers Canada
Cang Li Yu Qi 藏礼于 器 51
Canguilhem, Georges 521, 526, 529
Capurro, Raphael 717
care 461–474; and modernity 461–467; politics of 467, 469–470; and sustainability 467–473
caring 467–468, 470–474, 714
Carel, Havi 528–529
Carnap, Rudolf 116, 725
Carnegie Foundation for the Advancement of Teaching 666
Carnot, Sadi 14, 111–112, 114, 121
Carrara, M. 251
Carroll, N. 509–512
Carson, Rachel 691, 702
Catalano, George D. 466, 700–709
Cavoukian, Ann 599
Cech, E.A. 652
Cepeda Porras, G. 427
certainty 374–375, 380; *see also* uncertainty
Chakma, Bhumitra 56
Chakraborty, Partha 54
Challenger space shuttle 266, 561, 664, 667
Chalmers, David 267
change 141–144, 147; as driver of problems and solutions *140*; and cycle of replication–incrementation–disruption *144*
Channel Tunnel (Britain) 69
Chapman, Fredrik Henrik af 67
Charisi, V. 562
Chatterjee, A. 515
Chavez, Aura Lolita 683
Checkland, P. 185
Chennai, India 56
Chen, Changshu 51
Cheng 程 50–51
Chernobyl 664
China 50–53; Cultural Revolution 34
Chisholm, Roderick 211
Christman, J. 559
Chromosome Model 253, *254*, 255
CIB approach *see* Cross-Impact-Balancing Analysis

circular economics 351
citizenship 610; responsibility of 667, 669, 671
civil-aviation system 226, 228–231
civilising hypocrisies 466
Clancy, K.B.H. 649
Clarkson, John 321, 390
climate: engineering 586; global 33, 473; simulations 410; uncertainty 170; warming of 139
climate change *140*, 151, 670, 740; adapting to 157; catastrophic 171; environmental effects 153; geoengineering for 515; injustice of 691; modeling and measuring 377; Paris Agreement 638; professional response to 594, 596; reality of 346; and social justice 601; water contamination and 469
climate disruption 461, 468, 472
climate modeling 165
climate policy 161–162, 596; engineer's role in 631, 635
Cockburn, A. 255, *256*
code of conduct 588
code of ethics 584–585, 588–590, 594–595; in the United States 661–662; *see also* professional engineering societies
Code, L. 243
Codex Alimentarius 571
Coeckelbergh, M. 561
Cohen, Benjamin R. 464, 674, 687–699
Colby, A. 666–667, 671
Cold War 162, 663, 665
Cole, Luke 689
collaboration 324, 329–335, **337**, 338; and social justice 675; as value 739
collaborative creativity 330, 332; *see also* creativity
collaborative community development project 669
collaborative curiosity 330, 332; *see also* curiosity
colonialism 26, 678, 465, 688, 705; British 33–34, 50; European 31, 33, 345; Western 58
coloniality 465–467
Collberg, C. 430
communication networks 13
community college 14
community/ies: bonds 484; business 663; church 610; collaborations with 675; development 668, 670; empowered 597; engagement 682, 684, 687; and environmental injustice 688–696; of knowledge 450, 637; local 669; machine learning 198; marginalized 677, 679, 704; nurturing 706–707; philosophical 395; political 609; professional 371, 561; research 404, 430; resistance 683; and social ontology 21; systems engineering 181, 186–187; utopian socialist 18; whole 185; writing 575; *see also* engineering community
companies and corporations by name: Amazon 273; British Aerospace 179; Dupont 56; Facebook 273; Lucas Group 179; Nakajima Aircraft Company 60; RAND 162; Systems Engineering Company (SEC) 179; Union Carbide 56; Volvo 512

company 45, 98; automotive 171; and consumer 512; development 167; human resources 501; investment in 163; jet engine 387; market domination by 571; management board 168; mining 278, 675–676, 681–683; social responsibility programs 669–671; standardization 575; technological innovation 273; secrets 534; tobacco 452

company profit 56

comparative experiment 425

complexity 181–184, 187, 426, 431; computational 443; of design tasks 105, 315; of engineering problem-solving process 142; extrapolation as adding 109; nonlinear 463; of optimization 98; of prescriptive engineering knowledge 123; reducing 466, 490; sociotechnical 501; trust and 495, 498, 503; *see also* system complexity *183*

complexity barrier 415–417

comprehensive methods 324

compromise 308–310

compromise versus transformation 348, 350

computer 82; digital 266; personal 599; quantum 121, 222; tablet 328, 335

computer-aided design 324

computer-based tools 195, 209, 320, 332

computer engineering 33, 127; experimental method in 424–425, 429–432

Computer Ethics Institute 585

computer fluid dynamics (CFD) 323

computer models *see* models

computer mouse 426

computer networks 533, 577

computer programs/programmer 70, 73, 83, 134; ideal 319; Mississippi Basin Model 398; source code 536; weakness in 534, 538

computer science 32, 133, 423; verification and validation in 435

computer scientists 586, 602

computer simulation (CS) 89, 90, 408–418; complexity barrier 415–417; definitions 408–410; other methods, relation to 412–414; process 410–412

computer system 315–316; vulnerabilities in 540

computer systems design 329

Condorcet, J.-A. 464

conflict negotiation 347

conflict of interest 41, 43, 443, 574, 609, 621

Confucianism 51–53; neo- 59, 61, 346

Conga project 683

Conly, Sarah 560

consequentialism 44–47

Constance School 723, 725–727

Constant, Edward II 15

constructivism 537; critical 711; methodical 722–732

context 234–242, 302–306, 342–346; biological and medical 409; changing 641; of codes and principles for engineering 589, 661; of complex practice 583, 741; constraints 668; cross-cultural 625–626; cultural 18, 714; decision-making

in 312; designed 495–496; of discovery 415; discursive 466; economic 473, 710; of ecological resilience 483; of ecosystems 482; embedding into 410; and emergence, template for 265; of engineering 162, 728; of an ethics of relevance 688; globalized 620, 622–624, 691; of healthcare engineering 483; historic 349; of human social life 731; human-technology relations 719; of justification 415; multinational 694, 696; narrative 654–655; omnium/engineering 700, 706; ontological 486; operational 474; of order and orderability 711; physical 236, 239; of physics 730; political 281, 613, 667, 738; of political deliberation 109; of political order 279; of power differences and structures 609, 640, 666–667; professional 449; "regenerative design" 351; research and non-research 383–387, 598; risk assessment 450; scenario thinking in 170–171; scientific 454, 641; security 538, 540; social and socially-responsible 235, 239, 329, 462–464, 670; social experiment 715–716; social justice 675, 682–683, 690–691; structural 742; of technology 614–615; theoretical 224; of uncertainty 499; values in 302–303

context-dependent problems 342, 414, 562

context-sensitive ethics 328

context-specific design praxis 323

contextualization (and decontextualization) 115, 612, 654

contextual reasons 501

Continental philosophy 223, 507, 510, 726; *see also* Erlangen school; philosophy of technology

control 461–474; of behavior 594; cost 103; of disease 349; of the environment 133; excessive 19, excess of 330; failures of 473; five classes of 535; flood 28; of health 525, 716; hierarchies of 471; of information 303; of internet 576; knowledge of 375; of local weather 33; of means of production 348; of motion 24, 31; of movement 29; over nature 713; physical-technical means to 80–86; political 465, 610; process 157; of professional licensing 32; regulatory 577; and repeatability 413; safety and security 537, 539; self-56; social 496; and technology 564–565; *see also* air traffic controllers; data controllers; discontinuous automatic control theory; over-control argument; quality control; out of control

control-delusion 469

controlled conditions 641

controlled distortion 196

controlled environment 440

controlled experience 421

controlled experiment 421, 424–425, 428–429, 431–432

control measures 457

control subjects 715

control theory 228

control volume 150, 156

convergent thinking 144–145, 147; versus divergent thinking **145**
cookbook engineering 633–634, 642
cooking 727–729; use of artefacts in 241; utensils 229
cooking pot as closed system 182, 210–211
Cooper, Theodore 494
Copernicus 20
Corotis, R. 154
corporate social responsibility (CSR) 670–671, 679–684; "business case" for 671, 681
Corporation of the Seven Wardens, The 598
corporations 18, *496*, 512; and engineers 558, 561, 664, 668, 670; industrial 662; international 334; and the military 663; modern 664; nineteenth-century 345; as a person 45; transnational 466; *see also* companies and corporations by name
cosmos 344–345, 350, 707
Coulomb, Charles-Augustin de 67
cowardice 40, 330
Craig, C.H. 632
creative destruction 277
creative thinking **146**
creativity 138–147: and change 138–140, *140*; cooperative 331; and design 296–297; diminishing returns 142–143, *143*; need for 140–142; problem-solving 143–144, *144*, 297; *see also* innovation
critical thinking 144–145
Cropley, David H. 138–149
Cross-Impact-Balancing (CIB) Analysis 165, 168
Cukier, K. 716
culturalism 722–732
culture 469; 'of assessment' 352; Australian Aboriginal 139; Chinese 50, 52; of control 471–472; of depoliticization 656; design 197; Eastern 61–62; engineering 461, 464–466, 474, 590, 620–626, 652, 657; Harappan 27; high 22; Indian 50; as individualism 20; infraculture 464, 468; Japanese 50; material 13, 277, **283**, 361, 363; monoculture 541n2; nature and 212; organizational 561; and philosophy 38–40; police 333; relations of 471; safety 489; 'two cultures' 134; weed-out 649
Cummins, R. 239
Cunya, Elizabeth 683
curiosity 119; collaborative and cooperative 330–332, **337**, 338
cyborgs 352

Darwin, Charles 214, 268, 344, 378
data analytics 194, 197–198
data chain *192*, 197
data controllers 201, 502
datafication 716
datafied subject 716–717
data management tools 324
data visualization 191
David Taylor Model Basin 397–398

Davis, Michael 633; codes of ethics 44, 561, 580–591, 696, 700; norms and values 624, 627n8
Day, Rosie 690
De Bruijn, N.G. 444
Dechesne, Francien 195, 435–446
decision-making: algorithmic 108, 196–199; automated 323; data and 191, 193; deliberative 17; discriminatory 195; ethical 57; laws and 121; multidisciplinary clinical 498; multiple criteria 309; organizational 162; of organizations 161, 319; political 312; power of 34; process 334, 547, 551–557; and risk 455; scenarios for 162, 164, 168, 170–172; societal 454, 496; strategies of 144; systems engineering and 179; and systems thinking 185; techniques 39
degradation 213, 359, 379; environmental 465, 473, 512, 516, 681
degrowth hypothesis 351
de Jouvenel *see* Jouvenel, Bertrand de
De Léon, D. 241
democratisation 464–465
Denning, P.J. 425
derivation 268
Derrida, Jacques 332–334
Descartes, René 21, 39, 183, 264
design: ancient practice of 352; as activity 290–293; automated 259; conceptual *248*; and creativity 296–297; epistemology of 316–318; human-centered 332–334, **337**; ideal theory of 234; privacy by design 300; and values 294–296, 300–321; *see also* human-centered design (HCD); use plan; sustainable design; value sensitive design (VSD)
Design for Values (DfV) 300
designing preferred conditions 352
Design Methods Movement 319–320
design methods 16–17, 315–325; contemporary process models 320–321; history of 317–320; testing as 322–323
design paradigm 360–363
design parameters **106**
design process *322*
design tools 323–324
Dessauer, Friedrich 22
development: elementary 436–437; evolutionary (design) 439–440; multi-stage 437–438
De Vries, G. 610
de Vries, Mark J. *see* Vries, Mark J. de
Dewey, John 22, 149, 311; on inquiry and imagination 46–47, 335–337; on problem resolution 178; on *unmodern* 352
dialogue 242, 669; with customers 98; with designers 317; with engineers 722; in human centered design 335; philosophical 17, 38, 583; with public 602; responsibility of 667, 669, 671
Didier, C. 662
Dieckhoff, Christian 160–175, 342
Dieter, G.E. 138, 145, **146**

digital age 599, 602
digital art and music 507–508, 514
digital computer 33, 266
digital currency 716
digital data 737
digital devices 328, 717
Digital Divide 599
digital engineering 270
digital environment 497, 501
digital information 71
digitalization 736–737
digital native 498
digital networks 128, 135, *275*, 280
digital representation 738
digital simulation *see* computer simulation
digital space 736
digital system 270
digital technologies 633, 731, 736
Dingler, H. 724–725, 730
Dipert, Randall 212, 214, 359
directions of fit 383–391; in modeling relations
    384–388; target-to-vehicle 385–387; vehicle-to-
    target 385–388
discontinuous automatic control theory 227
discovery 138–147, 301; of black swans 152;
    discrimination 196; context of 415; cost 640;
    versus justification 119; of knowledge 191; 'lucky'
    139; of laws of lift 14; of proxies 197; of values
    302
discrimination 196–199; anti-198; in data 191, 195;
    in decision-making rules 195; in health 521;
    linguistic 576; sociocultural 599
discrimination analysis 196
discrimination discovery 196
discrimination prevention 196, 202
disenchantment 133, 464
disrupter 740
disruption 275–276; climate 461, 468, 474; cultural
    345; political 350; vulnerability to 576, 691;
    see also replication-incrementation-disruption
documentation 437
DOD-STD-216 437
Dongeun Huh *see* Huh, Dongeun
Doorn, Neelke 1–8, 736–744; resilience in
    engineering 351, 482–493
Douglas, Heather 453–454, 459n9
Drummond, C. 430
dualism 263; Cartesian 211; classical 269; critiques
    of 650; psychophysical 263–264; technical-social
    652
dual nature theory 71–72, 84, 117
Dupuit, Jules 45
Durand-Lesley Propeller Tests 119–120
Durkheim, Emile 582
duty ethics 42–44; *see also* ethics
Dwork, C. 203n8
Dyer, Henry 59–60
Dyson, Esther 648

Eckert, Claudia 321
Ecola, L. 168
École de Beaux-Arts (France) 128, 131
École Polytechnique (France) 18, 31, 67–68, 128,
    131
ecology 344–346, *348*, 350–351; agro- 472;
    behavioral 236; resilience and 483
economy, ecology, equity 344, 346, *348*, 350
ecosystems 185, 703; healthy 687; and resilience
    482–483; vulnerability of 691; *see also* Green
    engineering
ECPD *see* Engineering Council for Professional
    Development (ECPD)
Edenheim, S. 649
Eder, Wolfgang 252–255, 320
Edgerton, David 360, 362–364
Edwards, R. 675
Edwards, R.G. 564
Eggert's belt-and-pulley example *102*, 103–105
Egypt 26–31, 34; pyramids in 129, 178, 186
Ehrlenspiel, Klaus 320
Ehrlich, Paul 33, 346, 525
Eiffel, Gustav 132
Einstein, Albert 142, 268, 413
Eisenbart, Boris 233, 245–262
Elder, Crawford 215–216
electricity 32, 55, 131; Coulomb's work in 67; to
    electrons 184; into motion 240; piezo- 86; as
    technologically-produced phenomenon 81
electricity grid 632, 640–641; disruption of 536
electricity networks 176, 361
elementary development 436–437
eliminativism 113, 115, 122–123, 217, 225, 513;
    see also nihilism
Ellul, Jacques 711–712
emergence 263–270; beginnings of 263–264;
    engineering for 270; history of 264–265; modes of
    novelty 265–269; template for 265
empirical cycle 80
empirical investigation 16
'empirical turn' 517
empiricism 39, 186, 466
empowerment 331–332
Encyclopédie (Diderot and d'Alembert) 67
energy efficiency 347
enframing 20, 711–712
engineer 585–587; existential 739; future 738–740
engineering 11–23; autonomy in 558–565; *being*
    651; case studies, use of 741–742; climate
    586; computer 33; contemporary 32–35,
    714–718; as designing 15–16; discussion and
    definitions of 11–12, 21; and environmental
    justice 687–696; essence or defining function of
    587; ethics 547–557; ethics, professional codes
    of 580–590; feminist 651–654; financial 586;
    function in 245–259; future 736–738; genetic
    558, 564, 586, 729; Green 701–703; history
    of 25–35; humanitarian 625, 678–679, *703*,

704–705, *705*; imaginaries 651; meaning of
20–22; meritocracy in 652–653; nuclear 33, 35,
56; objects of 612–613; *omnium* 706–707; pre-
professional 27–31; philosophical inadequacy of
652; philosophy of (Eastern) 50–62; philosophy
of (Western) 38–47; as a practice 610–612;
prescriptive 111–123; problem definitions of 612;
as profession 17–18, 31–32; racial inequality in
652; social 586; socially responsible 661–671; as
society-building 613–615; sociotechnical 650;
standards 569–578; structural 157; and technology
691–696; textbooks 664–665; verification and
validation of 435–444; *see also* geoengineering;
resilience engineering; security engineering
engineering activities and methods *see* computer
simulation; experimentation; measurement;
models; scale models; verification and validation
engineering and social justice (ESJ) 675–676; *see also*
engineering for social justice
engineering as political practice 607–617
engineering community 651, 676; ethical
responsibilities 601–602; global 624; professional
self-understanding of 19; and Sustainability 469
Engineering Council for Professional Development
(ECPD) 663–665
engineering design 289–298; Arrow's theorem
103–105, 108; nominal requirements and
constraints 99–101; objective of 233; optimality
97–109; ordinal requirements 101–103; process
145, **146**; and the social world 297–298; trade-offs
105; use plans for 233–243
engineering design models, phases of **146**
engineering education 31–34; accreditation 44, 664;
aesthetics, role of 134; American 652; in China
53; convergent thinking, focus on 147; degree
programs 664; design, emphasis on 357; ethics and
standards in 577–578, 664; in France 67; future of
740–741, 743; "hidden curriculum" 742–743; in
India 57; in Japan 59–61; Kant's influence on 42;
militarized 649; in the military 18; practitioner's
movement in 318; racial inequality in 652
engineering ethics 676–677; education 741–743;
hidden curriculum 742–743; global 620–626
engineering for social justice 674–685; *see also*
Engineering, Social Justice, and Peace (ESJP)
engineering for policy 631–632; *see also* policy for
engineering
engineering in policy 632–633, 634–635
engineering knowledge 730–731; classification
72–73; dual nature theory of 71; epistemology
73–75; nature of 633–634; practical rule 75–76;
pragmatic 633, 635, 643, 651; systematic 74; tacit
72–73
engineering methodology 114; generic systems *180*
engineering methods *see* engineering activities and
methods
engineering narratives in policy 637–641
engineering objects 222–231; *see also* Quine, Willard
Van Orman

engineering, philosophy for 12
engineering philosophy 12, 57, 176–177; China
50–53; Eastern 50–62; India 54–57; Japan 57–61;
Western 38–47; *see also* Aristotle; Heidegger;
Kant; philosophy of engineering; systems
engineering
engineering policy advice 630–634, 641–642
engineering practice and policy 630–643, 722–732
engineering practitioners *see* practitioners
engineering problem-solving: five pathways *141*
engineering profession 663–664
engineering reasoning 144–145; *see also* reason in
engineering; reasoning
engineering responsibility 741; *see also* responsibility/
ies in engineering
engineering science 66–76; definition 80–84;
design-concepts 84–85; epistemology 72–75;
experimental tradition of 74; idealization 75–76;
metaphysics 71–72; methodology 80–91; models
87–91; as profession 70
Engineering, Social Justice and Peace (ESJP) 597,
674–676, 705–706
engineering societies 602–603; emergence of 31–34
engineering solutions, diminishing returns of *143*;
see also solutions
Engineers Australia 594, 600, 677–678
Engineers Canada 594, 596
Engineers Ireland 595
Engineers Without Borders (EWB) 598, 603
Enlightenment, the 20, 31, 67, 183–184, 318, 464
entropy 344–345
Environmental Impact Assessment (EIA) 684
environmental justice (EJ) 687–696
environmental racism 689
Environmental Protection Agency (EPA) 701
episteme 41, 48n5
epistemic aim 83
epistemic artifacts 82; *see also* artifact
epistemic building blocks 86
epistemic challenge 374
epistemic evaluation 376–377, 379
epistemic injustice *see* injustice
epistemic interpretation: of possibility *164*; of
scenarios 171–172
epistemic norms *see* norms
epistemic opacity 415–417
epistemic pluralism *see* pluralism
epistemic practices 39
epistemic processes 47
epistemic purposes 87, *89*, 91
epistemic tasks 85, 89
epistemic tools 383–384
epistemic uncertainty *see* uncertainty
epistemic values *see* values
epistemic virtues *see* virtues
epistemological challenges 191–192, 202
epistemological consequences 112
epistemological study 116
'epistemological turn' 214

epistemologies of ignorance 653
epistemology 40, 302; analytic ideal 242; of
  computer simulation 408–409, 414–417; of design
  316–318; of descriptive laws and theories 121; of
  engineering 72–75, 114, 176–177, 365; Kantian
  22, 42; Marxist 52; meaning and 21; new 350;
  and pragmatism 46; of prescriptive knowledge
  119; reductionism as 183; scientific 378; of
  systems engineering 186–188; unmodern 352; of
  verification and validity 435, 441, 450
equality 43; fairness and 198–200, 599; and
  inequality 461, 465, 473, 652; *see also* fairness;
  social justice
equity *348*, 350, 652, 687, 706; complex 349; desire
  for 676; inequity 653; and injustice 689; social
  344–347, 349, 461, 463, 466, 473
Eriugena, Johannes Scotus 66
Erlangen School 723, 726–727
established technology (ET) 120
ethical code *see* global ethical code
ethics 580–581; aspirational 625; feminist 654–656;
  Kantian 676; masculinist 654–655; *see also* code of
  ethics; data ethics; duty ethics; engineering ethics;
  professional code of ethics; virtue ethics
ethics-of-alterity 329, 332–333, **337**
etiology 485–486
EURON Special Interest Group on Good
  Experimental Methodology in Robotics Research
  428
European Network for Accreditation of Engineering
  Education (ENAEE) 44
evaluation 73, 115, 117, 119; during conceptual
  design *248*; of rationality of design 234; and
  convergent thinking 145, **146**, 147; of/by use
  plans 234–235, 237–239, 243; *see also* function-
  behavior-structure network
evaluation criteria 129, 292–293
evaluation strategies 144
Evetts, J. 496, 503
evidence 437
Evnine, Simon 210
evolution 345; and behavior of the device 256;
  co-evolution (in problem solving) 247, 252–253,
  258; co-evolution (of problem and solution) 317;
  co-evolution (of technology and social structures)
  301; Darwinian theory of 214, 344; human 47;
  of methods 440, 443; processes of 212; of product
  forms 17; of technology 46
evolutionary development paradigm *see* paradigm
evolutionary fitness 528
Evolutionary systems 235
expectation 268
experimental computer science and engineering
  (ECSE) 422–426; differing views on 424–426
Experimental Model Basin (Washington Navy Yard)
  398
experimental subject 715–716
experiments 421–432: explorative 432; *see also*
  controlled experiment; feasibility experiment;

laboratory experiments; social experiment;
  thought experiments; trial experiment
explanatory power 227
explanation: biological 119; causal 114; chain of
  292; conceptualizations of 303; defying 268;
  derivation or 269; and design-concepts 86;
  diagnostic theories 122; epistemic 415; functional
  291; mechanistic 416; narrative 630, 635–637,
  642; of phenomena 82, 87, 91; scientific 270, 394;
  systemic 42
Exxon Valdez 664

Fadel, G.M. 236, 240–241
FAES *see* Federation of American Engineering
  Societies (FAES)
fairness 191–201, 574; and justice 576; obligation of
  581; social justice and 599; unfairness 540; values
  of 501
fairness argument 540
Fairness, Accountability, and Transparency in
  Machine Learning (FAT-ML) 201
Fann, K.T. 158
Faulkner, Wendy 648, 650
Faustian aspect of innovation *see* innovation
FBS framework *see* Function-Behaviour-Structure
  (FBS) framework
feasibility experiment 425
Feinberg, Joel 559
Federation of American Engineering Societies
  (FAES) 662–663
Feenberg, Andrew 22, 52, 608, 711–712
Feldman, J. 424
Feldman, M. 196
Feldman report 424
feminism 467, 518; relational autonomy advocated
  by 560
feminist aesthetics *see* aesthetics
feminist engineering 47, 647–656
feminist philosophers 242
Ferguson, Eugene 15, 134
Feynman, Richard 725
fidelity 44
field experiment 425
finite element method (FEM) 323
finite element models 121, 158
firmness, commodity, delight 16, 130
fitness landscape 350, *351*
Flanagan, Mary 301
Fleck, L. 684–685
Fleddermann, Charles 742
Floridi, Luciano 535
Florman, Samuel C. 21, 739
Flügge-Lotz, Irmgard 227
Folke, C. 483
Food and Agriculture Organization (FAO) 571
Foot, Philippa 562
Forest de Bélidor, Bernard 68
Forlano, Laura 650
Forsythe, George 422

Fortun, Kim 693
Fortun, M. 654
Forum on Philosophy, Engineering and Technology (fPET) 736
framing: community and community-centric 484, 693; of design 320, 359; enframing 20, 711–712; of security concerns 540; social justice 674, 676, 684–685, 694; theoretical 689; user-centred design 677; *see also* problem framing
framing the scenario 168
framing of systems engineering 188
Frankenstein (Shelly) 132
Frankfurt School 52, 711
Franklin, Ursula 679
Franssen, Maarten 97–110, 305, 392n2, 722
Freidson, Eliot 561
French scenario school 162
Fricker, Miranda 381n10, 652
Friedler, S.A. 199
Friedman, Batya 300–302
Friedrich, Caspar David 132
Frigg, R. 414–415, 418n8
Fritzsche, Albrecht 38, 722–735
FRUX project 329, 332–334
Fuji Industries 60
Fujiki, Atsushi 50–65
Fulton, L. 161
Fulton, Robert 68–69
function 245–259; archetypes 251–251; as ontology 247–251; purpose 252; technical 252; teleological notion of 251
Functional Basis *250*
function-based design 256–259
Function-Behaviour-Structure (FBS) framework 252–253, 257, *258*
functionalist theory 623–624
functionalist stance 240, 510, 511–512
function carriers 248, *249*, 255, *256*, 257–258
functional taxonomies *250*
function structure 253
Funk, Michael 38, 722–735
future developments, assumptions of 163–164
future engineers, training of 5–6, 736
future of engineering 127, 135, 736–744; *see also* uncertainty
Future of Life Institute 603
future worlds 167

Gadamer, H.G. 521
GAIA hypothesis 352
Galileo Galilei 74, 413, 724
Galle, P. 386, 392n7
Gandhi (Mahatma) 55
Garzoni, Tommaso 12
Geddes, Patrick (Sir) 346
Gelernter, David 129
general problem solver (GPS) 319, 323
genetic editing *729*
genetic engineering *see* engineering

genetics, nanotechnology, and robotics (GNR) 602
geoengineering 33, 283, 515–516; aesthetics of 515–516; Anthropocene 469; technologies 468, 473
geometrical similarity 40
George IV (king) 12
Gericke, Killian 233, 245–262
German Association of Engineers *see Verein Deutscher Ingenieure* (VDI)
Gero, John S. 252, *253*, 257
Gertner, J. 142
Gewirth, Alan 561
Ghana 679
Giere, R.N. 82
Gilligan, Carol 654
Gilman, G.W. 178
global development 693–694
global engineering ethics 620–626
global ethical code 623–624
global ethics 623; *see also* ethics
Global Initiative on Ethics of Autonomous and Intelligent Systems 603
globalization 139, 142, 345, 620; economic 61; term 344
global navigation satellite systems (GNSS) 641
global positioning system (GPS) 192
global warming potential (GWP) *305*
Göteborg Bridge 132
Goethe, Johann Wolfgang von 378–379
Goldman, S.L. 633
Gong Cheng 工程 50–51
goods (of engineering practice) 611–612, 716
GPS *see* general problem solver
GPS *see* global positioning system
grand challenges 738, 740; *see also* problems: wicked
Gransche, Bruno 717
gratitude 44
Greece, ancient 30, 34, 38, 66, 130, 277, 281
Green engineering 701–703
greenhouse gas removal (GGR) 638–640
Greenpeace 170
greenwashing 349, 670
Grenon, Pierre 216, 219n7
Grewal, David 576
Gruenewald, D. 684
Grunwald, Armin 160–175, 342
Grüne-Yanof, T. 409
Guéhéneuc, Y.G. 427
Guilford, J.P. 145

Habermas, Jürgen 21, 149, 564
Hacker, Sally 648–649, 654
hackers 432, 577, 599, 614; ethical 537–538
Hacking, Ian 372
hacking science 424, 432n3
Haeckel, Ernst Heinrich Philipp August 344
Haiti 742
Hall, A.D. 178
Hamilton, John 573

handover dilemma 453, 455–456
Hansson, Sven Ove: action-guided experiments
  125n19; engineering science, definitions and
  discussions of 38–39, 66–79, 127, 130–131, 345,
  376, 450, 730; prediction typologies and scenarios
  164, 172; risk assessment *456*, 456–457, 458n3;
  'science,' definition of 450, 452; technical rule
  knowledge 124n1
Haraway, Donna 656
Harding, S. 651, 653, 656
Hardt, M. 203n6
hardware-and-software-in-the-loop (HiL/SiL) 323
Hare, R.M. 48n7
Harford, T. 155
Harris, Charles E. 40, 625, 741
Hartmann, D. 727
Hartmann, S. 409
Harvey, Bruce 680–682
Hashimoto, Takehiko 59
Haslanger, S. 650
HD method *see* hypothetical-deductive method
health 521–530; and disease 523–524; as engineering
  value 522–523; environmental; philosophical
  accounts of 527–530; and technology 524–526
Hegel, G.W.F. 21–22, 465
Heidegger, Martin 19–21, 730; *Being and Time*
  223, 723–724; on frustration 509; Riis on 712;
  on Stiegler, influence on 713; technology as
  "enframing" 711; on technology as "means to an
  end" 462; on technology, reductive nature of 39;
  on technology and *techne* 41
Hempel, Carl G.: argument from inductive risk 453;
  hypothetical-deductive method 87, *88*, 636
Herschel, John F.W. 378–379
Herkert, Joseph 32, 495, 592–606, 665–666, 687
heuristic(s) 12, 46–47; and analytic aesthetics 508;
  design 324; engineering 153; experiment 413;
  "general" 113; methods 427; of politics 609; of
  Simon 317–318; and Sustainability 470; systematic
  319, 323; and uncertainty 153, 156–158
Hevner, Alan 117–119
Heynen, H. 518
Hickman, Larry 47
high frequency (HF) communications 641
Hillier, J. 177
Hilpinen, Risto 212, 293, 359
Hinduism 54; nationalism 56
Hirota, Shigeru 61
Hirsch Hadorn, G. 172
Hirt, Sonia 349
Hobbes, Thomas 68
Hodge, Tony 680
Hoff, K.A. 499–500
holism 181, 183–185, 187–188
Holling, C.S. 351, 482–483
Hollnagel, Erik 485–490
Hooke's law 82, 87
Hoover Dam 132
Hoover, Herbert 662

Horenstein, M.N. 138
Horkheimer, Max 706
Houkes, Wybo 222–232, 233–243
Huber, Lara 371–382
Huber, M. 529
Hubka, Vladimir 252–255, 320
Hugh of Saint Victor 66
Huh, Dongeun 386
Hui, Yuk 51
human-centered design (HCD) 328–338
Human Genome Project 526
humanitarian engineering *see* engineering
human rights *683*; abuses 670; declarations of
  21; framework 683–685; *see also* engineering:
  humanitarian
human use and convenience 652–653
human subject 21, 715–719; experimenting on 562,
  590
human subjects research 598–599
Humboldt, Alexander 344
Hume, David 13, 39, 292; and Kant 42; on "use and
  convenience" 46
Humphreys, Paul 409, 413, 415
Hundal, M. *250*
Huning, A. 662
Husserl, Edmund 211, 723–727, 730
Hyatt Regency disaster 549, 664
hydraulics 13, 26–29, 129, *250*; of empire 32;
  infrastructure 148; models 398; pumping stations
  132
hydrogen economy 165
hydrogen gas 84, 183
hydrography 378
hypothetical deductive (HD) method 80, 87–89, *88*
Hypothetical Transparency Condition 502–503

identity: disciplinary 736; engineer's 662, 687, 693;
  experiences of 650; indigenous 34; professional
  661, 736, 740
ICE *see* Institution of Civil Engineers (ICE) (Britain)
idea 281
ideai 279–280, 284
IEEE *see* Institute of Electrical and Electronics
  Engineers
IEEE 829–2008 438
IEEE 1012–2004 437–438
IEEE Transactions of Software Engineering and *Software*
  427
Ihde, Don 223, 443, 712
Ilvari, Juhani 117–119
imaginaries: engineering 651; historical 352;
  sociotechnical 162
imagination(s) 362, 718; of control 463, 465–468,
  470, 474; engineer's 133, 466; human 129; joint
  337, 338; idealized 463; instrumental 463, 466;
  moral 336; in scenarios 163, 170; social 352; of
  Sustainability 471; *see also* Together Anywhere,
  Together Anytime (TA2)
imagining 714; *see also* reimagining engineering

Imhotep 129
Imperial Institute of Physics and Technology
 (Germany) 595
inaccuracy 377, 385
incendiaries 26, 30
INCOSE *see* International Council on Systems
 Engineering (INCOSE)
India 33–34, 54–57
individualism 16, 20–21, 242, 665
individualization 464–465
inductive risk 453–454
industrialization 464–465, 514; in Britain 345;
 capitalist 20, 31; in China 34; in Japan 59, 61
Industrial Revolution 17–19, 73, 111, 128,
 131–132
industrial sublime 133
information systems (IS) 116–119
information technology (IT) 117, 270, 737
infrastructure: carbon capture and storage 639;
 in China 53; community 693; corporate
 commitments to 681; data acquisition 201;
 design of 308; engineered 463, 640; of
 engineering 465; engineering's focus on 127–128;
 environmental impact of 573, 575; ethics of
 594; and geoengineering 468; great works of
 132; imaginaries and 352; laboratory 375; mass
 society and 133; modernist visions of 473–474;
 modern life and 614; needs 678; resilient
 631–633; overhauling 740; policy 639–641; and
 social justice 599; systems engineering and 176;
 technical 470, 484; telecommunications 53;
 transport and transportation 13, 16, 46, 461, 601;
 in the United States 356; weapons 472
Ingold, Tim 13, 358–359, 364
injustice 473, 597, 655, 700; of climate change 691;
 and corporate social responsibility (CSR) 681,
 684; distributive 694; and engineering and social
 justice (ESJ) 675–676; environmental 688–689,
 692; and environmental justice (EJ) 693–694;
 epistemic 379, 652, 740; global networks of 696;
 hermeneutic 653; institutional 742; procedural
 690; resisting 705–706; social 690, 717; *see also*
 justice
innovation 273–284; creative aspect 281–282;
 destructive aspect 277–281; Faustian aspect of 276,
 282, **283**; responsible; Schumpeter's influence on
 275–277, **283**; self-evident concept of **283**; *see also*
 Plato; Schumpeter
insight 171; epistemological 39; 'small' 91
Institute of Electrical and Electronics Engineers
 (IEEE) 33, 570; Code of Ethics 585, 588,
 593, 595, 600; *Ethically Aligned Design* 603;
 paramountcy principle 664
Institute of Electrical Engineers (Britain) 33
Institute of Radio Engineers 33
institution 19: codes of ethics of 588; and control
 464; coordinated 26; economic 16; educational
 59, 178; engineering 57, 357; financial 571;
 human 11; industrial 179; and innovation 281;

professional 593; and resilience 484, 489; social 38,
 41, 47, 191, 223, 230, 235–236, 242; specialized
 31; supra-national 623; and sustainability 468–471,
 473–474
Institution of Civil Engineers (ICE) (Britain) 12–13,
 32, 240, 372, 594; theory of function ascription
 235, 243
Institution of Engineers (AMIE) (India) 34
Institution of Engineering and Technology (Britain)
 593
Intergovernmental Panel on Climate Change (IPCC)
 161, 165, 469
International Council of Academies of Engineering
 and Technology 11
International Council on Mining and Minerals
 (ICMM) 680, 682
International Council on Systems Engineering
 (INCOSE) 179, 181, 187; Systems Engineering
 handbook 442
International Energy Agency (IEA) 161
International Union for the Conservation of Nature
 (IUCN) 343, 347
Internet of Things (IoT) 539, 599, 602, 637–638
invariants 404–405
IPAT equation 346–347
ISO/IEC 2008 437
ISO/IEC 15288 *436*

Jackson, D.C. 664
Jackson, Steven J. 362, 365
Jackson, W. 156
Jaggar, A.M. 242
Jainism 54, 57
Jamal, Karim 623
James, William 46, 335
Janich, Peter 723–728, 730
Janism 346
Japan 57–61
Japan Society of Civil Engineers 593, 595
Jensen, J.N. 138
Jerónimo, Helena 712
Jesiek, Brent K. 623–624
Jewson, N.D. 524
Jobs, Steve 443
Johnson, D. 668
Johnson, Gearold 736–744
Jonas, Hans 39, 710, 718
Jones, Sharon 678, 696
Joseph, M. 199
Jouvenel, Bertrand de 162
Joy, Bill 602
jurisdiction 456, 571, 690
Juristo, N. 427–428
justice 44, 598, 623; capabilities approach to
 690–691; distributive 312n2, *348*, 654, 689–691,
 695–696; energy 691; procedural 690–691,
 694–695; recognition 690, 692–696; safeguarding
 626; water 691; *see also* environmental justice;
 social justice

Kahneman, Daniel 549
Kahn, Herman 162, 171
Kambartel, Friedrich 723, 725
Kaminski, A. 411–412, 415–416
Kamiran, F. 196
Kamlah, Wilhelm 723–726
Kantian 213; representationalism 216; rights and
    representation 708
Kant, Immanuel 21–22; anti-functionalism of 511;
    on autonomy 558; on categories 211; duty ethics
    and virtue theory 42–44, 549, 676; systemic
    philosophy of 39
Kasher, A. 561
Kass, Leon 564
Kast, F.E. 181
Keinonen, T. 167
Keller, E. 653
Kelvin *see* Thomson, William (Lord Kelvin)
Kenny, A. 511
Kerckhoffs' principle 536
Khalil, H. 424
Kim, Daniel 185
Kingma, E. 528
Kleinberg, J. 199
Kline, Ronald 561
Knight, F.H. 152, 163
knowledge: descriptive 113–115, **115**; of physical-
    technological phenomena 85–86; propositional
    124n1, 242, 508; *see also* community/ies;
    engineering knowledge; prescriptive knowledge
Knowledge Discovery and Data Mining (KDD) 201
knowledge for 82–83
knowledge of 85–87
Knuuttila, T. T. 90, 93n5
Koen, Billy Vaughn 12, 46, 125n21, 361
Kojin Club 34
Koller, Stefan 506–520
Kondratieff, Nikolai 275
Korman, Daniel 210
Kosky, P. 132
Krauss, L. M. 167
Kripke, Saul 23
Kroes, Peter 53, 71, 407n2; engineering design
    289–299, 358–359, 392n2, 730; Nereda
    technology 125n17; prescriptive knowledge
    124n1
Kuhn, Thomas 15; "normal science" 413; paradigms
    in science 346, 637
Kuorikoski, J. 416
Kusner, M.J. 203n5
Kwakkel, Jan 191–206
Kyoto School 58

laboratory experiment (LE) 412
Lachman, Richard 135
ladder 71; brace of 236; as design problem 551–552
Lakshimi, Chitra 57
Lambrinidou, Yanna 742
Lampe, Hildrun 408–420

Latin America Mine Monitoring Program (LAMMP)
    679–680
Latour, Bruno 351–352, 380, 607
Lavoisier, Antoine Laurent 268
Layton, Edwin T., Jr. 15, 18–20, 662–663
Leadership in Energy and Environmental Design
    (LEED) 347, 573
Le Corbusier 133
Lee, J.D. 499
Leibniz, Gottfried Wilhelm 318
Lempert, R.J. 166
Lenhard, J. 412–413, 416
Leonardo da Vinci 12, 131
Lepri, B. 196
Leroi-Gourhan, André 13
Leuschner, A. 164
Levinas, Emmanuel 332–333
Levy, L.S. 424
Lewes, George Henry 264
Liang Zhi 良知 51
Li Bocong 51–52
Lincoln, Abraham 708
Liu Dachun 52
Living Building Challenge 347
Lo, Felix Hun Tan 719
logic 18, 114, **115**; *see also* convergent thinking
logos 40, 633
Longino, Helen E. 653
Lorenzen, Paul 723, 725–726, 731
Lorenz, K. 725–726, 731
Lotka-Volterra model 87
Love Canal, New York 689, 692, 702
Lovelock, James 352
Lowe, E. Jonathan 211, 223
Lower Mississippi River Physical Model 398
Lucena, Juan 693–694; engineering, socially
    responsible 661–673, 680, 682
Luegenbiehl, Heinz C. 561, 624, 662
Luhmann, Niklas 535
lung chip *see* models
Lynch, William T. 561

M&S *see* modeling and simulation
Mach, Ernst 147n1
Machiavelli, N. 281–282
MacIntyre, Alasdair 48n1, 295, 610
MacKay, B. 167
Mackey, Robert 39
Maema, Takanori 61
Maier, J.R. 236, 240–241
Maillart (Swiss engineer) 129, 132
maintenance 356–366; as conservation 357–360
Mai, T. 169
make-believe: games of 384; theory of 389–391
Mallet, Robert 483
Mameli, M. 564
management science 118–119
management theory (MT) 118
Manhattan Project 35

Mao Zedong 34
Marburg School 723, 727
Marcellus (general) 29
March, Salvatore T. 117–118
Marcuse, Herbert 52, 706, 711
Mariotte's law 112
Mark, Robert 129
Marres, Noortje 609
Mars Orbiter 549
Martin, Mike W. 562, 598, 716
Marullo, S. 675
Marxism: in China 50, 52–53; in Japan 58; neo- 22;
    on technological and socioeconomic relations 608
Marx, Karl 73, 564; Schumpeter, influence on 276;
    Stiegler, influence on 713
Masse, Pierre 162
mathematical complex systems 152
mathematical concepts 226
mathematical modeling/models 76, 82, 88, 90, 149,
    153; and approximate similarity 405; automobile
    engines designed by 154; computer simulations
    and 409–411, 415–417; explanatory power of 408;
    equation as 395; and fit, concept and direction of
    384; indirect views of representation and 388; of
    the target 410–411; verification and validation 435
mathematical modification 413
mathematical patterns 87
mathematical relations 324
mathematical representations 82
mathematical structures 91
mathematical theory 223
mathematicization 13
mathematics 17; in engineering 41, 74, 611,
    633–634, 736; in engineering curriculum (France)
    18, 31–32, 34, 67–68; equation solving 320;
    functions in 251; Kant's influence of 42; modern
    39; and mereology 218; see also STEAM; STEM
Maxwell-Boltzmann model 87
McAllister, J. 512
McCarthy, Natasha 630–644
McCarthy, Timothy G. 624
McDonough, W. 702–703
measurement 371–382; challenges of 376–379;
    diversity of 730; knowledge through 380;
    objectives of 372–375; as a practice 371–372;
    preconditions of 375–376; and trial systems 425
measurement theory 107, 372–374; accuracy 373;
    representational 381n1
Meijers, Anthonie 71
Meiji Restoration 34, 58–60
Meiksins, Peter F. 561
Melander, L. 168
mercury 374, 553, 555–556
Mercury (planet) 268
mereology 226; of artifacts 217–219; nihilism 231
meritocracy in engineering 652–653
Merleau-Ponty, Maurice 223
Metaphysics (Aristotle) 223

metaphysics: and aesthetics 517; analytic 517; of
    artifacts and artifact kinds 211, 214, 217; of
    engineering science 66, 71–72, 76; feminist 650;
    Kantian 42; Plato's 130; of Special Compromise
    Question 210; of systems engineering 188
methodology 183, 223, 239, 423–424; design
    115, 292–293, 702; engineering 106, 109, 114;
    experimental 427–428; 'function' and 245;
    generic systems engineering *180*, 321; meta- 176,
    186–188; and pragmatism 724–725; Q- 302;
    of scale modeling 394–395, 399, 402; scientific
    80–91, 122, 380; of scientific modeling 80; value
    sensitive design 301
methods: Agile 440, 442–443; analytic 409;
    comprehensive 324; computational 47; lost 27;
    heuristic 427; manufacturing 100; mathematical
    736; in mereology 217–218; model-based
    169–170; post-processing 197–198; research 653;
    in risk assessment 454–456, 458, 534; scenario
    160, 162–172; scientific 15, 450, 456; statistical
    149, 154; subsistence 139; of truth finding
    613; quality control 155; quantitative 38, 107;
    in value sensitive design (VSD) 301–302, 304;
    verification and validation 440–441; work 69;
    see also computer simulation (CS); design methods;
    engineering activities and methods; environmental
    justice (EJ); greenhouse gas removal
Meyer-Schoenberger, V. 716
Michelfelder, Diane P. 1–8, 444, 464, 710–721,
    736–744
Middleton, W.E. Knowles 377
Mietzner, D. 166
Millennium Development Goals 738
Miller, Glen: engineering, Eastern philosophical
    approaches to 47, 50–65; engineering, Western
    philosophical approaches to 38–49
Mill, John Stuart 39, 46, 264, 345
mining 676–677, 679–684; data 191, 196, 198;
    companies 278; industries 662, 669; in Japan 59;
    pollutants 675–676; in Soviet Union 52
Mining for Development 681
Mississippi Basin Model (MBM) 398
Missouri River flood 398
Mitcham, Carl 53, 666; on AIEE code 589; on
    ancient Egyptian engineering 36n1; design
    as essence of engineering 357–358, 652; on
    engineering, concepts and definitions of 11–24,
    38, 58, 687; engineering ethics 662–665,
    667–668; on epistemologies of ignorance 653;
    "Ethics is Not Enough 667–668; "humanitarian
    engineering" 625, 652, 678–679; on technology
    39, 176–177, 357, 363; *Thinking Through
    Technology* 176–177, 356
Mitchell, S.D. 381
Mittelstadt, Brent 191–206
Mittelstraß, Jürgen 723, 725–726, 731
model-based methods 169–170; *see also* methods;
    scenarios

model-building 91
model dynamics 417
modeling and simulation (M&S) 121
modeling relations 383–392
models 88, 91, 383–392, 611; 2D and 3D 324;
    architectural surrogate 386–387; bioengineering
    surrogate 385–386; computer 152, 154, 158,
    169, 408–410, 414, 601; contemporary process
    320–321, 323; designer's product 387–388;
    engineering 383–392; engineering design **146**;
    jet engine 387–388, 389–391; formal 169; lung
    chip 384–385, 391; multi-scale 86; "plant-in-the-"
    119; process and product 316, 320, 323, 390–391;
    propulsion 148n2; reference 431; scale 394–407,
    404; scientific 82–84, 86–91; simulation 408, 413,
    442, 444; triple-cycle 118; types 154; weekend
    cottage 386–387, 388–389, 391
model target system 169
model theory 169
Modernity 25, 464–467; care/control and
    461–474; challenges to 22; engineering
    as 19–21; Enlightenment notions of 464;
    infrastructure and 132; innovation and 134;
    nonmodern 352; precursors to 131; risk and
    safety assessment and 449; rethinking 62;
    technical rationality and 357; technological 692;
    unmodern 352
moha 55
Mohanani, R. 443
Mokyr, Joel 124n1
Möller, Niklas 152, 449–460
Monge, Gaspard 67
Monism 217, 263–264
Montfaucon, Bernard de 15
Moore, Steven A. 342–355
moral code 580, 588
moral concern 610, 708
moral goodness *305*
moral ideal 17, 583–584
morality: and engineering 739; engineering of 732;
    and ethics 580–581, 583–585, 588, 608, 731;
    mere-choice and 560; sexual 310; Socratic 583;
    standards and 569, 594
moral reasoning *see* reasoning
morals: petty 580; theory of 13
moral value *see* values
moral virtue *see* virtue
Morgan, C. Lloyd 264
Morgan, Mary 89, 92, 93n4, 630, 635–637, 642
Morgan, Robin 560
Morrison, Margaret 89, 92, 93n4
Müller, Johannes 319, 323
multibody simulation (MBS) 323
multiple criteria decision-making (MCDM) 309
multi-stage development 437–439, *438*, 443
Mumford, Lewis 13, 346
Murchison, Roderick Impey 69
murmurations 471, 473–474

Nabeshima, Naomasa 59
nanotechnology 128, 602, 729; *see also* genetics,
    nanotechnology, and robotics (GNR)
Narayan, Uma 559, 654
narrative: and engineering policy 630–643; of jet
    engines 389; received 620, 626; reductionist
    183; scenarios 163; of sustainability 461; of
    systems engineering origins 187–188; *see* context;
    explanation
narrative explanation 635–637
NASA *see* National Aeronautics and Space
    Administration
National Academy of Engineering (NAE) (United
    States) 20, 598
National Aeronautics and Space Administration
    (NASA) 178; projects 186; Space Pen 141; *Systems
    Engineering Handbook* 179
National Bureau of Standards (NBS) (United States)
    595
National Council on Systems Engineering (NCOSE)
    (United States) 179
National Defence Industrial Association (NDIA)
    (United States) 179
National Infrastructure Commission (UK) 632
National Institutes of Health (NIH) (United States)
    601
National Physical Laboratory (Britain) 595
National Society of Professional Engineers (NSPE)
    (United States) 44, 522, 585, 593, 664; Code of
    Ethics 569, 703; policy statements 603
nationalisation 465
nationalism 34, 62; Hindu 56
National Society of Black Engineers 593
National Society of Professional Engineers (NSPE)
    593, 664
national standards organizations 595
nation-forming 464–465
natural kinds 23
natural processes 352
natural sciences 13, 343; analytic approach of 317;
    applied 69; basic 80–81, 84; and computer science
    423–432; definition of 70; and disenchantment
    133; drawing in 17; vs. engineering sciences
    227; engineering's dependence on 41–42, 46,
    75–76; experiments/experimentation in 74,
    423–432; formalization and standardization in
    378; hypothetical deductive method *88*; and
    information technology (IT) 117; measurement
    in 380; and Monge 67; understanding 450; and
    ontology 225, 230
natural systems *see* systems
Nayar, Baldev Raj 56
Nedelsky, J. 560
Needham, Joseph 50
Needham-Schroeder protocol 538
net-zero 638–639
Neural Information Processing Systems (NIPS) 201
Newell, Allen 319, 423–424

Newmont Mining Corporation 683
Newtonian mechanics 115, 121
Newtonian physics 42, 87
Newton, Isaac (Sir) 115
Nickel, Philip J. 194, 494–505
Nietzsche, Friedrich 21, 284
Nieusma, Dean 694
nihilism 210, 217, 225–228; mereological 231;
 Platonic 284n3
Niiniluoto, I. 113, 118
Nissenbaum, Helen 717
Noble, David F. 663–664
noblesse oblige 616
Noë, A. 517
Nonaka, Ikujiro 72
non-maleficence 44, 495, 598
Nordenfelt, Lennart 528–529
Nordmann, Alfred 375
Norman, D.A. 236
norms: cultural 603; design 501; of engineering
 education 661; in engineering profession 551,
 592; epistemic 512; ethical 623; fairness 198–199;
 gender 648–649; vs. individualization 465; of
 practice (in engineering) 610, 611, 621–626,
 675; of privacy 498; of reproduction and survival
 528; scientific 636; social/societal 51, 198, 236,
 538, 561; subjective/objective 529; technical 113,
 118; trust and/in 495, 499; of use 230; user 501;
 values and 163, 295, 302–303, 306, *307*, 527; of
 Wittgenstein's lifeworld 725; workplace 743
Norström, Per 112–113
nous (understanding) 41, 48; *see also* understanding
Nozick, Robert 560
nuclear engineering *see* engineering: nuclear
nuclear power/reactor 152, 157, 458n7, 461,
 472–473, 558
nuclear waste 536, 562, 702
nuclear weapons 35, 56, 461, 533, 536, 601;
 thermonuclear war 162, 171

observational method 157
Oden, Michael 349
Ohashi, Hideo 61
Olsen, J. 157
Onondaga Nation 704
Ontological Commitment 224
ontological parsimony 240
ontology 40, 52, 486–; aesthetics and 511; analytic
 222, 225, 231; archetypal 251; definition
 of 223–225; of design 233; of engineering
 knowledge 177; and experience and value 514;
 feminist engineering 647–648, 650–651, 653–654,
 656; holistic 184; innovation, concept of **283**;
 meta- 223, 225–226, 231; reductionism as 183;
 social 21; of systems engineering 188; *see also*
 5-key terms; artifacts; emergence; engineering
 objects; function in engineering; health;
 innovation; use plans
open-endedness 6–7, 342, 470, 610

open system *182*
optimality 97–109, 573
order 279
Order of the Engineer 598
Organization for Economic Co-operation and
 Development (OECD) 273, **283**
organ structure 253
Orihuela, J.C. 681
Orlikowski, W.J. 614
Osborne, Alex 318
Oshana, M. 560
'out of control' 564–565
Ottinger, Gwen 692, 696
over-control argument 414
Owen, Robert 18

Pacey, Arnold 15
Pahl, Gerhard 247, 250–254, 257, 320
Pahl and Beitz approach 97, *250*, *258*
Panama Canal 34
paradigm 637, 642; in analytic aesthetics 513;
 Aristotelian 209; artifacts as 210, 213;
 automobility 614; being-in-the-world 177; of
 BUILD, TEST, FIX 435; causality 245; of control
 463; design 356, 360–363; engineering ethics
 625; ethical 707; evolutionary development 442;
 generic 186; of innovation 283, 695; input/output
 252; of natural kinds 214; of problem-solving 317;
 pro-business 666; poiesis- 728, 730; programming
 427; research 515; Resilience 482–485,
 487–488, 490; Safety 484–486, 489–490; systems
 engineering as187; techno-economic 284; of
 technological development 274; unjust 685;
 see also design paradigm
paradigm shift 147, 316, 323; circular economics
 351; Kuhnian 346
paramountcy clause 594–595, 600
paramountcy principle 664, 668
Pareto Improvement 312n3
Pareto optimality 97, 101–104, 108
Pareto, V.F. *see* Pareto optimality
Parfit, Derek 48n7
Paris Agreement 161, 631, 638, 640
Paris, France 57, 131–132, 364
Parker, W. 165
Parker, W.S. 418
Parsons, Glenn 357
partial methods 324
participants 437
participatory design (PD) 329–330, 332, 338
parts structure 253
Passino, Kevin 704
Patel, Sardar Vallabhbhai 57
Patočka, Jan 718
Paul, T.V. 56
PDF 571–572, 575, 577
Pearce, D. 234
Pedreschi, D. 196
Peirce, Charles Sanders 375

Pellow, David 693
Pérez, B. 194
Perrin, Noel 58
Perrow, C. 152
personhood 51, 500; moral 564
perspectives: agent 309–310; disciplinary 357;
   eliminativist 113; epistemological 2; on fairness
   198; feminist 674; on the human subject 715;
   male 651; marginalized 653; multiple 187;
   ontological 650–651; philosophical 38, 50, 366,
   537; reductionist 185; of success 199; on systems
   engineering 187–188; on technology 608,
   711–712; tool 332; White 651
perspectives, foundational *see* engineering;
   engineering science; methodology
Peru 26, 28, 33–34, 683
pesticides *307*, 467, 702
PETRAS research consortium 637
Petroski, Henry 47, 158
phenomenological laws 82
phenomenology 21, 39, 443, 485, 730; and analytic
   philosophy 725; of extended mind 507; and
   hermeneutics 723–724; Kamlah, influence on 726;
   models 227; of perception 223; post- 239, 719,
   741; of technology user 500
Phenomenology of Spirit (Hegel) 22
phenomenon 81–91, 264–269, 274, 375–377;
   artificial 431; assessment of 371, 380; bodily 527;
   compartmentalized 573; computer simulation as
   408, 426; concrete 637; contested 20; controlled
   experience of 421; creation of 416, 426, 431;
   data exploration as model of 408; disease as 525;
   emergent 267, 498; formative 471; health as
   521, 523–524, 528, 530; of inductive risk 454;
   lifeworld 731; of make-believe 389; natural 383,
   385, 431; objective/subjective 529; observations
   as 376; physical 414, 422, 736; of psychological
   flow 510; pluralistic 352; rhythmic 365; socially-
   constructed 647; transient 486; trust 497;
   underivable 268; unforeseen 266; ungenerable 269
Philon of Byzantium 112, 120
philosophy of engineering 422; six different
   approaches to 39
philosophy of technology 57–58, 289, 300;
   Continental 710–719
physicalism 263
Pickering, Andrew 365
Picon, Antoine 128, 352
Pielke, R.A., Jr. 630
Pieters, Wolter 533–544
Pippin, Robert 21–22
Pirtle, Zach 462, 631
Pissarskoi, E. 170
Pitt, Joseph C. 633, 642, 743
Planner's Triangle 347–350
Plato 130, 359; on innovation 277, 279–281; in
   Quine's thought 225; in Schumpeter's thought
   274
Platonic form 587

Platonism 22, 40; Neoplatonism 38
pluralism: epistemic 379; method 324;
   methodological 506; "model" 165; mutualistic
   471; radical 623, 627n3; rejection of 623; of
   Shintoism 61
pluralism of values 620–623, 627n2; *see also* values
poiesis 128; paradigm 728, 730
Polanyi, Michael 72
policy advice 630–643
policy for engineering 631; *see also* engineering for
   policy; engineering in policy; engineering policy
   advice; engineering practice and policy
policymaking 134, 162, 630, 635, 638
political control 465
political order 13, 20, 279–281, 284
political practice *see* engineering as political practice
political virtue *see* virtue
politics and the political *see* engineering as political
   practice
politics of care *see* care
Pols, Auke 233–244, 513
Pols, J. 525
Pool, R. 157
Poon, J. *248*
Popper, Karl 22, 156, 707
Poser, H. 728
Post-It notes 269–270
potentially discriminatory (PD) 196
Power, Michael 572, 574
Poznic, Michael 383–393, 414, 736–744
practical experiments (PE) 120
practical reasoning *see* reasoning
practice *see* engineering: practice; understanding
practitioners 246, 259, 371, 667, 688, 691;
   conception 304; dimensionless parameters **404**;
   human-centered design 333; individual 706;
   know-how of 375; of measurement/measuring
   372, 376–377, 379; of philosophy 583; and
   professional societies 593–594; and "profession"
   582; scientific 654; technical 696; trust in 379–380
pragmatic approach 82, 84, 416
pragmatism (philosophical) 39, 117, 185–186, 329,
   722; consequentialism and 44–47; methodology
   and 724–725; philosophers 311, 350
pragmatist ethics 329, 335, **337**
pragmatic order 727–728, 731
pragmatic truth 725, 731
praxis: cooking as 728; design 315–316, 319,
   323–324; human action as 516; human centered
   design as 332; humanitarian engineering as 679
precision 372–373; *see also* measurement
prediction *88*, 268; algorithmic 200; bias 198–199;
   design as 342; deterministic 164–165, 169, 171,
   640; experiments as 425; hoaxes and 451; models
   used for 91, 388, 398, 416; non-discriminatory
   196–198; possibilistic 163, 165; probabilistic 164,
   169; search for 383; and uncertainty 152–153;
   unforeseeable 268
prediction horizon 156

prescriptive knowledge 111–123, 124n1; vs. descriptive 113–115; in design science 116–118; emancipation or elimination of 121–123; and management science 118–119; reliability of 112–114, **115**, 119–121, 122

Preston, Beth 209–221, 233, 358, 507

Pritchard, Michael 741

Privacy by Design (PbD) 599

problem *180*: in bioethics 40; breaking down 318; of capitalism 18; code audit to detect 197; context-dependent 122; design 97, 318, 352, 361, 611; engineering 67, 76, 98; ethical 54; heuristics 47; multiple-criteria 115; of social equity 347; societal 283; 'tame' 320; technical 97, 121–123; 'wicked' 134, 320, 717, 738, 740

problem context *89*

problem definition 145, 612–613

problem finding 129, 134

problem framing 2, 6, 310, 357, 736, 738

problem-setting 335, **337**, 338

problem solving 111, 119, 138, 176; abstraction in 185; creative engineering 143, **146**; engineering design as a form of 319, 357; as engineer's task 611, 738; framework 186; rationality as 129; reductionist 184; stages of 145; technological 144, 352

problem solving techniques (pro-sol) 178

Product Data Management (PDM) 324

Product Life-cycle Management (PLM) 324

profession 582–584

professional codes/codes of ethics 580–591

professional engineering societies 592–603; codes of ethics 594–595; and social responsibility 594

progressivism 662–664

Progressive Era 662

Project Management Institute (PMI) 437

proof: of existence 426; of concept 426; performance 426

Protestant Reformation 20

Pugh, Stuart **146**, 309

quality control 154–155, 666

Quality Function Deployment (QFD) 309

Quebec Bridge Disaster 494

Quine, Willard Van Orman 222–231

Rabins, Michael 741

radiation *250*; ionising 467, 469

Rasmussen, Nicolas 377

Rasmussen Reports 564

Rawls, John 242; autonomy 558, 565n1; basic structure of society 609; fairness 199; overlapping consensus 310, 624–625; pluralism 621–622

Raz, J. 559

realism: Aristotelian 211; versus constructivism 537; instrumental 724; moral 625; of simulations 411; "world as it is" 216

reality 19–22; of aesthetic production 517; of artifacts 215, 230; assumptions found in *164*; concepts

of 176, 183, 654; constructions of 188; datasets reflecting 191; designs in 121; of engineering objects 228–230; historical emergence of 738; holistic ontology of 183–184; human 712, 722–723; of LGTBQ students 649; models as imperfect representations of 153–154, 158, 242; in natural sciences 117; objective 650, 653; ontology of 225, 227; reconstruction of 725; of scientific practices 372; transforming 133; and uncertainty 158

reasoning in engineering 138, 140, 144–147; *see also* art; Big Data; creativity; data analytics; engineering design; optimality; prescriptive knowledge; problem solving; scenarios; uncertainty

reasoning 41; based on alternative notions of function 252–256; consequentialist 46; designing as mode of 317, 320, 324; epistemic 724; about functional properties 292; hypothetical–deductive (HD) 87; logical 426; means-end 293; models involving 387–388; moral 43, 731; non-deductive 306; practical 233–234, 239; proportional **403**; scientific 424; from similarity 400–401, **403**; use case-based *258*; utilitarian 45; from user goals to function carriers *256*; "what if" 537; *see also* 5-key-term approach; *knowledge for*; praxis

reimagining engineering *see* constructivism; culturism; environmental justice (EJ); feminist engineering; future of engineering; philosophy of technology; social justice; social responsibility

Reiser, Stanley 524–525

reliability 14; of autonomous systems 564; of engineering objects 230; epistemic 377; externalist justification strategies based on 415, 417; of humans 500, 564; Indian engineers' commitment to 57; and maintenance 362–363; and measuring 375; as metric 270; of model building principles 414; of outputs 191; policy related to 638; of prescriptive knowledge 112–114, **115**, 119–121, 122; quality and 600; of simulation 410; and trust 503; as value 451; verification and validation of 435, 443

Rentsch, T. 725

Resch, M.M. 409, 414

respect for persons 43, 558, 598

resilience 157, 351, 482–490; paradigm 486–488

resilience engineering 484–486; implementing 488–489; and traditional safety management, compared 489–490

responsibility/ies in engineering *see* autonomy; code of ethics; engineering as political practice; engineering ethics; global engineering ethics; engineering practice and policy; professional codes/codes of ethics; professional engineering societies; social responsibility; standards

Reuleaux, Franz 318

revolution 55–56; Energy 170; Green 55; knowledge, 55; social 139; White 55

Reydon, Thomas 214

Reynolds number 403
Richters, F. 489
Rielder, Alois 318
Riley, Donna 47, 242, 467, 507, 647–660, 694, 741
Ringland, G. 167
risk 38; acceptable 535; adversarial 4, 540;
    attunement towards 40; engineering 43, 179;
    intrinsic 273, **283**, 284; normalized 155
risk analysis 45; two-stage picture of *450*
risk and safety assessment 41, 45, 449–458, *456*;
    external influences 451–452; handover dilemma
    455–456; inductive risk 453–454; and science
    450–451; value boundaries of 452–453
risk-taking **145**, 147
Rittel, Horst 320
Ritual of the Calling of the Engineer 494
Robinson, J. 509
Robinson, J.G. 659
Robison, Wade L. 547–557, 736–744
robotics 522, 599, 650; and artificial intelligence
    602–603; autonomous 422, 426–429,
    431–432; social 716, 731–732; *see also* genetics,
    nanotechnology, and robotics (GNR); systems
Robust Decision Making 166
robustness 157–158, 171, 311; in measurement
    outcomes 372–373, 375, 380; philosophical
    constructions of 656
Rodriguez-Nikl, Tonatiuh
Roeser, Sabine 514
Romei, A. 195–196
Rondon, Glevys 679
Rosenberg, C.E. 524–525
Rosenberg, Nathan 15, 361
Rosenzweig, J. 181
Ross, W.D. 39, 43–44
Roth, P.A. 630, 635–637, 642
Rowling, J.K. 227
Royal Academy of Engineering (England) 11, 598,
    635, 637–648; joint report 642; policy report 640
Royal Charter (Institution of Civil Engineers,
    England) 12
Royal Dutch Shell 162
Royal Society (England) 642
Royal Geographical Society (England) 69
Ruggieri, S. 195–196
Russell, Andrew 363, 595
Russell, John Scott 121
Ryle, Gilbert 211

Saam, N.J. 412, 414
Saarinen, M.M.E. 177
Saaskilahti, M. 167
Sain, Barbara 736–744
safety: engineering and 529, 562; operationalization
    of *305*; paradigm 485–486, 490; public 18–19,
    44, 588, 603, 627, 700; design value of 304; and
    security 533
safety assessment 449–459, 679
safety devices 152

safety factor 73, 154, 157
safety features 306
safety management 482–483, 488–490
safety measures 56
safety regulation 99
safety risk 487
safety science 537, 539
safety standards 573, 575
Salvadori, Mario 360, 362
Salvarsan treatment 33
Samuel, book of 281
Sandel, Michael 564
Sano, Tsunetami 59
Sarewitz, Daniel 22
satisficing 151, 179, 319; *see also* Simon, Herbert
Saul/Paul 281
Savulescu, Julian 564
scale modeling 394–406; inherent limitations 405;
    invariants 404–405; misconceptions 406; in
    philosophy 394–395; in practice 395–399
scenarios 160–172; as arguments 164–165; as
    communication 170; as constructs 162–163;
    cooperative methods 167–169; creating 165–167;
    creative methods 167; for decision-making 164,
    168, 170–172; historical 162; for insight 171;
    for integration 170–171; model-based methods
    169–170; as possibilities 163–164; in technological
    development 160–162
Schatzberg, Eric 13
Scheele, M. 235, 242
Schiaffonati, Viola 352, 412, 421–434, 715
Schlosberg, David 689–691
Schlossberger, Eugene 558–568
Schmidt, Jon A. 40, 634, 641
Schmidt, L.C. 138, 145, **146**
Schneider, F.G.K. 351
Schoemaker, P.J.H. 167
Schomberg, René von 616
Schön, Donald A. 317, 357
Schön, Ludwig 378
Schrijver, Lara 127–137, 719, 730
Schumpeter, Joseph 274–277; business cycles *275*;
    creative destruction 276; innovation 280, **283**
Schwartz, Shalom H. 302
Schwemmer, O. 726, 731
Schyfter, P. 235, 242
science *see* engineering sciences; natural sciences;
    social sciences
science policy 630
science, technology, engineering, art and
    mathematics (STEAM) 127, 129, 134–135
science, technology, engineering and mathematics
    (STEM) 127, 129, 135
scooter 60
Searle, John 392n3, 647
security: secret security 4; transparent 4
security engineering 533–541; ethics 539–540;
    metrics and testing 537–539; ontologies of
    533–536

self-improvement 44

Sen, Amartya 62n1, 690

Senge, Peter M. 184

Serra, Montse 625

Sethy, Satya Sundar 50–65

Seymour, E. 648

shadow measurements *401*

Shaw, John 319

Shima, Hideo 61

Shintoism 61

Sider, Ted 225

Sidgwick, Henry 48n7

Siemens, C.W. 69

Sigma Xi 598

similarity 399–400: concept of 394; generalizing 402–404; geometric 4, 397, 399, 400–402, 406; material 414; physical 405; as verification or validation method 441, 444

similarity relationship 91

Simon, Herbert 47, 116–122; "bounded rationality" 319; on computer science as experimental science 423; on design problem-solving 738; "design without final goals" 156, 343; "satisficing" 151, 179, 319; on science of design 317; *Sciences of the Artificial* 116, 133

Simon, J. 501

Simondon, Gilbert 285n4, 712–714, 719

Simons, Peter 159, 217; emergence in engineering 263–272

Simpson, Thomas 67

Singer, Peter 48n7

Singh, Baldev 55

Slaton, Amy E. 650, 652, 656

smart city 352

smartphone 222, 226–248, 328, 506, 737; and innovation 572; privacy issues and 599; trust in 512

smart world 637–638

Smeaton, John 27

Smith, Adam 16, 73

Smith, Barry 216

Smith, Gerald F. 117

Smith, Jessica M. 661–673, 676, 680, 682

Smith, Ralph J. 357–358

Snow, C.P. 134

Sobel, Dava 573

social contract 345

social engineering 497, 740

social equity *see* equity

social experiment 715

social inequality 465

social justice 469, 674–685; *see also* Brundtland

social media 1, 222, 273, 497–498, 715; data collection and 599; manipulation of 577; and politics 610; privacy-invasive 614; and totalitarianism 565; underdetermination of 608

social networking 328

social opacity 415

social organization 17

social responsibility: enacting 668–670; three pillars of 666–668; *see also* corporate social responsibility (CSR)

social sciences 1, 70, 198, 230; and engineering 626, 633, 665; naturalistic 213; and verification 44

social transformation 129, 471–473

Society for Technical Education (India) 593

Society of Women Engineers 593

Sōetsu, Yanagi 14

software engineering body of knowledge (SWEBoK) 443

Solano, Lina 683

solutionism 463–464

solution-finding (finding solutions) 253, 319, 331, 335, **337**, 338, 611

solutions 46–47, 76, 97–98, 138–144, **146**, 147, *247–249*; algorithmic 106; change as driver of *140*; comparative 425; compromise as 309; counterintuitive 226; design 108, 117–188, 293, 295–296, 357; engineered 461, 470–471; engineering 130, 134, *143*, 253, 439; ideal *180*; integrated 309–310; function-oriented 129, 131; in human-centered design 328–329, 331, 333, 335–338; modeling of 133; of societal challenge 274, 276–277, **283**; standard 444; steps toward 257; technical 245–246, 258; transition from problem to 252, 255–256; workable 83

solution theories 122–123

sophia (wisdom)

South Africa: Institute of Electrical Engineers 593

space weather 640–641

Special Compromise Question 210

specification 306

Sperber, D. 213–214

Standards Administration of China (SAC) 570

standards developing organizations (SDOs) 570–571, 574, 577

standards setting organizations (SSOs) 570, 577

standards 569–578; *de facto* 576–577; and engineering education 577–578; ethical evaluation of 576–577; ethical significance of 574–575; and power and social justice 575–576; and professional/social responsibilities 595–596; thin or thick 622; *see also* national standards organizations; safety standards

standards, technical 304–305, 437, 441, 569–571; four types of 572; logical properties of 572–574

STEAM *see* science, technology, engineering, art and mathematics

steam engine 131–132, 276; as artifact 297; Carnot's work on 14, 111; in Japan 59; new world order associated with 280, 285n4; and thermodynamics 14; Watt's condenser for 32; working of 291

steam rail *275*

Steen, Marc 239, 319, 328–341

STEM *see* science, technology, engineering and mathematics

Steinmueller, Edward 15

Sternberg, R.J. 144, 148n2

Sterrett, Susan G. 4, 394–407

Stiegler, Bernard 712, 713–714
Stirling, Andy 342, 461–471, 482, 654
Stone Age 123
Stone, R.B. *250*, 251
Stone, Taylor 735–744
strategic policy 637, 642
strategic reliance 495, 498–501
strategy/ies: adaptive 157, 311; anticipatory 311,
    488; biomimicry as 84; design 311; degrowth
    hypothesis 351; of democratization 615; for
    discrimination prevention 196; eliminativist 513;
    engineering 482; in engineering science 69, 86;
    evaluation 144; experimental 416; internalist
    justification 415, 417; manipulative 472;
    methodological 323; military 58, 162; mitigation
    555; of multiple-value conflict 308–310; of
    prescriptive knowledge 119–121; prevention
    56; problem-solving 178, 188, 191; scenario
    166, 170–172; self-determined 59; for socially
    responsible engineering 661; validation and
    verification 440
Strien, P.J. van 118–119, 122–123
Suchman, Lucy 241
Suez Canal 69
Sullivan, S. 653
Sullivan, W. M. 666–667, 671
Surgrue, Noreen M. 624
Sussman, Leonard 564
sustainability 342, 461–74; defining 346–348;
    Planner's Triangle 348–350; precedents of *343*,
    343–346
sustainable design 342–353
sustainable development 665–666
Sustainable Development Goals 463
Swan-Raven experiment 121
Swimme, Brian 707
systematic design *321*
systematic engineering knowledge 73
Systematic Parameter Variation (PV) 119–120
system complexity *183*
system development: function, centrality of 251;
    general process model for 247
system hierarchy 181
systems 149–150: artificial intelligence (AI) 562,
    603; Aristotelian 41; autonomous 563–564,
    602–603; behaviors 441; bias assessment 201;
    biological 84–85, 374; category 213; chaotic
    267; classification 574; closed 344; complex 264,
    270, 319, 417, 425, 442, 450, 489; complex
    and complex adaptive 135, 150–157, 168;
    complicated 150–151; computer 316, 329, 424,
    430, 540; computer operating 548, 571–572;
    data, large scale 599; and data 194–195, 633;
    decision-making 198; defensive 26; disrupting
    740; earth 470; economic 669; empirical claims
    regarding 122; energy 169–170, 310; engineered
    465, 474, 495, 569–570; engineering 25, 632;
    engineering education 60–61; and/of engineering
    objects 223, 227; and engineering practice 633;

entertainment 737; four types of 149–150;
    government 32; harmful 455; harm to 533;
    information 238, 300–301, 331, 333; infotainment
    713; infrastructure 631; intelligent transport 162;
    interoperable 216; irrigation 27–28; kinship 669;
    knowable 352; knowledge 214; large equation
    169; large scale 149, 231, 270; life cycle of **436**;
    living 691; machine learning 202; models and 388,
    611; modern weapons 19; monitoring 533; natural
    142, 642; ontological perception of 252; open
    738–740; physical 490, 736; physically similar 397,
    399, 402–406; product-service 258; pulley 129;
    rational 133; realist and non-realist 211; resilience
    in 351, 485–486; robot(ic) 428–429, 431, 563,
    731; satellite 640–641; sewage 129; simple
    150–151; "smart" 495; social 483; socioeconomic
    62; software 200, 422; sub- 256, 383; "substitute"
    317; supersystems 181, 222; sustainability 347;
    sustainable 342, 350; symbol 318; target 81, 388,
    391, 417; technical 245; technical, theory of
    234, 253–254; technological 80, 87, 463, 498,
    595, 664, 687, 692, 694, 718; transportation 171,
    222, 715; ultra-large-scale 270; unjust 675; value
    310–311; ventilation 33; verifying and validating
    435, *436*, *439*; water delivery 32; waste disposal
    688; weapons 19, 562, 602; *see also* civil-aviation
    system; ecosystems; information systems (IS);
    information technology (IT)
systems approach (policymaking) 633, 636, 638
systems, automated 500–501, 503; decision-making
    198; and discriminatory criteria, detection of 197;
    feedback 352; rise of 4
systems engineering 33, 176–189; concepts in
    180–183; an engineering philosophy 187–188;
    history 178–179; objectives and methods 179–180,
    180; origins of 186; reductionism and holism
    183–184; representations of 186–187
systems engineering body of knowledge (SEBoK)
    440
systems expert 739–740
systems realms 47
systems, sociotechnical 38, 47, 120, 179, 185–186,
    188, 235; complex 361, 483, 500–501; design of
    290, 298, 503, 539; literature on 235; models of
    383; resilience and 484, 489; safety reserves and
    490; values of 310–311
systems of systems *439*, 444
systems theory 518, 535
systems thinking 184–185
Sze, Julie 689, 691, 693

Tallant, J. 500
Tanaka, Hisashige 59
task 251
Taylor, Charles 621
Taylor, David *see* David Taylor Model Basin
Taylor, Frederick Winslow 73, 572
Taylor, J.S. 562
Team Data Management (TDM) 324

techne 40–41, 128, 633
technical standards *see* standards, technical
technics 13
technology assessment 608
technology/ies 311; digital 633, 731, 736; emerging 468, 527, 595–596, 602–603, 647, 667, 729; 'open' 712–713; as a process 363–365; rethinking 711–714; *see also* nanotechnology
technology, philosophy of *see* philosophy of technology
Tedri, Matti 425
Teilhard de Chardin, Pierre 707
Tejero, Debora 708
Telford, Thomas 15–16
tensile testing machine *248*
Tetri, P. 535
Thackara, John 332
Theory of Domains *258*
theory of function ascription 235, 243
Theory of Technical Systems 234
thinking, convergent versus divergent **145**
Thomasson, Amie 210, 215–216, 230, 297, 358–359
Thompson, Paul B. 201, 351, 441, 569–579
Thomson, William (Lord Kelvin) 158, 372
thought experiment (TE) 412, 416
Three Gorges Dam, Yangtze River, China 35
Three Mile Island 152, 157, 664–665
Tichy, W. 427
Tiefel, H.O. 564
Together Anywhere, Together Anytime (TA2) 334–337
Tonkens, Ryan 563
Tonso, Karen 649
tool perspective 332
Tormey, R. 742
Tosaka, Jun 58
Toshiba 59
toxicity *305*
toxic load 571
toxic materials 553–554, 667
toxic pollution 461
toxic waste 467, 667, 689–690, 702; *see also* Love Canal
toxicological effects 453, 455
toxicological standard 572, 574
training 489, 495; academic 31–33; and artificial intelligence 715; automation and 503; behavior and 539; data 195–199, 201, 715; datasets 201; in engineering for social justice 706; of engineers 570, 592, 694, 729; of future engineers 5–6, 736; in green engineering 702; in humanitarian engineering 704; in *omnium* engineering 707; preparatory 678; requirements 186; through practice 34; statistical 427; technical 25; use plans and 236; vocational 55; of women 669
training exercise 408
training facility 122
Transmission Control Protocol (TCP) 570, 572, 576–577

transformation: alchemical 30; of basic inputs 247; of buildings 364; causal 248; in China 39, 52; compromise versus 348, 350; computer 17; of culture 469; of data 192, 194, 196; in design 290–291; in Deweyan inquiry 46, 335; digital 497, 731, 737; in engineering education 42; of human condition 20; and innovation 282; of knowledge 375; of operands 250–251, 257; of politics of 472; science 414; social 13, 127–129, 131, 684; technological 164, 344, 508
transformation process structure 253–255, 258
transformations to Sustainability 470–474
transparency 4, 170, 172; algorithmic 197; in artificial intelligence 715; in China 53; in computer simulation (CS) 415–417; epistemic 413, 416; hypothetical 502–503; of methodological constraints 325; in risk assessment 458; in safety 563; in security 536; in social relations of knowledge 653; *see also* Fairness, Accountability and Transparency in Machine Learning (FAT-ML)
transsubjectivity 6, 725, 732
Tredgold, Thomas 13, 19, 46, 652
Treusch, Pat 650
trial experiments 425
Trolley Problem 562–563
Tronto, Joan 654
trust in automation 498–503
trust in engineers and engineering 494–503
truth: affirming or denying 40; criterion 724–725; descriptive knowledge and 111; discrimination and 195; duty of fidelity to 44; ethical 623; ground 195, 200–201; of knowledge 652; objective reality 651; observation of 377; partial 653; pragmatic 725, 731; technology as 19; theoretical 516; transcendental 42, 280; understanding of 41; and usability 117
truth-finding 613
truth-value 114, **115**, 118, 122
Tuana, N. 653
Turcotte, J. 497

Uchida, Hoshimi 60
ultra-large-scale systems 269; *see also* systems
Umbrello, Steven 736–744
uncertainty 149–158: aleatory 149; epistemic 149, 152, 343; living with 156–158; mitigation of 153–156; sources of 150–152
under construction 444
understanding: and action 471; aesthetic experience 509–510, 514, 516–516; behavior 395; of being an engineer 718; of closed systems 182; in codes of ethics 587, 595, 597; of complex reality 183; computations 424; computer simulation 408, 410, 412, 416–417; of creativity, importance of 147; design philosophies 426; design process as means of 316–317, 319, 321; of engineering as art and science 127–131, 134; engineering claims 107; of engineering education 666; of engineering in China 53; of engineering design

587; of engineering ethics 625; of engineering knowledge, nature of 632–634; of engineering practices 665, 722; of engineering's contributions 643; engineering's philosophical traditions 39, 47; of engineering's power 25; of engineering profession 18, 623; of engineering, responsibilities of 462, 671, 684–684, 707–708; of *engineering science* 80; epistemological 46; of evidence 456; of feminist ontology 650–651; of flight 14; of function 250; the function of an object 71; of general patterns 75; governance 632; of health and disease 522–523, 529–530; in human centered design 328, 333; humanitarian engineering 704; of Husserl 724–725; of ignorance, practices of 651, 653; injustice 688, 695; innovation 275–278, **283**; of laws of physics 444; of maintenance practices 356–361, 363–366; in measuring and measurement 51, 376, 378; models and material entities 611; of narrative structure 630; as *nous* 41; in participatory design 332; of phenomena 81, 83, 85, 117; philosophical 20; of policymaking 642; of politics 608, 613–614; practice-based 621; problems 440; professional 13, 19; professions 582; public 623; reductionism as means of 183–184; resilience as means of 351; Resilience paradigm 487, 489; in robotics 429; of risk and risk assessment 449; scale model experiments 399; scenarios 160, 162–165; of science 450; within science 636; scientific 83–84; scientific models 91; self- 226, 230; shared 387; simulation 408; of social justice 706; of sociotechnical environment 577; Sustainability 474; structural conditions 668; systems, goals of 255; systems thinking as means of 185; of *techne* 40; technology 177; of technology and society, co-construction of 700; theoretic 15; users' needs 437; of value conflict 310; and well-being 304; of world 274, 465, 506, 635

UNESCO 34

unexpected, unforeseen, underived, unexplained 269–270

Union of Concerned Scientists (UCS) 597, 601, 603

United Church of Christ (UCC) 600

United Nations Sustainable Development Goals 738

United States Green Building Council (USGBC) 347

UN WCED *see* World Commission on Environment and Development

"use and convenience" 16, 46

use case-based function reasoning 255–256

use plan 233–243; communicating 236–237; criticism of 240–242; design of 234–236; evaluating 237–233; executing 237; functions and affordances 239–240

utilitarian 14, 45; mining as 681

utilitarianism 14, 16, 39, 45, 345; and autonomy 560–561; and engineering ethics 676; "greatest good…" 574; in US Constitution 349; *see also* Bentham; Hume; Locke; Mill

utilitas 16, 130

utilities 13, 157; electric 664

utility 13–14; in analytic aesthetics 511; in engineering 128, 133; pluralist outlook on 185; and pragmatism 46–47, 117; of systems engineering 187–188

utility data *321*

Vajpayee, Ata Bihari 56

validation **146**; in/of computer simulation 408, 410, 414; in data science 194; of design 145, 246; and design methods 315–325; experimental 427; of function 259; in system development *247*; verification and 435–445

Valkenburg, Govert 27, 310, 607–619, 688, 712

Vallero, Daniel 693

value boundary 452–453

value conflict 306–308

value hierarchy 306, *307*

value judgement 306, 317

value-ladenness 115, 118, 300–301, 311, 449, 452; internal and external 451

values 300–312; compromise 309; conceptualization of 303–304; definition 301–303; engineering, held by 626, 739; epistemic 409, 414, 451–458; fact-value distinction 114; integrated solution 309–310; moral 57, 239, 295–296, 311, 452, 508; non-epistemic 451–458; operationalization of 303–305, *305*; pluralism of 620–623, 627n2; in risk and safety assessment 449–458; specification of 306; *see also* aesthetic value; truth-values

Values at Play 300

value sensitive design (VSD) 295, 300–302, 329, 338, 498; and value change 310–311; *see also* human-centered design (HCD)

values in Design (VID) 300

van Aken, Joan E. *see* Aken, Joan E. van

van den Hoven, Jeroen 301, 308

Van de Poel, Ibo 300–314, 328, 615, 621, 742

Vandersteen, J.D.J. 679

Vandewalle, P. 430

Van Gogh, Vincent 509

van Inwagen, Peter 210, 217, 225, 559

van Loosdrecht, Mark 119

variation theory 685

Veblen, Thorstein 18

VDI *see Verein Deutscher Ingenieure* (VDI)

Venable, J. 125n18

Verbeek, Peter-Paul 53, 223, 328

Verein Deutscher Ingenieure (VDI) *247*, 320, *321*, 588, 598

verification: formal 423, 435, 441–442, 444, 538; and validation *436*, *438–439*; *see also* validation

Vermaas, Pieter E.: 5-key-terms approach 256–258; on capability as a concept 251; in China 53; engineering, future of 736–744; on 'instrumental kinds' 230; use plan approach 233, 235–243

Vertov, Dziga 133

Veslind, Aarne P. 693

Vincenti, Walter 14, 46, 111–112, 120; *What Engineers Know* 423, 429, 633

Vinsel, Lee 363

virtue ethicists 48n1
virtue ethics 56, 329–332, **337**; and Aristotelian philosophy 16, 33, 39–42; Western 55
virtue 44, 54, 540, 695; of classical models 416; courage 42; of engineers 611, 630, 693, 741; epistemic 225, 451, 739; in human-centered design 329–330, **337**, 338; intellectual 40–41; moral 41; political 31; practical 377–379; and vices 740
virtue theory 676
visualization 151, 335–336, 410, 413, 416; computer 154; data 191; techniques 427
Vitek, B. 156
Vitruvius 12, 16, 128, 130, 318
Voigt, C. 166
Vogel, Steven 219n4, 366
Voland, G. 138
von Kármán, Theodore 15
von Wright, G.H. *see* Wright, G.H. von
Vorms, M. 411, 418n6
Vuorinen, J. 535
Vries, G. de *see* De Vries, G.
Vries, Marc J. de 124n1, 617n3

Wack, Pierre 167
Walker, Margaret Urban 654
Wallas, G. 145
Waller, A.A. 651
Walls, Joseph 116, 118
Walport, M. 632
Waltzer 349
Wang, Dazhou 47
Wang, Xiaowei (Tom) 50–65
Warren, Karen 654
Wasson, Charles S. 437, 443
Waterways Experiment Station 398–399
Watt, James 32
WCED *see* United Nations
Webber, Melvin M. 320
Weber, M. 464, 582
Weisberg, Michael 394, 397
well-being 41, 451; artificial intelligence's impact on 603; autonomy as element of 560; community 690; economic 670; illness and 524; of marginalized populations 625, 669, 700, 708; as paradigm 695; promoting 626; right of 690; social 521; sustainability and 461, 463, 466, 473; technology's impact on 576; three theories of 303–304; WeCare project and 329
well-being, engineering's role in 662, 718; challenges to 740; as ethics of profession 625–626; resilience as goal in 484; responsibilities to the public 592, 595; as value of profession 621–622
white people 651, 678, 692; norms of 649; wealthy 617n2, 684, 689
White Revolution 55
Whitney, Eli 571
wicked problems *see* problems
Wieringa, Roel 122–123
Wiggins, David 228
Wilson, Richard L.
Winner, Langdon 300, 607, 652, 666–668
Winsberg, E. 411–412, 445
Wittgenstein, Ludwig 182, 214, 725, 728, 730
Wolff, Christian 318
Women's Engineering Society (UK) 593
Wong, Pak-Hang 51, 352, 620–629, 696
Wood, K. L. 251
Woods, David 564
World Commission on Environment and Development (WCED) 343, 346
World Energy Outlook 161
World Federation of Engineering Organizations (WFEO) 593, 596
World War I 33
World War II 33, 663, 692; and China 50; and environmental concerns, rise of 702; and Japan 58, 61; nuclear weapons 35; problem-solving techniques 178; and research subjects, "drafting" of 715; spread of 'system' during 181, 186
Wulf, William 20
Wright brothers 112, 266
Wright, G. 168
Wright, G.H. von 113, 292
Wynn, David 390

Xenophon 278–279, 283

Yamao, Yozo 59
Yanacocha mine 683
Yanagi, Sōetsu 14
Young, Iris Marion 654
Young, Mark Thomas 356–368
Young, Robert 346

Zhang Wei 51
Zhu, Qin 39, 623–624
Zimmerman, J.B. 702
Zussman, R. 652
Zwart, Sjoerd 111–126, 466, 724, 730
Zwicky, Fritz 318